T0389458

Innovations as Key to the Green Revolution in Africa

Andre Bationo · Boaz Waswa ·
Jeremiah M. Okeyo · Fredah Maina ·
Job Kihara

Editors

Innovations as Key to the Green Revolution in Africa

Exploring the Scientific Facts

Volume 1

Springer

Editors

Andre Bationo
Alliance for a Green Revolution in Africa
   (AGRA)
Soil Health Program
6 Agostino Neto Road
Airport Residential Area
PMB KIA 114, Airport-Accra
Ghana
abationo@agra-alliance.org

Boaz Waswa
Tropical Soil Biology and Fertility Institute
   of the International Centre for Tropical
   Agriculture (TSBF-CIAT)
Nairobi, Kenya
bswaswa@yahoo.com

Jeremiah M. Okeyo
Tropical Soil Biology & Fertility (TSBF)
African Network for Soil Biology
   and Fertility (AfNet)
c/o ICRAF, Off UN Avenue
P.O. Box 30677-00100
Nairobi, Kenya
jmosioma@gmail.com

Fredah Maina
Kenya Agricultural Research Institute
Socio-economics and Biometrics
P.O. Box 14733-00800
Nairobi, Kenya
fredah.maina@yahoo.com

Job Kihara
Tropical Soil Biology & Fertility (TSBF)
African Network for Soil Biology
   and Fertility (AfNet)
c/o ICRAF, Off UN Avenue
P.O. Box 30677-00100
Nairobi, Kenya
j.kihara@cgiar.org

Please note that some manuscripts have been previously published in the journal 'Nutrient Cycling in Agroecosystems' Special Issue "Innovations as Key to the Green Revolution in Africa: Exploring the Scientific Facts". (Chapters 13, 14, 19, 20, 23, 36, 42, 57, 59, 78, 80 and 113)

Printed in 2 volumes
ISBN 978-90-481-2541-8   e-ISBN 978-90-481-2543-2
DOI 10.1007/978-90-481-2543-2
Springer Dordrecht Heidelberg London New York

Library of Congress Control Number: 2011930869

Springer is part of Springer Science+Business Media (www.springer.com)

Preface

Africa remains the only continent that did not fully benefit from the effects of the Green Revolution experienced in the 1960s. With the 2015 deadline for the millennium development goals (MDGs) rapidly approaching, the number of hungry in Africa is increasing again. Africa accounts for half of the 12 million children under the age of 5 years dying each year as a consequence of chronic hunger. Food production has not been able to keep pace with the ever growing human population in sub-Saharan Africa. The low and declining productivity can be attributed to Africa's impoverished agricultural resource base, unfavourable socioeconomic and policy environments for investment in agricultural sector development as well as the emerging challenges associated with unfavourable weather and climate change.

Over the last few years, various local, regional and international forums have been held to discuss how Africa's Green Revolution can be achieved. The African heads of state and governments have developed the Comprehensive African Agricultural Development Program (CAADP) as a framework for agricultural growth, food security and rural development. CAADP has set a goal of 6% annual growth rate in agricultural production to reach the UN's millennium development goal of halving poverty and hunger by 2015. The African Heads of State Fertilizer Summit held in Abuja Nigeria in June 2006 led to the Abuja Declaration on Fertilizer for the African Green Revolution. The Summit identified three most critical issues that need to be addressed if millions of African farmers are to increase utilization of fertilizer. These are access, affordability and the use of incentives. The Summit recognized that given the strategic importance of fertilizers in achieving the African Green Revolution, there is need to increase the level of use of fertilizer from the current average of 8 kg ha^{-1} to an average of at least 50 kg ha^{-1} by 2015. Similar sentiments were echoed at the African Green Revolution Conference in Oslo where it was resolved to take concrete and concerted action towards the development of self-sustaining changes in African agricultural growth through the use of enhanced approaches to public–private partnerships. Achievement of the desired growth in agricultural production calls for deliberate effort to increase access and affordability of inorganic fertilizers, seed, pesticides and profitable soil, water and nutrient management technologies by the smallholder farmers in Africa. All these components can best be explained under the integrated soil fertility management (ISFM) approach.

Crop diversification is an important instrument for economic growth. Through the use of biotechnology, high-yielding crop varieties have been bred with potential to significantly increase production. NERICA, "New Rice for Africa", for example, is a new rice variety that has been bred through the application of biotechnology and offers great potential for transforming agriculture in the continent. Other high-yielding crop varieties such as maize, sorghum, millet, cowpea, soya bean, cassava, and cotton with additional benefits of being disease and insect resistant have also been bred and these have the potential for increasing food production and incomes if accessed by smallholder farmers in the continent. Smallholder farmers should be empowered to confront the rapidly evolving production, consumption and marketing systems in the global systems. Farmers need to be linked to input–output markets and supported in order to access the required seed, fertilizer, and pesticides and also access market information and better prices for their produce. Further, there is need for change in paradigms in development practice where participation, diversity and self-reflection are incorporated in agricultural research and development. There is need therefore to build strong institutions among all actors in the natural resource management (NRM) sector as basis for influencing change.

Whereas numerous investments have been made in agriculture research in the continent, little impact has been seen especially with wide adoption of the promising soil fertility and food production technologies. There is need for a shift in paradigm from the linear model of research-to-development to the systems approach. This calls for agricultural innovation, which is the application of new and existing scientific and technological (S&T) knowledge to achieve the desired growth in agricultural production and overall economic development in Africa.

It is against the above backdrop that the African Network for Soil Biology and Fertility (AfNet) in collaboration with the Soil Fertility Consortium for Southern Africa (SOFECSA) organized this international symposium entitled *Innovations as Key to the Green Revolution in Africa: Exploring the Scientific Facts*. The overall goal of this symposium was to bring together scientists, agricultural extension staff, NGOs and policy makers from all over Africa to deliberate on the scientific facts and share knowledge and experiences on the role of innovation in soil fertility replenishment as a key to the Green Revolution in Africa.

The specific objectives of the symposium were the following:

1. To assess the potential and feasibility of use of external input and improved soil and crop management to achieve the African Green Revolution
2. To identify and learn about innovative approaches needed to build rural input market infrastructure
3. To review the main policy, institutional, financial, infrastructural and market constraints that limit access to innovations by poor farmers
4. To evaluate strategies for scaling out innovations to millions of poor farmers in the continent

The symposium was organized under four main themes, namely the following:

1. Constraints and opportunities towards the African Green Revolution
2. Potential and feasibility of use of external input and improved soil and crop management to achieve the African Green Revolution
3. Factors that limit access to and adoption of innovations by poor farmers
4. Innovations and their scaling-up/out in Africa

The symposium held in Arusha, Tanzania (17–21 September 2007), was attended by over 230 participants drawn from 20 African countries (Benin, Botswana, Burkina

Faso, Cameroon, Democratic Republic of Congo, Ethiopia, Ghana, Niger, Uganda, Ivory Coast, Kenya, Malawi, Mali, Namibia, Nigeria, Rwanda, Senegal, South Africa, Tanzania and Zimbabwe), Europe (Belgium, France, Netherlands, Norway, Scotland and Sweden), North and South America (Canada, Colombia and USA), Asia (Japan) and Australia. The symposium was also attended by representatives from the Bill and Melinda Gates Foundation; CG centres (The World Agroforestry Centre – ICRAF, International Centre for Research in Semiarid Tropics – ICRISAT, International Institute of Tropical Agriculture – IITA, Africa Rice Center – WARDA, International Maize and Wheat Improvement Centre – CIMMYT, International Livestock Research Institute – ILRI, International Centre for Tropical Agriculture – IITA, CIAT); advanced research organizations (Norwegian Institute of Agriculture, JIRCAS); international NGOs (Catholic Relief Services, IFDC, UNDP, AFRICARE, AVRDC – The World Vegetable Centre, United Nations Economic Commission for Africa – UNECA); universities (KTH University, Cornell University, Wageningen University, Columbia University, University of Aberdeen and La Trobe University) and the private sector (YARA, IFA, Chemplex Corporation Ltd).

This book presents papers of the symposium organized under the above four themes. It is worth noting that a selection of 12 papers at this symposium have been published in the special issue of *Nutrient Cycling in Agroecosystem Journal* (Volume 88, No. 1) titled: *Innovations as Key to the Green Revolution in Africa: Exploring the Scientific Facts*.

It is the our hope that the knowledge and wealth of experiences presented in this book and the special issue will enlighten the reader and other development partners in SSA to make informed choices that will result in the desired growth in the agricultural sector.

Nairobi, Kenya   Nteranya Sanginga
Nairobi, Kenya   Akin Adesina

Acknowledgements

The organizers would like to thank the Alliance for a Green Revolution in Africa (AGRA), the Canadian International Development Agency (CIDA), the International Development Research Centre (IDRC), the Ford Foundation (FF), International Foundation for Science (IFS), the Technical Centre for Agricultural and Rural Cooperation (CTA), the Rockefeller Foundation, Syngenta Foundation, Forum for Agricultural Research in Africa (FARA), the International Centre for Tropical Agriculture (CIAT) and the Tropical Soil Biology and Fertility (TSBF) for their financial contributions towards the organization of the symposium. We would also like to thank the Ministry of Agriculture Food Security and Cooperatives, Tanzania, for hosting this symposium and the local organizing committee for the logistical support.

Contents

Volume 1

Volume 2

About the Organizers

The African Network for Soil Biology and Fertility (AfNet)

The African Network for Soil Biology and Fertility (AfNet) was established in 1988 as a pan-African network of researchers in sub-Saharan Africa. AfNet is the single most important implementing agency of Tropical Soil Biology and Fertility Institute of the International Centre for Tropical Agriculture (TSBF-CIAT) in Africa. More recently, a Memorandum of Understanding (MOU) was signed between the Forum for Agricultural Research in Africa (FARA) and The International Centre for Tropical Agriculture (CIAT) for hosting AfNet under the umbrella of FARA. Since its inception, AfNet has grown steadily and the current membership stands at over 400 scientists. The network aims at strengthening and sustaining stakeholder capacity to generate, share and apply soil fertility management knowledge and skills to contribute to the welfare of farming communities in the Africa. This is achieved through the adoption of the integrated soil fertility management (ISFM), a holistic approach to soil fertility that embraces the full range of driving factors and consequences, namely biological, physical, chemical, social, economic and policy aspects of soil fertility.

The main activities of AfNet are the following:

(i) *Research and development activities*: Network trials are scattered in more than 100 sites across the continent. The research is undertaken in collaboration with national agricultural research systems (NARS), scientists, farmers, non-governmental organizations (NGOs), local and foreign universities and advanced research institutes (AROs). Other partners include the CGIAR centres, system-wide programmes (SWPs), challenge programmes (CPs) and other networks. The main research themes include soil fertility management, nutrient use efficiency, conservation agriculture, targeting of recommendations to farmers and scaling-up success stories, among others.

(ii) *Capacity building*: AfNet's capacity building agenda is achieved through degree-oriented training (M.Sc. and Ph.D. research) in the domain of ISFM as well as through short courses. Over the years, AfNet offered several training courses on topics such as participatory research and scaling-up, decision support systems (DSSAT), proposal and scientific writing, presentation skills, soil erosion and carbon sequestration and nutrient monitoring (NUTMON) in agro-ecosystems, markets and agroenterprise development.

(iii) *Information dissemination*: In an effort to facilitate exchange of information among all stakeholders, AfNet has published several books, newsletters, brochures and posters. AfNet has successfully organized nine international symposia where researchers from across the continent were able to share their research experiences. AfNet has also established The Essential Electronic Agricultural Library (TEEAL) to facilitate information dissemination to researchers and students.

The AfNet Coordination Unit is comprised of the coordinator, two research assistants and one administrative assistant. The AfNet Steering Committee consists of a multi-disciplinary and gender-balanced team of African scientists drawn from the eastern, southern, central and western Africa regions.

Soil Fertility Consortium for Southern Africa (SOFECSA)

The Soil Fertility Consortium for Southern Africa (SOFECSA) is a multi-institutional and interdisciplinary regional organization founded in 2005 to develop and promote technical and institutional innovations that enhance contributions of integrated soil fertility research and development to sustainable food security and livelihood options in southern Africa. SOFECSA is an impact-oriented consortium operationalized through a 15-member technical management/steering committee in collaboration with the host institution (CIMMYT, southern Africa), a regional coordinator and support staff, and country-level teams drawn from diverse stakeholders.

Contributors

R.C. Abaidoo Soil Research Laboratory, International Institute of Tropical Agriculture, Ibadan, Nigeria; C/O LW Lambourn & Co., Carolyn House, Croydon, UK, r.abaidoo@cgiar.org

G.S. Abawi Department of Plant Pathology, Cornell University, Ithaca, NY, USA, gsa1@cornell.edu

G.O. Abayo Agronomy Programme, Crop Development Department, Kenya Sugar Research Foundation (KESREF), Kisumu, Kenya, abayogo@yahoo.com

M. Abba Faculty of Science and Biotechnology Centre, University of Yaoundé I, 812, Yaoundé, Cameroon, maimounaabba@yahoo.co.uk

A. Abdou International Crops Research Institute for the Semi-Arid Tropics (ICRISAT) Sahelian Center, Niamey, Niger, a.abdou@cgiar.org

T. Abdoulaye JIRCAS, Tsukuba, Ibaraki, Japan; INRAN, Niamey, Niger, t.abdoulaye@cgiar.org

S. Abdoussalam International Crops Research Institute for the Semi-Arid Tropics (ICRISAT) Sahelian Center, Niamey, Niger, s.abdoussalam@cgiar.org

C. Achieng Department of Environmental Science, Kenyatta University, Nairobi, Kenya; School of Environmental Studies and Human Sciences, Kenyatta University, Nairobi, Kenya, cellineoduor@yahoo.com

M.A. Adamu Green Sahel Agro Venture, Gumel, Jigawa State, Nigeria, muhddansista@yahoo.com

A. Adesina Alliance for a Green Revolution in Africa (AGRA), Nairobi, Kenya, aadesina@agra-alliance.org

B.D.K. Ahiabor CSIR-Savanna Agricultural Research Institute, Tamale, Ghana, bahiabor@yahoo.com

I. Akintayo Africa Rice Center (WARDA), Cotonou, Benin, i.akintayo@cgiar.org

B. Akpatou Centre Swisse de Recherche Scientifique en Côte d'Ivoire (CSRS), Abidjan, Côte d'Ivoire; University of Cocody, Abidjan, Côte d'Ivoire, bertin.akpatou@csrs.ci

D. Allou University of Cocody, Abidjan, Côte d'Ivoire; Centre National de Recherche Agronomique (CNRA), Lamé, Côte d'Ivoire, desire.allou@gmail.com

T.E. Alweendo Division Plant Production Research, Ministry of Agriculture, Water and Forestry, Government Office Park, Windhoek, Namibia, Alweendot@mawf.gov.na; alweendotwewaadha@yahoo.com

B. Amadou FAO Projet Intrants, Niamey, Niger, malbachi61@yahoo.fr

A.R. Anuar Faculty of Agriculture, Department of Land Management, Universiti Putra Malaysia, Serdang, Selangor, Malaysia, anuar@Agri.upm.edu.my

H. Anyanzwa Department of Soil Science, Moi University, Eldoret, Kenya, hellenanyz@yahoo.com

J. Ashby Alianza Cambio Andino (Andean Change Program), International Potato Center (CIP), Cali, Colombia, jacashby@gmail.com

E. Atsu CSIR-Savanna Agricultural Research Institute, Tamale, Ghana, Ericatsu@yahoo.com

A. Aw Africa Rice Center (AfricaRice), Sahel Regional Station, Saint-Louis BP 96, Senegal, aaw@usaid-yaajeende.org

A. Ayemou Centre Suisse de Recherche Scientifique en Côte d'Ivoire (CSRS), Abidjan, Côte d'Ivoire; University of Abobo-Adjamé, Abidjan, Côte d'Ivoire, amandine.sopie@gmail.com

M. Azuba Kampala City Council District Urban Agriculture Office, Kampala, Uganda, azubam@yahoo.com

A. Babirye International Institute of Tropical Agriculture (IITA-Uganda), Kampala, Uganda, annetbabirye2@yahoo.com

A. Babu Agricultural Research Institute Ukiriguru, Mwanza, Tanzania, adventinababu@yahoo.com

B.V. Bado Sahel Regional Station, Africa Rice Center (AfricaRice), BP 96, Saint-Louis, Senegal; Institute of Environment and Agricultural Research (INERA), BP 910, Bobo-Dioulasso, Burkina Faso, V.Bado@cgiar.org

A. Bala School of Agriculture and Agricultural Technology, Federal University of Technology Minna, Minna, Niger State Nigeria, Abdullahi_bala@yahoo.com

K. Balozi Kenya Forestry Research Institute, Gede Regional Research Centre, Malindi, Kenya, balozibk@hotmail.com

T. Basamba Department of Soil Science, Makerere University, Kampala, Uganda, ateenyitwaha@hotmail.com

A. Bationo Alliance for a Green Revolution in Africa (AGRA), Accra, Ghana, abationo@agra-alliance.org

C.L. Bielders Tropical Soil Biology and Fertility Institute of the International Centre for Tropical Agriculture (TSBF-CIAT), Nairobi, Kenya, charles.bielders@uclouvain.be

M. Bonzi Institute of the Environment and Agricultural Research (INERA), Ouagadougou, Burkina Faso, bouabonzi@yahoo.fr

S. Boubacar Sasakawa Global 2000 (SG 2000), Bamako, Mali, b_sandinan@yahoo.fr

G. Burpee Catholic Relief Services, Baltimore, MD, USA, gaye.burpee@crs.org

J. Byalebeka Kawanda Agricultural Research Institute (KARI), National Agricultural Research Organization (NARO), Kampala, Uganda, jbyalebeka@yahoo.co.uk

P. Cattan Agricoles et de Formation de Kamboinsé, Institute of the Environment and Agricultural Research Institute (INERA), Centre de Recherches Environnementales, Ouagadougou, Burkina Faso; Station de Neufchateau-Sainte-Marie, CIRAD, Capesterre-Belle-Eau, France, philippe.cattan@cirad.fr

M.P. Cescas FSSA, Université Laval, Québec City, QC, Canada, Michel.Cescas@fsaa.ulaval.ca

A.L. Chek Kenya Agricultural Research Institute (KARI), National Agricultural Research Laboratories, Nairobi, Kenya, kss@iconnect.co.ke

H.K. Cheruiyot Desert Margins Programme, Kenya Agricultural Research Institute, Nairobi, Kenya, hkcheruiyot@kari.org

J. Chianu Tropical Soil Biology and Fertility Institute of the International Centre for Tropical Agriculture (TSBF-CIAT), UN Avenue, Gigiri, Nairobi, Kenya, jchianu@yahoo.com

R. Chikowo Department of Soil Science and Agricultural Engineering, University of Zimbabwe, Mt Pleasant, Harare, Zimbabwe, rchikowo@agric.uz.ac.zw; regischikowo@yahoo.co.uk

D. Chikoye International Institute of Tropical Agriculture (IITA), Ibadan, Nigeria, d.chikoye@cgiar.org

P. Chivenge Department of Plant Sciences, University of California, One Shields Ave., Davis, CA, USA, pchivenge@gmail.com

G. Cisse Swiss Tropical and Public Health Institute (Swiss TPHI), Bâle, Switzerland, gueladio.cisse@unibas.ch

M. Cisse Institut Sénégalais de Recherches Agricoles (ISRA), Saint-Louis, Senegal, sbamand@yahoo.com

E. Compaore LSE/ENSAIA, Vandoeuvre-lès-Nancy Cedex, France; Station de Recherches Agricoles de Farako-Bâ, Environment and Agricultural Research Institute (INERA), Bobo-Dioulasso, Burkina Faso, ecompaorez@hotmail.com

M. Corbeels Département Persyst, Centre de Coopération Internationale en Recherche Agronomique pour le Développement (CIRAD), Av Agropolis TA B-102/02, Montpellier Cedex 5, France, corbeels@cirad.fr

F.D. Dakora Science Faculty, Tswane University of Technology, Pretoria, South Africa, dakorafd@tut.ac.za

O.G. Dangasuk Department of Biological Sciences, Moi University, Eldoret, Kenya, georgedangasuk@yahoo.com

S. Danso Tropical Soil Biology and Fertility Institute of the International Centre for Tropical Agriculture (TSBF-CIAT), Nairobi, Kenya, danso@libr.ug.edu.gh

O. Deji Department of Agricultural Extension and Rural Development, Obafemi Awolowo University, Ile-Ife, Osun, Nigeria, odeji2001@yahoo.com

R.J. Delve Tropical Soil Biology and Fertility Institute of the International Centre for Tropical Agriculture (TSBF-CIAT), Mt Pleasant, Harare, Zimbabwe, rdelve@earo.crs.org

A. Deubel Institute of Soil Science and Plant Nutrition, Martin-Luther University Halle-Wittenberg, Halle, Germany, deubel@landw.uni-halle.de

M.E. Devries Sahel Regional Station, Africa Rice Centre, BP 96 Saint Louis, Senegal, michielerikdevries@gmail.com

M.K. Diallo International Crops Research Institute for the Semi-Arid Tropics (ICRISAT), Niamey, Niger, m.diallo@icrisatne.ne

A. Diallo African Livelihoods Program, CIMMYT, Nairobi, Kenya, a.o.diallo@cgiar.org

J. Diels Division of Soil and Water Management, Department of Land Management and Economics, KU Leuven, Leuven, Belgium, jan.diels@ees.kuleuven.be

J. Dimes International Crops Research Institute for the Semi Arid Tropics (ICRISAT), Bulawayo, Zimbabwe, j.dimes@cgiar.org

M. Diouf TROPICASEM/TECHNISEM, Km 5,6 Bd du Centenaire de la commune de Dakar, BP 999 Dakar, Sénégal, meissa.diouf@tropicasem.sn

P. Dugue CIRAD TERA, Montpellier Cedex 01, France, patrick.dugue@cirad.fr

J.W. Duindam International Institute of Tropical Agriculture, Yaoundé, Cameroon, jelleduindam@hotmail.com

W. Ego Kenya Agricultural Research Institute, Kiboko Research Centre, Makindu, Kenya; Desert Margins Programme, Kenya Agricultural Research Institute (KARI), Nairobi, Kenya, egowk@yahoo.com

F. Ekeleme Michael Okpara University of Agriculture, Umudike, Nigeria, fekeleme@yahoo.co.uk

I. Ekise Tropical Soil Biology and Fertility Institute of the International Centre for Tropical Agriculture (TSBF-CIAT), Nairobi, Kenya, iekise@hotmail.com

A.O. Esilaba Desert Margins Programme, Kenya Agricultural Research Institute (KARI), Nairobi, Kenya, aoesilaba@kari.org; aesilaba@gmail.com

F.-X. Etoa Department of Biochemistry, University of Yaoundé I, 812, Yaoundé, Cameroon, fxetoa@yahoo.fr

A. Fall Institut Sénégalais de Recherches Agricoles (ISRA), Route des Hydrocarbures Bel-Air, BP 3120 Dakar, Senegal, fallalio@refer.sn

H. Fankem Department of Plant Biology, Faculty of Science, University of Douala, P.O. Box 24157 Douala, Cameroon, fankemhenri@yahoo.fr

J.-C. Fardeau Département Environnement et Agronomie, INRA, Versailles, France, fardeau@versailles.inr.fr

A. Farrow International Center for Tropical Agriculture, Kampala, Uganda, a.farrow@cgiar.org

D. Fatondji International Crops Research Institute for the Semi-Arid Tropics (ICRISAT) Sahelian Center, Niamey, Niger, d.fatondji@cgiar.org; d.fantondji@gmail.com

A.O. Fatunbi Agronomy Department, Agricultural and Rural Development Research Institute (ARDRI), University of Fort Hare, Eastern Cape, South Africa, afatunbi@ufh.ac.za

C. Feller ENGREF:DFRT/UR IRD 179 SeqBio, Montpellier Cedex, France, feller@ird.fr

A.M. Fermont International Institute of Tropical Agriculture (IITA-Uganda), Kampala, Uganda, a.fermont@cgiar.org

S. Ferris Agriculture and Environment, Catholic Relief Services (CRS), Baltimore, MD, USA, shaun.ferris@crs.org

R. Fogain Centre Africain de Recherche sur le Bananier et Plantain (CARBAP), Njombé, Cameroon, rfogain7@yahoo.fr

M. Fosu CSIR-Savanna Agricultural Research Institute, Tamale, Ghana, mathiasfosu@yahoo.co.uk

B. Freyer Division of Organic Farming, University of Natural Resources and Applied Life Sciences, Vienna, Austria, bernhard.freyer@boku.ac.at

C.K.K. Gachene University of Nairobi, Nairobi, Kenya, gachene@uonbi.ac.ke

G.N. Gachini Kenya Agricultural Research Institute (KARI), National Agricultural Research Laboratories, Nairobi, Kenya, kss@iconnect.co.ke

M. Gafishi Institut des Sciences Agronomiques du Rwanda (ISAR), Musanze, Rwanda, mkgafishi@yahoo.fr

M. Gandah The Regional coordinator of the AGRA funded microdosing project, (ICRISAT), Niamey, Niger, m.gandah@cgiar.org

K.E. Gathoni Department of Environmental Science, Kenyatta University, Nairobi, Kenya, gathoni_edith@yahoo.com

S. Gaye Sahel Regional Station, Africa Rice Centre, BP 96 Saint Louis, Senegal, soulgaye@hotmail.com

R. Gentile Department of Plant Sciences, University of California, One Shields Ave., Davis, CA, USA; Departments of Agronomy and Range Science, University of California, Davis, CA, USA, rgentile@ucdavis.edu

B. Gerard International Livestock Research Institute (ILRI), Addis Ababa, Ethiopia, b.gerard@cgiar.org

E.W. Gikonyo Institute of Tropical Agriculture, Universiti Putra Malaysia, Serdang, Selangor, Malaysia, estgikonyo@yahoo.com

P. Gildemacher International Potato Center, Nairobi, Kenya, p.gildemacher@kit.nl

K.E. Giller Plant Production Systems, Department of Plant Sciences, Wageningen University, Wageningen, The Netherlands, ken.giller@wur.nl

H.K. Githinji Department of Soil Science, Moi University, Eldoret, Kenya, harrikag@yahoo.com

C.M. Githunguri Katumani Research Centre, Kenya Agricultural Research Institute, Machakos, Kenya, cgithunguri@kari.org; cyrusgithunguri@yahoo.com

S.O. Gudu Department of Soil Science, Moi University, Eldoret, Kenya, samgudu2002@yahoo.com

T. Gueye Ecole Nationale Supérieure d'Agriculture (ENSA)/Université de Thiès, BP A296, Thiès, Sénégal, tgueye@univ-thies.sn

B.K. Gugino Penn State Cooperative Extension, University Park, PA 16802, Ithaca, NY, USA, bkgugino@psu.edu

M.M. Hanafi Institute of Tropical Agriculture, Universiti Putra Malaysia, Serdang, Selangor, Malaysia, mmhanafi@Agri.upm.edu.my

O. Hassane International Crops Research Institute for the Semi-Arid Tropics (ICRISAT), Niamey, Niger, o.hassane@icrisatne.ne

S. Hauser International Institute of Tropical Agriculture, Kinshasa, Democratic Republic of Congo, s.hauser@cgiar.org

K. Hayashi JIRCAS, Tsukuba, Ibaraki, Japan; ICRISAT West & Central Africa, Niamey, Niger, khayash@jircas.affrc.go.jp

G. Heinrich Agriculture and Environment, Catholic Relief Services (CRS), Baltimore, MD, USA, gheinrich@saro.crs.org

E. Hien SVT Department, University of Ouagadougou, Ouagadougou, Burkina Faso; Université de Ouagadougou, UFR/SVT, Ouagadougou 03, Burkina Faso, edmond.hien@ird.fr

V. Hien INERA/CREAF, Kamboinse, Burkina Faso, vhien@ird.bf; vhien@fasonet.bf

L. Hitimana Secretariat of the Sahel and West Africa Club (SWAC/OECD), 2 rue André Pascal, 75775 Paris, Cedex 16, France, leonidas.hitimana@oecd.org

T. Hongo Kenyatta University, Nairobi, Kenya, tamollo@yahoo.com

L.N. Horn Division of Plant Production Research, Ministry of Agriculture, Water and Forestry, Government Office Park, Luther Str. Windhoek, Namibia, lnhorn@yahoo.com

P. Houngnandan Faculté des Sciences Agronomiques (FSA), Université d'Abomey-Calavi (UAC), Recette Principale, Cotonou, Benin

L. Hove International Crops Research Institute for the Semi Arid Tropics (ICRISAT), Bulawayo, Zimbabwe

J. Huising Tropical Soil Biology and Fertility Institute of the International Centre for Tropical Agriculture (TSBF-CIAT), Nairobi, Kenya, j.huising@cgiar.org

T. Hyuha Department of Agricultural Economics, Makerere University, Kampala, Uganda, thyuha@yahoo.com; thyuha@mak.ac.ug

O.J. Idowu Department of Extension Plant Sciences, New Mexico State University, Las Cruces, NM 88011, USA, jidowu@ad.nmsu.edu

S.T. Ikerra Mlingano Agricultural Research Institute, Tanga, Tanzania, susikera@yahoo.com

E.C. Ikitoo Department of Soil Science, Moi University, Eldoret, Kenya, ikitoo.caleb@gmail.com

J.K. Itabari Katumani Research Centre, Kenya Agricultural Research Institute, Machakos, Kenya, itabarijustus@yahoo.co.uk

O. Ito JIRCAS, Tsukuba, Ibaraki, Japan, osamuito@jircas.affrc.go.jp

M. Jemo International Institute of Tropical Agriculture, Humid Forest Ecoregional Centre (HFC), Yaounde, Cameroon, m.jemo@cgiar.org; mjemo2001@yahoo.com

A.S. Jeng Soil & Environment Division, Bioforsk – Norwegian Institute for Agricultural and Environmental Research, Fredrik A Dahls vei 20A, N-1432 Aas, Norway, Alhaji.Jeng@bioforsk.no

B. Junge University of Oldenburg, Germany, birte.junge@uni-oldenburg.de

V.H. Kabambe Bunda College of Agriculture, Lilongwe, Malawi, kabambev@yahoo.com

W.T. Kabore Université de Ouagadougou, UFR/SVT, Ouagadougou 03, Burkina Faso; IRD, UR SeqBio, DMP Program, Ouagadougou, Burkina Faso, kathewin@yahoo.fr

A. Kabuli Soil Fertility Consortium for Southern Africa, Bunda College of Agriculture, Lilongwe, Malawi, amonmw@yahoo.com

H.J. Kabuli Department of Agricultural Research, Chitedze Research Station, Ministry of Agriculture and Food Security, Lilongwe, Malawi, hjinazali@yahoo.com

E. Kaganzi CIAT/Enabling Rural Innovation (ERI), Kampala, Uganda, e.kaganzi@cgiar.org

K.C. Kaizzi Kawanda Agricultural Research Institute (KARI), National Agricultural Research Organization (NARO), Kampala, Uganda, kckaizzi@gmail.com

M.C. Kalumuna Agricultural Research Institute, Mlingano, Tanga, Tanzania, M.C.Kalumunakokwijuka@yahoo.co.uk

D. Kamalongo Chitedze Research Station, Department of Agricultural Research Services, Ministry of Agriculture and Food Security, Lilongwe, Malawi, dkamalongo@hotmail.com

A.Y. Kamara International Institute of Tropical Agriculture, Ibadan, Nigeria, akamara@cgiar.org

J. Kamau Kenyatta University, Nairobi, Kenya, joycekamau28@yahoo.com

W.M.H. Kamiri Tropical Soil Biology and Fertility Institute of the International Centre for Tropical Agriculture (TSBF-CIAT), Nairobi, Kenya, wangechikamiri@yahoo.com

R. Kamugisha African Highland Initiative, Kampala, Uganda, rkamu2000@yahoo.co.uk

F. Kanampiu CIMMYT, Nairobi, Kenya, f.kanampiu@cgiar.org

N.K. Karanja University of Nairobi, Nairobi, Kenya, nancy.karanja@cgiar.org

R. Karega School of Environmental Studies and Human Sciences, Kenyatta University, Nairobi, Kenya, rkarega@yahoo.co.uk

A.N. Kathuku Kenya Agricultural Research Institute, National Agricultural Research Centre, Nairobi, Kenya; Desert Margins Programme, Nairobi, Kenya, angelandan2000@yahoo.com

P. Kathuli Katumani Research Centre, Kenya Agricultural Research Institute, Machakos, Kenya, peterkathuli@yahoo.com

A.A. Kauwa (Deceased) Chitedze Research Station, Department of Agricultural Research Services, Ministry of Agriculture and Food Security, Lilongwe, Malawi

A. Kavatha Land O' Lakes Regional Office, Nairobi, Kenya, agnes@landolakes.co.ke

A. Kavoo Tropical Soil Biology and Fertility Institute of the International Centre for Tropical Agriculture (TSBF-CIAT), Nairobi, Kenya, agneskavoo@yahoo.com

G.A. Keya Desert Margins Programme, Kenya Agricultural Research Institute, Nairobi, Kenya, gakeya@kari.org

Z. Khan ICIPE, Mbita, Kenya, zkhan@icipe.org

M.J. Khaemba Department of Crops, Horticulture and Soil Sciences, Egerton University, Egerton, Kenya, khaemba03@yahoo.com

G. Khisa Ministry of Agriculture, Kakamega, Kenya, khisagodrick@yahoo.co.uk

I.D. Kiba Institute of the Environment and Agricultural Research Institute (INERA), Ouagadougou, Burkina Faso, ikiba@yahoo.fr

C.N. Kibunja Kenya Agricultural Research Institute, NARL-KARI, Nairobi, Kenya, catherine.kibunja@yahoo.com

J.N. Kigomo Kenya Forestry Research Institute, Nairobi, Kenya, kigomo2@yahoo.com

F.M. Kihanda Kenya Agricultural Research Institute (KARI), Embu Regional Research Centre, Embu, Kenya, kihandafm@yahoo.com

J. Kihara Zentrum für Entwicklungsforschung (ZEF), University of Bonn, Bonn, Germany; Tropical Soil Biology and Fertility Institute of the International Centre for Tropical Agriculture (TSBF-CIAT), Nairobi, Kenya, jkiharam@yahoo.com; j.kihara@cgiar.org

S.K. Kimani Kenya Agricultural Research Institute, National Agricultural Research Centre, Nairobi, Kenya, skimani@africaonline.co.ke

P.K. Kimani Kenya Soil Survey (KSS), Kenya Agricultural Research Institute (KARI), Nairobi, Kenya, pkkims@yahoo.com

J. Kimiywe Kenyatta University, Nairobi, Kenya, jokimiywe@yahoo.com

L.M. Kimotho Katumani Research Centre, Kenya Agricultural Research Institute, Machakos, Kenya, klmkimotho@yahoo.com

D.M. Kinfack Centre Africain de Recherche sur le Bananier et Plantain (CARBAP), Njombé, Cameroon; Faculty of Science and Biotechnology Centre, University of Yaoundé I, 812, Yaoundé, Cameroon, kinfack29@yahoo.fr

B. King'olla Tropical Soil Biology and Fertility Institute of the International Centre for Tropical Agriculture (TSBF-CIAT), Nairobi, Kenya, b.wawaka@cgiar.org

Z.M. Kinyua National Agricultural Laboratories, Kenya Agricultural Research Institute, Nairobi, Kenya, kinyuazm@gmail.com

M.J. Kipsat Department of Marketing and Economics, Moi University, Eldoret, Kenya, mjkipsat@yahoo.com

J. Kirui Land O' Lakes Regional Office, Nairobi, Kenya, j.kirui@cgiar.org

E.G. Kirumba Department of Environmental Science, Kenyatta University, Nairobi, Kenya, gathoni_edith@yahoo.com

M. Kisaka-Lwayo Agricultural Economics Discipline, School of Agricultural Sciences and Agribusiness, University of KwaZulu-Natal, Pietermaritzburg, South Africa, maggiekisaka@yahoo.com

P.O. Kisinyo Department of Soil Science, Moi University, Eldoret, Kenya, kisinyopeter@yahoo.com

E. Kituyi Department of Chemistry, University of Nairobi, Nairobi, Kenya, ekituyi@uonbi.ac.ke

S. Koala AfNet-TSBF, International Center for Tropical Agriculture (CIAT), Nairobi, Kenya, s.koala@cgiar.org

I. Kone Centre Swisse de Recherche Scientifique en Côte d'Ivoire (CSRS), Abidjan, Côte d'Ivoire; University of Cocody, Abidjan, Côte d'Ivoire, inza.kone@csrs.ci

A. Kone Institute of the Environment and Agricultural Research Institute (INERA), Ouagadougou, Burkina Faso, Kone.adama@reseaucrepa.org

J.B. Kung'u Department of Environmental Sciences, School of Environmental Studies, Kenyatta University, Nairobi, Kenya, kungu_james@yahoo.com

T.K. Kwambai National Agricultural Research Centre, Kenya Agricultural Research Institute, Kitale, Kenya, tkkwambai2003@yahoo.com

J.D. Kwari Department of Soil Science, University of Maiduguri, Maiduguri, Nigeria, jdkwari@yahoo.co.uk

K. Kwena Kenya Agricultural Research Institute, Katumani Research Centre, Machakos, Kenya, kwenakizito@yahoo.com

P.K. Kyakaisho District Agriculture and Livestock Office, Muheza, Tanzania, kassian03@yahoo.co.uk; kennemma@hotmail.com

M. Larwanou Faculté d'Agronomie, Université Abdou Moumouni de Niamey, Niamey, Niger, m.larwanou@coraf.org

J.J. Lelei Department of Crops, Horticulture and Soils, Egerton University, Njoro, Kenya, Joycendemo@yahoo.com

J.K. Lelon Kenya Forestry Research Institute, Nairobi, Kenya, jklelon@yahoo.com

M. Lepage IRD, UR SeqBio, DMP Program, 01 BP 182, Ouagadougou, Burkina Faso, lepage@ird.bf

I. Ligowe Chitedze Research Station, Department of Agricultural Research Services, Ministry of Agriculture and Food Security, Lilongwe, Malawi, ivyligowe@yahoo.co.uk

F. Lompo Institute of Environment and Agricultural Research (INERA), Ouagadougou, Burkina Faso, lompoxa1@yahoo.fr

B.A. Lukuyu Kenya Agricultural Research Institute, Nairobi, Kenya, b.lukuyu@cgiar.org

L. Lunze Centre de Recherche de Mulungu, INERA, D.S. Bukavu, D.R. Congo, llunze@yahoo.fr

C.M. Lusweti Kenya Agricultural Research Institute (KARI), Kitale Centre, Kitale, Kenya, lusweticharles@yahoo.com

S. Lwasa Kampala City Council District Urban Agriculture Office, Kampala, Uganda, s.lwasa@cgiar.org

P.N. Macharia KARI-Kenya Soil Survey, Nairobi, Kenya, kss@iconnect.co.ke

A. Macharia Department of Environmental Science, School of Environmental Studies and Human Sciences, Kenyatta University, Nairobi, Kenya, amacharia@nema.go.ke

E.J. Maeda Ministry of Agriculture Food Security and Cooperatives, Tanzania, elizabeth.maeda@kilimo.go.tz

P. Mahposa International Crops Research Institute for the Semi Arid Tropics (ICRISAT), Bulawayo, Zimbabwe

F. Mairura Tropical Soil Biology and Fertility, Institute of the International Centre for Tropical Agriculture (TSBF-CIAT), Nairobi, Kenya, F.Mairura@cgiar.org

P. Makhosi National Agricultural Research Laboratories (NARL), Kampala, Uganda, landuse@infocom.co.ug

E. Makonese International Fertilizer Industry Association (IFA), Paris, France; Chemplex Corporation Ltd., Harare, Zimbabwe, makonesee@chemplex.co.zw

W. Makumba Department of Agricultural Research, Chitedze Research Station, Ministry of Agriculture and Food Security, Lilongwe, Malawi, w.makumba@africa-online.net

A. Malmer Department of Forest Ecology and Management, Swedish University of Agricultural Sciences (SLU), Umea, Sweden, Anders.Malmer@slu.se

M. Mamo Department of Agronomy and Horticulture, University of Nebraska Lincoln, Lincoln, NE, USA, Mmartha@unlnotes.unl.edu

A. Mando Division of Afrique, An International Center for Soil Fertility and Agricultural Development (IFDC), Lome, Togo, amando@ifdc.org

N. Mangale Kenya Agricultural Research Institute, Katumani Research Centre, Machakos, Kenya, nmangale2005@yahoo.com

R.J. Manlay ENGREF:DFRT/UR IRD 179 SeqBio, Montpellier Cedex, France, raphael.manlay@agroparistech.fr

P. Mapfumo Department of Soil Science and Agricultural Engineering, University of Zimbabwe, Mount Pleasant, Harare, Zimbabwe; Soil Fertility Consortium for Southern Africa (SOFECSA), CIMMYT, Southern Africa, Mount Pleasant, Harare, Zimbabwe, p.mapfumo@cgiar.org

A.E.T. Marandu Mlingano Agricultural Research Institute, Tanga, Tanzania, atanasiom@yahoo.co.uk

D. Marchal FAO Projet Intrants, Niamey, Niger, Paule.marchal@laposte.net

H.K. Maritim (Deceased) Department of Soil Science, Moi University, Eldoret, Kenya

C. Martius Zentrum für Entwicklungsforschung (ZEF), University of Bonn, Bonn, Germany, c.martius@cgiar.org

N. Mashingaidze International Crops Research Institute for the Semi Arid Tropics (ICRISAT), Bulawayo, Zimbabwe

D. Masse IRD Institut de Recherche pour le Développement, UR 179 SeqBio, Montpellier Cedex, France; IRD, UR SeqBio, DMP Program, Ouagadougou, Burkina Faso, dominique.masse@ird.fr

K.F.G. Masuki African Highland Initiative, Kampala, Uganda, k.masuki@cgiar.org

E.N. Masvaya Department of Soil Science and Agricultural Engineering, University of Zimbabwe, Mt Pleasant, Harare, Zimbabwe, e.masvaya@cgiar.org

B.M. Mati Improved Management of Agricultural Water in Eastern & Southern Africa (IMAWESA), Nairobi, Kenya, b.mati@cgiar.org

F.M. Matiri Kenya Agricultural Research Institute (KARI), Embu, Kenya, francis_matiri@yahoo.com

R. Matsunaga JIRCAS, Tsukuba, Ibaraki, Japan; ICRISAT West & Central Africa, Niamey, Niger, ryoichi_matsunaga@affrc.go.jp

L.W. Mauyo Masinde Muliro University of Science and Technology, P.O. BOX 190-50100, Kakamega, Kenya, lmauyo@yahoo.com

G. Mbagaya Moi University, Eldoret, Kenya, gmbagaya@mu.ac.ke

D. Mbithe   Kenyatta University, Nairobi, Kenya, dorcusmbithe@yahoo.com

G.N. Mbure   Kenya Agricultural Research Institute, Nairobi, Kenya, mbureg@yahoo.com

M.W.K. Mburu   Kenya Agricultural Research Institute (KARI), Embu, Kenya, mwambui2011@gmail.com

W. Merbach   Institute of Soil Science and Plant Nutrition, Martin-Luther University Halle-Wittenberg, Halle, Germany, merbach@landw.uni-halle.de

R. Merckx   Department of Earth and Environmental Sciences, Katholieke Universiteit Leuven, Kasteelpark Arenberg 20, 3001 Heverlee, Belgium, roel.merckx@ees.kuleuven.be

J.N. Methu   Land O' Lakes Regional Office, Nairobi, Kenya, j.methu@asareca.org

C. Milaho   District Agriculture and Livestock Office, Kilosa, Tanzania, cmilaho@yahoo.com

J.M. Miriti   Desert Margins Programme, Kenya Agricultural Research Institute, Nairobi, Kenya, jmmiriti@yahoo.co.uk; angelandan2000@yahoo.com

M. Misiko   Tropical Soil Biology and Fertility Institute of the International Centre for Tropical Agriculture (TSBF-CIAT), Nairobi, Kenya, m.misiko@cgiar.org

R. Miura   Kyoto University, Kyoto, Japan, miurar@kais.kyoto-u.ac.jp

C.Z. Mkangwa   Ilonga Agricultural Research Institute, Kilosa, Tanzania, mkangwa@yahoo.co.uk

P.N.S. Mnkeni   Faculty of Science and Agriculture, University of Fort Hare, Eastern Cape, South Africa, pmnkeni@ufh.ac.za

B.N. Moebius-Clune   Department of Crop and Soil Sciences, Cornell University, Ithaca, NY, USA, bnm5@cornell.edu

A.U. Mokwunye   United Nations University (UNU), Institute for Natural Resources in Africa, Accra, Ghana, uzo.mokwunye@alumni.illinois.edu

T. Mombeyarara   Tropical Soil Biology and Fertility Institute of the International Centre for Tropical Agriculture (TSBF-CIAT), Harare, Zimbabwe, t.mombeyarara@cgiar.org

J.-L. Morel   LSE/ENSAIA, Vandoeuvre-lès-Nancy Cedex, France, Jean-Louis.Morel@ensaia.inpl-nancy.fr

J.G. Mowo   African Highland Initiative, Kampala, Uganda, j.mowo@cgiar.org

M. Moyo   International Crops Research Institute for the Semi Arid Tropics (ICRISAT), Bulawayo, Zimbabwe

J.P. Mrema   Mlingano Agricultural Research Institute, Tanga, Tanzania; Department of Soil Science, Sokoine University of Agriculture, Morogoro, Tanzania, jmrema@suanet.ac.tz

F. Mtambanengwe   Department of Soil Science and Agricultural Engineering, University of Zimbabwe, Mount Pleasant, Harare, Zimbabwe, fmtamba@agric.uz.ac.zw

R.M. Muasya Department of Seed, Crops and Horticultural Sciences, Moi University, Eldoret, Kenya, rmuasya@africaonlinc.co.ke

M. Mucheru-Muna Department of Environmental Sciences, School of Environmental Studies, Kenyatta University, Nairobi, Kenya, moniquechiku@yahoo.com

R.J. Mugabo Institut des Sciences Agronomiques du Rwanda (ISAR), Musanze, Rwanda, mugabojosa@yahoo.fr

D.N. Mugendi Department of Environmental Sciences, School of Environmental Studies, Kenyatta University, Nairobi, Kenya, dmugendi@yahoo.com

J. Mugwe Department of Agricultural Resource management, School of Agriculture and Enterprise Development (SAED), Kenyatta University, Nairobi, Kenya, jaynemugwe@yahoo.com

L. Muhammad Kenya Agricultural Research Institute, Katumani Research Centre, Machakos, Kenya, luttam2002@yahoo.com

J. Mukalama Tropical Soil Biology and Fertility Institute of the International Centre for Tropical Agriculture (TSBF-CIAT), Nairobi, Kenya, jmukalama@yahoo.com

A. Mukuralinda World Agroforestry Centre (ICRAF), Rwanda, mukuratha@yahoo.com

J. Mulatya Kenya Forestry Research Institute, Nairobi, Kenya, director@kefri.org

G. Muluvi Kenyatta University, Nairobi, Kenya, muluvi.geoffrey@ku.ac.ke

L.M. Mumera Department of Crops, Horticulture and Soil Sciences, Egerton University, Egerton, Kenya, lmmumera@africaonline.com; dvcaf@egerton.ac.ke

G.S. Mumina Kenya Agricultural Research Institute, National Arid Lands Research Centre, Marsabit, Kenya; Egerton University, Njoro, Kenya, muminalu@yahoo.com

J.M. Mungatu Tropical Soil Biology and Fertility Institute of the International Centre for Tropical Agriculture (TSBF-CIAT), c/o ICRAF, UN Avenue, Gigiri, Nairobi, Kenya, kmungatu@yahoo.com

J.W. Munyasi Kenya Agricultural Research Institute, Kiboko Research Centre, Makindu, Kenya, munyasijoseph@yahoo.com

S.W. Munyiri Department of Crop, Horticulture and Soils, Egerton University, Egerton, Njoro, Kenya, wanja_munyiri@yahoo.co.uk

W. Mupangwa International Crops Research Institute for the Semi Arid Tropics (ICRISAT), Bulawayo, Zimbabwe

P. Mureithi Department of Environmental Studies and Community Development, Kenyatta University, Nairobi, Kenya, petmukariuki@yahoo.co.uk

J.G. Mureithi KARI Headquarters, Nairobi, Kenya, JGMureithi@kari.org

J. Muriuki District Agricultural Office, Meru South District, Ministry of Agriculture, Chuka, Kenya, justinmuriuki@yahoo.co.uk

E. Murua Tropical Soil Biology and Fertility Institute of the International Centre for Tropical Agriculture (TSBF-CIAT), c/o ICRAF, UN Avenue, Gigiri, Nairobi, Kenya, libbymurua2002@yahoo.com

H.K. Murwira Tropical Soil Biology and Fertility Institute of the International Centre for Tropical Agriculture (TSBF-CIAT), Harare, Zimbabwe, h.murwira@cgiar.org

D.K. Musembi Kenya Agricultural Research Institute (KARI), Kiboko Research Centre, Makindu, Kenya, dkmusembi@yahoo.com

D. Mushabizi Institut des Sciences Agronomiques du Rwanda (ISAR), Musanze, Rwanda, mushabizi@yahoo.fr

C. Musharo Department of Soil Science and Agricultural Engineering, University of Zimbabwe, Mount Pleasant, Harare, Zimbabwe, cmusharo@yahoo.com

J.K. Mutegi World Agroforestry Centre (ICRAF), Nairobi, Kenya, mutegijames@yahoo.com

J.M. Muthamia Department of Environmental Sciences, Kenyatta University, Nairobi, Kenya, muthamiajoses@yahoo.com

L.M. Mutuku Katumani Research Centre, Kenya Agricultural Research Institute, Machakos, Kenya, mwangagilawrence@yahoo.com

G.M. Muturi University of Nairobi, Nairobi, Kenya, gmuturi@kefri.org

E.M. Muya National Agricultural Research Laboratories, Kenya Agricultural Research Institute (KARI), Nairobi, Kenya, edwardmuya@yahoo.com; kss@iconnect.co.ke

R. Muzira National Agricultural Research Organization, Mbarara, Uganda, nrmuzira@yahoo.com

M. Mwala School of Agricultural Sciences, University of Zambia, Lusaka, Zambia, mmwala@yahoo.com

C.D. Mwale Chitedze Research Station, Department of Agricultural Research Services, Ministry of Agriculture and Food Security, Lilongwe, Malawi, cyprianmwale@yahoo.com

D. Mwangi Kenya Agricultural Research Institute, Nairobi, Kenya, DMMwangi@kari.org

F.B. Mwaura University of Nairobi, Nairobi, Kenya, fbmwaura@uonbi.ac.ke

S.W. Mwendia Kenya Agricultural Research Institute, Nairobi, Kenya, smwendia@une.edu.au

S.M. Mwonga Department of Crops, Horticulture and Soil Sciences, Egerton University, Egerton, Kenya, smwonga@yahoo.com

N.L. Nabahungu ISAR-Rwanda, Butare, Rwanda, nabahungu@yahoo.com

L. Nakhone Department of Crops, Horticulture and Soil Sciences, Egerton University, Egerton, Kenya, lenahnakhone@yahoo.com

I. Nalukenge Department of Agricultural Economics, Makerere University, Kampala, Uganda, imeldanalukenge@yahoo.com; inalukenge@mak.ac.ug

B. Ncube WATERnet, Department of Civil Engineering, University of Zimbabwe, Harare, Zimbabwe

P.A. Ndakidemi Research & Technology Promotion, Cape Peninsula University of Technology, Keizersgracht, Cape Town, South Africa, ndakidemip@cput.ac.za

M. Ndiaye Africa Rice Center (AfricaRice), Sahel Regional Station, Saint-Louis BP 96, Senegal, a_malick_nd@yahoo.fr

J. Ndjeunga International Crops Research Institute for the Semi-Arid Tropics (ICRISAT) Sahelian Center, Niamey, Niger, ndjeungajupiter@gmail.com

J.K. Ndufa Kenya Forestry Research Institute, Gede Regional Research Centre, Malindi, Kenya, jndufa@africaonline.co.ke

A.O. Nekesa Department of Soil Science, Moi University, Eldoret, Kenya, amarishas@yahoo.com

H. Nezomba Department of Soil Science and Agricultural Engineering, University of Zimbabwe, Mount Pleasant, Harare, Zimbabwe, hnezomba@agric.uz.ac.zw

M.N. Ng'ang'a Department of Seed, Crops and Horticultural Sciences, Moi University, Eldoret, Kenya, marionnduta@yahoo.com

J.K. Ng'ang'a Kenya Agricultural Research Institute, National Agricultural Research Centre, P.O. Box 14733, Nairobi, Kenya, joxkiarie@yahoo.com

G.N. Ngae Kenya Agricultural Research Institute, Nairobi, Kenya, gnngae@gmail.com

M. Ngongo Centre de Recherche de Mulungu, INERA, D.S. Bukavu, D.R. Congo, nmulangwa@yahoo.com

M. Ngutu Kenya Agricultural Research Institute, National Arid Lands Research Centre, Marsabit, Kenya; Egerton University, Njoro, Kenya, mnthiani@yahoo.com

A. Niang Africa Rice Center (WARDA), Cotonou, Benin, abibou.niang1@cgiar.org

A. Nikiema International Crops Research Institute for the Semi-Arid Tropics (ICRISAT) Sahelian Center, Niamey, Niger, a.nikiema@cgiar.org; a.nikiema@gmail.com

P.M. Njingulula Socio-economist INERA-Mulungu, Kivu, DR Congo, pnjingulula@yahoo.fr

L. Ngo Nkot Department of Plant Biology, University of Douala, 24157, Douala, Cameroon, lnkot@yahoo.fr

C. Nolte FAO, Plant Production and Protection Division (AGP), Rome, Italy, cknmail@yahoo.co.uk; Christian.nolte@fao.org

E. Nsengumuremyi ISAR-Rwanda, Butare, Rwanda, nsenguemile@yahoo.fr

D. Nwaga Faculty of Science and Biotechnology Centre, University of Yaoundé I, 812, Yaoundé, Cameroon, dnwaga@yahoo.fr

O.C. Nwoke Department of Agronomy, Osun State University, Osogbo, Nigeria,
c.nwoke@cgiar.org; o.chik@yahoo.com

I. Nyagumbo Department of Soil Science and Agricultural Engineering, University
of Zimbabwe, Mount Pleasant, Harare, Zimbabwe, inyagumbo@agric.uz.ac.zw

A.S. Nyaki Mlingano Agricultural Research Institute, Tanga, Tanzania,
adolfnyaki@yahoo.com

J. Nyamangara Department of Soil Science and Agricultural Engineering,
University of Zimbabwe, Mount Pleasant, Harare, Zimbabwe; Chitedze Research
Station, Tropical Soil Biology and Fertility Institute of the International Centre for
Tropical Agriculture (TSBF-CIAT), Lilongwe, Malawi, jnyamangara@yahoo.co.uk;
j.nyamangara@cgiar.org

D.M. Nyariki University of Nairobi, Nairobi, Kenya, Dicksonnyariki@yahoo.com

R.W. Nyawasha Department of Soil Science and Agricultural Engineering,
University of Zimbabwe, Mt Pleasant, Harare, Zimbabwe,
rwnyawasha@yahoo.co.uk

G. Nyberg Department of Forest Ecology and Management, Swedish University of
Agricultural Sciences (SLU), SE-901 83, Umea, Sweden, Gert.Nyberg@slu.se

G. Nziguheba Soil Research Laboratory, International Institute of Tropical
Agriculture, Ibadan, Nigeria, gnziguheba@iri.columbia.edu

J.N. Nzomoi Central Bank of Kenya, Nairobi, Kenya, nnzomoi@yahoo.co.uk

G. Obare Department of Agricultural Economics and Agri-Business Management,
Egerton University, Njoro, Kenya, ga.obare@yahoo.com; g.obare@hotmail.com

A. Obi Department of Agricultural Economics and Extension, University of Fort
Hare, Alice, Eastern Cape, South Africa, aobi@ufh.ac.za

J. Obua Makerere University, Kampala, Uganda, j.obua@vicres.net

J.O. Ochoudho School of Agriculture and Biotechnology, Moi University, Eldoret,
Kenya, juliusochuodho@yahoo.com

M. Odendo Kenya Agricultural Research Institute (KARI), Regional Research
Centre, Kakamega, Kenya, Odendos@yahoo.com

J.A. Odhiambo Department of Crops, Horiculture and Soils, Egerton University,
Egerton, Kenya, otishanan@yahoo.com

A.J. Odofin School of Agriculture and Agricultural Technology, Federal University
of Technology Minna, Minna, Niger State Nigeria, ayoodofin@yahoo.com

O.V. Oeba Kenya Forestry Research Institute, Gede Regional Research Centre,
Malindi, Kenya, voeba@yahoo.co.uk

O. Ohiokpehai Tropical Soil Biology and Fertility Institute of the International
Centre for Tropical Agriculture (TSBF-CIAT), Nairobi, Kenya,
oohiokpehai@yahoo.com

S.O. Oikeh Africa Rice Center (WARDA), Cotonou, Benin,
S.Oikeh@aatf-africa.org

J.R. Okalebo Department of Soil Science, Moi University, Eldoret, Kenya, rokalebo@yahoo.com

M.M. Okeyo Kenya Forestry Research Institute (KEFRI), Nairobi, Kenya; Londiani Regional Research Centre, Londiani, Kenya, mikemairura@yahoo.com

P.F. Okoth Tropical Soil Biology and Fertility Institute of the International Centre for Tropical Agriculture (TSBF-CIAT), UN Avenue, Gigiri, Nairobi, Kenya, p.okoth@cgiar.org

M. Okoti Desert Margins Programme, Kenya Agricultural Research Institute (KARI), Nairobi, Kenya, michaeldominion2003@yahoo.com

G. Olukoye (Deceased) Kenyatta University, Nairobi, Kenya

E. Omami Department of Seed, Crops and Horticultural Sciences, Moi University, Eldoret, Kenya, elizabethomami@yahoo.com

L. Omoigui University of Agriculture, Makurdi, Nigeria, lomoigui@yahoo.co.uk

J.O. Omollo Agronomy Programme, Crop Development Department, Kenya Sugar Research Foundation (KESREF), Kisumu, Kenya, jac.omollo@gmail.com

W.O. Omondi Kenya Forestry Research Institute (KEFRI), Nairobi, Kenya, williamomondi2004@yahoo.co.uk

M.E. Omunyin Department of Seed, Crops and Horticultural Sciences, Moi University, Eldoret, Kenya, omunyinem2712@hotmail.com

R.N. Onwonga Department of Land Resource Management and Agricultural Technology, University of Nairobi, Nairobi, Kenya, ronwonga@yahoo.com

J.W. Onyango Irrigation and Drainage Research Programme, Kenya Agricultural Research Institute (KARI), Nairobi, Kenya, joabwamari@yahoo.com

C. Opondo African Highland Initiative, Kampala, Uganda, opondo08@yahoo.com

A.O. Osunde School of Agriculture and Agricultural Technology, Federal University of Technology Minna, Minna, Niger State Nigeria, akimosunde@yahoo.co.uk

L. Oteba Kenyatta University, Nairobi, Kenya, oteba-lawrence@yahoo.com

C.O. Othieno Department of Soil Science, Moi University, Eldoret, Kenya, cotieno19@yahoo.com

S. Otor Department of Environmental Science, School of Environmental Studies and Human Sciences, Kenyatta University, Nairobi, Kenya, cjotor@yahoo.com

N. Ouandaogo Institute of the Environment and Agricultural Research (INERA), Ouagadougou, Burkina Faso, ouandaogo_noufou@yahoo.fr

K. Ouattara Institute of Environment and Agricultural Research (INERA), 04 BP 8645 Ouagadougou 04, Burkina Faso, korodjouma_ouattara@hotmail.com

B. Ouattara Department of Natural Resource Management, Institute of Environment and Agricultural Research (INERA), 04 BP 8645 Ouagadougou 04, Burkina Faso, badiori.ouattara@coraf.org

T.G. Ouattara Bureau National des Sols (BUNASOLS), Ouagadougou, Burkina Faso, ouattarag@yahoo.fr

O. Owuor Moi University, Eldoret, Kenya, jbokeyo@yahoo.com

D. Pasternak International Crops Research Institute for the Semi-Arid Tropics (ICRISAT) Sahelian Center, Niamey, Niger, d.pasternak@cgiar.org

R.S. Pathak Department of Crop, Horticulture and Soils, Egerton University, Egerton, Njoro, Kenya, ramspathak@yahoo.com

M.A.R. Phiri Faculty of Development Studies, Bunda College of Agriculture, Lilongwe, Malawi, marphiri1996@yahoo.com

C.J. Pilbeam Cranfield University School of Management, Bedford, UK, colin.pilbeam@cranfield.ac.uk

P. Pote Department of Agricultural Economics and Extension, University of Fort Hare, Eastern Cape, South Africa, PPote@ufh.ac.za

P. Pypers Tropical Soil Biology and Fertility Institute of the International Centre for Tropical Agriculture (TSBF-CIAT), Nairobi, Kenya, p.pypers@cgiar.org

C. Quiros International Center for Tropical Agriculture (CIAT), Cali, Colombia, cquiros@cgiar.org

I. Rahimou Faculté des Sciences Agronomiques (FSA), Université d'Abomey-Calavi (UAC), Recette Principale, Cotonou, Benin

H. Recke Kenya Agricultural Research Institute Headquarters, European Union Coordination Unit, Nairobi, Kenya, h.recke@cgiar.org

C. Reij Vrije Universiteit Amsterdam, Amsterdam, The Netherlands, CP.Reij@cis.vu.nl

T. Remington Agriculture and Environment, Catholic Relief Services (CRS), Baltimore, MD, USA, tom.remington@crs.org

D. Rohrbach World Bank, Lilongwe, Malawi

D.L. Rowell Department of Soil Science, The University of Reading, Reading, UK, d.l.rowell@reading.ac.uk

J. Rurinda Department of Soil Science and Agricultural Engineering, University of Zimbabwe, Harare, Zimbabwe, jairurinda@yahoo.co.uk

J. Rusike Chitedze Research Station, International Institute for Tropical Agriculture, Lilongwe, Malawi

L. Rusinamhodzi Tropical Soil Biology and Fertility Institute of the International Centre for Tropical Agriculture (TSBF-CIAT), Harare, Zimbabwe, l.rusinamhodzi@cgiar.org

E.J. Rutto Department of Soil Science, Moi University, Eldoret, Kenya, emyruto@yahoo.com

S.M. Rwakaikara Department of Soil Science, Makerere University, Kampala, Uganda, mcrsilver2002@yahoo.co.uk

I.H. Rwiza Agriculture Research Institute Ukiriguru, Mwanza, Tanzania, rwizaih@hotmail.com

W.D. Sakala (Deceased) Chitedze Research Station, Department of Agricultural Research Services, Ministry of Agriculture and Food Security, Lilongwe, Malawi

G.M. Sakala Zambia Agriculture Research Institute, Mount Makulu Research Station, Chilanga, Zambia; Department of Soil Science, The University of Reading, Reading, UK, godfreysakala@yahoo.co.uk

K. Sako European Development for Rural Development (EUCORD)-Former Winrock International, Bamako, Mali, skaramoko@yahoo.fr

B. Salasya Kenya Agricultural Research Institute (KARI), Regional Research Centre, Kakamega, Kenya, beatsakwa@yahoo.com

N. Sanginga Tropical Soil Biology and Fertility Institute of the International Centre for Tropical Agriculture (TSBF-CIAT), UN Avenue, Gigiri, Nairobi, Kenya, n.sanginga@cgiar.org

P. Sanginga CIAT-Africa, Kawanda Agricultural Research Institute, Kampala, Uganda, Psanginga@idrc.or.ke

R.R. Schindelbeck Department of Crop and Soil Sciences, Cornell University, Ithaca, NY14853, USA, rrs3@cornell.edu

P.M. Sédogo Department of Natural Resource Management, Institute of Environment and Agricultural Research (INERA), 04 BP 8645 Ouagadougou 04, Burkina Faso, michel_sedogo@yahoo.fr

Z. Segda Institute of Environment and Agricultural Research (INERA), Ouagadougou, Burkina Faso, zacharie.segda@yahoo.fr

O. Semalulu National Agricultural Research Laboratories (NARL), Kampala, Uganda, o.semalulu@gmail.com

E. Semu Mlingano Agricultural Research Institute, Tanga, Tanzania; Department of Soil Science, Sokoine University of Agriculture, Morogoro, Tanzania, esemu@suanet.ac.tz; semu@yahoo.com

D. Senbeto International Crops Research Institute for the Semi-Arid Tropics (ICRISAT) Sahelian Center, Niamey, Niger, d.senbeto@cgiar.org

C. Serrem Department of Soil Science, Moi University, Eldoret, Kenya, cserrem@yahoo.com

R. Sheila African Farm Radio Research Initiative (AFFFRI), Developing Countries Farm Radio Network, Ottawa, ON, Canada, shrao@farmradio.org

M.G. Shibia Kenya Agricultural Research Institute (KARI), National Arid Lands Research Centre, Marsabit, Kenya, schibier@yahoo.com

H. Shinjo Kyoto University, Kyoto, Japan, shinhit@kais.kyoto-u.ac.jp

C.A. Shisanya Kenyatta University, Nairobi, Kenya, shisanya@yahoo.com

S.N. Silim Kenya Agricultural Research Institute (KARI), Embu, Kenya, s.silim@cgiar.org

J. Six Department of Plant Sciences, University of California, One Shields Ave., Davis, CA, USA; Departments of Agronomy and Range Science, University of California, Davis, CA, USA, jwsix@ucdavis.edu

D. Sogodogo Institut d' Economie Rurale (IER), Cinzana, Mali, sdiakalia@yahoo.fr

A. Sow Sahel Regional Station, Africa Rice Centre, BP 96 Saint Louis, Senegal, a.sow@cgiar.org

K. Stahr Institute of Soil Science and Land Evaluation, University of Hohenheim, Stuttgart, Germany, karl.stahr@uni-hohenheim.de

L. Stroosnijder Erosion and Soil & Water Conservation Group, Wageningen University, Wageningen, The Netherlands, leo.stroosnijder@wur.nl

K. Sukalac Information and Communications, IFA, Paris, France, ksukalac@fertilizer.org

I. Sumaila CSIR-Savanna Agricultural Research Institute, Tamale, Ghana, sumdanib@yahoo.com

R. Tabo Forum for Agricultural Research in Africa (FARA), Accra, Ghana, rtabo@fara-africa.org

I.M. Tabu Department of Crops, Horiculture and Soils, Egerton University, Egerton, Kenya, immtabu@yahoo.com

S. Tamani Bureau National des Sols (BUNASOLS), Ouagadougou, Burkina Faso, tamani.sohar@yahoo.fr

U. Tanaka Kyoto University, Kyoto, Japan, uerutnk@kais.kyoto-u.ac.jp

J. Tanui African Highland Initiative, Kampala, Uganda, j.tanui@cgiar.org

J.-B.S. Taonda Institut de l' Environnement et de Recherches Agricoles, INERA, Kamboinse, Ouagadougou, Burkina Faso, staonda2@yahoo.fr

T.P. Tauro Department of Soil Science and Agricultural Engineering, University of Zimbabwe, Mount Pleasant, Harare, Zimbabwe; Department of Research and Specialist Services (DR&SS), Chemistry and Soil Research Institute, Causeway, Harare, Zimbabwe, phirilani2@yahoo.co.uk

M. Tchienkoua Institut de la Recherche Agricole pour le Développement (IRAD), Yaounde, Cameroon, mtchienko@yahoo.com

A.J. Tenge The University of Dodoma, Dodoma, Tanzania, ajtenge@yahoo.co.uk

A. Tenkouano International Institute of Tropical Agriculture (IITA), Humid Forest Ecoregional Center, Yaoundé, Cameroon, a.tenkouano@cgiar.org

J.S. Tenywa Makerere University, Kampala, Uganda, jtenywa@agri.mak.ac.ug

C. Thierfelder CIMMYT Zimbabwe, Mount Pleasant, Harare, Zimbabwe, c.thierfelder@cgiar.org

M.N. Thuita Department of Soil Science, Moi University, Eldoret, Kenya, thuitam@yahoo.com

I. Tibo CSIR-Savanna Agricultural Research Institute, Tamale, Ghana, ibrahimtibo@yahoo.com

P. Tittonell Plant Production Systems, Department of Plant Sciences, Wageningen University, Wageningen, The Netherlands; Tropical Soil Biology and Fertility Institute of the International Centre for Tropical Agriculture (TSBF-CIAT), Nairobi, Kenya, Pablo.Tittonell@wur.nl

S. Tobita JIRCAS, Tsukuba, Ibaraki, Japan, bita1mon@jircas.affrc.go.jp

E. Tollens Faculty of Bioscience Engineering, Center for Agricultural and Food Economics, Catholic University of Leuven, Leuven, Belgium, Eric.Tollens@ees.kuleuven.be

K. Tomekpe Centre Africain de Recherche sur le Bananier et Plantain (CARBAP), Njombé, Cameroon, tomekpe@carbap-africa.org

A. Touré Africa Rice Center (WARDA), Cotonou, Benin

K. Traore Sahel Regional Station, Africa Rice Centre, BP 96 Saint Louis, Senegal, k.traore@cgiar.org

M. Traoré Bureau National des Sols (BUNASOLS), Ouagadougou, Burkina Faso, madouchef@yahoo.fr

G. Tsané International Institute of Tropical Agriculture (IITA), Humid Forest Ecoregional Center, Yaoundé, Cameroon; Faculty of Science and Biotechnology Centre, University of Yaoundé I, 812, Yaoundé, Cameroon, tsanegodefroy@yahoo.fr

A. Tschannen Centre Swisse de Recherche Scientifique en Côte d'Ivoire (CSRS), Abidjan, Côte d'Ivoire, andres.tschannen@gmail.com

S. Twomlow International Crops Research Institute for the Semi Arid Tropics (ICRISAT), Bulawayo, Zimbabwe, s.twomlow@cgiar.org

A. Uwiragiye ISAR, Kiruhura District, Butare, Rwanda, uwiragiyeambroise@yahoo.fr

H.M. van Es Department of Crop and Soil Sciences, Cornell University, Ithaca, NY, USA, hmv1@cornell.edu

B. Vanlauwe Tropical Soil Biology and Fertility Institute of the International Centre for Tropical Agriculture (TSBF-CIAT), Nairobi, Kenya, b.vanlauwe@cgiar.org

P. van Straaten School of Environmental Sciences, University of Guelph, Guelph, ON, Canada, pvanstra@uoguelph.ca

L.V. Verchot World Agroforestry Centre (ICRAF), Nairobi, Kenya, l.verchot@cgiar.org

P.L.G. Vlek Center for Development Research–ZEF, University of Bonn, Bonn, Germany, p.vlek@uni-bonn.de

P. Wakaba Kenya Agricultural Research Institute Muguga South, Nairobi, Kenya, petermwakaba@yahoo.com

C. Walela Kenya Forestry Research Institute, Gede Regional Research Centre, Malindi, Kenya, cwalela@une.edu.au

P.C. Wall CIMMYT, Harare, Zimbabwe, p.wall@cgiar.org

D.K. Wamae Kenya Agricultural Research Institute (KARI), Nairobi, Kenya, kdwamae@yahoo.com

J. Wamalwa Kenya Agricultural Research Institute, Nairobi, Kenya, jwamalwa60@yahoo.com

S.W. Wanderi Kenya Agricultural Research Institute (KARI), Embu, Kenya, wanderi_susan@yahoo.com

J.M. Wanyama Kenya Agricultural Research Institute (KARI), Kitale Centre, Kitale, Kenya, jmasindektl@yahoo.com

G.P. Warren Department of Soil Science, The University of Reading, Reading, UK, g.p.warren@reading.ac.uk

B.S. Waswa Tropical Soil Biology and Fertility Institute of the International Centre for Tropical Agriculture (TSBF-CIAT), Nairobi, Kenya, bswaswa@yahoo.com

F. Waswa Department of Environmental Planning, Management and Community Development, Kenyatta University, Nairobi, Kenya, wfuchaka@yahoo.com

L. Wekesa Kenya Forestry Research Station, Kibwezi, Kenya, weknus@yahoo.com

M. Welimo Kenya Forestry Research Institute, Gede Regional Research Centre, Malindi, Kenya, mwelimo@yahoo.com

G. Were Moi University, Eldoret, Kenya, gmwere@yahoo.com

K. Wilson Fletcher School of Law and Diplomacy, Tufts University, Boston, MA, USA, kimberley.wilson@tufts.edu

D.W. Wolfe Department of Horticulture, Cornell University, Ithaca, NY, USA, dww5@cornell.edu

L. Woltering International Crops Research Institute for the Semi-Arid Tropics (ICRISAT) Sahelian Center, Niamey, Niger, lennartwoltering@yahoo.com

P.L. Woomer FORMAT Kenya, Nairobi, Kenya, plwoomer@gmail.com

C.S. Wortmann Department of Agronomy and Horticulture, University of Nebraska Lincoln, Lincoln, NE, USA, cswortman@unlnotes.unl.edu

O. Yombo Faculty of Science and Biotechnology Centre, University of Yaoundé I, 812, Yaoundé, Cameroon, yombono@yahoo.fr

S. Youl IFDC-Ouaga 11 BP 82 CMS, Ouagadougou 11, Burkina Faso, syoul@ifdc.org

A.R. Zaharah Faculty of Agriculture, Department of Land Management, Universiti Putra Malaysia, Serdang, Selangor, Malaysia, Zaharah@Agri.upm.edu.my

S. Zingore Chitedze Research Station, Tropical Soil Biology and Fertility Institute of the International Centre for Tropical Agriculture (TSBF-CIAT), Lilongwe, Malawi, szingore@ipni.net

E. Zongo Bureau National des Sols (BUNASOLS), Ouagadougou, Burkina Faso, Zongoedouard60@yahoo.fr

R. Zougmoré Institute for Environment and Agricultural Research (INERA), Ouagadougou 04, Burkina Faso, robert.zougmore@messrs.gov.bf

S.J. Zoundi Secretariat of the Sahel and West Africa Club (SWAC/OECD), 2 rue André Pascal, 75775 Paris, Cedex 16, France, sibirijean.zoundi@oecd.org

Part I
Constraints and Opportunities
for the African Green Revolution

New Challenges and Opportunities for Integrated Soil Fertility Management in Africa

A. Bationo and B.S. Waswa

Abstract Sub-Saharan Africa (SSA) can increase food production at household level through wide-scale adoption of integrated soil fertility management (ISFM). Past and ongoing agricultural research shows that it is possible to double or even triple yields by improving soil, nutrient and water management at farm level. However, adoption of ISFM technologies in Africa remains low due to various biophysical and socioeconomic challenges. Supporting the ISFM with sound policy, financial and institutional support can stimulate the much needed increase in food production. This chapter explores some of the opportunities for increasing agricultural production in SSA. These opportunities include the innovative application of science and technology to arrest challenge of soil degradation and ensure improved soil fertility, promoting use of new and improved crop varieties through plant breeding and biotechnology, adoption of the value chain to ensure that investments in agriculture are profitable and facilitating farmers' access to credit and financing.

Keywords Land degradation · Integrated soil fertility management (ISFM) · Innovative financing · Value chain approach · Climate change

Introduction

Sub-Saharan Africa (SSA) has been identified as a future hotspot for food shortage due to low agricultural yields, high dependence on agriculture, costly agricultural inputs, weak economies and high population increase. Moreover, climate change is expected to negatively impact food security in the region. The United Nations Millennium Summit singled out eradication of extreme poverty and hunger as a key development goal to be achieved by all countries by the year 2015. Achievement of this goal will remain a mirage if no deliberate effort is made by the African countries to address the challenges contributing to the stagnant and declining agricultural production.

Food decline in Africa is manifested in increased food insecurity and declining export earnings from agricultural produce. Several African countries are food insecure and have persistently been unable to feed their population (NEPAD, 2003). Haggblade et al. (2004) note that over the past 40 years or so, agriculture production has increased at a rate of 2.5% per year in Africa compared to 2.9% in Latin America and 3.5% in developing Asia. As a result of the situation, Africa is a net food-importing region. Food imports in Africa rose from USD 88 billion in 2006 to about USD 119 billion in 2007. The number of chronically under-nourished people increased from 168 million in 1990–1992 to 194 million in 1997–1999 (NEPAD, 2003). Whereas indicators show that performance of agricultural GDP has been on the increase since 1995, this growth has not been sustained due to variations in weather, conflict, diseases and insect as well as lack of good technology.

A. Bationo (✉)
Alliance for a Green Revolution in Africa (AGRA), Accra, Ghana
e-mail: abationo@agra-alliance.org

A. Bationo et al. (eds.), *Innovations as Key to the Green Revolution in Africa*, DOI 10.1007/978-90-481-2543-2_1, © Springer Science+Business Media B.V. 2011

Currently, yield levels in SSA commonly are below 1 t ha^{-1} compared with 5 t ha^{-1} levels elsewhere. This apparent yield gap is partly related to mismanagement of water and nutrients, due to inherently low-fertility soils, droughts and dry spells in the sub-humid and semi-arid zones.

Africa suffers a loss between 30 and 60 kg of nutrients per hectare each year, while using only one-tenth of the average application of fertilizer if compared with the rest of the world. Africa loses equivalent of US $4 billion per year due to soil nutrient mining. An estimated US $42 billion in income is lost and 6 million ha of productive land threatened every year due to land degradation. In terms of productivity, 55% of the land is classified as unsustainable for crop production, while 28.3% is classified as medium and low potential. Only about 9.6% is prime land and another 6.7% is of high potential (Eswaran et al., 1996; 1997). More than 93% of agriculture in SSA is rainfed. On average, 9 out of 10 years offer rainfall that is sufficient to produce adequate crops in the dry sub-humid and semi-arid zones, but rainfall is erratic and often unevenly distributed over the cropping season.

In the last four decades in Africa, less than 40% of the gains in cereal production came from increased yields. The rest was from expansion of the land devoted to arable agriculture (Runge, 2008). With a rapidly increasing human population, opportunities for expansion of land for agriculture have declined. In future, Africa must depend more on yield gains than land expansion to achieve food security.

The African Heads of State and governments have developed the Comprehensive African Agricultural Development Program (CAADP) as a framework for agricultural growth, food security and rural development. CAADP has set a goal of 6% annual growth rate in agricultural production to reach the UN's Millennium Development Goal of halving poverty and hunger by 2015. The African Heads of State Fertilizer Summit held in Abuja Nigeria in June 2006 led to the Abuja Declaration on Fertilizer for the African Green Revolution. The Summit emphasized the need to ensure access, affordability and incentives to fertilizer. Given the strategic importance of fertilizers in achieving the African Green Revolution, there is need to increase the level of use of fertilizer from the current average of 8 kg ha^{-1} to an average of at least 50 kg ha^{-1} by 2015. Similar interventions are being spearheaded by major multilateral and bilateral donors such as the Bill and Melinda Gates and the Rockefeller Foundations through the Alliance for a Green Revolution in Africa (AGRA).

The above initiatives recognize that there is no single or "silver bullet" to the problem of declining soil fertility and low agricultural production in Africa. On-farm experiences show that it is possible to double or even triple yields by improving soil, nutrient and water management by adoption of integrated soil fertility management (ISFM). Unfortunately, adoption of ISFM technologies in Africa still remains low due to various biophysical and socioeconomic challenges. Supporting the ISFM with sound policy, financial and institutional support can guarantee a turnaround of the situation for the better of the continent. This chapter explores the emerging new challenges and opportunities for wider adoption of ISFM technologies in Africa.

Agriculture and African Economies

Around 65% of Africans live in rural areas. By 2030, more than 600 million Africans will live in rural areas (World Development Report, 2008). The majority will rely on agriculture for their livelihoods. Agriculture accounts for 70% of full-time employment, 33% of national income and 40% of total export earnings. About 80% of Africans depend on agriculture in one way or another – most of these through small farms (World Development Report, 2008). More than 80% of African farms are smaller than 2 ha. In the poorest countries, such as Malawi, more than 90% of the population depends on small-scale farming for their survival (Toenniessen et al., 2008). Most of these grow some form of staple crops such as rice, maize and sorghum, which represent around 70% of total agricultural output in Africa.

Agriculture plays a vital role in reducing poverty. Economic analysis shows that agricultural growth is more pro-poor (Ravallion and Chen, 2003; Kraay, 2003) than non-agriculture-based growth. The poorest households get up to four times more benefit from a 1% GDP increase if this increase is based on agricultural rather than non-agricultural growth (Ligon and Sadoulet, 2007).

Agricultural productivity in Africa lags all other continents. While cereal crop yields in Asia have doubled or quadrupled since the 1960s, they have

Fig. 1 Per capita food production trends in Asia, Latin America and SSA. *Source*: Haggblade and Hazell (2010)

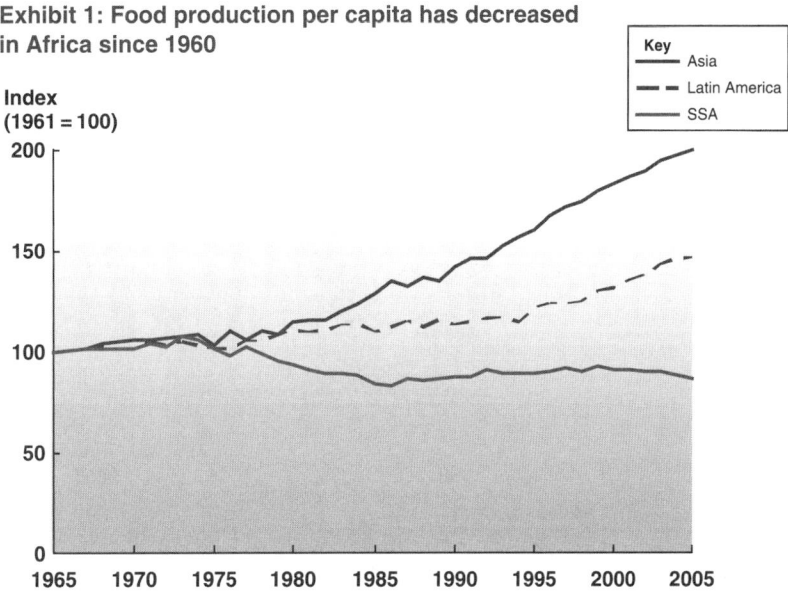

Exhibit 1: Food production per capita has decreased in Africa since 1960

Index (1961 = 100)

Key
— Asia
– – Latin America
— SSA

stagnated in Africa (Haggblade et al., 2004) and as populations have increased, food production per capita has been declining in Africa for the past three decades. This leaves African families with ever less opportunity to feed themselves and their children. Malnutrition remains shockingly common in Africa.

The Food Production Crisis

With the 2015 deadline for the millennium development goals (MDG) rapidly approaching, the number of hungry in Africa is increasing again (FAO, 2008) and Africa accounts for half of the 12 million children <5 dying each year as a consequence of chronic hunger (FAO, 2008). Food production is not keeping pace with population growth in Sub-Saharan Africa. An already low food production per capital faces continued decline. This happens while the doubling of yields obtained in the major cereals in developing countries from 1961 to 1997 (Dixon et al., 2001) also seems to have reached a plateau. In fact, the Green Revolution, which combined improved seeds, inorganic fertilizers, plant protection products with irrigation, has largely bypassed Africa. This problem is rooted in Africa's poor agricultural resource base that, together with socioeconomic and policy environments unfavourable to investments in development of

the agricultural sector, explains why the use of external inputs is generally unprofitable (Fig. 1).

The Poverty Trap

More than 50% of Africans live in extreme poverty – almost 400 million people live on less than "a dollar a day" (World Bank, 2009 PovCalNet). From 1981 to 2005 the share of Africa's population living in extreme poverty went from 53 to 51% (World Bank, 2009 PovCalNet) putting over 150 million more Africans under absolute poverty. This trend contrasts the happenings in other continents. For example, in East Asia, the share of its population living in extreme poverty fell from 78% in 1981 to 17% in 2005, while in South Asia, poverty incidence fell from 59 to 40% (World Bank, 2009 PovCalNet). This rapid decline in poverty incidence in Asia can be linked to the region's success in increasing agricultural productivity, rural employment and income growth (Fig. 2).

The Green Revolution: A Miss for Africa

In the 1960s and early 1970s, a number of Asian countries were facing severe food shortages, following a

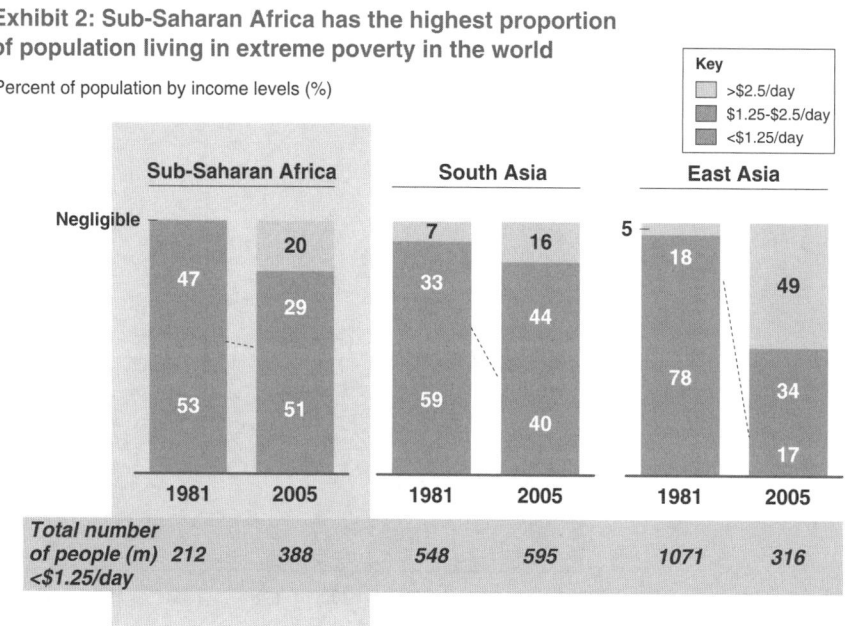

Fig. 2 Poverty levels in SSA, South and East Asia. *Source*: World Bank "PovCalNet" database, 2009

period of rapid population growth and low agricultural productivity. The Asian countries managed to become self-sufficient in food through a series of interventions that focused on smaller farmers. In India, for example, the government invested in irrigation, roads, education and subsidies for fertilizers, energy and credit. Targeted government interventions, combined with vibrant private sector activity, and the commitment of development partners like USAID and the Rockefeller Foundation led to a strong increase in agricultural productivity which made India self-sufficient in staple foods and laid the groundwork for later economic growth.

Unfortunately, Africa missed out on this revolution. Several factors have been postulated as contributing to this failure (de Janvry, 2009; Djurfeldt et al., 2005). First, wheat and rice (the major crops of the Asian Green Revolution) are not the major food crops in Africa, where farming systems are dominated by root and tuber crops, sorghum, millet, maize and pulses. The diversity of the crops makes priority setting for improvement more challenging. Second, Africa has more diverse agro-ecologies than Asia, which means technological change with continent-wide implications is much harder to achieve. Third, while the Asian Green Revolution largely arose in homogenous,

irrigated farming systems, African agriculture is dominated by diverse rainfed systems, with less than 5% of the arable land irrigated – compared to over 60% of irrigated arable land in Asia (BANR, 2008; de Janvry and Byerlee, 2007; FAO, 2006) (Fig. 3).

In addition to these natural barriers, other constraints included the following:

- *Negative effects of structural adjustment.* The withdrawal of the state from the agricultural sector since the 1980s resulted in a lack of agricultural extension and reduced capacity for agricultural research.
- *Considerable infrastructure challenges.* Africa today has only a fraction of the infrastructure that Asia had in the 1950s. Road density in countries like Ethiopia in 2000–2005 ranged from 3 to 4 km/1000 km^2, whereas India at the start of its Green Revolution had a road density of 388 km/1000 km^2. As a result, transportation costs in Africa today are far higher than in Asia – average transportation prices in the Douala-Ndjamena corridor, for example, are 11 cents km^{-1}, compared to 5 cents km^{-1} in China or 2 cents km^{-1} in Pakistan.
- *Disempowerment of the rural population.* African leaders for many years have not shown the

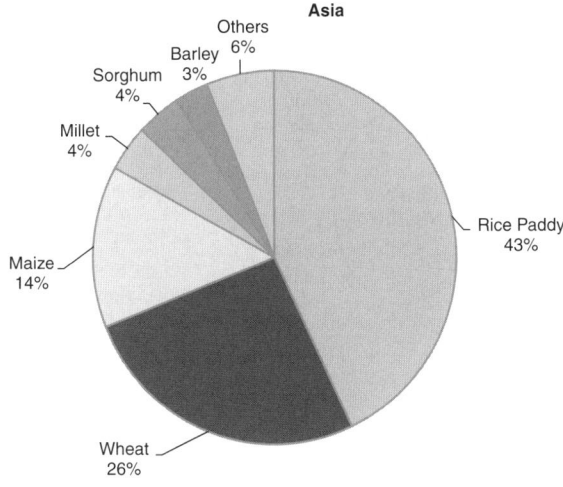

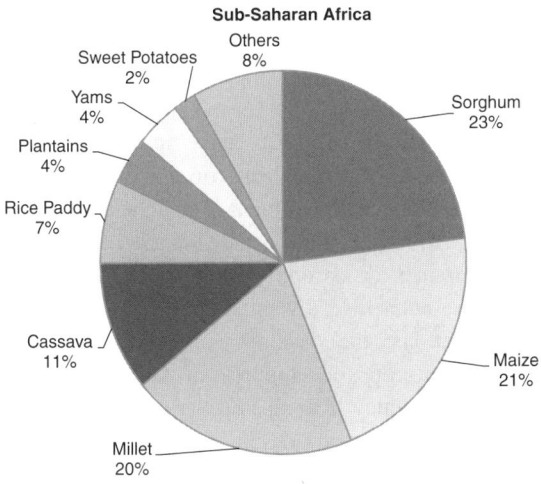

Fig. 3 Major food crops of Asia and Sub-Saharan Africa. *Note*: Percentages refer to hectares harvested, averaged for 2000–2004. *Source*: BANR, 2008; de Janvry and Byerlee, 2007; FAO, 2006

Soil Fertility and Land Degradation in Africa

Although significant progress has been made in research in developing principles, methodologies and technologies for combating soil fertility depletion, soil infertility still remains the fundamental biophysical cause for the declining per capita food production in SSA over the last 3–5 decades (Sanchez et al., 1997). This is evident from huge gap between actual and potential crop yields (FAO, 1995).

During the last 30 years, soil fertility depletion has been estimated at an average of 660 kg N ha^{-1}, 75 kg P ha^{-1} and 450 kg K ha^{-1} from about 200 million ha of cultivated land in 37 African countries (Sanchez et al., 1997). Stoorvogel et al. (1993) estimated annual net depletion of nutrients in excess of 30 kg N and 20 kg K ha^{-1} of arable land per year in Ethiopia, Kenya, Malawi, Nigeria, Rwanda and Zimbabwe (Table 1). Nutrient balances are negative for many cropping systems indicating that farmers are mining their soils of nutrient reserves.

The low and declining soil fertility can be attributed to low inherent fertility, nutrient depletion, weakened ability to maintain fertility and low returns on investment in raised soil fertility (Fig. 4). The inherent constraints in some soils have been exacerbated by their over-exploitation for agricultural production. Large areas of soils of high production potential in SSA have been degraded due to continuous cropping without replacement of nutrients taken up in harvests

necessary political will that is needed to achieve a Green Revolution. Part of the problem is that farmers are politically powerless. In Asia, the powerful rural population exerted pressure on politicians to support agriculture, but until now this has not been the case in Africa.

• *Weaker nature of African institutions.* African countries did not invest early in the development of strong food crop research institutions. Post-colonial states focused on export-oriented cash crops. Research, extension finance and marketing institutions were developed to serve primarily the export-oriented sector.

Table 1 Average nutrient balance of N, P and K (kg ha^{-1} yr^{-1}) for the arable land for some Sub-Saharan African countries (average of 1982–1984)

Country	N	P	K
Botswana	0	1	0
Mali	−8	−1	−7
Senegal	−12	−2	−10
Benin	−14	−1	−9
Cameroon	−20	−2	−12
Tanzania	−27	−4	−18
Zimbabwe	−31	−2	−22
Nigeria	−34	−4	−24
Ethiopia	−41	−6	−26
Kenya	−42	−3	−29
Rwanda	−54	−9	−47
Malawi	−68	−10	−44

Source: Stoorvogel et al., 1993

Fig. 4 Biophysico-chemical and socio-economic factors contributing to low soil fertility in Africa (adapted from Bationo et al., 2007)

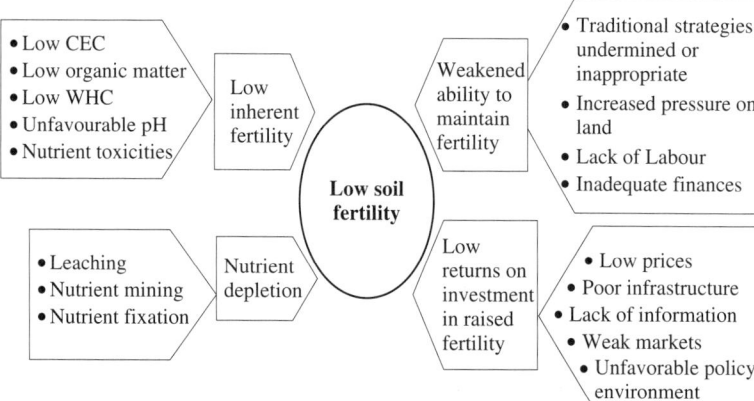

(Murwira, 2003). Increasing population pressure of up to 1200 persons per square kilometre (Shepherd et al., 2000) has necessitated the cultivation of marginal lands that are prone to erosion and other environmental degradation and it is also no longer feasible to use extended fallow periods to restore soil fertility. The shortened fallow periods cannot regenerate soil productivity leading to non-sustainability of the production systems (Nandwa, 2001) (Fig. 4).

Climate Change and Food Production

Climate change is another challenge faced by millions of poor farmers in Africa. It is expected that severe droughts will occur more frequently, especially in the dry semi-arid regions of the Sahel, which already suffer from low rainfall and high inter- and intra-seasonal variabilities in rainfall and lengths of the growing season. In the coastal areas, it is expected that rising sea temperature levels will lead to greater levels of flooding. It is estimated that 75–250 million people in the Sahel will be at risk of droughts as the region is expected to be drier, which will be more severe than the severe droughts of the 1970s in the Sahel. Flooding in southern Africa is expected to increase, bringing to mind the severe floods of 2000 that wiped out one-third of the crops in Mozambique, killed many and led to massive displacement of populations (Fleshman, 2007). The net loss due to climate change in Africa could be as high as US $133 billion with agriculture bearing the most of the brunt – an estimated loss of US $132 billion (Mendelsohn et al., 2000). In Mali, it is estimated that because of projected yield declines and

loss of forages the country will suffer economic losses in the range of US $70–$142 million (Butt et al., 2005).

Among some of the ways of adjusting to climate (Dinar et al., 2008) change includes (a) production practices: delayed planting; shifting into more drought-tolerant crops; shifting the composition of livestock between large animals such as cattle and smaller ruminants that better tolerate heat; investment in water harvesting and better water management, and changing land uses; (b) market approaches: the use of crop and flood insurance, catastrophic bonds, weather market derivatives, enterprise diversification; (c) technological innovations: the development of drought- and flood-tolerant crops, development of crops with better water use efficiency; increased reliance on biological nitrogen fixation, incorporation of organic matter and use of conservation tillage; and (d) policy interventions: support of farmers to purchase crop insurance, price stabilization, strategic grain reserves and use of market-hedging instruments.

Unfortunately, adaptation to climate change is constrained by lack of access to information on weather, lack of predictive information on climate change and expected impacts, lack of information on appropriate adaptation methods, cost of adjustments to adaptation, etc. Lack of access to credit by smallholder farmers makes them more vulnerable to impacts of climate change (Dinar et al., 2008). Access to finance reduces adjustment costs and allows farmers to move to more efficient and optimal adaptation pathways, invest in crop insurance, improve investments in appropriate land, water and soil fertility management strategies and reduce the effects of shocks on their overall incomes and assets, and smooth consumption demands (Adesina, 2009).

Public policies are needed to support the development of more reliable weather and climate forecasting. Africa will need more drought- and flood-tolerant crops for its farmers. More investment in breeding research is needed to develop new strains of crops with tolerance to drought, floods, diseases and pests. Major investments will be needed in irrigation and water management across Africa. Climate change also needs to be fully integrated into national agricultural sector planning strategies in Africa. Greater support is needed to broaden the genetic base of crops; promote diversification of farmers into more drought-tolerant crops; invest in irrigation and water management; and develop local capacity in climate science to improve climate predictions, preparedness and disaster management. Greater efforts must also be put into developing effective programs for reducing vulnerability of at-risk populations through social safety net programs.

Evolution of Soil Fertility Paradigm in Africa

Since the 1960s, the paradigms underlying soil fertility management research and development efforts have undergone substantial change because of experiences gained with specific approaches and changes in the overall social, economic and political environment faced by various stakeholders. During the 1960s and 1970s, an external input paradigm characterized by increased use of improved germplasm and fertilizer significantly led to a rapid increase in food production, commonly referred to as the 'Green Revolution', especially in Asia and Latin America. This paradigm put little if any significance to the use of organic resources as sources of nutrients or soil health. The impacts of the 'Green Revolution' strategy resulted only in minor achievements in SSA. The environmental degradation resulting from massive and injudicious applications of fertilizers and pesticides observed in Asia and Latin America between the mid-1980s and early 1990s (Theng, 1991) and the abolition of the fertilizer subsidies in SSA (Smaling, 1993) imposed by structural adjustment programs led to a renewed interest in organic resources in the early 1980s. The balance shifted from mineral inputs to low-input sustainable agriculture (LISA) where organic resources were believed to enable sustainable agricultural production

(Vanlauwe, 2004). The adoption of LISA technologies such as alley cropping and live mulch systems was constrained by both technical (e.g. lack of sufficient organic resources) and the socio-economic factors (e.g. labour-intensive technologies) (Vanlauwe, 2004). This led to the second paradigm, the integrated nutrient management (INM), which emphasized on the need for the judicious use of both mineral and organic inputs to sustain crop production (Sanchez, 1994).

A further shift in paradigm in the mid-1980s to the mid-1990s advanced the combined use of organic and mineral inputs accompanied by a shift in approaches towards involvement of the various stakeholders in the research and development process, mainly driven by the 'participatory' movement. One of the important lessons learnt was that the farmers' decision-making process was not merely driven by the soil and climate but by a whole set of factors cutting across the biophysical, socio-economic and political domains. The integrated natural resource management (INRM) research approach was thus formulated aimed at developing interventions that take all the above aspects into account (Izac, 2000).

Past paradigms of soil fertility management focused on fertilizer or 'low-input' methods but rarely on both and ignored the essential scientific fact that fertilizers are most effective and efficient in the presence of soil organic matter and well-conserved soil structure. This dichotomy is resolved by the integrated soil fertility management (ISFM) framework. ISFM is the application of locally adapted soil fertility management practices to optimize the agronomic efficiency of fertilizer and organic inputs in crop production. ISFM necessarily includes locally appropriate fertilizers and organic resources, the knowledge needed to conduct local experimentation and testing, and locally adapted grain and legume varieties. The inorganic fertilizer provides most of the nutrients and the organic fertilizer increases soil organic matter status, soil structure and buffering capacity of the soil in general. Use of both inorganic and organic fertilizers has proven to result in synergy, improving efficiency of both nutrient and water use. Considering the ISFM is a knowledge-intensive process, significant increases in farmer knowledge and investments in continued capacity building and generating new technologies are a prerequisite under ISFM.

The ISFM concept also considers other socioeconomic factors such as land tenure, input–output markets, access to credit and institutional support among others in a value chain approach. ISFM therefore seeks to develop competitive commodity chains by strengthening the technical and managerial competencies of the various factors involved, in particular, the farmers and local entrepreneurs (including inputs dealers, processors, stockists and traders) at the grassroots.

Opportunities for Soil Fertility Management

This section discusses the opportunities for improving food production in Africa by looking at adoption of ISFM technologies, the role of new cultivars, plant breeding and biotechnology, value chain approach, innovative financing and scientific knowledge.

Adoption of ISFM Technologies

Inorganic Fertilizer

There is ample evidence that increased use of inorganic fertilizers has been responsible for an important share of worldwide agricultural productivity growth. Van Keulen and Breman (1990) and Breman (1990) stated that the only real cure against land hunger in the West African Sahel lay in increased productivity of the arable land through the use of external inputs, mainly inorganic fertilizers. Surveys in the dry lands show that few small-scale farmers in these drought-prone regions use any fertilizer to increase crop production. For instance, surveys in southern Zimbabwe showed that less than 5% of the farmers used chemical fertilizers even when the deficiencies are evident. The low use is attributed to the perception that fertilizers applied under low soil moisture will scorch the plants. Second, the potential losses of crop and low crop yields and returns also are a disincentive to fertilizer use.

Pieri (1989) reporting on fertilizer research conducted from 1960 to 1985 confirmed that inorganic fertilizers in combination with other intensification practices had tripled cotton yields in West Africa from 310 to 970 kg ha^{-1}. There are numerous cases of strong fertilizer response for maize in East, South and West Africa. Maize yield increase over the control due to NPK fertilizer application from 6 AEZ and averaged

over 4 years was 149%, but when the soil was amended with lime and manure yield, response over the control increased to 184% (Mokwunye et al., 1996). Similarly higher yield improvements have been observed in East (Qureish, 1987) and southern (Mtambanengwe and Mapfumo, 2005) African countries. Despite the above response to fertilizer application, studies have shown that there are great on-farm soil fertility gradients and the yields are bound to vary greatly even on the same production unit. Prudencio (1993) observed such fertility gradients between the fields closest to the homestead (home gardens/infields) and those furthest (bush fields/outfields). Soil organic carbon contents of between 11 and 22 g kg^{-1} have been observed in home gardens compared to 2–5 g kg^{-1} soil in the bush fields. Similarly, higher total organic nitrogen, available phosphorus and exchangeable potassium have also been observed in the home gardens compared to the bush fields.

Despite the fact that deficiency of N and P is acute on the soils of Africa, local farmers use very low P fertilizers because of high cost and problems with availability. The use of locally available phosphate rock (PR) could be an alternative to imported P fertilizers. For example, Bationo et al. (1987a) showed that direct application of local PR may be more economical than imported water-soluble P fertilizers. Bationo et al. (1990) showed that Tahoua PR from Niger is suitable for direct application, but Parc-W from Burkina Faso has less potential for direct application. The effectiveness of local phosphate rock depends on its chemical and mineralogical composition (Chien and Hammond, 1978). Phosphate rocks can be used as a soil amendment and the use of water-soluble P can be more profitable. The efficiency of fertilizer use by plants is dependent on mode of application, with hill placement being the most efficient method.

In order to overcome the risks associated with fertilizer use under conditions of low soil moisture, farmers have adopted the 'microdose' technology that involves strategic application of small doses of fertilizer 4 kg P ha^{-1} and seed (Tabo et al., 2011, 2007). This rate of fertilizer application is only one-third of the recommended rates for the areas. In all the project study sites in the three West African countries (Burkina Faso, Mali and Niger), grain yields of millet and sorghum increases were up to 43–120% when using fertilizer 'micro-dosing'. The incomes of farmers using fertilizer 'micro-dosing' and inventory credit system or

Table 2 Effect of microdose on millet grain yield in Karabedji

Treatments	Millet grain yield (kg ha^{-1})
1. Farmers' practices	487
2. NPK HP	1030
3. DAP HP	924
4. PRT+NPK HP	1325

NPK, 15-15-15 compound fertilizers; DAP, diammonium phosphate; HP, hill placement at 4 kg P ha^{-1}; PRT, Tahona phosphate

'warrantage' increased by 52–134%. Small amounts are more affordable for farmers, give an economically optimum (though not biologically maximum) response, and if placed in the root zone of these widely spaced crops rather than uniformly distributed result in more efficient uptake (Bationo and Buerkert, 2001) (Table 2).

Crop Residue Management

Crop residue (CR) management can play an important role in improving crop productivity in SSA. Bationo et al. (1995) in a pearl millet (*Pennisetum glaucum* L.) experiment at Sadoré, Niger, observed that grain yields could be increased to 770 kg ha^{-1} with mulch application of 2 t ha^{-1} crop residue (CR) and to 1030 kg ha^{-1} with 13 kg P as SSP plus 30 kg N ha^{-1}. The combination of CR and mineral fertilizers resulted in a grain yield of 1940 kg ha^{-1}. Availability of organic inputs in sufficient quantities and quality is one of the main challenges faced by farmers especially in the dryland areas where the stover is used for multiple uses such as dry season feed for livestock. In an inventory of crop residue availability in the Sudanian zone of central Burkina Faso, Sedga (1991) concluded that the production of cereal straw can meet the currently recommended optimum level of 5 t ha^{-1} every 2 years. However, McIntire and Fussell (1986) reported that on fields of unfertilized local cultivars, grain yield averaged only 236 kg ha^{-1} and mean residue yields barely reached 1300 kg ha^{-1}. These results imply that unless stover production is increased through application of fertilizers and or manure, it is unlikely that the recommended levels of crop residue could be available for use as mulch.

Manure Management

Manure, another farm-available soil amendment, is an important organic input in African agro-ecosystems. Palm (1995) concluded that for a modest yield of 2 t ha^{-1} of maize, the application of 5 t ha^{-1} of high-quality manure can meet the N requirement but this cannot meet the P requirements in areas where P is deficient. The quality of manure, however, is highly variable depending on the quality of feed to livestock, manure collection and storage methods (Mueller-Saemann and Kotschi, 1994; Mugwira, 1984; Ikombo, 1984; Probert et al., 1995; Kihanda, 1996). Availability of sufficient manure is equally a challenge. Probert et al. (1995) estimated manure production levels of 1 t per livestock unit per year for unimproved local cattle breed in maize–livestock system in the semi-arid parts of eastern Kenya. De Leeuw et al. (1995) reported that with the present livestock systems in West Africa, the potential annual transfer of nutrients from manure is 2.5 kg N and 0.6 kg P ha^{-1} of cropland. Although the manure rates are between 5 and 20 t ha^{-1} in most of the on-station experiments, quantities used by farmers are very low and range from 1.3 to 3.8 t ha^{-1} (Williams et al., 1995). This is due to insufficient number of animals to provide the manure needed and the problem becomes more pronounced especially in post-drought years (Williams et al., 1995). Depending on rangeland productivity, it will require between 10 and 40 ha of dry season grazing land and 3–10 ha of rangeland of wet season grazing to maintain yields on 1 ha of cropland using animal manure (Fernandez, et al. 1995).

Cropping Systems

Cropping systems in Africa are characterized by mixed cropping of cereals and legumes in various temporal and spatial arrangements (Bationo et al. 1987b). These can be sole, rotations, intercropping or relay cropping systems. This practice provides the farmer with several options for returns from land and labour, often increases efficiency with which scarce resources are used and reduces dependence upon a single crop that is susceptible to environmental and economic fluctuations. The legumes not only fix nitrogen and thus replace fertilizer but also produce an economic crop that small farmers can sell. Many legumes can be used for food (grain), feed (forage) or dual purposes (grain and forage). The residues, being returned to the soil directly or indirectly (green manure), benefit the following crop, regenerate fertility and contribute to sustainability.

Cereal–Legume Intercropping

Significant advances have been made in the development and promotion of dual-purpose cowpea intercropped with sorghum in West Africa. The most attractive part of this system is availability of cowpea grain in August which is normally the off-season for cowpea when prices are high. In Senegal, millet intercropped with groundnut has given 75–195% yield increases in systems that include, in addition to the green manures, improved use of rock phosphate, water harvesting, composting and stall-fed livestock. Intercropping maize with groundnut and pigeon pea is widespread in East and southern Africa, justifying further ISFM investment (Woomer et al., 2004). Pigeon pea–maize intercropping is a common farmers' practice in southern Malawi, and parts of Mozambique and Tanzania. Pigeon pea is also capable of accessing sparingly soluble phosphate (P) and can efficiently utilize residual P remaining in the soil from fertilizer applied to the maize crop (Bahl and Pasricha, 1998). In East Africa, large-scale adoption of systems of cereal intercropped with forage legumes has provided pathogen resistance and soil fertility benefits (Khan et al., 2000). ICIPE has demonstrated that intercropping maize with the fodder legume *Desmodium uncinatum* reduced infestation of parasitic weed *Striga hermonthica* by a factor of 40 compared to a maize mono-crop.

Cereal–Legume Rotations

Rotation of cereals with legumes is another positive soil health intervention. Over several years, IITA, ICRISAT, CIAT and their partners have developed and implemented sustainable beans, groundnut, pigeon pea and soybean/maize rotations with proven success. In rotational maize/soybean systems where legume residues are retained and mineral fertilizer is applied (15 kg P ha^{-1} applied to soybean and 45 N kg ha^{-1} applied to maize), soybean yields were 2.5 t of grain per hectare with N fixation contributing about 50 kg N ha^{-1} (Sanginga et al., 1997). The maize following soybean had 75% greater yield than maize following maize. Furthermore, fertilizer utilization by maize was improved by 100% because applying soybean residues and 45 kg N ha^{-1} resulted in maize yields equal to those obtained by applying the 90 kg fertilizer N ha^{-1} without soybean.

Combination of Organic and Inorganic Nutrient Sources

Past research on ISFM technologies has shown that there is substantial gains in crop productivity from nutrient additions through mixtures of organic and inorganic nutrient compared with either input alone (Swift et al., 1994). Combination of animal manure with inorganic fertilizers, for instance, is a common practice among smallholder farmers in Zimbabwe. Supplementation of 5 t ha^{-1} with 40 kg N ha^{-1} (inorganic fertilizer) in Zimbabwe resulted in a statistically higher yield than sole manure treatment (Murwira et al., 2002). These responses are advantageous to smallholder farmers who cannot afford to purchase and apply fertilizers at recommended rates. Studies have also shown that there is higher synchrony between N release and crop uptake can be achieved when organic and inorganic nutrient sources are combined (Murwira and Kirchmann, 1993).

Interaction Between Water and Nutrients

African cropping systems, especially in the arid and semi-arid areas, continue to suffer from unreliable rainfall which often results in droughts and crop failures. Unlocking the potential of small-scale, rainfed agriculture in SSA in the sub-humid and semi-arid zones calls for the implementation of a combination of water and nutrient management measures. Dry spells can be bridged with technologies that more efficiently use the available soil moisture by decreasing non-productive evaporative losses or augment the soil moisture by irrigation. Water management also includes conservation agriculture, mulching and water harvesting to enable supplementary irrigation. Several reports (Camara and Heinemann, 2006; Rockstrom et al., 2007) emphasize that yields can be increased by a factor of 2–4 in many parts of SSA through better water management practices, such as adding organic matter to soils, preventing soil erosion, using water harvesting technologies and increasing water retention with tied ridges, bunds and terraces. Planting pits (zai system), ridges and stone bounds are examples of water harvesting technologies widely used in dry-land areas.

Practised mainly in Mali, Burkina Faso and Niger, where it is known as *tassa*, zaï is a traditional technique

Table 3 Effect of planting technique on millet grain and dry matter yield in un-amended and amended plots with organic matter; Damari and Kakassi, rainy seasons 1999 and 2000

| Sowing technique | Grain yield (kg ha^{-1}) | | | | Total dry matter (kg ha^{-1}) | | | |
| | 1999 | | 2000 | | 1999 | | 2000 | |
	Damari	Kakassi	Damari	Kakassi	Damari	Kakassi	Damari	Kakassi
Flat, un-amended	0.9	118	6	94	96	752	101	768
Flat, amended	416	366	292	389	1881	2382	1346	1704
Zaï, un-amended	17	434	19	388	303	2125	213	1938
Zaï, amended	662	628	488	637	3096	3800	1824	3593

for conserving water and rehabilitating degraded land. The zaï system is a series of man-made pits, or holes, dug on abandoned or unused land. The purpose of creating the holes is to capture run-off precipitation, because the land is typically less permeable to water. The zaï pits are often filled with organic matter so that moisture can be trapped and stored more easily. The pits are then planted with annual crops such as millet or sorghum. The zaï pits extend the favourable conditions for soil infiltration after run-off events, and they are beneficial during storms, when there is too much water. The compost and organic matter in the pits absorb excess water and act as a form of water storage for the planted crops.

The success of zaï planting pits has been documented all over the Sahel region. In a study by Fatondji et al. (2006) in Damari and Kakassi in Niger, higher millet total dry matter and grain were observed in zaï treatments compared to flat-planted plots (Table 3). In the zaï plots the grain yields were three to four times higher, illustrating the contributions of the water harvesting effects of the zaï techniques. Similarly, cattle manure significantly increased the positive effect of the zaï technique expressed in terms of grain yield and total dry matter (Table 2).

Reij and Thiombiano (2003) have also reported higher sorghum grain yields when the planting pits were amended with organic and/or inorganic nutrient sources (Table 4). This indicates the importance of nutrient management in improving the performance of the zaï technology.

Table 4 Effect of planting pits (zaï) and nutrient application on sorghum grain yield

Technology	Sorghum yield (kg ha^{-1})
Only planting pits	200
Pit + cow dung	700
Pit + mineral fertilizers	1400
Pit + dung and fertilizers	1700

New Cultivars, Plant Breeding and Biotechnology

Breeding crop varieties with adaptation to drought, high temperatures, salinity, aluminium toxicity and low nutrient conditions can significantly help increase food production in Africa (Umezawa et al., 2006; Sunkar et al., 2007). Because drought is a prevailing concern, improving the water use efficiency of crops (biomass produced per unit of water transpired) is of particular interest. Breeding for early flowering and developing crops that can be planted at a higher density to minimize soil water evaporation are two strategies to cope with water deficits.

Value Chain Approach

An agricultural transformation depends upon all parts of the agricultural system working together. This system relies on science and technology, farmer productivity, market access and an enabling environment (including policy, regulatory and infrastructure).

Improving farmers' access to better science (e.g. through improved seed varieties) and the knowledge on effective soil fertility management technologies and inputs required to drive farmer productivity constitute "supply side" interventions. Building robust post-harvest value chains to reliable demand sources to enable farmers to capture value from their crops constitutes "demand side" measures. Supply and demand side interventions must work together in a sequenced manner to be sustainable. They also need to be supported by an appropriate enabling environment, including infrastructure (transportation, power, financial services, etc.) and proper regulatory institutions and policies. The right policy environment will ensure there are incentives and appropriate institutions to encourage the uptake of technologies, while sustaining income growth from technological change and stabilizing food prices for consumers.

Innovative Financing: Filling a Financing Gap in Smallholder Agriculture

Lack of access to finance is a major constraint to unlocking the potential of agriculture in Africa. Whereas there exists significant excess financial liquidity in local financial institutions, commercial banks do not lend to agriculture for several reasons. These include the high dispersion of farmers, which increases lending and recollection costs; high level of covariate risks as farmers in a given location are often subject to the same sets of risks – both climatic and price; lack of acceptable collaterals by applicants; seasonality and low profitability of smallholder agriculture; lack of risk mitigating instruments to lower risk of lending to poor farmers; and high costs of borrowing money on capital markets which leads to higher interest rate charges for farmers (Adesina, 2009).

Informal financial support systems such as the warrantage system, otherwise called the inventory credit system (ICS), have been demonstrated in many parts of Africa with mixed successes. Microfinance institutions are another alternative for accessing finance for farmers. Despite the spread of microfinance institutions in Africa, they do not lend much to agriculture. Interest rates charged by microfinance institutions are extremely high well beyond the reach of poor farmers. Repayment schedules do not synchronize well with seasonal nature of agriculture and the loan durations are also too short (6–8 months) to be useful for farmers (Adesina, 2009).

New institutional innovations and policies are needed to leverage commercial financing into smallholder agriculture (Adesina, 2009). These should include (a) reducing the high perceived risks of lending to agriculture, especially through the use of loan guarantees; (b) encouraging banks to develop commercial lending operations in the rural space to mobilize savings and provide credit; (c) developing more appropriate loan products that can serve the needs of farmers and the entire agricultural value chain; (d) improving credit policies that synchronize credit needs with the seasonal nature of agriculture, especially disbursements and repayment schedules; and (e) providing financial literacy to farmers in managing farming as a business, as well as credit management.

Innovative financing is a loan guarantee scheme being spearheaded by AGRA to leverage commercial banks to lend to agriculture in Africa. With the use of US $16 million in loan guarantees for commercial banks, AGRA has been able to leverage US $170 million in market-based and affordable loans for smallholder farmers and agricultural value chains that support smallholder farmers in Tanzania, Uganda, Kenya, Mozambique and Ghana. What is needed now is to scale up commercial bank lending to the agriculture sector through the development of markets for risk sharing instruments so that commercial banks can leverage substantial flows of funds in support of agriculture across the value chain and allow African farmers to participate more effectively in meeting national, regional and global food needs.

Scientific Knowledge is a Cornerstone

Successful implementation of ISFM interventions above needs to be supported by a sound scientific background using analytical tools including models. Scientific support can help identify risks of given production systems and technologies and also to provide recommendation domains for similar technologies in other areas. Modelling can help generate scenarios including changes to climate and identify actions such as new agricultural management practices. In addition to scientific evidence, the quest for yield increases requires efforts in regard to knowledge creation and networking, capacity development of decision makers and implementers, stakeholder interaction, efficient and supportive policies and institutions, financial resources and all of the above embedded in a solid communications strategy.

Conclusion

In order for Sub-Saharan Africa to meet its future food security and development requirements, it must radically transform its agricultural system and ensure that the benefits of such transformation lead to both growth and poverty reduction. There exist numerous successful ISFM technologies capable of increasing agricultural productivity at farm level while at the same time ensuring sustainability of the farming systems. Access to credit to purchase fertilizers, improved seed and other production factors as well as climate change will remain a major challenge to the success of any ISFM technologies. As discussed, wide-scale and sustained adoption of ISFM technologies is knowledge intensive and must be supported by investment in research, capacity building, institutional strengthening and policy support.

References

Adesina AA (2009) Africa's food crisis: conditioning trends and global development policy. Keynote paper presented at the international association of agricultural economists conference, Beijing, China, 16 Aug 2009

Bahl GS, Pasricha NS (1998) Efficiency of P utilization by pigeon pea and wheat grown in a rotation. Nutr Cycling Agroecosyst 51:225–229

BANR (Board on Agriculture and Natural Resources) (2008) Emerging technologies to benefit farmers in sub-Saharan Africa and South Asia. Committee on a study of technologies to benefit farmers in Africa and South Asia. The National Academies Press, Washington, DC, pp 19–24. [Online] Available: http://www.nap.edu/catalog/12455.html. Accessed 8 Mar 2011

Bationo A, Buerkert A (2001) Soil organic carbon management for sustainable land use in Sudano Sahelian West Africa. Nutr Cycling Agroecosyst 61:131–142

Bationo A, Buerkert A, Sedogo MP, Christianson BC, Mokwunye AU (1995) A critical review of crop residue use as soil amendment in the West African semi-arid tropics. In: Powell JM, Fernandez Rivera S, Williams TO, Renard C (eds) Livestock and sustainable nutrient cycling in mixed farming systems of sub-Saharan Africa, vol II. Technical papers. Proceedings of an international conference held in Addis Ababa, 22–26 Nov 1993. ILCA (International Livestock Center for Africa), Addis Ababa, Ethiopia, 569 pp

Bationo A, Chien SH, Christianson CB, Henao J, Mokwunye AU (1990) A three year evaluation of two unacidulated and partially acidulated phosphate rocks indigenous to Niger. Soil Sci Soc Am J 54:1772–1777

Bationo A, Chien SH, Mokwunye AU (1987a) Chemical characteristics and agronomic values of some phosphate rocks in West Africa. In: Menyonga JM, Beguneh T, Youdeowwei A (eds) Food grain production in semi-arid Africa. SAFGRAD Coordination office. DAU/SAFGRAD, Essex

Bationo A, Christianson CB, Mokwunye AU (1987b) Soil fertility management of the millet-producing sandy soils of Sahelian West Africa: The Niger experience. Paper presented at the workshop on soil and crop management systems for rainfed agriculture in the Sudano-Sahelian zone,

International Crops Research Institute for the Semi-Arid Tropics (ICRISAT), Niamey, Niger

Bationo A, Kihara J, Vanlauwe B, Kimetu J, Waswa BS, Sahrawat KL (2007) Integrated nutrient management – concepts and experience from Sub-Saharan Africa. In: Aukland MS, Grant CA (eds) The Hartworth Press, New York, NY

Board on Agriculture and Natural Resources (BANR) (2008) Emerging technologies to benefit farmers in Sub-Saharan Africa and South Asia. National Academy of Sciences, Washington, DC

Breman H (1990) No sustainability without external inputs. Sub-Saharan Africa; beyond adjustment. Africa Seminar. Ministry of Foreign Affairs, DGIS, The Hague, pp 124–134

Butt TA, McCarl BA, Angerer J, dyke PT, Stuth JW (2005) The economic and food security implications of climate change in Mali. Climate Change 68(3):355–378, DOI: 10.1007/s10584-005-6014-0

Camara O, Heinemann E (2006) Overview of the fertilizer situation in Africa. Background paper presented at the African fertilizer summit, Abuja, Nigeria, 9–13 June

Chien SH, Hammond LL (1978) A simple chemical method for evaluating the agronomic potential of granulated phosphate rock. Soil Sci Soc Am J 42:615–617

de Janvry A (2009) Agriculture for development: New paradigm and options for success. Elmhirst Lecture, IAAE conference, Beijing, 16–22 August 2009

de Janvry A, Byerlee D (2007) Agriculture for development: World Development Report 2008. The World Bank, Washington, DC

De Leeuw PN, Reynolds L, Rey B (1995) Nutrient transfers from livestock in West African agricultural systems. In: Powell JM, Fernandez-Rivera S, Williams TO, Renard C (eds) Livestock and sustainable nutrient cycling in mixed farming systems of Sub-Saharan Africa, vol II. Technical papers. Proceedings of an international conference held in Addis-Ababa, 22–26 November 1993. ILCA (International Livestock Center for Africa): Addis Ababa, Ethiopia, 569 pp

Dinar A, Hassan R, Mendelsohn R, Benhin J et al (2008) Climate change and agriculture in Africa. Impact assessment and adaptation strategies. Earthscan, London

Dixon J, Gulliver A, Gibbon D (2001) Farming systems and poverty. Improving farmers' livelihoods in a changing world. FAO and World Bank, Rome and Washington, DC

Djurfeldt G, Holmen H, Jirström M, Larsson R (eds) (2005) The African food crisis: lessons from the Asian green revolution. CABI Publishing, Wallingford, CT

Eswaran H, Almaraz R, van den Berg E, Reich P (1996) An assessment of the soil resources of Africa in relation to productivity. World Soil Resources, Soil Survey Division, USDA Natural Resources Conservation Service, Washington, DC, 20013. Received 28 February, 1996

Eswaran H, Almaraz H, van den Berg E, Reich P (1997) An assessment of the soil resources of Africa in relation to productivity. Geoderma 77(1):1–18

FAO (1995) The state of food and agriculture. FAO, Rome, Italy

FAO (2006) Statistical yearbook 2005–2006. Food and Agriculture Organization of the United Nations, Rome, Italy

FAO (2008) The state of food insecurity in the world 2008. Food and Agricultural Organisation of the United

Nations. Available at: ftp://ftp.fao.org/docrep/fao/011/i291e/i0291e00.pdf

Fatondji D, Martius C, Bielders CL, Vlek PLG, Bationo A, Gerard B (2006) Effect of planting technique and amendment type on pearl millet yield, nutrient uptake and water use on degraded land in Niger. Special issue: Advances in integrated soil fertility management in Sub-Saharan Africa: challenges and opportunities (eds Bationo A, Waswa BS, Kihara J, Kimetu J). Nutr Cycling Agroecosyst 76(2–3): 203–217

Fernandez-Rivera S, Williams TO, Hiernaux P, Powell JM (1995) Livestock, feed and manure availability for crop production in semi-arid West Africa. In: Powell JM, Fernandez Rivera S, Williams TO, Renard C (eds) Livestock and sustainable nutrient cycling in mixed farming systems of Sub-Saharan Africa. Proceedings of the international conference ILCA, Addis Ababa, pp 149–170

Fleshman M (2007) Climate change and Africa: stormy weather ahead. http://www.un.org/ecosocdev/geninfo/afrec/newrels/climate-change-1.html

Haggblade S, Hazell PBR (2010) Successes in African agriculture: lessons for the future. IFPRI Issue Brief 63, May 2010. http://www.ifpri.org/sites/default/files/publications/ib63.pdf. Accessed 8 Mar 2011

Haggblade S, Hazell P, Kirsten I, Mkandawire R (2004) Building on successes in African agriculture: African agriculture – past performance, future imperatives, 2020 Vision Focus 12 Brief No. 1. IFPRI, Washington, DC

Ikombo BM (1984) Effect of farmyard manure and fertilizer on maize in a semi-arid area of eastern Kenya. East Afr Agric Forestry J (Special issue: Dryland Farming Research in Kenya) 44:266–274

Izac A-MN (2000) What paradigm for linking poverty alleviation to natural resources management? Proceedings of an international workshop on [Integrated natural resource management in the CGIAR: approaches and lessons, 21–25 August 2000, Penang – Malaysia]

Khan ZR, Pickett JA, Van den Berg J, Wadhams LJ, Woodcock CM (2000) Exploiting chemical ecology and species diversity: stem borer and Striga control in maize in Kenya. Pest Manage Sci 56:957–962

Kihanda FM (1996) The role of farmyard manure in improving maize production in the sub-humid highlands of Central Kenya. PhD thesis, The University of Reading, UK, 236 pp

Kraay A (2003) 'When is growth pro-poor? Evidence from a panel of countries,' Mimeo, Development Research Group, World Bank

Ligon E, Sadoulet E (2007) Estimating the effects of aggregate agricultural growth on the distribution of expenditures. Background paper for World Development Report 2008

McIntire J, Fussel LK (1986) On-farm experiments with millet in Niger. III. Yields and economic analyses. ISC (ICRISAT Sahelian Center), Niamey, Niger

Mendelsohn R, Dinar A, Dalfelt A (2000) Climate change impacts on African agriculture. Background paper. http://www.ceepa.co.za/Climate_Change/pdf/(5-22-01)afrbckgrnd-impact.pdf. Accessed 8 Mar 2011

Mokwunye AU, de Jager A, Smaling EMA (1996) Restoring and maintaining the productivity of West African soils: key to sustainable development. IFDC-Africa, LEI-DLO and SC-DLO

Mtambanengwe F, Mapfumo P (2005) Effects of organic resource quality on soil profile N dynamics and maize yields on sandy soils in Zimbabwe. Plant Soil 281(1–2):173–190

Mueller-Samann KM, Kotschi J (1994) Sustaining growth: soil fertility management in tropical smallholdings. Margraf Verlag, Germany

Mugwira LM (1984) Relative effectiveness of fertilizer and communal area manures as plant nutrient sources. Zimb Agric J 81:81–89

Murwira HK (2003) Managing Africa's soils: approaches and challenges. In: Gichuru MP, Bationo A, Bekunda MA, Goma PC, Mafongoya PL, Mugendi DN, Murwira HM, Nandwa SM, Nyathi P, Swift MJ (eds) Soil fertility management in Africa: a regional perspective. Academy Science Publishers: Nairobi, Kenya

Murwira HK, Kirchmann H (1993) Carbon and nitrogen of cattle manures, subjected to different treatments, in Zimbabwean and Swedish Soils. In: Mulongoy K, Merckx R (eds) Soil organic matter dynamics and sustainability of tropical agriculture. Wiley-Sayce, Chichester, pp 189–198

Murwira HK, Mutuo P, Nhamo N, Marandu AE, Rabeson R, Mwale M, Palm CA (2002) Fertiliser equivalency values of organic materials of differing quality. In: Vanlauwe B, Diels J, Sanginga N, Merkx R (eds) Integrated nutrient management in Sub-Saharan Africa. CAB International, Wallingford, CT, pp 113–122

NEPAD (New Partnership for African Development) (2003) Comprehensive Africa agricultural development programme. NEPAD, Pretoria

Nandwa SM (2001) Soil organic carbon (SOC) management for sustainable productivity of cropping and agro-forestry systems in Eastern and Southern Africa. Nutr Cycling Agroecosyst 61:143–158

Palm CA (1995) Contribution of agroforestry trees to nutrient requirements of intercropped plants. Agroforest Syst 30:105–124

Pieri C (1989) Fertilité des terres de savane. Bilan de trente ans de recherche et de développement agricoles au Sud du Sahara. Ministère de la Coopération. CIRAD, Paris, 444 pp

Probert ME, Okalebo JR, Jones RK (1995) The use of manure on smallholder farms in semi-arid eastern Kenya. Exp Agric 31:371–381

Prudencio CY (1993) Ring management of soils and crops in the West African semi-arid tropics: the case of the Mossi farming system in Burkina Faso. Agric Ecosyst Environ 47: 237–264

Qureish JN (1987) The cumulative effects of N–P fertilizers, manure and crop residues on maize grain yields, leaf nutrient contents and some soil chemical properties at Kabete. Paper presented at the national maize agronomy workshop, CIMMYT, Nairobi, Kenya, 17–19 February

Ravallion M, Chen S (2003) Measuring pro-poor growth. Econ Lett 78(1):93–99

Reij C, Thiombiano T (2003) Développement rural et environnent au Burkina Faso: la réhabilitation de la capacité productive des terroirs sur la partie nord du Plateau Central entre 1980 et 2001. Free University of Amsterdam, GTZ and USAID, Amsterdam, 80 pp

Rockstrom J, Lannerstad M, Falkenmark M (2007) Assessing the water challenge of a new green revolution in developing countries. PNAS 104(15):6253–6260

Runge J (2008) Dynamics of forest ecosystems in Central Africa during the Holocene: past–present–future. Taylor and Francis Group, London

Sanchez PA (1994) Tropical soil fertility research: towards the second paradigm. In: Inaugural and state of the art conferences. Transactions 15th world congress of soil science, Acapulco, Mexico, pp 65–88

Sanchez PA, Shepherd KD, Soule MJ, Place FM, Buresh RJ, Izac AMN, Mokwunye AU, Kwesiga FR, Ndiritu CG, Woomer PL (1997) Soil fertility replenishment in Africa: an investment in natural resource capital. In: Buresh RJ, Sanchez PA, Calhoun F (eds) Replenishing soil fertility in Africa. SSSA special publication No. 51. Soil Science Society of America, Madison, WI, pp 1–46

Sanginga N, Dashiell K, Okogun JA, Thottappilly G (1997) Nitrogen fixation and N contribution in promiscuous soybeans in the southern Guinea savanna of Nigeria. Plant Soil 195:257–266

Sedga Z (1991) Contribution a la valorization agricole des residus de culture dans le plateau central du Burkina Faso. Invetaire de disponibilites en metiere organique et e'tude des effets de l'inoculum, micro 110 IBF, Memoire d'ingenieur des sciences appliqués, IPDR/Katibougou, 100 pp

Shepherd K, Walsh M, Mugo F, Ong C, Hansen ST, Swallow B, Awiti A, Hai M, Nyantika D, Ombalo D, Grunder M, Mbote F, Mungai D (2000) Linking land and lake, research and extension, catchment and lake basin. Final technical report start-up phase, July 1999 to June 2000. Working paper 2000–2002. ICRAF

Smaling EMA (1993) An agro-ecological framework of integrated nutrient management with special reference to Kenya. Wageningen Agricultural University, Wageningen

Stoorvogel JJ, Smaling EMA, Janssen BH (1993) Calculating soil nutrient balances in Africa at different scales. I. Supra-National Scale. Fertilizer Res 35:227–235

Sunkar R, Chinnusamy V, Zhu J, Zhu J-K (2007) Small RNAs as big players in plant abiotic stress responses and nutrient deprivation. Trends Plant Sci 12(7):301–309

Swift MJ, Seward PD, Frost PGH, Qureshi JN, Muchena FN (1994) Long-term experiments in Africa: developing a database for sustainable land use under global change. In: Leigh RA, Johnson AE (eds) Long-term experiments in agricultural and ecological sciences, CAB International, Wallingford, UK, pp 229–251

Tabo R, Bationo A, Amadou B, Marchal D, Lompo F, Gandah M, Hassane O, Diallo MK, Ndjeunga J, Fatondji D, Gerard B, Sogodogo D, Taonda J-BS, Sako K, Boubacar S, Abdou A, Koala S (2011) Fertilizer microdosing and "warrantage"

or inventory credit system to improve food security and farmers' income in West Africa. In: Bationo A, Waswa BS, Okeyo JM, Maina F, Kihara J (eds) Innovations for a green revolution in Africa: exploring the scientific facts, vol 1. Springer, The Netherlands, pp 113–121

Tabo R, Bationo A, Bruno G, Ndjeunga J, Marcha D, Amadou B, Annou MG, Sogodogo D, SibiryTaonda JB, Ousmane H, Diallo MK, Koala S (2007) Improving the productivity of sorghum and millet and farmers income using a strategic application of fertilizers in West Africa. In: Bationo A, Waswa BS, Kihara J, Kimetu J (eds) Advances in integrated soil fertility management in sub-Saharan Africa: challenges and opportunities. Springer, The Netherlands

Theng BKG (1991) Soil science in the tropics – the next 75 years. Soil Sci 151:76–90

Toenniessen G, Adesina A, DeVries J (2008) Building an alliance for a green revolution in Africa. Ann NY Acad Sci 1136:233–242

Umezawa T, Fujita M, Fujita Y, Yamaguchi-Shinozaki K, Shinozaki K (2006) Engineering drought tolerance in plants: Discovering and tailoring genes to unlock the future. Curr Opin Biotechnol 17:113–122

Van Keulen H, Breman H (1990) Agricultural development in the West African Sahelian region: a cure against land hunger? Agric Ecosyst Environ 32:177–197

Vanlauwe B (2004) Integrated soil fertility management research at TSBF: the framework, the principles, and their application. In: Bationo A (ed) Managing nutrient cycles to sustain soil fertility in Sub-Saharan Africa. Academy Science Publishers, Nairobi

Williams TO, Powell JM, Fernandez-Rivera S (1995) Manure utilization, drought cycles and herd dynamics in the Sahel: Implications for crop productivity. In: Powell JM, Fernandez-Rivera S, Williams TO, Renard C (eds) Livestock and sustainable nutrient cycling in mixed farming systems of Sub-Saharan Africa. Vol 2: Technical papers. Proceedings of an international conference, 22–26 November 1993. International Livestock Centre for Africa (ILCA), Addis Ababa, Ethiopia, pp 393–409

Woomer PL, Lan'gat M, Tungani JO (2004) Innovative maize–legume intercropping results in above- and below-ground competitive advantages for understorey legumes. West Afr J Appl Ecol 6:85–94

World Bank (2008) World Development Report 2008: agriculture for development–response from a slow trade–sound farming perspective. World Bank, Washington, DC

World Bank (2009) "PovCalNet"; Online poverty analysis tool

Meeting the Demands for Plant Nutrients for an African Green Revolution: The Role of Indigenous Agrominerals

A.U. Mokwunye and A. Bationo

Abstract Africa is a vast continent with a tremendous resource endowment and offers great potential for increased agricultural productivity. Africa occupies 20% of the world's land mass but only 21% of this land is suitable for cultivation. Africa's soils have an inherently poor fertility because they are very old and lack volcanic rejuvenation. With increased cultivation without adequate application of external nutrients, soil fertility levels decline further leading to low per capita food production especially among the majority of the smallholder farmers in the continent. Fortunately, nature has been kind enough to provide the African continent with abundant supplies of local sources of P, calcium (Ca), magnesium (Mg) and sulphur (S) and many rocks that contain other nutrients needed by crops. Agricultural intensification would require harnessing these natural resources to complement good agronomic practices, improved seeds, improved input and output markets and vibrant agricultural research programs. This chapter discusses opportunities for Africa to tap into this natural wealth of agrominerals to improve the fertility of the soils and to promote agricultural development.

Keywords Africa soils · Agrominerals · Rock phosphate · Phosphorus recapitalization

Introduction

The inability to feed a projected population of more than 1 billion by 2025 is one of the major challenges facing Africa as the first decade of the twenty-first century slides by. There was a time not so long ago when South Asia and Latin America were the basket cases with respect to their inability to feed their growing populations. All that changed some 40 years ago through a remarkable process whereby sizable increases in agricultural productivity were achieved through a combination of high-yielding varieties of rice and wheat, inorganic fertilizers, good management practices, appropriate institutions and a conducive policy environment resulting in what has been called the Green Revolution. Unfortunately, the Green Revolution bypassed Africa. In the meantime, Africa's lands have become more degraded. Water erosion, wind erosion and chemical and physical processes have resulted in more than 17% of Africa's land area being degraded. Land degradation manifests itself in reduced soil fertility and increased water stress. In fact, only 14% of Africa's land is relatively free of water stress. These adverse conditions have reduced agricultural production. African farmers knew that their lands were chemically and physically fragile and used the practice of shifting cultivation to regenerate the fertility of cropped lands. But, increasing population has meant a reduction in the length of fallow periods, and this in turn has meant that marginal lands and lands of lower fertility have been brought under cultivation. This has further exacerbated the land degradation problem. From the national to the continental level, low agricultural productivity has resulted in stagnating and worsening economies. How to increase soil fertility thereby improving agricultural productivity was the issue that faced the African Presidents and Heads of Governments that met at the Fertilizer Summit at Abuja, Nigeria's capital, during 9–13 June 2006. The Presidents and Heads of Governments recognized the

A.U. Mokwunye (✉)
United Nations University (UNU), Institute for Natural Resources in Africa, Accra, Ghana
e-mail: uzo.mokwunye@alumni.illinois.edu

A. Bationo et al. (eds.), *Innovations as Key to the Green Revolution in Africa*,
DOI 10.1007/978-90-481-2543-2_2, © Springer Science+Business Media B.V. 2011

importance of fertilizer in achieving the African Green Revolution and therefore declared fertilizer from both inorganic and organic sources "a strategic commodity without borders". It was recommended that African Governments should aim to raise fertilizer consumption from the present average of 8 kg ha^{-1} to an average of at least 50 kg ha^{-1} by 2015.

Africa is endowed with many natural resources that can be used to improve the productivity of agriculture. Africa has over 70% of the world's known deposits of phosphate rock. There are accumulations of hydrocarbons and coal that can be used to produce nitrogen fertilizers, and there are well-documented deposits of potash, limestone, dolomite and gypsum. In this presentation, we argue that the time is ripe for Africa to utilize its abundant natural resources to improve the fertility of the soils and promote agricultural development.

Tropical Africa's Land Resources

Characteristics of Soils of the African Tropics

The parent materials of the soils of Tropical Africa are varied. They are primarily Pre-Cambrian materials, volcanic ashfalls and rocks that have been reworked by processes of erosion and denudation. These parent materials are not uniformly distributed, and the conditions under which they weathered to form soils have been diverse. The result is a tremendous variation in the inherent fertility of the soils. For example, soils that have been formed over basic materials such as basalts and dolerites tend to have inherent high soil fertility. In contrast, soils that have developed over granites which are acidic parent materials are usually sandy and vary in chemical fertility depending on the content of dark minerals such as olivine, pyroxenes and amphiboles. A great part of the soils of west and central Africa are complex basement soils developed over quartzite and other quartzite-like rocks such as sandstones. Soils developed over sandstones and quartzite are sandy and generally poor in fertility. Clay soils of poor inherent fertility are generally formed over mudstones. On the whole, the amount of rainfall determines whether soils developed over these parent materials are acidic or basic in reaction. The amount of rainfall

also determines the nature of the clay materials as well as the climax vegetation. Thus, the climax vegetation varies from tropical rainforest where rainfall exceeds 1500 mm in 5–10 months of the year to the tropical desert where annual rainfall of less than 200 mm occurs in less than 1 month. In forest zones, the high rainfall and high temperatures promote intense leaching and weathering. The clay fraction of these soils is dominated by kaolinite, halloysites and/or iron and aluminium oxides. The Oxisols and Ultisols of this region are devoid of basic cations. The clays are characterized by low and variable cation exchange capacity. Where the rainfall is less, as is the case with the savanna areas, bases are not easily leached and soils will have a relatively basic reaction. The clays may also contain substantial amounts of montmorillonites. Alfisols, Entisols and Andisols are the dominant soils of such environments. Overall, the largest category of land in tropical Africa (55%) is chemically fragile, easily degradable and of low inherent soil fertility. The main constraints of the soils are summarized below.

Generally low levels of organic matter: In tropical African soils dominated by Oxisols, Ultisols and Alfisols, the organic matter content varies from a high of 10% for soils derived from volcanic ashfalls and basic amphibolites to a low of less than 0.1%. Since organic matter is the main source of cation exchange capacity (CEC) and especially nitrogen, soils with low organic matter have low inherent fertility. The low levels of organic matter are exacerbated by intensive cultivation, deforestation, annual bush burning and the removal of crop residue from the land after harvest. Soil texture affects the rate of decline of organic matter. Reduced levels of organic matter result in poor stability of micro-aggregates making the soils more susceptible to erosion and surface sealing. These in turn lead to reduced moisture-holding capacity. Low organic matter results in low buffering capacity of the soils. Low buffering capacity reduces the ability of the soils to complex the iron and aluminium oxides and hydroxides that dominate the clay complex and influence the availability of such nutrients as phosphorus.

Low nutrient holding capacities: The kaolinites and halloysites that dominate the clay complex of the Oxisols, Ultisols and Alfisols are characteristically low-activity clays. These clays have variable charges, and hence, changes in pH of the soil solution result in drastic changes in the ability of the soils to hold on to exchangeable cations. Such cations are thus leached to lower levels where plant roots have no access to them. The combined effects of leaching of nutrients, losses through erosion and removal by crops leave the soils with a negative nutrient balance (Buresh et al., 1997). As we shall note later on, the presence of variable charge clays affects the reaction of phosphorus in these soils.

Aluminium toxicity: In Africa south of the Sahara and north of the Limpopo, approximately 550 million ha suffer from both acidity and the presence of free oxides of aluminium (Sanchez et al., 1991). The ability to render available soil phosphorus unavailable to crops is characteristic of these soils. In addition to phosphorus "fixation", the presence of free oxides of aluminium is toxic to crops.

Soil acidity: Nineteen percent or 481 million ha of soils in tropical Africa suffer from acidity even in the absence of toxic levels of the oxides of aluminium and iron. Characteristically, these soils are short of bases, and only specially adapted crops can grow. Such leguminous crops as groundnuts and cowpea which are excellent sources of protein cannot perform well in these soils because of limited capacity for nodulation processes to occur.

Poor infiltration and moisture retention: Excessive weathering and leaching result in shallow soils. Shallow soils and soils full of gravel make up more than 27% (645 million ha) of tropical Africa. These conditions promote poor water infiltration and retention. In such soils, moisture stress at critical periods in the growth cycle of crops is a common feature. Studies have shown that poor crop yields are often related more to poor soil moisture management than absolute insufficiency of rainfall. As was noted earlier, low levels of organic matter aggravate the situation. This is a vicious cycle since lack of moisture reduces biomass production that, in turn, lowers the amount of organic matter. In the absence of adequate organic matter and moisture, biological activity in soils is drastically reduced. The agro-ecological zones of tropical Africa and the characteristic soils are illustrated in Table 1.

Table 1 Agro-ecological zones and main soils in tropical Africa

Ecosystem	Main soils	Countries included
Miombo acid savannas	Ultisols	Zambia, Malawi, Southern DRC, Angola
	Oxisols	Mozambique, Tanzania and Madagascar
Tropical rainforest	Ultisols	Liberia, Sierra Leone, Eastern Nigeria
	Oxisols	Cameroon, Central African Republic, Gabon, DRC Congo, Madagascar
Guinea savanna	Alfisols	Senegal, Guinea, Cote d'Ivoire, Togo, Benin
Eutrophic East African savanna	Alfisols, Vertisols	Kenya, Tanzania, Uganda, Mozambique
Sahel	Entisols	Mauritania, Mali, Burkina Faso, Nigeria, Chad, Sudan
Sudan savanna	Alfisols	Senegal, Mali, Burkina Faso, Nigeria, Ghana, Cameroon, Chad, Sudan
Highlands	Oxisols, Andisols	Kenya, Uganda, Rwanda, Burundi, Eastern DRC, Ethiopia
Wetlands	Aquapts	Mali, Chad, DRC, Zambia, Gambia, Sudan
High veldt savannas	Alfisols	Southern Zambia, Zimbabwe
Low veldt savannas	Alfisols, Entisols	Zimbabwe, Botswana
Kalahari	Entisols	Angola, Namibia, Botswana

Source: Sanchez et al. (1991)

Impact on Agricultural Productivity

With the formidable array of limitations we have noted, it is not surprising to find that only 21% of Africa's land is suitable for cultivation. In better circumstances, this figure would not have been such a disastrous piece of news. However, stagnation and decline in food and cash crops are occurring in almost every region of sub-Saharan Africa. For example, the average cereal yield in sub-Saharan Africa, especially the Sudano-Sahelian region, is the lowest among the developing regions of the world (Henao and Baanante, 2006). We have noted that by their nature, the soils of tropical Africa, if not well managed, are highly susceptible to land degradation. One of the huge elements of land degradation in tropical Africa is human-induced depletion of soil fertility caused by a continuous and steady depletion of plant nutrients due to farming without the addition of nutrients that can replenish what the crops have taken. This process has become known as nutrient mining (Bationo et al., 1998; Buresh et al., 1997; Sanders et al., 1996; Smalling et al., 1997). Estimates by country show that nutrient mining is most severe (above 60 kg NPK/ha/year) in the Oxisols of the humid highlands of Central Africa (Eastern DRC, Uganda, Rwanda and Burundi). Nutrient losses due to erosion are highest in this region. Henao and Baanante (2006) conclude by noting that if erosion continues unabated, yield reductions by 2020 could be from 17 to 30% with an expected decrease of about 10 million tonnes of cereals, 15 million tonnes of tubers and 1 million tonnes of pulses. The consequences of low agricultural production caused by poor soil fertility can be summarized as follows:

- More than 50% of Africans live on less than US $1/day. Sub-Saharan Africa is the only region of the world where the population that lives on less than US $1/day has risen in the last two decades.
- Over 200 million Africans are chronically hungry.
- In 2002, Africa received 2.8 million tonnes of food aid.
- While the population has increased at the rate of about 3% per year, increases in grain production have averaged 1%.
- In terms of sustenance, cereal production has actually declined from 150 to 130 kg/person/year.
- In a continent where the greatest number of the population is engaged in agriculture, low agricultural production has meant low returns to labour and increase in poverty.

Where Do We Go From Here? Tried and Tested Solutions

When the African Presidents and Heads of Governments met in Abuja in June 2006, they were aware that eradicating extreme poverty by 2015 (Millennium Development Goal 1) would be daunting if not an impossible task unless agricultural production in sub-Saharan Africa is drastically improved. The population of Africa will top the 1 billion mark by 2025. If tropical African countries are to feed this population, the challenge of building the fertility of the soils to levels never before attained must be met. A tried and tested option for replenishing plant nutrients leached beyond the root zone, lost by erosion or removed by crops, is the application of fertilizers. Unfortunately, the use of inorganic fertilizers has come into disfavour among seemingly well-meaning

people who neither understand the magnitude of the food problem in Africa nor the history and geology of the continent. Incidentally, while for the past 30 years, 200 million ha of cropped land in tropical Africa has lost an average of 1,125 kg NPK/ha, over the same period, commercial farms in North America (the source of precious food aid) increased their nutrient capital by 3,700 kg NPK/ha (Pieri, 1989). The Abuja summit recommended the use of inorganic and organic sources of nutrients. When efficiently managed, inorganic fertilizers can rapidly boost soil fertility status. However, we must remember the very important roles played by organic matter in soils of tropical Africa. The dilemma that we face is the inadequate quantity and quality of organic sources of nutrients. There are simply not enough organic materials and besides, what organic matter there is has been produced in soils that are already very low in nutrients. As was aptly noted by Dennis Greenland (1992), *Recycling poverty levels of nutrients can only condemn most of Africa to continuing poverty*. It is however important to always remember that proper management of plant materials on the farm such as crop residues (that can be fed to animals to produce manure) is essential for sustainable agricultural production. A substantial body of evidence has accumulated (Dudal and Roy, 1995; Palm et al., 1997) to demonstrate that inorganic fertilizers alone cannot sustain high yields in the long term in most of the dominant ecosystems of tropical Africa. In a region where manure from livestock is scarce, crop residue is an important source of nutrients for soil fertility improvement.

In 1961, under the Freedom from Hunger Campaign, the Food and Agriculture Organization (FAO) of the United Nations initiated its Fertilizer Programme with the goal to "improve crop production and farmer's incomes through the efficient use of fertilizers" (Dudal, 2002). Fertilizer Programme activities, which included demonstrations and on-farm trials, were carried out in 23 countries in sub-Saharan countries from Senegal in extreme West Africa to Madagascar in East Africa. From the results obtained from the agricultural research stations, we know that there is a yield gap between what the potential yield in a given environment is and the amount of production by farmers. Biophysical factors that we have already discussed as well as the absence of improved crop varieties are partially responsible. Socio-economic and policy limitations are also partially responsible.

Through the work by FAO and that by numerous scientists since then, one can safely say that African farmers today know the value of NPK fertilizers. The average fertilizer application rates in kilograms per hectare in Africa have actually declined from nearly 20 kg nutrients/ha in 1983 to just over 16 kg of fertilizer nutrients/ha. The major obstacle is the high cost of inorganic fertilizers. For example, in 2005, it was observed that in the landlocked country of Zambia, approximately 230 million metric tonnes of fertilizers were imported. The import bill for fertilizers was only second to the import bill for petroleum products (K. Munyinda – personal communication). It was with this knowledge of the immense financial problems that face governments (who pay for and in cases subsidize fertilizers) and farmers (who have to scurry for funds to purchase fertilizers sometimes at full cost where subsidies do not exist) that the Presidents and Heads of Governments declared that *given the strategic importance of fertilizer in achieving the African Green Revolution to end hunger, the African Union Member States resolve to increase the level of the use of fertilizer from the current average of 8 kilograms per hectare to an average of at least 50 kilograms per hectare by 2015.* The rest of this document would attempt to show how Africa's abundant natural resources, especially agricultural minerals or agrominerals, can be used to supply the much needed nutrients for increased agricultural production.

Africa's Natural Resource Base

Apart from carbon, oxygen and hydrogen, the nutrients required by crops originate from the rocks that formed our soils. From the ancient Egyptians to the Greeks and the Romans, the value of rocks with regard to soil fertility has been known. However, the use of whole rocks for agriculture started in the late nineteenth century when Hensel (van Straaten, 2007) used finely ground rocks and minerals as low-cost, locally available geological nutrient resources for agricultural development. If the cost of manufactured inorganic fertilizers constrains their use to improve the productivity of Africa's impoverished soils, perhaps the time has come to find ways to use Africa's abundant natural resources as locally available sources of plant nutrients. Apart from potash, Africa is self-sufficient in fertilizer raw

materials. Ammonia synthesis is the first stage of nitrogen fertilizer production. The energy required for ammonia synthesis is usually provided by the use of coal, liquid hydrocarbons or natural gas. Africa has over 8, 9 and 6%, respectively, of the world's known reserves of these resources (van Kauwenbergh, 2006). Estimates are that Africa possesses in excess of 70% of the world's phosphate rock deposits. Almost as extensive is the occurrence of gypsum which supplies both calcium and sulphur. Substantive deposits of gypsum are found in at least 17 African countries from Angola to Zimbabwe. Limestone, another important source of calcium, is prevalent in many countries. Most of the gypsum and limestone deposits are currently exploited for the cement industry. Although Africa cannot boast of large sources of potassium, there are workable potassium deposits in Egypt, Ethiopia, Madagascar, Morocco, Niger and Tunisia. The production of fertilizers is capital intensive. In fact, fertilizer plant capacity utilization in sub-Saharan Africa is the lowest in the world, and production of N, P and K in this region has actually declined between 1990/1991 and 2002/2003. If the African Green Revolution is to become a reality, these local sources of plant nutrients should provide suitable and affordable alternatives to expensive manufactured fertilizers.

Increasing Africa's Natural Capital

Concept of "Capital"

Serageldin (1995) distinguished four categories of capital on the basis of value. *Natural capital* is defined as the stocks of resources generated by natural biogeochemical processes and solar energy that yield flows of useful services and amenities into the future (Daly, 1994). Examples of natural capital include forests, wetlands, soil and water. *Manufactured capital* constitutes what is normally considered in economic and financial terms. Examples are houses, factories, money, roads and equipment. *Human capital* represents the education, health and nutrition of individuals within a society. *Social capital* reflects the institutional and cultural basis by which a society functions. Natural capital can be either renewable or non-renewable, and soil fertility is an important form of renewable natural capital. For a more detailed discussion of this topic, the reader is referred to Izac (1997). All categories of capital play a key role in promoting human welfare. In fact, sustainable growth and poverty alleviation will require investment in all four forms of capital. Although, as we have stated, soil fertility is an important form of renewable natural capital, plant nutrients, the major indicator of soil fertility, are among the least resilient components of sustainability (Buresh et al., 1997).

The Ideal Entry Point: Boosting the Phosphorus Capital in Soils

If we must attain the goal of reducing abject poverty by 50% by 2015, the logical starting point is to build the fertility of our soils to levels never before attained. In other words, it is insufficient to talk of replenishing nutrients that have been lost through erosion and plant uptake. Because of the character of the parent materials of tropical African soils, the base of our soil fertility (natural capital) is extremely low. Therefore, for effective and sustained growth in agricultural production, it is imperative that we improve this low capital base. We recommend that this process should begin by boosting the P capital in our soils. Phosphorus plays key roles in plant and animal metabolism. It has functions of a structural nature in macromolecules such as nucleic acids and of energy transfer in metabolic pathways of biosynthesis and degradation. The irony is that although Africa accounts for over 70% of the world's known phosphate rock deposits, about 80% of the soils of tropical Africa have inadequate amounts of this critical nutrient element. In some of these soils with inadequate levels of P such as the sandy soils of the Sahel region, seedlings die when the phosphorus supply in the seeds is used up (Bationo et al., 1986). Nitrogen is also very essential for plant growth, and African soils are also notoriously low in this nutrient. However, as important as this nutrient is for crop production in tropical Africa, Mughogho et al. (1986) concluded that profitable returns to nitrogen addition to soils, especially in the drier regions, are possible only in the presence of threshold levels of P. In other words, a threshold level of P is required in the soils for there to be uptake of applied nitrogen.

There is another reason for choosing phosphorus as our entry point. Africa's indigenous phosphate rock deposits range from the highly reactive sedimentary deposits in Tanzania (Minjingu) and in the Tilemsi Valley of Mali to the relatively "unreactive" igneous deposits in Zaimbabwe (Dorowa) and Chilembwe (Zambia) (McClellan and Notholt, 1986). The literature has grown (and continues to grow) on the effect of the application of ground phosphate rock (PR) on crop production in sub-Saharan Africa (Bationo et al., 1986; Buresh et al., 1997). While this chapter will not review all the available agronomic information relating to the value of "direct application of PR" to crops, it is pertinent to point out that there is sufficient evidence to recommend the direct application of such phosphate rocks as the Minjingu, the Tilemsi, the Matam (Senegal) and the Tahoua (Niger) phosphate rocks (Bationo et al., 1986). Additional advantages of the direct application of PR to crops in tropical Africa can be summarized as follows:

☐ Many of the PR deposits in Africa are small and may not be used for industrial purposes. However, direct application enables the use of these deposits.

☐ Based on the unit cost of P, indigenous or locally available PR is usually the cheapest.

☐ PRs have a variable and complex chemical composition. As a result, PRs are sources of other nutrients than P. Such nutrients as calcium and magnesium can be released into the soil through the dissolution of PRs. The presence of higher levels of these exchangeable bases helps to neutralize the harmful effects of free iron and aluminium oxides in the Oxisols and Ultisols of tropical Africa.

☐ PRs are natural products or minerals requiring minimum metallurgical processing. The direct application of PRs avoids the traditional wet acidification process for the production of commercial water-soluble P fertilizers and circumvents the production cycle of polluting wastes such as phospho-gypsum and greenhouse gases. Use of PR therefore results in energy conservation and the protection of the environment from industrial pollution.

☐ Because they are natural products, PRs can be used in organic agriculture.

There are some limitations that must be borne in mind. These are as follows:

☐ Not all PRs are suitable for direct application especially to short-season crops such as cereals if the intention is to immediately increase crop yield. Several processes have been developed to increase the agronomic effectiveness of the unreactive PRs. The production of partially acidulated phosphate rocks (PAPR) using mineral acids has been perfected by the International Center for Soil Fertility and Agricultural Development (IFDC).

☐ Although a lot has been done to increase our knowledge of the main factors and conditions affecting the agronomic effectiveness of PRs, in Africa, there is lack of knowledge of the long-term effectiveness of PRs, the financial benefits as well as the socioeconomic and policy factors that affect the efficient use of PRs.

☐ The low grade (content of P) of some PRs compared with high-grade commercial P fertilizers makes them more expensive at the point of application. This economic evaluation is critical and should be made at the time of exploitation of the PR deposit.

☐ Because of their variable chemical composition, some sedimentary PRs may contain elements such as heavy metals and even radionuclides that upon dissolution of the PR in the soils may be harmful at some concentrations. For example, the phosphate rock from Togo is reputed to have high levels of cadmium.

Use of PR as an Investment in Natural Resource Capital in Tropical Africa

Sanchez et al. (1997) noted that there is an exact congruence between the concepts of capital stocks and service flows in economics and those of nutrient pools and nutrient fluxes in soil science. Addition of nutrients using either inorganic or organic forms increases the nutrient pools (natural capital). The P pools and fluxes are illustrated in Fig. 1. Total P in the soil can be grouped into three groups: The first group is the P that is in soil solution and P that is very lightly held on the clay surfaces (labile P). These two components constitute P that is readily available to plants. We have designated them as "Agricultural P" in the figure. The P that is held tightly by the clay minerals (the oxides and hydroxides of iron and aluminium) is not immediately available to plants. Soil organic matter

contains P. This P is not readily available to plants until the organic matter is mineralized. These two components, organic P and strongly sorbed P, constitute what we have called Capital P. It might be equated to money in the savings account. Parent materials from which soils have formed have unweathered apatite (the P-bearing mineral) in their structure. In addition to this "structural" P, the soil also has P that is irreversibly fixed by the clay minerals. The structural P (inert P) and P that is irreversibly fixed constitute the third group of P in the soil. The first two classes are very important in plant nutrition. It is important to note that there is equilibrium between Agricultural P and Capital P.

In Fig. 1, the P inflows define the various sources of P that are added to the soil. Thus, when water-soluble inorganic fertilizers such as Triple Superphosphate are added to soils, the Agricultural P is immediately boosted. As reactions with the soil system occur, some of this added P becomes transformed into Capital P. Of immense significance to us is the fact that when PR is added to soils, the "Agricultural" P is also boosted. The amount by which this fraction is boosted depends on the reactivity of the rock, the soil conditions (such as pH) and the characteristics of the plants growing in the soil (Sale and Mokwunye, 1993). As stated earlier, organic P contributes to Agricultural P upon mineralization of organic matter. In Fig. 1, the P fluxes are the "outflows" of P which can be the result of crop uptake, erosion, runoff, etc. During the cropping season, the farmer tries to maximize the service flows. P fluxes or outflows reduce the total P in the soil system. As the nutrient flows decrease (because of the fluxes) during the growing season, the P stock decreases and this decrease can be compared to capital depreciation.

Agricultural P is analogous to "liquid capital" or "money in a checking account" in the bank. As plants take up P from the Agricultural P component, more P is released from the Capital P to restore the equilibrium. The concept that we are attempting to promote is that the addition of PR to soils increases both Agricultural P and Capital P. Usually, it is the Capital P that is increased the greatest. Because the level of Capital P is greatly increased when PR is added and recognizing the equilibrium between Capital P and Agricultural P, this practice ensures that low Agricultural P levels are "a thing of the past" in such soils. While the degree to which the Agricultural P can be increased when PR is

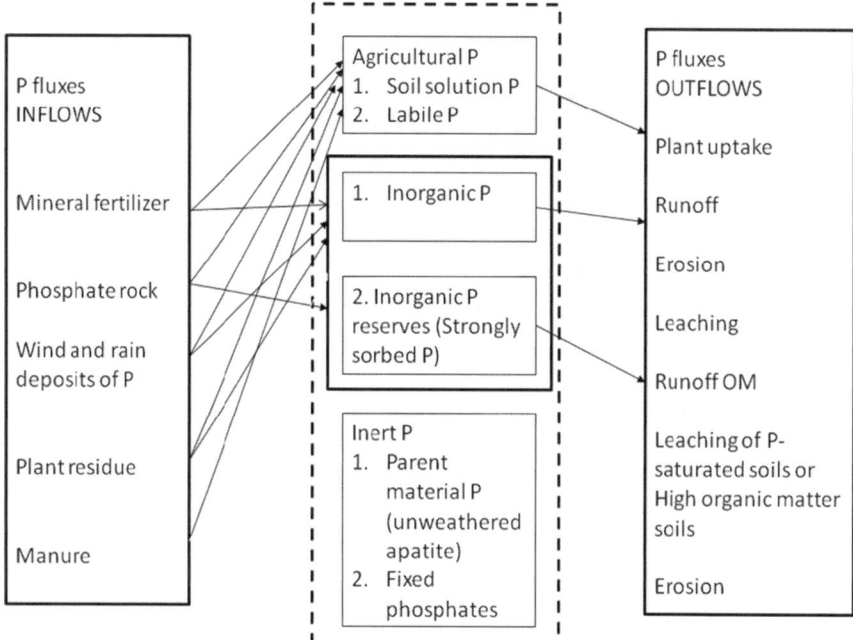

Fig. 1 Phosphorus pool, inflows and outflows in soils

added is dependent on the characteristics of the rock, such is not the case for Capital P. In soils with low native P, both Agricultural P and Capital P are low. The addition of locally available PR sources therefore boosts both Agricultural P and Capital P in Africa's soils.

What about the use of organic sources of P to boost both Agricultural P and Capital P? The use of manure or cow dung has been advocated, and the soil fertility literature in tropical Africa is replete with trials involving the use of manure (see the review by Ssali et al., 1986). It must not be forgotten that the use of manure in the African context is part of a farm's internal flow of nutrients. It does not involve the addition of nutrients from outside the farm. In West Africa, livestock can potentially transfer 2.5 kg N and 0.6 kg P per hectare of cropped land. To restore the nutrient levels removed by crops would require the application of 6–20 tonnes of manure per hectare. This would require about 10–40 ha of grazing land to maintain the yields of 1 ha of cropland. At best, about 700 kg of manure is available to the best managed farm in semi-arid West Africa (Bationo, 2004). This quantity is sufficient to raise yields by about 2% per year. If our goal is to raise the fertility of our soils to levels never before attained, other avenues must be tried.

Crystallizing the Concept of P "Recapitalization" Using PRs

We have seen how the P pool in the soil system can be built up, especially the Agricultural P and the Capital P. The reason that this is possible is the ability of all P sources to continue to supply P to crops beyond the initial year of application. This "residual effectiveness" of P fertilizers and of phosphate rocks in particular is a significant property which enables PRs to be used for the long-term improvement of the productive capacity of African soils (Bationo et al., 1995). The principle of "recapitalization" thus involves the build-up of the Capital P in tropical African soils. It is important, at this juncture, to remind ourselves that what is desirable is not to maintain the fertility of Africa's degraded lands. What is needed is to build the fertility to levels hitherto unknown to the African farmer. Drastic situations require drastic measures. It is also important to note that the quality of the PR is not important when we are considering its use for P recapitalization. Therefore, all local PR sources in Africa can be used to boost Capital P.

Because the P in PR can be released to the crop over a long period of time, the use of PR can be considered an investment in natural capital (Cescas M.P., Ph.D Thesis, University of Illinois, Urbana-Champaign, 1969). However, it should be noted that the duration of crop yield responses is dependent on the amount of P applied, the soil's P sorption capacity and the cropping intensity. In general, the larger the P applied, the longer the residual effect. PR applied to soils with high P sorption capacity has longer residual effectiveness and the higher the number of crops harvested during the year, the shorter is the residual effect. Strategies for P capitalization will therefore differ according to these variables. The strategy may involve a one-time application of a large dose, of the order of 150–500 kg P/ha, or annual applications of smaller doses using the local phosphate rock. Whatever approach is adopted, the principle is to ensure that the service flows are maximized for a given crop or cropping system.

What we are advocating is a two-pronged approach. There are reactive sedimentary PRs such as Minjingu PR and Tilemsi PR that can readily be used for direct application. These reactive phosphate rocks can readily increase the Agricultural P. The second approach is to use any locally available PR. This can be applied either as a one-time large dose or in smaller annual applications for boosting the Capital P. In 1994, the World Bank in consultation with the centers supported by the Consultative Group on International Agricultural Research (CGIAR) and other interested parties launched an initiative titled "Development of National Strategies for Soil Fertility Recapitalization in sub-Saharan Africa" (Buresh et al., 1997; Baanante, 1998). The goal was to encourage the development of Soil Fertility Action Plans by the different countries in sub-Saharan Africa. One of the outcomes was the document, Framework for National Soil Fertility Improvement Action Plans. Recognizing that the soils of tropical Africa were poor in phosphorus while the region has numerous deposits of phosphate rock, the Framework called for the use of local phosphate rocks as a capital investment in natural resources in Africa. The World Bank's plan called for a one-time massive application of PR which would be expected to overcome the high P sorption problems associated with the soils of the humid region of tropical Africa (World Bank, 1996).

We had earlier alluded to the fact that the main issue in tropical Africa is not the *maintenance* of soil fertility because simply maintaining the low fertility levels would not improve agricultural production. Where soil fertility has been increased through boosting the Capital P, it would make sense to develop technologies that would maintain the fertility levels of the soils. Therefore, unless the PR is highly reactive, its use, whether in small or large doses, will not immediately increase the levels of Agricultural P for purposes of obtaining optimum yields. To make up for this shortcoming, the use of small amounts of water-soluble P sources to immediately augment the amount of Agricultural P is recommended. What should be remembered is that because PR has already been applied to the soil, a greater portion of P applied through the water-soluble source remains as Agricultural P much longer in the soil and hence is more available to crops. Currently, African farmers apply less than optimal doses of NPK to their farms. A combination of this less than adequate amounts of NPK and the large dose of PR ensures that the service flows are maximized for the given crop or cropping system.

As enthusiastic as we are about the use of PR for boosting Capital P, this practice will not result in increased soil productivity unless additional steps are taken. Runoff and erosion are the main mechanisms of P loss from the crop/soil system. If the investment on Capital P is to pay off, soil erosion control technologies must be in place. In fact, the application of high amounts of PR when erosion control measures are not in place does more harm to the ecosystem than good as erosion promotes the pollution of rivers, lakes and other water bodies. There are both physical and biological methods of controlling erosion. Growing leguminous hedges or vegetative filter strips along the contour (Kiepe and Rao, 1994) are tried and tested biological methods to control erosion. These biological methods provide useful products for the household and are thus attractive to adopt. In addition to erosion control, the adoption of sound agronomic practices is essential to the success of the investment programme. Some of these practices involve the use of appropriate techniques of land clearing, the adoption of tillage practices that take into consideration the fragile nature of the soils of tropical Africa and the use of crop mixtures or rotations that include leguminous species. For the investment to pay off, the use of improved crop varieties is also essential.

Sharing Responsibilities

During the implementation of the World Bank project, case studies were carried out in Burkina Faso, Madagascar and Zimbabwe. The International Center for Soil Fertility and Agricultural Development (IFDC) assisted in setting up the National Soil Fertility Action Plan for Burkina Faso. The results of the case studies in the three countries showed that locally available PR could be used to boost Capital P. However, the thorny issue was "Who is to pay for this practice?" Sustainable agricultural development in sub-Saharan Africa will depend on strong partnerships between national governments, their agencies, the farming population and Africa's development partners. African Union (2003) demonstrated that African Governments recognize that agriculture is a way of life for the vast majority of people in tropical Africa. By declaring their intention to engineer an African Green Revolution before 2015, African Presidents and Heads of Governments meeting at the Abuja Summit gave teeth to the Maputo Declaration. But African Governments must go beyond the expression of what is desirable. Governments must put in place appropriate reform measures to ensure open, competitive markets for both inputs and outputs. In tropical Africa, a poorly defined land tenure system provides little incentive for farmers to invest their meagre resources on soil improvement. For more than two decades, we have been told that subsidies on fertilizers constituted an intolerable drain on government's resources. The recommendation of the Abuja Summit that *With immediate effect, the African Union Member States must improve farmers' access to fertilizer, by granting, with the support of Africa's Development partners, targeted subsidies in favour of the fertilizer sector, with special attention to poor farmers* must erase the practices of the 1980s and 1990s. Research scientists must not conduct research simply to obtain outputs. They must adopt an approach that enables them to transform the outputs to outcomes and eventually impact. The technologies that ensure the increase in soil fertility capital already set out in this chapter are not being used by the farmers because they have not been involved in their development and testing. The cost of the use of PR as an investment on natural capital must not be borne by the resource-poor farmers of sub-Saharan Africa. National Governments and their development partners

must come to their aid. This practice is not simply an investment on "natural capital"; it is an investment on "national capital". We recommend that the following actions be taken:

1. Study the feasibility of mining, grinding and spreading the local PR.
2. Use a crop duster to fly over the entire countryside and spread the PR on every homestead.
3. Develop, with local communities, appropriate soil conservation measures to prevent runoff and erosion.
4. Support the development and use of improved crop varieties.
5. Take necessary measures that would promote the development of input and output markets.
6. Monitor crop production increases and the regeneration of vegetation in erstwhile barren lands.

Conclusion

Do African soils only sustain subsistence agriculture? This was relevant given the fact that low soil fertility has resulted in the current mode of agricultural production in tropical Africa being subsistence agriculture. Africa occupies 20% of the world's land mass but only 21% of this land is suitable for cultivation. Low soil fertility is the major problem. Nature dealt the African farmer a bad hand. But we can reverse this hand. Investing in soil fertility capital is one sure way to compensate for the bad hand dealt by nature. But, as we have noted, nature has been kind enough to provide the African continent with abundant supplies of local sources of P, calcium (Ca), magnesium (Mg) and sulphur (S) and many rocks that contain other nutrients needed by crops. Agricultural intensification traditionally requires higher levels of plant nutrients. In Africa, agricultural intensification would require harnessing these natural resources to complement good agronomic practices, improved seeds, improved input and output markets and vibrant agricultural research programmes. The task of improving the *long-term* productivity of the soils through an investment in natural resource capital must not be left to those members of the society least able to afford it.

The benefits from the use of PR to increase the Capital P in the soils accrue to the "national society", the "global society" and to the farming households (Izac, 1997). Given these conditions, an equitable mechanism would be for all the beneficiaries to share the costs of implementing the boosting of Capital P in soils. Making sure that erosion control measures are put in place and that appropriate crop management practices are adopted are the responsibilities of the farmers. The cost of mining, grinding and spreading the PR must be shared between the national governments and their development partners.

References

African Union (2003) The Maputo declaration: assembly of the African Union (Second Ordinary Session, 10–12 July 2003). Declaration on agriculture and food security in Africa, Maputo, Mozambique

Baanante CA (1998) Economic evaluation of the use of phosphate fertilizers as a capital investment. In: Johnston AE, Syers JK (eds) Nutrient management for sustainable agriculture in Asia. Wallingford, CT, CAB International, pp 109–120

Bationo A (eds) (2004) Managing nutrient cycles to sustain soil fertility in Sub-Saharan Africa. Academy Science Publishers, Nairobi, Kenya, pp 1–23

Bationo A, Buerkert A, Sedogo MP, Christianson BC, Mokwunye AU (1995) A critical review of crop residue use as soil amendment in the West African semi-arid tropics. In: Powell JM, Fernandez Rivera S, Williams TO, Renard C (eds) Livestock and sustainable nutrient cycling in mixed farming systems of sub-Saharan Africa. Proceedings of the International Conference ILCA, Addis Ababa, pp 305–322

Bationo A, Lompo F, Koala S (1998) Research on nutrient flows and balances in West Africa: State-of-the-art. Agric Ecosyst Environ 71:19–35

Bationo A, Mughogho SK, Mokwunye AU (1986) Agronomic evaluation of phosphate fertilizers in tropical Africa. In: Vlek PLG, Mokwunye AU (eds) Management of nitrogen and P fertilizers in sub-Saharan Africa. Martinus Nijhoff Publ, Dordrecht, pp 283–318

Buresh RJ, Smithson PC, Heliums DT (1997) Building soil phosphorus capital in Africa. In: Buresh RJ et al (eds) Replenishing soil fertility in Africa. SSSA Spec. Publ. 51. SSSA, Madison, WI, pp 111–149

Daly HE (1994) Operationalizing sustainable development by investing in natural capital. In: Jansson A et al (eds) Investing in natural capital. Island Press, Washington, DC, pp 22–37

Dudal R (2002) Forty years of soil fertility work in sub-Saharan Africa. In: Vanlauwe B, Diels J, Sanginga N, Merckx R (eds) Integrated plant nutrient management in sub-Saharan Africa. From concept to practice. CABI Publishing, Wallingford, pp 7–21

Dudal R, Roy RN (1995) Integrated plant nutrition systems. FAO fertilizer and plant nutrition bulletin 12. FAO, Rome, Italy, pp 426

Henao J, Baanante C (2006) Agricultural production and soil nutrient mining in Africa: Implication for resource conservation and policy development. IFDC Tech. Bull. International Fertilizer Development Center. Muscle Shoals, AL

Izac A-MN (1997) Ecological economics of investing in natural resource capital in Africa. In: Buresh RJ et al (eds) Replenishing soil fertility in Africa. SSSA Spec. Publ. 51. SSSA, Madison, Wl, pp 237–251

Kiepe P, Rao MR (1994) Management of agroforestry for the conservation and utilisation of land and water resources. Outlook Agric 23:17–25

McClellan GH, Notholt AJG (1986) Phosphate deposits of Sub-Saharan Africa. In: Mokwunye AU, Vlek PLG (eds) Management of nitrogen and fertilisers in sub-Saharan Africa. Martinus Nighoff, Dordrecht, pp 173–223

Mughogho SK, Bationo A, Christianson B, Vlek PLG (1986) Management of nitrogen fertilizers for tropical African soils. In: Mokwunye AU, Vlek PLG (eds) Management of nitrogen and phosphorus fertilizers in sub-Saharan Africa. Martinus Nijhoff Publishers, Dordrecht, pp 117–172

Palm CA, Myers RJK, Nandwa SM (1997) Combined use of organic and inorganic nutrient sources for soil fertility maintenance and replenishment. In: Buresh RJ et al (eds) Replenishing soil fertility in Africa. SSSA Spec. Publ. 51. SSSA, Madison, WI, pp 193–217

Pieri CJMG (1989) Fertility of soils. A future for farming in the West African savannah. Springer, Berlin

Sale PWG, Mokwunye AU (1993) Use of phosphate rocks in the tropics. Fert Res 35:33–45

Sanchez PA, Shepherd KD, Soule MJ, Place FM, Buresh RJ, Izac A-MN, Mokwunye AU, Kwesiga FR, Ndiritu CG, Woomer PL (1997) Soil fertility replenishment in Africa: An investment in natural resource capital. In: Buresh RJ, Sánchez PA, Calhoun F (eds) Replenishing soil fertility in Africa. SSSA Spec. Publ. 51. SSSA, Madison, WI, pp 1–46

Sanchez PA et al (1991) Soils of tropical Africa. Report to the Rockefeller Foundation, New York, NY

Sanders JH, Shapiro BI, Ramaswamy S (1996) The economics of agricultural technology in semi-arid sub-Saharan Africa. Johns Hopkins University Press, Baltimore, MD

Serageldin I (1995) Sustainability and the wealth of nations. World Bank, Washington, DC

Smalling EMA, Nandwa SM, Janssen BH (1997) Soil fertility in Africa is at stake. In: Buresh RJ, Sánchez PA, Calhoun F (eds) Replenishing soil fertility in Africa. Special Publication no. 51 American Society of Agronomy and Soil Science Society of America, Madison, WI, pp 46–61

Ssali H, Ahn P, Mokwunye AU (1986) Fertility of soils in tropical Africa: a historical perspective.. In: Mokwunye AU, Vlek PLG (eds) Management of nitrogen and phosphorus fertilizers in sub-Saharan Africa. Martinus Nijhoff, Dordrecht, pp 59–82

Van Kauwenbergh SJ (2006) Fertilizer raw material resources of Africa. International Fertilizer Development Centre (IFDC), Washington, DC

Van Straaten P (2007) Agrogeology: the use of rocks for crops. ICRAF, Nairobi

World Bank (1996) Natural resource degradation in sub-Saharan Africa: Restoration of soil fertility. Africa Region. World Bank, Washington, DC

The Geological Basis of Farming in Africa

P. van Straaten

Abstract Soil fertility is largely a function of climate, parent material and management of the soils. Parent materials are derived from various rock types. Most of the rocks in Africa are of Precambrian age, more than 544 million years old. The oldest rocks in Africa, Archean granites and granite gneisses as well as old volcanic rocks, called greenstones, are more than 3.2 billion years old. They form the stable granite–greenstone nuclei of the continent. Folded Precambrian metamorphic belts, the so-called mobile belts, accreted around these nuclei. While the granite and granite gneiss-dominated areas are mineralogically and chemically more homogenous, the areas underlain by deeply eroded Precambrian fold belts vary widely in composition. The African post-Precambrian (<544 million years) history is marked by the development of sedimentary basins with varying sedimentary rocks at the perimeter of the igneous and metamorphic stable continent. Extensive parts of the interior of Africa are covered by Mesozoic to Cenozoic sandstone-dominated basins. Soil rejuvenating volcanic rocks extruded from lower parts of the earth crust and mantle mainly during Mesozoic and Cenozoic times in linear zones. Massive volcanic outpourings took place since about 30 million years, in Tertiary Rift environments. Soils derived from these rock types vary. Soils derived from silica over-saturated igneous and metamorphic rocks as well as sandstone-dominated lithologies form sandy soils with limiting, inherently low nutrient concentrations and low water-holding capacities. In contrast, soils derived from volcanic, silica-saturated or silica-undersaturated igneous rocks, young or old, contain higher concentrations of total Ca, Mg, P and micronutrients. Under suitable climatic conditions they weather into more clay-rich, fertile soils with high water-holding capacities. Examples of strong inherent soil differences related to the underlying parent materials are given from Zimbabwe and Uganda. Inherent differences in soil properties can spatially vary over short distances. The main nutrient constraints of African soils are N and P deficiencies. Most soils are also low in organic matter. There are many soil fertility restoration management strategies, making use of local organic and imported and local inorganic nutrient resources, such as phosphate rocks and fertilizers. Africa is endowed with large deposits of naturally occurring agrominerals and fertilizer raw materials. Natural gas, the principal feedstock for industrial N fertilizer production, is found in coastal zones of Africa, mainly at the northern and western sides of the continent. Extensive sedimentary phosphate rock resources occur along the west coast of Africa and in North Africa. Igneous phosphate rocks are found mainly in eastern and Central Africa and along linear zones in West and southwest Africa. Large gypsum deposits occur in Mesozoic strata in coastal sedimentary basins. Calcium–magnesium carbonates, in various forms, occur in almost every country of the continent.

Keywords Africa · Agrominerals · Fertilizer raw materials · Geology · Soil rejuvenation

P. van Straaten (✉)
School of Environmental Sciences, University of Guelph,
Guelph, ON, Canada
e-mail: pvanstra@uoguelph.ca

A. Bationo et al. (eds.), *Innovations as Key to the Green Revolution in Africa*,
DOI 10.1007/978-90-481-2543-2_3, © Springer Science+Business Media B.V. 2011

Introduction

African food security is at risk as per capita food production in large parts of Africa has not kept pace with population growth. The quality of the soil the farmers cultivate depends largely on environmental conditions, especially rainfall, temperature and length of growing period, and on the inherent soil fertility characteristics, including chemical and physical parent material attributes. It is well known that in Africa, like in many parts of the world, the most fertile soils are those derived from young volcanic rocks (volcanic lavas and tuffs), provided they occur in areas with suitable climatic conditions. Unlike soils where the interventions of the Green Revolution were successful, on alluvial and the young volcanic soils in humid climates in Asia, the majority of soils in Africa are derived from Precambrian igneous and metamorphosed rocks with differing chemical compositions, mainly aluminium silicate-rich igneous and metamorphosed rocks, and sandstone-dominated sediments with mostly low levels of weatherable minerals and generally low inherent geological nutrient content.

Nutrient depletion through continuous cropping ('nutrient mining') in smallholder farms is seen as fundamental root cause for declining food production in many parts of sub-Saharan Africa (Sanchez et al., 1997; Sanchez, 2002). Researchers have recognized that low soil fertility, especially N and P deficiencies, as well as low soil organic matter contents forms the main biophysical constraint to sustainable farming (Sanchez et al., 1997). In addition, wind and water erosion can cause major physical losses of topsoils.

The productivity of soils depends largely on climatic conditions and inherent physical and chemical status of soils as well as soil management by farmers. The chemical characteristics of soils are largely functions of parent materials (including the amount of weatherable minerals present), climatic conditions, weathering rates, as well as soil nutrient replenishment practices such as the use of organic and inorganic nutrient inputs, e.g. manures and fertilizers.

Where soil fertility and organic matter is low, either inherently low or anthropogenically degraded, nutrients and organic carbon have to be replenished to improve livelihoods of smallholder farmers in Africa. Among other practices to improve soil productivity (e.g. application of crop residue mulch, better organic residue and manure management, intercropping or rotation of cereals with legumes, agroforestry practices, improved legume fallow systems) it is important to replenish soil nutrients by nutrient inputs, either through imported fertilizers or locally available mineral and organic nutrient inputs.

This chapter looks at the basic geology of Africa with the distribution of various rock types in space and time. It discusses the relationship between inherent soil fertility and geological parent materials. In addition, Africa's large geological nutrient resources, agrominerals such as phosphate rocks, are seen as nutrient assets to enhance and restore inherently infertile or anthropogenically degraded soils in Africa.

Geology of Africa

The Oldest Rocks of Africa: The Precambrian (3,200–544 Million Years)

The simplified geological map of Africa (Fig. 1) shows the general distribution of rocks and structures in Africa. It shows that large parts are underlain by granites and metamorphic rocks, while other parts are either underlain by continental sedimentary rocks, mainly sandstones, and only very small parts underlain by marine sediments in coastal areas. The distribution of rocks is related to the development of the continent over a geologically very long time period.

Africa is one of the oldest continents on Earth. Africa has been a stable continent over most of its history. Only the northern and southern tips of the continent (the Atlas and the Cape Fold Belts) have been subjected to young mountain-building processes. The oldest rocks, found in southern Africa, were formed 3,600–3,200 million years ago, in the oldest Precambrian, the Archean. These Archean rocks consist of mainly granites and granite gneisses, igneous and metamorphic rocks formed at great depth and high temperatures. Enclosed in these largely igneous rocks are narrow belts of so-called greenstone belts, formed predominantly from volcanic and sedimentary rocks. The granite–greenstone terrains together form the so-called cratons. The granites and granite gneisses of these cratons make up relatively homogenous rock massifs. As seen in Fig. 2, more than 2,000 million

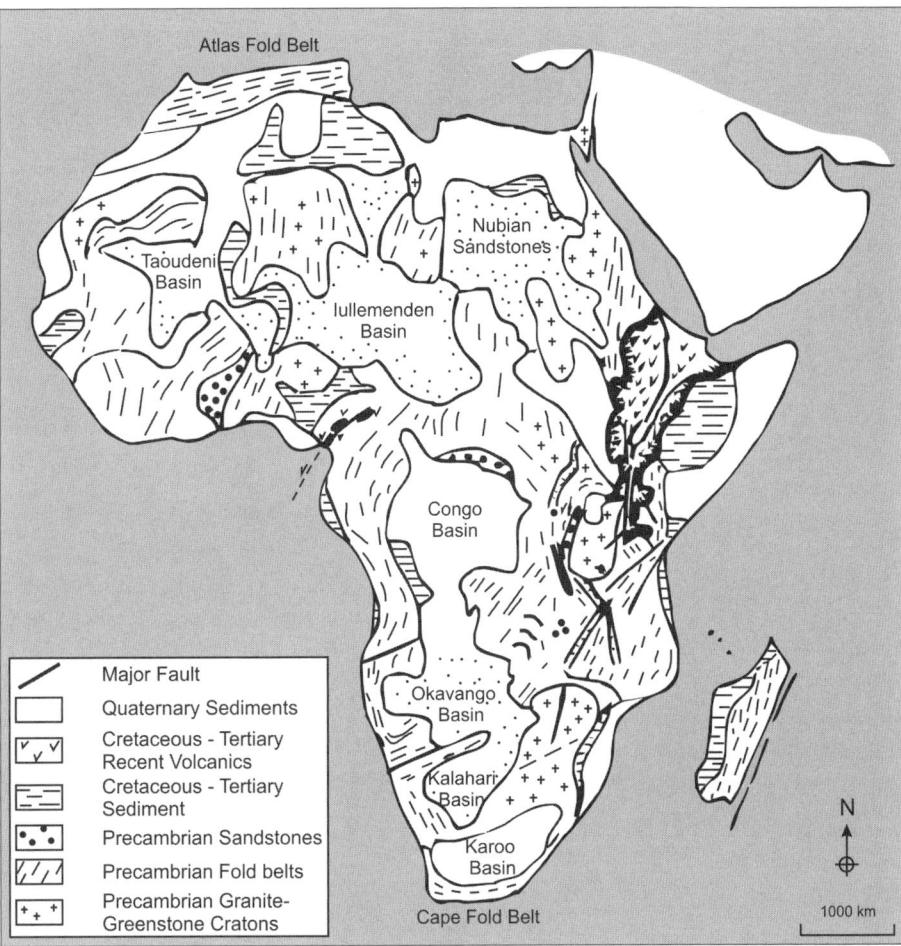

Fig. 1 Simplified geological map of Africa

years old, consolidated cratons underlie extensive parts of Africa (Kampunzu and Popoff, 1991; Schlueter, 2006).

The old granite–greenstone cratons are surrounded by 'younger' Precambrian rocks, made up of folded and metamorphosed sedimentary rocks as well as igneous rocks, the so-called mobile belts. These folded and metamorphosed Proterozoic (2,500–544 million years old) rock suites formed as a result of mountain-building processes along extended narrow belts and basins. However, the mountain ranges are not high mountain ranges any longer; they have been eroded to the crystalline core over millions of years. While the Archean granitic rocks are relatively homogenous in composition, the rock types in the folded and metamorphosed rock suites of the Proterozoic are largely heterogenous. They include rock types like mica-schists (metamorphosed mudstones), amphibolites (metamorphosed volcanic rocks), quartzites (metamorphosed sandstones), marbles (metamorphosed limestones and dolostones) and gneisses (highly metamorphosed sedimentary and igneous rocks).

At the end of the Precambrian and at the beginning of the Phanerozoic, around 500 million years ago, almost all of present Africa had consolidated into a stable continent, part of the super-continent 'Gondwanaland' which included South America, India, Australia and Antarctica. In large parts of the southern hemisphere this time period was a period of very cold climates. Rock types of glacial origin indicate that large parts of the earth were covered in ice and snow. Some scientists (Hoffman et al., 1998) claim at that time that the whole earth was entombed in ice; the earth was 'snowball earth'.

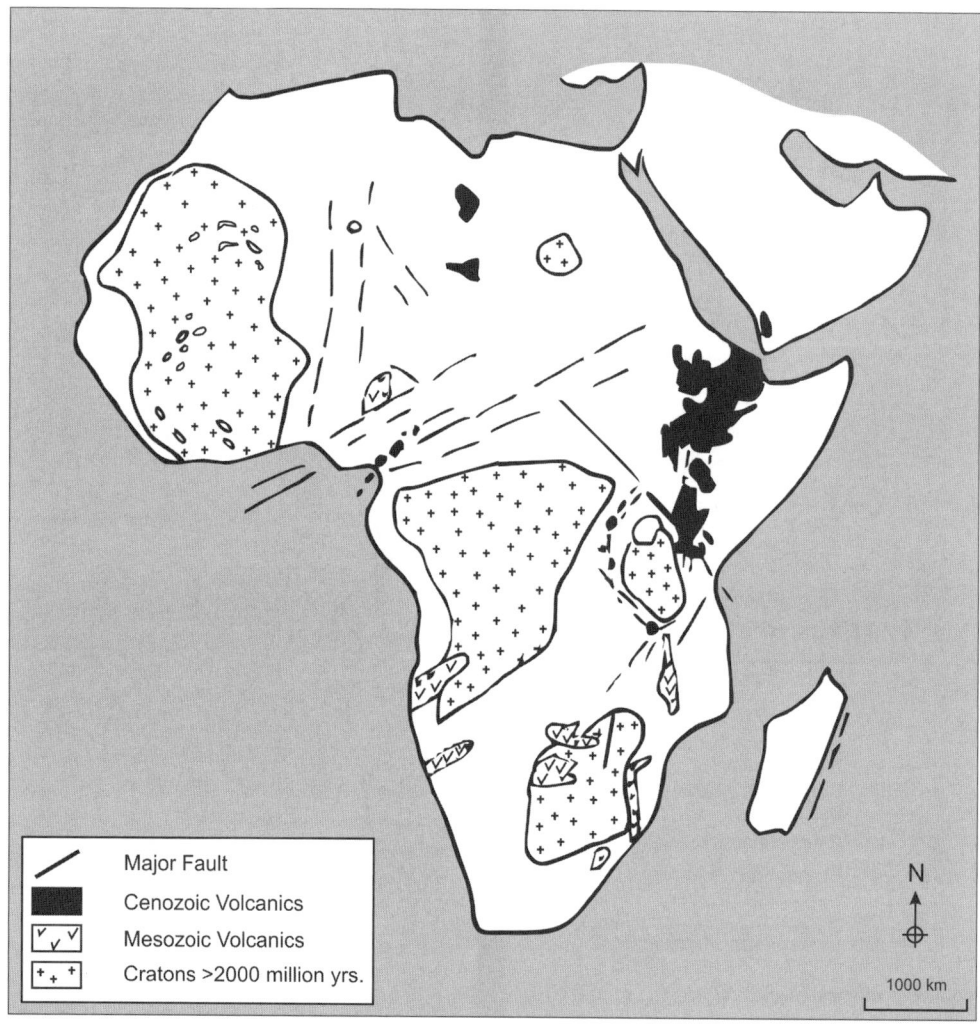

Fig. 2 Distribution of >2,000 million-year-old, consolidated cratons and Mesozoic and Cenozoic volcanic rocks in Africa (modified after Kampunzu and Popoff 1991)

Rocks of Africa: The Phanerozoic – 544 Million Years and Younger

In contrast to the metamorphosed and igneous Precambrian rocks in Africa, most rocks formed during the Phanerozoic, 544 million years and younger, are of sedimentary origin. They include limestones, sandstones, shales, and other sedimentary rocks. Volcanic rocks occur only along a few linear zones. They extruded mainly in the Mesozoic and Tertiary. Some of the volcanoes, along the rift valleys, are still active today.

At the end of the Precambrian and the beginning of the Phanerozoic, southern equatorial Africa

was in the grip of a very cold climate. While in the early Phanerozoic, large parts of North and West Africa were covered by marine and sediments, large parts of southern and central Africa remained elevated land under continental conditions, covered by ice-derived tillites, coal beds and deltaic, fluvial and aeolian sandy rocks (Schlueter, 2006). With the exception of voluminous outpourings of volcanic rocks in the Drakensberg area of southern Africa in the Lower Jurassic (190–155 million years) large parts of southeastern South Africa are underlain by continental sandstones and shales of the Karoo Supergroup.

Between the late Jurassic and early Cretaceous (140–130 million years), the super-continent Gondwanaland started to break apart, and at about

130 million years ago Africa started to separate from South America and the Atlantic formed. During the early phases of the fragmentation of the super-continent, the so-called sag basins and rifts developed as a result of crustal extension. Rifts like the Luangwa valley in Zambia and Karoo basins filled with continental sediments, and coal beds formed in southern Africa (Schlueter, 2006). During the early phases of the separation of Africa and South America extensive fault systems developed, along which carbonatites and alkaline rocks intruded.

From the Mesozoic onwards, continental Africa was surrounded by oceans and marine coastal basins developed with the deposition of marine sediments, such as shales, sandstones, limestones and gypsum deposits in the coastal basins. Phosphate-rich sediments were deposited mainly in the late Cretaceous to mid-Tertiary (75–55 million years) in coastal basins, mainly in western Africa (see below). Large parts of the interior of Africa were covered by Mesozoic to Cenozoic sandstone-dominated basins, for example, the Taoudeni, Illumenden, Okavango and Kalahari Basins (Fig. 1).

A major structural and magmatic event started in the early Tertiary, some 30 million years ago (Baker et al., 1972), when the East African Rift began to develop as a result of regional updoming and subsequent crustal extension with major faulting and rifting. The modern Rift Valley extends from the Red Sea southward and splits into two rift branches in Central East Africa, the Eastern Rift and the Western Rift. The formation of the Ethiopian part of the Rift Valley and the Eastern Rift was accompanied by extensive outpourings of massive volcanic lavas (Fig. 2). The updoming of the rift area and the accumulation of massive volcanic piles, totalling 220,000 km^3 of nutrient-rich volcanic rocks (Williams and Chapman, 1986), created highlands with elevations well above 1500 m. The volcanic extrusive rocks are mainly basalts, phonolites and trachytes as well as nephelinites. The Western Rift is less volcanic in nature and is dominated by deep lakes (e.g. Lake Tanganyika). Among the volcanic rocks found along the Western Rift are K-rich volcanics and carbonatites especially in intersecting rift structures (e.g. in the Mbeya area of Tanzania; van Straaten, 1989).

Another zone of structural weakness and subsequent volcanism is the Cameroon Fault Line (66–30 million years) which followed the Central African mobile zone (Kampunzu and Popoff, 1991) (Fig. 2).

The main Phanerozoic volcanic events took place in the Mesozoic (probably linked to the break-up of Gondwanaland) and in the Tertiary along extensional structures linked to the development of the Rift Valley System, as well as along the Cameroon Volcanic Line.

The volcanic eruptions that are associated with these large-scale structural zones brought young fresh rock material derived from great depth in the crust and upper mantle to the surface of the earth. These extrusive rocks are rich in easily weatherable minerals, such as olivines, pyroxenes, feldspars and feldspathoids.

From the foregoing description of the simplified geology of Africa it can be seen that extensive parts of the continent are underlain by Archean granite–greenstone terranes, Proterozoic mobile belts and Mesozoic and younger sandstone-dominated sedimentary inland basins. Marine sedimentary basins developed only along the edges of the continent and in some specific intercratonic 'straits', for example, the Gao Strait in West Africa. Only small areas are covered by young volcanic rocks associated with structurally active zones, e.g. the Rift Valley and the Cameroon volcanic line.

In general terms, the African continent formed a stable craton since the Precambrian. Except for the Paleozoic, Mesozoic and Cenozoic fold belts in Morocco (the Atlas fold belt) and South Africa (the Cape fold belt), Africa has not been subjected to large-scale and active subduction-related plate tectonics like in South America, Europe and Asia. Africa remained largely stable over the last 500 million years and only Mesozoic to Cenozoic rift and extensive fault structures lead to soil rejuvenating volcanism.

Geological Soil Rejuvenation

Fyfe et al. (1983) illustrated the importance of the geological concept of plate tectonics in geological nutrient cycling and distribution of fertile soils. They showed that geological subduction processes at the margin of continents, as well as formation of magmas along linear rift zones and 'hot spots', have given rise to new magmatic products with an abundance of nutrient elements. Some of the most fertile regions of the Earth, New Zealand, California, Japan, the Philippines and Java, are located at continental plate margins. Other regions of the Earth with 'hot spot' volcanic activities and volcanic activities related to rift formations are also known to be covered with fertile soils.

Other processes that lead to soil rejuvenation include glaciation, and redistribution of unweathered rock and mineral debris through fluvial processes, such as the redistribution of freshly exposed rocks into river plains and alluvial fans, like the Nile river plains and delta, as well as airborne dust redistribution, e.g. in West Africa (Fyfe et al., 1983; van Straaten, 2007).

In Africa, like in many parts of the world, volcanic activity provides one of the most effective and extensive basis for soil rejuvenation.

Soils of Africa

Some soils of Africa are very old. Two Mesozoic to Tertiary lateritic crusts, representing horizons in an ancient soil profile, occur in wide parts of Africa (de Swardt, 1964; King 1967). The 'African' erosion surface, found widely all over West and East Africa, is probably mid-Miocene in age and reflects prolonged periods of tectonic still-stand and extensive weathering. Paleosoils of similar age as this erosion surface, found in volcanic regions of the East African Rift system, have been dated as 14 million years (mid-Miocene). These mid-Miocene fossil soils supported wooded grassland vegetation (Retallack et al., 1990). Tertiary paleosoils found in Nigeria illustrate mineralogical and chemical differences and associated differing Zr/TiO_2 ratios in saprolites derived from different parent materials (Zeese et al., 1994).

Today, Africa has a great variety of soils. The notion that most of Africa is covered by Ultisols and Oxisols is incorrect. Bationo et al. (2006) showed that 40.3% of the soils of Africa are shallow Lithosols subjected to droughts and salinization. Some 18.7% of African soils are Regosols and Arenosols, quartz-rich soils with low water and nutrient-holding capacities. Ferralsols and Acrisols (similar to Oxisols and Ultisols in the USDA Soil Taxonomy) make up 16.2% of African soils and are found mainly in high rainfall regions. Fertile, high P-fixing soils are Andosols and Nitosols, mainly derived from volcanic parent material. They make up only 3.8% of the total land. In general, only about 16% of Africa's land surface is considered of high potential for cultivation, but as much as 55% of the land is unsuitable for cultivation except for grazing (Bationo et al., 2006).

Fertile and productive soils are predominantly found in areas where reliable and sufficient rain falls on nutrient-rich and weatherable parent materials. Fertile areas are where soils derived from volcanic parent material prevail and young alluvial plains. In contrast, very extensive areas of the continent are covered by soils derived from quartz-dominated igneous, metamorphic and sedimentary rocks, exposed to long periods of weathering and leaching.

Soils vary spatially as a function of differing parent material. Some of the areas underlain by a variety of rock types, for example, in metamorphic mobile belts, show a great spatial variability of soils on a relatively small area.

The granite areas of Africa as well as the extensive sandstone-dominated inland basins are generally weathered into soils with sandy texture and poor in weatherable minerals; they are quartz-rich, kaolinite clay dominated and Ca and Mg as well as micronutrient poor. Volcanic rocks, on the other side, are generally silica undersaturated to saturated and largely free of quartz. These Ca and Mg and micronutrient-rich rock types weather into more clay-rich loamy soils, whereby the clay fraction varies. Some of the deeply eroded mobile belts of Africa contain a variety of different metamorphosed rocks containing minerals with differing nutrient concentrations. Accordingly, the inherent soil fertility over these rock types varies strongly. Other Precambrian fold belts, for example, the extensive 1,600–1,200 million-year-old Kibaran fold belt, consist of mainly silicate-rich clastic rock types, without amphibolites and marbles, and provide for relatively infertile soils.

Strong geological contrasts are reported from the granite–greenstone terrain of Zimbabwe. Nyamapfene (1991) illustrated the 'strikingly close relationship between geology and soils, often creating very sharp boundaries between different soil types' in extensive areas of Zimbabwe, with annual precipitation rates of 700–1,000 mm. Nyamapfene (1991), Asumadu and Weil (1988) and Zingore et al. (2005) showed the inherent relationships between the underlying geology and soil fertility. The red, quartz-free clay soils are derived from the 2,600 million-year-old mafic volcanic rocks from the so-called greenstone belts, remnants of volcanic oceanic crust and dolerite dykes of basaltic composition, now exposed on the surface of the earth. Soils derived from these parent materials form the most fertile soils in Zimbabwe. They exhibit not only favourable chemical characteristics (high extractable Ca and Mg concentrations, high extractable P, high soil organic matter content) but also good soil structure with good water-holding capacities and good response to farm inputs. In contrast, the sandy soils derived from granites have much lower soil fertility, high quartz sand content and consequently lower water-holding capacities (Nyamapfene, 1991; Zingore et al., 2005). The contact between these two principal soils and rock types is at very sharp places.

In some parts of central Zimbabwe, strongly differing soils on strongly differing parent materials exhibit high spatial variability in areas where dolerite dykes (with high soil fertility) cut through granites (with low soil fertility). Several meters wide dolerite dykes and dyke swarms cutting through various rocks are common in some areas of southern, eastern, northwestern and West Africa.

Provided the climate is suitable, soils derived from rock sequences rich in mafic rocks types, e.g. amphibolites or dolerite dykes, are generally more fertile than soils on sandstone- or quartzite-dominated silica-rich clastic rock sequences. Examples are rock sequences in Uganda, where parts of the 2,200–1,800 million-year-old Buganda-Toro fold belt are amphibolite-rich, considerably different from the amphibolite-free, quartzite- and shale-rich 1,400–1,200 million-year-old Karagwe-Ankolean fold belt. Soils and soil fertility are correspondingly different on these rock suites.

Sharp contrasts between soils and underlying rocks are also documented from eastern Uganda (van Straaten, JAMA, in prep.). Over a distance of a few hundred meters, soil characteristics change rapidly from infertile sandy soils derived from granitic parent material ($P = 3$–5 mg kg^{-1}, $K = 0.1$ cmol(+) kg^{-1}, $Ca = 1$–3 cmol(+) kg^{-1}, $Mg = 0.3$–0.9 cmol(+) kg^{-1} and carbon $= 0.5$–1%) to very fertile soils ($P = 40$–80 mg kg^{-1}, $K = 1.15$ cmol(+) kg^{-1}, $Ca = 8$–14 cmol(+) kg^{-1}, $Mg = 3$–5 cmol(+) kg^{-1} and $C = 2.6$–3.5%). The latter soils are derived from the Tertiary volcanic agglomerate associated with the Bukusu alkaline complex.

However, not everywhere are geological contrasts sharp and differences in soils clearly outlined. In areas where precipitation rates are very high, such as in the humic tropics, the influence of the parent material on soil properties is considerably lower than in sub-humid or semi-arid areas due to high and sustained leaching rates.

Poor land management, including soil nutrient depletion ('soil mining'), lack of nutrient replenishment, soil erosion by wind and water, salinization and loss of vegetation, has exacerbated the situation. These man-made processes lead to soil degradation and pose serious threats to food production, food security and the natural environment.

In order to enhance and restore soil fertility on inherently or degraded soils it is necessary to add mineral and organic nutrient resources. Some of the mineral nutrient resources are locally or regionally available in Africa, others have to be imported.

Geological Fertilizer Raw Materials of Africa

All nutrients required by plants and fertilizers, with the exception of N, are derived from geological resources. And even industrially processed N requires geological feedstocks, for example, natural gas, or other sources of methane, all geological resources.

Nitrogen

About 79% of the atmosphere by volume is composed of the inert N_2 gas. This gas has to be converted into plant available form, ammonium or nitrate. Nitrogen can be 'fixed' from the air either biologically with the aid of various biological nitrogen-fixing organisms or by the industrial N-fixing process to produce nitrogen fertilizers, the Haber–Bosch process, based on the simplified reaction: $N_2 + 3H_2 \rightarrow 2NH_3$. The principal feedstock for this reaction is natural gas. In the past, also coal was used.

At present, 97–99% of all industrially manufactured N fertilizers used in agriculture are produced by the Haber–Bosch process. The N products from this industrial process include anhydrous ammonia and products like urea, ammonium nitrate, ammonium sulphate, as well as other N fertilizers and NP fertilizers like mono-ammonium and di-ammonium phosphates (MAP and DAP).

Natural gas and other hydrocarbon resources needed for the industrial N conversion of N from the air into ammonia, and other N fertilizer products are not distributed evenly over the surface of the earth. Natural gas resources are concentrated in specific regions of the world, mainly in the Middle East, and North and West Africa, as well as regions of the former Soviet Union. In fact, approximately 71% of the natural gas reserves are located in the Middle East and countries of the former Soviet Union. The Middle East accounts for 40% of the world's natural gas reserves followed by the Russian Federation (27% of the world's total).

As of 1 January 2007, proved world natural gas reserves, as reported by the international journal *Oil & Gas Journal*, were estimated at 6,183 trillion cubic feet, which is 71 trillion cubic feet (about 1%) higher than the estimate for 2006. Worldwide, the reserves-to-production ratio of natural gas is estimated at 65 years. In Central and South America the reserves-to-production ratio is about 52 years and in Africa 88 years.

Africa is well endowed with extensive natural gas resources. As can be seen from Fig. 3, most natural gas (and oil) resources of Africa are located in North and West Africa. Most proven natural gas reserves in Africa are located in Nigeria, the remainder being concentrated in Algeria, Egypt, Libya, Angola, Mozambique, Namibia and Tanzania. In recent years, several new oil resources were discovered in the continental basins of Chad and Sudan.

New, still untapped resources of methane gas related to coal beds, the so-called coalbed methane (CBM), have been proven in Botswana and Zimbabwe (Tromp et al., 1995; Botswana Department of Geological Survey 2003). A pilot drilling program in the Matabola Basin in west Zimbabwe showed promising results (Tromp et al., 1995). The study of the Geological Survey of Botswana indicated 60 trillion cubic feet (Tcf) of methane in the study area. It is expected that these resources could become a significant supply source of methane in the future. Reservoir modelling suggests that 15–20% of the known coalbed methane reserves of Botswana could potentially be developed

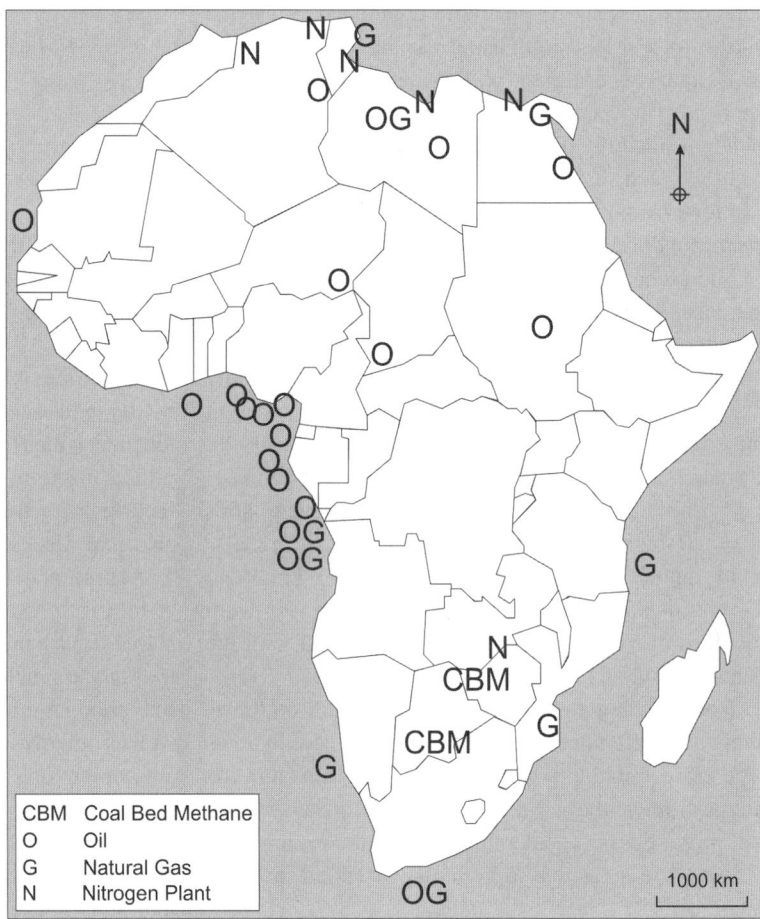

Fig. 3 Distribution of natural gas and oil in Africa

at a gas price of US $2.00 per thousand cubic feet at the wellhead. Coalbed methane resources have a potential as 'new' source of environmentally friendly energy, petrochemicals and N fertilizer feedstock.

In Africa, industrial N fertilizer production has been limited to only a few plants in North Africa. Since the industrial production of N fertilizers is largely linked to the availability of natural gas, the price of N fertilizers is fluctuating with the price of natural gas. In recent years the price of natural gas has risen and so has the price of N fertilizers. In 2000, the price of urea was about US $95 per tonne, in early October 2007 the price of 1 t of urea ex-factory was about US $341. The price of natural gas at the beginning of 2000 was about US $2.4 per mbtu (million British thermal units); in early October 2007, the price was about US $6.96 per mbtu.

Natural gas is a finite non-renewable resource and more efforts have to be directed to finding either additional natural gas resources or alternative feedstocks for industrial N production. Obviously, 'fixing' N from the air by biological means (biological nitrogen fixation) remains an extremely important process of supplying N to plants.

Phosphorus

While N can be 'fixed' from the air either biologically or industrially with the aid of natural gas feedstocks, P is derived predominantly from phosphate-bearing agrominerals. Rocks with significant amounts of phosphate minerals are called phosphate rocks. While most phosphate rocks are transformed industrially into P fertilizers some phosphate rocks can also be used directly without major chemical transformations.

Africa's phosphate rock resources are vast. Although the 2006 phosphate production in Africa is third after America and Asia, with 36% of the world production, 42% of the world's total phosphate reserves are derived from Africa (US Geological Survey 2007). Despite these huge phosphate resources most African countries are still importing phosphate fertilizers from all over the world to combat the serious P deficiencies in Africa's soils.

Figure 4 illustrates the distribution of major known phosphate rock (PR) deposits in Africa. While there are many PR deposits in Africa, it can be seen that the phosphate deposits with the highest production

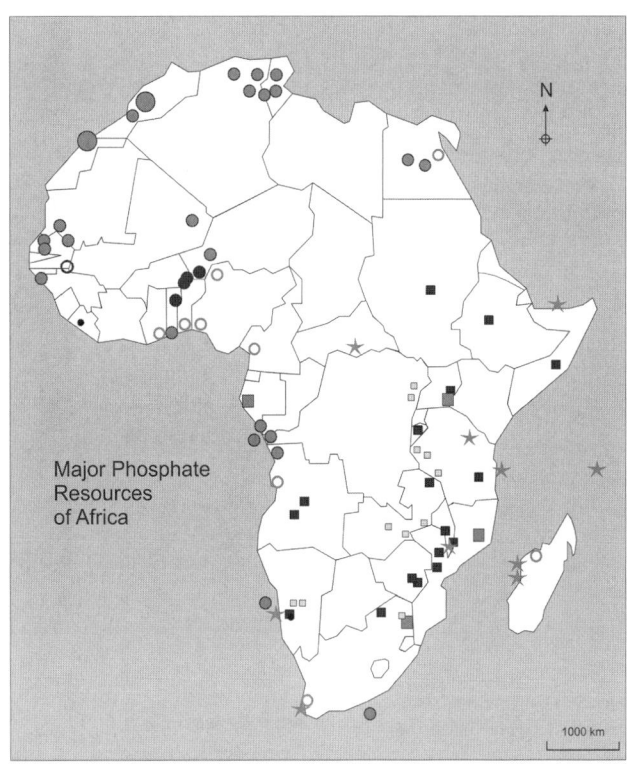

Fig. 4 Distribution of major phosphate rock deposits in Africa

rate come from North Africa (Morocco and Western Sahara, Tunisia, Egypt). Other large resources of phosphate rocks are located in South Africa, Senegal and Togo. Details of African phosphate deposits are shown by van Straaten (2002) and Van Kauwenbergh (2006).

There are several genetic types of phosphate rocks, sedimentary phosphates, also called phosphorites (Slansky, 1986), igneous phosphate rocks, as well as metamorphic, biogenic and residual phosphate rocks.

When plotting the distribution of the various types of phosphate rocks it becomes apparent that the main types of phosphate rocks, sedimentary and igneous, are spread unequally over the African continent. Sedimentary PR deposits are mainly found in the western and northern parts of the continent (Fig. 5). This distribution in the west of the continent is consistent with models of phosphate genesis related to ocean circulations. According to the genetic model of Sheldon (1980), the most suitable environments for phosphate precipitation are in areas of East–West equatorial currents and in trade wind-induced cold water upwelling zones along Africa's west coast and not along the east coast of the continent.

Igneous PR deposits are associated mainly with igneous carbonatite complexes. They are located mainly in the Central, East and South Africa, along a zone spatially related to the eastern and western rift structures. Clusters of igneous PR deposit-associated carbonatites are found at intersecting fault zones. Numerous Mesozoic igneous phosphate resources are located in western and southwestern Africa, in Gabon,

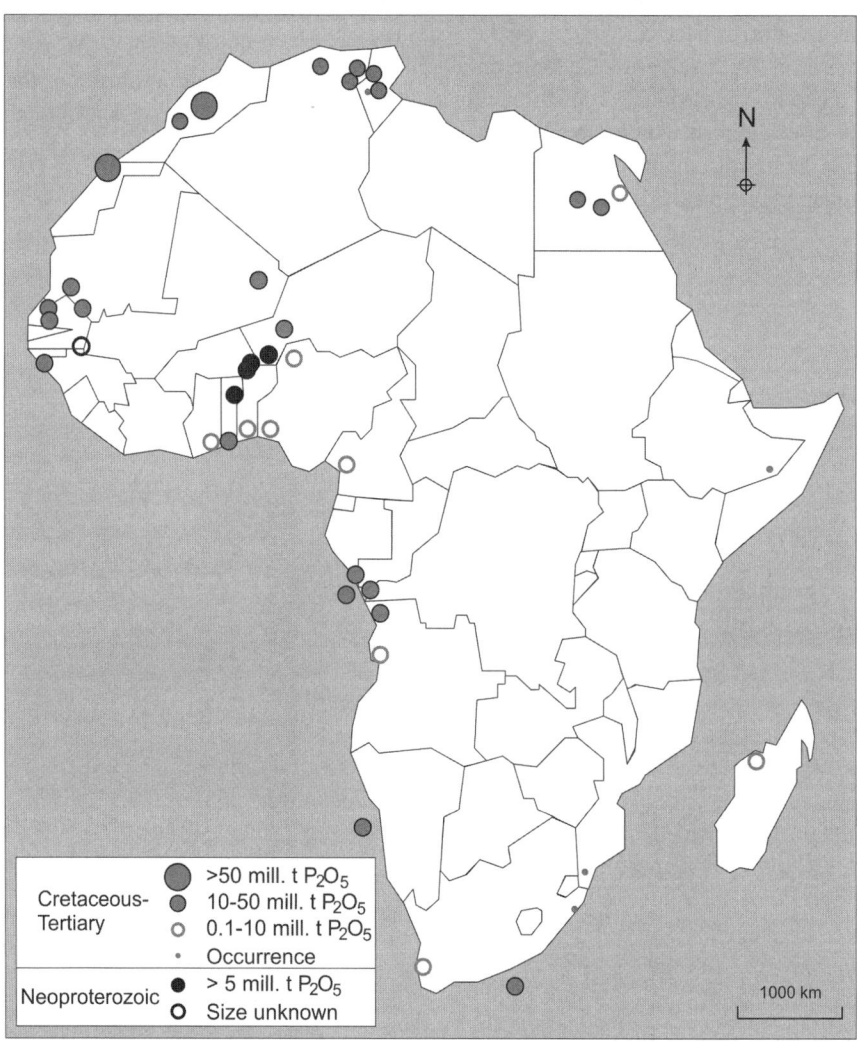

Fig. 5 Distribution of sedimentary phosphate rock deposits

Fig. 6 Distribution of igneous phosphate rock deposits in Africa

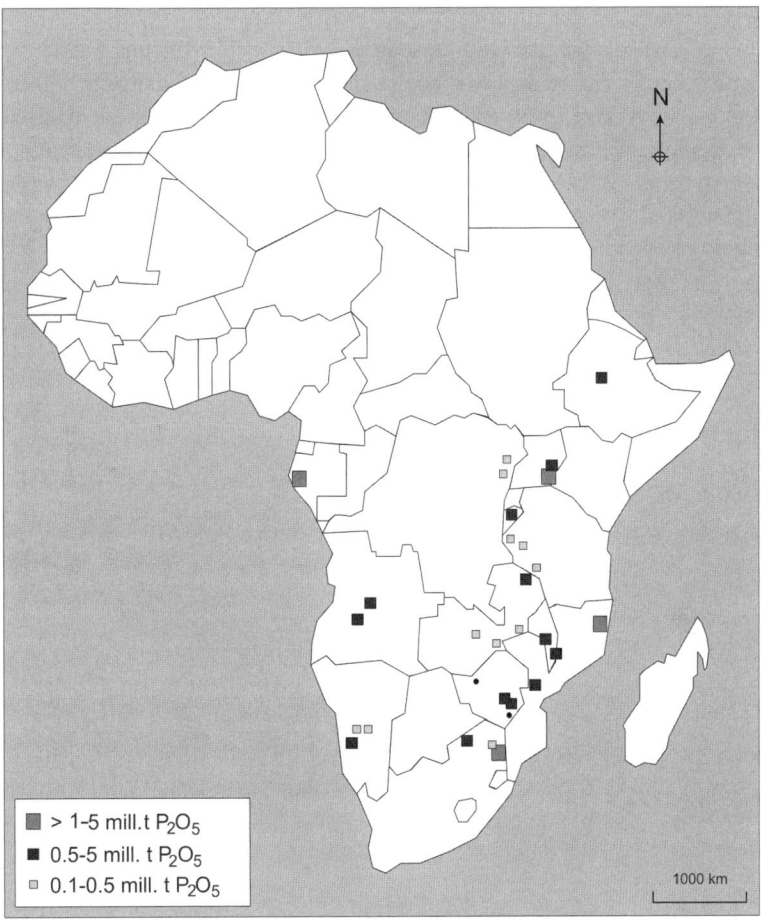

Angola and Namibia (Fig. 6). They occur along structural zones of faulting related to the opening of the Atlantic in the Cretaceous. The occurrences of igneous PR deposits are related to deep-seated structural zones of rifting, crustal extension and faulting.

A few exceptionally reactive biogenic phosphate rock deposits, Tilemsi in Mali and Minjingu in Tanzania, contain a high content of reactive bones (mainly hydroxyapatites). They have been used successfully in direct application in East and West Africa.

Although the igneous and most sedimentary phosphates have a low reactivity, there are several modification techniques that can make these phosphates agronomically more effective. Modification techniques of phosphate rocks, compiled in van Straaten (2002, 2007), include mechanical activation, fusion and calcination, partial acidulation, mixing with sulphur and organic acid-producing microorganisms, heap leaching, blending and compacting with acidulating fertilizers, phosphocomposting, biosolubilization and other biological methods. Promising modification techniques include partial acidulation with industrial acids (PAPR), blending with acid-producing fertilizers, blending with sulphur and *Acidithiobacillus* (Stamford et al., 2007) and biological solubilization processes using low molecular weight organic acids, e.g. citric acid, produced by microorganisms.

Potassium

It is reported that most soils of Africa are not K deficient. In many parts of Africa, the inherent K concentrations in soils are sufficiently high for low K-demanding crops and low crop yields. However, K fertilizers are used on plantation crops and on K-demanding plants like bananas, as well as in areas where continuous cropping has lead to K deficiencies.

Currently, Africa imports all K fertilizers from abroad. Africa's soluble K salt resources are restricted to two known major deposits, one in the Republic of Congo (de Ruiter, 1979) and the other in the Danakil depression of Ethiopia and Eritrea (Holwerda and Hutchinson, 1968) (Fig. 7). While the K deposit of the Republic of Congo is currently being reassessed by a mining company, the deposit of the Danakil depression has not been investigated in recent years. An extensive potash deposit has been discovered in oil well logs in Egypt (Michalski, 1997). Potash beds also occur in Morocco but have not been exploited (Van Kauwenbergh 2006). A few small K salt deposits are found in restricted inland basins of Africa.

K-bearing agromineral resources of low solubility include glauconite deposits (found in several countries, e.g. Angola, Madagascar, Mozambique and Togo) and K silicates such as K feldspathoids and biotite and phlogopite resources. The ultrapotassic pyroxenite of Mlindi, Malawi (Laval and Hottin, 1992), is one of

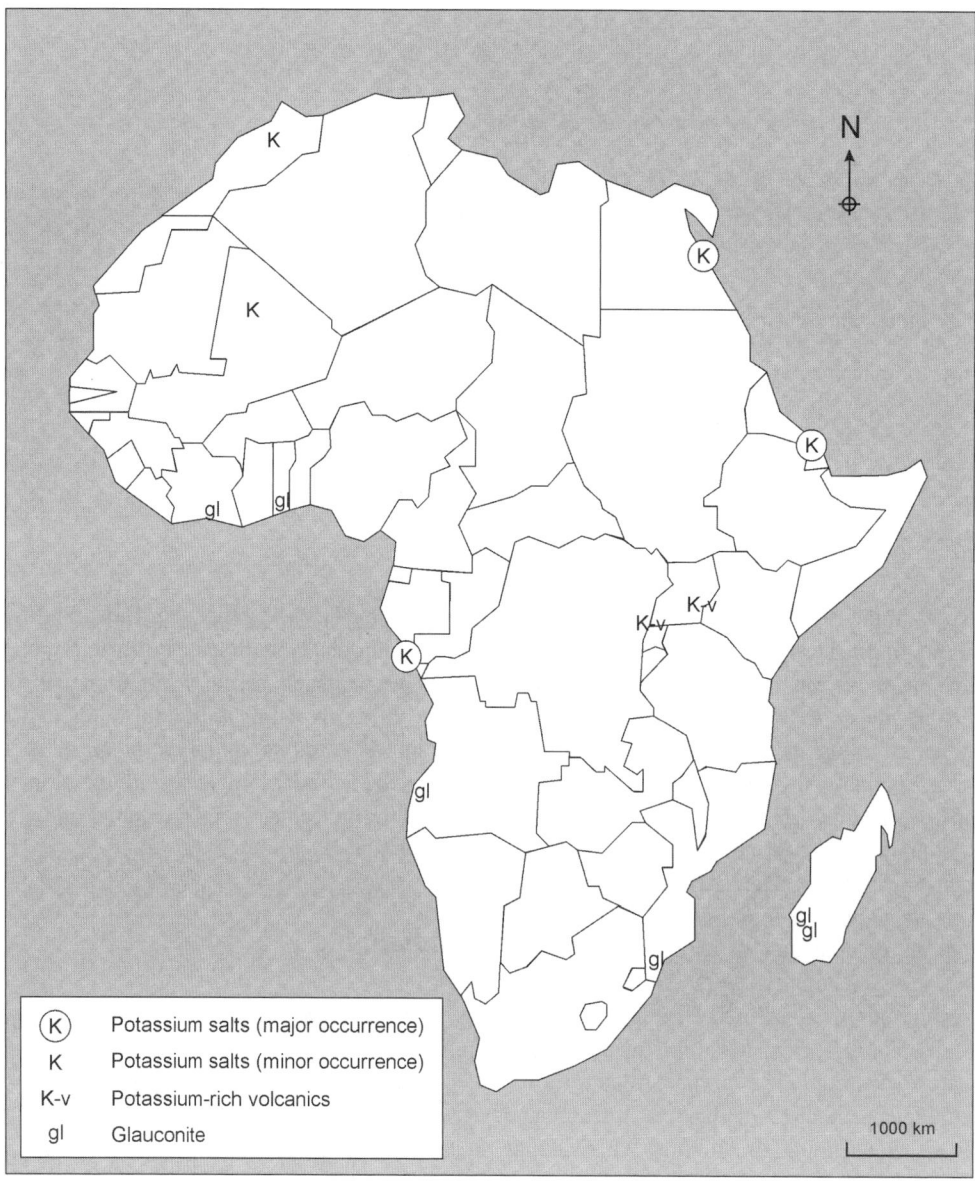

Fig. 7 Distribution of K deposits in Africa

the complexes that contain considerable amounts of weatherable mica. Large volumes of ultrapotassic lavas occur in N Rwanda, SW Uganda and parts of the Democratic Republic of Congo (Fig. 7).

The extensive low-grade K-silicate resources have not been tested on their agronomic effectiveness for reasons of low K concentrations and low solubility of these resources. However, there are efforts underway in many countries to increase the release rate of K from K-silicate minerals, specifically biotite/phlogopite resources as well as from K-rich volcanics with high nepheline and leucite components. In Brazil, for example, major efforts are being made to search for alternative mineral K resources to reduce the dependence on K fertilizer imports from abroad, especially for fertilization of crops used for biofuel production (Dr. Antonio Ramalho Filho, EMBRAPA-SOLOS, Rio de Janeiro, personal communication, 3 August 2007).

Sulphur

Sulphur deficiencies are increasingly reported from Africa's soils, although comprehensive data on the distribution of S deficiencies are still missing. The shift to low-S, high-analysis fertilizers, like urea, DAP and others, together with continuous cropping, and the planting of high-yield crop varieties could have increased the extent of S deficiencies in African soils (Weil and Mughogho, 2000).

Sulphur occurs naturally in organic compounds and in three mineral forms: elemental S, sulphates and sulphides. The resources of elemental S in Africa are very limited. Elemental S occurs in small amounts as yellow masses or crystals near hot springs and fumaroles in young volcanic areas. It also occurs in layers associated with limestone and gypsum in salt domes and bedded deposits, for example, in Mauritania (Van Kauwenbergh, 2006). Approximately 4,500 t/year by-product S from the oil and gas industry is recovered in Egypt and Libya (Bermúdez-Lugo, 2003; Mobbs, 2004). In South Africa, elemental S is also produced from pyrite by-products from the coal and metal industry.

Sulphates, in the form of gypsum, on the other hand, occur in Africa mainly in marine Mesozoic sediments, for example, in Egypt, Sudan, Ethiopia, Somalia, Kenya, Tanzania, Mozambique, South Africa, Namibia, Angola, Niger, Nigeria and Mauritania, and in smaller amounts in inland basins, for example, in Mali and Chad (Fig. 8). In Tanzania and Mozambique, the sedimentary gypsum resources are very extensive but covered by thick sedimentary overburden.

Phosphogypsum, a by-product of the phosphate fertilizer industry, is produced in Morocco, Senegal, South Africa and Zimbabwe. Instead of disposing the phosphogypsum, especially the ones derived from igneous phosphate rock resources, these resources can be used as inexpensive S source, for example, for groundnut fertilization.

Sulphides in the form of pyrite and chalcopyrite and other metal sulphides can be found in association with metal deposits. Sulphides occur as pyrite associated with gold or metal deposits or as 'waste' from the coal industry (coal-pyrites). Sulphides occur in many countries of Africa, but they are hardly extracted as S source or for the production of sulphuric acid for P fertilizer production. Availability, transport costs and potential metal contamination are the reasons for not using sulphides as S sources.

All in all, there are considerable S resources in Africa, but almost none of them are used for agricultural purposes.

Calcium and Magnesium

The Ca- and Mg-bearing agromineral resources of Africa are vast. Calcium and/or magnesium carbonates occur in many forms, including limestones and dolostones (rocks predominantly made up of dolomite), in sedimentary, metamorphic and igneous (carbonatite) rocks. The main Ca and Mg carbonate resources are marine limestones and dolostones, marbles, carbonatites and travertine and calcrete. Some of these agromineral resources are extremely large and have a wide range of industrial applications, including uses in agriculture. Chemical specifications dictate the suitability of some limestones, for example, the cement industry requires low Mg limestones. In contrast, Mg-rich varieties of limestones are preferred for agricultural purposes.

Limestones and dolostones and their metamorphic equivalents, marbles, occur in every country

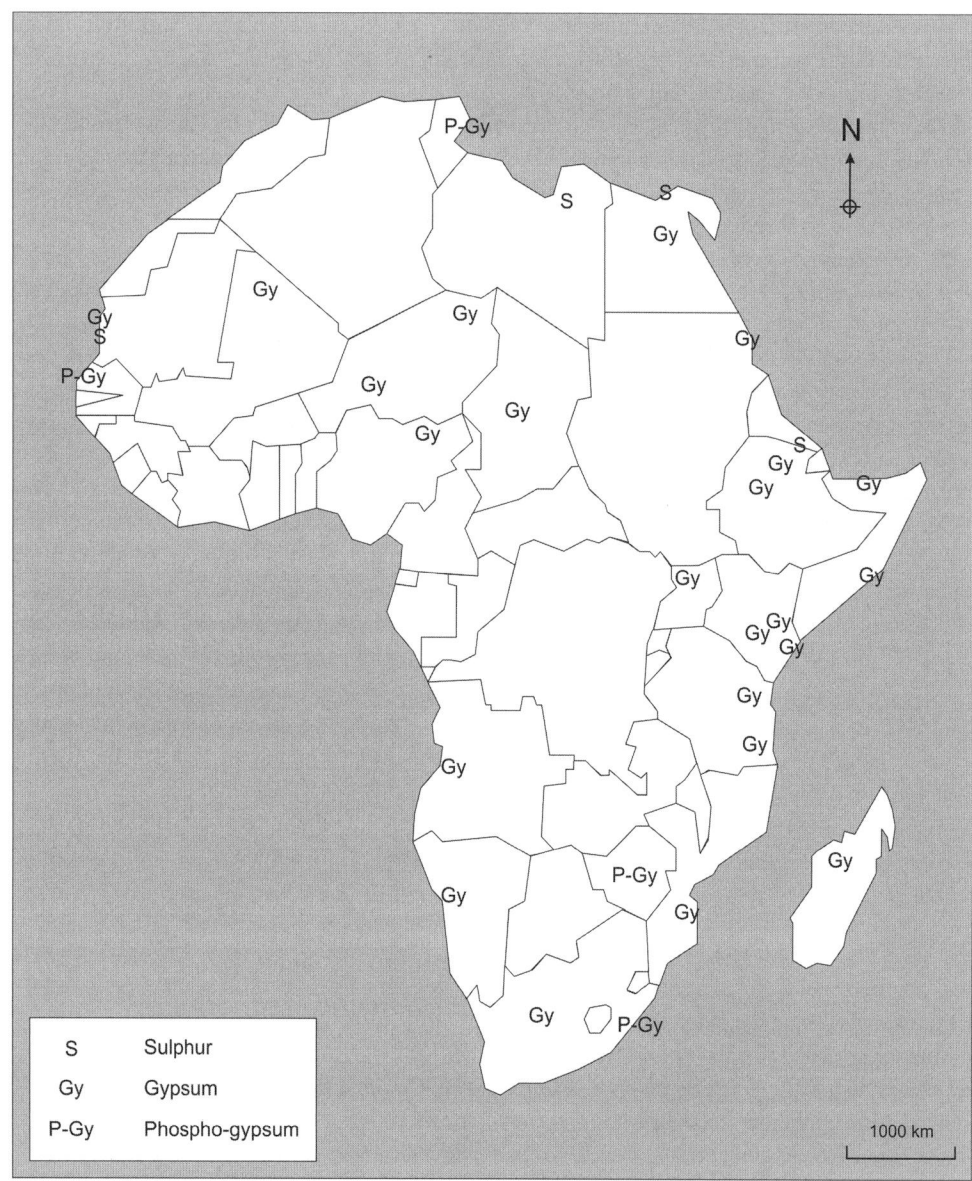

Fig. 8 Sulphur, gypsum and phosphogypsum resources in Africa

on the continent, except the Gambia, Guinea-Bissau, Equatorial Guinea, Liberia and Sierra Leone (Fig. 9). Large parts of Africa, where no limestone, dolostone or marble deposits were detected, are overlain by deserts, e.g. the Sahara and the Kalahari, or underlain by rock suites like the largely terrestrial Karoo Supergroup or granitic cratons.

A detailed compilation of the main Ca and Mg resources in all countries of Africa has been presented by Bosse et al. (1996). Limestones, dolostones and marbles occur in rocks of Archean, Paleoproterozoic and Neoproterozoic age. Few Ca and Mg carbonate resources have been found in the predominantly clastic Mesoproterozoic. Marine limestones and, to a much smaller extent, limestones developed in intracratonic basins are found in Jurassic and younger sedimentary successions of Africa. Travertine deposits are found mainly along fault systems, for example, in rift-related environments. Calcretes, usually $CaCO_3$-rich crusts of various thicknesses with a range of impurities, occur

Fig. 9 Distribution of Ca–Mg carbonate deposits in Africa (compiled from Bosse et al., 1996)

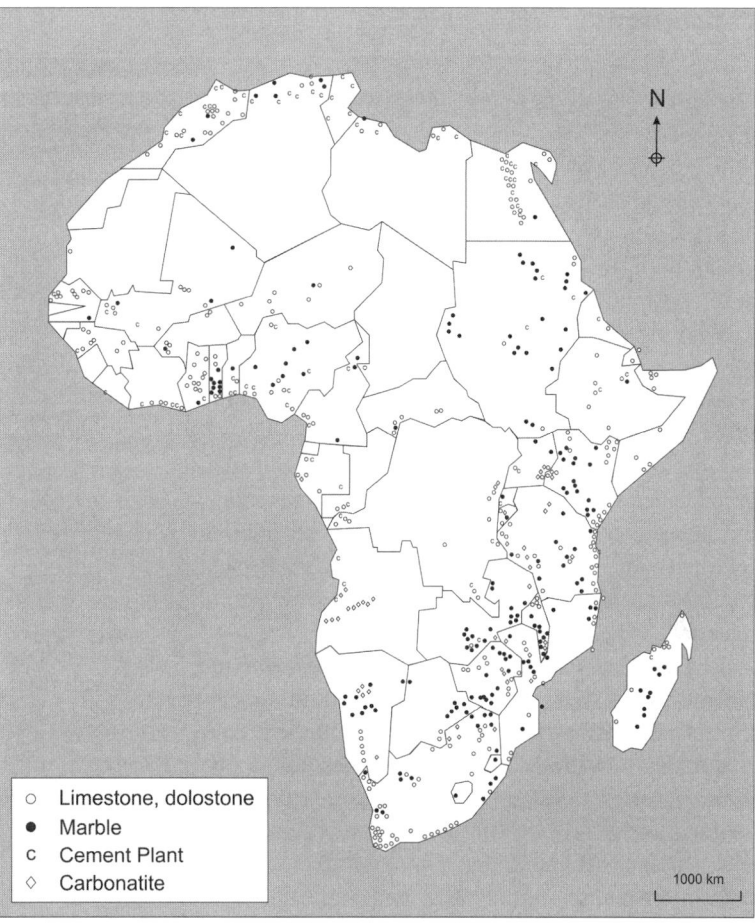

N

○ Limestone, dolostone
● Marble
c Cement Plant
◇ Carbonatite

1000 km

widely in Africa, predominantly in areas with annual precipitation rates of 400–600 mm (Bosse et al., 1996).

In addition, there are large Ca and Mg silicate resources, mainly volcanic rocks, either old 'greenstones' or younger volcanics, e.g. along the Eastern and Western Rift Valley or the Cameroon Volcanic Line. Some of these resources, especially along the Western Rift, contain also high K concentrations. The application of Ca, Mg, K, P-rich ultrapotassic, ultramafic rocks in agriculture has shown to be successful in Brazil (Theodoro and Leonardos 2006), but their use has not been explored in Africa as yet.

These vast resources of Ca and Mg resources as nutrient resources or as liming materials have not used extensively in Africa. The main areas where liming is being practiced is on plantation farms and in some farm operations with acid soils in southern Africa. Only rarely are liming methods being used on smallholder farms. In Zambia, the advantages of low cost mining and processing of dolomitic limestones ('farmlime')

for application on smallholder farms have been demonstrated (Mitchel et al., 2003).

Micronutrients

Comprehensive micronutrient data from Africa's soils are rare. Many of the micronutrient data are stored in national soil survey archives and/or in geological surveys. Except for some areas where soils were also analysed for micronutrient there are only few published comprehensive geochemical surveys from Africa. Presently, very few data exist on the distribution of selenium in African soils. There is a clear need to survey and analyse more soils on micronutrients as well as compile data from existing geochemical and soil files to provide some supporting information for future micronutrient fertilizing campaigns and environmental health surveys.

Conclusions

The inherent soil fertility is largely a function of the underlying rocks, climate and weathering history. Minerals and rocks have inherently differing mineralogical and chemical characteristics and differ widely in their nutrient release during weathering. The distribution of rocks is a function of the geological development of a given area.

With the exception of the volcanism that accompanied the formation of the Rift Valley and some fault zones, the African continent is underlain mainly by old Precambrian igneous and metamorphic rocks as well as sandstone-dominated lithologies. While granite and granite gneiss-dominated areas are more homogenous, the areas underlain by deeply eroded Precambrian fold belts vary widely in chemical and mineralogical composition and so does the fertility on these parent materials.

In general, soils derived from silica oversaturated igneous rocks and sandstone-dominated lithologies form sandy soils with limited inherent nutrient concentrations and low water-holding capacities. Large parts of Africa are underlain by these rock types. In contrast, soils derived from volcanic, silica saturated or silica undersaturated igneous rocks, young or old, contain higher concentrations of total Ca, Mg, P and micronutrients. Given suitable climatic conditions they weather into more clay-rich, fertile soils with high water-holding capacities.

To enhance the low nutrient status of inherently low soil fertility and to restore nutrient-depleted soils, nutrients and organic matter have to be added. This can be done by external nutrient inputs, such as imported fertilizers, or locally or regionally available agromineral resources. Africa's natural resource endowment of fertilizer raw materials of geological providence is large.

References

Asumadu K, Weil RR 1988. A comparison of the USDA soil taxonomy and the Zimbabwe soil classification system with reference to some red soils of Zimbabwe. Proceedings of an international symposium, Harare, Zimbabwe, 24–27 Feb 1986, IDRC-MR 170e, International Development Research Centre, Ottawa, ON, Canada, 413–445

Baker BH, Mohr PA, Williams LAJ (1972) Geology of the eastern rift system of Africa. Geol Soc Am Spec Paper136:1–67

Bationo A, Hartemink A, Lungu O, Naimi M, Okoth PF, Smaling E, Thiombiano L (2006) African soils: their productivity and profitability of fertilizer use. Background paper prepared for the African Fertilizer Summit June 9–13, 2006, Abuja, Nigeria, 25p. http://www.africafertilizersummit.org/Background_Papers/05%20Bationo-African%20Soils%20-%20Their%20Productivity.pdf

Bermúdez-Lugo O (2003) The mineral industry of Egypt. US Geol. Surv. Minerals yearbook 2002. http://minerals.usgs.gov/minerals/pubs/country

Bosse HR, Gwosdz W, Lorenz W, Markwich H, Roth W, Wolff F (1996) Limestone and dolomite resources of Africa. Geol Jb D102, Hannover, Germany, 532p

Botswana Department of Geological Survey (2003) The Department of Geological Survey Coalbed Methane Study. www.gov.bw/docs/dgscbmstudy.pdf

De Ruiter PAC (1979) The gabon and congo basins salt deposits. Econ Geol 74:419–431

De Swardt AMJ (1964) Lateritisation and landscape development in parts of equatorial Africa. Zeitschr Geomorph (N S) 3:313–333

Fyfe WS, Kronberg BI, Leonardos OH, Olorufemi N (1983) Global tectonics and agriculture: a geochemical perspective. Agric Ecosyst Environ 9:383–399

Hoffman PF, Kaufman AJ, Halverston GP, Schrag DP (1998) A neoproterozoic snowball Earth. Science 281:1342–1348

Holwerda JG, Hutchinson RW (1968) Potash-bearing evaporites in the Danakil area, Ethiopia. Econ Geol 63:124–150

Kampunzu AB, Popoff M (1991) Distribution of the main African rifts and associated magmatism. In: Kampunzu AB, Lubala RT (eds) Magmatism in extensional structural settings – the phanerozoic African plate. Springer, New York, NY, pp 2–10

King LC (1967) The morphology of the Earth, 2nd edn. Oliver and Boyd Ltd, Edinburgh

King BS (1978) Structural and volcanic evolution of the Gregory Rift. In: Bishop WW (ed) Geological background to fossil man. Geol Soc London. Scottish Academic Press, Edinburgh, pp 29–54

Laval M, Hottin AM (1992) The Mlindi ring structure: an example of an ultrapotassic pyroxenite to syenite differentiated complex. Geol Rdsch 81:737–757

Michalski B (1997) The mineral industry of Egypt. US Geol Surv Minerals Yearbook 1997. http://minerals.usgs.gov/minerals/pubs/country

Mitchel CJ, Simukanga S, Shitumbanuma V, Banda D, Walker B, Steadman EJ, Muibeya B, Mwanza M, Mtonga M, Kapindula D (2003) http://www.mineralsuk.com/britmin/farmlime_summary.pdf

Mobbs PM (2004) The mineral industry of Libya. US Geol Surv Minerals Yearbook 2004. http://minerals.usgs.gov/minerals/pubs/country

Nyamapfene K (1991) Soils of Zimbabwe. Nehanda Publishers, Harare

Retallack GJ, Dugas DP, Bestland EA (1990) Fossil soils and grasses of a Middle Miocene East African grassland. Science 247:1325–1328

Sanchez PA (2002) Soil fertility and hunger in Africa. Science 295:2019–2020

Sanchez PA, Shepherd KD, Soule MJ, Place FM, Buresh RJ, Izac AN (1997) Soil fertility replenishment in Africa:

an investment in natural resource capital. In: Buresh RJ, Sanchez PA, Calhoun F (eds) Replenishing soil fertility in Africa. SSSA Spec Publ 51, Madison, WI, pp 1–46

Schlueter T (2006) Geological Atlas of Africa. Springer, New York, NY

Sheldon RP (1980) Episodicity of phosphate deposition and deep ocean circulation – a hypothesis. Soc Econ Paleont Mineral, Spec Publ 29:239–247

Slansky M (1986) Geology of sedimentary phosphates. North Oxford Academic Press, London

Stamford NP, Santos PR, Santos CRS, Freitas ADS, Dias SHL, Lira MA Jr (2007) Agronomic effectiveness of biofertilizers with phosphate rock, sulphur and *Acidithiobacillus* for yam grown on a Brazilian tableland acidic soil. Biores Techn 98:1311–1318

Theodoro SHa, Leonardos OH (2006) The use of rocks to improve family agriculture in Brazil. Ann Braz Acad Sci 78:721–730

Tromp PL, O'Hare AM, Martin A (1995) Coalbed methane exploration in developing countries: Integration of proven technology and sub-Saharan coal. In: Blenkinsop TG, Tromp PL (eds) Sub-Saharan economic geology. Geol Soc Zimbabwe, Spec Publ 3, Balkema, Rotterdam/Brookfield, pp 55–83

U.S. Geological Survey (2007) Mineral commodity summaries 2007: U.S. Geological Survey. United States Government Printing Office: Washington, 195p

Van Kauwenbergh SJ (2006) Fertilizer raw material resources of Africa. Reference Manual IFDC R-16, IFDC, Muscle Shoals, AL

van Straaten P (1989) Nature and structural relationships of carbonatites from Southwest and West Tanzania. In: Bell K (ed) Carbonatites, genesis and evolution. Unwin Hyman, London, pp 177–199

van Straaten P (2002) Rocks for crops – Agrominerals of sub-Saharan Africa. International Centre for Research in Agroforestry – ICRAF, Nairobi

van Straaten P (2007) Agrogeology – the use of rocks for crops. Enviroquest Ltd, Cambridge, ON

Weil RR, Mughogho SK (2000) Sulfur nutrition of maize in four regions of Malawi. Agr J 92:649–656

Williams LAJ, Chapman GR (1986) Relationships between major structure, salic volcanism and sedimentation in the Kenya Rift from the equator northward to Lake Turkana. In: Frostick LE, Renault RW, Reid I, Tiercelin JJ (eds) Sedimentation in the African Rifts. Geol Soc Spec Publ, 25, 59–74

Zeese R, Schwerdtmann U, Tietz GF, Jux U (1994) Mineralogy and stratigraphy of three deep lateritic profiles of the Jos plateau (Central Nigeria). Catena 21:195–214

Zingore S, Manyame C, Nyamufagata P, Giller KE (2005) Long-term changes in organic matter of woodland soils cleared for arable cropping in Zimbabwe. Eur J Soil Sci 56: 727–736

The Challenges Facing West African Family Farms in Accessing Agricultural Innovations: Institutional and Political Implications

S.J. Zoundi and L. Hitimana

Abstract The challenge to provide producers with access to agricultural innovation has been at the centre of numerous reforms within research and extension (R&E) institutions since the 1970s. This challenge has also been the cause for the evolution in paradigms in the research to development to include aspects of participatory approaches and users assuming more responsibility. The recent past has seen a shift in system from innovation led by researchers' and extension practitioners' agenda to "demand-driven" innovation. As the continent strives to promote wider adoption of technologies, in particular those regarding soil and water management there is need to address the political challenges faced by especially smallholder farmers in accessing productivity enhancing technologies and information. The chapter presents a case study analyses on strategies for access and use of agricultural innovation in some West African countries: Burkina Faso, The Gambia, Ghana, Mali and Nigeria.

Keywords Agricultural innovations · Agricultural policies · Input-output markets · Research and extension (R&E) · Family farm

Introduction

This analysis of producers' access to agricultural innovation has been carried out within the global framework of a strategic thinking series on agricultural transformation that the Sahel and West Africa Club launched with regional actors in 2002. It concerns notably (i) West African agricultural transformation and the role of family farms; (ii) technological innovation within the structural change process of West African family farms: the role research and agricultural extension can play; (iii) support to the West African Network of Peasant Organisations and Producers (ROPPA) in the implementation of the West African Monetary and Economic Union's (UEMOA) policy.

The key question guiding the analysis is *how can access to and use of agricultural innovation by producers be improved thus enabling family farms (FF) to be able to benefit from the liberalised global context but in particular to respond to the food demands of an ever-growing, increasingly urbanised West African population?*

The working hypothesis is that there are other determining factors, sometimes forgotten and still noticed, in addition to the recriminations of poor performance facing research and extension (R&E) institutions and accusations of producers' "passivity" and "inertia". Thus, the goal of the study is to analyse the institutional, socio-economic and political environment linked to access and use of innovation by family farms with a view to stimulating decision making and action aimed at improving the livelihoods of family farms.

Background and Analysis Framework

Main Characteristics of Family Farms

In addition to numerous concepts and definitions, as a production model family farms entwine structure,

S.J. Zoundi (✉)
Secretariat of the Sahel and West Africa Club (SWAC/OECD),
2 rue André Pascal, 75775 Paris, Cedex 16, France
e-mail: sibirijean.zoundi@oecd.org

A. Bationo et al. (eds.), *Innovations as Key to the Green Revolution in Africa*,
DOI 10.1007/978-90-481-2543-2_4, © Springer Science+Business Media B.V. 2011

Table 1 Several comparisons between family farming and commercial agriculture

Characteristics	Family farm	Commercial agriculture
Role of household labour	Major	Little or none
Community linkages	Strong: based on solidarity and mutual help between household and broader group	Weak: often no social connection between entrepreneur and local community
Priority objectives	Consume	Sell
	Stock	Buy
	Sell	Consume
Diversification	High: to reduce exposure to risk	Low: specialisation on very few crops and activities
Flexibility	High	Low
Size of holding	Small: averaging 5–10 ha	Large: may exceed 100 ha
Links to market	Weak: but becoming stronger	Strong
Land access	Inheritance and social arrangements	Purchase

Source: Toulmin and Guèye (2003)

activities, household composition as well as the capital for production. This relationship is important and has implications on the way in which decisions are made with regard to types of production, organisation and allocation of resources namely household labour, capital, land management and inheritance issues (Belière et al., 2002). In terms of the social organisation of work, family farms use mainly unpaid family labour, even if, increasingly they are resorting to a salaried workforce as is the case on cotton and cocoa tree farms.

On the *socio-economic* level and in comparison with commercial agriculture, also called agribusiness, social and cultural values are still important with regard to the family farm. Managing risks is very costly and family farms use few agricultural inputs relying generally on a wide range of products including staple food and cash crops, livestock, fisheries, logging and other non-agricultural economic activities such as handicrafts, small trade and some family members may even take seasonal jobs via seasonal migration (Zoundi, 2003b).

Family farms are much smaller in comparison to commercial farms. Some studies carried out in Ghana in 1997 (Owusu et al., 2002) counted 800,000 family farms (FF) growing cocoa with an average area of 3 ha per farm, among which 80% had less than 4 ha. In Benin, the average size was 3.3 ha (Minot et al., 2001). In Mali, cotton production is carried out by more than 200,000 agricultural households with fewer than 15 people and 10 ha (Toulmin and Guèye, 2003). Even if production is primarily for FF home consumption, there is an increase in the sale of production due to the growing need for liquidity. Increasingly, grains provide households with both food and income in addition to other activities such as trade, livestock, handicrafts, and fisheries.

Thus, the basic element of family farms is the link between the economic, social as well as cultural aspects and the various objectives sought through a balance between individual and collective goals as well as risk management through the diversification of revenue sources (Table 1).

- *Socio-cultural*: Based on household labour, with many family relations, objectives and strategies combining both individual and collective concerns while emphasising solidarity. Due to their diverse activities, FF can easily adapt to the context's evolution. One of the adjustments, for example, is the reduction in cereal planted surface area or even ceasing production if the economic environment becomes unfavourable, such as a drop in price.
- *Economic*: Integrating or combining a diversified range of activities relative to the priority objectives (consumption, storage, sale) in order to minimise risk.
- *Technical*: Based on not only the desire to keep and improve the land on which it depends but also the concern to innovate technically and economically (to modernise) in order to respond to the context's evolution as well as to current and future challenges.

Access to Agricultural Innovation by Family Farms

In response to challenges regarding food security and poverty reduction, many countries have placed their hope in the agricultural innovation system even if research is not well implicated in the policy decision-making process in these countries (Butare and Zoundi, 2005). With regard to this expectation expressed to research and extension (R & E) institutions in West Africa and throughout Africa, the key issue is *how to promote access and effective use of agricultural innovation by producers with a view to increasing agricultural production, leading to economic growth and contributing to food security and poverty reduction.*

This challenge to provide producers with access to agricultural innovation has been at the inception of numerous reforms within R&E institutions since the 1970s. Without providing an exhaustive history, recent evolutions have included the promotion of participatory approaches and users assuming more responsibility. The system has gone from innovation led by researchers' and extension practitioners' agenda to "demand-driven" innovation.

This vision is explained notably by the establishment of new financing mechanisms with the aim of responding to the paradigm illustrated in Fig. 1 (Zoundi, 2004a).

There have been many organisational and institutional evolutions observed in the region among which are the regional research and agricultural extension committees (CRRVA) of Mali, and the Research and Extension Liaison Committee (RELC) in Ghana and the Gambia. In some situations, this has led to the establishment of semi-private institutions where the holding of social capital allows producers the right to voice their opinion and make decisions, such as the *Centre National de Recherche Agronomique* (CNRA) and the *Agence Nationale de Développement Rural* (ANADER) in Côte d'Ivoire, the *Agence Nationale de Conseil Agricole et Rural* (ANCAR) in Senegal. There are even some liberalised cotton sub-sectors in Burkina

Faso where the *Union Nationale des Producteurs de Coton du Burkina* (UNPC-B) holds shares in cotton companies giving them the right to voice their *desiderata* with regard to research and agricultural advice (Zoundi, 2004b). These reforms, still underway in countries, have been well documented in the work carried out by Coraf/Wecard (ODI/CIRAD/ITAD, 1999) and the Sahel and West Africa Club (Zoundi, 2003b). At the regional and international levels, these reforms have also benefited from the contribution of much strategic thinking such as that of the Neuchâtel initiative (www.neuchatelinitiative.net et www.lbl.ch/int) and the "Research–Extension and Producers' Organisations Partnership Network in West and Central Africa (REPO-Net)" (Zoundi, 2003c), as well as the partnership between the Netherlands' Royal Tropical Institute (KIT) and the World Bank on the participatory approaches of development (PAD).

Another important component in family farm's access to agricultural innovation has been the structural adjustment policies. International financing institutes imposed budgetary measures on the States relegated R&E institutions to second place. These institutions no longer have the means needed to respond to the requests made by the agricultural and rural environment. This situation seems contradictory because commercial trade liberalisation and globalisation in general raises other challenges for West African agriculture such as productivity, competitiveness, and sustainability (Zoundi, 2003a) – *How can access and use of agricultural innovations be promoted when faced with such challenges?* This question can be more specifically asked for FF. Thus, many FF are ever more dependent on the international market. Such is the case for producers of fruits and vegetables, cotton, coffee, cocoa, etc. Moreover, these FF should take into account market requirements related to sanitary, phyto-sanitary and quality standards.

Method – The Study's Structure

The methodology is based on case study analyses and strategies for access and use of agricultural innovation in some West African countries: Burkina Faso, The Gambia, Ghana, Mali, and Nigeria. This exercise was supplemented by an electronic discussion. Based on the preliminary results of the analysis, a regional consultation with actors was organised with

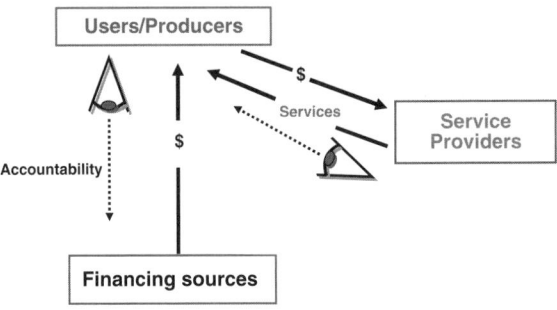

Fig. 1 New mechanisms for research and development and agricultural advice strengthening producers' power (Zoundi, 2004a)

a view to sharing and compiling other experiences so as to facilitate decision making with regard to forming development partnerships at the regional level.

Results and Key Lessons of the Study

In addition to producers' technical expertise with regard to agricultural innovation, strategic thinking carried out with actors on various case studies revealed the importance of taking into account the economic environment linked to the use of innovation. It involves notably factors needed in order to apply agricultural innovations such as inputs and other production factors, but in particular market opportunities that are able to justify or encourage producers to invest in agricultural innovation. In some situations, the implementation of projects and programmes has allowed such a favourable environment to emerge – but for the most part, this environment has remained "artificial", fading away after the end of the project or programme which raises the question of sustainability. *Is there a way to develop adequate policies enabling an environment to be created encouraging producers to use innovation?* The various documented case studies tend to provide responses to this question and some success stories which demonstrate that FF can access agricultural innovations.

The Key Role Played by Upstream and Downstream Production Services

Some case studies analysed revealed that access and use of agricultural innovation can be improved when there is an upstream and downstream favourable environment. Since the implementation of structural adjustment policies in the 1990s, several West African countries are currently experiencing (i) State withdrawal from production sectors, (ii) a reduction in all types of production support and (iii) private sector service providers are in a transitory period and for various reasons have difficulty filling the void left by public services.

However, experiences of rice producers of the Office du Niger in Mali (Box 1) show that the private sector's role in providing necessary services for the use of agricultural innovation is essential. This experience also reveals what role the State played, notably in (i) setting up needed reforms to create a favourable environment for emerging private operators and (ii) creating better land security conditions which was also a determining element for investment in agricultural production, and hence resorting to agricultural innovation.

Thus, access to the necessary inputs (seeds, fertilizer, small equipment, etc.) in order to use agricultural innovation as well as access to a remunerative market presents a favourable environment for family farms to invest in agricultural production through the implementation of agricultural innovations.

Box 1 Upstream and Downstream Services and Producers' Access to Agricultural Innovation: the Case of Rice Production within the *Office du Niger* in Mali

Background

Created in 1932, the *Office du Niger* (ON) was restructured in 1994. Some of the important reforms were (i) liberalising the marketing of threshed unmilled rice and abolishing the "economic police," (ii) securing land by establishing a *farming permit*/Permis d'exploitation agricole (PEA), (iii) signing of a three-party (state–ON–farmers) agreement.

One of the direct consequences of the restructuration was the emergence of a wide range of services offered by the private sector providing inputs, processing and marketing.

Availability of a Wide Range of Upstream and Downstream Services Favourable to Investment in Agricultural Innovation

Inputs and Other Production Factors

Access to production factors was greatly facilitated by various financing institutions including classic banks like the *Banque Nationale de Développement Agricole*/National Agricultural Development Bank (BNDA), but particularly many micro-finance institutions: *Centres d'assistance aux réseaux des caisses rurales*/Rural Credit and Savings Bank (CAREC), *Fédération des caisses rurales mutualistes du Delta*/Federation of Mutualist Rural Banks of the Delta (FCRMD), etc. These private institutions are essentially involved in access to inputs (seeds, fertilizer, pesticides, etc.) needed in order to use agricultural innovations. In 2003, the FCRMD alone provided 80% of producers with credit access to finance inputs for one agricultural campaign of which close to 100% was reimbursed. Similarly, access to agricultural equipment was encouraged with the emergence of private entities like the *Coopérative artisanale des forgerons de l'Office du Niger*/Blacksmith's cooperative at the Office du Niger (CAFON).

Services Facilitating Market Access

In addition to private infrastructure such as threshers, huskers and rice mills, several private rice harvesting and processing institutions have been set up. Such is the example of the producer association "*Jɛ ka fere*" ("Marketing Together" in the *bamana* language), which, with the support of other organisations such as the NGO *Afrique Verte* and the *Centre de prestation de services*/Provision of Services Centre (CPS), carries out processing and remunerative market research for rice sales.

Applying technological packages recommended to producers requires inputs and other production factors such as small equipment. The availability of campaign credit is possible from private financial institutions such as FCRMD, enabling producers to access various factors required for applying these technological packages. Similarly, facilitating market access assured by private organisations such as "*Jɛ ka fere*" allows producers to benefit from remunerative negotiated prices. This return on investment encouraged producers to improve productivity, thus to innovate.

This enabling environment for the use of agricultural innovation has led to spectacular rice yields, increasing from 3 t at the beginning of the restructuration to close to 6.1 t/ha on average in 2003. This illustrates as well that "*anything is possible when there is incentive to use agricultural innovations.*"

Private Companies, Agribusiness Actors and Family Farms' Access to Services Needed to Use Agricultural Innovations

Experiences presented in Boxes 2, 3 and 4 show the key role played by private operators in facilitating access and use of agricultural innovations.

Box 2 Innovation in Cereal Crops: Role of Private Companies and Operators

Background

Most production yield from cereal food crop production systems in West Africa is used for building up home consumption stocks. Aside from this general trend there are other essential cereal production systems focusing mainly on the market such is the case of mixed cotton–cereal systems. Whatever the system, one of the bottlenecks remains market uncertainty and in particular remunerative prices.

Market Opportunities for Food Crops: A Catalyst for Investment in Agricultural Innovation

Facilitating Market Access

The NGO *Afrique Verte* has been working in some countries in West Africa (Burkina

Faso, Mali and Niger) since 1990. Its experience in facilitating market access to food crop producers consists of (i) strengthening capacities of professional agricultural organisations (PAO) with regard to marketing techniques, stock management and trade negotiation, (ii) establishing a security system for PAOs to access credit to market cereals and (iii) putting producers and buyers into contact through the organisation of the "cereal stock market".

Associating *production–market* thus enables food crop producers to benefit from remunerative prices favourable to investing in innovation.

Setting Up Processor–Producer Contracts

Establishing agro-food producer–processor contracts has led to the Millet–Sorghum Initiative (MIS) framework in West and Central Africa (Burkina Faso, Chad, Mali, Niger, Senegal) or "*Downstream-driven*" has again illustrated the role played by the market in encouraging the use of agricultural innovation. This initiative implemented by the NGO Sasakawa Global 2000 has consisted of establishing contracts between producers and processors. This helps guarantee (i) delivery of a known quality product and in sufficient quantity to processors, (ii) a market and a remunerative price to producers known before seeding and providing a "price premium," (iii) producer access to agricultural inputs through credit provided by processors.

In *Niger*, for example, setting up contracts has involved the "Bunkasa Iri" collectif des groupements de producteurs privés de semences/collective of private seed producers (CGPPS) of Maradi and two processing entities: the Société de Transformation Alimentaire/Food processing company (STA) and the "ALHERI" Women's Processing Group. In *Senegal*, entering into contracts has involved local cereal processing economic interest groups (GIE TCL) made up of 11 processing companies and the Dramé Escale production economic interest group.

With a view to fulfilling the contract in terms of quality and quantity and motivated by guaranteed market access, family farms are investing in innovations, including improved varieties

(IKMP1, IKMP5, ZATIB and Souna III for millet; Framida for sorghum), fertilisation systems including chemical fertilizers, cultivation techniques and other post-harvest conservation techniques.

Applying these technological packages has led to significant yield improvement: 886 kg/ha as compared to 500 kg/ha on average for traditional crop millet; 1560 kg/ha as compared to 700 kg/ha on average for traditional crop sorghum.

In the Sahel, experiences of the NGO Sasakawa Global 2000 (Millet–Sorghum Initiative or "Downstream-Driven") and *Afrique Verte* (Box 2) and TIVISKI in Mauritania (Box 3) with regard to milk highlight the role played by private operators in creating an enabling market environment for investment in innovation. This underlines the key role that agro-food processing actors could play in agricultural development policies.

Box 3 The Private Sector and Producers' Access to Upstream and Downstream Production Services: The TIVISKI Dairy and the *Association des Producteurs Laitiers Transhumants*/Association of Transhumant Dairy-Farmers (APLT) in Mauritania

Background

The TIVISKI dairy (which means "Spring" in *Hassaniya*, a *Moor* dialect), created in 1989 in Nouakchott, started off processing dromedary milk and then cow milk in 1990 followed by goat milk in 1998. Daily production capacity is at 45 t and includes numerous products: pasteurised milk, UHT milk, buttermilk, crème fraîche, yogurt, unfermented cheese, camel cheese.

A Partnership with Livestock Breeders Fostering Access to Animal Feed and Medicines as well as Veterinary Inputs

Milk is provided to the TIVISKI dairy by a network of nomad livestock breeders located within a 300 km radius from the capital. In 2006, delivery of milk averaged 14,000 L/day, with peaks of 20,000 L.

Facing stiff competition from imported milk and in order to guarantee regular supply and the quality of raw material, the dairy initiated the establishment of a milk producers association, which in 2003 became an autonomous NGO (the "*Association des producteurs laitiers transhumants*" – APLT). The APLT is thus an innovative partnership between milk producers and a private company (the dairy).

Through the APLT, the dairy offers milk-supplying livestock breeders opportunities to access veterinary care and cattle feed on credit. More than 1000 nomadic livestock breeding families make up the network of milk suppliers.

The beginning was difficult due to many factors including consumer preference for imported goods. However, high-quality fresh dairy products gradually won over imported sterilised milk, and sales gradually increased. The dairy provides veterinary care, vaccination and feed, on credit, as well as instruction on hygiene. The milk quality has improved and it is so good that raw cow milk is now easily processed in the new UHT plant.

This has created an incentive for innovation in order to meet quantity and quality. There is a double incentive to invest in innovation:

- Facilitated access to inputs on credit, essential elements in order to use certain innovations such as feed formulations or health management of dairy cattle.
- Guaranteed market access for the TIVISKI dairy – With 140 MRO/L, livestock breeders have a source of permanent, substantial revenue. During the average production period, a livestock breeder can deliver 14–15 L of milk per day to the dairy earning 50,000 MRO/month (approximately 149 €/month).

Sources: TIVISKI Dairy (www.tiviski.com); Vétérinaires sans frontières (http://www. vsf-belgium.org/docs/info2002/aug_fr_2002. pdf); Hanak et al. (2002); Nouakchott info n° 734 of 24 February 2005 (http://www. akhbarnouakchott.com/mapeci/734/breves.htm)

In Ghana and Nigeria, the initiatives analysed (Box 4) highlight the place of private companies and agribusiness actors in supporting family farms in the access and use of agricultural innovation. The agribusiness/family farm partnership is in all regards one of the major issues in capitalising on agricultural innovation. This enables the FF to respond to concerns regarding productivity, competitiveness and quality standards required for products to be sold on the international market.

Box 4 Family Farms' Access to Agricultural Innovation: the Role of Agribusiness and Other Private Operators

Background

Besides the majority of family farms, more commercial farms (agribusiness) as well as private companies exporting agricultural products are being set up in countries. Thus quality issues need to be taken into account. *How can the existence of these private actors constitute an environment conducive to better access and use of agricultural innovations by family farms?*

Partnership Between Producers–Agribusiness/Private Companies: An Impetus Facilitating Access and Use of Agricultural Innovations

In *Ghana*, the Horticulturalists' Association of Ghana (HAG) was created in 1985. In 2003 it brought together 30 agribusiness actors working with a network of 600 family farms to produce pineapples for export. This agribusiness/family farm partnership, while at the same time allowing agribusiness to respect their commitments of regularly exported quantity, helps producers access inputs, credit, etc., necessary factors in order to use agricultural innovation required to better respond to the external market. This is in addition to other services offered by HAG: (i) information and facilitation in respecting EUREPGAP (Euro Retailer Working Group – Good Agriculture Practice) quality standards and (ii) research of vegetal material better responding to the market (variety MD-2).

In *Nigeria*, the "Okomu Oil Palm Company PLC", created in 1977 by the Federal Government and farming 8,000 ha, privatised in 1990, an approach geared towards establishing contracts with family farms for oil palm fruit production. This agribusiness/family farm partnership, while at the same time providing guaranteed market access, also assures producers of access to inputs and credit.

In *Ghana*, the Sea-freight Pineapple Exporters of Ghana (SPEG), a private company bringing together the majority of pineapple exporters, has developed solid partnerships with producers to better respond to the external market. Through this partnership, the SPEG offers producers a guaranteed market and remunerative environment through (i) facilitating access to market and (ii) facilitating compliance to EUREPGAP standards.

Whether the contract is between agribusiness/family farms or private companies/family farms, a guaranteed remunerative market and facilitating access to necessary production services (inputs, credit, etc.) constitute an incentive for using agricultural innovation. Pineapple producers' use of innovations has involved improved varieties such as the MD-2, cultural techniques such as *Plastic-Mulching*. The success achieved through these "win–win" partnerships has been a determining factor in the increase of SPEG members from 15 to 42 and an increase in production exported from 3,000 to 45,000 t from 1995 to 2003.

In all cases examined (IMS-SG2000, Afrique Verte, SPEG, HAG, etc.), the main lesson

learned is that "*. . . access and use of agricultural innovations can be improved, even in the food crop sub-sector, if there is a favourable market environment. . .*".

Professional Agricultural Organisations and Producers' Access to Agricultural Innovation

With the withdrawal by the state and the low level of development of private operators in some countries, many professional agricultural organisations (PAO) have developed their own initiatives aiming to facilitate access and use of agricultural innovations. These initiatives combine agricultural advice and in particular other strategies focusing on producers' access to necessary factors to use technological packages (Zoundi, 2003c).

The experience of the *Fédération des Paysans du Fouta Djallon/Federation of Farmers of Fouta Djallon* (FPFD) in Guinea is an illustration of the opportunities that PAO can offer their members in terms of access to inputs and the market. Two essential elements encourage agricultural innovation (Box 5).

Box 5 Producers' Access to Upstream and Downstream Production Services Through their PAO: the *Fédération des Paysans du Fouta Djallon* (FPFD) of Guinea

Background

The *Fédération des Paysans du Fouta Djallon* (FPFD) is a producers organisation created in November 1992. In 2005, the organisation had 440 groups within 21 unions, with a total of more than 15.000 members. The Federation focused in particular on some agricultural sub-sectors, notably potatoes, onions, tomatoes, etc.

How has the FPFD Facilitated Conditions for its Members to Access and Use Agricultural Innovation?

During the first years of the FPFD's creation, the producer members faced the major problem of high imports of potatoes and onions. Thus, the federation decided to take action in two ways: (i) negotiate with political authorities to regulate imports and impose taxes during the local production period and (ii) organise product marketing within the federation: negotiating prices and signing contracts with wholesalers at the local level and in the capital.

Having created the necessary foundation to ensure market access and remunerative prices, the federation decided to improve production in order to respond to demand, notably through capitalising on agricultural innovations. This twofold action enabled its members to access the following:

- Agricultural innovation and acquire technical skills – The FPFD signed cooperation frameworks with the *Institut de Recherches Agronomiques* of Guinea – IRAG and the *Service National de Promotion Rurale et de la Vulgarisation* – SNPRV. These partnerships include agricultural development and training for producers in order to use agricultural innovations related to potatoes and onions.
- Inputs needed to use agricultural innovations. For marketing, the federation sets up an internal mechanism assuring the ordering and distribution of inputs (seeds, fertilizer, etc.) to its members. Figure 2 presents a general overview of the inputs mobilised in 2001–2003.

The federation has been able to ensure part of its financing for charges linked to contracts with R & E institutions through product marketing and input supplies.

Access to inputs and the market has enabled most FPFD producers to access and use agricultural innovations needed to respond to local market demand for potatoes and onions. Average potato yields have increased from 10 t ha^{-1} during the 1993–1994 campaign to 18 t ha^{-1} during the 1996–1997 campaign. In 2003, 95% of the potatoes on the market was supplied by national production of which 50% was from producers' organisations (Camara, 2003).

FPFD's experience is not unique in the region and there are other success stories in West Africa such as the *Fédération provinciale des producteurs agricoles de la Sissili* (FEPPA-SI) in Burkina Faso for corn and cowpea crops and the *Unions sous-préfectorales des producteurs/Sub-Prefecture Producers Union* (USPP) in Benin with the establishment of the *Centrale d'achat et de gestion des intrants agricoles/Cooperative for Supply and Management of Agricultural Inputs* (CAGIA).

This illustrates the essential role played by professional agricultural organisations in the creation of an enabling environment for producers to access and use innovations.

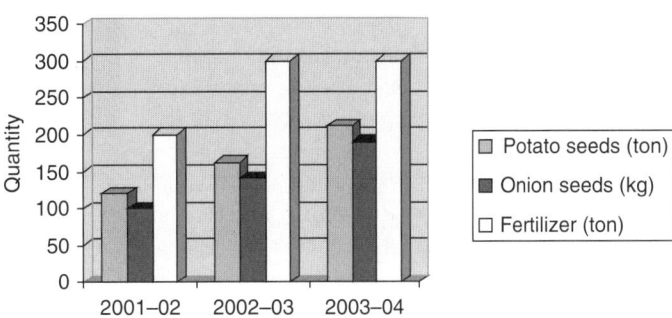

Fig. 2. Evolution of input quantities supplied by FPFD to its members

What are the Policy Implications at the National as well as at the Regional Level?

Can the Conception and Implementation of Agricultural Policies be Re-examined?

One of the major lessons learned from various experiences is that it is possible to improve family farms' access and use of agricultural innovations when they are able to access the necessary inputs and the market (local, national, regional and international). Besides cash crops, experiences related to cereals indicate that there could be a revolution even in the food crop sector if there is an incentive to innovate. This can be illustrated through the success of the sorghum variety SK 5912 in Nigeria (Box 6).

The veritable challenge is thus *how to realign agricultural policy conception with a greater and integrated vision*, that is to say taking into account all the other agricultural training or support sectors as is the case of agro-industry. Currently this concern is not sufficiently taken into account and the persistence of development sectoral approaches at the country level constitutes a less favourable environment to implement such an integrated vision.

Box 6 A Political Decision Leading to a Positive Effect on Agricultural Innovations' Access and Utilisation: *Nigeria Sorghum Success Story*

Context

In 1979, a Nigerian-released sorghum variety, SK 5912, which had been improved by the Institute for Agricultural Research (IAR), Samaru, Nigeria, was found unsatisfactory by farmers as food. So, the search for alternative uses for this very productive variety was on. In 1982, the collaborative efforts to evaluate suitable varieties of Nigerian sorghum have expanded to include other partners in industry.

The Political Decision and Its Impact

During the same period, the Nigerian government changed its policy of gradually substituting imported industrial raw materials to one of total and immediate substitution. IAR and the Federal Institute for Industrial Research, Oshodi, capitalised on the situation and began collaborative pilot and industrial scale brewing research and development for Lager beer. A series of tests were undertaken with Trophy Breweries, Double Crown Breweries, Premier Breweries and later Nigeria Breweries Ltd. By year 1983, appropriate malting and brewing procedures for sorghum were established and confirmed, finally using 100% substitution. This research for development and industrial success in the use of sorghum and sorghum malt for brewing Lager saves the country more than US $100 million annually. Malting companies have mushroomed, ever more grain is used in the poultry industry, and academic training and education for degrees and diplomas in the areas of food science and technology have increased significantly.

Specific Impacts on Innovation's Utilisation and Sorghum Production

In addition to the money saved, this success has raised the commercial production of Nigerian sorghum variety SK 5912, and two other later-released ICRISAT-bred varieties ICSV 400 and ICSV 111. This commercial use of sorghum is expected to increase the estimated sorghum requirements from the initial 67,000 t/annum in 1989 to 225,000 in 1995 and 1,500,000 in 2005.

Spillover success stories are spreading across Africa: (i) Sorghum as an adjunct by the Bralirwa (Heinneken) Brewery in Rwanda; for Lager beer in Uganda; (ii) ICSV 111 for Guiness Stout and malt drink in Ghana.

Source: SATrends – ICRISAT's monthly newsletter, SAT Trends Issue 30, May 2003 (http://www.icrisat.org/satrends/may2003.htm). For more information: *Dr A B Obilana* (a.obilana@cgiar.org)

More Favourable Trade Policies to Better Connect Family Farms to the Market

Most of the analysed case studies highlight the major importance of market opportunities in prompting investment in agricultural innovation. The FPFD's potato experience shows the relevance of greater implementation of liberalisation policies – market openings. In addition to the potato case study in Guinea, several other regional products are particularly involved by local market competitiveness such as rice, meat, and milk. For most of the products cited, extra-African imports seem to inhibit local production although production potential exists. This paradoxical situation can be illustrated through the case concerning milk (Box 7) where local production is suffering due to an unfavourable economic environment. *Can family farms truly be urged to invest in innovations in such an environment where it is difficult to sell products even at the local level?* This issue is elaborated in the frozen chicken case (GRET, 2004).

Whether it is the success story of milk in Kenya (Box 7) or that of sorghum in Nigeria (Box 6), the major changes having had impacts on incentives to invest in agricultural innovation indicate the relevance and need to take into account some measures or instruments aiming to accelerate the use of agricultural technologies. These two examples illustrate the determining role of the vision and political commitment in the creation of an economic environment needed for investment in agricultural innovation.

Box 7 Trade Policies having Impacts on Local Production and Investment in Agricultural Innovation

The European Union's (EU) dairy production subsidies are estimated at close to US $2/dairy cow/day (16 billion €/year). Among the instruments used are direct pricing support, production quotas, import restrictions and export subsidies. At the regional level, trade policies were implemented within the framework of liberalisation. The Common External Tariff (CET) has been set up and customs duties applied to imports are 5% for powdered milk and 20% for processed products.

All of these policies together present harmful consequences:

- *West African markets are invaded*: In Senegal, for example, in 2002 dairy product imports were close to 211,000 t, of which 75% was powdered milk in particular from the EU (80%). In monetary means, this is close to 22 billion CFA francs (33.5 million €).
- *Undermining development of processing factories*: In Koudougou (Burkina Faso), for example, women produce milk foods (from small millet flour diluted in curd cheese) with imported powdered milk. A kilogram of powdered milk retails for up to 1.700 CFA francs (approximately 2.59 €), from which 1 L of milk could be reconstituted for 200 CFA francs (0.30 €). A litre of fresh local milk would cost 300 CFA francs (0.46 €). Obviously, local milk is less competitive (price) than imported milk which is supported by the exporting countries, in particular, European countries. Processors cannot compete with the imported products.
- *Producers pushed aside*: A livestock breeder from Burkina Faso's Koudougou region attests: *"When I was young, my father and I supplied the French Government. Today we can produce milk all year long if it is worthwhile (which would mean being sold at around 300 CFA francs, or 0.46 € per litre of milk delivered). But we can't afford to feed our cows if we can't sell our milk."*

These policies greatly contrast the enabling environment in some countries such as *Kenya*, where success in the dairy sector has benefited from a national support policy that includes the following instruments:

- Establishment of a regulatory body, the Kenya Dairy Board (KDB), created in 1958 to regulate, promote and develop the national milk industry

- Strict control of imports by imposing a 60% customs tariff
- Subsidies by the state: since independence to the end of the 1980s 80% of the small producers' fees for artificial insemination were paid for in addition to providing veterinary services and medication (before their progressive withdrawal as from 1988).

This policy instituted in Kenya encouraged access and use of agricultural innovation by small milk producers. It has had effects on productivity; average production per cow increased from 462 to 507 kg between 1985and 1998, as compared to 192–209 and 350–350 kg, respectively, in Ethiopia and Uganda for the same period. Kenyan exports of milk, cream, butter and ghee increased from KES 117.5 million (1.29 million €) to KES 140.6 million (1.55 million €) in 1998–2002, respectively, while total value of import (only dry milk) decreased from KES 353 million (3.88 million €) to KES 135 million (1.48 million €) in the same period, respectively.

Sources: SWAC (2007); Haggblade (2004); EPZA (2005); Karanja (2003); Oudet (2005)

This issue raised with policy decision makers in various countries on their roles and responsibilities with regard to necessary support provided to the agriculture sector also concerns policy decisions at the regional level on the stakes and acceptable limits of international trade negotiations that are currently underway within the World Trade Organisation (WTO) and the Economic Partnership Agreements (EPA).

More Action-Oriented Commitment in Relation to Development of Private and Public Institutions Supporting the Agriculture Sector

The Private Sector

In relation to private sector actors, certain reforms resulting from liberalisation thus enabling private operators to play a leading role in economic development

strategies and policies had certain positive effects but there were also negative aspects. Experiences presented in this chapter highlight the importance of these actors notably in supporting the agricultural innovation process for FF. Currently in many countries, policies have not sufficiently emphasised the promotion of upstream and downstream support services in the food crop production sector. Worse yet, in some cases, selective policies have been implemented targeting products which present a so-called economic advantage to the detriment of other highly strategic products for food security. *How can investment in agricultural innovation be promoted in such contexts in order to improve agricultural production and meet the challenges of food security?*

Overall and with reference to the persistence of the changing situation where states are withdrawing and where private initiatives take a long time to materialise, the question is *how to go beyond political discourse and promote investment necessary to strengthening private sector capacities at the national level? How to improve the development of private trans-national operators and strengthen the dynamisation of the regional agricultural product market and thus stimulate investment for agricultural innovation?*

More specifically, in relation to the PAO, the issue of investment to strengthen their capacities and enable them to support family farms in innovation remains a major challenge. There is a large gap between dialogue and political will on the one hand and having it turned into action and investment on the other. *How to make radical changes at this level, notably in terms of concrete commitment?*

What is the State's Position as well as that of Public Institutions?

Reforms undertaken within the framework of the structural adjustment and liberalisation policies were perhaps necessary, but we can also question the role that public institutions should play in support of the agriculture sector. Analysis reveals that in many situations, public institutions seem dismantled, lacking capacities and means to accomplish the state directive and regulation missions, etc. With regard to access and use of agricultural innovation, the analysis shows that the public sector plays a crucial role in the implementation of structuring investments, notably those geared towards agricultural vulnerability (water management, etc.) facilitating market access (routes and other infrastructure, etc.), bearing in mind other forms of support to the agriculture sector. In some cases also, the state's role and above all policy decision makers have been decisive in the creation of an enabling economic environment as has been the case in decisions made by the Nigerian government in the promotion of local sorghum processing.

How to reaffirm that the state and public institutions have their part to play in support of producers in innovation in a complementary and synergistic approach with the private sector as well as for the creation of a favourable economic environment?

Conclusion

This analysis demonstrates that in order to "*create a green revolution through the accelerated wide-spread use of agricultural innovation*" there is a need to go beyond the recriminations made with regard to research and extension institutions. Thus, an upstream and downstream enabling production environment seems ever more crucial. However, the analysis reveals that the creation of this economic environment requires not just simple political will but in particular a vision and commitment with concrete actions and decisions. This is the main challenge facing policy decision makers at the national and regional levels. Although in some cases there seems a long way to go for such political commitment to emerge, the analysis reveals that it is possible to undertake other actions in order to accelerate the use of agricultural innovation. This constitutes at the same time calling upon research development (R&D) institutions to give priority to the analyses presented as simple messages facilitating decision making and action by policy decision makers. These analyses should demonstrate the interest of setting up an enabling environment to invest in agricultural innovation.

References

Belières J-F, Bosc P-M, Faure G, Fournier S and Losch B (2002) What future for West Africa's family farms in a world market economy? IIED Drylands Programme Issue Paper, No.113, 42 p

Butaré I, Zoundi JS (2005) Éclairer la prise de décision politique en Afrique subsaharienne: nouvelle donne pour la recherche agricole et environnementale. 2e éd. Butare et Zoundi, Dakar, Sénégal. 96 p. ISBN 2-9525390-0–6

Camara M (2003) Des formes d'organisation pour la commercialisation: L'expérience de la Fédération des paysans du Fouta Djallon, Guinée, 2 p. (http://ancien.inter-reseaux.org/publications/graindesel/gds24/dossier/formesdorganisation.pdf)

Export Processing Zones Authority (EPZA) (2005) Dairy industry in Kenya 2005. EPZ, Nairobi, 10 p

GRET (2004) Exportations de poulets: L'Europe plume l'Afrique – Campagne pour le droit à la protection des marchés agricoles, 20 p. (www.sosfaim.org/pdf/fr/poulets_brochure.pdf) (www.sosfaim.be/pdf/fr/dp/DP4.pdf)

Haggblade S (2004) Building on successes in African Agriculture, IFPRI Focus 12. Brief 1 of 10 Apr 2004, IFPRI, Washington, DC, USA, 24 p. http://www.ifpri.org/2020/focus/focus12/focus12.pdf

Hanak E, Boutrif E, Fabre P, Pineiro M (éditeurs scientifiques) (2002) Gestion de la sécurité des aliments dans les pays en développement. Actes de l'atelier international, CIRAD-FAO, 11-13 décembre 2000, Montpellier, France, CIRAD-FAO. Cédérom du CIRAD, Montpellier, France. http://wwww.cirad.fr/colloque/fao/pdf/12-abeiderrahmane-vf.pdf

Karanja AM (2003) The dairy industry in Kenya: Post-liberalization agenda. Working Paper no1, 2003, Tegemeo Institute/Egerton University http://www.aec.msu.edu/fs2/kenya/o_papers/dairy_sector_color.pdf

Minot N, Kherallah M, Soulé GB and Berry PH (2001) Impact of agricultural market reforms on smallholder farmers in Benin and Malawi. Results of survey of smallholder farmers, communities and villages, vol. 1, IFPRI, Washington, DC, 299 p

Nouakchott (2005) Info No 734 of 24 Feb 2005 http://www.akhbarnouakchott.com/mapeci/734/breves.htm

ODI/CIRAD/ITAD (1999) Strengthening Research-Extension-Farmers' Organisations linkages in West and Central Africa. Overview paper. A study prepared for CORAF, the Department for Development and the French Ministère de la Coopération. CORAF. CORAF (Dakar), 47 p. + Appendices. Voir Rapport sur le site: www.odi.uk/rpeg/coraf/overview.pdf

Oudet M (2005) La révolution blanche est-elle possible au Burkina Faso et plus largement en Afrique de l'Ouest. Misereor, Germany, 30 p. http://www.abcburkina.net/ancien/documents/filiere_lait_burkina.pdf

Owusu JGK, Osei Y and Baah F (2002) Current issues in agriculture in Ghana: the future of family Farming. Paper prepared for IIED Sahel, Dakar, Senegal

SWAC (2007) Livestock in the Sahel and West Africa: a series of policy notes http://www.oecd.org/document/53/0,3343,en_38233741_38246915_38402165_1_1_1_1,00.html

TIVISKI Dairy .www.tiviski.com

Toulmin C and Guèye B (2003) Transformation in West African Agricultures and the role of family farms. Sahel and West Africa Club (SWAC/OECD), SAH/D(2003)541, Paris, France, 144 p

Vétérinaires sans frontières (2002) http://www.vsf-belgium.org/docs/info2002/aug_fr_2002.pdf

Zoundi SJ (2003a) L'évolution du développement rural au sud: quels défis pour l'agriculture sub-saharienne? In Agridoc –

Revue Thématique, no 6 Octobre 2003, BDPA, Paris (France), pp 13–15 (voir document dans: www.agridoc.com/resdoc/revuethem/pdf/revue_6/Zoundi.pdf)

Zoundi SJ (2003b) Innovation technologique dans le processus de changement structurel de l'agriculture familiale en Afrique de l'Ouest: Quel rôle pour la recherche et la vulgarisation agricole? Club du Sahel et de l'Afrique de l'Ouest, Paris (France), 46 p. Voir document sur le site: www.sahel-club.org/fr/agri/index.htm

Zoundi SJ (2003c) Adapting agricultural institutions to the changing rural development context in West Africa: A participating framework offered by the REPO-Net. In Agricultural Research and Extension Network (AgReN) Newsletter no 47, pp 14 (Pour plus d'informations sur REPO-Net sont disponibles dans le bulletin Agricultural Research and Extension Network (AgREN) No 47 de Janvier 2003: www.odi.org.uk/agren/papers/newsletter47.pdf)

Zoundi SJ (2004a) Quels mécanismes de financement durable de la recherche agricole en Afrique Sub-Saharienne? In Grain de Sel, no 29, décembre 2004, Inter-Réseaux, Paris (France), pp 23 (voir document dans: http://www.inter-reseaux.org/IMG/pdf/5.13_dossier_zoundi.pdf)

Zoundi SJ (2004b) Processus d'innovation dans le secteur coton en Afrique de l'Ouest: Enjeux et défis pour les producteurs dans un contexte de libéralisation/privatisation de la filière coton, ROPPA, 18 p

Achieving an African Green Revolution: A Perspective from an Agri-Input Supplier

E. Makonese and K. Sukalac

Abstract Discussions about increasing agricultural production in Africa often either focus on farmers alone or rest on simple assumptions that producing fertilizers in Africa is a magic-bullet solution. Drawing on his experience, both as a fertilizer executive and as a farmer in Zimbabwe, the speaker discusses some of the major economic and policy issues that must be addressed for a sustainable, market-driven agriculture to take root in Africa.

Keywords Fertilizers · Seeds · Crop protection · Private sector · Policy and economic frameworks · Market-driven agriculture

Introduction

In recent years, a lot of ink has flowed to make the case for an African Green Revolution. Africa is the only region where agricultural productivity has actually fallen over the past three decades; yet World Bank research in countries with significant poverty reductions suggests that agricultural productivity growth may be responsible for as much as 40–70% of the improvement (World Bank, 2006). Both hunger and malnutrition remain persistent across the continent, and large swathes of Africa's agricultural lands have become significantly degraded, with some estimates of annual nutrient losses running into billions of dollars. That being said, there is growing evidence that important incremental progress is being made with regard to many development issues. At a continental level, momentum is building as Africa recovers from the shock of the structural adjustment programmes put in place in the 1980s. The picture at the national level is increasingly diverse, making it harder to generalize.

Many lessons have been learned about what should and should not be done to move forward: in some ways, there is too much information out there, making it difficult to maintain continuity. Using both personal perspectives and case studies from across Africa, this chapter will argue that the right enabling framework is vital and that it is in the interest of farmers, researchers, the agri-food industries and agricultural NGOs alike to be vocal advocates for the key measures that governments must enact in order to provide a solid and predictable framework to stimulate agricultural development. Success in this sector will support wider rural development as well as the growth of the entire continent.

Agriculture and Fertilizers: A Historical Perspective

During the past 45 years, the world population has more than doubled. The combined use of high-yielding varieties, fertilizers and improved crop management practices has allowed the world's farmers to keep pace with the fast growing demand for food and feed. However, the situation varies from one region to another, and access to these innovations has been particularly limited in sub-Saharan Africa, where the average application rate of manufactured fertilizers (excluding South Africa) is around 8 kg/ha, compared with a global average around 90–100 kg/ha.

K. Sukalac (✉)
Information and Communications, IFA, Paris, France
e-mail: ksukalac@fertilizer.org

A. Bationo et al. (eds.), *Innovations as Key to the Green Revolution in Africa*,
DOI 10.1007/978-90-481-2543-2_5, © Springer Science+Business Media B.V. 2011

In France, which is typical of mature markets, total cereal production has tripled since the early 1960s. The cultivated area actually declined somewhat, meaning that the increase is due solely to yield gains (of some 190%). This result correlates closely with increased fertilizer use (particularly nitrogen).

With regard to developing countries, India is one country that has successfully introduced high-yielding varieties, fertilizers, irrigation and mechanization. This has made it possible to greatly increase food security without encroaching on fragile lands. The cultivated land area grew only marginally from 1961 to 2001, but cereal yields went up some 180%. The government's strong policy commitment was a key factor in this remarkable result: measures include the development of an effective distribution network for fertilizers and other inputs, a well-developed transport infrastructure, appropriate and effective fertilizer quality assurance, widespread extension services and measures to increase the affordability of using modern agricultural technologies. A critical element of their strategy was the Retention Price Scheme and controlled retail prices for fertilizers, which kept fertilizer and crop prices low while ensuring a viable return to the fertilizer industry (Isherwood, 2004).

Challenges Facing African Agriculture

Because of the fragility of Africa's soils and the relative abundance of land and scarcity of labour, farmers have traditionally relied on long fallow periods to maintain soil fertility. However, the rapidly growing population and related food requirements have led to shorter fallow periods and to the cultivation of marginal lands, a situation that Norman Borlaug, the 1970 Nobel Laureate for Peace, has called "an environmental, social and political time bomb" (Borlaug, 2003). Contrary to France or India, production gains in Africa have largely come through the expansion of cultivated land, translating into 38% more land being cultivated to produce only 25% more cereal yields.

The vast majority of African agriculture is rainfed, increasing the inherent risk of being a farmer. Although water is fairly plentiful on a continental scale, it is unevenly distributed and irrigation has barely been developed, except for the cash crop sector, where higher output prices have helped compensate for the necessary investments.

As well as diverse agro-climatic conditions, African farmers over the past several decades have operated in an environment marked by heterogeneous and inconsistent policies. In the 1980s, under pressure from the donor community, most African governments eliminated their direct involvement in agriculture. While the transition to a market-based system was probably for the best in the long term, the brutality with which the government withdrawals took place was catastrophic in many countries because the private sector did not have time to develop enough to fill the gap. As a result, agricultural output plunged in many countries.

Add to these factors the demographic challenges: agricultural labour is relatively scarce in Africa, and this has worsened in recent years as many working-age people have migrated to cities or have become afflicted with debilitating or deadly diseases. The rural workforce is disproportionately made up of women, children and elders, who struggle to eke a living for their families out of small plots, let alone feeding city dwellers. The expansion of cities has hurt agriculture in another way: because most cities are built in fertile areas, their expansion reduces the average quality of the remaining land available for agriculture.

Good Technologies Fail with the Wrong Economics and Policies

Governments and donors alike have renewed their interest in African agriculture over the past decade, but the approach is now largely different: there is general agreement that governments need to support the development of rural agribusinesses and to provide support to poor farmers in a way that does not undermine fragile market structures. Furthermore, support systems should be designed with the government's exit strategy built in. When government intervention is deemed necessary, a growing number of voices are calling for "smart" subsidies. One example is a system of vouchers that work through the private sector. This fosters more attractive input prices for farmers (as retailers compete for their voucher business) and a greater, more varied supply of agricultural inputs.

In Malawi, a credit insurance programme put in place in 2001 greatly expanded the number of rural agrodealers. By 2005, the majority of the country's farmers purchased their inputs from these entrepreneurs, whose sales reached more than 1 million US dollars (plus a significant amount not underwritten by the credit insurance). As the number of dealers grew, the distances farmers travelled to obtain inputs decreased, sometimes quite dramatically. For the farmers, this meant a savings in both time and travel costs. As business boomed, retailers began hiring staff, thus boosting local economies, raising government tax receipts and increasing the provision of non-agricultural services. In 2005, Malawi was hit with a food crisis, spurring the government to provide subsidized seeds and fertilizers in order to prevent the situation worsening. The 2006 maize crop rebounded significantly, climbing to 1.60 million (mn) tonnes from 1.25 mn a year earlier. (The 2004 harvest was 1.73 mn tonnes.) However, the impact on private sector agrodealers was devastating: commercial sales of fertilizers slumped by 60–70%. A coalition from the private sector immediately began a dialogue with the government in order to find a way to transform the support programme into a private–public partnership. The private sector was involved in the 2006/2007 cycle and is heavily involved in planning for the coming year (Chapweteka, 2007).

Physical access to seeds, fertilizers and crop protection products is just part of the equation. Government programmes to foster their use can be important catalysts, but at the end of the day, the development of output markets is the key for agricultural intensification. Farmers only have an incentive to invest in their crops if they can sell their products profitably. A major difficulty at early stages of agricultural development is the boom and bust cycle: high food prices mean that many poor people cannot afford to eat, yet they encourage investment in agricultural technologies (often subsidized to stem potential food crises) in order to increase the food supply. The resulting harvest is likely to be significantly higher, thus driving down crop prices and undermining the ability to repay the credit for any inputs used. This discourages technology investments the following year, unless they are again government supported (which is unsustainable in the long term), leading to another food crisis, and so on.

Farmers, governments and the private sector need to work together to break the boom and bust cycle, thus putting agriculture and the wider economy on a more sustainable path. In this context, it is important to note that food prices have a much stronger impact (some 25% higher) on farmers' decisions to invest in fertilizers than do fertilizer prices themselves (Poulisse, 2007 and Isherwood, 2004). However, in countries with large, politically active urban populations and substantial poverty, increasing basic food prices is very sensitive. Furthermore, the very farmers who benefit from high food prices as entrepreneurs often suffer from high food prices themselves as poor consumers who spend a large percentage of their income buying food. This dilemma is perfectly illustrated by current debates over whether African farmers should grow biofuel crops in order to tap into a lucrative new cash crop: while this would provide new market opportunities and could help increase Africa's own energy supplies with all of the positive knock-on effects that would entail, it is likely that food–fuel tensions would persist during a transition period.

With regard to fertilizers in particular, which contribute to soil stewardship and conservation as well as yields and crop quality, it is also important to keep in mind that farmers are discouraged from making the necessary investments in long-term soil health if they do not enjoy secure land tenure. Depending on the context, policy reforms may be necessary to give farmers a reason to invest in the soil rather than trying to extract as much as possible out of it in the short term if we are to halt and reverse soil degradation. However, the management of land assets is a complex affair, where community needs and individual needs may differ significantly. Zimbabwe is a case in point. Formerly proof of agricultural qualifications was necessary to purchase a small-scale farm. As the original owners died, their children often preferred to pursue economic opportunities in the city. In many cases, this left prime farmland idle. If the children returned to the family farm, it was either for emotional reasons or because they had not succeeded in town, not because they were qualified or capable of running a successful farm business. The situation has been compounded by the deterioration of research and extension services so that the new generation of landholders is not gaining the skills needed to increase agricultural productivity.

Because of the market imperative, African farmers have, until now, largely used agricultural inputs on cash crops only. This has effectively meant that most small farmers have had little or no access to modern agricultural technologies. This is why there is a growing emphasis on helping subsistence farmers make the transition to agricultural entrepreneurship. Development projects that focus on agriculture now often incorporate a focus on value-chain linkages or helping match farmers to customers for their output.

The affordability of fertilizers is further hampered by a number of additional factors. As Jan Poulisse of the Food and Agriculture Organization (FAO) of the United Nations notes

> Many problems can arise when importing, storing, transporting and distributing a bulky input that is sensitive to heat and humidity. An adequate supply at the farm level is essential for maintaining reasonable fertilizer costs, even when domestic prices reflect real costs. Improved marketing systems, particularly through private marketing and better infrastructure, will reduce farm-gate prices of fertilizers regardless of their origin (Poulisse, 2007).

Underdeveloped transport infrastructure has a significant impact on the farm-gate costs of fertilizers, which are bulky, heavy and with seasonal demand. Just 5% of the rural population lives within 2 km of an all-season road in Chad and Uganda, although this rises to over 50% in countries such as Madagascar, Mali, Niger and Zambia. Intra-African transport costs are estimated by the World Bank to be nearly twice the level in other developing regions (World Bank, 2006). "Inland transportation costs, excluding duty, losses, bags and bagging, loading or unloading alone account for 15, 19 and 21% of the farm-gate price in Nigeria, Malawi and Zambia respectively", according to Camara and Heinemann (2006).

This is a major brake on agricultural development, especially in landlocked countries. Not only do the costs of agricultural inputs soar as they are moved overland but farmers' returns on crop sales fall: Rwandan coffee farmers only receive about 20% of the price of their coffee as it is loaded onto ships, according to the World Bank. The other 80% is eroded by transport and administrative costs (World Bank, 2006).

Improved transport networks should be designed in a cooperative manner among African policymakers in order to foster intra-regional trade, increase economies of scale and optimize investments. This physical investment should be bolstered by administrative and regulatory measures, including the reduction or elimination of tariffs and bureaucratic fees, simplified administrative procedures and regulatory harmonization. These are exactly the sort of steps that African Union (AU) Heads of State and Government agreed to at the June 2006 Africa Fertilizer Summit (see Annex) (African Union Heads of State and Government, 2006). Progress has been made, but not

at the extremely ambitious rate foreseen in the Abuja Declaration. In addition, reporting fatigue already seems to have set in, with the second progress report (scheduled for July 2007) still under preparation. Farmers, researchers, extension agents, agribusiness, NGOs and rural communities need to maintain positive pressure on AU governments to insist on their accountability for completing these critical milestones.

Another element of high agricultural input prices in Africa is the risk premiums built in by retailers and by their agri-input suppliers, thus raising transaction costs. The Malawi credit insurance scheme mentioned above is one initiative to address this. It guarantees agri-input suppliers that they will receive at least 50% of their payment should a retailer default. The programme then provides capacity building to help village retailers succeed in their business, including training on how to repackage inputs into smaller sizes without compromising the quality of what farmers buy. After the first 4 years, the default rate was less than 1% (Chapweteka, 2007). Such innovations help keep down final costs and credit costs, increasing the effective demand for agricultural technologies by making them more affordable. Building on this success, the Alliance for a Green Revolution in Africa (AGRA) has just provided a 3-year grant of some 13 mn US dollars to implement a comprehensive Agrodealer Support Program in Kenya, Malawi and Tanzania (CNFA, 2007).

Another useful tool for managing agricultural risk is crop insurance, but traditional insurance policies often exclude systemic risks such as drought, which is recurrent in much of Africa. When weather is included, insurance schemes are often too costly for the farmers who most need them. A policy brief by Bryla and Syroka (2007) describes how new index-based insurance products represent an attractive alternative for managing weather risk affordably. This can help to strengthen the resilience of farmers and agribusinesses to weather shocks, thus fostering greater investment in advanced agricultural technologies.

Irrigation is, of course, a well-known technical strategy to help reduce weather-related risks by decreasing the variability of water supplies. Yet only a very small share of sub-Saharan Africa's cropland is irrigated, on average less than 4%. Until the number of people with access to clean drinking water increases significantly or unless this goal is explicitly tied to new irrigation projects, it may prove politically sensitive to focus significant resources on irrigation rather than access to clean water. However, it is nonetheless imperative to increase the irrigated area and water use efficiency to make the investment in fertilizers and other agricultural inputs worthwhile.

Fertilizer Supplies – Imports or Local Production?

On the flip side of increasing effective demand for agricultural inputs is the issue of how to satisfy that demand. There are essentially two choices: importing products or producing them locally. With regard to fertilizers, it is clear that the only effective approach is a mixture of the two, for the simple reason that no country possesses all the raw materials necessary to furnish even the three macronutrients (nitrogen (N), phosphorus (P) and potassium (K)), let alone to provide a full range of products tailored to the needs of different soils, climates and crop rotations. The choice of products and overall nutrition management strategy should optimize the recycling of organic nutrient sources, but especially in Africa, there are not enough organic sources to meet existing, let alone future, requirements.

On the other hand, domestic fertilizer production, a tactic integrated into the food security plans of many Asian countries, does correlate positively to local food production. This is probably because it helps shield fertilizer prices from exchange rate variations, thus favouring price stability. For this reason and because of poor transport systems, local production may increase the security of supplies.

There is, however, an enormous caveat: domestic fertilizer production will not improve farmer access to agricultural inputs unless it is accompanied by the development of the necessary market infrastructure including vital transport. Eleven countries in Africa currently produce nitrogen fertilizers, and the continent is home to the largest phosphate reserves in the world, yet sub-Saharan Africa imports over 90% of the fertilizer it uses (Maene and Heffer, 2004; Van Kauwenbergh, 2006). Furthermore, large quantities of what is manufactured in Africa are sold in world markets, rather than within the region. In 2004/2005, sub-Saharan Africa's consumption of phosphate fertilizers equalled just 26% of what was produced in this continent (IFADATA, 2007). If North Africa is included, the

number rises to about 39%. More than half of African phosphate production goes to better endowed regions, despite the needs of farmers at home. Building more fertilizer plants without roads to agricultural areas would only exacerbate this phenomenon.

Certain conditions favour the choice for local production. First, a distance from the sea of about 400 km or more tends to make imports expensive. In addition to expense, poor logistics can mean that imported fertilizers arrive at the farm gate after the growing season.

Second, the availability of cheap raw materials improves the viability of local production. For example, in Zimbabwe, an ammonium nitrate plant based on the electrolysis process was built in 1971 to utilize abundant hydroelectric power. Prior to the plant's commissioning, energy was literally running down the Zambezi River.

Fortunately, there are some recent developments that favour the right investment environment to improve the feasibility of local production and the likelihood that locally manufactured fertilizer products are actually sold in Africa rather than in world markets. The first is a growing awareness of the importance of agriculture to help tackle other issues that have greater political capital. Possibly the most important of these is public health. There is a new appreciation among those fighting Africa's endemic disease burden that it makes little sense to provide relatively costly treatments to people whose immune systems are already weakened by hunger and malnutrition. If this synergy is properly harnessed, the use of agricultural technologies could receive an important political boost. From a purely agronomic perspective, significant gains in raising yields could be made in the short term by increasing supplies of just N, P and K (although even more could be accomplished by providing any deficient secondary and micronutrients). From a public health perspective, there is also an important argument to use micronutrient fertilizers that improve food's nutritive value. Without the additional policy support rallied by the link with health objectives, such products would probably remain out of reach for most African farmers for decades to come.

Concerns about global climate change have also improved the investment context for African agriculture. The first reason is the expected negative impact global warming could have on farming conditions in Africa. But there is also a positive side to the story, including the potential of African farmers to capture large amounts of solar energy through bioenergy crops.

Finally, the growing interest of large developing countries, such as China, in Africa's mineral resources has provided new partners and an external impetus for the necessary infrastructure projects. (Those countries are also potentially important customers for African agricultural output as their food self-reliance is undermined by the expansion and urbanization of their own populations.)

Market Innovations Bring New Opportunities

The situation on the ground is constantly evolving, and it would be wrong to imply that policymakers are the only actors who can make a difference. A number of initiatives from multiple sources have already had a huge impact on the lives of many Africans. Sometimes introduced by NGOs and sometimes as a result of solidarity among members, the creation of associations of farmers or agri-input retailers has been beneficial. Larger groups have more leverage to negotiate better terms, especially when facing powerful economic players. Groups may also pool together their credit needs, thus reducing the individual exposure to risk from defaults (but introducing the problem of free riders). And such associations serve as a platform for sharing experiences and expertise. They may also reach a critical threshold for sharing services whether hiring mechanical equipment together or providing a cost-effective audience for training courses.

Credit and finance systems themselves are evolving (IFAD, 1996). The 2006 Nobel Peace Prize recognized the positive influence of microcredit on the lives of many of the world's poorest people over the past 30 years.

More recently, mobile telephone technology has begun to make banking facilities accessible in remote locations, without the overhead of branch offices. Since March, some rural Kenyans have been able to withdraw cash and make payments or send money using their mobile phones. As in other sectors of the economy, the ability to make convenient and secure transfer of even small sums of money could go a long way towards greasing the wheels of the rural economy. Mobile phones have already been part of a revolution that lets farmers be better informed about input and

output markets. This has helped them make decisions whether to carry out transactions locally or invest in a trip to another market. Armed with knowledge about fair current prices, farmers with mobile phones have also become stronger negotiators. The internet has also contributed to these changes, but a computer cannot be carried around in a pocket as easily as a handset, and few African farmers have access to a computer or the internet without visiting a commercial cybercafé.

Making Knowledge Grow on Trees

Agriculture has always been a knowledge-intensive activity. In the past, the pace of change was slow enough that the right practices could be handed down from generation to generation. In our days, the introduction of improved varieties and modern inputs, coupled with the need to meet rapidly growing demands for agricultural goods and services, has overwhelmed the old system of transmitting knowledge. So-called traditional knowledge remains critical, but it now needs to be integrated with more precise scientific understanding of plant physiology and the biochemical properties of soil. Together these underpin the development of site-specific farming practices that maximize yields (and economic returns), optimize the use of natural resources and inputs and minimize unintended impacts.

Governments provide the foundation for this ongoing progress through basic schooling and appropriate investments in research. Governments and agribusinesses are natural partners in fostering a system of soil and crop testing. Industry efforts can also help develop improved crop management practices and can lead to the introduction of new products adapted to local commitments. Extension systems and crop advisors help farmers learn about such innovations. Agrodealers help deliver the solutions to the farm gate and are increasingly being trained to contribute to the spread of enhanced technologies and techniques.

Conclusions

Africa faces a long list of development challenges, but many of them would benefit from an interrelated set of solutions. By concentrating on some key actions and enlisting partners such as farmers themselves, NGOs and the private sector, governments can prime the well that will produce a flood. Progress will be incremental at first, but should accelerate when thresholds or tipping points are reached, particularly because of the stimulating effect that agricultural development has on the wider economy, as well as the spillover benefits of transport and other aspects of the enabling environment.

Farmers, the private sector, researchers, extension agents and NGOs should all advocate that governments support the development of all rural agricultural enterprises – those working the land, their input suppliers and buyers for their harvested crops – through targeted subsidies channelled via the retail network as well as through initiatives to make credit more affordable and less risky. Governments and donors should also build regionally coherent transport and communication networks to integrate rural communities into national and regional economies. These physical improvements should be supported by the simplification of custom arrangements and other bureaucratic formalities, reduced tariff burdens and regulatory harmonization.

Such a positive enabling environment would allow farmers and other rural entrepreneurs to blossom. As well as the immediate food security and rural development benefits this would entail, it would free resources for other economic activities and increase government tax revenues, thus enhancing the ability to address other policy priorities.

Annex: Action Points Excerpted from the Abuja Declaration on Fertilizer for the African Green Revolution (Emphasis Added)

1. Given the strategic importance of fertilizer in achieving the African Green Revolution to end hunger, the African Union Member States resolve to **increase the level of use of fertilizer** from the current average of 8 kg/ha **to an average of at least 50 kg/ha by 2015**.

2. By mid-2007, the African Union Member States and the Regional Economic Communities should take appropriate measures to reduce the cost of fertilizer procurement at national and regional levels especially through the **harmonization of policies and regulations to ensure duty- and tax-free movement across regions, and the development**

of capacity for quality control. As an immediate measure, we recommend the elimination of taxes and tariffs on fertilizer and on fertilizer raw materials.

3. By mid-2007, the African Governments must take concrete measures to **improve farmers' access to fertilizers, by developing and scaling up input dealers' and community-based networks** across rural areas. The Private Sector and Development Partners are hereby requested to support such actions.

4. By 2007, the African Union Member States must take concrete measures to specially **address the fertilizer needs of farmers, especially women, and to develop and strengthen the capacity of youth, farmers' associations, civil society organizations, and the private sector**.

5. With immediate effect, the African Union Member States must improve farmers' access to fertilizer, by granting, with the support of Africa's Development Partners, **targeted subsidies** in favor of the fertilizer sector, with special attention to poor farmers.

6. The African Union Member States should take immediate steps to **accelerate investment in infrastructure, particularly transport, fiscal incentives, strengthening farmers' organizations, and other measures to improve output market incentives**.

7. The African Union Member States should establish **national financing facilities for input suppliers** to accelerate access to credit at the local and national level, with specific attention to women.

8. The African Union Member States, hereby request the **establishment of Regional Fertilizer Procurement and Distribution Facilities** with the support of the African Development Bank, the Economic Commission for Africa, the Regional Economic Communities and the Regional Development Banks, through strategic public-private partnerships by the end of 2007.

9. Given the extensive fertilizer raw material resources in Africa and the fact that they are underutilized in many parts of the continent, the African Union Member States undertake to **promote national/regional fertilizer production and intra-regional fertilizer trade** to capture a bigger market and take advantage of economies of scale through appropriate measures such as tax incentives and

infrastructure development. This should be supported by the African Development Bank, the Economic Commission for Africa, the Regional Development Banks, the Regional Economic Communities, other Development Partners, and the Private Sector.

10. The African Union Member States should take specific action to **improve farmer access to quality seeds, irrigation facilities, extension services, market information, and soil nutrient testing and mapping** to facilitate effective and efficient use of inorganic and organic fertilizers, while paying attention to the environment.

11. The African Development Bank, with the support of the Economic Commission for Africa and the African Union Commission, is called to establish, by 2007, an **Africa Fertilizer Development Financing Mechanism** that will meet the financing requirements of the various actions agreed upon by the Summit. We, the African Union Member States, undertake to support the establishment of this facility and will pledge resources for its immediate operation.

12. The African Union Member States request the African Union Commission and the New Partnership for Africa's Development to set up a **mechanism to monitor and evaluate the implementation of this resolution**. This should be done in collaboration with the Economic Commission for Africa and the African Development Bank. The African Union Commission should give progress report to the African Heads of State at every sixth-monthly African Union Summit, starting in January 2007.

References

African Union Heads of State and Government (2006) Abuja declaration on fertilizer for the African green revolution. Final declaration from the Africa fertilizer summit, Abuja, 9–13 June 2006

Borlaug N (2003) Feeding a world of 10 Billion people: the TVA/IFDC legacy. IFDC – An International Center for Soil Fertility and Agricultural Development, Muscle Shoals, AL

Bryla E and Syroka J 2007. Developing index-based insurance for agriculture in developing countries. Sustainable development innovation briefs, Volume 2, March 2007. United Nations Department of Economic and Social Affairs, New York

CNFA (2007) CNFA announces Kenya, Malawi and Tanzania AGRA-funded programs. Press Release. Nairobi, Kenya, 13 Aug 2007

Camara O, Heinemann E (2006) Overview of the fertilizer situation in Africa. Background paper for the Africa fertilizer summit, Abuja, 9–13 June 2006

Chapweteka R (2007) Private correspondence. CNFA Malawi/RUMARK. Multiple dates in late 2005, early 2006 and August 2007

IFADATA Statistics: 1973/1974 to 2004/2005 (2007) International Fertilizer Industry Association, Paris, France. http://www.fertilizer.org/ifa/ifadata/search

International Fund for Agricultural Development (1996) Smallholder credit, input supply and marketing project: country portfolio evaluation. http://www.ifad.org/english/operations/pa/gha/i247gh/documents.htm. Cited 21 August 2007

Isherwood K (2004) Task force on fertilizer use constraints: Ghana. International Fertilizer Industry Association, Paris

Maene L, Heffer P (2004) Current situation and prospects for fertilizer use in Sub-Saharan Africa. Paper presented at the symposium on Nitrogen and crop production in Uganda, Makerere University, Kampala, Uganda, 14 Jan 2004

Poulisse J (2007) Increased fertilizer use opportunities and challenges for food security in Sub-Saharan Africa. Paper presented at the 13th International Annual Conference and Exhibition. Arab Fertilizer Association, Sharm El-Sheikh, 6–8 Feb 2007

Van Kauwenbergh SJ (2006) Fertilizer raw material resources of Africa. International Fertilizer Development Center (IFDC), Muscle Shoals, AL

World Bank (2006) Africa Development Indicators 2006 http://web.worldbank.org/WBSITE/EXTERNAL/COUNTRIES/AFRICAEXT/EXTPUBREP/EXTSTATINAFR/0,contentMDK:21102598~menuPK:3083981~pagePK:64168445~piPK:64168309~theSitePK:824043,00.html. Cited 20 August 2007

The African Green Revolution and the Role of Partnerships in East Africa

J.R. Okalebo, C.O. Othieno, S.O. Gudu, P.L. Woomer, N.K. Karanja, C. Serrem,
H.K. Maritim, N. Sanginga, A. Bationo, R.M. Muasya, A.O. Esilaba, A. Adesina,
P.O. Kisinyo, A.O. Nekesa, M.N. Thuita, and B.S. Waswa

Abstract Sub-Saharan African (SSA) region continues to experience perennial hunger, poverty and poor health of its people. Agricultural production has remained low over decades and is declining to extremely low staple maize yields below 0.5 t ha^{-1} season^{-1} at the smallholder farm scale, against the potential of 4–5 t ha^{-1} season^{-1} given modest levels of inputs and good crop husbandry. Constraints contributing to low productivity are numerous, but the planting of poor-quality seed, declining soil fertility, poor markets and value addition to products significantly contribute to poor productivity. Partnerships for development are weak even though there are numerous technologies to improve and sustain agricultural production arising from extensive research and extension in SSA. But, technology adoption rates have been extremely slow, and in some cases we find no adoption. In this chapter we highlight constraints which are bottlenecks for achievement of a green revolution in Africa. Success efforts are reported, but we moot a focus on efficient utilization of abundant and affordable African natural resources, such as phosphate rocks to replenish depleted phosphorus in soils. We argue that to achieve an African green revolution, partnerships with concerned global communities and national institutions, including universities, NGOs, CBOs and farming communities, need to be strengthened. Specifically, human capacity at all levels should be built through training. Without private sector's strong participation on acquisition of inputs and marketing proven products, it will be difficult to achieve a green revolution.

Keywords African green revolution · Bottlenecks · Partnerships · Food security · Poverty alleviation · Accept change

Introduction

In the past four decades, the green revolution has had a remarkable impact on food security and improved livelihoods in Asia and South America but very little impact on the continent of Africa. Africa has a very large food deficit, contributing to large-scale hunger, poverty and human malnutrition (FAO, 1996; Schaffert, 2007). In Asia, with a positive change to green revolution, interventions such as the cultivation of improved rice varieties, adopting the recommended agronomic practices, efficient irrigation have resulted in adequate food stocks for both local and export markets. In fact East African countries import rice from Asia at the moment. In Brazil, and other South American countries, the minimum tillage practice with fertilizer and lime applications gives grain yields of 10 and 4 t ha^{-1} of maize and soybeans, respectively.

Africa is well-known for its vast resources, which include minerals, phosphate rocks, oil reserves, diverse forests, crops and wildlife. But full benefits of these resources have not been exploited. Wide variations of climate, soils and farming practices significantly influence agricultural productivity and hence the livelihoods of African people. As an example, the yields of maize (staple) vary from near 0 to 10 t ha^{-1} in the cooler, high altitude areas on the continent (Ayaga,

J.R. Okalebo (✉)
Department of Soil Science, Moi University, Eldoret, Kenya
e-mail: rokalebo@yahoo.com

Maritim (Deceased)

A. Bationo et al. (eds.), *Innovations as Key to the Green Revolution in Africa*,
DOI 10.1007/978-90-481-2543-2_6, © Springer Science+Business Media B.V. 2011

2003), but with an average yield of <1 t ha^{-1} commonly obtained at smallholder farm level in the sub-Saharan Africa (SSA) (Sanchez et al., 1997; Nekesa et al., 1999). Low and unsustained agricultural production is partly attributed to the lowest usage of fertilizers, 9–15 kg ha^{-1} in Africa (A. Bationo, pers. commun.), which are necessary to add nutrients to the heavily fertility-depleted soils, for enhanced production. Thus, low and unstable agricultural productivity largely contributes to food insecurity, poverty and malnutrition in the continent, but particularly so in rural areas (YARA International, 2006).

Indeed, there are global, continental and national efforts that have and are being made to combat misery in Africa; these will be highlighted in this chapter, but the current endeavours seem to be focused on the declaration and implementation of the United Nations Millennium Development Goals (MDGs) of the year 2000 on efforts that targets reduction of hunger and extreme poverty, achievement of primary education, promotion of maternal health care, combating HIV/AIDS pandemic, controlling malaria and other diseases and sustaining environment and development of global partnerships for development. In endeavours to pursue these goals, global and national collaborations from development partners have been established, particularly in SSA. This chapter summarizes strategies from such partnerships. The senior author shares his experience on food security issues in East Africa from colonial and post-independence eras. The authors suggest the way(s) forward towards achievement of African green revolution in the context of food security and wealth and better livelihoods, targeting the agricultural sector and focusing the East African region.

The Constraint of Food Security with Reference to East Africa

It is well known that agriculture is the backbone for sustenance of livelihoods for the people in developing countries. Thus in the current East African Community countries (Burundi, Kenya, Rwanda, Uganda and Tanzania), agricultural production over many decades has consisted of the large-, medium- and small-scale categories of farmers, whereby the large-scale producers concentrate on cash and a few food crops (tea, coffee, sugarcane, rice, wheat) which usually fetch good global markets. These cadres of farmers usually have large parcels of land, have machinery and can readily purchase inputs.

On the other hand, the medium and smallholder farmers grow mainly food crops (maize, sorghum, millets, pulses, vegetables) basically to feed themselves and their families, with little produce to sell (Acland, 1971). To a certain extent, cotton is grown within less fertile soils and drier and hotter environments as a source of income. However, observations currently indicate that diversified agricultural enterprises now cut across the three categories of farmers, with livestock incorporation among all cadres of farmers in SSA.

As indicated above, agricultural production constraints continue to limit productivity, particularly so among smallholder farmers, who constitute about 75% of the world's poorest people who live on $1 a day or less. Most of these farmers cannot grow enough food; hence hunger and food aid are prevalent, particularly in the developing world (FAO, 1996; YARA International, 2006). Before constraints contributing to food insecurity are presented, it is perhaps useful to mention them as components of a maize production package developed for the highlands of Kenya (Allan et al., 1972). In this package, a good seedbed, early planting of high-quality seed, application of fertilizer (60 kg N plus 26 kg P ha^{-1}), timely weeding and pests/disease control will result in seasonal maize yields of 8–10 t ha^{-1}. In this presentation, we highlight the constraints and measures taken in East Africa to address the problems of poor-quality and low-yielding seed and soil fertility depletion.

High-Quality Seed

There is need to breed crops widely cultivated by smallholder farmers (including cash crops) towards the production of high-quality and high-yielding crops, with tolerance to harsh soil fertility, soil moisture, pests and disease stresses. It is generally observed that few farmers in SSA have access to new, improved varieties of local food crops capable of producing good harvests. With reference to maize (staple for SSA), traditionally, the farmers planted their own selected seed from healthy plants at maturity, having large and well-filled ears in the field. The ears were dried by hanging on tree branches of big trees close to the households or under the roofs of the kitchens. These drying techniques protected the seed from pest damages prior to planting. Although yields from landraces were low, the farmers knew how to select and store planting seed.

Plant breeding programmes, focusing maize and sorghum, were initiated for the East African region in the 1960s; the hybrids yielding 8–10 t ha^{-1} were produced at KARI's Kitale Centre in Kenya which were taken up for multiplication and distribution by the Kenya Seed Company based in Kitale. Many farmers (about 80%, C. Nkonge, pers. commun.) have adopted the planting of hybrid maize, resulting in the presence of about 75 maize seed companies or distributors (including overseas entrepreneurs) in Kenya to date. Other countries, e.g. Uganda and Rwanda, are setting up their seed distribution systems. Although competition in seed sales is healthy, the farmer needs to be educated in relation to seed type to grow. Some distributors compare side by side the performance of different seed varieties, but along roadsides, where the rural farmer still cannot be reached. The issue of a production culture also seems to be a barrier towards selection of maize seed hybrids. For example, farmers in Uasin Gishu and Trans Nzoia districts in Kenya have a maize hybrid H 614D and diammonium phosphate (DAP) fertilizer culture. They hardly accept alternatives to these two inputs. This is a negative attitude towards "change".

However, regarding seed, we note the successful introduction of IR-maize seed (coated with imazyhyr herbicide) which is tolerant to Striga weed infestation to maize in western Kenya and eastern Uganda (FORMAT, Kenya). Also there are currently very few improved seed varieties for most crops in the East African region.

Soil Fertility Depletion

Africa has some of the most weathered soils in the world (e.g. ultisols, oxisols, alfisols, cambisols) that are highly weathered and have lost substantial quantities of nutrients through leaching process, enhanced by heavy rainfall in some countries (Woomer and Muchena, 1996; A. Bationo, unpubl.). Low fertility in these soils is one of the factors accounting for depressed crop yields on smallholder farms across Africa, with the resultant continent's low agricultural productivity. Soil erosion further accelerates soil degradation in Africa. In relation to this, Norman Borlaug states that "soil losses in SSA are an environmental, social and political time bomb. Unless we wake up soon and reverse these disastrous trends, the future viability of African food systems will indeed be imperiled".

From this statement, it can be visualized that poor crop yields are directly linked to soil health. This statement has further support from data obtained from substantial research work done in the SSA region over the past century on soil fertility and improved crop production, but particularly so in the East African region, as highlighted in reviews by Okalebo (2003, 2006). These reviews confirm widespread nitrogen (N), phosphorus (P) and organic matter limitations in arable soils for adequate crop production as also found from diagnostic surveys by Okalebo et al (1992) in eastern Kenya, Ikombo et al. (1994) and Woomer et al. (1999) in central Kenya and TSBF (1994) and Shepherd et al. (1996) in western Kenya. Soil data supporting the low nutrient status of farmland soils in western Kenya are presented in Table 1, showing on the average, the acidity, low P and low to moderate organic matter status of soils.

Table 1 Soil properties prior to the start of best-bets fertility trials in western Kenya, February 2002

Location	pH	Total C (%)	Total N (%)	Olsen P (mg kg^{-1})	FAO soil class (Kenya Soil Survey, KARI)
Bungoma	5.6	1.4	0.18	4.8	Orthic acrisols and ferralsols
Busia	5.1	1.0	0.27	4.0	Orthic ferralsols w/iron stones
Homa bay	6.8	2.3	0.43	19.8	Ferralic cambisols
Siaya	4.8	1.6	0.17	3.3	Orthic to rhodic ferralsols
Teso	5.7	0.7	0.09	6.6	Orthic ferralsols and acrisols
Trans Nzoia	5.3	2.0	0.27	3.5	Rhodic ferralsols
Vihiga	4.8	1.2	0.16	3.3	Ferralo acrisols/orthic ferralsols
LSD$_{0.05}$	0.4	0.4	0.01	4.8	

Source: Woomer et al. (2003)

Note: The acidity, low P status and low to moderate soil C and N (organic matter) in these soils

Proven Technologies to Restore Soil Fertility

Poor soils and human poverty go hand in hand, especially in the tropics. Many studies (as indicated above) have therefore been devoted to finding factors constraining tropical soil fertility in order to achieve or enhance sustainable agriculture (Cardoso and Kuyper, 2006). These efforts have yielded many useful findings, concepts and technologies, particularly for East Africa (Sanchez and Jama, 2002; Vanlauwe et al., 2002; Okalebo et al., 2006). However, stagnation of agricultural development in SSA is commonly attributed to the limited adoption of new and improved technologies (Roben et al., 2006). In this context, the demonstrated positive and economically viable use of inorganic fertilizers applied from 40 to 175 kg N ha^{-1} and from 10 to 500 kg P ha^{-1} to a wide range of crops over 60 years in East Africa (Okalebo et al., 2006) has not been widely adopted.

Smallholder farmers have always indicated that the prices of fertilizers are high and increase season after season. Thus in a bid to lower the quantities of fertilizers used, incorporation of organic resources (manures, composts, crop residues, shrubs from hedges, e.g. tithonia, lantana camara, agroforestry prunings) together with modest quantities of inorganics has resulted in positive crop yield increases (Jama et al., 1997; Thuita, 2007). Research in SSA has gone a step further to apply "low-cost" phosphate rocks (PRs) to improve crop production (e.g. Okalebo et al., 2007). Information on effectiveness of PRs in western Kenya is now summarized below.

Effectiveness of Phosphate Rocks, with Reference to Western Kenya

Phosphate rocks (PRs), characterized by local names, rock origins, reserves, hardness and other qualities (Buresh et al., 1997; van Kauwenburgh, 1991), are widely distributed in Africa (Fig. 1), accounting for about 75% of the world's total phosphate deposits (A. Bationo, pers. commun.). These PRs vary widely in their agronomic effectiveness in that the sedimentary or biogenic PRs, such as the Minjingu PR from Tanzania, may be incorporated directly into the soil for

fast or direct improvement of P availability for crop uptake and yield (Nekesa et al., 1999; Obura et al., 2001). On the other hand, igneous rocks, such as the Busumbu PR from eastern Uganda, need the removal of iron through demagnetization followed by blending with a soluble phosphate source, such as triple superphosphate before beneficial use (Ngoze, 2000).

Towards Development of Affordable Soil Fertility Restoration Products: The PREP-PAC Example

From day-to-day life, consumers tend to buy products repackaged in small and affordable quantities. Therefore, to promote the use of reactive and cheap phosphate rock to replenish the P status of poor soils in western Kenya, an affordable product, the PREP-PAC, was developed and packaged at Moi University in 1998/1999. PREP is an acronym standing for the Phosphate Rock Evaluation Project, while PAC is simply a packet. The pack consists of 2 kg of biogenic/reactive Minjingu (Tanzania PR), 200 g urea, the rhizobial inoculant (Biofix) and 120 g of suitable legume seed.

The Biofix smaller package is reinforced with lime pellets to raise the pH of the inoculant and soil environment and also with gum Arabic sticker to hold the Biofix on the surface of the seed over reasonable periods of time. The package consists of instructions for use written in English and Kiswahili, and one package replenishes the fertility of depleted soils in patches that are widespread in the fields of smallholder farmers (Obura et al., 2001).

PREP-PAC has been tested in the field whereby synergy in the performance of its three main components (above) has been demonstrated (Obura et al., 2001), and significant maize and bean yield increases from its use have been shown on 52 smallholder farms in western Kenya (Table 2). In addition, the PREP-PAC intervention compares favourably with established soil fertility amelioration options tested in western Kenya, such as the use of inorganic fertilizers (FURP, 1994), the use of fortified compost (Muasya, 1996), the planting of improved/short agroforestry fallows (Jama et al., 1997) and the planting of maize-dolichos *Lablab* relays (Woomer et al., 2003; Okalebo et al., 2006).

Fig. 1 Map of Africa showing distribution of phosphate rocks (van Kauwenburgh, 1991)

Table 2 Effect of PREP-PAC on maize and bean yields (kg ha^{-1}) in western Kenya

| Crop | Yield (kg ha^{-1}) | | % Increase |
	No PREP-PAC	With PREP-PAC	
Maize	708	1442	104
Bush beans	150	289	93
Flora climbing beans	210	476	127

Source: Nekesa et al. (1999)

Note: PREP-PAC significantly ($p < 0.05$) increased crop yields, with flora bean out yielding the commonly grown bush bean; means of 52 smallholder farms are given

Other Benefits from Use of Phosphate Rocks (PRs)

Many investigations on the use of PRs to raise crop yields in SSA have centred on the P supplying aspect of these materials. However, the PRs contain both secondary and micronutrients, but specifically Ca and Mg present as $CaCO_3$, CaO, $MgCO_3$, the liming materials (van Kauwenburgh, 1991), in rather large quantities. This reflects added advantages of using PRs to supply secondary nutrients (and possibly micronutrients), to raise the pH of acid soils and to provide P. Nekesa (2007) has reported increases in soil pH through Minjingu phosphate rock application to acid soils of western Kenya, comparable to increases found from the application of agricultural lime mined in Koru, near Kisumu, Kenya (Figs. 2 and 3). These soil pH increases are reflected in positive maize yield increases from both P and lime additions (Table 3).

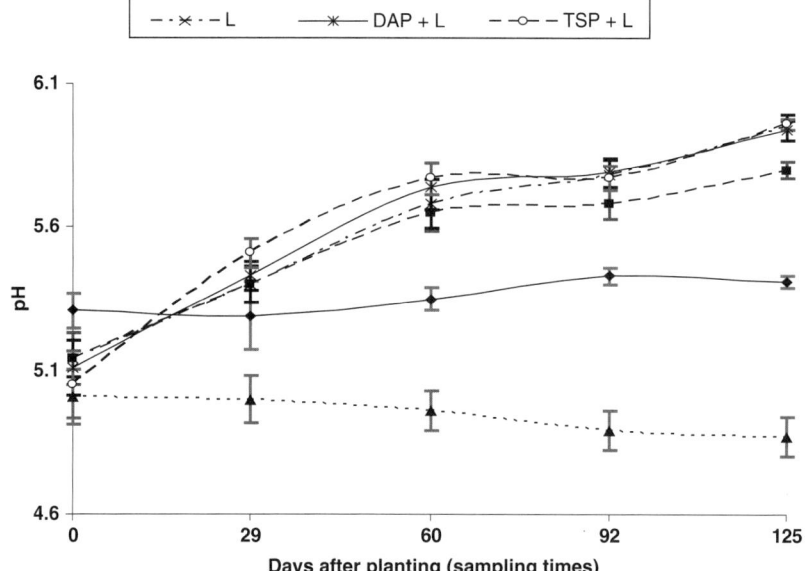

Fig 2 Changes in soil pH as affected by soil amendment material application during the 2005 LR at Mabanga (Bungoma) site (Nekesa, 2007)

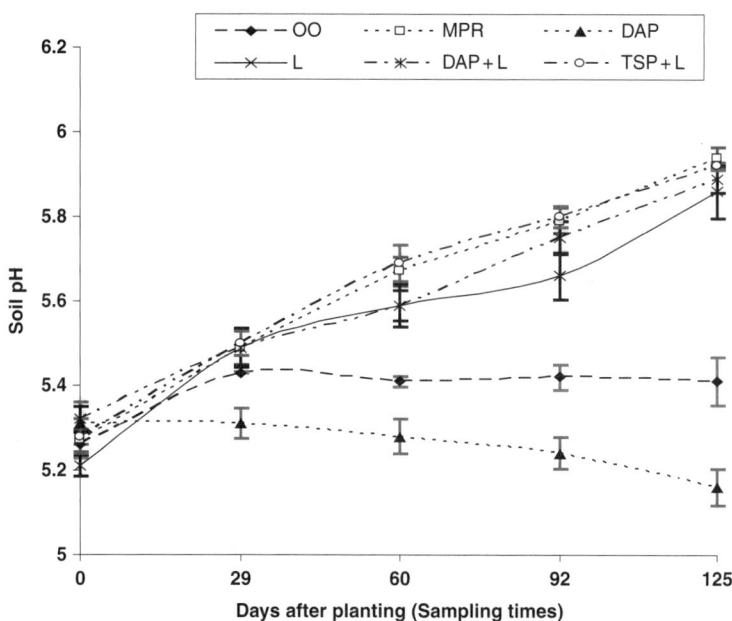

Fig 3 Changes in soil pH as affected by soil amendment material application during the 2005 LR at Sega (Siaya) site (Nekesa, 2007)

Table 3 Maize yields (t ha^{-1}) as affected by phosphate sources and agricultural lime in western Kenya, long rains 2005 (adapted from Nekesa, 2007)

Treatments	Mabanga (Bungoma)		Sega (Siaya)	
	Grain yield	Yield increase	Grain yield	Yield increase
Control	0.58	–	0.52	–
MPR	4.54	3.96	4.40	3.88
DAP alone	4.42	3.84	3.90	3.38
Lime alone	4.48	3.90	2.32	1.80
DAP + lime	5.40	4.82	4.52	4.00
TSP + lime	4.62	4.04	4.57	4.05
Means	(4.01)	–	(3.37)	–
SED trt. means	0.77	–	0.42	–
LSD ($p = 0.01$)	2.10		1.25	

Note: MPR = Minjingu phosphate rock (0–25/30–0), DAP = diammonium phosphate (18–46–0), TSP = triple super phosphate (0–46–0); MPR contains 38% CaO, while agricultural lime (Koru) contains 21% CaO

The message to the western Kenya farmers, with a DAP culture, is that they may reverse soil acidity and declining maize yields if they apply either Minjingu phosphate rock as an additional or alternative P source or DAP together with modest or affordable and available agricultural lime from Koru of 0.5–1 t ha^{-1} a season. On the use of PRs and agricultural lime, it is noted that economists tend to ignore the residual benefits arising from the use of these materials or any other nutrient inputs. These materials have positive long-term effects on crop yields (e.g. Ndungu et al, 2006) and are environmentally friendly as they raise soil pH to enhance long-term availability of P and Mo and possibly favour macrofauna and microfauna biodiversity in soils.

Maize Breeding Towards Acid Tolerance

There has been scientific thought to develop maize genotypes that are tolerant to soil acidity ($H^+ + Al^{+++}$). Specifically, soil acidity is associated with the release of Al that may be present to toxic levels in the soil solution. Retardation of plant root elongation is one of the indicators of Al toxicity (Kochian, 1995). A study in western Kenya compared the performance of five maize genotypes in acid and high-Al soils (Gudu et al., 2005). On the nutrient-depleted soils of western Kenya, both P and lime (L) additions were made to each of the genotypes (two from CIMMYT, Mexico, with one tolerant and the other susceptible to soil acidity) and the other two from Kenyan landraces, one tolerant and the other susceptible to soil acidity. The fifth genotype was the popular hybrid H 614D

moderately tolerant to soil acidity. Results (Table 4) showed significant maize yield increases from soil amendments, suggesting that in the nutrient-depleted soils, minimum nutrient inputs are required even for stress-tolerant genotypes.

The Influence of Policies, Market Forces, Value Addition to Products and Social Set-Up on Food Security, Poverty and Poor Health

It is common knowledge that without cash the constraints of poverty and its related turmoils will prevail among societies or communities, whether developed or developing. With regard to SSA, the cash constraint slows down and stops the purchases of inputs needed for sustained agricultural production; the communities cannot afford medical treatment and good health care, and overall, social and other financial demands cannot be met. Strategies to improve productivity are discussed.

Impact of Policies, Markets and Value Addition

During the colonial era (before 1960s for most of SSA), policies on crop production and marketing of raw materials existed and they were strongly enforced. Thus, large-scale farmers on the highlands cultivated mainly coffee, tea and pyrethrum for assured profitable markets. Meanwhile, the smallholder farmers

Table 4 Grain yields (t ha^{-1}) of maize cultivars receiving lime and phosphorus additions in Bumala, Busia, long rains 2003 (Gudu et al., 2005)

Cultivar	Treatments				
	Control	L	P	L + P	Means
Cimcali 97 BA Chap/ASA 4	0.38	2.10	1.14	1.88	1.37
Cimcali 97 ASA 3–1	1.53	1.09	1.90	2.24	1.65
2 A 1 – Vihiga	1.57	3.95	2.88	2.86	2.84
306 A – Kilifi	0.73	1.52	2.20	2.93	2.37
H 614D	2.27	2.64	4.73	4.39	3.54
Mean	1.30	2.27	2.57	2.86	2.36
SED	0.23	0.33	0.33	0.46	

Note: L = 4 t ha^{-1} agricultural lime (20% CaO), P = 26 kg P ha^{-1} TSP, blanket N added at 75 kg N ha^{-1} CAN. Cultivars 1 and 3 are acid tolerant, while 2 and 4 are susceptible to soil acidity

on lowlands, with poorer soils and climate, but with adequate land holdings, were forced to plant the minimum of 1 ha each of cotton (cash crop), groundnuts/beans, millet, cassava and sweet potatoes. The area chief made sure that each household planted the above crops and that after each harvest there was a reserve granary of millet/sorghum (staple) which could only be consumed upon his permission. This was at least an assurance of food security. The marketing of cotton and other food products was controlled by setting up seasonal raw product prices. In the case of cotton, the ginning (value addition) companies/firms distributed good-quality planting seed and bought all the raw cotton from farmers, providing transport themselves. In other words, the farmers recovered their labour costs with some minimum profits.

The results of such production and marketing system are that through immediate direct payments of their produce (cotton) they were able to enjoy their Christmas festivities, pay school fees and meet other family financial obligations at least for 4 months in a year.

They also received additional cash and food benefits from pulses maturing before cotton. After independence and the structural adjustment echo from the World Bank, a sizeable number of cooperatives to market a wide range of crops were formed in SSA. There were high expectations for better and competitive markets, which were a reality at the beginning. But most of these cooperatives collapsed due to mismanagement, particularly where producers waited for very long periods to receive payment for their commodities. The specific examples are the breakdowns of coffee, cotton and livestock products in East Africa.

Other contributing factors to these failures include poor management of natural resources. Thus in competitions for cash, the communities cut down trees with negligible replanting of "lost" trees; soil conservation structures became neglected or were even removed, thereby creating the widespread soil erosion. Dams constructed to conserve water for use during droughts became silted and neglected. Above all, human populations increased fast to the level of 500–1200 persons ha^{-1} (Swinkels et al., 1997). Certainly these issues had a role to play towards the recurrent hunger and poverty in SSA.

Social Set-Ups Among Rural Communities, Towards "Change"

Social and cultural issues significantly influence a "change" in development among rural communities. An example related to a change in agricultural production is given to support this statement. Thus, the SSA region has diversity in tribes and their populations, languages, cultural beliefs and social set-ups.

This is reflected in land tenures whereby families within a tribe and clan have continued to live on and subdivide their family land to very small units across generations. Certainly the current smallholdings of <0.2 ha per household in SSA highlands are a clear evidence of diminishing land sizes, with strong beliefs that one has to have a share of family land in spite of his/her capability to buy or hire land elsewhere. The immediate result is the continuous cultivation with negligible nutrient inputs of such small land parcels, resulting in low crop yields and perennial hunger. The practice of shifting cultivation, which could regenerate soil fertility, is now history in most of SSA.

Regarding the use of inputs, some beliefs hinder a change. For example, some communities believe that fertilizers applied to soils create thirst in soils; hence fertilizer use is discouraged. These could be reasons of soil acidity and application of very minimal amounts of fertilizers which do not make an impact on productivity. But a message such as this one will go a long way towards reduced fertilizer use. There are also preferences for specific products in the market in spite of products of greater potential introduced during research agenda.

For example, in the Kenyan granary in Uasin Gishu and Trans Nzoia districts (cited above), farmers believe that if they do not apply DAP and plant hybrid H 614D maize, they will incur losses in their annual maize production system. It is therefore difficult to introduce a change when the minds of farmers in these districts are firmly set. These anti-change beliefs are further compounded whereby specific tribes are not ready for change. They would rather watch a few individuals struggle for a change. In other words, they have given up in the area of development. They would also rather work for entrepreneurs and wait for fertilizer handouts from donors or their members of parliament or their employed relatives.

What Is the Way Forward Towards African Green Revolution?

Global Participation Towards an African Green Revolution

Arising from aspirations of the United Nations Millennium Development Goals (MDGs) proposed in 2000, various initiatives have been formed and are being implemented to reverse the African hunger, poverty and poor health miseries. These include the following.

Alliance for a Green Revolution in Africa (AGRA)

This is a broad-based partnership (alliance) dedicated to helping the hungry and poor millions of people in the continent to lift themselves out of the poverty and hunger traps through dramatic improvement of agricultural productivity, hence food security and improved livelihoods of smallholder farmers in particular, across Africa.

This alliance has visions for

☐ developing better and more appropriate seeds;
☐ fortifying depleted soils with responsible use of soil nutrients and better management practices;
☐ improving access to water and water use efficiency;
☐ improving income opportunities through better agricultural input and output markets;
☐ developing local networks of agricultural education;
☐ understanding and sharing the wealth of African farmer knowledge;
☐ encouraging government policies that support small-scale farmers; and
☐ monitoring and evaluating to ensure that the alliance efforts improve the lives of smallholder farmers and their households and help to build a sustainable future for all Africans.
☐ In addition, as livestock is often critical to small-scale farmers, the alliance will promote opportunities to strengthen mixed crop–livestock systems.

Most of the initiatives currently (and in the past) have targeted their visions, missions and activities towards the visions above. Thus in Africa, the Action Plan of the Environmental Initiative of the New Partnership for Africa's Development, NEPAD, in June 2003 states that most small-scale farmers across Africa realize that successful farming depends on a healthy environment, rich soils, quality seed and biodiversity on and off the farm, but substantial environmental degradation has already taken place. Hence sustainable farming conserves environment and natural resources.

In summary, NEPAD notes that for successful advancement of African agriculture the following should be initiated or strengthened:

(a) New, more productive technologies that lower costs and make farming more competitive and profitable.
(b) A focus on increasing Africa's domestic and export markets.
(c) High-level political commitment to agriculture.

Conclusion

It is a general view from the authors that the African green revolution is achievable but only through strong partnerships dedicated to building human capacity through training at different levels and increased agricultural production by use of the right inputs efficiently, but better still through the exploitation of the natural resources such as phosphate rocks. Effective markets, particularly of value-added products, will contribute to the continuity of interventions. Communities should be willing to accept a change, specifically in the context of their socio-cultural behaviour. Above all, partnerships including private sector, alliances, donors (such as The Rockefeller Foundation, McKnight, Bill and Melinda Gates Foundation and other nations) with national institutions are important in creating a positive change, loosely known as the African green revolution.

Acknowledgements The authors are grateful to The Rockefeller and McKnight Foundations in USA, RUFORUM in Uganda and TSBF-CIAT for funding the research and training activities which enabled the generation of data presented in this chapter. They also extend their gratitude to FORMAT (K) and SACRED-Africa NGOs for fine collaboration together with farmers in western Kenya, whose land was used for experimentation. We record our thanks to the technical staff and Mary Nancy (Secretary – for all typing and editing of this chapter) of Moi University for their willing participation in both field and laboratory activities. Finally, the senior author is grateful to TSBF-CIAT for funding to attend this symposium. He also records thanks to the vice chancellor, Moi University, for permission to attend the symposium.

References

Acland JD (1971) East African crops, an introduction to the production of field plantation crops in Kenya, Tanzania and Uganda. FAO, Rome

Allan AY, Were A, Laycock D (1972) Trials of types of fertilizer and time of nitrogen application. In: Kenya Department of Agriculture, N.A.R.S. Kitale. Ann Rpt. Part II: 90–94

Ayaga GO (2003) Maize yield trends in Kenya in the last 20 years. A keynote paper. In: Othieno CO, Odindo AO, Auma EO (eds) Proceedings of a workshop on declining maize yields trends in Trans Nzoia district. Kenya, Moi University, pp 7–13

Buresh RJ, Smithson PC, Hellums DT (1997) Building soil P capital in sub-Saharan Africa. In: Buresh RJ, Sanchez PA, Calhoun F (eds) Replenishing soil fertility in Africa. SSSA Special Publication No. 51, SSSA, Madison, WI, pp 111–149

Cardoso IM, Kuyper TW (2006) Mycorrhizas and tropical soil fertility. Agric Ecosyst Environ 116:72–84

FAO (Food and Agriculture Organization of United Nations) (1996) Soil fertility initiative. FAO, Rome

FURP (Fertilizer Use Recommendation Project) (1994) Fertilizer use recommendations Vols 1–2. Kenya Agricultural research Institute (KARI). Ministry of agriculture, Nairobi

Gudu SO, Okalebo JR, Othieno CO, Obura PA, Ligeyo DO, Santana DP, Schulze D, Johnston C, Kisinyo PO (2005) Maize response to phosphate and lime on marginal fertility soils of western Kenya. African Crop Sci. Proceedings, Entebbe, Uganda, December 2005

Ikombo BM, Esilaba AO, Kilewe AM, Okalebo JR (1994) Soil and water management practices in smallholder farms in Kiambu district, Kenya: a diagnostic survey. In Proceedings of the 4th KARI Annual Scientific Conference, Nairobi, Kenya, Oct 1994, p 30

Jama B, Swinkels RA, Buresh RJ (1997) Agronomic and economic evaluation of organic and inorganic sources of phosphorus in western Kenya. Agron J 89:597–604

Kochian LV (1995) Cellular mechanism of aluminum toxicity and resistance in plants. Ann Rev Plant Physiol 46:237–260

Muasya RM (1996) Wheat response to fortified wheat straw compost. M. Phil Thesis, Moi University, Eldoret, Kenya

Ndungu KW, Okalebo JR, Othieno CO, Kifuko MN, Koech AK, Kemenye LN (2006) Residual effectiveness of Minjingu phosphate rock and fallow biomass on crop yields and financial returns in western Kenya. Expl Agric 42:323–336

Nekesa AO (2007) Effect of Minjingu phosphate rock and agricultural lime in relation to maize, groundnut and soybean yields on acid soils of western Kenya. M. Phil Thesis, Moi University, Eldoret, Kenya

Nekesa P, Maritim HK, Okalebo JR, Woomer PL (1999) Economic analysis of maize-bean production using a soil fertility replenishment product (PREP-PAC) in western Kenya. Afr Crop Sci J 7:423–432

Ngoze S (2000) A comparison of the performance of Minjingu and Busumbu phosphate rocks in acid soils of Nyabeda, Siaya, Kenya. M. Phil Thesis, Moi University, Eldoret, Kenya

Obura PA, Okalebo JR, Othieno CO, Woomer PL (2001) Effect of PREP-PAC production on maize-soybean intercrops in acid soils of western Kenya. Afr Crop Sci Proc 5:889–896

Okalebo JR (2003) Inorganic resources management for sustainable soil productivity. Special Issue – E Afri Agric For J 69(2):119–129

Okalebo JR, Othieno CO, Woomer PL, Karanja NK, Semoka JRM, Bekunda MA, Mugendi DN, Muasya RM, Bationo A, Mukhwana EJ (2006) Available technologies to replenish soil fertility in East Africa. Nutr Cycling Agroecosyst 76:153–170

Okalebo JR, Othieno CO, Woomer PL, Maritim HK, Karanja NK, Mukhwana EJ, Kapkiyai JJ, Bationo A (2007) The potential of underutilized phosphate rocks for replenishment of soil fertility in Africa: experience from western Kenya. Proceedings of the 2nd rocks for crops workshop (Nairobi and Kisumu, Kenya), July 2007 (under preparation)

Okalebo JR, Simpson JR, Probert ME (1992) Phosphorus status of cropland soils in the semi-arid areas of Machakos and Kitui districts, Kenya. In: Probert ME (ed) A search for strategies for sustainable dryland cropping in semi-arid eastern Kenya. ACIAR, Canberra ACT Proceedings No. 41, pp 50–54

Roben R, Kruseman G, Kuyvenhoven A (2006) Strategies for sustainable intensification in East African highlands: labour use and input efficiency. Agric Econ 34:167–181

Sanchez PA, Jama BA (2002) Soil Fertility Replenishment takes off in East and Southern Africa. In: Vanlauwe B, Diels J, Sanginga N, Merekx R (eds) Integrated plant nutrient management in sub-Saharan Africa. Wallingford, CT, CABI, pp 23–46

Sanchez PA, Shepherd KD, Soule MJ, Place FM, Mokwunye AU, Buresh RJ, Kwesiga FR, Izac A-MN, Ndiritu CG, Woomer PL (1997) Soil fertility replenishment in Africa: an investment in natural resource capital. In: Buresh RJ, Sanchez PA, Calhoun F (eds) Soil fertility replenishment in Africa. SSSA Special Publication No. 51, Madison, WI, pp 1–46

Schaffert RE (2007) Integrated fertility management for acid Savanna and other low phosphorus soils in African and Brazil. A seminar presentation at KARI Headquarters, Nairobi, Kenya, Feb 2007

Shepherd KD, Ohlson E, Okalebo JR, Ndufa JK (1996) Potential impact of agroforestry on soil nutrient balances at farm scale in the East African Highlands. Fert Res 44:87–98

Swinkels RA, Franzel S, Shepherd KD, Ohlson E, Ndufa JK (1997) The economics of short-term rotation improved fallow: Evidence from areas of high population density in western Kenya. Agric Syst 55:99–121

Thuita MN (2007) A comparison between the "MBILI" and conventional intercropping systems on root distribution, uptake of N and P yields of intercrops in western Kenya. M. Phil Thesis, Moi University, Eldoret, Kenya

TSBF (Tropical Soil Biology and Fertility) (1994) Limiting nutrient studies conducted in Kabras division, Kakamega district, western Kenya. TSBF Ann. Rpt., 1994

Van Kauwenburgh SJ (1991) Overview of phosphate deposits in East and South Africa. Fert Res 30:127–150 (Special Issue)

Vanlauwe B, Diels J, Sanginga N, Merckx R (eds) (2002) Integrated plant nutrient management in sub-Saharan Africa. Wallingford, CT, CABI

Woomer PL, Karanja NK, Okalebo JR (1999) Opportunities for improving integrated nutrient management by smallholder farmers in the Central Kenya highlands. Afr Crop Sci J 7:441–454

Woomer PL, Muchena FN (1996) Recognizing and overcoming soil constraints to crop production in tropical Africa. Afr Crop Sci J 4:503–518

Woomer PL, Musyoka MW, Mukhwana EJ (2003) Best-Bet maize-legume intercropping technologies and summary of 2002 Network Findings. Best-Bets Bulletin No. 1, SACRED-Africa NGO, Bungoma, Kenya, pp 18

YARA International (2006) Proceedings of annual conference held in Oslo, Norway, September/October 2006

Optimizing Agricultural Water Management for the Green Revolution in Africa

B.M. Mati

Abstract The word "famine" nowadays applies almost exclusively to the African continent. Up till the end of the twentieth century, the "hunger belt" usually referred to the Horn of Africa countries, but the new millennium has seen several countries in southern Africa join the group requiring food aid. In all cases of famine, lack/shortage of rainfall is cited as the main cause. Yet, within the same continent there are countries which are food secure, but normally receive much less rainfall than the "drought" rainfall amounts of the hunger belts. Just how much is agriculture in Africa retarded by water scarcity? Or is it simply poor management of agricultural water? Or better still, by how much could African agriculture recoup from targeted management of water that is or could be made available to agriculture? The fact that the problem has persisted means that either not enough has been done or what has been done has had little impact. Thus, the innovations needed in agricultural water management so as to propel the African Green Revolution will require more than just exploring case studies of experimental trials. They have to include changing the perceptions and actions of vast cross sections of society, from farmers to decision makers, and even those in the non-farming sectors. This chapter therefore describes some project-scale water management initiatives relevant to smallholder agriculture in Africa. It also explores the wider options to enhance the prioritization of agricultural water management at decision-making levels and optimal management of water by practitioners.

Keywords Agricultural water management · Practices · Investment · Policy · Networking · Africa

Introduction

Nowhere in the world have the constraints associated with water management for agriculture been more acute than in Africa, specifically in sub-Saharan Africa (SSA). These problems are manifest in the vulnerability of smallholder farmers to weather-related events such as drought, prolonged dry spells, inadequate rainfall or occasionally floods, resulting in crop failure, loss of livestock, poverty and quite often famines. This vulnerability persists even where the risks of adverse weather are well known or form a regular occurrence. Already, about one-third of the people in Africa live in drought-prone areas. Paradoxically, Africa is also the cradle of early civilizations which were driven by water management for agriculture, as in Egypt. It is also in Africa, particularly North Africa, where advances in agricultural water management (AWM) have seen farmers enjoy high crop productivity. These disparities raise several questions, and there is growing evidence that the problem of high vulnerability of the smallholder farmer in SSA lies not with the climate or natural endowments, but with water management or more specifically lack of it.

At a glance, Africa appears to have abundant water with its 3,991 km^3 of renewable freshwater resources, 17 river basins having catchment areas larger than 100,000 km^2, more than 160 lakes larger than 27 km^2 and vast wetlands (UN Water/Africa, 2003; FAO, 2006a). However, most of these water resources are

B.M. Mati (✉)
Improved Management of Agricultural Water in Eastern & Southern Africa (IMAWESA), Nairobi, Kenya
e-mail: b.mati@cgiar.org

A. Bationo et al. (eds.), *Innovations as Key to the Green Revolution in Africa*,
DOI 10.1007/978-90-481-2543-2_7, © Springer Science+Business Media B.V. 2011

located around the equatorial region and the sub-humid East African Highlands. The rest of the continent comprises drylands which cover about 60% of the total land area. Moreover, rainfall distribution displays high spatial and temporal variability, ranging from less than 50 to over 1,700 mm, while mean annual rainfall is about 673 mm/year. As Africa straddles the Equator, the continent suffers high evaporation and tropical storms that yield high runoff losses, resulting in substantially lower percentage of precipitation contributing to renewable water resources. Thus, only 19.7% of rainfall becomes renewable water, which is equivalent to 149.9 mm/year (ECA, 1995; FAO, 1995). Moreover, large areas of the continent are underlain by low storage aquifers, thereby limiting the effectiveness of rainfall recharge. Compounded by the rainfall unpredictability and frequent droughts, sustainable agriculture in most countries of Africa has to be supported by targeted management of water. Otherwise, left to natural forces, agriculture and thus livelihoods will continue to suffer as is the current situation.

Importance of Water for Agriculture in Africa

Over the past three decades, agricultural production in Africa, growing at less than 2% per annum, has not matched population growth which has been rising at about 3%. In East and Southern Africa (ESA),

the number of people affected by food insecurity doubled from 22 million in the early 1980s to 39 million in the early 1990s (IFAD, 2000). In most cases, food insecurity is associated with drought and/or insufficient rainfall. To close the food gap, the FAO (2000) has estimated that between 1995 and 2030, about 75% of projected growth in crop production in SSA will have to come from land intensification with the remaining 25% coming from expansion of arable land. Water management, particularly irrigation (including partial irrigation from harvested rainfall), will play a major role. There is substantial underutilized potential for irrigation expansion (about 45 million hectares) since in the whole of Africa, only about 7% of the arable land is irrigated, while the proportion for SSA is about 3.7% (NEPAD-CAADP, 2003; FAO, 2006b; World Bank, 2005). Deliberate steps are therefore required to accelerate progress because the current growth in irrigation at 1% per annum (Fig. 1) in most countries is too slow (NEPAD-CAADP, 2003). Furthermore, as much as 70% of the water used for irrigation is lost and not used by plants. Thus, the proposed intensification will have to include improving the infrastructure and the water productivity of existing systems. The high levels of water wastage are attributed to inefficient technologies, lack of incentives to conserve water and poor operation and maintenance of installed facilities. More importantly, there is an even greater scope for improving agricultural productivity through managing the water in rain-fed systems, which cover about 95% of cultivated area. Moreover, subsistence-based agronomic practices have resulted in the "mining" of the natural

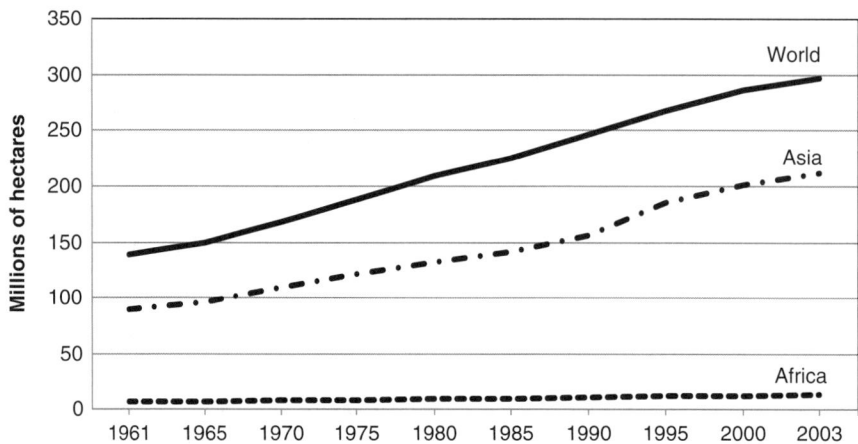

Fig. 1 Area equipped for irrigation in the world. *Source*: FAO (Food and Agriculture Organization) 2006b, FAOSTAT database, Rome (http://faostat.external.fao.org)

resource base in the process of crop production and livestock husbandry due to the need to produce more from the same land. Since land is inelastic, innovative ways that enable higher productivity will have to be adopted to meet the growing food gap. To achieve the Green Revolution in Africa will require water, without which all the other components of agricultural system cannot even begin to function.

Real Cost of Irrigation Development is Much Lower than Reported

There is divided opinion as to whether returns to investments for irrigation in SSA are really so low. Evidence on the ground suggests that the so-called poor returns are related to macro-economic and socio-political events of the 1970s to the 1990s which particularly affected large publicly funded irrigation schemes. Most of the countries were undergoing structural adjustment programmes (SAPs), which eroded the government's ability to cushion farmers against high input prices and aggressively competitive global markets. At the time, development of irrigation carried with it large expenditures on other infrastructure such as roads and administrative centres, with the burden of paying the full costs ending with the poor farmers. The inflated costs were also associated with the quality of appraisal and feasibility studies, implementation capacity, use of inappropriate technologies, limited competition among contractors and failure to realize the potential of alternatives to conventional irrigation in water management (Inocencio et al., 2003). Other reasons included higher proportion of investments allocated to appurtenant, higher mobilization costs due to remoteness of project areas and higher costs of input materials and equipment.

The neglect of water management utilizing rainfall or natural wetlands and informal systems failed to exploit the more positive aspects of quasi-irrigated agriculture, which holds even greater potential than conventional full-scale irrigation in the region (Mati, 2007). Meanwhile, private and smallholder group irrigation schemes have been thriving. This is partly because of lower and targeted initial investment costs, not tied to large loans, and therefore payable from the proceeds of the schemes. For instance, in Kenya, the Ministry of Water and Irrigation estimates that

the unit cost of small-scale irrigation project ranges between US \$2,000 and 3,000 per hectare (Mbatia, 2006). In comparison, World Bank analyses showed that between 1950 and 1993, irrigation investments in sub-Saharan Africa averaged more than US \$18,000 per hectare which was equivalent to over 13 times the South Asian average (World Bank, 1994). But since less than 5% of the potential in the region has been exploited, irrigation is too undeveloped to be ignored. Moreover, the most pressing constraint to smallholder agriculture in the region is not necessarily access to land, but access to water for crops and livestock (IFAD, 2002). Therefore, the existence of successful smallholder irrigation schemes in the region is an indicator that there is scope for profitable investment in irrigation.

The Need for Policy Support for AWM

Over the decades, governments and development partners in Africa have implemented many projects meant to improve the productivity of smallholder agriculture, which provides livelihoods to the rural poor. Many of these projects have involved interventions such as intensified use of inputs, land husbandry, microcredit, capacity building and various extension packages. However, as agriculture in sub-Saharan Africa deteriorated in the 1990s with the implementation of SAPs, attention was shifted from just technologies to the set of policies and institutional arrangements in use. It became evident that many countries continued to operate under the same policies inherited from the colonial era, or initiated soon after independence (OED, 2003). Increasingly, these policies and institutional frameworks were ineffective and rather than stimulate efficient economic activity and development, they were retarding it (Hunter, 1992). And what was true of economic activity in general was also true of resource management in particular. In general, the policies were driving mismatched resource use, exacerbating food insecurity and environmental degradation. Even as far back as the 1980s, the need for policy reform was expressed by stakeholders and summarized as follows: "the focus of African resource management is beginning to change. From seeing farmers and other resource users as the culprits of resource mismanagement, it is increasingly becoming evident that the culprit is to be found else-where" (Michanek, 1989). Thus, attention was shifting from too much emphasis on technological measures to addressing the farmer as the decision maker capable of controlling what happens on the land. Moreover, the farmer does not operate in a vacuum and affects, or is affected by, the prevailing economic, policy and institutional frameworks at local, national and international levels. Institutionalizing policies and frameworks that respond to the needs of smallholder farmers forms an entry point to the rural development process. Conducive policies for agricultural water management are an important ingredient to the whole agricultural sector revitalization. But the main question is whether these policies exist, and if so, to what extent they relate to enhancing management of agricultural water by smallholders in Africa.

A study conducted by IMAWESA (2007) covering nine countries of the ESA (Eritrea, Kenya, Madagascar, Malawi, Mauritius, Rwanda, Sudan, Tanzania and Zimbabwe) examined 78 policies that were deemed to have AWM implications in the target countries. The study found that all the countries had acknowledged water as a "social" good which needed to be shared equitably, although current IWRM concepts recognize water as an "economic" good and some countries are moving towards its commercialization. Furthermore, over the last 15 years, most countries have developed national and sector policies to address pertinent economic and development issues. Most of these policies are on the alleviation of poverty, achievement of economic growth, attainment of increased agricultural productivity and food security and securing sustainability in the environment. The water policies, however, place greater emphasis on drinking water or water used for sanitation systems and pay scant attention to water for agricultural purposes (IMAWESA, 2006). A critical review of the various policies (Fig. 2) showed that there is no specific policy document that addresses AWM in its broad sense in any of the nine countries. Instead, existing policies are scattered across different ministries or sectors. Moreover, the policies and legal instruments in most countries have been prepared on a sector basis without much regard to their cross-sectoral interactions or their impacts on customary laws. The scattering of AWM issues across several sectors has resulted in unavoidable overlapping of policies, duplication of efforts, and inefficient use of resources, as well as lack of clear ownership of AWM issues. Therefore, there is need for deliberate efforts towards creating awareness on AWM issues among policy makers and the general public as

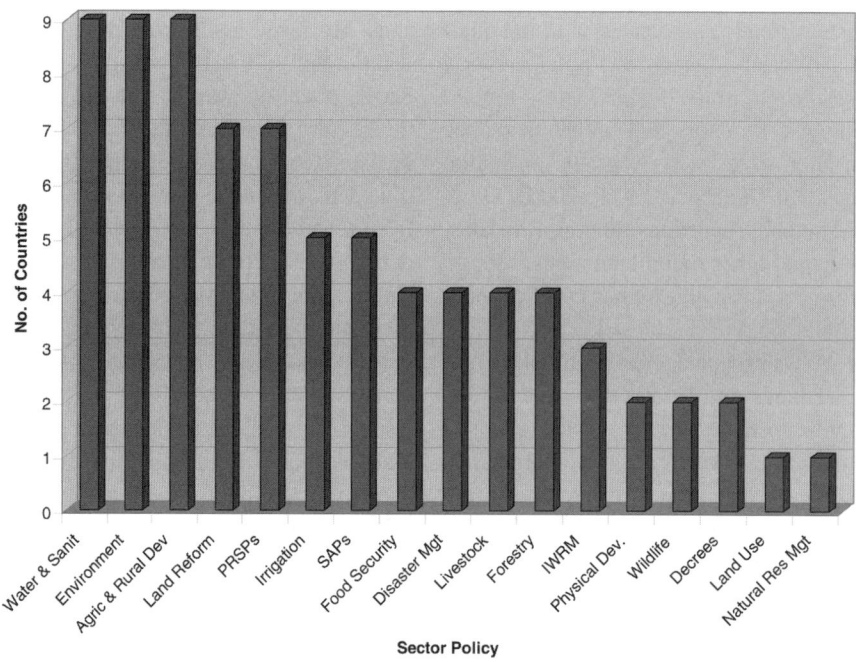

Fig. 2 Number of countries covering AWM issues in sector policies. *Source*: IMAWESA Policy Study, 2007

well as promoting the sharing of information between various institutions implementing AWM programmes.

Defining Agricultural Water Management

Until recently, water management for agriculture was synonymous with irrigation, and therefore other forms of water management beneficial to agriculture were downplayed. The current trend among professionals and practitioners is to adopt a more holistic approach to water for agriculture. Thus, agricultural water management (AWM) is here defined as "all deliberate human actions designed to optimize the availability and utilization of water for agricultural purposes. The source of water could include direct rain as well as water supplied from surface and underground sources. AWM is therefore the management of all the water put into agriculture (crops, trees and livestock) in the continuum from rainfed systems to irrigated agriculture and all relevant aspects of management of water and land". Therefore, AWM includes soil and water conservation, rainwater harvesting, irrigation (full and supplemental irrigation), agronomic management, rangeland rehabilitation, wetland management and utilization, drainage of waterlogged soils, water conservation for livestock, soil fertility management, conservation agriculture, agroforestry, climatic variability mitigation, use of low-quality or recycled water, water used in value addition and interventions such as integrated watershed management.

The Role of Networks

A network is any group of individuals and/or organizations that on a voluntary basis exchange information or goods or implement joint activities and organize themselves in such a way that the individual autonomy remains intact. In a network, members take part on a voluntary basis and they carry out joint activities that cannot easily be done alone. The network structure is often light and not very formal. Networks can have different forms and use different procedures depending on the specific situation. There is therefore a great diversity in networking experiences. Many development projects are nowadays promoted and implemented through networks. Farmer networks

have been particularly successful in promoting learning at the grassroots level, promoting the diffusion and adoption of innovative technologies especially among smallholder land users (Critchley et al., 1999; Haile et al., 2001; Kibwana, 2001). Networks have facilitated farmer participatory learning such as on-farm research, farming systems, training and visits, rapid and participatory rural appraisals, farmer participatory methods and farmer field schools, many of which have been tested with smallholder farmers in Africa (Norman et al., 1994; Chambers et al., 1989; Duveskog, 2001).

There are several types of networks and different criteria by which a typology can be based, for instance (i) the formal/centralized networks tend to have a strong secretariat, whereby most of the communication is initiated by or passes through the secretariat, and (ii) the more informal/decentralized networks have direct and systematic communication between the different members. Before a network is created, it is necessary to ensure that there is an atmosphere of openness among potential members which allows them to admit mistakes and learn from each other so that the co-ordination of the network can be assured, especially during the early stages. Moreover, the necessary activities will be performed by mobilized members or an assurance of sustainable source of funding, while enough commitment is required of initiators and/or supporting agents to overcome the organizational and establishment phase. These phases are particularly difficult from the point of view of resource availability. There should be a common vision and set of common goals among potential members (such as empowerment of marginal farmers, sustainability). It is important to understand the common problems faced by potential members, as well as create awareness of these problems and the importance of their influence on their work. There is need for a certain degree of maturity in management and sufficient organizational skills among potential members, especially NGOs. Relevant results/experiences are necessary on what could be shared and willingness by members to spend time and energy in sharing and networking at the expense of their own programmes. The role of host agencies and initiators as well as resource mobilization should always be clarified.

The place of networking for the African Green Revolution cannot be overstated, noting that it provides a means of knowledge sharing and learning. Africa is a continent fragmented by national boundaries, language and cultural barriers and especially techno-logical divides between scientists and farmers, to the level that there is a lot of duplication of efforts and missed opportunities. Networks provide one solution to break these barriers and to identify human resource pools for joint activities. Although most networks are thematic, bringing together scientists of a specific dis-cipline (e.g. soil scientists), there is a growing trend of multi-sectoral networks, e.g. IMAWESA (Improved Management of Agricultural Water in Eastern & Southern Africa), which bring together a wide cross section of stakeholders. The trend for more inclusive networking should be encouraged, especially to ensure messages from research have influenced policy more efficiently. A network is a means to an end; thus, if the benefits sought are achieved, then that network has served its purpose.

Project-Scale Interventions for Smallholder AWM in Africa

Africa has a long history with agricultural water man-agement, and neither water nor know-how is lacking. What is required is action. However, the need to adopt holistic approaches to AWM is based on the fact that in the past, sectorized approaches between what constitutes rain-fed as opposed to irrigated agricul-ture led to competition rather than complementarities for resources and services. Projects were sectorized as either soil and water conservation or irrigation, and implementation usually maintained this divide. Examples of these are scattered throughout Africa and have formed the foundation of many development projects with agriculture and land management on their agendas (Reij et al., 1996; Lundgren, 1993; Hurni and Tato, 1992; Penning de Vries et al., 2005). Addressing the continuum from rain-fed to irrigated agriculture is particularly pertinent and involves rainwater harvest-ing (RWH) systems, which are applicable in areas of insufficient rainfall, even with seasonal rainfall as low as 100–350 mm (Oweis et al., 2001; SIWI, 2001; Mati, 2007). There is a wide selection of AWM technologies and practices to choose from, which are practicable in smallholder agriculture in Africa and which have been tested and applied on the continent (Reij and Waters-Bayer, 2001; Mati, 2005, Negasi et al., 2000; Hatibu and Mahoo, 2000; IWMI, 2006; Mulengera,

1998). A more recent compilation of these technolo-gies and practices has put together a compendium of "100 ways to manage water under smallholder agricul-ture" (Mati, 2007), but this list is hardly exhaustive. As AWM under smallholder agriculture tends to be niche-relevant, there is need to identify what suits a certain section of the farm specifically, and any single farm may have multiple practices. At the continental level, identifying what AWM interventions will propel the Green Revolution is an even more daunting task. However, project-scale interventions can be identified from the multitude of possible technologies bringing down the options to some 12 broad-based approaches. These are summarized as follows.

Conserving Water on Steep Slopes

The African continent has large areas of agricultural land, where productivity is constrained due to slope steepness, causing soil erosion and loss of nutrients and water. Conservation of this water through vari-ous types of structures and agronomic management can improve water productivity by 50–100%. The main intervention is when some level of slope reduc-tion is achieved, normally by construction of soil and water conservation structures. The actual technologies may include stone bunds, ditches, earth bunds, fanya juu terraces, vegetative strips, trash lines, vegetative barriers, bench terraces, hedges and all manner of ter-racing land. These reduce surface runoff flows, soil erosion and water losses and thus achieve soil and water conservation by increasing infiltration and soil water storage. Most of the countries in Africa have various types of terraces (Christiansson et al., 1993; WOCAT, 1997; Wolde-Aregay, 1996; Lundgren and Taylor, 1993).

Stream Flow Diversion and Utilization for Irrigation

Africa is crisis-crossed by many rivers and streams which empty their waters to the lakes and oceans completely unutilized. Many of these streams are in dry areas that could benefit from irrigation. Generally, the irrigation potential in Africa (45 million hectares) is pegged to the availability of perennial rivers.

Moreover, in most countries, the potential irrigable areas are known, including the amounts of water that can be safely abstracted. What has been lacking has been political and financial capital to get it done. Thus, the diversion of stream flows and utilization in gravity-fed smallholder irrigated fields (furrow, basin, sprinkler, drip) offers a well-tested solution to water management for agriculture, in nearly all the countries of Africa.

Water-Lifting Devices

Use of low-head and small-powered petrol and electric pumps, including treadle pumps to lift water from rivers, small ponds and shallow water tables for irrigation (full or supplemental) has many examples all over Africa, especially in Malawi, South Africa, Swaziland, Kenya, Tanzania, Zambia, Ghana, Nigeria and Zimbabwe (Mangisoni, 2006; IWMI, 2006). In recent years, there have been many projects supported by governments and NGOs which have distributed treadle pumps and drip kits to small holder farmers. Some of these farmers are irrigating kitchen gardens or plots ranging from 20 m^2 up to several hectares. There is a wide scope for enhancing water management in rainfall deficit areas, constrained against gravity-based systems, through introduction of affordable water-lifting devices. However, there is a need to be aware of socio-economic and traditional norms of the people for the technology to be accepted and adopted by the community.

Small Individual Water Storages (Blue Water)

Rainwater harvesting and storage can provide drinking water for humans and livestock or is used for supplemental irrigation, to drought-proof crop production in dry areas. This can be achieved through (i) rooftop RWH and storages in either surface or underground tanks (Gould and Nissen-Peterssen, 1999); (ii) runoff harvesting from open surfaces and paths, roads, rocks and storage in structures such as pond or underground tanks (Guleid, 2002; Nega and Kimeu, 2002); (iii) flood flow harvesting from valleys, gullies, ephemeral

streams and its storage in ponds, weirs, small dams; and (iv) flood flow harvesting from ephemeral water courses and its storage within sand formations as sub-surface or sand dams (Nissen-Petersen, 2000). Pans and ponds are particularly easy to adopt. They are excavated where the catchment is appropriate and have been used for rainwater harvesting in many parts of East Africa, especially for livestock watering. The cost of construction can be relatively low since these structures utilize local knowledge in site selection, community labour through programmes such as food for work thereby reducing the costs even further (Natea, 2002; Mati and Penning de Vries, 2005). The main difference between ponds and pans is that ponds receive some groundwater contribution while pans rely solely on surface runoff. Thus, pans which range in size from about 10,000 to 50,000 m^3 (Bake, 1993) can be constructed almost anywhere as long as physical and soil properties permit. When properly designed and with good sedimentation basins, the water collected can be used for livestock watering or to supplement the irrigation of crops.

Runoff Harvesting and Storage in Soil Profile (Green Water)

This involves runoff harvesting from land, roads, paved areas and channelling it to specially treated cropped area for storage within the soil profile where the soil can hold water for crops. The cropped area may be prepared as planting pits, negarim, zai, ditches, bunded basins, demi-lunes, or simply ploughed land (Hai, 1998; Mati, 2005; Ngigi, 2003). Storing rainwater in the soil profile for evapotranspiration is sometimes referred to as "green water" and forms a very important component for plant growth. The design of run-on facility (e.g. semi-circular bund, negarim, zai pit) depends on many factors including catchment area, volume of runoff expected, type of crop, soil depth, and availability of labour (Hatibu and Mahoo, 2000). The source of water could be small areas or "micro-catchments" or larger areas such as external catchments. The latter involve runoff diversion from larger external catchments such as roads, gullies, open fields into micro-basins for crops, ditches or fields (with storage in soil profile) including paddy production where the profile can hold water relatively well.

Valley Bottom Utilization and Management

Another category of water management involves utilization of valley bottoms and areas where the natural water table is high. It can also be regarded as flood recession agriculture or wetland cultivation, involving cultivation of flood plains as flood water recedes. It involves appropriate management of lowlands such as *olonaka* and *olombanda* in Angola, *bas fonds* in Madagascar and *dambos* in Malawi, Zambia and Zimbabwe (IFAD, 2002), with or without irrigation, and by a combination of drainage and water storages to permit crop production including fisheries. For example, in Malawi, fish ponds are constructed and the same water used for supplemental irrigation of paddy and other field crops. Cultivation of inland valley lowlands which are seasonably saturated with water and retain a high water table even during the dry season is common in Tanzania, Uganda, Madagascar, Rwanda, Malawi, Mozambique, Swaziland, Zimbabwe, Zambia and indeed most countries (Critchley et al., 1999; Mati, 2007; IWMI, 2006; McCartney et al., 2005).

Spate Flow Diversion and Utilization

Spate irrigation or floodwater diversion includes techniques which force the water to leave its natural course, for the water to be used for agriculture, especially in arid areas. The runoff water is diverted through canals or *wadis* (wide valleys in arid areas that carry flush floods) for supplemental irrigation of low-lying lands, sometimes far away from the source of runoff. Spate irrigation has a long history in the Horn of Africa and still forms the livelihood base for rural communities in arid parts of Eritrea, Ethiopia, Kenya, Somalia and Sudan (SIWI 2001; Negasi et al., 2000; Critchley et al., 1992). In Tanzania, spate irrigation has been used to increase yield of rice in RWH systems (majaluba) from 1 to 4 t ha^{-1} (Gallet et al., 1996). Although spate flow irrigation has high maintenance requirements, its applicability is valid for large areas of the Sahel and the Horn of Africa, where other conventional irrigation methods may not be feasible.

Drainage of Waterlogged Soils

Generally, the word "drainage" is usually used to refer to the removal of excess water from waterlogged soils and also from irrigated fields. Traditionally, many communities in Africa have used waterlogged soils to grow sugarcane, vegetables and fodder for livestock. In addition, waterlogging, especially temporary waterlogging, is responsible for loss of crop production in these areas. Many development projects in recent years have shied away from large-scale drainage operations due to the environmental furore this might raise. However, it is important to note that as land space decreases, more waterlogged soils will have to be drained for agriculture. The actual size of land requiring this targeted drainage remains largely unknown, and there is scope for research in this area. Some of the techniques of achieving drainage include use of surface drains and flood control to overcome waterlogging and salinity problems (FAO, 2002). In areas where surface drains do not have a natural drainage outlet, low-head high-discharge pumps can be used to dispose the drainage effluent into appropriate outlets. Horizontal subsurface drainage is usually preferred but vertical drainage in the form of shallow tube-wells can also be applied. However, the density of the tube-wells should vary with the groundwater quality and recharge to avoid salinity buildup. Temporary drainage techniques such as mole drains and deep tillage are also advisable, as well as the management of shallow water tables.

Conservation Agriculture

Conservation agriculture (CA) or conservation tillage is a land cultivation technique which tries to reduce labour, promotes soil fertility and enhances soil moisture conservation. It is considered as one of the most efficient systems for harnessing green water. Conservation agriculture has come to be recognized as the missing link between sustainable soil management and reduced cost of labour, especially during land preparation, and it holds the potential for increased crop production and reduced erosion (Biamah et al., 2000). Conservation agriculture can be achieved through (i) minimum or zero soil turning, (ii) permanent soil cover, (iii) stubble mulch tillage and

(iv) crop selection and rotations. It may also include pot-holing, infiltration pits, strip tillage and tied ridging. CA has been gaining acceptance in Africa in countries such as Tanzania, Madagascar, Zambia and Zimbabwe (Nyagumbo, 2000; Biamah et al., 2000).

Recycling Waste Water and Use of Low-Quality Water

There are many areas in Africa which could benefit from the use of recycled water or low-quality water for irrigation. Many of these areas are sub-urban, some of them already utilizing raw untreated waste water. The niche of the Green Revolution would be to improve the quality and safety of the water so used, as well as its quantity and to streamline the policy and legal aspects for use of waste water for agriculture. Other sources include kitchen waste water, grey water and low-quality water. For instance, the reuse of irrigation effluents can supplement scarce irrigation water supplies and also help to alleviate disposal problems (FAO, 2002).

Water for Livestock

Water for livestock sometimes gets ignored in development projects targeting crop-based systems. It is quite common to see livestock walk very long distances to water, within an area that is benefiting from soil and water conservation initiatives. By providing livestock with water close to home, time and resources wasted in searching for water can be saved and livestock productivity enhanced. Water shortage for livestock is particularly acute in the Horn of Africa countries covering Eritrea, Ethiopia, Kenya, Sudan and Somalia, as well in many Sahelian countries, and this is associated with frequent drought occurrence. For instance, in a study covering 160 pastoralists in different locations of northern Kenya, drought was perceived by 94% of the sampled population to be the principal livelihoods challenge (Nyamwaro et al., 2006) in contrast to lack of pasture which was listed by 63% of the respondents. Thus, to enhance drought resilience to the poorest households, interventions could include

(a) enhancing the robustness of current drought coping strategies/mechanisms and (b) decreasing households' vulnerability to future drought-related shocks. Water provision for both drinking and pasture production comes top of the list of interventions required.

Soil Fertility Inputs and Management

Water for agriculture is well managed if the capacity of the soil to produce a good crop is not limited by fertility or any other constraint. Thus, the concept of "poor" and "fertile" soil is not just about nutrient availability, but water availability as well. Thus, a soil is considered fertile if it can supply adequate nutrients, resulting in sustained high crop yields, and if it has good rooting depth, good aeration, good water-holding capacity, the right pH balance, and no adverse soil-borne pests or diseases (Landon, 1991). Thus, sustainable soil management entails replenishment of chemical soil nutrients removed or lost; improvement and maintenance of soil physical conditions; prevention of accumulation of toxic elements and augmentation of soil pH; limiting the buildup of weeds, pests and diseases; and conservation of soil and water resources (Young, 1976; Greenland, 1975). Therefore, soil fertility interventions, fertilizers, manures, mulches and low external input sustainable agriculture (LEISA) systems should accompany any of the "water management" interventions mentioned above and are also inherently managing water for agriculture, and this applies to all the countries in Africa.

Conclusions and Recommendations

There is no doubt that water constitutes a key ingredient to the success of the African Green Revolution. Unlike the Asian revolution which found in place a relatively well-developed irrigation infrastructure, the African agricultural sector is predominantly rain-fed, even in ecological zones which should of necessity be irrigated or partially irrigated. Again, unlike fertilizers and improved seeds which can be imported, purchased and distributed in small packets, water has to be sourced locally, and a higher technological input is required. Such services as surveying, design, layout and construction of water supply and application equipment require some level of specialized expertise and cost. This is not to say that the farmer is completely helpless. There are many technologies where with just some minimum training the farmer can survey, lay out and construct water management infrastructure using own labour, especially soil conservation and runoff harvesting systems.

The interventions presented in this chapter therefore constitute broadly categorized water management interventions that can form the basis for extensive project-scale interventions across multiple countries. Within each category, there are several technologies to allow local adaptability and cost-effectiveness. Another aspect of these interventions is their applicability for smallholder agriculture and micro-irrigation projects. The success of these interventions will depend to a large extent on the operational framework against which they are implemented, especially the inclusion of the farmers in planning, implementation and management of the systems. Other supportive aspects include implementation of AWM interventions as part of a more inclusive integrated watershed management, and thus the institutionalization of management structures such as water users associations (WUAs). Whenever possible, AWM interventions should target to provide water for multiple purposes and enhance cost-effectiveness. Capacity building for all cadres of stakeholders and local ownership are necessary for success.

The major threats to enhancing the AWM for the African Green Revolution include negative perceptions about the returns to investment from irrigation, and as explained earlier, it has been proved that smallholder water management, especially where the farmer has some level of individual autonomy in decision making, is not only profitable but quite sustainable. Another constraint is the high initial investment required, as sometimes, supporting infrastructure such as roads and stores may have to be constructed first. Moreover, the poorest and most vulnerable communities tend to be located in the driest and remotest (far from roads, towns) part of the country where transaction costs of any activity tend to be high. This therefore poses a challenge as to where to allocate resources, especially given the slim chances of payback from such vulnerable groups as the poorest. The trial and error tendencies of farmers exposed to irrigation and/or water harvesting for the first time can lead to many mistakes which could discourage both farmers and investors. However, even with these limitations, the benefits of managing water for agriculture optimally far outweigh the threats, especially as there is increased food security, wealth creation and improved livelihoods for beneficiaries.

References

Bake G (1993) Water resources. In: Herlocker DJ, Shaaban SB, Wilkes S (eds) Range management handbook of Kenya, vol II, 5, Isiolo District. Republic of Kenya, Ministry of Agriculture, Livestock Development and Marketing, Nairobi, pp 73–90

Biamah EK, Rockstrom J, Okwach GE (2000) Conservation tillage for dryland farming. Technological options and experiences in eastern and Southern Africa. RELMA, Nairobi

Chambers R, Pacey A, Thrupp LA (eds) (1989) Farmer first. Farmer innovation and agricultural research. Intermediate Technology Publishers, London, Collaborative program partners. Pretoria: IWMI

Christianson C, Mbegu CN, Yrgard A (1993) The hand of man: soil conservation in Kondoa eroded area, Tanzania. SIDA/RSCU Report No. 12. Nairobi

Critchley W, Cooke R, Jallow T, Njoroge J, Nyagah V, Saint-Firmin E (1999) Promoting farmer innovation: harnessing local environmental knowledge in East Africa. Workshop Report No. 2. UNDP – Office to Combat Desertification and Drought and RELMA, Nairobi

Critchley W, Reij C, Seznec A (1992) Water harvesting for plant production. Volume II: case studies and conclusions for sub-Saharan Africa. World Bank Technical Paper Number 157. Africa Technical Department Series

Duveskog D (2001) Water harvesting and soil moisture retention. A study guide for farmer field schools. Ministry of Agriculture and Farmesa, Sida, Nairobi

ECA (Economic Commission for Africa) (1995) Water resources: policy assessment, Report of UNECA/WMO. International conference. Addis Ababa

FAO (Food and Agriculture Organization) (1995) Irrigation in Africa in figures. World Water Report 7. Rome

FAO (Food and Agriculture Organization) (2006a) AQUASTAT database. FAO, Rome. http://www.fao.org/ag/aquastat

FAO (Food and Agriculture Organization) (2006b) FAOSTAT database. FAO, Rome. http://faostat.external.fao.org

FAO (2000) Agriculture towards 2015/30. Technical Interim Report, Rome

FAO (2002) Agricultural drainage water management in arid and semi-arid areas. Irrigation and drainage paper, 61. FAO, Rome

Gallet LAG, Rajabu NK, Magila MJ (1996) Rural poverty alleviation: experience of SDPMA. Paper presented at a workshop on "Approaches to rural poverty alleviation in SADC countries". Cape Town, South Africa, Jan 31–Feb 3 1996

Gould J, Nissen-Peterssen E (1999) Rainwater catchment systems for domestic supply. Design, construction and implementation. Intermediate Technology Publications, London

Greenland DJ (1975) Bringing green revolution to the shifting cultivator. Science 190:841–844

Guleid AA (2002) Water-harvesting in the Somali National Regional State of Ethiopia. In: Haile M, Merga SN (eds) Workshop on the experiences of water harvesting in drylands of Ethiopia: principals and practices. Dryland Coordination Group (DCG Report No.19), Ethiopia, pp 45–49

Hai MT (1998) Water harvesting: an illustrative manual for development of microcatchment techniques for crop production in dry areas. Technical Handbook No. 16. RELMA, Nairobi

Haile M, Abay F, Waters-Bayer A (2001) Joining forces to discover and celebrate local innovation in land husbandry in Tigray, Ethiopia. In: Reij C, Waters-Bayer A (eds) Farmer innovation in Africa. A source of inspiration for agricultural development. Earthscan, London, pp 58–73

Hatibu N, Mahoo HF (2000) Rainwater harvesting for natural resources management. A planning guide for Tanzania. Technical Handbook No. 22. RELMA, Nairobi

Hunter JP (1992) Promoting sustainable land management: problems and experiences in the SADC region. Splash Vol. 8 No. 1. Maseru, Lesotho

Hurni H, Tato K (eds) (1992) Erosion, conservation and small-scale farming. Geographisca Bernesia, Walsworth Publishing Company, Missouri

IFAD (2000) Knowledge management. A thematic review. IFAD support for water management and irrigation in East and Southern Africa. IFAD, Rome

IFAD (2002) Assessment of rural poverty in eastern and southern Africa. International Fund for Agricultural Development, Rome

IMAWESA (2006) Preliminary assessment of policy and institutional frameworks with bearing on agricultural water management in eastern and southern Africa. Report of a baseline study, SWMnet Working Paper 11, Nairobi. www.asareca.org/swmnet/imawesa

IMAWESA (2007) Policies and institutional frameworks for agricultural water management in eastern and southern Africa. Synthesis report of a rapid appraisal covering nine countries in the ESA. A research report of IMAWESA, Nairobi. www.asareca.org/swmnet/imawesa

IWMI (2006) Agricultural water management technologies for small scale farmers in Southern Africa: an inventory and assessment of experiences, good practices and costs. Final report produced by the International Water Management Institute (IWMI) Southern Africa Regional Office Pretoria, South Africa. For Office of Foreign Disaster Assistance, Southern Africa Regional Office, United States Agency for International Development

Inocencio A, Sally H, Merrey D, de Jong I (2003) Irrigation capital investment costs in sub-Saharan Africa: an overview of issues and evidence, review of literature report submitted to the World Bank and the Collaborative Program Partners. IWMI, Pretoria

Kibwana OT (2001) Forging partnerships between farmers, extension and research in Tanzania. In: Reij C, Waters-Bayer A (eds) Farmer innovation in Africa. A source of inspiration for agricultural development. Earthscan, London, 51–57

Landon JR (1991) Booker tropical soil manual. A handbook for soil survey and agricultural land evaluation in the tropics and subtropics. Longman, New York, NY

Lundgren L (1993) Twenty years of soil conservation in Eastern Africa. RSCU/Sida. Report No. 9. Nairobi, Kenya

Lundgren L, Taylor G (1993) From soil conservation to land husbandry. Guidelines based on SIDA's experiences. Natural Resources Management Division, SIDA, Stockholm

Mangisoni J (2006) Impact of treadle pump irrigation technology on smallholder poverty and food security in Malawi: a case study of Blantyre and Mchinji Districts. Report written for IWMI. IWMI, Pretoria, South Africa

Mati BM (2005) Overview of water and soil nutrient management under smallholder rain-fed agriculture in East Africa. Working Paper 105. Colombo, Sri Lanka: International Water Management Institute (IWMI). www.iwmi.cgiar.org/pubs/working/WOR105.pdf

Mati BM (2007) 100 ways to manage water for smallholder agriculture in eastern and Southern Africa. SWMnet Working Paper No. 13, Nairobi, Kenya. www.asareca.org/swmnet/imawes

Mati BM, Penning de Vries FWT (2005) Bright spots on technology-driven change in smallholder irrigation. Case studies from Kenya. In: Bright spots demonstrate community successes in African Agriculture. International Water Management Institute. Working Paper 102:27–47. www.iwmi.cgiar.org/pubs/working/WOR102.pdf

Mbatia ED (2006) Irrigation sector in Kenya – status and challenges. Paper presented during the green water credits workshop, KARI, Nairobi, 11–12 October 2006

McCartney MP, Masiyandima M, Houghton-Carr HA (2005) "Working wetlands" classifying wetland potential for agriculture. IWMI Research Report No. 90. IWMI, Colombo

Michanek GLLM (1989) The legal context of conservation in Lesotho. SPLASH. 5 No. 1. Maseru, Lesotho

Mulengera MK (1998) Workshop for innovative farmers and researchers on traditional ways of soil conservation and land and water management. Ministry of Agriculture and Cooperatives, Dar es Salaam, Tanzania

Natea S (2002) The experience of CARE Ethiopia in rainwater harvesting systems for domestic consumption. In: Haile M, Merga SN (eds) Workshop on the experiences of water harvesting in drylands of Ethiopia: principles and practices. Dryland Coordination Group (DCG Report No.19), Ethiopia, pp 19–28

Nega H, Kimeu PM (2002) Low-cost methods of rainwater storage. Results from field trials in Ethiopia and Kenya. Technical Report No. 28. RELMA, Nairobi

Negasi A, Tengnas B, Bein E, Gebru K (2000) Soil conservation in Eritrea. Some case studies. Technical Report No. 23. RELMA, Nairobi

NEPAD (New Partnership for Africa's Development) (2003) Comprehensive Africa agriculture development programme. New Partnership for Africa's Development (NEPAD) and African Union, Midrand

Ngigi SN (2003) Rainwater harvesting for improved food security. Promising technologies in the Greater Horn of Africa. Kenya Rainwater Association, Nairobi

Nissen-Petersen E (1996) Groundwater dams in sand rivers. UNDO/UNHCHS (Habitat), Myanmar

Nissen-Petersen E (2000) Water from sand rivers. A manual on site survey, design, construction and maintenance of seven types of water structures in riverbeds. Technical Handbook No. 23. RELMA, Nairobi

Norman DW, Siebert JD, Modiakogotla E, Worman FD (1994) Farming systems research approach: a primer for eastern and southern Africa. F.S. Program. UNDP, Gaborone, Botswana

Nyagumbo I (2000) Conservation technologies for smallholder farmers in Zimbabwe. In: Biamah EK, Rockstrom J, Okwach J (eds) Conservation tillage for dryland farming. Technological options and experiences in Eastern and Southern Africa. RELMA, Nairobi, pp 70–86

Nyamwaro SO, Watson DJ, Mati B, Notenbaert A, Mariner J, Rodriguez LC, Freeman A (2006) Assessment of the impacts of the drought response program in the provision of emergency livestock and water interventions in preserving pastoral livelihoods in northern Kenya. Report of an ILRI multidisciplinary scientific team of consultants assessing the emergency drought response project in northern Kenya. ILRI

OED (Operations Evaluation Department) (2003) OED Review of the Poverty Reduction Strategy Paper (PRSP) process. Approach paper. World Bank, Washington, DC

Oweis T, Prinz P, Hachum A (2001) Water harvesting. Indigenous knowledge for the future of the drier environments. International Centre for Agricultural Research in the Dry Areas (ICARDA), Aleppo, Syria

Penning de Vries FWT, Mati B, Khisa G, Omar S, Yonis M (2005) Lessons learned from community successes: a case for optimism. In: Penning de Vries FWT (ed) Bright spots demonstrate community successes in African agriculture. Working paper 102:1–6. International Water Management Institute (IWMI), Colombo, Sri Lanka

Reij C, Scoones I, Toulmin C (eds) (1996) Sustaining the soil. Indigenous soil and water conservation in Africa. Earthscan, London

Reij C, Waters-Bayer A, (eds) (2001) Farmer innovation in Africa. A source of inspiration for agricultural development. Earthscan, London

SIWI (2001) Water harvesting for upgrading of rain-fed agriculture. Problem analysis and research needs. Report II. Stockholm International Water Institute. Stockholm International Water Institute, Stockholm, Sweden

UN (United Nations) Water/Africa (2003) Africa Water Vision 2025. Economic Commission for Africa. Addis Ababa

WOCAT (World Overview of Conservation Approaches and Technologies) (1997) WOCAT, World overview of conservation approaches and technologies – a program overview. May 1997. Bern, Switzerland

Wolde-Aregay B (1996) Twenty years of soil conservation in Ethiopia. A personal overview. Technical Report No. 14. SIDA/RSCU, Nairobi

World Bank (1994) A review of World Bank experiences in irrigation. Report No. 13676. World Bank, Washington, DC

World Bank (2005) Reengaging in agricultural water management. Challenges, opportunities and trade-offs. Water for food team. Agriculture and Rural Development Department (ARD). The World Bank, Washington, DC, 146 pp

Young A (1976) Tropical soils and soil survey. Cambridge University Press, Cambridge

Ex-ante Evaluation of the Impact of a Structural Change in Fertilizer Procurement Method in Sub-Saharan Africa

J. Chianu, A. Adesina, P. Sanginga, A. Bationo, and N. Sanginga

Abstract In June 2006, the African Heads of State made a declaration to support increase in use of fertilizers in the farming systems of sub-Saharan Africa from the present average of about 8 kg ha^{-1} to about 50 kg ha^{-1}. One route to attain this goal is to engender regional joint fertilizer procurement to reduce farm gate price and increase fertilizer demand and use. A review of fertilizer use in Africa has shown that structural changes in fertilizer procurement can reduce farm gate price by 11–18%. Using an average of these figures (15%), this study compares the effect of structural changes in fertilizer market (reducing farm gate price by 15%) on total fertilizer demand, total farm income, and additional farm income with the base situation (using FAO data) under three own fertilizer price elasticity of demand scenarios (low –0.38; medium –1.43; and high –2.24) for 11 sub-Saharan Africa countries. Data were analyzed using Microsoft Excel. Result shows that compared with the base level, structural change in fertilizer procurement arrangement (reducing farm gate price by 15%) led to 6% additional farm income (US $125 million) under low elasticity; 22% (US $472 million) under medium elasticity; and 34% (US $730 million) under high elasticity. Switching from one scenario to another indicates the potential to further increase farm income from 20 to 32%. The chapter concludes with the support for structural interventions that reduce farm gate price of fertilizers and other inputs. Such interventions increase farmer productivity, total production, and total farm income and lead to improved livelihoods.

Keywords Farm gate price · Fertilizer use · Joint procurement · Price elasticity scenarios · Sub-Saharan Africa

Introduction

Green revolution and high agricultural productivity have been attained in most regions of the world, except Africa. African agriculture is still characterized by low productivity and food insecurity – a situation most clearly observed in the sub-Saharan Africa (SSA), especially if South Africa is excluded. Among the farm families, this results in low returns on investments, poor livelihoods (low income, poor nutrition and lack of food security, vulnerability to risks, low life expectancy, etc.), and extractive coping behavior, leading to environmental and natural resource degradation (Chianu et al., 2006). This confirms that food security and availability is closely tied to agricultural productivity (Todd, 2004).

While agreeing that agricultural growth is central to winning the battle against hunger and poverty in SSA, scholars have largely blamed the rampant low agricultural productivity and widespread land degradation [65% of all Africa (Scherr (1999)] on the abysmally low use of farm inputs, especially mineral fertilizers. About 75–80% of Africa's farmland is degraded at an annual rate of 30–60 kg nutrients per hectare (Roy, 2006). Without adequate inputs, farmers often cannot meet the food needs of their families and those of the rapidly growing population. Land degradation is affecting up to two-thirds of Africa's productive land and nearly all of Africa's land is vulnerable to degradation (European Commission, 2007).

J. Chianu (✉)
Tropical Soil Biology and Fertility Institute of the International Centre for Tropical Agriculture (TSBF-CIAT), UN Avenue, Gigiri, Nairobi, Kenya
e-mail: jchianu@yahoo.com

Table 1 Population, cropped land, and fertilizer use (1961 and 1997)

Region	1961			1997		
	Population (million)	Crop land (million ha)	Fertilizer use (kg ha^{-1})	Population (million)	Crop land (million ha)	Fertilizer use (kg ha^{-1})
World	3136	1352	23.00	5823	1501	90.00
Developing countries	978	654	42.00	1294	640	87.00
SSA	*219*	*120*	*0.15*	*578*	*154*	*9.00*
DR Congo	16	7.0	0.04	48	8	0.80
Kenya	9	28.8	2.80	28	5	29.00
Nigeria	38	0.6	0.50	104	31	4.50
Egypt	*29*	*2.6*	*93.00*	*65*	*3*	*313.00*
France	*46*	*21.4*	*113.00*	*58*	*20*	*260.00*
India	452	160.9	21.00	966	170	95.00
USA	189	182.5	41.00	272	177	114.00

Source: FAOSTAT (2003)

Yet fertilizer use in Africa is the lowest in the world. In other regions, it is the excessive use of mineral fertilizers that is causing environmental degradation. However, Africa suffers from the opposite problem where lack of or limited use of mineral fertilizers in the farming systems is the main cause of environmental degradation due to soil nutrient mining.

In 2002/2003, fertilizer use (kg ha^{-1}) in various regions and countries of the world was as follows: 8 (for SSA), 80 (Latin America), 95 (India), 98 (North America), 114 (USA), 175 (Western Europe), 202 (East Asia), 260 (France), and 314 (Egypt) (Roy, 2006). Average fertilizer use in SSA is about 8–9 kg ha^{-1} compared to 260 kg ha^{-1} in France and 114 kg ha^{-1} in USA (Table 1).

At about 2 million tons per year, SSA's fertilizer consumption is less than the 3.4 million tons annually consumed by Bangladesh (Roy, 2006; Versi, 2006). A comparison of fertilizer consumption trend (1980–1989 and 1996–2000) in SSA and developing countries of Asia shows that while average annual fertilizer consumption increased by 182% in the latter, it increased by only 16% in the former (FAOSTAT, 2003). With 9% of the world's population, SSA accounts for <2% of global fertilizer use and <0.1% of global fertilizer production. It is estimated that since the 1950s, Africa has lost about 20% of its soil fertility irreversibly due to degradation (Dregne, 1990). The continent loses the equivalent of over US $4 billion worth of soil nutrients per year, severely eroding its ability to feed itself. Yield and production losses due to land degradation in SSA have been reported to range from 2 to 50% (Lal, 1995; Scherr, 1999). When soil nutrient inputs are lower than soil nutrient removals (due to water

and wind erosion, nutrient removal by crops, leaching, etc.), soil depletion and degradation occurs. Nutrient balances for cropping systems in most SSA countries are negative, implicating soil nutrient mining (Fig. 1).

Annual nutrient loss from cultivated land is estimated (in million tons) at 4.4 for nitrogen (N), 0.5 for phosphorus (P), and 3.0 for potassium (K). These rates are several times higher than SSA's (excluding South Africa) annual fertilizer consumption (in million tons) of 0.8 for N, 0.26 for P, and 0.20 for K. Nutrient loss (in kg ha^{-1}) during the last three decades is equivalent to 1400 of urea, 375 of triple superphosphate (TSP), and 896 of potassium chloride (KCl).

The above anomalies exist even though Africa (including North Africa) has the highest endowments of the principle ingredients (phosphates, nitrates, etc.) used in manufacturing fertilizers. With fallow periods getting shorter in many African countries, the absence

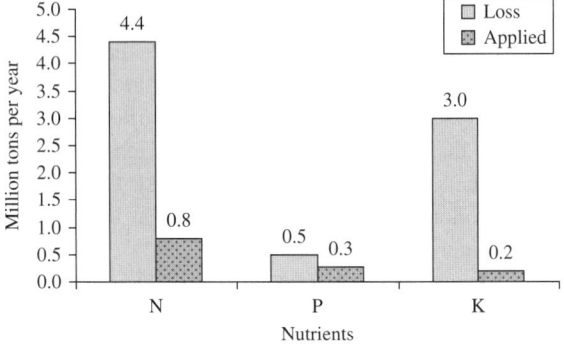

Fig. 1 Macronutrient applications versus losses in sub-Saharan Africa (Source: Sanchez et al., 1997)

of fertilizer has meant that soil is being leached of essential minerals. While land is being degraded, soil fertility is declining to levels insufficient to sustain and support economically feasible agricultural production.

Given that several alternative options for increasing soil fertility for enhanced agricultural productivity has been tried (without much success) in the past, a debate on the way forward concluded on the urgent need to increase the average rate of use of mineral fertilizers at farm level. This, unfortunately, is coming when the world fertilizer free on board (FOB) cost (US\$ t^{-1}) is trending upward, reaching historic highs. To follow up on the conclusion on the way forward, top-level lobbying and policy dialogue are being used, based on the conviction that a move toward reducing hunger on the African continent must begin by addressing its severely depleted soils.

As a result, in June 2006, following the Africa Fertilizer Summit (AFS), the African Heads of State and Government made a declaration to support increase in the use of fertilizers in the farming systems of SSA from the present average (about 8 kg ha^{-1}) to about 50 kg nutrient ha^{-1} by 2015. How this relates to fertilizer use in different regions of the world is shown in Fig. 2.

Following the political declaration, the question of how best to attain this goal became critical in the minds of policy and economic analysts, especially given the high farm gate price of mineral fertilizers in most parts of SSA (e.g., about US \$482 t^{-1} for NPK in Malawi: Saudi Arabia to Blantyre via Beira) compared with the NPK fertilizer FOB of US \$289 t^{-1} (Malawi delivery) or about US \$513 t^{-1} for NPK in Rwanda (Black Sea–Russia–Kigali via Dar es Salaam) compared with the NPK fertilizer FOB of US \$207 t^{-1} (Rwanda delivery) of mineral fertilizers in most parts of SSA. Using the examples of Malawi and Rwanda, the rather high farm gate price is partitioned and attributed (in %) as shown in Table 2.

Table 2 shows that FOB price accounts for only 40–60% of the farm gate price. Other key items of cost that add up to explain the high farm gate price are trucking (especially from sea port to in-country warehouse) which accounts for 15–34% of the farm gate price. Next comes sea shipping that accounts for 10–12% of the farm gate price.

Objectives of the Study

This chapter aims at contributing to answering the question: 'How best to attain the goal of increasing fertilizer use from about 8 kg ha^{-1} to 50 kg ha^{-1} by 2015' by carrying out an ex-ante analysis of the

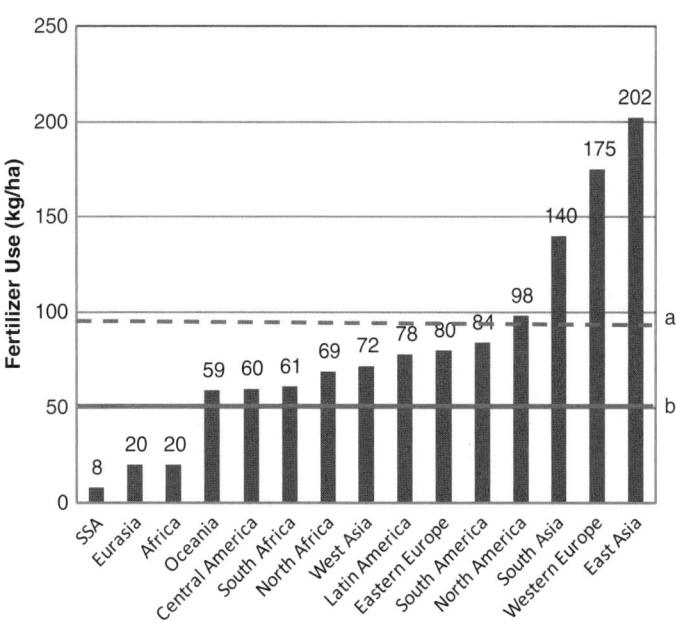

Fig. 2 Fertilizer use in different regions of the world. *a* Global mean fertilizer use (93 kg ha^{-1}); *b* African Fertilizer Summit recommendation (50 kg ha^{-1})

Table 2 Percent attribution of the high farm gate prices of fertilizers: Malawi and Rwanda

Farm gate price, fertilizer FOB price, and items of attribution	Malawi	Rwanda
Farm gate price (US$/ton)	482	513
Fertilizer FOB price (US$/ton)	289	207
Item of attribution		
Fertilizer FOB (%)	60	40
Sea shipping (%)	12	10
Stevedoring, shore handling, ad valorem,[a] bagging, warehousing, customs (on value), insurance and clearing (%)[b]	5	8
Trucking from port to in-country warehouse (+ border fees) (%)[c]	15	34
In-country trucking and warehousing (%)	8	8
Total (%)	100	100

[a]Various duty charges, taxes, and port fees based on FOB dollar value of shipment
[b]In the case of Rwanda, other items listed within this sub-group are demurrage, bags, port storage (indirect discharge), C+F ad valorem wharfage, private in-port security, three tallies, loading on truck, sundries, and losses from B/L
[c]This is made up of gasoline, overhead, taxes and tolls, depreciation, tire, and driver costs
Source: Adapted from Kumar (2007)

potential benefits of joint regional fertilizer procurement to reduce farm gate price and provide incentive for increased fertilizer demand and use in SSA? It compares the effect of structural changes in fertilizer market on total fertilizer demand, total farm income, and additional farm income with the base condition under three own (fertilizer) price elasticity of demand scenarios and provides information on potential benefits of regional joint fertilizer procurement to reduce farm gate price of fertilizers, creating incentives for increased fertilizer use in SSA. Increase in volume of order through consortia also leads to a fall in FOB price.

Justification for the Study

To move one metric ton of fertilizer to a distance of 1000 km costs an average of US $15 (in USA), US $30 (India), US $100 (SSA), and US $163 (Rwanda). Similarly, while it costs about US $50 to move one metric ton of maize from Iowa (USA) to Mombasa (Kenya), a distance of about 13,600 km, it costs about US $100 to move the same quantity of maize from Mombasa to Kampala (Uganda), a distance of about 900 km. These implicate a structural problem. Improvements in the structural supply issues [e.g., ordering trucks in advance, actively negotiating, use of large ships that stop in few locations, ordering 'generic' mass-produced NPK product, ordering in bulk, sourcing from the lowest cost plants, and

selecting low-cost (but 'right') fertilizers] have been found to lead to 11–18% decrease in the farm gate prices of fertilizer (Kelly et al., 2003; Kumar, 2007). These could reduce farm gate price of fertilizer in Malawi by 14.5%, from US $482 t^{-1} to US $412 t^{-1} (Kumar, 2007). As a result, Kumar (2007) indicated that improving the overall procurement process could reduce the price of mineral fertilizers in SSA.

Theoretical Framework and Elasticity Scenarios

Elasticity (*e*) is a measurement used by economists to assess the responsiveness of demand to changes in price. Technically, *e* refers to proportionate change in a dependent variable of a function (*Y*) divided by proportionate change in an independent variable (*X*) at a given value of the independent variable and is a product of ratios (independent of units of variables) (Bannock et al., 2003).

Observation of market behavior is important for the calculation of *e*, explaining why *e* is usually estimated using historical data from statistics. Alternatively, *e* can be derived from an econometric model expressing demand as a function of price in equation form. The formula is as follows:

$$e = \frac{\% \text{ change in } y}{\% \text{ change in } x}$$

e is a measure of the sensitivity of one thing (e.g., demand for a commodity) to another (e.g., the price of it). In the case of price elasticity, *Y* would be the quantity demanded, and *X* would be price. If elasticity has an absolute magnitude numerically <1, the quantity demanded is price inelastic [i.e., that if the price is increased (marginally), the quantity demanded will not fall proportionately as much and, therefore, the total expenditure on the good will increase]. If the good is price elastic (i.e., elasticity is numerically >1), demand will be reduced more than price, and therefore less will be spent on the good than before price was increased. We can have elasticity of anything with respect to anything else, not just in consumer demand theory. An own price *e* of say –0.35 for fertilizer means that a 10% increase in price would result in 3.5% decrease in fertilizer demand. Using the estimated price elasticity of a commodity (and other factors sometimes), percent increase or decrease in its usage in future years can be projected.

Materials and Methods

FAO data for 11 SSA countries (Burkina Faso, Ghana, Mali, and Nigeria in West Africa; Democratic Republic of Congo, Kenya, Rwanda, and Uganda in East and Central Africa; and Malawi, Mozambique, and Zambia in southern Africa) were used as the base data for this study. Crops used in the case study were maize, millet, sorghum, and cassava. The areas (ha), yields (kg ha^{-1}), outputs (metric tons, MT), prices (US\$/MT), and output values (US\$) of these crops were employed in various computations.

Literature information indicating that structural changes in fertilizer procurement can reduce the farm gate price of fertilizers by 11–18% (Kelly et al., 2003) was employed. An approximated middle value (15%) of this range was used in computations. Estimations were carried out under three-fertilizer own price elasticity of demand scenarios: 0.38, –1.43, and –2.24 earlier estimated in Ethiopia, Cote d'Ivoire, and Ghana, respectively. These were further classified as low elasticity for –0.38 (representing situations in less endowed countries), medium elasticity for –1.43 (actually high in own respect), and high elasticity for –2.24 (typical of highly endowed agriculture). The elasticity figures are generally interpreted as in the example given earlier.

Data analysis was carried out using Microsoft Excel. The increases in demand (even under the different elasticity scenarios) due to the reduction in farm gate prices were multiplied by the current level of use of fertilizers to arrive at new levels of demand for fertilizers. We used the estimate of the increase in yield per kilogram of fertilizer applied to estimate the overall increase in production from the expanded fertilizer use. Thereafter, we multiply by world market price for the test crops (maize, millet, sorghum, and cassava) to obtain the additional farm incomes or the value of additional food production.

For each of the elasticity estimates, we assumed an average decline in farm gate price of 15% [approximated middle value of the range given by Kelly et al. (2003)]. The demand for fertilizer will increase depending on the own price elasticity of demand for fertilizers. Based on the 15% decline in farm gate price, the percent increase in demand for fertilizer was computed under each elasticity scenario and as follows:

(i) Low own fertilizer price elasticity of demand:

$$\frac{3.8 \times 0.15}{0.10} = 5.7\%$$

(ii) Medium own fertilizer price elasticity of demand:

$$\frac{14.3 \times 0.15}{0.10} = 21.5\%$$

(iii) High own fertilizer price elasticity of demand:

$$\frac{22.4 \times 0.15}{0.10} = 33.6\%$$

With the above, all we need is to multiply the increase in demand by the current level of use of fertilizers to get the new level of demand for fertilizers. Then, we use the estimate of the increase in yield per kilogram of fertilizer used to estimate the overall increase in production from the expanded fertilizer use. We then multiply by the world market price for the crops to get the additional farm incomes or the value of additional food production.

Results and Discussions

Some selected characteristics (total land area cultivated; land area under maize, millet, sorghum, and
cassava; proportion of total land area accounted for by
maize, millet, sorghum, and cassava; and the current
average rate of fertilizer application) of the countries
of study are listed in Table 3. Among others, Table 3
shows that current mineral fertilizer consumption for
these countries ranges from a low value of 1 kg ha^{-1}
(Democratic Republic of Congo and Uganda) to 39 kg
ha^{-1} (Malawi) with a mean of 9.8 kg ha^{-1} across the
11 countries.

Table 4 shows the current total fertilizer marketed in
the 11 countries and the effect (under the three elasticity scenarios) on total fertilizer marketed of a structural

change in fertilizer marketing arrangement that leads
to a 15% reduction in farm gate price of fertilizer.
Without the intervention of 15% farm gate price reduction due to joint fertilizer procurement, an own price
elasticity of say –0.35 would mean that a 10% increase
in fertilizer price would result in a 3.5% decrease in
demand for and use of fertilizer. With this type of information, we can project how far down the percent usage
of fertilizer can go in the future. Table 4 shows how the
structural change that brings about a 15% reduction
in farm gate price of fertilizer actually results in an
increase in demand and use of fertilizer under the three
own price elasticity scenarios.

Results in Table 5 show that compared with base
situation, a structural change in fertilizer procurement
arrangement that resulted in a 15% reduction in farm

Table 3 Selected characteristics of study countries

Country	Total land cultivated (× 1000 ha)	Land area devoted to maize, millet, sorghum, and cassava (× 1000 ha)	Land area devoted to the four commodities as percentage of total land area	Average rate of fertilizer application (kg ha^{-1})
Burkina Faso	4250	2998	71	3
Congo, Dem R	7800	3507	45	1
Ghana	6214	2027	33	4
Kenya	5127	1842	36	29
Malawi	2340	1706	73	39
Mali	4691	2274	48	9
Mozambique	4268	2658	62	5
Nigeria	31683	20387	64	6
Rwanda	1265	411	32	4
Uganda	7187	1712	24	1
Zambia	5288	724	14	8
Mean	*7283*	*3659*	*46*	*9.8*

Table 4 Structural change in fertilizer market (15% price fall)

Country	Total fertilizer (× 1000 MT)			
	Current	Scenario 1 (−0.38)	Scenario 2 (−1.43)	Scenario 3 (−2.24)
Burkina Faso	12.4	13.1	15.1	16.9
DRC	4.4	4.7	5.4	5.9
Ghana	24.6	26.1	29.9	32.9
Kenya	146.1	154.5	177.5	195.2
Malawi	90.1	95.2	109.4	120.4
Mali	41.3	43.6	50.1	55.2
Mozambique	21.4	22.6	26.0	28.5
Nigeria	191.6	202.5	232.7	255.9
Rwanda	5.3	5.6	6.4	7.1
Uganda	7.2	7.7	8.8	9.7
Zambia	44.3	46.8	53.8	59.2
Across countries	*588.8*	*622.4*	*715.1*	*786.7*

Table 5 Farm (total and additional) income changes from 15% farm gate fertilizer price reduction

Current and different elasticity scenarios	Farm income (US$ × 1,000,000)		Percentage increase in income following switches		
	Total	Additional	C to 1,2,3	1–2, 3	2–3
Current status (C)	2198	–	C to 1,2,3	1–2, 3	2–3
Scenario 1 (1)	2324	125	6	–	–
Scenario 2 (2)	2670	472	22	20	–
Scenario 3 (3)	2920	730	33	32	27

gate price led to 6% additional farm income (US $125 million) under the low own price elasticity of fertilizer demand scenario, 22% (or US $472 million) additional farm income under the medium own price elasticity of fertilizer demand scenario, and 34% (or US $730 million) additional farm under the high own price elasticity of fertilizer demand scenario. Switching from one scenario to another also indicates a potential to further increase farm income from 20 to 32% (see Table 5).

Conclusion

This chapter notes the strong need to support structural changes in fertilizer market and any other interventions that can reduce farm gate price of fertilizers (and other inputs). Such interventions increase farm-level use of inputs, farmer productivity, and total production and farm income, thereby leading to improved livelihoods.

The fertilizer FOBs that are being reported are from the developed world where the cost of inputs (e.g., labor) is high. This presents a strong argument for fertilizer production in Africa where, apart from the existence of natural deposits of the key raw materials (e.g., over 75% of rock phosphate deposits in the world is found in Africa) for producing fertilizers, labor is also relatively cheap. Improving the overall fertilizer procurement process could reduce the price of fertilizer in sub-Saharan Africa.

References

Bannock G, Baxter RE, Davis E (2003) The Penguin dictionary of economics, 7th edn. Penguin, UK

Chianu J, Tsujii H, Awange J (2006) Environmental impact of agricultural production practices in the savannas of northern Nigeria. J Food Agric Environ 4(2):255–260

Dregne HE (1990) Erosion and soil productivity in Africa. J Soil Water Conserv 45(4):432–436

European Commission (2007) Advancing African agriculture. Proposal for continental and regional level cooperation on agricultural development for Africa. Unit B2 – Policies for sustainable management of natural resources

FAOSTAT (2003) http://apps.fao.org/page/collections?subset=agriculture

Kelly V, Adesina AA, Gordon A (2003) Expanding access to agricultural inputs in Africa: a review of recent market development experience. Food Policy 28:379–404

Kumar S (2007) Dynamics of the global fertilizer market. PowerPoint presentation. Institute of Agriculture and Environment Research, Oslo, Norway, 31 August

Lal R (1995) Erosion–crop productivity relationships for soil of Africa. Soil Sci Soc Am J 59(3):661–667

Roy A (2006) Africa fertilizer crisis: summit background and process. Presented at the technical session: high-level dialogue, Africa fertilizer summit, Abuja, Nigeria, 9 June

Sanchez PA, Shepherd JD, Soule MJ, Place FM, Buresh RJ, Izac AMN, Mukwonye AU, Kwesiga FR, Ndiritu CG, Woomer PL (1997) Soil fertility replenishment in Africa: an investment in natural capital. In: Buresh RJ, Sanchez PA, Calhoun F (eds) Replenishing soil fertility in Africa. Soil Sci Soc Am 51:1–46

Scherr SJ (1999) Soil degradation: a threat to developing-country food security by 2020? International Food Policy Research Institute (IFPRI). Food, agriculture and environment discussion paper 27

Todd B (2004) Africa's food and nutrition security situation: where are we and how did we get here? International Food Policy Research Institute (IFPRI). 2020 discussion paper 37

Versi A (2006) Fertilizers – food for as hungry earth, Africa launches green revolution. African business, 18 Dec

Preparing Groups of Poor Farmers for Market Engagement: Five Key Skill Sets

J. Ashby, G. Heinrich, G. Burpee, T. Remington, S. Ferris, K. Wilson, and C. Quiros

Abstract This is the second of two chapters that present the case for a portfolio of basic skills to prepare poor farmers for market engagement. This type of skill formation receives little attention in the current debate about how to overcome wealth-differentiated barriers to market entry in poor societies. The chapter discusses the findings of a multi-country Study Tour of three countries, Uganda, India, and Bolivia, organized by the Catholic Relief Services (CRS), a relief and development agency, and the Rural Innovation Institute (RII) of the International Center for Tropical Agriculture (CIAT), an international research organization to explore how their support to farmer groups could be improved and expanded to reach more of the rural poor and prepare them for agro-enterprise development. The Study Tour discovered that a common feature of the farmer groups visited was a drive to acquire and combine five basic "skill sets" that even the poorest groups were incorporating. The observed skill sets were classified as follows: (1) group management; (2) financial skills (usually developed through participation in internal savings and lending groups); (3) marketing skills; (4) experimentation and innovation skills for accessing new technology; and (5) sustainable production and natural resource management skills. Most groups proactively sought to develop multiple skill sets even in the absence of external support for this purpose. Although no single skill set is new in and of itself, the novel discovery was the expressed demand by farmer groups to *combine* several skill sets. The Study Tour participants concluded that combining skill sets has considerable promise for improving current group development approaches and preparing farmer groups to engage with markets.

Keywords Farmers · Markets · Skills · Social capital · Poverty

Introduction

Expanding and accelerating access to product markets for millions of poor farmers is critically important to reducing rural poverty (Dorward et al., 2003; Hulme, 2003; Altenberg and von Drachenfels, 2006). This chapter addresses the question of how to improve strategies for organizing poor farmers into groups and enhancing their capacity to access dynamic markets on a large scale. It argues that there is a need to include the formation of a portfolio of basic "market readiness" skills that will prepare the poor for market engagement.

Skill formation receives a relatively low level of attention in the current debate about how to overcome wealth-differentiated barriers to market entry in poor rural societies. Several recent studies call attention to the decline in public investment in skill development for the rural poor and the failure of supply-driven training programs to meet actual livelihood skill needs of rural people, in particular subsistence farmers and others in survival enterprises (Simmons and Supri, 1999; Grossman and Poston, 2003; Bingen et al., 2003; Shaw, 2004; Mayoux, 2006). In these studies, enterprise training and provision of business development services is faulted for being too heavily oriented to segments of the rural community that are already better endowed with human capital. The rural poor are

G. Heinrich (✉)
Agriculture and Environment, Catholic Relief Services (CRS),
Baltimore, MD, USA
e-mail: gheinrich@saro.crs.org

A. Bationo et al. (eds.), *Innovations as Key to the Green Revolution in Africa*,
DOI 10.1007/978-90-481-2543-2_9, © Springer Science+Business Media B.V. 2011

significantly handicapped by the deficit of a minimum set of skills that go beyond conventional business or technical competencies and are required to organize collectively and engage equitably in transactions with external agencies (Rigg, 2006: 194). Several authors emphasize the need for a more comprehensive redefinition of skills and learning for the rural poor, the need for training to improve group functioning and "readiness" for market engagement (Grosh and Somolekae, 1996; Evans et al., 1999; Krishna, 2003; Shaw, 2004; Thorpe et al., 2005: 918; Edgcomb, 2002; Grossmann and Poston, 2003; Altenberg and von Drachenfels, 2006; Schiferaw et al., 2006).

This chapter addresses this issue with an analysis of the findings of a multi-country Study Tour organized by a relief and development agency and an international research organization to explore how their support to farmer groups could be improved and expanded. The Study Tour discovered that a common feature of the farmer groups visited was a drive to acquire and combine five basic "skill sets" that even the poorest farmer groups were readily incorporating. The skill sets were classified as follows: (1) group management skills; (2) financial skills (usually developed through internal savings and lending); (3) marketing skills; (4) experimentation and innovation skills for accessing new technology; and (5) sustainable production and natural resource management (NRM) skills. Most groups proactively sought to develop most if not all of these skill sets even in the absence of external support.

Small agro-enterprise development for poor farmers needs to be closely (and simultaneously) linked with rural credit systems, technical innovation, improved information, better physical infrastructure, and enhanced human capital formation (Rigg, 2006; Barrett et al., 2001; Ruben and Pender, 2004; Altenberg and von Drachenfelds, 2006). However, at an early stage of development markets have inherent difficulties coordinating investments in these types of complementary activities with positive spillovers. The resulting "coordination failure" simultaneously depresses investments by the same set of mutually dependent investors and deters them from acting in a complementary way toward a common goal. A minimum threshold of support for successful rural micro-enterprise development among the very poor requires explicit links among different types of support (Shaw, 2004; Mayoux, 2006; Holovoet, 2005).

Coordination is especially important in promoting market linkages because there is a threshold level of complementary investments in codependent activities (e.g., input delivery, finance, trading, and skill training) below which separate investments are all ineffective. Pro-poor market-based growth occurs only if sufficient coordination mechanisms develop to create a virtuous cycle of increased investment in all the necessary interventions beyond a threshold level. Thus, increasing complementarities among interventions is a critical factor in "making markets work for the poor" (Poulton et al., 2006).

The formation of groups among the poor is widely practiced as a strategy for overcoming market failures (Thorpe et al., 2005). Groups help to overcome high transaction risk for their members in two ways: first, group organization increases capability for market engagement and income generation. Second, groups improve coordination with service providers, providing members with the means to pursue their interests or defend their assets with agencies such as banks, agricultural extension, input suppliers, marketing boards, or supermarkets (Thorp et al., 2005; Poulton et al., 2006; Kydd and Dorward, 2004; Holovoet, 2005; Krishna, 2000; Cleaver, 2005).

The development of farmer groups is of particular relevance to the issue of coordination failure (Kydd and Dorward, 2004: 964). Development practitioners emphasize the need for NGOs supporting farmer groups to facilitate or broker strong, active external linkages in order to foster initial conditions for market development (Clark et al., 2003; Kindness and Gordon, 2001; Edwards, 1999; Stringfellow et al., 1997). Farmer groups contribute to the growth of endogenous, local mechanisms necessary for overcoming coordination failure and making the inter-sectoral linkages needed to ensure that different types and sources of support aimed at making markets work for the poor complement each other.

However, group formation among poor farmers is not automatically associated with success in collective marketing (Evers and Walters, 2000; Krishna, 2003; Thorp et al., 2005; Cleaver, 2005; Barham, 2006). This may be a specific instance of coordination failure, where the expected return to investment in organizing poor producers into groups is not realized because other essential, complementary investments are insufficient. A gap remains therefore, in understanding how best to form and support farmer groups to successfully sustain engagement with dynamic and changing markets. This is the question the Study Tour addressed.

Our findings suggest that part of the answer to this questions lies in the importance of forming capacity that goes beyond the usual foundation business skills, such as numeracy and collective marketing, to a broader set of competencies identified in the literature and that include competency to visualize future goals; plan for transition from survival and subsistence; find sources of information; solve problems independently; speak up in public meetings; communicate, negotiate, and network; identify and demand services; assert rights; resolve conflict; lead collective action; and advocate for change (Krishna, 2003; Bingen et al., 2003; Mayoux, 2006; Edgcomb, 2002; Thorpe et al., 2005). These types of competencies are described as *"agency capacity."* Studies of how group formation contributes to poverty reduction find that the very poor benefit as much from initiatives that improve their agency capacity in terms of group cohesiveness and sense of self-worth, as from economic support (Thorpe et al., 2005). Higher agency capacity in rural communities is positively correlated with local development while low agency capacity can mean that abundant social capital is underutilized (Krishna 2003, 2004). The findings of the Study Tour show that poor farmers organized in groups to engage with markets have a felt need for multiple sets of competencies and skills. Understanding this need can provide important insights for supporting these groups.

Methodology

In 2002, a global relief and development organization, Catholic Relief Services (CRS), and an international agricultural research organization, the Rural Innovation Institute (RII) of the International Center for Tropical Agriculture (CIAT), formed an "Agro-enterprise Learning Alliance." Both agencies were applying different approaches to farmer group formation across the globe, with varied success. In order to improve approaches, a joint Study Tour was designed to analyze five different strategies for forming farmer groups. These were farmer field schools (FFS), farmer research committees (CIALS), savings-led microfinance self-help groups (SHGs), watershed management committees, and producer marketing (or

agribusiness) groups. The Study Tour's objective was to seek common elements in group development that could be combined to improve agro-enterprise development with poor farmers. Broadly defined for this purpose, agro-enterprise refers to the collection or production and marketing of an agricultural product that when conducted by a group can help poor households to improve the quality, volume, and thus the prices of their produce. A basic assumption for the design of the Study Tour was that linking poor farmers to markets in an equitable manner through agro-enterprise development is critical if they are to increase the financial assets central to reducing their economic poverty.

The Study Tour, an action research methodology used to identify and familiarize decision makers with successful field practice in order to facilitate its mainstreaming (Reason and Bradbury, 2001), was used because both organizations identified a pressing need to rapidly introduce new ideas into field programs. The eight-member Study Tour team assembled by the two organizations included a development sociologist, three agronomists/soil scientists, a socio-economist, and three microfinance/micro-enterprise advisors. Three members of the team were women. Three 1-week trips were organized to assess farmer groups in three countries – Uganda in September 2005; India in October 2005; and Bolivia in November 2005. In each country, staff from local partner organizations also participated in orientation, field visits, and discussions of findings. The field visits were followed by two 1-week "writeshops," one for systematization and analysis of findings immediately after the field visits and a second to produce an internal working document and draft a Field Guide for practitioners (Aldana et al. 2007; CRS, 2007).

Selection of field sites targeted comparison of different approaches to group formation within a given culture and socio-economic environment. The focus was on farmer groups that had already initiated successful collective marketing, but were in an early transition stage from subsistence to semi-commercial production, typically involving a mix of traded and non-traded products. In these countries, this type of farmer overlaps with, but does not automatically include, the chronically or ultra-poor. The Study Tour sought to compensate for bias by concentrating on local (in contrast to specialty, supermarket, or export) market linkages that included both livelihood and business types of small agro-enterprise.

Table 1 Number of groups of each type at group initiation visited during the CRS-RII CIAT agro-enterprise Study Tour, 2005

Country	Group type at initiation					
	Self-help[a]	Producer Agri-business	Technology		Watershed/NRM	Total
			FFS	FRC		
Uganda	2	6	3	0	0	11
India	15	0	0	0	7	21
Bolivia	0	2	0	4	1	7
Total	*17*	*8*	*3*	*4*	*8*	*40*

[a]Self-help group (SHG) models differ, and the SHGs in Uganda were established using a different approach from the one used by CRS to establish SHGs in India

Each country visit included a 1-day review of the country context and main group organization types, two to three sub-team visits to different areas to interview groups, and a final in-country debriefing with all local participants. Semi-structured interviews collected information on group background, history, gender, structure, leadership, and legal status; services provided to members; current and planned group activities; successes and challenges; common equity; training, financial and technical support received from external agencies, understanding of markets; improvement of products for market demand; and ability to negotiate prices.

Table 1 summarizes the number of groups of each type visited in each country. Table 2 shows the total number of members by group type and average group size. In all, the Study Tour interviewed 40 groups with a total membership of 993 members and an average membership of 25. The most common type visited was the self-help group organized with internal savings and lending as a primary purpose, and most of these were interviewed in India. Group maturity was classified as "Short" = 1–2 years, "Medium" = 3–5 years, "Long" = >5 years. Most of the groups visited were of medium or long maturity (Aldana et al. 2007). Producer groups tended to be larger, while self-help groups tended to be smaller in size.

After the field visits, a selective literature review was conducted of two to four seminal impact studies of each of the five group types visited, in parallel with the analysis of field notes. The literature review search used electronic databases, libraries, and specialized electronic resources including the Microfinance Gateway Library on the CGAP web site, FAO web sites for farmer field schools, and CIAT for CIALs. The team wanted to formulate observations in the field independently of the literature. Although these two forms of analysis were conducted interactively, for the sake of clarity, the ideas culled from the literature review are presented first, the analysis of field observations follows next, and finally the relationship between the two is discussed.

Results

Comparison of Farmer Group Formation Strategies from the Literature Review

A selective review of published studies of the impact of the five strategies for farmer group formation was undertaken to complement field observations. Two key findings emerged: first, the five contrasting strategies

Table 2 Number of members and average size of groups of different types visited during the CRS-RII CIAT agro-enterprise Study Tour, 2005

Country	Group type									
	Self-help		Producer		FFS		FRC[a]		Watershed	
	No. of members	Average	No. of members	Average	No. of members	Average	No. of members	Average	No. of members	Average
Uganda	40	*20*	263	*44*	72	*24*	0	*0*	0	*0*
India	167	*11*	0	*0*	0	*0*	0	*0*	120	*17*
Bolivia	0	*0*	150	*75*	0	*0*	91	*23*	90	*90*
Total/average	207	*12*	413	*52*	72	*24*	91	*23*	210	*26*

[a]FRC = Farmer Research Committee

for farmer group formation assessed by the Study Tour are typically implemented in isolation from each other and in ways that epitomize the coordination failure discussed earlier in the chapter. All of the farmer field schools (FFS), farmer research committees (CIALS), savings-led microfinance self-help groups (SHGs), and producer marketing or agribusiness groups observed by the Study Tour had been formed without any coordinated relationship to any other strategy. Second, among all five group types, the review of impact studies identified three common elements for group formation that contributed to the viability of groups and their development of market linkages and that were convergent with observations from field visits (discussed next). The commonalities were (a) the expansion into new activities beyond their initial purpose that depended on internal social capital and the capacity for collective action; (b) the attention of facilitating organizations to developing a sustained capacity for learning within the groups; and (c) the role played by facilitators in brokering linkages between groups and other service providers while building group capacity to manage these external relationships.

Findings from Field Visits to Farmer Groups

Analysis of field observations focused on identifying common aspects of the different group formation strategies that help the poor to engage successfully with markets. Independently of group formation strategy and the initial purpose of group formation, we found that all the groups visited were diversifying their activities in ways that engaged them in acquiring elements of a constellation of five critical skill sets. The Study Tour team classified these skill sets as follows: (1) group management skills; (2) financial management skills (usually developed through internal savings and lending groups); (3) marketing skills; (4) experimentation and innovation skills for accessing new technology; and (5) sustainable production or NRM skills. The five skill sets observed reflect the types of new activities that the groups were branching into, regardless of the initial purpose of group formation, a trend noted in the impact studies reviewed above. For example, some groups that started as self-help groups (with savings and lending) were seeking to improve

their marketing skills; some groups that started as FFS were trying to gain marketing skills and learn how to do internal saving and lending. Based on information for 36 of the 40 groups interviewed, we found that a total of 28 groups (78%) had acquired at least three or more of the five skill sets, while the other 8 groups (22%) had acquired two of the skill sets.

Most groups observed sought proactively to develop additional skill sets even in the absence of external support for this purpose. We concluded that this is evidence of a felt need by farmer groups to add critical skill sets. Moreover, although all groups were receiving assistance in developing or strengthening at least one skill set in a formal way through the efforts of their respective facilitating external organizations, *no one group was receiving facilitation in all five skill sets*. Field staff of facilitating organizations sometimes experienced bottleneck when pushed beyond levels of existing expertise to meet this demand for new skill sets (Aldana et al. 2007). Regardless of their original purpose, all groups were struggling to add the skill sets they were lacking in this minimum set of five, often without the knowledge of their facilitators.

Table 3 presents the finding that the majority of the farmer groups interviewed had evolved from the original purpose for which they were organized and were currently engaging in a different primary activity that involved acquiring at least one additional skill set. A primary activity was defined by a group as its current main purpose. As Table 3 shows, while 12 groups of the total of 40 visited did not adopt a new primary activity, 28 (70%) had defined a new purpose.

The Five Skill Sets and their Contributions

This section examines in detail the composition of the five key skill sets identified from field observations. We define a skill set as knowing *how to* undertake a specific activity. The skill sets discussed here include group management; financial management; market engagement; experimentation and innovation for accessing new technology; and sustainable production (including NRM). A profile for each of the five skill sets was developed from field observations and this produced a total of 27 skill indicators. The indicators are listed by skill set in the "Field Guide" (CRS, 2007).

Table 3 Change in primary activities of groups over time (observations from the CRS-RII CIAT agro-enterprise Study Tour, 2005)

Current primary activity	Group purpose at initiation					Total with this new primary activity
	Self help (internal savings)	Producer (marketing)	Farmer field school	Farmer research committee	Watershed or NRM	
Internal savings and lending	(India = 4)					0
Marketing	Uganda = 2 India = 2	(Uganda = 6)	Uganda = 3	Bolivia = 3	India = 6	16
Natural resource management	India = 9				(India = 1)	9
Technology experimentation		Bolivia = 2		(Bolivia = 1)	Bolivia = 1	3
Percent with new activity (Total N groups)	76.5% (17)	25% (8)	100% (3)	75% (4)	87.5% (8)	70% (28)

Note: Groups in parentheses: purpose at initiation was still their primary activity at the time visited by the Study Tour

Basic Group Management Skills

The group management skill set observed in the field relates to internal social capital, specifically procedures for internal democratic management, including the capacity to resolve internal conflicts and enforce compliance with internal rules. Two additional dimensions of this skill set include the development of *external* social capital or knowing how to build effective relationships with essential service providers and to take the initiative in problem solving. Being cohesive and able to act efficiently as a group provides a strong foundation for the acquisition and utilization of the other skill sets.

Basic Financial Skills

Financial skills are based on the SHG activities and involved understanding how to save regularly, how to manage savings so that they are protected, how to lend at a reasonable interest rate, and how this increases capital. They depend primarily on the group management skills described earlier required for transparent and responsible management and utilization of joint resources. Reciprocal financial responsibility built competency for collective action, while particularly in India the savings groups provided a platform for learning and knowledge sharing on other subjects and for building external relationships with external providers of credit and banking services.

There is a tendency among development professionals with non-financial technical backgrounds to expect that increasing the incomes of the poor will be sufficient to end poverty. We forget that the capacities to *save*, and to *protect assets*, are also essential for asset accumulation and escape from poverty. Internal savings and lending groups are an excellent approach for increasing the capacity of the poor in both of these areas. The savings build assets and both savings and access to additional credit help to protect assets when cash is suddenly needed in an emergency. Having savings and access to even small amounts of credit may significantly reduce the vulnerability of poor households. Coincidentally, participation in savings groups also helps the poor to learn about financial management, investment, and "growing your money."

The Study Tour participants concluded that the internal savings and lending skills could be extremely important to the durability and long-term success of groups because they provided an immediate, direct, and enduring benefit for participating in the groups and they reduced the vulnerability of both the group and its individual members (through access to credit and other social support in times of crisis). They also strengthened group cohesiveness because they met frequently and regularly.

Basic Business Skills

Collective marketing is often the central purpose around which external organizations develop farmer agri-business groups. Collective marketing can be very important because it can increase the leverage of poor farmers in the market and increase the overall efficiency of the market in rural areas. However, there are several other components of the business skill set that are important to the long-term effectiveness of farmer groups and which often receive much less attention in "market-oriented" development projects (CRS, 2007). The capacity to identify authentic market opportunities, to understand the basics of value chains, and to recognize viable points of entry for market engagement are especially important for the rural poor. This is because if they are assisted only to market collectively into one specific market, but do not have these additional skills, they will not be able to adjust and adapt when markets change. Ensuring that groups understand the concepts of profit and loss and are able to keep basic records is also very important in areas of low literacy. The Study Tour participants concluded that a basic but well-rounded set of business skills is crucial to the long-term success of groups that plan to participate in markets.

Basic Experimentation and Innovation Skills

Farmer experimentation with new technology was observed in essentially all farmer groups that were engaging with markets. Activities included evaluation of new or different crops and varieties; production methods; pest control measures; product processing; and packaging. However, when the innovation process was not assisted by an external supporting organization, it was often unsystematic or inefficient.

The Study Tour participants concluded that this was a vital capacity for smallholder farmers for several reasons. First, to efficiently engage with markets, farmers need to increase their production and productivity and often the quality of their product as well. This usually requires new technology (new methods, systems, and/or physical materials). Second, markets and consumer demands change constantly. If farmers are not able to obtain new technical knowledge and skills, and adapt and apply these to their production systems, they will not be able to adapt and compete. Thus learning how and where to acquire new technology and skills, and how to test, adapt and apply these, is critical to long-term success.

Sustainable Production and NRM Skills

Natural resources are the basis of rural productivity and the majority of rural livelihoods, especially in Africa. In many areas of the developing world, these resources are being rapidly degraded – soil, water, and forest resources in particular.

Basic sustainable production and NRM skills observed involved groups in learning and sharing knowledge about how on-farm practices impact in-field productivity, their neighbors, and the wider landscape. Some groups also displayed negotiation skills required to manage collective resources and relationships necessary for collective action and for interacting with outsiders. This skill set was the most developed in groups that were already using natural resources (e.g., the watershed committees in India). Demand for this skill set was weaker than for the other skill sets among the other groups.

Protecting natural resources is crucial to maintaining rural livelihoods in the medium and long term, as well as to the provision of essential ecosystem services for external communities (e.g., clean water for cities). Equally importantly, increasing the productivity of local natural resources can significantly increase the potential for local production of food and income. For example, one of the watershed groups visited in India had used the additional water collected through the watershed management project to irrigate an additional 400 ha of rice – radically altering the economy of the community (previously based largely on rain-fed agriculture). The Study Tour participants concluded that it was important to build this skill set early, before environmental degradation became a major limitation, and that it might warrant the use of subsidies such as the "Food For Work" that use relief aid to develop community assets.

Discussion

The major finding from the Study Tour was the characterization of five basic skill sets and the felt need by farmer groups to combine these. While none of the skill sets is new in and of themselves, the novel discovery was the consistent expression of a bottom-up drive to *combine* all five skill sets that emerged across different cultures, continents, and strategies for initial group formation. This drive can be interpreted as an expression of farmers' felt need to overcome inadequate or missing complementary investments in group development that are important for sustained and successful market engagement. In other words, farmer groups are attempting to avoid "coordination failure," or to reduce the likelihood that the group's investment in one aspect of their development, such as microfinance, will not pay off because of their lack of capability in a different area, such as enterprise development or technology innovation.

A second possible reason for the demand for multiple skill sets may be related to their contribution toward developing "agency capacity" as defined in the literature. As noted earlier, agency capacity contributes directly to the empowerment of the poor by increasing their capacity to act and effect change in their local socio-economic environment. Agency capacity is positively correlated with development (Krishna, 2003, 2004). Groups with strong agency capacity would also be more resilient and better able to deal with change and adversity.

Beyond strengthening individual groups, building agency capacity may be particularly important to the process of scaling-out to large numbers of farmers. This is because empowerment of the rural poor (through building agency capacity) would allow change to happen independently from specific project interventions and allow it to spread spontaneously within and across communities. The Study Tour participants believe that the development of groups with multiple skill sets could thus be an important component of effective exit strategies for agro-enterprise projects.

Two other important benefits could potentially flow from establishing large numbers of groups with these multiple skill sets across a broad geographic area. First, such a network of groups would provide a solid foundation for the development of specific commodity markets because large numbers of farmers would already be organized and able to respond quickly and positively to new market opportunities. Second, it would be relatively easy to disseminate new skills or knowledge (e.g., messages about health and nutrition) because a network to receive and apply the information would already exist.

Group formation would not necessarily need to start from scratch. Large numbers of groups already exist in most rural areas, and many of these would be happy to receive support in gaining new skill sets. The indicators in the "Field Guide" (CRS, 2007) could be used to assess current skill levels in existing groups, and only missing skill sets would need to be added. CRS

is currently testing this new approach to strengthening groups that wish to engage in agro-enterprise. Existing groups in agro-enterprise projects being implemented by CRS and its partners are being "retro-fitted" with savings and lending skills in several countries in East Africa (including Ethiopia, Kenya, Tanzania, and Uganda). CRS programs in India are starting to support a limited number of existing SHGs in gaining basic business and experimentation skills. These processes are being systematically monitored in both regions.

Conclusion

In conclusion, it is true that many factors need to come together for market-based approaches to succeed in reducing chronic rural poverty. These include the need for improved infrastructure, technology, markets, and market chains themselves. But the authors argue that building the capacity of the very poor to engage equitably and sustainably with markets is an equally important part of this process. The clear "bottom-up" demand for increased capacity and skills being expressed by the farmer groups supports this argument. Establishing large numbers of groups with multiple skill sets will help to overcome some of the coordination failures that have been common in the past, build strong agency capacity among the poor in rural communities, and contribute directly to enabling large numbers of the rural poor to exit from poverty through sustainable market participation. CRS has started using a multi-skills approach and expects to mainstream this approach in all future agro-enterprise development programs.

References

Aldana M, Burpee G, Heinrich G, Remington T, Wilson K, Ashby J, Ferris S, Quiros C (2007) The organization and development of farmer groups for Agroenterprise: conclusions from a CRS & RII-CIAT study tour in Asia, Africa and Latin America. Catholic Relief Services, Baltimore, MD

Altenberg T, von Drachenfels C (2006) The new minimalist approach to private sector development: a critical assessment. Dev Policy Rev 24(4):387–411

Barham J (2006) Collective action initiatives to improve marketing importance:lesons from farmer groups in Tanzania. Paper presented at the CAPRI research workshop on collective action and market access for smallholders, Cali, Colombia, 2–5 October

Barrett CB, Reardon T, Webb P (eds) (2001) Income diversification and livelihoods in rural Africa: cause and consequence of change. Food Policy 26(4 Special Issue)

Bingen J, Serrano A, Howard J (2003) Linking farmers to markets: different approaches to human capital development. Food Policy 28:405–419

Clark N, Hall A, Sulaiman R (2003) Research as capacity building: the case of an NGO facilitated post-harvest innovation system for the Himalayan hills. World Dev 31(11):1845–1863

Cleaver F (2005) The inequality of social capital and the reproduction of chronic poverty. World Dev 33(6):893–906

CRS (2007) Preparing farmer groups to engage successfully with markets. A field guide for five key skill sets. Catholic Relief Services & RII-CIAT Agroenterprise Study Tour Group Working Paper, Catholic Relief Services, Baltimore, MD

Dorward A, Poole N, Morrison J, Kydd J, Urey I (2003) Markets, Institutions and technology: missing links in livelihoods analysis. Dev Policy Rev 21(3):319–332

Edgcomb E (2002) What makes for effective micro enterprise training? J Microfinanc 4(1):99–114

Edwards M (1999) NGO performance- what breeds success? New evidence from South Asia. World Dev 27(2):361–374

Evans T, Adams A, Mohammed R (1999) Demystifying nonparticipation in micro credit: a population-based analysis. World Dev 27(2):419–430

Evers B, Walters B (2000) Extra-household factors and women farmers' supply response in sub-Saharan Africa. World Dev 28(7):1341–1345

Grosh B, Somolekae G (1996) Mighty oaks from little acorns: can micro enterprise serve as the seedbed of industrialization? World Dev 24(12):1879–1890

Grossman M, Poston M (2003) Skill needs and policies for agriculture-led, pro-poor development. Queen Elizabeth House Working Paper Number 112, University of Oxford, Oxford, UK

Holovoet N (2005) The impact of microfinance on decision-making agency: evidence from south India. Dev Change 36(1):75–102

Hulme D (2003) Conceptualizing chronic poverty: an introduction. World Dev 31(3):399–402

Kindness H, Gordon A (2001) Agricultural marketing in developing countries: the role of NGOs and CBOs. London: University of Greenwich, Natural Resources Institute Policy Series 13

Krishna A (2000) Creating and harnessing social capital. In: Dasgupta P, Serageldin I (eds) Social capital a multifaceted perspective. The World Bank, Washington, DC

Krishna A (2003) Understanding, measuring and utilizing social capital; Clarifying concepts and presenting a field application from India. CAPRI Working paper No 28, CGIAR Systemwide Program on Collective Action and Property Rights.

Krishna A (2004) Escaping poverty and becoming poor: who gains, who loses and why? World Dev 32(1):121–136

Kydd J, Dorward A (2004) Implications of market and coordination failures for rural development in least developed countries. J Int Dev 16:951–970

Mayoux L (2006) Learning and decent work for all. New directions in training and education for pro-poor growth. Draft discussion paper. http://www.lindaswebs.org.uk/Page2_Livelihoods/BDS/BDS_Docs/Mayoux_trainingoverview_2006.pdf

Poulton C, Kydd J, Dorward A (2006) Overcoming market constraints on pro-poor agricultural growth in sub-saharan Africa. Dev Policy Rev 24(3):243–277

Reason P, Brandbury H (eds) (2001) Handbook of action research. Participative enquiry and practice. Sage, London

Rigg J (2006) Land, farming, livelihoods and poverty: rethinking the links in the rural south. World Dev 34(1):180–202

Ruben R, Pender J (2004) Rural diversity and heterogeneity in less-favoured areas: the quest for policy targeting. Food Policy 29:303–320

Schiferaw B, Obare G, Muricho G (2006) Rural institutions and producer organizations in imperfect markets: experiences from producer marketing groups in semi-arid eastern Kenya. Paper presented at the CAPRI research Workshop on collective action and market Access for Smallholders, Cali, Colombia, 2–5 October

Shaw J (2004) Microenterprise occupation and poverty reduction in microfinance programs: evidence from Sri Lanka. World Dev 32(7):1247–1264

Simmons C, Supri S (1999) Failing financial and training institutions: the marginalization of rural house hold enterprises in the Indian Punjab. J Econ Iss 33(4):951–972

Stringfellow R, Coulter J, Lucey T, McKone C, Hussain A (1997) Improving access of small holders to agricultural services in sub-Saharan Africa. London: Natural resource perspectives, Number 20. Overseas Development Institute, London

Thorp R, Stewart F, Heyer A (2005) When and how is group formation a route out of chronic poverty? World Dev 33(6): 907–920

Fertilizer Microdosing and "Warrantage" or Inventory Credit System to Improve Food Security and Farmers' Income in West Africa

R. Tabo, A. Bationo, B. Amadou, D. Marchal, F. Lompo, M. Gandah, O. Hassane, M.K. Diallo, J. Ndjeunga, D. Fatondji, B. Gerard, D. Sogodogo, J.-B.S. Taonda, K. Sako, S. Boubacar, A. Abdou, and S. Koala

Abstract The fertilizer microdosing technology deals with the application of small quantities of fertilizers in the planting hole, thereby increasing fertilizer use efficiency and yields while minimizing input costs. In drought years, microdosing also performs well, because larger root systems are more efficient at finding water, and it hastens crop maturity, avoiding late-season drought. Recent research found that solving the soil fertility problem unleashes the yield potential of improved millet varieties, generating an additional grain yield of nearly the same quantity. Recognizing that liquidity constraints often prevent farmers from intensifying their production system, the warrantage or inventory credit system helps to remove barriers to the adoption of soil fertility restoration. Using a participatory approach through a network of partners from the National Agricultural Research and Extension Systems (NARES), non-governmental organizations (NGOs), farmers and farmer groups and other international agricultural research centres, the microdosing technology and the warrantage system have been demonstrated and promoted in Burkina Faso, Mali, and Niger during the past few years with encouraging results. Sorghum and millet yields increased by up to 120%, and farmers' incomes went up by 130% when microdosing was combined with the warrantage system. This chapter highlights the outstanding past results and the ongoing efforts to further scale up the technology using Farmer field schools (FFS) and demonstrations, capacity and institutional strengthening, private sector linkages and crop diversification amongst other approaches.

Keywords Farmer field schools · Fertilizer microdosing · Millet · Participatory approach · Sorghum · Warrantage or inventory credit system

Introduction

Poverty and food insecurity continue to create suffering across the semi-arid Sudano-Sahelian zone of West Africa. Unpredictable droughts cause food shortages for both humans and the livestock on which they depend. The predominantly sandy soils in this zone are of very low fertility, particularly in phosphorus (P) and nitrogen (N), with phosphorus being more limiting to crop growth and yield than is nitrogen (Bationo et al., 1998a, b). It was reported that crop response to nitrogen was minimal when crop phosphorous requirements were not met (Traore, 1974).

Sivakumar (1992) reported that the little arable land in the Sudano-Sahelian zones is being gradually reduced due to the southward creep of the 400-mm isohyet as a consequence of land degradation, drought and other human activities. In addition, the high population growth rate (3.4% per annum) and the increasing population density have put a lot of pressure on the cultivated lands, which leads to a significant decrease and disappearance, in some cases, of fallow lands. Because of this, farmers are increasingly being forced to cultivate marginal and degraded lands where moisture and nutrient stress significantly constrain crop production, which results in low yields and further land

R. Tabo (✉)
Forum for Agricultural Research in Africa (FARA), Accra, Ghana
e-mail: rtabo@fara-africa.org

A. Bationo et al. (eds.), *Innovations as Key to the Green Revolution in Africa*,
DOI 10.1007/978-90-481-2543-2_10, © Springer Science+Business Media B.V. 2011

degradation. Stoorvogel and Smaling (1990) reported that because of these problems, increases in crop production have resulted more from the expansion of cultivated areas than from increased crop productivity.

It is widely believed that the only real cure against land hunger in the West African Sahel lies in the intensification of agriculture and in increased productivity of the arable land through the use of external inputs, mainly inorganic fertilizers (Van Keulen and Breman, 1990; Breman, 1990). Although soil fertility enhancement technologies have been developed over the years for the main staple food crops in West Africa, such as sorghum and millet, these technologies have not been adopted by resource-poor farmers due to the high costs and unavailability of the inputs as well as the inappropriateness of the fertilizer recommendations made which are very high and not affordable by farmers. As a consequence of this, the productivity of the major staple food crops such as sorghum and millet has continued to decrease.

To address these constraints and increase the productivity of these major staple food crops, various national and international research institutions working in the Sahel joined forces through a collaborative research program and developed an effective technique to increase fertilizer use efficiency and reduce investment costs for resource-poor, small-scale farmers, thereby increasing crop growth and productivity (Bationo et al., 1998a, b; Buerkert and Hiernaux, 1998). This strategic application of fertilizer, also known as fertilizer microdosing, is based on applying small doses of fertilizer in the hill of the target grain crop at planting rather than broadcasting it all over the field. The microdosing technology is affordable to the poor because of the reduced investment cost, and it gives a quick start, thus avoiding early season drought, and an earlier finish, avoiding end-of-season drought while increasing crop yields (Tabo et al., 2006, 2007).

It is not enough to grow more. What is even more important is that farmers should get the right price for their product (i.e. grains) so as to increase their income and improve their livelihoods. Rather than selling their grain into a glutted market for low prices at harvest time, in the Warrantage or inventory credit system, farmers (or producer organizations) stock their produce at harvest in the warehouses of the farmers' associations and are issued cash loans. These loans enable them to meet immediate family cash needs, participate in collective fertilizer (and other input) purchases and carry out their income-generating activities during the off-season like fattening of small ruminants, vegetable gardening and trading. They then sell the stored grains at higher prices when the market supply begins to decline 4–5 months after harvest and pay back their loans with interest.

The warrantage system is used as a link between credit and cereal grain markets. This credit facility removes the barriers to the adoption of soil fertility restoration technologies. In order to make inputs accessible to farmers, sustainable farmer-based enterprises and cooperative organizations are developed, storage facilities and inputs shops (boutique d' intrants) are built, and credit and savings schemes are also developed. These facilities are managed by members of these cooperatives. Linking farmers to input or product markets and the vertical integration between these become prerequisites to the uptake of agricultural technologies. Efforts to develop institutional arrangements likely to improve the linkages of rural households to major markets are often major developmental challenges. The combination of fertilizer microdosing with the complementary institutional and market linkages, through the warrantage system, offers an excellent option for improving crop productivity and increasing farmers' incomes in the semi-arid Sudano-Sahelian zone of West Africa. The warrantage credit facility was initiated in Niger in the late 1990s to remove barriers to the adoption of soil fertility restoration inputs.

ICRISAT is working closely with Projet Intrants, FAO, several NGOs, national and international research organizations, development agencies, extension services and other stakeholders to help farmers develop and strengthen cooperatives. In the past few years, USAID assisted ICRISAT to complement FAO's efforts for demonstrating and promoting the fertilizer microdosing technology and the warrantage system in Burkina Faso, Mali and Niger. Recently we won a competitive grant from the West and Central African Council for Agricultural Research and Development (CORAF/WECARD), with funds from the African Development Bank to pursue our efforts to disseminate the technology in West Africa. We are also exploring other sources of funding to scale up and out this promising technology to millions of farmers across the West and Central African regions. This chapter reports the encouraging and good results that were obtained from on-farm evaluation trials and demonstrations of the technology in three selected countries in West Africa, namely Burkina Faso, Mali and Niger, and also discusses the future perspectives for wider dissemination across the region.

Materials and Methods

Demonstrations and on-farm trials involving microdosing technology were conducted in Burkina Faso, Mali and Niger between 1998 and 2006. These field experiments were designed by the researchers but were managed by the farmers themselves, with training and technical backstopping from extension agents, NGOs and scientists. Experimental plots and types of fertilizers used varied between study sites depending on the local conditions and the availability of these inputs.

On-farm Field Experiments

These on-farm field tests included demonstration plots and farmer field schools (FFS') using the fertilizer microdosing technology. The demonstration tests consisted of three plots per farmer, each plot measuring approximately 300 m^2. Three treatments consisted of the farmers' practice, the earlier recommended broadcasting system of fertilizer application (about 100 kg NPK (15:15:15) per ha) and the fertilizer microdosing at 4–6 g per hill of compound fertilizer (NPK) (40–60 kg NPK per ha) or 2 g of diammonium phosphate (DAP) per hill (20 kg DAP per ha). The test crops used were millet and sorghum. Plant densities under farmer conditions varied between 5,000 and 6,000 hills per ha, while the recommended densities in the microdosing plots varied from 10,000 to 20,000 hills per ha.

Farmers were given the option to plant their fields whenever they felt that the soil was moist enough for germination of seeds. They used their own densities in the control plots but were requested to follow the recommended densities in the microdosing plots, with guidance from the field technicians. They also weeded when it is time to do so, in some cases, on the advice of field technicians. Harvesting is done by farmers under the supervision of field technicians. Data collection was done by the field technicians.

In Burkina Faso, 30 villages and 210 farmers in the northern, central north zone were involved in these studies. In Mali the on-farm trials were carried out in 44 villages in the regions of Mopti, Segou, Koulikoro, Mande and Beledougou with 321 farmers. In Niger, approximately 1,536 demonstrations and field experiments were established in 254 villages in five departments in southern Niger, namely Tillabery, Dosso, Tahoua, Maradi and Zinder.

Socio-economic Assessment

In addition to the field trials, a socio-economic evaluation was carried out to assess the economic performance of the fertilizer microdosing technology. Net gain was calculated as the difference between the total revenues from the grain and the total cost of fertilizer as follows:

$$NG = R - C$$

where NG is the net gain; R is the revenue from grains; and C is the cost of fertilizer.

Net gain was expressed in Franc Communauté Financière Africaine (FCFA) per hectare. The cost of labour was not used as the data were collected from plots that are not large enough and the data were not reliable.

In November/December 2004, surveys were conducted to assess the effect of input shops on fertilizer use and crop yield. These surveys involved 10 villages and 10 input shops.

Capacity-Building Activities

Field technicians, extension agents and farmers in all the three participating countries were trained in the laying out of the demonstration plots and farmer field schools and the appropriate method of using the fertilizer microdosing technology. These training sessions demonstrated to them how to measure the recommended rate of fertilizer (microdose) in the field, how to apply it correctly in the field and how to manage the field after sowing. Emphasis was also put on the best way of collecting agronomic as well as socio-economic data from the trial set up. Several training sessions were given to farmers' organizations on the warrantage system.

Results

Microdosing Performance by Agro-ecological Zones

In all the three agro-ecological zones – Sahelian (400–600 mm), Sudano-Sahelian (600–1000 mm) and Sudano-Guinean (>1000 mm), sorghum under the microdose yielded higher than under the earlier recommended rates of broadcasting and the farmers' traditional practice (Fig. 1). The yield advantage of the microdose over the farmers' practice varied from 50 to 100%. As it is expected, yields were generally higher in the wetter Sudano-Guinean zone (1500 kg ha^{-1}) than in the drier Sahelian zone (750 kg ha^{-1}). This is due to the higher rainfall amount and to the better distribution of the rainfall in the Sudano-Guinean zone during the growing season, which reduces the risk of crop failure.

Burkina Faso

Grain Yields

Throughout the study period covering the cropping seasons from 2002 to 2006, the microdose treatments yielded, on average, higher than did the farmers' traditional practice. Millet grain yields ranged from 44% in 2002 to 101% in 2005, while sorghum grain yields

under microdose were 47 and 106% higher than the control in 2002 and 2005, respectively (Figs. 2 and 3). Sorghum and millet performed better in 2005 due to a better rainfall distribution during the growing period. In 2006, millet and sorghum grain yields were 64 and 90% higher, respectively, under microdose than with the farmers' practice.

Net Gains from Microdosing

Farmers obtained returns from their millet with microdose that were three times higher than the revenue from the broadcasting method (12575 FCFA ha^{-1} as compared to 5175 FCFA ha^{-1}). The net gains for sorghum were approximately 2.5 times higher with microdose (22780 FCFA ha^{-1} vs 9255 FCFA ha^{-1}).

Mali

Grain Yields

Sorghum and millet performed better under microdose than with the broadcasting method and farmers' traditional practice. In 2002, millet and sorghum grain yields with the microdose were, on average, 61 and 107% higher, respectively, than the control (Figs. 4 and 5). In 2003, millet and sorghum grain yields from the microdosing treatments were 90 and 69% higher, respectively than the farmers' practice.

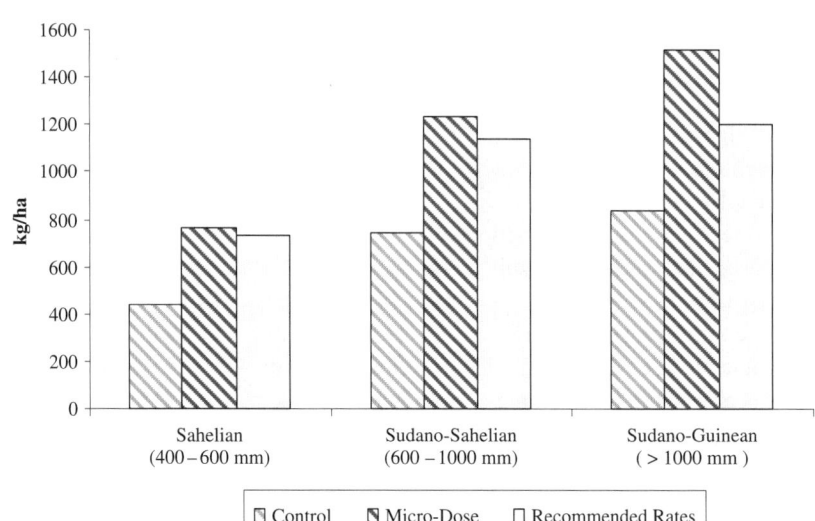

Fig. 1 Sorghum grain yield (kg ha^{-1}) by agro-ecological zones, WCA

Fig. 2 Millet grain yield (kg ha^{-1}) as affected by microdosing, Burkina, 2006

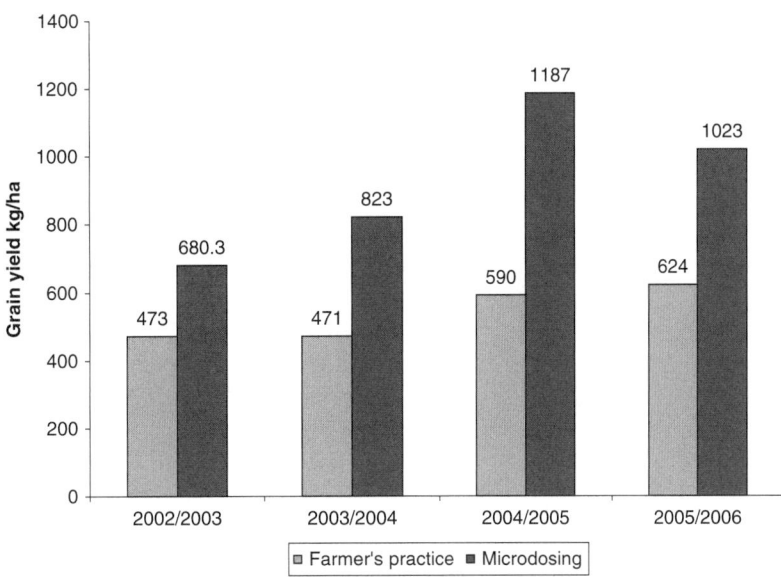

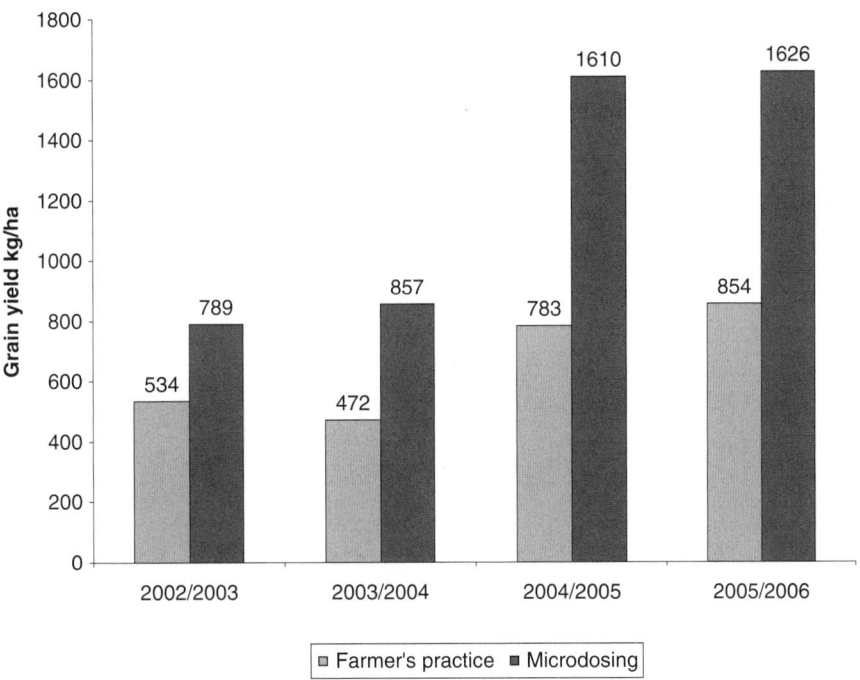

Fig. 3 Sorghum grain yield (kg ha^{-1}) as affected by microdosing, Burkina, 2006

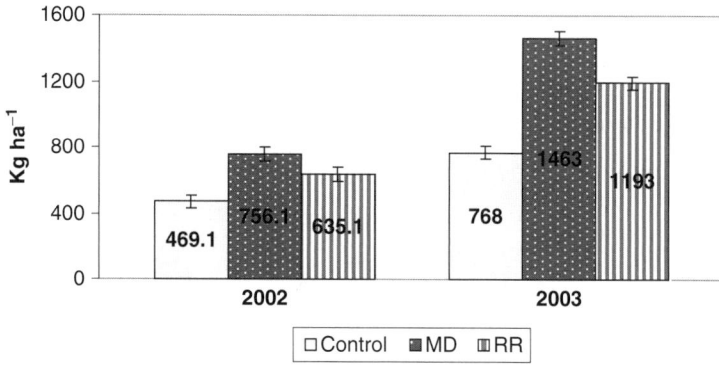

Fig. 4 Millet grain yields (kg ha^{-1}) for demonstration trials (control, microdose and recommended rates (RR)) in Mali, 2002 and 2003

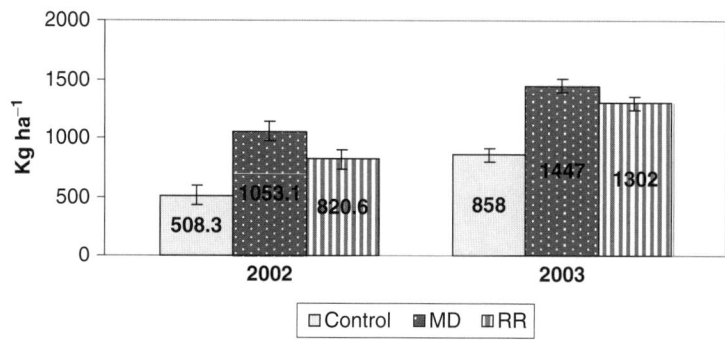

Fig. 5 Sorghum grain yields (kg ha^{-1}) for demonstration trials (control, microdose and recommended rates (RR)) in Mali, 2002 and 2003

Net Gains from Microdose

Millet under microdose gave net monetary gains of 119690 FCFA ha^{-1} which were 68% higher than the net returns from the traditional practice (71167 FCFA ha^{-1}) and 33% higher than the net gain from the broadcasting technique (89959 FCFA ha^{-1}).

Net Returns from Microdosing

In 2002, net returns were 74650 FCFA per ha for DAP + urea, 65642 FCFA per ha for DAP, 62619 FCFA per ha for NPK and 51745 FCFA per ha for the control. Net profits were, on average, 44 and 121% higher with the microdose than under the control plots in 2002 and 2003, respectively.

Niger

Grain Yields

In all agro-ecological zones, microdosing resulted in significant increase in grain yield (Fig. 6). There is significant yield increase at individual farmer's level due to microdosing of fertilizers. Grain yield increment from microdosing treatment over the control was as high as 89% with an average of 44% or about 300 kg ha^{-1}. Approximately half of the farmers (44%) reported yield increase of at least 50%, which is double the yield obtained from the farmers' practice.

Effect of the Presence of Input Shops on Fertilizer Use and Crop Performance

It was observed that the presence of input shops in a village has a positive effect on the intensity of fertilizer use as well as the crop yields. Figure 7 shows that on average, 5.52 kg of fertilizer per ha was used by farmers where input shops are established as compared to only 3.32 kg per ha in areas with no input shops. This translates to a higher grain yield from millet (541 kg ha^{-1}) where there are input shops, whereas

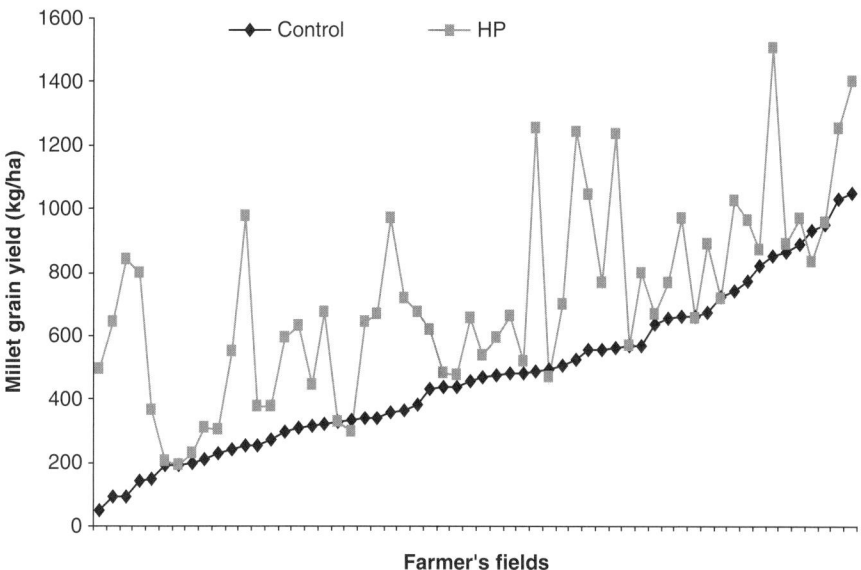

Fig. 6 Pearl millet grain response (kg ha^{-1}) to microdose hill placement (HP) management, Niger, 1998–2000

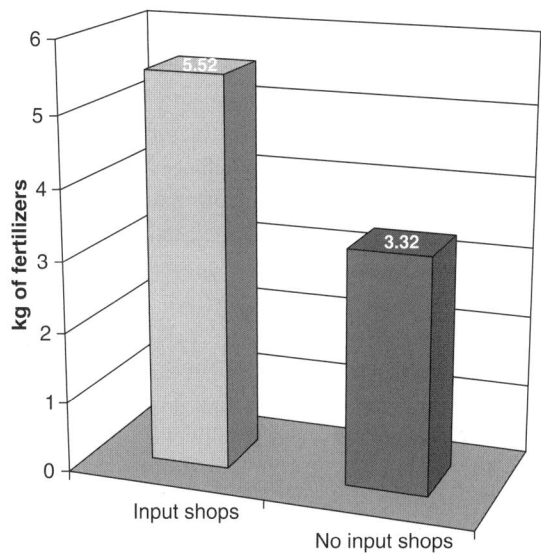

Fig. 7 Fertilizer use (kg ha^{-1}) affected by the presence of input shop

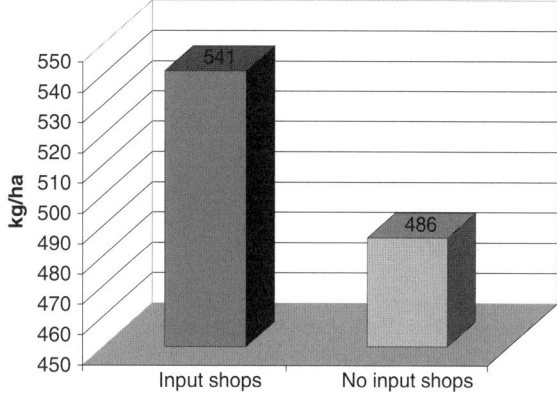

Fig. 8 Millet grain yield (kg ha^{-1}) as affected by the presence of input shop

'Warrantage' or Inventory Credit System

The warrantage scheme enables the establishment of a link between credit and cereal grain markets. This credit facility removes the barriers to the adoption of soil fertility restoration technologies. Farmers can have access to credit to enable them purchase external inputs such as fertilizers and invest in income-generating activities like fattening of small ruminants, horticulture and trading while using the stored grains to get higher prices at a time when the market supply begins

grain yields were lower (486 kg ha^{-1}) (Fig. 8). The presence of input shops where small packs of fertilizers (1, 2, or 5 kg of small pack) are sold enables farmers with limited resources to afford these small packs instead of trying to purchase 50-kg bags of fertilizers that are out of their reach financially.

Table 1 Example of warrantage performance in Mali, 2002/2003

NGO partner	Villages	Crops	Quantity of grain stored (kg) × 1000	Credit received under "warrantage" (FCFA) × 1000	Management fees (FCFA) × 1000	Net benefits (FCFA)
SG 2000	Kondogola	Millet	4	360 (US $720)*	4 (US $8)	236 (US $472)
	Niamabougou	Millet/sorghum	28.5	4246.5 ($8493)	1320 ($2640)	922.5 ($1845)
		Paddy rice	2	580 ($1160)	79 ($158)	Consumed
	Sélinkégny	Maize/millet	3.8	482 ($964)	–	29.4 ($59)
		Paddy rice	4.2	420 ($840)	100 ($200)	Consumed
	Tioribougou	Sorghum	4.1	619.5 ($1239)	10.5 ($21)	196 ($392)
ADAF/Gallé	Kénioroba	Millet/Sorghum	6.9	1141 ($2282)	–	91.4 ($183)
Winrock	Tissala	Millet/Sorghum	6.2	620 ($1240)	46.5 ($93)	35 ($70)
International	Sofara	Paddy rice	13.1	638.4 ($1277)	–	215.6 ($431)
		Sorghum	0.9	126.3 ($252)	49.9 ($100)	Consumed

Equivalence in US dollars (1 US$ = 500 FCFA)

to decline. In order to make inputs accessible to farmers, sustainable farmer-based enterprises and cooperative organizations are developed, storage facilities and inputs shops (boutique d' intrants) are built and credit and savings schemes are also developed. These facilities are managed by members of these cooperatives. Table 1 shows that farmers were able to make substantial benefits by practising the warrantage system in Mali in 2002/2003.

Discussion

In this chapter, it was shown that the fertilizer microdosing technology has great potential to improve crop productivity. Overall grain yield increases using the microdosing technology were double the yields obtained from the farmer's traditional practice. Although it is believed that fertilizer microdosing gives the plant a quick start thereby enabling it to escape drought, higher yields are generally achieved under assured rainfall conditions such as in the Sudano-Guinean zone. Net gains were obtained by farmers using this technology which is economically viable.

The fertilizer microdosing technology coupled with the warrantage system (credit scheme and input shops) is an entry point for the green revolution in Africa. This is a simple but efficient technology that is readily accessible to farmers. The role of input shops where small packs of fertilizers are sold is significant in making inputs available to and affordable by farmers. As

farmers experiment with the small packs and are convinced of the benefits of the microdosing technology, they are willing to invest more and more in purchasing fertilizers, thereby increasing the intensity of fertilizer use.

An issue that requires further investigation is the possibility of soil mining arising from using the fertilizer microdosing technology. As grain yields increase per unit area and very little organic matter (OM), including crop residues, are returned into the soil, there is the risk that nutrient imbalances will inevitably develop with time. There is therefore a need to ensure that OM is added and incorporated into these soils to improve their structure so that their capacity to store adequate moisture and nutrients even after crops are harvested is enhanced.

Labour could also be a major constraint to the wide adoption of the fertilizer microdosing technology. To further reduce the cost–benefit ratio, efforts should be made to develop labour-saving equipment to complement the farmers' efforts. The precise application of the fertilizer microdose in the hill of the plant requires that appropriate technology be developed and used.

The warrantage system offers an excellent opportunity to farmers to get better prices for their grain products like sorghum and millet, to have access to cash credit and to purchase the needed inputs for increasing their agricultural productivity. The example from Mali given in this chapter showed clearly that farmers can obtain great benefits by practising the warrantage system. There is, however, a need to strengthen farmers' organizations and assist them to establish effective linkages with financial stakeholders (commercial banks, etc.) for additional funding. Income-generating activities and options during the dry season should be made available to farmers so that they can make better use of the cash loan that they obtained from stocking their grains in the warrantage stores.

Conclusions

The fertilizer microdosing technology has shown its potential in all the three countries, namely Burkina Faso, Mali and Niger where it was tested, demonstrated and promoted. Overall millet and sorghum grain yields were 50–120% higher with microdosing than with the earlier recommended fertilizer broadcasting rates and farmers' traditional practices. Microdosing coupled with the warrantage system (credit scheme and input shops) is an entry point for the green revolution in Africa. It is a simple but efficient technology that is readily accessible by farmers. Farmers achieved net profits greater than 130% from microdosing than with their traditional practice or the broadcasting method.

In spite of the encouraging results that were presented in this chapter, there is a need to address some issues that could make the technology more robust and increase its adoption rate by resource-poor farmers in sub-Saharan Africa. Research should be pursued to investigate the possible soil mining issue, the interaction between improved varieties and microdosing, the water × microdosing interaction, the effect of input shops on the intensity of fertilizer use and crop productivity, and the mechanization of microdosing as a strategy to reduce labour costs. In our efforts to scale up and out the technology, we will emphasize capacity-building activities to strengthen farmers' associations, link farmers' organizations to decentralized financial systems and banks, improve on the various infrastructures for the warrantage system and facilitate exchange visits between farmers and across countries. Monitoring and evaluation of these activities will be intensified.

Acknowledgements The authors express their gratitude to USAID and CORAF/AfDB for providing funding to implement the project activities in Burkina Faso, Mali and Niger. We thank all the farmers, farmers' organizations and research and development partners and NGOs for their active involvement in the execution of the field activities.

References

Bationo A, Lompo F, Koala S (1998a) Research on nutrient flows and balances in West Africa: state-of-the art. Agric Ecosyst Environ 71:19–35

Bationo A, Ndjeunga J, Bielders C, Prabhakar VR, Buerkert A, Koala S (1998b) Soil fertility restoration options to enhance pearl millet productivity on sandy Sahelian soils in southwest Niger. In: Lawrence P, Renard G, von Oppen M (eds) Proceedings of an international workshop on the evaluation of technical and institutional options for small farmers in West Africa, University of Hohenheim, Stuttgart, Germany. Margraf Verlag, Weikersheim, Germany, pp 93–104, 21–22, April

Breman H (1990) No sustainability without external inputs. Sub-Saharan Africa; beyond adjustment. Africa seminar. Ministry of Foreign Affairs, DGIS, The Hague, The Netherlands, pp 124–134

Buerkert A, Hiernaux P (1998) Nutrients in the West African Sudano-Sahelian zone: losses, transfers and role of external inputs. J Plant Nutr Soil Sci 161:365–383

Sivakumar MVK (1992) Climate changes and implications for agriculture in Niger. Climate Change 20:297–312

Stoorvogel JJ, Smaling EMA (1990) Assessment of soil nutrient depletion in sub-Saharan Africa: 1983–2000. Report 28. Wageningen. The Winand Staring Center for Integrated Land, Soil and Water Research (SC-DLO), Wageningen, The Netherlands

Tabo R, Bationo A, Bruno G, Ndjeunga J, Marcha D, Amadou B, Annou MG, Sogodogo D, SibiryTaonda JB, Ousmane H, Diallo MK, Koala S (2007) Improving cereal productivity and farmers, income using a strategic application of fertilizers in West Africa, pp 201–208. In: Bationo A, Waswa BS, Kihara J, Kimetu J (eds) Advances in integrated soil fertility management in sub-Saharan Africa: Challenges and opportunities, 1091 pp

Tabo R, Bationo A, Diallo MK, Hassane O, Koala S (2006) Fertilizer micro-dosing for the prosperity of small-scale farmers in the Sahel: Final report. Global theme on agrocosystems report No. 23. P.O. Box 12404, International Crops Research Institute for the Semi-Arid Tropics, Niamey, Niger, 28pp

Traore F (1974) Etude de la fumure azotée intensive des céréales et du rôle spécifique de la matière organique dans la fertilité des sols du Mali. Agron Trop 29:567–586

Van Keulen H, Breman H (1990) Agricultural development in the West African Sahelian region: a cure against land hunger? Agric Ecosyst Environ 32:177–197

African Green Revolution Requires a Secure Source of Phosphorus: A Review of Alternative Sources and Improved Management Options of Phosphorus

A.S. Jeng

Abstract An African Green Revolution cannot succeed without a secured supply of mineral fertilizers. This is particularly true of phosphorus, one of the key essential macronutrients. In most tropical soils, P is one of the main limiting plant nutrients and its deficiency is a major constraint for better crop production. This is mainly attributable to (i) the low total P content in soil, (ii) the relative unavailability of inherent soil P for plant uptake, and lastly (iii) the relative speed at which applied soluble sources of P such as inorganic P fertilizers and manures become fixed or changed to unavailable forms. It is clear that mining P minerals and spreading P fertilizers over the landscape is not sustainable in the long run. Cultural practices which can secure P sources and which conserve P should be made use of. Some of the measures necessary to adequately address the P problem can be listed as follows:

- nutrient cycling through the recycling of crop residues, green manures, animal manures, domestic and industrial wastes;
- the integration into the cropping system of P-mobilizing plant species which show the ability to improve P uptake even from less labile P forms and store P in the aboveground biomass even in excess of their needs; and
- biological means making use of mycorrhiza and other soil fauna to help extract fixed P from deep soils under low pH conditions.

African Green Revolution must put a lot of emphasis on integrated soil fertility management (ISFM), which combines the use of plant residues and inorganic P fertilizers exploiting their high potential for increasing crop production and ensuring sustainability. Increased production and productivity should never be based on addressing the constraints surrounding inorganic (mineral) fertilizers alone.

Keywords Integrated soil fertility management (ISFM) · Mineral fertilizers · Mycorrhiza · Phosphorus mobilization

Introduction

Africa, especially sub-Saharan Africa, has the world's lowest per capita agricultural production causing widespread chronic food insecurity. The latest statistics indicate that some 200 million people in African are chronically hungry. The situation is aggravated by the high population growth rate (3%) relative to the continually declining cereal (food) production. We have in other words a serious decline in per capita agricultural production (Fig. 1).

Negligible quantities of phosphorus and nitrogen and low organic matter contents have combined with the lack of bases to make low soil fertility the major biophysical constraint affecting agriculture, generally, and causing declining per capita food production in SSA. Food production cannot be increased because of nutrient-poor (infertile) soils. The rate at which nutrients are being extracted from the soil has also accelerated dramatically to the point that replacement rates are often being exceeded.

A.S. Jeng (✉)
Soil & Environment Division, Bioforsk – Norwegian Institute for Agricultural and Environmental Research, Fredrik A Dahls vei 20A, N-1432 Aas, Norway
e-mail: Alhaji.Jeng@bioforsk.no

A. Bationo et al. (eds.), *Innovations as Key to the Green Revolution in Africa*,
DOI 10.1007/978-90-481-2543-2_11, © Springer Science+Business Media B.V. 2011

Fig. 1 Agricultural production. Indices: total production per capita index. Units: Percent (%) of 1999–2001 average agricultural production per capita. Source: FAO (2006): FAO online statistical service, http://apps.fao.org

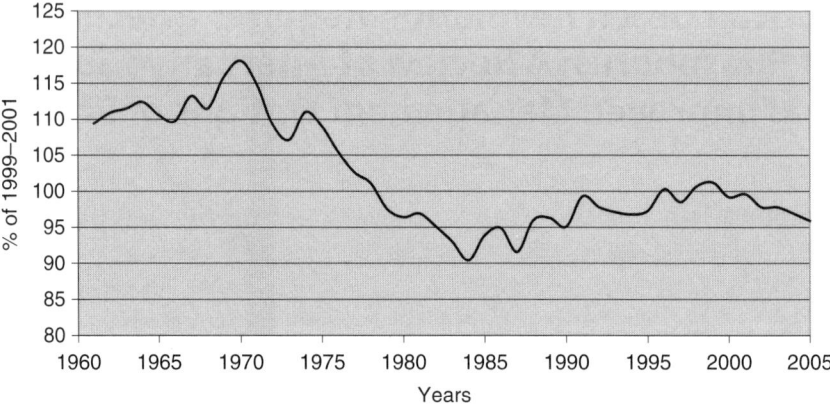

Of the essential macronutrients, phosphorus (P) is the most limiting nutrient for acceptable yields in African agriculture. While nitrogen inputs can be naturally available through biological N fixation (BNF) and the decomposition of crop residues and other organic compounds in the soil, P inputs need to be applied in order to improve the soil P status and ensure satisfactory yields. Many African soils are rich in sesquioxides with high levels of active Al and therefore have high P fixation capacities. As such large amounts of phosphate fertilizers are needed to overcome this high P fixation capacity (FAO, 2004) and to correct crop P deficiency and obtain reasonable yields and adequate food and fiber production (Sanchez and Buol, 1975; Date et al., 1995). Large amounts of inorganic fertilizers are however unaffordable to the African subsistence farmer. A Green Revolution aiming to reverse the trends of food insecurity and poverty must therefore make use of integrated practices and technologies for sustainable agricultural production.

The aim of this chapter is to highlight the importance of phosphorus as a plant nutrient, its place in an African Green Revolution, and to emphasize the need to develop and adopt technologies for a better conservation of P and better management of the available P sources (organic and inorganic).

Importance of Phosphorus as Plant Nutrient

Phosphorus is essential to all living organisms. It is irreplaceable in those compounds on which life processes depend. It is an important component of

enzymes which are the key players in energy transfer processes which drive growth processes in all living things. In plants, photosynthesis, which converts CO_2 to sugars, is such a process. Phosphorus is also an essential component of nucleic acids in which complex DNA and RNA structures carry and translate genetic information which controls all living processes, such as the production of proteins and vitamins.

It is clear from the above that plants that do not receive optimal amount of P will have their growth severely retarded. Deficiency does affect not only plant growth and development (vegetative) but also the generative/reproductive aspect of plant growth, decreasing fruit and seed formation and delaying ripening. In a recent communication in the Norwegian Teknisk Ukeblad, Prof. Dag Hessen of the University of Oslo, stated that if mankind exhausts phosphorus, he also exhausts the building blocks of RNA and DNA (Hessen, 2011). Plants will be the first victims to be closely followed by the animal kingdom. We must bear in mind that phosphorus is a limited resource and we must devise ways to economize its consumption.

The World's Phosphorus Reserves

Almost all phosphate fertilizers are derived from phosphate rock. The "reserves" of phosphate rock, i.e., deposits which are or could be profitably mined under prevailing costs, market prices and technology are rather limited. The reserve base is the in-place demonstrated (measured plus indicated) phosphorus resource from which reserves are estimated. It includes those resources that are currently economic (reserves),

Fig. 2 World phosphate rock reserves and reserve base. *Source*: US Geological Survey (1998)

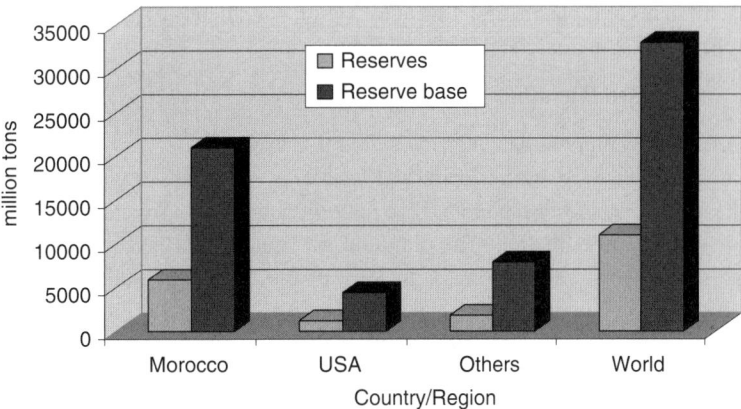

marginally economic (marginal reserves), and sub-economic (sub-economic resources). The "resources" which are at present not economically exploitable, but which could potentially become so, are much larger than the "reserves" and "reserves base."

In 1998, the US Geological Survey estimated that world phosphate rock reserves amounted to about 11 billion tonnes (Fig. 2), with a larger reserve base of about 33 billion tonnes. Over 60% of these reserves are concentrated in Morocco, the remainder being found in the USA, Jordan, South Africa, and Senegal.

These inorganic sources of the nutrient are limited and many estimates have indicated that the inexpensive rock phosphate reserves could be depleted in as little as 60–80 years (Council for Agricultural Science and Technology, 1988; Runge-Metzger, 1995). A potential phosphate crisis therefore looms for agriculture in the twenty-first century.

In view of the predicted future scarcity of mineral P fertilizers, other sources must be found and utilized effectively and in a sustainable manner. Sustainable management of P in agriculture requires that plant biologists discover mechanisms in plants that enhance P acquisition and exploit these adaptations to make plants more efficient at acquiring P, develop P-efficient germplasm, and advance crop management schemes that increase soil P availability.

P in Soil–Plant Continuum

The element phosphorus does not occur by itself in nature; it is always combined with other elements to form many different phosphates, some of which are very complex. Low availability of phosphorus is a major constraint on agricultural productivity in highly weathered tropical soils. Such soils have a significant capacity to sorb large amounts of phosphorus, taking them out of the soil solution. This limits the availability of inorganic phosphorus for plants, whether it is already contained in the soil or added as fertilizer. Further, some tropical soils contain only small amounts of total phosphorus, with a relatively large proportion of this present in organic forms (Nziguheba and Bünemann, 2005; Oberson and Joner, 2005).

Phosphorus (P) deficiency is a factor limiting crop production on tropical and sub-tropical soils (Fairhust et al., 1999; Mokwuny et al., 1986; Sanchez and Salinas, 1981). Correcting P deficiency with applications of P fertilizer is not possible for the mostly resource-poor farmers in the tropics and subtropics, especially on soils with high P-fixing capacity.

In a continent with relatively rapid population growth, declining per capita food production, endowed for the most part with highly weathered, acidic soils with low nutrient status, and where agriculture is mainly subsistence undertaken mainly by resource-poor farmers, sustainable management of soil nutrients is absolutely crucial for sustained increased agricultural production. P conservation for enhanced crop productivity is a must.

P Conservation for Enhanced Crop Productivity

Sustainable management of P in agriculture requires that plant biologists discover mechanisms in plants that enhance P acquisition and exploit these adaptations to make plants more efficient at acquiring P, develop

P-efficient germplasm, and advance crop management schemes that increase soil P availability.

(a) Increased P availability by biological means

Organic P compounds range from readily available, un-decomposed plant residues and microbes within the soil to stable compounds that have become part of soil organic matter. The amount of organic P present in soils varies from 20 to 80% of the total P (Prasad and Power, 1997; Schachtman et al., 1998). Phosphorus cycling and availability in soils is controlled by a combination of biological (mineralization–immobilization) and chemical (adsorption–desorption and dissolution–precipitation) processes (Frossard et al., 2000). Biological processes in the soil, such as microbial activity, tend to control the mineralization and immobilization of organic P. Phosphorus immobilization by microorganisms, turnover of microbial P, and mineralization of microbial by-products seem to be the major processes regulating P cycling and P availability from plant residues in soils (McLaughlin et al., 1988). Arbuscular mycorrhizal (AM) association plays an important role for plant P nutrition (Lopez-Gutierrez et al., 2004) and yield improvements (Young et al., 1986; Quilambo et al., 2005). This kind of mutualistic symbiosis between plant and fungus could be one of the most important and poorly understood processes for nutrient acquisition and plant growth in agriculture. The symbiotic interaction is localized in a root or a root-like structure in which energy moves primarily from plant to fungus and inorganic resources move from fungus to plant (Allen, 1991). AM contribution, particularly to P acquisition, may be direct or indirect. The direct effect is the consequence of the production of extracellular phosphatases and the access to distant P sources otherwise not available to plants (Joner and Johansen, 2000). The indirect effect is due to its extraradical hyphae that are capable of absorbing and translocating nutrients and of exploring more soil volume (Joner and Jacobsen, 1995; Singh and Kapoor, 1998). Both effects can contribute greatly to plant P nutrition even at P-deficient condition (Joner and Jacobsen, 1995). For soils that are low in P or have high P fixation capacity as is the case in most of the soils of SSA, AM will contribute to the efficient use of the nutrient for increased crop production. In addition to plant nutrition, AM fungi play an important role in enhancing physical and biological soil quality (soil structure), the interactions with other beneficial soil organisms (nitrogen-fixing rhizobia), and in improved protection against pathogens (Cardoso and Kuyper, 2006).

(b) P-mobilizing plant species

One of the most promising agronomic measures for P management is the integration of P-mobilizing plant species into the cropping system (Vanlauwe et al., 2000; Horst et al., 2001). These plant species show the ability to uptake P even from less labile P forms and store P in the aboveground biomass even in excess of their needs (Kahm et al., 1999). Culture of such species enhances uptake of soil P from deeper layers and by returning the plant residues, surface soil profiles could be enriched with P.

Upon decomposition, organic P in the incorporated green manure tissues could provide a relatively labile or available P to succeeding crops, thus providing a larger pool of mineralizable soil organic P to supplement soluble inorganic P pools (Tiessen et al., 1994). Along with the mineralized P, organic acids are released, which may help dissolve soil mineral P (Sharpley and Smith, 1989). Carbon dioxide (CO_2) released during decomposition of green material forms H_2CO_3 in the soil solution, which ultimately dissolves P minerals in soils, thus increasing P availability for plants (Tisdale et al., 1985). Cavigelli and Thien (2003) reported from a pot study that incorporation of green manure crops into soil may increase P bioavailability for succeeding crops. Pypers et al. (2005) reported from a field core incubation experiment that both P deficiency and Al toxicity can be amended through green manuring.

Studying P-deficient soils of the West African Sahel, Alvey et al. (2001) could show that legume–cereal rotations had significant positive effects on phosphate availability (Table 1), which in turn triggered improved cereal growth.

Li et al. (2007) reported that intercropping, which grows two or more crop species on the same piece of land at the same time, can increase grain yields greatly (Table 2). Their investigations indicated that the large increases of maize, a crop species that has a high requirement for P, resulted from a rhizosphere effect of faba bean on maize, even when P was provided in an insoluble form. This was attributable to, what was termed, the interspecific facilitation between the intercropped species (maize and faba beans).

Table 1 Phosphorus availability in bulk and rhizosphere soils of continuous sorghum and rotation sorghum soils from Fada without N application at 57 days after sowing (*Source*: Alvey et al., 2001)

Soil type	Total (P_t)	H_2SO_4	Organic (P_o)	Bray (P_b)
	Phosphorus fraction(mg P kg^{-1} soil)			
Rotation without plant	44.1	19.7	24.7	6.3
Continuous without plant	37.0	16.0	21.0	4.9
Rotation rhizosphere	45.0	15.3	29.7	5.8
Continuous rhizosphere	38.8	13.8	25.0	2.7
Rotation bulk soil	37.8	15.2	22.7	3.9
Continuous bulk soil	33.8	12.7	21.2	2.9
SED[a]	**1.08**	**1.12**	**1.42**	**0.31**
P>F[b]: System	0.009	0.390	< 0.001	< 0.001
P>F[b]: Location	0.001	0.296	0.029	0.015
P>F[b]: System × location	0.378	0.512	0.871	0.004

[a]Standard error of the difference
[b]Probability of a treatment effect (significance level)

Table 2 Biomass and grain yields (kg/ha) of intercropped maize and faba bean with and without root barriers under field conditions (field study 2) (*Source*: Li et al., 2007)

Crop	Solid barrier	Mesh barrier	No barrier	$F(2,30)$	P
Grain yield					
Maize	5,311[b]	5,341[b]	6,722[a]	20.32	<0.0001
Faba bean	4,176[b]	4,392[b]	5,527[a]	10.81	0.0003
Aboveground biomass					
Maize	12,560[b]	12,525[b]	15,783[a]	21.04	<0.0001
Faba bean	9,873[c]	11,027[ab]	12,468[a]	6.10	0.006

Values for grain yield are averages of all inoculations with rhizobium (two treatments), mycorrhiza (two treatments), and four replicates ($n = 16$), because there was no significant response to the inoculations

Values in the same row followed by different superscript letters are significantly different ($P < 0.05$)

(c) Integrated Soil Fertility Management (ISFM)

ISFM farming practices involve the integration of a range of actions that will result in raising productivity levels while maintaining the natural resource base. The basic focus of ISFM is sustainability, which creates a system that is able to provide adequate, affordable food, feed, and fiber supplies in a profitable manner without being detrimental to the environment. Key aspects of the approach include the following:

- Replenishing soil nutrient pools
- Maximizing on-farm recycling of nutrients

- Reducing nutrient losses to the environment
- Improving the efficiency of external inputs

Much of the above interventions for nutrient P management are aspects of ISFM. Several scientists have reported that the effect of organic amendments on crop yield increases is partly due to effects of SOC. Bationo et al. (2007) reported a large positive and additive effect of crop residue and mineral fertilizer application on pearl millet yield from a long-term experiment carried out in the West African Sahel (Table 3). The results also indicated that for these soils the potential

Table 3 Effect of crop residue and fertilizer on pearl millet grain and stover yields at Sadore, Niger (*Source*: Bationo et al., 2007)

Treatment	Grain yields (kg ha^{-1})				Stover yield (kg ha^{-1})			
	1983	1984	1985	1986	1983	1984	1985	1986
Control	280	215	160	75	NA	900	1100	1030
Crop residue (no fertilizer)	400	370	770	745	NA	1175	2950	2880
Fertilizer (no crop residue)	1040	460	1030	815	NA	1175	3540	3420
Crop residue plus fertilizer	1210	390	1940	1530	NA	1300	6650	5690
LSD$_{0.05}$	260	210	180	200		530	650	870

NA, not available

for continuous millet production is very limited in the absence of soil amendments. Gangwar et al. (2006) recorded increased soil organic carbon content and P availability and concluded that reduced tillage and in situ incorporation of crop residues at 5 Mg ha^{-1} along with 150 kg N ha^{-1} were optimum to achieve higher yield of wheat after rice in sandy loam soils.

Conclusion

In the context of an African Green Revolution, sustainability is the keyword. Slash and burn agriculture, which is still practiced in many places in SSA, is not sustainable. It counteracts most of the objectives of an integrated management system.

With the continuing decline in the world's phosphorus reserves, it is essential that production systems that encourage nutrient recycling be adopted making greater use of the small amounts of soil P in the soils of SSA. Crop residues, other organic wastes, green manures, and P-enhancing soil organisms must be given greater emphasis compared to P acquisition through mineral fertilizer alone.

References

Allen MF (1991) The ecology of Mycorrhiza. Cambridge University Press, Cambridge

Alvey S, Bagayoko M, Neumann G, Buerkert A (2001) Cereal/legume rotations affect chemical properties and biological activities in two west African soils. Plant Soil 231:45–54

Bationo A, Kihara J, Vanlauwe B, Waswa BS, Kimetu J (2007) Soil organic carbon dynamics, functions and management in west African agro-ecosystems. Agric Syst 94:13–25

Cardoso IM, Kuyper TW (2006) Mycorrhizas and tropical soil fertility. Agric Ecosyst Environ 116(1–2):72–84

Cavigelli MA, Thien SJ (2003) Phosphorus bioavailability following incorporation of green manure crops. Soil Sci Soc Am J 67:1186–1194

Council for Agricultural Science and Technology (1988) Report no. 114, long term viability of US agriculture. Council for Agricultural Science and Technology, Ames, IA

Date RA, Grundon NJ, Rayment GE, Probert ME (eds) (1995) Plant–soil interactions at low pH: Principles and management. Developments in plant and soil sciences, vol 64. Kluwer, Dordrecht, The Netherlands, 822 pp

Fairhust T, Lefroy R, Mutert E, Batjes N (1999) The importance, distribution and causes of phosphorus deficiency as a constraint to crop production in the tropics. Agroforest For 9:2–8

FAO (2004) Use of phosphate rocks for sustainable agriculture. FAO Fertil Plant Nutr Bull 13:172

Food and Agricultural Organization of the United Nations (FAO) (2006) FAO online statistical service. FAO, Rome. Available online at: http://apps.fao.org

Frossard E, Condron LM, Oberson A, Sinaj S, Fardeau JC (2000) Processes governing phosphorus availability in temperate soils. J Environ Qual 29:15–23

Gangwar KS, Singh KK, Sharma SK, Tomar OK (2006) Alternative tillage and crop residue management in wheat after rice in sandy loam soils of indo-gangetic plains. Soil Tillage Res 88(1–2):242–252

Hessen D (2011) Peak fosfor mye verre enn peak oil. Teknisk Ukeblad 0211:16

Horst WJ, Kahm M, Jibrin JM, Chude VO (2001) Agronomic measures for increasing P availability to crops. Plant Soil 237:211–223

Joner EJ, Jacobsen I (1995) Growth and extracellular phosphatase activity of arbuscular mycorrhizal hyphae as influenced by soil organic matter. Soil Biol Biochem 24:897–903

Joner EJ, Johansen A (2000) Phosphatase activity of external hyphae of two arbuscular mycorrhizal fungi. Mycol Res 104:81–86

Kahm M, Horst WJ, Amer F, Mostafa H, Maier P (1999) Mobilization of soil and fertilizer phosphate by cover crops. Plant Soil 211:19–27

Li L, Li SM, Sun JH, Zhou LL, Bao XG, Zhang HG, Zhang FS (2007) Diversity enhances agricultural productivity via rhizosphere phosphorus facilitation on phosphorus-deficient soils. Proc Natl Acad Sci USA 104(27):11192–11196

Lopez-Gutierrez JC, Toro M, Lopez-Hernandez D (2004) Seasonality of organic phosphorus mineralization in the rhizosphere of the native savanna grass, Trachypogon plumosus. Soil Biol Biochem 36:1675–1684

McLaughlin MJ, Alston AM, Martin JK (1988) Phosphorus cycling in wheat–pasture rotations. II. The role of microbial biomass in phosphorus cycling. Aust J Soil Res 26:333–342

Mokwunye ASH, Chien SH, Rhodes ER (1986) Reactions of phosphate with tropical African soils. In: Mokwunye A, Vlek PLG (eds) Management of nitrogen and phosphorus fertilizers in sub-Saharan Africa. Martinus Nijhoff, Dordrecht, The Netherlands, pp 253–282

Nziguheba G, Bünemann E (2005) Organic phosphorus dynamics in tropical agroecosystems. In: Turner BL, Frossard E, Baldwin DS (eds) Organic phosphorus in the environment. CAB International, Wallingford, UK, pp 243–268

Oberson A, Joner EJ (2005) Microbial turnover of phosphorus in soil. In: Turner BL, Frossard E, Baldwin DS (eds) Organic phosphorus in the environment. CAB International, Wallingford, UK, pp 133–164

Prasad R, Power JF (1997) Soil fertility management for sustainable agriculture. CRC, Lewis, Boca Raton, NY, p 356

Pypers P, Verstraete S, This CP, Merckx R (2005) Changes in mineral nitrogen, phosphorus availability and salt-extractable aluminum following the application of green manure residues in two weathered soils of South Vietnam. Soil Biol Biochem 37:163–172

Quilambo OA, Weissenhorn I, Doddema H, Kuiper PJC, Stulen I (2005) Arbuscular mycorrhizal inoculation of peanut in low-fertile tropical soil. II. Alleviation of drought stress. J Plant Nutr 28(9):1645–1662

Runge-Metzger A (1995) Closing the cycle: obstacles to efficient P management for improved global security. In: Tiessen H (ed) Phosphorus in the global environment. Wiley, Chichester, UK, pp 27–42

Sanchez PA, Buol SW (1975) Soils of the tropics and the world food crisis. Science 188:598–603

Sanchez P, Salinas JG (1981) Low input technology for managing Oxisols and Ultisols in tropical America. Adv Agron 34:280–406

Schachtman DP, Robert JR, Ayling SM (1998) Phosphorus uptake by plants: from soil to cell. Plant Physiol 116:447–453

Sharpley AN, Smith SJ (1989) Mineralization and leaching of phosphorus from soil incubated with surface-applied and incorporated crop residue. J Environ Qual 18:101–110

Singh S, Kapoor KK (1998) Effects of inoculations of phosphate solubilizing microorganisms and arbuscular mycorrhizal fungus on mung bean grown under natural conditions. Mycorrhiza 7:249–253

Tiessen H, Stewart JWB, Oberson A (1994) Innovative soil phosphorus availability indices: assessing organic phosphorus. In: Havlin JL, Jacobsen JS (eds) Soil testing: prospects for improving nutrient recommendations. SSSA Spec. Pub. No. 40. SSA, Madison, WI, pp 143–162

Tisdale SL, Nelson WL, Beaton JD (1985) Soil fertility and fertilizers, 4th edn. Macmillan, New York, NY, p 754

USGS (1998) Phosphate Rock, Mineral Commodity Summaries (1998) Reston, Virginia. Online linkage: http://minerals.usgs.gov/minerals/pubs/mcs

Vanlauwe B, Diels J, Sanginga N, Carsky RJ, Deckers J, Merckx R (2000) Utilization of rock phosphate by crops on a representative toposequence in the Northern Guinea savanna zone of Nigeria: response by maize to previous herbaceous legume cropping and rock phosphate treatments. Soil Biol Biochem 32:2079–2090

Young CC, Juang TC, Guo HY (1986) The effect of inoculation with vesicular–arbuscular mycorrhizal fungi on soybean yield and mineral phosphorus utilization in subtropical–tropical soils. Plant Soil 95:245–253

Part II
Potential and Feasibility of Use of External Input and Improved Soil and Crop Management to Achieve the African Green Revolution

Soybean Varieties, Developed in Lowland West Africa, Retain Their Promiscuity and Dual-Purpose Nature Under Highland Conditions in Western Kenya

B. Vanlauwe, J. Mukalama, R.C. Abaidoo, and N. Sanginga

Abstract Entry points that give farmers immediate benefits are required to reverse the ever-declining soil fertility status of a substantial area in sub-Saharan Africa. In West Africa, dual-purpose, promiscuous soybeans (*Glycine max* (L.) Merr.) that produce a substantial amount of grains and leafy biomass and do not require inoculation with specific *Rhizobium* spp. strains were developed and have increased resilience of farming while providing income to farmers. These crops could be a potential entry point for soil fertility improvement in western Kenya, provided they retain their promiscuity and dual-purpose character in this new environment. The major objective of this work was to quantify nodulation, biomass production and grain yield characteristics of a set of best-bet, dual-purpose varieties relative to a locally available variety at two sites (Vihiga and Siaya districts) in western Kenya. In the presence of P, most promiscuous soybean varieties showed substantial improvements in nodulation (19–165 nodules per 0.5 m of soybean) than did the local variety (3–13 nodules per 0.5 m of soybean). While grain yield was for all but one variety as good as the local control (845 kg ha^{-1}, on average), nearly half of the varieties produced significantly higher amounts of biomass at 50% podding than did the local variety (865 kg ha^{-1} in Siaya and 1877 kg ha^{-1} in Vihiga). Increases in nodulation, biomass production and grain yield were mainly observed after application of P fertilizer; in the absence of P, almost none of the varieties performed better than the local control for any of the measured characteristics. To fully exploit the potential soil fertility-improving characteristics of these varieties, it will be necessary to facilitate availability of P fertilizer and to foster demand for soybean at the farm, community and national levels.

Keywords Biomass production · N harvest index · Nodulation · Soil fertility

Introduction

Farmers in sub-Saharan Africa (SSA) usually cite declining soil fertility as one of their major constraints to crop production (De Groote et al., 2003). Negative N, P, and K balances in most farming areas in SSA are due to soil nutrient mining and soil fertility decline. (Smaling et al., 2002). Today, integrated soil fertility management (ISFM) is the accepted paradigm for devising and disseminating technologies alleviating soil fertility decline in SSA (Vanlauwe et al., 2011). Technically, ISFM advocates the use of mineral and organic nutrient inputs to enhance and sustain agricultural productivity. While the availability of mineral inputs is largely restricted by economic and policy constraints, producing a significant amount of organic resources in existing cropping systems also has technical challenges.

In implementing an ISFM research and development agenda, legumes have been attracting substantial attention because of their potential to fix atmospheric N through symbiosis with N-fixing bacteria, harboured in their root nodules. This phenomenon is often translated in substantial yield gains in a cereal that follows the legume crop. However, when using grain

B. Vanlauwe (✉)
Tropical Soil Biology and Fertility Institute of the International Centre for Tropical Agriculture (TSBF-CIAT), Nairobi, Kenya
e-mail: b.vanlauwe@cgiar.org

legumes that are traditionally grown in large parts of SSA, a substantial part of the legume N fixed from the atmosphere is usually removed from the field through harvested grains and/or stover, often resulting in marginally positive or even negative net N balances. Understandably, researchers aiming at improving the soil fertility status have traditionally opted for so-called green manure legumes, of which no or little harvested products are removed from the field. For instance, in West Africa, substantial efforts have been made to use *Mucuna pruriens* (L.) DC. to reverse soil degradation (Versteeg et al., 1998). In East Africa, *Sesbania sesban* (L.) Merr. tree fallows were proven to substantially enhance the soil N status (Sanchez and Jama, 2002). Unfortunately, although such technologies aiming solely at improving the soil fertility status were proven to work under on-station conditions, farmers have been and still are very reluctant to adapt such technologies, mainly because they do not derive any immediate benefits such as edible grains or fodder for animals (Vanlauwe et al., 2003).

Soybean is one of the most popular pulses in the world and its success stems from a number of factors related to its composition and productivity. Soybean yields more than other common pulses and has relatively few pest and disease problems, and a good grain storage quality. Although 'traditionally bred' soybean varieties do not contribute to the soil N status because most of the N fixed is removed by harvesting the grains, 'dual-purpose', promiscuous soybean varieties were developed by the International Institute of Tropical Agriculture (IITA) between the mid-1970s and early 1990s (Sanginga et al., 2003). Their dual-purpose nature stems from the fact that these varieties produce a substantial amount of grain and leafy biomass, resulting in a relatively low N harvest index. The promiscuous nature allows these varieties to nodule freely with the indigenous *Bradyrhizobium* spp. population, avoiding inoculation, a technique that has often failed in SSA (Mpepereki et al., 2000). When these new improved varieties are planted in rotation with maize or sorghum, the productivity and sustainability of crop production is enhanced (Sanginga et al., 2001; 2003). Having soybean in rotation with maize also has the potential to reduce *Striga hermonthica* parasitism (Carsky et al., 2000) and enhance the soil P status (Sanginga, 2003).

Although these varieties were bred under West African lowland conditions, their potential role as an entry point to reverse soil fertility decline in East Africa cannot be underestimated. Especially in densely populated areas such as western Kenya (Crowley and Carter, 2000), farmers required immediate benefits, e.g. legume grains, of any soil fertility-improving technology due to acute shortage of land. The current dual-purpose soybean could be such technology offering immediate benefits if the promiscuity and the low N harvest index feature are retained under western Kenyan conditions. The objectives of this chapter, therefore, were (i) to evaluate the nodulation and biomass potential of a set of best-bet, dual-purpose varieties from the 2001 International Trials of IITA under western Kenyan conditions, relative to a commercially available, local variety, and (ii) to identify some varieties that could be used for large-scale testing in subsequent initiatives aimed at improving the soil fertility status of farmland in that area.

Materials and Methods

Varieties Used in the Screening Work

Several varieties, developed at the International Institute of Tropical Agriculture (IITA), Ibadan, Nigeria, based on the selection of progenies from crosses between Asian and American soybean varieties with good nodulation in local soils without external rhizobia application (Kueneman et al., 1984), were used in the screening work (Table 1). These comprised 10 early and 10 medium-to-late varieties with corresponding dates to maturity of 85–100 and 101–120 days, respectively. One commercially available, short-duration, local variety (X-Baraton) was included in the screening trial.

Site Characteristics

The different soybean varieties were screened during the 2001 short rainy season (between 30 August 2001 and 19 December 2001/6 February 2002, depending on the variety) in four sites in Vihiga district and during the 2002 long rainy season (between 19 March 2002 and 1 July 2002/26 August 2002, depending on the

Table 1 Soybean varieties screened in western Kenya. All varieties except SB21 belong to the soybean international trial 2001 set of IITA. R2 means the full bloom growth stage, while R8 means the full maturity growth stage. The SED for the days to maturity in 2001 is 3, while the SED for the days to flowering and maturity in 2002 are 2 and 3, respectively

Label	Variety	Duration	Days to full maturity (R8); IITA Ibadan, Nigeria	Days to full maturity (R8) (Vihiga, 2001)	Days to flowering (R2) (Siaya, 2002)	Days to full maturity (R8) (Siaya, 2002)
SB1	TGx1830-20E	Early	95–100	122	54	112
SB2	TGx1831-32E	Early	95–100	131	50	107
SB3	TGx1835-10E	Early	85–95	115	49	102
SB4	TGx1871-12E	Early	85–95	111	52	103
SB5	TGx1876-4E	Early	95–100	128	49	105
SB6	TGx1895-4F	Early	95–100	130	52	110
SB7	TGx1895-6F	Early	95–100	131	57	112
SB8	TGx1895-33F	Early	95–100	131	52	117
SB9	TGx1895-49F	Early	95–100	138	57	118
SB10	TGx1844-18E	Medium to late	115–120	147	57	118
SB11	TGx1866-12F	Medium to late	115–120	161	72	141
SB12	TGx1869-31E	Medium to late	105–110	129	52	111
SB13	TGx1873-16E	Medium to late	110–115	150	60	118
SB14	TGx1878-7E	Medium to late	115–120	139	55	117
SB15	TGx1889-12F	Medium to late	115–120	135	51	112
SB16	TGx1893-7F	Medium to late	105–110	138	59	115
SB17	TGx1893-10F	Medium to late	105–110	130	52	109
SB18	TGx1894-3E	Medium to late	110–115	133	52	110
SB19	TGx1740-2F	Early	95–100	104	51	105
SB20	TGx1448-2E	Medium to late	115–120	141	53	112
Local	X-BARATON	Early	–	116	53	101

variety) in four sites in Siaya district. Both districts are in Western Kenya. The sites in Vihiga lie at an altitude of 1540 masl and have an average minimum and maximum temperature of 14 and 28°C, respectively, while the sites in Siaya lie at an altitude of 1330 masl and have an average minimum and maximum temperature of 16 and 29°C, respectively. The total rainfall was 526 mm during the short rainy season of 2001 and 937 mm during the long rainy season of 2002 season (Fig. 1).

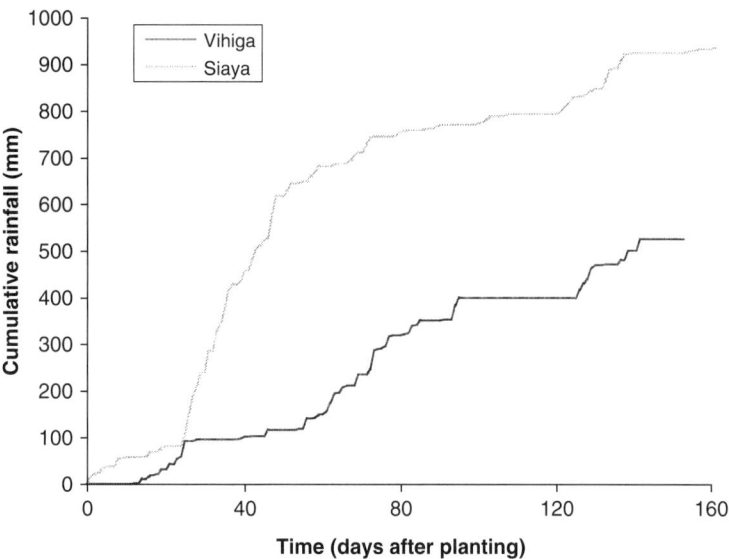

Fig. 1 Cumulative rainfall distribution during the short rainy season of 2001 in Vihiga and the long rainy season of 2002 in Siaya

Table 2 Topsoil characteristics of the trial sites

District – Farmer	Soil organic C (g kg⁻¹)	Total soil N (g kg⁻¹)	Available P (mg kg⁻¹ Olsen-P)	pH		Sand (g kg⁻¹)	Silt (g kg⁻¹)	Clay (g kg⁻¹)
				Water	KCl			
Vihiga								
– S Khaynaje	14.2	1.39	2.7	5.4	4.7	435	315	350
– P Omulema	11.2	1.07	1.6	5.4	4.4	420	195	385
– N Muhando	16.6	1.49	2.8	5.4	4.6	335	285	380
– A Akero	11.2	1.12	1.2	5.5	4.7	420	220	360
Siaya								
– J Oloo	17.7	1.59	2.1	5.3	4.4	156	295	549
– F Odinga	17.4	1.85	0.9	5.4	4.6	196	270	534
– J Aketch	16.4	1.43	1.5	5.5	4.8	216	240	544
– A Meso	17.6	1.54	2.6	5.4	4.6	283	220	497

The four farms were randomly chosen within the districts but care was taken to include both relatively poor and fertile sites. The Vihiga soils contained between 1.07 and 1.49 g N kg⁻¹, while the Siaya soils were heavier and contained between 1.43 and 1.85 g N kg⁻¹ (Table 2). Available P content was below 3 mg kg⁻¹ at all sites (Table 2).

Field Trial Establishment and Management History

In each of the above sites, all varieties were planted, one line of 4 m per variety. Either all varieties were treated with 22 kg P ha⁻¹ or not. Each site contained two replicates. Planting density was 75 cm (between lines) × 5 cm (between plants within the line). The plants were regularly weeded till harvesting.

Observations

In 2001, at 50% podding, the aboveground biomass of 0.5 m of the soybeans at the side of the row was harvested and the dry matter of the leaves, stems and pods was determined. The roots systems were dug out down to the level where roots were no longer obvious (often up to 50 cm depth) and the number of nodules and their fresh weight were recorded. All nodules were sliced with a blade and the colour determined. Active nodules were pink, red or brown, while non-active nodules were white or light green (Somasegaran and Hoben, 1994). In 2002, the same observations were taken at

50% flowering and 50% podding. In both years, final yield data (grains, husks and haulms) and days to full maturity were recorded.

Mathematical and Statistical Analyses

The MIXED procedure of the SAS system (SAS, 1992) using 'variety' and 'P application' as fixed factors following a split plot design ('P application' as main plot and 'variety' as sub-plot) and 'replicate within site' and 'variety * replicate within site' as random factors was used to determine the significance of the treatment effects. Significantly different means were separated with the PDIFF option of the LSMEANS statement. Regression analysis (SAS, 1985) was used to explore potential linear relationships between various attributes measured in this work.

Results

Days to Maturity

The total number of days to full maturity varied from 104 to 161 days in Vihiga and from 101 to 141 days in Siaya (Table 1). Variety SB 11 showed the longest duration to maturity in both sites. Varieties SB9, SB10, SB13 and SB14 expressed long-duration characteristics at both sites, while varieties SB1, SB3, SB4, SB5 and SB19 retained their short-duration characteristics (Table 1).

Total Biomass and Grain Production

Varieties SB9, SB10, SB11, SB13, SB14, SB15 and SB20 showed significantly higher total biomass production at 50% podding than did the local variety at both sites in response to P fertilizer application (Fig. 2). In the absence of P fertilizer, no variety showed higher total biomass accumulation at both sites at 50% podding relative to the local variety. Application of P increased total biomass production for all varieties at both sites, although in Vihiga, the differences were not significant for varieties SB1, SB4, SB6, SB7, SB8, SB9, SB16, SB18, SB19 and the local variety (Fig. 2a). In Siaya, the difference was not significant

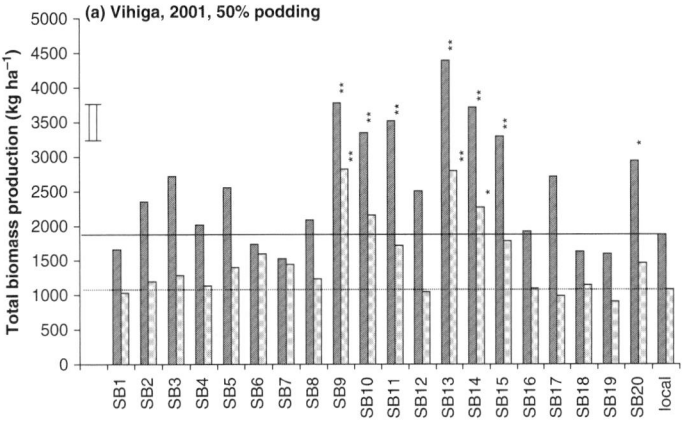

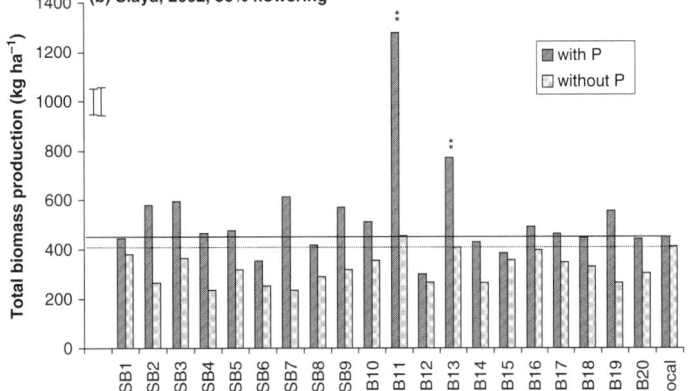

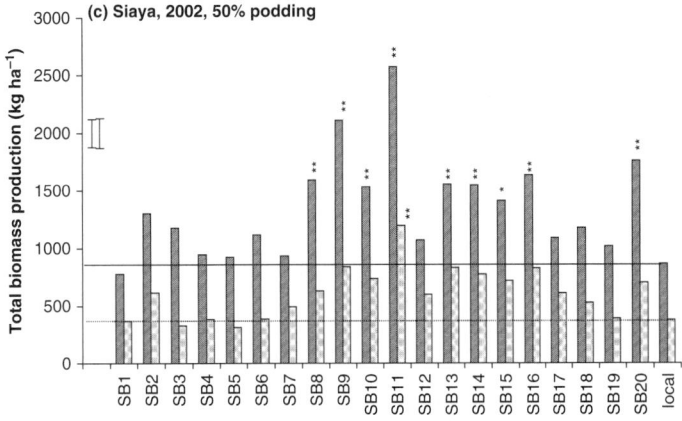

Fig. 2 Total shoot biomass (leaves, stems and pods) production at 50% podding in Vihiga in 2001 (**a**) and at 50% flowering (**b**) and 50% podding (**c**) in Siaya in 2002. The *horizontal full line* shows the total biomass yield of the local variety in the presence of P and the *horizontal dashed line* in the absence of P. Above each bar, the level of significance is shown relative to the local variety only if the treatment is significantly different from the local variety at least at the 5% level. The *left error bar* is the standard error of the difference (SED) to compare between varieties in the presence or the absence of P, while the *right error bar* is the SED to evaluate the effect of P for a single variety

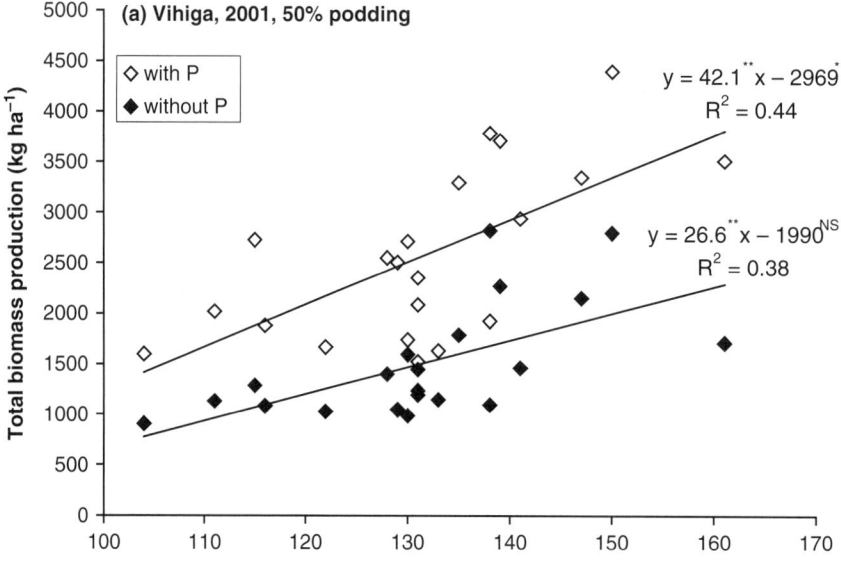

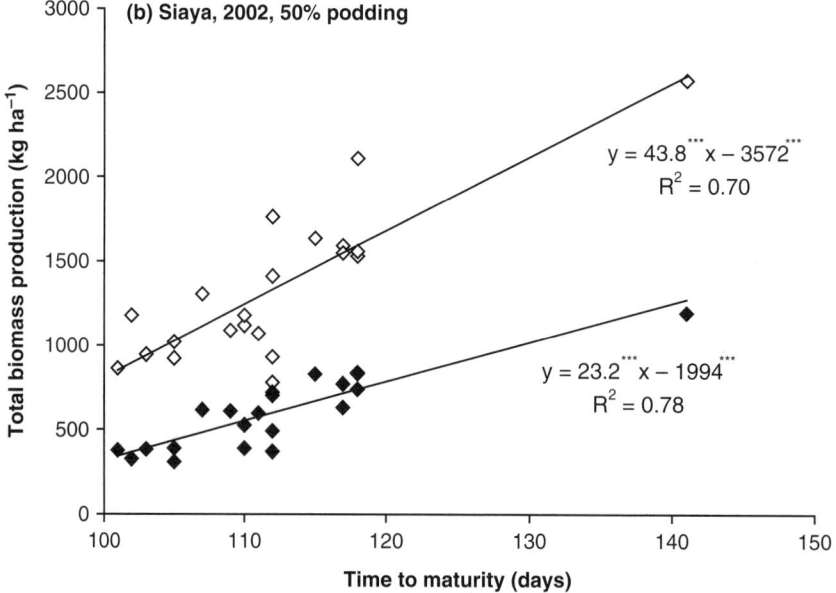

Fig. 3 Relationships between the time to maturity and the total shoot biomass production at 50% podding in the presence or the absence of P in Vihiga in 2001 (**a**) and in Siaya in 2002 (**b**)

for varieties SB1, SB7, SB12, SB17 and the local variety (Fig. 2b). Generally, varieties with a longer time to maturity produce a larger amount of biomass at 50% podding, both in the presence and in the absence of P (Fig. 3). Varieties SB8, SB9 and SB17 produce a significantly higher amount of grains than did the local variety at both sites after application of P, while

SB11 showed a significantly lower grain yield (Fig. 4). Although grain yields were higher for all varieties in Vihiga, response to P was significant only for varieties SB3, SB4, SB5, SB6, SB8, SB15, SB17 and SB19 (Fig. 4a). In Siaya, all varieties except SB1 produced significantly higher grain yields after application of P (Fig. 4b).

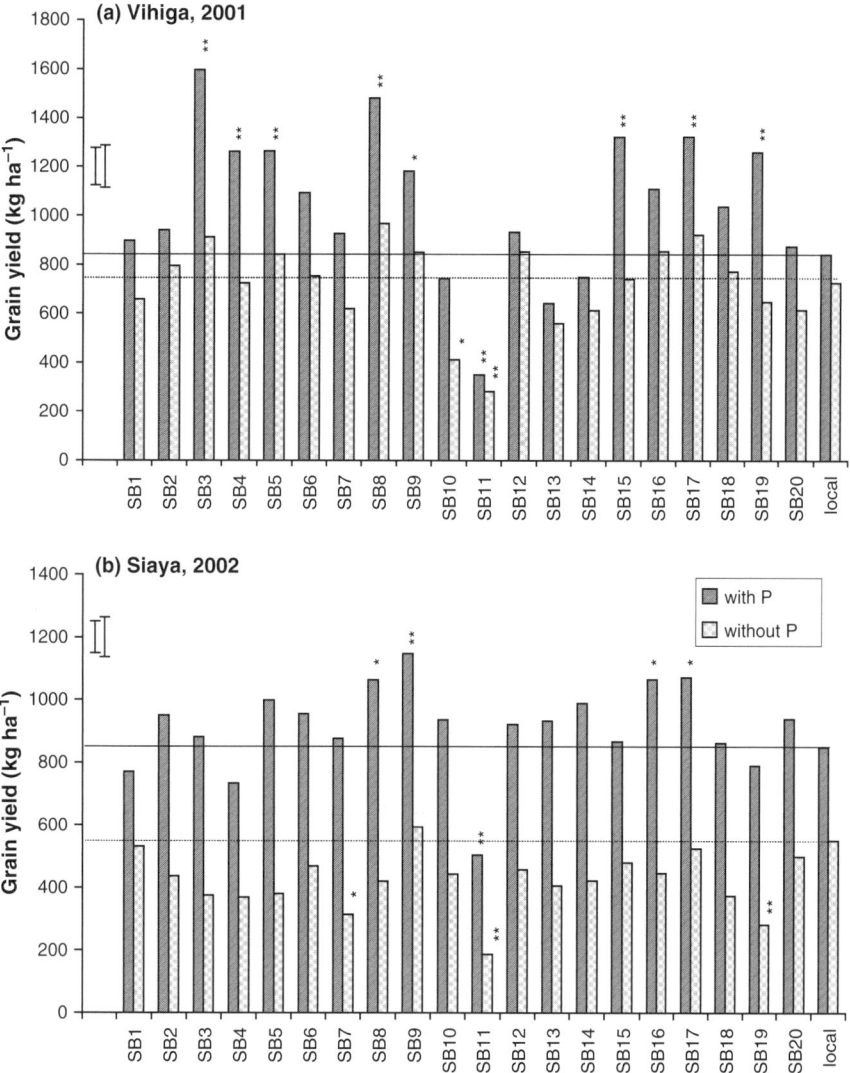

Fig. 4 Grain yield in Vihiga in 2001 (**a**) and in Siaya in 2002 (**b**). The *horizontal lines*, levels of significance and error bars are described in the caption of Fig. 1

Nodulation

In Vihiga, in the presence of P, all improved varieties had a significantly higher number of nodules (ranging from 33 to 165 per 0.5 m of soybean) than did the local variety (13 per 0.5 m of soybean), except for SB4, SB6 and SB18 (Fig. 5a). In the absence of P, all varieties except SB4, SB6, SB17 and SB18 had a higher number of nodules (ranging from 14 to 76 per 0.5 m of soybean) than did the local variety (5 per 0.5 m of soybean) (Fig. 5b). Of all nodules, $78 \pm 26\%$ were active in the presence of P and $73 \pm 29\%$ in the absence of

P (*data not shown*). In Siaya at 50% podding, varieties SB2, SB8, SB9, SB13, SB14, SB15, SB19 and SB20 had a significantly higher number of nodules (ranging from 19 to 36 per 0.5 m of soybean) than did the local variety (3 per 0.5 m of soybean) in the presence of P (Fig. 5c). In the absence of P, only varieties SB8, SB9 and SB 19 had more nodules (ranging from 2 to 3 per 0.5 m of soybean) than did the local variety (0 per 0.5 m of soybean) (Fig. 5c). Nodule fresh weight was linearly related to nodule number at both sites and in the presence or the absence of P with R^2 values ranging from 0.42 to 0.67 and slopes from 0.052 to 0.129 g

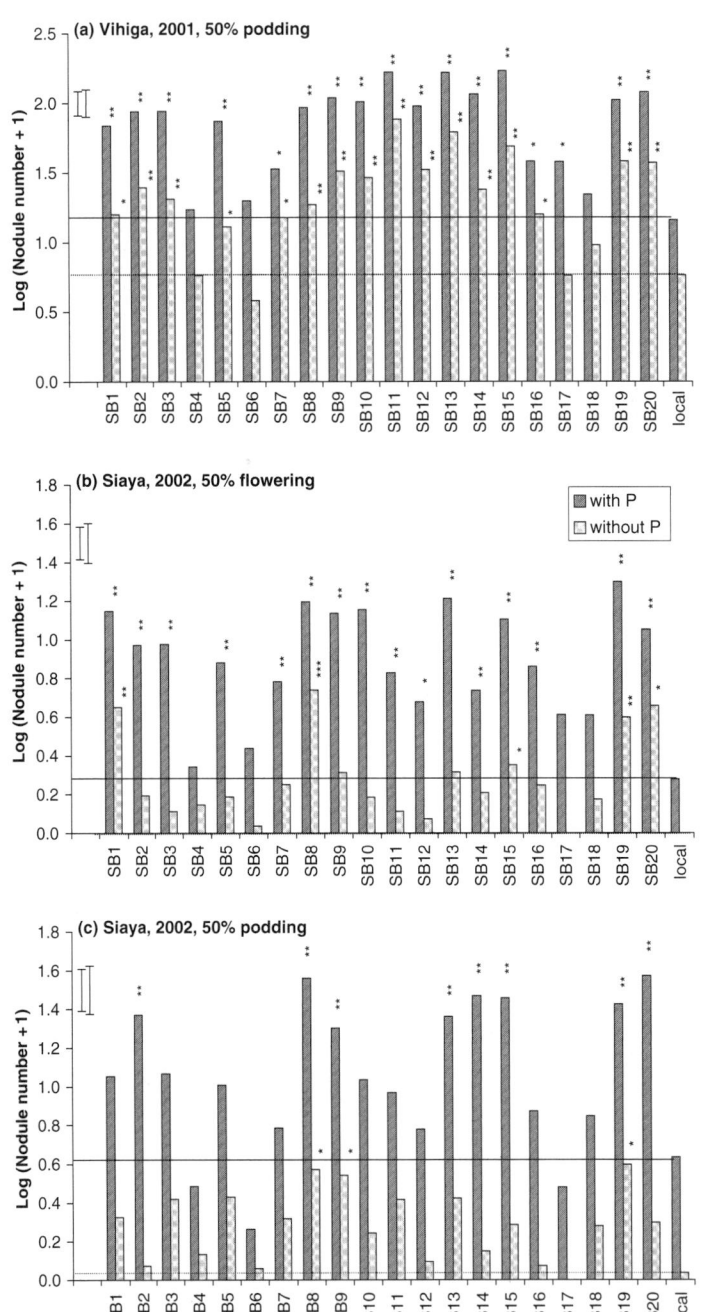

Fig. 5 Nodule numbers (total number of nodules from all plants in a 0.5-m-long line) at 50% podding in Vihiga in 2001 (**a**) and at 50% flowering (**b**) and 50% podding (**c**) in Siaya in 2002. The number of nodules found in 0.5 m of soybean was log transformed before statistical analysis. The *horizontal lines*, levels of significance and error bars are described in the caption of Fig. 1

Fig. 6 Relationships between the nodule fresh weight and the nodule numbers in the Vihiga (**a**) and Siaya (**b**) sites. The regression lines were forced through 0. '*', '***', '****' and 'NS' signify 'significance at 5, 1, 0.1% levels' and 'not significant', respectively

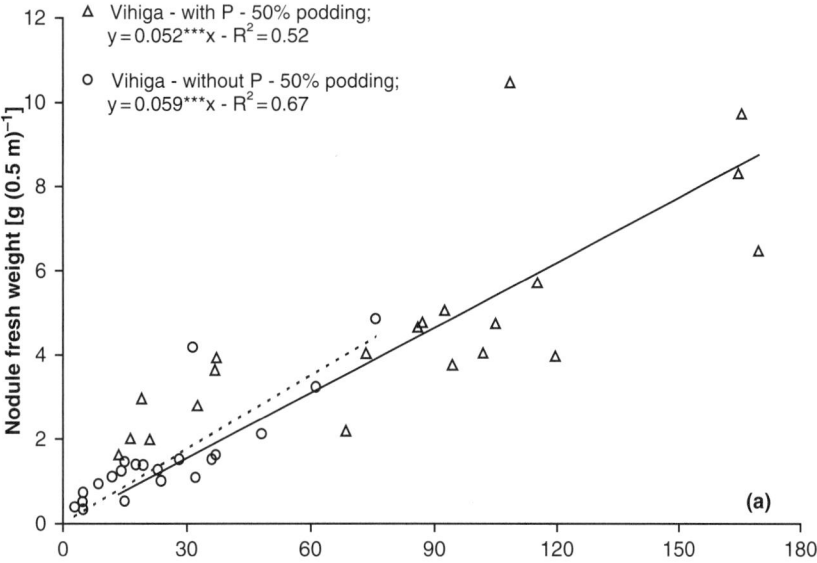

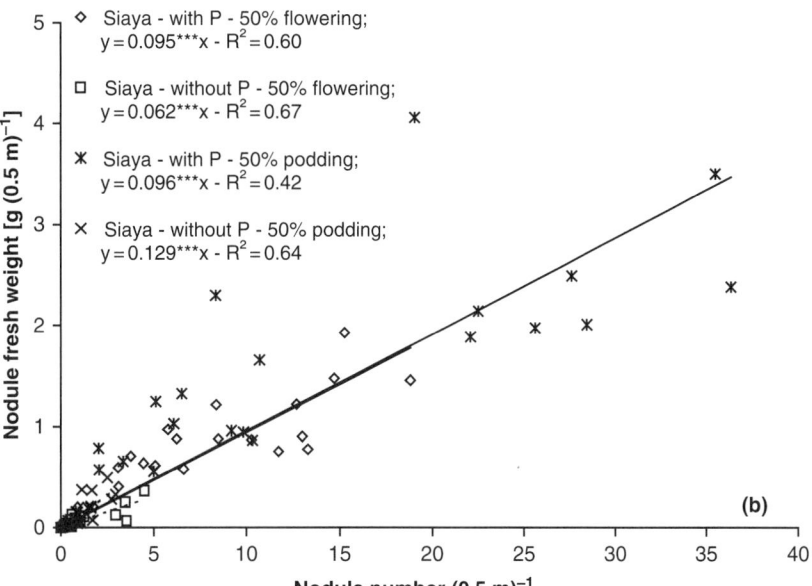

(Fig. 6). In the presence of P, total biomass production at 50% podding was linearly related to the total fresh weight of the nodules in both Siaya and Vihiga (Fig. 7).

Discussion

In the presence of P, nearly all improved varieties had more nodules than did the local variety at both sites and most of the nodules found contained leghaemoglobin, indicating active N_2 fixation. These varieties were developed under West African conditions with the aim to nodulate with indigenous *Bradyrhizobium* spp. (cowpea rhizobia) as well as *Bradyrhizobium japonicum* and *Bradyrhizobium elkanii* strains (Abaidoo et al., 1999). The fact that these varieties nodulate equally well under western Kenyan conditions indicates that the breeding for effective promiscuous nodulation had been successful and that these varieties

Fig. 7 Relationship between the total shoot biomass production at 50% podding and the nodule fresh weight, in the presence of P. '***' signifies significance at the 0.1% level

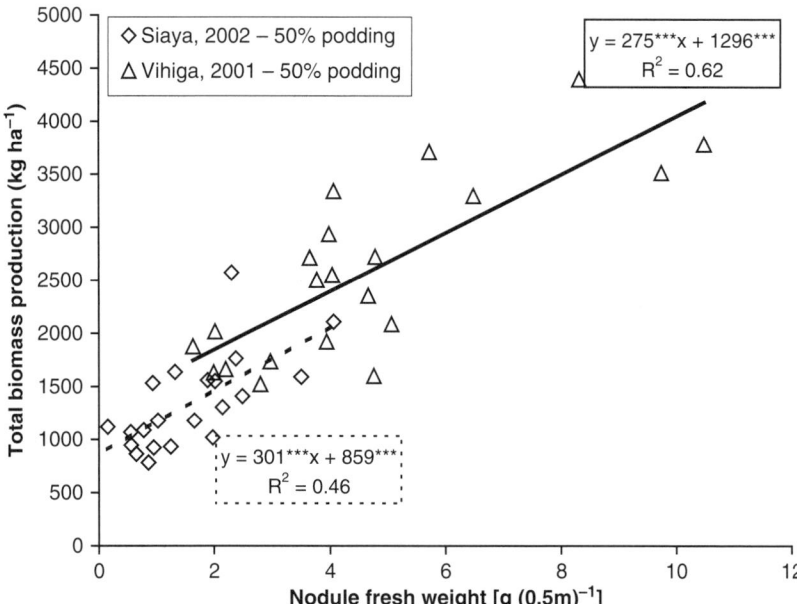

are expected to perform creditably in most tropical soil environments without inoculation, provided adequate bradyrhizobia populations exist in such soils. Important activities to follow up on the present observations are (i) identification of the exact strains that are nodulating with the improved varieties, (ii) quantification of the minimal population required to have proper nodulation and (iii) direct quantification of the total amount of N_2 fixed from the atmosphere as nodule numbers are usually not a conclusive indicator for appropriate N_2 fixation (Sanginga et al., 1997).

At least seven of the tested varieties also produced substantially more biomass at 50% podding than did the local variety while producing similar amounts of grain. Most of these varieties had the longest duration to full maturity. These observations indirectly indicate that these varieties are likely going to have a lower N harvest index and consequently leave a relatively higher amount of N in the soil compared with the local variety (Giller, 2001) as nearly all leaves have fallen at full maturity. Exact quantification of the proportion of N fixed and the total N uptake at peak biomass production would then allow to judge whether a net amount of N is retained in the soil after harvesting the beans. Sanginga et al. (2001) reported net soybean N inputs for dual-purpose varieties varying between 17 and 27 kg N ha^{-1}. For Samsoy 2, a local variety commonly used in northern Nigeria, this value was −8 kg

N ha^{-1} (Sanginga, Unpublished Results). Variety SB11 that showed an exceptionally long time to full maturity produced high amounts of biomass but much less grains than did the local variety, thus likely having the lowest N harvest index. Although this may result in the largest net N inputs in the cropping system, the low grain yields are likely going to make this variety unacceptable to the farmer community. This variety, however, could be useful as parent for future breeding activities.

It needs to be stressed that both the promiscuous nodulation and low N harvest index characteristics hold only after applying P fertilizer. Ogoke et al. (2003) previously observed that the N balance of dual-purpose varieties was − 9 kg N ha^{-1} without P and 5 kg N ha^{-1} with P. Applying P is especially important in western Kenya due to the common occurrence of soils with high P fixation (Nziguheba et al., 2000). In the current trials, P was applied at a single rate of 22 kg P ha^{-1}. This may not be the optimal application rate, so further fine-tuning of actual P requirements could enhance the economic benefits of inclusion of soybeans into existing cropping systems. Moreover, P fertilizer usually shows some residual effects that may lessen the need to apply P to soybean, specifically when grown in rotation with maize that received fertilizer. Currently, varieties are also being screened for relative tolerance to low available P conditions, although

Table 3 Ranking of the various soybean varieties in terms of grain yield, total biomass production and nodulation [log(number+1)]. These traits were scored from 1 to 21 following the relative size of each of these parameters and the averages of these scores are presented in the current table. Varieties indicated in bold have positions within the first 10 in both sites

Label	Variety	Vihiga		Siaya	
		Score	Position	Score	Position
SB1	TGx1830-20E	17	19	16	18
SB2	TGx1831-32E	8	6	12	14
SB3	**TGx1835-10E**	11	**10**	7	**3**
SB4	TGx1871-12E	19	21	13	15
SB5	TGx1876-4E	12	11	9	6
SB6	TGx1895-4F	14	15	15	17
SB7	TGx1895-6F	16	18	17	20
SB8	**TGx1895-33F**	4	**1**	8	**4**
SB9	**TGx1895-49F**	4	**1**	5	**2**
SB10	TGx1844-18E	10	9	11	11
SB11	TGx1866-12F	12	11	9	6
SB12	TGx1869-31E	15	17	11	11
SB13	**TGx1873-16E**	8	**6**	8	**4**
SB14	**TGx1878-7E**	5	**4**	9	**6**
SB15	**TGx1889-12F**	9	**8**	4	**1**
SB16	TGx1893-7F	7	5	13	15
SB17	TGx1893-10F	12	11	9	6
SB18	TGx1894-3E	14	15	16	18
SB19	TGx1740-2F	13	14	11	11
SB20	**TGx1448-2E**	4	**1**	9	**6**
Local	X-BARATON	18	20	18	21

it is unlikely that varieties will be identified that result in net N inputs in the soil without applying any P. Besides the need for P, no other important constraints for grain and biomass production were observed. The occurrence of root rot nematodes was minimal and no major leaf diseases were seen. Flowers, pods and grains were also quasi-disease free. It will be important, however, to continue to monitor the occurrence of pest and diseases as soybean is a relatively new crop in the area and enlarging its production area may result in the build-up of pest and disease pressure.

In this work, several improved varieties were screened with varying nodulation, biomass and grain production characteristics. When using these traits to score and identify the best-bet varieties, SB3, SB8, SB9, SB13, SB14, SB15 and SB20 appear amongst the best 10 in both sites (Table 3). The local variety scores low in both sites. It is also interesting to note that SB 20 (TGx-1448-2E) is one of the varieties taking off in Nigeria (Sanginga et al., 2003). Obviously, the current scoring technique has many flaws as various traits are not included (e.g. grain quality and ease of processing) and the traits included may not be ranked with equal weight by the end-users. A holistic evaluation phase is required involving all potential beneficiaries (farmers, local processors, large-scale millers, etc.).

In order to realize the potential of dual-purpose soybean in enhancing the soil fertility status in SSA, it is necessary to improve the demand for soybean products. In Nigeria, for instance, this was achieved by including soybean products in local meals and supporting community-based processing of soybean (Sanginga et al., 1999). Soybean is an emerging industrial crop with high market potential in Nigeria, although linkages between producers, processors and consumers, hindering production, expansion and utilization of soybean, could substantially enhance large-scale processing (Kormawa and Von Oppen, 2001). It would also be essential to link soybean production with a credit scheme for P fertilizer as without such inputs the potential in terms of N_2 fixation, biomass production, grain yield and soil fertility improvement is likely not going to be achieved in most of western Kenya.

Acknowledgements The authors gratefully acknowledge the Bundesministerium für Wirtschaftliche Zusammenarbeit und Entwicklung (BMZ) and the Rockefeller Foundation for providing the necessary financial support in the framework of the projects on 'Improving integrated nutrient management practices on small-scale farms in Africa' and on 'Exploring the multiple potentials of soybeans in enhancing rural livelihoods and small industry in East Africa'.

References

Abaidoo RC, Dashiell KE, Sanginga N, Keyser HH, Singleton PW (1999) Time-course of dinitrogen fixation of promiscuous soybean cultivars measured by the isotope dilution method. Biol Fertil Soils 30:187–192

Carsky RJ, Berner DK, Oyewole BD, Dashiell K, Schulz S (2000) Reduction of *Striga hermonthica* parasitism on maize using soybean rotation. Int J Pest Manage 46:115–120

Crowley EL, Carter SE (2000) Agrarian change and the changing relationships between toil and soil in Maragoli, Western Kenya. Hum Ecol 28:383–414

De Groote HJ, Okuro O, Bett C, Mose L, Odendo M, Wekesa E (2003) Assessing the demand for insect resistant maize varieties in Kenya combining participatory rural appraisal into a geographic information system. In: System wide program on participatory research and gender analysis (PRGA) and Centre de Coopération Internationale en Recherche Agronomique pour le Développement (CIRAD) (eds) Participatory plant breeding and participatory genetic resource enhancement: an Africa-wide exchange of experiences. Centro International d'Agronomia Tropical International, Cali, Colombia, pp 145–158

Giller KE (2001) Nitrogen fixation in tropical cropping systems, 2nd edn. CABI, Wallingford, UK

Kormawa PM, von Oppen M (2001) Linking potential supply and demand for soybean in West Africa: a location analysis of a developing industry in Nigeria. Q J Int Agric 40:211–226

Kueneman EA, Root WR, Dashiell KE, Hohenberg J (1984) Breeding soybean for the tropics capable of nodulating effectively with indigenous *Rhizobium* spp. Plant Soil 82:387–396

Mpepereki A, Javaheri F, Davis P, Giller KE (2000) Soyabeans and sustainable agriculture promiscuous soyabeans in southern Africa. Field Crops Res 65:137–149

Nziguheba G, Merckx R, Palm CA, Rao MR (2000) Organic residues affect phosphorus availability and maize yields in a Nitisol of western Kenya. Biol Fertil Soils 32:328–339

Ogoke IJ, Carsky RJ, Togun AO, Dashiell K (2003) Effect of P fertilizer application on N balance of soybean crop in the guinea savanna of Nigeria. Agric Ecosyst Environ 100:153–159

Sanchez PA, Jama BA (2002) Soil fertility replenishment takes off in East and Southern Africa. In: Vanlauwe B, Diels J, Sanginga N, Merckx R (eds) Integrated plant nutrient management in sub-Saharan Africa: from concept to practice. CABI, Wallingford, UK, pp 23–46

Sanginga N (2003) Role of biological nitrogen fixation in legume based cropping systems: a case study of West Africa farming systems. Plant Soil 252:25–39

Sanginga PC, Adesina AA, Manyong VM, Otite O, Dashiell KE (1999) Social impact of soybean in Nigeria's southern Guinea Savanna. IMPACT, IITA, Ibadan, Nigeria

Sanginga N, Dashiell K, Diels J, Vanlauwe B, Lyasse O, Carsky RJ, Tarawali S, Asafo-Adjei B, Menkir A, Schulz S, Singh BB, Chikoye D, Keatinge D, Rodomiro O (2003) Sustainable resource management coupled to resilient germplasm to provide new intensive cereal–grain legume–livestock systems in the dry savanna. Agric Ecosyst Environ 100:305–314

Sanginga N, Dashiell K, Okogun JA, Thottappilly G (1997) Nitrogen fixation and N contribution by promiscuous nodulating soybeans in the southern Guinea savanna of Nigeria. Plant Soil 195:257–266

Sanginga N, Okogun JA, Vanlauwe B, Diels J, Carsky RJ, Dashiell K (2001) Nitrogen contribution of promiscuous soybeans in maize-based cropping systems. In: Tian G, Ishida R, Keatinge JDH (eds) Sustaining soil fertility in West Africa. SSSA Special Publication Number 58. SSSA, Madison, WI, pp 157–178

SAS (1985) SAS user's guide: statistics, 5th edn. SAS Institute Inc., Cary, NC

SAS (1992) The MIXED procedure. SAS technical report P-229: SAS/STAT software: changes and enhancements. SAS Institute Inc., Cary, NC, pp 287–366

Smaling EMA, Stoorvogel JJ, de Jager A (2002) Decision making on integrated nutrient management through the eyes of the scientist, the land-user and the policy maker. In: Vanlauwe B, Diels J, Sanginga N, Merckx R (eds) Integrated plant nutrient management in sub-Saharan Africa: from concept to practice. CABI, Wallingford, UK, pp 265–284

Somasegaran P, Hoben HJ (1994) Handbook for *Rhizobia*. Springer, New York, NY

Vanlauwe B, Bationo A, Carsky RJ, Diels J, Sanginga N, Schulz S (2003) Enhancing contribution of legumes and biological nitrogen fixation in cropping systems: experiences from West Africa. In: Waddington S (ed) Proceedings of the SoilFertNet meeting, Vumba, Zimbabwe. SoilFertNet, Harare, Zimbabwe, pp 3–13

Vanlauwe B, Bationo A, Chianu J, Giller KE, Merckx R, Mokwunye U, Ohiokpehai O, Pypers P, Tabo R, Shepherd K, Smaling E, Woomer PL, Sanginga N (2010) Integrated soil fertility management: operational definition and consequences for implementation and dissemination. Outlook Agric 39:17–24

Versteeg MN, Amadji F, Eteka A, Gogan A, Koudokpon V (1998) Farmers' adoptability of *Mucuna* fallowing and agroforestry technologies in the coastal savanna of Benin. Agric Syst 56:269–287

Nutr Cycl Agroecosyst (2010) 88:133–141
DOI 10.1007/s10705-010-9355-7

RESEARCH ARTICLE

Long-term effect of continuous cropping of irrigated rice on soil and yield trends in the Sahel of West Africa

B.V. Bado · A. Aw · M. Ndiaye

Received: 13 May 2009 / Accepted: 23 February 2010 / Published online: 9 March 2010
© Springer Science+Business Media B.V. 2010

Abstract The effects of 18 years continuous cropping of irrigated rice on soil and yields were studied in two long-term fertility experiments (LTFE) at Ndiaye and Fanaye in the Senegal River Valley (West Africa). Rice was planted twice in a year during the hot dry season (HDS) and wet season (WS) with different fertilizer treatments. Soil organic carbon (SOC) under fallow varied from 7.1 g kg^{-1} at Fanaye to 11.0 g kg^{-1} at Ndiaye. Rice cropping maintained and increased SOC at Ndiaye and Fanaye, respectively and fertilizer treatments did not affect SOC. Soil available P and exchangeable K were maintained or increased with long-term application of NPK fertilizers. Without any fertilizer, yields decreased by 60 kg ha^{-1} (1.5%) and 115 kg ha^{-1} (3%) per year at Fanaye and Ndiaye, respectively. The highest annual yield decreases of 268 kg ha^{-1} (3.6%) and 277 kg ha^{-1} (4.1%) were observed at Fanaye and Ndiaye, respectively when only N fertilizer was applied. Rice yields were only maintained with NPK fertilizers supplying at least 60 kg N, 26 kg P and 50 kg K ha^{-1}. It was concluded that the double cropping of irrigated rice does not decrease SOC and the application of the recommended doses of NPK fertilizer maintained rice yields for 18 years.

B. V. Bado (✉) · A. Aw · M. Ndiaye
Africa Rice Center (AfricaRice), Sahel Regional Station,
Saint-Louis BP 96, Senegal
e-mail: V.Bado@cgiar.org

Keywords Fertilizer · Organic carbon · *Oryza sativa* · Phosphorous · Potassium · Soil

Introduction

Weak soil buffering capacity due to low soil organic carbon (SOC) and clay content, low cation exchange capacity (CEC) and P deficiency are the main limiting factors to agricultural productivity of the upland soils of West Africa. Data from many long-term experiments in upland soils usually show yield declines over time as a consequence of a decrease in soil organic carbon (SOC), soil acidification and a decrease of nutrient use efficiency (Bationo and Mokwunye 1991; Bado et al. 1997; Bationo 2008). In contrast to the poor upland soils, lowland soils of the inland valleys have generally higher organic carbon and clay contents, a better CEC and water retention capacity, offering better conditions for crop production.

As with Asia, rice has become the most important cereal crop of the inland valleys of West Africa. Rice is cultivated as a staple food by farmers in the small inland valleys, or as a cash crop in many irrigated schemes. Consequently, the cropping systems are becoming more and more intensified and, whenever possible, many farmers grow two rice crops (during the wet and hot dry seasons) per year. While the simulated potential yield of irrigated rice is

🖄 Springer

This article has been previously published in the journal "Nutrient Cycling in Agroecosystems" Volume 88 Issue 1.
A. Bationo et al. (eds.), Innovations as Key to the Green Revolution in Africa: Exploring the Scientific Facts. © 2010 Springer.

8–12 tonnes ha^{-1} (Dingkuhn and Sow 1997), the average yields in farmers' fields vary from 4 to 6 tonnes ha^{-1} (Haefele et al. 2002; Kebbeh and Miezan 2003). This means that there is scope for increasing rice yields with the sustainable intensification of the existing cropping systems.

The long-term and intensive cultivation of rice with periodical flooding can affect the dynamics of SOC, soil pH, cation exchange capacity and nutrient use efficiency (Kenneth et al. 1997). A key research question is whether long-term intensive lowland rice-rice cropping in the irrigated schemes is sustainable. Some results from long-term experiments in Asia showed yield declines of 70 to more than 200 kg ha^{-1} with best management practices over a period of 10–24 years (Flinn and De Datta 1984; Cassman et al. 1995; Kenneth et al. 1997). There is very little information on soil changes and rice crop yields in long-term intensive rice cultivation in Africa. Working with only 10 years data on two long-term fertility experiments (LTFE) in West Africa, Haefele et al. (2002, 2004) found a slight but not significant yield decline (−27 kg ha^{-1} per season) at one site and a significant yield increase (+86 kg ha^{-1} per season) at other site. A non-significant decrease in SOC was also observed. The authors recognized that because of the short duration of the experiments (10 years) and climatic influences, the observed yield trends could not give an accurate indication of the biophysical sustainability of the cropping system.

The overall objective of this study was to assess the sustainability of rice cropping in terms of soil fertility and yield stability using more data (18 years or 36 cropping seasons) from the same LTFE. This research is mainly focussed on soil P and K, SOC and yield trends under long-term intensive cropping of rice.

Materials and methods

Experimental sites

Two long-term fertility experiments (LTFE) were carried out over a period of 18 years (1991–2008) at AfricaRice's research farms in Ndiaye and Fanaye, both located in the Senegal River Valley. Ndiaye (16°14′ N, 16°14′ W) is located close to the coast (about 40 km inland) in the Senegal River delta. Soil

salinity at the site and in the river delta is generally high, due to the occurrence of marine salt deposits in the subsoil (Haefele et al. 2002). The soil profile at Ndiaye corresponds to a typical Orthithionic Gleysol (FAO 1998). The original soil contained at least 10 mg C kg^{-1} of soil and 5 mg P kg^{-1} (P-Bray1). Fanaye (16°33′ N, 15°46′ W) is located in the middle valley of the Senegal River, ∼240 km inland, where natural soil salinity is low or absent. The soil profile in Fanaye belongs to a Eutric Vertisol (FAO 1998). The original soil contained at least 6.5 mg C kg^{-1} of soil and 4 mg P kg^{-1} (P-Bray1). The climate of the two sites is characterized by a wet season (WS) with ∼200 mm rainfall per year from July to October, a cold dry season from November to February and a hot dry season (HDS) from March to June (Dingkuhn and Sow 1997; Haefele et al. 2002).

Agronomic experiments

The LTFE was established at Ndiaye during the hot dry season (HDS) of 1991. The first rice crop was cultivated to homogenize soil fertility variability with the rice cultivar SIPI 692033, planted with a uniform application of NPK fertilizer supplying 120, 26 and 50 kg ha^{-1} of N, P and K, respectively. At Fanaye, soil fertility variability was homogenized by growing different cultivars of rice during the cold dry season with the same uniform dose of NPK fertilizer used at Ndiaye. From 1991 to 1997, the medium duration cultivar Jaya was used in the WS and the short duration cultivar IR 50 in the HDS. From the WS of 1997 (Ndiaye) and the HDS of 1998 (Fanaye), both cultivars were replaced by Sahel 108, a short duration cultivar (Miézan and Diack 1994).

Six fertilizer treatments with four replications were laid out in a randomized complete block design. Experimental plots measured 25 m^2 (5 × 5 m) and were separated by small 30 cm high dikes. The six fertilizer treatments were: a control (without any fertilizer); the recommended dose of N without PK fertilizer (120 kg N ha^{-1}, 0 kg P ha^{-1}, 0 kg K ha^{-1}); the recommended dose of NPK fertilizer (R_NPK) supplying 120 kg N ha^{-1}, 26 kg P ha^{-1} and 50 kg K ha^{-1}); low dose of N-medium PK fertilizer (L_NPK: 60 kg N ha^{-1}, 26 kg P ha^{-1} and 50 kg K ha^{-1}); high dose of N-medium PK fertilizer (H_NPK: 180 kg N ha^{-1}, 26 kg P ha^{-1} and 50 kg K ha^{-1}); and medium dose of N-high PK fertilizer (M_NPK: 120 kg N ha^{-1}, 52 kg P ha^{-1} and 100 kg K ha^{-1}). Urea (46% N), ammonium phosphate (18% N and 20% P) and potassium chloride (47% K) were used.

Rice seedlings were transplanted at the rate of 25 hills m^{-2}. Crop management was adjusted to the farmers' practice. For all fertilizer treatments, 50% N, 100% P and 100% K were broadcast at 21 days after transplanting. The remaining N dose was split-applied at panicle initiation (25%) and 10 days before flowering (25%). Herbicide (6 l ha^{-1} Propanyl) and manual weeding were used for weed control. Herbicide was applied once at 21 days after sowing, 1 day before the first N application; thereafter, plots were kept weed-free by manual weeding. Insecticides (Furadan) were sometimes used at 25 kg ha^{-1} for pest control at the start of tillering, maximum tillering, panicle initiation and flowering. A constant water layer of 10-15 cm was maintained during the whole cropping season and the harvested straw was removed from the plots every season. Rice grain yields were determined from a 4 m^2 harvest area in each plot at maturity and reported at a standard water content of 140 g water kg^{-1} fresh weight.

Soil sampling and analysis

Soil from each of the six fertilizer treatments were sampled at the end of the dry season of 2007 (after 17 years of cultivation). Data of the original soils of the two sites are already published (Haefele et al. 2002). However, in the absence of the original database with repeated values, it was not possible to conduct an analysis of variance in comparison with soil samples from the experimental plots. We have selected some uncultivated areas (native fallow) around the experiments to serve as a reference. These areas also received the same water regimes (flooding and aeration) as the experimental plots. The sub-samples were mixed thoroughly to get a composite sample, air-dried immediately after collection and dried in the laboratory at 60°C for 1 week. The samples were analyzed in the soil laboratory at AfricaRice. Soil organic C was determined using the wet digestion method (Walkley and Black 1934). Available soil phosphorous was determined by the Bray-P1 method using the extractant 0.03 N NH$_4$F in 0.025 N HCl (Bray and Kurtz 1945). Soil exchangeable K (K$_{exch}$) was extracted with 1 M NH4OACc solution (Helmke and Sparks 1996).

Statistical data analysis

The statistical analyses of the rice yields were firstly done using the four factors: year, site, season and fertilizer treatments (Gomez and Gamez 1983). The SAS software was used for data calculations (SAS Inst. 1995). The relationships between rice yield variations and years of cultivation were tested with different models before selecting the simple linear model which better described the relationship between the two factors (Eq. 1):

$$Y\left(\text{Mg ha}^{-1}\right) = aX + b \quad 1 \leq X \leq 18 \tag{1}$$

where Y = rice grain yields (dependent variable), X = number of years of cultivation (independent variable), a and b = the coefficient and the intercept of the linear regression, respectively (Gomez and Gamez 1983).

Results and discussion

Soil organic carbon, phosphorous and potassium

The mean soil organic carbon (SOC) of the uncultivated soil was higher ($P < 0.01$) at Ndiaye (11.0 g kg^{-1}) than Fanaye (7.1 g kg^{-1}) (Table 1). Data of the original soil also indicated that the site of Ndiaye contained more carbon (10.0 g kg^{-1}) than Fanaye (6.6 g kg^{-1}) (Haefele et al. 2004). At Ndiaye, SOC varied from 10.6 g kg^{-1} in the control (without fertilizer) plots to 12.3 g kg^{-1} in the high N fertilizer-medium PK treatment, with no significant differences between the fertilizer treatments. At Fanaye, SOC ranged from 12.9 g kg^{-1} in the control plots to 13.6 g kg^{-1} in the plots receiving the recommended NPK treatment. SOC was not decreased by long-term cultivation at either site but was even increased at Fanaye.

Soil extractable P-Bray 1 of the uncultivated soil was very low (3.8 mg P kg^{-1} at Fanaye and 4.9 mg P kg^{-1} at Ndiaye). All NPK fertilizer treatments increased soil extractable P at the two sites (Table 1). The soil available P ranged from 5.3 mg P kg^{-1} (control without any fertilizer) to 19.7 mg kg^{-1} in the medium N fertilizer-high PK treatment at Ndiaye and from 3.8 (control) to 7.1 mg kg^{-1} in the medium N fertilizer-high PK treatment at Fanaye.

While soil exchangeable K (K$_{exch}$) was affected by fertilizer application at Fanaye, K$_{exch}$ was not affected by fertilizer treatments at Ndiaye. In general,

Table 1 Effects of 18 years of continuous lowland irrigated rice mono cropping with NPK chemical fertilizer application on soil K, organic carbon and extractable Bray 1-P) (0–20 cm surface layer), at two experimental sites (Ndiaye and Fanaye)

Sites	N–P-K fertilizer (kg ha^{-1})	Organic carbon (g kg^{-1})	Extractable P (mg P kg^{-1})	Exchangeable K (cmol kg^{-1})
Ndiaye	Control (0 N-0 P-0 K)	10.6	5.3^{c}	0.56
	60 N-26 P-50 K	11.4	9.9^{b}	0.59
	120 N-0 P-0 K	11.7	4.2^{c}	0.64
	120 N-26 P-50 K	11.0	9.7^{b}	0.53
	120 N-52 P-100 K	10.5	19.7^{a}	0.66
	180 N-26 P-50 K	12.3	8.9^{b}	0.49
	Uncultivated soil	11.0	4.9^{c}	0.61
Fanaye	Control (0 N-0 P-0 K)	12.9^{a}	4.2^{b}	0.58^{a}
	60 N-26 P-50 K	13.3^{a}	6.2^{a}	0.55^{ab}
	120 N-0 P-0 K	13.5^{a}	2.8^{c}	0.45^{c}
	120 N-26 P-50 K	13.6^{a}	6.8a	0.48^{bc}
	120 N-52 P-100 K	13.2^{a}	7.1^{a}	0.63^{a}
	180 N-26 P-50 K	13.3^{a}	5.2^{b}	0.45^{c}
	Uncultivated soil	7.1^{b}	3.8^{c}	0.57^{ab}

Values followed by the same letters on the same site and the same column are not significantly different ($P < 0.001$) according the Fisher test. Significant differences were not observed for soil organic carbon at Ndiaye

soil K_{exch} was increased at the two sites with the application of K fertilizers. The lowest levels of K_{exch} were also noted at the two sites with the high doses (180 kg N ha^{-1}) of N fertilizers (Table 1).

Yield variations and trends

Grain yields were highly affected by fertilizer application ($P < 0.001$), year ($P < 0.001$) and site ($P < 0.0001$). However, grain yields were not affected by the season (Table 2). Significant simple interactions (site-season; site-year; site-fertilizer; year-site; season-site and year-fertilizer), triple interactions (site-year-season; site-season-fertilizer; year-season-fertilizer and site-year-fertilizer), and quadruple interaction (site-year-season-fertilizer) were observed on grain yields.

The parameters of the linear regression models used to analyze grain yield trends over the 18 years are presented in Table 3. The intercepts of the linear regression summarize the mean grain yields during the first year of cultivation. Considering the yield trends (Table 3; Figs. 1, 2), two main groups of fertilizer treatments were identified: (1) the control treatment and the recommended N dose alone without PK, and (2) the NPK treatments.

Long-term cultivation without fertilizer application decreased grain yields—by 115 kg ha^{-1} per year, with a confidence interval of 59–259 kg ha^{-1} ($P < 0.05$) at Ndiaye during the WS (Table 3). The same practice induced an annual yield loss of 60 kg ha^{-1} at Fanaye during the same WS. Thus, long-term cultivation without fertilizer decreased rice yields by 1.5% per year at Fanaye and 3% at Ndiaye during the WS. While a high yield of 7 Mg ha^{-1} was obtained in the first year of cultivation, N-fertilizer application alone induced the highest annual yield decreases at both sites during the two seasons (Table 1; Figs. 1, 2). The rice yield decreased by 268 kg ha^{-1} (−3.6%) at Fanaye and 277 kg ha^{-1} (−4.1%) at Ndiaye during the WS. A significant yield decrease of 147 kg ha^{-1} was observed during the HDS at Ndiaye (Table 3). With the NPK treatments, there was no significant linear relationship ($P > 0.05$) between yield variations and years. Otherwise, no significant yield decline was observed at the two sites when NPK fertilizers were applied (Table 3; Figs. 1, 2). However, slight (and nonsignificant) yield decreases ($P > 0.05$) were always observed during the WS at both sites while NPK fertilizers seemed to increase rice yields ($P > 0.05$) during the HDS.

Table 2 Influence of year, site, season and fertilizer on rice grain yields at Ndiaye and Fanaye between 1991 and 2008

Source of variations	Degree of freedom	F value	Probability level
Site	1	41.84	<.0001
Year	17	38.95	<.0001
Season	1	0.02	0.8939
Fertilizer	5	1002.47	<.0001
Site × Season	1	207.32	<.0001
Site × Year	17	23.55	<.0001
Site × Fertilizer	5	10.19	<.0001
Season × Fertilizer	5	20.30	<.0001
Year × Fertilizer	85	4.55	<.0001
Site × Year × Season	13	18.81	<.0001
Site × Season × Fertilizer	5	3.47	0.0041
Year × Season × Fertilizer	75	3.13	<.0001
Site × Year × Fertilizer	85	3.41	<.0001
Site × Year × Season × Fertilizer	65	2.57	<.0001

Discussion

In comparison with the uncultivated soil, SOC was maintained at the same level at Ndiaye or even increased at Fanaye with rice cropping. Working on the same two LTFE, Haefele et al. (2004) also reported a higher SOC in cultivated soils at one site (Fanaye), while a slight but nonsignificant decreasing trend of SOC was observed at the other site (Ndiaye). Probably because of the limited amount of data (three sample dates) and special variability, the authors recognized that they could not conclude that SOC did not change. Unfortunately, soil samples were not taken every year in the LTFE and there are insufficient data to assess the trends of SOC over years. Our data also confirm higher SOC in the cultivated soil at Fanaye and treatments with NPK fertilizer applications always had higher SOC. Consequently, we can at least conclude that SOC did not decrease with rice cropping and this is true at the two sites but was rather increased at one site (Fanaye).

The recycling of crop residues and roots in the flooded conditions of irrigated rice systems can explain the status of SOC. The decomposition of plant residues is typically slower in submerged than in aerated soil (Powlson and Olk 2000; Regmi et al. 2002; Sahrawat 2004; Zhang and He 2004; Mirasol et al. 2008). Crop residues are continuously recycled in the soil twice a year as a source of carbon, which are incorporated into young soil organic matter fraction

(Cassman et al. 1995; Olk et al. 1996, 2000; Bronson et al. 1997; Witt et al. 2000). A significant increase of SOC was observed at Fanaye, probably because of the initial low level of SOC. After 17 years of cultivation, there was no significant difference in SOC between the cultivated soils at the two sites. The SOC had increased in the low C content soil at Fanaye to the same level as at Ndiaye. The buildup of SOC by periodical anaerobic conditions seamed to raise SOC to an equilibrium of 11–13 mg C kg^{-1}. More significant increases of SOC would probably not occur during the forthcoming years.

Low yields and significant decreases in yields over the years in the control treatment and N fertilizer without PK plots may be explained by the initial poor soil fertility status. The low levels of extractable P in the original soil confirmed the need for P-fertilizer applications to overtake the critical limit of soil P for rice (Bado et al. 2008). The high yields obtained with N-fertilizer alone without PK treatment during the first 3 years, followed by the quick yield decline, showed the limiting effect of these nutrients with time of cultivation.

In general, most of the irrigated lowland valleys in the Sahel and Sudan savannah have considerable soil K reserves (Buri et al. 1999; Wopereis et al. 1999; Haefele et al. 2004). Probably because of the relatively high soil K status, the exchangeable K of the LTFE did not show any significant depletion with the recommended doses of NPK fertilizer as already

 Springer

Table 3 Rice grain yields in the first year of cultivation, annual variations in yield and correlation coefficient of the linear equations between years of cultivation for each season (hot dry and wet season) and rice grain yields during 18 years (1991–2008) in the delta (Ndiaye) and middle (Fanaye) valley of the Senegal River

Sites	Seasons	NPK fertilizer (kg ha^{-1} nutrients)	Intercept Mg ha^{-1} (b)	Yield trends Kg ha^{-1} (a)	Yield variations % per year	Coefficient of determination (R^2)	Number of values (n)
Ndiaye	Hot dry season	Control (0-0-0)	3.3	−49	−1.5	0.15	13
		120-0-0	6.1	−147 (−258; −37)[a]	−2.8	0.37*	16
		60-26-50	5.1	92	1.8	0.15	17
		120-26-50	6.9	19	0.3	0.01	17
		120-52-100	6.6	32	0.5	0.01	17
		180-26-50	7.3	13	0.2	0.03	17
	Wet season	Control (0-0-0)	3.9	−115 (−173; −59)	−2.9	0.58**	16
		120-0-0	6.8	−277 (−362; −188)	−4.1	0.77***	16
		60-26-50	6.1	−59	−0.1	0.14	16
		120-26-50	7.3	−78	−1.1	0.20	16
		120-52-100	7.3	−100	−1.4	0.20	16
		180-26-50	7.1	−54	−0.8	0.08	16
Fanaye	Hot dry season	Control (0-0-0)	2.2	−47	−2.1	0.14	16
		120-0-0	4.7	−46	−0.9	0.06	16
		60-26-50	4.9	−44	−0.9	0.05	16
		120-26-50	6.1	24	0.4	0.01	15
		120-52-100	6.0	60	1.0	0.06	16
		180-26-50	6.8	70	1.0	0.08	16
	Wet season	Control (0-0-0)	4.0	−60 (−124; 3)	−1.5	0.22*	17
		120-0-0	7.4	−268 (−372; −166)	−3.6	0.67***	17
		60-26-50	5.9	−23	−0.4	0.02	17
		120-26-50	7.3	−68	−0.9	0.10	17
		120-52-100	7.3	−42	−0.6	0.07	17
		180-26-50	7.9	−62	−0.8	0.06	17

Each n value represents the mean yield of 4 values of the 4 replications

[a] Confidence interval of yield variations at 95% probability level

*, **, *** Significant at 0.05, 0.01 or 0.001 probability levels according the Fisher test

noted by Haefele et al. (2004). However, the highest doses of N fertilizers (180 kg N ha^{-1} or more) can probably induce K depletion with long-term cropping.

As important factor of soil fertility, the maintenance or buildup of SOC in flooded conditions, coupled with the improvement of soil chemical properties (Sahrawat 2004) and better use of nutrients both from soil and fertilizers, can explain the yield trends during the 18 years of cropping. Except for the control treatment (without any fertilizer) and the control without PK fertilizer, rice grain yields were maintained during the 18 years with the seasonal applications of chemical NPK fertilizers. Using the yields of the best NPK treatment, Haefele et al. (2002) observed a slight and nonsignificant yield decline (−27 kg ha^{-1} per season) at Ndiaye and significant yield increase at Fanaye (+86 kg ha^{-1} per season). Otherwise, no yield declines were observed when the three NPK nutrients were applied. The authors recognized that because of the short duration of the experiments (10 years) and extreme conditions of the Sahelian climate, the observed yield trends could not give an accurate indication of the

Fig. 1 Influence of fertilizer application on rice grain yields during the hot dry season (HDS) and wet season (WS) at Ndiaye in the delta valley of the Senegal River over 18 years (1991–2008)

Fig. 2 Influence of fertilizer application on rice grain yields during the hot dry season (HDS) and wet season (WS) at Fanaye in the middle valley of the Senegal River over 18 years (1991–2008)

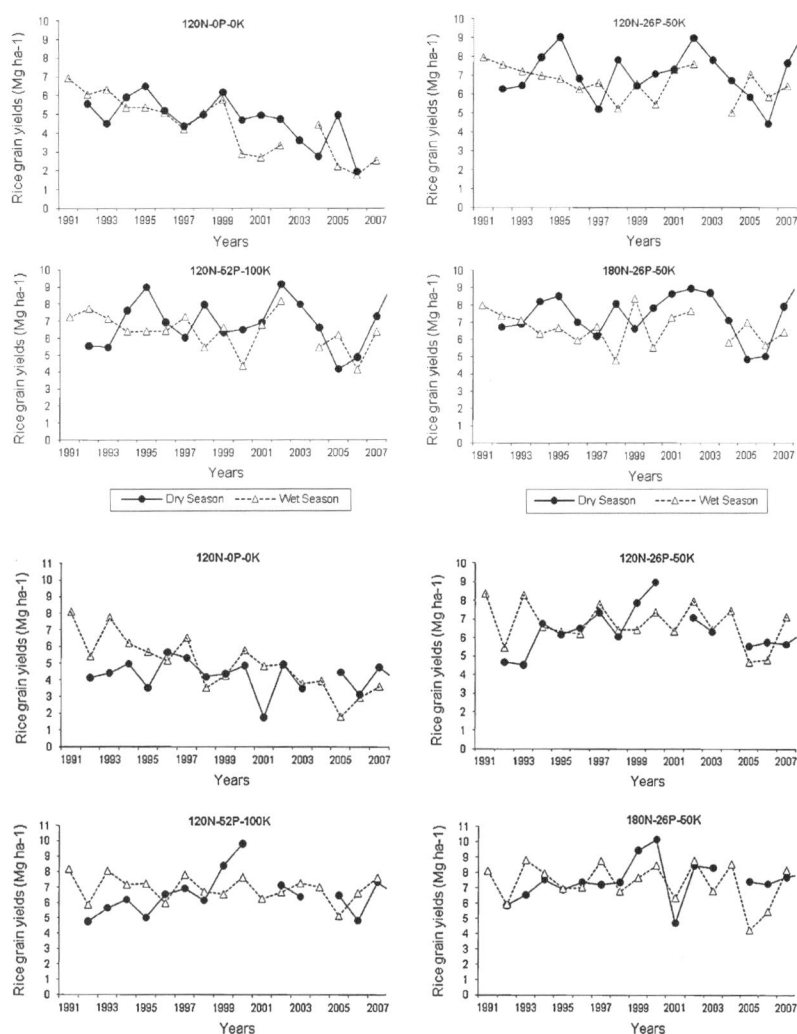

biophysical sustainability of the cropping system. They probably did not observe the interactions within the factors site, season and fertilizer as we did with 36 cropping seasons (Table 2). The climatic variability and site-specific crop management at the two sites may explain the influence of sites and season on grain yield (Dingkuhn and Sow 1997). Compared to the continental zone (Fanaye), the hot dry season was found to be most productive in the coastal zone of the delta (Ndiaye) because of the moderate air temperatures in the proximity of the sea. Cold-induced spikelet sterility that frequently affects rice yield

(Dingkuhn and Sow 1997), climatic variations and harvest operation may explain the diverse effects of site and season on biomass and grain yields.

The data confirmed that rice yields were maintained over 18 years (36 seasons) of continuous cropping, probably because of the improvement of soil chemical fertility with flooding (Sahrawat 1998) and the maintenance of SOC and NPK nutrient status with the seasonal applications of NPK fertilizers. This is confirmed for all the three treatments of NPK fertilizer and not only for the best treatment per season as observed by Haefele et al. (2002).

Conclusions

Soil organic carbon was always maintained or increased irrespective of fertilizer application and rice yields declined only when rice was cultivated without NPK fertilizer or when only N-fertilizer was used. However, these data suggest research axes to develop alternative and better management options of chemical fertilizers to improve irrigated rice productivity and profitability. While N should be applied each season, it is probably not necessary to apply P and K each season. For example, seasonal applications of N (each cropping season) and annual applications (one season per year) of P and K can be an option to improve fertilizer management, rice production and profitability.

References

Bado BV, Sedogo MP, Cescas MP, Lompo F (1997) Effet à long terme des fumures sur le sol et les rendements du mais au Burkina Faso. Cah Agric 6:571–575

Bado BV, DeVries ME, Haefele SM, Marco MCS, Ndiaye MK (2008) Critical limit of extractable phosphorous in a Gleysol for rice production in the Senegal River valley of West Africa. Comm Soi Sci Plant Anal 39:202–206

Bationo A (2008) Organic amendments for sustainable agricultural production in Sudano-Sahelian West Africa. In: Bationo A (ed) Integrated soil fertility management options for agricultural intensification in Sudano-Sahelian zone of West Africa. Academy Science Publishers, Nairobi

Bationo A, Mokwunye AU (1991) Role of manure and crop residues in alleviating soil fertility constraints to crop production: with special reference to the Sahelian and Sudanian zone of West Africa. Fert Res 29:117–125

Bray RM, Kurtz LT (1945) Determination of total, organic and available forms of phosphorous in soils. Soil Sci 59:39–45

Bronson KF, Cassman KG, Wassmann R, Olk DC, van Noordwijk M, Garrity DP (1997) Soil carbon dynamics in different cropping systems in principal ecoregions of Asia. In: Lal R et al (eds) Management of carbon sequestration in soil. CRC Press, Boca Raton

Buri MM, Ishida F, Kubota D, Masunaga T, Wakatsuki T (1999) Soils of flood plains of West Africa: general fertility status. Soil Sci Plant Nutr 45:37–50

Cassman KG, De Datta SK, Olk DC, Alcantara J, Samson M, Descalsota J, Dizon M (1995) Yield decline and the nitrogen economy of long-term experiments on continuous, irrigated rice systems in the tropics. In: Lal R, Stewart BA (eds) Soil management: experimental basis for sustainability and environmental quality. CRC Press, Boca Raton

Dingkuhn M, Sow A (1997) Potential yields of irrigated rice in the Sahel. In: Miezan KM, Wopereis MCS, Dingkuhn M, Deckers J, Randolph TF (eds) Irrigated rice in the Sahel: prospects for sustainable development. West Africa Rice Development Association, Bouake

FAO (1998) World references base for soil resources. World soil resource 84. FAO, Rome

Flinn JC, De Datta SK (1984) Trends in irrigated-rice yields under intensive cropping at Philippine research stations. Field Crops Res 9:1–15

Gomez KA, Gamez AA (1983) Statistical procedure for agricultural research, 2nd edn. IRRI, Los Banos

Haefele SM, Wopereis MCS, Wiechmann H (2002) Long-term fertility experiments for irrigated rice in the West African Sahel: agronomic results. Field Crop Res 78:119–131

Haefele SM, Wopereis MCS, Schloebohm AM, Wiechmann H (2004) Long-term fertility experiments for irrigated rice in the West African Sahel: effects on soil characteristics. Field Crop Res 83:61–77

Helmke PA, Sparks DL (1996) Lithium, sodium, potassium, rubidium and cesium. In: Bigham JM (ed) Method of soil analysis. Part 3. Chemical methods. Soil Science Society of America/American Society of Agronomy, Madison, WI, pp 551–574

Kebbeh M, Miezan K (2003) Ex-ante evaluation on integrated crop management options for irrigated rice production in the Senegal River Valley. Field Crop Res 81:87–94

Kenneth G, Cassman Shaobing, Peng DobermannA (1997) Nutritional physiology of the rice plant and productivity decline of irrigated rice systems in the tropics. In: Ando T et al (eds) Plant nutrition for sustainable food production and environment. Kluwer Academic Publishers, Japan

Miézan KM, Diack S (1994) Senegal releases three new cultivars selected by WARDA. West Africa rice development association, Annual report 1994, Bouaké

Mirasol FP, Eufrocino VL, Gines HC, Buresh JR (2008) Soil carbon and nitrogen changes in long-term continuous lowland rice cropping. Soil Sci Soc Am J 72:798–807

Olk DC, Cassman KG, Randall EW, Kinchesh P, Sanger LJ, Anderson JM (1996) Changes in chemical properties of organic matter with intensified rice cropping in tropical lowland soil. Eur J Soil Sci 47:293–303

Olk DC, Brunetti G, Senesi N (2000) Decrease in humification of organic matter with intensified lowland rice cropping: a wet chemical and spectroscopic investigation. Soil Sci Soc Am J 64:1337–1347

Powlson DS, Olk DC (2000) Long-term soil organic matter dynamics. In: Kirk GJD, Olk DC (eds) Carbon and nitrogen dynamics in flooded soils. Paper presented at the Workshop on carbon and nitrogen dynamics in flooded soils. IRRI, Los Baños, 19–22 April 1999

Regmi AP, Ladha JK, Pathak H, Pasuquin E, Bueno C, Dawe D, Hobbs PR, Joshy D, Maskey SL, Pandey SP (2002) Yield and soil fertility trends in a 20-year rice-rice-wheat experiment in Nepal. Soil Sci Soc Am J 66:857–867

Sahrawat KL (1998) Flooding soil: a great equalizer of diversity in soil chemical fertility. Oryza 35:300–305

Sahrawat KL (2004) Organic matter accumulation in submerged soils. Adv Agron 81:169–201

SAS Institute (1995) SAS system for windows. Release 6.11. SAS Inst, Cary, NC

Walkley A, Black JA (1934) An examination of the Detjareff method for determining soil organic matter and a

Nutr Cycl Agroecosyst (2010) 88:133–141

proposed modification of the chromatic acid titration method. Soil Sci 37:29–38

Witt C, Cassman KG, Olk DC, Biker U, Liboon SP, Samson MI, Ottow JCG (2000) Crop rotation and residue management effects on carbon sequestration, nitrogen cycling and productivity of irrigated rice systems. Plant Soil 225:263–278

Wopereis MCS, Donovan C, Nebié B, Guindo D, Ndiaye MK (1999) Soil fertility management in irrigated rice systems in the sahel Savanna region of West Africa. Part 1. Agronomic analysis. Field Crops Res 61:125–145

Zhang M, He Z (2004) Long term changes in organic carbon and nutrients of an Ultisol under rice cropping in southeast China. Geoderma 118:167–179

 Springer

Nutr Cycl Agroecosyst
DOI 10.1007/s10705-011-9423-7

ORIGINAL ARTICLE

Conservation tillage, local organic resources and nitrogen fertilizer combinations affect maize productivity, soil structure and nutrient balances in semi-arid Kenya

J. Kihara · A. Bationo · D. N. Mugendi ·
C. Martius · P. L. G. Vlek

Received: 25 April 2008 / Accepted: 13 January 2011
© Springer Science+Business Media B.V. 2011

Abstract Smallholder land productivity in drylands can be increased by optimizing locally available resources, through nutrient enhancement and water conservation. In this study, we investigated the effect of tillage system, organic resource and chemical nitrogen fertilizer application on maize productivity in a sandy soil in eastern Kenya over four seasons. The objectives were to (1) determine effects of different tillage-organic resource combinations on soil structure and crop yield, (2) determine optimum organic–inorganic nutrient combinations for arid and semi-arid environments in Kenya and, (3) assess partial nutrient budgets of different soil, water and nutrient management practices using nutrient inflows and outflows. This experiment, initiated in the short rainy season of 2005, was a split plot design with 7 treatments involving combinations of tillage (tied-ridges, conventional tillage and no-till) and organic resource (1 t ha^{-1} manure + 1 t ha^{-1} crop residue and; 2 t ha^{-1} of manure (no crop residue) in the main plots. Chemical nitrogen fertilizer at 0 and 60 kg N ha^{-1} was used in sub-plots. Although average yield in no-till was by 30–65% lower than in conventional and tied-ridges during the initial two seasons, it achieved 7–40% higher yields than these tillage systems by season four. Combined application of 1 t ha^{-1} of crop residue and 1 t ha^{-1} of manure increased maize yield over sole application of manure at 2 t ha^{-1} by between 17 and 51% depending on the tillage system, for treatments without inorganic N fertilizer. Cumulative nutrients in harvested maize in the four seasons ranged from 77 to 196 kg N ha^{-1}, 12 to 27 kg P ha^{-1} and 102 to 191 kg K ha^{-1}, representing 23 and 62% of applied N in treatments with and without mineral fertilizer N respectively, 10% of applied P and 35% of applied K. Chemical nitrogen fertilizer application increased maize yields by 17–94%; the increases were significant in the first 3 seasons ($P < 0.05$). Tillage had significant effect on soil macro- (>2 mm) and micro-aggregates fractions (<250 µm >53 µm: $P < 0.05$), with aggregation indices following the order no-till > tied-ridges >

J. Kihara (✉) · A. Bationo
Tropical Soil Biology and Fertility Institute of CIAT,
30677-00100 Nairobi, Kenya
e-mail: jkiharam@yahoo.com

J. Kihara · C. Martius · P. L. G. Vlek
Zentrum für Entwicklungsforschung (ZEF), University
of Bonn, Walter-Flex-Strasse 3, 53113 Bonn, Germany

A. Bationo
Alliance for Green Revolution in Africa
(AGRA-Alliance), CSIR Office Complex, #6 Agostino
Neto Road, Airport Residential Area, PMP KIA 114
Airport-Accra, Ghana

D. N. Mugendi
Kenyatta University, 43844-00100 Nairobi, Kenya

Present Address:
C. Martius
Inter-American Institute for Global Change Research
(IAI), Avenida dos Astronautas 1758, São José dos
Campos, SP 12227-010, Brazil

Published online: 29 January 2011

 Springer

This article has been previously published in the journal "Nutrient Cycling in Agroecosystems" Volume 88 Issue 1.
A. Bationo et al. (eds.), Innovations as Key to the Green Revolution in Africa: Exploring the Scientific Facts. © 2010 Springer.

conventional tillage. Also, combining crop residue and manure increased large macro-aggregates by 1.4–4.0 g 100 g^{-1} soil above manure only treatments. We conclude that even with modest organic resource application, and depending on the number of seasons of use, conservation tillage systems such as tied-ridges and no-till can be effective in improving crop yield, nutrient uptake and soil structure and that farmers are better off applying 1 t ha^{-1} each of crop residue and manure rather than sole manure.

Keywords Crop residue · Manure · Soil aggregation · Nutrient balance · Tillage

Introduction

Farming in sub-Saharan Africa (SSA) traditionally depended upon natural fallows to restore soil fertility (Maroko et al. 1998). Due to factors such as increased population pressure and demand for food, natural fallows are not feasible hence the need for sustainable land management systems (Pascual and Barbier 2006). Maintenance and or improvement of soil organic matter (SOM) is one of the key components of sustainable management of agricultural lands due to the associated system productivity (Bationo and Vlek 1997; Martius et al. 2001; Bationo et al. 2007). One promising approach to manage SOM and promote activities of soil micro-organisms is the concomitant use of organic (such as crop residue and manure) and inorganic nutrient resources (Goyal et al. 1999). The extent of organic amendments on SOM, soil structure and soil micro-organisms vary according to the crop residue type and quantity (Scopel et al. 1998).

In many cropping systems within poor smallholder farms, little or no agricultural residue is returned to the soil due to low productivity in these areas and competing uses such as fodder and fuelwood (Erenstein 2003). Also, and especially in Kenyan drylands, manure is limited due to free range system of livestock management practiced here that leaves large amounts of manure in the bushlands, and only small amounts dropped in the shed at night are available for use in crop production. The challenge is optimizing the limited resources available to the

small-scale farmers for improved productivity of the systems. Quite often though, the needed or recommended levels of organic amendments used by scientists in experimental set-ups are too high for any practical use by majority of farmers. It is needed to investigate the effect of the small quantities of organic resources available to the farmers on both crop productivity and soil structure, and how these effects can be improved by a judicious choice of amendments of varying quality. In this study, we assess soil aggregation and crop yield under different tillage and organic resource systems. Village-level residue management study in West African agro-ecosystems showed that 21–39% of residue from previous crops is available on the farm at planting (Bationo et al. 2007). Thus, given the low productivity observed in the dryland agro-ecosystems (3–4 t ha^{-1} of maize stover Miriti et al. 2007), farmers can retain about 1 t ha^{-1} of maize stover residue in their fields and this quantity is tested in this study.

Conservation tillage practices such as reduced tillage have been observed to result in better soil structure (Six et al. 2000) and have higher soil organic carbon compared to conventional tillage practice, especially at the top soil depths (Madari et al. 2005; McCarty and Meisinger 1997). Combined with crop residue applied as mulch, reduced tillage also conserves available rainwater important for crop growth. Such rainwater is currently lost in the magnitude of 70–85% from cropping systems in sub-Sahara Africa through soil evaporation, deep percolation and surface runoff (Rockstrom et al. 2003). The additional mulch produced by the crops could play a key role in reducing runoff and direct evaporation from the soil (Erenstein 2003), and often reduces the emergence of weeds (Erenstein 2003). In low rainfall seasons, substantial increases in crop yields above conventional practices are expected where tied-ridges are used for in situ water harvesting (Gebrekidan 2003).

We hypothesized that combining conservation tillage with the limited quantities of organic and inorganic nutrient resources available and affordable by smallholder farmers would result in increased maize productivity and higher efficiency of resources through improvements in soil structure. The objectives were to (1) determine effects of different tillage systems and low quantities of organic resources on soil structure and crop yield, (2) determine optimum organic–inorganic nutrient combinations for arid and semi-arid environments in Kenya and, (3) assess partial nutrient budgets of different soil, water and nutrient management practices using nutrient inflows and outflows.

Materials and methods

Location

This study was conducted in a semi-arid zone at Machang'a in Mbeere, Eastern Kenya. The site is located at 0°45′S and 37°45′E, at an altitude of 1,050 m.a.s.l. The study was conducted for 2 years (from September 2005 to August 2007) which constituted four seasons of experimentation (two seasons per year). The two seasons in a year are the short rainy season from September to March and the long rainy season from April to July. Much of the seasonal rainfall often falls in a few storms distributed over the season. Figure 1 shows cumulative rainfall recorded in the study site during the four seasons of experimentation. Short rains 2005 (SR2005, season one) had lower rainfall (245 mm) compared to long rains 2006 (LR2006, season two) with 285 mm, although there was better distribution in the former. During short rains 2006 (SR2006, season 3) up to 825 mm rainfall was recorded, while 374 mm of rainfall, very poorly distributed, was recorded during long rains 2007 (LR2007, season 4). Average temperature was 23.4 ($\pm$1.53) as observed during the four seasons.

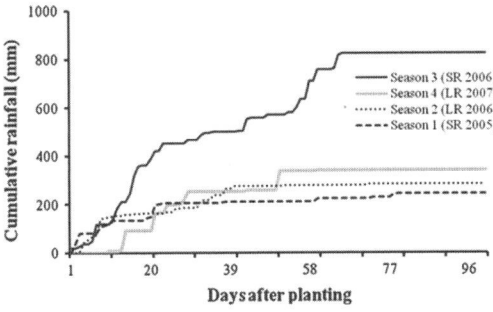

Fig. 1 Seasonal cumulative rainfall recorded at Machang'a in Mbeere, eastern Kenya during the four seasons of study (September 2005 to July 2007). SR = short rainy season, LR = long rainy season

Soils characteristics

The soils, classified as chromic cambisols (Warren et al. 1997), had 65% sand, 22% clay and 13% silt. Chemical analysis was done at the main plot level because the field was observed to exhibit some variability. The Analysis showed that SOC content was 0.61 ($\pm$0.07)%, total N 0.05 ($\pm$0.01)%, pH water 6.5 ($\pm$0.46), extractable phosphorus 6.3 ($\pm$7.74) mg P kg^{-1}, exchangeable potassium 0.56 ($\pm$0.14) me 100 g^{-1} of soil, exchangeable magnesium 0.78 ($\pm$0.077) me 100 g^{-1} of soil, and exchangeable calcium 2.29 ($\pm$1.04) me 100 g^{-1} of soil. The site was characterized by C/N ratio varying between 9.0 and 22.1. The data showed high nutrient variability (Fig. 2) that is common in farmers' fields due to local land use histories or biological activity (termite mounds that concentrate nutrients).

Treatments and crop management

The experiment, initiated in the short rainy season of 2005, was a split plot design with seven (7) tillage-organic resource treatments in main plots as follows:

(1) Conventional tillage without organic residue (control)
(2) Conventional tillage + crop residue (1 t ha^{-1}) + manure (1 t ha^{-1})
(3) Conventional tillage + manure (2 t ha^{-1})
(4) No-Tillage + crop residue (1 t ha^{-1}) + manure (1 t ha^{-1})
(5) No-Tillage + manure (2 t ha^{-1})
(6) Tied-Ridges + crop residue (1 t ha^{-1}) + manure (1 t ha^{-1})
(7) Tied-Ridges + manure (2 t ha^{-1})

Split plots comprised 0 and 60 kg N ha^{-1} as chemical nitrogen fertilizer. The experiment was replicated three times. Tillage in the conventional system was done to about 15 cm soil depth using hand hoes as commonly done by farmers. Tied-ridges were prepared by digging, using hand hoes, during trial initiation and they were maintained throughout the experiment, with tillage restricted to refreshment of the ridges. Ridge tops were not hoed but retained some fresh soils as a result of the ridge maintenance work. The main ridges on which maize was planted were 90 cm apart while the ties were made at distances of 2 m. In the no-tillage system, tillage

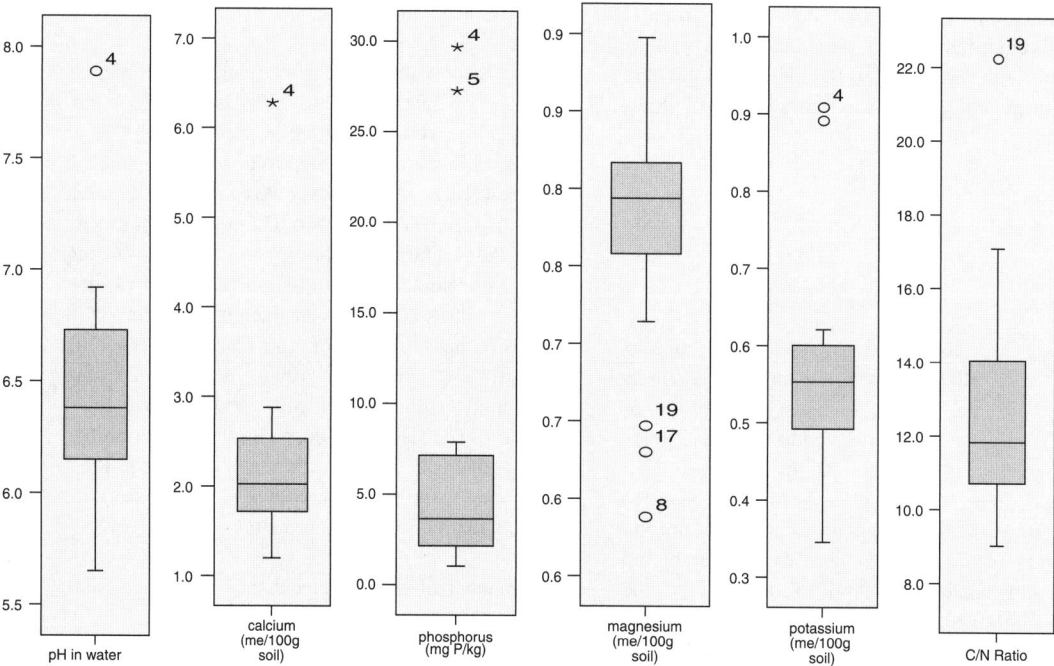

Fig. 2 *Box plots* describing initial characteristics for selected soil parameters ($n = 21$) at Machang'a in Mbeere, Eastern Kenya in 2005

for land preparation was done using hand hoes with no subsequent use of hand hoes in between the season. Weeding in no-till system was done by hand pulling.

Organic resources applied were goat manure (on average 2.67, 0.52 and 4.17% of N, P and K, respectively) broadcast at planting at 1 and 2 t ha^{-1} and maize stover (0.65, 0.08 and 1.78% of N, P and K, respectively) applied at 1 t ha^{-1}. The manure was imported from one farmer at all times while maize stover was from the harvested crop, except in the first season when it was sourced from an immediate farm. The stover was spread over the plots prior to land preparation. Each of the treatments was split into two subplots that received 0 and 60 kg N ha^{-1}, respectively. The chemical nitrogen fertilizer was split applied with $^1/_3$ at planting and $^2/_3$ at 4 weeks after planting.

Test crop was maize (*Zea mays*) Katumani composite variety that takes 72 days to mature. The seeds were planted at 90 and 60 cm inter and intra-row spacing, respectively and thinned to two plants per

hole after germination. Harvesting was done leaving one border row and end plants making a harvest net plot area of 4.5 m by 1.8 m from the initial subplot plot of 6 m by 3 m.

Plant nutrient uptake

Nutrient inputs from fertilizer were calculated based on application amounts and fertilizer formulation. Inputs from organic resources were derived from nutrient contents of the manure and maize stover applied in each season. At the end of season 4, maize grain and stover at subplot level were analyzed for %N, P and K for calculations of seasonal plant nutrient uptake. Sampling for the grain and stover was done at harvest by selecting five (5) plants from each plot. The stover was chopped into small pieces and both grain and the stover air-dried to constant weight. A subsample was then ground, separately for grain and stover, before analysis. The results were used to calculate partial nutrient balances for N, P and K as:

Partial nutrient balance = input (fertilizers + organics) − uptake (removal in grain and stover)

The estimates excluded inputs in wet and dry deposition, sedimentation, and nutrient accessions by deep roots from subsoil layers, and outputs by leaching, erosion, run-off, and gaseous losses.

Separating aggregate fractions

Soil sampling for aggregation assessment was done at the end of the 4th season in all −N plots at 0–15 cm depth at five spots (along the two diagonals). A sampling core with 5 cm diameter was used and the soil was mixed before taking a total sample of 1 kg. Within tied-ridges, soil sampling was at sloping part of the ridge (mid-way between furrow and ridge-top). The samples were air dried for 2 days to constant weight before passing through an 8 mm sieve from where 80 g was weighed for wet sieving, and an additional 10 g for moisture content determination. The 80 g soil was submerged in water over a 2 mm sieve for 5 min to allow slaking, followed by sieving for 2 min. Soil that passed through the 2 mm sieve was sieved again for 2 min using a 250 μm sieve to obtain small macro-aggregates. The aggregates not captured by the 250 μm sieve were then sieved for 2 min using a 53 μm sieve to obtain micro aggregates. The filtrate obtained after sieving with a 53 μm sieve was shaken using a dispenser and 250 ml sampled into a beaker. The four different samples were dried at 60°C for 48 h before weighing. Sand correction was not done. Mean weight diameter (MWD) of the aggregate fractions was calculated as MWD=$\Sigma X_i W_i$, where X_i is the diameter of the ith sieve size and W_i is the proportion of the total aggregates in the ith fraction. Higher MWD indicate higher proportions of macro-aggregates. Geometric mean diameter (GMD) was calculated as, GMD = $\exp\left(\frac{\Sigma W_i \ln X_i}{\Sigma W_i}\right)$, where W$i$ is the weight of aggregates in size class i and Xi is the mean diameter of that size class. GMD estimates the size of the most frequent aggregate size class (Filho et al. 2002).

Chemical soil and plant analysis

Plant and soil analysis were conducted at ICRAF laboratories in Nairobi, Kenya, and all analysis were done according to the ICRAF laboratory procedures as detailed by (ICRAF 1995). Analysis for plant N, P and K was by wet oxidation based on Kjeldahl digestion with sulphuric acid (Parkinson and Allen 1975). Potassium was determined through flame photometry, and N and P were determined by colorimetric determination. Soil total N was also based on wet oxidation using Kjeldahl digestion while soil total carbon was extracted using acidified dichromate. Analysis for extractable inorganic phosphorus and exchangeable potassium were based on a modified Olsen extractant. Soil pH was determined in water while exchangeable Ca and Mg were extracted using 1 N KCl extractant and determined using a spectrophotometer.

Data analysis

Maize germination and yield data (grain and stover), were analyzed in Genstat Discovery edition 3, using split plot procedure with initial soil parameters values (available Phosphorus, exchangeable calcium, magnesium and potassium and C/N ratio) as covariates to take care of the variability observed in the experimental field. Soil aggregation data was obtained at main plot level and was analyzed using generalized linear model (GLM procedure) in Statistical Analysis System (SAS) version 9.1 and treatment least square means obtained by the LSmeans statement. Pearsons correlation coefficients were used to establish relationship between different soil parameters using SAS software but first, the outliers and extreme values, identified by box plots in SPSS version 14.0 for windows, were removed.

Results

Soils across the experimental field were heterogeneous. Excluding outliers and extreme values shown in Fig. 2, Pearson's correlations showed pH to be largely correlated to exchangeable calcium ($R^2 = 0.8$).

Effect of treatment on germination

Germination was significantly affected by tillage-organic resource system ($P < 0.01$) during the last two seasons with treatments under tied-ridge systems having lower germination than no-till and conventional tillage systems (Table 1). Germination was highest in the 2nd season but declined in the subsequent seasons. The decline was highest in tied-ridges where 88, 62 and 29% germination was observed in the 2nd, 3rd and 4th seasons respectively compared to 91, 81 and 77% in the other systems, during the same period. Chemical nitrogen fertilizer application did not affect germination in any of the seasons.

Effect of treatment on maize yield

Tillage-organic resource system had no significant effect on maize grain yield in any of the season. However, important trends were observed when comparing the different tillage and organic residue treatments. For example, treatments under no-till system had the lowest yields during the first two seasons but improved to achieve the highest yield by the fourth season (Table 2). During the fourth season, and without N application, no-till + manure achieved significantly higher maize grain yields than control ($P < 0.05$). These data show continual progression of crop yield improvement under no-till system over the 4 seasons. Regardless of tillage-organic resource system, the highest seasonal yield was observed in season 3 (SR2006).

Combining CR and manure resulted in additive maize yields over manure only treatments in all the seasons of study. Over the four seasons for example, combining 1 t ha^{-1} of crop residue and 1 t ha^{-1} of manure, in plots not applied with chemical N, increased average maize grain yield above manure only treatments by 52, 17 and 27% in conventional, no-till and tied-ridge treatments, respectively (Table 2). During the same period, average maize stover yield in CR + manure treatments was more than manure only treatments by 22–36%, for treatments not applied with chemical nitrogen fertilizer. We observed that under tied-ridges, conventional and no-till systems, there were $^6/_8$, $^6/_8$ and $^4/_8$ cases respectively in which organic resource combination (1 t ha^{-1} each of crop residue and manure) had additive maize grain yields over manure only (applied at 2 t ha^{-1}).

Chemical nitrogen fertilizer application had a positive and significant effect on maize yield in the first three seasons while its interaction with tillage-organic resource system was significant during the fourth season. On average, application of chemical nitrogen fertilizer increased yield by 39, 59, 91 and 13% over plots not applied with the nitrogen fertilizer in season 1, 2, 3 and 4 respectively. Averaged over the four seasons, maize grain increases due to N application in CR + manure and manure only treatments were 23 and 34% in conventional tillage, 18 and 59% in no-till and, 17 and 35% in tied-ridges, respectively. Greater response to fertilizer N in manure only treatments over manure + CR was also observed with maize stover.

Table 1 Effects of tillage and organic resource on percent germination of maize at Machang'a in Mbeere, Kenya during the September 2005 to July 2007 period

Tillage system	SR2005		LR2006		SR2006		LR2007	
	−N	+N	−N	+N	−N	+N	−N	+N
Conventional (control)	83.3^b	85.8^{bc}	93.0^{bc}	89.0^{ab}	81.7^b	81.7^b	66.7^b	78.3^{bc}
Conventional + CR + manure	83.3^b	75.3^a	91.7^{abc}	93.0^b	83.3^b	83.3^b	80.0^c	71.7^b
Conventional + manure	91.7^c	81.7^{ab}	95.0^c	91.3^{ab}	78.3^b	85.0^b	78.3^c	71.7^b
No-till + CR + manure	80.0^{ab}	86.7^{bc}	89.7^{ab}	92.3^{ab}	73.0^b	85.3^b	76.7^c	76.7^{bc}
No-till + manure	80.3^{ab}	91.7^c	87.0^a	89.7^{ab}	80.0^b	80.0^b	81.7^c	80.0^c
Tied-ridges + CR + manure	75.0^a	85.0^{bc}	89.0^{ab}	88.3^{ab}	73.0^b	56.7^a	33.3^a	28.3^a
Tied-ridges + manure	83.3^{ab}	88.3^c	88.3^{ab}	87.3^a	57.7^a	60.0^a	28.3^a	26.7^a
SED	3.2	(4.1)	2.3	(2.8)	6.3	(7.7)	3.4	(6.3)

CR = crop residue, SR2005 = short rainy season of 2005, LR2006 = long rainy season of 2006, SR2006 = short rainy season of 2006, LR2007 = long rainy season of 2007, Values in the same column followed by the same letter are not significantly different ($P = 0.05$), SED = standard error of the differences of means. The values in bracket are SED to compare across different tillage treatments and nitrogen levels

Table 2 Effect of tillage, manure, crop residue and chemical nitrogen fertilizer on maize grain and stover yield at Machang'a in Mbeere, eastern Kenya during the September 2005 to July 2007 period

Tillage system	SR2005		LR2006		SR2006		LR2007	
	−N	+N	−N	+N	−N	+N	−N	+N
Maize grain yield (t ha⁻¹)								
Conventional (control)	1.14[ab]	1.62[ab]	1.03[ab]	1.52[ab]	1.22[ab]	3.10[a]	1.00[a]	2.29[a]
Conventional + CR + manure	2.05[b]	2.62[b]	1.07[ab]	1.57[ab]	3.91[b]	4.89[a]	2.83[ab]	3.09[a]
Conventional + manure	1.30[ab]	1.87[ab]	1.36[ab]	1.68[ab]	2.12[ab]	3.40[a]	1.73[ab]	1.74[a]
No-till + CR + manure	0.55[a]	0.75[a]	0.61[ab]	0.92[ab]	1.73[ab]	3.10[a]	2.77[ab]	1.91[a]
No-till + manure	0.76[ab]	1.73[ab]	0.24[a]	0.65[a]	0.76[a]	2.02[a]	3.09[b]	3.29[a]
Tied-ridges + CR + manure	2.13[b]	1.92[ab]	1.53[b]	2.20[b]	2.35[ab]	3.56[a]	1.81[ab]	1.48[a]
Tied-ridges + manure	1.53[ab]	1.64[ab]	1.15[ab]	1.43[ab]	1.84[ab]	3.65[a]	1.65[ab]	1.59[a]
SED	0.60	(0.65)	0.52	(0.52)	1.24	(1.21)	0.80	(0.77)
Maize stover yield (t ha⁻¹)								
Conventional (control)	1.23[a]	2.09[a]	1.59[a]	2.30[a]	1.43[a]	2.68[b]	1.07[a]	1.72[a]
Conventional + CR + manure	2.95[ab]	3.63[a]	1.26[a]	1.69[a]	2.20[a]	2.83[b]	2.43[a]	1.66[a]
Conventional + manure	2.55[ab]	2.81[a]	1.80[a]	2.55[a]	1.20[a]	2.32[ab]	1.64[a]	1.18[a]
No-till + CR + manure	2.11[ab]	2.33[a]	1.34[a]	1.82[a]	1.52[a]	3.12[b]	1.24[a]	1.34[a]
No-till + manure	1.53[a]	2.56[a]	0.68[a]	1.27[a]	0.90[a]	1.23[a]	2.00[a]	2.12[a]
Tied-ridges + CR + manure	4.15[b]	2.33[a]	1.84[a]	2.80[a]	2.06[a]	2.36[ab]	1.66[a]	1.06[a]
Tied-ridges + manure	2.84[ab]	3.51[a]	1.37[a]	1.35[a]	1.54[a]	2.61[b]	1.39[a]	1.13[a]
SED	0.81	(0.90)	0.70	(0.72)	0.58	(0.68)	0.70	(0.68)

CR = crop residue, SR2005 = short rainy season of 2005, LR2006 = long rainy season of 2006, SR2006 = short rainy season of 2006, LR2007 = long rainy season of 2007, Values in the same column followed by the same letter are not significantly different ($P = 0.05$), SED = standard error of the differences of means. The values in bracket are SED to compare across different tillage treatments and nitrogen levels. Minus N (−N) refers only to chemical fertilizer N and not N contained in the organic resources used. As such, the −N treatments contain some N from these organic resources

Partial nutrient balances

Summed up over the four seasons of this study, nutrient inputs of up to 455 kg N ha⁻¹, 200 kg P ha⁻¹ and 560 kg K ha⁻¹ were applied as fertilizer, manure and crop residue (Table 3). Highest total nutrient uptake contained in the above ground plant parts was 196 kg N ha⁻¹, 27 kg P ha⁻¹ and 191 kg K ha⁻¹. Of the total uptake, uptake by stover ranged from 21.6 kg N ha⁻¹ in control to 47.2 kg N ha⁻¹ in tied ridge, 3.0 kg P ha⁻¹ in no-till to 6.6 kg P ha⁻¹ in tied ridge and 87 kg K ha⁻¹ in no-till to 165 kg K ha⁻¹ in conventional tillage (data not shown). This represented 25–35, 20–33 and 86–92% of total N, P and K uptake respectively. Generally, total N uptake, in treatments not applied with chemical N fertilizer for example, was in the order conventional tillage + CR + manure > tied-ridges + CR + manure > tied-ridge + manure > conventional tillage + manure > no-till + CR +

manure > no-till + manure > control (Table 3). As with harvested yield, combination of crop residue and manure resulted in an additional uptake of 79, 16 and 9 kg N ha⁻¹ in conventional tillage, no-till and tied ridge systems, respectively, for plots not applied with chemical N fertilizer. Also, plots applied with inorganic fertilizer N had more K and N taken up than non-N plots. Grain/stover nutrient ratio varied from 1.5 to 4.0 for N, 2.0 to 6.1 for P and 0.15 to 0.25 for K. Highest ratios for N and P were observed in no-till treatment.

Nitrogen partial nutrient balances were positive (+6 to +350 kg N ha⁻¹) except for the control (−57 kg N ha⁻¹) and conventional tillage + CR + manure (−48 kg N ha⁻¹; Table 3). Negative balances indicate nutrient mining. Phosphorus and potassium not used by crops ranged between 141 to 193 kg P ha⁻¹ and 75 to 462 kg K ha⁻¹. Although apparent P and K recoveries (uptake/input) were similar between N and non chemical N fertilizer plots, apparent

Table 3 Effects of tillage and organic resource on total N, P and K uptake by maize and partial balances observed over four seasons (September 2005 to July 2007) at Machang'a in Mbeere, eastern Kenya

Tillage system	Organic resource	Chemical N Fertilizer	Input$^{\pm}$ (kg nutrient ha^{-1})			Uptake (kg nutrient ha^{-1})			Partial balance (kg nutrient ha^{-1})		
			N	P	K	N	P	K	N	P	K
Control	None	−N	0	160	240	67	14	120	−67	146	120
Conventional	CR + manure	−N	133	186	470	183	22	185	−50	164	285
Conventional	Manure	−N	214	206	557	104	18	143	110	188	414
No-till	CR + manure	−N	133	186	470	97	12	122	36	174	348
No-till	Manure	−N	214	206	557	81	15	102	133	191	455
Tied-ridges	CR + manure	−N	133	186	470	127	17	164	6	169	306
Tied-ridges	Manure	−N	214	206	557	118	21	159	96	185	398
Control	None	+N	240	160	240	137	20	168	103	140	72
Conventional	CR + manure	+N	373	186	470	196	27	181	177	159	289
Conventional	Manure	+N	454	206	557	137	18	191	317	188	366
No-till	CR + manure	+N	373	186	470	111	18	182	262	168	288
No-till	Manure	+N	454	206	557	109	18	153	345	188	404
Tied-ridges	CR + manure	+N	373	186	470	160	20	178	213	166	292
Tied-ridges	Manure	+N	454	206	557	144	20	172	310	186	385

CR = crop residue, $^{\pm}$ applied to maize through manure, crop residue and fertilizer over four season

recoveries of N were 0.76 for non-N treatments and 0.38 for treatments applied with N respectively. In all cases, lower uptake and higher nutrient balances were observed in 'manure only' treatments than treatments involving combination of manure and CR.

In most cases, partial N, P and K balances were significantly and negatively correlated with maize yield (Table 4), indicating that increased crop productivity led to lower partial nutrient budgets. Taking partial P balance as an example, treatments combining both CR and manure had more significant correlations with grain and stover compared to treatments with manure as the only organic resource applied. For the different organic resources and N fertilizer application combinations, there was positive influence of initial soil total N and exchangeable P and K on total N, P and K uptake by maize (data not shown).

Soil aggregation

Soil aggregation was assessed at the main plot level, allowing to determine the separate effects of tillage and organic resources. Tillage had significant effect on large macro-aggregate fraction (>2 mm) and micro-aggregates fraction (<250 μm >53 μm), with the highest macro-aggregates and aggregation indices in no-till followed by tied-ridges and least in conventional tillage (Table 5). In treatments combining CR and manure, large macro-aggregates were up to 2 times higher in conservation systems (no-till and tied ridges) compared to conventional tillage system, with resultant reductions in micro-aggregates. Also, combining crop residue and manure increased large macro-aggregates by 1.4–4.0 g 100 g^{-1} soil above manure only treatments and simultaneously decreased micro-aggregates by 6.2 g 100 g^{-1} soil in tied-ridges and 4.0 g 100 g^{-1} soil in no-till below manure only treatments. Aggregate mean weight diameter (MWD) was significantly affected by tillage ($P > 0.05$) being greatest in no-till followed by tied-ridges and least in conventional tillage. As with large macro-aggregates, MWD was higher (not significant) in treatments combining CR and manure compared to those with only manure as the organic resource. Particulate organic matter (POM > 2 mm), small macro-aggregates (<2 mm >53 μm), silt and clay (<53 μm) and geometric mean diameter (GMD) were not affected either by tillage or organic resource.

Table 4 Pearson correlation coefficients between maize yield and partial N, P and K balances after four seasons of cropping (September 2005 to July 2007) at Machang'a in Mbeere, eastern Kenya

Fertilization	Partial N balance		Partial P balance		Partial K balance	
	Grain	Stover	Grain	Stover	Grain	Stover
CR + manure −N	−0.95***	−0.67*	−0.92***	−0.81**	−0.85**	−0.72*
Manure −N	−0.78*	−0.63	−0.75*	−0.49	−0.81**	−0.73*
CR + manure +N	−0.89**	−0.50	−0.82**	−0.69*	−0.56	−0.73*
Manure +N	−0.90***	−0.45	−0.66	−0.54	−0.66	−0.85**

CR = crop residue, * significant at $P = 0.05$, ** significant at $P = 0.01$, *** significant at $P = 0.001$

Table 5 Soil aggregation as affected by tillage and organic resources at 0–15 cm at Machang'a in Mbeere, eastern Kenya in 2007

Tillage system	Organic resource	POM (>2 mm) g 100 g^{-1} soil	Large macro (>2 mm) g 100 g^{-1} soil	Small macro (<2 mm >250 μm) g 100 g^{-1} soil	Micro (<250 >53 μm) g 100 g^{-1} soil	Silt + clay (<53 μm) g 100 g^{-1} soil	MWD	GMD
Conventional	Zero	0.05[a]	2.6[a]	27.4[a]	59.9[bc]	8.8[a]	0.60[a]	0.25[a]
Conventional	CR + manure	0.09[a]	3.7[a]	24.7[a]	62.3[c]	8.0[a]	0.62[a]	0.24[a]
Conventional	Manure	0.05[a]	2.4[a]	27.1[a]	60.7[c]	8.3[a]	0.59[a]	0.24[a]
No-till	CR + manure	0.11[a]	8.3[b]	27.9[a]	53.7[a]	7.9[a]	0.88[b]	0.30[a]
No-till	Manure	0.09[a]	4.3[ab]	29.6[a]	57.7[abc]	7.9[a]	0.71[ab]	0.29[a]
Tied-ridges	CR + manure	0.08[a]	6.4[ab]	29.1[a]	54.0[ab]	8.4[a]	0.80[ab]	0.28[a]
Tied-ridges	Manure	0.07[a]	4.4[ab]	26.9[a]	60.2[c]	7.4[a]	0.68[ab]	0.27[a]
SE		0.032	1.344	1.631	2.024	0.854	0.073	0.026

CR = crop residue, SE = standard error, POM = particulate organic matter, MWD = mean weight diameter of aggregates, GMD = geometric mean diameter of aggregates, Values in the same column followed by the same letter are not significantly different ($P = 0.05$)

Discussion

No-till and reduced tillage systems have been reported to perform better (Mrabet 2002), similar and sometimes poorer than conventional tillage systems in terms of agronomic yields (Blaise and Ravindran 2003; Diaz-Zorita 2000). In our case, agronomic performance in no-till system improved throughout the four seasons. Reduced yield under no-till compared to conventional system, especially during initial seasons, has been reported elsewhere in Africa (Hoogmoed 1999; Rawitz et al. 1986) and attributed to crusting, surface sealing off the pores for water percolation and runoff (Osunbitan et al. 2005; Rosolem et al. 2002). We observed leaf closing (wilting) and drying off plants indicating low water availability in the rooting zone of no-till during the first seasons. The amount of applied crop residue in our case was lower than the amount used in many conservation tillage trials and this could have

contributed to the low yields under no-till during the initial seasons. The absence of residue during the fallow periods (between seasons as in our case) likely decreased agronomic performance in the no-till system, and future no-till trials should maintain residue cover throughout the year. High performance in no-till by the 4th season could be attributed to improvements of soil structure and therefore rooting and soil water relations, or simply response to the rainfall regime. Residue retention is necessary in no-till systems and Govaerts et al. (2009) and Verhulst et al. (2009) have shown that similar improvements in soil quality (increased direct infiltration in the soil) and crop yields can be achieved with partial or with full residue retention. Recently, in order to improve water relations and crop performance in no-till systems, modifications have emerged including ripping and sub-soiling and the results are promising (Busscher et al. 1995; Diaz-Zorita 2000; Motavalli et al. 2003), especially where soil compaction is

Fig. 3 Soil and water sinks formed by Tied ridges in Mbeere, Eastern Kenya. Picture taken on 12th April 2007 after 77 mm of rainfall 15 h earlier. *Rings* around the basins represent silt and SOM retained rather than washing away in runoff

evident (Evans et al. 1996) but also depending on the rainfall regime (Gameda et al. 1994). The next challenge in no-till farming is weed management in systems where chemical control is not viable.

Similar yield between conventional tillage and tied-ridges as observed during the first three seasons, even though tillage under tied-ridge was restricted to ridge maintenance, is attributed to soil moisture improvement since the basins of tied-ridges harvest rainwater and allow it to percolate rather than runoff (Fig. 3). Rainfall was often stormy and in season two for example, 75% of the seasonal rainfall was received only in 7 days. Tied-ridge systems have been observed to increase yield over conventional tillage by up to 85% (Nyakatawa et al. 1996) but the increase depends on soil type, ridge flow (closed or open) and planting position (in furrow or ridge top—as in our case Belay et al. 1998). Negative yields under tied ridges have also been reported and were attributed to high rainfall (>700 mm Jensen et al. 2003) and water logging (Olufayo et al. 1994). Sandy soils in eastern Kenya allow quick percolation of water into subsurface horizons and no water-logging has been observed. Decreased yield in tied-ridges compared to conventional tillage and no-till systems especially during the last season is attributed to lower germination. As also reported by Twomlow and Bruneau (2000), poor crop germination in tied ridges, especially for season four that had lowest rainfall during the first weeks, is attributed to incomplete wetting up of the tied-ridges at beginning of season. In Zimbabwe, the current practice to counter low germination on ridges is to use post emergence tied-

ridges (Nyagumbo, personal communication), but this requires rebuilding the ridges every season rather than maintaining old ones across seasons.

System performance under different organic resources shows clearly that farmers are better off combining different organic resources (manure and crop residue) for better yields. The benefits maybe derived from a synchrony mechanism where manure releases nutrients more readily meeting plant demand at early stages, while crop residues release nutrients slowly meeting requirements at later stages when plant demand is also lower. Also, in manure-only treatments, greater amounts of N could be released and get leached out of the system while presence of crop residue in CR + manure treatment could lead to immobilization of such readily mineralizable manure N by micro-organisms during decomposition of the low quality crop residue. In northern France, for example, low quality straw retained in treatments reduced average mineralization over 4 months from 45 to 21 kg N ha^{-1} through immobilization (Beaudoin et al. 2005). Such immobilized N could be released at later crop stages thereby enhancing crop performance. Greater losses of N in manure only than in CR + manure treatments could explain the greater maize response to chemical N fertilizer application observed in manure only treatments, compared to CR + manure treatments. Sandy environments, as is the case in our study site, are reported to experience excess drainage below the rooting zone accompanied by nitrogen leaching (Cameira et al. 2003) and up to 154 kg N ha^{-1} could be leached annually (Beaudoin et al. 2005). Higher yields with than without crop residue within conservation tillage systems (Dam et al. 2005) could be attributed to reduction in surface runoff. In Nigeria, Erenstein (2003) observed that runoff was reduced from 75.4 to 43.4% using 2 t ha^{-1} of surface mulch. Combining crop residue and manure could be a good strategy for semi-arid environments such as Mbeere, eastern Kenya, where poor rainfall distribution could lead to occasional flushing of plant available nutrients down the soil profile. It is such technologies that enhance synchrony between nutrient release and uptake by crops, and that consider farmers' organic resource access that could increase farm productivity and profitability.

Significant response to chemical N fertilizer indicates the importance of application of fertilizer in the cropping systems. Cropping system management should thus entail combinations of an appropriate tillage system, application/retention of organic resources, even though in small quantities, and the use of chemical nutrient sources. Nevertheless, the additional benefits following fertilizer application vary with seasonal rainfall. Lowest response to fertilizer in season 4 (LR 2007) is attributed to lack of adequate rainfall during the first 2 weeks after planting. This was likely through negative effects on root development for the already germinated crop (there was no effect of chemical N fertilizer on germination percentage). On the contrary, highest response to fertilizer was observed during the season with highest, well-distributed rainfall (SR 2006). In general, highest yields achieved when chemical N fertilizer was combined with CR and manure demonstrate the importance of integrated soil fertility management (ISFM). This is well demonstrated by Vanlauwe et al. (2010) through a meta-analysis involving 90 ISFM-based per-reviewed publications that revealed greater agronomic performance when fertilizer was combined with manure or compost.

For the apparent nutrient recovery, the values observed in this study are similar to those reported by Saidou et al. (2003). Leaching is a prevalent problem in sandy soils and 25–33% of applied N could be leached as observed in a sandy soil in England (Salazar et al. 2005), and values could even be higher. The large partial balance observed in manure only treatments compared to those combining CR and manure could constitute more of nutrients lost by leaching or runoff. The correlations of NPK balances with maize yield further show that there were some manure only treatments where partial balance was not related with yield and this could be due to such leaching losses. Low nutrient uptake in no-till system compared to conventional and tied-ridge systems is attributed to the lower yields observed during the initial seasons. Our observation of higher N uptake following application of fertilizer N agrees with the observation of reduced aboveground N uptake in N-stress treatments in West Africa (Oikeh et al. 2003).

Greater soil macro-aggregation in no-till systems due to reduced disturbance normally caused by plowing has been reported by several authors (Filho et al. 2002; Pinheiro et al. 2004). The presence of crop residue within no-till improved soil aggregation, agreeing with Lal (1984) who observed that crop

residue and reduced tillage drastically reduced soil detachment and its transport through runoff. Similarly, Ogunremi et al. (1986) reported increased proportion of aggregates exceeding 2 mm diameter in no-till compared to ploughed plots, and the differences were more pronounced when no-till was combined with crop residue application. In another study, Ley et al. (1989) found soil passing through 125-μm sieves at Samaru to be 10 times less in no-till compared to tilled treatment. Aggregation indices also showed that soil structure was improved under tied-ridges compared to conventional tillage plots, again due to less disruptions by tillage and enhanced soil moisture. But aggregation in tied-ridges was lower than in no-till because the restricted tillage for ridge maintenance contributed to aggregate disruptions.

Conclusions

Since combining 1 t ha^{-1} manure and 1 t ha^{-1} crop residue result in improvement of maize yield and soil aggregation compared to sole application of manure at 2 t ha^{-1}, we conclude that application of high quality manure combined with low quality crop residue is a good practice that could reduce N leakage and increase nutrient use efficiency within dryland maize cropping systems. No-till system can achieve better yield than conventional tillage but this is achieved after some cropping seasons. The best practice for farmers will be to use combination of conservation tillage (no-till or tied-ridges) and manure plus crop residue. Also, the application of chemical N fertilizer in the system is appropriate for further increases in yield. These results are applicable in farmers' fields and can be implemented immediately since the rates of organic resources used are within the access of majority of small-holder farmers in the dryland environments.

Acknowledgments This work was made possible by the funding from the International Foundation for Science (IFS) to which I am greatly indebted. Thanks also to Ivan Adolwa (TSBF-CIAT) and Samuel Njoroge (Kenyatta University) for assisting with literature searches and soil fractionation.

References

Bationo A, Vlek PLG (1997) The role of nitrogen fertilisers applied to food crop in the Sudano-Sahelian zone of West

Africa. In: Renard G et al. (ed) Soil fertility management in West African land use systems. Proc. regional workshop Univ. Hohenheim, ICRISAT Sahelian Centre and INRAN 4–8 March 1997, Niamey, Niger. Margraf Verlag, Germany

Bationo A, Kihara J, Vanlauwe B, Waswa B, Kimetu J (2007) Soil organic carbon dynamics, functions and management in West African agro-ecosystems. Agric Syst 94:12–25

Beaudoin N, Saad JK, Van Laethem C, Machet JM, Maucorps J, Mary B (2005) Nitrate leaching in intensive agriculture in Northern France: effect of farming practices, soils and crop rotations. Agric Ecosyst Environ 111:292–310

Belay A, Gebrekidan H, Uloro Y (1998) Effect of tied ridges on grain yield response of maize (Zea mays L.) to application of crop residue and residual N and P on two soil types at Alemaya, Ethiopia. S Afr J Plant Soil 15:123–129

Blaise D, Ravindran CD (2003) Influence of tillage and residue management on growth and yield of cotton grown on a vertisol over 5 years in a semi-arid region of India. Soil Tillage Res 70:163–173

Busscher WJ, Edwards JH, Vepraskas MJ, Karlen DL (1995) Residual effects of slit tillage and subsoiling in a hardpan soil. Soil Tillage Res 35:115–123

Cameira MR, Fernando RM, Pereira LS (2003) Monitoring water and NO_3-N in irrigated maize fields in the Sorraia Watershed, Portugal. Agric Water Manag 60:199–216

Dam RF, Mehdi BB, Burgess MSE, Madramootoo CA, Mehuys GR, Callum IR (2005) Soil bulk density and crop yield under eleven consecutive years of corn with different tillage and residue practices in a sandy loam soil in central Canada. Soil Tillage Res 84:41–53

Diaz-Zorita M (2000) Effect of deep-tillage and nitrogen fertilization interactions on dryland corn (Zea mays L.) productivity. Soil Tillage Res 54:11–19

Erenstein O (2003) Smallholder conservation farming in the tropics and sub-tropics: a guide to the development and dissemination of mulching with crop residues and cover crops. Agric Ecosyst Environ 100:17–37

Evans SD, Lindstrom MJ, Voorhees WB, Monscrief JF, Nelson GA (1996) Effect of subsoiling and subsequent tillage on soil bulk density, soil moisture, and con yield. Soil Tillage Res 38:35–46

Filho CC, Lourenco A, Guimaraes FM, Fonseca ICB (2002) Aggregate stability under different soil management systems in a red latosol in the state of Parana, Brazil. Soil Tillage Res 65:45–51

Gameda S, Raghavan GSV, McKyes E, Watson AK, Mehuys G (1994) Response of grain corn to subsoiling and chemical wetting of a compacted clay soil. Soil Tillage Res 29:179–187

Gebrekidan H (2003) Grain yield response of sorghum (Sorghum bicolor) to tied ridges and planting methods on entisols and vertisols of Alemaya Area, Eastern Ethiopian Highlands. J Agric Rural Dev Trop Subtrop 104:113–128

Govaerts B, Verhulst N, Sayre KD, Kienle F, Flores D, Limon-Ortega A (2009) Implementing conservation agriculture concepts for irrigated wheat based systems in Northwest Mexico: a dynamic process towards sustainable production. In: Innovations for improving efficiency, equity and environment. Lead papers for the 4th world congress on conservation agriculture held 4–7th February 2009 in New Delhi, India. 4th world congress on conservation agriculture, New Delhi, India

Goyal S, Chander K, Mundra MC, Kapoor KK (1999) Influence of inorganic fertilizers and organic amendments on soil organic matter and soil microbial properties under tropical conditions. Biol Fertil Soils 29:196–200

Hoogmoed WB (1999) Tillage for soil and water conservation in the semi-arid tropics. Doctoral Thesis, Wageningen University, Wageningen, the Netherlands

ICRAF (1995) International Centre for Research in Agroforestry: Laboratory methods of soil and plant analysis, Nairobi, Kenya

Jensen JR, Bernhard RH, Hanse S, McDonagh J, Moberg JP, Nielsen NE, Nordbo E (2003) Productivity in maize based cropping systems under various soil-water-nutrient management strategies in a semi-arid, alfisol environment in East Africa. Agric Water Manag 56:217–237

Lal R (1984) Mechanized tillage systems effects on soil erosion from an alfisol in watersheds cropped to maize. Soil Tillage Res 4:349–360

Ley GJ, Mullins CE, Lal R (1989) Hard-setting behavior of some structurally weak tropical soils. Soil Tillage Res 13:365–381

Madari B, Machado PLOA, Torres E, de Andrade AG, Valencia LIO (2005) No tillage and crop rotation effects on soil aggregation and organic carbon in a rhodic ferralsol from southern Brazil. Soil Tillage Res 80:185–200

Maroko JB, Buresh RJ, Smithson PC (1998) Soil nitrogen availability as affected by fallow-maize systems on two soils in Kenya. Biol Fertil Soils 26:229–234

Martius C, Tiessen H, Vlek PLG (2001) The management of organic matter in tropical soils: what are the priorities? Nutr Cycl Agroecosyst 61:1–6

McCarty GW, Meisinger JJ (1997) Effects of N fertilizer treatments on biologically active N pools in soils under plow and no tillage. Biol Fertil Soils 24:406–412

Miriti JM, Esilaba AO, Bationo A, Cheruiyot H, Kihumba J, Thuranira EG (2007) Tied-ridging and intergrated nutrient management options for sustainable crop production in semi-arid eastern Kenya. In: Bationo A et al (eds) Advances in integrated soil fertility management in sub-Saharan Africa: challenges and opportunities. Springer, Dordrecht

Motavalli PP, Stevens WE, Hartwig G (2003) Remediation of subsoil compaction and compaction effects on corn N availability by deep tillage and application of poultry manure in a sandy-textured soil. Soil Tillage Res 71:121–131

Mrabet R (2002) Stratification of soil aggregation and organic matter under conservation tillage systems in Africa. Soil Tillage Res 66:119–128

Nyakatawa EZ, Brown M, Maringa D (1996) Maize and sorghum yields under tied-ridges of fertilized sandy soils in semi-arid south-east lowveld of Zimbabwe. Afr Crop Sci J 4:197–206

Ogunremi LT, Lal R, Babalola O (1986) Effects of tillage and seeding methods on soil physical properties and yield of upland rice for an ultisol in southeast Nigeria. Soil Tillage Res 6:305–324

Oikeh SO, Carsky RJ, Kling JG, Chude VO, Horst WJ (2003) Differential N uptake by maize cultivars and soil nitrate dynamics under N fertilization in West Africa. Agric Ecosyst Environ 100:181–191

Olufayo A, Baldy C, Some L, Traore I (1994) Tillage effects on grain sorghum (Sorghum bicolor (L) Moench) development and plant water status in Burkina Faso. Soil Tillage Res 32:105–116

Osunbitan JA, Oyedele DJ, Adekalu KO (2005) Tillage effects on bulk density, hydraulic conductivity and strength of a loamy sand soil in southwestern Nigeria. Soil Tillage Res 82:57–64

Parkinson JA, Allen SE (1975) A wet oxidation procedure suitable for the determination of nitrogen and mineral nutrients in biological materials. Commun Soil Sci Plant Anal 6:1–11

Pascual U, Barbier EB (2006) Deprived land-use intensification in shifting cultivation: the population pressure hypothesis revisited. Agric Econ 34:155–165

Pinheiro EFM, Pereira MG, Anjos LHC (2004) Aggregate distribution and soil organic matter under different tillage systems for vegetable crops in a red latosol from Brazil. Soil Tillage Res 77:79–84

Rawitz E, Hoogmoed WB, Morin J (1986) The effect of tillage practices on crust properties, infiltration, and crop response under semi-arid conditions. In: Proceedings of international symposium on assessment of soil surface sealing and crusting, ISSS Ghent, Belgium, pp 278–284

Rockstrom J, Barron J, Fox P (2003) Water productivity in rainfed agriculture: challenges and opportunities for small holder farmers in drought prone tropical agro-ecosystems. CABI; IWMI, Wallingford, UK; Colombo, Sri Lanka

Rosolem CA, Foloni JSS, Tiritan CS (2002) Root growth and nutrient accumulation in cover crops as affected by soil compaction. Soil Tillage Res 65:109–115

Saidou A, Janssen BH, Temminghoff EJM (2003) Effects of soil properties, mulch and NPK fertilizer on maize yields and nutrient budgets on ferralitic soils in southern Benin. Agric Ecosyst Environ 100:265–273

Salazar FJ, Chadwick D, Pain BF, Hatch D, Owen E (2005) Nitrogen budgets for three cropping systems fertilised with cattle manure. Bioresour Technol 96:235–245

Scopel E, Muller B, Tostado JMA, Guerra EC, Maraux F (1998) Quantifying and modelling the effects of a light crop residue on the water balance: an application to rainfed maize in Western Mexico XVI world congress of soil science—Montpellier, France, August 1998

Six J, Elliott ET, Paustian K (2000) Soil macroaggregate turnover and microaggregate formation: a mechanism for C sequestration under no-tillage agriculture. Soil Biol Biochem 32:2099–2103

Twomlow SJ, Bruneau PMC (2000) The influence of tillage on semi-arid soil-water regimes in Zimbabwe. Geoderma 95:33–51

Vanlauwe B, Kihara J, Chivenge P, Pypers P, Coe R, Six J (2010) Agronomic use efficiency of N fertilizer in maize-based systems in sub-Saharan Africa within the context of integrated soil fertility management. Plant Soil 339:35–50

Verhulst N, Govaerts B, Verachtert E, Kienle F, Limon-Ortega A, Deckers J, Raes D, Sayre KD (2009) The importance of crop residue management in maintaining soil quality in zero tillage systems; a comparison between long-term trials in rainfed and irrigated wheat systems. In: Innovations for improving efficiency, equity and environment. Lead papers for the 4th world congress on conservation agriculture held 4–7th February 2009 in New Delhi, India. 4th world congress on conservation agriculture, New Delhi, India

Warren GP, Atwal SS, Irungu JW (1997) Soil nitrate variations under grass, sorghum and bare fallow in semi-arid Kenya. Exp Agric 33:321–333

 Springer

Long-Term Land Management Effects on Crop Yields and Soil Properties in the Sub-humid Highlands of Kenya

C.N. Kibunja, F.B. Mwaura, D.N. Mugendi, D.K. Wamae, and A. Bationo

Abstract The effect of continuous cultivation using inorganic and organic fertilizers on crop yields and soil agro-properties was studied in a 30-year-old long-term field experiment at Kabete, near Nairobi, in the highlands of Kenya. The area is sub-humid with an average bimodal rainfall of 980 mm and two cropping seasons per year. The soil is a dark red, friable clay classified as a Humic Nitisol and is considered to be moderately fertile. The main treatments consisted of three rates of inorganic fertilizers nitrogen (N) and phosphorus (P), farmyard manure with or without stover restitution. Maize and beans were planted during the long and short rains seasons, respectively. Results indicate that the use of chemical fertilizers alone increased maize grain yields by more than 50% during the first 6 years of experimentation but declined thereafter. Application of combined chemical fertilizers and farmyard manure proved superior to inorganic fertilizers alone and maintained maize yields at 3–5 t ha^{-1}. Farmyard manure also gave better yields than did chemical fertilizers. However, application of chemical fertilizers alone led to decreased maize yields, increased soil acidification from pH of 5.5 to 4.3 and raised bulk density from 1.04 to 10.8 g cm^{-3} soil. The total % N declined by 25% from 0.16%, while soil organic carbon decreased from 2 to 1.2% after 27 years. Fertilizer N utilization ranged from 25 to 33% but was higher in plots supplied with chemical fertilizers than in those with combined organic and inorganic inputs.

Keywords Long term · Chemical fertilizers · Farmyard manure · Crop residues · Maize · Soil properties

Introduction

The continuous cultivation of land accompanied by rising population pressure has led to serious soil fertility decline in sub-Saharan Africa (SSA). This situation is manifested by declining crop yields, decreasing vegetation cover and increasing soil erosion. As a result, farm productivity and agricultural incomes are falling, and migration to urban centres is on the increase, while both household and countrywide food securities are declining (FAO, 1995). In general, the SSA region faces two interrelated problems: declining per capita production and progressive deterioration of the environment (Scholes et al., 1994; FAO, 2002). While food insecurity occurs in most countries in the developing world, it is most acute in the SSA region. In contrast to sustained increases in other parts of the developing world, per capita food production continues to decrease in this region (World Bank, 1996a; Bationo et al., 2004).

Sanchez et al. (1997) contend that 'soil fertility depletion in smallholder farms is the fundamental biophysical root cause of declining per capita food production in Africa and soil fertility replenishment should be considered as an investment in natural resource capital'. Poor and inappropriate soil management practices are cited as the main causes of soil fertility decline of cultivated lands. The smallholder farming community in Kenya plays a major role in domestic food production and is responsible for

C.N. Kibunja (✉)
Kenya Agricultural Research Institute, NARL-KARI, Nairobi, Kenya
e-mail: catherine.kibunja@yahoo.com

A. Bationo et al. (eds.), *Innovations as Key to the Green Revolution in Africa*, DOI 10.1007/978-90-481-2543-2_15, © Springer Science+Business Media B.V. 2011

meeting up to 80% of the domestic food requirement in the country. Yet they are faced with many challenges due to lowland productivity as a result of declining soil fertility coupled with inadequate use of agricultural inputs to replenish the nutrients. Shortage of capital and poor access to credit facilities often hamper the use of chemical fertilizers to maintain soil fertility (Minae and Akyeampong, 1988; Makokha et al., 2001), while the economic policy reforms have led to declining output versus fertilizer prices (Heerink, 2005). Continuous cropping, with little or inadequate external inputs, causes heavy nutrient losses through the harvested food and cash crops which are exported out of the system (Stoorvogel et al., 1993). Continuous land sub-division on the other hand reduces the opportunity for fertility restoration through fallowing leading to stagnation in agricultural productivity.

Land management alters the pattern of decomposition and nutrient release and also affects the relative proportion of soil organic matter in cropped land. This may be due to changes in soil structure caused by continuous cultivation. The magnitude of this loss depends on the intensity of cultivation as well as the quality and quantity of fertilizers and organic residues returned to the soil. Soil fertility research in the east African region began in the early 1900s and recognized the fact that continuous cropping with the advent of permanent settlement led to decline of crop productivity accompanied by land degradation (Nye and Greenland, 1960). This led to the setting up of several long-term experiments in various parts of Africa to study appropriate interventions for the small-scale farming community. In Kenya, a long-term experiment was established in 1976 to investigate the effect of continuous cropping on soil chemical properties and crop yields (maize and beans) using various combinations of farmyard manure, crop residues and inorganic fertilizers (in particular nitrogen and phosphorus). It was designed specifically as an agronomic trial but has in addition to the crop yields provided data on soil chemical and physical properties (Qureshi, 1991; Swift et al., 1994) and biological parameters (Kibunja et al., 2005). It now serves as a unique resource to investigate the long-term influences of continuous cropping using various organic and inorganic inputs.

Materials and Methods

A long-term experiment was set up in 1976 at the experimental grounds of the National Agricultural Research Laboratories, Kabete, of the Kenya Agricultural Research Institute, at Kabete, near Nairobi. The centre is located about 8 km northwest of Nairobi at 36°41′E and 01°15′S and at an altitude of 1,737 m above sea level. The area is in the sub-humid agro-ecological zone with an annual average temperature of 23°C with the coolest temperatures in July (16°C) and the warmest in January (28°C). The rainfall is bimodal with two cropping seasons per year. The first and longer rainy season starts in March and ends in September, while the second rain falls from October to December (Jaetzold and Schmidt, 1983). The average annual precipitation is 980 mm ranging from 600 to 1,800 mm with about 70 rainy days with an extended dry period of 9 months (Kenya Meteorological Department, 1984).

The soil type is a well-drained, very deep dark reddish brown to dark red, friable clay classified as a Humic Nitisol according to the FAO-UNESCO (1974) Soil Map of the World and is known locally as the Kikuyu red clay loam (Sombroek et al., 1980). Siderius and Muchena (1977) classified the soils as clayey, kaolinitic, non-calcareous, acid, isothermic, very deep Paleustults according to the USDA-Soil Conservation Service (1975) system of classification. The soil is considered to be inherently fertile with a deep, well-drained profile with moderate amounts of available calcium and magnesium, and adequate potassium but low amounts of total phosphorus and available nitrogen. The soil type Nitisol represents an important agricultural zone in Kenya, i.e. the central Kenya and Kisii regions covering a total of about 55,000 ha of land in Kenya and about 228,000 km^{-2} within the east African highlands (Swift et al., 1994). The area enjoys a relatively good and reliable rainfall with fertile soils, which supports a rapidly increasing population.

The experimental procedures consisted of 18 treatments laid out in a randomized complete block design with four replications in various combinations of inorganic fertilizers, crop residues and farmyard manure to give a $2 \times 2 \times 3$ factorial arrangement as shown in Table 1. Phosphate fertilizer was applied as a blanket

Table 1 Experimental treatments applied in the long-term experiment at NARL, Kabete, Kenya

Factor	Level 1	Level 2	Level 3
N, P (kg ha^{-1})	0,0	60,26	120,52
FYM (t ha^{-1})	0	5	10
Crop residues	+	−	N/A

treatment at the rates of 0, 26 and 52 kg P ha^{-1} as triple super-phosphate (TSP). Nitrogen fertilizer was applied at three rates (0, 60 and 120 kg N ha^{-1} as calcium ammonium nitrate (CAN)) at the seventh week after planting maize. Farmyard manure was applied at three rates (0, 5 and 10 t ha^{-1}) in shallow furrows approximately 15 cm deep at least 2 weeks before planting. Both fertilizers and manure were applied once a year during the long rains season at the time of maize planting.

Land preparation was carried out by hand according to the practice of small-scale farmers. Stover, chopped into small pieces, was returned and incorporated in the appropriate plots during land preparation. Two crops are planted each year as is common for this agro-ecological zone. Maize hybrid variety '512' and beans cultivar *mwezi moja* as sole crops were grown in rotation during the first and second rains of each year, respectively. Maize was planted at the onset of the long rains at the spacing of 75 cm between the rows and 25 cm within the row. Two seeds were planted per hill and later thinned to one in order to give a plant density of 55,000 plants per hectare. Beans were planted at the beginning of the short rains at a spacing of 45 cm between the rows and 15 cm within the row. Gapping and thinning took place about 2 weeks after germination to ensure the right plant density. The crop was maintained till maturity observing all good husbandry practices recommended for small-scale farmers (Qureshi, 1991).

Soil chemical and physical properties were determined at the beginning of every year. Soil organic content, soil acidity, bulk density as well as the macronutrients N, P, K, Mg and Ca were determined using standard methodology used at the National Agricultural Research Laboratories (Hinga et al., 1980). Fertilizer N recovery was studied through the use of labelled ^{15}N fertilizer as calcium ammonium nitrate (CAN) at 10% atom excess (a.e.) for a period of 2 years (2000–2001). The fertilizer N was applied

to maize in 1×2 m^2 micro-plots at the rate of 60 kg N ha^{-1} year^{-1} at the seventh leaf stage. Maize from the micro-plots was harvested at physiological maturity and separated into cobs, stover and roots and weighed. The plant parts were shredded when still fresh and then oven-dried before fine grinding (Zapata and Hera, 1995). The ^{15}N content of the plant tissue was determined using an emission spectrometer (NO-I PC, GmHB, Germany).

Results and Discussion

Soil Properties

Changes in some selected physical and chemical properties that have occurred over the last 25 years are shown in Table 2. These results indicate that continuous cultivation had an adverse effect on both soil physical and chemical properties, that is, total N, pH in water, soil organic C and the macronutrients Ca, Mg, K and P over time. Soil organic carbon declined by 63% from 2% in the no-input control plots. According to studies using the ROTH-C-26.3 model, simulated SOC turnover supported the measured decreases in SOC with the model and further predicted lengthy and difficult amelioration process for such degraded soils (Ayaga, 2000). Further results showed increased soil acidity in plots receiving no inputs where soil pH dropped from 5.5 to 4.2 to date. Soil bulk density also seems to be on the rise, albeit slowly. The greatest changes were noted in the no-input control plots, but it is noteworthy that even the application of mineral nitrogen, crop residues and farmyard manure did not totally arrest the decline of soil nitrogen and carbon content over time (Kibunja et al., 2005).

Crop Yields

Some of the major highlights of the experiment, now in its 30th year, indicate that continuous cropping has led to marked decline in crop yields regardless of land management option used (Table 3). The maize grain

Table 2 Effect of continuous cropping with different land management options on soil chemical and physical properties of the top soil (0–20 cm) at Kabete, Kenya, for 25 years (Kibunja et al., 2005)

Properties	Initial 1976[a]	Treatments 2002				
		Nil	N_1P_1	FYM_1	$N_1P_1FYM_1$	$N_1P_1FYM_1+R$
pH (1:2 in H_2O)	5.50	4.20	4.27	4.45	4.35	4.40
SOC (%)	2.00	1.26	0.93	1.26	1.44	1.43
Total N	0.16	0.12	0.14	0.15	0.15	0.16
Ca (m.e./100 g)	3.65	2.50	4.05	2.75	3.95	4.05
Mg (m.e./100 g)	2.20	2.01	1.60	1.98	1.90	2.06
K (m.e./100 g)	0.73	0.56	0.62	0.54	0.52	0.64
Available P (mg kg^{-1})	15.00	13.50	15.00	15.00	16.00	16.00
CEC (m.e./100 g)	56.0	n.d.	n.d.	n.d.	n.d.	n.d.
Soil texture (%)	...........................67% clay:22% silt:11% sand.....................................					
Textural class	....Clay..					
Bulk density (g cm^{-2})	1.04	1.08	1.08	1.06	1.04	1.0

[a]1976 *Source* of data: Qureshi (1987)

Nil, control; N_1P_1, chemical fertilizer at 60 kg N ha^{-1} and 26.4 kg P ha^{-1}; FYM_1, farmyard manure at 5 t ha^{-1}; $N_1P_1 + FYM_1$, chemical fertilizer at 60 kg N ha^{-1} and 26.4 kg P ha^{-1} with farmyard manure at 5 t ha^{-1}; $N_1P_1 + FYM_1 + R$, chemical fertilizer at 60 kg N ha^{-1} and 26.4 kg P ha^{-1} with farmyard manure at 5 t ha^{-1} and stover returned; n.d., not determined

Table 3 Productivity of maize and fertilizer N utilization using either inorganic or combined organic and inorganic inputs

Treatments*	Grain yield (kg ha^{-1})	Total % N	% Ndff	% Fertilizer N utilization
N_1P_1	2,750	1.57	14.5	33
N_1P_1+R	3,538	1.49	12.5	27
$N_1P_1+FYM_1$	6,003	1.40	9.3	25
$N_1P_1+FYM_1+R$	5,692	1.38	10.5	29

*N_1, 60 kg N ha^{-1} (applied as calcium ammonium nitrate (CAN)); P_1, 26.4 kg P ha^{-1} (applied as triple super-phosphate (TSP)); FYM_1, 5 t ha^{-1} boma manure (supplying about 15 kg N and 4 kg P t^{-1} dry manure); R, all harvested stover returned back

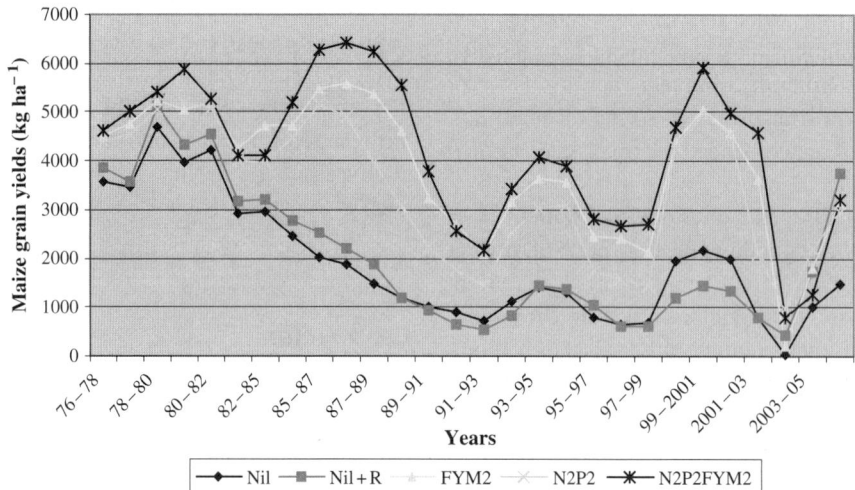

Fig. 1 Maize grain yields using various management options in the long-term trial at Kabete, Kenya, 1976–2006. Nil, no inputs control; Nil + R, all harvested stover returned; N_2, 120 kg N ha^{-1} (applied as calcium ammonium nitrate (CAN)); P_2, 52 kg P ha^{-1} (applied as triple super-phosphate (TSP)); FYM_2, 10 t ha^{-1} boma manure (supplying about 15 kg N and 4 kg P t^{-1} dry manure)

yields in the no-input control declined from 4.0 to 2.0 t ha^{-1} within the first 10 years (Fig. 1). During the first 5 years after the start of the experiment, application of chemical fertilizers alone increased maize yields by about 50% and out-yielded all other treatments (Qureshi, 1991). However, this trend changed soon after the sixth year and the combination of farmyard manure with chemical fertilizers gave higher yields than did all other treatments (Qureshi, 1987, 1991). To date, the combined use of chemical fertilizers and farmyard manure continues to consistently give higher yields than did all the other treatments (Qureshi, 1987, 1991; Swift et al., 1994; Kapkiyai et al., 1998; Kibunja and Gikonyo, 2000; Kibunja, 2007) as shown in Fig. 1. The yields of beans (not shown) declined from 950 kg ha^{-1} to about 300–500 kg ha^{-1} probably due to increasing soil acidity, declining soil fertility and high incidences of diseases.

Conclusions

Crop yields have declined steadily with continuous cropping, regardless of inputs. The system is also losing SOC at a rate higher than the current returns and its amelioration may be an expensive and time-consuming exercise. Application of combined organic and inorganic fertilizers gave the highest yields but there is need to investigate ways of reducing loss of N through leaching and thereby improve nutrient use efficiency. Liming may be an important strategy to deal with rising soil acidification, especially where inorganic fertilizers have been applied without any organic inputs. Though potassium fertilizers have not been applied in the past, it is now clear that the system is losing K probably through plant uptake and leaching and hence there is need to further investigate K dynamics in this cropping system. In general, better management strategies of available plant nutrients are needed in order to improve soil organic matter, nutrient use efficiency and crop productivity and hence arrest land degradation in continuously cultivated lands. There is further need to advise farmers and policy makers adequately on the maintenance and restoration of soil fertility of these more fertile African soils.

Acknowledgements The authors wish to thank the director, KARI for permission to publish this work and TSBF-CIAT for its continued support of the long-term trial during the last 4 years through its AfNet Programme.

References

Ayaga GO (2000) Improving the efficiency of phosphate utilization in low-input maize production in Kenya. PhD thesis, University of Nottingham, Nottingham, UK

Bationo A, Kimetu J, Ikerra S, Kimani S, Mugendi DN, Odendo M, Silver M, Swift MJ, Sanginga N (2004) The African network for soil biology and fertility: new challenges and opportunities. In: Bationo A (ed) Managing nutrient cycles to sustain soil fertility in sub-Saharan Africa. AfNet-CIAT. Academy Science Publishers, Nairobi, Kenya, pp 1–23

FAO (1995) Planning for sustainable use of land resources: towards a new approach. In: Sombroek WG, Sims D (eds) Land and water bulletin 2. FAO, Rome, Italy

FAO (2002) Food insecurity: when people must live with hunger and fear starvation. The state of 2002. Food and Agriculture Organization of the United Nations, Rome, Italy

FAO-UNESCO (1974) Soil map of the world: volume VI, Africa. UNESCO, Paris

Heerink N (2005) Soil fertility decline and economic policy reform in sub-Saharan Africa. Land Use Policy 22:67–74

Hinga G, Muchena FN, Njihia CM (1980) Physical and chemical methods of soil analysis report. National Agricultural Research Laboratories, Nairobi, Kenya

Jaetzold R, Schmidt H (1983) Farm management handbook for Kenya. Natural conditions and farm management information. Part C, East Kenya. Ministry of Agriculture, Nairobi

Kapkiyai JJ, Karanja NK, Woomer P, Qureshi JN (1998) Soil organic carbon fractions in a long-term experiment and the potential for their use as a diagnostic assay in highland farming systems of Central Kenya. Afr Crop Sci J 6(1):19–28

Kenya Meteorological Department (1984) Climatological statistics of Kenya. Kenya Meteorological Department, Nairobi, 61pp

Kibunja CN (2007) Impact of long-term application of organic and inorganic nutrient sources in a maize-bean rotation to soil nitrogen dynamics and soil microbial populations and activity. PhD thesis, University of Nairobi, Kenya

Kibunja CN, Gikonyo EW (2000) Maintenance of soil fertility under continuous cropping in a maize-bean rotation (long-term trial, NARL-Kabete). The biology and fertility of tropical soils. Report of the tropical soil biology fertility programme. TSBF. 1997–1998

Kibunja CN, Mugendi DN, Mwaura FB, Kitonyo EM, Salema MP (2005) Fate of applied nitrogen in a long-term maize – bean cropping system in Kenya. In: Mugendi DN, Kironchi G, Gicheru PT, Gachene CKK, Macharia PN, Mburu M, Mureithi JG, Maina F (eds) Proceedings of the 21st Conference of the Soil Science Society of East Africa (SSSEA). Eldoret, Kenya, pp 363–371

Makokha S, Kimani S, Mwangi W, Verkuijl H, Musembi F (2001) Determinants of fertilizer and manure use in maize production in Kiambu District, Kenya. CIMMYT/KARI report

Minae S, Akyeampong E (1988) Agroforestry potentials for the land use systems in the bimodal highlands of eastern Africa, Kenya. Report. AFRENA No. 3. ICRAF, Nairobi, 153pp

Nye PH, Greenland DJ (1960) The soil under shifting cultivation. Commonwealth Agric, Bureaux, Harpenden, England

Qureshi JN (1987) The cumulative effects of N–P fertilizers, manure and crop residues on maize grain yields, leaf contents and some soil chemical properties at Kabete. National maize agronomy workshop, Nairobi, 17–19th 1987, p 12

Qureshi JN (1991) The cumulative effects of N–P fertilizers, manure and crop residues on maize and bean yields and some chemical properties at Kabete. In: Proceedings of the 2nd KARI conference, Nairobi, Kenya, 5–7 Sept 1991

Sanchez PA, Shepherd KA, Soule MJ, Place FM, Buresh RJ, Izac A-MN, Mokwunye AU, Kwesiga FR, Ndiritu CG, Woomer PL (1997) Soil fertility replenishment in Africa: an investment in natural resource capital. In: Buresh RJ, Sanchez PA, Calhon F (eds) Replenishing Soil Fertility in Africa. SSSA special publication no. 51. Soil Science Society of America & American Society of Agronomy, Madison, WI, pp 1–46

Scholes MC, Swift MJ, Heal OW, Sanchez PA, Ingram JSI, Dalal R (1994) Soil fertility research in response to the demand for sustainability. In: Woomer PL, Swift MJ (eds) The biological management of tropical soil fertility. Wiley, England, UK, pp 1–14

Siderius W, Muchena FN (1977) Soils and environmental conditions of agricultural research stations in Kenya. Misc.

publication no. 5. Kenya Soil Survey, National Agricultural Laboratory, Nairobi, Kenya

Sombroek WG, Braun HMH, Van der Pouw BJA (1980) Exploratory soil map and agro-climatic zone map of Kenya. Report no. E 1. Ministry of Agriculture, Kenya Soil Survey, Nairobi, Kenya

Stoorvogel JJ, Smaling EMA, Janssen BH (1993) Calculating soil nutrient balances in Africa at different scales. 1. Supranational scale. Fertil Res 33:227–235

Swift MJ, Seward PD, Frost PGG, Qureshi JN, Muchena FN (1994) Long-term experiments in Africa: developing a database for sustainable land use under global change. In: Leigh RA, Johnson AE (eds) Long-term experiment in agricultural and ecological sciences. CAB International, Wallingford, CT, pp 229–251

USDA-Soil Conservation Service (1975) Soil taxonomy. A system for soil classification. Soil survey staff. SMSS technical monograph no. 19. United States Department of Agriculture, Washington, DC

World Bank (1996a) African development indicators 1996. World Bank, Washington, DC

Zapata F, Hera C (1995) Enhancing nutrient management through use of isotope techniques. In: Proceedings of an international symposium on nuclear and related techniques in soil-plant studies on sustainable agriculture and environmental preservation jointly organized by IAEA and FAO, Vienna, 17–21 October 1994, pp 83–105

Integrated Management of Fertilizers, Weed and Rice Genotypes Can Improve Rice Productivity

B.V. Bado, K. Traore, M.E. Devries, A. Sow, and S. Gaye

Abstract   The influence of weed control on fertilizer nitrogen use efficiencies (NUEs) by rice genotypes was studied in the Senegal River valley of West Africa with a field experiment during four rice growing seasons. It was hypothesized that integrated management of technologies could improve rice productivity. The objective was to develop integrated high-return technologies that improve irrigated rice-based systems productivity and profitability. Data indicated that rice grain yields were affected by N fertilizer, genotypes and plant densities. In good weed control conditions, optimum doses of recommended N fertilizer varied from 80 to 180 kg N ha^{-1}. Fertilizer N use efficiencies by genotypes were affected by weed control. Profitable management options of genotypes and N fertilizer recommendations have been identified. With a good control of weed, varieties and N fertilizer recommendations were suggested as integrated management options for farmers. But poor control of weed increased N lost, decreased grain yields and profitability. Two genotypes (WAS 55-B-B-2-1-2-5 and WAS 191-1-1-7 FKR) were found to be most competitive against weeds. However, no more than 60 kg N ha^{-1} should ever be recommended when weeds are poorly controlled. It was concluded that productivity and profitability of irrigated rice-based systems could be improved with integrated management options of genotypes, fertilizers and weed.

Keywords Fertilizer · Nitrogen · Weed · Varieties · Rice

B.V. Bado (✉)
Sahel Regional Station, Africa Rice Centre (Africa Rice), BP 96, Saint-Louis, Senegal
e-mail: V.Bado@cgiar.org

Introduction

Rice is a strategic crop in West Africa, an important source of food and farm income for most rural households in the region. In addition to its growing importance in rural diets, rice has rapidly become the major source of calories for most urban households in West Africa. While rice has proved to be a sustainable crop, preserving the natural resource base, West Africa currently only produces 40–60% of their total consumation in rice (Haefele et al., 2002). In 1995, Mali, Burkina Faso, Senegal and Gambia together imported 578,000 tonnes of milled rice, valued at US $154 million (WARDA, 2000).

Enormous efforts and resources have been invested in the development of irrigation schemes to enhance irrigated rice production as a means of promoting food security and tackling rural poverty in the sub-region. With favourable climatic conditions, the potential for increasing irrigated rice productivity and output in the Sahel is indeed high (Dingkuhn and Sow, 1995). The adoption of structural adjustment and market liberalization policies, with support from the World Bank and IMF, has further enhanced this potential. These factors together offer an opportunity for increasing the competitiveness of the rice sector in West Africa (Kebbeh and Miezan, 2003).

However, many factors limit farmer's capacities to exploit potential production and market opportunity offered by rice. Rice yields are mainly limited by low inputs such as fertilizer, weed and pests and inadequate management of inputs, water and cropping systems. Although technologies have been developed to address some of these constraints, an important problem is that improved technologies generated to facilitate the productivity and profitability of irrigated rice systems

A. Bationo et al. (eds.), *Innovations as Key to the Green Revolution in Africa,*
DOI 10.1007/978-90-481-2543-2_16, © Springer Science+Business Media B.V. 2011

have not been successfully adapted and integrated into farmers' production environments. For instance, improved varieties need appropriate fertilizer doses coupled with good management practices to exploit the performances of new varieties. In order to improve rice productivity, a major challenge is to develop integrated and cost-effective crop management technologies taking into account the biophysical and socioeconomic environments of medium- and small-scale irrigated rice producers. Meeting this challenge will result in an increase of the productivity and profitability of irrigated rice production systems, thereby contributing to higher food security and increased farm revenues (WARDA, 2000).

Nitrogen is the most limiting factor for irrigated rice in the Sahel. The main causes of yield variations are related to low doses of N applied by farmers, timing of N fertilizer application, low N recoveries from applied fertilizers, use of old seedling at transplanting, weed management and agronomic management practices (Wopereis et al., 1999).

This research aimed to study different management options of fertilizer, weeds and varieties to improve rice productivity and profitability. The goal is to suggest integrated management options that could improve inputs, use efficiency and rice productivity and profitability for farmers.

Materials and Methods

The study has been undertaken during four rice growing seasons of 2005 and 2006 (two seasons per year) on two sites of the agronomic research stations of WARDA in the Senegal River Valley of Senegal (West Africa). The first site is located at Ndiaye (16°14′N, 16°14′W) in the delta valley. Soil salinity at Ndiaye and in the River delta is generally high due to marine salt deposits in the subsoil (Raes and Deckers, 1993). The second site is located at Fanaye (16°33′N, 15°46′W) in the middle at 240 km inland. Natural soil salinity is low or absent in the middle valley (Haefele et al., 2002). The original soil of Ndiaye (clay 40%; pH water 6.5; organic C 10 g kg^{-1}; CEC 13.0 cmol$_c$ kg^{-1}) was a typical Orthithionic Gleysol, according to FAO, or a Sulfic Tropaquept, according to the American classification system. The soil of Fanaye was an Eutric Vertisol or Typic Haplustert (Haefele et al., 2002). The new improved irrigated rice varieties developed by WARDA were used. The main two varieties released

by WARDA Sahel station (SAHEL 108 and SAHEL 202) and cultivated by more than 75% of farmers in the Senegal River Valley and IR 64 from IRRI were used as check varieties. New intraspecific (cross breeding of *Oryza sativa* × *O. sativa*) and interspecific (*O. sativa* × *O. glaberrima*) varieties were used (Table 1).

Two field experiments were used on the two sites during the two seasons of years 2005 and 2006. In the first experiment, the responses of 27 new improved varieties to N fertilizer applications were evaluated using a factorial 4 × 27 in a split-plot arrangement (Gomez and Gomez, 1983). Five levels of N fertilizer (0, 80, 120, 160 and 200 kg N ha^{-1}) and a control without fertilizer were used in the main plots as first factor and the varieties were used in the sub-plots as second factor.

The effects of interactions of the three factors (fertilizer, weed and varieties) were studied in a second experiment. A factorial 2 × 4 × 6 experiment (two weeding treatments, four levels of N fertilizer and six varieties) in a split-split-plot arrangement with randomized block design in three replications was used. The experiment was conducted at Ndiaye during the two seasons of 2006 (dry and wet season). The weeding treatments (weeding or no weeding) were used as first factor in the main plots. The four levels of N fertilizer (0, 60, 120 and 160 kg N ha^{-1}) were used as second factor in the sub-plots. The six varieties were used as third factor in the sub-sub-plots. Weeding, nitrogen and variety treatments were randomized in the blocks, subplots and sub-sub-plots, respectively. The plant seeds were sown in nurseries each season and transplanted 25 days after sowing. The experiment was conducted at Ndiaye during the two seasons of 2006 (dry and wet season).

In the two experiments, the plant seeds were sown in nurseries each season and transplanted 25 days after sowing. A basal uniform PK fertilization (26 and 50 kg ha^{-1} of P and K, respectively) was applied 10 days after transplanting in the form of triple super phosphate and potassium chloride, respectively. Nitrogen fertilizer was applied in the form of urea in three applications: 40, 30 and 30% at 21, 45 and 65 days after sowing, respectively. Rice grain yields have been evaluated at 14% of humidity for all samples at the end of each season.

At the start of the first season, soil samples were taken in the first 0–20 cm layer for laboratory analysis. Phosphorous was extracted by the Bray P1 (0.03 N NH$_4$F in 0.025 N HCl) and Olsen bicarbonate (0.5 N

Table 1 Responses in kilogram per hectare of new irrigated rice varieties to N fertilizer applications during the dry season of 2006 at Ndiaye, Senegal

| Varieties | Fertilizer N-P_2O_5-K_2O (kg ha^{-1}) | | | | | |
	0-0-0	0-46-30	80-46-30	120-46-30	160-46-30	200-46-30
C15	2830	4130	5970	6930	8530	7930
ECIA-31	3370	5130	6270	7470	7500	9670
IR64	3770	4800	4600	7000	8700	9700
ITA 344	3130	5800	5770	5970	5870	7400
JAYA	4000	4970	6600	6530	7130	9470
SAHEL 108	2500	4200	5900	6500	8430	10,600
TOX 32	2070	3500	6630	7270	5600	6470
[a]WAS 105b IDESSA	3500	6270	5700	6770	6700	7070
[a]WAS 122 IDESSA10	4020	4850	8520	6600	7420	8780
[a]WAS 161	3400	3870	6500	5900	6500	6930
WAS 173B221	3170	3770	5900	5830	8470	8130
WAS 173BB53	3700	4500	3330	5370	6070	7100
WAS 19bb	4000	5030	6500	7470	8130	7500
[a]WAS 191 17fkr	4030	4900	6370	6430	7670	8770
[a]WAS 191 81fkr	3400	4970	5900	7630	6670	8570
WAS 21 bb	4030	4030	6000	6270	9270	8530
WAS 30	4200	3470	6030	7400	6900	9470
WAS 33	4870	4470	7700	6930	8100	7000
WAS 44	3700	5700	6730	8430	6330	7630
WAS 49 bb9134	4100	4430	4970	5400	6630	7970
WAS 49 bb9144	2870	4900	5630	6630	6600	7730
WAS 55	2470	6030	6300	8030	7770	9630
WAS 57bb31	4800	6170	6470	8570	9230	9070
WAS 62bb141	3650	4680	7070	8110	7400	8470
WAS 63 22 11 33	1570	3930	4570	5570	5800	7300
WAS 63 59101	1830	5330	4470	7700	6970	8300
WAS 161 b92[a]	3500	3030	6700	7330	7770	9070
Minimum	821	3030	3330	5370	5600	6470
Maximum	4870	6270	8520	8570	9270	10,600
Mean	3425	4699	6041	6890	7339	8306
SD	158	846	1038	893	1036	1052

[a]Crossing *O. sativa* × *O. glaberrima* or NERICA

$NaHCO_3$) methods (Sharpley et al., 1984). Soil pH was measured in 1 N KCl using a 2:1 solution to soil ratio. Organic carbon was measured by the Walkley and Black (1934) procedure.

The GLM procedure of SAS software has been used for statistical analysis of the variance.

The response curves to fertilizer N applications were used for fertilizer N use efficiencies by rice varieties. Fertilizer N use efficiency (NUE) was calculated as a ratio between total N mobilized in the grain and N fertilizer applied (1).

$$NUE\ (\%) = ((N_x - N_o) \times N_x)/100 \qquad (1)$$

N_o and N_x are grain N yields for the control (no N application) and for the applied dose of N, respectively.

Results and Discussion

Nitrogen Use Efficiency

Fertilizer N recoveries were only measured during the dry season of 2005 at low (80 kg N ha^{-1}) and medium (120 kg N ha^{-1}) level of N applications. Nitrogen recoveries were affected by varieties and the quantities of N fertilizer applied (Fig. 1). NUE varied from 9 to 43%. More N was generally mobilized with low levels of N applied. This indicated a better use of low quantities of N by varieties. A better use of low N fertilizer was particularly observed with two newly developed varieties (WAS 191-1-5 and WAS 57-B-B-3) that recovered more N than the check varieties

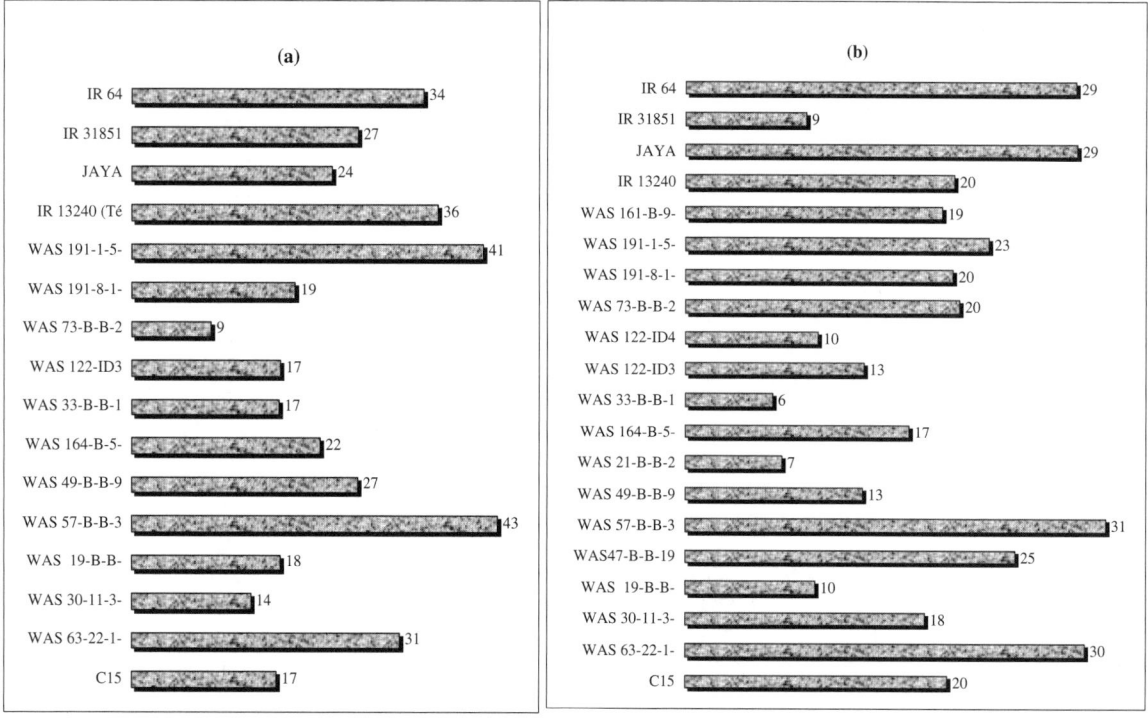

Fig. 1 Nitrogen use efficiency (NUE in % of N applied) by varieties for (**a**) 80 and (**b**) 120 kg N ha^{-1} during the humid season at Ndiaye

(IR 13240 and IR 64). For high level of N applied (120 kg N ha^{-1}), WAS 57 B-B-3 and WAS 63-22-1 recovered more N than the check IR 64.

Responses to N Fertilizer

Good responses to N application ($p<0.001$) and varieties ($p<0.001$) were observed on the two sites (Tables 2 and 3). But interaction was observed ($p<0.001$) between the two factors, indicating that N responses depended on varieties.

Ndiaye

Rice grain yields varied from 1.8 t ha^{-1} with zero fertilizer to 9.2 t ha^{-1} with an application of 160 kg N ha^{-1} on WAS 21-B-B or WAS 57-B-B (Table 2). Six new varieties (WAS 19-B-B, WAS 191, WAS 21, WAS 30, WAS 33 and WAS 57) produced more than 4 t ha^{-1} of paddy without fertilizer. Four new varieties (WAS

105 B, ITA 344, WAS 55 and WAS 57-B-B) produced at least 6 t ha^{-1} without N fertilizer applications. This indicated the ability of those varieties for internal N use efficiency. Four new varieties (WAS105 B, ITA 344, WAS 55 and WAS 57-B-B) produced at least 6 t ha^{-1} without N fertilizer applications (only PK fertilizer was applied). This indicated the ability of those varieties for internal N use efficiency. Statistically differences were observed between the 27 new varieties.

Combined management of varieties and N fertilizer increased grain yields from 2.5 to 10.1 t ha^{-1}. Differences were observed. Significant interaction existed between the two factors (N and variety), meaning that N recommendations should take into account the N use efficiency of each variety. Six varieties (C15, ECIA-31, JAYA, TOX 32, WAS 173-B and WAS 30-11) with efficient use of low quantities of N fertilizer (80 kg N ha^{-1}) and good yields (6.5–7.3 t ha^{-1}) were identified. Seven varieties (C15, ECIA-31, JAYA, TOX 32, WAS 161-B, WAS 173-BB and WAS 62-BB) efficiently used high quantities of N fertilizer (160 kg N ha^{-1}) and yielded (9.0–10.5 t ha^{-1}). The new varieties can be classified according to their responses

Table 2 Responses in kilogram per hectare of new irrigated rice varieties to fertilizer N applications during the dry season of 2005 at Fanaye

| Varieties | Fertilizer N-P_2O_5-K_2O (kg ha^{-1}) | | | | | |
	0-0-0	0-46-30	80-46-30	120-46-30	160-46-30	200-46-30
C15	4177	3781	6788	9760	6307	7283
ECIA-31	3945	3938	6969	10,051	9707	7031
IR64	3464	2496	6756	6485	6463	3708
ITA 344	2791	3779	7873	8367	8767	4267
JAYA	4390	4741	8442	9528	8919	6779
SAHEL 108	3041	3496	6422	7434	6143	4963
TOX 32	3148	4215	9512	8314	5806	6922
WAS 161-b-6[a]	3136	3744	8121	9066	8937	6172
WAS 122 ID1[a]	3042	2246	4325	5709	4555	3872
WAS 122 ID2[a]	2726	1853	5181	4230	6042	3147
WAS 122 IDES[a]	2376	2132	5064	4841	6867	3123
WAS 173BB53	2646	3461	6742	9217	8985	5258
WAS 19bb	219	417	6195	7290	8023	4485
WAS 191 81fkr[a]	2339	2554	6701	6050	8402	4839
WAS 44	2547	3500	6550	6750	7966	5628
WAS 49 bb9134	3699	2812	6721	7138	6469	4757
WAS 62bb141	3244	3188	7229	7707	7556	5138
WAS 62bB2	3522	2404	5628	7945	6245	4338
WAS 63 22 11 33	1974	2563	7163	4602	6732	4284
WAS 63 59101	2484	2316	6576	5490	6985	4545
WAS 105b IDESSA[a]	4031	4273	8373	8013	9667	5875
WAS 173B221	3425	3557	6790	7617	8659	6740
WAS 191 17fkr	2198	2554	6464	7644	7863	6059
WAS 21 bb	2713	2801	7533	6698	7704	5566
WAS 30	2621	3089	6178	6810	7124	6495
WAS 33	2645	2594	6175	8063	7240	5332
WAS 49 bb9144	3570	2752	7578	7769	9356	5166
WAS 55	2921	2937	7146	7912	5911	3932
WAS 57bb31	3272	4201	6784	7867	9301	5013
WAS 161 b92[a]	2813	3436	6750	6787	6938	3670
Minimum	219	417	4325	4230	4555	3123
Maximum	4390	4741	9512	10,051	9707	7283
Mean	2971	3061	6824	7372	7521	5146
SD	794	884	1045	1464	1338	1172

[a]Crossing *O. sativa* × *O. glaberrima* or NERICA

Table 3 Management options of varieties and fertilizers to improve rice productivity and profitability on the two sites (Ndiaye and Fanaye)

Sites	Fertilizer applied N-26P-50 K[a]	Recommended varieties	Yields (t ha^{-1})
Ndiaye	Less than 80 kg N ha^{-1}	WAS 57-B-B-3, WAS 55-B-B-2, WAS 105-B-IDESSA1, ITA 344, WAS 44-B-B-6	5–6
	80 kg N ha^{-1}	WAS 33-B-B-1, WAS 122-IDESA2 WAS 62B-B-1, WAS 122-IDESA1	7–8.5
	120 kg N ha^{-1}	WAS 57-B-B-3, WAS 55-B-B-2, WAS 62-B-B-1, WAS 44-B-B-6	8–9
Fanaye	Less than 80 kg N ha^{-1}	WAS 57, ECIA, JAYA, JAYA, C15, ECIA, TOX	4–6
	80 kg N ha^{-1}	JAYA, C15, ECIA, TOX, WAS 173	7
	120 kg N ha^{-1}	WAS 161-B, JAYA, WAS 105-B-IDESSA1, TOX 32	8–9

[a]N and basal uniform PK fertilization of 26 and 50 kg h^{-1} of P and K, respectively

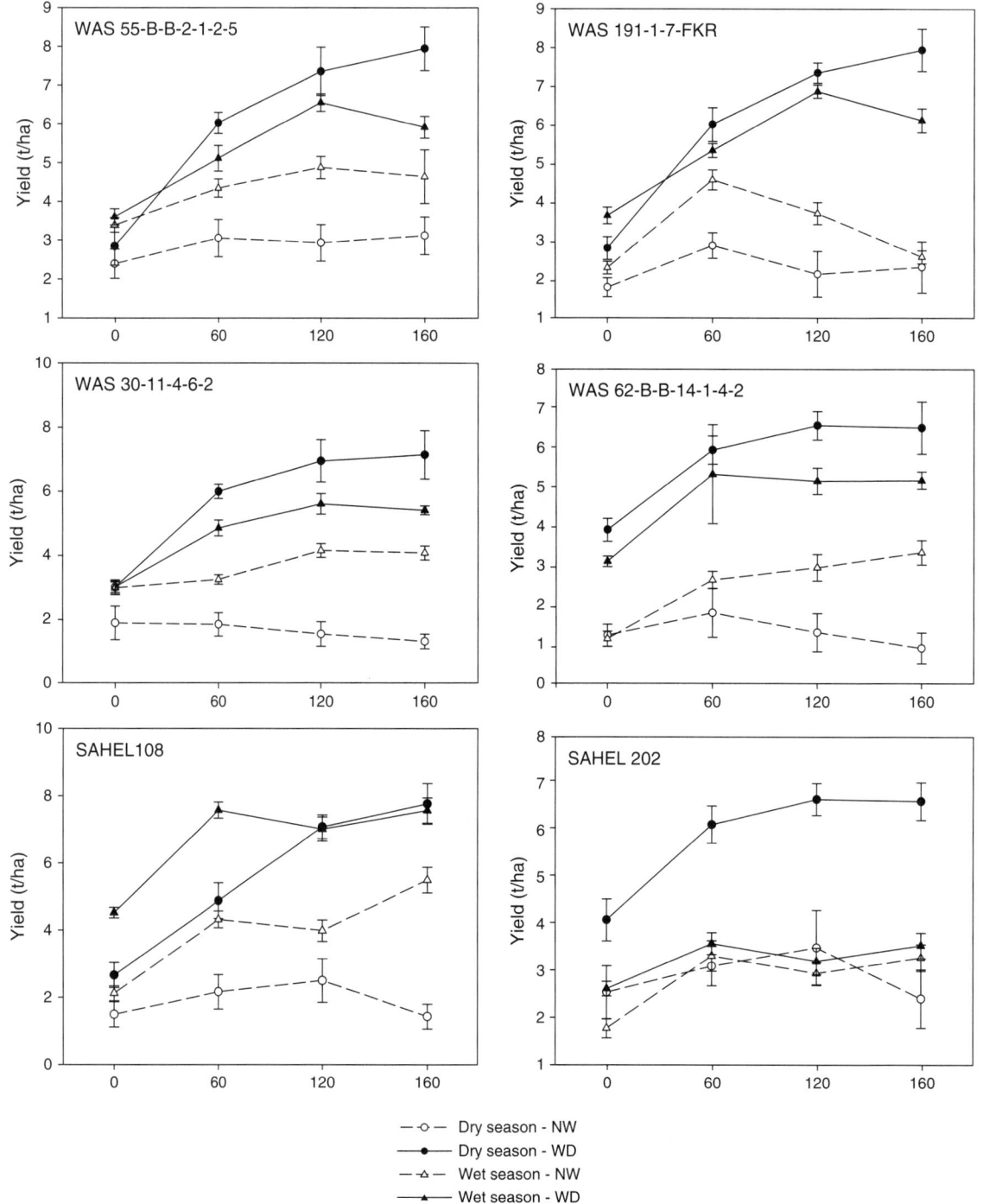

Fig. 2 Effects of weed control on the responses of six varieties to N fertilizer applications during the dry and wet season of 2006 at Ndiaye

to N fertilizer applications. Five varieties had highest yields without nitrogen application, yielding more than 5 t ha^{-1} of grain when only P fertilizer was used. With low application of N fertilizer (80 kg N ha^{-1}), four varieties (WAS 33, WAS 62, WAS 122 IDSA-1 and WAS 122 IDSA-2) were the best and produced 7–9 t ha^{-1} of grain. With medium level of N fertilizer (120 kg N ha^{-1}), four varieties (WAS 57, WAS 55, WAS 62 and WAS 44) were the best and produced 8 t ha^{-1} of grain. Two varieties (WAS 57-B-B-3 and WAS 21-B-B-2) produced 9 t ha^{-1} of grain with an application of 160 kg N ha^{-1}.

Fanaye

Rice grain yield varied from 2 t ha^{-1} without fertilizer to 10 t ha^{-1} with an application of 120 kg N ha^{-1} on ECIA-31 (Table 2). With no fertilizer application, JAYA and WAS 105-B-IDSA produced at least 4 t ha^{-1}. Without N applications ECIA-31, SAHEL 108, C15 and WAS 105-IDSA, WAS 105-B-IDSA and WAS 57-B-31 produced at least 4 t ha^{-1}. Good responses were observed with low N applications (80 kg N ha^{-1}) at Fanaye during the dry season of 2005. Twenty-six varieties produced from 6 to 9.5 t ha^{-1}. TOX 32 was the most productive (9.5 t ha^{-1}) with 80 kg N ha^{-1} applied.

Taking into account the financial capacities of farmers to invest on fertilizer, the results provide indications that could be used to suggest different management options to improve rice productivity depending on the site and good weed control conditions (Table 3).

Weed × N Interactions

Rice yield was affected by N and weed control during the two seasons. But interaction was not observed between the two factors, meaning that weeds affected rice response to N applications in the same way during the two seasons. Weeds reduced rice responses to N and NUE during the two seasons (Fig. 2). This indicated that weed control is essential to improve fertilizer use efficiency in irrigated rice system (Haefele et al., 2000; Johnson et al., 2004). During the two seasons, good responses to N fertilizer applications were particularly observed with six varieties. Without weed control, all varieties produced less than 3 t

ha^{-1} of paddy despite the application of N fertilizer. Otherwise, fertilizer applications cannot be recommended for the six varieties in poor weed control conditions. However, two new varieties (WAS 55-B-B-1 and WAS 191-1-7-FKR) performed well during the wet season without weeding. The two varieties produced an optimum yield of 4 t ha^{-1} with 60 kg N ha^{-1} applied.

Conclusion

Weed control is an essential factor to improve rice response to N fertilizer. In poor weed control conditions, no more than 60 kg N ha^{-1} should be recommended. But with a good weed control, many cost-effective management options could be used to improve rice productivity. Nitrogen use efficiencies can be improved with appropriate choice of varieties, and fertilizer recommendation should take into account the abilities of varieties for a better use of N applied. While individual technologies are less effective, integrated management technologies of fertilizer, weed and varieties are more appropriate to improve irrigated rice productivity.

References

Dingkuhn M, Sow A (1995) Potential yields of irrigated rice in the Sahel. In: Miézan KM et al (eds) Irrigated rice in the Sahel: prospects for sustainable development. Africa Rice Development Association, Bouaké

Gomez AK, Gomez AA (1983) Statistical procedures for agricultural research. Wiley, New York, NY

Haefele SM, Johnson DE, Diallo S, Wopereis MCS, Powers S, Jamin I (2000) Improved soil fertility and weed management is profitable for irrigated rice farmers in Sahelian West Africa. Field Crops Res 66:101–113

Haefele SM, Wopereis MCS, Wiechmann H (2002) Long-tem fertility experiments for irrigated rice in the West African Sahel: agronomic results. Field Crop Res 78:119–131

Johnson DE, Wopereis MCS, Mbodj D, Diallo S, Powers S, Haefele SM (2004) Timing of weed management and yields losses due to weeds in irrigated rice in the Sahel. Field Crops Res 85:31–42

Kebbeh M, Miezan KM (2003) Ex-ante evaluation of integrated management options for irrigated rice production in the Senegal River valley. Field Crops Res 81:87–94

Raes D, Deckers J (1993) Les sols du delta du fleuve Senegal. Propriétés physiques et chimiques. Bulletin technique No 8, KULeuven-SAED, Saint Louis, Sénégal

Walkley A, Black JA (1934) An examination of the Detjareff method for determining soil organic matter and a proposed modification of the chromatic acid titration method. Soil Sci 37:29–38

West Africa Rice Development Association (WARDA) (2000) Partners in development: Responding to the challenges of food security and poverty eradication in Africa. WARDA, Bouaké

Wopereis MCS, Donovan C, Nebié B, Guindo D, N'Diaye MK (1999) Soil fertility management in irrigated rice systems in the Sahel and Savanah regions of West Africa. Part I. Agronomic analysis. Field Crops Res 61:125–145

Integrated Soil Fertility Management for Increased Maize Production in the Degraded Farmlands of the Guinea Savanna Zone of Ghana Using Devil-Bean (*Crotalaria retusa*) and Fertilizer Nitrogen

B. D. K. Ahiabor, M. Fosu, E. Atsu, I. Tibo, and I. Sumaila

Abstract   The native N and P of soils of the Guinea Savanna Zone of northern Ghana are only about 20 and 10% of the crops' requirements, respectively, and organic matter content is usually below 1%. Hence, cereal yields without soil amendments are usually below 500 kg/ha. Organic residue and mineral fertilizer combinations are necessary to increase nutrient use efficiency. Devil-bean is a very promising leguminous cover crop for this agro-ecology. The best time to intercrop devil-bean in maize, effect of P on the maize, and the effect of incorporated devil-bean biomass on grain yield of N-fertilized maize were investigated. In 2003, devil-bean was drilled in maize at 1, 3, and 4 weeks after planting (WAP) the maize which received 0, 20, and 40 kg P/ha. Phosphorus enhanced maize growth and yield. The devil-bean biomass was incorporated into the soil in the 2004 growing season. Maize was planted, fertilized with 0, 20, and 40 kg N/ha, and intercropped again with devil-bean as before. About 40 kg N/ha fertilized maize grown on incorporated devil-bean intercropped at 1 WAP in 2003 had the highest grain yield of 1.59 t/ha. In 2005, the devil-bean intercropped in the 2004 40 kg N/ha maize at 3 WAP produced the highest biomass containing 42–88, 4–11, and 25–52 kg/ha of N, P, and K, respectively. Maize grain yield significantly increased with incorporated biomass with the highest biomass producing the highest grain yield. The cumulative effect of the biomass applications was significant in this study.

Keywords Devil-bean biomass · Fertilizer nitrogen · Intercropping · Maize production · Soil fertility management

Introduction

Soils in the Savanna zones of northern Ghana are generally poor in nutrient status, especially in available nitrogen and phosphorus and have low organic matter content. The native N and P of soils of the Guinea Savanna Zone of northern Ghana are only about 20 and 10% of the crops' requirements, respectively, and organic matter content is usually below 1% (Tiessen, 1989). Crop production in this area must always be backed by nutrient amendments for economic yield but unfortunately the mostly resource-poor peasant farmers in this agro-ecological zone cannot afford to buy the mostly expensive chemical fertilizers to ameliorate the effects of poor fertility of their farmlands. Even where the application of such chemical fertilizers is possible, nitrogen is the most frequently applied fertilizer and often the only nutrient element added to the soil. There are even some indications from literature (Hayman, 1982) that increasing levels of N fertilizers may inhibit arbuscular mycorrhizal (AM) formation and may negatively affect AM fungal population in the soil, thereby adversely affecting phosphorus uptake.

An alternative that has been used by such farmers for ages is biological nitrogen fixation (BNF) which involves the use of legumes (grain, herbaceous, and tree) as components of the farming systems found in this area. Such legumes are grown as sole crops in rotation, relay crops, or intercrops with cereals (sorghum, millet, or maize). The promotion of herbaceous legumes as relay or rotation crops with

B.D.K. Ahiabor (✉)
CSIR-Savanna Agricultural Research Institute, Tamale, Ghana
e-mail: bahiabor@yahoo.com

cereals is currently on the increase. These herbaceous nitrogen-fixing cover legumes have been excellent ways to supply substantial amounts of organic carbon, N, and recycled nutrients to annual crop rotation systems by returning the total biomass produced to the soil prior to planting cereals. The direct N benefit from a well-developed leguminous cover crop on the subsequent crop has been estimated by a number of authors, and it ranges between 50 and 100 kg N/ha (Greenland, 1985). Fosu (1999) also observed a range of 40–76 kg N/ha. *Canavalia ensiformis, Mucuna pruriens, Glycine max,* and *Vigna unguiculata* have been reported to potentially contribute considerable amounts of N to succeeding crops (Sanginga et al., 1996; Ravuri and Hume, 1992). The residual effects on the following crops are less certain and depend on both the qualities (decomposition rate) and time of incorporation of the organic matter as well as the soil conditions influencing N mineralization rate from the young organic matter (Cheruiyot et al., 2007). Apart from being a source of N and other nutrients, cover crops also provide protection against soil erosion, improve soil structure, and interrupt the cycles of diseases and insect pests.

The nitrogen fixed by these cover crops becomes available to the subsequent or companion cereal crops which are high nitrogen feeders. The nitrogen-fixing efficiency of the legume depends on the legume type, the degree of nodulation, and the nitrogen-fixing capacity of the nodules. In screening four legume species, Ahiabor et al. (2007) obtained relatively high values of dry matter, N_2 fixation, and N accumulation for devil-bean (*Crotalaria retusa*) which is a perennial legume with high tolerance to drought and bushfires.

Experiments were therefore conducted to determine the most appropriate time of intercropping maize with devil-bean and to determine the optimum P level required by the intercropped maize. The effect of the biomass of the incorporated devil-bean on growth and grain yield of maize when ploughed into the soil in the subsequent year and planted to maize and fertilized with mineral nitrogen was also studied.

Materials and Methods

Experimental Site

The experiment was set up on-station in the experimental field of the Savanna Agricultural Research Institute of the Council for Scientific and Industrial Research, Ghana, situated at Nyankpala (9°25″N and 0°58″W) in the Tolon–Kumbungu district of the Northern region of Ghana. This area experiences a monomodal rainfall pattern (April–October) with a mean annual rainfall of 1000 mm and a variability of between 15 and 20% (Kasei, 1988). The mean annual temperature is about 28°C with the daily maximum sometimes being around 42°C during the hottest months of February and March, and the lowest temperatures (about 20°C) are recorded in December and January when the area comes under the influence of the cold dry North-Easterly Trade winds ('Harmattan' winds) from the Sahara Desert.

The soil was classified as Gleyi-ferric Lixisol (FAO/UNESCO) with a pH of 4.5. Base saturation was 32% and CEC was below 4 cmol$^+$/kg. Organic matter was below 1% and total N was less than 0.4%. Available P was 14 mg/kg and exchangeable K was less than 40 mg/kg.

Experimental Procedures

In 2003, the best time to intercrop maize (Dorke SR) with devil-bean and also the optimum P level to apply to the maize were investigated. The devil-bean was drilled at 40 kg seed/ha in two rows 40 cm apart in between two maize rows on 6 m × 4 m plots at 1, 3, and 4 weeks after planting the maize. The planting distance for the maize (2 seeds per hill) was 80 × 40 cm. The maize received P application rates of 0, 20, and 40 kg P/ha as triple super-phosphate (TSP) in addition to 40 kg N/ha (as sulphate of ammonia) and 30 kg K/ha (as muriate of potash). Half of the nitrogen and all of the potassium were applied 2 weeks after emergence. The remaining nitrogen was applied 4 weeks later. The fertilizers were deposited in bands about 5 cm away from the maize stands. The devil-bean was also given 40 kg P/ha (as TSP) and 30 kg K/ha (as muriate of potash) which were applied the same way as for the maize. At full maturity, the maize was harvested minus the outermost rows. The cobs were de-husked, shelled, and the grains were spread on a concrete floor and sufficiently dried in the sun for 5 days and weighed. The devil-bean stands were, however, left to grow into the following year.

At the beginning of the 2004 growing season, the devil-bean biomass (stems plus leaves) was harvested by slashing at the base of the plant. Sub-samples of this matter were dried in a forced-air oven at 80°C for 48 h and weighed. The rest of the biomass was chopped into pieces by cutlass and evenly spread on the plots and then incorporated into the soil 20 cm deep by hoe. After 2 weeks, the triplicate plots were hoe-harrowed, and the same maize variety was planted at the same spacing and fertilized with 0, 20, and 40 kg N/ha (sulphate of ammonia) as well as with 30 kg/ha each of P (TSP) and K (muriate of potash).

In 2005, 12 treatments (in triplicates) comprising devil-bean, N, and P combinations were tested on-station at Nyankpala as was in 2003 and 2004. Devil-bean was either intercropped in the maize at 1, 3, or 4 WAP the maize or not, and the latter was fertilized with 0, 20, and 40 kg N/ha and all treatments received 40 kg P/ha and 30 kg K/ha.

The dry matter of the devil-bean biomass incorporated was determined as stated earlier whereas the nitrogen, phosphorus, and potassium contents were, respectively, determined by the Kjeldahl digestion method (Tel and Hagatey, 1984), Bray I method (Bray and Kurz, 1945), and ammonium acetate method (Nelson and Sommers, 1982).

In 2006, a follow-up experiment was conducted to confirm the cumulative effect of incorporating small quantities of devil-bean into maize plots on maize growth and grain yield. All treatments were replicated three times. The 2005 devil-bean stands still growing on the respective plots were slashed, and small quantities of between 2.1 and 4.4 t/ha were ploughed into the soils of the respective plots 2 weeks prior to

planting the maize variety Obatanpa (110-day maturity variety) which was fertilized with 60 kg N/ha, 30 kg P/ha, and 30 kg K/ha. No devil-bean was intercropped in the maize this time. The maize was grown to full maturity, and the cobs were harvested, shelled, and thoroughly sun-dried on a concrete floor for 5 days and then weighed.

Statistical Analysis

The data were subjected to analysis of variance using Statistix software (STATISTIX 7), and significance of treatment effects was tested at 5% level of Treatment means were separated using the least significant difference (LSD).

Results and Discussion

The yield of maize grain was not significantly increased by intercropping with devil-bean whose growth in the maize was not as good as expected (Table 1). Maize intercropped with devil-bean but without fertilizer P generally performed lower than those with devil-bean and applied P, or without devil-bean intercrop. Usually, an associated legume crop rarely has any beneficial effect on the cereal intercrop partner in the first season as any nitrogen that

Table 1 Effects of devil-bean intercrop and fertilizer P on grain yield, growth, shoot nutrient (N, P, and K) uptake, and arbuscular mycorrhizal formation of maize grown in a Gleyi-ferric Lixisol at Nyankpala in 2003

Treatment	Grain yield (t/ha)	Stover yield (t/ha)	AMF spore no. (/5 g a.d.s.)	AMF colonization (%)	N (kg/ha)	P (kg/ha)	K (kg/ha)
Maize/devil-bean 1 WAP + P0	1.57 b[a]	2.83 c	238	12.3 ab	99.6	6.8 ab	86.4 ab
Maize/devil-bean 1 WAP + P20	1.84 ab	3.78 a–c	125	10.8 ab	96.8	7.9 ab	71.5 ab
Maize/devil-bean 1 WAP + P40	1.77 ab	3.86 a–c	222	21.6 a	113.9	15.8 a	96.3 a
Maize/devil-bean 3 WAP + P0	1.71 b	3.02 c	174	4.9 b	103.8	6.9 ab	89.4 ab
Maize/devil-bean 3 WAP + P20	1.89 ab	4.23 ab	66	17.9 ab	85.7	8.1 ab	63.7 ab
Maize/devil-bean 3 WAP + P40	1.96 ab	4.22 ab	170	9.0 ab	83.9	11.1 ab	74.1 ab
Maize/devil-bean 4 WAP + P0	1.69 b	2.89 c	236	15.0 ab	76.1	6.3 b	64.2 ab
Maize/devil-bean 4 WAP + P20	1.87 ab	3.95 ab	125	10.2 ab	104.3	6.8 b	43.8 b
Maize/devil-bean 4 WAP + P40	2.16 ab	4.04 ab	124	13.1 ab	104	12.1 ab	96.2 a
Maize + P0	1.74 ab	3.04 bc	149	14.6 ab	67.9	5.6 b	66.0 ab
Maize + P20	1.74 ab	3.38 a–c	122	9.8 ab	100.1	12.0 ab	87.9 ab
Maize + P40	2.55 a	4.21 a	91	13.9 ab	70.8	10.3 ab	64.4 ab

[a]Means followed by the same letter(s) within a column are not significantly different at $P = 0.05$ by the least significant difference method

may have been fixed is not immediately accessible by the companion cereal crop. The intercrop partners therefore tend to compete with each other, and the degree of this competition largely depends on the intercrop arrangement and management practices adopted. The results here indicate that in the absence of an external supply of a vital nutrient like phosphorus to the maize, the effect of the competition on the growth and grain yield of the cereal was more pronounced than when P was present resulting in the decreased growth and lower grain yield in the former treatment. Maize fertilized with 40 kg P/ha and intercropped with devil-bean 1 week after planting (1 WAP) was generally the most colonized (22%) by arbuscular mycorrhiza (even though there was no significant effect of treatment on the AMF spore production) and tended to have the highest content of shoot P. Arbuscular mycorrhizal fungi (AMF) have been reported to enhance P nutrition in almost all agricultural crops (Atayese et al., 1993; Smith and Read, 1997). Mosse (1973) and Gerdemann (1975) reported increased nutrient uptake, especially that of phosphorus, in soils with low P content when plants were infected with arbuscular mycorrhizal fungi to produce mycorrhizas. Even though application of high rates of phosphate fertilizers to soil decreases the percentage of infection of roots with AMF (Hayman et al., 1975) and then inhibits the ameliorating effects of the fungi on plant growth (Thomson et al., 1986), the highest P rate of 40 kg/ha applied to the maize in this work rather produced the highest AMF colonization compared to the lower rates, thereby implying

that this level might be just about the optimum for the maize and the AMF in terms of P nutrition. The relatively high AMF colonization rate and the enhanced P nutrition were not accompanied by any significantly improved plant growth (stover dry weight) which is contrary to the findings of Daft and El-Giahmi (1974), Mosse et al. (1976) and Atayese et al. (1993) that improved P uptake associated with arbuscular mycorrhizal plants is largely the result of enhanced shoot growth of such mycorrhizal plants. Devil-bean dry matter estimated at the beginning of the following growing season was very low, ranging between 2 and 4 t/ha (Table 2).

Results obtained in 2004 showed that maize grown on incorporated biomass of devil-bean that had been intercropped at 1 WAP and fertilized with 40 kg N/ha was tallest at 31, 42, 56, and 77 DAP (Table 2). Maize grain yield was also highest with 40 kg N/ha without devil-bean or when devil-bean was intercropped at 1 WAP compared with all the integrated treatments whereas the least grain yield was obtained with 1 WAP + 20 kg N/ha treatment (Table 2).

The dry matter and estimated nutrient content of devil-bean incorporated in 2005 are presented in Table 3. Devil-bean planted 3 WAP in the 2004 maize which received 40 kg N/ha produced the highest dry matter. This is similar to devil-bean intercropped 1 WAP in maize that received 20 kg N/ha. These treatments also produced the highest amounts of biomass N. Organic (biomass) N applied with the cover crop generally ranged from 88 to 42 kg/ha. The P applied

Table 2 Effects of incorporated devil-bean biomass (t/ha) and fertilizer N (kg/ha) on plant height (cm) at 31, 42, 56, and 77 DAP and grain yield of maize (t/ha) grown in a Gleyi-ferric Lixisol at Nyankpala in 2004

Treatments (crop combinations in 2003)	Devil-bean biomass incorporated in 2004	Fertilizer N	31 DAP	42 DAP	56 DAP	77 DAP	Grain yield
Maize/devil-bean 1 WAP	3.01	0	31.2 bcd[a]	60.2 cd	83.0 e	99.6	0.35 de
	3.15	20	32.5 abc	65.6 bc	99.4 cde	96.8	0.22 e
	3.53	40	41.1 a	90.8 a	142.4 a	113.9	1.59 a
Maize/devil-bean 3 WAP	2.49	0	31.2 bcd	59.6 cd	89.8 de	103.8	1.24 abc
	2.49	20	29.9 bcd	65.8 bc	105.0 b–e	85.7	0.87 b–e
	4.16	40	36.6 ab	85.3 a	136.3 ab	83.9	1.21 abc
Maize/devil-bean 4 WAP	2.04	0	24.4 cd	49.0 d	76.5 e	76.1	0.51 de
	2	20	28.1 bcd	60.0 cd	94.3 de	104.3	0.71 cde
	2.19	40	31.6 bcd	78.2 ab	130.2 abc	104	1.40 ab
Maize	0	0	23.8 d	54.1 cd	75.4 e	67.9	0.70 cde
	0	20	26.6 cd	64.8 bcd	106.7 b–e	100.1	0.92 bcd
	0	40	24.4 cd	67.6 bc	117.2 cd	70.8	1.81 a

[a]Means followed by the same letter(s) within a column are not significantly different at $P = 0.05$ by the least significant difference method

Table 3 Total devil-bean biomass (t/ha) and estimated nutrients (kg/ha) incorporated into the soil at the beginning of 2005 cropping season and their effects on grain yield of maize (t/ha) grown in a Gleyi-ferric Lixisol at Nyankpala in 2005

Treatments (crop combinations in 2004)	Devil-bean biomass incorporated in 2005	Fertilizer N	N	P	K	Grain yield
Maize/devil-bean 1 WAP	2.5	0	50	6.2	29.5	1.1
	4.1	20	82	10.1	48.3	1.2
	3.6	40	72	8.9	42.4	2
Maize/devil-bean 3 WAP	2.4	0	48	5.9	28.3	1.4
	2.5	20	50	6.2	29.5	1.9
	4.4	40	88	10.8	51.9	2.7
Maize/devil-bean 4 WAP	2.1	0	42	5.2	24.8	1.7
	2.2	20	44	5.4	25.9	1.3
	2.6	40	52	6.4	30.6	2.3
Maize	0	0	nd[a]	nd	nd	0.7
	0	20	nd	nd	nd	0.9
	0	40	nd	nd	nd	1.1
LSD (5%)	1.8		36	4.4	21.2	0.8

[a]Not determined

was between 11 and 5 kg/ha, and the K was between 52 and 25 kg/ha.

The devil-bean intercropped in the maize at 1 and 3 WAP might have had enough time to establish before the maize got fully established compared to the 4 WAP treatment, thereby experiencing a lesser degree of competition from the maize which may have resulted in a better growth of the former. Competition for space, light, soil water and nutrients has been known to be a major factor that influences the relative performance of partner crops in intercrop and relay cropping systems.

The yield of maize grain was significantly influenced by incorporating the biomass of devil-bean that was intercropped in the maize the previous year. The treatments having the highest devil-bean biomass production also resulted in the highest maize grain yield (Table 3). This observation agrees with the findings of Cheruiyot et al. (2001) who evaluated some legumes for short-rain conditions and identified dolichos [*Lablab purpureus* (L.) Sweet] as the most suitable legume for the fallow periods because of its production of a larger and higher quality aboveground biomass than the other legumes. It is commonly recognized that the amount of biomass generated by a pre-cereal legume crop greatly influences the growth, mineral nutrition, and yield effects of the legume on the cereal as a result of the high amounts of organic matter and mineralized N made available to the cereal (Cheruiyot et al., 2001, 2003). The high N contents of the biomass of the treatments that resulted in the highest grain yields may have enhanced the

decomposition of the biomass leading to a relatively higher mineralization rate and more rapid nutrient release to the succeeding maize. Palm and Sanchez (1991), Palm et al. (2001), Wang et al. (2004), and Nziguheba et al. (2005) have linked the rate of nutrient release to biochemical properties, especially lignin, polyphenols, and N content from studies done on litter mineralization.

It should, however, be understood that the cumulative effect of devil-bean dry matter application also played a significant role. This can be deduced from the fact that in 2003, the treatments without devil-bean but with 40 kg P/ha (Table 1) gave the highest yield suggesting the presence of some competition of devil-bean with the maize. In the subsequent years, the cumulative effect of devil-bean biomass application on maize grain yield far exceeded the competitive effect, hence the increase.

The devil-bean intercropped in 2005 at either 1 or 3 weeks after planting (WAP) maize that received 40 kg N/ha and at 1 WAP maize supplemented with 20 kg N/ha tended to produce the highest shoot dry matter (biomass) when harvested in 2006 (Table 4), which confirms the conclusions drawn in 2005 on the effects of the same treatments. However, the growth and yield responses of the maize did not generally depend on the quantities of the devil-bean biomass ploughed in Table 4. This is explained by the fact that plots that had no devil-bean grown on them or incorporated into them for 2–3 years (especially those which even did not receive any N application) produced the same

Table 4 Cumulative effects of intercropping devil-bean and incorporating its biomass on stover dry weight and grain yield of maize grown in a Gleyi-ferric Lixisol at Nyankpala in 2006

Treatment combinations (in 2005)[a]	Devil-bean dry matter[b]	Stover yield	Grain yield
Maize/devil-bean 1 WAP + N0	2.6 a–d[c]	1.63 ab	4.13 ab
Maize/devil-bean 1 WAP + N20	4.1 a–c	1.63 ab	3.72 ab
Maize/devil-bean 1 WAP + N40	4.3 ab	1.78 ab	4.90 a
Maize/devil-bean 3 WAP + N0	2.3 bcd	2.16 a	3.73 ab
Maize/devil-bean 3 WAP + N20	2.5 a–d	2.05 a	4.70 ab
Maize/devil-bean 3 WAP + N40	4.4 a	1.92 ab	4.74 ab
Maize/devil-bean 4 WAP + N0	2.2 cd	1.59 ab	3.91 ab
Maize/devil-bean 4 WAP + N20	2.1 d	1.73 ab	4.53 ab
Maize/devil-bean 4 WAP + N40	2.2 cd	1.85 ab	4.68 ab
Maize + N0	nd[d]	1.63 ab	4.87 a
Maize + N20	nd	1.30 b	3.42 b
Maize + N40	nd	1.84 ab	4.21 ab

[a]In 2006, each plot was fertilized with 60 kg N/ha, 30 kg P/ha, and 30 kg K/ha
[b]Devil-bean biomass incorporated at the beginning of the 2006 growing season
[c]Means followed by the same letter(s) within a column are not significantly different at $P = 0.05$ by the least significant difference method
[d]Not determined

yields as the other treatments, thus indicating that the maize highly responded to the 60 kg/ha of N applied in 2006. The lack of dependency of the maize on the biomass applied might therefore be due probably to the sufficiency of the rate of fertilizer N applied for its growth and grain production under the conditions of this experiment. Even though supplementary N is required by maize when it follows legumes sequentially (Asibuo and Osei-Bonsu, 1999), it appears that the positive cumulative effects of the small quantities of the devil-bean biomass on growth and grain yield responses observed can be overshadowed when high rates of fertilizer N are introduced as the maize tends to depend on the fertilizer N rather than the organic N released from the decomposing legume biomass.

The use of devil-bean (as a short fallow herbaceous cover crop) as a nitrogen and organic matter source is therefore recommended in the absence of high doses of fertilizer nitrogen for enhanced maize production in the Guinea Savanna zone of Ghana.

Conclusions

Some farmers in the Guinea Savanna zone of Ghana express concern over the rotation of cover crops with maize as they obtain no economic gain from the land when it is under cover crop. This fear has been addressed in this study which has found out that

devil-bean (a cover crop) can be intercropped in maize consecutively, say for 2–3 years, and when the devil-bean is repeatedly ploughed into the soil, the organic matter buildup and the associated release of N into the soil enhance maize growth and grain yield. Better yields of maize can be obtained when the maize is supplied with moderate rates of fertilizer N. This integrated system allows the farmer to obtain economic benefits from the cereal while simultaneously improving the fertility level and structural stability of his farmland.

From this investigation, the use of devil-bean as a short fallow herbaceous cover crop for nitrogen and organic matter supply is therefore recommended in the absence of high doses of fertilizer nitrogen for enhanced maize production in the Guinea Savanna zone of Ghana.

Acknowledgements The authors are profoundly grateful to the Food Crops Development Project of the Ministry of Food and Agriculture (MoFA), Ghana, for providing the funds for these studies. We also appreciate the administrative, material, infrastructural, support given to the authors by Savanna Agricultural Research Institute of the Council for Scientific and Industrial Research during the conduct of this investigation.

References

Ahiabor BDK, Fosu M, Tibo I, Sumaila I (2007) Comparative nitrogen fixation, native arbuscular mycorrhiza formation and biomass production potentials of some grain legume species grown in the field in the Guinea Savanna zone of Ghana. West Afr J Appl Ecol 11:89–107

Asibuo JY, Osei-Bonsu P (1999) Influence of leguminous crops and fertilizer N on maize in the forest-savanna transition zone of Ghana. In: Carsky RJ, Etèka AC, Keatinge JDH, Manyong VM (eds) Cover crops for natural resource management in West Africa. Proceedings of a workshop organized by IITA and CIEPCA, Cotonou, Benin, pp 40–46, 26–29 Oct 1999

Atayese MO, Awotoye OO, Osonubi O, Mulongoy K (1993) Comparisons of the influence of vesicular-arbuscular mycorrhiza on the productivity of hedgerow woody legumes and cassava at the top and base of a hillslope under alley cropping systems. Biol Fertil Soils 16:198–204

Bray RH, Kurz LT (1945) Determination of total organic available forms of phosphorus in soils. Soil Sci 59:39–45

Cheruiyot EK, Mumera LM, Nakhone LN, Mwonga SM (2001) Rotational effect of legumes on maize performance in the Rift valley Highlands of Kenya. Afr Crop Sci J 9:667–676

Cheruiyot EK, Mumera LM, Nakhone LN, Mwonga SM (2003) Effect of legume-managed fallow on weeds and soil nitrogen in following maize (Zea mays L.) and wheat (*Triticum aestivum* L.) crops in the Rift valley highlands of Kenya. Aust J Exp Agric 43:597–604

Cheruiyot EK, Mwonga SM, Mumera LM, Macharia JK, Tabu IM, Ngugi JG (2007) Rapid decay of dolichos [*Lablab purpureus* (L.) Sweet] residue leads to loss of nitrogen benefit to succeeding maize (Zea mays L). Aust J Exp Agric 47:1000–1007

Daft MJ, El-Giahmi AA (1974) Effect of endogone mycorrhiza on plant growth. VIII. Influence of infection on the growth and nodulation in French bean (*P. Vulgaris*). New Phytol 73:1139–1147

Fosu M (1999) The role of cover crops and their accumulated N in improving cereal production in Northern Ghana. PhD thesis, Georg-August-Universität, Göttingen, Germany

Gerdemann JW (1975) Vesicular-arbuscular mycorrhizae. In: Torrey JR, Clarkson DT (eds) The development and function of roots. Academic Press, New York, pp 575–591

Greenland DJ (1985) Nitrogen and food production in the tropics: Contribution from fertilizer nitrogen and biological nitrogen fixation. In: Kang BT, van der Heide J (eds) Nitrogen management in farming systems in humid and subhumid tropics. Institute of Soil Fertility and the International Institute of Tropical Agriculture, Haren, Netherlands, pp 9–38

Hayman DS (1982) Influence of soils and fertility on activity and survival of vesicular-arbuscular mycorrhizal fungi. Phytopathology 72:1119–1125

Hayman DS, Johnson AM, Ruddlesdin I (1975) The influence of phosphate and crop species on *Endogone* spores and vesicular-arbuscular mycorrhiza under field condition. Plant Soil 43:489–495

Kasei CN (1988) The physical environment of semi-arid Ghana. In: Unger PW, Sneed TV, Jordan WR, Jensen R (eds) Challenges in dryland agriculture, a global perspective. Proceedings of international conference on dryland farming, Amarillo/Bushland, TX, pp 350–354

Mosse B (1973) Advances in the study of vesicular-arbuscular mycorrhiza. Ann Rev Phytopathol 11:171–196

Mosse B, Powell CLI, Hayman DS (1976) Plant growth responses to vesicular-arbuscular mycorrhiza. IX. Interactions between VA mycorrhiza, rock phosphate and symbiotic nitrogen fixation. New Phytol 76:331–342

Nelson DW, Sommers LW (1982) Total carbon and organic matter. In:Page AL, Miller RH, Keeney DR (eds) Methods of soil analyses, part 2, 2nd edn. No. 9 Soil Society of America Book, Madison, WI, pp 301–312

Nziguheba G, Merckx R, Palm CA (2005) Carbon and nitrogen dynamics in phosphorus-deficient soil amended with organic residues and fertilizers in western Kenya. Biol Fertil Soils 41:240–248

Palm CA, Giller K, Mafongoya PL, Swift MJ (2001) Management of organic matter in the tropics: translating theory into practice. Nutr Cycling Agroecosyst 61:63–75

Palm CA, Sanchez PA (1991) Nitrogen release from the leaves of some tropical legumes as affected by their lignin and polyphenolic contents. Soil Biol Biochem 23:83–88

Ravuri V, Hume DJ (1992) Performance of a superior *Bradyrhizobium japonica* and a selected *Sinorhizobium fredii* strain with soybean cultivars. Agron J 84:1051–1056

Sanginga N, Ibewiro B, Hougnandan P, Vanlauwe B, Okogun JK (1996) Evaluation of symbiotic properties and nitrogen contribution of *Mucuna* to maize growth in the derived savannas of West Africa. Plant Soil 179:119–129

Smith SE, Read DJ (1997) Mycorrhizal symbiosis, 2nd edn. Academic, San Diego, CA

Tel DA, Hagatey M (1984) Methodology in soil chemical analyses. In soil and plant analyses. Study guide for agricultural laboratory directors and technologists working in tropical regions. IITA, Nigeria, pp 119–138

Thomson BD, Robson AD, Abbott LK (1986) Effects of phosphorus on the formation of mycorrhizas by *Gigaspora calospora* and *Glomus fasciculatum* in relation to root carbohydrates. New Phytol 103:751–765

Tiessen H (1989) Assessment of soil fertility management in sub-sahelian savannahs. In: Unger PW, Sneed TV, Jordan WR, Jensen R (eds) Challenges in dryland agriculture: a global perspective. Proceedings of the international conference on dryland farming, Amarillo/Bushland, TX, pp 396–399, 15–19 Aug 1988

Wang WJ, Baldock JA, Dalal RC, Moody PW (2004) Decomposition dynamics of plant materials in relation to nitrogen availability and biochemistry determined by NMR and wet-chemical analysis. Soil Biol Biochem 36: 2045–2058

Effect of Organic Inputs and Mineral Fertilizer on Maize Yield in a Ferralsol and a Nitisol Soil in Central Kenya

M. Mucheru-Muna, D.N. Mugendi, P. Pypers, J. Mugwe, B. Vanlauwe, R. Merckx, and J.B. Kung'u

Abstract Declining land productivity is a major problem facing smallholder farmers in Kenya today. This decline primarily results from a reduction in soil fertility caused by continuous cultivation without adequate addition of external nutrient inputs. Improved fertility management combining organic and mineral fertilizer inputs can enable efficient use of the inputs applied and increase overall system's productivity. Field trials were established at three sites in distinct agro-ecological zones of central Kenya (one site at Machang'a and two sites at Mucwa with different soil fertility status) aiming to determine the effects of various organic sources (tithonia, lantana, mucuna, calliandra and manure) and combinations with mineral N fertilizer on maize grain yield during four consecutive seasons. In Machang'a site, sole manure recorded the highest maize grain yield across the four seasons. In Mucwa poor site, sole tithonia gave the highest maize grain yield during the four seasons, while in Mucwa good site, sole calliandra gave the highest maize grain yields. Generally, the maize grain yields were lower in the treatments with fertilizer alone compared to the treatments with organics across the three sites in the four seasons due to the poorly distributed rainfall. In Machang'a during the SR 2004 and SR 2005 seasons, the treatments with integration of organic and mineral fertilizer inputs were significantly higher than treatment with the sole organics; however, in Mucwa good and poor sites, generally the treatments with sole organics did better than the ones with integration of mineral N fertilizer and organics with the exception of the mucuna treatment which did significantly better in the integration compared to the sole application.

Keywords Manure · Tithonia · Lantana · Mineral fertilizer · Maize grain yields

Introduction

The central highlands of Kenya cover both areas with high potential for crop production on inherently fertile Nitisols and drier areas with lower potential on lighter, fragile soils, prone to quick degradation. The high potential areas of the central highlands (e.g. Meru South) are among the most densely populated regions in the country with an average of more than 700 person km^{-2}, leading to land fragmentations. This has eventually led to small farm sizes of about 0.5–1 ha per household.

The population pressure in the high potential areas of the central highlands has resulted in spill over of the population to the low potential areas of the highlands (e.g. Mbeere district). Mbeere district is characterized by frequent droughts due to low and erratic rainfall (Jaetzold and Schmidt, 1983). The soils are generally sandy loam, shallow and are low in organic matter (Warren, 1998); they are also characterized by physical soil loss from erosion, aluminium and iron toxicity, crusting and moisture stress (Place et al., 2003). These soils are also deficient in major plant nutrients such as nitrogen and phosphorus, a situation significantly influencing crop productivity (Warren, 1998).

M. Mucheru-Muna (✉)
Department of Environmental Sciences, School of Environmental Studies, Kenyatta University, Nairobi, Kenya
e-mail: moniquechiku@yahoo.com

A. Bationo et al. (eds.), *Innovations as Key to the Green Revolution in Africa*,
DOI 10.1007/978-90-481-2543-2_18, © Springer Science+Business Media B.V. 2011

The soil fertility in the central highlands has declined over time, with an annual net nutrient depletion exceeding 30 kg N (Smaling, 1993) as a result of continuous cropping with inadequate nutrient replenishment (Mwangi et al., 1996). In most smallholder farms, these deficiencies could be replenished through the use of mineral fertilizers and cattle manure. However, few farmers can afford the mineral fertilizers and the ones using them hardly use the recommended rates (60 kg N ha^{-1}) in the area, with most of them applying less than 20 kg N ha^{-1} (Adiel, 2004); on the other hand, the use of manure is also limited by its low quality (Ikombo, 1984; Kihanda, 1998). As a result, soil fertility has continued to decline as has the productivity of the land (Kapkiyai et al., 1998; Adiel, 2004). The situation in Mbeere district is further aggravated as the immigrants in these areas continue growing crops, which they used to grow in the high potential areas, consequently leading to environmental degradation and occasional crop failures.

Locally available organic inputs could be used to curb this problem. For instance, Kimani et al. (2004) reported a 92% increase in maize grain yields after applying manure compared to the control. Jama et al. (1999), Nziguheba and Mutuo (2000) and Mucheru-Muna et al. (2007) reported more than 50% increase in maize grain yields as a result of applying *Tithonia diversifolia* in the soil compared to the control. Mugendi et al. (1999) reported that application of *Calliandra calothyrsus* green biomass increased maize grain yield by 32%, while Kimetu et al. (2004) reported an increase of 48% compared to the control. Incorporating *Mucuna pruriens* biomass into the soil has been found to increase maize grain yields by 46% above farmer practice in the central highlands of Kenya (Gitari et al., 1998), while Gachene et al. (1999) reported maize yield of 88% higher than the control after incorporating mucuna.

Technologies that combine mineral fertilizers with organic nutrient sources can be considered as better options in increasing fertilizer use efficiency and providing a more balanced supply of nutrients (Donovan and Casey, 1998). Combination of organic and mineral fertilizer nutrient sources has been shown to result in synergistic effects and improved synchronization of nutrient release and uptake by crop (Palm et al., 1997) leading to higher yields, especially when the levels of mineral fertilizers used are relatively low as is the case in most smallholder farms of central Kenya

(Kapkiyai et al., 1998). Maize yields were increased with increasing rates of farmyard manure application; however, maize grain yields above 3.5 t ha^{-1} were obtained only when both farmyard manure and NP fertilizers were applied (Kihanda, 1996). Calliandra biomass combined with mineral fertilizer gave higher crop yields as compared to sole use of mineral fertilizer or sole calliandra biomass (Mugendi et al., 1999; Mucheru, 2003). The practice may hold the key to effective soil fertility management in the central highlands.

Trials using organic and mineral fertilizer inputs were established in the main maize-growing areas of the central highlands of Kenya in 2004 with the main objective of addressing the decline in soil fertility. The study aimed to determine the effects of different organic sources and combinations with mineral fertilizer inputs on maize grain yield.

Materials and Methods

The Study Area

The study was conducted in Meru South and Mbeere districts in the central highlands of Kenya. In Meru South, the experiment was conducted in Mucwa (00°18′48.2″S; 37°38′38.8″ E), which is located in upper midland 3 with an altitude of approximately 1373 m above sea level. The soils are Rhodic Nitisols (Jaetzold and Schmidt, 1983), which are deep, well weathered with moderate to high inherent fertility. The study was conducted in two farms: one that had relatively good (fertile) soils and another that had poor (unfertile) soils (Table 1).

Meru South is characterized by rapid population growth and low soil fertility (Government of Kenya, 2001). The main staple food crop is maize (*Zea mays* L.), which is commonly intercropped with beans (*Phaseolus vulgaris* L.). Other food crops include Irish potatoes, bananas, sweet potatoes, vegetables and fruits that are mainly grown for subsistence consumption. Livestock production is a major enterprise, especially dairy cattle of improved breeds. Other livestock in the area include sheep, goats and poultry. The main cash crops include coffee, tobacco and tea in that order. The rainfall is bimodal, falling in two seasons: the long rains (LR) lasting from March to June and

Table 1 Soil characterization in Mucwa good and poor sites, Meru South district, Kenya

	Site	
Soil parameters	Mucwa good	Mucwa poor
pH in water	5.0	4.6
Total N (%)	0.25	0.24
Total soil organic carbon (%)	2.1	1.8
Exchangeable P (ppm)	33.8	20.4
Exchangeable K (cmol kg^{-1})	0.36	0.21
Exchangeable Ca (cmol kg^{-1})	1.13	0.9
Exchangeable Mg (cmol kg^{-1})	0.20	0.20
Clay (%)	40	40
Sand (%)	32	32
Silt (%)	28	28

short rains (SR) lasting from October to December. The area receives an annual mean rainfall of 1400 mm. Rainfall for the four seasons in which the experiment was conducted is presented in Fig. 1.

In Mbeere district, the experiment was conducted in Machang'a (00°47′26.8″S; 37°39′45.3″E) with an altitude of approximately 1028 m above sea level and annual mean temperature of about 23°C. The soils are sandy clay loam, blackish grey or reddish brown, classified as the Nitro-rhodic Ferralsols (Jaetzold and Schmidt, 1983). The soils are shallow (about 1 m deep) and lose their organic matter, including nutrient-rich aggregates within 3–4 years of cultivation without adequate external organic material inputs and soil protection from water erosion (Jaetzold and Schmidt, 1983; Warren et al., 1998; Micheni et al., 2004). Table 2 shows the soil characteristics of the soils in Machang'a.

The major cropping enterprises include maize (Z. mays L.) and beans (P. vulgaris L.). Other food crops include cowpea, millet, sorghum, green grams and fruits (pawpaws and mangoes). Livestock (cows, goats, sheep and poultry) production is a major enterprise, and the farmers mainly keep the local breeds. Bee keeping is also a major enterprise in the area. Farmers plant food crops and keep livestock with a high staple and economic value as they do not grow any "cash crop"; therefore the crops they grow double as food crops and cash crops.

Machang'a lies at the transition between the marginal cotton (LM4) and main cotton (LM3) agro-ecological zones (Jaetzold and Schmidt, 1983). The rainfall is bimodal, falling in two seasons: the long rains (LR) lasting from March to June and short rains

(SR) from October to December. The total rainfall is however unreliable, with a mean annual rainfall of 900 mm (Government of Kenya, 1997). Total rainfall per season during the study period ranged between 209 and 731 mm. Rainfall for the four seasons in which the experiment was conducted is presented in Fig. 2.

Experimental Layout

The experiments were established in Meru South and Mbeere districts, and were laid out as randomized complete block design replicated thrice with the plots measuring 6 m × 4.5 m. In Meru South district, the test crop was maize (Z. mays L, var. H513) planted at a spacing of 0.75 and 0.5 m inter- and intra-row, respectively, while in Mbeere district, the test crop was maize (Z. mays L, var. Katumani) planted at a spacing of 0.90 and 0.6 m inter- and intra-row, respectively. Three (3) seeds were sown per hole and thinned 4 weeks later to two plants. External nutrient replenishment inputs were applied to give an equivalent amount of 60 kg N ha^{-1} (this is the recommended rate of N to meet maize nutrient requirement for an optimum crop production in the area; FURP, 1987) with the exception of the herbaceous legume treatment whereby the amount of N was determined by the biomass harvested and incorporated in the respective treatments. The treatments included (1) manure, (2) manure + 30 kg N ha^{-1}, (3) tithonia, (4) calliandra/lantana, (5) tithonia + 30 kg N ha^{-1}, (7) calliandra/lantana + 30 kg N ha^{-1}, (8) mucuna, (9) mucuna+ 30 kg N ha^{-1}, (10) fertilizer (60 kg N ha^{-1}) and (11) control.

Trial Management

The organic materials were harvested from nearby plots established for that purpose. A sample of each organic input was taken and N content determined, and then the amount of organic to be applied, equivalent to 30 or 60 kg N, was determined (for the treatments with sole organic and integration, an equivalent of 60 and 30 kg N ha^{-1}, respectively, was applied). The weight of mucuna biomass applied during the second, third and fourth seasons is presented in Table 3.

All organic inputs were harvested, weighed, chopped and incorporated into the soil to a depth of

Fig. 1 Rainfall distribution from 2004 to 2006 in Mucwa site, Meru South district, Kenya

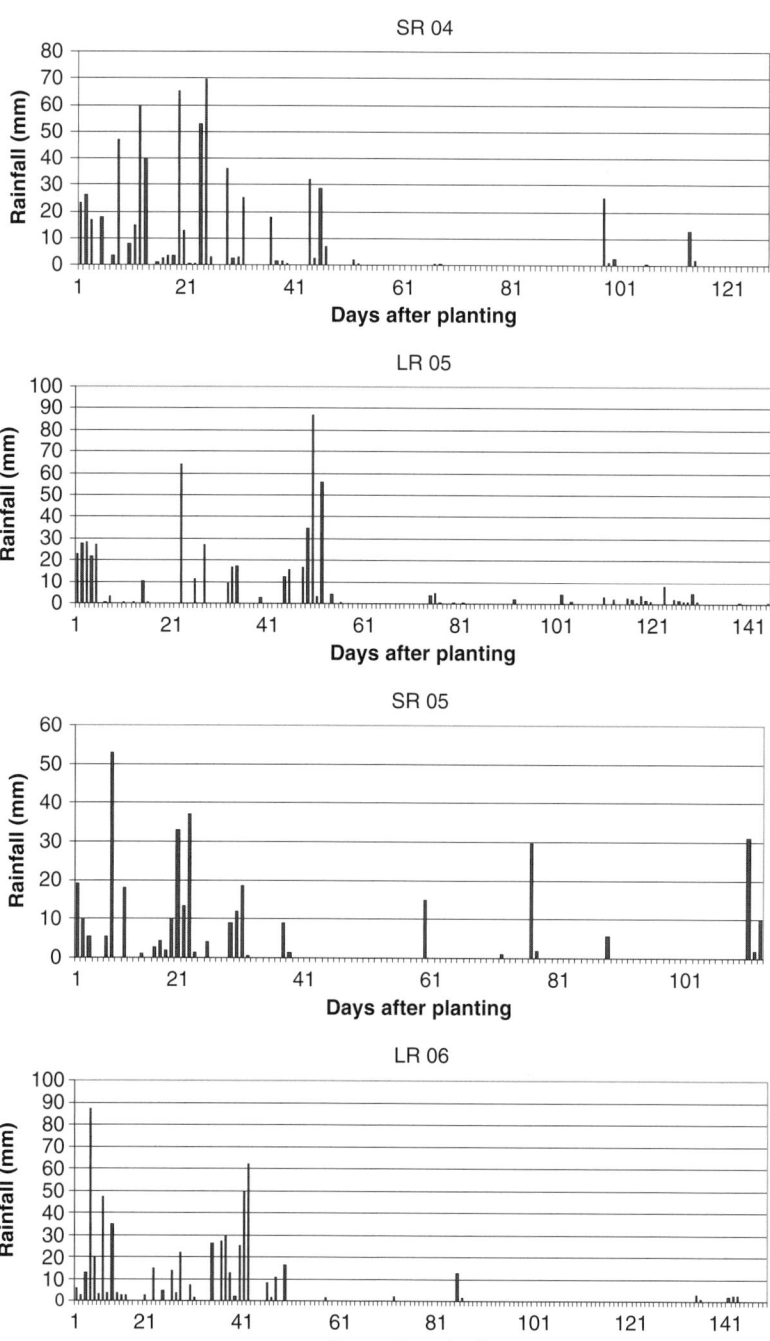

15 cm during land preparation. CAN was the source of mineral N and at all application rates, one-third was applied 4 weeks after planting and two-thirds 6 weeks after planting. Since organic inputs were being applied in this experiment and they (organic inputs) are often limited in their ability to increase P availability due to their low P content (Palm et al., 1997), P was applied in all plots at the recommended rate (60 kg P ha^{-1}) as triple super phosphate (TSP) to minimize the possibility of its confounding effects. This way it was assumed that nitrogen was the only macronutrient limiting maize yields. Other agronomic procedures

Table 2 Soil characterization in Machang'a site, Mbeere district, Kenya

Soil parameters	
pH in water	6.4
Total N (%)	0.09
Total soil organic carbon (%)	1.05
Exchangeable P (ppm)	12.9
Exchangeable K (cmol kg^{-1})	0.35
Exchangeable Ca (cmol kg^{-1})	1.0
Exchangeable Mg (cmol kg^{-1})	0.14
Clay (%)	22
Sand (%)	67
Silt (%)	11

for maize production were appropriately followed after planting.

Maize Harvesting

Maize grain and stover were harvested at maturity from a net area of 21.0 m^2 (out of the total area of 27 m^2) after leaving out one row on each side of the plot and the first and last maize plants on each row to minimize the edge effect. Cobs in each plot were separated from the stovers and their fresh weight was determined. At maturity, the maize and legumes were harvested and the fresh weight of both grain and stover was taken. Maize grains were dried and expressed in terms of dry matter content. Maize stovers were cut at ground level and the total fresh weight was determined. After harvesting, all the maize stovers were removed from the experimental plots to ensure that no nutrients were returned to the plots from the stovers that may confound the effects of adding a material of different quality into the experimental plots. Stover samples were oven dried at 70°C for 72 h to determine moisture contents, which were used to correct stover yields measured in the field to dry matter produced.

Statistical Analysis

Data were subjected to analysis of variance using GENSTAT programme, and the means were separated using Tukey's test at $p < 0.05$.

Results and Discussions

Results

There was a significant ($p < 0.001$) effect of seasons on maize grain yield in Mucwa good site. The mean maize grain yields were highest during the LR 05 season followed by SR 05, LR 06 and SR 04 seasons with 5.6, 3.1, 2.2 and 1.6 t ha^{-1}, respectively (Table 4). During the LR 05 season, the maize grain yield ranged between 3.3 and 6.5 t ha^{-1}, while during the SR 04 season, the maize grain yield ranged between 0.6 and 3.4 t ha^{-1} (Table 4).

In Mucwa good site, sole calliandra recorded the highest maize grain yields with 3.4, 4.8 and 3.7 t ha^{-1} during the SR 04, SR 05 and LR 06 seasons, respectively, while sole mucuna recorded the highest grain yields with 6.5 t ha^{-1} during the LR 05 season (Table 4). On the other hand, sole mucuna, sole calliandra, calliandra + 30 kg N ha^{-1} and manure + 30 kg N ha^{-1} recorded the lowest maize grain yields with 0.7, 5.2, 2.9 and 0.8 t ha^{-1} during the SR 04, LR 05, SR 05 and LR 06 seasons, respectively, among the treatments with inputs. Control recorded the overall lowest maize grain yield in all the four seasons.

The treatments with organic and mineral fertilizers increased the maize grain yield in comparison to the control. During the SR 04 season, the grain yields increased from 3% (sole mucuna) to 430% (sole calliandra) compared to the control, while during the LR 05, SR 05 and LR 06 seasons, the grain yields increased from 57% (sole calliandra) to 95% (sole mucuna), 81% (calliandra + 30 kg N ha^{-1}) to 193% (sole calliandra) and 132% (manure + 30 kg N ha^{-1}) to 988% (sole calliandra).

The treatments with organic and mineral fertilizers increased the maize grain yield in comparison to the sole mineral fertilizer (60 kg N ha^{-1}) in Mucwa good site. For instance, during the SR 04 season, the grain yields increased from 37% (tithonia + 30 kg N ha^{-1}) to 151% (sole calliandra) compared to the sole mineral fertilizer, while during the LR 05, SR 05 and LR 06 seasons the maize grain yields increased from 1% (manure + 30 kg N ha^{-1}) to 11% (sole mucuna), 1% (manure + 30 kg N ha^{-1}) to 57% (sole calliandra) and 32% (sole mucuna) to 119% (sole calliandra).

Generally, in Mucwa good site, treatments with sole organics performed significantly better than did the

Fig. 2 Rainfall distribution from 2004 to 2006 in Machang'a site, Mbeere district, Kenya

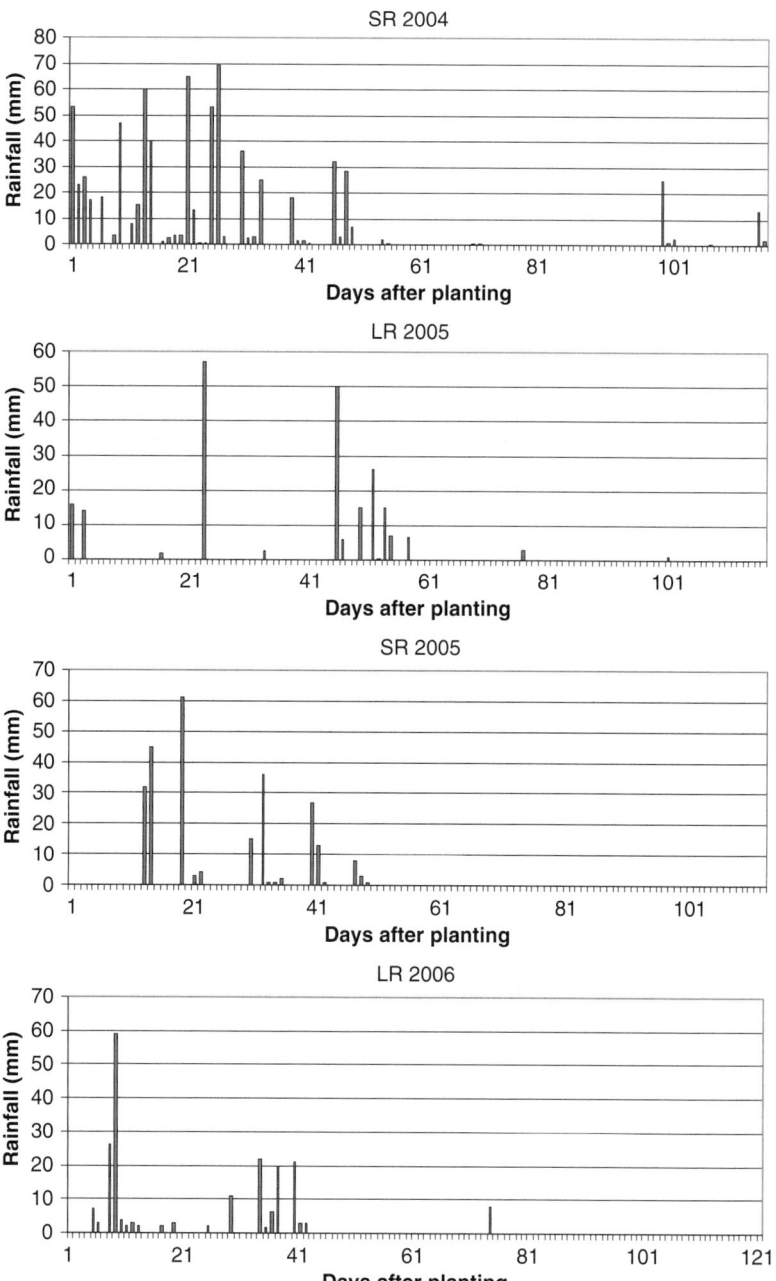

ones with integration of organic and mineral fertilizer across the four seasons. For instance, during the SR 2004 season, treatments with sole organics recorded significantly higher maize grain yields compared to the integration of organic and mineral fertilizer with the exception of mucuna treatment where the integration

was significantly higher than the sole application at $p < 0.001$.

In Mucwa poor site, there was a significant ($p < 0.001$) effect of seasons on maize grain yield. The mean maize grain yields were highest during the LR 05 season followed by SR 05, LR 06 and SR 04 seasons

Table 3 Mucuna incorporated in the 2005 long rain, 2005 short rain and 2006 long rain seasons in Mucwa good, Mucwa poor site and Machang'a

Treatment	2005 LR season		2005 SR season		2006 LR season	
	Biomass incorp. (t ha^{-1})	N equivalence (kg ha^{-1})	Biomass incorp. (t ha^{-1})	N equivalence (kg ha^{-1})	Biomass incorp. (t ha^{-1})	N equivalence (kg ha^{-1})
Mucwa good						
Mucuna	12.5	30	32.8	83	16.6	45.2
Mucuna + 30 kg N ha^{-1}	12.7	29.2	38.6	108.9	16.8	50
Mucwa poor						
Mucuna	9.2	23	32.3	111.4	13.2	37
Mucuna + 30 kg N ha^{-1}	6.3	16.4	34.4	115.9	16.2	45.5
Machang'a						
Mucuna	12.6	18.9	26.9	48.2	18.6	38.5
Mucuna + 30 kg N ha^{-1}	13	18.9	27.3	43.4	18.1	28.6

Table 4 Maize grain yields (t ha^{-1}) under different treatments during the 2004 short rain, 2005 long rain, 2005 short rain and 2006 long rain seasons at Mucwa good site

Treatment	SR 2004 season		LR 2005 season		SR 2005 season		LR 2006 season	
	−N	+N Fertilizer	−N	+N Fertilizer	−N	+N Fertilizer	−N Fertilizer	+N
Organic residue application								
Calliandra	3.39	2.01	5.21	5.39	4.75	2.93	3.70	2.60
Mucuna	0.66	0.82	6.47	6.09	3.04	3.00	2.23	2.67
Tithonia	2.97	1.85	6.04	6.05	3.07	3.51	2.86	2.75
Manure	1.35	1.01	5.45	5.89	3.51	2.97	2.28	0.79
Fertilizer (60 kg N ha^{-1})								
Control	0.64		3.32		1.62		0.34	
SED (N$_{level}$)			0.28*		ns			
SED (treatment)			ns		ns			
SED (N$_{level}$×treatment)	0.16***		ns		ns		0.39*	

*, ** and *** significant at $p < 0.05$, $p < 0.01$ and $p < 0.001$, respectively

ns, not significant

SED, standard error of differences between means

with 5.0 2.0, 1.6 and 1.4 t ha^{-1}, respectively (Table 5). During the LR 05 season, the maize grain yield ranged between 2.4 and 6.7 t ha^{-1}, while during the SR 04 season the maize grain yield ranged between 0.3 and 2.9 t ha^{-1} (Table 5).

Sole tithonia gave the highest maize grain yield during the SR 2004, LR 2005, SR 2005 and LR 2006 seasons with 2.9, 6.6, 2.8 and 3.1 t ha^{-1}, respectively (Table 5) in Mucwa poor site. Sole mucuna gave the lowest maize grain yield during the LR 2005 and the SR 2006 seasons, while mucuna + 30 kg N ha^{-1} and manure + 30 kg N ha^{-1} gave the lowest maize grain yield during the SR 2004 and LR 2006 seasons with 0.2, 4.9, 1.4 and 0.7 t ha^{-1}, respectively, in the treatments that received inputs. Overall, control gave

the lowest maize grain yields during the LR 2005, SR 2005 and LR 2006 seasons with 2.4, 0.8 and 0.5 t ha^{-1}, respectively.

The treatments with organic and mineral fertilizers increased the maize grain yield in comparison to the control during all the seasons. During the SR 04 season, the maize grain yields increased from 58% (manure + 30 kg N ha^{-1}) to 282% (sole tithonia), while during the LR 05, SR 05 and LR 06 seasons the maize grain yields increased from 104% (sole mucuna, calliandra + 30 kg N ha^{-1} and manure + 30 kg N ha^{-1}) to 185% (sole tithonia), 84% (sole mucuna) to 268% (sole tithonia and manure) and 30% (manure + 30 kg N ha^{-1}) to 474% (sole tithonia), respectively, compared to the control.

Table 5 Maize grain yields (t ha^{-1}) under different treatments during the 2004 short rain, 2005 long rain, 2005 short rain and 2006 long rain seasons at Mucwa poor site

Treatment	SR 2004 season		LR 2005 season		SR 2005 season		LR 2006 season	
	−N	+N Fertilizer	−N	+N Fertilizer	−N	+N Fertilizer	−N	+N Fertilizer
Organic residue application								
Calliandra	2.10	2.82	5.57	4.80	2.60	1.74	2.91	1.08
Mucuna	0.28	0.21	4.79	5.88	1.44	2.65	1.05	2.74
Tithonia	2.88	2.34	6.65	5.00	2.80	2.18	3.06	1.78
Manure	0.73	1.17	5.85	4.79	2.76	1.70	1.80	0.73
Fertilizer (60 kg N ha^{-1})								
Control	0.76		2.35		0.76		0.54	
SED (N$_{level}$)								
SED (treatment)								
SED (N$_{level}$×treatment)	0.62*		0.46*		0.46*		0.55**	

* and ** significant at $p < 0.05$ and $p < 0.01$, respectively
ns, not significant
SED, standard error of differences between means

The treatments with sole organics and organics integrated with mineral fertilizers increased the maize grain yield in comparison to the sole mineral fertilizer (60 kg N ha^{-1}). During the SR 04 season, the grain yields increased from 58% (manure + 30 kg N ha^{-1}) to 282% (sole tithonia), while during the LR 05, SR 05 and LR 06 seasons, the grain yields increased from 12% (sole mucuna, calliandra + 30 kg N ha^{-1} and manure + 30 kg N ha^{-1}) to 57% (sole tithonia), 1% (calliandra + 30 kg N ha^{-1} and manure + 30 kg N ha^{-1}) to 67% (sole tithonia and manure) and 35% (manure + 30 kg N ha^{-1}) to 496% (sole tithonia), respectively, compared to the sole mineral fertilizer.

Generally, in Mucwa poor site, treatments with sole organics performed significantly better than did the ones with integration of organic and mineral fertilizer across the four seasons. For instance, during the SR 2004 season, treatments with sole organics recorded significantly higher maize grain yields compared to treatments with integration of organic and mineral fertilizer with the exception of mucuna treatment where the integration was significantly higher than the sole application at $p < 0.001$. There was a significant ($p < 0.001$) effect of seasons on maize grain yield in Machang'a. The mean maize grain yields were highest during the LR 05 season followed by SR 04, LR 06 and SR 05 seasons with 2.5, 2.1, 2.1 and 1.6 t ha^{-1}, respectively (Table 6). During the LR 05 season, the maize grain yield ranged between 1.7 and 3.1 t ha^{-1}, while during the SR 05 season the maize grain yield ranged between 0.9 and 2.9 t ha^{-1} (Table 6).

In Machang'a site, sole manure recorded the highest maize grain yield during the SR 04, LR 05, SR 05 and LR 06 seasons with 3.0, 3.1, 2.9 and 2.7 t ha^{-1}, respectively (Table of 6). Sole mucuna recorded the lowest yield during the SR 04 and LR 05 seasons, whereas sole tithonia and mucuna + 30 kg N ha^{-1} recorded the lowest maize grain yields during the SR 05 and LR 06 seasons with 1.5, 2.3, 1.1 and 1.6 t ha^{-1}, respectively, among the treatments with inputs. Overall, control gave the lowest maize grain yields with 1.3, 1.9, 0.9 and 1.4 t ha^{-1} during the SR 04, LR 05, SR 05 and LR 06 seasons, respectively.

The treatments with sole organics and integrations of organics with mineral fertilizers increased the maize grain yield in comparison to the control. During the SR 04 season, the grain yields increased from 15% (sole mucuna) to 131% (sole manure), while during the LR 05, SR 05 and LR 06 seasons the grain yields increased from 24% (sole mucuna) to 67% (sole manure and tithonia + 30 kg N ha^{-1}), 20% (sole tithonia) to 215% (sole manure) and 11% (mucuna + 30 kg N ha^{-1}) to 88% (sole manure), respectively, compared to the control. The treatments with sole organics and combination of organics and mineral fertilizers increased the maize grain yield in comparison to the sole mineral fertilizer (60 kg N ha^{-1}) in Machang'a. During the SR 04 season, the grain yields increased from 18% (sole lantana) to 97% (sole manure), while during the LR 05, SR 05 and LR 06 seasons the grain yields increased from 36% (sole mucuna) to 83% (sole manure and tithonia + 30 kg N ha^{-1}), 14% (manure

Table 6 Maize grain yields (t ha^{-1}) under different treatments during the 2004 short rain, 2005 long rain, 2005 short rain and 2006 long rain seasons at Machang'a site

Treatment	SR 2004 season		LR 2005 season		SR 2005 season		LR 2006 season	
	−N	+N fertilizer	−N	+N fertilizer	−N	+N fertilizer	−N	+N fertilizer
Organic residue application								
Lantana	1.80	2.43	2.66	2.99	1.49	1.97	2.15	2.47
Mucuna	1.47	2.57	2.32	2.78	1.49	2.03	1.89	1.62
Tithonia	2.21	2.37	2.64	3.09	1.08	1.84	2.34	2.36
Manure	3.04	2.22	3.12	2.55	2.89	1.25	2.73	2.18
Fertilizer (60 kg N ha^{-1})								
Control	1.29		1.86		0.92		1.44	
SED (N$_{level}$)			ns				ns	
SED (treatment)			ns				ns	
SED (N$_{level}$×treatment)	0.30**		ns		0.38**		ns	

** significant at $p < 0.01$
ns, not significant
SED, standard error of differences between means

+ 30 kg N ha^{-1}) to 154% (sole manure) and 2% (mucuna + 30 kg N ha^{-1}) to 72% (sole manure), respectively, compared to the sole mineral fertilizer.

During the SR 04 and SR 05 seasons in Machang'a, there was a positive interaction between the mineral N fertilizer and organic residues ($p < 0.01$). There was also a significant difference between the treatments with sole organic and the treatments with organics integrated with mineral N fertilizer. For instance, treatments of lantana and sole mucuna with mineral fertilizer N were significantly higher than the treatments with sole lantana and sole mucuna, while sole manure treatment was significantly higher than treatment with manure integrated with mineral N fertilizer during the SR 04 and SR 05 seasons. In addition, during the SR 2005 season, tithonia integrated with mineral N fertilizer was significant higher than treatment with sole tithonia.

Discussions

The application of organic alone or in combination with mineral fertilizers led to increased maize yield compared to the control. On the one hand, in Mucwa good site, sole calliandra recorded a yield increase of up to 988%; on the other hand, sole tithonia recorded an increase of 474% in Mucwa poor site, while in Machang'a, manure recorded a yield increase of up to 215%. Several authors have reported increased

yields as a result of applying tithonia, calliandra and manure inputs in other areas (Kihanda, 1996; Gachengo et al., 1999; Jama et al., 1999; Mugendi et al., 1999; Mutuo et al., 2000; Kimani et al., 2004; Kimetu et al., 2004; Kihanda et al., 2006; Mucheru-Muna et al., 2007). In western Kenya, yield increase of up to 200% was reported following application of tithonia biomass (Jama et al., 2000), while in central Kenya Mucheru-Muna et al. (2007) reported yield increases of up to 267% following the application of sole tithonia biomass. Mucheru-Muna et al. (2007) reported an increase of 227% using calliandra biomass in Kenya, while Mtambanengwe et al. (2006) reported an increase of 525% following manure application in Zimbabwe.

Generally, the maize grain yields were lower in the treatments with mineral fertilizer alone compared to treatments with organic and organic combined with mineral fertilizers across the three sites in the four seasons. For instance, in Machang'a, increases of 154% were reported with the application of sole manure, in Mucwa good site, increases of 151% were reported with the application of sole calliandra and in Mucwa poor site, increases of 496% were reported with the application of sole tithonia. Mtambanengwe et al. (2006) reported a yield increase of 104% following manure application against sole mineral fertilizer. This implies an increased nutrient recovery in the organic and organic integrated with mineral fertilizer treatments compared to sole mineral fertilizer treatment.

The lower yields in the mineral fertilizer treatments could be as a result of the poorly distributed rainfall (Figs. 1 and 2) during the seasons. The timing of the application of the N from the organic treatments compared to the mineral fertilizer N could also explain the difference in yields. For the organics, all the 60 kg N ha^{-1} (sole organic) and 30 kg N ha^{-1} (treatments with integration) were applied at planting when there was adequate rainfall, while the mineral fertilizer N was applied in splits (0.33% was applied 4 weeks after planting, while the other 0.66% was applied 2 weeks later after which there was a long dry spell). Consequently, the growing maize crop may not have utilized this portion of the mineral N fertilizer, thus leading to the low maize grain yields. Other researchers have observed higher maize grain yields as a result of applying organic inputs like tithonia combined with mineral fertilizers as compared to sole application of mineral fertilizers (Mugendi et al., 1999; Mutuo et al., 2000; Nziguheba et al., 2000; Kimetu et al., 2004; Mucheru-Muna et al., 2007). Nutrients supplied in less soluble forms are less prone to loss and more suitable than mineral fertilizers when rainfall tends to be irregular and then heavy (Kihanda et al., 2006) like they were in this study.

Bekunda et al. (1997) reviewed information from selected experiments in Africa, and the results indicate that continuous application of mineral fertilizer without organic inputs eventually results in a decline in crop yields. They attributed such declining yields to soil acidification through continuous mineral fertilizer application and decline in soil organic matter. The soil organic matter in the Machang'a soils was very low as depicted by the carbon concentration (Table 2). According to Telalign et al. (1991), the soil organic carbon rated low (0.5–1.5%) in all the treatments at the beginning of the experiment, and it gradually reduced as cropping continued (data not shown). This low soil organic carbon could also have reduced the response of mineral fertilizers, agreeing with Greenland (1994), who reported that at low levels of soil organic matter, crop response to inputs is relatively poor and it is difficult to maintain yields with mineral fertilizers alone. The nutrient limitation may also be directly or indirectly related to the decline in soil organic matter with the multiple loss of soil physical condition.

Across the four seasons in Machang'a, sole manure was significantly higher than manure with mineral N fertilizer. This could be as result of the manure not decomposing as fast as the other organics applied (tithonia and lantana), thereby having more organic matter which was able to retain more water moisture that could be utilized by the crop during the growing season as depicted by the higher carbon concentration in the soil in this treatment compared to other treatments (data not shown).

Machang'a has a drier environment, and the most limiting factor to crop growth is soil moisture; during the study period, an average of 227 mm per season was received (Fig. 2). Interactions occur between water and the availability of nutrients because water increases the rate of release of nutrients from organic or insoluble forms and enables transport to roots and losses from soil to occur. Eghball (2002) reported that manure application led to a higher increase in soil organic carbon compared to other treatments. Sole manure treatment was also the only treatment that showed superiority in all nutrient provision, and it could also have supplied nutrients to the maize plant throughout the season, thereby giving higher yields. Some mineralization studies (Olsen, 1986) have shown that N release from manure is low but can persist throughout the maize growth period. These results, however, do not agree with Kihanda (1996) who reported that maize grain yields above 3.5 t ha^{-1} were obtained only when both farmyard manure and NP fertilizers were applied in acid soils in Embu.

In Mucwa good and poor sites, however, the manure treatment did not perform as well as in Machang'a site in comparison to the other treatments. Unlike in Machang'a, water is not the most limiting factor to growth in Mucwa (the area received an average of 556 mm per season during the study period; Fig. 1) but nutrients like N and P are the most limiting factors to growth. Hence, the lower rates of manure decomposition leading to low availability of nutrients to the maize crop could have led to the lower yields. Though the amount of N added via all these organic inputs was the same (60 kg N ha^{-1}), manure had lower N concentration than did all the other organic inputs and could have released the N slower due to the higher C:N ratio (Kimani et al., 2004).

During the short rain seasons in Machang'a, the treatments with integration of organic (lantana, mucuna and tithonia) and mineral N fertilizer inputs were significantly higher than the treatments with sole organics. This site was characterized by low soil fertility (Table 2), and the integration of the organic and mineral fertilizers could have probably led to an enhanced available N pool. The higher yields in the integration compared to the sole organic concur with results by Gachengo (1996), Mugendi et al. (1999) and Mutuo et al. (2000) on the integration of organic and mineral soil fertility inputs. Woomer and Swift (1994) demonstrated that integration of organics and mineral fertilizers can enhance the efficiency of mineral fertilizers. Integration of mineral and organic nutrient inputs can therefore be considered as a better option in increasing fertilizer use efficiency and providing a more balanced supply of nutrients (Vanlauwe et al., 2002). Kapkiyai et al. (1998) reported that combination of organic and mineral fertilizer nutrient sources has been shown to result in synergy and improved synchronization of nutrient release and uptake by plants leading to higher yields.

In Mucwa good and poor sites, however, the treatments with sole organics generally did better than did the ones with integration of mineral N fertilizer and organics with the exception of the mucuna treatment which did significantly better in the integration compared to the sole application. The rainfall during the four seasons was very unevenly distributed (Fig. 1), and the organic inputs could have conserved more soil moisture; hence there was more moisture made available to the growing maize in the organic treatments than in the treatments with integrations where there was less organic applied (in the sole organic treatments the organic input applied was double the one in the integration). The higher maize grain yields from the organic treatments could also be due to positive effects of the organic materials on soil physical and chemical properties (Murwira et al., 2002; Kimetu et al., 2004; Kihanda et al., 2006).

The consistently higher yields with sole tithonia biomass in the Mucwa poor site could be associated with the fast decomposition of tithonia, leading to rapid release of nutrients to the crop (Nziguheba et al., 1998; Gachengo et al., 1999). Tithonia contains high amounts of nutrients, especially N, and other nutrients such as phosphorus, potassium and magnesium, and may thus prevent other nutrient deficiencies such as micronutrients (Murwira et al., 2002). In addition, the P concentration of tithonia leaves is greater than the critical 2.5 g kg^{-1} threshold for net P mineralization (Palm et al., 1999), meaning addition of tithonia biomass to soil results in net mineralization rather than immobilization of P (Blair and Boland, 1978). The application of tithonia leaves would probably result in increased P availability by both net mineralization and decreased soil sorption (Nziguheba et al., 1998; George et al., 2001).

The consistently higher maize grain yields in the sole calliandra treatment in Mucwa good site could be due to build-up of soil organic matter by calliandra biomass which is of relatively low quality (Niang et al., 1996; Palm et al., 2001). The higher yields in calliandra treatment in Mucwa good site compared to other treatments however do not agree with results by other authors (Mwale et al., 2000; Kimetu et al., 2004) who reported lower yields with the addition of sole calliandra compared to other treatments like tithonia which decomposes very fast.

Conclusion

Sole tithonia biomass, sole calliandra and sole manure treatment recorded consistently higher maize grain yields in Mucwa poor site, Mucwa good site and Machang'a, respectively, across the four seasons. Generally, the treatments with application of organics resulted in higher maize grain yields compared to the treatments with sole mineral fertilizer, demonstrating the superiority of the organics in yield improvement due to their beneficial roles other than the addition of plant N like in the mineral fertilizer treatments. In Machang'a, the integrations of organics and mineral fertilizer recorded higher maize grain yields than did the sole organics, while in Mucwa good and Mucwa poor sites, sole organics performed better than the integrations of organic and mineral fertilizer across the four seasons.

Acknowledgements The authors wish to thank VLIR and IFS for providing financial support for the field experimentation. They also appreciate the contribution and collaborative efforts by the Tropical Soil Biology and Fertility Institute of CIAT (TSBF-CIAT), Kenya Agricultural Research Institute (KARI), Kenya Forestry Research Institute (KEFRI) and Kenyatta University (Department of Environmental Sciences) in administering field activities.

References

Adiel RK (2004) Assessment of on-farm adoption potential of nutrient management strategies in Chuka Division, Meru South, Kenya. MSc thesis, Kenyatta University, Kenya

Bekunda AM, Bationo A, Ssali H (1997) Soil fertility management in Africa: a review of selected research trials. In: Buresh RJ, Sanchez PA, Calhoun F (eds) Replenishing soil fertility in Africa. SSSA special publication No. 51. SSSA, Madison, WI, pp 63–79

Blair GJ, Boland OW (1978) The release of P from plant material added to soil. Aust J Soil Res 16:101–111

Donovan G Casey F (1998) Soil fertility management in sub-Saharan Africa. World Bank. Technical paper. 408

Eghball B (2002) Soil properties as influenced by phosphorus and nitrogen based manure and compost applications. Agron J 94:128–135

Fertilizer Use Recommendation Project (FURP) (1987) Description of first priority trial site in the various districts. Final report, vol 24. Embu District, National Agricultural Research Laboratories, Nairobi

Gachene CKK, Palm CA, Mureithi JG (1999) Legume cover crops for soil fertility improvement in the East African region. Report of an AHI workshop held at TSBF, Nairobi, Kenya, 18th–19th February 1999

Gachengo C (1996) Phosphorus release and availability on addition of organic materials to phosphorus fixing soils. MPhil thesis, Moi University, Eldoret, Kenya

Gachengo CN, Palm CA, Jama B, Otieno C (1999) Tithonia and senna green manures and inorganic fertilizers as phosphorus sources for maize in western Kenya. Agrof Syst 44:21–26

George TS, Gregory PJ, Robinson JS, Buresh RJ, Jama BA (2001) Tithonia diversifolia: variations in leaf nutrient concentration and implications for biomass transfer. Agrof Syst 52:199–205

Gitari JN, Karumba S, Gichovi M, Mwaniki K (1998) Integrated nutrient management studies at Embu site. In: Legume Research Network Project annual report

Government of Kenya (1997) Mbeere District Development Plan, 1997–2001. Ministry of Planning and National Development, Nairobi

Government of Kenya (2001) The 1999 Kenya national census results. Ministry of Home Affairs, Nairobi

Greenland DJ (1994) Soil science and sustainable land management. In: Syers JK, Rimmer DL (eds). Soil science and sustainable land management. CAB International, Wallingford, pp 1–15

Ikombo BM (1984) Effects of farmyard manure and fertilizers on maize in semi-arid areas of eastern Kenya. East Afr Agric J 44:266–274

Jaetzold R, Schmidt H (1983) Farm management handbook of Kenya. Natural conditions and farm information, vol 11/C. East Kenya. Ministry of Agriculture, Kenya

Jama B, Palm CA, Buresh RJ, Niang A, Gachengo C, Nziguheba G, Amadalo B (1999) Tithonia diversifolia green manure for improvement of soil fertility: a review from western Kenya. ICRAF, Nairobi

Jama B, Palm CA, Buresh RJ, Niang A, Gachengo C, Nziguheba G, Amadalo B (2000) Tithonia diversifolia as a green manure for soil fertility improvement in western Kenya: A Review. Agrof Syst 49:201–221

Kapkiyai JJ, Karanja NK, Woomer P, Qureshi JN (1998) Soil organic carbon fractions in a long-term experiment and the potential for their use as a diagnostic assays in highland farming systems of central Kenya. Afr Crop Sci J 6: 19–28

Kihanda FM (1996) The role of farmyard manure in improving maize production in the sub-humid central highlands of central Kenya. PhD thesis, UK

Kihanda FM (1998) Improvement of farmyard manure quality through composting with high quality organic residues. Research proposal funded by the African Science-Based Development Career Awards of the Rockefeller Foundation (1998–1999). KARI, Nairobi, Kenya

Kihanda FM, Warren GP, Micheni AN (2006) Effect of manure application on crop yield and soil chemical properties in a long-term field trial of semi-arid Kenya. Nutrient Cycling in Agroecosystems 76:341–354

Kimani SK, Macharia JM, Gachengo C, Palm CA, Delve RJ (2004) Maize production in the central highlands of Kenya using cattle manures combined with modest amounts of mineral fertilizer. Uganda J Agric Sci 9:480–490

Kimetu JM, Mugendi DN, Palm CA, Mutuo PK, Gachengo CN, Bationo A, Nandwa S, Kung'u JB (2004) Nitrogen fertilizer equivalencies of organics of differing quality and optimum combination with inorganic nitrogen source in central Kenya. Nutr Cycling Agroecosyst 68:127–135

Micheni A, Kihanda FM, Irungu J (2004) Soil organic matter: the basis for improved crop production in arid and semi-arid climates of eastern Kenya. In: Bationo A (ed) Managing nutrient cycles to sustain soil fertility in sub-Saharan Africa. Academy, Nairobi, pp 239–248

Mtambanengwe F, Mapfumo P, Vanlauwe B (2006) Comparative short-term effects of different quality organic resources on maize productivity under two different environments in Zimbabwe. Nutr Cycling Agroecosyst 76:271–284

Mucheru MW (2003) Soil fertility technologies for increased food production in Chuka, Meru South District, Kenya. Master's thesis, Kenyatta University, Nairobi, Kenya

Mucheru-Muna MW, Mugendi DN, Kung'u JB, Mugwe J, Bationo A (2007) Effects of organic and mineral fertilizer inputs on maize yield and soil chemical properties in a maize cropping system in Meru South District, Kenya. Agrof Syst 69: 189–197

Mugendi DN, Nair PKR, Mugwe JN, O'Neill MK, Woomer PL (1999) Calliandra and Leucaena alley cropped with maize. Part 1: soil fertility changes and maize production in the sub-humid highlands of Kenya. Agrof Syst 46:39–50

Murwira HK, Mutuo PK, Nhamo N, Marandu AE, Rabeson R, Mwale M, Palm CA (2002) Fertilizer equivalency values of organic materials differing quality. In: Vanlauwe B, Diels J, Sanginga N and Merckx R (eds) Integrated plant nutrient management in sub-Saharan Africa: from concept to practice. CAB International, Wallingford, CT, pp 113–122

Mutuo PK, Mukalama JP, Agunda J (2000) On-farm testing of organic and inorganic phosphorous source on maize in western Kenya. In: The biology and fertility of tropical soils, TSBF report, p 22

Mwale M (2000) Effect of application of S. sesban prunnings on maize yields. In: The biology and fertility of tropical soils, TSBF report 1997–1998

Mwangi JN, Mugendi DN, O'Neill KM (1996) Crop yield response to incorporation of leaf prunnings in sole and alley cropping systems. East Afr Agric For J 62:209–218

Niang A, Amadalo B, Gathumbi S, Obonyo CO (1996) Yield response to green manure application from selected shrubs and tree species in western Kenya: A preliminary assessment

Nziguheba G, Mutuo PK (2000) Integration of *Tithonia diversifolia* and inorganic fertilizer for maize production. In: The biology and fertility of tropical soils, TSBF report, p 23

Nziguheba G, Palm CA, Buresh RJ, Smithson PC (1998) Soil phosphorus fractions and adsorption as affected by organic and inorganic sources. Plant and Soil 198:159–168

Nziguheba G, Palm CA, Buresh RJ, Smithson PC (2000) Soil phosphorus fractions and adsorption as affected by organic and inorganic sources. In: The biology and fertility of tropical soils, TSBF report, p 22

Olsen SR (1986) The role of organic matter and ammonium in producing high corn yields. In: Chen Y, Avnimelech Y (eds) The role of organic matter in modern agriculture. Martinus Nijhoff, Dordrecht, pp 29–54

Palm CA, Gachengo CN, Delve RJ, Cadish G, Giller KE (2001) Organic inputs for soil fertility management in tropical agroecosystems application of an organic resource data base. Agriculture, Ecosystems and Environment 83:27–42

Palm CA, Myers RJK, Nandwa SM (1997) Organic–inorganic nutrient interactions in soil fertility replenishment. In: Buresh RJ, Sanchez PA, Calhoun F (eds) Replenishing soil fertility in Africa. Soil Science Society of America special publication 51. Soil Science Society of America, Madison, WI, pp 193–218

Palm CA, Nzuguheba G, Gachengo C, Gacheru E, Rao MR (1999) Organic materials as sources of phosphorus. Agroforestry Forum 9(4):30–33

Place F, Barret CB, Freman HA, Ramisch JJ, Vanlauwe B (2003) Prospects for integrated soil fertility management using organic and inorganic inputs evidence from small holder African agricultural systems. Food Policy 28:365–378

Smaling E (1993) Soil nutrient depletion in sub-Saharan Africa. In: Van Reuler H, Prins W (eds) The role of plant nutrients for sustainable food crop production in sub-Saharan Africa. VKP, Leidschendam, pp 367–375

Telalign T, Haque I, Aduayi EA (1991) Soil, plant, water, fertilizer, animal manure and compost analysis manual. Plant Science Division Working Document 13, ILCA, Addis Ababa, Ethiopia

Vanlauwe B, Diels J, Aihou K, Iwuafor ENO, Lyasse O, Sanginga N, Merckx R (2002) Direct interactions between N fertilizer and organic matter: evidence from trials with ^{15}N-labelled fertilizer. In: Vanlauwe B, Diels J, Sanginga N, Merckx R (eds) Integrated plant nutrient management in sub-Saharan Africa: from concept to practice. CAB International Wallingford, Oxon, pp 173–184

Warren GP (1998) Effects of continuous manure application on grain yields at seven sites of trial, from 1993 to 1996. In: Kihanda FM, Warren GP (eds) Maintenance of soil fertility and organic matter in dryland areas. The Department of Soil Science, University of Reading occasional publication No. 3. The University of Reading, Reading

Woomer PL, Swift MJ (eds) (1994) The biological management of tropical soil fertility. Wiley, New York, NY

Nutr Cycl Agroecosyst (2010) 88:39–47
DOI 10.1007/s10705-008-9210-2

RESEARCH ARTICLE

Effects of conservation tillage, crop residue and cropping systems on changes in soil organic matter and maize–legume production: a case study in Teso District

H. Anyanzwa · J. R. Okalebo · C. O. Othieno ·
A. Bationo · B. S. Waswa · J. Kihara

Received: 5 March 2008 / Accepted: 29 August 2008 / Published online: 13 September 2008
© Springer Science+Business Media B.V. 2008

Abstract The effects of conservation tillage, crop residue and cropping systems on the changes in soil organic matter (SOM) and overall maize–legume production were investigated in western Kenya. The experiment was a split-split plot design with three replicates with crop residue management as main plots, cropping systems as sub-plots and nutrient levels as sub-sub plots. Nitrogen was applied in each treatment at two rates (0 and 60 kg N ha^{-1}). Phosphorus was applied at 60 kg P/ha in all plots except two intercropped plots. Inorganic fertilizer (N and P) showed significant effects on yields with plots receiving 60 kg P ha^{-1} + 60 kg N ha^{-1} giving higher yields of 5.23 t ha^{-1} compared to control plots whose yields were as low as 1.8 t ha^{-1} during the third season. Crop residues had an additive effect on crop production, soil organic carbon and soil total nitrogen. Crop rotation gave higher yields hence an attractive option to farmers. Long-term studies are needed to show the effects of crop residue, cropping systems and nutrient input on sustainability of SOM and crop productivity.

Keywords Crop residue management ·
Maize yield · Particulate organic matter ·
Residue addition · Soil fertility and productivity

Introduction

Soil organic matter (SOM) is a useful indicator of soil quality since it determines the fertility, productivity and sustainability in tropical agricultural systems where nutrient poor and highly weathered soils are managed with little external input (Feller and Beare 1997; LaI 1997). The dynamics of SOM are influenced by practices such as tillage, mulching, removal of crop residues and application of organic and mineral fertilizers. Removal of crop residue from the fields is known to hasten soil organic carbon (SOC) decline especially when coupled with conventional tillage (Yang and Wander 1999; Mann et al. 2002). This is a common practice in most smallholder farms in western Kenya. Carbon sequestration in agricultural soils can be enhanced by addition of carbon inputs from crop residues. Locally available organic nutrient sources are often added in form of farmyard manure (Mugwira and Murwira 1997), leaf litter (Nyathi and Campbell 1993), crop residues (Campbell et al. 1998; Powell and Unger 1998), green manure (Giller et al. 1998) and agroforestry tree prunings (Mafongoya and Nair 1997; Chikowo et al. 2004). However, most of the available organic

H. Anyanzwa (✉) · J. R. Okalebo · C. O. Othieno
Department of Soil Science, Moi University,
P. O. Box 1125, Eldoret, Kenya
e-mail: hellenanyz@yahoo.com

A. Bationo
Alliance for a Green Revolution in Africa (AGRA), Accra, Ghana

B. S. Waswa · J. Kihara
Tropical Soil Biology and Fertility Institute of the International
Centre for Tropical Agriculture (TSBF-CIAT), Nairobi, Kenya,
bswaswa@yahoo.com

This article has been previously published in the journal "Nutrient Cycling in Agroecosystems" Volume 88 Issue 1.
A. Bationo et al. (eds.), Innovations as Key to the Green Revolution in Africa: Exploring the Scientific Facts. © 2010 Springer.

Ⓓ Springer

resources are from medium to poor quality and there is paucity on their potential to build up the fertility status of soils or influence short-term productivity (Mtambanengwe et al. 2006). Incorporation of crop residues favours short term nitrogen (N) immobilization because of their high C:N ratios but repeated additions has been known to increase SOM levels (Nicholson et al. 1997).

Tillage plays an important role in the manipulation of nutrient storage and release from SOM, with conventional tillage (CT) inducing its rapid mineralization and potential loss of C and N from the soil. Greater SOM loss is observed in coarse textured soils than in fine textured ones primarily due to lack of physical protection of organic matter in sandy soils (Hassink 1995; Feller and Beare 1997). Soil disturbance through tillage is a major cause of reduction in the number and stability of soil aggregates and subsequently organic matter depletion (Six et al. 2000). Tillage reduces SOM in all size fractions, but particulate organic matter (POM) is more readily lost than other fractions (Cambardella and Elliott 1992; Six et al. 1999). In continuously cultivated soils, decrease in SOC in sandy soils is primarily due to loss in POM (Feller and Beare 1997).

Type and length of tillage practice and soil texture influence the amount of SOC present in the soil, the rate of SOC turnover and its distribution among size fractions (Cambardella and Elliott 1992; Six et al. 2002).

The objectives of the study were: (i) to assess the effects of conservation tillage, crop residue and cropping systems on the changes in SOM, (ii) to determine and compare the overall maize and legume yields between different cropping systems in conservation tillage with and without crop residue.

Materials and methods

Site description

The field trial was established in 2005 long rain season (LR) on a smallholder farm in Teso (0°36′N; 34°20′E and 1,420 m above sea level) district in western Kenya. The mean annual rainfall is 1,800 mm with a bimodal distribution pattern. The long rain season begins from March to July where as the short rain season occurs from September to January.

Experimentation

The experiment was a split-split plot design arranged in a factorial combination of crop residue management (with and without crop residue) as main plots, cropping systems (cereal–legume intercropping, cereal–legume rotation, continuous cereal) as sub-plots and nutrient levels as sub-sub plots (Table 1). Main plots of 12 × 12 m were split into sub-plots of 5 × 4.5 m (separated by 0.5 m space) to accommodate different fertilizer combinations and N response. N fertilizers were applied at 0 and 60 kg N/ha (urea), P fertilizers at 0 and 60 kg P/ha triple superphosphate (TSP) with a blanket application of K fertilizer at 60 kg K/ha muriate of potash (MOP). Maize and soybean were planted as the test crops.

For each cropping system, the plots were divided into sub-plots measuring 5 × 4.5 m. For each of the rotation system two of the sub-plots were planted with legume at different N and P rates while the other two were planted with cereal and rotated every season. For the intercrops, all the plots were planted with legume and cereal at different rates of N and P. Four plots with two levels of N fertilizer (0 and 60 kg N/ha) and all receiving P were used in continuous cereal cropping system.

Initial land preparation and planting

Initial land preparation was by hand digging with a hoe at 15 cm depth in all plots; later, the practice changed to conservation tillage where minimal tillage operations were done and at least 30% of residue was

Table 1 Conservation tillage experimental treatments as applied at the study site (Asinge) in Teso district, western Kenya, 2005–2006

Treatment no.	Crop residue management	Cropping system (CR)
1	+ Crop residue	Legume–cereal ROTATION
2	+ Crop residue	Legume–cereal INTERCROP
3	+ Crop residue	CONTINUOUS cereal
4	−Crop residue	Legume–cereal ROTATION
5	−Crop residue	Legume–cereal INTERCROP
6	−Crop residue	CONTINUOUS cereal

Nutr Cycl Agroecosyst (2010) 88:39–47

left on the surface after planting. Maize variety that is striga tolerant (IR-CIMMTY hybrid, with seed coated with imazapyr chemical) was used as a test crop. It was planted at a spacing of 75 cm between rows and 30 cm within rows while soybean in rotation was planted at an inter-row and intra-row spacing of 75 and 10 cm, respectively.

Treatment application

Crop residue (maize stover) obtained from maize harvested each season on the field was applied at 2 t ha^{-1} before planting the plots. Fertilizers were placed at the time of sowing. Two levels of TSP (0 and 60 kg P/ha) were applied with a blanket application of 60 kg K/ha MOP. The rationale was to test whether reduced rates could give substantial results. The fertilizers were banded close to the seed row to increase contact between the fertilizer and the roots early in the growing season hence increased nutrient uptake by plants. Urea was applied at two rates (0 and 60 kg N/ha), respectively. This was split-applied; 1/3 at planting and 2/3 at 5 weeks after planting (WAP). Bulldock 025EC (active ingredient: Beta-cyfluthrin 25 g/l, manufactured by Bayer AG Germany) and gladiator 4TC (active ingredient: chlorpyrifos 480 g/l, manufactured by Dow Agrosciences Limited, Kingslynn, UK) were applied on maize at 5 WAP to control stem borers and termites, respectively. From this experiment sub-plots receiving 0 and 60 kg N/ha from both residue and no-residue treatments were sampled for analysis.

Crop management

All plots were kept weed free by hand pulling. Maize stover (C:N = 47)and soybean (C:N = 20) residues were left on the surface of the field after harvesting the crop. Maize stover was applied the following season as crop residue.

Soil sampling

Soil sampling was done before the start of the experiment for initial characterization at (0–15 cm) depth using an auger. A composite sample was made from 10 samples collected randomly from different parts of each plot, mixed, sub-sampled, air dried and passed through a 2 mm sieve for pH, particle size,

extractable phosphorus and through 60 mesh soils for organic carbon and total nitrogen analysis (Okalebo et al. 2002). Soil sampling was done each season immediately after harvesting the crop for three consecutive seasons to determine changes in SOM and soil total nitrogen.

Crop harvesting procedures

Maize was harvested from the net plots at maturity leaving two border plants (25 cm spacing) on either side and one row (75 cm spacing) from other ends to eliminate edge effects. Thus, the harvested effective area per plot was 14.0 m^2. In the harvested area, total weights of maize cobs (unshelled maize grain) were taken and sub-samples of 8–10 cobs taken from an arrangement of cobs into different classes (big, medium, small). For soybean, about 20 plants were taken as the plot sample. Maize was hand shelled and the grain weights recorded for each plot. Maize stover was cut at ground level and its fresh weights taken. Sub-samples of 8 stalks/plot from the stover were taken and cut into small pieces and mixed thoroughly. Fresh sub-samples of about 500 g of chopped stover/ plot were taken. These were air-dried and dry weights recorded.Yield calculations were done using the following expression:

$$\text{Dry matter factor} = (\text{sample dry weight} \times 100)/\text{sample fresh weight} \tag{1}$$

$$\text{Yield}\,(\text{kg ha}^{-1}) = (\text{total fresh weight} \times 10{,}000)/\text{Effective area}\,(14.0\,\text{m}^2) \tag{2}$$

Soil chemical analysis

Soil organic carbon was determined using the Nelson and Sommers (1975) oxidation method while soil total nitrogen was determined using the colorimetric method. Procedures are outlined in Okalebo et al. (2002).

Data analysis

All yields, organic carbon and total soil nitrogen were subjected to analysis of variance (ANOVA) using Genstat Discovery Edition 3.

 Springer

Rainfall

Cumulative rainfall amounts of 769, 453 and 661 mm were recorded during the three cropping seasons, i.e. 2005 long rain (LR), 2005 short rain (SR) and 2006 LR seasons, respectively. During 2005 SR season rainfall received was very low to sustain crop growth to maturity hence there was crop failure and only above ground biomass was harvested.

Results and discussion

Initial soil characterization

Initial properties of the top (0–15 cm) soils obtained before treatment additions are presented in Table 2.

The soil pH was within the recommended range of 5.5–6.5 for most food crops. Available P (bicarbonate extractable P) was very low below the critical level of 10 mg P/kg of soil below which fertilizer P responses are expected according to ratings given in Okalebo et al. 2002. The organic carbon and N contents in surface soils were very low. This was due to the sandy nature of the soil hence less carbon is protected within the soil particles. Soil aggregate dynamics strongly influence carbon sequestration and cycling (Six et al. 1999)

Effects of residue (maize stover) application on soil organic carbon (SOC) under the three cropping systems

Residue application significantly increased ($P < 0.001$) SOC in surface soils during the three cropping seasons (Fig. 1 and Table 3). Mean SOC contents of 1.09, 1.36, 1.57% C and 0.82, 0.79, 0.78% C were

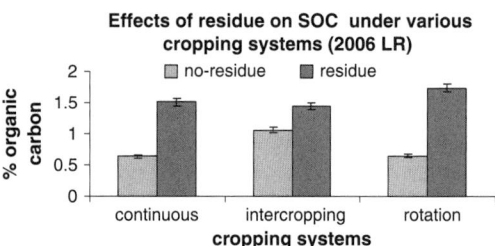

Fig. 1 Effects of residue application on SOC contents under various cropping systems during 2006 LR

obtained under residue and no-residue treatments during 2005 LR, 2005 SR and 2006 LR seasons, respectively. The higher SOC contents under residue treatments could be attributed to the beneficial effects of residue such as erosion control, nutrient cycling and soil quality enhancement. However, SOC analysis (Nelson and Sommers 1975) may have overestimated the contents. Dalal et al. (1991) observed that in the surface soil layers, the interactive effects of zero tillage and returning residues resulted in higher SOC contents compared with zero and conventional tillage with residue burning.

Effects of N fertilizer (urea) application on SOC contents under the various cropping systems

Fertilizer N addition significantly increased ($P < 0.001$) SOC in surface soils during the three cropping seasons (Table 3). Higher SOC contents were obtained in treatments receiving 60 kg N/ha. The increase in SOC because of fertilizer N addition agrees with findings of Paustian et al. (1992) who reported that fertilizer N addition increased SOC contents by 15–19% by increasing the net primary productivity and residue carbon input. Franzluebbers et al. (1994) also observed that the SOC of the 0–50 mm depth was 62% higher, in wheat cultivation, with fertilizer than without fertilizer, implying synergy in organic and inorganic resource inputs.

Effects of residue application and fertilizer N addition on soil total N content under various cropping systems

Residue application significantly increased ($P < 0.001$) soil total N contents in surface soils of all cropping seasons (Table 4). Total N contents of 0.083

Table 2 Soil characterization for the top (0–15 cm) before treatment application

Soil pH (H$_2$O)	5.50
% Organic carbon	0.83
Olsen P mg kg^{-1} of soil	6.04
% Total N	0.08
% Silt	7.45
% Clay	7.85
% Sand	84.7
Textural class	Loamy sand

 Springer

Table 3 % Organic carbon content during 2005 and 2006 LR seasons, respectively

Cropping systems	N rates	2005			2006		
		No-residue	Residue	Means	No-residue	Residue	Means
Continuous	0N	0.74	0.89	0.81	0.6	1.44	1.02
	60N	0.86	1.26	1.06	0.69	1.58	1.13
	Means	0.8	1.07	0.94	0.64	1.51	1.08
Intercropping	0N	0.88	0.94	0.91	1.01	1.35	1.18
	60N	0.73	1.33	1.03	1.11	1.55	1.33
	Means	0.81	1.14	0.97	1.06	1.45	1.26
Rotation	0N	0.81	0.91	0.86	0.58	1.6	1.09
	60N	0.88	1.24	1.06	0.72	1.88	1.3
	Means	0.85	1.08	0.96	0.65	1.74	1.2
l.s.d residue				0.23			0.15
l.s.d cropping system				0.08			0.04
l.s.d fertilizer				0.1			0.07
l.s.d residue * cropping system				0.18			0.17
l.s.d residue * fertilizer				0.18			0.16
l.s.d cropping system * fertilizer				0.14			0.08
l.s.d residue * cropping system * fertilizer				0.22			0.2

and 0.073% N were obtained during the first season (2005 LR) under residue and no-residue treatments and 0.096 and 0.080% N obtained during the third season (2006 LR) under residue and no-residue treatments, respectively. The high total N content under residue treatments could be attributed to the decomposition of the residues thus releasing nutrients to the soil particularly nitrogen. Application of organic amendments under conservation tillage systems is expected to provide a long-term source of N and reduce the need for N fertilization. In western Kenya, incorporation of crop residues (maize stover, wheat straw, bean trash and improved fallows) into soils have been associated with improved N levels in soils and subsequent high maize and bean yields (Palm et al. 1997; Okalebo et al. 1999; Kifuko 2002; Ndung'u et al. 2006; Waigwa 2002). Cropping systems had no significant effect on total N content in all cropping seasons. Fertilizer N addition significantly increased $P < 0.001$ total N contents in all cropping seasons. Total N contents of 0.072 and 0.083% N were obtained during the first season in treatments receiving 0 and 60 kg N/ha while 0.080 and 0.096% N were obtained during the third season also in treatments receiving 0 and 60 kg N/ha, respectively. The high total N content under fertilized treatments could be as a result of N application at

planting time since urea contains 45% N. The minimal N contents obtained under control treatments could be due to continuous crop uptake of N.

Maize yields

Effects of residue and fertilizer application on maize grain yields

Crop residue and fertilizer N application significantly increased $P < 0.001$ maize grain yields during the first and third seasons (Table 5). The high yields obtained under residue treatments could be as a result of weed suppression and moisture conserved under these plots. Experiments conducted in western Kenya have demonstrated that higher yields can be obtained when organic residues have been incorporated (Gachengo et al. 1999; Palm 1996). While the higher yields obtained under fertilized treatments indicate that the crop responded well to N fertilization due to the low organic matter content of the soil. In rotation cropping system, for example, the high yields obtained under treatments receiving N indicate that even where soybean had been planted as a previous crop, some mineral N may still need to be applied to the succeeding cereal.

 Springer

Table 4 % Soil total N content during LR in 2006

Cropping systems	N rates	(Residue)		
		No-residue	Residue	Means
Continuous	0N	0.067	0.09	0.078
	60N	0.086	0.1	0.095
	Means	0.077	0.096	0.086
Intercropping	0N	0.075	0.089	0.082
	60N	0.091	0.106	0.099
	Means	0.083	0.097	0.09
Rotation	0N	0.07	0.086	0.078
	60N	0.089	0.101	0.095
	Means	0.079	0.093	0.086
l.s.d cropping system				0.006
l.s.d residue				0.004
l.s.d fertilizer				0.003
l.s.d cropping system * residue				0.006
l.s.d cropping system * fertilizer				0.005
l.s.d residue * fertilizer				0.006
l.s.d cropping system * residue * fertilizer				0.008

Table 5 Maize grain yields (t/ha) during 2005 and 2006 LR seasons

Cropping systems	N rates	2005			2006		
		No-residue	Residue	Means	No-residue	Residue	Means
Continuous	0N	0.92	1.18	1.05	2.53	1.93	2.23
	60N	2.55	3.32	2.94	3.4	4.53	3.96
	Means	1.74	2.25	1.99	2.96	3.23	3.09
Intercropping	0N	1.3	1.09	1.2	1.51	2.04	1.78
	60N	1.88	1.9	1.89	2.57	2.5	2.54
	Means	1.59	1.49	1.54	2.04	2.27	2.15
Rotation	0N	1.36	3.14	1.68	1.69	2.78	2.23
	60N	1.99	3.63	3.38	5.03	5.44	5.23
	Means	2.25	2.81	2.53	3.36	4.11	3.73
l.s.d cropping system				0.15			0.22
l.s.d residue				0.05			0.72
l.s.d fertilizer				0.13			0.22
l.s.d cropping system * residue				0.17			0.57
l.s.d cropping system * fertilizer				0.2			0.32
l.s.d residue * fertilizer				0.13			0.57
l.s.d cropping system * residue * fertilizer				0.27			0.59

Effects of cropping systems on maize yields

Cropping systems had a significant effect $P < 0.001$ on yields during the first and third cropping seasons

(Table 5). During the first cropping season rotation system gave higher yields of (2.53 t ha^{-1}) followed by continuous cropping (1.99 t ha^{-1}) with intercropping giving lower yields of (1.54 t ha^{-1}) while during the third season also higher yields of (3.73 t ha^{-1}) were obtained under rotation plots (3.1 t ha^{-1}) under continuous cropping and (2.15 t ha^{-1}) in intercrop plots, respectively. The higher yields under rotation plots could be attributed to the beneficial effects of the previous legume crop. Giller (1999) found that maize planted following soybeans benefited more compared to continuous maize, largely due to effects of improved soil fertility. Kasasa et al. (1999) also found increased maize yields following the planting of promiscuous soybean varieties. However, yields under intercropping were lower compared to continuous and rotation plots. The low yields could possibly be as a result of interspecific competition, which might have occurred when the two crops were grown together hence, resulting to reduced yields (Van der meer 1992).

Soybean yields

Effects of residue and fertilizer N application on soybean yields under various cropping systems

Cropping systems had a significant effect $P < 0.001$ on soybean yields. The highest yields in both 2005 and 2006 LR seasons were obtained under rotation cropping systems while intercropping had lower yields (Table 6). The lower yields obtained under intercrop plots could be as a result of light and nutrient competition, soybeans are susceptible to intercropping (Van der meer 1992). Fertilizer N application at 60 kg N/ha significantly increased $P < 0.001$ yields. Higher yields in the two cropping seasons were obtained in fertilized plots. The higher yields obtained under fertilized treatments could be an indication that the crop needed some starter N at planting to ensure the young seedlings will have an adequate supply until the rhizobia can become established on their roots for N fixation Agboola and Fayemi (1972)

Conclusion

Primary plant production and soil microbial activity are the two main biological processes governing inputs and outputs of SOM. The balance between them determines SOM turnover and is controlled by biotic and abiotic factors. Appropriate levels of SOM ensure soil fertility and minimize agricultural impact on the environment through carbon sequestration, reducing soil erosion, and preserving soil biodiversity.

Conservation tillage combined with residue addition can lead to build up in soil organic carbon of surface soils.

Surface application of crop residue under conservation tillage is beneficial in trapping soil and water.

Higher yields were observed under rotation systems accompanied with crop residue and minimal

Table 6 Soybean grain yields in (kg/ha) during 2005 and 2006 (LR) season

Cropping system	N rate	2005			2006		
		No-residue	Residue	Means	No-residue	Residue	Means
Intercropping	0	131	146	139	143	140	142
	60	182	209	196	293	277	285
	Means	157	177	167	218	208	213
Rotation	0	257	257	257	370	379	374
	60	311	298	305	397	442	419
	Means	284	277	281	383	410	397
l.s.d cropping system				4309.91			9020.91
l.s.d residue				3011.21			4348.11
l.s.d fertilizer				5576.79			5334.88
l.s.d cropping system * residue				5258.48			10014.03
l.s.d cropping system * fertilizer				7047.38			10485.13
l.s.d residue * fertilizer				6334.37			6881.86
l.s.d cropping system * residue * fertilizer				9479.27			12541.41

 Springer

levels of nitrogen fertilizer, hence an attractive alternative to farmers.

Long-term studies are needed to show the effect of residue and nutrient inputs on the sustainability of SOM and crop productivity through improvement of soil properties.

Acknowledgements We acknowledge the support of John Mukalama for coordinating field activities, the farmer for availing the experimental site and casual labourers. Also thanks to the technicians at Moi University, Department of Soil Science for guidance in laboratory analyses. We are grateful for financial support from Tropical Soil Biology and Fertility (TSBF) institute of CIAT.

References

Agboola AA, Fayemi AA (1972) Fixation and excretion of nitrogen by tropical legumes. Legumes Agron J 64:409–412

Cambardella CA, Elliott ET (1992) Particulate soil organic matter changes across a grassland cultivation sequence. Soil Sci Soc Am J 56:777–783

Campbell BM, Frost PGH, Kirchmann H, Swift M (1998) A survey of soil fertility management in small-scale farming systems on north-eastern Zimbabwe. J Sustain Agric 11:19–39. doi:10.1300/J064v11n02_04

Chikowo R, Mapfumo P, Nyamugafata P, Giller KE (2004) Woody legume fallow productivity, biological N_2-fixation and residual benefits to two successive maize crops in Zimbabwe. Plant Soil 262:303–315. doi:10.1023/B:PLSO. 0000037053.05902.60

Dalal RC, Henderson PA, Glasby JM (1991) Organic matter and microbial biomass in a vertisol after 20 yrs of zero tillage. Soil Biol Biochem 23:435–441. doi:10.1016/0038-0717(91)90006-6

Feller C, Beare MH (1997) Physical control of soil organic matter dynamics in the tropics. Geoderma 79:69–116. doi: 10.1016/S0016-7061(97)00039-6

Franzluebbers AJ, Hons FM, Zuberer DA (1994) Long term changes in soil carbon and nitrogen pools in wheat management systems. Soil Sci Soc Am J 58:1639–1645

Gachengo CN, Palm CA, Jama B, Othieno CO (1999) Tithonia and Senna green manures and inorganic fertilizers as phosphorus sources for maize in western Kenya. Agrofor Syst 44:21–36. doi:10.1023/A:1006123404071

Giller KE (1999) Nitrogen mineralization: leaching tube protocol. In: Mutuo P, Palm C (eds) Combined inorganic and organic nutrient sources: experimental protocols for TSBF-AfNeT, SoilFert and SWNM. Tropical Soil Biology and Fertility, Nairobi, Kenya, pp 17–19

Giller KE, Gilbert R, Mugwira LM, Muza L, Patel BK, Waddington S (1998) Practical approaches to soil organic matter management for smallholder maize production in Southern Africa. In: Waddington SR, Murwira HK, Kumwenda JDT, Hikwa D, Tagwira F (eds) Soil fertility research for maize-based farming systems in Malawi and

Zimbabwe. Soil FertNet and CIMMYT-Zimbabwe, Harare, Zimbabwe, pp 139–153

Hassink J (1995) Density fractions of soil macroorganic matter and microbial biomass as predictors of C and N mineralization. Soil Biol Biochem 27:1099–1108. doi:10.1016/0038-0717(95)00027-C

Kasasa P, Mpepereki S, Musiyiwa K, Makonese F, Giller KE (1999) Residual nitrogen benefits of promiscuous soybean to maize under field conditions. Afr Crop Sci J 7:375–382

Kifuko MN (2002) Effect of combining organic residues with phosphate rock and triplesuperphosphate on phosphorus sorption and maize performance in acid soils of western Kenya. M.Phil Thesis, Dept of Soil science, Moi Univ, Eldoret

LaI R (1997) Residue management, conservation tillage and soil restoration for mitigating green house effect by CO_2 enrichment. Soil Tillage Res 43:81–107. doi:10.1016/S0167-1987(97)00036-6

Mafongoya PL, Nair PKR (1997) Multipurpose tree prunnings as sources of nitrogen to maize under semi-arid conditions in Zimbabwe: interaction of pruning quality and time and method of application on nitrogen recovery by maize in two soil types. Agrofor Syst 37:1–14. doi:10.1023/A:1005833528619

Mann L, Tolbert V, Cushmann J (2002) Potential environmental effects of corn (*Zea mays* L) stover removal with emphasis on soil organic matter and erosion. Agric Ecosyst Environ 89:149–166. doi:10.1016/S0167-8809(01)00166-9

Mtambanengwe F, Mapfumo P, Vanlauwe B (2006) Comparative short-term effects of different quality organic resources on maize productivity under two different environments in Zimbabwe. Nutr Cycl Agroecosyst 76:271–284. doi:10.1007/s10705-005-4988-7

Mugwira LM, Murwira HK (1997) Use of cattle manure to improve soil fertility in Zimbabwe: past and current research and future research needs. Network of working paper no. 2 soil fertility network for maize based cropping systems in Zimbabwe and Malawi. CIMMYT, Harare, Zimbabwe

Ndung'u KW, Okalebo JR, Othieno CO, Kifuko MN, Kipkoech AN, Kimenye LN (2006) Residual effectiveness of Minjingu Phosphate Rock and fallow biomass on crop yields and financial returns in western Kenya. Exp Agric 42:323–336

Nelson DW, Sommers LE (1975) A rapid and accurate method of estimating organic carbon in soil. Proc Indian Acad Sci 84:456–462

Nicholson FA, Chambers BJ, Mills AR, Stachon PJ (1997) Effects of repeated straw additions on crop fertilizer requirements, soil mineral nitrogen and nitrate leaching loses. Soil Use Manage 13:136–142. doi:10.1111/j.1475-2743.1997.tb00574.x

Nyathi P, Campbell BM (1993) The acquisition and use of miombo litter by small-scale farmers in Masvingo, Zimbabwe. Agrofor Syst 22:43–48. doi:10.1007/BF00707469

Okalebo JR, Palm CA, Gichuru M, Owuor JO, Othieno CO, Munyampundu A et al (1999) Use of wheat straw, soybean trash and nitrogen fertilizer for maize production in the Kenyan highlands. Afr Crop Sci J 7(4):433–440

Okalebo JR, Gathua KW, Woomer PL (2002) Laboratory methods of plant and soil analysis: a working manual, 2nd edn. TSBF-UNESCO, Nairobi

Palm CA (1996) Nutrient management: combined use of organic and inorganic fertilizers for increasing soil phosphorus availability. Annual report of Tropical Soil Biology and Fertility Programme (TSBF). Nairobi, Kenya

Palm CA, Myers RJK, Nandwa SM (1997) Combined use of organic and inorganic nutrient sources for soil fertility maintenance and replenishment. In: Buresh JR et al (eds) Replenishing soil fertility in Africa. SSA, Madison, pp 193–217

Paustian K, Parton WJ, Persson J (1992) Modelling soil organic matter in organic amended and nitrogen-fertilized long-term plots. Soil Sci Soc Am J 56:476–488

Powell JM, Unger PW (1998) Alternatives to crop residues for sustaining agricultural productivity and natural resource conservation. J Sustain Agric 11:59–84. doi:10.1300/J064 v11n02_07

Six J, Elliott ET, Paustian K (1999) Aggregate and SOM dynamics under conventional and no-tillage systems. Soil Sci Am J 63:1350–1358

Six J, Elliot ET, Paustian K (2000) Soil macroaggregate turnover and microaggregate formation: a mechanism for carbon sequestration under no tillage agriculture. Soil Biol Biochem 32:2099–2103. doi:10.1016/S0038-0717(00)00 179-6

Six J, Conant RT, Paul EA, Paustian K (2002) Stabilization mechanism for soil organic matter: implications for carbon saturation of soils. Plant Soil 141:155–176. doi: 10.1023/A:1016125726789

Van der meer JH (1992) The ecology of intercropping. Cambridge University Press, Cambridge

Waigwa M (2002) The use of manure and crop residues to improve the solubility of Minjingu phosphate rock for phosphorus replenishment in depleted acid soils of western Kenya. M.Phil Thesis. Moi University, Eldoret Kenya

Yang XM, Wander MM (1999) Tillage effects on soil organic carbon distribution and storage in silt loam soil in Illinois. Soil Tillage Res 52:1–9. doi:10.1016/S0167-1987(99) 00051-3

Nutr Cycl Agroecosyst (2010) 88:17–27
DOI 10.1007/s10705-008-9191-1

RESEARCH ARTICLE

Benefits of integrated soil fertility and water management in semi-arid West Africa: an example study in Burkina Faso

R. Zougmoré · A. Mando ·
L. Stroosnijder

Received: 5 March 2008 / Accepted: 9 July 2008 / Published online: 14 August 2008
© Springer Science+Business Media B.V. 2008

Abstract The synergistic effect of soil and water conservation (SWC) measures (stone rows or grass strips) and nutrient inputs (organic or mineral nutrient sources) was studied at Saria station, Burkina Faso. The reduction in runoff was 59% in plots with barriers alone, but reached 67% in plots with barriers + mineral N and 84% in plots with barriers + organic N, as compared with the control plots. Plots with no SWC measure lost huge amounts of soil (3 t ha^{-1}) and nutrients. Annual losses from eroded sediments and runoff reached 84 kg OC ha^{-1}, 16.5 kg N ha^{-1}, 2 kg P ha^{-1}, and 1.5 kg K ha^{-1} in the control plots. The application of compost led to the reduction of total soil loss by 52% in plots without barriers and 79% in plots with stone rows as compared to the losses in control plots. SWC measures without N input did not significantly increase sorghum yield. Application of compost or manure in combination with SWC measures increased sorghum grain yield by about 142% compared to a 65% increase due to mineral fertilizers. Yields increase did not cover annual costs of single SWC measures while application of single compost or urea was cost effective. The combination of SWC measures with application of compost resulted in financial gains of 145,000 to 180,000 FCFA ha^{-1} year^{-1} under adequate rainfall condition. Without nutrient inputs, SWC measures hardly affected sorghum yields, and without SWC, fertilizer inputs also had little effect. However, combining SWC and nutrient management caused an increase in sorghum yield.

Keywords Stone row · Grass strip · Nutrient input · Sorghum · Economic benefit

R. Zougmoré (✉)
Institute for Environment and Agricultural Research (INERA), 04 BP 8645, Ouagadougou 04, Burkina Faso
e-mail: robert.zougmore@messrs.gov.bf

A. Mando
Division Afrique, An International Center for Soil Fertility and Agricultural Development (IFDC),
BP 4483, Lome, Togo
e-mail: amando@ifdc.org

L. Stroosnijder
Erosion and Soil & Water Conservation Group,
Wageningen University, Nieuwe Kanaal 11,
6709 PA Wageningen, The Netherlands
e-mail: leo.stroosnijder@wur.nl

Introduction

Soil degradation is a major constraint to the sustainability of agricultural systems in the arid and semiarid tropics (Ryan and Spencer 2001) and particularly in the semiarid zone of West Africa (Laflen and Roose 1998; Lal 1998). Water erosion is a serious threat to sustainable agricultural land use as it affects soil productivity (Laflen and Roose 1998). In this region, erosion is

This article has been previously published in the journal "Nutrient Cycling in Agroecosystems" Volume 88 Issue 1.
A. Bationo et al. (eds.), Innovations as Key to the Green Revolution in Africa: Exploring the Scientific Facts. © 2010 Springer.

worsened by poor soil and crop mismanagement, which jeopardize the integrity of soil's self-regulatory capacity (Lal 1998). Indeed, erosion by runoff water is responsible for negative nutrient and carbon balances in most farming systems in West Africa (Stoorvogel and Smaling 1990) and for the reduction of crop rooting depth (Morgan 1995). Erosion influences several soil properties, such as topsoil depth, soil organic carbon content, nutrient status, soil texture and structure, available water holding capacity and water transmission characteristics.

In addition, plant nutrient use efficiency in cereal-based farming systems is often very low because of limited soil moisture conditions (Buerkert et al. 2002). The low soil quality combined with the harsh Sahelian climate leads to a low efficiency of fertilizers (Breman et al. 2001). Indeed, considering the importance of soil moisture for crop growth and for the uptake of plant nutrients in this zone, the effectiveness of soil fertility enhancing measures should be related to the rainfall regimes (FAO 1986).

What is responsible for water deficiency (i.e. more and/or longer periods of water stress), low water use efficiency and crop production is not primarily water shortage, but loss of water through runoff, soil evaporation and drainage below the root zone (Mando 1997). According to Lal (1997), one of the key conditions to increase soil productivity in the sub-Saharan zone is to ensure effective water infiltration and storage in the soil. In the last two decades, several water-harvesting technologies such as tillage, stone rows, hedgerows, earth bunds and dikes have been used to improve soil water infiltration and storage (Nicou and Charreau 1985; Perez et al. 1998; Zougmoré et al. 2000b). Moreover, alleviating erosion-induced loss will require the development and adoption of land use systems that are capable of replenishing or maintaining the nutrient status of the soil in addition to controlling water runoff and soil erosion (Sanchez et al. 1997; Quansah and Ampontuah 1999). Therefore, interactions of soil and water conservation measures with organic or mineral source of nutrients need to be examined. Indeed, combining soil and water conservation measures (SWC) with locally available nutrients inputs may optimize crop production and economic benefit in cereal-based farming systems. This can be best illustrated through work conducted at Saria agricultural station (INERA-Burkina Faso) on the combined use of runoff barriers (stone rows or grass strips) and organic or mineral nutrient sources.

Materials and methods

Site description and experimental design

The experimental field was at Saria Agricultural Research Station in Burkina Faso (12°16′ N, 2° 9′ W, 300 m altitude), characterised by a north-sudanian climate (Fontès and Guinko 1995). Over the last 30 years the average annual rainfall was 800 mm. Rainfall is mono-modal, lasts for 6 months (May–October) and is distributed irregularly in time and space. Mean daily temperatures vary between 30°C during the rainy season and may reach 35°C in April and May. The mean potential evapotranspiration is 2,096 mm in dry years and 1,713 mm in wet years. The site was previously under fallow for about 15 years. The soil type was a Ferric Lixisol (FAO-UNESCO 1994) with an average slope of 1.5% and a hardpan at 0.7 m depth. The textural class according to the USDA system is a sandy loam at 0.3 m (62% sand, 28% silt, 10% clay) with average gravel content of 30% at 0.1 m depth.

The trial was conducted over three seasons (2000–2002) and combined linear SWC measures with organic or fertiliser sources of nitrogen. The experimental design was a randomized Fisher bloc with nine treatments and two replications:

T_{SR}: stone rows, no N input	T_{GS}: grass strips, no N input
T_{SRC}: stone rows + compost-N	T_{GSC}: grass strips + compost-N
T_{SRM}: stone rows + manure	T_{GSM}: grass strips + manure
T_{SRU}: stone rows + urea-N	T_{GSU}: grass strips + urea-N
T_0: no SWC measures, no N input	

Taking into consideration that in 2000, results of treatments with compost or animal manure showed the same trend and because compost was more available than manure (Sédogo 1993), treatments T_{GSM} and T_{SRM} were replaced in year 2001 respectively by T_C (Compost-N, no SWC measure) and T_U (Urea-N, no SWC measure).

The plots, of 100 m by 25 m, were isolated from the surrounding area by an earth bund 0.6 m high.

 Springer

The first replication was instrumented with runoff collection devices and equipment to record runoff. Runoff and sediment in each plot were collected from a 100 m by 1 m subplot. A metal sheet was used to direct runoff into a 6 m^3 cement-lined pit. The covered pits were designed to cope with an exceptional 120 mm rain event. Each pit in one replicate only was equipped with a water level recorder (TD-divers, Eijkelkamp, Giesbeek, The Netherlands) that recorded the overland flow hydrograph. Rainfall intensity was recorded using an automatic rain gauge (tipping bucket). In each plot, 36 subplots of 10 m by 2 m were delimited. These subplots, which were used to record sorghum yield variation down the length of slope, were located in pairs at 99, 96, 83, 78, 70, 67, 65, 62, 50, 45, 37, 34, 32, 29, 17, 12, 4 and 1 m from the downslope border of each plot. Stone rows and grass strips had been installed during the preceding 1999 rainy season, spaced 33 m apart (i.e. 3 barriers per plot) along the contours (Zougmoré et al. 2000a, b). Each stone row consisted of two rows of stones placed in a furrow. The upslope row of large stones was stabilized by the downslope row of small stones. Stone rows were about 0.2–0.3 m high. Grass strips were made of three rows of grass, resulting in a thick barrier of 0.3 m width.

The 110-day sorghum (*Sorghum bicolor* (L.) Moench) variety Sariasso 14 was sown by hand in rows across the slope at 31,250 seedlings per hectare in all plots. The plots were weeded with hand hoes twice a year. Prior to sowing, plots were ploughed to 0.15 m depth using oxen power to incorporate manure, compost, and urea. Manure, compost and urea were applied each year at a rate of 50 kg N ha^{-1}. The amounts of compost or animal manure applied to attain this N-rate varied between 5 to 7 t ha^{-1}, and were about the minimal rates recommended in Burkina (Sédogo 1993; Berger 1996). Urea application was split into two rates (at 21 and 56 days after planting). All plots received a basal dressing of 20 kg ha^{-1} P in the form of triple superphosphate to eliminate phosphorus deficiency.

Data collection

Runoff was recorded for each rainfall event that generated overland flow using the TD-divers placed in the runoff collection pits. Before pumping out the water, runoff water was thoroughly stirred and sampled in plastic barrels (60 l) after each runoff-producing rain event for the determination of suspended sediment concentration. The amount of water that was pumped-out divided by the amount of well-stirred sample in the barrel is called sample fraction (SF). The fine sediments, after decantation and filtration of the barrel content, were dried and weighted. Coarse sediment in the pit after each runoff event was also dried and weighted. The total soil loss per each rain event was the sum of the dried fine sediment times SF and the total collected coarse sediment. Sorghum grain and straw yields were measured after sun drying at harvest from the 36 subplots in each plot.

Data analysis

Runoff was analyzed from 10 erosive rain events for the 2000 rainy season, 9 erosive rain events for the 2001 rainy season, and 16 erosive rain events for the 2002 rainy season. Cumulative runoff during the crop-growing period (from sorghum planting to its harvest) of each year was related to cumulative rainfall to assess the ratio of annual runoff. Cumulative soil loss was compared per treatment to assess the effect of treatment on erosion during the 3 years. STATITCF package (Gouet and Philippeau 1986) was used for statistical analyses of sorghum grain and straw yields. Newman-Keuls test was used for mean separation at $P < 0.05$.

Assuming $x_1 =$ Stone rows or grass strips, $x_2 =$ Application of compost-N or urea-N, $x_0 =$ Control (no SWC measures, no N input), $Y =$ yield, $(x_1 + x_2) =$ Combined SWC measure (x_1) and compost-N or urea-N (x_2). The interaction effect (IE) in crop yield is the benefit in crop yield (in comparison to the control treatment) of the combined application of both SWC measure and urea-N or compost-N ($\Delta Y(x_1 + x_2)$) minus the sum of the benefits from the two components (ΔYx_1 and ΔYx_2) when applied in isolation (Iwuafor et al. 2002).

$$\Delta Yx_1 = Y(x_1) - Y(x_0) \tag{1}$$

$$\Delta Yx_2 = Y(x_2) - Y(x_0) \tag{2}$$

$$\Delta Y(x_1 + x_2) = Y(x_1 + x_2) - Y(x_0) \tag{3}$$

$$IE = \Delta Y(x_1 + x_2) - (\Delta Yx_1 + \Delta Yx_2) \tag{4}$$

There is positive interaction between x_1 and x_2 when IE > 0, and negative interaction between x_1 and x_2 when IE < 0.

A minimum yield value, which corresponds to the minimum excess yield that supports the annual cost of the applied technology, was calculated per treatment to determine the economic benefit of single or combined N-input and SWC measures. To that end, the yield increase per kg N ($\Delta Y/\Delta N$) was calculated for the applied 50 kg ha^{-1} urea-N or compost-N. ΔY stands for yield increase and ΔN for applied N amount i.e. 50 kg N ha^{-1}.

In 2001 and 2002, the price of 1 kg urea-N was about 544 FCFA (1 USD = 650 FCFA in 2003). The price of 1 kg of sorghum in the region of Saria fluctuated between 100 FCFA and 180 FCFA. The average of 140 FCFA for 1 kg sorghum and a minimum of 3.9 kg sorghum per kg of urea-N were used in this paper. This corresponds to a minimum yield of 195 kg ha^{-1} for urea-N.

Several studies (Graaff 1996; Zougmoré et al. 2000b; Posthumus et al. 2001) have defined total costs for stone rows and grass strips installation in Burkina Faso to be 75,520 CFA ha^{-1} and 33,200 CFA ha^{-1}, respectively. The calculation of annual costs took into account the amortization, the opportunity cost of the capital and the repair and maintenance costs. The discounted average cost for stone rows using truck transport was 48,312 FCFA ha^{-1} year^{-1} while grass strips installation cost using root transplanting was 26,240 FCFA ha^{-1} year^{-1}. The minimum yield was therefore 345 kg ha^{-1} sorghum grain for stone rows and 187 kg ha^{-1} sorghum grain for grass strips.

The discounted annual cost for compost was evaluated to 37,900 FCFA ha^{-1} and the price of 1 kg of nitrogen deriving from compost was 758 FCFA. A minimum sorghum yield for compost-N was 5.4 kg kg^{-1}, which corresponds to a minimum yield of 271 kg ha^{-1}. This implies that for a technology to be beneficial, its excess yield, which is the difference between yield increase (ΔY) and the minimum yield, should be greater than zero.

Results and discussions

Rainfall characteristics

Figure 1 shows the cumulative rainfall patterns over the 3 years which were less than the regional average of 800 mm. The rainfall was 796 mm in 2000,

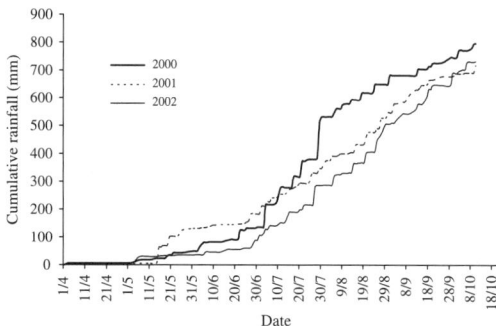

Fig. 1 Daily cumulative rainfall for the 2000, 2001 and 2002 rainy seasons at Saria, Burkina Faso

719 mm in 2001, and 733 mm in 2002. In 2000 there were 43 rain events, 4 of which were exceptionally heavy (53, 56, 81 and 127 mm during July); in 2001 there were 56 rain events, all less than 40 mm and well distributed in time. In 2002, 53 rain events occurred, 2 of which were greater than 50 mm and very influential on total runoff and soil loss. The rainfall during the sorghum cropping period (June–October) was more evenly distributed in 2001 and 2002 than in 2000; this contributed to the better crop performance in 2001 and 2002 (Table 2). A drought of 13 days occurred early in September 2000 (Fig. 1), coinciding with the sorghum maturation stage. The total rainfall in September 2000 was only 65 mm compared with 131 mm in September 2001 and 183 mm in September 2002. After a long period of drought during the whole month of June, rainfall was well distributed from July to October 2002. However, the delay of the rainy season in June postponed crop establishment in 2002.

Runoff

During the 3 years of study, all treatments reduced runoff compared to control plots (Table 1). In the 2000 rainy season, the runoff in treatments with stone rows was always less than in treatments with grass strips when compared in pairs (T_{SR}/T_{GS}; T_{SRC}/T_{GSC}; T_{SRM}/T_{GSM}; T_{SRU}/T_{GSU}). This difference in runoff reduction between stone rows and grass strips was only 2% in composted plots, 5% in manured plots, 7% in non-amended plots, and reached 19% in plots with fertilizer N. The same trend was observed in 2001 and 2002 with larger differences than in 2000 (Table 1). Overall in 2000, T_{SRC} and T_{GSC} had the

Nutr Cycl Agroecosyst (2010) 88:17–27   21

Table 1 Effect of treatments on runoff in the rainy seasons of 2000, 2001 and 2002 at Saria, Burkina Faso	Annual runoff (% of rainfall)			Runoff reduction (%)		
	2000	2001	2002	2000	2001	2002
T_0	15.9 (2.1)	12.2 (1.1)	17.6 (1.9)	0	0	0
T_{SR}	7.1 (2.0)	3.5 (1.3)	5.0 (0.9)	55	71	71
T_{SRU}	8.3 (2.7)	4.2 (1.6)	5.3 (0.9)	48	65	70
T_{SRC}	6.8 (1.6)	3.2 (1.6)	1.0 (0.6)	58	74	94
T_{GS}	8.3 (1.1)	5.9 (1.0)	8.2 (1.3)	48	51	53
T_{GSU}	11.4 (0.9)	9.5 (1.1)	7.6 (2.2)	29	22	57
T_{GSC}	7.1 (1.6)	4.5 (1.8)	2.8 (1.2)	56	63	84
T_{SRM}/T_U	7.5 (2.2)	6.6 (0.9)	9.0 (1.0)	53	46	49
T_{GSM}/T_C	8.2 (1.8)	8.2 (0.7)	2.4 (0.9)	48	32	87
Number of rain events	10	09	16	10	09	16

Values in brackets: ±standard deviation between runoff volumes measured in pits and recorded values of runoff. Treatments are explained in materials and methods section

least runoff, followed by T_{SR}, T_{SRM}, T_{GSM}, T_{GS}, T_{SRU}, T_{GSU} and T_0. In 2001 and 2002, except for T_{GSU}, the treatments without barriers (T_U, T_C, T_0) showed the greatest runoff, confirming the positive effect of stone rows and grass strips on runoff reduction. Under the climatic conditions of the study zone, it appears that as filtering barriers, stone rows had a greater effect in reducing runoff than grass strips; this was certainly because the stone rows were better able to slow down runoff water than vegetation bunds. Indeed, the stone row structure (The second line of stones, which supports the first row of big stones), allows to close all small gaps in the first row; thus, the runoff flow is uniformly dispatched throughout the contour line; the consecutive reduced velocity of runoff leads to an increased storage and infiltration of water within the field plot. Although the grass strips comprised three regular lines of plants, the resulted runoff barrier in this experiment remained more permeable to runoff water than stone rows. Contrarily to stone rows structure, some gaps were observed throughout the grass strips, which lead to the concentration of runoff water flow at these specific places and to little retention of water within the field plot. Grass strips took a few years (2 years in this experiment) to become thick enough to be fully effective. Moreover, grass strips have to endure and need about one month of re-growth after the inhospitable dry season before they can fully resume their anti-erosion role. This is consistent with the results reported by Lamachère and Serpentié (1991) and Zougmoré et al. (2000a) in similar climatic zones. It is however reported by numerous studies (Buldgen and François 1998; Sharma et al. 1999) that under

good climatic conditions, grass strips are very efficient soil and water conservation techniques that appreciably reduce runoff, leading to increased water infiltration into the land.

Compared to the control treatment, runoff was reduced more in the stone row and grass strip treatments with compost (by 74% and 63% respectively). Combining stone rows or grass strips with fertilizer N also resulted in runoff reduction: stone rows with urea reduced runoff by up to 65% whereas grass strips with urea reduced runoff by 22%. Applying compost (T_C) alone reduced runoff by 32%, which was almost as much as applying fertilizer N (T_U: 46%). Treatments without barriers (T_U, T_0) showed the highest runoff in both 2001 and 2002. Organic amendments were more effective than fertilizer N in reducing runoff but combining stone rows or grass strips with compost application induced the greatest reducing effect on runoff (Table 1). Application of compost for three successive years may notably improve soil physical properties, which could enhance water infiltration. This is in accordance with results of Cogle et al. (2002) in semiarid India who found that organic amendments, of farmyard and straw significantly reduced annual runoff, compared to non-amended treatments.

Soil and nutrient losses

All treatments reduced soil loss compared to the control treatment, which showed the average highest soil loss of 217 kg ha^{-1} in 2000, 236 kg ha^{-1} in 2001 and 32711 kg ha^{-1} in 2002 (Fig. 2a, b). In 2001, the effect of stone rows in soil loss reduction

 Springer

was much more noticeable than the effect of grass strips. Soil losses were very low and less than 70 kg ha^{-1} in the three stone row treatments (T_{SRC}, T_{SRU}, T_{SR}) and rose to more than 100 kg ha^{-1} in grass trips treatments (T_{GSC}, T_{GSU}, T_{GS}). Stone rows and grass strips alone (T_{SR}, T_{GS}) were able to reduce soil loss respectively by 70% and 58% as compared to the control treatment. Plots with stone rows or grass strips (T_{SR}, T_{GS}) were less erosive than those with compost-N or urea-N only (T_C, T_U). Indeed, the two latter treatments induced 51% soil loss reduction while treatments T_{SR} and T_{GS} induced respectively 70% and 58%, confirming the positive effect of permeable barriers on soil loss reduction.

The application of compost and urea in plots with soil conservation barriers (stone rows or grass strips)

improved the efficiency of the system to reduce soil loss (Fig. 2a, b). However, the efficiency of the water conservation system was better when compost is applied than when urea is applied. Applying compost on plots without barriers (T_C) enabled soil loss to be reduced by 52% while combining compost and stone rows (T_{SRC}) induced a 79% reduction in soil loss compared to the control (T_0). The difference in soil loss reduction rate was only 2% between T_C and T_{GSC} and was up to 26% between T_C and T_{SRC}. These trends were also observed in 2000 and 2002. Thus, in this study, application of organic amendment (compost) reduced soil loss more than application of mineral fertilisers (urea). Thanks to the better availability of macro-nutrients and micro-nutrients released along the cropping season from applied

Fig. 2 Cumulative soil loss as affected by SWC measures and nutrient inputs at Saria, Burkina Faso; (**a**) 2001, (**b**) 2002. Treatments are explained in materials and methods section

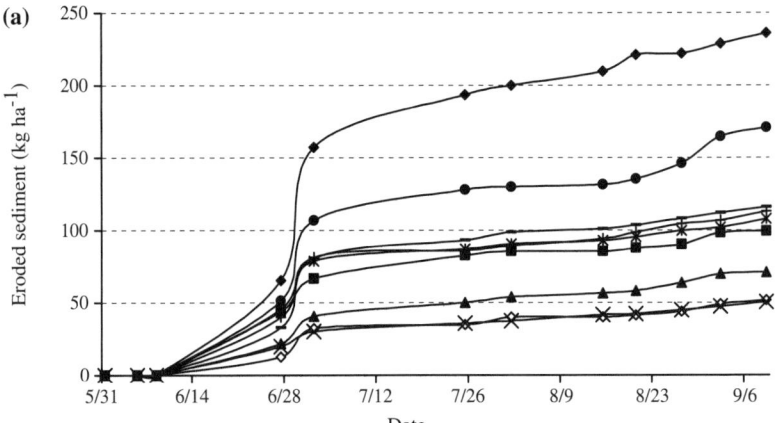

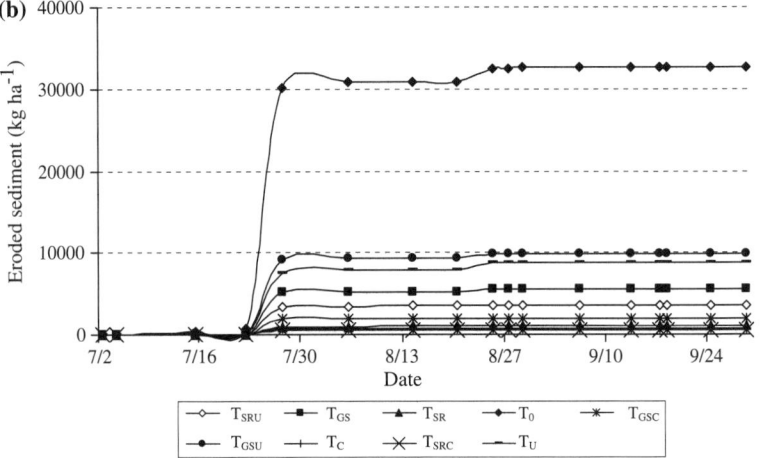

compost, sorghum biomass was more important in plots with application of organic amendment than in plots supplied with urea; this important biomass ensured a good crop canopy and soil cover, which may contribute slowing down runoff and reducing the displacement of soil particles, particularly the finest (Zougmoré et al. 2000a). This is consistent with Lal (1998) who reported that permanently protecting the soil surface with a dead or living cover is one of the most effective ways of controlling erosion.

Carbon and nutrient concentrations of generated sediments were very high and reached 14–29 g kg^{-1} for OC, 1.0–3.7 g kg^{-1} for N and 318–709 mg kg^{-1} for total P. Treatments with stone rows barriers had on average NO$_3^-$-N concentration in runoff water of 23 mg L^{-1} compared to only 2 mg L^{-1} with treatments without barriers (11 times greater). Annual losses of organic C, N, P and K were high and greatly dependant on the magnitude of soil loss. Combining runoff barriers with compost application induced the least organic C and plants nutrients losses. Integrated water and nutrient management is effective in alleviating total soil, carbon and nutrients losses by water erosion and therefore could play a major role in sustaining crop production in Sahelian smallholder farming systems.

Effects of SWC measures and nutrient management on sorghum performance

Sorghum yields were significantly different among treatments over the 3 years (Table 2). Except for T$_{GS}$

(grass strips alone), sorghum grain yield was lower in the control treatment than in the water and nutrient management treatments during the 3 years. In 2001, although not statistically different, sorghum yields with stone rows alone (T$_{SR}$) increased by 12% whereas with grass strips alone (T$_{GS}$) sorghum yield decreased by 18% when compared to the control (T$_0$). In 2002, single stone rows plots induced 12% grain yield increase whereas in single grass strips plots, grain yield decreased by 15% compared to the control. These results indicate that during well-distributed rainfall years in the Sahel, implementing water conservation measures without adding nutrients induced little or even negative influence on crop yields. This may be explained by the inherent low nutrient content of soils, mainly for N and P (Bationo et al. 1998). Also, thanks to the better water availability induced by the runoff barriers, nutrient needs of sorghum plants may increase, leading to greater deficits and competitions between sorghum plants and those from the grass strips. This is consistent with results of previous studies in the region (Hamer 1996).

In 2001, application of compost (T$_C$) or urea (T$_U$) alone significantly increased sorghum yield by respectively 107% and 92% compared to the control (T$_0$). Thus, applying nutrient inputs alone (T$_C$, T$_U$) induced much higher grain yields than laying SWC barriers without nutrient inputs: 80% compared to stone rows plots (T$_{SR}$) and 145% compared to grass strips plots (T$_{GS}$).

Table 2 Effect of treatments on sorghum performance for rainy seasons 2000, 2001 and 2002 at Saria, Burkina Faso

	Grain yield (kg ha^{-1})			Straw yield (kg ha^{-1})		
	2000	2001	2002	2000	2001	2002
T$_{SRC}$	2,308 (18) a	2,535 (43) a	2,766 (58) a	4,844 (244) ab	5,139 (208) a	4,598 (175) a
T$_{GSC}$	2,324 (75) a	2,338 (73) ab	2,536 (74) b	4,997 (401) a	4,742 (240) a	4,564 (212) b
T$_{GSM}$/T$_C$	1,558 (78) b	2,278 (68) ab	2,385 (79) b	3,591 (181) ab	4,570 (84) a	4,038 (147) b
T$_{SRU}$	1,444 (10) bc	1,796 (50) c	1,511 (39) c	3,891 (116) ab	4,024 (193) ab	3,023 (165) c
T$_{GSU}$	932 (13) cd	1,537 (22) c	1,411 (30) c	2,815 (169) ab	3,523 (204) ab	2,376 (76) c
T$_0$	838 (70) cd	1,099 (76) d	1,164 (75) d	2,623 (99) ab	2,857 (172) ab	1,967 (143) d
T$_{SR}$	739 (53) d	1,226 (74) d	1,308 (54) cd	2,439 (121) b	3,005 (118) ab	2,374 (87) cd
T$_{GS}$	664 (9) d	896 (72) d	983 (42) d	2,321 (215) b	2,056 (162) b	1,669 (159) d
T$_{SRM}$/T$_U$	1,692 (55) b	2,106 (14) b	1,403 (30) c	3,534 (68) ab	3,823 (220) ab	2,468(165) cd
Prob.	<0.001	0.022	0.021	0.020	0.021	0.026

Treatments with the same letter are not statistically different at $P = 0.05$. Values in brackets: $\pm$ are standard deviations. Treatments are explained in materials and methods section

In 2002, as in 2001, single application of N-input (T_C, T_U) induced significant higher grain yield than single application of SWC measures: applying urea (T_U) induced 7% and 43% yield increase compared to plots with stone rows (T_{SR}) or grass strips (T_{GS}) alone respectively, whereas application of compost alone (T_C) significantly increased yield by 82% and 143% compared to stone row (T_{SR}) and grass strip plots (T_{GS}), respectively. As the productivity of most soils in their native state in the study zone is very low (Bationo et al. 1998), applying plants nutrients (50 kg N ha^{-1}) in these poor soils induced great positive reaction in crop production (Table 2), particularly during good rainfall years (Fig. 1) when soil moisture constraint is small. In plots with compost application, the mineralization of compost releases not only the macronutrients such as nitrogen and phosphorus, but also considerable amounts of micronutrients for plants (Velthof et al. 1998). This explains also why in 2001, combining compost with stone rows (T_{SRC}) induced a significant yield increase of 106% when compared to plots with stone rows alone (T_{SR}). Similarly, combining grass strips and compost (T_{GSC}) induced a significant grain yield increase of 160% when compared to plots with grass strips alone (T_{GS}). Also, adding urea to plots with barriers (T_{SRU}, T_{GSU}) significantly increased grain yield by 46% and 71% respectively compared to plots with barriers only (T_{SR}, T_{GS}). In general, only slight differences were observed between treatments combining barriers with N-input (T_{SRC}, T_{GSC}, T_{SRU}, T_{GSU}) and receiving-N treatments without barriers (T_C, T_U). These results were confirmed in 2002: treatments combining barriers with compost (T_{SRC}, T_{GSC}) induced significant yield increase by respectively 138% and 118% compared to the control. Adding urea to plots with barriers (T_{SRU}, T_{GSU}) significantly increased grain yield by 30% and 21% respectively when compared to control plots. The above results suggest that under Sahelian conditions, SWC in

combination with nutrient management can be used to alleviate risks and to achieve production intensification. This attests that to develop new strategies of agricultural production in sub-Saharan West Africa, one should take into account local SWC technologies and improved practices of soil fertility replenishment (Dudal 2002).

Economic benefit of water and nutrient management

Positive interactions ($\Delta Y(x_1 + x_2)$) of combined SWC measures and N-inputs were observed apart for T_{GSU} in 2001 (-367 kg ha^{-1}) T_{SRU} in 2001 (-437 kg ha^{-1}) and 2002 (-36 kg ha^{-1}). The high response of sorghum yield to added N-inputs only (T_C, T_U) suggests that nutrient supply more than water retention by the filtering barriers (T_{SR}, T_{GS}) increased the yield in combined SWC and nutrient plots. Yield increases did not cover annual costs of stone rows or grass strips alone (Table 3).

Conversely, economic benefits of treatments in Table 4 showed that single application of compost-N or urea-N were cost-effective but supply of urea-N was less beneficial (6,160 FCFA) in 2002 compared to compost-N (133,040 FCFA).

The combination of SWC measures with urea-N (T_{SRU}, T_{GSU}) or compost-N (T_{SRC}, T_{GSC}) induced positive economic benefits in 2001 (Table 5), indicating that at least the annual costs for implementing SWC measures and applying compost-N or urea-N were covered by the excess yields in the combined SWC measure and N-input treatments.

Ganry and Badiane (1998) noted that in cultivated sandy soils of dry tropical zones, organic matter becomes more important in the soil surface layer, because of its effects on the water balance and the mobility of mineral elements. Easily decomposable organic material like compost may well make available

Table 3 Economic benefits of single stone rows or single grass strips in 2001 and 2002

	Stone rows		Grass strips	
	2001	2002	2001	2002
Annual cost (FCFA ha^{-1})	48,312	48,312	26,240	26,240
Sorghum average price (FCFA kg^{-1})	140	140	140	140
Minimum yield (kg ha^{-1})	345	345	187	187
ΔY (kg ha^{-1})[a]	127	144	-203	-181
Excess yield (kg ha^{-1})	-218	-201	-390	-368
Economic benefit (FCFA ha^{-1})	$-30,520$	$-28,140$	$-54,600$	$-51,520$

[a] ΔY stands for yield increase for stone rows or grass strips treatments compared to the control treatment

Table 4 Economic benefits of single urea-N or single compost-N in 2001 and 2002

Treatment	Urea-N		Compost-N	
	2001	2002	2001	2002
N-input cost (FCFA kg^{-1} N)	544	544	758	758
Sorghum average price (FCFA kg^{-1})	140	140	140	140
Minimum yield increase (kg kg^{-1})	3.9	3.9	5.4	5.4
$\Delta Y/\Delta N$ (kg kg^{-1})[a]	20.2	4.8	23.6	24.4
Excess yield increase (kg kg^{-1})	16.3	0.9	18.1	19.0
Excess yield (kg ha^{-1})	813	44	907	950
Economic benefit (FCFA ha^{-1})	113,820	6,160	127,020	133,040

[a] ΔY stands for yield increase and ΔN for applied N amount of 50 kg N ha^{-1}

Table 5 Economic benefits of combining stone rows or grass strips with compost-N or urea-N in 2001 and 2002 at Saria, Burkina Faso

	2001				2002			
	T_{SRU}	T_{GSU}	T_{SRC}	T_{GSC}	T_{SRU}	T_{GSU}	T_{SRC}	T_{GSC}
Minimum yield for N inputs (kg ha^{-1})	195	195	271	271	195	195	271	271
Minimum yield for SWC (kg ha^{-1})	345	187	345	187	345	187	345	187
Minimum yield for SWC + N input (kg ha^{-1})	540	382	615	457	540	382	615	457
Excess yield (kg ha^{-1})	158	54	821	782	−193	−135	987	915
Economic benefit (FCFA ha^{-1})	22,120	7,560	114,940	109,480	−27,020	−18,900	138,180	128,100

T_{SRC}: stone rows + compost-N; T_{GSC}: grass strips + compost-N; T_{SRU}: stone rows + urea-N; T_{GSU}: grass strips + urea-N

plant nutrients at crucial time of sorghum production (maturing phase), which, in combination with better water availability thanks to SWC measures, have improved sorghum productivity. Moreover, organic matter maintains soil physical, chemical and biological balance that would accelerate crop root formation in the soil profile (Piéri 1989). The best sorghum production resulting from this positive interaction effect of SWC measures and N-inputs explains their greatest economic benefits. Loss of available Urea-N through runoff or leaching could explain the lower interactive effect of urea-N that has resulted in lower economic benefits in comparison to organic-N (Ouédraogo et al. 2007).

Conclusions

Results of this study suggest that:

- The semi-permeable soil and water conservation barriers combined with compost application significantly reduced runoff and soil loss. Stone rows and grass strips increased soil moisture, especially upslope, and could thus play a major role in harvesting runoff water.

- Soil organic C, N, P and K concentrations of eroded sediments were very high, particularly for N, indicating severe losses of one of the more deficient element in West African soils.

- When annual rainfall is well distributed in time (as was the case in 2001 and 2002 at Saria, Burkina Faso), installation of stone rows only induced very limited sorghum yield increase while *Andropogon gayanus* grass strips induced sorghum yield decrease. These yields were not enough to support installation costs due to high labor, transport and material inputs.

- Application of the sole compost-N or urea-N induced significant greater sorghum yield increase than SWC measures only.

- Stone rows or grass strips combined with compost-N induced positive interaction effects while stone rows combined with urea-N showed negative interactions. A positive interaction of grass strips combined with urea-N was observed only after 2 years.

- Economic benefits when combining compost-N to both stone rows and grass strips were substantial (109,000–138,000 FCFA ha^{-1}) while the greatest amounts observed with added urea-N were small (7,560–22,120 FCFA ha^{-1}).

 Springer

- These results indicate that in the Sahel, opportunities do exist for making more efficient use of local sources of nutrients such as compost in combination with locally accepted SWC measures. This may empower farmers to invest for sufficient nutrient supply in the sub-Saharan soils characterized by poor fertility.

Acknowledgements This study was funded by the IFAD project in Burkina Faso (CES/AGF), the University of Wageningen, the International Foundation for Science and INERA. We are grateful to Mr. Silga Mathias (BUMIGEB laboratory), Zacharia Gnankambary, Noufou Wandaogo, and Martin Ramdé (Soil-Water-Plant laboratory, INERA) for samples analyses, to Zacharie Zida, Moctar Ouédraogo, Saidou Simporé and Martin Sanon at Saria agricultural station for their support.

References

Bationo A, Lompo F, Koala S (1998) Research on nutrient flows and balances in West Africa: state-of-the-art. Agric Ecosyst Environ 71:19–35. doi:10.1016/S0167-8809(98)00129-7

Berger M (1996) L'amélioration de la fumure organique en Afrique soudano-sahélienne. Fiches techniques. Agriculture et développement n° hors série. CIRAD, Montpellier

Breman H, Groot JJR, van Keulen H (2001) Resource limitations in Sahelian agriculture. Glob Environ Change 11:59–68. doi:10.1016/S0959-3780(00)00045-5

Buerkert A, Piepho HP, Bationo A (2002) Multi-site time-trend analysis of soil fertility management effects on crop production in sub-Saharan West Africa. Exp Agric 38:163–183. doi:10.1017/S0014479702000236

Buldgen A, François J (1998) Physiological reactions to imposed water deficit by *Andropogon gayanus cv. Bisquamulatus and Cenchrus ciliaris cv. Biloela* in a mixed fodder crop. J Agric Sci 131:31–38. doi:10.1017/S0021859698005462

Cogle AL, Rao KPC, Yule DF, Smith GD et al (2002) Soil management for Alfisols in the semi-arid tropics: erosion, enrichment ratios and runoff. Soil Use Manage 18:10–17. doi:10.1079/SUM2002094

de Graaff J (1996) The price of soil erosion. An economic evaluation of soil conservation and watershed development. TRMP 14. Wageningen University, Wageningen

Dudal R (2002) Forty years of soil fertility work in sub-Saharan Africa. In: Vanlauwe B (ed) Integrated plant nutrient management in Sub-Saharan Africa: From concept to practice. CAB International, New York, pp 7–21

FAO (1986) African agriculture: the next 25 years. Annex II. The land resource base. FAO, Rome

FAO-UNESCO (1994) Soil map of the world. ISRIC, Wageningen

Fontes J, Guinko S (1995) Carte de la végétation et de l'occupation du sol du Burkina Faso. Note explicative. Ministère de la coopération française, Toulouse

Ganry F, Badiane A (1998) La valorisation agricole des fumiers et des composts en Afrique soudano-sahélienne. Diagnostic et perspectives. Agric Dev 18:73–80

Gouet JP, Philippeau G (1986) Comment interpréter les résultats d'une analyse de variance. I.T.C.F, Paris

Hamer A (1996) Etude sur les bénéfices des cordons pierreux dans les régions de Kaya et Manga, Burkina Faso. Rap Etudiants 80 Antenne Sahélienne, Ouagadougou

Iwuafor ENO, Aihou K, Jaryum JS, Vanlauwe B et al (2002) On-farm evaluation of the contribution of sole and mixed applications of organic matter and urea to maize grain production in the Savannah. In: Vanlauwe B, Diels J, Sanginga N, Merckx R (eds) Integrated plant nutrient management in Sub-Saharan Africa: From concept to practice. CAB International, New York, pp 185–197

Laflen JM, Roose E (1998) Methodologies for assessment of soil degradation due to water erosion. In: Lal R, Blum WH, Valentin C (eds) Methods for assessment of soil degradation. Advances in Soil Science. CRC Press, Boca Raton, pp 31–55

Lal R (1997) Soil quality and sustainability. In: Lal R, Blum WH, Valentin C (eds) Methods for assessment of soil degradation. CRC Press, Boca Raton, pp 17–31

Lal R (1998) Soil erosion impact on agronomic productivity and environment quality. Crit Rev Plant Sci 17:319–464. doi:10.1016/S0735-2689(98)00363-3

Lamachère JM, Serpantié G (1991) Valorisation agricole des eaux de ruissellement et lutte contre l'érosion sur champs cultivés en mil en zone soudano-sahélienne, Bidi, Burkina Faso. In: Kergreis A, Claude J (eds) Utilisation rationnelle de l'eau des petits bassins versants en zone aride. John Libbey Eurotext, Paris, pp 165–178

Mando A (1997) The effect of mulch on the water balance of Sahelian crusted-soils. Soil Technol 11:121–138. doi:10.1016/S0933-3630(97)00003-2

Morgan RPC (1995) Soil erosion & conservation, 2nd edn. Longman, Addison

Nicou R, Charreau C (1985) Soil tillage and water conservation in semi-arid West Africa. In: Ohm W, Nagy JC (eds) Appropriate technologies for farmers in semi-arid West Africa. Purdue University, Lafayette, pp 9–39

Ouédraogo E, Stroosnijder L, Mando A, Zougmoré R et al (2007) Agroecological analysis and economic benefit of organic resources and fertiliser in till and no-till sorghum production after a 6-year fallow in semi-arid West Africa. Nutr Cycl Agroecosyst 77:245–256. doi:10.1007/s10705-006-9063-5

Perez P, Albergel J, Diatta M et al (1998) Rehabilitation of a semi-arid ecosystem in Senegal. 2. Farm-plot experiments. Agric Ecosyst Environ 70:19–29. doi:10.1016/S0167-8809(97)00157-6

Piéri C (1989) Fertilité des terres de savane. Bilan de trente ans de recherche et de développement agricoles au sud du Sahara. Min Coop-CIRAD, Paris

Posthumus H, Zougmoré R, Spaan W, de Graaff J (2001) Incentives for soil and water conservation in semi-arid zones: a case study from Burkina Faso. ZALF 47:91–97

Quansah C, Ampontuah EO (1999) Soil fertility erosion under different soil and residue management systems: a case study in the semi-deciduous forest zone of Ghana. Bull Eros 19:111–121

Nutr Cycl Agroecosyst (2010) 88:17–27

Ryan JG, Spencer DC (2001) Future challenges and opportunities for agricultural R&D in the semi-arid tropics. International Crops Research Institute for the Semi-Arid Tropics, Patancheru

Sanchez PA, Shepherd KD, Soule JM et al (1997) Soil fertility replenishment in Africa: an investment in natural resource capital. In: Buresh RJ, Sanchez PA (eds) Replenishing soil fertility in Africa. SSSA Special Publication N°51, Madison, pp 1–46

Sédogo PM (1993) Evolution des sols ferrugineux lessivés sous culture: incidence des modes de gestion sur la fertilité. Thèse de doctorat es sciences Université Nationale de Côte d'Ivoire, Abidjan

Sharma KD, Joshi NL, Singh HP, Bohra DN, Kalla AK, Joshi P (1999) Study on the performance of contour vegetative barriers in an arid region using numerical models. Agric Water Manage 41:41–56. doi:10.1016/S0378-3774(98)00114-0

Stoorvogel JJ, Smaling EMA (1990) Assessment of soil nutrient depletion in sub-Saharan Africa 1983–2000. Report 28; Winand Staring Centre for Integrated Land, Soil, and Water Research (SC-DLO). Wageningen

Velthof GL, Van Beuichem ML, Raijmakers WMF et al (1998) Relation between availability indices and plant uptake of nitrogen and phosphorus from organic products. Plant Soil 2000:215–226. doi:10.1023/A:1004336903214

Zougmoré R, Guillobez S, Kambou NF et al (2000a) Runoff and sorghum performance as affected by the spacing of stone lines in the semi-arid Sahelian zone. Soil Tillage Res 56:175–183. doi:10.1016/S0167-1987(00)00137-9

Zougmoré R, Kaboré D, Lowenberg-Deboer J (2000b) Optimal spacing of soil conservation barriers: example of rock bunds in Burkina Faso. Agron J 92:361–368. doi:10.1007/s100870050045

 Springer

Survival and Soil Nutrient Changes During 5 Years of Growth of 16 *Faidherbia albida* Provenances in Semi-Arid Baringo District, Kenya

O.G. Dangasuk, S.O. Gudu, and J.R. Okalebo

Abstract Farmers throughout Africa for improving soil fertility and crop yields have long used *Faidherbia albida*. Increase in yield from crops grown below the trees has been attributed to increased nutrients. Sixteen provenances of *F. albida* were planted in a randomized complete block design (RCBD) with five replications in April 1997 at Noiweit sub-location in semi-arid Baringo district of Kenya. The objectives were (1) to investigate the pattern of genetic variation and performance among the 16 provenances to determine their suitability for introduction in this environment and (2) to assess the soil fertility development under *F. albida*. Assessments of growth variables and soil properties were done in October 1997 (6 months after planting) and March 2002 (5 years after planting). There were significant differences in height and diameter growth among provenances at 6 months but not at 5 years. Survival percentage was higher among the Eastern and Southern African provenances, while four of the five West African provenances had 0% survival at 5 years. Soils data showed significant increase in soil pH, organic C, total N, available Olsen P and exchangeable Na and K; a significant decrease in exchangeable Ca and Mg; and no significant difference in exchangeable Al in 5 years. Eastern and Southern African provenances were found to be suitable for introduction, and *F. albida* has proved to be effective in soil fertility improvement in this region.

Keywords Arid and semi-arid lands · *Faidherbia albida* · Genetic variation · Soil nutrients

Introduction

Faidherbia albida (Del.) A. Chev., a leguminous tree species belonging to the mimosoideae subfamily, is very important for agro-silvo-pastoral as well as soil conservation and soil fertility improvement throughout the arid and semi-arid zones of Africa. The tree has unusual characteristics of shedding off its leaves during the rainy season. Due to this reverse phenology compared to the normal phenologic system observed for most other species, it does not compete with agricultural crops for nutrients, water and light during the rainy season (ICRAF, 1989). On the contrary, in the dry season it is in full leaf and thus provides valuable shade for domestic animals while its pods and foliage constitute an excellent fodder rich in protein for livestock during the dry season when grass is not available. *F. albida* is a nitrogen-fixing tree species; and it also improves soil fertility by recycling mineral elements to the surface owing to its deep root system (Le Houerou, 1980). Decomposition of shed foliage as well as animal droppings under the trees increases the organic matter and microbial activity in the soil, thus improving soil structure and permeability. *F. albida* also acts as an efficient windbreaker against wind erosion, offering notable resistance to the displacement of the surface layers of sand. This effect is strong when the trees are growing sufficiently close (Dancette and Poulain, 1969).

O.G. Dangasuk (✉)
Department of Biological Sciences, Moi University, Eldoret, Kenya
e-mail: georgedangasuk@yahoo.com

Recently, there has been renewed interest in this species particularly on its impact on soil characteristics, microclimate and crop yields. Indeed, there is a concentration of research efforts on genetic resources and stock improvement as well as on the relationship between soil nutrient availability and *F. albida's* establishment, growth and biomass production, particularly in the arid and semi-arid environments. Noiweit sub-location of Baringo district contains several native Acacia species, but not *F. albida*. Owing to its importance in agro-silvo-pastoral systems, it would be beneficial to introduce *F. albida* in this environment.

Studies by Dangasuk et al. (2001) on morphological evaluation and soil properties 6 months after planting reflected the results at juvenile developmental stage of *F. albida* which would definitely change with time as the plant continues to grow. In addition, the influence of 6 months seedlings on the soil would be negligible if any at all. Thus, there is a need to monitor the species' survival, performance and influence on soil properties at a 5-year interval in order to understand the genetic diversity of the species as well as its effects on soil properties in semi-arid Baringo district of Kenya. This could help evaluate the suitability of certain provenances of *F. albida* in this area. The study objectives were (1) to investigate the growth variation among 16 provenances of *F. albida* based on morphological characteristics namely height, diameter at ground level and survival percentage with the aim of selecting the best for introduction in this semi-arid Baringo district of Kenya and (2) to assess the soil fertility improvement under *F. albida* at 5 years after planting.

Materials and Methods

Comparison of Field Growth Performance in 16 Provenances of F. albida at Noiweit

Bulk seeds of 16 *F. albida* provenances from different geographical regions of Africa were used in this study (Table 1). The samples were representative of the entire natural distribution range of the species in Africa. The field trial was based at Noiweit, located at 0°10′N and 36°00′E and at an altitude of about 1500 m above sea level, in Mogotio division of Baringo district. The district is classified as semi-arid (Teel, 1984), with mean annual rainfall range between 500 and 800 mm, and has an average annual temperature of 21°C.

Four-month-old potted seedlings were transplanted into the field in April 1997, in a randomized complete block design (RCBD) with repeated observations. There were five blocks (replications) and in each block, there were 16 experimental units representing each of the provenances of *F. albida*. In each unit, there were

Table 1 Collection site data for 16 provenances of *F. albida* used in this study

CI	Code	Provenance	Country	Zone	Latitude	Longitude	Altitude (m)	Rain (mm)	Temp. °C
OFI	1	Bignona	Senegal	WA	12°45′N	16°25′W	10	1408	26.5
OFI	2	Mana Pools	Zimbabwe	SA	15°45′S	29°20′E	360	628	25.1
OFI	3	Chawanje	Malawi	SA	14°36′S	34°48′E	600	900	24.2
OFI	4	Bolgatanga	Ghana	WA	10°46′N	01°00′W	201	1057	28.2
OFI	5	Mwembe	Tanzania	EA	04°08′S	37°51′E	860	569	23.5
OFI	6	Dumisa	Zimbabwe	SA	22°13′S	31°24′E	280	438	24.7
OFI	7	Debre zeit	Ethiopia	EA	08°43′N	38°59′E	1850	730	–
OFI	8	Pongola River	South Africa	SA	22°20′S	30°03′E	540	340	–
OFI	9	Rama	Ethiopia	EA	14°23′N	38°48′E	1350	742	19.1
OFI	10	Lake Awassa	Ethiopia	EA	07°03′N	38°25′E	1650	961	–
OFI	11	Hoanib River	Namibia	SA	19°15′S	13°23′E	350	98	24.0
CDF	12	Moulvouday	Cameroun	WA	10°23′N	14°50′E	330	815	–
CDF	13	Tera	Niger	WA	14°00′N	00°45′E	240	458	–
CDF	14	Banbora	Burkina Faso	WA	10°34′N	04°46′W	280	800	–
KFI	15	Kainuk	Kenya	EA	01°14′N	35°09′E	500	230	25.0
KFI	16	Tot	Kenya	EA	01°30′N	35°45′E	750	800	21.0

CI = Collector's Identity; OFI = Oxford Forestry Institute; CDF = Cirad-Foret; KFI = Kenya Forestry Research Institute; WA = West Africa; SA = Southern Africa; EA = East Africa; – = missing data

two seedlings from the same provenance. In total there were 10 seedlings per provenance in the whole design at a spacing of 3×3 m. Height was measured using a meter rule; collar diameter and diameter at breast height (DBH) were measured using a vernier calliper; and survival percentage was assessed first at 6 months and later at 5 years after transplanting.

Soil Sampling and Analysis

Soils from a depth of 0–30 cm were sampled in between the trees at three randomly selected sites from each of the five blocks, first at 6 months and later at 5 years after transplanting the *F. albida* seedlings into the field. At each site samples were collected from top-soil (0–15 cm depth) and subsoil (15–30 cm depth). Thus, six soil samples were taken from each block (total of 30 samples for all the five blocks). The soil samples were air dried, crushed and sieved through a 2 mm sieve and a further sub-sample ground finely to pass through a 60-mesh screen for all the soil chemical property analysis.

Soil pH was measured in a 2.5:1 water to soil sus-pension using a glass electrode in a pH meter. Organic carbon was estimated by the dichromate oxidation method of Nelson and Sommers (1975), followed by back titration of residual chromate using ferrous ammonium sulphate solution. Total nitrogen was deter-mined by the wet digestion procedure, whereby the ammonium was distilled off and measured by titra-tion with hydrochloric acid (the semi-micro-Kjeldahl method), as described by Okalebo et al. (2002). Available phosphorus was extracted from soil using the sodium bicarbonate (0.5 M $NaHCO_3$; pH 8.5) procedure of Olsen et al. (1954), followed by the P determination in extracts using the colorimetric ascor-bic method described by Murphy and Riley (1962). Exchangeable cations (K^+, Na^+, Mg^{2+} and Ca^{2+}) were determined by extraction of soils using 1 M ammonium acetate solution buffered to pH 7.0, fol-lowed by the measurements of Na, K and Ca in the extracts on a flame photometer and Mg on atomic absorption spectrophotometer (Okalebo et al., 2002). Because of pH dependency on Al activities in soils (McLean, 1965), exchangeable acidity (H^+ and Al^{3+}) was determined titrimetrically using unbuffered neu-tral salt 1 M KCl. Thus, the soils were extracted with

1 M KCl solution followed by filtration and titration of the aliquots of the extracts using 0.05 M NaOH. The amount of this base used was equivalent to the total amount of acidity in the aliquot taken. The soil particle size analysis (clay, silt and sand) was estimated only at 6 months after planting. It was assumed that no sig-nificant change would take place within 5 years, since soil particle size does not change that easily. The anal-ysis was based on hydrometer method described by Bouyoucos (1962), using the sodium hexametaphos-phate solution (Calgon) as a soil-dispersing agent and making temperature corrections of the solution.

Statistical Analysis

Both plant growth parameter data and soil analysis data were subjected to analysis of variance (ANOVA) and significant treatment effects at $P < 0.05$ separated using Duncan's multiple range test (DMRT), using the Statistical Package for Social Sciences (SPSS) version 12.0.

Results

Growth Variability Among the 16 Provenances 6 Months and 5 Years After Planting

The 6 months field performance of *F. albida* in Noiweit showed that differences in plant height and collar diameter were significant ($P < 0.01$) between the 16 provenances. With regard to collar diameter and height growth, the three leading provenances were all of Southern African origin. The Eastern and Southern African provenances were completely integrated with no clear-cut variation in height and collar diameter growths, while the West African provenances formed a distinct group with shorter heights and smaller collar diameters. The trees planted showed 100% survival for all provenances at 6 months of age (Table 2).

Provenance evaluation 5 years after planting showed no significant differences ($P > 0.05$) in height and diameter at breast height (DBH) among the 12 surviving provenances. Mana Pools (Zimbabwe) had the highest mean height (2.72 m) followed by Tot

Table 2 Comparison between 6 months and 5 years growth performance of 16 *F. albida* provenances at Noiweit

Height (m)				Diameter (cm)				Survival (%)			
Prov.	Zone	6 months	5 years	Prov.	Zone	6 months	5 years	Prov.	Zone	6 months	5 years
2	SA	0.79^a	2.73^a	3	SA	1.30^a	2.58^a	15	EA	100^a	70^a
6	SA	0.78^{ab}	2.20^a	6	SA	1.30^a	2.42^a	6	SA	100^a	60^{ab}
11	SA	0.74^{ab}	2.05^a	2	SA	1.28^a	2.75^a	3	SA	100^a	60^{ab}
15	EA	0.68^{abc}	2.00^a	8	SA	1.24^{ab}	2.00^a	2	SA	100^a	60^{ab}
5	EA	0.63^{abcd}	1.88^a	9	EA	1.18^{abc}	2.25^a	16	EA	100^a	50^{abc}
3	SA	0.62^{abcd}	2.49^a	7	EA	1.18^{abc}	3.13^a	10	EA	100^a	50^{abc}
9	EA	0.62^{abcd}	1.74^a	10	EA	1.11^{abcd}	2.10^a	9	EA	100^a	40^{abc}
7	EA	0.61^{abcd}	2.19^a	15	EA	1.11^{abcd}	2.50^a	8	SA	100^a	40^{abc}
8	SA	0.61^{abcd}	1.83^a	11	SA	1.10^{abcde}	1.75^a	7	EA	100^a	40^{abc}
10	EA	0.60^{bcde}	1.71^a	5	EA	1.04^{abcde}	2.50^a	5	EA	100^a	40^{abc}
16	EA	0.59^{bcde}	2.57^a	16	EA	0.95^{bcde}	3.10^a	1	WA	100^a	30^{bc}
13	WA	0.49^{cdef}	0.00	1	WA	0.88^{cde}	2.67^a	11	SA	100^a	20^c
1	WA	0.46^{def}	1.68^a	14	WA	0.87^{cde}	0	14	WA	100^a	0
12	WA	0.43^{ef}	0	4	WA	0.85^{de}	0	13	WA	100^a	0
14	WA	0.39^f	0	13	WA	0.85^{de}	0	12	WA	100^a	0
4	WA	0.32^f	0	12	WA	0.79^e	0	4	WA	100^a	0

Prov. = Provenance, Codes for provenances and abbreviations for zones are given in Table 1. Means followed by the same letter in each column are not significantly different at $P < 0.05$. (DMRT = Duncan's Multiple Range Test)

(Kenya) with 2.57 m. There was no differentiation among the Southern and Eastern African provenances as regard to height growth. The least height growth was recorded from the only surviving West African provenance, Bignona (Senegal), with a mean height of 1.68 m (Table 2). The greatest mean diameter of 3.13 cm was observed in Debre zeit (Ethiopia), followed by 3.10 cm for Tot (Kenya) provenances. The least mean diameter of 1.75 cm was recorded from Hoanib River (Namibia) provenance. Survival percentage showed significant differences ($P < 0.05$) between the Eastern, Southern and the one surviving West African provenances. Kainuk (Kenya) had the highest survival percentage (70%), followed by three Southern African provenances. Hoanib River from Namibia had the least survival percentage (20%) while Bignona (Senegal) had 30% survival percentage. The rest of the West African provenances showed 100% mortality 5 years after planting (Table 2).

Correlation Analysis Within and Between Growth Variables and Seed Source Parameters

Correlation coefficient (r) within and between growth variables and seed source parameters are presented in Table 3. Highly significant ($P < 0.01$) correlation ($r = 0.828$) was observed between collar diameter and height growth. Similarly, survival percentage 5 years after planting showed significant ($P < 0.05$) correlation to collar diameter ($r = 0.797$) and height ($r = 0.777$) attained 6 months after planting. Highly significant correlation ($P < 0.01$) was observed between growth variables and seed source parameters such as longitude, latitude and rainfall (Table 3). For example, at 6 months, height and diameter growths were negatively correlated to latitude, but positively correlated to longitude. In addition, at 5 years survival percentage showed positive correlation to longitude of seed source.

Changes in Soil pH and Soil Nutrient Status at Noiweit 5 Years After Planting F. albida

Based on particle size distribution (Table 4), the soil was classified as clay loam and hence has low water retention capacity. Soil data (Table 4) obtained at 5 years after planting the 16 provenances of *F. albida* showed that replication had no effect on soil pH, organic C, total N, exchangeable Mg and Al and

Table 3 Correlation coefficients (r) among growth variables and seed source parameters (P values in parentheses)

	DBH (5 years)	Height (6 months)	Collar diameter (6 months)	Survival % (5 years)	Latitude	Longitude	Altitude
Height (5 years)	0.535 (0.073)	0.489 (0.107)	0.356 (0.256)	0.470 (0.123)	−0.416 (0.178)	0.360 (0.251)	−0.193 (0.548)
DBH (5 years)		−0.222 (0.489)	−0.188 (0.557)	0.336 (0.286)	0.370 (0.237)	0.167 (0.604)	0.150 (0.641)
Height (6 months)			0.828** (0.000)	0.777** (0.000)	−0.721** (0.002)	0.683** (0.004)	0.221 (0.410)
Collar diameter (6 months)				0.797** (0.000)	−0.713** (0.002)	0.708** (0.002)	0.357 (0.174)
Survival % (5 years)					−0.505* (0.046)	0.754** (0.001)	0.334 (0.205)

DBH = diameter at breast height
*significant at $P < 0.05$; **significant at $P < 0.01$

available P, but was significant for exchangeable Na, K and Ca. Five years after planting, only Al levels in the soil did not change significantly ($P > 0.05$), while the rest of the soil parameters, including soil pH, organic C, total N, Olsen P, Na, K, Ca and Mg, showed significant changes ($P < 0.05$) in the experimental plot (Table 4). Six parameters: pH, organic C, total N, Olsen P, Na and K increased significantly, while Ca and Mg showed significant decrease (Table 4).

Discussion and Conclusions

Growth Variability Among the 16 Provenances 6 Months and 5 Years After Planting

All the Southern and Eastern African provenances survived the 5 years of growth, while only one provenance from West Africa, Bignona (Senegal) survived. The

Table 4 Comparison of soil properties at 6 months (1997) and at 5 years (2002) after planting 16 provenances of *F. albida* at Noiweit

Soil parameter	Time (year)	Mean	SD	SE
pH (H_2O)	1997	4.72	0.23	0.059
	2002	>5.82**	0.11	0.029
Organic C%	1997	0.98	0.06	0.016
	2002	>1.36**	0.27	0.070
Total N%	1997	0.11	0.03	0.007
	2002	>0.28**	0.09	0.022
Olsen P (ppm)	1997	0.01	0.04	0.009
	2002	>0.96**	0.43	0.110
Na (cmol kg^{-1})	1997	6.97	1.93	0.499
	2002	>8.60**	0.45	0.116
K (cmol kg^{-1})	1997	24.13	2.62	0.677
	2002	>26.44**	2.21	0.570
Ca (cmol kg^{-1})	1997	9.81	2.56	0.662
	2002	<7.45**	1.26	0.326
Mg (cmol kg^{-1})	1997	3.83	0.72	0.185
	2002	<1.77**	0.54	0.140
Al (cmol kg^{-1})	1997	1.50	0.46	0.119
	2002	<1.47 ns	0.64	0.165
Sand%	1997	41.65	1.51	0.399
Clay%	1997	37.63	2.09	0.500
Silt%	1997	20.71	2.27	0.582

Sand, Clay and Silt percentages were determined only at 6 months and classified the soil as clay loam.
**significant increase or decrease at $P < 0.01$; ns = not significant change ($P > 0.05$)

poor survival of West African provenances in Baringo district of Kenya suggests that they are not adapted to this region and hence not suitable for introduction and domestication in this Eastern African region. Only about six Eastern and Southern African provenances whose survival percentage ranged between 50 and 70% are likely potential candidates for introduction and domestication in this region.

This result is comparable to those obtained from trial in Western and Southern Africa with the same species. The trials in Burkina Faso (West Africa) and Zimbabwe (Southern Africa) showed better initial growth among Eastern and Southern African provenances, but in the Burkina Faso trial, the Eastern and Southern African provenances were poorly adapted to the Sahelian environment and their mortality reached 100% by the second year (Joly, 1992). In Zimbabwe, the Eastern African provenances remain vigorous (Joly, 1992), but this provenance also failed in Mouda, Northern Cameroun (West Africa) (Joly 1992). This confirms the conclusions that there is genetic difference in *F. albida* particularly between Western and Southern African provenances (Vandenbelt, 1991; Dangasuk et al., 1997), which should be taken into account during any introduction and domestication program.

Significant variation in height and diameter growths was found among the 16 provenances of *F. albida* at 6 months of age. However, 5 years later no significant differences in height and diameter were detected among the surviving provenances. This observation seems to agree with the findings of Awang et al. (1994) in *Acacia auriculiformis*. The general lack of clear differentiation among the surviving provenances 5 years after planting provides any breeding program with a limited opportunity for selection between provenances. However, there was considerable variation within each provenance in these growth variables as evidenced by the relatively high standard deviation (not shown). The existence of individual trees with superior growth therefore offers an opportunity for selection within provenances.

Six months height and diameter growth data were highly correlated to seed source parameters (latitude and longitude). However, after 5 years there was no more significant relationship between the growth variables and seed source parameters, thus conforming to the report by Awang et al. (1994). The significantly

high correlation between survival percentage at 5 years and the 6 months height and diameter seems to agree with the suggestion of Burley and Wood (1976) that growth characteristics in the nursery explain the variation and vigour of the species, which is ultimately correlated to survival and production.

The significant inter-provenance variation in growth variables, observed at 6 months of age and its absence 5 years later, is indicative of the diminishing influence of seed sources and the increasing influence of the new environment on the species' genotypes. This was also confirmed by the significant correlation between the growth variables at 6 months and seed source parameters, and its absence between the growth variables at 5 years and the seed source parameters. Survival at 5 years was highly influenced by the seed sources as evident by the significant correlation between survival percentage at 5 years and seed source parameters (latitude and longitude), as well as with 6 months height and diameter data.

Changes in Soil Parameters Based on 6 Months and 5 Years Assessments

In semi-arid Baringo district of Kenya where the soil is generally infertile and the entire annual precipitation occurs in less than one quarter of the year, soil erosion is a serious problem. Planting of fast growing, drought tolerant and nitrogen-fixing tree species such as *F. albida* is, therefore, an important soil conservation and fertility restoration strategy.

Soil analysis conducted 6 months after planting showed that the soil was the clay-loam type of low water-holding capacity; strongly acidic; and low in total N, organic C and available Olsen P contents, typical of arid and semi-arid soils in the tropics. According to soil fertility rating by Tekaglin et al. (1991), it was noted that K, Mg, and Ca levels were adequate, which might have partly contributed towards the establishment of the species in this acidic soil (Fisher and Juo, 1995). In addition, soil fertility constraints may act only at certain stages of stand development, for example shortly after planting when root systems are poorly developed (Szott, 1995). This was observed in this study when 3 months after planting, the leaves of the seedlings turned pinkish to purplish, which is indicative of phosphorous deficiency, but later reverted to green.

Five years after planting the levels of soil pH, organic carbon (C), total N, Olsen P, exchangeable Na and K increased significantly, which could be due to the nitrogen-fixing ability of the species, decomposition of its fallen leaves and roots and its efficiency in recycling plant nutrients from deep underground soil horizons with its long taproot system (Le Houerou, 1980). This resulted in a general improvement in soil fertility. However, there was a significant decrease in the levels of Ca and Mg, which was attributed to possible depletion of these two nutrients as a result of increased uptake. This observation seems to agree with the findings of Fisher and Juo (1995). In general, trees are less sensitive to low Ca availability and their root growth is less affected by low levels of exchangeable Ca at pH 4.0 and above (Fisher and Juo, 1995). There was no significant change in the concentration of Al during the 5 year growth period. According to Tepper et al. (1989), the range of Al concentration (1.47–1.50 cmol kg^{-1}) observed in Noiweit is the ideal range for proper root growth. Higher concentration of Al^{3+} ions in the soil (>3 cmol kg^{-1}) could inhibit root growth directly by inhibition of cell division and elongation or indirectly by detrimental effect on the legume–rhizobium symbiosis needed for effective nodulation and N_2 fixation (Chong et al., 1984).

Both the chemistry of the soil and the biochemistry of the root are such that it is difficult for a plant or tree to obtain micronutrients from acidic soil (Fisher and Juo 1995). *F. albida* could be growing well in acid soils because it is better adapted to obtaining nutrients and water under acidic soil conditions coupled with its ability to improve soil fertility over time. Therefore, the nitrogen fixing ability of the species, its efficiency in recycling plant nutrients, its reverse phenology, its ability to thrive in acid soils and its soil fertility restoration and improvement capacity make *F. albida* the most important agro-silvo-pastoral tree species in the arid and semi-arid lands of Africa.

Acknowledgement International Foundation of Science (IFS) funded the research under grant number D/2539-2 for which we are extremely grateful.

References

Awang K, Nor Aini AS, Adjers G, Bhumibhamon M, Pan FJ, Venkateswarlu P (1994) Performance of *Acacia auriculiformis* provenances at 18 months in four sites. J Trop For Sci 7:251–261

Bouyoucos GJ (1962) Hydrometer method improved for making particle size analysis of soils. Agron J 53:464–465

Burley J, Wood PJ (eds) (1976) A manual on species and provenance research with particular reference to the tropics. Tropical Forestry Papers No. 10 and 10A. Commonwealth Forestry Institute, Oxford

Chong K, Wynne JC, Elkan GH, Schneeweis TJ (1984) Effects of soil acidity and aluminium content on Rhizobium inoculation, growth and nitrogen fixation of peanuts and other grain legumes. Trop Agric 64(2):97–104

Dancette C, Poulain JF (1969) Influence of *Acacia albida* on pedoclimatic factors and crop yield. Afr Soil 14:143–184

Dangasuk OG, Gudu S, Okalebo JR (2001) Early growth performance of sixteen populations of *Faidherbia albida* in semi arid Baringo district of Kenya. Paper presented at the 10th international soil conservation organization conference on sustaining the global farm, Purdue University, West Lafayette, Indiana, USA, 23–28 May 1999

Dangasuk OG, Seurei P, Gudu S (1997) Genetic variation in seed and seedling traits in 12 African provenances of *Faidherbia albida* (Del.) A. Chev. at Lodwar, Kenya. Agroforest Syst 37:133–141

Fisher RF, Juo ASR (1995) Mechanism of tree growth in acid soils. Paper presented at the nitrogen fixing tree association and the centro agronómico tropical de ynvestigación y Enseñanza, Turrialba, Costa Rica, 3–8 Jul 1994

ICRAF (International Center for Research in Agroforestry) (1989) The apple ring. Agroforest Today 1:11–12

Joly HI (1992) The genetic of *Acacia albida* (syn *Faidherbia albida*). Paper presented in the international *Faidherbia albida* workshop on *Faidherbia albida* in the West African semi arid tropics, Niamey, Niger, 22–26 Apr 1991

Le Houerou MN (1980) Chemical composition and nutritional value of browse in tropical West Africa. In: Le Houerou MN (ed) Browse in Africa the current state of knowledge. ILCA, Ethiopia, pp 261–289

McLean EO (1965) Aluminium. In: Black CA (ed) Methods of soil analysis, Part II, Agronomy No. 9. Lewis Publishers, Broca Raton, pp 986–994

Murphy J, Riley JP (1962) A modified single solution method for the determination of phosphate in natural waters. Anal Chem Acta 27:31–36

Nelson DW, Sommers LE (1975) A rapid and accurate method for estimating organic carbon in soil. Proc Indiana Acad Sci 84:456–462

Okalebo JR, Gathua KW, Woomer PL (2002) Laboratory methods of soil and plant analysis. A working manual. SSSEA, KARI, SACRED Africa, TSBF, Nairobi, Kenya

Olsen SR, Cole CV, Watanabe FS, Dean LA (1954) Estimation of available phosphorus in soils by extraction with sodium bicarbonate, USDA Circular No. 939

Szott LT (1995) Growth and biomass production of nitrogen fixing trees on acid soils. Paper presented at the nitrogen fixing tree association and the centro agronómico tropical de ynvestigación y Enseñanza, Turrialba, Costa Rica, 3–8 Jul 1994

Teel W (1984) A pocket directory of trees and seeds in Kenya. KENGO, Nairobi

Tekaglin T, Hague I, Aduayi EA (1991) Soil, plant, water, fertilizer, animal manure and compost analysis manual. Plant Science Division, Warring Documents 13, Ilk, Addis Ababa, Ethiopia

Tepper HB, Yang CS, Schaedle M (1989) Effect of aluminum on growth of root tips of honey locus and loblolly pine. Environ Exp Bot 29:165–173

Vandenbelt RJ (1991) Rooting systems of western and southern African *Faidherbia albida* (Del.) A. Chev. (syn. *Acacia albida* Del.): a comparative analysis with biogeographic implications. Agroforest Syst 14:233–244

The 'Secret' Behind the Good Performance of *Tithonia diversifolia* on P Availability as Compared to Other Green Manures

S.T. Ikerra, E. Semu, and J.P. Mrema

Abstract Use of organic materials to improve soil nutrient and increase crop production is well documented. Organic resource quality influences the effect of these organic materials. Although characterization of these organic materials has been done for most parameters, a comprehensive characterization in terms of type and concentration of organic anions is still lacking. An incubation experiment was conducted to characterize *Tithonia diversifolia, Lantana camara, Gliricidia sepium* and farmyard manure (FYM) and relate the parameters to P availability. The experiment was in a completely randomized design, four treatments replicated four times. *Tithonia* produced the highest concentration of basic cations and oxalic acid while *Gliricidia* had the lowest. Farmyard manure and *Lantana* were intermediate. At 5 t ha^{-1} these organic materials significantly ($p < 0.05$) increased P availability through reduction of P maximum on Chromic Acrisol. The order of this reduction was *Tithonia*>FYM>*Lantana*=*Gliricidia*. The influence of organic materials on P sorption was highly dependent on their pH, P, Ca and oxalic acid concentration. There was a significant ($p < 0.05$) negative correlation between P maximum and oxalic acid concentration of green manures ($r = -0.97$); this was attributed to the high Al complexation capacity of oxalic acid (log $k_{Al} = 6.1$). Selection of organic materials, therefore, besides being based on other conventional quality parameters, should consider the concentration of oxalic acid. It was concluded that the better performance of *Tithonia* on P availability through reduction of P sorption is due to its higher oxalic acid. The agronomic implication of these results is that proper handling of *Tithonia* is required to avoid loss of oxalic acid during its application.

Keywords Farmyard manure · Green manure · Oxalic acid · Phosphorus sorption

Introduction

Phosphorus deficiency caused by low inherent P stocks and P fixation is one of the major factors limiting crop production in Tanzania particularly in highly weathered soils (Ikerra and Kalumuna, 1992; Mnkeni et al., 1994). Most crops usually respond to P fertilization, but phosphate fertilizers' use by small-scale farmers is limited by the high fertilizer costs (Ikerra, 2004). Use of organic materials as alternatives to mineral fertilizers to improve P availability and increase crop production is well documented (Nziguheba et al., 2000). Organic materials that have been documented in increasing P availability when applied to soil include FYM (Ikerra, 2004) and agroforestry green manures (Nziguheba et al., 2002).

Organic resource quality influences the effect of these organic materials on P availability. For instance, the N, P, C/N, lignin and polyphenolic contents are important determinants of P release from any organic material (Palm et al., 2001). Organic matter and its decomposition products (organic anions) can markedly reduce P adsorption by soils and hence increase P availability. This happens because decomposing organic materials produce organic anions which complex Al and Fe responsible for P adsorption (Hue,

S.T. Ikerra (✉)
Mlingano Agricultural Research Institute, Tanga, Tanzania
e-mail: susikera@yahoo.com

A. Bationo et al. (eds.), *Innovations as Key to the Green Revolution in Africa*,
DOI 10.1007/978-90-481-2543-2_22, © Springer Science+Business Media B.V. 2011

1991). Organic anions also compete with P for the same adsorption sites, thereby reducing the chances of P being adsorbed. They also prevent precipitation of P by Fe and Al oxides/hydroxides through their liming effects. Organic materials increase the size of soil aggregates and decrease the specific surface area, which leads to decrease in P adsorption and hence increase in P availability (Phan Thi Cong, 2000).

However, the effect of organic materials on P adsorption is highly dependent on their quantity and quality, type and concentration of organic anions and the soil type (Singh and Jones, 1976; Iyamuremye et al., 1996a; Nziguheba et al., 2000). Although studies conducted in the region (Nziguheba et al., 2000) attributed the decrease in P adsorption upon the application of green manure to the production of organic acids, no work has been done to characterize the produced organic acids and relate their concentrations to P adsorption capacities. These data are important for selection of various organic materials intended to be used for increasing P availability in highly weathered soils.

An incubation study was carried out (i) to characterize FYM, *Lantana camara*, *Gliricidia sepium* and *Tithonia diversifolia* on their chemical characteristics and light molecular weight organic acids (LMWOA) and (ii) to find the influence of these organic materials on other soil parameters which affect P availability.

Materials and Methods

Study Site

The incubation experiment was conducted in 2002 at the Department of Soil Science of the Sokoine University of Agriculture (SUA), Morogoro, Tanzania.

Soil and Organic Materials Sampling and Analysis

Twenty spot samples were randomly collected from the 0 to 20 cm layer from a half hectare experimental plot at Magadu near SUA, Morogoro, Tanzania. This site is located at 6°51′S and 37°39′E, at an altitude of 568 m above sea level. The spot soil samples were thoroughly mixed, air-dried and ground to pass through a 2 mm sieve. The composite sample was subjected to physical and chemical analyses in the laboratory. This soil was an isohyperthemic ustic, fine clayey kaolinitic, kanhaplic Haplustults (Soil survey Staff, 1975) or Chromic Acrisol (WRB, 1998).

Leaves and twigs of *L. camara*, *G. sepium* and *T. diversifolia* were collected from 20 trees in the SUA botanical garden, thoroughly mixed, washed thoroughly with deionized water and chopped. Partially decomposed FYM was collected from the SUA farm cattle shed. These green manures (GMs) and FYM were oven-dried at 70°C for 24 h, ground to pass a 0.5 mm sieve and subjected to chemical analysis.

Laboratory Analyses

Soil pH was determined with a combination glass electrode in a 1:2.5 soil:water ratio according to McLean (1965). Exchangeable acidity and Al were extracted by 1 M KCl and determined titrimetrically using the method of McLean (1965). Organic C was determined by the wet oxidation method (Anderson and Ingram, 1993). Total nitrogen was determined by the Kjeldahl method (Bremner and Mulvaney, 1982). Nitrate-nitrogen was extracted using 2 M KCl and determined by cadmium reduction (Dorich and Nelson, 1984) while ammonium-nitrogen in the 2 M KCl extract was determined by the salicylate–hypochlorite colorimetric method (Anderson and Ingram, 1993). The sum of nitrate and ammonium constituted the total inorganic nitrogen. Exchangeable bases (Ca, Mg, K, Na) and CEC were determined after extraction with 1 M ammonium acetate (pH 7) (Rhodes, 1982). DTPA-extractable Fe, Zn and Mn were extracted (Lindsay and Novell, 1978) and determined using atomic absorption spectrophotometry (AAS). Particle size analysis was determined by the hydrometer method (Ghee and Bauder, 1986).

The organic amendments were analysed for pH in water at a manure:water ratio of 1:5, total N (Bremner and Mulvaney, 1982), P, K, and soluble and organic C (Anderson and Ingram, 1993). Lignin and polyphenols were determined by the acid detergent method and the revised Folin–Denis method, respectively (Anderson and Ingram, 1993). Extraction and measurement of LMWOA produced by different organic materials was done according to the method of Bolan et al. (1994). This method involved extraction of the organic materials with distilled water at a 1:100 organic material:water extraction ratio. The extracts were centrifuged at 6000 rpm for 30 min and filtered through 0.45 μm Millipore filters. The LMWOAs were determined at 210 nm on a Shimadzu LC 6A high-performance liquid chromatography (HPLC) equipped with a modified C18 column containing silicon. The organic acids were identified by comparing the retention times of the LMWOA peaks with those obtained for pure standard organic acids. Peak heights (recorded on a Shimadzu SPD-10AVP detector) were measured and used to quantify the organic acids.

Incubation of Organic Materials and Soil

Two hundred grams of soil were mixed with each of the organic materials at the rate of 2.27 or 5.4 g kg^{-1} (equivalent to 5 or 10 t ha^{-1}) on dry weight basis. Using distilled water, the mixture was moistened to 66–80% field capacity and incubated on benches in triplicates, at 25–26°C, for 12 weeks. Samples were aerated after every 2 days. After 2 and 12 weeks, soil sub-samples were withdrawn from the incubation vessels, air-dried at room temperature for 4 days and used for the determination of pH, Al and the Langmuir P adsorption parameters.

Determination of Phosphorus Adsorption Parameters

Three grams of the incubated soil + organic material samples were equilibrated in 30 ml of 0.01 M CaCl$_2$ solutions containing KH$_2$PO$_4$ – P at 0, 10, 20, 30, 40, 50 or 60 mg P l^{-1}, equivalent to 0, 100, 200, 300, 400, 500 or 600 mg P kg^{-1}. Three drops of toluene were added to each of the extraction bottles prior to incubation to arrest microbial activity. The samples were equilibrated at 26–27°C for 6 days. During this period they were shaken twice daily for half an hour, at 175 rpm. They were filtered through Whatman No. 42 filter papers. The filtrates were analysed for P according to the method of Murphy and Riley

(1962). Adsorbed phosphorus was determined by the difference between the applied P and the P that was in solution at equilibrium (Fox and Kamprath, 1970). The experiment was undertaken using the completely randomized design with five treatments, namely control (i.e. soil alone), soil + FYM, soil + *Lantana*, soil + *Gliricidia* and soil + *Tithonia*, all replicated three times.

Statistical Analysis

Analysis of variance was conducted using the general linear model procedure of the SAS program (SAS Institute, 1995). Mean separation was done according to the Duncan's new multiple range test. The raw data for adsorbed P and equilibrium P were fitted into the Langmuir, using the SAS program procedure of non-linear regression (SAS Institute, 1995) to generate the P adsorption parameters. The 95% level of probability was used to test for significance.

Results and Discussion

Characteristics of Soil and Organic Materials

The soil used was very acidic, very low in available P, had moderate aluminium saturation and was low in exchangeable Ca. It had low organic matter content and low cation exchange capacity (Table 1). The soil was classified as a Chromic Acrisol (WRB 1998). All these parameters indicate poor soil fertility status. The low pH, moderate Al and clay fraction dominated by 1:1 type of clay imply net positive surface charges favouring P adsorption (Asenga and Mrema, 1991), implying that there is a need for P application in order to improve crop yields.

All the organic materials used (Table 2) are generally classified as being high-quality materials. According to Palm et al. (2001), high-quality organic materials have N>2.5%, P>0.24%, lignin<15% and polyphenol<4%. Based on these critical values, manure had rather low N content but high P content. *Lantana* and *Gliricidia* had P contents less than the critical P value while their N content was above the

Table 1 Some physicochemical properties of the topsoil (0–15 cm) of the Chromic Acrisol used

Parameter	Value	Rating	Reference
pH (H_2O) (1:2.5)	4.8	Very acidic	Landon (1991)
pH $_{KCl}$ (1:2.5)	4.4		–
Organic C (g kg^{-1})	9.9	Low	Landon (1991)
Total N (g kg^{-1})	0.9	Low	Landon (1991)
Mineral N (mg kg^{-1})	6.8	–	–
Total P (mg kg^{-1})	0.3	Very low	Landon (1991)
Available P (mg kg^{-1})			
Bray-1	6.4	Low	Singh et al. (1977)
Resin	4.1	Low	Landon (1991)
Exch. Al (cmol$_c$ kg^{-1})	3.2	Moderate	Landon (1991)
Exch. bases (cmol$_c$ kg^{-1})			
Ca	1.80	Low	Landon (1991)
Mg	1.34	Medium	Landon (1991)
K	0.53	Medium	Landon (1991)
Na	0.05	–	–
CEC (pH 7)	8.40	Low	Landon (1991)
DTPA (mg kg^{-1})			
Zn	1.04	Low	Lindsay and Norvell (1978)
Fe	60.69	Very high	Lindsay and Norvell (1978)
Texture (%)			
Sand	32		
Silt	13		
Clay	55		

Table 2 Some characteristics of the organic materials used

Material	pH	N	P	K	Ca	Mg	T C	S C	Lig	Poly	L/P	C_s/P_t
						---%---						
FYM	8.5	1.76	0.43	1.87	0.89	0.59	34.6	1.21	11.5	–	–	2.6
Lantana	6.8	3.00	0.21	2.78	1.10	0.53	43.6	2.69	11.4	3.48	3.2	12.8
Gliricidia	5.7	3.70	0.22	2.61	1.16	0.53	47.0	3.37	11.9	3.13	3.8	15.3
Tithonia	6.9	4.00	0.46	4.06	2.03	0.64	41.4	2.65	7.2	3.17	2.2	5.8

C_s = total soluble carbon, P_t = total phosphorus, Lig. = lignin, Poly. = polyphenols

critical value. The P and N contents of *Tithonia* were all above the critical values. The lignin/polyphenol, soluble C/total P and carbon/nitrogen ratios were all below the critical values, indicating that these materials will undergo fast decomposition (Palm et al., 2001).

The concentrations and properties of the light molecular weight organic acids (LMWOA) contained in the experimental materials and their aluminium complexation constants are presented in Tables 3 and 4, respectively. A variety of organic acids were identified in the organic materials, whereas in the soil only small amounts of oxalic acid were detected. All types of the investigated LMWOA were detected in various concentrations in *Tithonia* and *Lantana* green manures. The concentrations of organic acids extracted

from the *Tithonia* green manure were relatively higher than those from the other green manures or FYM. Generally, the concentrations followed the trend: *Tithonia*>*Lantana*>*Gliricidia*>FYM. Formic, malonic, citric and tartaric acids were not detected in farmyard manure samples, while succinic acid was not detected in *Gliricidia*. Oxalic acid was the most dominant organic acid, followed by tartaric and formic acids in all GM materials, *Tithonia* having the highest concentrations of these acids. The oxalate concentration in the Chromic Acrisol is within the range of 4–22 mmol kg^{-1} reported by Hue et al. (1986) and Bolan et al. (1994) for mineral soils. However, this value was lower than the levels of oxalic acid reported by Fox et al. (1990), which ranged from 44 to 955 mmol kg^{-1} in a mineral soil under pine tree seedlings.

Table 3 Types and concentrations of light molecular weight organic acids released from the Chromic Acrisol and the organic materials

Material	Oxalic	Malic	Acetic	Formic	Succinic	Malonic	Citric	Tartaric
	--------------------------(mmol kg^{-1})-------------------------------------							
Soil	10	npi	npi	npi	npi	npi	npi	npi
FYM	47	14	2	npi	2	npi	npi	npi
Lantana	468	110	11	36	10	17	13	177
Gliricidia	50	10	20	177	npi	52	52	36
Tithonia	1620	93	28	210	20	84	7	209

npi = no peak identified. All values are means from two replications

Table 4 Formulae and aluminium–organic acid complexation constants of the LMWOA identified

Organic acid	Formula	Complexation constant, log k_{Al}
Acetic acid	CH_3COOH	1.57
Citric acid	$(COOH)CH_2C(OH)(COOH)CH_2COOH \cdot H_2O$	7.98
Formic acid	HCO_2H	1.36
Malic acid	$(CHOHCH_2)(CO_2H)_2$	5.34
Malonic acid	$CH_2(CO_2H)_2$	5.24
Oxalic acid	HO_2CCO_2H	6.1
Succinic cid	$(CH_2CO_2H)_2$	2.09
Tartaric acid	$HO_2CCH(OH)CH(OH)CO_2H$	5.62

Source: Fox et al. (1990)

P Adsorption Isotherms

At 2 weeks of incubation, the different organic materials incorporated into soil at 5 t ha^{-1} generally reduced P adsorption by the treated soil compared to that in the control (Fig. 1). The lowest P adsorption (meaning highest solution P) was encountered where *Tithonia* was used, and the trend of decrease in P adsorption was *Tithonia*<manure<*Lantana*<*Gliricidia*<control (Fig. 1). Increasing the application rate from 5 to

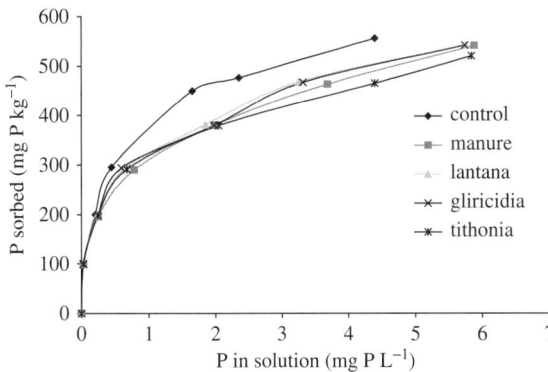

Fig. 1 Adsorption isotherms for the Chromic Acrisol incubated for 2 weeks with organic materials at 5 t ha^{-1}

10 t ha^{-1} at the same incubation period of 2 weeks had similar effects on the P adsorption.

Phosphorus Adsorption Characteristics

All organic materials incorporated into soil at 5 t ha^{-1} and incubated for 14 days resulted in decreases in the soil's adsorption maximum (*b*) compared to that of the control soil (Table 5). The adsorption maximum ranged from 585 mg kg^{-1} in control soil to the lowest value of 500 mg kg^{-1} in *Tithonia*-amended soil. The value for FYM was similar to that of *Lantana* and was significantly higher than that of *Tithonia* or *Gliricidia*. Increasing the application rate of the organic materials from 5 to 10 t ha^{-1} (with 14 days of incubation) resulted in generally similar trends, but the lowest adsorption maximum was associated with manure. Increasing the incubation period from 14 to 112 days for the 10 t ha^{-1} application rate of organic materials also reduced the adsorption maximum compared to that of the control, although the reduction was not significant in the case of manure and *Tithonia* (data not shown).

The decrease in adsorption capacity was not accompanied by a significant decrease in adsorption affinity

Table 5 Effects of organic materials on the P adsorption maximum (*b*), adsorption affinity (*k*) and SPR (*q*) based on the Langmuir equation

Week/rate	Variable unit	Control	Manure	*Lantana*	*Gliricidia*	*Tithonia*	R^2	CV (%)
2 wks/5 t	b (mg P kg^{-1})	585 a	552 b	552 b	524 c	500 c	0.82	2.48
	k (L mg^{-1})	2.95 a	1.63 c	2.2 b	2.74 a	1.73 bc	0.53	22.94
	q (mg P kg^{-1})	217	136	169	185	131		
2 wks/10 t	b (mg P kg^{-1})	606 a	451 d	540 c	562 b	523 c	0.96	2.25
	k (L mg^{-1})	2.2 a	1.80 c	2.00 b	1.93 b	2.22 b	0.81	9.20
	q (mg P kg^{-1})	185	119	154	157	161	–	–
12 wks/10 t	b (mg P kg^{-1})	589 a	573 ab	563 b	580 b	577 ab	0.50	1.65
	k (L mg^{-1})	2.12 a	1.59 c	1.98 ab	2.03 ab	1.77 cb	0.55	8.10
	q (mg P kg^{-1})	175	138	160	167	151	–	–

SPR = (*q*), P needed to obtain a soil solution concentration of 0.2 mg P L^{-1}, calculated from the Langmuir equation

Means in a row followed by the same letter are not significantly different (*p* < 0.05) according to the Duncan's new multiple range test

or bonding energy, '*b*', for the 5 t ha^{-1} at 14 days incubation treatment as compared to that in the control. When the application rate was increased from 5 to 10 t ha^{-1}, there was a significant reduction of bonding energy values compared to that of the control. However, there were no significant differences between different organic materials. After increasing the incubation period from 14 to 112 days, the organic materials did not change the adsorption affinities as compared to that in the control.

The decrease in P adsorption capacity of soil caused by the GMs was closely correlated to the oxalic acid concentration of the GMs (*r* = 0.97) (data for farmyard manure, whose organic acid concentration was not related to P sorption capacity, were omitted from the regression analysis). The higher the oxalic acid concentration, the lower was the P adsorption capacity of the soil. These data showed that 97% of the variation in phosphorus adsorption capacity from GMs could be explained by the oxalic acid content (Fig. 2).

The decreases in P adsorption maximum and affinity caused by organic materials can be attributed to several factors. First is the quenching of P sorption sites by the P released from the decomposing organic materials, as they contained substantial quantities of P (Table 2). This is supported by the observation that within the GMs the decrease in P sorption was greatest for *Tithonia*, which contained P levels that were greater than the critical value of 2.2 g P kg^{-1} (0.22% P). Singh and Jones (1976) reported that organic materials with P contents greater than 2.2 g P kg^{-1} reduced P sorption and vice versa.

This was confirmed in this study whereby the decrease in P sorption was in the order *Tithonia*< *Lantana*<*Gliricidia* <control (untreated soil), which agrees with the trend in the P content of the organic materials (Table 2). The second factor was the increase in pH, which was observed in the soil samples amended with the organic materials (Table 6). The effect of the organic materials on pH may be attributed to the self-liming effect due to the release of Ca and Mg. *Tithonia* had the highest Ca content, and it produced the highest increase in pH. Manure, though low in Ca and Mg contents, was ranked second in raising the soil pH, probably due to its initial high pH (Table 2). The reasons for increase in pH could be release of NH$_3$ during decomposition, as well as production of OH^{-1} upon dissolution of solid Mn and Fe oxides or upon ligand exchange, whereby the terminal OH^{-1} on Al and Fe hydroxyls is replaced by organic anions.

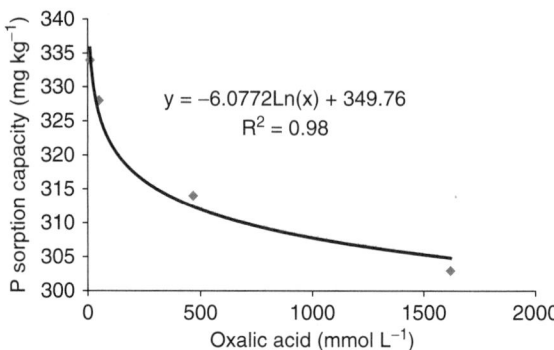

Fig. 2 Relationship between P adsorption capacity and oxalic acid concentration

Table 6 Effect of organic materials on pH and exchangeable Al

Organic material	pH (water)	Exch. Al (mol $(+)kg^{-1}$)
Control (soil alone)	4.60 d	1.4 0 a
Soil + *Tithonia*	5.25 a	0.62 d
Soil + manure	4.90 b	0.80 c
Soil + *Lantana*	4.85 c	0.98 b
Soil + *Gliricidia*	4.85 c	0.95 b

Means in a row followed by the same letter are not significantly different ($P < 0.05$) according to the Duncan's new multiple range test

Similar findings have been reported by other workers (Hoyt and Turner, 1975; Hue and Amien, 1989; Noble et al., 1996). Noble et al. (1996) reported that among 16 green manures tested that caused significant increase in soil pH, *Melia azedarach* had the highest Ca and Mg contents and gave the highest increase in pH. The increase in soil pH reduced the soil's Al^{3+} activity and contributed to the production of OH^- ions that competed with orthophosphate ions for the soil's adsorption sites, thereby reducing P adsorption. At the end of incubation all organic materials reduced exchangeable Al compared to that in the control, but the biggest decrease was caused by *Tithonia* (Table 6).

The decrease in exchangeable Al could be attributed to the binding of the Al by the LMWOA released during the decomposition of the organic materials (Table 2). This is partly supported by the observation that *Tithonia*, which had the highest contents of both oxalic and tartaric acids, resulted in the highest reduction in exchangeable Al. Oxalic and tartaric acids have been reported to form stable Al–organic complexes due to their high log k_{Al} or high complexation constants (Table 4) (Jones, 1998; Jones and Farrar, 1999). Although farmyard manure had low concentrations of LMWOA, it was second to *Tithonia* in reducing exchangeable Al, and this could be due to its high pH (pH = 8.5) (Table 2) and OH^- ions produced upon its decomposition. Similar findings of high pH values in manure have been reported by Mowo (2000). Organic anions produced during decomposition of farmyard manure could also have caused an increase in the pH of the soil through removal of the associated protons (Whalem et al., 2000). Phan Thi Cong (2000) similarly reported that *Tithonia* at 2.5 g kg^{-1} applied to a Ferralic Cambisol decreased exchangeable Al from 2.16 to 1.50 cmol(+) kg^{-1} after 49 days of incubation. At higher *Tithonia* rates of 40 g kg^{-1}, Phan Thi Cong (2000) observed that the concentration of

exchangeable Al was reduced to zero. The observed decrease in exchangeable Al, which otherwise fixes P, is, therefore, an important mechanism in reducing P adsorption and increasing P availability.

Third, complexation of exchangeable Al by the organic acids produced during decomposition of the organic materials may have prevented the precipitation of P by Al. This is supported by the decrease in the concentration of exchangeable Al in organic material-treated soils relative to that in the control soil (Table 6). Fourth, the organic acids produced during decomposition of the organic materials (Table 2) may have competed with orthophosphates for the same P adsorption sites. The decrease in standard phosphorus requirement (SPR) due to the presence of organic materials implies that less fertilizer P should be required for crop production upon amending the Chromic Acrisol with organic materials.

The more pronounced influence of *Tithonia* on P adsorption as compared to the other green manures is attributed to its higher content of P (Table 2) and of organic acids (Table 3). *Tithonia* contained higher concentrations both of oxalic and of tartaric acids than the other organic materials. These acids have higher Al complexing constants (greater than 4) (Table 4), and this normally leads to P desorption (Jones, 1998). This argument is supported by the observation that 97% of the variation in phosphorus adsorption capacities due to GM could be explained by the oxalic acid concentration (Fig. 2). Farmyard manure had low concentrations of these acids and its low depressing effect on P sorption was possibly caused by these low concentrations, as well as its tendency of increasing pH and its high P content (Table 2). The higher depressing effect on P adsorption that was caused by increase in application rate of the organic materials from 5 to 10 t ha^{-1} would be due to greater production of LMWOAs, PO_4^{3-} and OH^- ions at the higher application rate (Iyamuremye et al., 1996b). Incubating the organic materials for 12 weeks did not result in more intense effects than those from the 2-week incubation period. This was probably because with time some of the organic acids were transformed or neutralized as the pH increased (Krzyszowska et al., 1996).

The results obtained from this study are in conformity with those of Singh and Jones (1976), Iyamuremye et al. (1996b) and Nziguheba et al. (2000). However, these workers did not give a comprehensive characterization of the organic materials used and, hence, no correlation of P adsorption capacity with LMWOA was established, as has been in this study.

Conclusions

In conclusion, the higher concentrations of LMWOA particularly oxalic acid in *Tithonia* as compared to other agroforestry green manures used in this study is one of the reasons why *Tithonia* was more efficient in reducing P adsorption and increasing P availability when it was incorporated into the soil. Selection of organic materials for improving P availability should be based on oxalic acid concentration in addition to the other conventional quality parameters. Since oxalic acid is extracted by water, *Tithonia* should be properly handled before use to avoid losses of this LMWOA. Sun drying of *Tithonia* and keeping the already cut *Tithonia* in conditions where it is rained upon are some of the conditions which will decrease the concentration of LMWOA.

Acknowledgements We would like to thank the Tanzanian TARP II World Bank Research Program for providing research funds for this study. Financial support provided by the Rockefeller Foundation is highly acknowledged. We thank all the technical staff from the laboratories of the Department of Soil Science at SUA, the Tropical Soil Biology and Fertility (TSBF) at Nairobi, Kenya, and the Catholic University of Leuven in Belgium, for their support.

References

Anderson JM, Ingram JS (1993) Tropical soil biology and fertility. A handbook of methods, 2nd edn. CAB International, Wallingford, CT

Asenga RH, Mema JP (1991) Phosphorus adsorption by some Rhodic Ferralsols (Ultic haplustox) from Tanga, Tanzania. East Afr Agric For J 57(1):1–14

Bolan NS, Naidu R, Mahmairaja BS (1994) Influence of low molecular weight organic acids on the solubilization of phosphate. Soil Biol Fertil 18:311–319

Bremner JM, Mulvaney CS (1982) Nitrogen total. In: Page AL (ed) Methods of soil analysis, Part 2. Soil Science Society of America, Madison, WI, pp 595–624

Dorich RA, Nelson DW (1984) Evaluation of manual cadmium reduction methods for determination of nitrate in potassium chloride extracts of soil. Soil Sci Soc Am J 48:72–75

Fox TR, Comerford NB, McFee WW (1990) Phosphorus and aluminium release from spodic horizon mediated by organic acids. Soil Soc Am J 54:1763–1767

Fox RL, Kamprath EJ (1970) Phosphate sorption isotherms for evaluating the phosphate requirements of soil. Soil Sc Soc Am Proc 34:902–907

Ghee GW, Bauder JW (1986) Particle size analysis. In: Klute A (ed) Methods of soil analysis, physical and mineralogical methods, Agronomy monograph No. 9, 2nd edn. American Society of Agronomy Inc, Madison, WI, pp 383–412

Hoyt PB, Turner RC (1975) Effects of organic materials added to very acid soil on pH, aluminium, exchangeable NH_4 and crop yields. Soil Sci 119:227–237

Hue NV (1991) Effects of organic acids/anions on P sorption and phytoavailability in soils with different mineralogies. Soil Sci 152:463–471

Hue NV, Amien I (1989) Aluminium detoxification with green manures. Commun Soil Sci Plant Anal 20:1499–1511

Hue NV, Craddock GR, Fred A (1986) Effect of organic anions on Aluminium toxicity in subsoils. Soil Sci Soc Am J 50: 28–34

Ikerra ST (2004) Use of Minjingu phosphate rock combined with different organic inputs in improving phosphorus availability and maize yields on a Chromic Acrisol in Morogoro, Tanzania. PhD thesis, Sokoine University of Agriculture, Morogoro Tanzania

Ikerra ST, Kalumuna MC (1992) Phosphorus adsorption characteristics of soils and their influence on maize grain yield responses to P application along the Mlingano catena. Tanga, Tanzania. In: Zake JK et al (eds) Proceedings of the 11th AGM of the SSSEA, Mukono Kampala. December 1991, pp 85–94

Iyamuremye F, Dick RP, Boham J (1996a) Organic amendments and phosphorus dynamics. Phosphorus chemistry and sorption. Soil Sci 161:426–435

Iyamuremye F, Dick RP, Baham J (1996b) Organic amendments and phosphorus dynamics. Distribution of soil phosphorus fractions. Soil Sci 161:436–443

Jones L (1998) Organic acids in the rhizosphere – a critical review. Plant Soil 205:25–44

Jones D, Farrar J (1999) Phosphorus mobilization by root exudates in the rhizosphere: fact or fiction?. Agroforest For 9:20–25

Krzyszowska AJ, Blaylock MJ, Vance GF, David MB (1996) Ion-chromatographic analysis of low molecular weight organic acids in spodosol forest floor solutions. Soil Sci Soc Am J 60:1565–1571

Landon JR (1991) Booker tropical soil manual. A handbook for soil survey and agricultural and land evaluation in the tropics and subtropics. Longman, London

Lindsay W, Norvell WA (1978) Development of DTPA soil test for zinc, iron, manganese and copper. Soil Sci Soc Am J 42:421–428

McLean EO (1965) Soil pH and lime requirement In: Black CA et al (eds) Methods of soil analysis. Part 2. Agronomy 9. American Society of Agronomy, Madison, WI, pp 199–223

Mnkeni PSN, Semoka JMS, Kaitaba EG (1994) Effect of Mopogoro philipsite on the availability of Phosphorus in phosphate rocks. Trop Agric Trin 71(4):249–253

Mowo JG (2000) Effectiveness of phosphate rock on Ferralsols in Tanzania and the influence of within field variability. PhD thesis, Wageningen Agricultural University, The Netherlands, 164 pp

Murphy J, Riley JP (1962) A modified single solution method of the determination of phosphate in natural waters. Anal Chim Acta 27:31–36

Noble AD, Zenneck I, Randall PJ (1996) Leaf litter ash alkalinity and neutralization of soil acidity. Plant Soil 179: 293–302

Nziguheba G, Merckx R, Palm CA, Mutuo P (2002) Combining Tithonia diversifolia and fertilizers for maize production in phosphorus deficient soils in Kenya. Agroforestry Syst 55:165–174

Nziguheba G, Merckx R, Palm CA, Rao M (2000) Organic residues affect phosphorus availability and maize grain yields in a Nitisol of western Kenya. Biol Fertil Soils 32:328–339

Palm CA, Gachengo G, Delve RJ, Cadisch G, Giller KE (2001) Organic inputs for soil fertility management in tropical agroecosystem: Application of an organic data base. Agric Ecosyt Environ 83:27–42

Phan Thi Cong (2000) Improving phosphorus availability in selected soils from the uplands of south Vietnam by residue management. A case study: Tithonia diversifolia. PhD thesis Nr. 439. Katholieke Universiteit Leuven, Belgium

Rhodes JD (1982) Cations exchange capacity. In: Miller RH, Keeney PR (eds) Methods of soil analysis. Part 2. chemical and microbiological properties. Agronomy monographs No 9. American Society of Agronomy Inc, Madison, WI, pp 149–169

SAS Institute (1995) SAS/stat user's guide, 6th edn. SAS Institute, Cary, NC

Singh BB, Jones JP (1976) Phosphorus sorption and desorption characteristics of soil as affected by organic residues. Soil Sci Soc Am J 40:389–394

Singh BR, Uriyo AP, Mnkeni PNS, Msaky JJ (1977) Evaluation of indices for N and P availability for soils of Morogoro Tanzania. In: Proceedings of the conference on classification of tropical Soils, Kuala Lumpur Malaysia, pp 273–278

Soil Survey Staff (1975) Soil taxonomy United States, Department of Agriculture, McGraw-Hill, New York, NY

Whalem JK, Chi C, Clayton GW, Carefoot JP (2000) Cattle manure amendments can increase the pH of acid soils. Soil Sci Am J 64:962–966

WRB (1998) World reference base for soil resources. World soil resources report 84. FAO ISRIC, ISSS, Rome, Italy

Nutr Cycl Agroecosyst (2010) 88:49–58
DOI 10.1007/s10705-008-9187-x

RESEARCH ARTICLE

Biological nitrogen fixation potential by soybeans in two low-P soils of southern Cameroon

M. Jemo · C. Nolte · M. Tchienkoua ·
R. C. Abaidoo

Received: 5 March 2008 / Accepted: 16 June 2008 / Published online: 3 July 2008
© Springer Science+Business Media B.V. 2008

Abstract Biological nitrogen fixation (BNF) potential of 12 soybean genotypes was evaluated in conditions of low and sufficient phosphorus (P) supply in two acid soils of southern Cameroon. The P sources were phosphate rock (PR) and triple superphosphate (TSP). The experiment was carried out during two consecutive years (2001 and 2002) at two locations with different soil types. Shoot dry matter, nodule dry matter, and nitrogen (N) and P uptake were assessed at flowering and the grain yield at maturity. Shoot dry matter, nodule dry matter, N and P uptake, and grain yield varied significantly with site and genotypes ($P < 0.05$). On Typic Kandiudult soil, nodule dry matter ranged from 0.3 to 99.3 mg plant^{-1} and increased significantly with P application ($P < 0.05$). Total N uptake of soybean ranged from 38.3 to 60.1 kg N ha^{-1} on Typic Kandiudult and from 18 to 33 kg N ha^{-1} on Rhodic Kandiudult soil. Under P-limiting conditions, BNF ranged from -5.8 to 16 kg N ha^{-1} with significantly higher values for genotype TGm 1511 irrespective of soil type. Genotype TGm 1511 can be considered as an important companion crop for the development of smallholder agriculture in southern Cameroon.

Keywords Biological nitrogen fixation (BNF) · P-uptake · Soybean

Introduction

Soybean (*Glycine max* L. Merr) is a relatively new crop for smallholder farming communities in most African countries, gaining popularity as a consequence of the increasing need for food and fodder (Sanginga et al. 2002). In the humid forest zone (HFZ) of southern Cameroon, soybean is also being conserved as an important component of cropping systems through soil fertility restoration and provision of dietary proteins to small-scale farmers (Maesen and Somaatmadja 1992). However, many soils in humid forest ecosystems have low level of available P due to high phosphorus (P) sorption by Fe and Al oxides (Ssali et al. 1996; Menzies and Gillman 1997). Growth and biological nitrogen fixation (BNF) of legumes such as soybean are hampered by P deficiency (Giller 2001). Soil-P availability during plant seedling development is an important determinant for plant growth, N_2 fixation,

M. Jemo (✉)
International Institute of Tropical Agriculture, Humid Forest Ecoregional Centre (HFC), BP 2008 (Messa), Yaounde, Cameroon
e-mail: m.jemo@cgiar.org; mjemo2001@yahoo.com

C. Nolte
FAO, Plant Production and Protection Division (AGP), Rome, Italy

M. Tchienkoua
Institut de la Recherche Agricole pour le Développement (IRAD), Yaounde, Cameroon

R. C. Abaidoo
International Institute of Tropical Agriculture (IITA), Oyo Road, PMB 5320, Ibadan, Nigeria

This article has been previously published in the journal "Nutrient Cycling in Agroecosystems" Volume 88 Issue 1.
A. Bationo et al. (eds.), Innovations as Key to the Green Revolution in Africa: Exploring the Scientific Facts. © 2010 Springer.

Nutr Cycl Agroecosyst (2010) 88:49–58

and grain formation of legumes (Vance 2001). Low P availability in soils results in a decrease in shoot growth, affects the photosynthetic activity, and limits the transport of photosynthates to nodules (Jakobsen 1985) with a significant decline in N_2 fixation by the plant (Israel 1987).

Successful integration of grain legumes into the cropping system of humid forest ecosystems will depend on alleviation of the soil constraints such as those mentioned above. This may be difficult in areas where the supply of P is limited or where added P may be fixed into forms unavailable to plants (Sample et al. 1980). Selection of plant species or genotypes efficient in acquiring P from sparingly available sources or that make better use of P applied to the soil and fix N_2 from the atmosphere will benefit subsequent crops and represent keys elements of sustainable cropping systems in such regions (Horst et al. 2001).

Wide genotypic differences in BNF potential under P-deficient soil conditions have been documented among and within many legumes, including soybean (Alves et al. 2003; Sanginga 2003). However, no attempt has been made so far to select soybean genotypes for southern Cameroon that grow and fix N_2, and contribute to N inputs under low-P soil conditions or in response to limited P application.

Low-P-tolerant plants have been shown to develop several physiological mechanisms to acquire P from the soil system. Among the mechanisms described are the development of specific root morphological system that permit the exploration of a large soil volume (Krasilnikoff et al. 2003), mycorrhizal associations (Smith and Read 1997), acidification of the soil rhizosphere that helps solubilize less-labile P pools (Gahoonia et al. 1992), and excretion of organic acid anions to solubilize inorganic P (Raghothama 1999) or by secreting phosphatase enzymes making P bound to organic matter available (Li et al. 1997).

The present study was carried out to evaluate the potential of 12 soybean genotypes for BNF in two acid soils of southern Cameroon.

Materials and methods

Site characteristics

The field experiments were conducted during two consecutive years (2001 and 2002) at two locations of southern Cameroon. The first location Abang (3°24′ N, 11°47′ E) at 660 m asl was situated 50 km south of Yaounde. Mean annual rainfall was 1,513 mm with a bimodal distribution. The second location was Minkoameyos (3°51′ N, 11°25′ E), 10 km west of Yaounde at an altitude of 780 m asl with mean annual rainfall of 1,643 mm. The soils were classified as Typic Kandiudult and Rhodic Kandiudult (USDA soils classification system) for Abang and Minkoameyos, respectively. Selected physicochemical properties for surface horizons (0–10 cm) are shown in Table 1. The two sites were selected on the basis of their low P availability (between 3 and 5 μg by the Bray1-P method) and soil acidity (pH < 5.50). The soybean genotypes tested in the field were provided by the breeding program of the International Institute of Tropical Agriculture (IITA), Ibadan, Nigeria. Some morphological and growth characteristics of the genotypes are presented in Table 2.

Field experiment

The field experiment was carried out with a split-block design with four replications. Phosphorus treatments were applied on the main plots at rates of 0 and 30 kg P ha^{-1} in the form of triple superphosphate (TSP) and 90 kg P ha^{-1} with phosphate rock (PR). Subplots, measuring 4 × 4 m, comprised the soybean genotypes. The PR was a granulated powder from Hahotoe mine in Togo, sieved at 0.15 mm. The PR contained 17.5% P of which 3% was citrate soluble. No N fertilizer or *Bradyrhizobia* inoculation was applied. Soybean seeds were drilled along a ridge with 75 cm between rows and 5 cm separation within the row and were thinned to one plant 1 week after emergence. All the soybean plants were sprayed with the insecticide Thiodan® (endosulfuran organochlorine insecticide) at the rate of 0.33 mg l^{-1} (corresponding to approximately 2 kg ha^{-1}) at 14 and 28 days after sowing (DAS) and hand weeded at 14, 28, and 48 DAS.

Plant sampling and analysis

Plants were sampled at mid-pod fill stage, i.e., 56 DAS for shoot growth, nodulation, and shoot N and P concentrations. Six plants were chosen from the middle rows of each plot, their shoots cut at

Table 1 Physical and chemical properties of the topsoil (0–10 cm) used in the field experimental sites in southern Cameroon (Mean ± SE, $n = 36$)

Soil properties	Sand (%)	Silt (%)	Clay (%)	pH[a]	OC (g kg⁻¹)	TN (g kg⁻¹)	Ca (cmol (+) kg⁻¹)	Mg (cmol (+) kg⁻¹)	K (cmol (+) kg⁻¹)	Al (cmol (+) kg⁻¹)
Rhodic Kandiudult	40.8 (5.3)	10.2 (1.8)	48.9 (7.1)	4.5 (0.10)	22.5 (1.9)	2.1 (0.1)	1.1 (0.10)	0.33 (0.01)	0.08 (0.01)	0.65 (0.13)
Typic Kandiudult	50.6 (1.58)	9.7 (0.85)	39.7 (1.2)	5.4 (0.10)	16.0 (1.1)	1.6 (0.05)	1.3 (0.17)	0.64 (0.06)	0.06 (0.04)	0.12 (0.02)

	P sorption (%)[b]	FeOx (%)[c]	P (Bray 1) (mg kg⁻¹)	NaHCO₃–Pi (mg kg⁻¹)[d]	NaOH–Pi (mg kg⁻¹)[e]	HCl–P (mg kg⁻¹)[f]	H₂SO₄–P (mg kg⁻¹)[g]
Rhodic Kandiudult	88.3 (3.2)	0.69 (0.04)	2.50 (0.95)	1.7 (0.29)	25.4 (3.2)	1.0 (0.5)	154.1 (2.3)
Typic Kandiudult	77.5 (0.30)	2.3 (0.07)	5.0 (1.20)	3.2 (0.50)	29.2 (1.2)	0.86 (0.38)	190.9 (12)

Values in parentheses represent the standard error of the mean

[a] pH, measured in water

[b] P sorption, percentage of P sorbed into Fe and Al ions

[c] FeOx, oxalate-extractable Fe

[d] NaHCO₃–Pi, NaHCO₃-extractable inorganic P

[e] NaOH–Pi, NaOH-extractable inorganic P

[f] HCl–Pi, concentrated HCl-extractable inorganic P

[g] H₂SO₄–Pi, H₂SO₄-extractable inorganic P

Table 2 Characteristics of the soybean genotypes used for the field experiments in 2001 and 2002

Genotype	Mature pod color	Lodging score	Plant height (cm)	Days to 50% maturity	Origin geographic code
TGX-1456-2E	Brown	Unknown	Unknown	86	IITA
TGm 1511	Tan	Moderate	49	86	USA
TGm 1196	Brown	Moderate	65	89	Puerto Rico
TGm 1293	Brown	Severe	70	89	Unknown
TGm 1420	Brown	Slight	54	89	USA
TGm 1419	Tan	Severe	49	89	USA
TGm 1039	Brown	Slight	67	92	Taiwan
TGm 1251	Brown	Unknown	51	92	Unknown
TGm 1566	Brown	Moderate	85	94	USA
TGm 1576	Brown	Moderate	85	94	USA
TGm 944	Brown	Severe	72	96	Nigeria
TGm 1360	Brown	Moderate	62	96	USA
TGm 1540	Brown	Moderate	77	104	USA

0.05 m above ground level, and their fresh weight recorded. A representative subsample of about 500 g per plot was oven-dried at 70°C for 72 h for shoot biomass determination. To minimize the damage to the root system, the soil around the plant was loosened using a fork and the roots were carefully removed from the soil. Fresh nodules were cautiously removed from the roots, their number determined, and oven-dried at 70°C for 72 h for their dry matter determination. At grain maturity (84 DAS), a subplot of 1.5×1.5 m was harvested. The fresh weights of shoot biomass and grain yield were determined in the field. A subsample of about 500 g fresh biomass and 100 g fresh grain yield were retained and oven-dried at 70°C for 72 h and the respective dry weights were measured. Total P in plant samples was determined by the vanadomolybdate-yellow method (Motomizu et al. 1983) and total N was determined by the colorimetric method after sample ashing in concentrated sulfuric acid (Powers et al. 1981).

BNF of soybean genotypes

The BNF from soybean genotypes was estimated using the plant N uptake and nodule dry matter. We assumed that genotypes with nodules dry matter less than 20 mg plant^{-1} were a non-N$_2$ fixing control and their means were calculated, being 44 and 27 kg N ha^{-1} for the Typic and Rhodic Kandiudult, respectively. Differences in N accumulation between the mean of non-N$_2$ fixing controls and each genotype

with nodule dry matter more than 20 mg plant^{-1} were considered as the N contribution from BNF.

Statistical analysis

Statistical analyses of the data were carried out using the Statistical Analysis System (SAS) software version 8.2 (2001). Analysis of variance (ANOVA) was performed using the general linear procedure (GLM). Regression analysis was used to establish the relationship between pairs of variables.

Results

Nodulation, N uptake, BNF, and N contribution of soybean

Nodule dry matter and N uptake of soybean differed significantly in all soils (Table 3), with genotypes in the Typic having better values than those in the Rhodic Kandiudult soil.

Significant genotypic variation in nodule dry matter and N uptake among the soybean genotypes was observed on both soils ($P < 0.05$). Five soybean genotypes (TGm 1566, TGm 1511, TGm 1293, TGm 1540, and TGm 0944) produced higher nodule dry matter than other cultivars on both soils (Table 4). With respect to N uptake, soybean genotypes TGm 1511 and TGm 1566 had higher N uptake than other on both soils. The BNF and N contribution of soybean was

Table 3 Nodule dry matter and N uptake (kg N ha^{-1}) of the soybean genotypes grown on Typic and Rhodic Kandiudult soils of southern Cameroon at mid-pod filling (56 DAS)

P application	Nodule dry matter (mg plant^{-1})		N uptake (kg N ha^{-1})	
	Typic Kandiudult	Rhodic Kandiudult	Typic Kandiudult	Rhodic Kandiudult
0P	27.4 b	15.8 b	37.6 c	27.3 b
PR	30.8 ab	24.5 b	47.8 b	31.3 b
TSP	54.8 a	44.0 a	56.8 a	36.4 a
F value				
Year	ns		ns	
Soil type (S)	14.9***		97.6***	
P application (P)	3.0**	4.6**	18.6***	13.0***
Genotype (G)	4.5***	3.4***	2.0*	2.0*
S × P	ns		ns	
S × G	1.9*		ns	
P × G	ns	ns	ns	ns
S × P × G	ns		ns	

Means across genotypes, 2001 and 2002, $n = 96$

Numbers followed by the same lower-case letter are not significantly different between genotypes at $P > 0.05$, LSMEANS/PDIFF. ns, not significant; *, **, ***, significant at $P \leq 0.05$, $P \leq 0.01$ and $P \leq 0.001$, respectively

Table 4 Nodule dry matter, N uptake, and N$_2$ fixation indicator of the soybean genotypes grown on Typic and Rhodic Kandiudult soils of southern Cameroon at mid-pod filling (56 DAS)

Genotype	Nodule dry matter (mg plant^{-1})		N uptake (kg N ha^{-1})		Biological nitrogen fixation potential (kg N ha^{-1})	
	Typic Kandiudult	Rhodic Kandiudult	Typic Kandiudult	Rhodic Kandiudult	Typic Kandiudult	Rhodic Kandiudult
TGm 1196	0.3 c	2.5 b	40.7 c	29.6 b	−3.3	2.6
TGm 1360	3.3 c	0.0 b	47.1 b	31.2 b	3.1	4.2
TGm 1039	3.6 c	1.7 b	44.9 b	26.2 bc	1.9	−0.8
TGm 1251	6.5 c	8.6 b	41.1 c	30.1 b	2.9	3.1
TGx 1465-2E	7.6 bc	3.5 b	38.3 c	18.8 d	−5.8	−8.2
TGm 1419	13.5 bc	24.5 b	52.1 b	28.8 b	8.1	1.8
TGm 1420	15.0 bc	9.8 b	46.5 b	27.1 bc	2.5	0.1
TGm 1566	52.5 bc	28.8 b	51.5 b	33.0 b	7.5	6.0
TGm 1511	69.4 ab	25.5 b	60.1 a	41.4 a	16.1	14.4
TGm 1293	82.2 ab	64.6 ab	46.6 b	33.0 b	2.6	6.0
TGm 0944	98.6 a	78.8 ab	52.0 b	32.4 b	8.0	5.4
TGm 1540	99.3 a	88.7 a	48.1 b	23.3 cd	4.1	−3.7

Means across P rates, and 2001 and 2002, $n = 24$

Numbers followed by the same lower-case letter are not significantly different between genotypes at $P > 0.05$, LSMEANS/PDIFF. Forty-four and 27 kg N ha^{-1} were the mean N uptakes of genotypes with nodule mass less than 20 mg plant^{-1}, used to separate the N contribution efficiency of the soybean genotypes

higher in soybean genotypes TGm 1511, TGm 0944, and TGm 1293 than the others in the Typic and Rhodic Kandiudult soils, respectively, while the genotype Tg×1456-2E had a negative BNF and N contribution (−5.8 kg N ha^{-1}), suggesting poor N$_2$ input from the atmosphere. Under field conditions this genotype

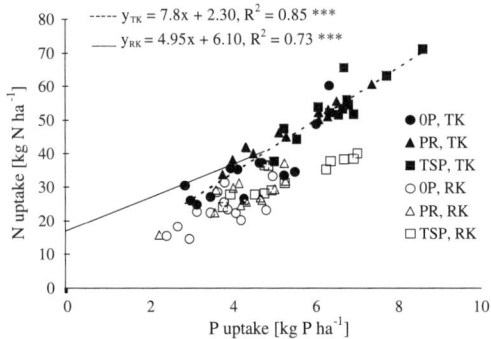

Fig. 1 Relationship between N and P uptake (in plants) of the soybean genotypes at mid-pod filling (56 DAS) grown on a Typic and Rhodic Kandiudult soils of southern Cameroon. $n = 8$. TK, Typic Kandiudult; RK, Rhodic Kandiudult; 0P, without P application; PR, phosphate rock (90 kg P ha^{-1}); TSP, triple superphosphate (applied at 30 kg P ha^{-1})

presented symptoms of N deficiencies in older leaves, which was recognized by a yellowish color in old leaves a few weeks after plant sowing.

Phosphorus application significantly increased nodule dry matter and N uptake ($P < 0.05$) on both soils, with TSP being more effective than PR in general (Table 3). The soil × phosphorus interaction

was significant for plant N uptake on the Typic Kandiudult soil. The soil × genotype interaction was only significant for the nodule dry matter in the Typic Kandiudult soil. However, the interactions phosphorus × genotype and soil × genotype × phosphorus were not significant for either nodule and plant N uptake in the Typic or the Rhodic Kandiudult soil.

The individually calculated regressions between plant P uptake and N uptake of the soybean genotypes were significant irrespective of the soil type (Fig. 1).

Shoot biomass production, P-uptake, and grain yield

The effects of soil types, P application, and genotype differences on soybean biomass, P-uptake, and grain yield differed significantly ($P < 0.001$). The soybean TGm 0944, TGm 1511, and TGm 1566 produced significantly higher shoot dry matter than other genotypes irrespective of the soil. With respect to the P-uptake, soybean genotypes TGm 0944, TGm 1511, TGm 1566, and TGm 1540 had significantly higher P uptake than the other (Tables 5 and 6). The genotypes TGm 1039, TGm 1566, and TGm 1511

Table 5 Shoot dry matter, P uptake at mid-pod filling (56 DAS), and grain yield at maturity (84 DAS) of the soybean genotypes grown on Typic and Rhodic Kandiudult soils of southern Cameroon

P application	Shoot dry matter (kg ha^{-1})		P uptake (kg P ha^{-1})		Grain yield (kg ha^{-1})	
	Typic Kandiudult	Rhodic Kandiudult	Typic Kandiudult	Rhodic Kandiudult	Typic Kandiudult	Rhodic Kandiudult
0P	1822 c	1284 c	3.9 c	3.7 c	715.0 b	237.7 b
PR	2074 b	1503 b	4.7 b	4.3 b	871.9 a	258.9 ab
TSP	2379 a	1678 a	5.5 a	5.1 a	904.4 a	286.3 a
F-value						
Year (Y)	ns		ns		8.7**	
Soil types (S)	198.7***		32.0***		666.9***	
P application (P)	27.5***	29.0**	19.6***	23.8***	10.9***	5.7**
Genotypes (G)	5.0***	8.0**	3.3*	5.3**	15.5***	8.9**
S × P	4.1**		ns		11.5**	
S × G	ns		2.3**		8.3**	
P × G	ns	ns	ns	ns	ns	ns
S × P × G	ns		ns		ns	

Means across genotypes, and 2001 and 2002, $n = 96$

Numbers followed by the same lower-case letter are not significantly different between genotypes at $P > 0.05$, LSMEANS/PDIFF. ns, not significant; *, **, ***, significant at $P \leq 0.05$, $P \leq 0.01$ and $P \leq 0.001$, respectively

Table 6 Shoot dry matter and P uptake at mid pod filling (56 DAS) of the soybean genotypes grown on a Typic and Rhodic Kandiudult soils of southern Cameroon

Genotype	Shoot dry matter (kg ha^{-1})		P uptake (kg P ha^{-1})	
	Typic Kandiudult	Rhodic Kandiudult	Typic Kandiudult	Rhodic Kandiudult
TGm 1196	1401 e	1375 cd	3.5 c	3.5 bc
TGm 1360	1748 cd	1748 ab	4.2 bc	4.0 b
TGm 1039	2223 bcd	1317 cd	4.6 bc	5.3 ab
TGm 1251	1353 e	1364 cd	3.4 c	3.1 c
TGx 1465-2E	1472 e	821 e	3.4 c	3.0 c
TGm 1419	2178 bc	1550 bc	4.3 b	4.2 b
TGm 1420	2269 bc	1343 cd	4.4 b	4.1 bc
TGm 1566	2776 a	1878 a	6.4 a	5.6 a
TGm 1511	2559 ab	1722 ab	6.7 a	4.6 b
TGm 1293	2085 cd	1360 cd	4.7 b	3.6 bc
TGm 0944	2574 a	1772 ab	5.3 ab	5.7 a
TGm 1540	2326 bc	1609 ab	4.9 bc	5.5 ab

Means across P rates, and 2001 and 2002, $n = 24$

Numbers followed by the same lower-case letter are not significantly different between genotypes at $P > 0.05$, LSMEANS/PDIFF

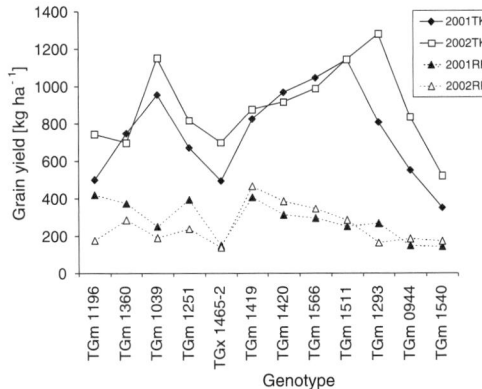

Fig. 2 Grain yield at maturity (84 DAS) of the soybean genotypes grown on Typic (TK) and Rhodic (RK) Kandiudult soils (southern Cameroon) in 2001 and 2002. Means across P rates, and 2001 and 2002, $n = 12$

were among the highest yielding on both soils across P rates.

Soybean grain yield of the genotypes significantly varied over the year and soil types with the highest values from the Typic Kandiudult soils (Fig. 2). There were significant correlations between grain yield and shoot dry matter (Fig. 3a), and plant N uptake (Fig. 3b) at low level of available P for both soils combined.

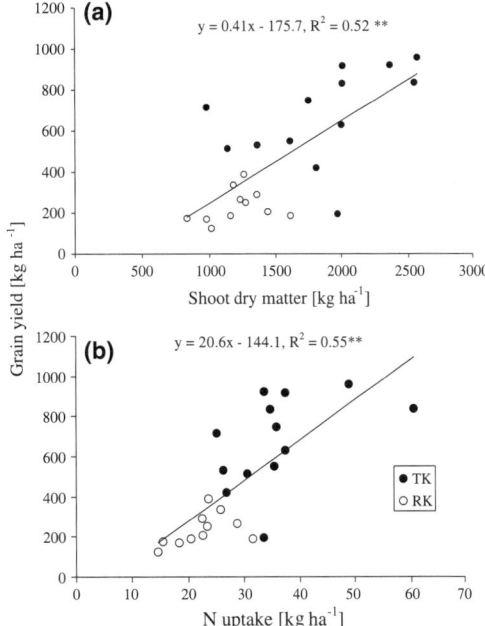

Fig. 3 Relationship between shoot dry matter (**a**), and N uptake (**b**) at mid-pod filling (56 DAS) and grain yield at maturity (86 DAS) on low-available P soils of southern Cameroon. Means at 0P and across soils. Open and filled symbols represent data points for the Typic (TK) and Rhodic Kandiudult (RK) soils, respectively. $n = 8$

Discussion

The present results show a positive response in shoot dry matter and P uptake in soybean genotypes to P application. The positive effects of P nutrition on plant growth were generally observable a few weeks after P application. Jakobsen (1985) related positive response to increased photosynthetic surface area, root growth, and nodulation. Results from our study are comparable with those obtained by several other authors on the effects of P application in bean (Christiansen and Graham 2002), in cowpea (Othman et al. 1991), and in pigeon pea (Adu-Gyamfi et al. 1989). Significant genotypic variations among the soybean genotypes were also observed. Assuming that only seeds of the soybean are to be removed from the plots, genotypes such as TGm 1511 and TGm 1566 will be of great interest to farmers because they will supply nutrients, particularly N and P, to the associated and/or rotational crops after decomposition and mineralization of shoot and leaf organs. In a parallel study, it was observed that maize grain yield production significantly increased after the genotypes TGm 1511 and TGm 1566 were grown as preceding crops (Jemo et al. 2006).

Significant genotypic variation for grain yield was observable depending on the soil used. Although such significant differences between the sites can be partly attributed to the variation in cropping history and the soil's chemical composition, such as high P fixation capacity and Al toxicity, the difference among genotypes is of practical interest in a low inputs agricultural system of southern Cameroon. In general grain yields of soybean were low compared to that produced in many other regions such as Argentina and Brazil (Alves et al. 2003). Although soybean cultivation in these areas is highly dependent on inputs, the low grain yield obtained in this present study might indicate that the rate of P application used was not optimal to allow sufficient nodulation and thus highest N_2 fixation of soybean. Additional research work will be conducted to support this hypothesis. Grain yield of the soybean genotypes TGm 1566, TGm 1511, and TGm 1039 was particularly higher compared to other genotypes previously grown on these soils (Wendt and Atemkeng 2004). Such findings provide useful information to farmers to increase yield production and to breeders for the possibility of improving soybean grain yield on soils with low available P in southern Cameroon.

P application significantly increased nodulation and grain yield, indicating that P was a limiting factor to nodulation and yield formation. Enhanced nodulation and plant growth after application of P have been reported by various authors (Othman et al. 1991; Adu-Gyamfi et al. 1989) with positive plant response to P applied under P-deficiency conditions.

The BNF potential of the soybean genotypes varied from −5.8 to 16.1 kg N ha^{-1} in the Typic Kandiudult and from −8.2 to 14.4 kg N ha^{-1} in the Rhodic Kandiudult soil with superior ability of the genotypes TGm 1566, TGm1511, and TGm 0944 over the others. Studies conducted in the southern Guinea Savanna zone of Nigeria with soybean-maize rotation indicated that the N contribution of soybean were in the range of −8 to 47 kg N ha^{-1} (Sanginga et al. 1997). However, the method of estimation of N_2 fixation was based on N isotope dilution. The superior ability of the genotypes TGm 1566, TGm1511, and TGm 0944 could be attributed to their efficient strategies for acquiring P under limited-P conditions of soils in southern Cameroon, a process which requires large amounts of P for each atom of N_2 fixed (Vance 2001). The possibility that these genotypes rapidly contact root symbiosis with the indigenous population of *Bradyrhizobium japonicum* strains at earlier stages of their growth than others could also not be excluded. It was generally observed that these genotypes were forming nodules only 8 DAS under the field conditions at both sites (data not shown). However, the question of whether the bacteria populations of these soils are large and diverse enough to allow these genotypes to optimize their BNF will need further investigation.

In most tropical areas, such as in southern Cameron, previous research studies have shown that soils supporting maize, the commonly grown cereal, must supply 50 to 60 kg N ha^{-1} for each ton of grain produced per hectare (Weber et al. 1995). From our study, it was observed that the BNF will only contribute to 15 and 16 kg N ha^{-1}, respectively from genotypes TGm 1511 and TGm 1566. Although the observed BNF contributions are still insufficient to cover the total maize N need, the obtained N contribution remains important to smallholder farmers in order to reduce part of the N inputs to agronomically and economically support maize yield in soils of southern Cameroon. In addition, the soybean genotypes produce grain important for food or fodder and

Nutr Cycl Agroecosyst (2010) 88:49–58

extra income for farmers in the local market or in neighboring countries. It was generally observed that, even when the stovers are exported, maize grain yield increased following soybean cultivation (Carsky et al. 1997), implying that the benefit from the incorporation of soybean into cropping system may go beyond the N contribution. Such benefits may include a suppression of root nematodes (Bagayoko et al. 2000), efficient P use by the subsequent maize crops (Carsky et al. 1997), and possibly the belowground-N contribution from legumes (Wichern et al. 2008).

Conclusion

The present study identified two soybean genotypes (TGm 1511 and TGm 1566) that thrive well under low-P conditions with potential to obtain part of their N from atmospheric fixation in the humid regions of southern Cameroon. These genotypes can be considered important and may be introduced into smallholder agriculture in Cameroon. However, at the present state of the study further work will need to be undertaken to improve their nodulation efficiency and effectiveness of *Bradyrhizobium* nodulating by the genotypes as well as the contribution from BNF to cropping systems in the humid forest of southern Cameroon.

Acknowledgements The research work was financially supported by the Australian Centre for International Agricultural Research (ACIAR) through contract grant no. SMCN2/1999/004. The breeding program of IITA-Ibadan is acknowledged for the grain legume seed supply.

References

Adu-Gyamfi JJ, Fujita K, Ogata S (1989) Phosphorus absorption and utilisation efficiency of pigeon pea (*Cajanus cajan* L. Millsp) in relation to dry matter production and nitrogen fixation. Plant Soil 119:315–324. doi:10.1007/BF02370424

Alves BJR, Boddey RM, Urquiaga S (2003) The success of BNF in soybean in Brazil. Plant Soil 252:1–9. doi:10.1023/A:1024191913296

Bagayoko M, Buerkert A, Lung G, Bationo A, Römheld V (2000) Cereal/legume rotation effects on cereal growth in Sudano-Sahelian West Africa: soil mineral nitrogen, mycorrhizae and nematodes. Plant Soil 218:103–116. doi:10.1023/A:1014957605852

Carsky RJ, Abaidoo R, Dashiell K, Sanginga N (1997) Effect of soybean on subsequent maize grain yield in the Guinea savanna zone of West Africa. Afr J Crop Sci 5:31–38

Christiansen I, Graham PH (2002) Variation in nitrogen (N_2) fixation among Andean bean (*Phaseolus vulgaris* L.) genotypes grown at two levels of phosphorus supply. Field Crops Res 73:133–142. doi:10.1016/S0378-4290(01)00190-3

Gahoonia TS, Claassen N, Junky A (1992) Mobilisation of phosphate in different soils by ryegrass supplied with ammonium and nitrate. Plant Soil 143:241–248. doi:10.1007/BF00010600

Giller KE (2001) Nitrogen fixation in tropical cropping systems, 2nd edn. CAB International

Horst WJ, Kamh M, Jibrin JM, Chude VO (2001) Agronomic measures for increasing P availability to crops. Plant Soil 237:211–223. doi:10.1023/A:1013353610570

Israel DW (1987) Investigation of the role of phosphorus in symbiotic dinitrogen fixation. Plant Physiol 84:835–840

Jakobsen I (1985) The roles of phosphorus in symbiotic nitrogen fixation by young pea plants (*Pisium sativum*). Physiol Plant 64:190–196. doi:10.1111/j.1399-3054.1985.tb02334.x

Jemo M, Abaidoo RC, Nolte C, Tchienkoua M, Sanginga N, Horst WJ (2006) Phosphorus benefits from grain-legume crops to subsequent maize grown on acid soils of southern Cameroon. Plant Soil 284:385–397. doi:10.1007/s11104-006-0052-x

Krasilnikoff G, Gahoonia T, Nielsen NE (2003) Variation in phosphorus uptake efficiency by genotypes of cowpea (*Vigna unguiculata*) due to differences in root and root hair length and induced rhizosphere processes. Plant Soil 251:83–91. doi:10.1023/A:1022934213879

Li M, Osaki M, Rao IM, Tadano T (1997) Secretion of phytase from the roots of several plant species under phosphorus-deficient conditions. Plant Soil 195:161–169. doi:10.1023/A:1004264002524

Maesen LJG, Somaatmadja PS (1992) Plant resources of southeast Asia (PROSEA). Pulses. Prosea Foundation, Bogor, Indonesia, pp 106

Menzies NW, Gillman GP (1997) Chemical characterization of soils of a tropical humid forest zone: a methodology. Soil Sci Soc Am J 62:1355–1363

Motomizu S, Wakimoto P, Toei K (1983) Spectrophotometric determination of phosphate in river waters with molybdate and malachite green. Analyst (Lond) 108:361–367. doi:10.1039/an9830800361

Othman WMW, Lie TA, Mannetje LT, Wassink GY (1991) Low level phosphorus supply affecting nodulation, N_2 fixation and growth of cowpea (*Vigna unguiculata* L. Walp). Plant Soil 135:67–74. doi:10.1007/BF00014779

Powers RF, van Gent D, Townsend RF (1981) Ammonia electrode analysis of nitrogen in microKjeldahl digests of forest vegetation. Commun Soil Sci Plant Anal 12:19–30

Raghothama KG (1999) Phosphate acquisition. Ann Rev Plant Physiol Plant Mol Biol 50:665–693. doi:10.1146/annurev.arplant.50.1.665

Sample EC, Soper RJ, Racz GJ (1980) Reactions of phosphate in soils. In: Khasawneh FE, Sample EC, Kamprath EJ (eds) The role of phosphorus in agriculture. Am. Soc. Agron, Madison, pp 263–310

Sanginga N (2003) Role of biological nitrogen fixation in legume based cropping systems; a case study of West Africa farming systems. Plant Soil 252:25–39. doi:10.1023/A:1024192604607

Sanginga N, Dashiell KE, Okogun JA, Thottappilly G (1997) Nitrogen fixation and N contribution by promiscuous nodulating soybeans in the southern Guinea Savanna of Nigeria. Plant Soil 195:257–266. doi:10.1023/A:1004207530131

Sanginga N, Okogun J, Vanlauwe B, Dashiell KE (2002) The contribution of nitrogen by promiscuous soybeans to maize based cropping the moist savanna of Nigeria. Plant Soil 241:223–231. doi:10.1023/A:1016192514568

SAS Institute (2001) SAS/STAT user's guide, vol 1. 4th SAS, Cary NC, USA

Smith SE, Read DJ (1997) Mycorrhizal symbiosis. Academic, San Diego, pp 126–160

Ssali H, Ahn PM, Mokwunye AU (1996) Fertility of soil tropical Africa: a historical perspective. In: Mokwunye AU, Vlek PLG (eds) Management of nitrogen and phosphorus fertilizers in Sub-Saharan Africa. Martinus Nijhoff, Dordrecht, pp 59–82

Vance CP (2001) Update on the state of nitrogen and phosphorus nutrition symbiotic nitrogen fixation and phosphorus acquisition plant nutrition in a world of declining renewable resources. Plant Physiol 127:390–397. doi:10.1104/pp.127.2.390

Weber GK, Chindo PS, Elemo KA, Oikeh S (1995) Nematodes as production constraints in intensifying cereal based cropping systems of the northern Guinea savannas. Resource and Crop Management Research, Monograph no. 17. IITA, Ibadan, Nigeria

Wendt JW, Atemkeng MF (2004) Soybean, cowpea, groundnut, and pigeon pea response to soils, rainfall, and cropping season in the forest margins of Cameroon. Plant Soil 263:121–132. doi:10.1023/B:PLSO.0000047731.35668.e0

Wichern F, Eberhardt E, Mayer J, Joergensen RG, Muller T (2008) Nitrogen rhizodeposition in agricultural crops: methods, estimates and future prospects. Soil Biol Biochem 40:30–48

Roles for Herbaceous and Grain Legumes, Kraal Manure, and Inorganic Fertilizers for Soil Fertility Management in Eastern Uganda

K.C. Kaizzi, J. Byalebeka, C.S. Wortmann, and M. Mamo

Abstract Grain sorghum [*Sorghum bicolor* (L.) Moenich] is an important food crop in semi-arid areas of sub-Saharan Africa. Crop yields are generally low and declining partly due to low soil fertility. Therefore on-farm research was conducted on 108 farms at three locations over 3 years to evaluate alternative low-input strategies for soil fertility improvement in sorghum-based cropping systems. The strategies were use of herbaceous legumes in improved fallow, a grain legume in rotation with sorghum, use of cattle manure, and application of low levels of N and P fertilizers. *Mucuna* (*Mucuna pruriens*) on average produced 7 t ha^{-1} of aboveground dry matter containing 160 kg N ha^{-1}. Application of 2.5 t ha^{-1} of kraal manure and a combination of 30 kg N and 10 kg P ha^{-1} both increased grain yield by a mean of 1.15 t ha^{-1}. A combination of 2.5 t ha^{-1} manure with 30 kg N ha^{-1} increased grain yield by 1.4 t ha^{-1} above the farmer practice (1.1 t ha^{-1} grain). The increase in sorghum grain yield in response to 30 kg N ha^{-1}, to a *Mucuna* fallow, and to a rotation with cowpea (*Vigna unguiculata*) was 1.0, 1.4, and 0.7 t ha^{-1}, respectively. These alternative strategies were found to be cost-effective in increasing sorghum yield in the predominantly smallholder agriculture where inorganic fertilizer is not used. Results of the study indicated that on-farm profitability and food security could be improved through integration of inorganic fertilizers, herbicides, manure, *Mucuna* fallow, and cowpea rotation into grain sorghum cropping systems.

Keywords Cowpea · Low input · *Mucuna pruriens* · Resource-poor farmers

Introduction

Cereals are important crops for smallholder farmers in sub-Saharan Africa. However, crop yields per unit area of production are declining (Greenland et al., 1994; Sanchez et al., 1996). The main contributing biophysical factors are low inherent soil fertility, particularly N and P deficiencies (Bekunda et al., 1997; Ssali and Keya, 1986), exacerbated by soil fertility depletion through nutrient removal as components of harvests and losses with runoff and soil erosion (Vlek, 1993; Sanchez et al., 1997). Many farmers are unable to compensate for these losses, resulting in negative nutrient balances at the national level for sub-Saharan Africa countries (Stoorvogel and Smaling, 1990) and at the regional scale for the farming systems of eastern and central Uganda (Wortmann and Kaizzi, 1998).

Nutrient availability can be improved through the use of inorganic or organic nutrient sources. However, the profitability of fertilizer use depends on agroclimatic and economic conditions at local and regional levels (Vlek, 1990). Infra-structural and other marketing constraints and lack of agricultural subsidies make the use of inorganic fertilizers very costly in Uganda, and real costs to farmers are two to six times as much as in Europe (Sanchez, 2002). Resource-poor farmers need large returns on the small investments that they can make, often requiring benefit to cost returns of 2 within a 6- to 12-month period (Wortmann and Ssali, 2001).

K.C. Kaizzi (✉)
Kawanda Agricultural Research Institute (KARI), National Agricultural Research Organization (NARO), Kampala, Uganda
e-mail: kckaizzi@gmail.com

A. Bationo et al. (eds.), *Innovations as Key to the Green Revolution in Africa*,
DOI 10.1007/978-90-481-2543-2_24, © Springer Science+Business Media B.V. 2011

Use of organic materials is constrained by lack of labor for collecting and applying them (Ruhigwa et al., 1995), limited quantities and variation in quality (Palm et al., 1997), and competing uses of crop residues as fuel and fodder (Palm, 1995). Green manure production requires land that could be used for food or cash crops (Giller et al., 1997), which is a luxury for smallholder farmers. Farmyard manure is available to many smallholder farmers but the supply is limited (Lekasi et al., 2001). Transfer of plant materials from field boundary areas, or near-by fallow or grazing areas, often has potential in sub-humid areas but less potential in semi-arid areas (Kaizzi and Wortmann, 2001; Wortmann and Ssali, 2001).

Biological nitrogen fixation (BNF) may contribute much N through better integration of legumes in farming systems. It also contributes to productivity both directly, when the fixed N is harvested in protein of grain or other food for human or animal consumption, or indirectly by adding N to the soil and thus contributing to the maintenance or enhancement of soil fertility (Giller and Cadisch 1995). Under favorable environmental conditions with a good supply of soil water and nutrients (Peoples et al., 1995; Giller and Cadisch, 1995; Ssali and Keya, 1986), BNF can meet N requirements of tropical agriculture (Giller et al., 1994; 1997). Economic constraints make BNF an attractive N source for resource-poor farmers in sub-Saharan Africa (Van Cleemput, 1995; Giller and Wilson, 1991).

Several low-input approaches to soil fertility management are feasible for sorghum production systems of resource-poor farmers in drought-prone areas of eastern Uganda, but requires verification and fine-tuning to production systems. The objectives of the study were to determine sorghum grain yield in response to inorganic N and P fertilizers, kraal manure, or a combination of inorganic N with kraal manure application and to *Mucuna* fallow and cowpea rotation.

Materials and Methods

Characteristics of the Study Sites

Farmer-managed trials were conducted at three sites namely Kadesok and Opwatetta parishes in the Southern and Eastern Lake Kyoga Basin and Kapolin parish in Usuk Sandy Farm Grasslands (Wortmann

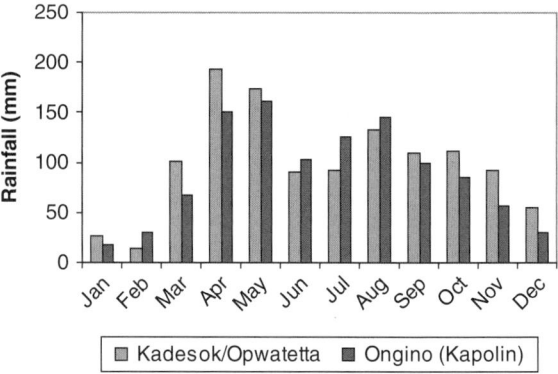

Fig. 1 Mean monthly rainfall at the research sites

and Eledu, 1999) at an altitude ranging from 1050 to 1150 m asl. The rainfall at the research sites allows for two cropping seasons per year with an annual mean of approximately 1150 mm for Kadesok and Opwatetta and 1000 mm for Kapolin (Fig. 1). The rainfall is more unreliable at Kapolin because of its location in the rain shadow of Mt. Elgon.

Soil samples from 0 to 20 cm depth were collected for each trial site, air-dried and ground to pass through a 2 mm sieve, and analyzed according to Foster (1971). Extractable P, K, and Ca were measured in a single ammonium lactate/acetic acid extract buffered at pH 3.8. Soil pH was measured using a soil to water ratio of 1:2.5. Soil organic matter was determined according to a modified Walkley–Black method (Foster, 1971). The dominant soil type in the area is Petroferric Haplustox.

Experimental Trials and Treatments

Four sets of trials were conducted. Each farm was a replication to minimize the probability of type I error in extrapolating results throughout eastern Uganda. The plot size was 100 m² for all trials.

Trial 1: Use of the herbaceous legume (*Mucuna pruriens*) in improved fallow and grain legumes (cowpea) in rotation with sorghum was evaluated on 39 farms. The treatments were (i) sorghum following sorghum (control), (ii) sorghum following cowpea produced during the short rains, (iii) continuous sorghum with 30 kg N ha⁻¹ applied to the long rains sorghum crop, and (v) *Mucuna* fallow during the short rains followed by sorghum.

Trial 2: Use of inorganic fertilizers and kraal manure was evaluated on 63 farms. The treatments were (i) no nutrients applied (control), (ii) 30 kg N ha^{-1}, (iii) 30 kg N plus 10 kg P ha^{-1}, (iv) 2.5 t kraal manure ha^{-1}, and (v) 30 kg N plus 2.5 t kraal manure ha^{-1}. The manure was collected from open kraals where farmers keep their cattle at night. Cowpea was planted on all plots during the short rains.

Trial 3: The N response curve for sorghum was determined using results from trials on 33 farms. Nitrogen levels of 0, 20, 40, and 60 kg ha^{-1} were used to determine the response curve. The experimental area received a blanket application of 40 kg P ha^{-1} and 60 kg K ha^{-1}.

Trial 4: Reduced tillage as an alternative to conventional tillage with hand hoes and ox-ploughs was evaluated on 12 farms. The treatments were (i) farmers' practice of conventional tillage, (ii) use of glyphosate herbicide (Round Up^{TM}) for early weed control, and (iii) a combination of glyphosate herbicide with 30 kg N plus 10 kg P ha^{-1}. Sorghum was planted immediately after applying the herbicide. Fertilizer P and 5 kg N ha^{-1} were applied at planting and the remaining 25 kg N ha^{-1} top dressed 6 weeks later.

Crop Management Practices

Manure and P fertilizers were applied at planting. Nitrogen was applied in two splits with 5 kg N ha^{-1} at planting and the remaining 25 kg N ha^{-1} 6 weeks later. However, for the N response curve, N and K were applied in two equal splits with half at planting and the remaining half 6 weeks later. The N and K fertilizers were incorporated into the soil immediately after their application.

Sorghum cv. "Sekedo" and "Epurpur" were planted at 60 by 20 cm during the long rains of 2004 (LR 2004) and 2005 (LR 2005), respectively. *Mucuna* was planted at 60 by 45 cm and cowpea at 45 by 20 cm during the short rains of 2003 (SR 2003) and 2004 (SR 2004). Weeds were controlled using hand hoes. Beta-cyfluthrin 0.05–2.5% (Bulldock^{TM}) was applied 3–4 weeks after sorghum had germinated to prevent damage by stem borers and chlorpyrifos 5% (Dursban^{TM}) was used for termite control. Sorghum stover and cowpea residues were left in the field.

Data Collection and Analysis

Sorghum grain yield was determined by harvesting the whole plot excluding the guard rows at physiological maturity. The panicles were dried in the open air and threshed. Sub-samples were collected for moisture determination and the grain yield was adjusted to 14% water content. *Mucuna* biomass production was determined after 22 weeks by harvesting a 3 m^2 area using a 1 m^2 quadrat placed randomly within a plot. All the materials within the quadrant including litter were collected and weighed. The *Mucuna* that remained in the plots continued growing. *Mucuna* seeds were not harvested but left in the fields. *Mucuna* germinated during the subsequent long rains and volunteer plants were controlled through weeding until after the second weeding when they were allowed to grow in competition with the sorghum crop. Cowpea grain yield was determined at physiological maturity by picking pods only, a common practice in the area. The pods were dried in open air and threshed. The grain was weighed and sampled for moisture determination. The grain yield was adjusted to 14% water content.

The data were analyzed by community and season using Statistix V. 8.0 (Statistix for windows, 1988). Differences were considered significant at the $p \leq 0.05$ level.

The Economic Analysis

The profitability of fertilizer use was estimated based on the following assumptions:

i. Opportunity cost, including risk allowance, was assumed to add 25, 50, and 75% to the cost of using fertilizer for wealthiest, poor, and very poor farmers, respectively.

ii. Farmgate crop prices were reduced by 10% to cover the cost of harvesting, processing, and marketing additional produce.

iii. Fertilizer costs were increased by 10% to cover transport and application costs.

iv. Plot yields were assumed to be highly relative to yields that small-scale farmers can achieve at a farm level. Therefore, plot yields were reduced by 10% in the economic analysis.

v. Farmgate prices were assumed to be Uganda shillings, 200 and 300 per kg for sorghum and cowpea, respectively.

vi. Prices for 50 kg bags of urea and triple superphosphate were assumed to be UGSh. 40,000.

Results and Discussion

Soil Characteristics

During the characterization and diagnosis processes, farmers recognized the different soils in the area, their location on the landscape, associated problems, and soil-specific coping mechanisms (Table 1). The soils have different names and the major constraints mentioned by farmers included low soil fertility, low water-holding capacity for sandy soils, water logging for the clay soils, and weeds. Farmers and researchers discussed potential solutions to the problems and farmers agreed to participate in the evaluation of possible solutions.

Soil texture class for the experimental sites ranged from sandy clay loam to sand, with sand content ranging from 58 to 92% (Table 2). Sandy loam and loamy sand were the predominant textural classes. The soils were generally of low fertility with the majority of the fields having soil chemical values below the critical low levels estimated for Uganda soils (Foster, 1971). Extractable K and Ca levels were low in 24–33% and 10–16% of the fields, respectively. The low K levels are partly due to K removal in past harvests of cassava and sweet potato. Soil organic matter varied widely but was often low. Available P was below the critical level in most fields. The results shown are in the ranges reported by Harrop (1970) and Ssali (2000).

Trial 1: Sorghum Response to Improved Fallow and Rotation with Cowpeas

The mean aboveground dry matter production by *Mucuna* across the three sites was 7 t ha^{-1} containing 160 kg N ha^{-1}. This agrees with other findings in the region where mean dry matter production was 7.3 t ha^{-1}, containing 180 kg N ha^{-1} with 103 kg N ha^{-1} derived from the atmosphere (Kaizzi, 2002; Kaizzi et al., 2004, 2006). Farmers usually either leave

Table 1 Local soil names, problems associated with the soil, and coping mechanisms by farmers

Local name	Textural class	Problems	Coping mechanism
Eitela, Akao, Nu-Ipokoras	Clay	Low fertility, difficult to work, water logging, weeds	Frequent weeding, early planting
Aputon, Nu-Isingekitos	Sand	Poor water-holding capacity, poor fertility	Fallow
Ecuicui, Luepusuk	Loam	Weeds	Frequent weeding
Ingooroi	Loam with stones	Low fertility, poor water-holding capacity, quickly, stones	Fallow

Table 2 The median values and range (values in parenthesis) of selected soil properties for farmers' fields at the research sites

Soil property	Community Kadesok	Opwatetta	Kapolin	Critical values[a]
pH[b]	6.1 (5.4–6.6)	6.0 (5.2–7.2)	6.1 (5.3–7.5)	5.2
SOM[c] (mg kg^{-1})	28 (19–42)	28 (26–41)	22 (17–35)	30
Extractable P (mg kg^{-1})[d]	1.3 (1–6)	1.3 (1–5)	2.80 (1–9)	5.0
Extractable K (cmol$_c$ kg^{-1})[d]	5.1 (2–10)	6.2 (3–10)	5.4 (2–10)	0.4
Extractable Ca (cmol$_c$ kg^{-1})[d]	41 (8–45)	36 (5–54)	31 (5–68)	0.9
Sand (%)	71 (66–88)	76 (58–82)	84 (62–92)	na
Silt (%)	7 (2.6–13)	5 (1.3–15)	5. (3–11)	na
Clay (%)	20 (6.5–27)	19 (14–29)	11 (5–29)	na

na = not applicable
[a]Below these values soils are deficient or poor (Foster, 1971)
[b]Measured in 1:2.5 (soil:water) suspension
[c]Soil organic matter measured with the Walkley–Black method, modified according to Foster (1971)
[d]Measured in single ammonium lactate/acetic acid extract (pH 3.8) (Foster, 1971)

their fields under fallow or grow cowpeas and rarely grow sorghum during the short rain season due to the uncertainty of the rains and extensive damage by birds. The mean cowpeas grain yield during the short rains was 0.82 t ha^{-1}, and the mean sorghum grain yield for the few farmers who harvested a short-season crop at Opwatetta and Kadesok was 1.05 t ha^{-1}.

All treatments resulted in a significant increase ($p \leq 5\%$) in sorghum grain yield in all sites in both years relative to the control under continuous sorghum with no fertilizer applied (Table 3). The mean increase in sorghum grain yield due to the effect of rotation with cowpea as compared to continuous sorghum was 0.72 t ha^{-1}, with a range of 0.2–1.4 t ha^{-1}, across sites and years. The effect of applying 30 kg N ha^{-1} had a mean increase in sorghum yield of 1.03 t ha^{-1} with a range of 0.4–1.7 t ha^{-1}. The effect of improved *Mucuna* fallow during the previous short rain season was a mean increase in sorghum yield of 1.43 t ha^{-1} with a range of 0.5–2.2 t ha^{-1}. Thus, inorganic N fertilizers, cowpea rotation, and *Mucuna* fallow served as effective N sources for sorghum at the three sites. The relatively greater increase in sorghum grain yield due to the improved fallow with *Mucuna* is in agreement with results reported for maize (Versteeg et al., 1998; Fischler and Wortmann, 1999; Wortman et al., 2000; Tian et al., 2000). The higher average response obtained for *Mucuna* compared to cowpea is expected as much fixed nitrogen was removed in the harvest of cowpea leaves and grain, while all *Mucuna* biomass was left in the field, but some was grazed by freely roaming livestock.

Including legumes in the rotation apparently improved N availability. The cowpea–sorghum rotation resulted in a significant increase in sorghum yield while providing a cowpea harvest the previous season. Using the short rain season to produce a *Mucuna* fallow resulted in the highest increase in sorghum yield. The application of N fertilizer resulted in a mean sorghum grain yield that was intermediate relative to the yields following the legumes. The use of N fertilizer means a cash expense to the farmer but allows flexibility in land use during the short rain season. The results of the economic analysis show that fertilizer use is profitable for all the farmers but less so for the poorest farmers because of the high greatest opportunity cost.

Trial 2: Sorghum Response to Applied Kraal Manure and Inorganic Fertilizer

Manure and fertilizer treatments resulted in a significant ($p \leq 5\%$) increase in sorghum grain yield in all sites in all years relative to the control of sorghum with no fertilizer applied (Table 4). The mean increases in sorghum grain yield, across sites and years, due to application of 30 kg N ha^{-1} and 2.5 t ha^{-1} manure were 1.3 and 1.2 t ha^{-1}, respectively. Application of 10 kg P ha^{-1} in addition to the 30 kg N ha^{-1} resulted in a mean additional yield increase of 0.27 t ha^{-1} with site–year mean increases ranging from <0.1 to 1.32 t ha^{-1}. Application of 2.5 t ha^{-1} of manure in combination with 30 kg N ha^{-1} resulted in a mean yield increase of 1.08 t ha^{-1} compared to N and manure used alone with site–year mean increases ranging from <0.30 to 1.47 t ha^{-1}. It was likely that manure supplied P and other nutrients required by sorghum. There was, however, no increase in the mean grain yield when P or manure was applied in addition to N during LR 2003 and LR 2004 seasons at Kapolin and Opwatetta, respectively.

The nutrient content of the manure on a dry matter basis was in the range 0.7–1.8%, 0.1–0.2%, and 0.8–2.4% for N, P, and K, respectively. A large proportion of the manure N was probably in organic form

Table 3 Sorghum grain yield (t ha^{-1}) during the long rains of 2004 and 2005

Treatment	Site						
	Kadesok		Opwatetta			Kapolin	
	LR 2004	LR 2005	LR 2004	SR 2005	LR 2005	LR 2004	LR 2005
Number of farmers	5	8	5	3	7	6	5
Control	1.03	1.76	1.17	1.00	1.39	0.67	0.96
Previous cowpea	1.84	3.19	1.61	2.00	2.17	1.43	1.18
30 kg N ha^{-1}	2.30	3.49	1.65	2.50	2.43	2.05	1.36
Previous *Mucuna*	2.64	3.98	1.76	2.67	3.00	2.81	1.47
LSD$_{0.05}$	0.27	0.27	0.29	0.56	0.39	0.53	0.15

Table 4 Sorghum grain yield (t ha^{-1}) during the long rains of 2003[a], 2004, and 2005

| | Site | | | | | | | |
| | Kadesok | | Opwatetta | | | Kapolin | | |
Treatment	LR 2004	LR 2005	LR 2004	SR 2005	LR 2005	LR 2003	LR 2004	LR 2005
Number of farmers	5	9	5	3	17	6	7	12
Control	0.98	1.66	0.61	1.50	1.39	1.15	0.79	0.98
30 kg N + 10 kg P ha^{-1}	2.62	3.78	1.74	2.92	2.69	2.15	2.50	1.48
30 kg N + 2.5 t M[b] ha^{-1}	3.02	3.93	1.60	3.80	2.85	2.26	2.47	1.88
30 kg N ha^{-1}	2.54	3.23	1.46	2.13	2.46	na	1.43	1.40
2.5 t kraal manure ha^{-1}	2.06	3.51	1.44	2.25	2.41	1.74	2.04	1.74
LSD$_{0.05}$	0.50	0.25	0.34	0.61	0.31	0.57	0.44	0.15

na = not applicable
[a]For Kapolin site only
[b]M = manure

rather than inorganic form which was slowly mineralized following application. Manure is a good source of P since it is not lost through volatilization and leaching, as for N and K. The advantage with manure application is that a wide range of nutrients and organic matter is applied which may contribute to sustainability of higher yield levels. However, manure use is a process of transfer of nutrients from one part of the farming system to another. The higher yields associated with manure application may lead to negative farm level nutrient balance if much of the harvested grain is marketed. Hence, there will be a need to bring nutrients to the farm to sustain the higher levels of productivity.

Trial 3: Nitrogen Response Curve for Sorghum

There was a significant increase ($p \leq 5\%$) in sorghum grain yield in response to increasing N levels across the sites, confirming that N is a limiting nutrient to

sorghum production in the area (Table 5). The shapes of the response curves were very similar for all locations. The response function determined using the mean yield by N rate averaged across all locations and years accounted for 99% of the variation in mean yields for the four N rates:

$$Yield = 1.050 + 0.044\,N^{0.9}$$

It is observed from Table 5 that the average increase was 0.74, 1.24, and 1.88 t ha^{-1} of grain in response to 20, 40, and 60 kg N ha^{-1}, respectively. The function indicates that N response will occur at N levels above 60 kg N ha^{-1}.

Trial 4: Sorghum Yield Under Reduced Tillage

There was a significant increase ($p \leq 5\%$) in sorghum grain yield with reduced tillage where early weed growth was controlled with glyphosate herbicides

Table 5 Sorghum grain yield response to increasing levels of inorganic N at the three sites

| | Site | | | | | | |
| | Kadesok | | Opwatetta | | | Kapolin | |
Treatment[a]	LR 2004	2003B	LR 2004	SR 2005	LR 2005	LR 2004	LR 2005
Number of farmers	3	5	5	4	5	6	5
0 kg N ha^{-1}	1.03	0.98	1.18	1.18	1.05	0.96	0.90
20 kg N ha^{-1}	1.87	1.44	2.15	1.70	1.87	1.91	1.34
40 kg N ha^{-1}	2.50	2.06	2.37	2.15	2.43	2.38	1.65
60 kg N ha^{-1}	3.27	2.40	3.34	2.60	3.25	2.91	2.14
LSD$_{0.05}$	0.65	0.42	0.51	0.43	0.32	0.49	0.37

[a]All treatments received a blanket application of 40 kg P and 60 kg K ha^{-1}

Table 6 Sorghum grain yield (t ha^{-1}) under reduced tillage during the short rains of 2004 and long rains of 2005

| | Site | | | |
| | Opwatetta | | Kadesok | |
Treatment	SR 2005	LR 2005	SR 2005	LR 2005
Number of farmers	3	3	3	3
Control (conventional tillage)	0.87	1.83	1.17	1.39
Glyphosate + (30 kg N + 10 kg P) ha^{-1}	1.80	4.00	1.61	2.43
Glyphosate	1.53	3.00	1.65	2.17
LSD$_{0.05}$	0.47	0.58	0.29	0.39

rather than plowing (Table 6). Yield with reduced tillage was further increased with N and P application. The response was most likely due to early planting and better weed control. During reduced tillage farmers weeded only once as compared to twice with conventional. Weeding of sorghum is typically done late as farmers give priority to weeding cash crops.

Reduced tillage is a potentially viable option for sorghum production in eastern Uganda. It enables early planting and better weed control early in the season resulting in increased yield. Any delay in field preparation will result in delayed planting which reduces the yield. Furthermore with the use of herbicides, farmers will weed their crops once, an advantage because labor for weeding and for other farm operations is scarce during the cropping season.

Economic Analysis

Most smallholder farmers operate at near-subsistence levels, selling the surplus only after their food needs are met. The results in the previous section show that the alternative strategies are effective in increasing sorghum production in eastern Uganda. The increase in sorghum production can contribute to food security,

which is the primary objective of the smallholder farmers. Since there are costs associated with the alternative strategies, it is important to subject the observed yield increment to economic analysis in order to determine the economic benefits associated with the alternative strategies. The benefit to cost ratio (B/C) of greater than 1 for alternative practices was an indicator of the profitability of a given practice, assuming the opportunity cost of money was included in the cost. The returns to application of low levels of N plus P and of manure are sufficient for these practices to be profitable as shown in Table 7.

The use of animal manure had a benefit to cost ratio of 2.67 when labor costs for collection, transport, and application of manure were not included. The profitability of fertilizer use was less for the poorer farmers whose access to money has a high opportunity cost due to the numerous demands on the poorest farmers' money. The wealthiest farmers normally have alternative income generating activities and in some cases even have access to credit. Combining inorganic N fertilizers with manure was the best strategy because manure is a good source of P and a substitute for inorganic P fertilizers. The yield response to inorganic P fertilizer was profitable but the benefit to cost ratio was lower than with the application of inorganic N fertilizer

Table 7 Increase in sorghum grain yield (t ha^{-1}) and net returns and benefit–cost ratios for opportunity costs of money of 25, 50, and 75%, averaged across all location and years

| Treatment | Increase in grain yield (t ha^{-1}) | Net returns to input use[a] (,000 UgSh ha^{-1}) | | | Benefit to cost ratio | | |
		25%	50%	75%	25%	50%	75%
30 kg N + 10 kg P ha^{-1}	1.27	186	162	137	2.52	2.10	1.80
30 kg N + 2.5 t M^a ha^{-1}	1.40	231	217	203	3.23	2.52	2.02
30 kg N ha^{-1}	0.93	154	140	126	3.15	2.63	2.25
[b]2.5 t kraal manure ha^{-1}	1.02	Labor	Labor	Labor			

[a]Conversion rate of 1,800 Uganda shillings per US dollar
[b]The benefit to cost ratio of kraal manure was 2.67

Table 8 Increase in sorghum grain yield (t ha^{-1}) and net returns and benefit–cost ratios for opportunity costs of money of 25, 50, and 75%, averaged across all location and years

Treatment	Increase in grain yield (t ha^{-1})	Net returns to input use[a] (,000 UgSh ha^{-1})			Benefit to cost ratios		
		25%	50%	75%	25%	50%	75%
20 kg N ha^{-1}	0.67	115	105	96	3.40	2.84	2.43
40 kg N ha^{-1}	1.11	174	155	135	2.82	2.35	2.01
60 kg N ha^{-1}	1.69	267	238	210	2.86	2.39	2.04

[a]Conversion rate of 1,800 Uganda shillings per US dollar

only. The use of legumes in rotation with sorghum is profitable; the value of the increased production was Uganda shillings 291,000 and 284,000 per hectare per year for cowpea and *Mucuna*, respectively.

Increasing the N rate to 60 kg N ha^{-1} was profitable at all three levels of opportunity cost (Table 8), but the benefit:cost ratio was greatest with the N rate of 20 kg ha^{-1}. The results indicate that higher N rates above 60 kg ha^{-1} were likely to be profitable when soil P, K, and other nutrient deficiency are less limiting than N.

Conclusion

Inorganic fertilizers, kraal manure, a combination of inorganic fertilizers with kraal manure, *Mucuna* fallow and cowpeas rotation, and reduced tillage were effective and economical strategies for increasing sorghum yield. In addition, *Mucuna* and cowpeas like inorganic N fertilizers also increased sorghum yield because N is limiting crop production in the study area. Application of P either as inorganic fertilizer or from organic sources like kraal manure resulted in additional yield increment and was also profitable at three opportunity costs for money. Therefore, any of these sources could be used for increased sorghum production in the region, in order to improve food security, which is a major concern of the smallholder farmers. Though use of green manures for soil fertility had declined in many countries where inorganic N fertilizers were widely used, they have a potential in the agriculture of the smallholder of sub-Saharan Africa since it is typically based on low external input, due to economic constraints limiting the use of inorganic fertilizers. Kraal manure also has a potential to increase crop production in the area. However, the use of a combination of inorganic and organic fertilizers is a better approach

considering the sustainability of the agro-ecosystem as whole.

Acknowledgments The authors are grateful to the farmers, Messrs Dennis Odelle, Bazil Kadiba, Patrick Odongo, and William Acoda, and the field assistants at the research sites for the good work done and the National Agricultural Research Organization (NARO), International Foundation for Science (IFS), and Sorghum/Millet Collaborative Research Support Program (INTSORMIL) for provision of the Research funds.

References

Bekunda MA, Bationo A, Ssali H (1997) Soil fertility management in Africa. A review of selected research trials. In: Buresh RJ, Sanchez PA, Calhoun F (eds) Replenishing soil fertility in Africa. SSSA spec. Publ. 51 SSA, Madison WI, pp 63–79

Fischler M, Wortmann CS (1999) Green manures maize-bean systems in eastern Uganda: agronomic performance and farmers' perceptions. Agroforest Syst 47:123–138

Foster HL (1971) Rapid routine soil and plant analysis without automatic equipment. I. Routine soil analysis. E Afr Agric For J 37:160–170

Giller KE, Cadisch G (1995) Future benefits from biological nitrogen fixation: an ecological approach to agriculture. Plant Soil 174:255–277

Giller KE, Cadisch G, Ehakuitusm C, Adams E, Sakala WD, Mafongoya PL (1997) Building soil nitrogen capital in Africa. In: Buresh RJ, Sanchez PA, Calhoun F (eds) Replenishing soil fertility in Africa. SSSA spec. Publ. 51 SSSA, Madison, EI, pp 151–192

Giller KE, McDonagh JF, Cadisch G (1994) Can biological nitrogen fixation sustain agriculture in the tropics?. In: Syers JK, Rimmer DL (eds) Soil Science and sustainable land management in the tropics. CAB International, Wallingford, CT, pp 173–191

Giller KE, Wilson KJ (1991) Nitrogen fixation in tropical cropping systems. CAB International, Wallingford, CT, 313p

Harrop J (1970) Soils. In: Jameson JD (ed) Agriculture in Uganda. Oxford University Press, Oxford, pp 43–71

Greenland DJ, Bowen G, Eswaran H, Rhoades R, Valentin C (1994) Soil, water, and nutrient management research, A new research agenda. IBSRAM Position Paper. Bangkok

(Thailand): International Board for Soil Research and Management, 60pp

Kaizzi CK (2002) The potential benefit of green manures and inorganic fertilizers in cereal production on contrasting soils in eastern Uganda. Ecology and Development series No. 4. Cuviller Verlag Göttigen, German, 102 pp

Kaizzi CK, Ssali H, Vlek PLG (2004) The potential of velvet bean (*Mucuna pruriens*) and N fertilizers in maize production on contrasting soils and agro-ecological zones of East Uganda. Nutr Cycling Agroecosyst 68:59–73

Kaizzi CK, Ssali H, Vlek PLG (2006) Differential use and benefits of velvet bean (*Mucuna pruriens*) and N fertilizers in maize production in contrasting agro-ecological zones of East Uganda. Agric Syst 88:44–60

Kaizzi CK, Wortmann CS (2001) Plant materials for soil fertility management in subhumid tropical areas. Agron J 93: 929–935

Lekasi JK, Tanner JC, Kimani SK, Harris PJC (2001) Managing Manure to Sustain Smallholder Livelihoods in the East African Highlands. HDRA Publications, Coventry, UK, 32 pp

Palm CA (1995) Contribution of agro forestry trees to nutrient requirements of intercropped plants. Agroforest Syst 30:105–124

Palm CA, Myers RJK, Nandwa SM (1997) Combined use of organic and inorganic nutrient sources for soil fertility maintenance and replenishment. In: Buresh RJ, Sanchez PA, Calhoun F (eds) Replenishing soil fertility in Africa. SSSA spec. Publ. 51. SSSA, Madison, WI, pp 193–217

Peoples MB, Herridge DF, Ladha JK (1995) Biological nitrogen fixation: an efficient source of nitrogen for sustainable agriculture? Plant Soil 174:3–28

Ruhigwa BA, Gichuru MP, Spencer DSC, Swennen R (1995) Economic analysis of cut-and carry and alley cropping systems of mulch production for plantains in South-eastern Nigeria. IITA Res 11:11–14

Sanchez PA (2002) Soil fertility and hunger in Africa. Science 295:2019–2020

Sanchez PA, Izac A-M N, Valencia I, Pieri C (1996) Soil fertility replenishment in Africa; a concept note. In: Breth SA (ed) Achieving greater impact from research investments in Africa. Sasakawa Africa Assoc, Mexico City, pp 200–207

Sanchez PA, Sheperd KD, Soule MJ, Place FM, Buresh RJ, Izac A-M N, Mokwunye AV, Kwesiga FR, Ndiritu CG, Woomer PL (1997) Soil fertility replenishment in Africa: An investment in natural resource capital. In: Buresh RJ,

Sanchez PA, Calhoun F (eds) Replenishing soil fertility in Africa. SSSA spec. Publ. 51. SSSA, Madison, WI, pp 1–46

Ssali H (2000) Soil Resources of Uganda and their relationship to major farming systems. Resource paper, Soils and Soil Fertility Management Programme, Kawanda, NARO

Ssali H, Keya SO (1986) The effects of phosphorus and nitrogen fertilizer level on nodulation, growth and dinitrogen fixation of three bean cultivars. Trop Agric 63:105–109

Statistix for Windows (1998) Analytical Software, Tallahassee, FL, USA

Stoorvogel JJ, Smaling EMA (1990) Assessment of soil nutrient depletion in sub-Saharan Africa, 1983–2000. Rep. 28 Win and Staring ctr, for Integrated Land, Soil and Water Res., Wageningen, The Netherlands

Tian G, Kolawole GO, Kang BT, Kirchhof G (2000) Nitrogen fertilizer replacement indexes of legume cover crops in the derived savanna of West Africa. Plant Soil 224:287–296

Van Cleemput O (1995) Fertiliser, Sustainable Agriculture and Preservation of the Environment. In: IAEA (ed.) Nuclear methods in Soil-Plant aspects of Sustainable Agriculture pp 7–16. Proceedings of an FAO/IAEA Regional Seminar for Asia and Pacific held in Colombo, Sri Lanka, 5-9 April 1993. IAEA-TECDOC-785. Vienna, Austria.

Versteeg MN, Amadji F, Eteka A, Gogan A, Koudokpon V (1998) Farmers' adoptability of Mucuna fallowing and Agro forestry technologies in the coastal savanna of Benin. Agric Syst 56:269–287

Vlek PLG (1990) The role of fertilizers in sustaining agriculture in sub-Saharan Africa. Fertil Res 26:327–339

Vlek PLG (1993) Strategies for sustaining agriculture in sub-Saharan Africa. In: Rogland J, Lal R (eds) Technologies for sustaining agriculture in the tropics. ASA Spec. Publ. 56. ASA CSSA, and SSSA, Madison, WI, pp 265–277

Wortman CS, Kaizzi CK (1998) Nutrient balances and expected effects of alternative practices in farming systems of Uganda. Agric Ecosyst Environ 71(115):129

Wortman CS, McIntyre BD, Kaizzi CK (2000) Annual soil improving legumes: agronomic effectiveness, nutrient uptake, nitrogen fixation and water use. Fields Crops Res 68:75–83

Wortmann CS, Eledu CA (1999) Uganda's agroecological zones: a guide for planners and policy makers. Centro Internacional de Agricultura Tropical, Kampala, Uganda

Wortmann CS, Ssali H (2001) Integrated nutrient management for resource-poor farming systems: A case study of adaptive research and technology dissemination in Uganda. Am J Alt Agric 16:161–167

The Effects of Integration of Organic and Inorganic Sources of Nutrient on Maize Yield in Central Kenya

A.N. Kathuku, S.K. Kimani, J.R. Okalebo, C.O. Othieno, and B. Vanlauwe

Abstract Low soil fertility is one of the major constraints to food production in the central highlands of Kenya. Majority of the farmers use the recommendations, made more than two decades ago, which were general for central Kenya, and others meander in the maze of guessing depending on crop performance or availability of manures or fertilizers. An experiment was set up in two districts of central Kenya with the main aim of checking the effects of manure on integration with mineral N fertilizer and the N response on increasing levels of addition of 5 t ha^{-1} manure. The experiment was an RCBD design, which consisted of manure as single applications and manure in combination with increasing rates of nitrogenous fertilizer application and replicated three times. The results showed that application of manure alone at 5 t ha^{-1} was not sufficient to give high crop performance, although higher yields above the control were reflected, but on addition of nitrogen, higher yields were obtained even at the lowest rates of 20 kg N ha^{-1}. Nitrogen response curves showed that application of 5 t ha^{-1} manure and N in the form of calcium ammonium nitrate up to 40 N kg ha^{-1} gave yield increases but in excess of that yields decreased. This indicated that the best quantity of nitrogen to add to 5 t ha^{-1} manure was 40 kg N ha^{-1}. Thus there is great potential in manure–nitrogen combinations.

Keywords Maize yield · Manure · Mineral fertilizer N fertilizer · N response

Introduction

Nitrogen (N) is one of the key nutrients for crop production. However, it is the most mobile and also the most easily exhausted nutrient in the soil due to its ability to exist in different forms and its easy leachability. To sustain high crop yields in intensive and continuous crop production system, nitrogen fertilizer input is required. Nitrogen is one of the most costly nutrients and its availability can also be a problem in smallholder farms (Palm et al., 1997). It is therefore essential to develop an integrated N management system that can fully exploit biological N fixation and other organic sources of N in various production systems.

This is because earlier research was more on inorganic fertilizers rather than a combination and the fact that the quality of organic sources of nutrients varies from place to place, source and handling methods, i.e. preparation and storage. Organic sources of nitrogen have the potential to supply large quantities of N required for growing crops; however, to obtain maximum production and for more sustainability, they should be supplemented with mineral fertilizers (Mugendi, 1997; Jama et al., 2000; Vanlauwe et al., 2001). The combination of mineral N fertilizers with organic N sources increases the rate of decomposition of low N and quality materials (Mugendi et al., 1999).

Manure or other organic inputs applied to the soil control the rate, pattern and extent of growth and activity of soil microorganisms. Manure can increase the humus content of soils by 15–50%, depending on soil type; this is in addition to increasing soil aggregate stability and root permeability (Klapp, 1967). Manure acts as a buffer, thus improving nutrient uptake for

A.N. Kathuku (✉)
Kenya Agricultural Research Institute, National Agricultural Research Centre, Nairobi, Kenya
e-mail: angelandan2000@yahoo.com

A. Bationo et al. (eds.), *Innovations as Key to the Green Revolution in Africa*,
DOI 10.1007/978-90-481-2543-2_25, © Springer Science+Business Media B.V. 2011

crops grown in acid soils; it also alleviates aluminium toxicity and improves the availability of the nutrients such as phosphorus (P), particularly in soils with high P and sulphur (S) fixation. Manure also supplies several nutrients such as magnesium (Mg) and trace elements which are not available in commonly used inorganic fertilizers (Simpson, 1986).

Due to high cost of mineral fertilizers, majority of farmers uses manure as a source of plant nutrient. According to a recent study in the central highlands of Kenya, farmers use mineral N fertilizers at rates between 15 and 25 kg N ha^{-1}, which are far below the recommended rate of 40 kg N ha^{-1} (Kimani et al., 2001). High costs, marketing problems and poor infrastructure were cited as the major reasons for the low use of inorganic fertilizers. The use of other resources (green manure, farmyard manure, crop residues and composts) available on-farm is therefore increasingly gaining importance. Of these resources, farmyard manure was found to be by far the most important (Kimani et al., 2001).

In most farms in central Kenya, the manure used is mainly cattle (65%) with the rest comprising other sources such as shoats (6%) and poultry (4%). Most of the manure is from own sources (83%) with a very small proportion of farmers (2%) purchasing manure (Kimani et al., 2000). In systems where manure is a major avenue for recycling nutrients, an important consideration is whether manure alone can meet and sustain the nutrient needs for a reasonable crop yield. There is sufficient evidence to indicate that N demand (and to some extent K) to produce a reasonable maize yield crop can be met by manure (Palm et al., 1997). Integration of manure and inorganic fertilizers may result in greater residual effect of the organic than the inorganic sources and also other added advantages of manure such as improved soil physical properties (Murwira, 1995).

Therefore, the main objective of this study was to determine the effects of manure on integration with mineral N fertilizer and N response on increasing levels of additional N on 5 t ha^{-1} manure.

Materials and Methods

The experiment was carried out in two districts of central Kenya, namely Kirinyaga and Maragua districts.

Kirinyaga District

Kirinyaga district occupies an area of 1437 km^2. Mukanduini (experimental site) is one of the divisions in the district and lies within S0° 34.678′ E37°16.215′ and at an altitude of 1303 m above the sea level (ASL). The soils are humic nitisols and 80% of the district is arable. The average farm size is 1.25 ha per family. Major agro-ecological zone (AEZ) includes upper humid (UH0), lower humid I (LH1), upper midland (UM) and lower midland III (LM3), while major enterprises include maize–bean, tomato and banana production. Mukanduini village is in Kerugoya division and lies within UM2 AEZ. The study area receives a mean annual rainfall of 1378 mm per annum with long rainy season occurring between March and June and short rainy season between July and December. Throughout the year, the area experienced rainfall, though low precipitation was experienced in January and February.

The organic carbon (%C) of the soil was low at the topsoil (0–20 cm) but lower than the one observed at the farmers' fields (Table 1). Percent N content of the soil (0.13% N) was moderate, according to Okalebo et al.'s (2002) rating (0.12–0.25% N), at the topsoil (0–20 cm) but low (0.11–0.06) at underneath soil (20–60 cm). The available Bray 2 P and cation exchange capacity (CEC) was lower than that at the farmers' fields. Generally the soil at the experimental site before treatment application was poor.

Maragua District

Maragua district covers an approximate area of 1065 km^2 and lies between 1100 and 2950 m above sea level (ASL). Kariti location (experimental site) of Maragua district lies within S0° 52.236′ E37°176′ and at an altitude of 1667 m ASL. The soils are mainly humic nitisols. The average farm size is 0.93 ha (IEA Kenya, 2002). Annual average rainfall ranges from 900 to 2700 mm p.a. (Government of Kenya, 2002). The district has four main agro-ecological zones, namely lower humid I (LH1), upper midland I (UM1), upper midland II (UM2) and upper midland III (UM3). Kariti location is in Kandara division, Maragua district and lies within UM2 AEZ in central Kenya.

Table 1 Soil characterization in Mukanduini site, Kirinyaga district and Kariti site, Maragua district

	Depth (cm)	pH	%C	Mineral N (ppm)	Total %N	P (Bray2) (mg kg^{-1})	K (mg kg^{-1})	Ca (mg kg^{-1})	CEC (cmol/kg)
Mukanduini	0–20	5.6	1.1	41.1	0.1	5.1	98.1	965.5	5.2
	20–40	5.6	0.8	37.9	0.1	4.3	61.1	915.4	4.7
	40–60	5.7	0.8	12.8	0.1	3.6	51.8	867.1	3.9
Kariti	0–20	5.0	1.1	14.7	0.2	0.8	376.7	1217.8	4.4
	20–40	5.0	1.2	12.7	0.2	0.5	327.1	1422.5	5.0
	40–60	5.3	1.1	10.2	0.2	0.2	354.0	1566.3	5.5

In Kandara division, the mean annual rainfall received was 1078 mm per annum (p.a.). The expected annual rainfall ranges between 900 and 2700 mm p.a. and this figure (1078 mm rainfall p.a.) lies within the average. Long rains were experienced between April and May, while the short rains occurred between October and December and January, February, March, June, July and August remained dry.

In Kariti, the topsoil (0–20 cm) was acidic with a pH of 4.98 (Table 1). Soil organic carbon was low (1.06% C) at the topsoil (0–20 cm) but total nitrogen (%N) was moderate (0.12–0.25% N).

The experiment was set up in a randomized complete block design replicated three times (Table 2). Plots of 6 m × 4 m were hand prepared and demarcated giving 1 m separation between plots and 2 m separation between blocks. Soils were sampled from each plot before application of any external source of nutrients and manure was collected from the farmers' farms and chemically analyzed using Okalebo et al.'s (2002) manual. Manure, which was previously chemically analyzed, and inorganic N fertilizer (calcium ammonium nitrate) were applied using 'per hole' method of application. The weight of manure was based on moist manure. Certified maize (*Zea mays*) seeds were planted at spacing of 75 cm × 30 cm and intercropped with common bean (*Phaseolus vulgaris*). All the crops were harvested at physiological maturity, fresh total and

sub-sample weight taken, dried and dry sub-sample weight taken. The yields per hectare were extrapolated based on the effective area harvested (7.56 m^2). The yield data were analyzed using Statistical Analysis System (SAS 10.0 for windows); this gave the ANOVA tables and comparison means.

Results and Discussions

Figure 1 shows the maize yields obtained on application of manure and mineral fertilizers at increasing levels and their corresponding yields. In Mukanduini, the highest maize grain yield was observed under manure 5 t ha^{-1} plus 60 kg N treatment, which showed significant differences from the other treatments ($P<0.05$), an indication that any treatment did equally well.

During the long rains, in Maragua, Kariti, the highest yield was observed under manure 5 t ha^{-1} plus 80 kg N ha^{-1} treatment. Application of manure 5 t ha^{-1} plus 60 kg N and manure 10 t ha^{-1} showed no significant differences, while application of manure 5 t ha^{-1} alone gave the lowest maize grain yield. This shows that 5 t ha^{-1} manure alone may not be adequate for crop production and farmers with large quantities of manure can apply 10 t ha^{-1} manure and cannot use any mineral fertilizers or alternatively where the manure is little, application of 60 kg N ha^{-1} would give equally good yield. This was applicable only to farmers who could afford to buy fertilizers as majority of the farmers were resource poor and could not afford to buy fertilizers.

During the short rains, in Mukanduini, Kirinyaga district, application of 5 t ha^{-1} manure plus 40 kg N ha^{-1} and 10 t ha^{-1} manure showed no significant differences ($P<0.05$), but in Kariti, 5 t ha^{-1} manure plus 80 kg N ha^{-1} gave the highest maize grain yield. The argument here is that a farmer can use either 10 t ha^{-1} manure or 5 t ha^{-1} manure on addition of 40 kg N ha^{-1}.

Table 2 Treatments applied in the two districts

Treatment	Description
1. Control	No external nutrient inputs
2. Manure	5 t ha^{-1}
3. Manure	10 t ha^{-1}
4. Manure	5 t ha^{-1} plus 20 kg N ha^{-1}
5. Manure	5 t ha^{-1} plus 40 kg N ha^{-1}
6. Manure	5 t ha^{-1} plus 60 kg N ha^{-1}
7. Manure	5 t ha^{-1} plus 80 kg N ha^{-1}

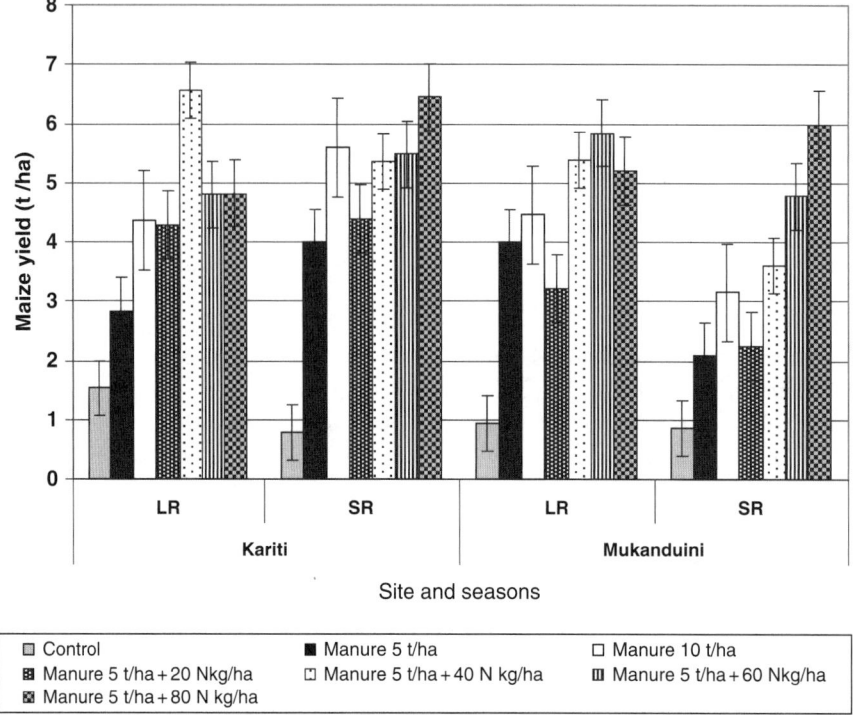

Fig. 1 Maize yields in kg ha^{-1} in two sites of central Kenya during two cropping seasons, 2003. LR, long rains; SR, short rains

In Kariti, Maragua district, application of 5 t ha^{-1} manure plus 40 kg N ha^{-1}, 5 t ha^{-1} manure plus 60 kg N ha^{-1} and 10 t ha^{-1} manure showed no significant differences ($P<0.05$) in maize grain yield. Accordingly, it would be more preferable to use lower rate of N (40 kg N ha^{-1}) than the higher rate as both were giving more or less the same yields. Similarly no significant differences were observed on application of 5 t ha^{-1} manure alone and 5 t ha^{-1} manure plus 20 kg N ha^{-1}. A general look at the results shows that in any combination of organic and inorganic sources of nutrients, there was at least an increase in yield above the control; particularly manure 5 t ha^{-1}, which was kept as the minimum manure application.

Nitrogen Response

To shed some light on the differences in crop yield under application of same treatments at the two sites, N response in these two soil types was assessed. Consider the law of minimum, which states that the amount of plant growth (maize yield in our case) is regulated by a factor present in the minimum amount. We assumed that N is the limiting factor and made a 'blanket' application of 40 kg P ha^{-1}. Therefore the graph rises or falls accordingly as this factor (N) is increased or decreased in amount or quantity. We found that this law did apply in both sites. This Liebig's law of minimum suggests that growth (as reflected in the results) is directly proportional to the nutrients applied to the value (quantity) where it was no longer limiting. However, this law is limited because if several factors are low, increases in one will in fact increase the yields to a slight extent. In many cases, although a plant contains a small amount of a compound, it would respond if fertilizer addition of this compound were made. According to an earlier scholar Hellriegel, a sigmoid curve would be obtained if an element limiting in a certain soil was increased systematically. This sigmoid curve is found in fields where the nutrient(s) are very deficient but in soils where nutrient (s) are less deficient, only the top part of the curve exists. The graphs (Figs. 2 and 3) were obtained from the two sites. Similarity in response was observed in both the long and the short rains.

In Maragua (Fig. 2) there was a rapid N response when N was increased from 0 to 40 kg ha^{-1} but

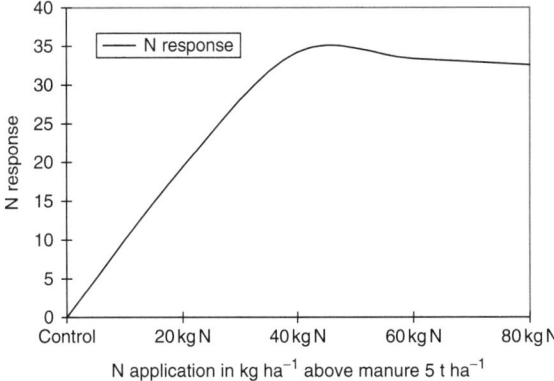

Fig. 2 Nitrogen response in Kariti, Maragua

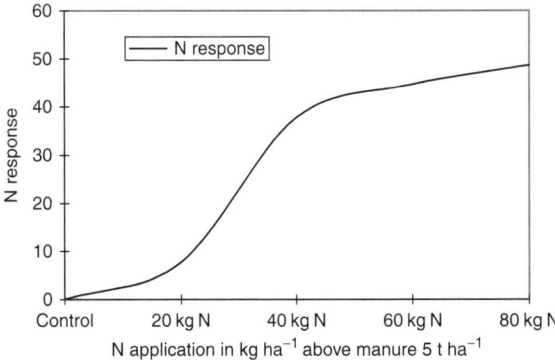

Fig. 3 N response in Mukanduini, Kirinyaga

application of N above 50 kg ha^{-1} showed a slow N response, thus low application of N (20 kg N ha^{-1}) would bring about a change in crop yield. In Kirinyaga (Fig. 3), the soils showed a better N response due to limiting levels of N and hence a sigmoid curve. There was a gradual increase in N response as the N rate was increased up to 40 kg N ha^{-1}. For any application of N above 40 kg ha^{-1} the N response showed a decrease. This showed that N was already sufficient in the soil at that level. This tells us that application of N could not be made blanket, only a site-specific recommendation would make a difference. Higher rates of N (40–60 kg ha^{-1}) were more likely to bring about yield increases in Kirinyaga.

Conclusions

The results have shown that application of manure is vital for higher crop yields, more importantly when inorganic fertilizers are added. High yields can be achieved on application of manure at higher rates when mineral fertilizers are not available; however, little manure addition of up to a maximum of 60 kg N ha^{-1} would be of great benefit to crop performance. From the curves in Figs. 2 and 3, the excess N application did increase maize grain yields, up to 60 kg N ha^{-1} application particularly in Mukanduini, but rates of N application above 60 kg N ha^{-1} plus manure 5 t ha^{-1} decreased maize grain yields in all the sites during the long rains. This shows that the site used in Kariti was less nutrient deficient as compared to Mukanduini site. This gives warning lights against regional fertilizer recommendation and advocates for site-specific fertilizer recommendations, but these recommendations will largely depend on the fertilizer availability and the economic status of the farmers in the area.

In conclusion, application of 5 t ha^{-1} manure plus rates of N fertilizers between 30 and 40 t ha^{-1} was recommended for Kariti, while application of 5 t ha^{-1} manure plus rates of N fertilizers between 40 and 60 t ha^{-1} was commended for Mukanduini.

Acknowledgement I am grateful to the Ministry of Agriculture field staff (extension officers in Kerugoya and Kariti) and KARI, Muguga Chemistry Laboratory technicians. My sincere gratitude goes to Dr S. K. Kimani of KARI, Muguga, for his critical scrutiny of this chapter, dedicated and collaborative assistance throughout the research period, all reviewers for their comments and correction and finally but not least to Rockefeller Foundation for financial support.

References

Government of Kenya (2002) National Development Plan (2002–2008) on effective management for sustainable economic growth and poverty reduction. Government Printer, Nairobi, Kenya

Jama B, Palm CA, Buresh RJ, Niang A, Gachengo CN, Nziguheba G, Amadalo B (2000) *Tithonia diversifolia* as a green manure for soil fertility improvement in western Kenya: a review. Agroforest Syst 49:201–221

Kimani SK, Mangale N, Gichuru M, Palm C, Wamuongo J (2001) Integrated use and effects of manures with modest application of inorganic fertilizers on soil properties and maize production in the central highlands of Kenya. Final technical report to the Rockefeller Foundation, May 2001

Kimani SK, Odera MM, Musembi F (2000) Factors influencing adoption of integrated use of manures and fertilizers in central Kenya. Proceedings of KARI scientific conference, Nov 2000

Klapp E (1967) Lehbuch des Acker-und pflanzebaus. 6. Aufl. Parey, Berlin. 603 pp, As cited in Karl et al (1994), pp 381–437

Mugendi DN (1997) Tree biomass decomposition, nitrogen dynamics and maize growth under agroforestry conditions in the sub-humid highlands of Kenya. PhD thesis, University of Florida Gainesville, Florida, USA

Mugendi DN, Nair PKR, Mugwe JN, O'Neill MK, Swift MJ, Woomer PL (1999) Alley cropping of maize with calliandra and leucaena in the sub-humid highlands of Kenya. Part. 2. Biomass decomposition, N mineralization and N uptake by maize. Agroforest Syst 46:51–64

Murwira HK (1995) Ammonia losses from Zimbabwean cattle manure before and after incorporation into soil. Trop Agric Trin 72:269–276

Okalebo JR, Gathua KW, Woomer PL (2002) Laboratory methods of soil and plant analysis: a working manual, 2nd edn. TSBF-CIAT and sacred Africa, Nairobi, Kenya

Palm CA, Myers RJK, Nandwa SM (1997) Combined use of organic and inorganic nutrient sources for soil fertility maintenance and replenishment. In: Buresh RJ, Sanchez PA, Calhoun F (eds) Replenishing soil fertility in Africa. Special publication no. 51. ASA, CSSA, SSSA, Madison, WI, pp 193–217

Simpson K (1986) Manures. In: Fertilizers and manures. Longman, London, pp 83–108

Vanlauwe B, Aihou K, Houngnandan P, Diels J, Sanginga N, Merkx R (2001) Nitrogen management in 'adequate' input maize-based agriculture in the derived savanna benchmark zone of Benin Republic. Plant Soil 228:61–71

Forage Legume–Cereal Double Cropping in Bimodal Rainfall Highland Tropics: The Kenyan Case

M.J. Khaemba, S.M. Mwonga, L.M. Mumera, and L. Nakhone

Abstract A 2-year study was conducted at two sites, Njoro and Mang'u, both in Nakuru district, in 1998–2000 to test the validity of double cropping of selected forage legumes with wheat and test their contribution to enhanced crop yields in a cereal-based cropping system. It was hypothesized that double cropping was possible in order to utilize the short rainy season, where the land is left under natural fallow. The main factor during the short rainy season was different fallow species: *Medicago sativa* L., *Vicia sativa* L., *Melilotus alba* Desr., *Crotalaria ochroleuca* G. Don, *Trifolium subterraneum* L., *Triticum aestivum* L. and a natural fallow. During the long rains, inorganic nitrogen supply to wheat test crop at 0, 20 and 40 kg N/ha was applied to make three sub-plots representing zero, full and double N rates, respectively. This factorial combination gave 21 treatments. The trial was laid as RCBD with split plot arrangement replicated three times. Double cropping was possible with some species with only *V. sativa* and *C ochroleuca* nodulating. Natural fallow plot recorded yield suppression probably due to wide C:N ratio and possibly allelopathic effects of the previous year's weeds. Wheat grown after *C. ochroleuca* produced the highest biomass and grain yields. This demonstrated that double cropping was possible using some species where the complementary legume crop could offer higher nutritive content to maize stover, the main livestock feed during the dry season.

Keywords Forage · Legumes · Double cropping · Wheat

M.J. Khaemba (✉)
Department of Crops, Horticulture and Soil Sciences, Egerton University, Egerton, Kenya
e-mail: khaemba03@yahoo.com

Introduction

Continuous cropping in sub-Saharan Africa (SSA) is a practice that has been known to lead to a decline in soil fertility and low crop yields (Bekunda et al., 1997). Though commercial fertilizers may be available, their use is limited due to poverty and variable response, which lead to their sub-optimal use (Mose et al., 1997; Mwaura and Woomer, 1999). Being in SSA, the Kenyan highlands face the same problem. They are characterized by low-resource farmers who practice mixed crop/livestock farming (Lekasi et al., 2001). The livestock component is important both as a nutritional supplement and in income generation. During the dry season of December to March, animal feeds are a constraint to livestock production, a period when they are fed mainly on cereal stover. FAO (1983) found that low-protein forage is a limiting factor to animal production in the tropics. The inclusion of forage legumes in the cereal stover fed to animals would improve the crude protein content and hence feed quality. Since similarities have been drawn between plant tissue decomposition in the soil and its digestion in the animal rumen (Palm et al., 2001), a plant that has good qualities for soil improvement is also good as an animal feed. Coincidentally, this region also suffers from shortage of wood fuel. A competition therefore exists between livestock and wood fuel uses for maize stover.

Forage legumes offer an opportunity in the intervention process. Their advantages have been cited by many researchers (Kariuki, 1998; de Jong and Mukisira, 1999; Palm et al., 2001). Whereas other researchers have used grain legumes in the region to demonstrate the possibility of double cropping, little information exists in the use of forage legumes (Onwonga, 1997; Cheruiyot et al., 2001). The need

A. Bationo et al. (eds.), *Innovations as Key to the Green Revolution in Africa*,
DOI 10.1007/978-90-481-2543-2_26, © Springer Science+Business Media B.V. 2011

therefore arises to use suitably adapted forage legumes as fallow species during the short rainy season for improving nutrient content of stover for dairy animals and the soil nitrogen pool through biological nitrogen fixation (BNF) and the general organic matter status of the soil. This improved soil nutrient status may be reflected in increased grain yields of succeeding cereal crop (Horst and Hardter, 1994).

The purpose of this study was to evaluate the tenability of forage legume species as fallow species during the short rainy season using wheat as a test crop in the long rains. It was hypothesized that double cropping in the wheat zone of the Kenyan highlands is possible and the use of forage legumes in the short rainy season would improve the soil and lead to a higher grain yield of the succeeding wheat.

Materials and Methods

Site Description

The experiment was carried out at two sites in Nakuru district, Rift Valley Province of Kenya: Egerton University (0° 23′ S and 35° 35′ E) and Mang'u (0° 10′ S and 36° 01′ E) in Rongai division. Altitude of Egerton is 2,250 m above sea level (m.a.s.l.) with mean annual rainfall of over 1,000 mm and mean temperatures of 14–16°C. Soils are well drained, sandy clay loam with a thick humic top classified as *Mollic Andosols*. Mang'u has an attitude of 1,945 m.a.s.l., mean temperature of 17°C and annual average rainfall of 900 mm. Soils are friable sandy clay loams classified as *Vitric Andosols*. The region has two rainfall patterns: the more reliable long rains of March/April to August (300–600 mm) and the unreliable short rains of October to December (250–400 mm). December to March is dry.

Soil Characterization

Soil samples from the sites were taken at 0–15 and 15–30 cm depths before sowing and analysed in the laboratory for total nitrogen (N) and phosphorus (P), organic carbon (OC), texture, pH and cation exchange capacity (CEC). Total N and P, OC, texture and pH

were determined as described by Okalebo et al. (2002), while CEC was determined by leaching with ammonium acetate method (Okalebo et al., 2002).

Experimental Design and Treatments

Short Rain Fallow Management

Randomized complete block design (RCBD) replicated three times was used. Plot size was 4 × 3 m. Fallow species (main factor) at seven levels were planted. These included *Medicago sativa* L. (lucerne) var. Hunter River, *Vicia sativa* L. (common vetch.), *Melilotus alba* Desr. (white sweet clover), *Crotalaria ochroleuca* G. Don (crotalaria), and *Trifolium subterraneum* L. (subclover). *Triticum aestivum* L. (wheat) var. Kenya Fahari and a natural fallow served as controls. Lucerne, sweet clover, crotalaria, subclover and wheat were drilled 20 cm apart, while the common vetch was drilled 30 cm apart. Seed rates were 20 kg/ha for lucerne and subclover, 25 kg/ha for crotalaria, sweet clover and the common vetch and 75 kg/ha for wheat. There was no rhizobia inoculation of the forage legume seeds in order to test the availability of compatible resident strains. The forage legumes were kept weed free in the field for 6 months with an 8-week harvesting interval and then incorporated in the soil.

Long Rain Test Crop (Wheat)

Inorganic nitrogen supply to wheat test crop at 0, 20 and 40 kg N/ha using calcium ammonium nitrate (CAN 26% N) to make three sub-plots representing zero, full and double N rates, respectively, was applied during the long rain as sub-plot treatment. These gave 21 treatments in split plot arrangement. The inorganic N was applied in two equal splits, at sowing and at 4 weeks after emergence. Phosphorus as TSP 20% P was uniformly applied at planting in all plots at the rate of 22 kg P/ha.

The first cycle of legumes was sown in August 1998, then incorporated in the soil in February 1999 and the test crop was sown in late April and harvested in September. Cycle two of legumes was sown in September 1999, then incorporated in the soil in February 2000 and the second wheat crop was sown in June and harvested in October 2000.

Sampling Procedures

Nodulating ability was assessed by randomly selecting three plants and then digging them out with the surrounding soil. The soil was carefully removed by hand and then the plants with all their roots were put in plastic bags and taken to the laboratory for removal of soil under a gentle stream of water. Nodules were detached, counted, and three of them were sliced into two pieces to observe the existence of pink colour, which signifies possible active N fixation. All nodules were dried in an oven at 65°C to constant weight. Vegetative biomass was harvested at 5 cm height (adapted from Akundabweni, 1986) to simulate grazing pressure at 8 weeks after planting and was maintained at eight weekly intervals from a quadrat of 0.5 m^2. Fresh weight was recorded and then dried in an oven at 65°C to constant weight.

Total Nitrogen and Phosphorus in Soil and Grains and Plant Tissue Organic Carbon and N Determination

Wheat grains were ground and passed through a 60-mesh sieve. Wet acid digestion of both grains and soil based on laboratory methods described by Okalebo et al. (2002) was used. N was determined via steam distillation followed by titration, while P was determined after extraction and colour development using ammonium molybdate and ammonium vanadate mixture and then measured using spectrophotometer at 400 nm wavelength. Plant tissue was analysed for carbon and N contents as described by Okalebo et al. (2002).

Wheat grain yield was determined by harvesting the three middle rows in each sub-plot, spikes threshed, winnowed and grains dried to 13% moisture content and then weighed.

Data Analysis

Collected data was subjected to analysis of variance (ANOVA) and general linear model (GLM) using SAS computer package and means separated by DMRT accordingly (SAS, 1996; Steel et al., 1997).

Results and Discussion

Soil Chemical and Physical Characteristics

Some soil chemical and physical characteristics are given in Table 1.

The top soil at Njoro is classified as clay loam, while that of Mang'u is sandy clay loam. Soils at Mang'u and Njoro have been classified as *Vitric Andosols* and *Mollic Andosols*, respectively (Jaetzold and Schmidt, 1983). From Table 1, Mang'u soils have very low P available levels and this can have a profound effect on

Table 1 Soil chemical and physical characteristics of Njoro and Mang'u experimental sites

Characteristics	Experimental sites			
	Njoro		Mang'u	
	Soil depths			
	0–15 cm	15–30 cm	0–15 cm	15–30 cm
Soil pH (H$_2$O)	6.2	5.4	5.6	5.4
Carbon (%)	3.14	2.13	2.05	0.76
Total N (%)	0.24	0.23	0.12	0.06
Phosphorus (ppm)	24.00	19.00	0.10	0.20
Organic matter (%)	5.46	3.70	3.56	1.32
C:N ratio	13.0	9.3	17.1	12.7
CEC	23.6	21.4	32.2	22.6
Sand (%)	37	66	68	86
Silt (%)	30	13	8	9
Clay (%)	33	21	24	5

the crop growth, while Njoro soils have adequate levels (Schulte and Kelling, 1996).

The apparent higher cation exchange capacity at Mang'u could be due to the fact that soils at this site are shallow yielding to cinders of volcanic nature that have high mineral contents (Jaetzold and Schmidt, 1983). These, if included during soil preparation, can give a high value when actually the cations are not available.

Nodulation

The level of plant nodulation is given in Table 2 where only two species (*V. sativa* and *C. ochroleuca*) were nodulated in 1998/1999.

In 1998/1999, only *V. sativa* and *C. ochroleuca* nodulated and the rest did not. There was no nodulation in 1999/2000 for all the five species at both sites. The non-nodulation of the three species in 1998/1999 could have been due to the unavailability of their specific strains in the soil, while the low rainfall received

in 1999/2000 (Figs. 1 and 2) may have affected nodulation of all the species.

For the nodulating species of *C. ochroleuca* and *V. sativa*, there was a significant ($p < 0.05$) difference in the nodule numbers for crotalaria at the two sites, with Njoro site recording a higher number for *V. sativa*. On the other hand, there was no significant difference in the nodule numbers at Mang'u. A comparison of nodule weights on site basis showed no significant differences, but the differences were observed between species at each site. It is possible that the slightly higher nodule numbers in Mang'u may not have translated to higher N since N fixation is not just about nodule numbers but also about the efficiency of the formed nodules and other favourable environmental factors. The wide difference in the soil P levels at the two sites did not affect nodulation since both *Crotalaria* and *Vicia* species nodulated despite that difference. The major factor for non-nodulation of the other species therefore could be unavailability of their specific strains.

Prolonged dry conditions have been reported to reduce the survival of rhizobium bacteria in the soil (Bogdan, 1977; Whiteman, 1980) and that could be one of the reasons for non-nodulation in the second short rainy season. On the same note, photosynthate, needed for root growth is in short supply during the dry season. This compromises on nodule formation, maintenance and subsequent dinitrogen fixation, are energy-requiring processes that if not supplied with enough photosynthate for root growth as in the dry season may fail (Peoples and Craswell, 1992; Fujita et al., 1992). Perhaps that explains why Mang'u with lower rainfall (Appendices 1 and 2) than Njoro was affected more. Sometimes, nodulation may also be interfered with when the population of ineffective strains of rhizobium is high compared to the effective ones (Wheeler and Jones, 1977; Fujita et al., 1992).

Table 2 Nodule number and weight of *V. sativa* and *C. ochroleuca* in 1998/1999

Germplasm	Nodule number per plant		Nodule weight (g) per plant	
	Njoro	Mang'u	Njoro	Mang'u
V. sativa	25[a]	24[ns]	0.077[a]	0.077[a]
C. ochroleuca	7[b]	18[ns]	0.023[b]	0.032[b]
M. alba	*	*	*	*
M. sativa	*	*	*	*
T. subterraneum	*	*	*	*
Weedy plot	–	–	–	–
Wheat	–	–	–	–
CV (%)	73.6	56.4	92.6	49.2
LSD	11.8	12.0	0.05	0.03

ns, not significant; *, no nodulation; –, not applicable. Means followed by the same letter within a column are not significantly different by LSD at $\alpha = 0.05$

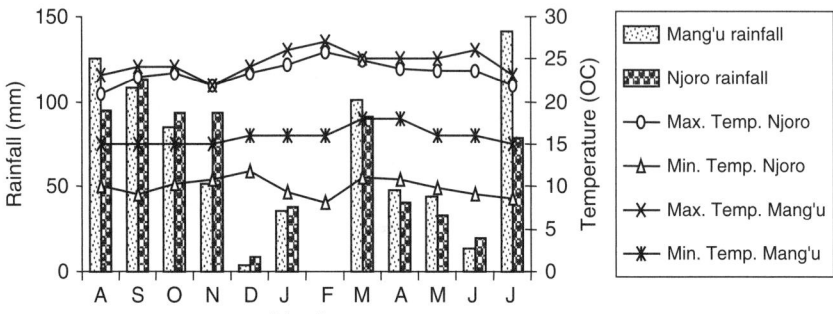

Fig. 1 The 1998/1999 rainfall and temperature for two sites

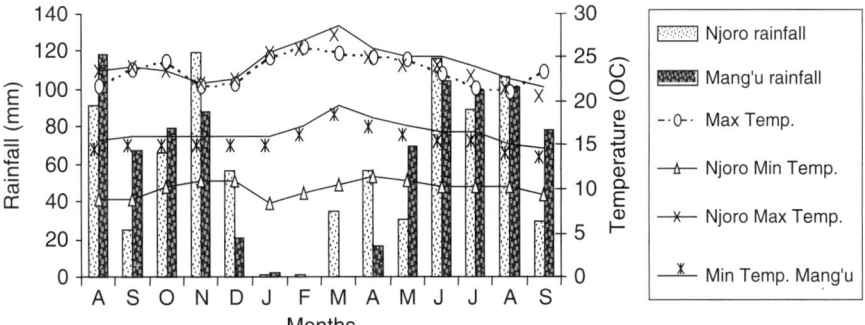

Fig. 2 The 1999/2000 rainfall and temperature for two sites

The observed erratic nodulation phenomenon is similar to observations made by McDonald (1935) for both exotic and indigenous forage legumes unless inoculated with the right strain of bacteria.

Among the species used, it is only crotalaria that is of tropical origin. It is locally adapted and may not have lost its 'primitive' association with the cross-inoculating rhizobium that it is known to be promiscuous with. The other legumes are temperate in origin and need specific strains to nodulate (Norris, 1967). This probably accounts for crotalaria's successful nodulation.

Legume Dry Matter Yield

The yields of the fallow species are shown in Table 3.

In general, more biomass was produced in 1998/1999 than in 1999/2000 at each site with some exceptions. In Njoro, *T. subterraneum* yielded more in

1999/2000 than in 1998/1999. *Vicia sativa* and the natural fallow plot produced similar amount of biomass in Njoro in 1998/1999, while *M. alba* produced consistently low biomass in both Njoro and Mang'u. In 1998/1999, *V. sativa*, *C. ochroleuca* and natural fallow plot registered the highest amount of biomass in Mang'u. In 1998/1999, all treatments in Njoro yielded more biomass than did treatments in Mang'u except *M. sativa* and *C. ochroleuca*, while in 1999/2000, the exception was *T. subterraneum*. This could be attributed to better rainfall distribution, higher soil N and P at Njoro and the differences between soil types at the two sites (Table 1). *Mollic* and *vitric* properties of *Andosols* dictate the amount of nutrients and water a soil can hold (Sombroek et al., 1982; Soil Survey Staff, 1992). This may have been a major contributing factor to the differential performance of the crops in the two soil types. Plants grown on a low-fertility soil with low and poorly distributed rainfall amounts yield lower than those grown on a higher fertility and better distributed rainfall conditions as happened in Mang'u and Njoro, respectively.

Table 3 Dry matter yield of the fallow species at Njoro and Mang'u in 1998/1999 and 1999/2000

| Fallow species | DM yield of the fallow species (kg/ha) | | | |
| | 1998/1999 | | 1999/2000 | |
	Njoro	Mang'u	Njoro	Mang'u
V. sativa	6,815[a]	4,237[a]	3,267[b]	2,336[b]
M. alba	289[c]	43[c]	132[d]	$
M. sativa	1,192[bc]	1,306[b]	1,070[c]	322[c]
C. ochroleuca	2,753[b]	5,000[a]	2,001[c]	694[c]
T. subterraneum	2,840[b]	1,141[b]	1,954[c]	3,735[a]
Wheat*	1,220	1,040	1,100	980
Weedy plot	5,541[a]	4,823[a]	5,155[a]	$
CV (%)	33.2	19.9	19.8	50.0
DMRT	1,959	1,002	816	1,339

Means followed by the same letter within a column are not significantly different by DMRT at $\alpha=0.05$; $, missing value due to the two samples being mixed. *, Wheat grain yield not statistically analysed

Some Chemical Properties of the Forage Legumes

Some chemical characteristics of the forage species are given in Table 4. From this data, *C. ochroleuca* gave the highest CP at both sites, while the natural fallow gave the lowest. Despite the high amount of biomass produced by the natural fallow plot, the percentage of N, hence crude protein content, was low. This suggests that natural fallows have a limitation in quality both as a soil N source and as an animal feed and have a potential of locking up N, thus making it unavailable for plant use. *Medicago sativa* had the lowest percentage of C, while the natural fallow had the highest in Njoro; in Mang'u it was *M. alba* and fallow plot, respectively. The calculated C:N ratio was lowest in *C. ochroleuca* and highest in the fallow plot.

Crotalaria can be classified as a better quality plant material for soil fertility improvement than the rest, based on its low C:N ratio (Table 4). Low C:N ratio has a positive effect on the plant decomposition rate in the soil. When C:N ratios are low, decomposition, and therefore mineralization, is faster. Crotalaria's decomposition is likely to be faster than the rest whose C:N ratios are higher. It may therefore provide a more immediate source of N to the crops grown after its incorporation compared to the other treatments. However, *Crotalaria* species are known to have toxic glucosides and a bitter taste, which interferes with their palatability, but this varies between species and varieties (Bogdan, 1977; Weber, 1996). The variety used in this case was not determined as to whether it was sweet or bitter. Mkiwa et al. (1990a, b) used 'Marejea' variety (sweet type) and found it to be having a high intake and digestibility. If C:N ratio and N and P contents are used as a standard, then this plant is good for soil improvement and possible animal use. *Vicia sativa* at Njoro had 2.3% N, thus being classified

as a marginal N source in the soil. For faster decomposition and nutrient release, the N content should be more than 2.5% N (Palm et al., 2001). The differences in the chemical properties of the species at the two sites may be attributed to site soil moisture differences. Stress to a plant can lead to a higher lignin and hence C at the expense of N and this may widen the C:N ratio. Cheruiyot et al. (2001) working at the same sites found that the weed biomass had a C:N ratio of 32 and 22 for Njoro and Mang'u, respectively. This makes natural fallow to have little impact on subsequent soil fertility improvement since it is likely to take too long to decompose. Thus, although the natural fallow plot had a high biomass, its value in soil fertility improvement is lower than the legume-managed fallows with lower biomass yield. As an animal feed source, it is similarly poor in quality.

Influence of Fallow Species on Wheat Dry Matter and Grain Yields

Wheat Dry Matter Yield

The first year (1998/1999) experienced crop failure and only biomass of the test crop was harvested to indicate biological yield at 104 days after planting and results are shown in Table 5. No biomass was harvested in 1999/2000 year due to the good rains experienced.

The fallow species impact on the wheat dry matter was significant ($p < 0.05$) at Njoro. Wheat planted in long rainy season of 1998/1999 after *C. ochroleuca* showed the highest biomass yield compared to that in weedy plot at both sites (farmer practice). No significant differences in wheat biomass yield were observed at Mang'u. This suggests that there was no improvement in soil fertility or else severe moisture stress as

Table 4 CP, C, P and N contents of bulked selected forage legumes in Njoro and Mang'u

Fallow species	Njoro					Mang'u				
	%C	%N	%CP	C/N	P (ppm)	%C	%N	%CP	C/N	P (ppm)
V. sativa	30.0	2.3	14.4	13:1	11.2	36.4	2.8	17.5	13:1	10.7
M. alba	38.4	2.6	16.8	14:1	14.8	30.4	2.7	16.9	11:1	16.1
M. sativa	29.6	2.8	17.5	11:1	7.3	35.2	2.7	16.9	13:1	7.7
C. ochroleuca	31.8	4.4	27.5	7:1	17.4	37.6	4.2	26.3	9:1	17.5
T. subterraneum	40.0	2.4	15	17:1	11.9	37.4	2.1	13.1	18:1	12.1
Weedy plot	39.6	1.2	7.4	34:1	$	40.2	1.2	7.4	34:1	$

$, missing value (data lost); CP, crude protein

Table 5 Wheat dry matter yield (kg/ha) in the 1998/1999 season

Fallow species	Njoro	Mang'u
V. sativa	1,035^c	1,091^{ab}
M. alba	1,023^c	1,131^{ab}
M. sativa	621^d	889^b
C. ochroleuca	1,436^a	1,378^a
T. subterraneum	1,206^b	1,373^a
Weedy plot	127^e	930^b
Wheat	627^c	1,105^{ab}
CV (%)	18.3	26.6
DMRT	157	332

Means within columns followed by the same letter are not significantly different by DMRT at $\alpha = 0.05$

Table 6 Wheat grain yields for the two sites for 1999/2000

	Grain yield (kg/ha)	
Fallow species	Njoro	Mang'u
V. sativa	1501^b	813^b
M. alba	2,279^a	1,118^a
M. sativa	2,111^a	785^{bc}
C. ochroleuca	2,077^a	626^{bc}
T. subterraneum	2,008^a	780b^c
Weedy plot	2,370^a	532^c
Wheat	2,020^a	830^b
CV (%)	20.9	34.0
DMRT	420	262

Means within columns followed by the same letter are not significantly different by DMRT at $\alpha = 0.05$

a result of poor rainfall received masked the effect of soil fertility improvement.

The poor wheat response in Njoro subsequent to the natural weedy plot could be attributed to the high weed biomass (Table 3) that could have used a lot of moisture and also locked up most of the available soil N due to its wide C:N ratio (Cheruiyot et al., 2001). As for the case of wheat after *M. sativa*, it could be due to low moisture amounts in the soil occasioned by the legume use and possibly allelopathic effects. Lucerne is known to be deep rooted, a feature that helps it to survive in dry seasons (Whiteman, 1980; Ivanov, 1988). With enough moisture, most of its roots are close to the surface, but they may reach up to 3 m deep under drought conditions. Drought interferes with the growth of the plants through the inhibition of root growth and the resultant limitation of the plants to explore new nutrient sites in the soil (van Duivenbooden, 1993).

Inorganic N supply had a significant ($p < 0.05$) effect on the yield of wheat biomass at both sites (data not shown). N is needed by the plant for purposes of vegetative material production and its deficiency can be reflected in the reduction in leaf area due to senescence of the older leaves and failure to grow (Gardner et al., 1985; Uhart and Andrade, 1995; Pandey et al., 2001). This significant increase in dry matter yield in Mang'u compared to Njoro could be due to inherently low N content in Mang'u which on addition of nutrients leads to quick response in dry matter yield (Gardner et al., 1985).

Wheat Grain Yield

Wheat grown after fallow treatments had a mixed response in 1999/2000 as given in Table 6.

The results of Table 6 contrast with those of Table 4, where the N content of the forage legumes is high but with no effect on the subsequent wheat yield. This is because the harvested biomass was not returned to the plots as it is assumed that the short rain legume biomass is used as fodder. The quantity of biomass incorporated therefore was limited to what was left at the end of the dry season. In Njoro, wheat yield after *V. sativa* gave the lowest yield compared to the other treatments. This suggests that the amount of N added was so low that it could not contribute significantly to the total N or the soils had enough N and what was added could not make any difference (except for *V. sativa*). The yield depression in the *V. sativa* treatment could be due to its annual growth habit where probably the only source of N could have been root remains and senesced parts only and these contain very little N. It is also possible that there was no synchrony between N release patterns from the incorporated legume biomass and crop demand. This is supported by the fact that incorporation of the biomass was done in February and planting of the wheat crop was done in June due to erratic rains. In Mang'u, however, treatment differences were noted. This could be due to poor soils at that site (Table 1) such that any little N added was felt and possible synchrony between crop demand and nutrient release, a condition where moisture is important. Wheat after *M. alba* yielded the highest, while that after *T. subterraneum* and the weedy plot yielded the lowest. *Melilotus alba* had a narrow C:N ratio and high P supply to the soil.

The benefit of growing these forage legumes could however be obtained from the quality improvement of the cereal stover that is otherwise fed to the animals and the resultant manure from these animals.

Conclusion

From the results of this study, it is evident that the growing of forage legumes can be a viable enterprise that can have an impact on the cereal production systems on smallholder farms. This is through the diversity at farm level resulting from the added crops and the increased productivity thereof. *Crotalaria ochroleuca*, an indigenous plant that withstands local environmental stress with its higher tissue N content, is suitable as a fallow management species. It can be used to improve the soil N, hence reducing the quantity of inorganic N required by crops. It is a promising species in increasing the grain and straw yields of the succeeding wheat crop. Second, sequential growing of *V. sativa* with wheat in smallholder mixed farming regions of the Kenyan highlands is viable since it caters to the farmers' soil and probable livestock needs especially during the dry seasons when the legume is used to improve the quality of the straw. Its benefits are best seen in cooler areas like Njoro. In addition, nodulation was affected by rainfall amounts and distribution and temperate legumes were poor in nodulation compared to tropical ones. *Vicia villosa* had the highest nodule numbers. In addition, the cooler environment of Njoro was more favourable for nodulation. Finally, biomass production may have been affected by the influence of rainfall amounts and distribution and soil type. This is why Njoro site yielded more than did the Mang'u one. The forage species took more time to flower in Njoro and produced more seed than in Mang'u.

References

Akundabweni LMS (1986) Forage potential of some native annual *Trifolium* species in the Ethiopian highlands. In: Haque I, Jutzi S, Neate PJH (eds) Potentials of forage legumes in farming systems of sub-Saharan Africa. Proceedings of workshop held at ILCA, Addis Ababa Ethiopia, 16–19 Sep 1985, pp 439–459

Bekunda MA, Bationo A, Ssali H (1997) Soil fertility in Africa: a review of selected research trials. In: Sanchez PA, Buresh RJ, Calhoun F (eds) Replenishing soil fertility in Africa. American Society of Agronomy and Soil Science Society of America Special publication 51, Madison, WI

Bogdan AV (1977) Tropical pasture and fodder plants (grasses and legumes). Longman, London

Cheruiyot EK, Mumera LM, Nakhone LN, Mwonga SM (2001) Rotational effect of grain legumes on maize performance in the Rift Valley highlands of Kenya. Afr Crop Sci J 9: 667–676

de Jong R and Mukisira EA (1999) National Dairy Cattle and Poultry Research Programme Technical Report. In: de Jong R, Mukisira EA (eds) National Agricultural Research Project II: National Dairy and Cattle and Poultry Research Programme. Technical Reports for discussion at NAHRC Naivasha, 7–10, June 1999. Document No. 1A/1999. KARI, Nairobi, Kenya, pp 43–54

FAO (1983) Towards sustainable agriculture; the conservation and rehabilitation of African lands: an international scheme. FAO ARC/90/4

Fujita K, Ofosu-Budu KG, Ogata S (1992) Biological nitrogen fixation in mixed legume–cereal cropping systems. Plant Soil 141:155–175

Gardner FP, Pearce RB, Mitchel RL (1985) Physiology of crop plants. Iowa State University Press, Ames, IA

Horst WJ, Hardter R (1994) Rotation of maize with cowpea improves yield and nutrient use of maize compared to maize monocropping in an alfisol in the northern Guinea Savannah of Ghana. Plant Soil 160:171–183

Ivanov AI (1988) Alfalfa. Oxonian Press Pvt Ltd, New Delhi, pp 254–257

Jaetzold R, Schmidt H (1983) Farm management handbook of Kenya, vol II/B Central Kenya. Ministry of Agriculture, Nairobi

Kariuki JN (1998) The potential of improving napier grass under smallholder dairy farmers' condition in Kenya. PhD thesis, Wageningen Agricultural University, The Netherlands

Lekasi JK, Tanner JC, Kimani SK, Harris PJC (2001) Managing manure to sustain smallholder livelihoods in the east African highlands. HDRA – the Organic Organization, Ryton-on-Dunsmore

McDonald J (1935) The inoculation of leguminous crops. East Afr Agric For J 1:8–13

Mkiwa FE, Sarwatt SV, Lwoga AB, Dzowela BH (1990a) Nutritive value of *Crotalaria ochroleuca*: I. Chemical composition and in vitro dry matter digestibility at different stages of growth. In: Dzowela BH, Said AN, Wendem-Agenehu AS, Kategile JA (eds) Utilization of research results on forage and agricultural by-product materials as animal feed resources in Africa. Proceedings of the 1st joint workshop held in Lilongwe, Malawi, 5–9 December 1988 by the Pastures Network for Eastern and Southern Africa (PANESA) and African Research Network for Agricultural By-products (ARNAB). International Livestock Centre for Africa, Addis Ababa, Ethiopia, pp 321–329

Mkiwa FE, Sarwatt SV, Lwoga AB, Dzowela BH (1990b) Nutritive value of *Crotalaria ochroleuca*: II. The effect of supplementation on feed utilization and performance of growing sheep. In: Dzowela BH, Said AN, Wendem-Agenehu AS, Kategile JA (eds) Utilization of research results on forage and agricultural by-product materials as animal feed resources in Africa. Proceedings of the 1st joint workshop held in Lilongwe, Malawi, 5–9 December 1988 by the Pastures Network for Eastern and Southern Africa (PANESA) and African Research Network for Agricultural By-products (ARNAB). International Livestock Centre for Africa, Addis Ababa, Ethiopia, pp 330–344

Mose LO, Nyangito HO, Mugunieri LG (1997) An economic analysis of the determinants of fertilizer use among smallholder maize producers in Western Kenya. In: Fungoh PO, Mbadi GCO (eds) Focus on agricultural research for sustainable development in a changing economic environment.

Proceedings of the 5th KARI scientific conference, 14–16 Oct 1996, Nairobi, pp 504–510

Mwaura FM, Woomer PL (1999) Fertilizer retailing in the Kenya highlands. Nutr Cycling Agroecosyst 55:107–116

Norris DO (1967). The intelligent use of inoculants and lime pelleting for tropical legumes

Okalebo JR, Githua KW, Woomer PL (2002). Laboratory methods of soil and plant analysis: a working manual. Tropical Soil Fertility and Fertility (TSBF)/Kenya Agricultural Research Institute (KARI), Nairobi

Onwonga RN (1997) Effect of chickpea incorporation on soil nutrient status and wheat grain yield in a wheat–chickpea rotation. MSc thesis, Egerton University, Njoro, Kenya

Palm CA, Gachengo CN, Delve RJ, Cadisch G, Giller KE (2001) Organic inputs for soil fertility management in tropical agroecosystems: application of an organic resource database. Agric Ecosyst Environ 83:27–42

Pandey RK, Maraville JW, Bako Y (2001) Nitrogen fertilizer response and use efficiency for three cereal crops in Niger. Commun Soil Plant Anal 32:1465–1482

Peoples MB, Craswell ET (1992) Biological nitrogen fixation: investments, expectations and actual contributions to agriculture. Plant Soil 141:13–39

SAS Institute (1996) SAS for Windows 6.12. Cary, NC

Schulte EE, Kelling KA (1996) Understanding plant nutrients: soil and applied phosphorus. University of Wisconsin—System Board of Reagents and University of Wisconsin—Extension Cooperative Extension

Soil Survey Staff (1992) Key to soil taxonomy, 5th edn. SMSS technical monograph No. 19. Pocahontas Press, Inc., Blacksburg, VI, 556p

Sombroek WG, Braun HMH, van der Pouw BJA (1982) Exploratory soil map and agro-climatic zone map of Kenya 1980. Scale 1:1,000,000. Exploratory soil survey report No. E1 Kenya Soil Survey Ministry of Agriculture – National Agricultural Laboratories, Nairobi

Steel RGD, Torrie JH, Dickey DA (1997) Principles and practices of statistics, a biometrical approach. McGraw-Hill, New York, NY

Uhart SA, Andrade FH (1995) Nitrogen deficiency in maize: 1. Effects on crop growth, development, dry matter partitioning and kernel set. Crop Sci 35:1375–1383

van Duivenbooden N (1993) Nitrogen, phosphorus and potassium relationships in five major cereals received in respect to fertiliser recommendations and land use planning. In: van Duivenbooden N (ed) Land use systems analysis as a tool in land use planning with reference to north and west Africa agroecosystems. Doctoral thesis, Wageningen Agricultural University, Wageningen, pp 65–79

Weber G (1996) Legume based technologies for sustainable African Savannas: challenges for research and development. Biol Agric Horticult 13:309–333

Wheeler JL, Jones RJ (1977) The potential of forage legumes in Kenya. Trop Grassl 11:273–282

Whiteman PC (1980) Tropical pasture science. Oxford University Press, Oxford

Effects of Conservation Tillage, Fertilizer Inputs and Cropping Systems on Soil Properties and Crop Yield in Western Kenya

H.K. Githinji, J.R. Okalebo, C.O. Othieno, A. Bationo, J. Kihara, and B.S. Waswa

Abstract An on-farm experiment was conducted in Western Kenya (Busia district) in the long rain season of 2005 to investigate the effects of conservation tillage on soil properties and the crop yields. The experiment based on a split–split plot design with three replicates and six core treatments arranged in a factorial combination of nitrogen application and cropping systems was adopted. Maize variety IR (striga resistant) was planted as a test crop, soybean (SB20) variety as an intercrop and for maize–legume rotation. Soil pH, moisture content and organic carbon were analysed in soil whereas total P and N were analysed in the plant tissue. Conservation and conventional tillage systems combined with cropping systems (intercropping, rotation and continuous) at 0 and 60 kg N/ha application were tested. Residue incorporation was done to all plots. The soil was sampled before and after harvesting to compare the effects of the treatments. Weeding for conservation tillage plots was by hand pulling. Combinations of conservation tillage, continuous and with application of 60 kg N/ha, for maize gave the highest yield of 2.8 t/ha. The combination of conservation tillage, rotation and application of 60 kg N/ha gave 2.5 t/ha maize grain. Combination of conservation tillage rotational cropping system at 60 kg N/ha application gave the highest soybean yield (1.23 t/ha). Soil carbon showed that there was significant difference between the conservation tillage and conventional tillage as well as the increase of the soil carbon from initial level of 1.44% to the highest percentage soil carbon of 1.9%.

Keywords Conservation tillage · Conventional tillage · Cropping system · Soil properties · Residue incorporation

Introduction

Conservation agriculture (CA) is now widely recognized as a viable concept for sustainable agriculture due to its comprehensive benefits in economic, environmental and social sustainability. Its principles are already widely adopted in the world (FAO, 2002). The US Conservation Technology Information Center developed the first widely accepted definition of conservation tillage as "any tillage and planting system that covers at least 30% of the soil surface with crop residue, after planting, in order to reduce soil erosion by water" (CTIC, 1996). It is estimated that at present no-tillage is practiced on more than 95 million hectares worldwide. Approximately 47% of the technology is practiced in South America, 39% is practiced in the United States and Canada, 9% in Australia and about 3.9% in the rest of the world, including Europe, Africa and Asia. Despite good and long-lasting research in this part of the world showing positive results for no-tillage, this technology has had only small rates of adoption worldwide (Derpsch, 2001).

Conservation tillage maintains a minimum coverage of soil surface that improves rainwater management through reduced erosion and runoff, and increased nutrient use efficiency as compared to conventional tillage that is largely practised by farmers. Conservation tillage has the potential of soil moisture conservation and mitigation of intra-season dry spells that often result in low productivity and

H.K. Githinji (✉)
Department of Soil Science, Moi University, Eldoret, Kenya
e-mail: harrikag@yahoo.com

A. Bationo et al. (eds.), *Innovations as Key to the Green Revolution in Africa*,
DOI 10.1007/978-90-481-2543-2_27, © Springer Science+Business Media B.V. 2011

crop failure especially in dry parts of East Africa (Rockstrom and Johnson, 1999). Thus, conservation tillage may be interpreted as "any system that promotes good crop yields while at the same time maintaining soil fertility, minimizing soil and nutrient loss, and saving energy/fuel inputs".

It is essential to increase food production by utilizing all the available resources. However, international donors have been reluctant to promote modern agricultural inputs in sub-Saharan African countries, arguing that agrochemicals are hazardous to the environment. In these circumstances, farmers in Africa are not able to get out of poverty (Findlay et al., 2002).

Conservation tillage has several positive effects on water productivity (Rockstrom et al., 2001) compared to traditional soil and water conservation systems. Besides enhancing infiltration and soil moisture storage, it improves timing of tillage operations and spatial distribution of soil moisture at field scale, which is crucial in semi-arid rainfed agriculture. In Kenya, promotion of animal-drawn conservation tillage tools such as rippers, ridgers and sub-soilers among smallholder farmers in semi-arid Machakos and Laikipia districts has resulted in improved water productivity and crop yields (Kihara, 2002; Muni, 2002).

Through a conservation tillage technology package that includes herbicides, improved seed and fertilizer, farmers can reduce the labour needed for ploughing and weeding, save time for other activities and expand the size of the cultivated area at little cost. In the long run, conservation tillage offers such advantages as increased organic matter in the soil, increased rainwater penetration and retention, prevention of soil erosion, soil fertility restoration and overall improvement of the physical soil structure (Ekboir et al., 2002).

Africa has the highest rate of population growth of 2.9% per year (Cleaver and Swiber, 1994). This results in increased demand for food production to cater to this population increase. However, the extra arable land to facilitate increased food production is not available, and hence leads to declining per capita food production of the African continent (World Bank, 1994). Many small-scale farmers who usually dominate in African farming lack the financial resources and incentives to purchase fertilizer, thus leading to decline of soil fertility. This leads to loss of soil nutrients that are removed mainly from harvesting crop products (Cooper et al., 1996). This has also resulted in the depletion of major soil nutrients, mainly nitrogen, phosphorus and potassium as well as the decline in the organic matter

content of the soil (Stoorvogel et al., 1993; Kapkiyai et al., 1998). This has contributed to the decline in per capita food production over the period from the 1960s to the 1990s to date (Ikombo et al., 1994).

Perennial crops are generally considered to improve soil structure and may help prevent land degradation. Conversion to conventional annual row cropping, however, may degrade soil structure and thereby negatively affect soil physical processes, crop growth potential and yield (Unger, 1975; Hermawan and Cameron, 1993). Use of conservation tillage is generally believed to reduce these negative aspects of row crop production by maintaining soil structure and limiting erosion.

Recent conferences (Tullberg and Hoogmoed, 2003; Wang and Gao, 2004) have illustrated large interest in conservation tillage. Yet, long-term research on conservation tillage has been carried out for at least 30 years, especially in the semi-arid and semi-humid regions with dry land farming, where the focus was crop production without supplemental irrigation. Several benefits from conservation tillage systems have been reported: (1) economical benefits (such as labour, energy, machinery cost and time saved) (Stonehouse, 1997; Uri et al., 1998; Uri, 2000), (2) positive effects from erosion protection and soil and water conservation and (3) increases in soil organic matter.

Much research has been carried out on surface water management in Kenya, on water management for dry land farmers, but there has been little impact on subsistence farmers. In semi-arid areas, moisture stress has been recognized as a major constraint to crop production, most of the times leading to famine relief (Rockstrom et al., 2001). Use of conservation tillage is generally believed to reduce these negative aspects of row crop production by maintaining soil structure and limiting erosion. The specific effects of conservation tillage practices on soil properties may nevertheless be contradictory due to variations in soil and environmental conditions. For example, yield difference between conventional and conservation tillage is known to vary greatly among soil types with fine-textured soils generally being less suitable for reduced and no-tillage (Cox et al., 1990).

On-farm research studies using conservation agriculture technologies in 2000–2002 in Eastern Province in Kenya showed potential to raise maize yields over three times, reduce labour by 75% and enhance soil fertility as a result of crop rotation. Further work in 2003–2004 building on the previous experiences on the same sites with cover crops planted between two rows of different-promising improved drought-tolerant maize varieties (WH 403, WS 202 and WS 103) and local variety also showed an increase in maize productivity compared to pure maize stand (with no cover crops). Results showed that when all plots were sub-soiled and cover crop managed, maize production increased to 4.5 t/ha during the short rain (Mwangi, 2003, 2004).

Conservation tillage has the potential of soil moisture retention and mitigation of intra-season dry spells that often result in low productivity and crop failure, especially in dry parts of East Africa. It also conserves available rainwater currently lost in the magnitude of 70–85% of rainfall from the cropping systems in sub-Saharan Africa, both through soil evaporation and through deep percolation and surface runoff (Rockstrom et al., 2001) and therefore makes it beneficial to the crops.

Materials and Methods

Site

Busia district lies between latitude $0°$ and $0°25'$ North and longitude $33°54'$ East. The altitude varies from 1,130 to 1,375 m above sea level. The area has bimodal rainfall distribution pattern, with long rains starting from March to May while short rains start from August to October. The annual rainfall ranges from 1,270 to 1,790 mm. The annual mean maximum temperature ranges from 26 to 30°C while the annual mean minimum temperature varies between 14 and 18°C. The soils are developed from various parent materials including intermediate and basic igneous rocks, sedimentary rocks and celevium. Most of the soils in the district are moderately deep, generally rocky and stony, consisting of well-drained red clays of low natural fertility (Republic of Kenya, 1997). The experimental site is located at Matayos location and division.

Experimental Treatments

The experiment recognized the N- and P-depleted status of the Busia agricultural soil (FURP, 1994). Thus, a blanket of 60 kg P/ha and two levels of N fertilizer, i.e. 0 and 60 kg N/ha, in the form of urea were applied. One-third of N was applied at planting time and two-thirds at knee height (maize) as top dressing. The experiment also compared the effects of the two tillage systems, namely conservation and conventional tillages on crop yields and soil properties. The experiment further compared the effects of cropping systems, i.e. continuous, intercropping and rotational cropping, in relation to maize and legume production. The conventional intercropping was that of soybean between maize rows. The plots were surrounded by border rows to which N and P were also applied.

Experimental Layout

The researcher-managed on-farm experiment of conservation tillage was conducted at Matayos site in Busia district in western Kenya. A $3×3×2$ factorial experiment with treatments applied in a randomized complete block design with three replicates was adopted. Two main plots (conservation and conventional tilled) of 18 m × 12 m were split into 12 m × 6 m plots (continuous, intercrop and rotational) and then subplots of 6 m × 6 m (with or without N fertilizer). Maize spacing was 75 cm × 25 cm inter-row and intra-row, respectively. Soybean spacing was 75 cm × 10 cm inter-row and intra-row, respectively. The N fertilizer was applied at 0 and 60 kg N/ha to come up with the optimal level, while basal application rate of 60 kg K/ha was applied in all plots. Since the soils of the area are highly P depleted, and according to the fertilizer use and recommendation by KARI 1994, a uniform amount of 60 kg P /ha was applied to all plots.

The main factors or plots were the cropping systems (cereal–legume intercropping, cereal–legume rotation and continuous cereal cropping). Tillage systems were subplots while sub-subplots were the fertilizer treatment.

Experiment Management

The experiment was set on an established farm where land preparation was done by hoe for conventional tillage and hand pulling of weeds for the conservation tillage. The organic material was applied before tilling to all plots. Striga-resistant maize IR hybrid with seeds coated with imazyphyrl chemical and soybeans (SB20) was done on the same day. The fertilizers were hill-placed at the time of sowing and N top dressing was done at maize knee height. The weeding was by hand pulling for conservation-tilled plots and by hoe for the conventionally tilled plots. Weeding was done thrice for the season. Soil sampling was done before and after the season to determine the effects of treatments. From each plot, a composite sample from eight sampling points taken at random (0–15 cm) was taken for laboratory analyses. The samples were air-dried and sieved through a 2-mm sieve for pH, particle size analysis and extractable P and bases analysis and 60 mesh soils for organic carbon and total N (Okalebo et al., 2002).

Plant samples were collected by first discarding two outer rows per plot and two plants at the end of each row giving a total effective area of 20.25 m^2 per plot. Total weight of unshelled maize cobs was taken and a sub-sample of cobs taken from big to small cob sizes proportionately, shelled and dried to obtain grain fresh and dry weights. For soybeans, 15 plants were taken as the plot sample determined fresh weight; the samples were shelled and dried to obtain fresh and dry grain weights. All plant tissues were ground for various analyses.

Statistical analysis was done by analysis of variance using the Statistical Analysis System (SAS Institute, 1994). Reference to statistical significance refers to a probability level of 0.05 unless otherwise noted.

Results and Discussion

Soil Carbon

In the first season, the percent (%) soil organic carbon varied significantly ($p \geq 0.05$) with treatment. Soil organic carbon accumulation in the soil was highest in the conservation tillage plots compared to the conventionally tilled plots (Table 1). Soil organic carbon in 0 N kg/ha fertilized soil was higher compared to 60 kg N/ha rate applied plots. Rotational cropping system gave the highest mean carbon accumulation in the soils, followed by continuous cropping system and then lastly the intercropping system. The interaction between the rotation cropping system by conservation tillage resulted in highest carbon accumulation at 60 kg N/ha rate. There is general increase of % carbon from the initial value of 1.44% C.

The lower level of organic carbon for conventional tillage was probably a result of high organic matter and its decomposition which is usually enhanced by disruption of soil aggregates (Hassink, 1995). This could have been enhanced by the removal of residue from the surface and mixing with the subsurface soil under conventional tillage compared to conservation tillage where residues are left on the surface, increasing organic matter inputs (Chivenge et al., 2004). The lower organic carbon for conventional tillage stimulates organic matter decomposition while conservation tillage enhances organic matter retention (Table 1) (Blair et al., 1997).

Soil pH

Soil pH had no significant difference as indicated in Table 2. This was attributed to the fact that no liming material was applied as part of treatments. However, minimal differences were observed between treatments, i.e. continuous cropping and conservation-tilled plots showed the highest value and the lowest was with rotational cropping and conventionally tilled plots. The overall pH value was higher than the initial value of 5.4.

Table 1 Effects of tillage systems and cropping systems on % soil organic carbon in a Busia soil

| | First season | | | | | |
| | 0 kg N/ha | | | 60 kg N/ha | | |
Cropping systems	NT	CT	Mean	NT	CT	Mean
Con	1.97[b]	1.63[b]	1.80	1.83[b]	1.56[a]	1.73
Rot	1.93[b]	1.72[b]	1.82	1.93[b]	1.57[a]	1.75
Int	1.93[b]	1.51[a]	1.72	1.72[b]	1.4[a]	1.56
Mean	1.95	1.71	1.78	1.89	1.51	1.71
Lsd$_{(0.05)}$	0.292					

NT, conservation tillage; CT, conventional tillage; Con, continuous cultivation; Rot, rotation cultivation; Int, intercropping
Means within each column followed by the same letter do not differ significantly ($p = 0.05$)

Table 2 Soil pH

| Cropping systems | Long rain season soil pH, 2005 | | | | | |
| | 0 kg N/ha | | | 60 kg N/ha | | |
	NT	CT	Mean	NT	CT	Mean
Con	5.8[a]	5.7[a]	5.8	5.9[a]	6.0[a]	6.0
Rot	5.7[a]	5.5[a]	5.6	5.8[a]	5.6[a]	5.7
Int	5.8[a]	5.6[a]	5.7	5.8[a]	5.9[a]	5.9
Lsd	0.6					

NT, conservation tillage; CT, conventional tillage; Con, continuous cultivation; Rot, rotation cultivation; Int, intercropping
Means within each column followed by the same letter do not differ significantly ($p = 0.05$)

Gravimetric Soil Moisture Content

Table 3 shows the effects of tillage systems, cropping systems and N application on gravimetric soil moisture content. In this season, the soil moisture (0–15 cm depth) content varied significantly between the tillage systems except in the continuous plots ($p \geq 0.05$). Conservation tillage registered the highest retained soil moisture content compared to conventional tillage. This could be probably due to the fact that the soil in the conventionally tilled plots was exposed to evaporation by the act of turning the subsoil through tilling. Also, the presence of the residue on the surface of the soil had mulching effects on the soil surface (Waddington et al., 2003) while in the conventionally tilled plots there was incorporation of residue into the soil which limits the residue to act as the mulch and hence more evaporation on the soil surface.

This agrees with the work done by Ohiri and Ezumah (1990) which reported increased soil drying rates and decreased water contents after tillage due to vapour movement being enhanced by increased macroporosity within the ploughed layer. There were significant differences among the cropping systems but

the trend of differences was that the highest moisture content was measured in the intercropped plots and at 60 kg N/ha fertilizer rate and the least moisture content was registered in the continuous cropping system plots and at 0 kg N/ha fertilizer rate of application. Probably the differences in the moisture content are caused by the amount of cover crop on the soil surface, thus reducing substantially the rate of evaporation. For instance, those plots of intercropping had a lot of biomass, thus giving shade effect on the soil surface and therefore leading to reduced evaporation rate unlike the plots of continuous cropping system. N fertilizer was also associated with increased biomass.

Effects of Tillage Systems, Cropping Systems and N Application on Maize and Soybean Grain Yield

Significant differences in crop yields were observed in both crops in different treatments across the cropping systems, tillage systems and the N levels (Table 4). At 0 kg N/ha, maize yield showed significant difference between rotation cropping system and both

Table 3 Effects of tillage systems, cropping systems and N application on gravimetric soil moisture content during the long rains of 2005

| Cropping systems | Gravimetric soil moisture content (%) | | | | | |
| | 0 kg N/ha | | | 60 kg N /ha | | |
	NT	CT	Mean	NT	CT	Mean
Con	14.5[a]	13.3[a]	13.9	16.4[c]	14.8[b]	15.6
Rot	14.9[b]	13.6[a]	14.3	16.4[c]	14.6[b]	15.5
Int	16.9[c]	14.4[a]	15.7	18.6[d]	16.2[c]	17.4
Mean	15.4	13.7	14.6	17.1	15.8	16.1
Lsd[(0.05)]	1.23					

NT, conservation tillage; CT, conventional tillage; Con, continuous cultivation; Rot, rotation cultivation; Int, intercropping
Means within each column followed by the same letter do not differ significantly ($p = 0.05$)

Table 4 The effects of tillage systems, cropping systems and N application on maize and soybeans grain yield (t/ha)

| | Maize yields (t/ha) | | | | | | Soybean yields (t/ha) | | | | | |
| | 0 kg N/ha | | | 60 kg N /ha | | | 0 kg N/ha | | | 60 kg N /ha | | |
Cropping systems	NT	CT	Mean	NT	CT	Mean	NT	CT	Mean	NT	CT	Mean
Con	1.26^a	1.00^a	1.13	2.89^d	2.24^c	2.57	–	–	–	–	–	
Rot	1.93^b	1.21^a	1.57	2.51^d	2.32^c	2.42	0.69^a	0.64^a	0.67	1.10^c	1.27^d	1.19
Int	1.02^a	0.98^a	1.00	2.18^c	1.62^b	1.90	0.70^a	0.61^a	0.66	1.07^c	1.19^d	1.13
Mean	1.40	1.06	1.23	2.53	2.06	2.30	0.70	0.63	0.67	1.09	1.23	1.16
$Lsd_{(0.05)}$	0.483						0.18					

NT, conservation tillage; CT, conventional tillage; Con, continuous cropping; Rot, rotational cropping; Int, intercropping; Lsd, least significant difference
Means within each column followed by the same letter do not differ significantly ($p = 0.05$)

intercropping and continuous systems. However, the highest yield was realized with the rotation, followed by the continuous and the intercrop systems. At 60 kg N/ha, the difference was significant between the tillage systems and cropping systems except between rotation and continuous cropping. Conservation-tilled plots collectively gave the higher yield mean than conventionally tilled plots and so did the use of nitrogen fertilizer at 60 kg N/ha. For soybean yield, there was no significant difference at 0 kg N/ha, neither between the cropping and tillage systems. However, at 60 kg N/ha, there was a significant difference between tillage systems, conservation-tilled plots being the highest. In both crops, the yields were better at 60 kg N/ha than at 0 kg N/ha and also with conservation tillage than with conventional tillage.

Nitrogen and Phosphorus Uptake by Maize and Soybean

The results revealed that nitrogen concentration in the grains of both maize and soybean differed significantly between the N treatment levels and between the tillage systems (Table 5). From the study it was noted that conservation-tilled plots with the 60 kg N/ha application had the highest N uptake in both crops while controls had the lowest. The combination which had the highest N uptake is conservation tillage, rotation and N at 60 kg N/ha with uptake of 80 kg N/ha and the one with the lowest was conventional tillage, 0 kg N/ha and continuous cropping for maize with uptake of 14.4 kg N/ha. Soybeans had highest N uptake of

Table 5 Nutrient uptake in both maize and soybeans in kg/ha

| Tillage system | N level | Cropping system | Maize | | Soybeans | |
			N uptake (kg N/ha)	P uptake (kg P/ha)	N uptake (kg N/ha)	P uptake (kg P/ha)
cons	0	Con	21.1^a	3.38^a		
cons	0	Rot	33.1^a	5.4^b	56.7^a	2.2^a
cons	0	Int	15.9^a	2.7^a	54.1^a	2.27^a
cons	60	Con	68.3^c	7.8^c		
cons	60	Rot	80.3^c	9.3^d	93.1^c	4.8^c
cons	60	Int	64.9^c	7.8^c	91.5^c	4.2^b
conv	0	Con	14.4^a	2.6^a		
conv	0	Rot	33.1^a	3.2^a	55.8^a	2.0^a
conv	0	Int	14.1^a	2.6^a	53.7^a	2.2^a
conv	60	Con	65.7^c	8.3^c		
conv	60	Rot	44.0^b	7.9^c	89.8^c	3.8^b
conv	60	Int	50.2^b	6.3^b	86.74^c	3.0^a
$Lsd_{(0.05)}$			24.17	2.14	27.8	1.4

cons, conservation tillage; conv, conventional tillage; Lsd, least significant difference; con, continuous cropping; rot, rotational cropping; int, intercropping
Means within each column followed by the same letter do not differ significantly ($p = 0.05$)

93.1 kg N/ha in conservation tillage, rotation and at 60 kg N/ha combination, and lowest of 53.7 kg N/ha in conventional tillage, 0 kg N/ha and rotational cropping combination (Table 5). The relatively high N uptake from the treatment could be attributed to the readily available N from urea. The highest N concentration in the grain in the non-tilled plots could have been attributed by the fact that there was reduced N leaching, hence little of the nitrogen was percolated down the soil profile in the form of nitrates in turn making it available to the plant (Di and Cameron, 2002). P uptake indicated the same trend as in the N uptake. The highest P uptake for maize grain was found to be with the combination of conservation tillage, rotational and at 60 kg N/ha of 9.3 kg P/ha and the lowest was conventional tillage, 0 kg N/ha and continuous cropping for maize of 2.6 kg P/ha. For the soybeans, the highest was observed to be 4.8 kg P/ha and the lowest was 2.0 kg P/ha. There was a significant difference in P uptake between the tillage systems, conservation tillage indicating to have the highest P uptake of 9.3 and 4.8 kg P/ha in both maize and soybeans grains, respectively. There was a significant difference in P uptake in both crops between the two levels of N application. This could be attributed to the P and N released after the decomposition of the residue making them available to the plant and this agrees with the work done by Gachengo (1996).

Conclusions and Recommendation

Crop yield under a tillage system is the integrated effect of changes in soil properties and associated plant response (Karunatilake and Schindelbeck, 2000). The highest average yield under the conservation tillage system suggests that surface shaping improves the performance of reduced tillage systems on these soils, as was also concluded by Cox et al. (1990). Arora et al. (1991) reported that better plant growth of conservation tillage maize occurred only in coarse-textured soils as a result of modified root growth. However, higher water losses from increased air circulation in the cloddy seedbeds under conventional tillage may be responsible for lower plant growth in the early growing season and in turn lower yield.

Conservation tillage emerges to be better than conventional tillage on both crop yield and soil improvement. In crop yields and nutrients quality i.e. phosphorus and nitrogen in the grain, conservation tillage not only saves labour cost but also increases production and improves the soil quality sustainably. Nitrogen application at 60 kg N/ha showed the best production over 0 kg N/ha. It is also observed that in order for other nutrients to work effectively and for residue decomposition there must be application of nitrogen to start off. This may not satisfactorily give the ultimate performance of conservation tillage, therefore I recommend the research can be tried for several seasons.

Acknowledgements This chapter acknowledges TSBF-CIAT for providing financial support through AfNet. The author also thanks Mr. John Mukhalama for his skilled assistance with the field activities and the laboratory staff of Department of Soil Science at Moi University.

References

Arora VK, Gajri PR, Prihar SS (1991) Tillage effects on corn in sandy soils in relation to water retentivity, nutrient and water management and seasonal evaporativity. Soil Tillage Res 21:1–21

Blair GJ, Singh BP, Till AR (1997) Development and use of a carbon management index to monitor changes in soil C pool size and turnover rate. In: Cadisch G, Giller KE (eds) Driven by nature: plant litter quality and decomposition. CAB International, Caen, p 114

CTIC (1996) Conservation technology information centre, CTIC Partners, April/May 1996, vol 14 N° 3

Chivenge PP, Murwira HK, Giller KE (2004) Tillage effects on soil organic carbon and nitrogen distribution in particle size fractions of a red clay soil profile in Zimbabwe. In: Bationo A (ed) Managing nutrient cycles to sustain soil fertility in sub-Saharan Africa. Academy, Nairobi, pp 113–126

Cleaver KM, Schreiber GA (1994) Reversing the spiral, the population, agriculture and environment nercus in sub-Saharan African world bank

Cooper PSM, Leaky RRB, Rao MR, Reynolds L (1996) Agroforestry and the mitigation of land. Degrading in the humid tropics of Africa. Exp Agric 32:235–290

Cox WJ, Zobel RW, van Es HM, Otis DJ (1990) Growth development and yield of maize under three tillage systems in the northeastern USA. Soil Tillage Res 18(2–3):107–310

Derpsch R (2001) Frontiers in conservation tillage and advances in conservation practice. In: Stott DE, Mohtar RH, Steinhardt GC (eds) Sustaining the global farm. Selected papers from the 10th international soil conservation organization meeting held May 24–29, 1999 at Purdue University and the USDA-ARS National Soil Erosion Research Laboratory, pp 248–254

Di HJ, Cameron KC (2002) Nitrate leaching in temperate agroecosystems; sources, factors and mitigating strategies. Nutr Cycling Agroecosyst 46:237–256

Ekboir J, Boa K, Dankyi AA (2002) Impacts of no-till technologies in Ghana. CIMMYT, Mexico, DF, pp 295–310

FAO (2002) The conservation agriculture working group activities 2000–2001. Food and Agriculture Organization of the United Nations, Rome, 25pp

Findlay JBR, Collins SC, Boa-Amponsen K, Mabuza S, Miheso V (2002) Conservation tillage programmes for small-holder farmers in Africa. In: American society of agronomy congress symposium on conservation agriculture for small-scale farmers in developing countries, Indianapolis, USA, 10–14 Nov

FURP (1994) Fertilizer use recommendation project. Fertilizer use recommendations volume II, Bungoma district. Kenya Agricultural Research Institute, Nairobi, 14 pp

Gachengo CN (1996) Phosphorus release and availability on addition of organic materials to phosphorus fixing soils. M.Sc. Thesis, Moi University, Eldoret Kenya

Hassink D (1995) Density fractions of soil macroorganic matter and microbial biomass as predictors of C and N mineralization. Soil Biol Biochem 27(8):1099–1108

Hermawan B, Cameron KC (1993) Structural changes in a silt loam under long-term conventional and minimum tillage. Soil Tillage Res 26:139–150

Ikombo BM, Esilaba AD, Kilewe AM (1994) A diagnostic survey of farming systems in Kiambu district Kenya. A report 1978 NARC Muguga, Karl

Kapkiyai JJ, Okalebo JR, Maritim HK (eds) (1998) Proceedings of the PREP phosphate rock explanatory project (PREP) workshop held at Moi University Chepkoilel Campus 12–15 May 1997, p 82

Karunatilake HM, Schindelbeck RR (2000) Soil and maize response to plow and no tillage after alfalfa-to-maize conversions on a clay loam soil in New York. Department of Crop and Soil Sciences, Cornell University, Ithaca, NY 14853-1901, USA.

Kihara FI (2002) Evaluation of rainwater harvesting systems in Laikipia District, Kenya. GHARP case study report. Greater horn of Africa rainwater partnership (GHARP), Kenya Rainwater Association, Nairobi, Kenya

Muni RK (2002) Evaluation of rainwater harvesting systems in Machakos District, Kenya. GHARP case study report. Greater Horn of Africa Rainwater Partnership (GHARP), Kenya Rainwater Association, Nairobi, Kenya

Mwangi HW (2003) Cover crops in conservation agriculture: a case study on D. lablab with farmer groups in Eastern Province

Mwangi HW (2004) Promoting soil moisture conservation through conservation agriculture and drought tolerant maize: Technical report

Ohiri AC, Ezumah HC (1990) Tillage effects on cassava (Manihot esculenta) production and some soil properties. Soil Tillage Res 17(3–4):221–229

Okalebo JR, Kenneth W, Woomer PL (2002) Laboratory methods of soil and plant analysis: a working manual, 2nd edn. TSBF_CIAT and SACRED Africa, Nairobi

Republic of Kenya (1997) Busia district development plan (1997–2001). Office of the vice president and minister of planning and national development. Government Printer, Nairobi

Rockstrom J, Barron J, Fox P (2001) Water productivity in rain-fed agriculture: challenges and opportunities for smallholder farmers in drought-prone tropical agro-systems. Paper presented at an IWMI workshop. Colombo, Sri Lanka, 12–14 Nov 2001

Rockstrom J, Johnson LO (1999) Conservation tillage systems for dry landform. On-farm research for extension experiences. East Afr Agric For J 65(2):101–114

SAS Institute (1994) SAS user's guide. SAS Institute, Cary, NC

Stonehouse DP (1997) Socio-economics of alternative tillage systems. Soil Tillage Res 43:109–130

Stoorvogel JJ, Smaling EMA, Jansen BH (1993) Calculating soil nutrient balances in Africa at different scales supranational scale. Fertil Res 35:227–235

Tullberg J, Hoogmoed WB (eds) (2003) Proceedings of the 16th international soil tillage research organization conference, CD-ROM, Brisbane, QLD, 13–18 Jul 2003

Unger PW (1975) Relationships between water retention, texture, density and organic matter content of west and south central Texas soils. Texas Agricultural Experiment Station Miscellaneous Publication, MP–1192C

Uri ND (2000) Perceptions on the use of no-till farming in production agriculture in the United States: an analysis of survey results. Agric Ecosyst Environ 77:263–266

Uri ND, Atwood D, Sanabria J (1998) The environmental benefits and costs of conservation tillage. Sci Total Environ 216:13–32

Waddington JM, Greenwood M, Petrone R, Price JS (2003) Mulch decomposition impedes recovery of net carbon sink function in a restored peatland. Ecol Eng 20:199–210

Wang ZC, Gao HW(eds) (2004) Conservation tillage and sustainable farming. Agricultural Science and Technology Press, Beijing, 560pp

World Bank (1994) Role of phosphorus in agriculture in feasibility of phosphate rock as a capital investment in sub-Sahara African issues and opportunity. World Bank Bay.1 IFA/MIGA, pp 01–37

Effect of Manure Application on Soil Nitrogen Availability to Intercropped Sorghum and Cowpea at Three Sites in Eastern Kenya

F. M. Kihanda and G.P. Warren

Abstract A trial was conducted where sorghum and cowpea were intercropped at three sites with different natural concentrations of soil phosphorus (1.0, 7.0 and 26.3 mg kg^{-1} Olsen-P) and with biannual cropping. There was a total of five field treatments, namely goat manure applied at rates of 0, 5 and 10 t ha^{-1} annually for 6 years and at 5 and 10 t ha^{-1} annually for 4 years followed by 2 years without manure. Continuous manuring at 10 t ha^{-1} created the same amount of soil N as 5 t ha^{-1} manure so the residual effects were the same for 5 and 10 t ha^{-1} manure and the effects were the same at all sites. In the season studied, between 1.8 and 4.1% of the native soil N was mineralized and taken up by the crops. Eleven percentage of the manure residual N (applied between 2 and 6 years previously and remaining in the soil) was taken up and this fraction was the same in all soils. Recently applied manure N contributed significantly to crop N in the most nutrient-deficient soil. On the soils with Olsen-P ≥ 7 mg kg^{-1}, cowpeas obtained a significant extra 32 kg N ha^{-1}, attributed to biological N fixation (BNF). If Olsen-P was < 6 mg kg^{-1}, BNF was negligible.

Keywords Intercropping · Manure · Nitrogen fixation · Phosphorus · Residual effect · *Sorghum bicolor* · *Vigna unguiculata*

F.M. Kihanda (✉)
Kenya Agricultural Research Institute (KARI), Embu Regional Research Centre, Embu, Kenya
e-mail: kihandafm@yahoo.com

Introduction

Intercropping of sorghum and cowpea gave up to 80% higher economic returns than sole crops in Nigeria (Andrews, 1972). Ofori and Stern (1986b) identified crop component density, plant arrangement, sowing time and applied nitrogen as important agronomic factors influencing productivity of cereal–legume intercropping systems. Despite the requirement for sufficient P for significant biological nitrogen fixation (BNF) by legumes (Giller and Wilson, 1991) and the obvious role of manure as a source of N and P, there has been little or no direct comparison of the effects of soil P or manure residual effects on BNF and the soil N cycle of the sorghum–cowpea system.

Nutrient inputs are required to sustain any cropping system where products are removed. Although much information is available on fertilizer requirements for sole cropping, relatively little is known about fertilizer requirements for intercropping with legumes in semi-arid climates (Bationo et al., 1991). In the better-studied maize–cowpea system, cowpeas were found to compete strongly for N (Wahua, 1983; Ofori and Stern, 1986a), reducing the yield advantage of intercropping, even though N was added via BNF (Eaglesham et al., 1981; Ofori et al., 1987). On the other hand, small additions of N increased yields of well-nodulated cowpeas, probably by helping to establish good root development (Kang et al., 1977). There is less information about manures as an alternative to fertilizer for intercropping, although manure application is one of the most effective ways of improving soil fertility in tropical African conditions (Watts-Padwick, 1983). Because of manure's multiple

A. Bationo et al. (eds.), *Innovations as Key to the Green Revolution in Africa*,
DOI 10.1007/978-90-481-2543-2_28, © Springer Science+Business Media B.V. 2011

benefits for soil chemical and physical properties, it is difficult to specify the most important benefit in any particular case.

Studies of N mineralization for organic additions in temperate climates indicate that in the "short" term (<1 year), a single-pool exponential model describes well the N release from organic material. But in the longer term, a multi-pool model is better, so Sluijsmans and Kolenbrander (1976) proposed that manure N should be considered in three components of different availability: the most recalcitrant could supply N for 20 years. Williams et al (1995) estimated that the annual breakdown of manure was in the ratio 50:40:10 over 3 years for Sahelian conditions and manure residual effects lasted three seasons in Botswana (Carter et al., 1992). But longer residual effects have been reported from other semi-arid experiments, such as over 9 years for millet and 13 years for cotton (Peat and Brown, 1962) in Tanzania, while residual manure increased grain yield, Olsen-P and soil organic C up to 9 years later in Kenya (Kihanda et al., 2005; Warren et al., 1997).

Thus, although the general benefits of intercropping and manure are well documented, Haque et al. (1995) concluded that further research is necessary to exploit the potential of crop–livestock systems for sustainable agriculture, particularly in the semi-arid tropics. These authors identified mineralization of N from organic matter, quantification of the effects of legume-based crop–livestock systems on soil and nutrient management and BNF by intercropped legumes as among the research priorities. The work described here was carried out to examine the following hypotheses:

- Manure supplies N to both sorghum and cowpea.
- The effectiveness of manure N is influenced by soil available P.
- Residues of earlier manure application supply available N.
- The effectiveness of residual manure N is influenced by soil properties.

Materials and Methods

Field Experiment

The sites were located in Mbeere and Tharaka-Nithi Districts and information on their location, climate and soil is given in Table 1. Rainfall was bimodally distributed with peaks in November and April and normally exceeding evaporation only in November and December. The soil classifications (WRB/FAO) were chromic Cambisol at Machang'a and Mutuobare and chromic Luvisol at Kajiampau. Machang'a and Mutuobare sites were cleared from native bush while Kajiampau had been cropped for some years.

Experimental work started in 1988 and the manure treatments described below commenced in 1989. The agronomic history to 1992 is given by Gibberd (1995a, b). Full agronomic history for Machang'a is given by Kihanda et al. (2005) and in the period under consideration the same treatments were applied at the other sites. In summary, there were five treatments. These were goat manure applied annually at 5 or 10 t ha^{-1} between 1989 and 1994 hereafter referred to as C1 and C2, respectively, and residual effects of goat's manure which had been applied at 5 and 10 t ha^{-1} between

Table 1 Characteristics of the sites and unmanured soils at Machang'a, Mutuobare and Kajiampau

	Machang'a	Mutuobare	Kajiampau
Latitude	0°47′S	0°47′S	0°20′S
Longitude	37°40′E	37°50′E	37°52′E
Altitude (m)	1050	900	750
Rainfall (mm)			
Annual	789[a]	803[b]	1040[c]
Oct. to Jan. season	437[a]	490[b]	683[c]
Oct. 1994 to Jan. 1995	439	616	862
pH (water)	6.55	6.84	7.10
pH (CaCl$_2$)	5.75	6.18	6.39
Olsen-P (mg kg^{-1})	1.0	26.3	7.0
Texture	Sandy clay loam	Sandy clay loam	Sandy loam

[a] 1989–2002 mean
[b] 1989–2000 mean
[c] 1989–1997 mean

1989 and 1992 hereafter referred as B1 and B2, respectively. The fifth treatment received no manure and was referred to as the control (C). The five treatments were arranged in a randomized complete block design, replicated three times within a site. The experiment was conducted in three sites, namely Machang'a, Mutuobare and Kajiampau. In the season under study here, the crops were (i) sorghum (*Sorghum bicolor,* var. 954066) planted in rows 70 cm apart at a spacing of 25 cm within rows and (ii) cowpea (*Vigna unguiculata,* var. M66) planted at the same spacings in rows placed midway between the sorghum rows in the second week of October. The cumulative amounts of this manure from 1989 to September 1994 were 30, 60, 20 and 40 t ha^{-1} in treatments A1, A2, B1 and B2, respectively. The amounts of N, P and K applied with 10 t ha^{-1} manure (treatment A2) applied in September 1994 were 188.7, 46.9 and 372 kg ha^{-1}, respectively, and half of these amounts were applied in treatment A1. In previous years, the annual nutrient rates were similar since manure was obtained from the same source.

Soil Sampling and Analysis

In September 1994, before cultivation, manuring and sowing, soil was sampled from the 0 to 20 cm horizon of each plot (as described by Kihanda et al., 2005) and air-dried. Soil total N (N_{soil}) was measured by the Kjeldahl method and expressed in kg N ha^{-1} using the bulk density measured at sampling. Extractable P was measured by Olsen's method (0.5 M NaHCO$_3$, pH 8.5).

Crop Harvests

Cowpeas were harvested in January 1995 and sorghum in February 1995. For each crop, the grains and above-ground residues (leaves, stalks and threshing residues) were collected separately. They were weighed and the residues were chopped into pieces >10 cm and mixed. Samples were finely ground and analysed by digestion in H$_2$SO$_4$/H$_2$O$_2$/Li$_2$SO$_4$ solution followed by automated colorimetric analysis for N and P. N taken up in grain plus above-ground residues was measured for N sorghum (N_{sg}) and N cowpea (N_{cp}).

Statistical Analyses

Initially, all data were analysed by one-way ANOVA or linear regression for each site individually. There was high variability for cowpea data at Machang'a, and this approach enabled some relevant differences to be identified. For N_{soil} and crop P/N data, similar effects were found at all sites, so the results were presented as a two-way ANOVA. Correlations were made between (i) N_{sg} or N_{cp} and (ii) N_{soil} with data for all sites pooled. Statistical calculations were made with INSTAT version 5.31 (Stern et al., 1990).

Results

Soil Total N

For each treatment, N_{soil} increased in the order Kajiampau < Machang'a < Mutuobare (Table 2). N_{soil} was always significantly more in the continuously manured soils (treatments A1 and A2) compared to the control soil (C). N_{soil} for the residual manure treatments (B1 and B2) was intermediate between treatments A and C, but the differences between treatments A and B were not large enough to be significant. The increases in N_{soil} caused by manure appeared larger at Mutuobare than at the other sites, but there was no significant interaction between manure treatments and sites, so it must be concluded that the treatments had the same effect at all sites.

Crop N Uptakes

Sorghum showed many significant responses to manure addition. Treatment A2 gave the highest N_{sg} and treatment C gave the lowest, always with a significant difference between them (Table 3). The other treatments gave intermediate results. Differences between residual and continuous manure treatments were largest at Machang'a, where continuously applied manure (treatments A1 and A2) gave higher N_{sg} than residual manure (treatments B1 and B2). At Mutuobare and Kajiampau, the differences between residual and continuous manure were not significant for N_{sg} (Table 3). Comparisons with treatment C

Table 2 Total soil N (%) for each site and treatment in September 1994

Treatment	Machang'a	Mutuobare	Kajiampau	Mean
C	0.067[a]	0.092[a]	0.059[a]	0.0724
B1	0.082[a, b]	0.121[b]	0.071[a, b]	0.0919
B2	0.088[a, b]	0.119[b]	0.067[a, b]	0.0913
A1	0.095[b]	0.116[b]	0.084[b]	0.0983
A2	0.094[b]	0.133[b]	0.087[b]	0.1047
ANOVA table				
Source of variation	d.f.	*F* ratio	Probability	SE
Site	2	38.9	0.000	0.0035
Treatment	4	7.0	0.000	0.0046
Site × treatment	8	0.5	0.837	0.0079
Error	30			

[a,b]Groups of results that are not significantly different ($P = 0.05$) within each site as identified by multiple *t*-tests

Table 3 Total nitrogen uptake in grain plus above-ground residues (kg N ha^{-1}), for sorghum and cowpea at each site and in each treatment with the standard error (SE, d.f. = 30) for a treatment

Treatment	N uptake (kg ha^{-1})		
	Machang'a	Mutuobare	Kajiampau
Sorghum			
C	22.4[a]	49.2[a]	32.0[a]
B1	62.9[b]	105.3[b, c]	49.0[a, b]
B2	59.4[b]	93.2[b]	74.6[b]
A1	113.2[c]	101.8[b, c]	68.0[b]
A2	128.4[c]	129.5[b, c]	78.2[b]
SE		*11.6*	
Cowpea			
C	4.8[a]	73.0[a]	50.3[a]
B1	22.4[a]	74.7[a]	48.1[a]
B2	11.9[a]	70.9[a]	52.2[a]
A1	7.9[a]	82.1[a]	62.7[a, b]
A2	28.9[a]	84.4[a]	69.9[b]
SE	*11.4*		*8.1*

[a, b, c, d]Groups of results that are not significantly different ($P = 0.05$) within each combination of soil and crop, identified by multiple *t*-tests

showed that residual manure significantly increased N_{sg} at Mutuobare and Kajiampau. Comparisons of the two continuous manure rates showed that treatment A2 gave higher N_{sg} at Mutuobare only. For the residual manure, the extra manure applied in B2 gave no increase in N_{sg}.

N_{cp} did not respond significantly to manure at Machang'a and Mutuobare (Table 3). At Kajiampau, continuous manuring (A1 and A2) gave higher N_{cp} than treatments C and B1, the lowest manure rates. At Machang'a, cowpea growth was very poor, with

N_{cp} being significantly less than at Kajiampau and Mutuobare for every treatment. Cowpea performed similarly at the latter two sites: grain yields were higher at Mutuobare but N_{cp} was not significantly more.

Grain yields were discussed by Kihanda et al. (2004) and showed almost the same pattern of responses.

Crop P/N Ratio

The crop P/N ratio was not altered significantly by the manure treatments at any site. There were highly significant differences between the sites in P/N ratios of every crop component (Table 4) and Machang'a always gave the lowest P/N ratio.

Correlations Between N_{sg}, N_{cp} and N_{soil}

The correlation between N_{sg} and N_{soil} increased in the order Machang'a ($r = 0.628^*$) < Mutuobare

Table 4 Mean P/N ratio over all manure treatments for grains and crop residues of intercropped sorghum and cowpea at each site

	Sorghum		Cowpea	
	Grain	Residues	Grain	Residues
Machang'a	0.124	0.062	0.061	0.067
Mutuobare	0.187	0.166	0.110	0.216
Kajiampau	0.256	0.213	0.119	0.103
SE (d.f. = 34)	0.0097	0.0126	0.0044	0.0098

($r = 0.657$**) < Kajiampau ($r = 0.760$***). The data for N_{sg}, N_{cp} and N_{soil} were then pooled over all sites. N_{sg} was significantly correlated with N_{soil} (Fig. 1). No significant improvement in overall joint fit was found if a separate line was allowed for any one site. The following equation sufficed to describe all the results (SE of each parameter, d.f. = 43):

$$N_{sg} = -9.9\ (13.7) + 0.0261\ (0.0060^{***})\ N_{soil}\ \text{(Variance accounted for} = 30.5\%).$$

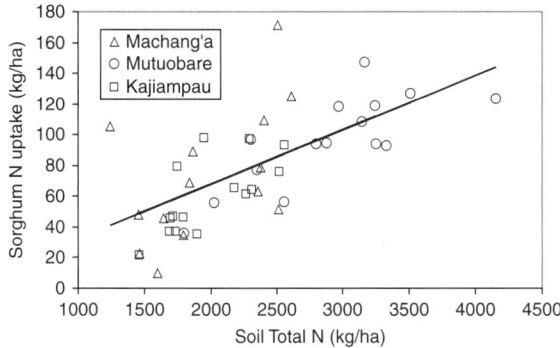

Fig. 1 Relationship between N uptake by intercropped sorghum (N_{sg}, kg N ha^{-1}) and soil total N (N_{soil}, kg N ha^{-1}) over all manure treatments and sites

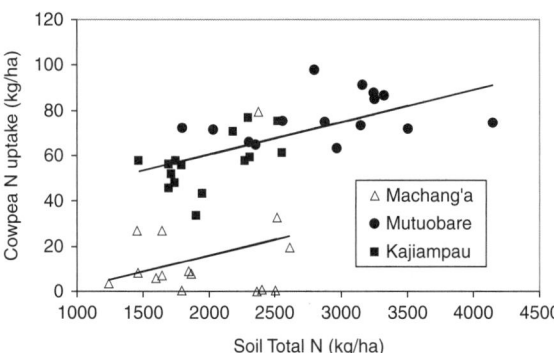

Fig. 2 Relationships between N uptake by cowpea (N_{cp}, kg N ha^{-1}) and soil total N (N_{soil}, kg N ha^{-1}) for (i) Machang'a site and (ii) Mutuobare and Kajiampau sites

For the cowpea data, the Machang'a results formed a distinct separate group (Fig. 2). The best fit to the data was obtained by the joint fitting of two parallel lines with separate intercepts for (i) Machang'a and (ii) Mutuobare plus Kajiampau (variance accounted for = 75.1%): Machang'a:

$$N_{cp} = -12.7\ (8.2) + 0.0143\ (0.0037^{***})\ N_{soil}$$

Mutuobare and Kajiampau:

$$N_{cp} = 32.0\ (6.1^{***}) + 0.0143\ (0.0037^{**})\ N_{soil}$$

Thus, at Mutuobare and Kajiampau only, cowpeas obtained a constant and highly significant extra 32 kg ha^{-1} of N in addition to the supply of N correlated with N_{soil}. As discussed below, this was attributed to BNF, and on that assumption, estimates were made of the contributions to N uptake from various sources (Table 5).

It can be argued that the above results are incorrect because the data include plots where the soil was sampled and analysed before the October 1994 manure application that must influence the subsequent crop N uptakes. Therefore, the above regression analyses were repeated excluding treatments A1 and A2. The fitted parameters were not significantly different for either sorghum (parameters: –21.0 and 0.0390) or cowpea (parameters: –11.8 or 31.3 and 0.0136), so the equations given above are considered appropriate.

Table 5 Estimates of the amounts of N (kg ha^{-1}) taken up from different sources by sorghum plus cowpea at each site, with standard error (d.f. = 30)

Source	Machang'a	Mutuobare	Kajiampau	SE (d.f. = 30)
Native soil	27.2	90.2	50.4	10.5
Biological nitrogen fixation	0.0	32.0	32.0	6.1
Residues of manure (1989–1992)	51.1	49.9	29.6	14.9
Recent manure (1993 and 1994), A1	42.8	11.8	18.7	14.9
Recent manure (1993 and 1994), A2	79.0	41.8	36.1	14.9

The contribution to the total N taken up by both crops from the native soil N was estimated from the N uptake from treatment C minus BNF. The contribution of manure residues (manure added in 1989–1992) to crop N (Table 5) could be estimated by the mean of the differences B1 – C and B2 – C (Table 3), because there were no significant differences between treatments B1 and B2. Similarly, the amount of manure residual N contained in N_{soil} was estimated by the mean of the differences B1 – C and B2 – C. Recently applied manure (in 1993 and 1994) was assumed to supply the remainder of the crop requirement and calculated from the differences A1 – B1 and A2 – B2.

Discussion

Differences Between Sites

Highly significant differences between sites were found in crop P/N ratios (Table 4), showing that the balance of available N and P differed between sites. The following observations showed that at Machang'a, P was more limiting than at the other sites: (i) the crop P/N ratios were significantly lower than at the other sites (Table 4) and (ii) without fresh manure (treatments C, B1 and B2), there was no significant difference between Machang'a and Kajiampau in N_{sg}, although N_{soil} was higher at Machang'a (Table 2), i.e. to benefit from the higher N at Machang'a, it was essential that more P be added. The limitation of crop growth by N deficiency was most pronounced at Kajiampau, as indicated by the following evidence: (i) the highest sorghum P/N ratios were at Kajiampau, (ii) under continuous manuring, N_{sg} was lowest for Kajiampau and (iii) for every manure treatment, Kajiampau had the lowest N_{soil} (Table 2). Furthermore, measurements of soil nitrate during the season indicated shortage at Kajiampau and surplus at Machang'a (Kihanda et al., 2004). Thus the principal benefit of manure was through P supply at Machang'a and N supply at Kajiampau.

Sources of N to Crops

The sources of available N were native soil N, residues of manure added in previous years (1989–1992),

manure applied recently (1993 and 1994) and biological N fixation (BNF) for cowpeas, plus other small sources such as atmospheric deposition, non-symbiotic N fixation and uptake from the subsoil. Recent and residual manure were clearly effective in supplying N to sorghum, because they increased grain yield and N_{sg} significantly in most cases (Table 3). Results for cowpea were less clear. Yield and N_{cp} were increased by the continuous manure treatments (A1 and A2) at Kajiampau (Table 3), the most N-deficient site, but not at Machang'a and Mutuobare. However, there was a significant general correlation of N_{cp} with N_{soil} over all sites. Since manure applied before September 1994 was the source of within-site differences in soil N, this result indicates that a part of the stabilized residues of manure N was available to cowpeas.

Biological Nitrogen Fixation

We suggest that the contribution of cowpea BNF can be estimated from the correlation equations between N_{cp} and N_{soil}. Cowpeas obtained a constant and significant extra 32 kg N ha^{-1} at Mutuobare and Kajiampau, but not at Machang'a. We attribute this to BNF because, if N_{soil} was the only source of available N, the intercepts of the relationships of N_{sg} and N_{cp} with N_{soil} should be close to zero. The non-significant intercept for sorghum (non-N fixing) at all sites (Fig. 2) agreed with this hypothesis, suggesting that other sources such as deposition and sub-soil N were small (BNF being impossible). The intercept would be greater than zero only if there were available N sources that were independent of both soil organic matter and treatment differences between plots. BNF is the only source that is available to cowpeas but not to sorghum, so we suggest that BNF contributed 32 kg N ha^{-1} at Mutuobare and Kajiampau only.

There was a clear difference in Olsen-P between Machang'a and the other two sites. BNF cannot take place without sufficient P (Giller and Wilson, 1991) and the lack of BNF by cowpea is in agreement with the low Olsen-P values at Machang'a, where the highest Olsen-P was 5.3 mg kg^{-1} in treatment A2. In contrast, the lowest Olsen-P at Kajiampau and Mutuobare was 7.0 mg kg^{-1} (Kajiampau treatment C). These data suggest that there is a critical value of Olsen-P, between 5.3 and 7.0 mg kg^{-1}, below which cowpea BNF does not take place.

The likely range of BNF values was compared with other results in Kenya. BNF was measured with the same cowpea variety in sole-crop experiments at nearby farmers' fields by both the difference method and ^{15}N labelling of soil N (Irungu et al., 2002). Measured BNF was about 4 kg N ha^{-1} (19% of N_{cp}) in soil without P fertilizer and about 10 kg ha^{-1} (30% of N_{cp}) in soil with P fertilizer at 20 kg P ha^{-1}. At Kiboko, Pilbeam et al. (1995) found BNF of 187 kg ha^{-1} (50% of N_{cp}) with P fertilizer at 40 kg P ha^{-1}. These results suggest that the higher the soil available P, the higher the proportion of N_{cp} supplied by BNF. An approximate range for BNF was estimated by multiplying N_{cp} by a percentage contribution from BNF observed in the work described by Irungu et al. (2002) and Pilbeam et al. (1995). The range was from 14.4 kg ha^{-1} (low values: Kajiampau treatment B1 $\times$ 30% from BNF) to 42.2 kg ha^{-1} (high values: Mutuobare treatment A2 $\times$ 50% from BNF). The range is in agreement with the significant (± 2 SE, $P = 0.05$) range of the regression intercept estimate described above, i.e. 19.2–44.8 kg ha^{-1}. The range is somewhat lower than the results for maize–cowpea intercrops, such as 81 kg ha^{-1} (Eaglesham et al., 1981) or 73–59 kg ha^{-1}, with and without fertilizer, respectively (Ofori et al., 1987). However, our results are consistent with the hypothesis that cowpeas fix similar amounts of N either as a sole crop or when intercropped with sorghum. Lower N fixation compared to the maize–cowpea system may be a consequence of local climate.

BNF is a key component for sustainability of many cropping systems, but its direct measurement is especially difficult under intercropping. The "difference" method depends on comparison of the N uptakes by N fixing and non-fixing crops with similar rooting systems. Methods using isotopic labelling of the soil N also depend on the assumption that root systems of both legume and non-fixing reference crop exploit equally the soil mineral N (Peoples et al., 1989). Under cereal/legume intercropping, the root systems of the partner crops are highly likely to be different in spatial distribution, so an essential assumption of both methods is violated. Alternative approaches to investigation of BNF under intercropping are therefore useful. The result obtained here is important because it does not depend on the assumption of similar rooting systems, but the value obtained is in broad agreement with other results. It would have been good to confirm the results in additional seasons, but the data presented here are the only complete set because of missing samples in some seasons and crop failure in others. However, in a subsequent season with complete data for Machang'a and Kajiampau sites, the cowpea data again fitted the model of two parallel lines, confirming the main results.

Native Soil N

The supplies from native soil N as a percentage of N_{soil} were 1.84, 4.08 and 2.88% for Machang'a, Mutuobare and Kajiampau, respectively, in approximate agreement with the general observation that between 1 and 3% of soil N can be mineralized in one season (Kihanda, 1996). It is suggested that the high value for Mutuobare was because soil organic matter in treatment C was not at a steady state and declining more quickly than at the other sites. Mutuobare soil was clearly the most fertile, as shown by the yields, N uptakes, Olsen-P and N_{soil}. It had been cleared from a dense thicket and was therefore undergoing the decline in SOM normally observed when tropical forest is cleared (Mueller-Harvey et al., 1985). In contrast, Kajiampau had been cropped for some years, while the natural vegetation at Machang'a was relatively sparse, indicating that at these sites soil organic matter was more depleted and resistant to mineralization.

Residual Manure N

The uptakes by sorghum plus cowpea of manure residual N were 51.1, 49.9 and 29.6 kg ha^{-1} at Machang'a, Mutuobare and Kajiampau, respectively (Table 5) and these values were not significantly different. This shows that in the fifth season after application, manure residues remained a significant source of N in all soils, but there was no difference between the soils. When expressed as a proportion of manure residual N in the soil, these uptakes were 17.1, 6.8 and 9.5%, respectively, but there was no significant difference between soils and the mean was 11.1%. For tropical agriculture, there are many reports of manure residual effects in the year following application, but specific information is scarce on subsequent years and apparently none on the N mineralization component. Our mean value for decomposition of SOM N derived from manure residues is for one season only, but it agrees with an annual mineralization fraction of 19.3% found by Mueller-Harvey et al. (1985) for forest SOM. If the mineralized fraction in each season is constant, this also agrees with the proposal of Sluijsmans and Kolenbrander (1976) that the most recalcitrant fraction could supply N for 20 years because it would take 20 years to reduce the residual N to 1% of that applied. Our results were statistically identical at three sites so we favour the idea that residual effects for N applied as manure are long lasting in the semi-arid climate and propose that 11% is mineralized each season.

Recent Manure N

The balance of total crop N was supplied from the recently applied manure. Its contribution to total crop N uptake was in the range 14–49% of the N applied in manure applied in 1993 and 1994. By comparison, 20–40% of N was mineralized from other farmyard manures (C:N<17) taken from the Embu region in incubation and pot experiments (Kihanda, 1996). A drawback of manures is their variable nutrient content, which makes it difficult to predict their effectiveness. Mineralization of N is likely in manure of C:N<15; immobilization for C:N>15 (Kirchmann, 1985; Castellanos and Pratt, 1981). The manure used here had C:N = 13.4 and so N mineralization appeared likely and was then found to be significant in field conditions. Although the primary constraint to crop nutrition at Machang'a was P shortage, the amount of N from recent manure was significantly higher than in the other soils. About twice as much recent manure N was taken up at Machang'a compared to the other two sites (Table 5), perhaps compensating for the lack of BNF. Thus, the uptake of recent manure N depended on pre-existing soil fertility.

Conclusions

In respect of the four hypotheses proposed, the following conclusions are made:

- Manure supplied N to both sorghum and cowpea.
- The effectiveness of recently applied manure N was influenced by soil properties. The contribution of

manure to available N was higher in soil with low available soil P.
- Residues of earlier manure application supplied available N in all soils. Manure had significant effects on N availability in the fifth season after the last application. Thus, by applying manure on different small areas each year labour may be saved.
- The amount of N supplied from residual manure did not differ between soils.

Furthermore, the results showed that for inter-cropped cowpea, BNF was negligible when Olsen-P was less than 5.3 mg kg^{-1}, and when Olsen-P was more than 7 mg kg^{-1}. BNF provided 32 kg N ha^{-1}, 38–67% of the cowpea N requirement. BNF provides only a part of the cowpea N requirement and cowpeas utilize mineralized soil N even under conditions favourable to N fixation.

Acknowledgements We thank the directors of KARI Embu Regional Research Centre and the National Agricultural Research Laboratories (NARL), Nairobi, for provision of field and laboratory facilities, and PI Mutwiri, P Njoroge and J Odhiambo for maintaining the field experiments and technical assistance. This publication is an output of research funded by the Department for International Development (DFID) of the UK Government, but the DFID accepts no responsibility for any information provided or views expressed.

References

Andrews DJ (1972) Intercropping with sorghum in Nigeria. Exp Agric 8:139–150

Bationo A, Ndunguru BJ, Ntare BR, Christianson CB, Mokwunye AU (1991) Fertilizer management strategies for legume based cropping systems in the West African semi-arid tropics. In: Johansen C, Lee KK, Sahrawat KL (eds) Phosphorus nutrition of grain legumes in the semi-arid tropics. ICRISAT, Patancheru, pp 213–226

Carter DC, Harris D, Youngquist JB, Persaud N (1992) Soil properties, crop water use and cereal yields in Botswana after additions of mulch and manure. Field Crops Res 30:97–109

Castellanos JZ, Pratt PF (1981) Mineralization of manure nitrogen: correlation with laboratory indexes. Soil Sci Soc Am J 45:354–357

Eaglesham ARJ, Ayanaba A, Rao VR, Eskew DL (1981) Improving the nitrogen nutrition of maize by intercropping with cowpea. Soil Biol Biochem 13:169–171

Gibberd V (1995a) A farmer-friendly research project in semi-arid Kenya. Tropical Sci 35:308–320

Gibberd V (1995b) Yield responses of food crops to animal manure in semi-arid Kenya. Tropical Sci 35:418–426

Giller KE, Wilson KJ (1991) Nitrogen fixation in tropical cropping systems. CAB International, Wallingford, CT

Haque I, Powell JM, Ehui SK (1995) Improved crop-livestock production strategies for sustainable soil management in tropical Africa. In: Lal R, Stewart BA (eds) Soil management: experimental basis for sustainability and environmental quality. CRC Press, Boca Raton, FL, pp 293–345

Irungu JW, Wood M, Okalebo JR, Warren GP (2002) Enhancement of the biological nitrogen input in smallholder farms of semi-arid Kenya through application of phosphorus fertiliser. In: Karanja NK, Kahindi JHP (eds) Challenges and imperatives for biological nitrogen fixation research and application in Africa for the 21st century. African Association for Biological Nitrogen Fixation, Nairobi, pp 88–91

Kang BT, Nangju D, Ayanaba A (1977) Effects of fertilizer on cowpea nodulation and nitrogen fixation in the lowland tropics. In: Ayanaba A, Dart PJ (eds) Biological nitrogen fixation in farming systems of the tropics. Wiley-Interscience, Chichester, pp 205–216

Kihanda FM (1996) The role of farmyard manure in improving maize production in the sub-humid highlands of central Kenya. PhD thesis, University of Reading, Reading

Kihanda FM, Warren GP, Atwal SS (2004) The influence of goat manure application on crop yield and soil nitrate variations in semi-arid eastern Kenya. In: Bationo A (ed) Managing nutrient cycles to sustain soil fertility in Sub-Saharan Africa. Academy Science Publishers in association with the Tropical Soil Biology and Fertility Institute of CIAT, Nairobi, pp 173–186

Kihanda FM, Warren GP, Micheni AN (2005) Effects of manure and fertilizer on grain yield, soil carbon and phosphorus in a 13-year field trial in semi-arid Kenya. Exp Agric 41: 389–412

Kirchmann H (1985) Losses, plant uptake and utilization of manure nitrogen during a production cycle. Acta Agric Scand Suppl 24. 69, 63–69

Mueller-Harvey I, Jou ASR, Wild A (1985) Soil organic C, N, S and P after forest clearance in Nigeria: mineralization rates and spatial variability. J Soil Soc 36:585–591

Ofori F, Pate JS, Stern WR (1987) Evaluation of N_2-fixation and nitrogen economy of a maize/cowpea intercrop system using ^{15}N dilution methods. Plant Soil 102: 149–160

Ofori F, Stern WR (1986a) Maize/cowpea intercrop system: effect of nitrogen fertilizer on productivity and efficiency. Field Crops Res 14:247–261

Ofori F, Stern WR (1986b) Cereal-legume intercropping systems. Adv Agron 41:41–90

Peat JE, Brown KJ (1962) The yield responses of rain-grown cotton, at Ukiriguru in the Lake Province of Tanzania. II. Land resting and other rotational treatments contrasted with the use of organic manure and inorganic fertilizers. Empire J Exp Agric 30:305–314

Peoples MB, Faizah AW, Rerkasem B, Herridge DF (eds) (1989) Methods for evaluating nitrogen fixation by nodulated legumes in the field. ACIAR, Canberra, ACT

Pilbeam CJ, Wood M, Mugane PG (1995) Nitrogen use in maize-grain legume cropping systems in semi-arid Kenya. Biol Fertil Soils 20:57–62

Sluijsmans CMJ, Kolenbrander GJ (1976) The nitrogen effect of farmyard manure in the short and the long term. Stikstof 7(83/84):349–354

Stern RD, Knock J, Burn RW (1990) INSTAT introductory guide. Statistical services centre, University of Reading, Reading

Wahua TAT (1983) Nitrogen uptake by intercropped maize and cowpeas and a concept of nutrient supplementation index (NSI). Exp Agric 19:263–275

Warren GP, Muthamia J, Irungu JW (1997) Soil fertility improvements under manuring in semi-arid lower Embu and Tharaka-Nithi. In: Fungoh PO, Mbadi GCO (eds) Focus on agricultural research for sustainable development in a changing economic environment. Proceedings of the 5th KARI Scientific Conference, 14 to 16 October 1996, Nairobi. Kenya Agricultural Research Institute, PO Box 57811, Nairobi, pp 151–163. ISBN 9966-879-21-8

Watts-Padwick G (1983) Fifty years of experimental agriculture. II. The maintenance of soil fertility in tropical Africa: a review. Exp Agric 19:293–310

Williams TO, Powell JM, Fernandez-Rivera S (1995) Manure utilization, drought cycles and herd dynamics in the Sahel: implications for cropland productivity. In: Powell JM, Fernandez-Rivera S, Williams O, Renard C (eds) Livestock and sustainable nutrient cycling in mixed farming systems of sub-Saharan Africa. Vol II: Technical Papers. Proceedings of an international conference held in Addis Ababa, Ethiopia, 22–26 Nov 1993. International Livestock Centre for Africa, Addis Ababa

The Effect of Organic-Based Nutrient Management Strategies on Soil Nutrient Availability and Maize Performance in Njoro, Kenya

J.J. Lelei, R.N. Onwonga, and B. Freyer

Abstract A field experiment based on the concept of organic nutrient management was conducted in Njoro, Kenya, to test the effect of improved legume fallows, crotalaria (CR), lablab (LB), garden pea (GP) and natural fallow (NF, as control) on available soil N and P and maize performance. The experimental design was a split plot fitted to a randomized complete block. The main plots were two cropping systems involving the improved legume fallows and NF preceding sole maize and maize–bean (M/B) intercrop. The sub-plots were two residue management types: residue incorporation and residue removal with farm yard manure (FYM) incorporated instead. Residue incorporation resulted in higher concentrations of N and P in soil than FYM in both cropping systems. Under sole maize, grain yield following LB was significantly higher than after CR, GP and NF. In the M/B intercrop, maize grain yield following LB was significantly higher than after GP and NF, with no significant differences in yields following CR and LB. Maize dry matter (DM) yields followed a similar trend. Overall, maize grain and DM yields were higher in sole maize cropping system than in M/B intercrop but an additional 0.5–0.6 kg ha^{-1} of bean grain yield was realized in the latter cropping system. The improved fallow legumes enhanced soil productivity, besides the seeds providing ancillary protein in diet locally, with resultant higher yields of the succeeding crop.

Keywords Biological nitrogen fixation · Farm yard manure · Improved legume fallow · Minjingu rock phosphate · Residue management

Introduction

Maize is the most important agricultural commodity and main staple food in Kenya (Kandji et al., 2003). The annual maize production is 2.6 million metric tonnes and it provides about 40% of the populations' caloric requirements (Hedin et al., 1998). Due to the ever growing human population, coupled with the declining soil fertility, demand for maize continues to outstrip supply (CBS, 1996). Nitrogen is undoubtedly the major nutrient limiting cereal growth (Snapp et al., 1998). Maize removes about 40 kg N ha^{-1} to produce 2–2.5 t of grain yield per hectare in the tropics (Mathews et al., 1992). Phosphorus is also a limiting nutrient in maize production due to the low native soil P and high P fixation by iron and aluminium oxides (Kwabiah et al., 2003). Nitrogen deficiency can be ameliorated through application of inorganic and organic fertilizers and biological nitrogen fixation (Hartemik et al., 2000). Phosphorus, unlike N, cannot be replenished through biological fixation. For many cropping systems in the tropics, application of P from organic and inorganic sources is essential to maximize and sustain high crop yield potential in continuous cultivation systems.

In most sub-Saharan African countries, fertilizer is not readily available and when available the cost is often a limiting factor to small-scale resource-poor farmers. Annual fertilizer applications on African agricultural soils are only 9 kg ha^{-1} compared to 83 kg

J.J. Lelei (✉)
Department of Crops, Horticulture and Soils, Egerton University, Njoro, Kenya
e-mail: Joycendemo@yahoo.com

A. Bationo et al. (eds.), *Innovations as Key to the Green Revolution in Africa*,
DOI 10.1007/978-90-481-2543-2_29, © Springer Science+Business Media B.V. 2011

ha^{-1} elsewhere in the developing world (Reardon et al., 2001). In Kenya, for instance, the estimated N and P fertilizer use was only 6 and 3 kg ha^{-1}year^{-1}, respectively, in 1981 (Smaling et al., 1992). In light of the aforementioned, other sustainable alternatives of enhancing soil fertility management become imperative in ensuring continued crop production and consequently food security.

Organic nutrient management (ONM), based on biodegradable material, is one such alternative. The ONM practices include the use of animal manure and compost, incorporation of crop residues, crop rotation, intercropping with legumes, green manuring, natural fallowing and improved fallows (Place et al., 2003). As a result of increasing human population, the ecosystem fertility functions of traditional fallows must now be improved via the use of managed fallows (Noordwijk, 1999). Short-term improved fallow technology is characterized by deliberate planting of fast growing nitrogen fixing legume species in rotation with cereal crops (Niang et al., 2002).

Efficient legume growth and nitrogen fixation are highly dependent on an adequate supply of phosphorus (Chiezey et al., 1992). To supplement P, low P release organic fertilizers such as Minjingu rock phosphate (MRP) can be applied. In Kenya, several studies conducted on the effectiveness and efficiency of MRP (Okalebo and Nandwa, 1997; Lelei, 2004) have revealed that it is on average 65% as effective as processed P fertilizers. MRP costs about 50% of processed P fertilizers, on elemental P basis, and hence makes it suitable for use under organic-based nutrient management systems.

The objective of this study was to determine the effect of organic-based nutrient management strategies involving improved legume fallows and residue management types in a legume–cereal cropping system on available soil N and P contents and maize performance.

Materials and Methods

Site Description

The study was conducted in Field 7 station of the Department of Agronomy, Egerton University, Njoro (longitude 35°23′ and 35°36′ East and latitude 0°13′ and 1°10′ South) during the short rain season (SRS) and long rain season (LRS) of the years 2003 and 2004, respectively. The LRS occurs between March/April to August and SRS from September/October to December with peaks in April and November, respectively. The average annual temperature is 15.9°C (Jaetzold and Schmidt, 1983). The total annual rainfall and mean temperature during the experimental period were 1443.9 mm and 19.6°C, respectively (Fig. 1). The soils are well drained, dark reddish in colour and classified as mollic Phaeozems (FAO/UNESCO, 1990).

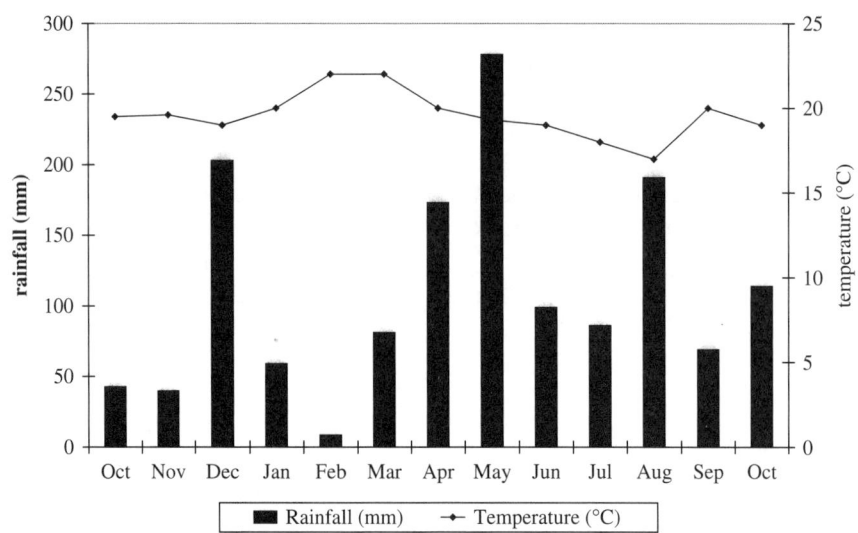

Fig. 1 Total monthly rainfall (mm) and mean temperature (°C) during the experimental period

Table 1 Selected chemical and physical properties of the top soil layer

| pH (H$_2$O) | Org C (g kg^{-1}) | Total N (g kg^{-1}) | Olsen P (mg kg^{-1}) | Exchangeable bases (cmol kg^{-1}) | | | Texture (%) | | | Textural class |
				Ca	Mg	K	Sand	Silt	Clay	
6.8	15	3.2	6.29	3.8	0.87	1.12	45	26	29	Clay

The analysed chemical and physical characteristics of the top (0–0.2 m) soil layer prior to the commencement of the experiment (Table 1) were neutral in pH, moderate in organic C, high in total N, low in Olsen extractable P and exchangeable bases and clay in texture (Okalebo et al., 1993).

Experimental Design and Field Practices

Treatments and Experimental Design

The treatments were improved fallow legumes: crotalaria (CR), lablab (LB), garden pea (GP) and natural fallow (NF, as control) with maize as the test crop. The experimental design was a split plot fitted to a randomized complete block. The main plots were two cropping systems involving improved legume fallows and NF preceding (i) sole maize (M): NF-M, CR-M, LB-M and GP-M and (ii) maize–bean (M/B) intercrop: NF-M/B, CR-M/B, LB-M/B and GP-M/B. The sub-plots were two residue management types: residue incorporation and residue removal with manure (FYM) incorporated instead. The treatments were replicated four times.

Field Practices

The experimental field was previously under a 2-month weedy fallow. It was cleared of vegetation, tilled and raked using hand tools prior to the experimental setup. MRP, at the rate of 290 kg ha^{-1} (40 kg P ha^{-1}), was broadcasted on all plots and mixed well with the top soil, prior to planting of the legumes. The legumes were planted on 10 October 2003 during the SRS at a spacing of 60 cm $\times$ 30 cm within rows. Two seeds were planted per hill.

To estimate biological nitrogen fixation (BNF), a reference crop, barley, was sown at a rate of 80 kg ha^{-1} in furrows spaced 20 cm apart. The amounts of nitrogen fixed by LB, CR, GP and NF weeds were then estimated using the difference method (Hauser, 1987):

$$\text{BNF (kg ha}^{-1}) = (\text{shoot N}_{\text{leg}} + \text{root N}_{\text{leg}}) \, \text{kg ha}^{-1}$$
$$- (\text{shoot N}_{\text{ref}} + \text{root N}_{\text{ref}}) \, \text{kg ha}^{-1}$$
$$+ (\text{N}_{\text{in}} \text{ in soil} - \text{N}_{\text{in}} \text{ in soil}_{\text{ref}}) \, \text{kg ha}^{-1}$$

where leg = legume; ref = reference crop; in = mineral N.

This method assumes that the uptake of soil-derived N is the same in the legume and reference crop and hence barley, a crop with similar N uptake characteristics as legumes, was chosen (Jensen, 1986; Reining, 2005).

Immediately after grain harvest, the aboveground residues of all crops were either completely removed and FYM added, at the rate of 5 t ha^{-1}, instead (to mimic the farmers' practice of recycling legume residue through livestock as FYM) or chopped into 5–20 cm pieces, spread across plots and incorporated during land preparation for the subsequent maize crop. Weed biomass in NF plots was handled similarly as the legume residues.

All plots were planted with H624 maize at the beginning of the LRS of 2004 at a spacing of 75 cm $\times$ 30 cm within hills. In the M/B intercrop, beans were sown between two rows of maize. Weeds were controlled by hand hoeing three times during the 2004 LRS.

Soil Sampling and Analysis

Soil samples for determination of soil available N and P were obtained from the upper soil surface layer using a 5 cm diameter soil auger. The soil samples were collected at 2 weeks after planting, seedling, tasselling, cobbing and maturity stages of maize growth, between the plants within a row in every plot at random. Four augerings were done in every plot and the soil bulked together to get one composite sample. The soil samples were then put in polythene bags and kept in a portable cool box. This approach was undertaken in order to portray actual field conditions of N concentration and moisture regimes, as of sampling, during analysis.

The samples for analysis of available N (NO_3–N^- + NH_4–N^+) were refrigerated before analysis. For the analysis of available P, soil samples were air dried by placing them in a shallow tray in a well-ventilated area. Soil analysis was done according to the methods described by Okalebo et al. (1993).

Plant Sampling and Analysis

The aboveground legume and weed biomass were obtained from two internal rows of each plot and 1 m^2 area, respectively. The biomass was weighed fresh in the field. Sub-samples were taken to the laboratory and oven-dried at 70°C for 48 h for determination of dry weight. Oven-dried biomass samples of legumes and weeds were ground to pass through a 0.5 mm sieve and analysed for total N and P using the methods described by Okalebo et al. (1993).

At harvest, maize samples were obtained from three centre rows for grain (adjusted to 13% moisture content) and stover dry matter (DM) yield (70°C) determination. The legume grain and DM yields were determined from two internal rows of each plot at 13% moisture content and expressed on hectare basis.

Statistical Analysis

The measured soil and plant parameters were subjected to analysis of variance (ANOVA) appropriate for a split plot design. The SPSS software (SPSS Incorporated, 1999) was used for statistical analysis.

Results and Discussion

Dry Matter (DM) Yields of NF Weeds and Improved Fallow Legumes

The DM yield (t ha^{-1}) of the NF weeds (6.9) was significantly higher ($P \leq 0.05$) than LB (5.7), CR (4.9) and GP (2.7). CR and LB had significantly higher DM yield than GP (Fig. 2).

The high DM yield of the NF weeds may partially be due to the highly vegetative nature of the weeds present (*Amaranthus hybridus*, *Commelina benghalensis*, *Galinsoga parviflora*, *Digitaria scalarum*, *Cynodon dactylon*, *Pennisetum clandestinum*) coupled with their rapid growth and establishment in comparison with the legume species. Gathumbi et al. (2004) studying short-term fallows reported highest biomass production in weedy fallow and observed that the type of fallow species greatly influences biomass production. The differences in legume DM yields may be attributed to increased available soil P supplied through solubilization of MRP and varietal differences in P uptake and P use efficiency. Chiezey et al. (1992) pointed out that phosphorus application enhances nodulation, total dry matter and grain yield of legumes.

Grain Yield of the Improved Fallow Legumes

Grain yield (t ha^{-1}) of CR (1.25) was significantly higher ($P \leq 0.05$) than that of LB (0.98) and GP (0.83). The grain yields did not differ significantly between

Fig. 2 Aboveground DM yield of NF weeds and improved fallow legumes and grain yield of improved fallow legumes. *Note*: Means of pre-maize treatments followed by the same letter are not significantly different according to Tukey mean separation procedure at $P < 0.05$

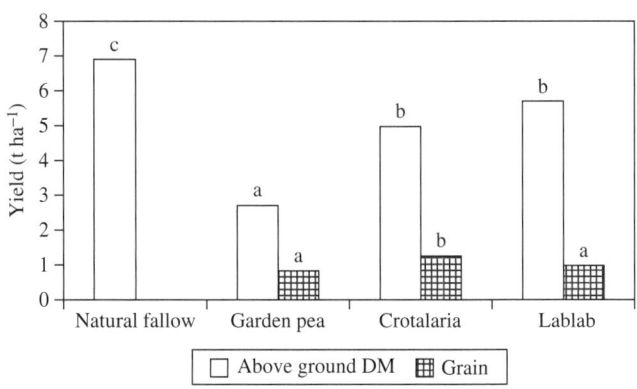

LB and GP (Fig. 2). The differences in yields could be in light of the pronounced differences in phenotypes and genotypes of the legumes involved with resultant variations in the amounts of nitrogen supplied to the crops through BNF. Nodulation and N_2 fixation in legumes are generally thought to be quantitatively inherited traits (Bliss, 1993). In terms of DM yields, LB had relatively higher DM yields than CR (Fig. 2). This observation indicates that LB plant tissue developed early in the growing season is not an effective N source for the later reproductive sink that exists during seed development (McConnell et al., 2002).

N and P Contents of Improved Fallow Legumes and NF Weed Residues

The total N content of LB and CR was significantly higher than that of the NF weeds and GP (Table 2). The same trend was observed for P. The high N and P contents of LB and CR residues may have been due to the high aboveground DM produced (Fig. 2) and consequently higher nutrient accumulation. Gathumbi et al. (2004) reported that CR fallows recorded the greatest total N yield due to fast establishment and higher total biomass production. Similar observations were reported by Niang et al. (2002).

The natural fallow weeds had the highest foliage biomass yield, but a correspondingly lower N yield (Table 2). This may be due to their low N concentration (1.5%) in comparison with GP (2.5%), CR (3.4%) and LB (3.2%). This is in agreement with the findings of Niang et al. (2002) who found that natural fallow weeds contained low N contents but high phosphorus and potassium contents.

N Fixation by the Improved Fallow Legumes

CR and LB fixed significantly higher ($P \leq 0.05$) amounts of nitrogen than GP (Table 2). This may be attributable to their better genetic potential, coupled with phosphate availability released by MRP, and hence had a comparative advantage for enhanced N fixation.

Low amounts of N were fixed by the NF weeds (Table 2) compared to the legumes. This may be due to the inherent poor N fixing ability of the weeds present. The improved fallow legumes with high amounts of N fixed (LB and CR) also had high aboveground biomass (Fig. 2). The correlation (r^2) between amounts of N fixed and legume DM yield for these legumes was 0.94. This agrees closely with the findings of Kumar and Goh (2000) who reported strong correlations between the amounts of N fixed for both legume DM yield ($r^2 = 0.96$) and N accumulation ($r^2 = 0.97$–0.99). Ladha et al. (1996) had also reported that larger amounts of N_2 fixed in broad bean resulted from better growth and higher biomass accumulation.

Effect of Improved Fallow Legumes on Soil Nutrient Availability and Maize Performance

Soil Available N

Maize growth stages: There was a decline in soil available N across maize growth stages for all pre-maize and residue management treatments (Table 3). The observed decline could be due to increased crop uptake during development. Berger (1962) stated that

Table 2 N and P contents and BNF by legumes and NF weeds

Pre-maize treatment	Nutrient content in residue (kg ha^{-1})		Biological nitrogen fixation (BNF, kg ha^{-1})
	N	P	
Natural fallow	104[b](28)	8[b](2)	4[a](2.5)
Garden pea	60[a](4)	4[a](3)	126[b](3)
Crotalaria	169[c](17)	12[c](1)	175[bc](31)
Lablab	180[c](14)	12[c](1)	196[c] (30)

Means in a column followed by the same letter are not significantly different according to Tukey mean separation procedure at $P < 0.05$. Values in parentheses are standard deviations

Table 3 Available nitrogen (kg ha^{-1}) measured at different stages of maize growth

Pre-maize and residue treatments		Cropping sequence and stages of maize growth									
		Sole maize					Maize/bean intercrop				
		Two weeks after planting	Seedling	Tasselling	Cobbing	Maturity	Two weeks after planting	Seedling	Tasselling	Cobbing	Maturity
NF	RI	36.80 (1.56)	15.40 (1.52)	17.80 (0.93)	19.30 (2.35)	17.20 (1.46)	34.05 (1.00)	14.20 (1.00)	12.20 (3.36)	10.20 (2.00)	9.90 (2.00)
	FYM	37.14 (2.18)	18.40 (2.00)	14.20 (1.26)	11.60 (2.62)	9.80 (1.50)	34.55 (1.23)	19.50 (2.20)	15.90 (2.26)	9.41 (0.46)	7.80 (2.40)
LB	RI	65.98 (2.56)	22.58 (2.24)	18.80 (1.47)	14.49 (0.93)	12.30 (2.30)	62.94 (2.36)	17.35 (1.28)	14.30 (0.58)	12.70 (2.42)	11.60 (1.62)
	FYM	52.62 (1.80)	21.43 (2.30)	12.60 (1.18)	11.80 (1.28)	9.20 (1.68)	50.56 (1.42)	16.33 (1.18)	14.80 (2.60)	10.70 (2.60)	8.60 (1.42)
CR	RI	49.22 (2.36)	20.51 (2.90)	16.70 (1.24)	11.10 (1.18)	9.38 (2.40)	48.18 (2.10)	16.42 (0.62)	13.70 (2.96)	10.62 (0.69)	8.60 (2.23)
	FYM	46.70 (2.32)	24.80 (3.36)	26.30 (1.18)	19.70 (1.70)	18.56 (1.16)	43.34 (2.10)	29.60 (2.36)	18.90 (1.23)	14.60 (2.36)	12.50 (3.80)
GP	RI	41.26 (1.18)	14.23 (3.60)	14.90 (4.20)	17.30 (1.90)	14.80 (2.50)	39.79 (2.24)	28.74 (1.00)	23.60 (0.89)	20.08 (1.37)	18.58 (2.00)
	FYM	40.89 (1.00)	15.39 (0.62)	12.70 (1.00)	14.49 (1.62)	12.60 (2.28)	36.70 (1.12)	17.70 (3.26)	20.20 (1.14)	15.60 (1.24)	4.30 (1.24)

RI = residue incorporation, FYM = farm yard manure
Figures in parentheses are standard deviations

N is taken up by maize throughout the growing season with maximum uptake 10 days before tasselling to 25–30 days after tasselling. Losses of N would also be due to leaching, as a result of heavy rains, microbial immobilization and denitrification (Richter et al., 1982; Scherer et al., 1992).

There were low levels of soil available N observed 2 weeks after maize was planted in the NF treatment (Table 3). This suggests depletion of soil available N reserves due to the low N content and poor N fixing ability of the weeds. N uptake therefore depended on the residual mineral N reserve in the soil and the rate of N mineralization. Gallagher (1984) had reported that the rate of N uptake is greater than the rate of mineralization during the growth period and consequently mineral N reserve in the soil is depleted quickly.

Legume residues vs. FYM incorporation: The soil available N was higher in the legume residue than FYM incorporated treatments, except for the NF treatment (Table 3). The FYM treatment had significantly low mineral N in soil at all stages of maize growth (Table 3), for both sole maize and M/B intercrop. This could be attributed to the manure's slow release of nutrients. Murwira and Kirchman (1993) studying

mineralization from organic fertilizers found that mineralization of manure N increased to a maximum of 46% in the third season of application. Kimble et al. (1972) hypothesized that NO_3^- from manure may be more prone to denitrification especially with above average rainfall. This was the case as above average rainfall was received close to tasselling and cobbing stages of maize growth (Fig. 1) during the experimental period.

Sole maize vs. maize/bean cropping systems: The available N in the soil was significantly higher ($P \leq 0.05$) in sole maize than in M/B intercrop (Table 3). The relatively low soil available N in the M/B intercrop compared to sole maize may be due to the competition between beans and maize for limited N and sharing of fixed N_2 with maize crop (Vandermeer, 1989).

Legume vs. legume: There were differences in soil available N following the different legumes with significantly higher ($P \leq 0.05$) amounts following CR and LB. The variation among the legumes may be attributed to differences in total N yield (Table 2) and mineralization rates after incorporation. Apart from input to the system through BNF, the higher N content following LB and CR may partly be explained by their deep root systems that captured nitrate from the sub-soil. This is also consistent with observations made by Wortmann et al. (2000) and Smestad et al. (2002).

Legume vs. NF: Soil available N content was significantly higher ($P \leq 0.05$) after legumes compared to the NF and this may be attributable to the superior mineralization of the incorporated N-rich legume residues. Szott et al. (1991) observed higher soil N and P levels in fallows improved with legumes than in NF. Microbial tie-up of soil inorganic N may have occurred during the decomposition of low-quality weed biomass with wide C/N ratios (27:1) and concomitantly nutrient release was slow.

Kumar and Goh (2000) reported lower soil N mineralization with incorporation of non-leguminous residues and attributed the same to greater immobilization of N because of high C/N ratios. Added organic material with a high C/N ratio (e.g. >20 kg C $(kg N)^{-1}$) provides adequate C substrate, but if the C/N ratio of the microbial biomass is lower ($3–14$ kg C $(kg N)^{-1}$), N is taken up from the mineral N pool in the soil to meet the shortfall (Kamukondiwa and Bergstorm, 1994; Palm et al., 1997).

Soil Available P

Soil available P did not vary significantly following the pre-maize treatments at the different maize growth stages (Table 4). This is attributable to the blanket application of P as MRP on all plots and the fact that P is an immobile element. Furthermore, Niang et al. (2002) had reported that phosphorus recycling by fallow species through deep uptake is likely to be negligible owing to the very low concentration of available P in the subsoil. Conversely, increased biological immobilization of P may have been responsible for the lack of differences (Palm et al., 1997; Lelei et al., 2006).

There were no significant differences in the concentration of soil available P for the pre-maize and residue management treatments. However, fluctuations in soil available P from planting to maturity stages of maize growth in all the treatments were observed (Table 4).

General decrease of available P with growth period could be as a result of plant uptake which is continuous throughout growth. Syre (1955) and Seeling and Zasoski (1993) reported that P uptake is continuous throughout the growing season.

Table 4 Available P (ppm) measured at different stages of maize growth

		Cropping sequence and stages of maize growth									
		Sole maize					Maize/bean intercrop				
Pre-maize and residue treatments		Two weeks after planting	Seedling	Tasselling	Cobbing	Maturity	Two weeks after planting	Seedling	Tasselling	Cobbing	Maturity
NF	RI	1.60 (0.20)	2.00 (0.20)	1.50 (0.12)	1.20 (0.26)	0.90 (0.52)	3.20 (0.56)	1.70 (0.20)	2.00 (0.75)	1.00 (0.26)	0.80 (0.10)
	FYM	4.40 (0.17)	1.70 (0.26)	4.00 (0.36)	3.00 (0.10)	2.80 (0.42)	2.70 (0.44)	2.10 (0.40)	1.40 (0.17)	1.10 (0.26)	0.70 (0.17)
LB	RI	5.00 (0.28)	2.60 (0.14)	3.80 (0.42)	2.80 (0.20)	2.50 (0.30)	2.80 (0.26)	1.70 (0.20)	2.20 (0.10)	1.20 (0.34)	1.00 (0.41)
	FYM	2.70 (0.24)	2.00 (0.10)	2.90 (0.17)	1.90 (0.21)	1.60 (0.10)	2.20 (0.70)	2.00 (0.26)	1.60 (0.17)	1.20 (0.52)	0.90 (0.26)
CR	RI	2.70 (0.10)	2.10 (0.36)	2.20 (0.10)	1.20 (0.26)	1.00 (0.46)	3.80 (0.58)	2.10 (0.10)	1.80 (0.12)	1.40 (0.28)	1.20 (0.56)
	FYM	1.90 (0.23)	1.60 (0.50)	1.00 (0.26)	0.70 (0.46)	0.40 (0.33)	2.70 (0.24)	2.00 (0.10)	2.90 (0.17)	1.90 (0.21)	1.60 (0.10)
GP	RI	2.50 (0.36)	1.50 (0.10)	1.70 (0.20)	1.20 (0.43)	0.70 (0.10)	2.40 (0.20)	2.10 (0.10)	2.90 (0.46)	1.80 (0.35)	1.50 (0.41)
	FYM	2.70 (0.44)	2.10 (0.10)	2.00 (0.18)	1.70 (0.10)	1.20 (0.36)	3.20 (0.20)	1.50 (0.38)	2.60 (0.20)	2.40 (0.10)	2.00 (0.26)

RI = residue incorporation, FYM = farm yard manure
Figures in parentheses are standard deviations

Maize DM and Grain Yield in the Sole Maize and M/B Intercrop Systems

Sole Maize Yield

Sole maize grain and DM yields were significantly higher ($P < 0.05$) following improved fallow legumes than NF with sole maize yields following LB being the highest (Table 5). Significant differences were, however, not observed after CR and GP residue and FYM incorporation (Table 4).

The higher maize grain and DM yields following LB may be attributed to higher biomass produced by LB coupled with its superior BNF, thus leading to high N content in the legume residues. In spite of the higher weed biomass produced by the NF weeds, the N content was low implying that even upon mineralization the N release was equally low and hence the associated low maize yields.

Higher maize and DM yields following leguminous fallow may be attributed to the higher N additions through mineralization of the high-quality legume residues, as amounts of available N in soil were greater after legumes than after natural fallow (Table 3). Fischler and Wortmann (1999) reported 50% higher maize grain yields following one season fallow with LB than maize following maize. Cheruiyot et al. (2003) in a study on the effect of legume-managed fallows on soil N in the Rift valley highlands of Kenya reported that among the legume species, lablab showed outstanding positive effect on succeeding maize. This was attributed to increased soil nitrate levels. In addition to nutrient release, increase in moisture conservation and weed suppression by the residues probably contributed to the increased maize yields (Barrios et al., 1998).

There were no significant differences between FYM and legume residue incorporation in terms of maize grain and DM yields even though the yields in the latter were relatively higher. This may be due to the higher available soil N resulting from BNF by the legumes. This is further supported by the low yields realized in the NF treatment in both the FYM and the residue incorporated treatments and sole maize and M/B intercrop.

Maize and Bean Yield in the Maize/Bean Intercrop Treatment

Maize yield: Intercropping beans with maize obviously reduced the pre-maize fallow crop effect compared to the sole maize system as shown by the higher maize yields in NF-M/B than NF-M treatments (Table 5). This may be explained by an increase in soil N availability to maize due to the BNF by the beans, mainly in the low-N fallow treatment.

Maize grain and DM yields in the M/B intercrop were relatively lower, with no significant differences, compared to sole maize in both the residue and the FYM incorporated treatments. This may be as a result of the competition between beans and maize for limited N and sharing of fixed N_2 with the maize. According to Vandermeer (1989), intercropping advantage depends on net effect in tradeoff between interspecific competition and facilitation (positive facilitation) in which one plant species enhances the survival, growth or fitness of another. In both cropping sequences and residue management

Table 5 Maize and common bean yield (kg ha^{-1}) in sole and maize/bean intercrop system

Pre-maize and residue treatments		Sole maize		Maize/bean intercrop		Bean yield	
		Grain yield	DM yield	Grain yield	DM yield	Grain yield	DM yield
NF	RI	2.20 (0.05)	4.20 (0.04)	2.80 (0.10)	3.45 (0.05)	0.50 (0.08)	1.00 (0.29)
	FYM	2.00 (0.06)	4.23 (0.05)	2.84 (0.05)	3.50 (0.28)	0.51 (0.19)	1.01 (0.03)
LB	RI	4.60 (0.10)	6.00 (0.22)	3.90 (0.36)	5.60 (0.17)	0.60 (0.13)	1.60 (0.10)
	FYM	3.67 (0.04)	4.78 (0.08)	3.13 (0.20)	4.50 (0.04)	0.48 (0.05)	1.28 (0.10)
CR	RI	3.50 (0.03)	5.50 (0.04)	3.10 (0.04)	5.50 (0.06)	0.60 (0.11)	1.30 (0.15)
	FYM	3.32 (0.04)	5.22 (0.02)	2.79 (0.05)	4.95 (0.05)	0.54 (0.07)	1.17 (0.18)
GP	RI	3.30 (0.04)	4.25 (0.17)	2.40 (0.45)	4.50 (0.05)	0.50 (0.14)	1.20 (0.02)
	FYM	3.27 (0.05)	4.20 (0.04)	2.21 (0.05)	4.15 (0.33)	0.46 (0.09)	1.10 (0.26)

RI = residue incorporation, FYM = farm yard manure
Figures in parentheses are standard deviations

systems, the yields following the order LB, CR, GP and NF were reflective of the amounts of N fixed and N content of residue incorporated (Table 2).

Bean: Bean grain yield ranged between 0.5 and 0.6 t ha^{-1} (Table 5) for all pre-maize treatments. Although yields of maize were lower in the M/B intercrop (Table 5), the fact that two crops could be harvested from the same plot more than compensated for the higher yields realized in the sole maize cropping system. According to Helenius and Jokinen (1994), the potential advantages of intercropping include overyielding, i.e. improved utilization of growth resources by the crop and improved reliability from season to season.

Conclusion

In this study, LB was most effective in improving soil productivity as indicated by the performance of subsequent maize crop. Though lower maize yields were realized in the M/B intercrop than sole maize cropping system, the M/B intercrop system may be better and preferable to the small-scale farmers due to the dual purposes of ensuring food and nutritional security, as two crops are harvested in one season from the same land, and improving soil fertility through BNF.

As the yields obtained with the incorporation of legume residues were not significantly different from those of FYM, farmers are at liberty to either directly incorporate the legume residues or recycle them as livestock feed and be brought back to the farm as FYM.

Since LB was superior in performance compared to other improved legume fallows, scaling up the technology to farmers' fields is necessary as it provides a feasible means of boosting maize production and consequently food security. From the results of this study, it is evident that legumes have the potential to substitute for inorganic fertilizers. This is in view of the fact that, the maize yields realized in the current study is comparable to those obtained in other studies involving use of inorganic fertilizers. This further attests to the fact that ONM strategy is a feasible technology that could easily fit into the circumstances of the resource-poor farmers within the region.

Acknowledgements We are grateful to the Agronomy and Soil Science Departments, Egerton University, for respectively providing the experimental site and the laboratory facilities. We also thank the ÖAD (Österreichischer Akademischer Austauschdienst) for funding the research.

References

Barrios E, Kwesiga F, Buresh RJ, Sprent JI, Coe R (1998) Relating pre season soil nitrogen to maize yield in tree legume–maize rotations. Soil Sci Soc Am J 62:1604–1609

Berger JE (1962) Maize production and the manuring of maize, 2nd edn. Cornzette and Herber, Zurich, p 520

Bliss FA (1993) Breeding common bean for improved biological nitrogen fixation. Plant Soil 152:71–79

CBS (Central Bureau of Statistics) (1996) Economic survey 1996. Office of the vice president and ministry of planning and national development, Nairobi, Kenya

Cheruiyot EK, Mumera LM, Nakhone LN, Mwonga SM (2003) Effect of legume managed fallows on soil nitrogen in following maize (Zea mays L.) and wheat (Triticum aestivum) crops in the Rift Valley highlands of Kenya. Eur J Exp Agr 43:597–604

Chiezey UF, Yayock JY, Shebayan JAY (1992) Response of soybean (Glycine max(L.) Merill) to Nitrogen and fertilizer levels. Trop Sci 32:360–368

FAO-UNESCO (1990) FAO-UNESCO Soil map of the world. Revised legend. World resources. Report 60, FAO, Rome

Fischler M, Wortmann CS (1999) Green manures for maize-bean system in Eastern Uganda: agronomic performance and farmers' perceptions. Agroforest Syst 9:123–138

Gallagher EG (1984) The role of nitrogen in yield formation of cereals especially winter wheat. In: Piertz et al. (eds) Cereal production. Butterworth and Company, London, p 121

Gathumbi SM, Cadisch G, Giller KE (2004) Improved fallows: effects of species interaction on growth and productivity in monoculture and mixed stands. For Ecol Manage 187:267–280

Hartemink RJ, Buresh PM, van Bodegom AR, Braun C, Jama BI, Janssen BH (2000) Inorganic nitrogen dynamics in fallows and maize on an Oxisol and Alfisol in the highlands of Kenya. Geoderma 98:11–33

Hauser S (1987) Schatzung der symbotisch fixierten Stickstoffmenge von Ackerbohne (Vicia faba L.) mit erweiterten Differenzmethoden. Dissertation, Georg-August- Universitat, Gottingen

Hedin PA, Williams WP, Buckley PM (1998) Caloric analysis of the distribution of energy in corn plants. J Agric Food Chem 46:4754–4758

Helenius J, Jokinen K (1994) Yield advantage and competition should therefore be recommended in intercropped oats (Avena sativa L.) and faba bean (Vicia faba L.), application of the hyperbolic yield-density model. Field Crops Res 37:85–94

Jaetzold R, Schmidt H (1983) Farm management handbook of Kenya. Volume 2 part B. central Kenya. Ministry of agriculture in co-operation with GAT and GTZ, pp 397–400

Jensen ES (1986) Symbiotic N$_2$ fixation in pea and field bean estimated by ^{15}N fertilizer dilution in field experiments with barley as a reference crop. Plant Soil 92:3–13

Kamukondiwa W, Bergstorm L (1994) Leaching and crop recovery of ^{15}N from ammonium sulphate and labelled maize material in lysimeters at a site in Zimbabwe. Plant Soil 162:193–210

Kandji ST, Ogol CKPO, Abrecht A (2003) Crop damage by nematodes in the improved fallow fields in Western Kenya. Agroforest Syst 57:51–57

Kimble JM, Bartlett RJ, McIntosh JL, Varney KE (1972) Fate of nitrate from manure and inorganic nitrogen in a clay soil cropped to continuous corn. J Environ Qual 1:413–415

Kumar K, Goh KM (2000) Biological nitrogen fixation, accumulation of soil nitrogen and nitrogen balance for white clover (Trifolium repens L.) and field pea (Pisum sativum L.) grown for seed. Field Crops Res 68: 49–59

Kwabiah AB, Stoskopf NC, Palm CA, Voroney RP, Rao MR, Gacheru E (2003) Phosphorus availability and maize response to organic and inorganic fertilizer inputs in a short term study in Western Kenya. Agric Ecosyst Environ 95: 49–59

Ladha JK, Kundu DK, Angelo-Van Coppenolle MG, Peoples MB, Carangal VR, Dart PJ (1996) Legume productivity and soil nitrogen dynamics in lowland rice-based cropping systems. Soil Sci Soc Am J 60:183–192

Lelei NJJ (2004) Impact of soil amendments on maize performance and soil nutrient status in legume–maize intercropping and rotation systems in central Rift Valley province of Kenya. Dissertation, University of natural resources and applied life sciences (BOKU) Vienna, Austria

Lelei JJ, Onwonga RN, Mochoge BO (2006) Interactive effects of lime, manure, N and P fertilizers on maize (Zea mays L.) yield and N and P uptake in an acid Mollic Andosol of Molo Kenya. Egerton J: Sci Technol Series 6(1):141–156

Mathews RB, Holden ST, Lungu S, Volk J (1992) The potential of alley cropping in improvement of cultivation systems in the high rainfall areas of Zambia. Agrofor Syst 17:219–240

McConnell JT, Miller PR, Lawrence RL, Engel R, Nielsen GA (2002) Managing inoculation failure of field pea and chickpea based on spectral responses. Can J Plant Sci 82:273–282

Murwira HK, Kirchmann H (1993) Nitrogen dynamics and maize growth in a Zimbabwean sandy soil under manure fertilization. Commun Soil Sci Plant Anal 24:2343

Niang AI, Amadalo BA, Wolf de J, Gathumbi SM (2002) Species screening for short-term planted fallows in the highlands of Western Kenya. Agroforest Syst 56:145–154

Noordwijk M (1999) Production of intensified crop-fallow rotations in the Trenbath model. Agroforest Syst 9:223–237

Okalebo JR, Gathua KW, Woomer PL (1993) Laboratory methods of soil and plant analysis. A working manual. Printed by Marvel E.P.Z (Kenya) LTD, Nairobi, Kenya, 88pp

Okalebo JR, Nandwa SM (1997) Effect of organic resources with and without inorganic fertilizers on maize yields, mainly on P deficient soils in Kenya. National soil fertility plant nutrition research programme technical report series No. 12, KARI, Nairobi

Palm CA, Myers RJK, Nandwa SM (1997) Organic-inorganic nutrient interactions in soil fertility replenishment. In: Buresh RJ, Sanchez PA, Calhoun F (eds) Replenishing soil fertility in Africa. Soil Science Society of America. Madison WI, pp 193–218

Place F, Christopher B, Barrett H, Freeman A, Ramisch JJ, Vanlauwe B (2003) Prospects for integrated soil fertility management using organic and inorganic inputs: evidence from smallholder African agricultural systems. Food Policy 28:365–378

Reardon T, Berdegué J, Escobar G (2001) Rural non-farm employment and incomes in Latin America: overview and policy implications. World Dev 29(3):395–409

Reining E (2005) Assessment tool for BNF of Vicia faba cultivated as spring main crop. Eur J Agron 23:392–400

Richter J, Nuske A, Hebenicht W, Bauer J (1982) Optimised N-mineralization parameters of Loess soils from incubation experiments. Pant Soil 68:379–388

Scherer HW, Werner W, Rossbach J (1992) Effects of pretreatment of soil samples on N mineralization in incubation experiments. Biol Fertil Soils 14:135–139

Seeling B, Zasoski RJ (1993) Microbial effects in maintaining organic and inorganic solution phosphorus concentration in a grassland topsoil. Plant Soil 148:277–284

Smaling EMA, Nandwa SM, Prestele H, Roetter R, Muchena FN (1992) Yield response of maize to fertilizers and manure under different agro-ecological conditions in Kenya. Agric Ecosyst Environ 41:241–252

Smestad BT, Tiessen H, Buresh RJ (2002) Short fallows of Tithonia diversifolia and Crotalaria grahamiana for soil fertility improvement in western Kenya. Agroforest Syst 55: 181–194

Snapp SS, Mafongoya PL, Wadington S (1998) Organic matter technologies to improve nutrient cycling in small holder cropping systems in southern Africa. Agric Ecosyst Environ 71:187–202

SPSS Incorporated (1999) Applications Guide, SPSS Base 9.0. SPSS Inc, Chicago, IL

Syre JD (1955) Mineral nutrition of corn and corn improvement. Academic, New York, NY, p 369

Szott LT, Palm CA, Sanchez PA (1991) Agroforestry in the acid soils of the humid tropics. Adv Agron 45:275–301

Vandermeer JH (1989) The ecology of intercropping. Bridge University Press, Cambridge

Wortmann CS, McIntyreb BD, Kaizzi CK (2000) Annual soil improving legumes: agronomic effectiveness, nutrient uptake, nitrogen fixation and water use. Field Crops Res 68:75–83

Using Forage Legumes to Improve Soil Fertility for Enhanced Grassland Productivity of Semi-arid Rangelands of Kajiado District, Kenya

P.N. Macharia, C.K.K. Gachene, and J.G. Mureithi

Abstract A two-phase study was conducted in the semi-arid rangelands of Kajiado District, Kenya, to determine the effect of forage legumes on soil fertility improvement and grassland productivity of natural pastures. During legume evaluation phase, *Neonotonia wightii* (Glycine), *Macroptilium atropurpureum* (Siratro), *Lablab purpureus* cv. Rongai (Dolichos), *Mucuna pruriens* (Velvet bean) and *Stylosanthes scabra* var. *seca* (Stylo) were screened for adaptability and growth performance. Results of soil analysis showed that soil pH, organic carbon, nitrogen and potassium significantly increased after 2 years of study due to the large amounts of organic residues produced by the legumes (particularly the perennials). The calcium content decreased significantly (which was attributed to plant uptake) while the decrease of phosphorus was not significant. After integration of Glycine, Siratro and Stylo into natural pastures during the second phase of the study, the crude protein content of grasses intercropped with legumes increased from 7.1 to 14.3, 11.9 and 10.2%, respectively. Grasses intercropped with legumes also had higher digestibility contents than grasses in monoculture stands. The study concluded that addition of organic residues by the introduced forage legumes improved the soil fertility status and hence the crude protein and digestibility of grasses.

Keywords Forage legumes · Grass/legume intercrops · Grassland productivity · Semi-arid rangelands · Soil fertility

P.N. Macharia (✉)
KARI-Kenya Soil Survey, Nairobi, Kenya
e-mail: kss@iconnect.co.ke

Introduction

In semi-arid regions of eastern Kenya, the rapid increase in population densities, continuous cultivation and overgrazing have contributed to the depletion of soil fertility resulting in low yields from croplands and pastures (Njarui et al., 2004). The infertility of the soils, particularly lack of N, is the main contributory factor to the low productivity of such tropical soils (Giller, 2001). In addition, the frequent fires that occur especially during the dry seasons in the rangelands lead to considerable losses of N in gaseous form to the atmosphere (Brady, 1984; Grace et al., 2006). According to Nyathi et al. (2003), some forms of tillage, particularly in arid and semi-arid areas, encourage oxidation of organic matter throughout the tilled profile, resulting in release of carbon to the atmosphere rather than its build-up in the soil. This leads to reduced biomass production from crops or pastures and lower carbon inputs to the soil in subsequent periods because less root matter, leaf litter and crop residues are returned to the soil.

The low pasture availability in terms of quantity and quality in semi-arid rangelands of Kajiado District remains a challenge to range managers and scientists (Too, 1985; de Leeuw, 1991). This is because livestock production based on natural pastures remains the key livelihood strategy to the inhabitants of the district. The current lifestyle of semi-nomadic pastoralism, where livestock and herders move during the dry season in search of pasture and water, encounters problems as the areas for free-range grazing become limited. Thus, ways have to be found to improve the productivity (quality and quantity) of the natural pastures in the semi-arid rangelands of Kenya.

A. Bationo et al. (eds.), *Innovations as Key to the Green Revolution in Africa*,
DOI 10.1007/978-90-481-2543-2_30, © Springer Science+Business Media B.V. 2011

One way of improving the productivity of the natural pastures is to integrate forage legumes into the natural pastures, especially in the smallholder livestock production systems. This will enhance forage production, increase amount of protein available for the grazing animals, and also prolong grazing periods as legumes and other pasture dicots remain green long after grasses have dried up during the dry season (Skerman et al., 1988; Kinyamario and Macharia, 1992). Legumes have the potential to improve the quality and quantity of pasture swards through fixation of atmospheric nitrogen (Guretzky et al., 2004). Legumes also contain higher crude protein content than grasses and therefore their integration with grasses improves the dry matter yields and forage quality of natural pastures.

Regarding soil fertility improvement, legumes have the potential to improve soil fertility through the release of nitrogen from decomposing leaf residues, roots and nodules which results in increased sward productivity after nitrogen uptake by the companion grasses (Guretzky et al., 2004; Cherr et al., 2006). The slow release of nitrogen may be better synchronized with plant uptake than other sources of inorganic N, thereby increasing nitrogen uptake efficiency and crop yields while reducing nitrogen loss through leaching (Cherr et al., 2006). Thus, integration of forage legumes into natural pastures is an option to improve soil fertility through addition of organic residues and soil nutrients (especially nitrogen) in the semi-arid rangelands of Kajiado District.

This study was therefore carried out with the aim of introducing forage legumes into natural pastures as a way of improving soil fertility and also the quality and quantity of livestock feed. The results obtained on the effect of legumes on dry matter productivity of grasses have been presented in Macharia et al. (2007) and hence will not be presented in this chapter.

Materials and Methods

Study Site

The study was carried out in Mashuru Division of Kajiado District (altitude of 1,280 m above sea level). Most parts of the division lie in an agro-climatic zone V which is semi-arid (Sombroek et al., 1980) with an average annual rainfall of 600–800 mm per year

(GoK, 1991). According to Sombroek et al. (1980), rainfall is the major limiting factor for maximum primary production in the area. The rains are received in two distinct rainfall seasons, with the long rainfall season occurring between March and May while the short rainfall season occurs between October and January of each year. The dominant farming system is agro-pastoralism.

Legume Screening Experiments

The first phase of the study was legume screening to identify suitable forage legumes for integration into natural pastures for enhancement of soil fertility and productivity (quantity and quality) of grasses. The forage legumes selected were *Neonotonia wightii* (Arn.) Lackey (Glycine), *Macroptilium atropurpureum* (DC) Urb. (Siratro), *Lablab purpureus* cv. Rongai (L.) Sweet (Dolichos), *Mucuna pruriens* (L.) DC (Velvet bean) and *Stylosanthes scabra* var. *seca* Vog. (Shrubby Stylo). These legumes were selected because they had been identified as green manure legumes for soil fertility improvement in cultivated smallholder agriculture in Kenya (LRNP, 1999). The legumes were sown as monoculture stands and soil samples collected at the beginning and the end of the 2-year study period. The soil pH was determined in a 1:2.5 soil:water suspension. The soil organic carbon was determined through the Walkley and Black method as described by Hinga et al. (1980). The soil nitrogen was determined by the macro-Kjedahl method while the exchangeable cations (phosphorus, potassium and calcium) were determined through the Mehlich Double Acid Method following procedures described by Hinga et al. (1980). The leaf litter yields of the legumes were determined at the peak of the long dry season by collection of all litter drops within an area of 1 m^2 in each experimental plot.

Grass/Legume Integration Experiments

Results from the legume screening experiment showed that Glycine, Siratro and Stylo were perennial, deep rooted and yielded high dry matter and organic matter than Dolichos and Velvet bean. They were therefore integrated into natural pastures through sowing along 20-cm-wide bands cleared of vegetation.

The treatments were natural pasture (NP), monoculture stands of Glycine, Siratro and Stylo, and intercrops of NP+Glycine, NP+Siratro and NP+Stylo. Soil samples were collected at the beginning and the end of the 2-year study period for soil fertility assessments. Data on the effect of legumes on the nutritive content of grasses were collected during the legume's fourth season of growth at three stages of growth. The plant samples were analysed for crude protein (AOAC, 1995), in vitro dry matter digestibility (Tilley and Terry, 1963) and fibre content (Van Soest, 1963).

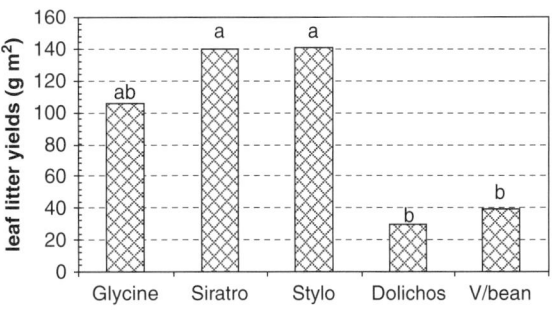

Fig. 1 Mean leaf litter yields (g m^2) of the legumes at the peak of the long dry season. *Bars* with the same letter are not significantly different at the 0.05 level of probability

Results and Discussion

Soil Fertility During Screening Phase

The results of soil analysis at the beginning and the end of the experimental period showed that soil pH, carbon (C), nitrogen (N) and potassium (K) significantly ($P \leq 0.05$) increased by the end of the experiment (Table 1). However, calcium (Ca) significantly ($P \leq 0.05$) decreased while P had a non-significant ($P \geq 0.05$) decrease from the soil by the end of the experiment.

The increase in soil pH was attributed to the addition of organic residues from the legumes in the form of leaf litter drops and from the decay of roots and nodules. Figure 1 shows that Glycine, Siratro and Stylo contributed the highest amounts of leaf litter which was significantly ($P \leq 0.05$) higher than that of Dolichos and Velvet bean. These results are in conformity with those of Njunie et al. (1996) who conducted a leaf drop study at coastal KARI-Mtwapa Research Station for 18 herbaceous legumes. The authors concluded that *Macroptilium lathyroides* (L.) Urb, *L. purpureus* and *M. atropurpureum* were capable of improving soil fertility by providing leaf mulch in situ and hence improving the soil organic matter content.

During decomposition and mineralization of organic residues, nutrients are released to the soil

due to increased soil microbial activities which are activated by more favourable soil conditions and availability of C (Landon, 1984; Muriuki and Qureshi, 2000) after an initial N immobilization by soil microorganisms (Giller and Wilson, 1991). Therefore, it may be argued that with an increased soil organic matter due to addition of plant residues by the legumes, the soil microbial activity increased. In addition, the shading effect by the legumes may have favourably improved the micro-environment of the soil in terms of soil temperature and moisture conditions, thus favouring increased microbial activity. This was reflected in the rise of total soil N by 29% by the end of the experiment, which indicated the positive contribution by legumes as regards enhancement of soil fertility.

By the end of the experiment, the P content in the soil decreased. This decrease was non-significant and was attributed to uptake by the legumes after it became more available as a result of the rise in soil pH. According to Rowell (1994), a rise in soil pH results in soil P becoming more available for uptake by plants. Once P contained in organic matter is released through mineralization into the soil solution, it can be immobilized by soil microorganisms, be taken up by plants or become fixed in the soil (Muriuki and Qureshi, 2000). Phosphorus is required for growth and development of plant roots and leaves,

Table 1 Effect of legumes on soil fertility during the screening phase

Period	pH	C (%)	N (%)	P (ppm)	K (me %)	Ca (me %)
Beginning	4.92^b	1.17^b	0.17^b	178.8	1.23^b	7.97^a
End	5.36^a	2.57^a	0.22^a	177.0	1.68^a	4.50^b
SE (means)	0.08	0.12	0.01	5.52	0.06	0.79

Means with different letters in a column are significantly different at the 0.05 level of probability

and additionally in legumes, the growth in number and density of nodules is greatly stimulated by P (Crowder and Chheda, 1982; Skerman et al., 1988). The authors added that the rate of fixation of N by the legume nodules resulted in better nodulation and improved plant metabolism.

Available soil Ca content decreased significantly ($P \leq 0.05$) by 44% at the end of the experiment. Muriuki and Qureshi (2000) stated that as the soil pH rises from 5.0 to 8.0, Ca becomes more available to plants. Results from the current study showed that the pH rose from 4.92 to 5.36 by the end of the experiment, a rise that may have led to increased Ca uptake by the legume plants therefore leading to a decrease from the soil. Calcium also raises the pH level of the soil, which in turn enables successful nodulation of the legumes by their associated *Rhizobium* bacteria (Skerman et al., 1988). Therefore, it implied that the high demand of Ca for the growth of legume meristematic tissue and the nodules may have led to a decreased Ca content in the soil.

Soil Fertility During Grass/Legume Integration

The results showed that soil pH significantly ($P \leq 0.05$) increased from 5.23 at the beginning of the experiment to 5.31 by the end of the experiment (Table 2). This increase in pH was similar to that obtained in the screening experiment. The results further showed that percent C, N, K and Ca significantly ($P \leq 0.05$) decreased while the decrease of P was not significant ($P \geq 0.05$).

The increase in soil pH after addition of legume residues is in agreement with Bationo et al. (1995) who stated that application of organic residues to the soil is one way of increasing soil pH. As a result aluminium toxicity which inhibits plant growth is

alleviated. Therefore, it can be argued that addition of organic residues and their decomposition and mineralization through action of soil microorganisms may have resulted in increased soil pH. Miles and Manson (2000) added that a change of soil pH to less acidic conditions causes the release of major nutrients like C, N, P, K, Ca, Mg and S which becomes available for plant uptake, thereby decreasing in the soil.

The results showed a decrease of organic C, N, P and K from the soil in contrast to results of the screening experiment where these nutrients significantly ($P \leq 0.05$) increased in the soil, except P. These different results may be attributed to the higher amounts of dry matter (hence more residues) produced by the legumes during screening as compared to the dry matter produced during the grass/legume integration study. The highest dry matter yield produced by Glycine, Siratro and Stylo during screening was 10.31, 7.81 and 3.52 t ha^{-1}, respectively, compared to 8.43, 3.46 and 3.29 t ha^{-1} produced by the same legumes, respectively, during the grass/legume integration study. Therefore, the results indicated that more organic residues from the legumes were returned to the soil during the screening study than during the grass/legume integration study.

According to Landon (1984), N-rich plant materials like those from legumes decompose more rapidly than those usually low in N. This statement is in conformity with the findings of Crespo et al. (2005) who reported that litter from *Desmodium ovalifolium* and *Stylosanthes guianensis* legumes decomposed after 180 days compared to 210 days for litter from *Pueraria phaseoloides* and *N. wightii/M. atropurpureum*. However, the rate of litter disappearance was low in grasses with 30, 15 and 10% of the original litter deposition remaining after 1 year in the case of *Cynodon nlemfuensis*, *Panicum maximum* and *Brachiaria decumbens*, respectively. Skerman et al. (1988) stated that the main source of available N in natural pastures is from soil organic matter; therefore, the mineralization of soil N is the most important

Table 2 Effect of forage legumes on soil fertility at the beginning and the end of the integration experiment

Time	pH	C (%)	N (%)	P (ppm)	K (me%)	Ca (me%)
Beginning	5.23[b]	1.37[a]	0.18[a]	170.3	1.32[a]	6.70[a]
End	5.31[a]	1.30[b]	0.13[b]	165.5	1.09[b]	3.08[b]
SE (means)	0.04	0.04	0.01	7.35	0.06	0.53

Means with a different letter in a column are significantly different at the 0.05 level of probability

source of N for the growth of unimproved grasslands. However, in improved pastures containing legumes, soil N was supplemented by N fixed by the legume–*Rhizobium* symbiosis. The authors added that the fixed N was used first for the growth of legume plants which later contributed to growth of other plants in the pasture.

By the end of the experiment, organic C, N and K significantly ($P \geq 0.05$) decreased in the soil while the decrease of P was not significant. These results are in conformity with those of Mureithi et al. (1994) who reported that after opening up virgin land for cultivation, C, N, P and K decreased up to the second year. However, during the third year of cropping, C, N and P increased (though non-significantly) due to application of mulch from *Leucaena leucocephala* (Lam) de Wit. The authors concluded that in the long term, improved soil nutrients will likely result not from direct addition of mulch from outside sources, but from increased return of crop residues to the soil after crop production.

The significant ($P \leq 0.05$) decrease in Ca content from the soil was similar to the decrease obtained during the screening experiment. The results from the integration experiment implied that as soil pH rose from 5.23 at the beginning of experiment to 5.31 by the end of the experiment, soil calcium became more available for plant uptake which is in agreement with Muriuki and Qureshi (2000). Thus the high demand of calcium for the growth of grass and legume meristematic tissues and also during nodulation of legumes may have led to the 118% significant ($P \leq 0.05$) decrease of the nutrient from the soil. According to Marschner (1995), dicotyledons such as legumes required higher levels of calcium than monocotyledons such as grasses for optimum growth. In addition, the process of nodule formation in legumes was dependent on calcium

which is required by the associated *Rhizobium* bacteria (Skerman et al., 1988).

Effect of Improved Soil Fertility on Crude Protein Content of Grasses

Grasses intercropped with forage legumes contained significantly ($P \leq 0.05$) higher crude protein (CP) content than grasses in the natural pasture at all stages of growth (Table 3). These results implied that grasses grown in mixtures with legumes benefited from the association, thereby increasing their CP contents.

A number of authors (Crowder and Chheda, 1982; Steele and Vallis, 1988; Miles and Manson, 2000) stated that one of the possible mechanisms of N transfer from legumes to grasses revolved around direct transfer of N between legume plants and grass roots or after a phase of nodule and litter decomposition in the soil. Miles and Manson (2000) added that nodule senescence was stimulated by defoliation and adverse environmental conditions and was characteristically more rapid than that of the remaining root material. After mineralization of plant litter, between 10 and 30% of the N becomes available to pasture plants within the first year after deposition (Steele and Vallis, 1988). Further, Crowder and Chheda (1982) stated that underground N transfers during growth of perennial legumes may not exceed 1 or 2% in the short term, but over a longer period of 2 years, the amounts varied from less than 5 to more than 30%. The N transfer processes mentioned above may have resulted in the increased CP content of grasses in mixed pastures, especially during the vegetative stage. The same processes may have resulted in the decreased

Table 3 Crude protein content (%) of monoculture and intercrops of grass/legume pastures at three stages of maturity

Treatments	Vegetative		Flowering		Senescent	
	Grass	Legume	Grass	Legume	Grass	Legume
Natural pasture (NP)	7.1[i–l]	–	5.9[j–l]	–	2.8[l]	–
Glycine	–	25.3[a–b]	–	17.6[c–f]	–	6.6[i–l]
Siratro	–	28.3[a]	–	21.9[j–l]	–	7.7[i–l]
Stylo	–	14.5[e–h]	–	14.0[e–h]	–	10.4[g–j]
NP + Glycine	14.3[e–h]	19.0[c–e]	7.5[j–l]	21.0[b–c]	4.0[k–l]	7.6[i–l]
NP + Siratro	11.9[g–j]	21.5[b–c]	8.5[h–l]	20.0[c–d]	3.1[l]	9.8[g–k]
NP + Stylo	10.2[g–k]	15.6[d–g]	6.1[i–l]	12.6[f–i]	3.1[l]	11.1[g–j]

Means with the same letter in a column or row are not significantly different at the 0.05 level of probability

CP contents in some legume plants after they were mixed with grasses. Therefore, the plant litter shedded by legumes during the long dry seasons every year must have decomposed after four seasons (2 years) and thus may have been a source of soil N that was taken up by grasses that were mixed with legumes, thereby raising their CP contents. These results are in conformity with those of Zemenchick et al. (2002) who studied the effect of Kura Clover (*Trifolium ambiguum*) on legume/grass intercrops in north-central USA. The authors reported that addition of legumes to cool-season grass monocultures improved the DM yield and CP content of the companion grasses.

Effect of Improved Soil Fertility on Digestibility of Grasses

The in vitro dry matter digestibility (IVDMD) of the treatments was significantly ($P \leq 0.05$) different (Table 4). In addition, digestibility of the legumes whether in monocultures or as intercrops with grasses was in majority of cases significantly ($P \leq 0.05$) higher than that of grasses at all stages of growth. Grasses grown as intercrops with legumes were more digestible than grasses in the natural pasture, indicating that inclusion of the legumes into natural pastures improved the digestibility of the associated grasses. This was attributed to increased CP contents of the grasses intercropped with legumes as it was shown in Table 3. For instance, Glycine, Siratro and Stylo raised the crude protein content of grasses from 7.1 to 14.3, 11.9 and 10.2%, respectively, at the vegetative stage.

The effect of CP contents on digestibility of forages is important for the maintenance of rumen microbial activities (Crowder and Chheda, 1982). If the crude protein content in the forage fell below 7%, the microbial activity in the rumen was depressed by lack of nitrogen, thus significantly reducing digestibility of the forage. Kinyamario and Macharia (1992) stated that when microbial activity in the rumen was depressed due to lack of nitrogen, the end result was that nitrogen excretion exceeded the nitrogen intake. Legumes contain higher amounts of N, Ca, P, K and Mg than grasses (Meissner et al., 2000). Therefore, when litter from legumes decomposed in the soil, the released nutrients may have become available for uptake by the grasses intercropped with legumes, thereby improving their mineral contents, hence the improved digestibility. Thus, livestock fed on fodder from such grass/legume mixed pastures were liable to feed on higher-quality fodder than livestock fed on natural pasture alone.

Conclusions

Addition of organic residues to the soil by the legumes improved the soil fertility by enhancing the soil pH, carbon, nitrogen and potassium. However, phosphorus and calcium levels in the soil decreased possibly due to utilization by the legumes during the nodulation and growth processes. The amount of legume residues returned to the soil determined the amount of soil nutrients released to the soil for subsequent uptake by the companion grasses. Glycine, Siratro and Stylo sown into natural pastures improved the soil fertility and hence the crude protein content and digestibility of grasses. This improvement was attributed to the

Table 4 Percent in vitro dry matter digestibility of monoculture and grass/legume mixed pastures at three stages of growth

Treatments	Vegetative		Flowering		Senescent	
	Grass	Legume	Grass	Legume	Grass	Legume
Natural pasture (NP)	54.6[e–i]	–	49.7[h–i]	–	49.5[h–i]	–
Glycine	–	59.7[a–h]	–	70.0[a–b]	–	51.2[g–i]
Siratro	–	71.4[a]	–	65.5[a–e]	–	52.4[f–i]
Stylo	–	65.7[a–e]	–	66.3[a–e]	–	58.1[b–h]
NP + Glycine	43.5[i]	56.2[c–h]	58.8[a–h]	68.4[a–d]	53.7[e–i]	59.5[a–h]
NP + Siratro	59.3[a–h]	65.4[a–e]	62.9[a–g]	64.5[a–f]	51.1[g–i]	55.7[d–h]
NP + Stylo	59.7[a–h]	68.8[a–c]	50.1[h–i]	67.8[a–d]	53.7[e–i]	59.1[a–h]

Means with the same letter in a column or row are not significantly different at the 0.05 level of probability

addition of organic matter by the legume residues. Therefore, livestock fed on fodder from grass/legume mixed pastures benefit more than those fed on natural pastures alone.

Acknowledgements The authors are grateful to the Rockefeller Foundation for funding the research work through the Legume Research Network Project of the Kenya Agricultural Research Institute. The authors are also very grateful to the coordinator of TSBF-CIAT and the conference organizers for facilitation to the first author to attend the conference and present this chapter.

References

AOAC (1995) Official methods of analysis. Association of Official Analytical Chemists, Washington DC,

Bationo A, Buekert A, Sedogo MP, Christianson BC, Mokwunye AU (1995) A critical review of crop-residue use as soil amendment in the West African semi-arid tropics. In: Powell JM, Fernandez-Rivera S, Williams TO, Renard C (eds) Livestock and sustainable nutrient cycling in mixed farming systems of sub-Saharan Africa. Volume II: Technical papers. Proceedings of an International Conference. 22–26 November, 1993. ILCA, Addis Ababa, Ethiopia

Brady NG (1984) The nature and properties of soils, 9th edn. Macmillan Publishing Company, New York, NY

Cherr CM, Scholberg JMS, McSorley R (2006) Green manure approaches to crop production: a synthesis. Agron J 98: 302–319

Crespo G, Rodriguez I, Dias MF, Lok S (2005) Accumulation and decomposition rates and N, P, and K returned to the soil by the litter of tropical legumes and grasses. In: Proceedings of the XX International Grassland Congress: Offered papers. Wageningen Academic Publishers, The Netherlands

Crowder LAV, Chheda HR (1982) Tropical grassland husbandry. Longman, London

de Leeuw PN (1991). The study area: biophysical environment. In: Bekure S, de Leeuw PN, Grandin BE, Neate PJH (eds) Maasai herding: an analysis of the livestock production system of Maasai pastoralists in eastern Kajiado District, Kenya. International livestock centre for Africa, Addis Ababa, Ethiopia

Giller KE (2001) Nitrogen fixation in tropical cropping systems, 2nd edn. Department of Soil Science and Agricultural Engineering, University of Zimbabwe, Harare

Giller KE, Wilson KJ (1991) Nitrogen fixation in tropical cropping systems. CAB International, UK

GoK (1991) National Atlas of Kenya, 4th edn. Survey of Kenya, Government of Kenya, Nairobi

Grace J, Jose JS, Meir P, Miranda HS, Montes RA (2006) Productivity and carbon fluxes of tropical savannas. J Biogeogr 33:387–400

Guretzky JA, Moore KJ, Burras CL, Brummer EC (2004) Distribution of legumes along gradients of slope and soil electrical conductivity in pastures. Agron J 96:547–555

Hinga G, Muchena FN, Njihia CM (eds) (1980) Physical and chemical methods of soil analysis. National Agricultural Laboratories, Ministry of Agriculture, Nairobi

Kinyamario JI, Macharia JNM (1992) Aboveground standing crop, protein content and dry matter digestibility of a tropical grassland range in the Nairobi National Park, Kenya. Afr J Ecol 30:33–41

Landon JR (ed) (1984) Booker tropical soil manual: a handbook for soil survey and agricultural land evaluation in the tropics and subtropics. Booker Agriculture International Ltd. Longman, New York, NY

LRNP (1999) Legume research network project newsletter. Issue No.1. Kenya Agricultural Research Institute, Nairobi

Macharia PN, Kinyamario JI, Ekaya WN, Gachene CKK, Mureithi JG (2007) Forage legumes for improvement of grassland productivity in semi-arid smallholder agro-pastoral systems in Kenya. In: Hare MD, Wongpichet K (eds) Proceedings of an international forage symposium 'forages: a pathway to prosperity for smallholder farmers'. Ubon Ratchathani University, Thailand, pp 187–202

Marschner H (1995) Mineral nutrition of higher plants, 2nd edn. Academic, London

Meissner HH, Zacharias PJK, O'Reagain PJ (2000) Forage quality (feed value). In: Tainton NM (ed) Pasture management in South Africa. University of Natal Press, Pietermaritzburg, South Africa

Miles N, Manson AD (2000) Nutrition of planted pastures. In: Tainton NM (ed) Pasture management in South Africa. University of Natal Press, Pietermaritzburg, South Africa

Mureithi JG, Tayler RS, Thorpe W (1994) The effects of alley cropping with *Leucaena leucocephala* and of different management practices on the productivity of maize and soil chemical properties in lowland coastal Kenya. Agroforest Syst 27:31–51

Muriuki AW, Qureshi JN (2000) Fertiliser use manual. Kenya Agricultural Research Institute, Nairobi

Njarui DMG, Beattie WM, Jones RK, Keating BA (2004) Evaluation of forage legumes in the semi-arid region of eastern Kenya. I. Establishment, visual bulk rating, insect pests and diseases incidences of a range of forage legumes. Trop Subtrop Agroecosyst 4:33–55

Njunie MN, Reynolds L, Mureithi JG, Thorpe W (1996) Evaluation of herbaceous legume germplasm for coastal lowland East Africa. In: Ndikumana J, de Leeuw P (eds) Sustainable feed production and utilization for smallholder livestock enterprises in sub-Saharan Africa. Proceedings of the Second African Feed Resources Network (AFRNET). 6–10th December, 1993. Harare, Zimbabwe

Nyathi P, Kimani SK, Jama B(2003) Soil fertility management in semi-arid areas of East and Southern Africa. In: Gichuru MP, Bationo A, Bekunda MA (eds) Soil fertility management in Africa: a regional perspective. TSBF-CIAT, Nairobi

Rowell DL (1994) Soil science: methods and applications. Department of Soil Science, University of Reading, United Kingdom

Skerman PJ, Cameron DG, Riveros F (1988) Tropical forage legumes, 2nd edn. FAO, Rome

Sombroek WG, Braun HMH, van der Pouw BJA (1980) Exploratory soil map and agro-climatic zone map of Kenya. E1 Report. Kenya Soil Survey, Nairobi

Steele KW, Vallis I (1988) The nitrogen cycle in pastures. In: Wilson JR (ed) Advances in nitrogen cycling in agricultural ecosystems. CAB International, UK

Tilley JMA, Terry RA (1963) A two-stage technique for the in vitro digestion of forage crops. J B Grassland Soc 18: 104–111

Too DK (1985) Effects of defoliation frequency and intensity on production of four burned and unburned bushed grassland communities in south-central Kenya. MSc thesis, Texas A&M University, USA

Van Soest PJ (1963) The use of detergents in the analysis of fibrous feeds. II. A rapid method for determination of fibre and lignin. J Assoc Off Agric Chem 46:829

Zemenchick RA, Albrecht KA, Shaver RD (2002) Improved nutritive value of Kura clover- and birds-foot trefoil-grass mixtures compared with grass monocultures. Agron J 94:1131–1138

Potential of Cowpea, Pigeonpea and Greengram to Contribute Nitrogen to Maize in Rotation on Ferralsol in Tanga – Tanzania

A.E.T. Marandu, J.P. Mrema, E. Semu, and A.S. Nyaki

Abstract A glasshouse incubation of soil from field experiment where cowpea, pigeonpea and greengram were grown in rotation with maize was carried out at Mlingano Agricultural Research Institute. The objective was to study the N mineralization of the soil and its potential to supply N to the maize crop following legume rotation. The experiment comprised of eight treatments which included the three legumes in rotation with maize, with the maize stover removed or retained on the plots and continuous maize with the stover removed or retained. Soil sampling for the incubation was carried out before maize planting from treatment plots at 0–20 cm depth, the soil sieved through a 6-mm screen while fresh, and then 250 g of the soil incubated in 500 ml volumetric flasks at 60% water-holding capacity for 42 days. Destructive samplings were done at 14-day intervals and analysed for mineral N. The N mineralization increased with incubation time, with cowpea and pigeonpea having significantly higher quantities at the 42nd day sampling compared to those from greengram and continuous maize plots. The contributions of cowpea, pigeonpea and greengram to the soil N at this sampling time were equivalent of 47, 40 and 13 kg ha^{-1}, respectively. Out of the total N mineralized during the entire period of incubation, 67, 51, 86 and 80% were on the 14th day for the cowpea, pigeonpea, greengram and continuous maize plots, respectively. Such early mineralization is not in synchrony with maize plants demand necessitating top dressing of N fertilizer as supplement for increased maize yields.

Keywords Legume residues · Maize stover · Nitrate leaching · Nitrogen mineralization

Introduction

Nitrogen is one of the major elements required by plants for growth and production. The nutrient is depleted in most soils in sub-Saharan Africa to a level that adversely affects soil productivity. It has been reported that nutrient depletion in this region has reduced the potential yields of some crops two to four times (Bationo, 2003). Nutrient balance studies in Tanga region of Tanzania have shown that the rate of N loss from arable land is 32 kg N ha^{-1} year^{-1} (Mbogoni et al., 2004). Continuous cropping without appropriate land management practices to replenish the N taken and harvested with crops is the primary cause of this negative N balance (CP-URT, 2000). Such situation contributes much to declining land productivity leading to increased food insecurity of the farming communities.

Cereal–legumes cropping system is among the strategies to alleviate the N depletion in cereal monocropping practice. The system enriches the soil with nitrogen through above- and below-ground residues, particularly for legumes with low N harvest index (Jeranyama et al., 2000). Part of the N is being derived from the soil and part from atmospheric N$_2$ fixation.

Maize grain yield increase when grown in rotation with legumes has been reported in various studies and was attributed to improvement in the soil N status (Horst and Hardter, 1994; Adetunji, 1996; Rao and Mathuva, 2000; Bloem and Barnard, 2001). Cowpea,

A.E.T. Marandu (✉)
Mlingano Agricultural Research Institute, Tanga, Tanzania
e-mail: atanasiom@yahoo.co.uk

A. Bationo et al. (eds.), *Innovations as Key to the Green Revolution in Africa*,
DOI 10.1007/978-90-481-2543-2_31, © Springer Science+Business Media B.V. 2011

for example, has been reported to increase maize grain yields equivalent to application of mineral fertilizer more or equal to 32 kg N ha^{-1} (Dakora et al., 1987; Adetunji, 1996; Bloem and Barnard, 2001), while pigeonpea has been reported to contribute equivalent of 30–70 kg N ha^{-1} to the succeeding maize (Peoples and Herridge, 1990). The yield increase has also been attributed to some positive rotational effects (Johnson et al., 1992; George et al., 1994; Ortas et al., 1996; Vaughan and Evanylo, 1999; Bagayoko et al., 2000; Rao and Mathuva, 2000).

Availability of N from crop residues depends on the decomposition rates and their synchronization with crop demand (Palm et al., 1997; Muza and Mapfumo, 1998; Whitbread et al., 2002). The ammonium (NH_4^+) formed during the decomposition process is relatively immobile and is subject to nitrification under aerobic conditions, leading to the formation of nitrate (NO_3^-), which is a relatively mobile form (Mekonnen et al., 1997; Purnomo et al., 2000). The NO_3^- has great potential for loss from the soil through denitrification, particularly under anaerobic condition and leaching.

Both the NH_4^+ and NO_3^- forms are taken by plants but the NO_3^- is normally the dominant form. Its concentration in the top soil changes due to plant uptake, microbial immobilization, leaching and gaseous losses (Warren et al., 1997). Nitrogen flush at the onset of rains following a dry season has been reported to substantially increase the NO_3^- concentration of the top soil (Wong and Nortcliff, 1995; Weber et al., 1995; Warren et al., 1997). Hagedorn et al. (1997) observed a constant mineral N in the 0–20 cm during a dry season. However, the mineral n content doubled within 5 days at the onset of rainfall due to N flush and then decreased by 50–70% within the first 2 weeks through leaching caused by heavy rains.

Pre-season mineral N is among the tests used to estimate the N availability to plants and has been reported to correlate well with maize grain yield (Weber et al., 1995; Barrios et al., 1998).

Farmers in Muheza district grow maize in rotation with cowpea, pigeonpea or greengram. However, the advantages of such cropping system have not been optimized as maize plants show N deficiency symptoms and yields as low as 1.3 t ha^{-1} are common (Ministry of Agriculture and Cooperatives, 1998). This situation indicates that either there is a minimal contribution of N by legumes to the subsequent rotational maize or N is lost through various processes during and between rotation cycles. This research was therefore designed to investigate (i) nitrogen mineralization pattern of incubated soil sampled 2 weeks before the onset of rains and (ii) pre-season mineral N of soil sampled at maize planting following legume rotation. The objective was to assess the potential of soils to supply N to maize in such cropping system.

Materials and Methods

Field Experiment

This research was carried out at Mlingano Agricultural Research Institute in Muheza district, Tanga region, Tanzania. The institute is located at 39°52′E and 5°10′S at an altitude of 183 m.a.s.l. The area receives a bimodal rainfall pattern with long rains between March and June and short rains between October and December. Cowpea, pigeonpea and greengram are grown during short rains followed by maize during long rains.

A rotational field experiment was established in 2002 during short rains with eight treatments which included the three legumes in rotation, with the maize stover removed or retained in the plots, and continuous maize, with the maize stover removed or retained on the plots. The treatments were arranged in a randomized complete block design with three replications. During legume rotation, the continuous maize plots were not grown with a crop, but maintained weed-free during weeding operations of the legume plots. At legume harvesting, the above-ground residues were weighed and left on respective plots. During long rains in 2003, maize was grown on treatment plots without application of nitrogen fertilizer. At maturity, maize cobs were harvested and the above-ground stover weighed and then removed or uniformly spread on respective plots. The second cycle of rotation started in October 2003 with planting of the legumes as before. Two weeks before maize planting during the second cycle in 2004, soil sampling for incubation study was carried out as detailed below. Further, at maize planting time, soil for pre-season mineral N determination was sampled from 0 to 20 cm using an auger and packed in polythene bags before being subsequently analysed for NH_4^+ and NO_3^- following the procedure under incubation experiment. Before the experiment was laid out, a composite soil sample for site characterization was collected from 0 to 20 cm depth, air dried, ground and then passed through a 2-mm sieve before the analysis.

Incubation Experiment

Two weeks before maize planting in 2004 season, soil samples for incubation were collected from 0 to 20 cm depth using a shovel at four positions in each treatment plot. The fresh soil samples were passed through a 6-mm sieve and portions equivalent to 250 g oven dry weight were placed in 500 ml volumetric flasks. These samples were brought to 60% water-holding capacity and incubated in a glasshouse at room temperature for 42 days. The moisture content of incubated soil was maintained at this level by constant addition of distilled water as necessary. Destructive sampling was carried out at 14-day intervals for mineral N determination as follows: Ten gram portions of fresh soil were weighed into plastic bottles, and 100 ml of 2 M KCl was added and shaken in a horizontal position on a shaker for 1 h. The suspensions were filtered through pre-washed Whatman no. 42 filter paper. Ammonium N (NH_4^+–N) and nitrate N (NO_3^-–N) were determined from the filtrates by steam distillation method according to Okalebo et al. (1993). The mineral N was subjected to analysis of variance using Mstat-C statistical package and means separated by Duncans New Multiple Range Test.

Results and Discussion

Soil Characterization

The chemical and physical characteristics of the experimental site are presented in Table 1. The field experimental soil is classified as Rhodic Ferralsol. According to Landon (1991), the soils' reaction was medium acid. The total N was very low, indicating a need for external N input for high maize yields. The organic carbon was very low, whereas the C:N ratio indicated presence of a good-quality soil organic matter. The site had low available P and exchangeable Ca. The exchangeable K and Mg were medium whereas the CEC was low.

Incubation Experiment

The quantities of mineral N of incubated soils are presented in Table 2.

The mineral N at the time of soil sampling in the field (0-day incubation) ranged from 5.7 mg kg^{-1} (continuous maize with stover left on the plots) to 18.1 mg kg^{-1} (greengram–maize rotation with stover left on the plots). During incubation, the quantities of mineral N at 14, 28 and 42 sampling days increased with incubation time. Higher mineral N was obtained from soils where the maize stover was left on the plots than where it was removed. During the 14th day sampling in treatments where the pigeonpea was grown, significantly lower mineral N was observed compared to those of cowpea or greengram regardless of the stover management practice. This observation was attributed to immobilization of soil N during decomposition of the low N pigeonpea residues (Table 3) which was harvested at the end of January 2004 whereas the cowpea and greengram were harvested in mid-December 2003. (sampling for the incubation was carried out in March 2004). The pigeonpea residue therefore was at earlier stages of decomposition compared to those of cowpea and greengram. The differences were however not observed beyond the 14th day of sampling. At the end of incubation among the legumes, significantly lower mineral N was determined in samples from plots where greengram was grown compared to those under the cowpea or pigeonpea, which reflects numerically the lowest quantity of residue added to the soil and the expected amount of N to be released upon complete decomposition (Table 3).

The increase in mineral N with incubation was not the same between sampling intervals. The highest

Table 1 Physical and chemical characteristics of the experimental site

Texture	pH	OC	Total N	C:N	Available P	Exchangeable				CEC
						Ca	Mg	K	Na	
	H$_2$O	(%)			.(mg kg^{-1})	(cmol (+) kg^{-1})				
Sandy clay	5.6	1.22	0.09	13	3	2.0	0.5	1.02	0.15	6.33

Table 2 Cumulative quantities of mineral N (mg kg^{-1}) at 0, 14, 28 and 42 days of incubated soils

Treatment	Sampling intervals (days)			
	0	14	28	42
Cowpea/maize minus stover	17.2	34.3 cd	42.3ab	56.4ab
Pigeonpea/maize minus stover	11.1	28.4e	37.4b	52.0bc
Greengram/maize minus stover	15.0	33.9 cd	39.3b	41.1d
Continuous maize minus stover (control)	6.7	26.7e	28.1c	35.2d
Cowpea/maize plus stover	16.3	44.1a	48.9a	60.2ab
Pigeonpea/maize plus stover	11.2	30.6de	41.5b	64.6a
Greengram/maize plus stover	18.1	39.9ab	42.8ab	44.3 cd
Continuous maize plus stover	5.7	36.1bc	37.3b	42.9 cd
F-test	Na	**	**	**
CV%		5.4	6.8	8.2

Na, not statistically analysed

Means followed by the same letter within the column are not significantly different according to Duncan's New Multiple Range Test

**Significant at p $\leq$ 0.05

Table 3 Legumes above-ground residues and total N content

Treatment	Residue (kg ha^{-1})	N (%)	Total N (kg ha^{-1})
Cowpea/maize minus stover	2,284	1.68	38.4
Pigeonpea/maize minus stover	1,715	1.49	25.6
Greengram/maize minus stover	1,441	1.32	19.0
Cowpea + maize plus stover	2,192	1.68	36.8
Pigeonpea/maize plus stover	1,755	1.49	26.2
Greengram/maize plus stover	1,576	1.32	20.8

increase was obtained between 0 and 14th day. In treatments where the stover was removed, the proportions of mineral N at the 14th day sampling in plots where the cowpea, pigeonpea or greengram had been grown and continuous maize plots were 61, 55, 82 and 76% of the cumulative quantities at the end of incubation, respectively. For the soils from the treatment under stover left on the plots, the proportions were 73, 47, 90 and 84% for the soil sampled from cowpea, pigeonpea or greengram in rotation and from continuous maize plots, respectively.

The observed lower mineral N determined at the time of soil sampling in the field (0 day) relative to that obtained during incubation indicated that decomposition rates at the time of sampling in the field were lower than those during incubation period. This could be attributed to increased microbial activities of the incubated soil as a result of increased soil moisture content. During the time of soil sampling in the field, the soil was fairly dry. At the beginning of incubation, the moisture status of the soil was raised and maintained at 60% water-holding capacity

throughout the incubation period. This might have activated the soil microorganisms, particularly bacteria, leading to faster decomposition and release of the organically bound N from the crop residues and the native soil organic matter. Soil moisture is among the important factors, which influences aerobic decomposition processes (Scholes et al., 1994; Myers et al., 1994). The highest proportions of cumulative mineral N determined at 14th day sampling could be attributed to increased soil microbial activities accompanied by release of N from rapidly decomposable fractions of the legume residues and soil organic matter. The decreased rate of mineralization with incubation time could be attributed to slow decomposition of the more resistant organic fractions. Such phenomenon has also being reported in other studies (Palm, 1995; Trinsoutrot et al., 2000).

The high cumulative mineral N at 14th day incubation is of practical importance to N availability to maize under this cropping system. The quantities of N determined (when converted on hectare using the 2.2 × 10^6 kg of furrow slice soil) at this period in the treatments where the maize stover was removed were equivalent to 75, 62, 75 and 59 kg N ha^{-1} for the soils from the cowpea, pigeonpea or greengram rotation and that from continuous maize, respectively. In treatments where maize stover was left on the plots, the quantities of N were equivalent to 97, 67, 88 and 79 kg N ha^{-1} for the soils from the cowpea, pigeonpea or greengram rotation and that from continuous maize plots, respectively. These quantities of mineral N are released when the maize plants are at early stages

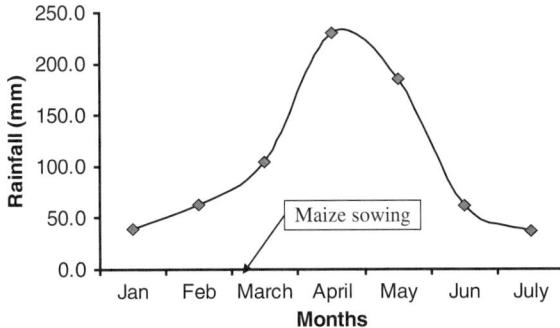

Fig. 1 Mean 10 year (1995–2004) rainfall distribution during maize rotation

of growth (14 days) with relatively low N demand. In most cases, such early stages of maize growth coincide with heavy rains (Fig. 1) which may subject the N to leaching and/or losses in gaseous forms resulting from denitrification.

The quantities of mineral N at 42nd day of incubation of soils from plots under cowpea, pigeonpea or greengram rotation and that from continuous maize with corresponding amounts per hectare are presented in Table 4. The quantities of mineral N per hectare, which could be attributed to the effects of the three legume rotation, were of the order greengram < pigeonpea < cowpea under the maize stover removed treatment and greengram < cowpea < pigeonpea where the stover was left on the plots. Theoretically, the quantities of N from the cowpea and pigeonpea rotation plots are about the same as the 50 kg N ha^{-1} that is recommended for continuous monocrop maize at the experimental site (Mowo et al., 1993). In practice, not all of this amount of N would become available to the maize plants due to lack of synchrony with the maize

crop N demand as already stated above. This could account for the maize plants' N deficiency symptoms and lower grain yields obtained in this cropping system compared to maize under continuous monocropping with application of the recommended 50 kg N ha^{-1}.

The potential of soils to supply mineral N under aerobic mineralization is also reported in other incubation studies. Horst and Hardter (1994), for example, observed that net N mineralization after 42 days incubation of top soil sampled from plots previously grown with cowpea in rotation with maize was higher by 33% as compared to that from maize monocropped plots. They attributed this to the N released from above- and below-ground cowpea residues. The study, however, was on a different soil type sampled from 0 to 30 cm, air-dried and sieved through a 2-mm sieve before incubation (as compared to a 6-mm sieve in our study). The deeper soil sampling accompanied by the finer grinding and sieving might have eliminated a substantial quantity of organic matter, whereas air-drying of the samples before incubation might have killed some of the microorganisms. In our study, the soil was sampled from 0 to 20 cm depth, sieved through a 6-mm screen, and then incubated while fresh. In so doing, the soil microbial environment was less disturbed. This could account for the relatively higher mineral N obtained.

Pre-season Mineral N

The quantities of pre-season mineral N for soil samples taken at the time of maize planting are shown in Table 5. As was the case for mineral N from the above incubation study, mineral N was higher in plots where legumes were grown compared to the

Table 4 Mineral N at 42nd day of incubation and corresponding values per hectare

	Mineral N	Net N[a]	N equivalent #
Treatment	(mg kg^{-1})		(kg ha^{-1})
Cowpea/maize minus stover	56.4	21.2	46.6
Pigeonpea/maize minus stover	52.0	16.8	40.0
Greengram/maize minus stover	41.1	5.9	13.0
Continuous maize minus stover (control)	35.2	Rd	Rd
Cowpea/maize plus stover	60.2	25.0	55.0
Pigeonpea/maize plus stover	64.6	29.4	64.7
Greengram/maize plus stover	44.3	9.1	20.0
Continuous maize plus stover	42.9	7.7	16.9

[a]Net N is the difference between mineral N of treatments in rotation and continuous maize with stover removed (control); # equivalent N ha^{-1} was obtained by converting the net N on hectare using the 2.2 × 10^6 kg of furrow slice soil; Rd = data for mineral N was the reference

Table 5 Quantities of mineral N and NO_3^- proportions at maize planting

Treatments	Total	NH_4^+	NO_3^-	NO_3^- proportion
		(mg kg^{-1})		(%)
Cowpea/maize minus stover	28.15	10.1	18.05	64
Pigeonpea/maize minus stover	42.5	12.15	29.85	71
Greengram/maize minus stover	30.4	7.55	22.85	75
Continuous maize minus stover	21.65	6.71	14.94	69
Cowpea/maize plus stover	30.35	8.65	21.7	72
Pigeonpea/maize plus stover	42.8	10.95	31.85	74
Greengram/maize plus stover	37.4	9.4	28.0	75
Continuous maize plus stover	24.45	5.35	19.07	78

continuous maize plots which could be attributed to mineralization of N from legume residues from the previous season (Table 3) and/or native soil organic matter.

The contribution of mineral N from the decomposition of the maize stover that was left on the respective plots (Table 6) could also account for the observed increased mineral N. This was shown by the relatively higher mineral N in treatments where the stover was left on the plots than where the stover was removed.

More than 69% of the mineral N was in NO_3^- form. Higher proportions of pre-season NO_3^- relative to those of NH_4^+ were also reported in other field experiments (Weber et al., 1995; Maroko et al., 1998; Barrios et al., 1998). Presence of such high proportion of NO_3^- at this time of maize planting indicates high possibility of some of the NO_3^- being leached by heavy rains usually received at this time of the season (Fig. 1). These rains may have also caused anaerobic pockets in the soil leading to denitrification of the NO_3^- to gaseous forms, thereby reducing the N which would have been available to the maize plants as stated under the incubation study.

Conclusions

There is a potential to increase the soil N following legume rotation which could contribute to the maize N requirement in this cropping system. Legumes grown in rotation with maize contribute to the maize N requirement. However, there is lack of synchrony between N mineralization and maize plants demand as much of the N is released at about 2 weeks of maize growth when its requirement is low, subjecting it to leaching. Increased maize plants N demand is therefore coupled with low N availability and hence maize plants N deficiency symptoms and low yields. To increase and sustain high maize yields in this cropping system, supplementation with mineral N fertilizer about 1 month after maize planting is necessary.

Acknowledgements This research was funded by the Ministry of Agriculture and Food Security, Tanzania, through the Tanzania Agricultural Research Project Phase 11 (TARP 11) for which we are most grateful. The first author would like to appreciate the financial support from AfNet to participate in this symposium.

Table 6 Quantities of maize stover and N removed or added to the soil in the field

Treatment	Stover	N[1]
	(kg ha^{-1})	
Cowpea/maize minus stover	1,923	−14.8
Pigeonpea/maize minus stover	1,614	−12.4
Greengram/maize minus stover	1,889	−14.5
Continuous maize minus stover	1,236	−9.5
Cowpea/maize plus stover	2,472	+19.0
Pigeonpea/maize plus stover	2,056	+15.8
Greengram/maize plus stover	1,683	+13.0
Continuous maize plus stover	1,387	+10.7

− = N removed in maize stover; + = N added in maize stover; N[1], content of stover = 0.77%

References

Adetunji MT (1996) Nitrogen utilization by maize in maize – cowpea sequential cropping of an intensively cultivated tropical Ultisol. J Ind Soc Soil Sci 44:85–88

Bagayoko M, Buerkert A, Lung G et al (2000) Cereal/legume rotation effects on cereal growth in Sudano – Sahelian West Africa: soil mineral nitrogen, mycorrhizae and nematodes. Plant Soil 218:103–116

Barrios E, Kwesiga F, Buresh RJ et al (1998) Relating pre-season soil nitrogen to maize yield in tree legume–maize rotations. Soil Sci Soc Am J 62:1604–1609

Bationo A (2003) Introduction. In: Gichuru MP, Bationo A, Bekunda MA et al (eds) Soil fertility management in

Africa: a regional perspective. Academy Science Publishers, Nairobi, Kenya, pp xiv–xvi

Bloem AA, Barnard RO (2001) Effect of annual legumes on soil nitrogen and on the subsequent yield of maize and grain sorghum. South Afric J Plant Soil 18(2):56–61

CP-URT (2000) Tanzania soil fertility initiative: concept paper. Report No.00/081. Ministry of Agriculture and Cooperatives/FAO-Investment Center. FAO/World Bank Cooperative Programme

Dakora FD, Aboyinga RA, Mahama Y et al (1987) Assessment of N_2 fixation in groundnuts (*Arachis hypogaea* L.) and cowpea (*Vigna unguiculata* L.Wasp) and their relative contribution to a succeeding maize crop in Northern Ghana. J Appl Microbiol Biochem 3:389–399

George E, Romheld V, Marschner H (1994) Contribution of mycorrhizal fungi to micronutrient uptake by plants. In: Manthey JA, Crowle DE, Luster DG (eds) Biochemistry of metal micronutrients in the rhizosphere. CRC Press, Boca Raton, FL, pp 93–109

Hagedorn F, Sterner KG, Sekayange L et al (1997) Effect of rainfall pattern on nitrogen mineralization and leaching in a green manure experiment in South Rwanda. Plant Soil 195:365–375

Horst WJ, Hardter R (1994) Rotation of maize with cowpea improves yield and nutrient use of maize compared to maize monocropping in an alfisol in northern Guinea Savanna of Ghana. Plant Soil 160:171–183

Jeranyama P, Hesterman BO, Waddington SR et al (2000) Relay-intercropping of sunnhemp and cowpea into a smallholder maize system in Zimbabwe. Agron J 92:239–244

Johnson CN, Copeland PJ, Crookston RK et al (1992) Mycorrhizae: possible explanation for yield decline with continuous corn and soybean. Agron J 84:387–390

Landon JR (ed) (1991) Booker tropical soil manual. A handbook for soil survey and agricultural land evaluation in the tropics and subtropics. Wiley, New York, NY

Maroko JB, Buresh J, Smithson PC (1998) Soil nitrogen availability as affected by fallow-maize systems on two soils in Kenya. Biol Fertil Soils 26:229–234

Mbogoni JDJ, Wickama JM, Kiwambo BJ et al (2004) Nutrient flows and their implication for sustainability of farming systems in the Eastern Zone of Tanzania. Technical report, Mlingano Agricultural Research Institute, Tanga, Tanzania

Mekonnen K, Buresh RJ, Jama B (1997) Root and inorganic distribution of sesbania fallows, natural fallow and maize. Plant Soil 188:319–327

Ministry of Agriculture and Cooperatives (1998) Basic Data, Agricultural and Livestock Sector. Dar es Salaam, Tanzania

Mowo JG, Floor J, Kaihura FBS et al (1993) Review of fertilizer recommendations in Tanzania. Soil Fertility Report F6, Mlingano, Tanga, Tanzania

Muza L, Mapfumo P (1998) Constraints and opportunities for legumes in the fertility enhancement of sandy soils in Zimbabwe. In: Proceedings of Eastern and Southern Africa Region Maize Conference, Adis Ababa, Ethiopia, 21–25 Sept 1998

Myers RJK, Palm CA, Cuevas E et al (1994) The synchronisation of nutrient mineralization and plant nutrient demand. In: Woomer PL, Swift MJ (eds) The biological management of tropical soil fertility. Wiley, Chichester, UK, pp 81–116

Okalebo JR, Gathua KW, Woomer PL (1993) Laboratory methods of soil and plant analysis. A working manual. KARI, SSEA, TSBF, UNESCO-ROSTA, Nairobi

Ortas I, Harris PJ, Rowel DL (1996) Enhanced uptake of phosphorus by mycorrhizal sorghum plants as influenced by forms of nitrogen. Plant Soil 184:255–264

Palm CA (1995) Contribution of agroforestry trees to nutrient requirements of intercropped plants. Agroforest Syst 30:105–124

Palm CA, Myers RJK, Nandwa SM (1997) Combined use of organic and inorganic nutrient sources for soil fertility maintenance and replenishment. In: Buresh RJ, Sanchez PA, Calhoun F (eds) Replenishing soil fertility in Africa. Soil Sci Soc Am Special Publication No 51, Madison, WI, USA, pp 193–217

Peoples MB, Herridge DF (1990) Nitrogen fixation by legumes in tropical and subtropical agriculture. Adv Agron 44:155–223

Purnomo E, Black AS, Conyers MK (2000) The distribution of net nitrogen mineralization within surface soil. 2. Factors influencing the distribution of net N mineralization. Aust J Soil Res 38:643–652

Rao MR, Mathuva MN (2000) Legumes for improving maize yields and income in semi-arid Kenya. Agric Ecosyst Environ 78:123–137

Scholes RJ, Dala R, Singer S (1994) Soil physics and fertility: the effects of water, temperature and texture. In: Woomer PL, Swift MJ (eds) Biological management of tropical soil fertility. Wiley, Chichester, UK, pp 117–136

Trinsoutrot I, Recous S, Bentz B et al (2000) Biochemical quality of crop residues and carbon and nitrogen mineralization kinetics under nonlimiting nitrogen conditions. Soil Sci Soc Am J 64:918–926

Vaugham JD, Evanylo GK (1999) Soil nitrogen dynamics in winter cover crop-corn system. Commun Soil Sci Plant Anal 30:31–52

Warren GP, Atwal SS, Irungu JW (1997) Soil nitrate variations under grass, sorghum and bare fallow in semi arid Kenya. Exp Agric 33:321–333

Weber B, Chude V, Pleysier J et al (1995) On-farm evaluation of nitrate nitrogen dynamics under maize in the northern Guinea savanna of Nigeria. Exp Agric 31:333–344

Whitbread A, Jiri O, Maasdorp B et al (2002) The movement and loss of soil nitrate and labile C in a tropical forage legume/ maize rotation in Zimbabwe. http://www.sfst.org/proceedings/17WCS-CD/Abstracts/0719.pdf site visited on 28/4/2004

Wong MTF, Nortcliff S (1995) Seasonal fluctuations of native available N and soil management implications. Fertil Res 42:13–26

Model Validation Through Long-Term Promising Sustainable Maize/Pigeon Pea Residue Management in Malawi

C.D. Mwale, V.H. Kabambe, W.D. Sakala, K.E. Giller, A.A. Kauwa, I. Ligowe, and D. Kamalongo

Abstract In the 2005/2006 season, the Model Validation Through Long-Term Promising Sustainable Maize/Pigeon Pea Residue Management experiment was in the 11th year at Chitedze and Chitala, and in the 8th year at Makoka and Zombwe. The experiment was a split-plot design with cropping system as the main plot and residue management as the sub-plot. All treatments were subjected to two fertilizer regimes. In the first regime, there was no addition of inorganic fertilizer and in the second, there was addition of inorganic fertilizer at area-specific fertilizer recommendation rate. The evaluation was done at Chitala, Chitedze, Makoka and Zombwe. Significant differences ($P \leq 0.05$) were observed in maize grain yield among sites and cropping systems. Highest grain yields were recorded at Chitedze (5,342 kg/ha). However, the response trend in grain yield to different cropping systems remained the same in all sites. Best yields were recorded in maize grown following pigeon pea in rotation system followed by maize intercropped with pigeon pea. The addition of inorganic fertilizer increased maize yield significantly. Removal or retention of crop residue in the field did not contribute any significant yield increase of maize across sites. For resource-poor smallholder farmers, growing maize/pigeon pea in rotation and maize intercropped with pigeon pea seems to be more profitable in terms of resource utilization and soil fertility improvements.

Keywords Maize/pigeon pea cropping systems · Long-term sustainability · Maize grain yield · Model validation

Introduction

In order to clearly understand short- and long-term nitrogen dynamics in the cereal/legume system, such as maize/pigeon pea intercropping, there is a need to obtain information on soil and crop yields from the field experiments on a long-term basis. One way of describing the sustainability of a cropping system would be to follow nitrogen and carbon pools in the system. Several ways for describing carbon and nitrogen pools in agro-ecosystems exist through the use of mechanistic models (Rao et al., 1982; Parton et al., 1987; Young and Muraya, 1991; Kumwenda et al., 1997). Models are simplified, formal representations of relationships between defined quantities or qualities in physical and mechanical terms (Jeffers, 1982). Although models are widely used in agro-ecosystems, soil spatial variability is the main limiting factor to model development and verification under field conditions (Rao et al., 1982). In this study, the CENTURY model (Parton et al., 1987) was used to determine the long-term effects of retaining and removing crop residues in a maize/pigeon pea intercropping system. This model was chosen because it allows the study of both carbon and nitrogen pools at the same time. The dynamic nature of this model is an essential component for simulating plant residue decomposition with time.

C.D. Mwale (✉)
Chitedze Research Station, Department of Agricultural Research Services, Ministry of Agriculture and Food Security, Lilongwe, Malawi
e-mail: cyprianmwale@yahoo.com

Sakala and Kauwa (Deceased)

A. Bationo et al. (eds.), *Innovations as Key to the Green Revolution in Africa*,
DOI 10.1007/978-90-481-2543-2_32, © Springer Science+Business Media B.V. 2011

SCUAF is a simulation model developed at the International Centre for Research in Agroforestry (ICRAF) Nairobi, Kenya. The model consists of a plant compartment, which deals with the processes occurring with the plant material before entering into the soil, and a soil compartment, which simulates what happens in the soil. The model encompasses two plant components, tree and crop, which can be present in a rotation or spatial system. It simulates carbon and nitrogen cycling and predicts a number of variables, which include changes in soil organic carbon (C) and nitrogen (N), biomass and grain yields and system nitrogen balance. The model programme initially requires information on soil conditions, plant growth parameters, organic and inorganic additions, crop removals and soil plant processes. In this study, maize and pigeon pea are grown under an intercropping system. In the model, pigeon pea is declared as the tree component and maize as the crop component growing in a 1:1 spatial mixture.

The CENTURY Agro-Ecosystem Version 4.0 model was developed as a project of the United States National Science Foundation Ecosystem Studies Research Projects. The model simulates the long-term dynamics of carbon, nitrogen, phosphorus and sulphur for different plant–soil systems. The primary purposes of the model are to provide a tool for ecosystem analysis, to test the consistency of data and to evaluate the effects of changes in management and climate on ecosystems. In this study, only the soil organic matter subcomponent of the model is used to predict long-term soil organic matter and maize yield changes under different maize/pigeon pea management regimes on a long-term basis.

Maize and pigeon pea intercropping under low input is common and seems like a promising way of growing maize and pigeon pea together in the southern parts of Malawi (Sakala et al., 1996; Kumwenda et al., 1997) where average land-holdings are small. Despite this common practice, the long-term effects on maize yields and soil organic C and N under such low input system have not been measured. The rationale behind this long-term experiment is that by comparing soil changes in organic C, N and maize yield in sole maize with that of intercropped maize and pigeon peas, a prediction of the long-term benefits or disadvantages of these management options could be determined and appropriate advice could be given for the overall maize/pigeon pea intercropping system. In the simulations, several possible options are considered.

The general goal of the experiment was to predict long-term changes in organic C, N and texture when residues of different amounts were added or removed from the field during crop harvest and the effect of continuous maize cropping on soil changes. The specific objectives of the study were (1) to follow soil changes, mainly C, N, texture and maize yields under long-term maize/pigeon pea systems; (2) to compare simulations predicted by the CENTURY model based on different soil types, compositional soil and litter characterization; and (3) to assess model short- and long-term fitness to the measured maize yields, soil organic C, N and texture.

Materials and Methods

The trial was initially established at five sites, namely Chitala, Chitedze, Lisasadzi, Makoka and Zombwe. Chitala is located at latitude 13°40′ South and longitude 34°15′ East in the Low-Altitude Plain (Salima Lakeshore Plain). It has an altitude of 606 m above sea level. There is high variability in mean annual rainfall (800 mm) which is poorly distributed, occurring within 3 months (normally between December and March). The mean annual maximum temperature is 28°C and with a minimum of 16°C. The soils are Ferruginous Latosol (sandy–clay–loam). Chitedze is located at latitude 13°59′ South and longitude 33°38′ East with an elevation of 1,146 m above sea level. It lies in the Mid-Altitude Plateau of Central Africa (Lilongwe Plain). It has a mean maximum temperature of 24°C and a minimum of 16°C. The station has medium variability rainfall with a long-term annual mean of 892 mm. Rainfall occurs between November and March. The soils are Alfisols/Oxisols (sandy–clay–loam) which are well supplied with organic matter and exchangeable cations. Lisasadzi lies at latitude 13°05′ South and longitude 33°29′ East in the Mid-Altitude Plateau (Kasungu-Lilongwe Plain). Means of annual temperature and rainfall range are similar to Chitedze. The soils are Ferruginous Latosol (loam–sand).

Makoka lies at latitude 15°32′ South and longitude 35°11′ East with an elevation of 1,029 m above sea level. The site has a mean maximum temperature of 25°C and a mean minimum temperature of 15.6°C. The site has low variability rainfall with a mean annual of 1,044 mm, most of which occurs from November to March. The soils are Ferruginous

Latosol (sandy–loam). Zombwe lies at latitude 11°19′ South and longitude 33°49′ East at an elevation of about 1,233 m above sea level. The site mean maximum temperature is 24°C and the minimum temperature is 15°C. The mean annual rainfall is 890 mm occurring between December and April. The soils are Ferruginous Latosol (sandy–clay–loam).

Chitedze, Chitala and Lisasadzi sites were established in the 1995/1996 cropping season while Makoka and Zombwe were established during the 1999/2000 cropping season. Lisasadzi was discontinued and put under natural fallow starting from 2005/2006 season after the 2004/2005 maize/pigeon pea intercropping system results showed that there were competitions for growth resources between maize and pigeon pea. The experiment was a split-plot design with three factors: fertilizer addition, cropping system and residue management. Factors studied were sole maize, maize/pigeon pea rotation, maize/pigeon pea intercropping systems with fertilizer or without fertilizer application and crop residue management options (removing or retaining residue on plots) as outlined in Table 1. The main plot sizes were 14 ridges × 9 m long × 0.9 m between ridges × 0.9 m between planting stations and subplot sizes were 7 ridges × 9 m long × 0.9 m between ridges × 0.9 m between planting stations. Four maize seeds were planted per station and later thinned to three seeds. Hybrid DK8031 was planted substituting MH18, the initial variety index of the model.

Inorganic fertilizers were applied using area-specific fertilizer recommendation rate of 92-21-0 + 4S kg/ha N-P_2O_5-K_2O for Chitedze, Chitala, Lisasadzi, Zombwe and 69-21-0 + 4S kg/ha N-P_2O_5-K_2O for Makoka. Fertilizer sources were 23:21:0 + 4S (NPK) basal dressed at planting or soon after germination and top dressed with urea (46% N) 3 weeks after planting.

Data were collected on site characteristics (soil, temperature, rainfall and cultivation practices), plant characteristics (plant genetic potential, chemical composition), external additions (fertilizers, organic additions) and grain and stover yields.

Results and Discussion

Results presented in Figs. 1, 2, 3, 4, 5 and 6 are for maize grain yields. Initial soil physical and chemical characteristic results are presented in Table 2 while final results on soil texture, soil organic carbon (OC), soil nitrogen (N) and nitrogen and phosphorus percentages in grain and stover will be presented at the final stage of the trial.

Maize grain yield (Fig. 1) differed significantly at $P \leq 0.05$ among sites due to treatment effects. The highest mean grain yield was recorded from Chitedze (5,342 kg/ha) and the lowest yield was recorded from Zombwe (2,188 kg/ha). Makoka site recorded the highest maize yield (7,108 kg/ha) while the lowest yield was recorded at Zombwe (977 kg/ha). Differences in site performance on maize grain yield could be due to site differences in soil types and climatic conditions. Despite differences in the sites, the response trend in maize yields to different cropping systems remained the same. The highest grain yield across sites was obtained when maize was grown in rotation with pigeon pea followed by maize/pigeon pea intercropping. The maize/pigeon pea rotation treatment gave

Table 1 Main and subplot treatment description for the experiment

| Plot | Fertility | Cropping system | Residue management | |
		Main plot	Subplot 1	Subplot 2
1	No fertilizer added	Sole maize	Minus stover	Plus stover
2		Maize/pigeon pea rotation	Minus pigeon pea stems + pigeon pea leaves	Plus pigeon pea stems + pigeon pea leaves
3		Maize/pigeon pea intercropping	Minus p/pea stems + p/pea leaves	Plus maize residues + p/pea leaves + p/pea stems
4	Fertilizer added	Sole maize	Minus stover	Plus stover
5	Area-specific recommendation	Maize/pigeon pea rotation	Minus pigeon pea stems + pigeon pea leaves	Plus pigeon pea stems + pigeon pea leaves
6		Maize/pigeon pea intercropping	Minus p/pea stems + p/pea leaves	Plus maize residues + p/pea leaves + p/pea stems

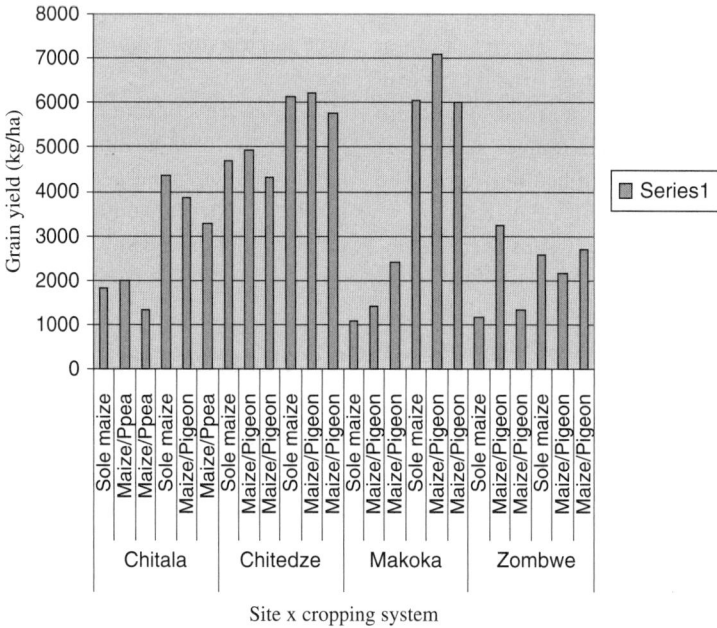

Fig. 1 Mean grain yield of maize at Chitala, Chitedze, Makoka and Zombwe in the 2005/2006 season

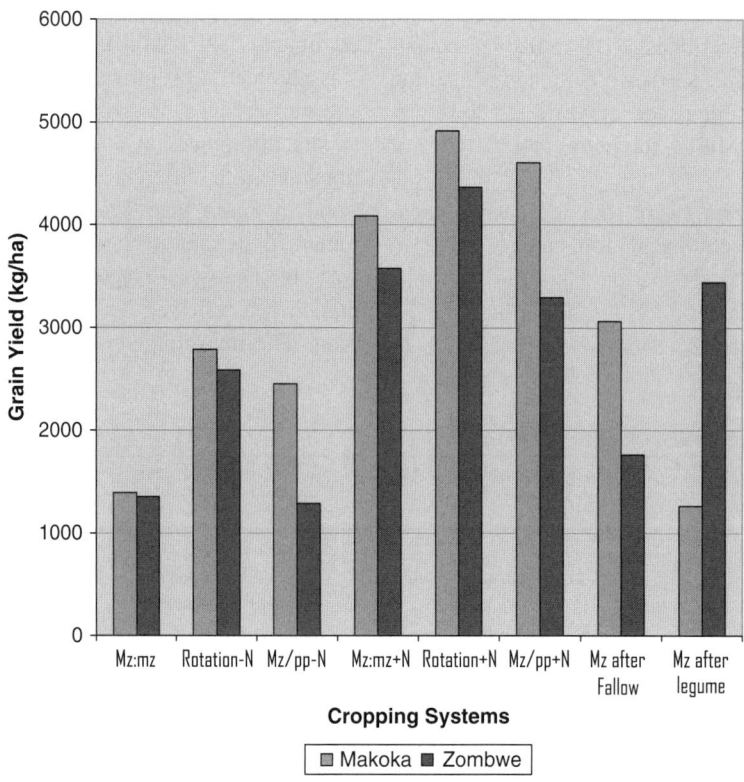

Fig. 2 Maize yield response to different cropping systems over a period of 6 years (2000–2005) at Makoka and Zombwe

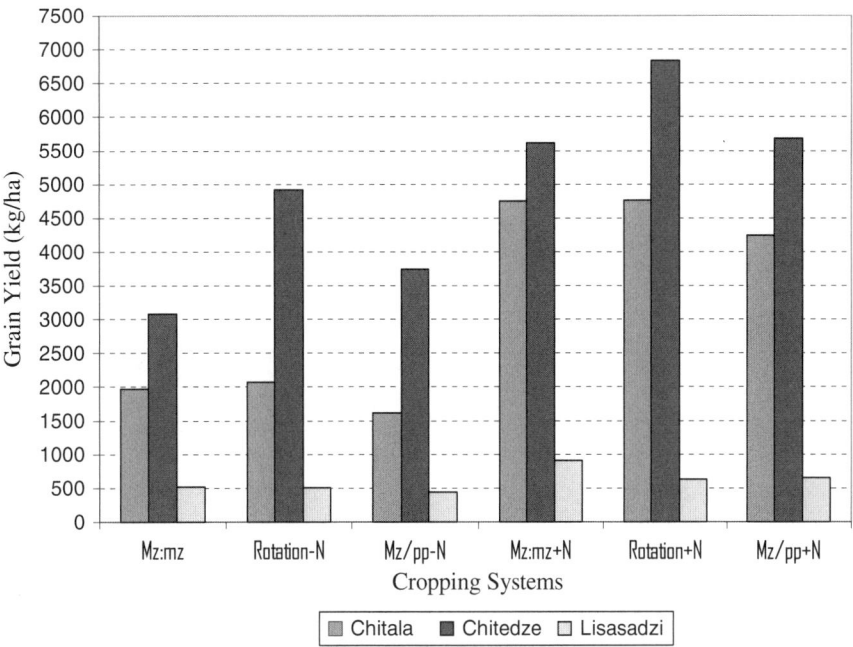

Fig. 3 Maize yield response to different cropping systems over a period of 6 years (2000–2005) at Chitala, Chitedze and Lisasadzi

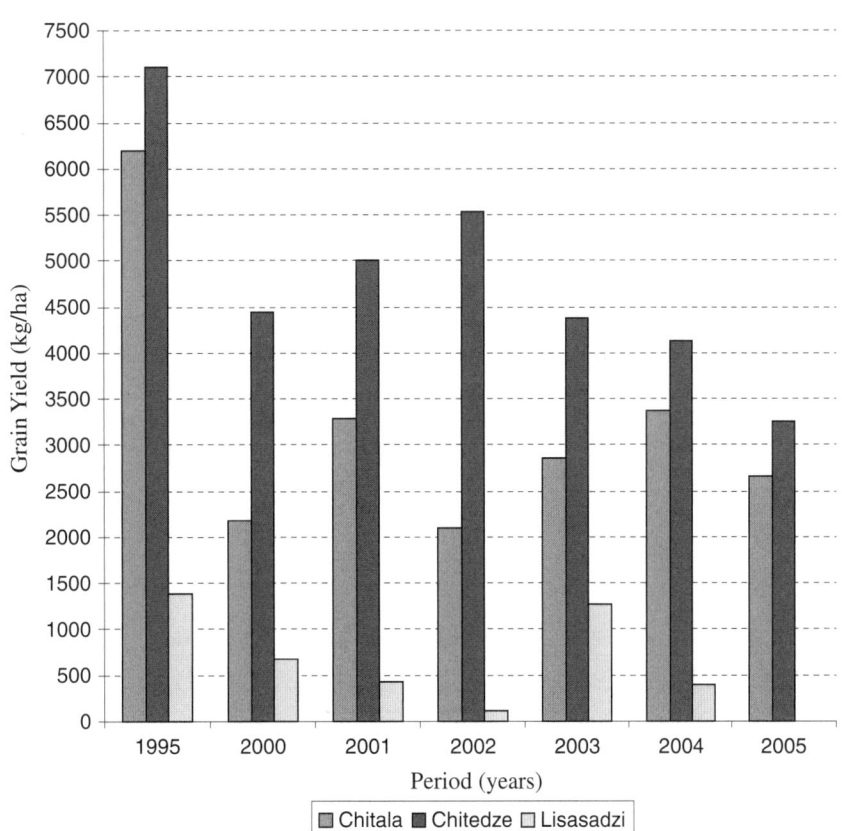

Fig. 4 Maize yield trend over a period of 7 years (1995–2005) at Chitala, Chitedze and Lisasadzi

Fig. 5 Maize yield trend for 6 years (2000–2005) at Makoka and Zombwe

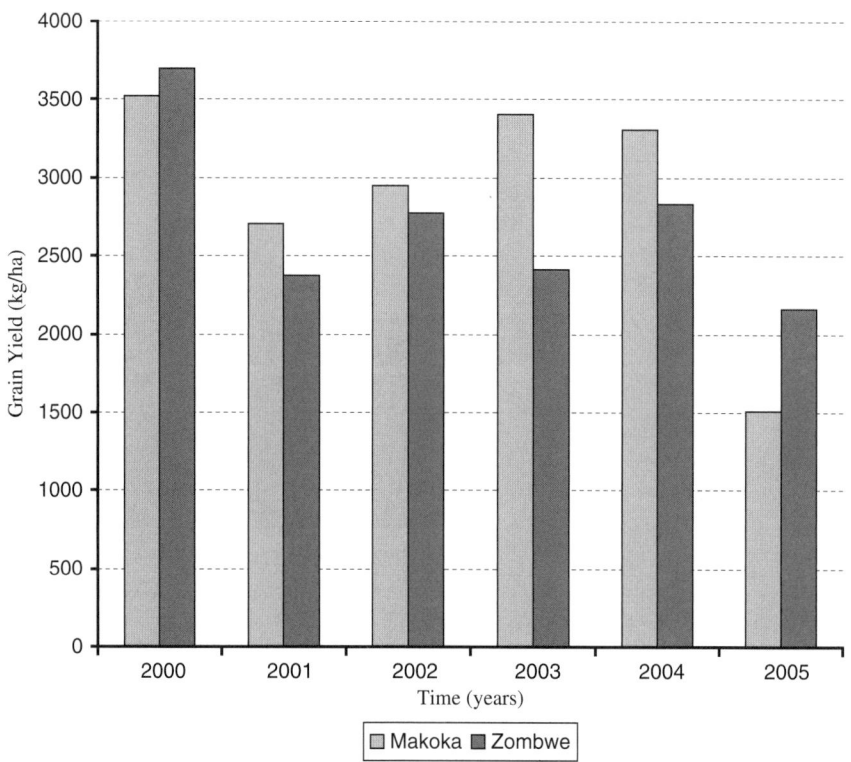

Fig. 6 Mean stover yield of maize at Chitala, Chitedze, Makoka and Zombwe in the 2006 season

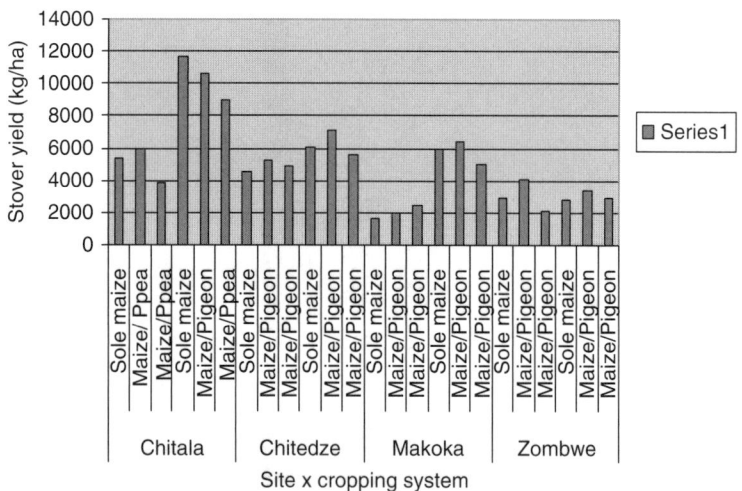

Table 2 Initial soil physical and chemical characteristics of experimental sites

Soil origin	pH (H$_2$O)	OC (%)	N (%)	Clay (%)	Silt (%)	Fine sand (%)	Coarse sand (%)	Soil texture
Chitala	5.6	1.07	0.07	26	9	25	40	SCL
Chitedze	5.4	2.39	0.16	28	21	21	30	SCL
Lisasadzi	5.8	0.59	0.04	4	4	36	56	SL
Makoka	5.3	0.30	0.02	16	6	28	50	LS
Zombwe	5.3	0.68	0.05	22	4	30	44	SCL

similar yields in all the sites when compared at the same fertility level. This shows that pigeon pea had the largest nitrogen residual effect to benefit the subsequent maize crop (Kumwenda, 1996, Kumwenda et al., 1997).

The lowest mean grain yield was 2,194 kg/ha (sole maize) and 2,344 kg/ha (maize/pigeon pea intercropping). In continuous maize cropping system, there was depletion of soil nitrogen, low depositions of soil organic matter due to poor quality of maize residues and high levels of nitrogen immobilization. There was also limited root depth and cultivation depth creating hardpan soil that minimizes soil permeability resulting in water floods within the root zone. Parasitic witch weed *Striga asiatica* was prevalent in sole maize plots, adversely affecting the maize crop and reducing its yield.

Maize yields were significantly higher in plots which had received fertilizer (area-specific fertilizer recommendations) compared to plots which had not received fertilizer in all the four sites. The semi-arid tropics have unimodal rainfall pattern, such that there was prolonged immobilization caused by yearly retention of maize residue whose decomposition is impeded by the unimodal rainfall pattern. In monocropping system, an addition of sufficient nitrogen fertilizer is important to overcome the immobilization process that can deprive the crop of nitrogen at the early growth stage.

There were no significant differences in maize grain yield due to crop residue management (removal or addition of crop residues). This implies that removing or retaining the residues across cropping systems does not affect maize yield. However, soil analytical data will verify whether soils in the plots where residues were retained were improved. The results also show that the farmer is given an opportunity on how best to manage crop residues in the farm.

Maize Yields Trend over Years, Sites and Across Cropping Systems

Highest maize yields were obtained in the initial year across sites of the trial as indicated in Figs. 3 and 4. Average maize yields decreased with time. Deviation in maize yields in some years and sites was attributed

to drought conditions and change of the trial variety index from MH18 to DK3031. The initial maize variety index MH18 was changed because the former variety became scarce on the market due to its high susceptibility to foliar diseases such as Grey Leaf Spot. Normally, DK8031 has a higher yield potential than MH18.

Amount of Stover Retained or Removed from Different Sites

The analysis of variance showed no significant differences ($p \leq 0.05$) across years and sites. Crop residue management treatments were placed in subplots to measure their precision on soil condition improvements over time in relation to maize grain yield. The amounts of maize stover yield were again higher in plots which had received fertilizer compared to plots which had not received any fertilizer (Fig. 5). Maize stover yield tended to be higher in intercropped maize than in sole maize plots. In general, maize stover yield followed a similar trend to that of grain yield.

Site Characteristics

Initial site characteristics are indicated in Table 2. Soil pH ranged from 5.3 (Zombwe and Makoka) to 5.6 (Chiatala). Soil pH is in the good range to support crop production. All sites had low initial soil nitrogen that made all sites suitable for this study. Initial soil nitrogen ranged from 0.002 to 0.16%. Sites had different levels of soil organic carbon ranging from 0.30 to 2.07%. At the end of the 2005/2006 season, soil samples were collected in all the five sites. However, due to financial problems and end of the project, our laboratory ran out of chemicals and soil analyses were not carried out.

Rainfall

Average monthly total rainfall varied across years and sites (Fig. 7). In drought years, it was observed that

Fig. 7 Mean monthly rainfall (mm) across sites over 10 years (1995–2005)

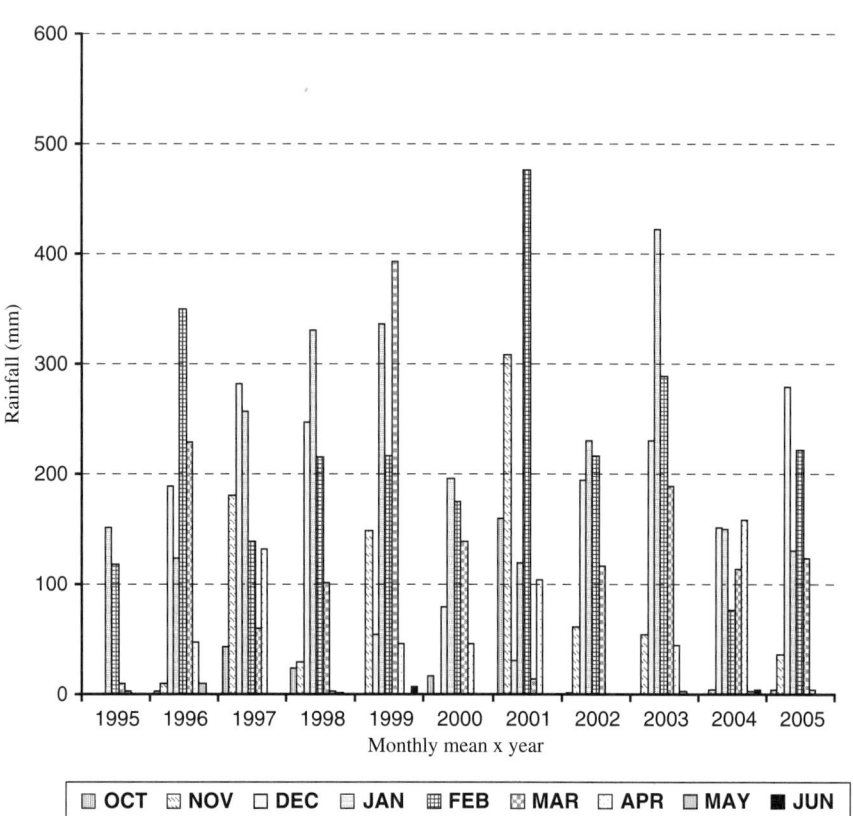

there were less amounts of total rainfall during the planting time in December and less total monthly rainfall during the flowering time (January and February) which is a critical time in the growth stage of the maize plant. During this time, more water is required to enhance the process of grain filling. High monthly total rainfall recorded does mean it was a good year; it might appear that more rains came after February when the crop demand for water had already passed. Dry years were observed in 1995, 1999, 2000, 2001, 2004 and 2005 with 282, 1,200, 652, 1,214, 661, and 801 mm of monthly rainfall totals, respectively.

fertilizer performed better than continuous maize cropping system. Maize grain yield reduction was more stable in maize/pigeon pea cropping system due to sustainability of the system while in continuous maize cropping system, grain yield reduction was so eminent due to the inferiority of the system. Over time, crop residue management did not affect maize grain yield across the sites. However, since soil analyses were not conducted, there is no soil data for comparing the removal or retention of crop residues within plots on soil improvement. Modelling has never been done since the trial started, and there is a need to conduct modelling on yields obtained over the past 10–12 years.

Conclusion

Results show that sites perform differently due to their ecological position and soil status. Chitedze had the best mean maize grain yields (6,405 kg/ha). Cropping systems affected average maize grain yield differently. Maize/pigeon pea cropping systems with or without

Acknowledgements The excellent technical assistance of Drs. (late) Webster D. Sakala and John D.T. Kumwenda, maize agronomists who started the experiment, and all technical support staff of maize commodity for good data collection are acknowledged. The Rockefeller Foundation is acknowledged for financial support and Malawi Government for financial and infrastructural support.

References

Jeffers J (1982) Modelling, outline studies in ecology. Chapman and Hall, London

Kumwenda JDT (1996) Interplanting legume manure crops in maize in Malawi. Preliminary results. In: Waddington SR (ed) Research results and network outputs 1994 and 1995. Soil Fertility Network for Maize-Based Farming Systems, CIMMYT, Harare, Zimbabwe, pp 35–40

Kumwenda JDT, Saka AR, Snapp SS, Ganunga R, Benson TB (1997) Effects of organic legume residues and inorganic fertilizer nitrogen on maize yield. In: Benson TB, Kumwenda JDT (eds) Maize Commodity Team. Annual Report for 1996/97. Chitedze Agricultural Research Station, Lilongwe, Malawi, pp 77–83

Parton WJ, Schimel DS, Cole CV, Ojima DS (1987) Analysis of factors controlling soil organic matter levels in Great Plains grasslands. Soil Sci Soc Am J 51:1173–1179

Rao PSC, Jessup RE, Hornsby AG (1982) Simulation of nitrogen in agroecosystems: criteria for model selection and use. Plant Soil 67:35–43

Sakala WD, Cadisch G, Giller KE (1996) Interaction of *Zea mays* and *Cajanus cajan* residue mixtures on nitrogen mineralization. Target News (7):4–5.CIMMYT, Harare, Zimbabwe

Young A, Muraya P (1991) SCUAF: soil changes under agroforestry. Agric Syst Inf Technol Newslett 3:20–23

Use of *Tithonia* Biomass, Maize Residues and Inorganic Phosphate in Climbing Bean Yield and Soil Properties in Rwanda

N.L. Nabahungu, J.G. Mowo, A. Uwiragiye, and E. Nsengumuremyi

Abstract Lack of adequate nutrient supplies from both organic and inorganic sources is the principal constraint to bean production in Rwanda. Field experiments were conducted in two sites, Rubona and Butare, Rwanda, in 2005/2006 to assess the effect of *Tithonia diversifolia* biomass and maize residues in the yield of climbing bean (Var. G2331) and soil chemical properties. The soil used for this experiment is Ultisols with contrasting properties; the soil of site 1 has high nutrient content compared to the soil of site 2. Seven treatments were studied including a control and either *T. diversifolia* biomass or maize residues applied alone, together or in combination with triple superphosphate. There was a significant positive effect ($P < 0.01$) in the yield of climbing beans in the two sites. Increased yields compared to the control were higher in site 1 than in site 2. The highest yield increase was achieved from the treatment of *Tithonia* plus maize residues and TSP ($3.3 \, t \, ha^{-1}$ for site 1 and $1.74 \, t \, ha^{-1}$ for site 2). The combination of tithonia and TSP increased soil pH_w compared to control from 5.45 to 6.05 and 4.81 to 5.46, respectively, for sites 1 and 2. The high performance of *Tithonia* application in bean yield confirms its capacity to increase bean production compared to inorganic fertilizers at equal rates of P. It is therefore concluded that *Tithonia* and maize residues are a more effective source of nutrient resources for climbing beans apart from their positive attributes in positively influencing the chemical properties of the soils.

Keywords Acidic soils · Aluminium toxicity · Climbing beans · Decomposition · Inorganic fertilizers

Introduction

The acid soils in Rwanda like in many sub-Saharan African countries are characteristically nitrogen (N) and phosphorus (P) deficient, high in aluminium (Al) toxicity and poor in calcium (Ca) and magnesium (Mg) (Nabahungu et al., 2007). Although the fertility of these soils can be restored through application of organic and/or inorganic fertilizers, the use of mineral fertilizers in Rwanda is limited by several socio-economic factors such as high prices, untimely availability and lack of credit. Moreover, most of the mineral fertilizers available in Rwanda have acidifying effect and therefore not suitable for acid soils (Nyabyenda and Rutunga, 1985). As an alternative to these inorganic fertilizers, many farmers apply farmyard manure and compost in their fields. However, the amount of organic manure available is low and of poor quality. Therefore, other locally available nutrient sources such as green manures and crop residues need to be evaluated for use in crop production. Although most organic soil amendments are low in P (Palm et al., 1997), they can be used to improve soil properties such as soil pH and exchangeable Al, K, Ca and Mg which are closely related to P fixation (Warren, 1992; Nziguheba et al., 2000). However, despite the potential of using organic nutrient resources to improve soil fertility, many cropping systems in Rwanda do not utilize crop residues and green manures. This has led to decline in soil fertility, biomass production and crop yields (Bationo and Vlek, 1997).

N.L. Nabahungu (✉)
ISAR-Rwanda, Butare, Rwanda
e-mail: nabahungu@yahoo.com

A. Bationo et al. (eds.), *Innovations as Key to the Green Revolution in Africa*,
DOI 10.1007/978-90-481-2543-2_33, © Springer Science+Business Media B.V. 2011

Given that average land per household in Rwanda is 0.5 ha, there is a need for affordable and appropriate nutrient replenishment strategies that can maximize crop production per unit area. Adoption of combined application of organic and inorganic nutrient resources has been emphasized by many researchers including Beernaert (1999), Nabahungu and Ruganzu (2000) and Scroth (2003) as an effective strategy for improving soil fertility and crop production.

Due to the low decomposition rate of maize residues and inherent low soil fertility status, especially N and P, Mtambanengwe et al. (2006) recommended that maize residues be applied together with fast decomposing biomass (e.g. *Tithonia*) and inorganic P. An understanding of the nutrient release dynamics from organic materials will enable smallholder farmers to manage their organic resources in a manner that will optimize nutrients availability in the soil and crop productivity. Furthermore, beans is an important food and cash crop in Rwanda, it is grown by 88% of rural households and is the major source of protein. Climbing bean yield is two to three times compared to that of bush beans. Rwanda produces more than 350,396 MT per year of beans. Therefore, the improved climbing bean production system will contribute to nutritional status and income of rural households in Rwanda. The aim of this study was to investigate the effects of *Tithonia*, maize residues and inorganic P fertilizers on soil properties and climbing bean yield. It was hypothesized that a combination of maize residues with *Tithonia* green biomass and inorganic P will increase the rate of decomposition of the maize residues and hence release of nutrients and improve P availability, leading to improved bean yields.

Materials and Methods

Study Site

The trial was conducted at two sites in the Institut des Sciences Agronomiques du Rwanda (ISAR) – Rubona agricultural station in Butare ($2°29'S$ and $29°47'E$, and 1630 m above sea level) – in the Southern Province of Rwanda during the short rains of 2005/2006 (December 2005 to March 2006) and the long rains of 2006 (March – June). The mean annual temperature ranges between 19.8 and 21.7°C and mean

Table 1 Chemical properties of soils at the experimental sites (top soil, 0–20 cm)

Parameter	Site 1	Site 2
Sand (%)	45	45
Silt (%)	7	7
Clay (%)	48	48
pH_{water}	5.5 (0.2)	4.7 (0.2)
pH_{KCl}	5.0 (0.1)	3.8 (0.2)
Exchangeable acidity	0.05 (0.01)	0.65 (0.07)
H^+ (cmol (+) kg^{-1})	0.03 (0.01)	0.11 (0.03)
Al^{3+} (cmol (+) kg^{-1})	0.02 (0.004)	0.55 (0.02)
Organic C (%)	3.14 (0.5)	1.38 (0.1)
N (%)	0.35 (0.05)	0.15 (0.04)
C/N ratio	8.97 (0.42)	9.39 (0.23)
Bray-1 P (mg kg^{-1})	48.48 (0.9)	8.79 (0.6)
K (cmol (+) kg^{-1})	0.58 (0.05)	0.44 (0.05)
Ca^{2+} (cmol (+) kg^{-1})	8.63 (0.7)	2.47 (0.5)
Na^+ (cmol (+) kg^{-1})	0.35 (0.03)	0.13 (0.01)
Mg^{2+} (cmol (+) kg^{-1})	2.37 (0.2)	1.94 (0.2)
CEC (cmol (+) kg^{-1})	14.07 (0.5)	9.69 (0.3)

Figures in parentheses indicate standard error

annual precipitation is 1271 mm. The site receives an average of 681.5 mm rainfall during both the growing seasons. The properties of the study soils are shown in Table 1.

Organic Resource Characterization and Experimental Treatments

Tithonia diversifolia and maize residue organic materials were used according to the Decision Guide for organic N management (Giller, 2000). Samples from *Tithonia* and maize residues were thoroughly washed and rinsed using distilled water, dried, ground and analysed for N, P, K Ca, OC, lignin and polyphenols. The samples were analysed for P, K and organic (Anderson and Ingram, 1993). Total N was determined by Kjeldahl digestion (Foster, 1995). Lignin and polyphenols were determined by the acid detergent method and the revised Folin–Denis method, respectively (Anderson and Ingram, 1993). The chemical composition of quality of *Tithonia* and maize residues is given in Table 2.

Tithonia green manure and maize residues were applied at 2.5 and 7.5 t ha^{-1}, respectively. Phosphorus was applied at 50 kg P ha^{-1} as triple superphosphate (TSP). Maize residues and the fresh biomass of

Table 2 Quality of organic resources at the time of application

Organic resources types	DM (%)	OC (%)	N (%)	Total P (%)	K (%)	Na (%)	Lignin (%)	Polyphenol (%)	C:N (%)
Tithonia	14.0	38.0	4.0	0.36	2.5	0.51	4.7	2.2	9.5
Maize residues	26.5	46.0	1.33	0.12	1.2	0.09	3.1	1.1	34.6

Table 3 The major nutrient contents of applied treatments

Treatments	N	P (kg ha$^{-1)}$	K
T0: control	0	0	0
T1: TSP	0	50.0	0
T2: *Tithonia*	100	9.0	75.0
T3: maize residues	100	9.0	90.0
T4: *Tithonia*+TSP	100	50.0	75.0
T5: maize residues+TSP	100	50.0	90.0
T6: maize residues+*Tithonia*+TSP	100	50.0	66.3
T7: maize residues+ *Tithonia*	100	9.0	66.3

Tithonia were cut into small pieces and then incorporated into the soil (10–20 cm) using hand hoes 2 weeks before planting. Triple superphosphate was banded along the rows at the time of planting beans. Table 1 shows the amounts of N, P and K contents in each treatment studied. The experimental design was a randomized complete block design with eight treatments (Table 3), replicated three times.

Climbing beans (variety G2331) was sown at the recommended spacing of 40 cm × 20 cm.

The agronomic parameters monitored were (i) number of pods per plant (based on a random sample of 10 plants per plot), (ii) grain yield and (iii) 1000-grain weight. Grains were sun dried, winnowed and weighed. The 1000-grain weight was obtained using electronic grain counter and an electronic balance.

Soil Sampling and Analysis

Topsoil (20 cm) samples were taken three times: (i) before sowing for soil characterization, (ii) at flowering stage and (iii) after harvest using a soil auger. The soils were sieved through 0.5 mm for N and OC analysis and 2 mm sieves for other soil parameters. Particle size analysis was done using the hydrometer method (Gee and Bauder, 1986). Soil pH was determined in water and KCl using pH meter at a ratio of 1:2.5 soil–water suspension and KCl suspensions (Page et al., 1982). Cation exchange capacity was determined by the ammonium acetate saturation method (Rhoades,

1982). Exchangeable Ca and Mg in the ammonium acetate leachate and Al + H were determined using absorption spectrophotometer. Exchangeable K and Na were determined using flame spectrophotometer. Total N was determined by Kjeldahl digestion (Foster, 1995) and the organic carbon was measured by Walkley and Black method (Page et al., 1982). Samples were digested in perchloric acid, and Bray-1 P content was determined by the molybdenum blue method (Page et al., 1982).

Statistical Analysis

Data on climbing bean yield and soil properties were analysed statistically by analysis of variance for each site using Genstat for windows Discovery Edition 3.2. Mean comparisons were done using LSD. Mention of statistical significance refers to $P < 0.05$ unless otherwise stated.

Results and Discussion

Influence of Organic Resource Quality and TSP on Bean Productivity

The yields of climbing beans (number of pods, grain yields and 1000 grains weight) for two sites are summarized in Table 3. In site 1, *Tithonia* combined with maize residues and TSP increased yield by 1.3 compared to the control while *Tithonia* combined with maize residues increased yield by 1.2. The corresponding yields of similar treatment in site 2 were 2.9 and 2.0 times, respectively. Neither maize residues alone nor TSP alone had significant effect on bean grain yield in site 1. In site 2, the corresponding yields of similar treatments were significantly ($P < 0.05$) increased by 1.4 and 1.6, respectively. In site 1, the grain yield of *Tithonia* was 115% that of the check treatment and 102% of the maize residue treatments. The corresponding yields of similar treatment in site 2 were 149 and 140%, respectively. The high yield associated with *Tithonia* green manure can be attributed to the faster decomposition and release of the nutrients from *Tithonia* than from maize residues (Nabahungu et al., 2007). The possible short-term immobilization of P and slower provision of N and K resulting from application of the low-quality maize residues could have negatively affected the early growth of climbing beans in the treatment with maize residues alone. Bean yields from *Tithonia* in combination with maize residues were higher than either *Tithonia* or maize residues applied alone by 138 and 147%, respectively, in site 1, while in site 2 they were 101 and 114%, respectively (Table 4). These observed yields may be due to the synchronization of release of nutrients of different quality organic resources. *Tithonia* may have enhanced the decomposition of maize residues when used together. The synergy between different green manures (*Tithonia* and maize residues) is high during flowering and fructification stage to ensure good growth of crop (Trutmann and Graf, 1981). Therefore, the yield increased with the increase in pod number and the yield diminished significantly during dry season because of the decrease of pod quantity due to abortion of flowers and pods and this result in decrease in yields. The mean average of pods in this study is very small, and this is tightly related to unfavourable crop production factors especially soil fertility. The high improvement of yield associated with TSP applied alone in site 2 compared to site 1 is attributed to improved P supply in soils of site 2 which was lower in P content. The average grain yield results obtained from site 1 were higher than from site 2. However, the responses of beans to the treatments were higher in site 2 than in site 1. This difference is related to the different soil fertility status of these two sites. Site $\times$ treatment interaction was highly significant ($P < 0.01$) meaning that different soil types reacted differently to the same treatments. The type of soil and nutrient status are important in determining the response of crop to input. In general, the organic materials applied had significant influence on crop growth and yield. The more efficient organic input in increasing the number of pods was *Tithonia* followed by maize residues, which means that this parameter is dependent on the treatments than environmental factors.

Effect of Treatments on Soil Chemical Properties After Harvest in Sites 1 and 2

The treatment which combined both *Tithonia* and maize residue biomasses and TSP has increased more soil pH and Bray-1 P and therefore gave the highest grain yield. This can be explained by the synchronization of release of nutrients, different quality of organic matter and improved plant nutrient uptake through cellular root membrane (Mukesimana, 1999). The weight of grains depended on the type of input used. Except for the control, other treatments had increased the weight at the medium-sized (250–400 g) grain. Grain weight determines the density of grain, low weight is below 250 g, medium weight is between 250 and 400 g and high weight is above 400 g.

Table 4 Effects of different treatments on climbing bean yields

Treatments	Site 1			Site 2		
	Grain yields (t ha^{-1})	Ponds/plant	Weight of 1000 grains (g)	Grain yields (t ha^{-1})	Ponds/plant	Weight of 1000 grains (g)
Control	2.57	14.07	438.7	0.597	8.16	241.7
TSP	2.64	15.20	496.7	0.945	6.34	331.7
Tithonia	2.96	18.83	532.7	0.888	10.84	365.0
Maize residues	2.64	13.73	485.7	0.833	10.84	300.0
Tithonia +TSP	2.83	20.07	496.7	1.600	9.65	275.0
Maize residues+TSP	2.86	12.07	484.3	1.000	10.09	325.0
Tithonia+maize residues+TSP	3.30	19.87	456.0	1.738	13.11	325.0
Tithonia+maize residues	3.00	19.30	465.3	1.222	10.81	383.3
LSD	0.13	1.34	–	0.15	2.42	60.25
F value	11.49***	52.86***	2.226^{ns}	23.77**	3.26**	5.34**
C.V. (%)	4.34	4.59	6.96	12.64	13.96	10.81

**significant at 0.01, ***significant at 0.001, ns = non significant

However, the phosphate inputs were not successful in both sites. Aluminium toxicity and low N in site 2 may have caused the P fixation resulting in low bean grain yield. Therefore, application of lime and organic materials is crucial to enable successful effect of phosphate inputs in soils with aluminium toxicity (Nabahungu et al., 2007). However, the TSP application was important in supplying P which was limiting bean growth.

The nutrients status in the soil 45 days after sowing was analysed in order to determine nutrients release from the inputs applied (Table 5). Soil pH after harvest increased from 5.4 to 6.05 and 4.8 to 5.56 in sites 1 and 2, respectively. In general, all treatments increased soil pH from 0.01 to 0.55. This small increase is acceptable because the sudden increase of pH counterblocks the micronutrient uptake and reduce organic matter in soil (Morel, 1989). High pH values were found in plots which received the *Tithonia* + maize residues + TSP. According to Iyamuremye and Dick (1996), the organic acids released during decomposition of organic matter compete with acidifying elements such as H^+, Fe^{2+}, Al^{3+} and Mn^{2+} on soil adsorbent complex thus reducing soil acidity. Total acidity at crop harvest varied between 0.25 and 0.48 cmol (+) kg^{-1} of soil meaning that soil acidity effect was due to H^+ ion and not Al^{3+} which was found to be in small quantities in the soils (<0.09 cmol (+) kg^{-1}). However, Al being toxic in site 2 (0.55 cmol (+) kg^{-1} of soil) and fixed P thus reduces Bray-1 P in soil. However, combining TSP and *Tithonia* improved P availability. Mbonigaba (2002) reported that micro-organisms

can also immobilize P in a short period and release it slowly during mineralization.

The soil acidity decrease has resulted in Bray-1 P increase and pH after bean harvest ranged between 47.5 and 98 mg kg^{-1}. In general, Bray-1 P content in soil after harvest increased in all plots except in control when compared to soil status. Treatments with TSP (i.e. TSP applied alone, TSP+ maize residues and TSP + *Tithonia*) have increased Bray-1 P significantly compared with other treatments. Thus organic manure increased the TSP solubility due to organic acids released during micro-organism activity (Soltner, 1981; Iyamuremye and Dick, 1996). That organic matter reduces anion adsorption sites for Bray-1 P and therefore enhances Al and Fe phosphates solubility and slowly Bray-1 P can be released in soil solution for crop uptake (Hue and Sobieszczyk, 1999; Nabahungu et al., 2007). This probably explains the increase in Bray-1 P at 45th days after sowing (Table 6).

The organic carbon and total N ranged between 2.87–3.12% and 0.196–0.26%, respectively, for site 1 and between 1.23–1.96% and 0.171–0.317% fo2ite 2 (Table 5). These values indicate that the soils at the two sites contain optimum organic matter and the mineralization is likely to be very fast to release the nutrients considering the C/N ratio ranged between 5.31 and 13.25 (Table 5). Sanchez and Miller (1986) reported that organic carbon in tropical soil at 50 cm deep varies from 0.95 to 1.36% and the total N from 0.09 to 0.131% such that the C/N is between 10.56 and 12.38 depending on soil moisture. This means

Table 5 Effect of *Tithonia*, maize residues and inorganic P on soil chemical properties at 45 days after sowing

Treatments	Site 1				Site 2			
	OC (%)	N (%)	C/N	Bray-1 P (mg kg^{-1})	OC (%)	N (%)	C/N	Bray-1 P (mg kg^{-1})
Control	3.03	0.25	12.35	39.34	1.55	0.24	6.52	12.15
TSP	3.17	0.24	14.07	53.18	1.50	0.27	5.59	27.60
Tithonia	3.03	0.23	13.20	40.44	1.44	0.24	6.39	11.80
Maize residues	3.11	0.25	12.47	44.76	1.43	0.27	5.31	14.30
Tithonia+TSP	3.04	0.25	12.89	57.46	1.53	0.24	6.42	38.00
Maize residues+TSP	3.08	0.27	11.63	68.34	1.44	0.22	6.71	35.10
Tithonia+maize residues+TSP	3.06	0.25	12.49	72.81	1.52	0.21	7.46	22.90
Tithonia+maize residues	3.19	0.24	13.25	44.30	1.58	0.19	8.75	14.35
LSD	–	–	–	–	–	–	–	6.38
F value	1.02[ns]	3.25[ns]	2.99[ns]	2.98[ns]	0.57[ns]	4.00[ns]	1.59[ns]	30.52***
C.V. (%)	2.82	3.94	4.70	19.83	7.03	8.0	18.17	12.26

***significant at 0.001, ns = non significant

Table 6 Effect of *Tithonia*, maize residues and inorganic P on pH and Bray-1 P at bean harvest

Treatments	Site 1			Site 2		
	pH_{water}	pH_{KCl}	Bray-1 P (mg kg^{-1})	pH_{water}	pH_{KCl}	Bray-1 P (mg kg^{-1})
Control	5.45	4.95	47.50	4.81	3.91	17.71
TSP	5.51	5.01	83.79	4.95	4.01	49.76
Tithonia	5.71	5.21	52.81	5.31	4.41	16.90
Maize residues	5.54	5.04	44.79	5.09	4.19	15.45
Tithonia+TSP	5.54	5.04	88.45	5.50	4.61	37.08
Maize residues+TSP	5.71	5.21	61.98	5.42	4.52	27.99
Tithonia+maize residues+TSP	6.05	5.45	98.76	5.46	4.56	52.12
Tithonia+maize residues	5.77	5.27	54.71	5.11	4.21	11.89
LSD	0.20	0.17	33.21	0.42	0.44	5.52
F value	10.89*	12.07*	4.34*	4.03*	4.33*	39.13***
C.V. (%)	1.47	1.38	21.09	3.43	4.15	15.96

*significant at 0.05, ***significant at 0.001

C/N ratio is less than 25 which allowed organic matter mineralization.

The value of C/N ratio is adequate for SOM mineralization indicating that the process was dominated by mineralization. The C data from site 2 show how this element content in soil was not generally changed in all treatments considering the value before the trial setting.

In general, Mg^{2+}, K^+ and Ca^{2+} contents in soil were in optimum range. In both sites these cations are well balanced (from 1.961 to 2.15) which is sufficient for tropical crops (Boyer, 1982). The base saturation rate being more than 50% (varying from 50 to 67%) indicates the soil is well saturated in cations (Gobat et al., 1998). However, considering the constraints related to organic materials availability and to soil fertility replenishment, a combination of *Tithonia* + maize residues + TSP seems to be an appropriate alternative for nutrient addition in soil.

Conclusion

The results from this study show that *Tithonia* green manure and maize residues improve soil fertility and the effectiveness of TSP as P source for beans by improving its availability. The two organic nutrient sources, therefore, are appropriate alternatives for soil nutrients replenishment. Where P content is low as in the soils of site 2, response of beans to TSP application is higher than in soils with high P content (site 1). This study complements others which have shown that combined use of organic and inorganic inputs is a better option for soil fertility improvement and more so when the soils are acidic and low in nutrient content. Judicious combination of organic resources and mineral inputs is necessary to improve crop production and maintain soil properties level. A combination of *Tithonia* green manure + maize residues + TSP is proposed as an appropriate option for soil fertility improvement in poor acidic soils under bean production in Rwanda.

Acknowledgements We thank the ECABREN and the "Institut des Sciences Agronomiques ud Rwanda (ISAR)" for providing financial support. We also thank the technical staff at ISAR, Rubona, for the excellent field experiments and laboratory work.

References

Anderson JM, Ingram JSI (1993) Tropical soil biology and fertility. A handbook of methods, 2nd edn. CAB International, Wallingford, CT

Bationo A, Vlek PLG (1997) The role of nitrogen fertilizers applied to food crops in the Sudano-Sahelian zone of West Africa. In: Renard G, Neef A, Becker K, von Oppen M (eds) Soil fertility management in West African land use . Margraf Verlog, Weikersheim, pp 41–51

Beernaert FR (1999) Etude de faisabilité d'un projet de production de chaux et/ ou de travertins broyés pour la gestion des sols acides au Rwanda. Rapport agro-pédologique, mission du 19/06 au 15/08/1999

Boyer J (1982) Les sols ferrallitiques Tome X: facteurs de fertilité et utilisation des sols. In: Initiations-documentations techniques no 52. ORSTOM, Paris, France

Foster JC (1995) Soil nitrogen. In: Alef K, Nannipieri P (eds) Methods in applied soil microbiology and biochemistry. Academic, London, pp 79–81

Gee GW, Bauder JW (1986) Particle size analysis. In: Klute A (ed) Methods of soil analysis part 1, agronomy monograph No. 9, 2nd edn. Soil Science Society of America, Madison, WI, pp 383–412

Giller KE (2000) Translating science into action for agricultural development in the tropics: an example from decomposition studies. Appl Soil Ecol 14:1–3

Gobat JM, Aragno M, Mattey W (1998) Le sol vivant. Base de pédologie-biologie des sols. Lausanne Presses polytechniques et universitaires romandes

Hue NV, Sobieszczyk BA (1999) Nutritional values of some biowastes as soil chemical and microbiological properties. http://www.pfa.ca/agrenv/fertorg.html

Iyamuremye F, Dick RP (1996) Organic amendments and phosphorus sorption by soil. Adv Agron 56:134–186

Mbonigaba JJ (2002) Essaie de compostage de déchets verts et évaluation des effets des composts obtenus sur des sols acides du Rwanda. Université de Gembloux. Thèse de Maîtrise, 154p

Morel R (1989) Les sols cultivés. Colléction technique, Paris, 373p. ISBN 2 – 85206 – 513 – 4

Mtambanengwe F, Mapfumo P, Vanlauwe B (2006) Comparative short term effects of different quality organic resources on maize productivity under two different environments in Zimbabwe. Nutr Cycling Agroecosyst 76:271–284

Mukesimana G (1999) Evaluation de l'effet de quelques sources organiques et inorganiques de phosphore et d'azote ainsi que leur interaction sur le rendement du maïs à Rubona. Mémoire, Faculté d'agronomie, U.N.R Butare 78p

Nabahungu NL, Ruganzu V (2000) Effet du travertin et de *Tithonia diversifolia* sur la productivité du haricot volubile en sols acides du Rwanda. Symposium, ECABREN, 2000, 5p

Nabahungu NL, Semoka JMR, Zaongo C (2007) Limestone, Minjingu phosphate rock and green manure application on improvement of acid soils in Rwanda. In: Bationo A et al (eds) Advances in integrated soil fertility in Sub Saharan Africa: challenges and opportunities. Springer, Dordrecht, pp 691–703. ISBN 13978-1-4020-5759-5 (HB)

Nyabyenda P, Rutunga V (1985) Fertilisation à l'ISAR. In Minagri: Premier séminaire National sur la fertilisation des sols su Rwanda, du 17 au 20 Juin 1985, Kigali, 116–135 pp

Nziguheba G, Merckx R, Palm CA, Rao M (2000) Organic residues affect phosphorus availability and maize yields in a Nitisols of Western Kenya. Biol Fertil Soils 32: 328–339

Page JR, Miller RH, Keeney DR, Baker DE, Roscoe Ellis JR, Rhoades JD (1982) Methods of soil analysis. II. Chemical and microbiology properties, 2nd edn. American Society of Agronomy, Madison, WI, 1159 pp

Palm CA, Myers RJK, Nandwa S (1997) Combined use of organic and inorganic nutrients sources for soil fertility maintenance and replenishment. In: Buresh RJ, Sanchez PA (eds) Replenishing soil fertility in Africa. Soil science society of America No 51. American Society of Agronomy and Soil Science Society of America, Madison, WI, pp 193–218

Rhoades JD (1982) Cation exchange capacity. In: Page AL, Miller RH, Keeney PR (eds) Methods of Soil Analysis Part 2. Chemistry and mineralogical properties, agronomy monograph No. 9. American Society of Agronomy, Madison, WI, pp 149–169

Sanchez A, Miller RH (1986) Organic matter and soil fertility management in acid soils of the tropics. Trans. 13th international congress of soil science, Hamburg, vol 6, pp 609–625

Scroth S (2003) Trees, crops and soil fertility. CABI Publishing, Cambridge, MA, 437p. http://www.pfa.ca/agrenv/matorg.html consulté le 08/03/2005

Soltner D (1981) Phytotechnique Générale, les bases de la production végétale. Tome 1: le sol. 10ème Edition, 287p

Trutmann P, Graf W (1981) Les facteurs Agronomiques limitant la production du haricot commun au Rwanda et les stratégies de leur maîtrise. In: Séminaire sur maladies et ravageurs des les cultures vivrières d'Afrique centrals, 157–161 pp

Warren GP (1992) Fertilizer phosphorus sorption and residue value in tropical African soils (NRI, Bulletin 37). Natural Resource Institute, Chatham

The Potential of Increased Maize and Soybean Production in Uasin Gishu District, Kenya, Resulting from Soil Acidity Amendment Using Minjingu Phosphate Rock and Agricultural Lime

A.O. Nekesa, J.R. Okalebo, C.O. Othieno, M.N. Thuita, A. Bationo, and B.S. Waswa

Abstract In Kenya, soil acidity is a major contributor to declining soil fertility and 20% of the soils are acidic and are considered to be of low fertility. Most farmers are unaware of the benefits of liming acid soils. A study was carried out during the 2005 and 2006 long rain seasons at Kuinet in Uasin Gishu District of the Rift Valley Province in Kenya to delineate the effects of Minjingu phosphate rock (MPR) and agricultural lime as liming materials on yields of soybeans intercropped with maize. The maize responded to application of soil amendment materials for the first season with the diammonium phosphate and lime (DAPL) treatment giving the highest maize yields of 6.19 t ha^{-1} compared to the control which gave 1.36 t ha^{-1}. Soybean yields were low in the first season with the DAPL treatment and control treatment giving yields of 0.32 and 0.14 t ha^{-1}, respectively. This, however, changed significantly after the variety was changed in the second season, with yields going up to 0.68 t ha^{-1} for the triple superphosphate and lime (TSPL) treatment. From the study, it was concluded that there is potential for growing soybean in Uasin Gishu District of Kenya. However, a study and/or research is recommended to screen and identify a suitable variety for increased soybean yields in this district.

Keywords Declining soil fertility · Liming · MPR · Soil acidity · Soybean yields

Introduction

Most farmers in sub-Saharan Africa (SSA) face urgent problems of food security for their families for the whole year as well as earning sufficient income (Rienke and Joke, 2002). In Kenya, production levels of maize (staple) and legumes indicate deficits. There is concern of declining legume yields in western Kenya even with the application of fertilizers (Nekesa et al., 1999). Indeed, yields as low as 500 and 200 kg grain/ha/season for maize and beans, respectively, have been reported in western Kenya. To a farmer, these yields are negligible compared to the demands and pressures from his/her family to provide enough food and other household requirements (Nekesa, 2007).

In an attempt to increase food production in SSA, many options ranging from restoring lost soil fertility to introduction of improved crop varieties have been tested. Intercropping of legumes with a cereal, for instance maize, has been considered one of the options in increasing food production as well as providing a better diet in terms of protein by the legume.

Soybean is one of the most important food legumes grown in the SSA due to its high nutritive value. This legume grows in both sandy and heavy textured soils and over a wide range of soil pH 5.5–8.5. For rainfed production, good commercial yields obtained range from 1.5 to 2.5 t ha^{-1} as a sole crop but average lower yields range from 0.15 to 1.6 t ha^{-1} in subsistence farming (Rienke and Joke, 2002). In Brazil, yields as high as 4 t ha^{-1} have been reported (J.R. Okalebo, *personal communication*).

Major constituents of soybean seed are protein (40–45%) and oil (20–22%). The protein is higher in lysine and tryptophan than in common cereals.

A.O. Nekesa (✉)
Department of Soil Science, Moi University, Eldoret, Kenya
e-mail: amarishas@yahoo.com

A. Bationo et al. (eds.), *Innovations as Key to the Green Revolution in Africa*, DOI 10.1007/978-90-481-2543-2_34, © Springer Science+Business Media B.V. 2011

Soybean protein equals that of milk, meat and eggs and can thus be used as a health food since it is high in fibre and low in fat and cholesterol (Riaz, 1999). Its production is increased by soil fertilization. Production of soybean is currently being promoted by TSBF-CIAT in western Kenya. Hence, in this chapter, we report part of a research carried out in Kuinet area, Uasin Gishu District, Kenya, to determine the potential of growing soybeans intercropped with maize in terms of grain yields in Uasin Gishu District, Kenya, where legumes are seldom grown. The findings of this study are expected to benefit the local farmer in Uasin Gishu District in terms of crop diversification and increased food production and security.

Materials and Methods

The experiment was carried out in Uasin Gishu District, on Walter Kimwatan's farm at Kuinet. This district lies between longitudes 34°50′ and 35°37′ East and latitudes 0°30′ South and 0°55′ North, some 300 km NNW of the capital Nairobi. Uasin Gishu District lies in Uasin Gishu plateau, which is one of Kenya's largest wheat producing areas, and most of it falls under the agroecological zone commonly known as wheat/barley zone (LH 3) (Republic of Kenya, 1997). Its terrain varies greatly with altitude, which ranges between 1500 m above sea level (asl) at Kipkaren in the west to 2100 m asl at Timboroa in the east. Eldoret (Kenya) town has an altitude of 2085 m asl and marks the boundary between the highest and the lowest altitudes of the district (Republic of Kenya, 1997). The general landscape is that of undulating plateau with no significant mountains or valleys. The land is higher in the east and declines in its western borders. The average rainfall is between 800 and 1000 mm, with temperatures of about 22°C. The soils are underlain by murram, are well drained, moderately deep, dark red friable clay of petroplinthite. The soils are classified as Rhodic Ferralsols with pH 4.5–5.0 (Republic of Kenya, 1997).

The experiment consisted of two cropping seasons: 2005 long rains (LR) and 2006 LR. In the first season (2005 LR), the liming effects of Minjingu phosphate rock (MPR) (from Tanzania) (Buresh et al., 1997) and agricultural lime (L) (quicklime – CaO – from Koru, Kenya) were compared. The liming effects of agricultural lime were tested individually as well as

diammonium phosphate (DAP) without lime. Triple superphosphate (TSP) in combination with lime was also tested. The highest DAP level of 90 kg P ha^{-1} provided the nitrogen (N) level of 81 kg N ha^{-1} to which other phosphorus (P) rates of 30 and 60 kg P ha^{-1} were adjusted to, or to all treatments apart from the control, through urea application at vegetative stage of maize growth. DAP, TSP, urea, agricultural lime and MPR were broadcasted and incorporated evenly within 15 cm depth using hand hoes for the plots receiving these inputs. This was to remove any N deficiency that would limit crop growth. For the second (2006 LR) season, the residual effect of the liming and P materials added in the first season was tested hence no liming or P material was applied. However, the plots received a blanket rate application of 75 kg N ha^{-1} applied as urea which was split into two applications: 30 kg N ha^{-1} at planting and 45 kg N ha^{-1} as top dressing to all treatments apart from the control. The experiment was laid down in a randomized complete block design (RCBD) replicated three times in plots measuring 5 × 4.5 m each which gave plot areas of 22.5 m^2 with the treatments described in Table 1.

Planting of Maize and Soybean Intercrops

During the 2005 LR, maize (H 614 D variety) and soybean (market variety) were planted using the "MBILI" intercropping system (Tungani et al., 2002) where two maize rows are spaced at 50 cm pairs that are 100 cm apart (the gap) and two rows of legumes planted within the gap of 33 cm row spacing, giving the maize plant population of 44,000 plants ha^{-1} and a legume population of about 88,000 plants ha^{-1} (Tungani et al., 2002). This allows adequate light and heat radiation to reach the legumes (Tungani et al., 2002). The market variety for the soybean was chosen because of the possibility for adoption by farmers. However, due to the poor performance of this variety in the first season, a known variety (TGx 14482E) was chosen and planted in the second season. Two seeds of maize and four seeds of the soybean were planted per hill and later thinned to one and two, respectively, 2 weeks after germination. The crops were sprayed to control pests using the pesticides Ogor (acts by both systemic and contact means and was used to control aphids which cause leaf miner in legumes) and Buldock (used to control stalk borer in maize).

Table 1 Experimental treatments as applied at Kuinet site in Uasin Gishu District of Kenya during the first season (2005 LR)

Treatment no	Treatment description	Treatment code
1	Control. No nutrient inputs	00
2	0.25 t ha^{-1} MPR; 96 kg CaO ha^{-1} + 30 kg P ha^{-1}	MPR 1
3	0.5 t ha^{-1} MPR; 192 kg CaO ha^{-1} + 60 kg P ha^{-1}	MPR 2
4	0.75 t ha^{-1} MPR; 287 kg CaO ha^{-1} + 90 kg P ha^{-1}	MPR 3
5	DAP; 30 kg P ha^{-1}	DAP 1
6	DAP; 60 kg P ha^{-1}	DAP 2
7	DAP; 90 kg P ha^{-1}	DAP 3
8	Agricultural lime (L); 96 kg CaO ha^{-1}	L 1
9	Agricultural lime (L); 192 kg CaO ha^{-1}	L 2
10	Agricultural lime (L); 287 kg CaO ha^{-1}	L 3
11	DAP + L; 30 kg P ha^{-1} + 96 kg CaO ha^{-1} in agricultural lime	DAP 1 L 1
12	DAP+ L; 60 kg P ha^{-1} + 192 kg CaO ha^{-1} in agricultural lime	DAP 2 L 2
13	DAP+L; 90 kg P ha^{-1} + 287 kg CaO ha^{-1} in agricultural lime	DAP 3 L 3
14	TSP; 30 kg P ha^{-1} + 78 kg CaO ha^{-1} in agricultural lime	TSP 1 L 1
15	TSP; 60 kg P ha^{-1} + 156 kg CaO ha^{-1} in agricultural lime	TSP 2 L 2
16	TSP; 90 kg P ha^{-1} + 233 kg CaO ha^{-1} in agricultural lime	TSP 3 L 3

MPR = Minjingu phosphate rock, DAP = diammonium phosphate, TSP = triple superphosphate, L = agricultural lime (as CaO)

The crops were weeded three times and harvested at maturity for all the seasons planted. Yields of the crops grown were reported on fresh and dry weight basis.

Soil Sampling and Preparation

For each experimental plot of 22.5 m^2, a composite surface soil (0–15 cm) sample was collected from 10 random sampling points, mixed and sub-sampled, air dried and passed through a 2 mm sieve, to determine particle size, soil pH, soil available P, exchangeable bases (Ca^{2+}, Mg^{2+}, K$^+$ and Na$^+$) and exchangeable acidity (H$^+$ and Al^{3+}). The soil was further passed through a 0.02 mm sieve for determination of organic carbon and total N. These parameters were used to characterize the soil at the beginning of the experiment for each site. After application of the treatments, soil samples, following the procedure described above, were taken at different intervals during the growing period to monitor changes in pH and available P which were determined in the laboratory according to methods described by Okalebo et al. (2002).

Crop Harvesting Procedures

Harvesting of maize and soybean intercrops was done on centre rows of each plot at final harvests by discarding two outer rows per plot and two plants/row at the ends of each plot. Thus, four inner rows for both maize and soybean per plot were harvested giving an effective harvest area of 11 and 14 m^2 for maize and soybeans, respectively. In the harvest area, total grain fresh weights were taken and sub-samples of 200 g for soybean (shelled) and 8 ears for maize taken and recorded. The maize and soybeans were sundried and then later hand-shelled (maize) and their dry weights recorded and used to compute grain yields per plot. Yield was calculated using the relationships:

$$\text{Yield/plot} = \frac{\text{Total fresh weight} \times \text{Sample dry weight}}{\text{Sample fresh weight}} \tag{1}$$

$$\text{Yields (kg ha}^{-1}) = \frac{\text{Yields/plot 10,000}}{\text{Effective area harvested}} \tag{2}$$

Statistical Analysis

The soil amendment materials included some that had two components of different origins or compositions within themselves. For instance, MPR provided both CaO and P, whereas DAP provided both N and P. Similarly, TSP treatment provided both CaO and P. However, the CaO in the TSP was low (18, 36 and 54 kg ha^{-1}) and had to be topped to the required (96, 192 and 287 kg ha^{-1}) rates used in this study. In order to be able to assess the degree of variation due to each component (CaO and P) in the materials, nesting was done. Hence, statistical analysis was done considering the nesting design with the following model:

$$Y_{ijk} = \mu + \alpha_i + \beta_{j(i)} + \Sigma_{k(ij)}$$

where Y_{ijk} represent plot observations, μ is the mean of plot observations, α_i the effect due to soil amendment material (fertilizer), $\beta_{j(i)}$ the effect due to CaO within the fertilizer, $\Sigma_{k(ij)}$ the experimental error.

Crop yields, soil and plant data were subjected to analysis of variance (ANOVA) with the general linear model (GLM) described above using the Statistical Analysis System (SAS) for windows, Version 8 computer software package. To make comparisons between the means of the soil amendment materials, single degree of freedom contrasts were used.

Results and Discussions

Soil Characterization of the Study Site

Initial soil characterization of the study site indicated strong acidity (pH 4.97) as classified by the National Agricultural Research Laboratories (NARL) – Kabete in Nairobi (Kanyanjua et al., 2002). The low amounts of soil available P (less than 10 mg kg^{-1}) considered as critical for the Olsen-extractable P in soils (Landon, 1984) suggest the need for supplemental P addition (Ndung'u et al., 2006). The soil from Kuinet had moderate levels of organic C and N giving a C:N ratio of 10:1. The site had exchangeable acidity (Al^{3+} + H$^+$) of 1.90 cmol kg^{-1}. According to the soil particle size analysis, the soil was classified as sandy clay loam. Table 2 gives the soil physical and chemical characteristics of the soil taken from the study site before planting.

Crop Yields as Affected by Soil Amendments

Amendment of soil acidity with lime and P additions gave significant increases ($p < 0.01$) above the control for the maize grain yields during the 2005 long rains

Table 2 Soil physical and chemical characteristics of surface (0–15 cm) soils taken before planting (start of 2005 LR) at the Kuinet experimental site

Soil property	Amount
%Sand	51
%Clay	25
%Silt	24
Textural class	Sandy clay loam
pH (1:2.5 soil:water)	4.97
Exchangeable acidity (cmol kg^{-1})	1.9
%N	0.21
%C	2.13
C:N ratio	10:1
Olsen P (mg kg^{-1})	4.94
Exchangeable bases (cmol kg^{-1})	
K	0.30
Na	0.66
Ca	0.82
Mg	0.31
Micronutrients (cmol kg^{-1})	
Fe	0.31
Mn	0.78
Zn	0.005
Cu	0.002

(LR) with a mean of 5.30 t ha^{-1}. The L (lime alone) treatment gave significantly ($p < 0.01$) lower yields than the other soil amendment materials. Generally, materials with both P and CaO gave higher yields compared to materials with either CaO alone or P alone. Orthogonal contrast tests gave significant differences in maize yields ($p < 0.05$) due to the soil amendment materials. When all the materials were compared, no significant differences were observed when L was compared with either DAP or MPR and when MPR was compared with TSPL in terms of maize grain yields. Table 3 gives maize yields obtained at Kuinet during the 2005 LR. Generally, maize grain yields increased due to liming of the acid soils and improved nutrition of added N and P (Nekesa, 2007).

No maize yields were obtained for the second season (2006 LR) due to poor rainfall distribution at the early stage of growth. Due to infestation by stalk borer pests in the first season, we decided to plant early during the second season which was a big risk because there occurred a dry spell of 2 months that led to the total destruction of the maize crop. However, the soybean crop survived giving better yields than the first season. This underlies the importance of crop

Table 3 Maize yields (t ha^{-1}) per site as affected by treatment application at Kuinet during the 2005 LR

Treatment	Maize grain yields (t ha^{-1})	Percent increase above the control
OO	*1.36*	–
MPR1	4.65	242
MPR2	5.70	319
MPR3	5.79	326
MPR mean	*5.38*	*296*
DAP1	5.60	312
DAP2	5.14	278
DAP3	5.53	307
DAP mean	*5.42*	*299*
L1	5.24	285
L2	6.04	344
L3	4.81	254
L mean	*5.36*	*294*
DAP1L1	5.74	322
DAP2L2	5.83	329
DAP3L3	6.99	414
DAPL mean	*6.19*	*355*
TSP1L1	5.80	326
TSP2L2	5.54	307
TSP3L3	5.01	268
TSPL mean	*5.45*	*300*
Overall mean	*5.30*	*309*
SED	*0.53*	–
LSD$_{(0.01)}$	*1.45*	–

P source – rates 1, 2 and 3 contain 30, 60 and 90 kg P ha^{-1}, respectively; lime (CaO) source – rates 1, 2 and 3 contain 96, 192 and 287 kg CaO ha^{-1}, respectively; combined P and CaO source – rates for both P and CaO as above
MPR = Minjingu phosphate rock, DAP = diammonium phosphate, TSP = triple superphosphate, L = agricultural lime

diversification especially in those areas where monoculture of cereals is prevalent, for instance, Uasin Gishu District. In such situations where one crop totally fails due to unforeseen circumstances, a farmer can benefit from the second crop. Farmers should avoid total dependence on planting of one crop due to such unforeseen circumstances that could lead to total crop failure, hence impacting negatively on the per capita food production.

Figure 1 presents soybean yields as affected by treatment application at Kuinet for both the first (2005 LR) and the second (2006 LR) seasons. During the 2005 LR, the soybean legume responded to lime and P additions giving higher yields for the soil amendment materials compared to the control treatment. The DAPL treatment gave the highest yields of 0.32 t ha^{-1}, whereas the control treatment gave the lowest soybean yields of 0.14 t ha^{-1}. Orthogonal contrast tests showed no significant differences when control was compared with the soil amendment materials probably due to the very low yields obtained. The low yields in this season could be attributed to the variety rather than the conditions of the soil. This is because, despite the soil having improved conditions in terms of soil pH and soil available P, the yields were still very low suggesting that the problem in low soybean production in this area was partly due to the choice in variety used for planting.

The second season (2006 LR), however, showed an improvement in the soybean grain yields due to a change in the variety used for planting. For this season, yields obtained varied significantly ($p < 0.01$) due to treatments ranging from 0.2 t ha^{-1} for the control to 0.68 t ha^{-1} for the TSPL treatment. However, in this season, the soybean plant seemed to invest more in the trash at the expense of the grain yield. This could be due to the residual effect of the P fertilizer as well as the effect of the blanket N fertilizer application to all treatments apart from the control. Phosphorus is required for the formation of strong stalks/stems while excess nitrogen results in rapid

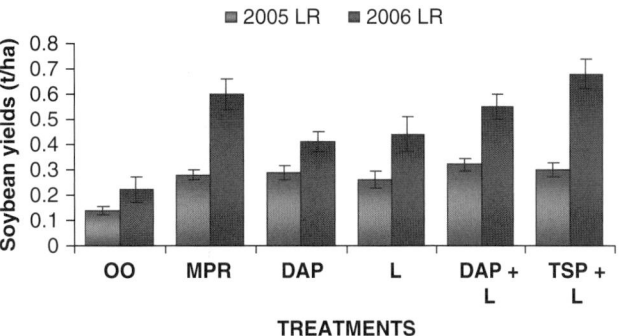

Fig. 1 Soybean yields (t ha^{-1}) for the Kuinet site as affected by lime and P additions in the soil

plant growth and excessive plant biomass, also at the expense of grain formation and development (Obura et al., 2001). Orthogonal contrast tests showed significant differences ($p < 0.05$) when the DAP treatment was compared to the L treatment and when MPR was compared to either L or DAPL treatments. Plots receiving both CaO and P nutrients gave higher soybean yields compared to plots receiving either P or CaO (lime) alone indicating that the soybeans responded well to residual effects of both P and CaO (lime) from the previous season. According to Rienke and Joke (2002), soybean yields as high as 1.5 and 2.5 t ha^{-1} in Nigeria and South America, respectively, have been realized in commercial production. This suggests that if the right variety was identified and planted in soil that is not limited by either N or P deficiencies and the constraints due to soil acidity removed, then high soybean yields in Uasin Gishu would be realized.

Conclusions and Recommendations

There were significant maize and soybean yield increases above the control for all the soil amendments applied. However, the lime alone and DAP alone treatments gave lower crop yields compared to either fertilizer (TSP or DAP) combined with lime or with MPR applied alone. Hence, it is concluded that to achieve high to maximum yields from acidic soils, combination of a fertilizer with agricultural lime or use of a material with both components is necessary. There is economic potential for soybean crop in western Kenya, particularly in Uasin Gishu and the neighbouring districts where cropping is majorly monoculture of maize and wheat. Liming of acid soils in this region is paramount to realize increased yields. In this era of frequent droughts and total crop failures, especially of maize, farmers should move away from the culture of monocropping and engage in more diversified cropping especially the intercropping of maize or wheat with legumes such as soybeans which have high economical and nutritive values. There is, however, a need to identify and screen a soybean variety suitable to the agroecological conditions found in Uasin Gishu District for increased soybean yields.

Acknowledgements We are greatly indebted to RUFORUM for funding this study as well as giving us an opportunity to explore our areas of study interest. Our most appreciation goes to the RUFORUM Coordinator, Prof. Adipala Ekwamu, and the Programme Assistant, Dr. Patricia Masanganise who took time to come and visit our experimental sites. We would like to acknowledge all the participating collaborators and Moi University as a whole for appreciating our work. Specific thanks go to our field and laboratory technicians, Ruth Njoroge, Scholastica Mutua and David Ndung'u, our Project Accountant, Mr. John Emongole, and our Project Secretary, Ms. Mary Nancy Wairimu, who have enabled the completion of this study research in various ways. To all we say, "May God add where you took."

References

Buresh RJ, Smithson PC, Hellums DT (1997) Building soil P in Sub-Saharan Africa. In: Buresh RJ, Sanchez. PA (eds) Replenishing soil fertility in Africa. SSSA special publication 51 SSSA and ASA, Madison, WI, pp 111–149

Kanyanjua SM, Ireri L, Wambua S, Nandwa SM (2002) KARI technical note No. 11; acidic soils in Kenya: constraints and remedial options. KARI headquarters, Nairobi

Landon RJ (ed) (1984) Booker tropical soil manual: a handbook for soil survey and agricultural land evaluation in the tropics and subtropics. Longman Group UK, Essex, pp 137–140

Ndung'u KW, Okalebo JR, Othieno CO, Kifuko MN, Kipkoech AK, Kimenye LN (2006) Residual effectiveness of Minjingu phosphate rock and fallow biomass on crop yields and financial returns in western Kenya. Exp Agric 42: 323–336

Nekesa AO (2007) Effect of Minjingu Phosphate Rock and agricultural lime in relation to maize, groundnut and soybean yields on acid soils of western Kenya. MPhil thesis, Department of Soil Science, Moi University, Eldoret

Nekesa PO, Maritim HK, Okalebo JR, Woomer PL (1999) Economic analysis of maize bean production using a soil fertility replenishment product (PREP-PAC) in western Kenya. Afr Crop Sci J 7:157–163

Obura PA, Okalebo JR, Othieno CO, Woomer PL (2001) Effect of Prep-Pac product on maize-soybean intercrop in the acid soils of western Kenya. Afr Crop Sci Proc 5:889–896

Okalebo JR, Gathua KW, Woomer PL (2002) Laboratory methods of soil and plant analysis: a working manual, 2nd edn. TSBF-CIAT and SACRED Africa, Nairobi

Republic of Kenya (1997) Uasin Gishu district development plan (1997–2001) office of the vice-president and minister of planning and national development. Government Printer, Nairobi

Riaz MN (1999) Soybeans and functional foods. Cereal Foods World 44(2):88–92

Rienke N, Joke N (2002) Cultivation of soya and other legumes. Agromisa Foundation, Wageningen

Tungani JO, Mukhwana EJ, Woomer PL (2002) MBILI is Number 1: a handbook for innovative maize-legume intercropping. SACRED Africa, Bungoma, 20pp

Residual Effects of Contrasting Organic Residues on Maize Growth and Phosphorus Accumulation over Four Cropping Cycles in Savanna Soils

O.C. Nwoke, R.C. Abaidoo, G. Nziguheba, and J. Diels

Abstract Management options to improve phosphorus (P) availability in the West African savanna include the application of organic residues, separately or with inorganic fertilizers. The quality and quantity of organic resources affect their contribution to nutrient availability. The quality of residues influences decomposition and nutrient release in the short term but its importance in the long term is unclear. A greenhouse study with six different soils assessed the residual effect of contrasting organic residues (maize stover, farmyard manure, *Senna siamea*, *Mucuna pruriens*, *Leucaena leucocephala*, *Gliricidia sepium*, *Pueraria phaseoloides*, and *Lablab purpureus*) and triple superphosphate (TSP) on maize growth during four cropping cycles of 7 weeks. For the first cropping, the average shoot dry matter yield (DMY), 14 g pot^{-1} obtained with residues of carbon to phosphorus (C:P) ratio ≤ 200, was similar to the TSP treatment and higher than the yield obtained with maize stover (C:P ratio = 396). From the second cropping, the shoot DMY of the maize stover treatment (15 g pot^{-1}) was higher than those of the other treatments. Both high- and low-quality residues had similar effects on the cumulative DMY. Cumulative shoot P accumulation in the maize stover treatment (53 mg pot^{-1}) was significantly higher than in the control (30 mg pot^{-1}) and similar to those from high-quality residue treatments. Significant interaction ($P < 0.05$) between soil and organic resource for DMY occurred only during the first two croppings. This indicates that the effects of soil type and residue quality on nutrient availability dwindle with time.

Keywords Cumulative dry matter yield · Phosphorus accumulation · Phosphorus availability · Residual effect · Residue quality

Introduction

Soil fertility problems impose considerable limitations on crop production in the moist savanna zone of West Africa. These problems range from land deterioration as a result of the poor status of soil organic matter (SOM) and the declining physical and chemical properties of the soils (Pieri, 1992) to the inadequate use of fertilizers for various reasons. Consequently, crop yields have remained low on farmers' fields (Tian et al., 1995), although there is tremendous potential for increased and sustainable crop production (Pieri, 1992).

The surface layer of most soils in the moist savanna zone of West Africa is often shallow and sandy. The need to incorporate organic resources in management strategies aimed at improving the fertility of these soils has been stressed (Buresh et al., 1997; Vanlauwe et al. 2001). Organic resources include plant or crop residues, prunings and root biomass from trees, cover crops (mainly herbaceous legumes) used as live mulch, and animal manure. These are particularly important for the maintenance of the physical and chemical conditions (Vanlauwe et al., 2001) of the soils. Besides soil maintenance, organic inputs can supply nutrients to growing crops, even though they usually contain low concentrations of nutrient elements. Organic inputs can

O.C. Nwoke (✉)
Department of Agronomy, Osun State University, Osogbo, Nigeria
e-mail: c.nwoke@cgiar.org; o.chik@yahoo.com

A. Bationo et al. (eds.), *Innovations as Key to the Green Revolution in Africa*, DOI 10.1007/978-90-481-2543-2_35, © Springer Science+Business Media B.V. 2011

also improve the availability of nutrients such as P upon decomposition, enhance microbial activity, and replenish the SOM pool (Palm et al., 1997). However, the contribution of organic resources to the availability of nutrients, particularly N and P, to the growing crop depends, to a large extent, on their quality and quantity. The quality of organic residues considerably influences their decomposition and nutrient release pattern in the short term but its importance in the long term is unclear.

Assessment of changes in P availability following the incorporation of organic residues and crop growth warrants the separation of soil P into pools with different levels of functional significance. This is necessary, considering the probable contribution of organic P fractions to P availability in tropical soils (Tiessen et al., 1992; Beck and Sanchez 1994). The labile or biologically meaningful pools can be separated into organic (Po) and inorganic (Pi) components by sequential fractionation (Hedley et al., 1982) and could be helpful in assessing changes in soil P availability following organic amendments. This study evaluated the residual effect of contrasting organic amendments on maize (in terms of dry matter yield and nutrient accumulation) during four cycles and the changes in labile soil P fractions during three cycles of maize cropping in savanna soils.

Materials and Methods

Collection of Soils and Organic Resources

Six soils (of varying levels of available P and P sorption capacity) were taken at a depth of 0–10 cm in experimental fields in Ibadan, Kasuwan Magani,

and Danayamaka, Nigeria; Niaouli, Bénin Republic; Sarakawa and Davié, Togo. The soils were air dried and passed through a 4 mm sieve. The agroecological zone, classification, and selected soil characteristics have been presented by Nwoke et al. (2004). Briefly, the soils had organic C 5–10 g kg^{-1}, total N 0.4–0.8 g kg^{-1}, clay 56–167 g kg^{-1}, and pH 5.1–5.8. The organic resources comprised of tree prunings (*L. leucocephala*, *G. sepium*, and *S. siamea*), herbaceous legume biomass (*M. pruriens*, *L. purpureus*, and *P. phaseoloides*), maize stover (*Zea mays*), and farmyard manure (FYM). These were obtained from experimental fields within the International Institute of Tropical Agriculture, Ibadan, Nigeria; fresh twigs (of diameter less than 6 mm) with leaves were picked from several stands of each plant species, chopped (about 5 mm long), and thoroughly mixed together. FYM was included as a different kind of organic resource. The characteristics of the residues have been given by Nwoke et al. (2004) and are shown in brief (Table 1).

Experimental Setup

All the organic resources and mineral P fertilizer (triple superphosphate, TSP) were mixed with 5 kg of dry soil in separate pots at the rate of 10.7 mg P kg^{-1}. The control treatment had no external P source. All treatments were replicated three times. The arrangement of the pots followed a randomized complete block design. The soil was wetted with distilled water and four pregerminated maize (hybrid Oba Super 2) seeds were sown into each pot, and then thinned to two after emergence. All the treatments received N (64 mg N kg^{-1} soil, amounting to 320 mg N pot^{-1}) as KNO$_3$ solution. This solution also supplied about 179 mg K kg^{-1} soil. The plants were watered daily with distilled

Table 1 Selected quality parameters of the organic resources and triple superphosphate (TSP)

Resource	Total N (%)	Total P (%)	Lignin (%)	C:N	C:P	N:P
Pueraria phaseoloides	3.82	0.23	11.1	12	194	16
Lablab purpureus	3.45	0.28	8.5	12	153	13
Mucuna pruriens	2.68	0.20	8.7	16	219	14
Leucaena leucocephala	4.06	0.22	9.2	10	176	18
Gliricidia sepium	3.75	0.20	9.4	11	211	19
Senna siamea	2.67	0.16	8.5	17	292	17
Zea mays (maize stover)	0.67	0.11	13.5	65	396	6
Farmyard manure	1.30	0.31	18.1	22	93	4
TSP	0	19.57	0	0	0	0

water. The maize was harvested after 7 weeks, considered as one cropping cycle in this study, by cutting the aboveground biomass at soil surface (harvest 1). The roots were separated by passing the soil through a 4 mm sieve and washed with water. The plant samples were dried (65°C), weighed, and ground. The soil was returned to the pot and maize was replanted (four pregerminated seeds per pot). The maize plants were thinned to two after emergence, watered daily with distilled water, and harvested after 7 weeks (harvest 2). The planting and harvesting of the third (harvest 3) and fourth (harvest 4) maize crop followed in a similar manner. All subsequent maize crops received the same amount and form of N and K as the first maize crop and soil samples were taken at all maize harvests.

Soil and Plant Analyses

Soil P fractions were determined, on selected treatments, after sequential extractions (Tiessen and Moir 1993). Harvest 4 soil samples were not included. The

anion exchange resin was used in the chloride form; the P in extracts was measured by the method of Murphy and Riley (1962). The ground plant samples were analyzed for total N and P contents (IITA 1982).

Statistical Analysis

Analysis of variance (ANOVA) was conducted with the MIXED model of the SAS system (SAS 1992) to assess treatment effects. Soil and treatment were regarded as fixed effects while replication was taken as a random effect.

Results and Discussion

Dry Matter Yield

Organic resources with a high percentage of N, a low C:N ratio, and a low lignin content are regarded as

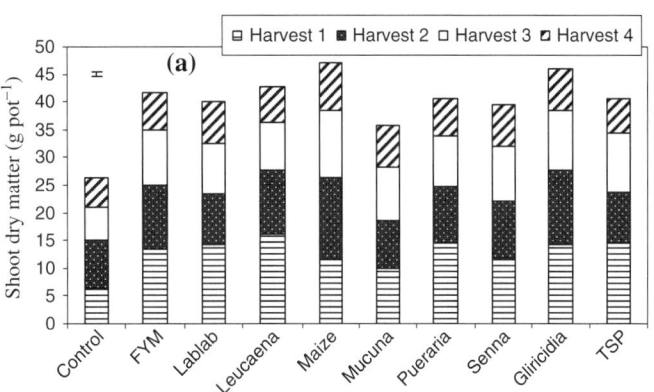

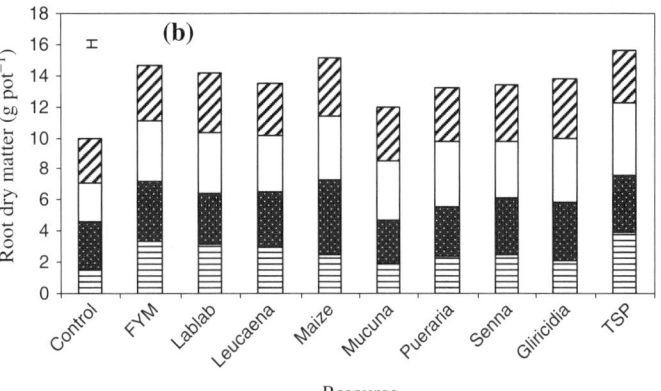

Fig. 1 The shoot (**a**) and root (**b**) dry matter yields of maize grown with organic sources of nutrients and TSP in savanna soils under greenhouse conditions. Values plotted are means across soils. *Bars* represent standard error of difference (SED)

high-quality materials, whereas those with low N, a
high C:N ratio, and a high level of lignin belong to
the low-quality group. The growth of maize was significantly enhanced by the addition of both high- and
low-quality organic resources and TSP during the four
growth periods, as evidenced by the DMY achieved
compared with the control. Although soil and treatment means were significantly ($P < 0.05$) different for
shoot DMY of the four maize crops, significant soil ×
treatment (resource) interaction occurred only with the
first and second crops. The cumulative biomass yields
are presented (Fig. 1) to depict the average effect of the
organic resources and TSP on DMY during the four
growth periods. The addition of maize stover resulted
in a yield slightly higher than or similar to those
obtained with the addition of the other resources (e.g.,
Gliricidia and *Pueraria*). The maize stover used was
low in N and high in lignin and also had high C:N and
C:P ratios (Table 1). Thus, it was of low quality while
Gliricidia and *Pueraria* were of high quality. For the
high-quality residue treatments, the largest proportion

of the cumulative shoot DMY came from the first crop,
whereas for the maize stover treatment it was from
the second crop. These results point to the huge residual effects of maize stover compared with the other
organic resources. The trend of the root DMY was
similar to that of the shoots but root growth was better during the second cropping than during the first in
most of the residue treatments (Fig. 1). With respect to
the soils, a plot of treatment means versus soil means
(Fig. 2) showed wide variability in the productivity of
the soils and also revealed that shoot DMY from the
maize stover treatment was among the highest in nearly
all the soils.

The contribution of organic resources to nutrient
availability is dependent on their decomposition pattern which has been shown to be strongly influenced
by some quality indicators (Palm and Sanchez, 1991;
Vanlauwe et al., 1996). For the first maize crop, shoot
DMY was highly influenced by the nature of the
organic resource and the soil but this influence faded
with time. For example, for harvest 1, treatments that

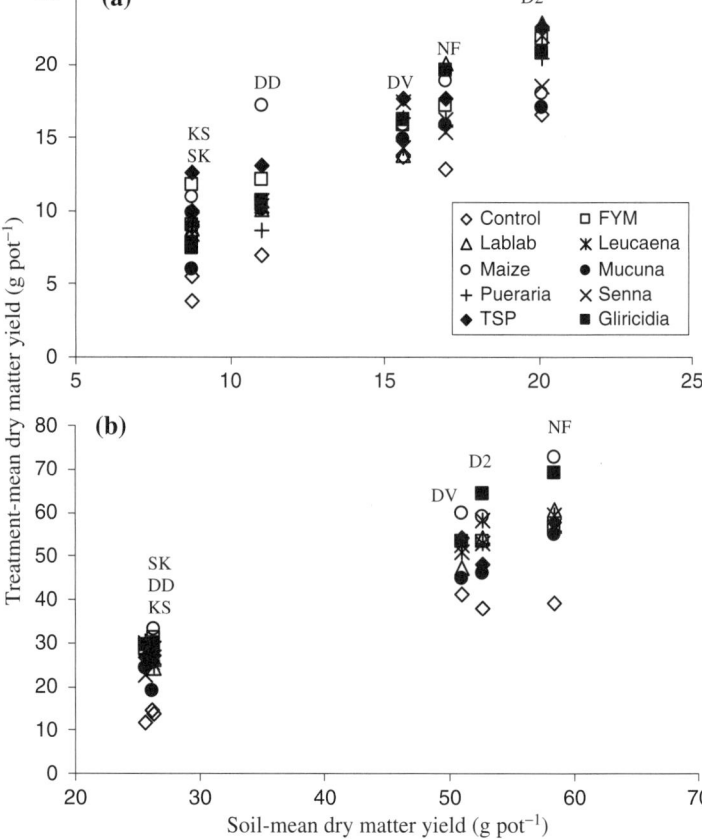

Fig. 2 Plots of treatment
means versus soil means for
root (**a**) and shoot (**b**) dry
matter yields of four maize
crops grown in six soils
amended with organic sources
of nutrients and TSP under
greenhouse conditions. *Labels*
represent soil codes:
KS = Kasuwan Magani,
SK = Sarakawa,
DD = Danayamaka,
DV = Davié, NF = Niaouli,
and D2 = Ibadan

Fig. 3 The accumulation of phosphorus (**a**) and nitrogen (**b**) in the shoot biomass of maize grown with organic sources of nutrients and TSP in savanna soils under greenhouse conditions. Values plotted are means across soils. *Bars* represent standard error of difference (SED)

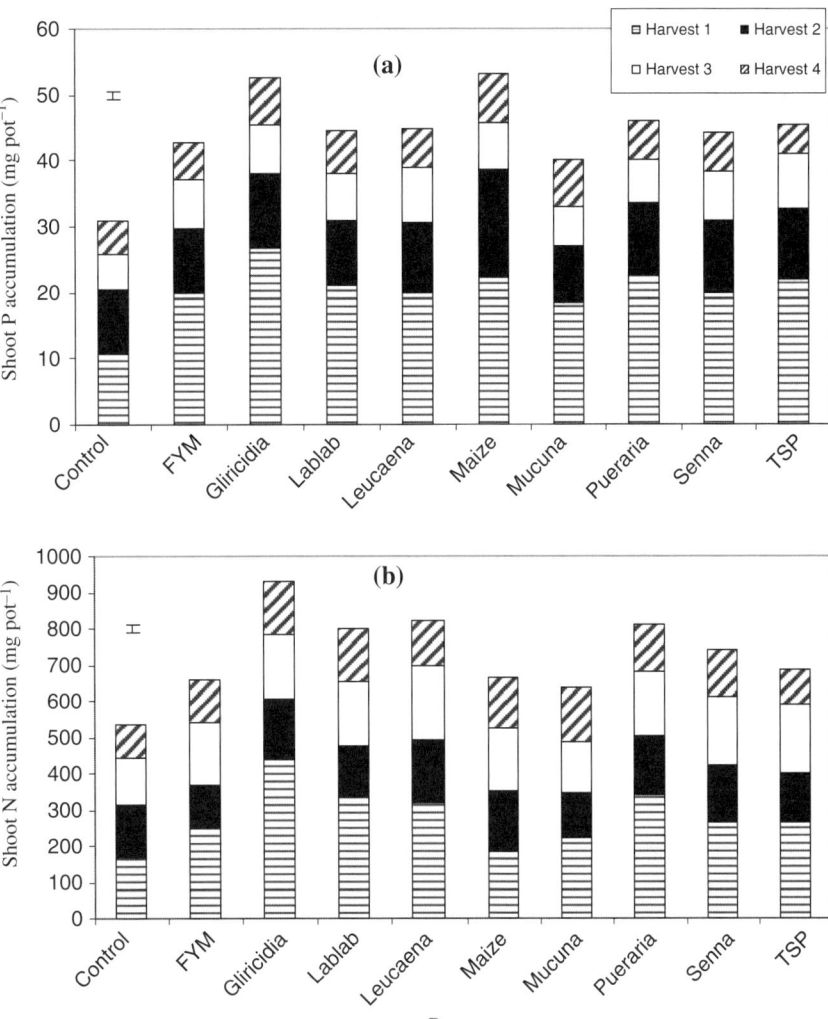

received high-quality residues had higher DMY than the maize stover treatment, and there was a significant soil × organic resource interaction. However, for harvests 2 and 3, the DMY from the maize stover treatment became slightly higher. These results signify that the high-quality residues decomposed and released a large proportion of their nutrients rapidly and thus benefited the first more than the subsequent crops, whereas maize stover, a low-quality resource, appeared to have decomposed and released nutrients more slowly than the high-quality sources to the benefit of the subsequent maize crops. For harvest 4, all the treatments had similar DMY with no significant soil × organic resource interaction, indicating that the contribution of the organic resources had been exhausted. Although the addition of organic resources can improve soil

conditions by enhancing the SOM pool, the influence of the contrasting residues on SOM was not assessed. Nevertheless, the results of this study imply that the quality of organic resources used for soil improvement may be less important under long-term management strategies.

Nutrient Accumulation in the Biomass

Generally, the addition of organic resources increased the accumulation of P in the shoot biomass of maize during the four growth cycles. Increased accumulation of N was also recorded. A combined analysis of the data for all the growth cycles showed that treatment (resource) means as well as soil means were

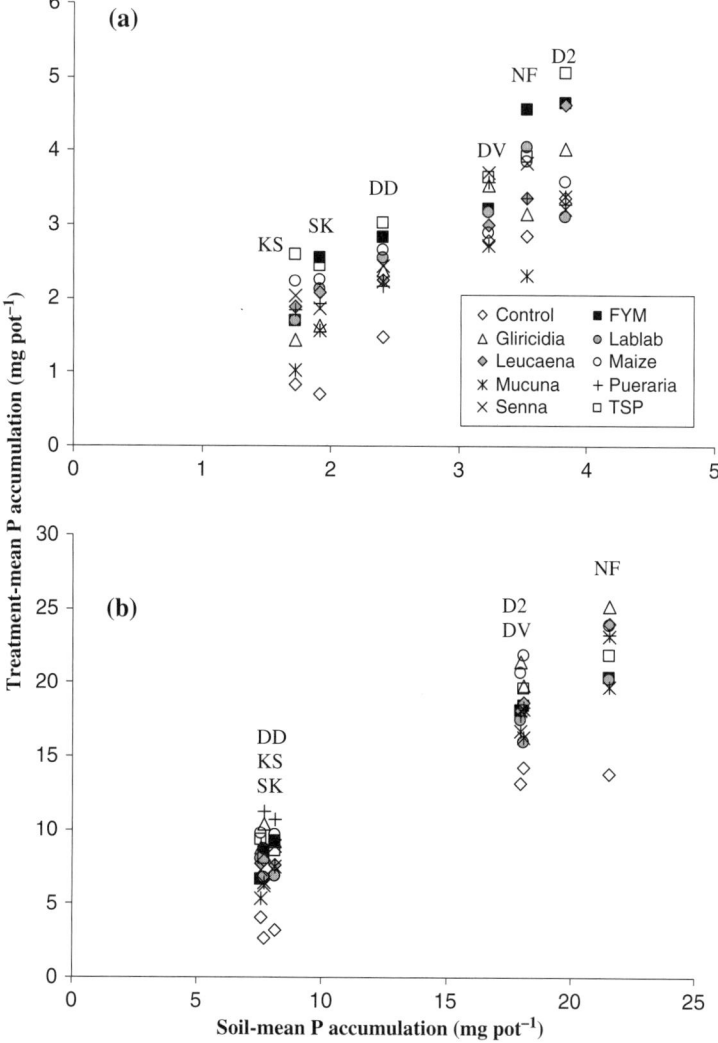

Fig. 4 Plots of treatment means versus soil means for P accumulation in roots (**a**) and shoots (**b**) of four maize crops grown in six soils amended with organic sources of nutrients and TSP under greenhouse conditions. *Labels* represent soil codes: KS = Kasuwan Magani, SK = Sarakawa, DD = Danayamaka, DV = Davié, NF = Niaouli, and D2 = Ibadan

significantly different ($P < 0.05$) but no soil × treatment interaction occurred for either nutrient parameter. The addition of high-quality *Gliricidia* and low-quality maize stover, on average, resulted in the accumulation of similar amounts of P in the shoot biomass of maize. These were also higher than the amounts of P that resulted from the addition of the other residues and TSP (Fig. 3). The amount of P accumulated in the shoots of maize grown with maize stover could not be attributed entirely to P released from the decomposing residue as it had the lowest P concentration (Table 1). The maize stover might have improved soil P availability through other mechanisms. Such mechanisms may include the reduction of the P sorption capacity (Iyamuremye and Dick, 1996) and improvement of the availability of native soil P by decomposition products

(Nagarajah et al., 1970). Plotting treatment means against soil means for P accumulation in both roots and shoots (Fig. 4) showed wide variability among the soils and among the organic resources within a soil. For shoot N accumulation, differences from the addition of the various residues were more pronounced than differences in shoot P accumulation (Fig. 3). As in shoot P accumulation, shoot N accumulation was highest with the addition of *Gliricidia* residues. The addition of *Lablab*, *Leucaena*, and *Pueraria* residues resulted in the accumulation of similar amounts of N which were higher than those from the addition of maize stover, *Mucuna* residues, or TSP. These differences occurred with the first crop that received the organic inputs and P fertilizer (when shoot N accumulation in some residue treatments was larger than the amount supplied by the inorganic N source) and were clearly visible even with the cumulative data, and may be attributed to the disparity in the amounts of N supplied to the different treatments. While all the treatments received the same amount of inorganic N, the organic resources applied differed in their N content and hence contributed different amounts of N which resulted in dissimilar N:P ratios to the maize crop.

Labile and Moderately Labile Soil P Pools

The effects of the organic resources and TSP on the measured soil P pools were minimal. The treatment

(resource), soil, and harvest (cropping) means are given in Table 2. The anion resin-extractable (resin–Pi) and the sodium bicarbonate-extractable (HCO$_3$–Pi) inorganic P fractions were slightly increased by the addition of the residues and TSP and the magnitude of the effects was similar. While resin–Pi represents the readily exchangeable P fraction in soil (Tiessen and Moir, 1993), which can be of immediate benefit to the growing crop and thus a good indicator of P availability, HCO$_3$–Pi is also considered labile and therefore a part of the available P pool. The labile organic fraction (HCO$_3$–Po) was not significantly different ($P = 0.19$) among the various treatments within the soil but, on average, tended to increase with each cropping cycle

Table 2 The effect of organic resources (herbaceous legume residues and farmyard manure, FYM) and phosphate fertilizer (TSP) on the labile and moderately labile inorganic (Pi) and organic (Po) soil P pools extracted from six soils at the first three maize harvests under greenhouse conditions

	Labile pool			Moderately labile pool	
	Resin–Pi (mg kg^{-1})	HCO$_3$–Pi (mg kg^{-1})	HCO$_3$–Po (mg kg^{-1})	NaOH–Pi (mg kg^{-1})	NaOH–Po (mg kg^{-1})
Resource					
Control	2.71	3.70	7.72	15.67	39.65
FYM	3.53	4.18	7.54	16.55	40.76
Lablab	3.22	4.26	7.44	16.02	42.11
Maize	3.97	4.13	7.67	16.23	41.74
Mucuna	3.35	3.88	8.14	16.38	42.03
Pueraria	3.37	4.66	7.69	16.38	44.96
TSP	3.75	4.90	6.97	17.53	39.85
SED	0.134	0.209	0.412	0.435	1.208
Soil					
Ibadan	2.80	5.44	10.66	23.60	52.83
Danayamaka	1.50	3.89	6.96	16.35	34.47
Davié	9.66	4.26	5.72	9.37	33.19
Kasuwan Magani	1.30	4.06	9.70	26.01	60.48
Niaouli	3.63	5.45	7.33	15.12	37.14
Sarakawa	1.61	2.37	5.22	7.90	31.38
SED	0.124	0.194	0.381	0.403	1.118
Cropping (harvest)					
1	4.14	4.68	6.92	16.44	49.37
2	3.28	3.83	7.08	16.45	37.06
3	2.82	4.21	8.80	16.28	38.32
SED	0.087	0.137	0.270	0.285	0.791
P > F					
Soil (S)	<0.0001	<0.0001	<0.0001	<0.0001	<0.0001
Resource (R)	<0.0001	<0.0001	0.1906	0.0022	0.0003
S × R	0.0090	0.0060	0.1231	0.0115	0.0197
Harvest (H)	<0.0001	<0.0001	<0.0001	0.8017	<0.0001
S × H	<0.0001	<0.0001	0.0276	<0.0001	<0.0001
R × H	<0.0001	0.0301	<0.0001	0.3152	0.0039
S × R × H	0.0532	0.0616	<0.0001	0.5069	<0.0001

SED represents standard error of the difference

(Table 2). Generally, the moderately labile (NaOH–Pi and NaOH–Po) pools from the residue treatments were not significantly different from the control. However, among the soils, the extracted amounts of the various P pools differed widely due to differences in inherent P status. The effect of P removal by harvested crops could be seen on the resin–Pi fraction which, on average, decreased slightly with each cropping (Table 2). Similarly, the average amount of NaOH–Po extracted after harvest 1 (49 mg kg^{-1}) was significantly ($P <$ 0.0001) higher than the amounts extracted after harvest 2 (37 mg kg^{-1}) and harvest 3 (38 mg kg^{-1}), indicating that part of this pool has entered into other pools, probably through mineralization processes. No treatment differences were observed in the quantity of P in most of the bioavailable (labile and moderately labile) pools extracted in any of the soils. Because the organic resources were applied once and cropped several times, their impact on soil P pools may have been low. Moreover, as different amounts of P were removed from the various treatments by the maize crop (Fig. 3), the contribution of the organic resources may have been obscured.

In conclusion, the data presented show clearly that the relative importance of the quality of applied organic resources decreases as cropping times increase considering that organic resources, both high and low quality, resulted in similar cumulative dry matter yields and P accumulation. In addition, there was no significant soil × resource × harvest interaction at the 5% probability level for the inorganic soil P pools (Table 2). Thus, the choice of organic resources for soil amendment can better be guided by the desired outcome. However, further studies are warranted under field conditions where maize is grown to maturity in less artificial conditions to corroborate these findings.

Acknowledgments The Belgian Directorate General for Development Cooperation (DGDC) is gratefully acknowledged for sponsoring this work under the collaborative project between KU Leuven and IITA on 'Balanced Nutrient Management Systems for Maize-based Farming Systems in the Moist Savanna and Humid Forest Zone of West Africa.'

References

Beck MA, Sanchez PA (1994) Soil phosphorus fraction dynamics during 18 years of cultivation on a Typic Paleudult. Soil Sci Soc Am J 58:1424–1431

Buresh RJ, Sanchez PA, Calhoun F (eds) (1997) Replenishing soil fertility in Africa, SSSA special Publication 51. SSSA, ASA, Madison, WI

Hedley MJ, Stewart JWB, Chauhan BS (1982) Changes in inorganic and organic soil phosphorus fractions induced by cultivation practices and by laboratory incubations. Soil Sci Soc Am J 46:970–976

IITA (International Institute of Tropical Agriculture) (1982) Automated and semi-automated methods for soil and plant analysis, manual series no. 7. IITA, Ibadan, p 33

Iyamuremye F, Dick RP (1996) Organic amendments and phosphorus sorption by soils. Adv Agron 56:139–185

Murphy J, Riley JP (1962) A modified single solution for the determination of phosphorus in natural waters. Anal Chem Acta 27:31–36

Nagarajah S, Posner AM, Quirk JP (1970) Competitive adsorption of phosphate with polygalacturonate and other organic anions on kaolinite and oxide surfaces. Nature 228:83–85

Nwoke OC, Vanlauwe B, Diels J, Sanginga N, Osonubi O (2004) Impact of residue characteristics on phosphorus availability in West-African moist savanna soils. Biol Fertil Soils 39:422–428

Palm CA, Myers RJK, Nandwa SM (1997) Combined use of organic and inorganic nutrient sources for soil fertility maintenance and replenishment. In: Buresh RJ, Sanchez PA, Calhoun F (eds) Replenishing soil fertility in Africa, SSSA special publication 51. SSSA, ASA, Madison, WI, pp 193–217

Palm CA, Sanchez PA (1991) Nitrogen release from the leaves of some tropical legumes as affected by their lignin and polyphenolic contents. Soil Biol Biochem 23:83–88

Pieri CJMG (1992) Fertility of soils: a future for farming in the West African savannah. Springer, New York, NY

SAS (1992) The mixed procedure. SAS technical report P-229: SAS/STAT software: changes and enhancements. SAS Institute, Cary, NC, pp 287–366

Tian G, Kang BT, Akobundu IO, Manyong VM (1995) Food production in the moist savanna of West and Central Africa. In: Kang BT, Akobundu IO, Manyong VM, Carsky RJ, Sanginga N, Kueneman EA (eds) Moist Savannas of Africa: potentials and constraints for crop production. Proceedings of the international workshop held at Cotonou, Republic of Benin, 19–23 Sept 1994. IITA, Ibadan, pp 107–127

Tiessen H, Moir JO (1993) Characterization of available P by sequential extraction. In: Carter MR (ed) Soil sampling and methods of analysis. Canadian Society of Soil Science (Lewis Publishers), Boca Raton, FL, pp 75–86

Tiessen H, Salcedo IH, Sampio EVSB (1992) Nutrient and soil organic matter dynamics under shifting cultivation in semi-arid northeastern Brazil. Agric Ecosyst Environ 38:139–151

Vanlauwe B, Ahiou K, Aman S, Iwuafor ENO, Tossah BK, Diels J, Sanginga N, Lyasse O, Merckx R, Deckers J (2001) Maize yield as affected by organic inputs and urea in the West African moist savanna. Agron J 93:1191–1199

Vanlauwe B, Nwoke OC, Sanginga N, Merckx R (1996) Impact of residue quality on the C and N mineralization of leaf and root residues of three agroforestry species. Plant Soil 183:221–231

Nutr Cycl Agroecosyst (2010) 88:103–109
DOI 10.1007/s10705-009-9282-7

RESEARCH ARTICLE

Interactive effects of selected nutrient resources and tied-ridging on plant growth performance in a semi-arid smallholder farming environment in central Zimbabwe

J. Nyamangara · I. Nyagumbo

Received: 5 March 2008 / Accepted: 21 April 2009 / Published online: 15 May 2009
© Springer Science+Business Media B.V. 2009

Abstract Crop production in sub-Saharan Africa is constrained by numerous factors including frequent droughts and periods of moisture stress, low soil fertility, and restricted access to mineral fertilisers. A 2 year (2005/6 and 2006/7) field study was conducted in Shurugwi district, central Zimbabwe, to determine the effects of different nutrient resources and two tillage practices on the grain yield of maize (*Zea mays* L.) and soybean (*Glycine max* (L.) Merr). Six nutrient resource treatments (control, pit-stored manure, leaf litter, anthill soil, mineral fertiliser, mineral fertiliser plus pit-stored manure) were combined with two tillage practices (conventional tillage and post-emergence tied ridging). Basal fertilisation was done with 0 kg ha^{-1} as control, 240 kg ha^{-1} PKS fertiliser, 18 t ha^{-1} manure, 10 t ha^{-1} manure plus 240 kg ha^{-1} PKS fertiliser, 35 t ha^{-1} leaf litter, 52 t ha^{-1} anthill soil. About 60 kg N/ha was applied to fertiliser only and fertiliser plus manure treatments as top dressing in the form of ammonium nitrate (34.5%N). A split-plot design was used with nutrient resource as the main plot and tillage practice as the subplot, and five farmers' fields were used as

replicates. Grain yield was determined at physiological maturity (140 and 126 days after planting for maize and soybean, respectively) and adjusted to 12.5% moisture content for maize and 11% for soybean. In the first season (2005/06), addition of different nutrient resources under conventional tillage increased ($P < 0.05$) maize grain yield by 102–450%, with leaf litter and manure plus fertiliser treatments, giving the lowest (551 kg ha$^{-1)}$) and highest (3,032 kg ha^{-1}) increments, respectively, compared to the control. For each treatment, tied-ridging further increased maize grain yield. For example, for leaf litter, tied-ridging further increased grain yield by 96% indicating the importance of integrating nutrient and water management practices in semi-arid areas where moisture stress is frequent. Despite the low rainfall and extended dry spells in the second season, addition of the different nutrient resources still increased yield which was further increased by tied-ridging in most treatments. Besides providing grain, soybean had higher residual effects on the following maize crop compared to *Crotalaria gramiana*, a green manure. It was concluded that the highest benefits of tied-ridging, in terms of grain yield, were realised when cattle manure was combined with mineral fertiliser, both of which are available to resource-endowed households. Besides marginally increasing yield, leaf litter and anthill which represent resources that can be accessed by very poor households, have a positive effect of the soil chemical environment.

J. Nyamangara (✉) · I. Nyagumbo
Department of Soil Science & Agricultural Engineering,
University of Zimbabwe, P.O. Box MP167, Mount
Pleasant, Harare, Zimbabwe
e-mail: jnyamangara@yahoo.co.uk;
jnyamangara@agric.uz.ac.zw

Springer

This article has been previously published in the journal "Nutrient Cycling in Agroecosystems" Volume 88 Issue 1.
A. Bationo et al. (eds.), Innovations as Key to the Green Revolution in Africa: Exploring the Scientific Facts. © 2010 Springer.

Keywords Conventional tillage ·
Maize · Nutrient resources ·
Post-emergence tied ridging · Soybean

Introduction

Soil fertility is the most important factor limiting crop productivity in sub-Saharan Africa (SSA) (Smaling et al. 1997) and this is often exacerbated by inadequate moisture content in semi-arid areas (Cleaver and Schreiber 1994). The majority of soils in SSA are light-textured and acidic, and have critically low concentrations of soil organic carbon (SOC) and nutrients (Grant 1981; Nyamangara et al. 2000) due to a combination of high rainfall that promote leaching (Kamukondiwa and Bergström 1994; Twomlow 1994; Vogel et al. 1994) and long-term cultivation with sub-optimal or no nutrient inputs (Stoorvogel and Smaling 1990). For example, in 1990 the average fertiliser use in SSA was only 8.4 kg ha^{-1} compared to 81 kg ha^{-1} in other developing countries, and 75% of the fertiliser use was limited to only six countries (Gerner and Harris 1993). In extreme cases of nutrient mining, multiple nutrient deficiencies have been reported (Mugwira and Nyamangara 1998; Mukurumbira and Nemasasi 1998), and the soils hardly respond to NPKS mineral fertiliser additions and will only respond to continual application of high rates of animal manure after several seasons (Zingore et al. 2007). The dominant smallholder cropping systems of southern and eastern and southern Africa are based on maize, the staple food, which accounts for about 50% of the calories consumed (Mugwira et al. 2002). However, soybean has also become an important source of cash income and nutritious diet with the latter being critical in reducing the negative impacts of HIV/AIDS ravaging the region.

In the semi-arid areas of Zimbabwe (rainfall < 600 mm per annum) crop failure occurs in three out of every five years and poor rainfall distribution (extended dry spells) within a relatively wet season often reduces yield potential. Under such marginal agro-ecological conditions crop yields can be enhanced by harvesting the rain which is often in the form of high intensity storms. However, some studies have shown that sandy soils may not benefit from extra water supplies unless if soil nutrients are also added (Anshuttz et al. 1997).

Soil water regime represents a balance among four processes; evaporation, transpiration, infiltration and internal drainage. These processes are in turn controlled in part by water retention and transmission properties of the soil. Tillage practices such as tied-ridging affect hydraulic properties of the soil (Singh et al. 1998). Tied-ridging involves construction of mounds of soil of 15–25 cm high along crop rows using a hoe, a mouldboard plough or a locally available high-wing ox-drawn ridger (Elwell and Norton 1988). Cross ties, half to two-thirds the height of the ridges, are made in the furrows at 1.5–3 m intervals to capture run-off and enhance water infiltration. Tied-ridging increases water infiltration reduces run-off and controls soil erosion (Nyagumbo 2002). Vogel (1993) reported high soil loss (3.3 t/ha) and run-off (90.4 mm) under conventional tillage compared to only 0.7 t/ha soil erosion and 30.1 mm run-off under tied-ridging in a sub-humid area in Zimbabwe. Subsequent studies at Hatcliffe and Domboshawa in Zimbabwe also showed a 5-year mean annual run-off of 22 and 9% at Hatcliffe and 20 and 4% at Domboshawa of the seasonal rainfall from conventional tillage and tied-ridging, respectively (Nyagumbo 2002). However, the water harvesting effects of tied ridging have been reported to increase loading of nitrates and herbicides at the bottom of the root zone through increased drainage (Hatfield et al. 1998), a situation more relevant in high fertiliser input systems as opposed to the cropping systems in SSA where sub-optimal levels of fertiliser are used. Moyo and Hagmann (1994) reported a higher drainage (174 mm) and much lower run-off (22 mm) under tied-ridging compared to conventional tillage (147 and 118 mm, respectively) over a 5-year study conducted on a sandy soil in a semi-arid area in southern Zimbabwe.

Integrated tillage and nutrient resource management practices need to be developed for sustainable crop production, especially in semi-arid areas where smallholder farmers use a wide range of locally available nutrient resources and crop response to fertilisation is often limited by moisture stress. Pre-planting tied-ridging has been reported to cause poor germination and establishment (Meijer 1992; Vogel 1993). This study aimed to determine the effects of

 Springer

selected sources of nutrients commonly used by smallholder farmers in SSA in combination with two tillage practices, conventional and post-planting tied-ridging, on the growth and grain yield of maize and soybean. In the second season the residual effect of soybean and *Crotalaria gramiana* on maize growth and grain yield were also measured. It was hypothesized that tillage practices have no effect on maize and soybean growth, and there is no interaction between tillage practices and nutrient resources used.

Materials and methods

Site description and soil characterisation

The field experiments were conducted in Ward 5 of Shurugwi smallholder area from 2005 to 2007 in central Zimbabwe. The area is classified as semi-arid, receiving low and erratic rainfall (450–650 mm per annum). The area is covered by coarse-grained sandy soils derived from granitic parent material with low soil organic matter content (<1%), water retention capacity, and also deficient in N, P and S (Grant 1981). The dominant farming system is mixed cropping and livestock-based with maize as the staple crop. Arable fields are individually owned and cropped but grazing is communal in open-access areas, and also in arable fields during the dry season. Cattle-owners collect crop residues from their fields and store as dry season animal feed.

Manure is a key source of plant nutrients although mineral fertilisers are also used by resource-endowed farmers but at low rates (<35 kg/ha N) for fear of perceived crop burn. Higher fertiliser rates are only used in vegetable gardens situated in wetlands where irrigation water is assured. However, higher N fertiliser rates (up to 100 kg N ha^{-1}) are recommended in seasons where seasonal rainfall distribution is predicted to be good.

Soil sampling and analysis

Soil samples were taken from the top 15 cm soil layer in each field (before ploughing) using a soil auger at the start of the experiment. Ten sub-samples were randomly taken from each field and thoroughly mixed to make a composite sample. After harvesting, soil samples were taken from each plot. Soil samples were taken from intra-row spaces between the planting stations. The composite samples were air-dried and passed through a 2-mm sieve before analysis. The samples were analysed for soil organic C (modified Walkely-Black), total N (micro-Kjeldahl), total P, available P (Olsen) and pH [water and 0.01 M CaCl$_2$] (Anderson and Ingram 1993).

Experimental design and planting

Five sites were selected across the ward with the participation of farmers and ploughed using an ox-drawn mouldboard plough. Two tillage treatments (conventional and post-emergence tied ridging) were used as main plots and five nutrient resource treatments as sub-plots in a split-plot design. The five experimental sites were used as replicates and each site was managed by a group of nearby farmers and each group consisted of at least ten farmers. The nutrient resource treatments were pit-stored cattle manure (18 t ha^{-1} manure), leaf litter (35 t ha^{-1}), anthill soil (52 t ha^{-1}), mineral fertiliser (240 kg ha^{-1} PKS containing 14% P 13% K 5% S + 60 kg ha^{-1} ammonium nitrate (34.5% N), mineral fertiliser plus pit-stored manure (10 t ha^{-1} manure plus 240 kg ha^{-1} PKS plus 60 kg ha^{-1} ammonium nitrate). The application rates were based on recommendations from extension agents for manure, leaf litter and mineral fertiliser, and farmer practice for anthill soil. Planting rows were marked out using ox-drawn ploughs without the mouldboard attached and nutrient resources were applied in the rows. The nutrient materials were partially covered with soil before sowing the maize to avoid seed burn. All the mineral N fertiliser was applied as top dressing at 5 weeks after planting. Post-emergence tied ridges were constructed at 4 after maize germination.

For the soybean and *Crotalaria gramiana* experiment in the first season, a similar experimental design was used but only one fertiliser treatment (240 kg ha^{-1} PKS—14% P 13% K 5% S) was used at planting. Seed for both crops was dribbled along the planting furrows, covered with soil and thinned to 10 cm intra-row spacing 2 weeks after germination. Soybean grain was harvested at its physiological maturity, and *Crotalaria* was cut after senescence, and the residues of both crops were left in the field. At the beginning of the second season, the crop residues were evenly spread in the respective plots

106

Nutr Cycl Agroecosyst (2010) 88:103–109

and ploughed in. Maize was then planted in order to assess the residual effects of the legume crops.

In each case tied ridges were made on the relevant plots using a mouldboard plough followed by cross-ties using a donkey-drawn tie-maker 4 weeks after planting. In both experiments plot sizes were 100–500 m^2 depending on the size of the field and what the participating farmers at each site perceived was an adequate size to enable them objectively assess crop performance. Weeding was done by hand-hoeing when required. Grain yield was determined at physiological maturity and adjusted to 12.5% moisture content for maize and 11% for soybean.

Results

Effect of nutrient resources on soil properties

All the soil nutrient resources reduced the initial soil pH by 3.3–13.5% (Table 1). The lowest pH (pH 4.74, 0.01 M CaCl$_2$) was recorded in the fertiliser only treatment while the least reduction (0.18 units) was recorded in the anthill soil treatment. Even in the control treatment, cultivation induced the lowering of soil pH. Available soil P was only increased (5.8–40.6%) by leaf litter, manure only and manure plus fertiliser treatments when compared to the initial soil P status (Table 1). Available soil P decreased by 18.6, 7.9 and 1.9% in the control, anthill soil and fertiliser only treatments, respectively, compared to the initial status. Table 2 shows that the C-to-N ratios of the organic resources were much lower for leaf litter and anthill soil, although manure had higher N

concentration. Of the three organic resources, cattle manure was a better source of P, K and Mg, and leaf litter for Ca (Table 2).

Effect of nutrient resources and tied-ridging on maize grain yield

All nutrient resources significantly ($P < 0.05$) increased maize grain yield without tied-ridging compared to the control, and highest yields (>3 t/ha in year 1; >1.5 t/ha in year 2) were realised when manure and fertiliser were applied in combination (Figs. 1, 2). Post-emergence tied-ridging further increased maize grain yields, even in the control treatments where yield was less than 1 t/ha, but this effect was more consistent in the first year (Fig. 1). In the second year tied-ridging depressed yield in leaf litter and mineral fertiliser only treatments. Maize grain yield in the control was similar to that in the leaf litter treatment with or without tied-ridging (Fig. 2). There was no interaction between the nutrient resources and tied-ridging.

Table 2 Selected properties of the soil amendments applied to experimental field plots

	%						C-to-N ratio
	Org C	N	P	K	Ca	Mg	
Cattle manure	16.8	0.70	1.10	0.30	0.60	0.56	24.0
Leaf litter	6.90	0.49	0.47	0.13	0.79	0.47	14.1
Anthill soil	3.34	0.25	0.29	0.11	0.50	0.31	13.4

NB Org.—Organic

Table 1 Some initial soil properties of the experimental sites and after one season of application of different nutrient resources

	pH		SOC (%)	Total N (%)	Available P (mg kg^{-1})
	H$_2$0	CaCl$_2$			
Initial	6.04	5.48	0.82	0.060	7.20
Control	5.72	5.18			5.86
240 kg PKS + 100 kg N	5.30	4.74			7.06
10 Mg manure + 240 kg PKS + 100 kg N	5.70	5.14			10.12
18 Mg manure	5.32	4.90			7.62
35 Mg leaf litter	5.64	4.96			10.12
52 Mg Anthill soil	5.86	5.30			6.63

NB Resource applications per ha; *PKS* 14% P 13% K 5% S; *N* top-dressed as ammonium nitrate at 5 weeks after planting; Soil samples taken from conventional tillage treatments

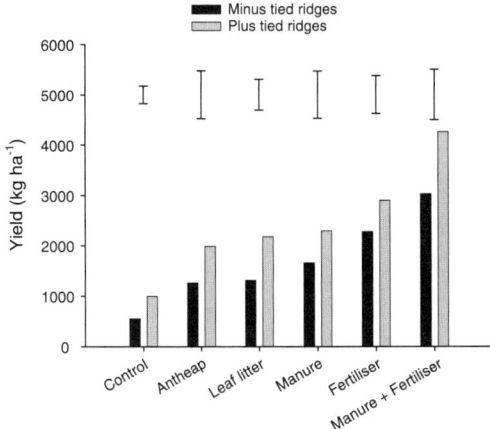

Fig. 1 Effect of different nutrient resources and tied-ridging on maize grain yield in season 1 (2005/06). Error bars represent least significant difference ($P < 0.05$)

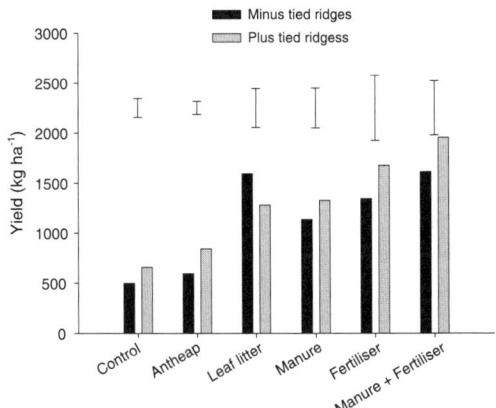

Fig. 2 Effect of different nutrient resources and tied-ridging on maize grain yield in season 2 (2006/07). Error bars represent least significant difference ($P < 0.05$)

Effect of tied-ridging on legume growth and effect of legume residue on maize grain yield

Tied-ridging increased soybean grain yield by up to 35% over the conventional tillage in the first season but the difference was not significant (data not shown). There was evidence of Zn, Mn and Mg deficiency (interveinal chlorosis, necrotic lesions on older leaves and chlorotic spotting) during the vegetative stage at all the five sites. Dry matter yield and N content of *Crotalaria gramiana* was not determined.

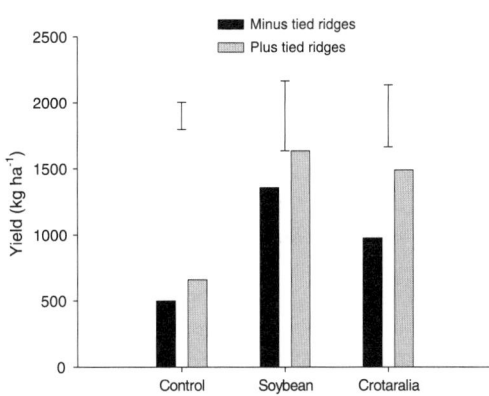

Fig. 3 Effects of soybean and *Crotalaria gramiana* residues on maize grain yield. Error bars represent least significant difference ($P < 0.05$)

Both legumes had a significant ($P < 0.05$) residual soil fertility effect on the yield of maize grown in the second year without tied-ridging (Fig. 3). Tied-ridging further enhanced the residual effect of both legumes, and the higher enhancement (500 kg/ha) was by *Crotalaria gramiana* (Fig. 3). Soybean had a higher residual effect than *Crotalaria gramiana*.

Statistical analysis

Comparisons on the effect of the different fertility treatments on grain yield were carried out using analysis of variance (ANOVA) (Genstat 2002). Separation of means were done using the least significant difference procedure ($P < 0.05$).

Discussion

Sandy soils in Zimbabwe are inherently deficient in N, P, S and also low in organic C (Grant 1981; Nyamangara et al. 2000), and therefore yields in control treatments were expected to be low (<500 kg ha^{-1}) (Figs. 1–3). The apparent lowering of pH (acidification) after just one year of cultivation, especially in mineral fertiliser treatments, indicates the poor buffering capacity of these largely sandy soils (Grant 1981). However, the decrease did not affect yield because the pH was still above threshold limits below which is reduced (Nyamangara et al. 2000). Though conducted in a semi-arid area, the study underscores the need for periodic liming if acidifying nitrogenous fertilisers are continually applied at similar (35 kg/ha) or higher rates in poorly buffered soils. Soil acidification due to the use of nitrogenous fertilisers has been reported in higher rainfall areas where lime application is recommended approximately every third year (Nyamangara et al. 2000). The frequency of application would be lower in semi arid areas where leaching of basic cations is relatively low and also as farmers apply lower fertiliser rates for fear of crop burn. The pH changes observed also seem to suggest that organic fertilisers buffer soils from acidification better than mineral fertilisers, and anthill soil had the highest pH buffering effect. Although the nutrient value of anthill soil was very low, farmers who use it would benefit from the pH moderation effect which will in turn ensure the availability of nutrients like P that become locked up when the soil becomes acidic. Organic fertilisers also improve the soil physical environment, especially the water-holding capacity which is key in semi arid areas (Nyamangara et al. 2001).

The reduction in available P concentration in just one season shows the inherent low status of the nutrient, and thus the soils are vulnerable to nutrient mining when rainfall is adequate. Soil nutrient mining has been widely reported in smallholder areas in SSA (Smaling et al. 1997). However, the decrease in soil pH due to cultivation and mineral fertiliser application could have also contributed to the reduction in the soil available P.

Besides supplying N, P, K and S, anthill soil, leaf litter and cattle manure also supply Ca, Mg and micronutrients to crops. These extra nutrients, which were not contained in NPKS fertiliser, could have had some growth effect in the manure, leaf litter and anthill soil treatments. However, Ca, Mg and micronutrient deficiencies have been reported on maize in high rainfall areas where leaching under acid soil conditions is high (Mugwira and Nyamangara 1998).

Increase in yield induced by all nutrient resources compared to the control treatment implied that smallholder farmers can apply any of these resources available to them in order to enhance yield and improve food security as well as to counter nutrient mining. Although the yield increases were small for anthill soil and leaf litter, these resources are often used by the most resource-poor farmers who do not have access to any other better sources of nutrients. However, the results showed that highest yields, and

hence food security, can be achieved through combined application of cattle manure and mineral fertiliser. The higher residual fertility of soybean than *Crotalaria gramiana* suggested that incorporation of soybean into the cropping system will be relatively more attractive to farmers given its dual role.

The enhanced yield under tied-ridging compared to conventional tillage, especially in the first year, implied that tied-ridging can increase rain water use efficiency. When rainfall is low, as in year 2, the benefits are diminished or absent due to extended dry spells. Lack of interaction between soil nutrient resources and tillage practices implied that these treatments acted independent of each other. Besides moisture conservation, the practice of constructing post-emergence tied-ridges has also been found to enhance weed control (Nyagumbo 1993). The benefits of tied-ridging were also evident on soybean yield and on maize grown on legume residual fertility implying that the practice can be applied to legume-maize-based cropping systems.

Conclusions

The integration of soil fertility with moisture conserving techniques such as tied-ridging can potentially increase productivity under semi-arid conditions, suggesting increased water use efficiency. The highest yield benefits occur when livestock manure and mineral fertilisers are applied in combination but such benefits are diminished when prolonged dry spell conditions occur. Nutrient-poor fertility resources such as leaf litter and anthill soil are worthwhile investments for resource-constrained farmers as they also improve the soil chemical and possibly physical, environment. Inclusion of soybean in farming systems can enhance crop productivity through its residual fertility effects though its yield remains low.

Acknowledgments We thank the Rockefeller Foundation for funding this work through the Soil Fertility Consortium for Southern Africa (SOFECSA).

References

Anderson JM, Ingram JSI (1993) Tropical soil biology and fertility: a hand book of methods. CAB International, Wallingford

Nutr Cycl Agroecosyst (2010) 88:103–109

Anshuttz J, Kome A, Nederlof M, de Neef R (1997) Water harvesting and soil moisture retention. Agrodok series 13, CTA, Agromisa, Wageningen, The Netherlands

Cleaver K, Schreiber G (1994) Reversing the spiral: the population, agriculture and environment nexus in sub-Saharan Africa. World Bank, Washington

Elwell HA, Norton AJ (1988) No-till tied ridging: a recommended sustainable crop production system. Institute of Agricultural Engineering, Ministry of Agriculture, Harare 85pp

Genstat (2002) Genstat release 6.1. VSN International Ltd, Oxford

Gerner H, Harris G (1993) The use and supply of fertilisers in sub-Saharan Africa. In: van Reuler H, Prins WH (eds) The role of plant nutrients for sustainable food crop production in sub-Saharan Africa. Dutch Association of Fertiliser Producers (VKP), The Netherlands, pp 107–125

Grant PM (1981) The fertilisation of sandy soils in peasant agriculture. Zim Agric J 78:169–175

Hatfield JL, Allmaras RR, Rehm GW, Lowery B (1998) Ridge tillage for corn and soybean production: environmental quality impacts. Soil Tillage Res 48:145–154. doi: 10.1016/S0167-1987(98)00141-X

Kamukondiwa W, Bergström L (1994) Nitrate leaching in field at an agricultural site in Zimbabwe. Soil Use Manage 10:118–124. doi:10.1111/j.1475-2743.1994.tb00471.x

Meijer RA (1992) Ridging as a primary tillage practice, results of three years of applied research. Animal draught Power Research and Development Programme, Magoye Research Station, Magoye, Zambia 14pp

Moyo A, Hagmann J (1994) Growth-effective rainfall in maize production under different tillage systems in semi-arid conditions and granitic sands of southern Zimbabwe. Proceedings of the 13th international conference, International Soil Tillage Research Organisation (ISTRO), Aalborg, Denmark, pp 475–480

Mugwira LM, Nyamangara J (1998) Organic carbon and plant nutrients in soils under maize in Chinamhora communal area. In: Bergström L, Kirchmann H (eds) Carbon and nutrient dynamics in natural and agricultural tropical systems. CAB International, Wallingford, pp 15–21

Mugwira LM, Nyamangara J, Hikwa D (2002) Effects of manure and fertilizer on maize at a research station and in smallholder (peasant) area of Zimbabwe. Commun Soil Sci Plant Anal 33:379–402. doi:10.1081/CSS-120002752

Mukurumbira LM, Nemasasi H (1998) Micronutrient in the maize-based system of the communal areas of Zimbabwe. In: Waddington SR, Murwira HK, Kumwenda JDT,

Hikwa D, Tagwira F (eds) Soil fertility research for maize-based farming systems in Malawi and Zimbabwe. CIMMYT/Soil FertNet, Harare, Zimbabwe, pp 229–232

Nyagumbo I (1993) Farmer participatory research in conservation tillage: practical experiences with no-till tied ridging in communal areas lying in the sub-humid north of Zimbabwe. Proceedings of the 4th Annual Scientific Conference, 11–14 October 1993, SADC Land and Water Management Research Programme, Gaborone, Botswana, pp 236–249

Nyagumbo I (2002) Effects of three tillage systems on seasonal water budgets and drainage of two Zimbabwean soils under maize. Doctor of Philosophy thesis, Department of Soil Science and Agricultural Engineering, University of Zimbabwe, Harare. 270pp

Nyamangara J, Mugwira LM, Mpofu SE (2000) Soil fertility status in the communal areas of Zimbabwe in relation to sustainable crop production. J Sustain Agric 16:15–29. doi:10.1300/J064v16n02_04

Nyamangara J, Gotosa J, Mpofu SE (2001) Effects of cattle manure on the structure and water retention capacity of a granitic sandy soil in Zimbabwe. J Soil Tillage Res 62:157–162. doi:10.1016/S0167-1987(01)00215-X

Singh B, Chanasyk DS, McGill WB (1998) Soil water regime under barley with long-term tillage-residue systems. Soil Tillage Res 45:59–74. doi:10.1016/S0167-1987(97)00067-6

Smaling EMA, Nandwa SM, Janssen BN (1997) Soil fertility is at stake. In: Buresh RJ, Sanchez PA, Calhoun F (eds) Replenishing soil fertility in Africa. SSSA Special Publication Number 51, Madison, WI, pp 47–61

Stoorvogel JJ, Smaling EMA (1990) Assessment of soil nutrient depletion in sub-Saharan Africa, 1983–2000. Report 28 DLO Winand Staring Ctr for Integrated Land, Soil and Water Research, Wageningen, The Netherlands

Twomlow SJ (1994) Field moisture characteristics of two fersilliatic soils in Zimbabwe. Soil Use Manage 10:168–173. doi:10.1111/j.1475-2743.1994.tb00481.x

Vogel H (1993) Evaluation of five tillage systems for smallholder agriculture in Zimbabwe. Tropenlandwirt 94:21–36

Vogel H, Nyagumbo I, Olsen K (1994) Effect of tied-ridging and mulch ripping on water conservation in maize production on sandveld soils. Tropenlandwirt 95:33–44

Zingore S, Murwira H, Delve RJ, Giller KE (2007) Soil type, historical management and current resource allocation: three dimensions regulating variability of maize yields and nutrient use efficiencies on smallholder farms. Fld Crop Res 101:296–305. doi:10.1016/j.fcr.2006.12.006

In Vitro Selection of Soybean Accessions for Induction of Germination of *Striga hermonthica* (Del.) Benth Seeds and Their Effect on *Striga hermonthica* Attachment on Associated Maize

J.A. Odhiambo, B. Vanlauwe, I.M. Tabu, F. Kanampiu, and Z. Khan

Abstract Production of maize in western Kenya is adversely affected by *Striga hermonthica*. Integrating legumes as intercrops is one way of reducing the density of *S. hermonthica* in the soil and improving the livelihood of subsistence farming communities. Legume species and varieties, however, vary in the ability to stimulate suicidal germination of *S. hermonthica* seeds. A study was conducted to select soybean (*Glycine max*) accessions with ability to stimulate germination of *S. hermonthica* seeds from western Kenya. The cut-root technique was used to screen 32 soybean accessions with *Desmodium*, *Mucuna* and maize varieties Nyamula, KSTP92 and WH502 as checks. Fourteen soybean accessions (selected from the cut-root experiment), *Desmodium* and *Mucuna* were grown in association with maize variety WH502 in pots inoculated with *Striga* seeds. There was a significant variation among soybean accessions in inducing germination of *Striga*. The relative germination of *Striga* seed by soybean accessions ranged from 8 to 66% compared to 70% for synthetic germination stimulant Nijmegen 1®. Accessions TGx1448-2E, Tgm1576, TGx1876-4E and Tgm1039 had the highest relative germination percent. Most accessions that stimulated high germination of *Striga* seeds increased *Striga* attachment by 6–95%. There was a negative correlation ($R^2 = 0.7$) between maize shoot dry weight and intercrop shoot dry weight. Accessions TGx1831-32E, Tgm944, Tgm1419 and Namsoy 4m had high stimulation but low attachment, hence making them potentially important trap crops.

Keywords Maize · Soybean · Suicidal germination · *S. hermonthica*

Introduction

Striga hermonthica, one of the major constraints limiting maize production in western Kenya, causes 30–100% loss of maize yield annually (Hassan, 1998; Khan et al., 2001; Gressel et al., 2004). Studies by Khan et al. (2002) have shown that the *Striga* problem is serious in areas where continuous cultivation is practised and soil fertility is low. The high genetic diversity and fecundity (Ejeta et al., 1992), viability of seeds and practice of continuous maize cultivation often result in a large seed bank and the seeds can remain dormant in the soil for many years until stimulated by the host plant. Management of the *Striga* weed should therefore aim at restraining development and seed production, depleting the soil of the *Striga* seeds and improving soil fertility. Many technologies aimed at restraining *Striga* seed development have been developed but adoption has remained low because of the cost, environmental implications and limited knowledge by the farmers (Joel, 2000).

Depletion of the soil *Striga* seed bank deemed as low cost and fitting within the cropping system can be achieved by stimulation of *Striga* seeds to germinate in the absence of a host plant (Berner et al., 1996). The practice is also referred to as trap cropping or suicidal germination, because it results in death of the *Striga* seedling and depletion of the seed bank in the soil.

J.A. Odhiambo (✉)
Department of Crops, Horiculture and Soils, Egerton University, Egerton, Kenya
e-mail: otishanan@yahoo.com

A. Bationo et al. (eds.), *Innovations as Key to the Green Revolution in Africa*,
DOI 10.1007/978-90-481-2543-2_37, © Springer Science+Business Media B.V. 2011

Chemical stimulants such as strigol have been identified and synthesized into analogues such as GR24, GR7 and Nijmegen 1[®] (Gerard et al., 1997). The analogues are, however, too expensive and therefore used mainly under research condition and in developed countries.

Integrating trap crops whose exudates induce suicidal germination of *Striga* seeds but are themselves not parasitized could go a long way to control *Striga*. Use of legume trap crops to reduce *Striga* seed bank is a priority because in addition, soil fertility and the livelihood of farmers could be improved. A number of legumes have been identified as potential trap crops for *Striga*. Cowpea, pigeon pea, cotton, soybean and groundnut induce abortive germination of *Striga* seeds with a consequent reduction in infestation (Parkinson et al., 1988). In Kenya, *Desmodium uncinatum,* which is a forage legume, has been found to reduce infestation by allelopathic root exudation that stimulates germination of *S. hermonthica* seeds but inhibits growth of its radicle (Khan et al., 2002; Tsanuo et al., 2003). Isoflavanones uncinanone B and uncinanone C have been isolated and identified as being responsible for the stimulation of *S. hermonthica* seed germination and inhibition of radicle growth, respectively. The process has subsequently been modelled into the 'push-and-pull' concept for control of *Striga* and stem borers (Khan et al., 2000). Though novel, adoption of this system is low probably because of the low viability of *Desmodium* seeds and its use as a fodder and not a food.

Soybean is one of the important legumes being promoted for use as alternative source of proteins, cooking oils and improvement of soil fertility in Kenya (Nassiuma and Wafula, 2002). Several soybean varieties accessions have been recommended for diverse agro-ecological zones of Kenya (Nassiuma and Wafula, 2002). Diversifying the use of soybean would increase potential for adoption and integration into cropping systems.

Preliminary studies in the savannah zones of West Africa have shown that some soybean accessions induce germination (Carsky et al., 2000). Berner and Williams (1998) found variability among non-host crops and within crop cultivars in the ability to stimulate *Striga* biotypes of different populations to germinate. A study was therefore carried out to identify and select in vitro, soybean accessions with high ability to stimulate *S. hermonthica* germination and to determine the effect of selected soybean trap crops on *S. hermonthica* emergence in a maize–legume intercrop.

Materials and Methods

Study Site

Laboratory screening and pot experiments were carried out at the International Center of Insect Physiology and Ecology (ICIPE) Thomas Odhiambo Campus, Mbita point, Kenya, located at an altitude of 1240 m above sea level, a longitude between $34°10'E$ and $34°15'E$ and a latitude between $0°52'S$ and $0°30'S$. The soil at the experimental site is well-drained alluvial and sandy loam classified as *Chromic Vertisols* (Jaetzold and Schimdt, 1983). The area receives annual rainfall between 700 and 1200 mm and has a temperature range of 17–35°C.

Screening Experiment

Treatments consisted of 32 soybean accessions, velvet bean (*Mucuna pruriens*), silverleaf desmodium (*D. uncinatum*) and maize (varieties WH502, KSTP94 and Nyamula a local cultivar). A synthetic stimulant Nijmegen 1[®] and distilled water were used as positive and negative controls, respectively. The 32 soybean accessions included ten early maturing (TGx1830-20E, TGx1831-32E, TGx1835-10E, TGx1871-12E, TGx1876-4E, TGx1895-4F, TGx1895-6F, TGx1895-33F, TGx1895-49F and TGx1740-2F), eight medium to late maturing (TGx1844-18E, TGx1869-31E, TGx1878-7E, TGx1889-12F, TGx1894-3F, TGx1893-10F, TGx1893-7F and TGx1448-2E), two local varieties (X-Baraton and Nyala) and twelve varieties whose phonological information is least known (Tgm1420, Tgm1511, Tgm1293, Tgm1360, Tgm944, Tgm1196, Tgm1419, Tgm1039, Tgm1576, Marksoy 1a, Namsoy 4m and J-499).

Petri dish screening technique as described by Berner et al. (1997) was used to screen the ability of test plant roots to stimulate *S. hermonthica* germination. Two pieces of regular filter papers were put at the bottom of the 9 cmdiameter Petri dish and then moistened with 5 ml distilled water. Aluminium foil ring of 2 cm diameter and 1.5 cm height was placed at the centre of Petri dish. The test plants seeds were surface sterilized with 1% sodium hypochlorite and then planted in the 625 ml plastic pots that had been filled with sterilized sand. At 21 days after sowing, roots were obtained from the test plants, washed free from soil, cut into pieces of 1 cm long and 1 g of root pieces was placed in the aluminium foil ring at the centre of the Petri dish lined with moistened double-layer filter paper. Fiber discs each containing 50 conditioned striga seeds were placed in four lines to form across radiating from the central aluminium foil ring (Fig. 1). Each line had four discs with the first disc in each line touching the central aluminium ring and subsequent discs touching one another edge to edge, i.e. 1, 2, 3 and 4, with the disc closest to the well taken as distance 1. Sterile distilled water (300 µl) was then pipetted over the roots in the centre of the well. For controls, 300 µl of 10^{-4} M Nijmegen 1[®] and 300 µl of sterile water were substituted for the root pieces in the aluminium well as positive and negative controls, respectively. The experiment was arranged in a completely randomized design (CRD) with three replications.

The Petri dishes were covered with lids, sealed with parafilm, wrapped in aluminium foil and then incubated at 30°C for 48 h. The number of *S. hermonthica* seeds germinating after 48 h on each glass fibre disc out of the total number of seeds on that disc was counted using dissecting microscope. Relative germination percentage (the absolute germination percentage adjusted for viability of *S. hermonthica* seeds) and the specific effectiveness index (mean germination of *S. hermonthica* seeds when exposed to exudates of potential trap crop divided by the mean germination when exposed to Nijmegen 1[®]) for each potential trap crop were then calculated.

Pot Preparation

Fourteen soybean accessions selected based on the ability to induce germination of *S. hermonthica* seeds (screening experiment) were used in a pot experiment to determine their effects on *Striga* emergence and attachment to the associated maize. Accessions comprised of the high-stimulant producers (TGx1448-2E, TGx1740-2F, Tgm1576, TGx1876-4E, Tgm1039, TGx1831-32E, Namsoy 4m, TGx1871-12E, Tgm944 and Tgm1419), medium-stimulant producers (Nyala and Tgm1293) and low-stimulant producers (TGx-1869-31E and TGx1895-6F). *Mucuna* and *Desmodium* were included for comparison.

Filter papers were placed at the bottom of each perforated 5 l plastic pot measuring 23 cm diameter at the top and 30 cm high. This was to avoid loss of *S. hermonthica* seeds through drainage. *Striga*-free soil from the ICIPE field station was mixed thoroughly with sand at the ratio of 4:1, sterilized at 121°C for 2 h and then 4 kg of it was filled in each pot. Each pot was inoculated with 0.075 g of *Striga* seeds whose viability and germination had been determined earlier. The pots were first filled to three-fourth level with clean soil, *Striga* seeds were sprinkled and the remaining soil was added, i.e. placing the seeds at about 8 cm below the soil surface. The pots were watered carefully to avoid leakages from the pots on the first day of infestation and then later after 4 days in order to condition the *Striga* seeds. Maize and the intercrops were planted

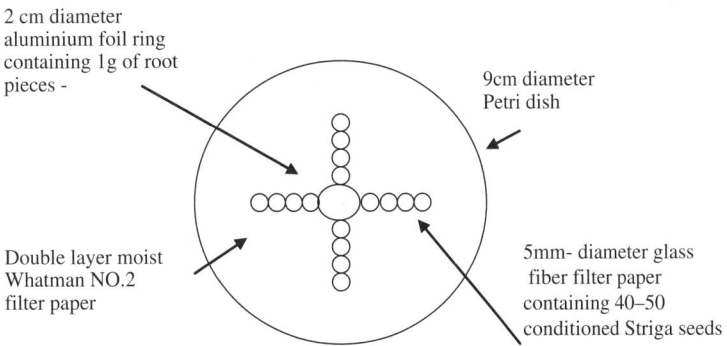

Fig. 1 Set-up of the test plant roots in the Petri dish

2 cm diameter aluminium foil ring containing 1g of root pieces -

9cm diameter Petri dish

Double layer moist Whatman NO.2 filter paper

5mm- diameter glass fiber filter paper containing 40–50 conditioned Striga seeds

7 days later to allow conditioning of *Striga* seeds. Each pot was provided with 0.25 g of N and P fertilizers at equivalent rates of 60 kg ha^{-1} in the form of CAN and TSP, respectively. Other weeds were controlled by hand removal.

Hybrid maize (WH502) which is recommended for the *Striga*-prone areas was planted together with each selected legumes in *Striga*-inoculated pots at a depth of about 2–3 cm. The experiment was arranged as a completely randomized design (CRD) with three replications.

Treatment combinations were as follows:

(a) Maize monocrop in infested soil
(b) Maize intercropped with *Mucuna* in infested soil
(c) Maize intercropped with *Desmodium* in infested soil
(d) Maize intercropped with TGx1448-2E in infested soil
(e) Maize intercropped with TGx1740-2F in infested soil
(f) Maize intercropped with Tgm1576 in infested soil
(g) Maize intercropped with TGx1876-4E in infested soil
(h) Maize intercropped with Tgm1039 in infested soil
(i) Maize intercropped with TGx1831-32E in infested soil
(j) Maize intercropped with Namsoy 4m in infested soil
(k) Maize intercropped with TGx1871-12E in infested soil
(l) Maize intercropped with Tgm944 in infested soil
(m) Maize intercropped with Nyala in infested soil
(n) Maize intercropped with Tgm1293 in infested soil
(o) Maize intercropped with TGx1869-31E in infested soil
(p) Maize intercropped with TGx1895-6F in infested soil
(q) Maize monocrop in non-infested soil

Three maize seeds were planted in each pot and thinned to one plant per pot, while soybeans were thinned to eight plants per pot. The pots were watered with equal amounts of water every day and plants were allowed to grow for 8 weeks after which the soil was washed off the roots of the plants for observation of *Striga* attachment on the roots of maize.

Results and Discussions

Viability and Germination

The percent viability of *S. hermonthica* seeds ranged from 53.9 to 66 with a mean viability of 58.6. The absolute germination percent ranged from 49.1 to 56.7 with a mean absolute germination percent of 50.4. This indicated that 86% of the viable seeds were able to germinate. Gbehounou and Adango (2003) similarly found the absolute germination of *S. hermonthica* seeds to vary from 3 to 55% depending on source of seeds and time of testing.

The Influence of Soybean Accessions on S. hermonthica Seed Germination

The test plants stimulated *Striga* seeds to germinate. The number of *S. hermonthica* seeds germinated, however, varied significantly (Table 1). Distilled water, the negative control, did not elicit any stimulatory effect, while strigol (Nijmegen 1®) at 10^{-4} M showed the highest (70.1%) relative germination. Maize varieties (WH502, Nyamula and KSTP 94) showed relative germination potentials of 61, 56.5 and 54.9% and SEI of 0.87, 0.81 and 0.78, respectively. The stimulation potential is a kin to that of *Desmodium* and *Mucuna* whose abilities have been documented in Kenya (Tsanuo et al., 2003; Khan et al., 2002; Ndungu, 1999). Maize varieties WH502 and KSTP94 are recommended for the *Striga*-prone areas. The stimulation implies that maize may be using a different mechanism to tolerate *Striga* stress.

The *Striga* germination induction of soybean (Table 1) ranged from 7.9 to 65.7%. Some accessions had high stimulation (relative germination of more than 50%), while others had medium stimulation (relative germination stimulation between 30 and 49%) and yet others had low stimulation (relative germination less than 30%). Soybean accessions TGx1448-2E, TGx1740-2F, Tgm1576, TGx1876-4E, Tgm1039, TGx1831-32E, Namsoy 4m, TGx1871-12E, Tgm944 and Tgm1419 caused relative germination of conditioned *S. hermonthica* seed ranging from 53.4 to 65.7%. These levels were comparable to the stimulation caused by *Desmodium*, *Mucuna* and

Table 1 Stimulation of *S. hermonthica* seed germination by root pieces of different plant species and accessions

Treatment	Actual germination %[1]	Relative germination %[2]	SEI[3]
TGx1830-20E	19.3[hijk]	32.7[hijk]	0.47
TGx1831-32E	37.0[abc]	62.8[abc]	0.90
TGx1835-10E	19.6[hijk]	33.2[hijk]	0.47
TGx1895-4F	15.2[kl]	25.7[kl]	0.37
TGx1895-6F	12.7[klm]	21.5[klm]	0.31
TGx1844-18E	13.9[kl]	23.5[kl]	0.34
TGx1869-31E	8.2[lm]	14.5[lm]	0.20
TGx1878-7E	4.7[lm]	07.9[lm]	0.11
TGx1889-12F	15.8[jkl]	26.8[jkl]	0.38
Tgm1420	20.6[hijk]	35.0[hijk]	0.50
Tgm1511	26.4[defgh]	44.8[defgh]	0.64
Tgm1293	22.0[ghijk]	37.3[ghijk]	0.53
Tgm1360	16.5[jkl]	27.9[jkl]	0.40
Tgm944	33.8[abcdef]	57.3[abcdef]	0.82
J-499	26.0[efghi]	44.0[efghi]	0.63
Nyala	25.6[efghi]	43.4[efghi]	0.62
Marksoy 1a	25.1[fghij]	42.6[fghij]	0.61
Namsoy 4m	35.0[abcdef]	59.4[abcdef]	0.85
TGx1871-12E	34.2[abcdef]	57.9[abcdef]	0.83
TGx1876-4E	38.0[ab]	64.3[ab]	0.92
TGx1895-33F	28.6[bcdefgh]	48.5[bcdefgh]	0.69
TGx1895-49F	26.7[defgh]	45.3[defgh]	0.65
TGx1893-7F	14.6[kl]	24.8[kl]	0.35
TGx1893-10F	27.9[cdefgh]	47.2[cdefgh]	0.67
TGx1894-3F	27.1[cdefgh]	46.0[cdefgh]	0.66
TGx1740-2F	38.7[a]	65.6[a]	0.94
TGx1448-2E	38.8[a]	65.7[a]	0.94
Tgm1196	21.4[hijk]	36.3[hijk]	0.52
Tgm1419	31.5[abcdefg]	53.4[abcdefg]	0.76
Tgm1039	37.9[ab]	64.2[ab]	0.92
Tgm1576	38.3[ab]	64.8[ab]	0.93
Mucuna	34.0[abcdef]	56.6[abcdef]	0.81
Desmodium	35.3[abcde]	59.8[abcde]	0.85
Nyamula	33.3[abcdef]	56.5[abcde]	0.81
WH502	36.0[abcd]	61.0[abcd]	0.87
KSTP 92	32.4[abcdef]	54.9[abcdef]	0.78
Nijmigen 1®	41.3[a]	70.1[a]	1.00
Distilled water	0.1[n]	0.1[n]	0.00

[1] Means followed by the same letter(s) in the column are not significantly different at $p = 0.05$
[2] The actual germination adjusted for viability of the *S. hermonthica* seeds used, i.e. (actual germination/mean viability of *S. hermonthica*)*100
[3] The SEI (specific effectiveness index) is the mean germination of *S. hermonthica* seeds when exposed to different test plant root pieces divided by mean germination when exposed to Nijmegen

the positive controls, maize hybrids and Nijmegen 1®. Soybean accessions TGx1895-33F, TGx1893-10F, TGx1894-3F, TGx1895-49F, Tgm1511, J-499, Nyala Marksoy 1a, Tgm1293, Tgm1196, Tgm1420, TGx1835-10E and TGx1830-20E caused medium stimulation (relative germination of between 32.7 and 48.5%) while accessions Tgm1360, TGx1898-12F, TGx1895-4F, TGx1893-7F, TGx1844-18E, TGx1895-6F, TGx1869-31E and TGx1878-7E caused low stimulation (relative germination of 7.95–27.9%). Therefore 31% of soybean accessions were high-stimulant producers, while 41% were medium and 25% were low-stimulant producers. Cotton and sorghum genotypes have also been found to vary in the ability to stimulate *S. hermonthica* seed germination (Buttler, 1995; Botanga et al., 2003). The differences in relative germination of *S. hermonthica* seeds induced by different soybean accessions may probably be due to different levels of germination stimulants.

Striga hermonthica *Attachment and Parasitism on the Associated Maize*

Maize grown in soil which was free from *Striga* seeds had no emerged *Striga* seedlings. Cropping system (intercropping) had a significant effect on *Striga* attachment. Intercropping maize and legumes that had previously been selected for high *Striga* induction led to variation in attachment on associated maize ranging from 6 to 95%. Total *S. hermonthica* seedlings that emerged and attached to maize roots were highest when maize was grown together with *Mucuna* and soybean accessions TGx1448-2E, TGx1740-2F, Tgm1576, TGx1876-4E, Tgm1039, TGx1871-12E and TGX1895-6F (Fig. 2). Except for TGx1895-6F, these accessions had earlier been identified as having

high *Striga* germination induction. They were expected to have resulted in higher stimulation and ultimately higher attachment on the associated maize.

There was, however, no significant difference in *S. hermonthica* attachment when it was planted alone on infested soil and when maize was intercropped with *Desmodium* and soybean accessions TGx1831-32E, Namsoy 4m, Tgm944, Tgm1419, Nyala, Tgm1293, Tgm944 and TGx1869-31E. In the laboratory screening experiment, the legumes showed medium to low suicidal germination potential except for *Desmodium*, TGx1831-32E and Namsoy 4m. In addition to germination stimulation, *Desmodium* also produce a radicle growth inhibition chemical, which hinders attachment of the parasite radicle onto the associated maize host plant (Tsanuo et al., 2003). This could explain why there was low attachment on maize associated with *Desmodium*. The mechanism behind the low attachment of *Striga* on maize intercropped with some soybean accessions such as TGx1831-32E and Namsoy 4m needs to be investigated further.

S. hermonthica infestation reduced maize height by 65%. There was, however, no significant effect of *Striga* on maize height when it was intercropped

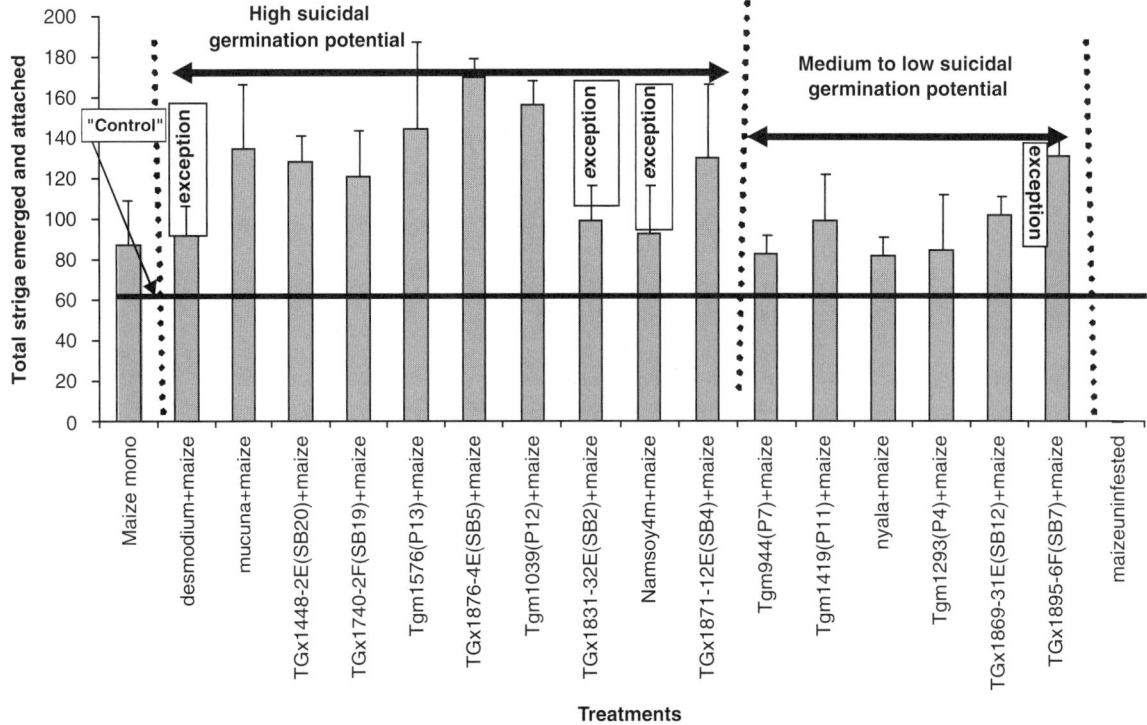

Fig. 2 Effects of legumes on the total emerged and attached *S. hermonthica* seedlings on maize plant roots

with *Desmodium*, *Mucuna* and soybean accessions, TGx1448-2E, Tgm1576, TGx1876-4E, Tgm1039, TGx1831-32E, Namsoy 4m, Nyala and Tgm1293. This suggests that the intercrops had mechanisms of suppressing the parasitic effect of *Striga* on the associated maize.

Most of the soybean accessions that had low *Striga* seed stimulation also resulted in low *Striga* attachment on the associated maize and also had a significant reduction on maize height. *Striga* reduced maize shoot weight by over 80% (Table 2). Mumera and Bellow (1993) and Aflakpui et al. (2002) also found reduction in shoot weight and attributed it to *Striga* effect on partitioning of photosynthates to the vegetative parts of the host plant. Intercropping maize with *Mucuna* and soybean accessions TGx1740-2F, Tgm944, Tgm1419 and TGx1869-31E resulted in significantly low maize shoot weight. The accessions had low stimulation of *S. hermonthica* seed germination in the laboratory with the exception of *Mucuna* and TGx1740-2F. These intercrops also had high legume shoot dry weight. The treatments did not have a significant effect on maize root dry weight and there was no correlation between maize/legume root weight and number of *Striga* attached. Aflakpui et al. (2002) also found no significant difference in root biomass between infected and uninfected maize plants. Thus combinations of

factors determine *Striga* attachment on maize roots under intercropping.

Conclusion

The laboratory germination experiment results showed that some soybean accessions grown in Kenya have high potential of producing *S. hermonthica* seed germination stimulants. Soybean accessions TGx1448-2E, TGx1740-2F, Tgm1576, TGx1876-4E, Tgm1039 and TGx1831-32E had higher germination potential than did maize varieties used in the study. The varieties have much potential as *Striga* trap crops. From the pot experiment, there was a positive correlation between the laboratory and the pot experiments. Most of the soybean accessions (TGx1448-2E, Tgm1576, TGx1876-4E and Tgm1039) which stimulated high *Striga* germination also caused high *Striga* attachment and emergence on associated maize crop. They also had no significant negative effect on the growth of the associated maize height and shoot dry weight. The soybean could therefore be used in intercropping with susceptible crop species or in crop rotations to reduce the amount of *S. hermonthica* seed bank in the field. Further work is needed to determine the possibility of

Table 2 Effect of soybean intercrops on growth of maize infested with *Striga*

Treatment	Maize height (cm)	Maize shoot weight (g)	Maize root weight (g)	Legume shoot weight (g)	Legume root weight (g)
Maize mono infested	92.0^b	36.7^b	27.6		
Desmodium +maize	86.3^{bc}	35.3^{bc}	24.4	17.0^{de}	13.3^a
Mucuna +maize	70.7^{bcd}	20.7^d	18.3	35.3^a	9.7^{bc}
TGx1448-2E+maize	65.3^{cd}	27.3^{bcd}	23.0	23.7^{bcde}	9.0^{bc}
TGx1870-2F+maize	53.3^d	24.7^d	22.2	22.0^{bcde}	9.1^{bc}
Tgm1576+maize	75.3^{bcd}	29.3^{bcd}	20.4	23.3^{bcde}	9.3^{bc}
TGx1876-4E+maize	77.3^{bcd}	27.3^{bcd}	18.8	19.3^{cde}	8.8^{bc}
Tgm1039+maize	77.3^{bcd}	26.4^{bcd}	22.0	23.0^{bcde}	9.3^{bc}
TGx1831-32E+maize	73.0^{bcd}	26.7^{bcd}	21.1	19.3^{cde}	8.8^{bc}
Namsoy 4m+maize	86.0^{bc}	30.7^{bcd}	21.6	21.7^{bcde}	9.0^{bc}
TGx1871-12E+maize	63.3^{cd}	26.3^{bcd}	19.2	25.7^{abcd}	10.7^{bc}
Tgm944+maize	65.7^{cd}	24.3^d	18.8	27.0^{abcd}	9.0^{bc}
Tgm1419+maize	56.7^d	25.7^{cd}	21.7	21.0^{bcde}	9.3^{bc}
Nyala+maize	76.7^{bcd}	29.0^{bcd}	22.5	22.7^{bcde}	9.0^{bc}
Tgm1293+maize	73.3^{bcd}	27.7^{bcd}	22.2	27.7^{abc}	11.0^b
TGx1869-31E+maize	62.0^{cd}	22.0^d	22.2	30.7^{ab}	9.3^{bc}
TGx1895-6F+maize	63.7^{cd}	35.3^{bc}	23.4	16.0^e	8.6^c
Maize mono uninfested	143.0^a	52.0^a	14.2		

Means followed by the same letter(s) within a column are not significantly different at 5%

radicle attachment inhibition by accession TGx1831-32E and Namsoy 4m that caused high stimulation with low attachment on associated maize for their inclusion into cereal cropping systems.

Acknowledgements The authors thank the African Agricultural Technology Foundation, which funded the research activities. International Centre of Insect Physiology and Ecology is thanked for providing space in their laboratory and screenhouse. Assistance provided by George Oriyo and Nelson Otieno is greatly acknowledged.

References

Aflakpui GKS, Gregory PJ, Froud-Williams RJ (2002) Growth and biomass partitioning of maize during vegetative growth in response to *Striga hermonthica* infection and nitrogen supply. Exp Agric 38:265–276

Berner DK, Carsky R, Dashiel K, Kling J, Manyong V (1996) A land management based approach to integrated *Striga hermonthica* control in sub Sahara Africa. Outlook Agric 25:157–164

Berner DK, Williams OA (1998) Germination stimulation of *Striga gesnerioides* seeds by hosts and non-hosts. Plant Dis 82:1242–1247

Berner DK, Winslow WD, Awad AE, Cardwek KF, Mohanraj DR, Kim SK (1997) *Striga* research methods, 2nd edn. International Institute for Tropical Agriculture, Ibadan

Botanga CJ, Alabi SO, Echekwu CA, Lagoke STO (2003) Genetics of suicidal germination of *Striga hermonthica* (Del.) Benth by cotton. Crop Sci 43:483–488

Buttler LG (1995) Chemical communication between parasitic weed *Striga* and its host crop, a new dimension on allelochemistry. In: Inderjig K (ed) Allelopathy organisms, process and applications, ACS symposium series 582. American Chemical Society, Washington, DC, pp 158–168

Carsky RJ, Berner DK, Oyewole BD, Dashell K, Schulz S (2000) Reduction of *Striga hermonthica* parasitism on maize using soybean rotation. Int J Pest Manage 46(2):115–120

Ejeta G, Butler LG, Babiker AGT (1992) New approaches to control of *Striga*: *Striga* research at Purdue University. Agricultural Experimental Station, Purdue University, West Lafayette, IN

Gbehounou G, Adango E (2003) Trap crops of *Striga hermonthica*: in vitro identification and effectiveness in situ. Crop Protect 22:395–404

Gerard HLN, Jan Willem JFT, Marco FMB, Zwanenburg B (1997) Synthesis of a phthaloylglycine-derived strigol analogue and its germination stimulatory activity toward seeds of the parasitic weeds *Striga hermonthica* and *Orobanche crenata*. J Agric Food Chem 45:2273–2227

Gressel J, Hanafi A, Head G, Marasa W, Obilana AB, Ochanda J, Souissi T, Tzotzos G (2004) Major heretofore intractable biotic constraints to African food security that may be amenable to novel biotechnological solutions. Crop Protect 23:661–689

Hassan RM (1998) Maize technology and transfer application for research planning in Kenya. CAB International, Wallingford, CT

Jaetzold R, Schimdt H (1983) Farm management handbook of Kenya: Natural and farm management information, vol II/B. Ministry of Agriculture, Livestock Development and Marketing, Nairobi

Joel MD (2000) The long approach to parasitic weeds control: manipulation of specific developmental mechanism of the parasite. Crop Protect 19:753–758

Khan ZR, Hassanali A, Overhot W, Khamia TM, Hooper AM, Pickett JA, Wadhams LJ, Wood-cock C (2002) Control of the witch weed *Striga hermonthica* by intercropping with *Desmodium* spp. and the mechanism defined as allelopathic. J Chem Ecol 28:1871–1885

Khan ZR, Picket JA, Van den Berg J, Wadhams LJ, Woodcock CM (2000) Exploiting chemical ecology and species diversity: stemborers and *Striga* control for maize and sorghum in Africa. Pest Manage Sci 56:957–962

Khan ZR, Picket JA, Wadhams L, Muyekho F (2001) Habitat management strategies for the control of cereal stem borers and *Striga* in maize in Kenya. Insect Sci Appl 21(4):375–380

Mumera LM, Bellow FE (1993) Role of nitrogen in resistance to *Striga* parasitism of maize. Crop Sci 33(4):758–763

Nassiuma D, Wafula W (2002) Stability assessment of soybean varieties in Kenya. Afr Crop Sci 2:139–144

Ndungu DK (1999) Effect of fodder legume species on germination, infestation and parasitism of *Striga hermonthica* (*Del*) *Benth* on maize. MSc thesis, University of Nairobi

Parkinson V, Kim SK, Efron Y, Bello L, Dashiel K (1988) Potential of trap crops as a cultural measure of *Striga* control in Africa. FAO Bull No 96

Tsanuo KM, Hassanali A, Hooper AM, Khan Z, Kaberia F, Pickett J, Wadhams L (2003) Isoflavanones from allelopathic aqueous root exudates of *Desmodium uncinatum*. Phytochemistry 64:265–273

Innovations in Cassava Production for Food Security and Forest Conservation in Western Côte D'ivoire

A. Ayemou, A. Tschannen, I. Kone, D. Allou, B. Akpatou, and G. Cisse

Abstract To support food security and reduce human pressure on the Taï National Park, Côte d'Ivoire, we introduced two improved cassava varieties, innovative farming techniques, and processing technologies at the western fringe of the park in 2004. The strategy was to (i) increase cassava productivity on reduced land area (0.5 ha), (ii) limit conflicts for access to arable land, (iii) increase the added value of cassava, and (iv) form a new generation of producers. After the in situ multiplication of introduced germplasm using the mini-cutting technique with three groups of producers, communal fields were established following multiple stem harvestings. After these two community-based multiplication steps, farmers transferred the improved varieties to their individual fields, and average yield was estimated at 20 t/ha compared to 5–12 t/ha observed in this area with the local variety. The experience resulted in a massive and rapid distribution of improved varieties in the Taï region. Parallel to this, two women were trained in "attiéké" production and entrusted with transferring this skill to their peers. "Attiéké," a cassava semolina obtained after fermentation, is a widely consumed and commercialized food in Côte d'Ivoire but often not available in the Taï region because of poor production standards. The induced diversified sources of income may reduce poaching and land-use conflicts, but this remains to be properly evaluated.

Keywords Cassava · Côte d'Ivoire · Farming innovations · Food security · Improved varieties

Introduction

Côte d'Ivoire is one of the tropical countries that have experienced the highest deforestation rates. Since 1960, the country has lost approximately 67% of its original forest cover (Tockman, 2002). Today, less than one-quarter of its primary forest remains, totaling approximately 24,000 km^2. While large amounts of timber were removed to supply tropical hardwood markets, agriculture was prized by the authorities as a cornerstone of the country's economic development, and the replacement of old forests by both commercial and subsistence farming was actively encouraged. As a result of that policy, Côte d'Ivoire is the world leader in cocoa production and the Africa's largest exporter of coffee. Today, the majority of the remaining forest consists of relatively small fragments, and farmers increasingly encroach on protected forests. Authorities perceive these encroachments often as a solution to forestall the outbreak of latent land conflicts between communities. The effects of deforestation and illegal hunting on the country's wildlife have been devastating. Most national parks and forest reserves contain dwindling populations of animals and several species have been hunted to extinction or near extinction.

The Taï National Park (TNP), in western Côte d'Ivoire, is viewed as an oasis in the midst of this alarming situation. The TNP, which covers 4570 km^2, is the largest block of continuous forest under protection in West Africa and is considered as a world heritage site. The park is situated within a bio-diversity

A. Ayemou (✉)
Centre Suisse de Recherche Scientifique en Côte d'Ivoire (CSRS), Abidjan, Côte d'Ivoire; University of Abobo-Adjamé, Abidjan, Côte d'Ivoire
e-mail: amandine.sopie@gmail.com

hot spot, and its preservation has been assigned a top conservation priority in Africa (IUCN, 1996; Hacker et al., 1998; Myers et al., 2000). Nevertheless, human pressure over the park is far from being nil. Indeed, human population size at the periphery of the TNP increased from 113,000 in 1992 to 527,000 in 1998 (Caspary et al., 2001). The burgeoning human population in that region, which is among the least developed regions in Côte d'Ivoire (IUCN, 1996), has been increasing the need for more agricultural lands, non-timber forest products, and animal protein resulting in a growing threat to the park. Poaching was identified as the major threat for the survival of wildlife within the TNP (Caspary et al., 2001; Refisch and Koné, 2005). The annual takeoffs by professional hunters were estimated at 56–78 t, consisting primarily of duikers and monkeys (Caspary et al., 2001). The local people generally consider poverty and the lack of alternatives as the factors favoring most poaching and other forms of encroachment of the park (Caspary et al., 2001). The situation worsened when Côte d'Ivoire plunged in a sociopolitical crisis in September 2002 when violent conflicts opposed the government loyalists to a rebel movement, dividing the country into two. Parallel to the military crisis, conflicts over land property rights broke out along the so-called cocoa belt in south-western Côte d'Ivoire, and the risk for flaring land conflicts built up in most other regions. The resulting insecurity in the Taï region provoked the temporary cessation of research and surveillance activities, and those of most development agencies and NGOs at its western periphery. The subsequent intensification of poaching in the TNP then dictated an urgent resumption of surveillance and development activities to rapidly reduce peoples' dependency on forest resources. In mid-2003, researchers from the "Centre Suisse de Recherches Scientifiques" (CSRS) resumed the surveillance of habituated groups of chimpanzees and monkeys near the town Taï, and until 2006, they implemented an ambitious project on the reconciliation of communities through shared values, such as protecting natural resources and improving livelihoods. Sensitization campaigns, multi-stakeholder sessions, and small-scale development projects were initiated to improve communication among TNP stakeholders. Rearing of chickens and pigs, on the one hand and cultivation of rice and cassava, on the other hand, were identified as top priorities by the local people (Goh et al., 2003).

In this chapter, we describe the introduction of two improved cassava varieties combined with innovative agricultural techniques and processing technologies to the western fringe of the TNP. Our goal was to contribute to food security and to the reduction of human pressure on the park. The approach consisted of the following elements: (i) increasing cassava productivity on reduced land area, (ii) limiting the risks for encroachment upon the park, and (iii) increasing the added value of cassava notably through transformation.

Materials and Methods

Study Site

The project was carried out at the north-western fringe of the TNP in south-western Côte d'Ivoire over a band of 60 km comprising 21 villages (Fig. 1). Annual rainfall in the Taï region averages 1800 mm and daily temperature averages 24°C. The climate is characterized by two rainy seasons (March to June and September to November) and two dry seasons (December to February and July to August) (Collinet et al., 1984). The indigenes of the study zone belong to the Oubi, the Dao, and the Gneho tribes. They represent only 10% of the population of the study area. The remaining 90% consist of a mix of ethnic groups from other regions of Côte d'Ivoire and immigrants from Liberia, Burkina Faso, Mali, and Guinea. Cassava roots and leaves are frequently consumed in various forms in the study area by both indigenes and immigrants.

Improved Cassava Varieties

Two improved cassava varieties (TME7 or "Yavo" and TME9 or "Olekanga") were introduced in the study area in November 2003. Both varieties had been received from the IITA (International Institute of Tropical Agriculture, Ibadan, Nigeria) and had previously been introduced with great success in the northern, central, and eastern regions of Côte d'Ivoire. They are characterized by (i) high productivity, up to 40 t/ha (Béhi et al., 2002), (ii) good tolerance to the African cassava mosaic virus and other diseases, and (iii) good processing and culinary qualities for various

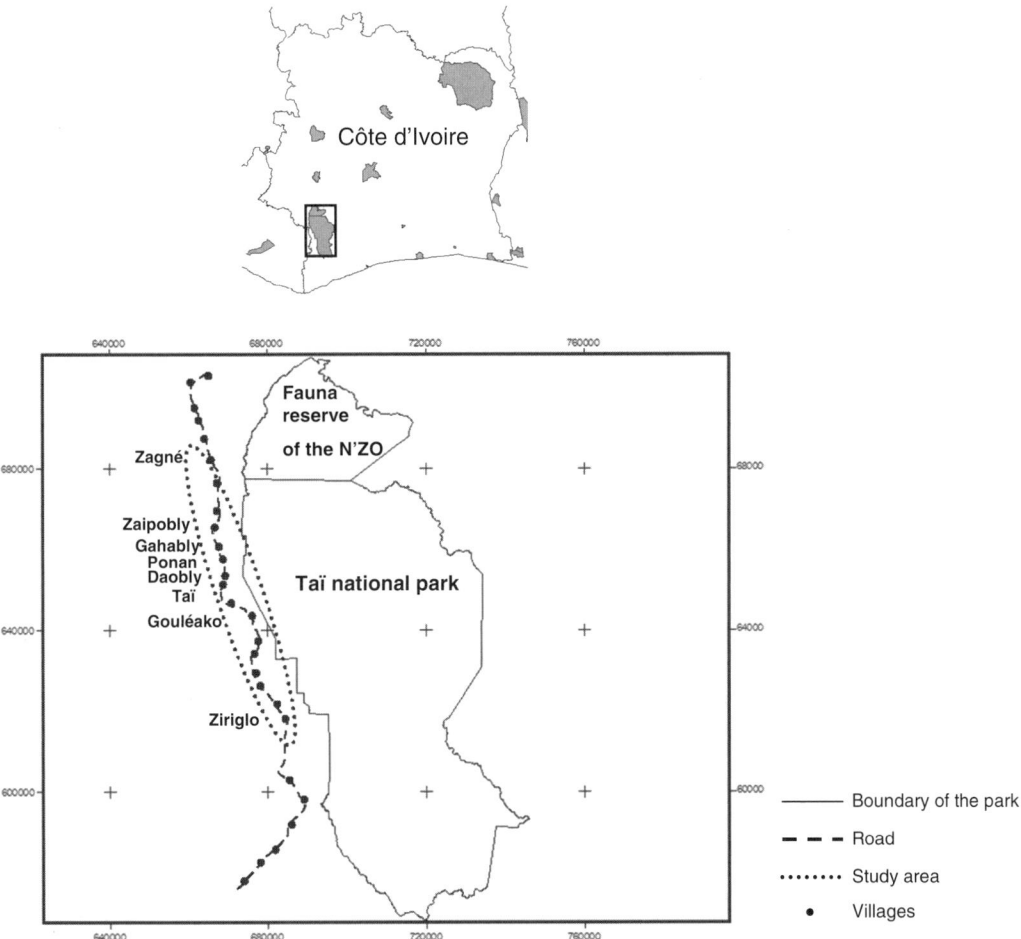

Fig. 1 Location of the project area at the western fringe of the Taï National Park

food forms (Bakayoko, 2007; Ayémou-Allou et al., 2007). The planting materials were obtained from the "CSRS" field station in Bringakro (central region) to Taï (western region).

Selection and Training of Pilot Farmers

Ten farmers per village were selected from three villages for a total of 30 farmers (Djidoubaye, Zaïpobly, and Gouléako) based on (i) their commitment to the protection of the TNP as ascertained by their previous or planned collaboration with the park managers and conservation NGOs, (ii) their organizational performance, (iii) their willingness to test new cassava varieties, and (iv) their acceptance of innovative farming techniques. Linkages were also established with two non-governmental organizations, "Vie et forêt" and "SOS Taï," to promote technology transfer in cassava production. The training of the three selected groups of farmers included theoretical aspects closely followed by field demonstrations. The theoretical training took place in Taï; the advantages of the improved varieties over the local ones were highlighted, and strategies for a cost-effective diffusion of the improved varieties in the region were also discussed. These strategies were based on the local rapid multiplication of the improved cassava stems using the mini-cutting technique (Otoo, 1996). Further training was given on good cultural practices for cassava cropping. Field demonstrations were conducted in the three selected villages on farmers' fields. In each village, a multiplication plot (3 m^2) was established with the project team and another one (100 m^2) without them. Six months later, the first harvest of cassava stems was carried out as outlined by

N'Zué (2006), followed by subsequent stem harvests. No fertilizer and no pesticide were used.

Monitoring of the Multiplication and Production Fields

The multiplication and the production plots were visited three times during the vegetative phase of the plant by the researchers and farmers. Shoot emergence, plant height and sturdiness, presence of diseases, and pests were recorded. The yield of fresh tuber was measured at 11 months after planting in one village (Gouleako 1).

Data Collect and Analysis

Three years after the introduction of innovations into three pilot villages (2006), a survey was carried out among the farmers in six villages including the three pilot ones. A sample of 12 farmers by villages was questioned. The survey was carried out by direct interview, based on an individual questionnaire, composed of closed questions. The main questions were related (i) to the farming system (crops, surfaces used, purchase, adoption of innovations, etc.) and (ii) to the post-harvest of cassava (consumption, processing, marketing, etc.). A household survey was also carried out in the six surveyed villages. Forty heads of household were interviewed on the quality and the consumption of attiéké. The collected data were controlled and codified. Descriptive statistics of the data was carried out using XLSTAT-Pro version 7.1 released by Addinsoft in 2004. Concerning the yield of cassava tubers, a descriptive analysis (means and standard errors) was carried out with the same software.

Selection and Training of Women in Cassava Processing

Two women were selected for training on attiéké production from Taï (west), the most populous town in the study area, where cassava has a high commercial value. The selection was based on attiéké production as their primary activity, willingness to improve their skills, and dedication to wildlife conservation by discouraging the consumption of bushmeat. They were trained for 7 days at Azito (south), by their peers renowned for the quality of their attiéké. After the training, the trainees received a kit of attiéké production from the project and a starter culture (rotting cassava tubers with the right mix of microorganisms to trigger the fermentative process) from their trainers.

Results and Discussion

Cassava leaves and fresh roots are traditionally used to make highly appreciated and common dishes for both natives and foreigners at the western periphery of the TNP as well as in many regions of Côte d'Ivoire. In addition, cassava cultivation is relatively easy to practice since it is moderately tolerant to drought and not very demanding in soil quality (Ayémou, 2007). The project's approach was to increase the socioeconomic productivity of cassava with the help of a rapid spread of two improved cassava varieties, the use of innovative multiplication techniques, and better post-harvest processing standards. This should contribute to reduce the use and conflict over arable lands and to limit encroachment of the TNP.

The Diffusion of Cassava Varieties: A Contribution to Sustainable Development in the Taï Region

In November 2003, we first introduced the TME9 or "Olekanga" variety in three villages (Djidoubaye, Zaïpobly, and Gouléako 1) and the TME7 or "Yavo" variety in one village (Keïbly) where multiplication plots were created using the mini-cutting technique (Otoo, 1996). In total, the farmers harvested twice stems from the multiplication plots in Djidoubaye and Gouléako and thrice in Zaïpobly. The very first stem harvesting allowed the creation of production fields of 0.5 ha in Zaïpobly, Gouléako, and Keïbly and 1 ha in Djidoubaye. A one-point estimate of cassava yield was measured in Gouléako 1 at 25 t/ha, 11 months after planting. In comparison, the yield of the local variety "Bonoua" was estimated at 5 t/ha (Table 1). The

Table 1 Average tuber yield at Gouleako 11 months after plantation

Varieties	Average yield (t/ha)	Standard error
Olekanga (improved variety)	25.07	7.1
Bonoua (local variety)	9.44	3

stems from subsequent harvests of the multiplication plots as well as those of the production farms were used to introduce the cassava varieties to other villages and to create individual production farms (four villages: Gahably, Amanikro, Taï, and Paule-Oula).

In the beginning (2003), the distribution of the planting materials was controlled and monitored. But 3 years later (2006), less than half of the surveyed farmers had obtained the new varieties through controlled channels (project and NGOs), whereas almost two-thirds had received it from relatives or friends (Fig. 2). Three years after introducing the varieties, interviews and field observations revealed that at least 167 farmers grow these varieties in 12 villages but this number is certainly underestimated. Already in 2007, three more villages had obtained the improved varieties from "Vie et Forêt," one of our local partner NGOs, and were establishing plots to multiply the planting material.

The future of these improved varieties at the fringe of the TNP seems to be bright. Indeed, the park managers expressed interest in extending these varieties to the eastern side. The success of the improved varieties in the region may be explained by their performances and culinary qualities. Indeed, during the first tuber harvest, tasting of fresh and boiled roots at Gouléako 1 and Djidoubaye was witnessed, and although no suitable assessment was performed, the enthusiasm of the participants illustrated the exceptional qualities of the introduced varieties.

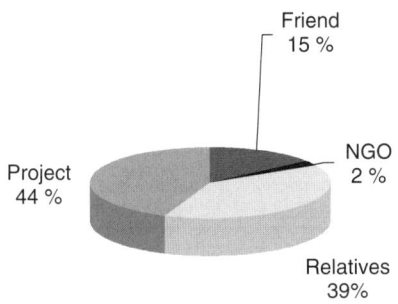

Fig. 2 Mode of purchasing the improved varieties from 61 farmers in 2006

The Role of the Transmission of Acquired Skills in the Diffusion of Innovative Cropping Techniques

Initially, 30 men and women learnt about the characteristics of improved cassava varieties and then adopted the innovative cropping techniques. These farmers can now produce cassava on reduced surfaces and with unprecedented yields. The project could not afford to train any more farmers interested in the improved varieties. Nevertheless, the varieties continued to spread based on local resources and on local initiatives. For example, interested farmers did not just request stems from the pilot farmers, but they also requested training on the innovative cropping techniques. We witnessed the training of the farmers of Keïbly by those of Zaïpobly and were satisfied with the ease and the result of this peer-to-peer training. Visits to some of the third-generation farms confirmed that the technique taught to the pilot farmers had been well transmitted and largely adopted by their trainees. Some farmers were testing the nursery technology to multiply the local variety.

Improved Standards for Attiéké in Taï as an Asset for the Reduction of Bushmeat Consumption

Attiéké is one of the most common and most appreciated cassava-based dishes not only in Ivory Coast but also in the neighboring countries. The food originated from southern Cote d'Ivoire, where women have acquired very high production standards to satisfy the growing need of urban centers. As part of a classical diner "out" in Abidjan and other towns, attiéké is frequently consumed with grilled fish or chicken, and almost never with bushmeat. Conversely, bushmeat is frequently consumed with traditional sauces. We hypothesized, therefore, that raising attiéké consumption through higher product quality in the Taï region could result in a reduced frequency of bushmeat consumption. Consequently, the quality of the attiéké produced in the region needs to be improved. The training of two female attiéké producers from Taï by their peers in Azito, southern Cote d'Ivoire, proved to be relevant in that respect. The women improved their skills in all steps of the attiéké production process including

Table 2 Attiéké consumption, quality, and supply in Taï

	Modality	Frequency	Percentage
Consumption of attiéké	Yes	30	75
	No	10	25
Time of consumption per week	1	10	25
	2	0	0
	3	20	50
	4	10	25
Improvement in attiéké quality	Yes	40	100
	No	0	0
Way of purchasing attiéké	Trainee	30	75
	Other	10	25

(i) starter preparation, (ii) peeling and crushing of the fresh roots, (iii) incorporation of the starter for fermentation, (iv) pressing of fermented paste, (v) production of semolina, (vi) winnowing of semolina, and (vii) cooking.

Once back home, one of the trained women, who is a restaurant keeper, modernized her restaurant into what is allegedly the best place to eat out among the rare places in Taï when one wants to avoid bushmeat. She acknowledged that the quality of her attiéké is the major reason for the current success of her restaurant. She claims that even people with low income can afford a meal in her restaurant. In a rapid assessment, we interviewed 40 consumers at that restaurant (Table 2) and all of them declared that the attiéké quality in Taï has improved, and therefore they eat attiéké more often than in the past (1–4 times per week compared to 1–4 times per month). We did not measure as to how far these new trends had an impact on bushmeat consumption in the Taï region, but the success of the very first restaurant in Taï to shun bushmeat is symbolic. The experience is worth repeating in other densely populated parts of the region, and its impact on bushmeat consumption properly assessed.

Anticipated Impact on Reduction of Need for Arable Land and Protection of the PNT

The project did not have sufficient means to soundly assess the impact of the introduced varieties and of consecutive improvements in cassava production on reserves and use of arable land and on encroachment of the TNP. Also, this would be premature at the present stage, merely 4 years after the first introduction of improved varieties with 30 farmers, because the

area presently occupied by improved varieties is still minuscule compared to the total land under cultivation. Indeed, the wider agricultural system, including the major cash crops, cocoa and rubber, must be taken into account. Not only the prevalent rotation patterns but also land property and usufruct rights must be included in a more comprehensive survey to validate the hypothesis. Yet, limiting ourselves to the cassava share of land use, there are good indications that the land under cassava cultivation might indeed decrease due to better yields. This follows from a point survey of objectives pursued by farmers producing cassava. Consumption at home, alone or in addition to sale to meet daily expenses, was the main motivation to plant cassava (Table 3), and out of the 61 interviewed cassava farmers, only 15 stated that they would use the sale of cassava to invest or save money (Table 4). This indicates that field sizes might be a function of consumption power and not of market forces. Whereas in a production scheme aiming to supply the market, the individual producer has incentives to produce as much as possible, and consequently maximize the field size, in subsistence farming it makes more sense to adjust the field size to the estimated needs of food. In particular, in areas with well-developed cash crops and consequential limitation of manpower, such as around the TNP, this logic might well be validated.

Table 3 Evolution over time of participating villages, farmers, and acreage cultivated with improved cassava varieties

Year	Total number of villages	Total number of farmers	Acreage (ha)
2003	3	30	0.03
2004	7	70	2.0
2005	8	ns	ns
2006	12	167	7.0

ns, no estimation

Table 4 Objectives pursued by 61 cassava farmers interviewed in 2006

Modalities	Frequency	Percentage
Marketing for investor and save money	2	3
Consumption	12	20
Consumption and marketing for daily needs	34	56
Consumption and marketing for investor	13	21
Total	61	100

Conclusion

The adopted approach for introducing innovation in cassava production to farmers had a major impact on the spread of two new cassava varieties at the western fringe of the TNP. It secured the supply of planting material and gave farmers the opportunity to accept or reject the improved varieties. These innovations in cassava production, improved varieties, mini-cuttings, multiple stem harvesting, establishment, and good cultural agronomic practices were simple, inexpensive, and proved sustainable in peer-to-peer training. The rapid spread of the new varieties witnessed that these innovations answered farmer's felt needs. In the town of Taï, the improvement of attiéké quality resulted in the success of a restaurant that does not sell bushmeat.

The results obtained with these two interventions in the value chain of cassava at the fringe of an important national park are certainly very encouraging. We recommend that further work is done to assess the impact of these innovations on the reduction of pressure on the TNP and to expand the experience around the entire TNP.

Acknowledgments We acknowledge the funding of the above research through the Swiss Development Cooperation program "Contribution du CSRS au Processus de la Réconciliation Nationale en Côte d'Ivoire" (2003–2006). In kind and financial contributions of the CSRS, the Max Planck Institute for Anthropology in Leipzig, Germany, and the University of Cocody, Abidjan, have contributed significantly to the research. We are particularly thankful to the participating farmers and communities, their authorities, and to the NGOs and TNP managers who have uncompromisingly supported the protection of this last continuous area of rainforest in west Africa.

References

Ayémou SA (2007) Etude des facteurs influençant les caractéristiques agronomiques et sensorielles du manioc (*Manihot esculenta Crantz*) dans la zone de transition forêt-savane de Côte d'Ivoire. Dissertation, University of Cocody, Abidjan

Ayémou-Allou A, Nindjin C, Dao D, Girardin O, Assa A (2007) Effet du type de sol sur les attributs sensoriels de trois produits de transformation à base de manioc (*Manihot esculenta* Crantz). Paper presented at the 1st international workshop on cassava processing in West Africa, Abidjan, Côte d'Ivoire, 4–7 June 2007

Bakayoko S (2007) Amélioration de la productivité du manioc (*Manihot esculenta,* Crantz) en Côte d'Ivoire: cas des variétés améliorées et influence de la fertilisation organique. Dissertation, University of Cocody, Abidjan

Béhi YEN, Diallo SS, Ayémou SA, Kouadio KKH, Abega J, Kouamé P, Ayémou A, Lehmann B, Girardin O (2002) Evaluation agronomique et technologique de variétés améliorées de manioc en Côte d'Ivoire en vue de leur diffusion. Bioterre (spécial):271–285

Caspary H-U, Koné I, Prouot C, de Pauw M (2001) La chasse et la filière viande de brousse dans l'espace Taï, Côte d'Ivoire. Tropenbos Côte d'Ivoire série 2. Tropenbos Côte d'Ivoire, Abidjan

Collinet J, Monteny B, Poutaud B (1984) Le milieu physique. In: Guillaumet J, Couturier G, Dosso H (eds) Recherche et aménagement en milieu forestier tropical humide: le Projet Taï de Côte d'Ivoire. UNESCO, Paris

Goh D, Acka-Douabelé C, Essé-Diby C, Akpatou AN (2003) Diagnostic participatif à la périphérie ouest du Parc National de Taï, Project report for CSRS and DDC. CSRS, Abidjan

Hacker JE, Cowlishaw G, Williams PH (1998) Patterns of African primate diversity and their evaluation for the selection of conservation areas. Biol Conserv 84:251–262

IUCN (1996) African primates: status survey and conservation action plan, revised edition. IUCN, Gland

Myers N, Mittermeier RA, Mittermeier CG, da Fonseca GAB, Kent J (2000) Biodiversity hotspots for conservation priorities. Nature 403:853–858

N'Zué B (2006). An innovative ratooning technique for rapid propagation of cassava in Côte d'Ivoire. Paper presented at the 1st international meeting on cassava breeding, biotechnology and ecology on "Cassava improvement to enhance livelihoods in sub-Saharan Africa and northeastern Brazil", Brasilia, 11–15 Nov 2006

Otoo JA (1996) Rapid multiplication of cassava. IITA research guide 51. International Institute of Tropical Agriculture, Ibadan, 22p

Refisch J, Koné I (2005) Impact of commercial hunting on the monkey populations in the Taï, Côte d'Ivoire. Biotropica 37(1):136–144

Tockman J (2002) Côte d'Ivoire: IMF, Cocoa, coffee, logging and mining. *World Rainforest Mov Bull* 54 (January). Available on http://www.wrm.org.uy/bulletin/54/CoteIvoire.html. Accessed 12 May 2011

Promoting Uses of Indigenous Phosphate Rock for Soil Fertility Recapitalisation in the Sahel: State of the Knowledge on the Review of the Rock Phosphates of Burkina Faso

M. Bonzi, F. Lompo, N. Ouandaogo, and P.M. Sédogo

Abstract   The widespread phosphorus deficiency in tropical soils constitutes one of the main factors limiting crop production in sub-Saharan Africa. This review documents the experiences in the use of Burkina rock phosphate in the recapitalisation of soil P fertility in order to increase agricultural production. The research found that the rates of rock phosphate recommended for major crops were (i) 400 kg ha^{-1} in first year and 100 kg ha^{-1} year^{-1} for the following years or 200 kg ha^{-1} every year for sorghum, maize, cotton, peanut and cowpea; (ii) for the pluvial rice it is recommended to apply 500 kg ha^{-1} in first year and 200 kg ha^{-1} year^{-1} for the following years; and (iii) 600 kg ha^{-1} in first year and 300 kg ha^{-1} year^{-1} the following years for the irrigated rice. The research developed a formula for partially acidulated rock phosphate as 4.22 N – 24.55 P$_2$O$_5$ – 6.26 S – 25.52 CaO – 0.16 MgO. This product is practically equivalent to TSP in terms of the production of cereals and is better in terms of increasing the nutrient content in soil; it is also economically profitable. Another finding was that phosphate rock solubility could be improved by adding 80 kg of PR per tonne of organic residue at the beginning of the composting of organic residues. Mixed formulas combining 75% rock phosphate to 25% TSP or 50% rock phosphate + 50% TSP for the rotation with cowpea increased sorghum production. Thus, rock phosphate can be effective in the managing soil degradation, in the recuperation of degraded soils, and in the stabilisation of crop production for a lasting agriculture.

Keywords   Tropical soils · Burkina rock phosphate · Rate · Mixed formula

Introduction

Nitrogen (N), phosphorus (P), and potassium (K) are major plant nutrients because of their quantitative and qualitative importance in plant growth. Phosphorus (P), an essential molecular component for energy transporting (ATP) and information (DNA) for living organisms, is essential for plant nutrition. In West African agro-systems P deficiency is the major constraint for crop production and can be one limiting factor for the plant's response to other nutrients such as nitrogen when water is not limiting (Bationo and Mokwunye, 1991). The total P content of non-cultivated soils is low and generally less than 200 mg kg^{-1}. It is the same for available P (Compaoré et al., 2003). Phosphorus deficiency in arid and semi-arid areas is explained by (i) the low amount of total P in the soil, (ii) the low amount of P that provides inorganic P, and (iii) the low content of organic P, due to low soil organic matter content. Organic matter normally contributes at least 50% to the pool of available P in soils with low fixing capacity and is also able to reduce P sorption in soil with high fixing capacity (Lompo, 2007, unpublished data). Dupont et al. (1967) also showed that P deficiency can result from excessive farming pressure on the land. According to Nwoke et al. (2003) and Kwabiah et al. (2003), the low availability of phosphorus also depends on factors

M. Bonzi (✉)
Institute of the Environment and Agricultural Research (INERA), Ouagadougou, Burkina Faso
e-mail: bouabonzi@yahoo.fr

such as (i) the weak activity and reduced surface area of clays such as kaolinite, (ii) the high content of aluminium and iron oxides, (iii) their low cation exchange capacity (CEC), (iv) and the soil acidity. Soil fertility management, crop type, as well as rainfall also play a part in affecting P supply (Bationo and Mokwunye, 1991). Inadequate soil fertility management and social and economic factors in Africa limit food security (Thiombiano et al., 1996; Marjatta, 2006). The low level of soil fertility, the climate adversity (arid and semi-arid climate), inadequate systems of production (traditional farming system), and the inadequacy of the policies are constraints to the development of agriculture in Africa (Lompo et al., 2000; Bationo et al., 2006).

Burkina Faso has important phosphate rock deposits (BP). Earlier researchers have shown that their use can result in the long-term improvement of the productive capacity of agro-ecosystems of the country. This chapter reviews the results of past phosphate rock research and demonstrates how the utilisation of indigenous phosphate rocks can restore and maintain productivity of soils with low agricultural production due to their P deficiency.

Materials and Methods

The earlier research work used different materials and methodologies according to the specific experimental approaches. A number of crops including sorghum, millet, maize, groundnuts, rice, cowpea, and cotton were used. The experiments could be divided into two groups: the first involved on-farm research, using dispersed blocks, while the second involved research station studies, using classical experimental designs such as the complete randomised block, the split plot, and the Fischer block design. The studies were carried out in three agro-climatic zones in Burkina which were in the 500–600 mm, the 600–800 mm, and the higher than 800 mm annual rainfall zones. The main trials were carried out at the agronomic research stations of Saria, Gampela, and Farako-Bâ in Burkina Faso. Fertilisers used in the comparison with phosphate rock (BP) or partially acidulated phosphate rock (BPA) were TSP (46% P_2O_5) and NPK (14-23-14-6S-1B). The organic manures used were cow manure and phosphocompost.

Results

Knowledge on the Agronomic Value of the Burkina Rock Phosphate (BP) and Its Effects on Cultivated Soils

The characteristics of Burkina rock phosphate (BP) are compared with rock phosphates from other West

Table 1 Comparison of the chemical characteristics of BP, Gasfa, and other West African rock phosphates

Country	Total P%[a]	Total Ca%[a]	P solubility[b]	Silica %[c]
Burkina Faso				
Arly	13.4	34.0	49.29	10.03
Kodjari	13.1	32.0	48.48	6.21
Senegal				
Matan	12.8	35.5	–	4.52
Taiba	15.9	32.0	41.54	2.24
Niger				
Tahoua	15.0	32.0	36.10	3.31
Mali				
Tilemsi	12.2	30.8	61.21	3.64
Togo				
Kpeme	15.4	26.0	40.93	0.55
Tunisia				
Gasfa[d]	13.2	22.8	91.58	1.60

Source: Bulletin UGFS (1999)
[a]The equivalent of P_2O_5 content is obtained by multiplying the total P content by 2.29 and that of CaO by multiplying the total Ca content by 1.4
[b]Solubility expressed in % of total P in 2% formic acid
[c]Silicon content mainly related to quartz
[d]Very good quality

African countries and also with the best quality Gasfa rock phosphate from Tunisia (Table 1). These results show that Burkina rock phosphates (BP) contain average concentrations in P and Ca but their P solubility in 2% formic acid is low (less than 50%). Various approaches were examined to improve the agronomic effectiveness of rock phosphates and to make the products more economically attractive. The following factors were identified by different authors. Bikienga (1980) suggested that the soil must be acid soil (pH lower than 6.5); it must have low P fixing capacity, have high organic matter content, be P deficient, and have high water-holding capacity. The same author indicated that Burkina phosphates must be spread and incorporated in the soil by ploughing such that the contact surface between soil and rock phosphate is high. The response time of Burkina phosphate is an important factor because the dissolution of rock phosphate extends over a long time period, and this gives a high residual value to these products (Bumb and Teboh, 1996). Phosphorus supply during this period of initial dissolution can be increased by applications of more soluble P sources depending on the cropping system.

On amended soils, many authors Hien et al. (1992), Bado (1985), Lompo et al. (1994b), and Sedogo et al. (1995 and 2001) showed that in spite of their low solubilities, Burkina rock phosphate improves the physical, chemical, and biological properties of soil (Tables 2 and 10).

An improvement in the total P and Ca concentrations occurred in soil treatments that received BP (Table 2). Local and annual tests were carried out between 1976 and 2007 where BP was compared with other phosphorus sources used in Burkina Faso including TSP, SSP, DAP. The results obtained with these tests are of a great interest and are summarised below.

Effects of Burkina Rock Phosphate on Rice in Burkina Faso

In irrigated valleys in west of Burkina Faso, Hien et al. (1992) compared the responses of rice to BP with those of TSP over three cropping cycles. These rice yields (Table 3) indicated that rates less than 500 kg of BP ha^{-1} were not effective on irrigated system. Hien et al. (1992) showed that on non-irrigated rice production results are best as Table 4 indicates on slightly acidic ferrasoil. The analysis of yields and amounts of BP showed that BP is effective as TSP in ferrasoil. The optimal amount is 600 kg ha^{-1} at first year plus 300 kg ha^{-1} as annual complement, in the following years.

Effects of BP on the Production of Maize, Millet, Sorghum, Cotton, Groundnut, and Soybean

A summary of results of tests carried out between 1976 and 1980, on the tropical lixisoil in the centre and southwest of Burkina Faso (Bikienga and Sedogo, 1983), is given in Table 5. The application of 400 kg BP ha^{-1} the first year plus a further 100 kg BP ha^{-1} in annual applications the following years in lixisoils

Table 2 Beneficial effect of Burkina rock phosphate mixed with manure on chemical characteristics of lixisoil from Saria, Burkina Faso (1991)

Treatments	pH water	Total C (%)	Total N (%)	Total P (ppm)	P Bray1 (ppm)	Ca (meq/100 g)	Mg (meq/100 g)
Control	4.5	1.17	0.42	197	1.42	1.67	0.34
NPK	4.2	0.21	0.20	198	3.40	1.29	0.23
Annual dose BP	5.0	0.21	0.19	252	4.78	2.15	0.34
Annual dose BP + manure	4.8	0.28	0.17	241	6.79	2.42	0.38
Correction dose BP + annual dose BP	5.0	0.20	0.14	215	5.71	2.16	0.38
Correction dose BP + annual dose BP + manure	5.2	0.25	0.09	222	6.30	2.16	0.31
Soil at beginning (1982)	5.5	0.82	0.21	–	–	1.70	0.68

Source: Lompo et al. (1994b)

Table 3 Effect of increasing rates of BP on rice production (kg ha^{-1}), Vallée du Kou, Burkina Faso

Treatments	Burkina phosphates (BP)			TSP		
P (kg ha^{-1})	HS88	SS88	HS89	HS88	SS89	HS89
0	4975	944	1020	4975	994	1020
13	5122	1216	1023	4634	1726	1676
26	5131	1461	1214	4979	1390	1389
39	4807	1817	1133	4556	1707	1268

Source: Hien et al. (1992)

HS = humid season, SS = dry season

Table 4 Effect of graduate doses of phosphates on rice yields in ferrasoil of Farako-Bâ (kg ha^{-1}), Burkina Faso

Treatments	BP		TSP	
P (kg ha^{-1})	1988	1989	1988	1989
0	2533	1643	2533	1643
13	2735	2269	2799	2065
26	3301	2562	3054	1593
39	3014	2387	3128	2471

Source: Hien et al. (1992)

increased grain yields by 600 kg ha^{-1} for sorghum, 450 kg ha^{-1} for cotton grain, and 1500 kg ha^{-1} for maize.

The on-farm research from Hien et al. (1992) gives site responses of maize, millet, and sorghum to BP applications compared with the mineral fertiliser (NPK). The major results are shown in Table 6. These results confirm the effectiveness of BP on maize in most rainy zones and on the millet. The responses by

sorghum to BP were variable depending on the sites and could be rather a function of P deficiency and soil pH.

Development of Technologies for Better Improvement of the Quality of Rock Phosphates: Development of Formula from BP Partially Solubilised to Water (BPA)

Small quantities of minerals acids (less than those used in manufacturing of TSP fertilizer) were used to partially solubilize BP. The result gives the product named BPA with the formula 4.22 N–24.55 P$_2$O$_5$–6.265 S-25.52 CaO–0.16 MgO (Hien et al., 1992). This product was then evaluated and the results are shown in Tables 7, 8, and 9.

Table 5 A comparison of effects of P from BP with P from inorganic fertiliser from 1976 to 1980 in Saria and Farako-Bâ, Burkina Faso, on crops yields (kg ha^{-1})

	Saria (Lixisol)				Farako-Bâ (Ferrasoil)		
	Cotton (76–79)	Sorghum (77–80)	Groundnut (76–79)	Millet (77–80)	Cotton (76–79)	Maize (77–80)	Soya (78–79)
Control	834	549	1041	673	285	329	480
NK	906	615	1085	623	360	444	596
P from mineral fertiliser (NPK)	1260	1310	1178	906	732	1748	1338
P from BP (correction)	1253	1138	1162	867	699	1873	1115

Source: Bikienga and Sedogo (1983)

Table 6 Response of maize, millet, and sorghum to BP applications compared with the mineral fertiliser (NPK): the sites effects through Burkina Faso (grains yield, kg ha^{-1})

	Maize 50 sites		Millet 52 sites		Sorghum 127 sites	
Treatments	Yield	Coeff. eff. (%)	Yield	Coeff. eff. (%)	Yield	Coeff. eff. %
Control	1020	–	542	–	812	–
BP	1759	79	723	55	1095	63
NPK	1973	100	869	100	1260	100

Source: Hien et al. (1992)

Coeff. eff. = effectiveness coefficient

Table 7 Response of irrigated and rain-fed rice to increasing rates of BPA in the Vallée du Kou, Burkina Faso (kg ha^{-1})

| | Irrigated rice | | | | | | Rain-fed rice | | | |
| | BPA | | | TSP | | | BPA | | TSP | |
Amounts (kg ha^{-1})	88 HS	88 SS	89 HS	88 HS	88 SS	89 HS	88	89	88	89
0	4975	944	1020	4975	994	1020	2355	1643	2533	1643
26	5157	1177	1568	4979	1390	1389	3049	2053	3054	1593
39	4493	1660	1030	4556	1707	1268	3318	2444	3128	2471

Source: Hien et al. (1992)
HS = raining season, SS = dry season

Table 8 Response of the sorghum, millet, and maize with the BPA and coefficient of effectiveness induced by the contributions of the BPA compared to the NPK (grain yield in kg ha^{-1})

| | Maize | | Millet | | Sorghum | |
Treatment	Yield	Coeff. eff.	Yield	Coeff. eff.	Yield–1	Coeff. eff.
Control	1020	–	542	–	812	–
BPA	1716	74	757	66	1153	76
NPK (P = TSP)	1973	100	869	100	1260	100

Source: Hien et al. (1992)

Table 9 Response of cotton and groundnut to the BPA (kg ha^{-1}), Burkina Faso

	Cotton	Groundnut
Mineral fertiliser NPK	821	1233
BPA	1059	1591

Source: Kamboulé (1984)

The agronomic results contained in Tables 7, 8, and 9 show that sometimes the partially acidulated rock phosphates are practically equivalent to superphosphates. The mineral composition established by Lompo (1993) indicated higher concentrations of P$_2$O$_5$, K$_2$O, and CaO than TSP. The BPA can be used to improve the levels of acidity in the soil by the contributions from CaO and K$_2$O but this fertiliser is unfortunately not available.

Development of Techniques of Solubilisation of Rock Phosphates by Phosphocomposting

The composting of the organic substrates is a biological process of fermentation which results in the production of organic acids. These acids at low concentrations will dissolve some of the BP during composting and will thus release P and the other elements. Research work undertaken at the Saria research station (Bado, 1985; Bonzi, 1989) gave the following results (Table 10). The action of organic acids on

Table 10 Effects of the addition of Burkina rock phosphates on the dry matter losses after 6 months of aerobic composting in Saria, Burkina Faso

	Starting quantity (kg MS)	Quantity at the end of 6 months (kg)	Losses (%)	pH water
Straw	60.0	51.0	15	8.2
Straw + BP	62.4	49.9	20	6.9
Straw + urea + BP	62.4	47.0	25	8.2
Litter	60.0	41.4	31	8.2
Litter + BP	62.4	42.4	32	8.5
Litter + urea	60.0	36.0	40	7.4
Litter + urea + BP	62.4	31.1	50	6.8

Source: Bado (1985)

phosphate results in a stimulation of microbial activity favourable to the decomposition of the organic substrates. The percentages of losses of organic substrates are increased when the BP is added which indicates an induced intense biological activity due to BP addition. Bonzi (1989) indicates that one needs 80 kg of BP per ton of farming residues to be added at the beginning of composting.

Effects of the Combining Soluble P and Rock P for Best Efficiency of BP in Burkina Faso

The recent studies made by Bonzi et al. (2007), in the north zone of Burkina Faso, were based on combinations of soluble P with rock phosphate P to increase in situ the efficiency of rock phosphate. They include also water harvesting technologies to make the soil more humid in order to increase BP solubilisation. The results are presented in Tables 11, 12, and 13.

The results in these tables show that after 2 years of recent experimentations there were differences between treatments for sorghum grain and straw yields. There was an increase in the effect of the treatment with 50% RP and 50% SP compared to 100% SP, with an additional 12% of yield. The combination 75% RP gave a bit lower yield than 100% SP, although not significantly. This is interesting given the high cost of the mineral P fertilisers. The same results were obtained with cowpea. The differences between treatments were statistically significant with a least significant difference (LSD) of 110.8 kg. The comparison of the combinations with manure applications indicated 75% RP closed to soluble phosphorus. On the cowpea the

Table 11 Effects of the combining soluble P and rock P on sorghum grain and straw yields

Treatment	Yields (kg ha^{-1})				Treatment effects on grain yield: % compared to 100% SP	
	Grain_year 1	Grain_year 2	Straw_year 1	Straw_year 2	Year 1	Year 2
Control	910.0	671.9	1850	1812.5		
Control NK	750.0	671.9	1800	1937.5		
100SP	1260.0	1281.3	2470	3312.5		
25RP	1260.0	1984.4	2530	4015.6	0.0	54.9
50RP	1380.0	1437.5	2510	3218.8	9.5	12.2
75RP	1160.0	1203.1	2560	3031.3	−7.9	−6.1
100RP+NK	750.0	687.5	1950	2203.1	−40.5	−46.3
100% RP	630.0	468.8	1360	2078.1	−50.0	−63.4

RP = rock phosphate, SP = soluble phosphate

Table 12 Effects of the combinations of P sources, organic matter, and zaï pit on sorghum yields

	Yields (kg ha^{-1})				Effect of treatment compared to control [(treat–Te)/Te]× 100 (%)	
	Grain_year 1	Grain_year 2	Straw_year 1	Straw_year 2	Year 1	Year 2
Direct sowing + NPK	300.0	99.4	810	607.2		
Zaï + 100SP+OM	1500.0	683.7	2870	2070.1	400	587.6
Zaï + 100RP+OM	760.0	373.1	1500	1479.2	153	275.2
Zaï + 75RP + OM	1200.0	585.2	2470	2015.2	300	488.6

SP = soluble phosphate, RP = rock phosphate, OM = manure, (treat–Te)/Te = (treatment yield – control yield)/control yield

Table 13 Effects of the combinations of P sources, organic matter, and half-moon on sorghum yields ($n = 10$)

Treatment	Yields (kg ha^{-1})				Effect of treatment compared to control [(treat–Te)/Te]× 100 (%)	
	Grain_year 1	Grain_year 2	Straw_year 1	Straw_year 2	Year 1	Year 2
Direct sowing + NPK	210.0	156.3	710	713.1		
HM + 100 SP+OM	820.0	539.8	1900	1735.8	290	245
HM + 100 RP+OM	300.0	250.0	780	1213.1	43	60
HM + 75 RP+OM	850.0	366.5	1680	1488.6	305	135

SP = soluble phosphate, RP = rock phosphate, HM = half-moon, OM = manure, (treat–Te)/Te = (treatment yield – control yield)/control yield

ranking of treatments was similar to that observed with the sorghum, and the results confirmed the first year finding. The results are presented in Table 6.

The combined application of 75% RP and 25% SP gave higher grain yields than the treatment 100% SP, in the first year but lower yields in the second year. The results with the cowpea were similar to those of the sorghum in the half-moon plots. The applications of 25% SP and 75% RP increased the yields compared to the application of 100% RP.

Discussions

Rock Phosphate as a Fertiliser

According to the data collected in Table 1, BP is relatively insoluble in formic acid (about 50% of total P). This means that BP is very insoluble in water; indeed, according to Lompo (1993) the dissolution rate of BP in water is approximately 0.03%. To increase BP solubility, the first finding is to reduce the size of granules, from 1.18 to 3.35 mm diameter and powdering from size 90 to 112 μm (Roy and McCallan, 1986). Grinding BP can increase its solubility but it remains very low as the molar ratio PO_4/CO_3 of BP, which is a main indicator of phosphate solubilisation, is 23.0 compared to <5 for more reactive rock phosphates (Gachon, 1977; Mokwunye, 1994).

Soil pH is an important factor for the release of P from rock phosphates. Indeed the soil's capacity to provide ions H^+ is a major condition for rock phosphate effectiveness (Kashewneh and Doll, 1994; Kouma, 2000). The soil must have a pH lower than 6.5. Rock phosphate application on soil with higher pH values is not recommended and particularly not in calcareous soils. Fixing capacity of soil is an important factor in the solubilisation of rock phosphates. The higher it is, the less available is the P from rock phosphates. The lower the P fixing capacity, the higher is the availability of the P that is solubilised from the BP. According to Morel (1996) the clay content can increase the fixing capacity while organic matter (MO) decreases it.

The inefficiency of BP at low rates in an irrigated system depends on soil proprieties, as in hydromorphic soils, which are very rich in clays and hydroxides. A proportion of any solubilised P would very quickly be fixed in soil and would not be available to plants.

Thus the small quantities of solubilised P from the BP are completely fixed. At high rates of BP there is enough solubilised P to saturate the fixing sites and some of the soluble P remains available for plants. Indeed, in this hydromorphic acid soil conditions with the presence of sulphurous ions and other hydroxides combined with air, continual dissolution of carbon dioxide in water creates good conditions of BP solubilisation. BP is effective in acid hydromorphic soils at the rates of 500 kg ha^{-1} in the first year and then at 200 kg ha^{-1} in annual applications in the following years.

The effectiveness of BP in ferrasoil with rice can be related to two factors. The first is that the soil was very deficient in P such that P is the major limiting factor which must be corrected to increase production (Hien et al., 1992). The second factor is that there was sufficient acidity in the soil to result in the solubilisation of the phosphate (Truong, 1984). It is also thought that the P nutrition would be improved by the roots developed by rice.

From the results obtained and the climatic conditions, in particular the rainfall limitation (500–1300 mm year^{-1}), one can speculate that BP could be effective in the major climatic zones of Burkina Faso. The favourable zones would be (i) the sub-sahelian zone with 500–700 mm of rainfall, (ii) the northern sahelian zone with 700–950 mm, (iii) the central Sudanese zone with 950–1100 mm, and (iv) the south Sudanese with 1100–1400 mm (Lompo (1995). On this basis the potential area of production in Burkina Faso would be 160,000 ha of irrigable soils; 3,765,000 ha of lixisoil for sorghum, millet, groundnut, corn, and cotton; and 4,560,000 ha of ferrasoil. However, the poor initial effectiveness of BP is considered to be the reason for low adoption of BP by farmers. This is why research needs to develop the new technologies and formulas containing BP that have an immediate response when used in the field.

Rock Phosphate and Soluble P Fertiliser Combinations

The results showed that phosphorus plays a key role in the production of the two crops if one considers the yield differences resulting from the treatments including more available P (100% SP and 50% SP) compared to those with less available P (100% RP).

This indicates the low solubility of rock phosphates and then the low availability of the P in Burkina phosphate as stated by Sedogo et al. (2001).

It is interesting with this experiment to notice that one can solve the problem of low solubility of natural phosphate by combining it with soluble P and using crop rotation with a legume crop like cowpea. These results support our conclusions from earlier years indicating that RP which is not readily available can result in increased yields if 25–50% of the P is supplied in a soluble form, increasing the amount of available P for plants. The 2 years of results are similar to earlier findings. Indeed, the direct application of natural phosphates in the region can be an economical alternative to the use of imported and expensive mineral fertilisers for the production of certain crops. West Africa has many sources of natural phosphates (The Park W of Kodjari, the valley of Tilemsi Hahotoe, Tahoua, and Sokoto) but there are differences in terms of solubility and availability of P when applied to crops. These results with BP can assist us to make better use of these RPs.

Rock Phosphates and Manure Combinations in Zaï Pits

The zaï used in this experiment is considered as a water harvesting technique and has probably favoured crops in terms of water and nutrient supply. Furthermore it will most likely increase RP solubility due to the increase in soil water content in the zaï pits, which in turn resulted in higher yields than with direct sowing. The highest yields recorded in zaï plots were partly due to the addition of organic matter (5 t ha^{-1}) which is associated with the zaï practice. The organic matter increases the availability and the use of nutrients by crops and may increase the solubility of natural phosphate due to the organic acids.

It is interesting, in this experiment, to notice that combining 25% soluble P with 75% natural phosphate partly solves the problem of P availability because it enables the crop to use readily available P during its first stage of growing, before the P in the natural phosphate becomes available later during the growing season. The additions of animal manure also improve the efficiency of the combinations.

Rock Phosphates and Manure Combinations in Half-Moon Pits

The half-moon, like the zaï, enables water harvesting and probably increases water and nutrient use by the crops. The increase in soil water content due to the water harvested with the half-moons has probably slightly improved the solubility of the natural phosphates. The application of 25% SP and 75% RP partially solves the problem of P availability and significantly increases the yields compared to RP application and the farmers' practice. Moreover, these farmers thought that it was quite impossible to grow crops on these degraded soils.

From these experiments it appears that the rock phosphate (Burkina phosphate) is of low solubility and this problem can be solved by combining it with soluble P. This enables crop to use the readily available P. According to the high cost of soluble P fertilisers (1800–2000 F CFA kg^{-1} P in Burkina in 2007), the combination with the rock phosphate (1000 F CFA kg^{-1} P of Burkina phosphate) can be an alternative. To make profit with the production we recommend for sorghum and cowpea, the combinations 75% RP + 25% SP or 50% RP + 50% SP particularly in a cropping system that includes crop rotations involving legume crops like cowpea. One can also consider a scenario with water harvesting techniques (zaï, half-moon) and organic matter (manure, compost) additions for the dry areas.

General Conclusion

The agronomic effectiveness of the BP was proven by many authors. BP is effective for (1) general management of low P soils, (2) plant nutrition by increasing physical, chemical, and biological properties of the soils, and (3) in stabilising crop yields for sustainable agriculture. This effectiveness is obtained under agro-pédo-climatic conditions which improve solubility of the BP. These conditions include acid soils with pH lower than 6.5, having a weak P fixing capacity, a strong phosphorus deficiency, and high levels of soil organic matter. The amounts of BP recommended for crops are # rain-fed rice: 600 kg ha^{-1} the first year followed by 300 kg ha^{-1} year^{-1} the following years in acid ferrasoil; # irrigated rice: 500 kg ha^{-1} the first year followed by 200 kg ha^{-1} year^{-1} following years; # sorghum, millet, corn, cotton, groundnut, soybean: 400 kg ha^{-1} the first year followed by 100 kg ha^{-1} year^{-1} following years – for zones whose annual rainfall lies between 500 and 1300 mm year^{-1}.

Work was also completed with an aim of improving the solubility of Burkina phosphate and reached the following conclusions: the solubility of the BP is improved when it is added in compost at a rate of about 80 kg t^{-1} of organic substrate, the solubility of the BP is improved in situ when it is added together with the organic manure in the fields, the solubility of the BP is improved in the process of phosphocomposting.

We recommend that the BP be used like an amendment for the improvement of soil fertility. To improve solubilisation and the agronomic effectiveness of the BP it will be necessary to consider the industrial production of phosphocomposts or the use of BP in a granulated partially solubilised BPA. In order to make profits with the sorghum and cowpea, we recommend using the combinations of 75% RP + 25% SP, but 50% RP + 50% SP in a cropping system that includes a crop rotation involving cowpea. One edge also considers a scenario with toilets harvesting technical (zaï, half-moon) and organic matter (manure, compost) additions for the dry areas.

References

Bado BV (1985) Improvement of rock phosphate by the use of the organic matter. Memory of end of study IDR/UO, 107p

Bationo A, Hartemink A, Lungu O, Naimi M, Okoth PF, Smaling E, Thiombiano L (2006) African soils: their productivity and profitability of fertilizer uses. Background paper prepared for the African fertilizer summit, Abuja, Nigeria, Jun 9–13 2006, 20p

Bationo A, Mokwunye AU (1991) Alleviating soil fertility constraints to increase crop production in West Africa: the experiment in the Sahel. Fertil Res (Special Exit) 20(1): 95–115

Bikienga IM (1980) Use of rock phosphate of Kodjari for the manufacture of phosphate-enriched fertilizers, 29p

Bikienga I, Sedogo PM (1983) Use of rock phosphate of Upper Volta: synthesis of work of experimentation on the improved volta phosphate and phosphates. IRAT/HV, 62p

Bonzi M (1989) Study of the techniques of composting and evaluation of the quality of perforate: effects of the organic matter on the cultures and the fertility of the grounds. Memory of engineer of the rural development, ISN/IDR-UO, 66p

Bonzi M, Ouattara K, Lompo F, Sedogo MP, Bationo A (2007) CORAF Project report on the use of rock phosphate in the recapitalization of soil fertility in the Sahel region of Burkina Faso, 19p

Bulletin UGFS (1999) Durable agriculture. Bull UGFS 4 & 5:10p

Bumb BL, Teboh JF (1996) Use of rock phosphates for the restoration and the maintenance of the fertility of the grounds in Burkina Faso. Background document. IFDC-A, 35p

Compaoré E, Frossard E, Sinaj S, Burden JC, Morel JL (2003) Influence of land-uses management one isotopically exchange phosphate in soils from Burkina Faso. Commun Ground Sci Anal Seed 34(1 & 2):201–223

Dupont OF, Dinechin B, Dumont C (1967) Phosphatic manure of the food crops in Upper Volta. Report/ratio IRAT, 23p

Gachon L (1977) Utility of a good level of reserves phosphates of the ground. Phosphorus Agric 70:27–33

Hien V, Youl S, Sanon K, Traoré O, Kaboré D (1992) Review article on the activities of the shutter experimentation of the project food manure 1986–1991. Agronomic results and economic evaluation of the formulas of manure at lower cost for cereals, 1984p

Kamboulé YP (1984) Back effects of rock Phosphates and phosphates improved with Gampela of 1981 to 1984. Mém Fine Cycle IDR 67

Kashewneh FE, Doll EC (1994) The uses of phosphate direct rock'n'roll for application to soils. Adv Agron 30:159–206

Kouma K (2000) Study of the effects of Burkina phosphates on the fertility of the grounds and the production of the sorghum in the western zone of Burkina Faso. Mém End Study IPB/IDR 81

Kwabiah AB, Stoskopf NC, Palm AC, Voroney RP (2003) Soil availability have affected by the chemical composition of seedling materials: implications for P-limiting agriculture in tropical Africa. Agric Ecosyst Environ 100:53–61

Lompo F (1993) Contribution to the valorization of rock phosphates of Burkina Faso. Study of the effects of the interaction phosphates organic matter. PhD thesis, Ing. University National of Côte.d'ivoire, 249p

Lompo F (1995) Case study in Burkina Faso of the initiative rock phosphates. Provisional report, 36p

Lompo F, Bonzi M, Zougmoré R, Youl S (2000) Rehabilitating soil fertility in Burkina Faso In: Hilhorst T, Muchena F (eds) Nutrients one the move. Soil fertility dynamic in African farming systems. IIED, pp 103–118

Lompo F, Sedogo MP, Assa A (1994a) Long-term effects of rock phosphates of Kodjari (Burkina Faso) on the production of the sorghum and the assessments mineral. Rev LMBO Amélior Agr Arid Medium 6:163–178

Lompo F, Sedogo MP, Hien V (1994b) Agronomic impacts of rock phosphate and the dolomite of Burkina Faso. Act of the seminar "use of rock phosphate for a durable agriculture in West Africa, pp 60–72

Marjatta F (2006) Achieving year African green revolution: with vision for sustainable agriculture growth in Africa. Background paper prepared for the African summit, Abuja, Nigeria, 9–13 June 2006, 29p

Mokwunye AU (1994) Recapitalizing west Africa' S soil fertility: of role local phosphate rock'n'roll. Agro-Minerals: News in Brief 19(4):10–12

Nwoke OC, Vanlauwe B, Diels J, Sanginga N, Osonubi O, Merckx R (2003) Unstable Assessment of phosphorus fractions and adsorption characteristics in relation to soil properties of West African savanna soils. Agric Ecosyst Environ 100:285–294

Roy AH, McCallan GH (1986) Processing phosphates into fertilizers. In: Mokwunye AU, Vlek PLG (eds) Management of nitrogen and phosphorus fertilizer in sub-Saharan Africa. Martinus Nijhoff, Dordrecht, pp 244–281

Sedogo MP, Lompo F, Bonzi M, Ouedraogo S, Bikienga IM (2001) Synthesis of work on Burkina Phosphates, Doc. of work UGFS, Mini Agri. Burkina Faso, 160p

Sedogo MP, Lompo F, Hien V, Assa A (1995) Study of the modes of management of the fertility on polysaccharides of the washed ferruginous grounds of the Central Plate of Burkina. Sci Technol 21(2):77–86

Thiombiano L, Lompo F, Ouédraogo S (1996) Natural resources and durable agriculture. In: Acts of the 2nd edition of the national forum of scientific research and the technological innovations, Apr 9–13, 1996, Ouagadougou, Burkina Faso, pp 99–101

Truong B (1984) Study of partially attacked rock phosphates, Upper Volta and Togo. Report of the experimental results of 1983. Report/ratio IRAT/MPL, 20p

Selecting Indigenous P-Solubilizing Bacteria for Cowpea and Millet Improvement in Nutrient-Deficient Acidic Soils of Southern Cameroon

H. Fankem, M. Abba, L. Ngo Nkot, A. Deubel, W. Merbach, F.-X. Etoa, and D. Nwaga

Abstract A trial of a screening and selection strategy for phosphate-solubilizing bacteria based on phosphate solubilization ability, and the subsequent effect of these bacteria on plant growth promotion under in situ conditions, was conducted. Of the 277 (187 from soils and 90 from roots) microorganisms tested, 10 bacteria (BOR$_8$, LEJ$_{14}$, DR$_5$, DR$_9$, EDJ$_4$, EDJ$_6$, EDJ$_8$, SR$_7$, EMJ$_5$, LR$_7$) were selected. All the bacteria were able to show P dissolution halo zone particularly on agar plates containing sparingly soluble iron phosphate as well as they were able to mobilize important amount of P in liquid media supplemented with either Ca$_3$(PO$_4$)$_2$ or AlPO$_4$·H$_2$O or FePO$_4$·2H$_2$O. Calcium phosphate (Ca-P) solubilization resulted from combined effects of pH decrease and carboxylic acids synthesis. However, the synthesis of carboxylic acids was the main mechanism involved in the process of aluminium phosphate (Al-P) and iron phosphate (Fe-P) solubilization. Both nutrients were mobilized at pH 4 corresponding to their natural occurrence by citrate, malate, tartrate, on much lower level by gluconate and *trans*-aconitate. Subsequently, a greenhouse trial using cowpea and millet inoculated with selected bacteria showed a significant improvement of plant phosphorus uptake as well as root and shoot dry weight. However, the selection of phosphate-solubilizing bacteria as possible inoculation tools for phosphate-deficient soils should focus on the integral interpretation of laboratory assays, greenhouse experiments, and field trials.

Keywords Cowpea · Phosphorus uptake · Phosphate-solubilizing bacteria · Organic acids · Millet

Introduction

After nitrogen, phosphorus (P) is the most plant growth-limiting nutrient despite being abundant in soils in both organic and inorganic forms (Gyaneshwar et al., 2002). The concentration of soluble phosphorus in soil is usually very low (Goldstein, 1994). Both P fixation and precipitation occur in soil, because of the large reactivity of phosphate ions with numerous soil constituents (Rodriguez and Fraga, 1999; Fernandez et al., 2007), calcium (Ca) in calcareous soils, aluminium (Al) or iron (Fe) in acidic ones. The greater quantity of applied phosphorus is fixed quickly in soil into fractions that are poorly available to plant roots (Yadav and Dadarwal, 1997). Moreover, unfavourable pH and high reactivity of Al and Fe in soils decrease P availability as well as P-fertilizer efficiency also with high total P contents (Börling et al., 2001; Hao et al., 2002).

While the use of mineral phosphate fertilizers is the best means to combat phosphate deficiency in Cameroon, their use could be limited by availability and high cost for the farmers' means. Interest has been focused on the inoculation of phosphate-solubilizing microorganisms into the soil in order to increase the availability of native, fixed P and to reduce the use of fertilizers (Illmer and Schinner, 1992). Microorganisms are involved in a range of processes that affect the transformation of soil phosphorus and are thus an integral component of the soil P cycle (Deubel and Merbach, 2005). In particular,

H. Fankem (✉)
Department of Plant Biology, Faculty of Science, University of Douala, P.O. Box 24157 Douala, Cameroon
e-mail: fankemhenri@yahoo.fr

A. Bationo et al. (eds.), *Innovations as Key to the Green Revolution in Africa*,
DOI 10.1007/978-90-481-2543-2_40, © Springer Science+Business Media B.V. 2011

391

soil microorganisms are effective in releasing P from organic pools of total soil P by mineralization (Abd-Alla, 1994; Bishop et al., 1994) and from inorganic complexes through solubilization (Richardson, 1994; Narula et al., 2000). Moreover, the microbial biomass in soil also contains a significant quantity of immobilized P that is potentially available to plants (Oberson et al., 2001). Phosphate solubilization occurs by carboxylic acids synthesized and released by microorganisms; this release also decreases pH (Puente et al., 2004; Rodriguez et al., 2006).

Therefore, the objective of the present study was to assess the efficiency of 10 selected bacteria in solubilizing sparingly soluble phosphates in laboratory media and by monitoring cowpea (*Vigna unguiculata*) and millet (*Pennisetum americanum*) plants grown under greenhouse-limiting conditions.

Methodology

Soil Characteristics of the Investigated Areas

The soils investigated are ferralitic, yellowish or brownish soils (Littoral and Southwest, Cameroon), ferralitic, yellowish or reddish soils (Centre and South). The locations include soils with high total P contents (500–1700 mg·kg^{-1}) in Southwest, but very low P availability because of high iron and aluminium contents. Furthermore, soils in the Centre and Littoral provinces are very poor in total phosphorus (140–410 mg·kg^{-1}). Most soils are acidic, with the most acidic ones being in the South province (pH 3.69–4.12), while some are alkaline (pH 7.40) at Bokito in the Centre province (Fankem et al., 2006).

Isolation and In Vitro Preselection of Microorganisms

The 277 microorganisms used were obtained from soil and root fragment samples collected in oil palm rhizospheres located in two humid agro-ecological zones of Cameroon and representing as far as possible various levels of acidity, aluminium, and iron

toxicity (Fankem, 2007). Isolation and purity testing of microorganisms were performed on agar plates containing nutrient agar supplemented with one of the three sparingly soluble phosphates $Ca_3(PO_4)_2$, $AlPO_4 \cdot H_2O$, $FePO_4 \cdot 2H_2O$ plus 0.5% Bromo-Cresol-Green (Fankem et al., 2006). Twenty isolates with halo zone (showing P solubilization) not only on calcium phosphate but particularly on iron and aluminium phosphates (characteristic of acidic soils) and showing high z/n values were preselected for liquid culture experiments. The phosphate solubilization activity of the tested microorganisms was performed in Reyes basal medium (Reyes et al., 1999) with the inorganic phosphate sources at the concentration of 30 mM P $Ca_3(PO_4)_2$, or $AlPO_4 \cdot H_2O$ or $FePO_4 \cdot 2H_2O$ (Fankem et al., 2006). Acidic compounds were purified and separated by ion chromatography (Johnson et al., 1996; Deubel et al., 2000; Fankem et al., 2006) and the effect of pH and carboxylic acids on P solubilization was assessed (Fankem et al., 2008). Ten isolates with high phosphate solubilization activity were selected for green house assays.

Characteristics and Preliminary Identification of Selected Microorganisms

The shape, colour, and diameter of colonies were assessed on 2-day-old culture plates filled with nutrient agar and inoculated with each of the selected isolates. Bacterial morphology was determined using Gram staining under a light microscope (objectives, ×100 with immersion oil). To assess the growth characteristics of each selected phosphate-solubilizing isolate, the bacterial colony was thoroughly mixed in 50 ml nutrient broth. The homogenate was incubated at 28°C on a rotary shaker for 6 h. At each sampling time (1 h for the first 2 h and 30 min for the remaining 4 h), 5 ml sub-samples were aseptically withdrawn from the flask for optical density measurement and for cfu determination. The optical density of the homogenate was measured twice at 650 nm using a spectrophotometer. To determine the number of bacterial cells per millilitre, the remaining 1 ml of the homogenate was serially diluted until 10^{-7}. The dilutions 10^{-5}, 10^{-6}, and 10^{-7} were then plated (3 plates/dilution) on nutrient agar and incubated at 28°C for 2 days. Bacterial colonies were counted on the plates and the mean value of bacterial cells of the homogenate was evaluated. Growth curves of the isolates were obtained by plotting the optical density of the homogenate against time.

Green House Trials with the Selected Phosphate-Solubilizing Microorganisms

Green house experiment was carried out in Leonard jar filled with a mixture of sterilized sand in which 1 g of Tanzania rock phosphate from Mbolea (P_2O_5: 31%; CaO: 40%; Na_2O: 13%; MgO_3: 2%; SiO_2: 9.4%; CS P_2O_5: –5.6%) was added before cultivation. The bottom part of the jar was filled with INAGROSA nutrient solution made of total nitrogen (3.8%), ammoniac (2.1%), nitrite (1.4%), organic nitrogen (0.3%), soluble phosphorus (6.0%), free amino acids (3.75 $g·l^{-1}$). During the growth period, the nutrient solution was added when necessary. Cowpea (*Vigna unguiculata*) and millet (*Pennisetum americanum*) seeds were sterilized with sodium hypochlorite (25 $g·l^{-1}$) and then finely washed before sowing. All the sowed seeds received 1 ml of a 2-day microbial culture containing approximately one to two·107 $cfu·ml^{-1}$. Uninoculated control pots were also set up. All pots were irrigated after sowing and randomly deposited in a greenhouse subjected to ambient light. The experiment included a control without inoculation and 10 inoculated treatments (BOR_8, $LEJ1_4$, DR_5, DR_9, EDJ_4, EDJ_6, EDJ_8, SR_7, EMJ_5, LR_7) performed in 12 replications. Growth characteristics in terms of plant height and leaf number were evaluated 42 days after planting (DAP). Thereafter, all the plants were harvested and parameters such as root and shoot dry weight and shoot P content were evaluated.

Statistical Analysis

Classic experimental designs were performed with representative replicates. Mean data were compared after statistic analysis using ANOVA tests performed with SPSS 10.1 software. When comparing the data, the risk of 5% was considered for the evaluation of the signification.

Results

In Vitro Phosphate-Solubilizing Efficiency of Microorganisms

Of the 277 isolates obtained, 65 (23.5%) were able to show P-solubilizing halo zone either on Ca-P, Al-P, or Fe-P agar plates. Among the 65 isolates, 36 showed halo zone on Ca-P agar plates, 28 and 45 on Al-P and Fe-P, respectively. Twenty isolates showed halo zone on both Al-P and Fe-P while only 15 showed halo zone on the three phosphate types. The z/n varied from one phosphate type to another. On agar plates containing sparingly Ca-P, z/n varied from 1.10 to 2.37 with an average of 1.74, while those values ranged from 1.12 to 2.86 and from 1.35 to 3.13 with an average of 1.19 and 2.24 on Al-P and Fe-P agar plates, respectively. Figure 1 shows an overview of halo zone on agar plate inoculated with bacterial isolate (EMJ_5).

All the 20 tested bacterial isolates showed good aptitude in mobilizing important amounts of P from insoluble sources, regardless of the phosphate type. The average amounts of solved P were 40.69, 60.04, and 90.13 mg $P·l^{-1}$ in media supplemented with Fe-P, Al-P, and Ca-P, respectively. Hence, the phosphate solubilization decreases in the order: Fe-P<Al-P<Ca-P.

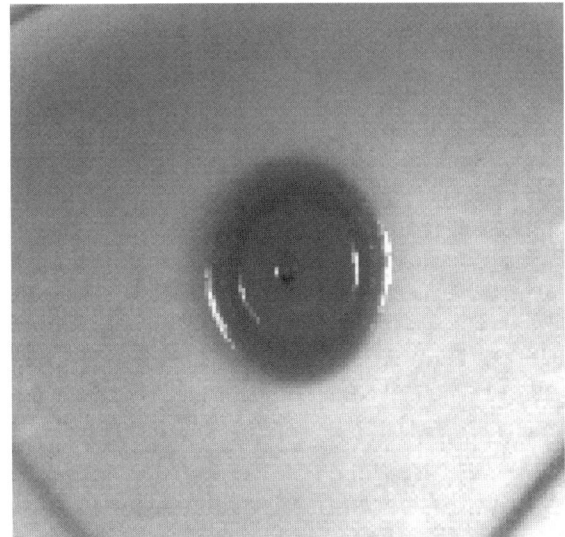

Fig. 1 Halo zone surrounding EMJ_5 colony on 5-day agar plates culture containing $FePO_4·2H_2O$

The amount of the solved P is much more important when the 10 selected bacterial isolates were considered. In the medium supplemented with Ca-P, the highest activity is recorded with the strain EDJ_6 (308.4 mg $P \cdot l^{-1}$) while the lowest is for DR_9 (59.5 mg $P \cdot l^{-1}$), with an average activity of 117.7 mg $P \cdot l^{-1}$. In the medium supplemented with Al-P, the highest activity is recorded with the strain EMJ_5 (94 mg $P \cdot l^{-1}$) while the lowest is for SR_7 (43.9 mg $P \cdot l^{-1}$), with an average activity of 64.3 mg $P \cdot l^{-1}$. In the medium supplemented with Fe-P, the highest activity is recorded with the strain EDJ_8 (64.8 mg $P \cdot l^{-1}$) while the lowest is for DR_5 (38.1 mg $P \cdot l^{-1}$), with an average activity of 46.8 mg $P \cdot l^{-1}$. All the average values are greater than those obtained with the 20 bacterial isolates.

Various molecules such as oxalate, tartrate, succinate, *trans*-aconitate, gluconate, lactate, oxaloacetate, xylonate, fumarate, citrate, and malate were purified from different bacterial growing media. The type and concentration of each carboxylic acid varied from one bacterial strain to another and were independent of P type. However, concerning the carboxylic acid patterns of the bacterial isolates, it appears that bacteria that regularly produced citrate, malate, tartarate, and gluconate were also those that mobilized great amounts of phosphate in liquid media.

Acidification improves the solubilization of tertiary Ca-P; its solubility increases exponentially with decreasing pH between 5.5 and 4.5. Al-P and Fe-P showed the lowest solubility at that pH range. While the solubilization of Fe-P was negligible in the full range of pH, Al-P showed strong solubility increase at pH level below 3 that is not important for natural soil conditions. At pH 7, where sparingly soluble Ca-P naturally occurs, only citrate solubilized remarkable P amounts. Conversely, at pH 4 Ca-P is mobilized by most of the organic acids used including citrate, tartrate, malate, *trans*-aconitate, and succinate. This suggested that acidification contributes to Ca-P mobilization by organic acids. Al- and Fe-P were mobilized at pH 4 (corresponding to their natural occurrence) mainly by citrate, malate, tartrate, and gluconate in the order: citrate > malate > tartrate > gluconate.

Characteristics of the Preselected Bacterial Isolates

Colony size generally varied from 0.5 to 5 mm. The small size colonies were observed with isolates BOR_8 (1.0–1.6 mm), EDJ_8 (0.5–2.3 mm), DR_9 (1.2–2.3 mm), DR_5 (1.5–2.0 mm), and LR_7 (1.5–2.0 mm). Meanwhile, isolates SR_7 (3.0–5.0 mm) and LEJ_{14} (3.0–4.0 mm) had large size colony. Based on the colony shape, three distinguishable groups were observed. These three groups were spherical for LEJ_{14}, DR_5, EDJ_6, EDJ_8, and SR_7; irregular for EMJ_5 and LR_7; round shape for BOR_8, DR_9, and EDJ_4. Most colonies were domed, effuse, or flat with colour varying from cream to white (DR_5, EMJ_5), whitish, brown (DR_9), orange (EDJ_8), or red (LR_7) (Table 1).

Bacterial isolates shapes vary from rod (bacilli) to spherical (cocci). Cell shape and size include cocci with diameters between 0.5 and 1 μm and bacilli between 1 and 4 μm, which had large size (3–4 μm) to DR_5 (Table 2). Seven bacterial isolates are Gram-negative (BOR_8, LEJ_{14}, DR_9, EDJ_4, EDJ_6, SR_7, and EMJ_5) while three are Gram-positive bacteria (DR_5, EDJ_8, and LR_7).

A growth curve of the 10 phosphate-solubilizing bacterial isolates shows three main phases (Fig. 2):

Table 1 Morpho-cultural characters of colonies of the 10 selected phosphate-solubilizing bacterial isolates

Isolate	Size (mm)	Shape	Aspect	Colour
BOR_8	1.0–1.6	Round shape	Effuse or flat	Cream colour
LEJ_{14}	3.0–4.0	Spherical	Domed	Whitish
DR_5	1.5–2.0	Spherical	Effuse or flat	White
DR_9	1.2–2.3	Round shape	Domed	Brown
EDJ_4	2.0–4.5	Round shape	Domed	Cream colour
EDJ_6	2.0–3.1	Spherical	Effuse or flat	Whitish
EDJ_8	0.5–2.3	Spherical	Domed	Orange
SR_7	3.0–5.0	Spherical	Effuse or flat	Whitish
EMJ_5	2.0–3.0	Irregular	Effuse or flat	White
LR_7	1.5–2.0	Irregular	Effuse or flat	Red

Table 2 Microscopic description of the 10 selected phosphate-solubilizing bacterial isolates

Isolate	Morphology	Gram	Size (μm)
BOR_8	Rod shape cells, straight	–	1.0–1.5
LEJ_{14}	Cocci (isolated, heaped, or in chains)	–	0.5–1.0
DR_5	Long bacilli (capsule shape)	+	3.0–4.0
DR_9	Rod shape cells, straight	–	1.0–1.3
EDJ_4	Rod shape cells, straight	–	1.0–1.5
EDJ_6	Cocci	–	1.0
EDJ_8	Cocci	+	1.0
SR_7	Cocci	–	1.0
EMJ_5	Short bacilli	–	0.5
LR_7	Cocci (heaped)	+	1.0

Fig. 2 Growth curves of the 10 phosphate-solubilizing bacterial isolates during 6 h of cultivation

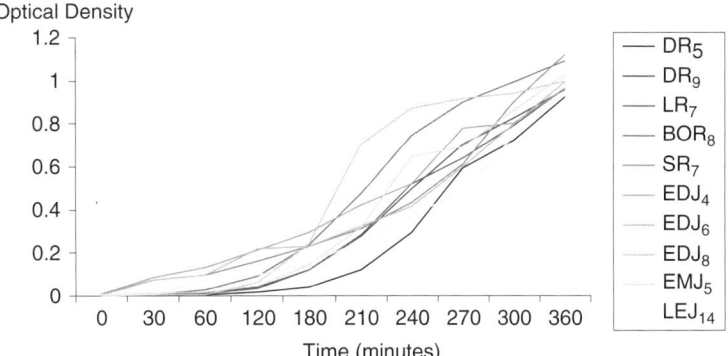

the lag phase, followed by the logarithmic or exponential phase (log phase), both are observed in all the curves. Last, the stationary phase (post-log phase) which in 6 h of growth is not really observed in all the curves. The duration the lag phase varied according to bacterial isolate is 2 h for some and 3 h for others (Fig. 2).

Bacterial Activity on the Growth and Phosphorus Uptake of Cowpea and Millet

When plant height is considered, 60% of bacterial isolates showed significant results for cowpea and 40% for millet. The highest value for cowpea is 105.89 cm performed by isolate EDJ$_8$ while LEJ$_{14}$ showed the highest value (26.35 cm) for millet (Table 3). Root

dry weight was improved by bacterial inoculation. The values obtained by 90% of bacteria were significantly different from the controls for cowpea and 70% for millet. The highest score was performed by isolate EDJ$_4$ (0.50 g) for cowpea and EDJ$_8$ (1.82 g) for millet (Table 3). Like root dry weight, bacterial inoculation significantly improved the shoot dry weight of both cowpea and millet, at 70 and 80%, respectively. The highest score was performed by isolate SR$_7$ (1.46 g) for cowpea and EDJ$_8$ (2.50 g) for millet (Table 3).

The P uptake of cowpea 42 days after planting (DAP) was significantly improved by bacterial inoculation. All the selected bacterial isolates contributed to P uptake of cowpea while 70% contributed to P uptake of millet. According to P uptake effect, the bacterial isolates can be classified as follows: EDJ$_8$ (+196%) > LEJ$_{14}$ > EMJ$_5$ > EDJ$_4$ > EDJ$_6$ > LR$_7$ > BOR$_8$ > DR$_5$ > DR$_9$ > SR$_7$ (+33%) (Table 3).

Table 3 Effect of inoculation of phosphate-solubilizing bacteria on the growth, phosphorus uptake, root and shoot dry weight of cowpea and millet under green house limiting conditions

Bacteria code	Plant height (cm)		Root dry weight (g)		Shoot dry weight (g)		P uptake (mg)		Mean P uptake effect (%)[a]
	Cowpea	Millet	Cowpea	Millet	Cowpea	Millet	Cowpea	Millet	
Control	75.75a	21.43a	0.10a	0.87a	0.70a	1.14a	1.35a	3.44a	–
BOR$_8$	76.05a	21.56a	0.21b	1.03a	0.76a	1.70c	2.94d	4.48b	+73
LEJ$_{14}$	nd	26.35b	nd	1.64c	nd	2.44e	3.72f	8.58e	+163
DR$_5$	nd	22.00a	0.32c	1.35b	0.91b	1.63c	2.32c	4.64b	+53
DR$_9$	nd	20.72a	0.26b	1.12b	0.61a	1.48b	1.89b	4.57b	+35
EDJ$_4$	88.31b	23.55a	0.50d	1.28b	1.51d	2.04d	5.7i	5.84c	+126
EDJ$_6$	102.78d	24.58b	0.45d	1.24b	1.06b	1.76c	4.61 h	nd	+108
EDJ$_8$	105.89d	25.43b	0.31c	1.82d	1.31c	2.50e	3.00d	8.25e	+196
SR$_7$	93.87c	22.48a	0.46d	0.68a	1.46d	1.34a	2.77d	nd	+33
EMJ$_5$	85.09b	21.58a	0.24b	1.12b	1.25c	1.40b	4.45 g	6.38d	+158
LR$_7$	94.07c	24.38b	0.31c	0.96a	1.22c	1.16a	3.46e	3.84a	+84

Note: Values in the same column followed by different letters are significantly different
[a]$(T - C) \times C^{-1} \times 100$ where T = treatment and C = control

Table 4 Global evaluation of phosphate-solubilizing bacterial isolates according to four parameters

Bacteria code	Plant height (cm)		Root dry weight (g)		Shoot dry weight (g)		P uptake (mg)		Synthesis[a]
	Cowpea	Millet	Cowpea	Millet	Cowpea	Millet	Cowpea	Millet	
BOR_8	NS	NS	S	NS	NS	S	S	S	4
LEJ_{14}	NS	S	NS	S	NS	S	S	S	5
DR_5	NS	NS	S	S	S	S	S	S	6
DR_9	NS	NS	S	S	NS	S	S	S	5
EDJ_4	S	NS	S	S	S	S	S	S	7
EDJ_6	S	S	S	S	S	S	S	NS	7
EDJ_8	S	S	S	S	S	S	S	S	8
SR_7	S	NS	S	NS	S	NS	S	NS	4
EMJ_5	S	NS	S	S	S	S	S	S	7
LR_7	S	S	S	NS	S	NS	S	NS	5

NS, non-significant; S, significant

[a]Number of significant results over 8

According to significant result of each parameter tested for both cowpea and millet, bacterial isolates can be classified in the following order EDJ_8 (8) > EDJ_4 = EDJ_6 = EMJ_5 (7)> DR_5 (6)> LEJ_{14} = DR_9 = LR_7 (5) > BOR_8 = SR_7 (4) (Table 4).

Discussion

On agar plate containing sparingly soluble phosphate, the activity of bacteria isolates was associated with a pH decrease of the medium, shown by the yellow zone surrounding bacterial colonies (halo zone). Bromo cresol green was associated to nutrient agar for a better observation (Mehta and Nautiyal, 2001) and as pH indicator (Gadagi and Sa, 2002). According to the obtained results, it could be suggested to improve such simple plate tests by using iron phosphate for the first selection step, because a halo zone on this P source clearly indicates other mobilizing mechanisms than acidification (Fankem et al., 2008).

The solubility of the different sparingly soluble phosphates in liquid media decreased in the following order: Ca-P > Al-P > Fe-P. This result is in agreement with that obtained by Ahn (1993) working on Ca-, Fe-, and Al-phosphates solubilization in tropical soils. Similar results were also obtained by Gadagi and Sa (2002) using *Penicillium oxalicum* as phosphate solubilizer. Unlike the results obtained on agar plates, some bacterial isolates such as EDJ_6 that were not able to show halo zone on agar plate containing sparingly soluble Ca-P were the best P solubilizer in liquid media containing the same phosphate source.

These contradictory results between plate halo detection and phosphate solubilization in liquid cultures were found by Deubel and Merbach (2005). This indicates that halo zone as criteria is not enough for phosphate-solubilizing bacteria selection, as many isolates that did not produce any visible halo zone on agar plates could conversely mobilize significant amount of phosphate in liquid media (Gupta et al., 1994). However, while screening a large number of microorganisms, this method can be regarded as generally reliable for isolation and preliminary characterization of phosphate-solubilizing microorganisms (Rodriguez and Fraga, 1999). Ca-P, Al-P, and Fe-P solubilization is associated with a pH decrease of the media, but this decrease is not proportional to the amount of P solved.

Ryan et al. (2001) stated that, among the carboxylic acids identified, dicarboxylic (oxalic, tartaric, malic, fumaric, malonic acids) and tricarboxylic (citric) acids were more effective for P mobilization. Illmer and Schinner (1992) reported that gluconic acid might be the most frequent agent for mineral P solubilization. According to Deubel and Merbach (2005), Ca-P appears under neutral or alkaline conditions where soil phosphates are fixed mainly in the form of apatite and their solubility is increasing with a decrease of soil pH. However, a strong pH decrease is impossible under well-buffered soil conditions. Acidification cannot be the explanation for P mobilization in bacterial culture containing Al-P and Fe-P. Therefore, the low pH recorded in their respective media is always the result of the carboxylic acids synthesis. May be, there are interactions between different carboxylates too. Furthermore, we can only find carboxylates in solutions, some of them may be insoluble, for instance, calcium citrate (Deubel et al., 2005). Because rhizosphere bacteria probably are not able to change the pH of the rhizosphere largely, the production of carboxylic acids will be more important as a mechanism for P mobilization. Previous studies have shown the importance of those carboxylic acids in the process of P solubilization. The presence of organic acids as anions in bacterial cultures may help to displace adsorbed P through ligand exchange reactions (Whitelaw, 2000; Egle, 2003).

The morpho-cultural and microscopic characters indicate that the microbial isolates are all bacteria. Among the phosphate-solubilizing bacteria, *Pseudomonas* and *Bacillus* are well studied. The genus *Pseudomonas* is a very large and important group of Gram-negative and nonfermenting bacteria, which is found in substantial numbers as free-living saprobes in soil, fresh water, and many other natural habitats (Bisen and Verma, 1994). The basic morphological characters common to *Pseudomonas* are the rod-shaped cells either straight or slightly curved in one plane, usually motile, with flagella and no spore production. *P. pudica* and *P. fluorescens* are plant growth-promoting bacteria. The *Bacillus* genus is a rod-shaped, spore-bearing, usually Gram-positive bacteria and catalase producing. Some *Bacillus* species are recognized as good phosphate solubilizers including *Bacillus megatherium* var *phosphaticum* that act in increasing the growth and yield of wheat plants cultivated in desert soils of Egypt (Abdel-Azeem and Hoda, 1994). Furthermore, *Bacillus circulans* is known as good mineral iron-phosphate fertilizer. Among the last 10 bacteria isolates, EDJ_4, DR_9, and BOR_8 are strains of *Pseudomonas fluorescens* (Fankem et al., 2006), while DR_5, EDJ_8, and LR_7 are very close to *Bacillus* genus.

Cowpea and millet inoculation using the selected bacterial isolates showed positive effects on their growth and P uptake (Table 4). Differences obtained in the fertilizer treatments can be attributed to the nutrients being readily available from the insoluble sources (rock phosphate). Thus, inoculation with phosphate-solubilizing bacteria made more soluble phosphates available to the growing plants. This may be the reason for improved growth and P uptake of plants and could have stimulated microbial growth and activity. Many bacteria (Rodriguez and Fraga, 1999) and fungi (Whitelaw, 2000) are able to improve plant growth by solubilizing sparingly soluble inorganic phosphates in the soil. However, the beneficial effects of bacteria on plant growth varied significantly depending on environmental conditions, bacterial strains, and plant and soil conditions (Sahin et al., 2004).

Conclusion

African soils such as those from oil palm rhizosphere in Cameroon constituted a good reservoir for phosphate-solubilizing bacteria, which could be very useful for improving crop productivity in poor fertile soils. In order to put this potential into activity, it is very important to carry out a succession of screenings from isolation to selection of the best-adapted isolates for specific environment targeted. It is thus necessary to confirm the good response of millet and cowpea improvements obtained under sterilized sand on green house conditions by assessments on farm in acidic soils with the competition of other soil microorganisms.

References

Abd-Alla MH (1994) Use of organic phosphorus by *Rhizobium leguminosarum* biovar. *viceae* phosphatases. Biol Fertil Soils 18:216–218

Abdel-Azeem AA, Hoda H (1994) Utilization of biofertilizers to improve garbage compost properties for increasing wheat yield in desert soil. MSc Thesis, Institute of Environmental Studies and Research. Ain Shams University, Cairo, Egypt

Ahn PM (1993) Tropical soils and fertilizer use. Longman Scientific & Technical, Malaysia, 264pp

Bisen PS, Verma K (1994) Handbook of microbiology. CBS Publishers and Distributors, Delhi

Bishop ML, Chang AC, Lee RWK (1994) Enzymatic mineralization of organic phosphorus in a volcanic soil in Chile. Soil Sci 157:238–243

Börling K, Otabbong E, Barberis E (2001) Phosphorus sorption in relation to soil properties in some cultivated Swedish soils. Nutr Cycling Agro Ecosyst 59:39–46

Deubel A, Fankem H, Nwaga D, Antoun H, Merbach W (2005) Characterization of phosphorus-mobilizing ability of rhizosphere bacteria of African oil palm tree (Elaeis guinensis). In: Merbach W, Beschow H, Augustin J (eds) Wurzelfunktionen und Umweltfaktoren. Grauer, Beuren Stuttgart, pp 60–63

Deubel A, Gransee A, Merbach W (2000) Transformation of organic rhizodepositions by rhizosphere bacteria and its influence on the availability of tertiary calcium phosphate. J Plant Nutr Soil Sci 163:387–392

Deubel A, Merbach W (2005) Influence of microorganisms on phosphorus bioavailability in soils. In: Buscot F, Varma A (eds) Microorganisms in soils: roles in genesis and functions. Springer, Berlin Heidelberg, pp 177–191

Egle K (2003) Untersuchungen zum Phosphor-, Kupfer-, Zink- und Cadmium-Aneignungsvermögen von drei Lupinenarten und Weidelgras unter Berücksichtigung wurzelbürtiger organischer Säuren, PhD thesis. Georg-August-Universität Göttingen, Shaker Verlag, Aachen. ISBN: 3–8322–1310–4

Fankem H (2007) Occurrence and potentials of phosphate solubilizing microorganisms associated with oil palm (*Elaeis guineensis*) rhizosphere in Cameroon. PhD thesis, University of Yaoundé I, Cameroon, 116p

Fankem H, Ngo Nkot L, Deubel A, Quinn J, Merbach W, Etoa FX, Nwaga D (2008) Solubilization of inorganic phosphates and plant growth promotion by strains of *Pseudomonas fluorescens* isolated from acidic soils of Cameroon. Afr J Microbiol Res 2(7):171–178

Fankem H, Nwaga D, Deubel A, Dieng L, Merbach W, Etoa FX (2006) Occurrence and functioning of phosphate solubilizing microorganisms from oil palm tree (*Elaeis guineensis*) rhizosphere in Cameroon. Afr J Biotechnol 5(24):2450–2460

Fernandez LA, Zalba P, Gomez MA, Sagardoy MA (2007) Phosphate-solubilization activity of bacterial strains in soil and their effect on soybean growth under greenhouse conditions. Biol Fertil Soils 43:805–809

Gadagi RS, Sa T (2002) New isolation method for microorganisms solubilizing iron and aluminium phosphates using dyes. Soil Sci Plant Nutr 48:615–618

Goldstein AH (1994) Involvement of the quinoprotein glucose dehydrogenase in the solubilization of exogenous phosphates by Gram-negative bacteria. In: Torriani-Gorini A, Yagil E, Silver S (eds) Phosphate in microorganisms: cellular and molecular biology. ASM Press, Washington, DC, pp 197–203

Gupta R, Singal R, Shankar A, Kuhad RC, Saxena RK (1994) A modified plate assay for screening phosphate solubilizing microorganisms. J Gen Appl Microbiol 40:255–260

Gyaneshwar P, Naresh Kumar G, Parekh LJ, Poole PS (2002) Role of soil microorganisms in improving P nutrition of plants. Plant Soil 245:83–93

Hao X, Cho CM, Racz GJ, Chang C (2002) Chemical retardation of phosphate diffusion in an acid soil as affected by liming. Nutr Cycling Agroecosyst 64:213–224

Illmer P, Schinner F (1992) Solubilization of inorganic phosphate by microorganisms isolated from forest soil. Soil Biol Biochem 24:389–395

Johnson JF, Vance CP, Allan DL (1996) Phosphorus deficiency in *Lupinus albus*. Plant Physiol 112:31–41

Mehta S, Nautiyal SC (2001) An efficient method for qualitative screening of phosphate-solubilizing bacteria. Curr Microbiol 43:51–56

Narula N, Kumar V, Behl RK, Deubel A, Gransee A, Merbach W (2000) Effect of P-solubilizing *Azotobacter chroococcum* on N, P, K uptake in P-responsive wheat genotypes grown under greenhouse conditions. J Plant Nutr Soil Sci 163:393–398

Oberson A, Friesen DK, Rao IM, Bühler S, Frossard E (2001) Phosphorus transformations in an oxisol under contrasting land-use systems: the role of the microbial biomass. Plant Soil 237:197–210

Puente ME, Bashan Y, Li CY, Lebsky VK (2004) Microbial populations and activities in the rhizoplane of rock-weathering desert plants. Root colonization and weathering of igneous rocks. Plant Biol 6:629–642

Reyes I, Bernier L, Simard RR, Tanguay P, Antoun H (1999) Characteristics of phosphate solubilization by an isolate of a tropical *Penicillium rugulosum* and two UV-induced mutants. FEMS Microbiol Ecol 28:291–295

Richardson AE (1994) Soil microorganisms and phosphorus availability. In: Pankhurst CE, Doube BM, Gupta VVSR, Grace PR (eds) Soil Biota management in sustainable farming system. CSIRO, Melbourne, VIC, pp 50–62

Rodriguez H, Fraga R (1999) Phosphate solubilizing bacteria and their role in plant growth promotion. Biotechnol Adv 17:319–339

Rodriguez H, Fraga R, Gonzalez T, Bashan Y (2006) Genetics of phosphate solubilization and its potential applications for improving plant growth promoting bacteria. Plant Soil 287:15–21

Ryan PR, Delhaise E, Jones DL (2001) Function and mechanism of organic anion exudation from plant roots. Ann Rev Plant Physiol Plant Mol Biol 52:527–560

Sahin F, Cakmakci R, Kantar F (2004) Sugar beet and barley yield in relation to inoculation with N2-fixing and phosphate solubilizing bacteria. Plant Soil 265(7):123–129

Whitelaw MA (2000) Growth promotion of plants inoculated with phosphate solubilizing fungi. Adv Agron 69:99–151

Yadav KS, Dadarwal KR (1997) Phosphate solubilization and mobilization through soil microorganisms. In: Dardawal KR (ed) Biotechnological approaches in soil microorganisms for sustainable crop production. Scientific Publishers, Jodhpur, pp 293–308

Evaluation of Human Urine as a Source of Nitrogen in the Co-composting of Pine Bark and Lawn Clippings

A.O. Fatunbi and P.N.S. Mnkeni

Abstract The introduction of urine diversion toilets in South Africa has created opportunities for the recycling of human urine in agriculture. One of such possibilities is the use of human urine as a source of nitrogen during the composting of organic wastes. This study evaluated the possibility of (i) replacing urea fertilizer with human urine as a source of N in the co-composting of pine bark and lawn clippings, (ii) minimizing the possible N loss during composting by encouraging struvite ($MgNH_4PO_4 \cdot 6H_2O$) precipitation with the addition of magnesium oxide (MgO) + single super phosphate (SSP) fertilizer or rock phosphate (RP) to the composting mixtures. Results showed that composting progressed faster where urine rather than urea was used as the source of nitrogen, as reflected by early attainment of peak thermophilic temperature (65°C). The faster composting of the urine-treated mixtures translated into a 22% greater degree of degradation relative to the urea-enriched compost. After 84 days of composting, inorganic N (NO_3 + NH_4) was 45% higher in the urine-treated compost than the urea-treated materials possibly as a result of the greater degradation observed in urine-enriched compost mixtures. Struvite crystals were observed in the MgO, SSP, RP and urine-treated composts but the elemental composition of the struvite crystals varied with the composting mixtures. However, none of the precipitated struvite crystals contained ammonium but potassium suggesting that the precipitated struvites were not effective in conserving nitrogen in the compost mixtures.

Keywords Composting · Human urine · Inorganic N · Pine bark · Struvite

Introduction

Recycling nutrients in human wastes (faecal and urine) for plant use has been accorded considerable attention in the recent times (Bo-Bertil et al., 2000). Human urine is responsible for over 94% of the nitrogen (N), phosphorus (P), potassium (K) and abundant micronutrients found in toilet wastewater (Wolgast, 1993; Bo-Bertil et al., 2000). Eco-cycling of this nutrient base is facilitated by source separation of urine from faecal waste in a specially designed urine separation toilet seats (Larsen et al., 2001; Udert, 2003). Direct use of human urine in crop production systems was reported to contribute substantially to crop growth (Mnkeni et al., 2005; Sridhar et al., 2005), yet it is not widely accepted due to issues pertaining to handling, transportation, storage, hygiene and health (Daughton and Ternes, 1999; Hanaeus et al., 1996). Bo-Bertil et al. (2000) suggested that the transformation of nutrients in human urine into solid form could solve the identified problems.

Co-composting human urine with other organic waste appears to be an attractive option that could yield end products with higher value. Composting is the microbial-mediated conversion of organic waste into humus-like material that is relatively stable and could be used as an organic fertility amendment in soil. Compost is known to improve soil structure, stimulate

A.O. Fatunbi (✉)
Agronomy Department, Agricultural and Rural Development Research Institute (ARDRI), University of Fort Hare, Eastern Cape, South Africa
e-mail: afatunbi@ufh.ac.za

A. Bationo et al. (eds.), *Innovations as Key to the Green Revolution in Africa*, DOI 10.1007/978-90-481-2543-2_41, © Springer Science+Business Media B.V. 2011

the activities of soil organism and serve as source of nutrient to cultivated crops by its contribution to the soil organic matter. Composting of organic materials is associated with substantial loss of nitrogen through volatilization. Loss of gaseous ammonia during composting of domestic waste has been reported to range between 50 and 60% of the original N content (Brink, 1995). In some farmyard manure, gaseous N loss during composting was over 70% (Martins and Dewes, 1992), while it was up to 68% in sewage sludge (Witter and Lopez-Real, 1988). This extensive loss of N is the major cause of reduced fertilizer value in composts. Ammonia volatilization has been reported as the major way in which N is lost during composting (Witter and Lopez-Real, 1988), losses may also occur as N_2, NO and N_2O (Mahimairaja et al., 1994; Martins and Dewes, 1992; Sibbesen and Lind, 1993). Ammonia formation and loss have been found to be influenced mainly by C/N of the original materials, composting temperature, ammonia and ammonium-adsorbing capacity of the added materials and availability of energy sources that may lead to immobilization (Morisaki et al., 1989; Witter and Kirchmann, 1989; Kirchmann, 1985).

Efforts at reducing ammonia losses during composting have involved the use of clay, peat and zeolite as covering materials over the composting pile. This was found to minimize loss of the volatilized ammonia and increase N content of the zeolite, peat and clay (Witter and Lopez-Real, 1987; Bernal et al., 1993; Liao et al., 1997). A novel approach to reduce ammonia loss is the precipitation of struvite ($MgNH_4PO_4·6H_2O$) during composting (Jeong and Kim, 2001). Struvite has been known to form and adhere to the surface of pipes in wastewater treatment (Ohlinger et al., 1999) and it has been used to retrieve and control N and P in waste water treatment facilities (Schulze-Rettmer, 1991). Struvite is reported to be a valuable slow release fertilizer, as the N content must dissolve in water before nitrification reactions could commence on it (Jeong and Kim, 2001). Bo-Bertil et al. (2000) demonstrated the recovery of 65–80% N content of human urine, through struvite precipitation when small amount of MgO was added, under laboratory condition. Jeong and Kim (2001) also demonstrated the precipitation of struvite in a laboratory-scale composting experiment, when salts of magnesium and phosphorus were applied on molar basis of 20% of the total N content of the composting mixture. This crystallization

resulted in substantial reduction in ammonia loss. The addition of magnesium and phosphorus salt to composting mixture for struvite formation will introduce extra cost, although the cost to benefit implication of such practice is yet to be ascertained.

In this study, we considered the composting of pine bark because of its availability and uses in South Africa. Composted pine bark is mainly used as growing media for vegetable transplants (Mnkeni et al., 2006). Pine bark-based growing media has many advantages but its high C:N ratio (200–400) makes it unsuitable for direct composting (Smith, 1992). Usually, urea fertilizer is added as a supplementary source of N in its composting which takes about 18 months; this has direct implication on the cost of the final product. Finding a cost-effective alternative source of N for pine bark composting is imperative.

The objectives of this study are (i) to evaluate the effects of replacing urea fertilizer with human urine in the composting of pine bark for use as a horticultural growing medium; (ii) to evaluate the possibility of struvite ($MgNH_4PO_4·6H_2O$) precipitation as a way to reduce N loss under aerobic composting condition.

Materials and Methods

Composting Process

A composting experiment was carried out indoors, in wooden boxes of 1 m^3 volume, during the summer of the year 2006. The boxes were constructed with wooden slabs having slits of air space between each slab. Pine barks, collected from Rance Timber in Stutterheim, South Africa, were sun dried and crushed to pass through a 15 mm metallic mesh fitted to a motorized crusher. Lawn clippings were obtained from the cricket field of the University of Fort Hare, Alice campus. Human urine used was obtained as voluntary donations from the male hostel of the University of Fort Hare, Alice campus. The chemical characteristics of the materials used in composting are presented in Table 1.

The composts were made by combining 81 kg pine bark with 50 kg lawn clipping (dry weight basis). This combination gives a mixture with a C:N ratio of 40:1. Additional materials were added to this mixture to reduce the C:N ratio to 30:1; these constituted the

Table 1 Chemical characteristics of the materials used in composting

Characteristics	Pine bark	Lawn clippings	Human urine
pH (H_2O)	4.03	5.86	9.04
Electrical conductivity (EC)	1007 $\mu S\ cm^{-1}$	6.06 ms cm^{-1}	92.2 $\mu S\ cm^{-1}$
Total C (%)	48	29.9	–
Total N (%)	0.28	2.21	0.74
Total P (g kg^{-1})	0.29	1.99	0.13
Total K (%)	0.259	0.826	–
Total Ca (%)	0.080	0.109	0.23
Total Mg (%)	0.337	0.135	0.003
Total Na (%)	0.112	0.160	0.593
Total Fe (mg kg^{-1})	749	1289.3	5.511
Total Cu (mg kg^{-1})	10.43	9.69	0.402
Total Mn (mg kg^{-1})	2.82	2.56	0.085
Polyphenol (%)	14.7	0.70	–
Density (g ml^{-1})	–	–	1.14

treatments. Treatment 1 (control) was the addition of 0.96 kg of urea fertilizer. Treatment 2 was the addition of 59.45 l of human urine. Treatment 3 was the addition of 59.45 l of human urine, 1.012 kg magnesium oxide (MgO) and 6.860 kg of single super phosphate (SSP). Treatment 4 was the addition of 59.45 l of human urine, 1.012 kg of MgO and 3.272 kg of rock phosphate (RP). The Mg and P salt used in treatments 4 and 5 correspond to 20% molar mass of the total N content of the mixture (Jeong and Kim, 2001). Each composting mixture was thoroughly mixed and water was added to bring the moisture content up to 60% on weight basis. The experiment was laid out in randomized complete block design with four replicates. The compost mixtures were turned and watered at the fourth and eighth weeks of composting after moisture determination.

Analytical Methods

The temperature of the compost was monitored daily during the first week of the study and thereafter on weekly basis using the A-type Fluke 52 thermocouple probe. Compost samples were collected on weekly basis for chemical analysis. The samples were oven dried at 70°C until constant weight was achieved, afterwards they were grinded in a hammer mill fitted with 1 mm mesh. The compost pH and electrical conductivity (EC) were determined in water extract (1:10 by weight) after shaking for 30 min (Okalebo et al., 2002). Volatile matter was determined from quadruplet samples per treatment, by ashing oven-dried sample in a

muffle furnace at 550°C for 4 h (Jeong and Kim, 2001). The degree of degradation was calculated from the percentage volatile matter, based on the premise that the total amount of fixed solid in the compost remains constant, while only the volatile matter will diminish by the composting process:

$$\text{Degree of degradation} = \frac{(a_0 - b_x)}{a_0} \times \frac{100}{1}$$

where a_0 is the percentage volatile matter at inception of composting and b_x the percentage volatile matter at sampling time × (other sampling time).

The milled materials were then digested at 36°C for 2 h with the selenium powder, lithium sulphate, hydrogen peroxides and sulphuric acid digestion mixture (Anderson and Ingram, 1993). Total P was determined from the digest using the colorimetric method without pH adjustment as described by Okalebo et al. (2002). Total K, Mg and Na were determined in the digest using the atomic absorption spectrometry. Mineral N (NO_3-N and NH_4^+-N) was determined colorimetrically in fresh compost samples extracted with 0.5 M K_2SO_4 (Okalebo et al., 2002).

Struvite formation was investigated on the final compost samples (12 weeks) dried at 105°C for 24 h and milled to pass through 1 mm mesh. The samples were mounted with carbon glue on carbon stubs used in scanning electron microscopy (SEM) and energy-dispersive X-ray spectroscopy (EDS), using the INCA X-sight Oxford instrument. To ascertain the crystal structure of the struvite under SEM pure struvite was precipitated in the laboratory as described by Kristell et al. (2005). Equal volumes (100 ml) of $MgCl_2 \cdot 6H_2O$

and $NH_4H_2PO_4$ stock solution were diluted in 250 ml flask so that the final concentration of Mg:N:P equals 1:2:2 prior to mixing. The pH of the mixture was adjusted to 9 by the addition of 0.01 N NaOH with continuous stirring and monitoring to ensure stability over 25 min. The precipitates were later filtered through Whatman 41 filter paper and dried at room temperature. Data was subjected to analysis of variance (ANOVA) using the General Linear Model (GLM) procedure (SAS, 1999).

Results

Compost Formation and Maturity

Composting reactions proceeded in all the treatments as shown in Fig. 1. There was a high rise in temperature within 24 h, leading to the first thermophilic phase (temperature > 45°C) which spans over the first 7 days. Treatments with human urine had rapid temperature development with peaks that ranged between 52.7 and 65.4°C in the first 3 days of composting. Their temperatures were significantly ($P < 0.05$) higher compared with composting mixtures where urea fertilizer was used as a supplementary source of N. The three composting mixture with human urine had higher degree of degradation on the long run (Fig. 1). At the fourth week of composting, the degree of degradation was 47.2, 38.8 and 26.5% in treatments 2, 3 and 4 where human urine was applied compared to 17.5% in treatment 1 with urea fertilizer (Fig. 1). The C:N ratio of the composting mixtures was not significantly affected by the treatments at all sampling time. However, gradual reduction was observed in all composting treatments with increasing composting days.

There were significant differences ($P < 0.05$) in the NH_4^+-N concentration of the composting mixtures at day 28, 56 and 84 of composting (Fig. 2). The NH_4^+-N concentration ranged between 20.0 and 80.0 mg kg^{-1} at day 14 of composting and was relatively stable with gradual reduction across the sampling period. There was a substantial reduction in the NH_4^+-N concentration in treatments 1, 2 and 3 from the 56th day of composting; this corresponds with the second mesophilic phase of the composting experiment. The NO_3-N concentration was initially low, while substantial rise in concentration occurred after the thermophilic phase in

all treatments (Fig. 2). Nitrate N concentration was higher in the human urine-treated compost compared to the urea fertilizer-treated compost across the sampling period.

Struvite Formation

The presence of struvite crystals was confirmed with EDS and SEM analysis in the three human urine-treated composts (Fig. 3). The crystals were observed under high magnification (>×100). The chemical structure of the struvite found on the dry compost samples was comparable to what was obtained in the laboratory-precipitated struvite. However, the chemical composition of the observed struvites varied with each composting mixtures (Table 2). None of the observed struvite structure contained nitrogen but potassium and manganese.

Chemical Characteristics of the Final Compost

The chemical properties of the four composts are presented in Table 3. Compost pH was significantly ($P < 0.01$) different with a range of 4.66–6.24 across the treatments. The form of P salt applied to treatments 3 and 4 for struvite precipitation (SSP and RP, respectively) could be responsible for the observed higher pH. The EC was also significantly different among the treatments; treatments 3, 4 and 5 had higher EC values. A significantly higher ($P < 0.05$) EC value was observed in treatment 3 compared to other treatments. The NH_4^+-N concentration was significantly ($P < 0.01$) different among the four composts. Treatment 4 had a significantly ($P < 0.05$) higher NH_4^+-N concentration compared to other composts. The NO_3^-N concentration was not significantly affected by the composting treatments, although treatment 5 had the least value, which may be due to its corresponding higher NH_4^+-N value. Total N content of the final composts ranged from 10.1 to 11.5 g kg^{-1} and it was not significantly affected by the difference in the supplementary source of N. Similar trend was observed with the C:N ratio which ranged between 26 and 29 and was not significantly affected by the treatment.

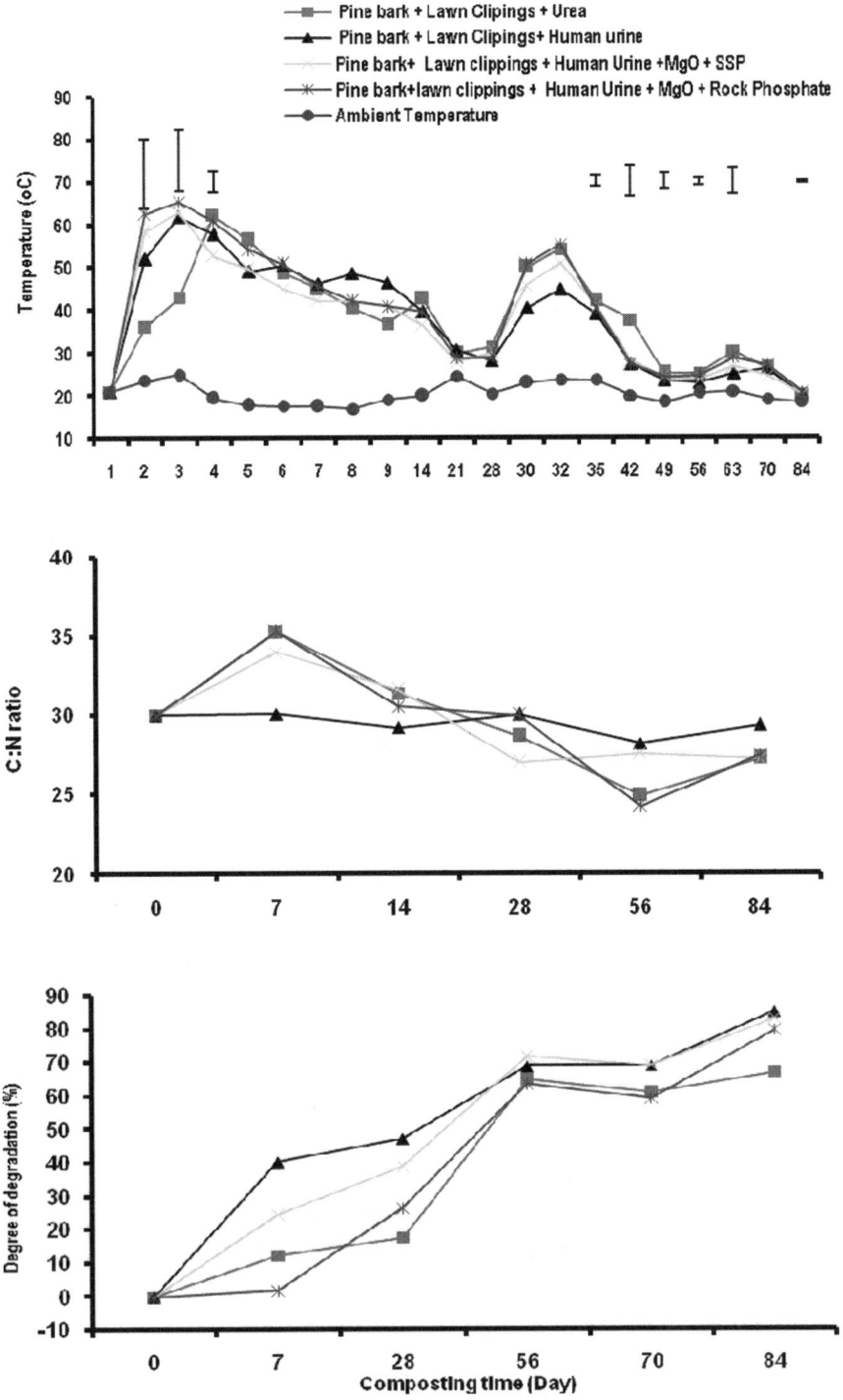

Fig. 1 Effects of composting treatment on temperature (°C), C:N ratio and the degree of degradation. *Bars* represent LSD (0.05)

Fig. 2 Effects of composting treatments on the dynamics of inorganic nitrogen

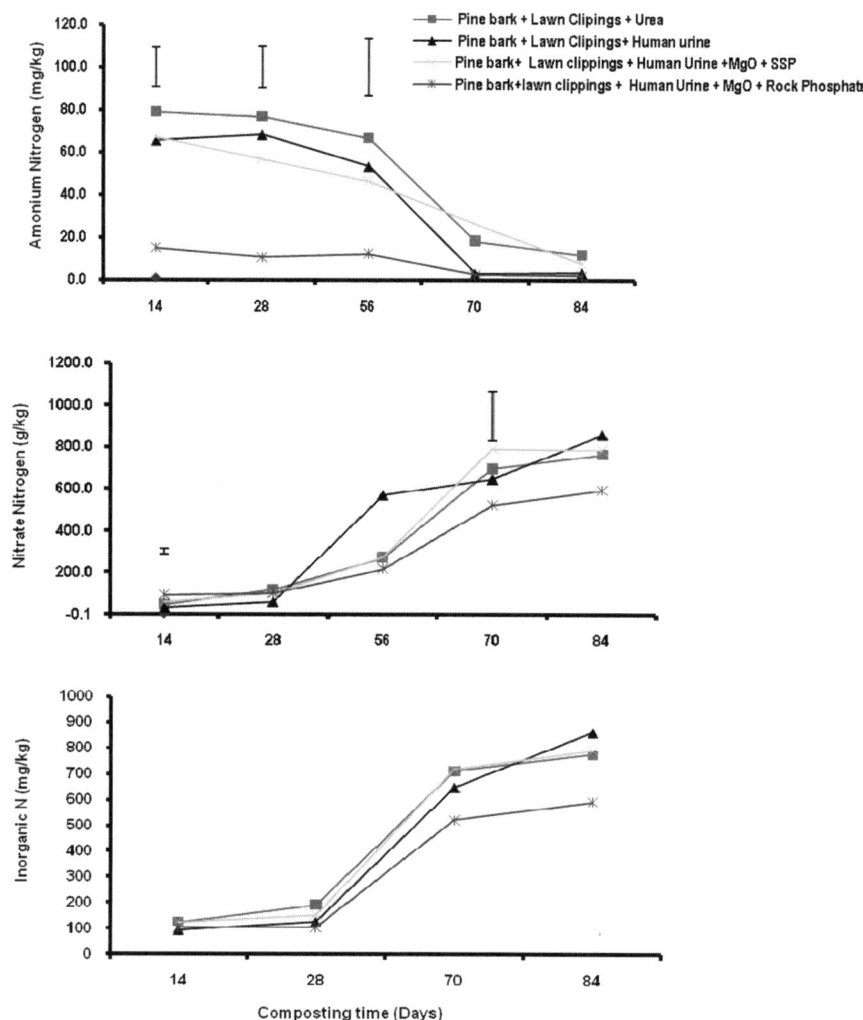

Discussion

Co-composting human urine with other organic wastes has the potential to solve most of the problems which are known to affect the adoption of urine as source of nutrient for crop production. It could also lead to the production of compost with higher agronomic value. In this study compost formation was enhanced with the addition of human urine. This was established by the observed rapid attainment of peak thermophilic temperature range in compost treated with human urine. This observation could partly be attributed to the supply of nutrients required by the thermophilic microorganism in a readily available form (Riddech et al., 2002). This would enhance the thermophilic phase by rapid proliferation of the microorganisms

(Ryckeboer et al., 2003) and subsequent degradation of organic materials. Optimum decomposition of organic materials has been reported to occur at around 55°C (Ryckeboer et al., 2003). The pH trend in this study is also consistent with what was reported in convectional composting (Day et al., 1998). A slight increase in pH is a common characteristic of composting of agricultural wastes; pH is known to increase slightly following a drop in the mesophilic phase (Day et al., 1998; Shin and Jeong, 1996). High pH during composting is also associated with the ammonification process which is the precursor of N loss in aerobic composting (Sanchez-Monedero et al., 2001). The observed initial increase in the EC values of the compost could be attributed to the release of mineral salts of phosphates and ammonium through the decomposition of organic

Fig. 3 Struvite crystals observed with scanning electron microscope (SEM) and X-ray analysis of struvite crystals with energy-dispersive spectroscopy (EDS). (a) Laboratory struvite (mag. ×2400). (b) Pine bark + lawn clippings + human urine (mag. ×100). (c) Pine bark + lawn clippings + human urine + MgO + single super phosphate (mag. ×100). (d) Pine bark + lawn clippings + human urine + MgO + rock phosphate (mag. ×100)

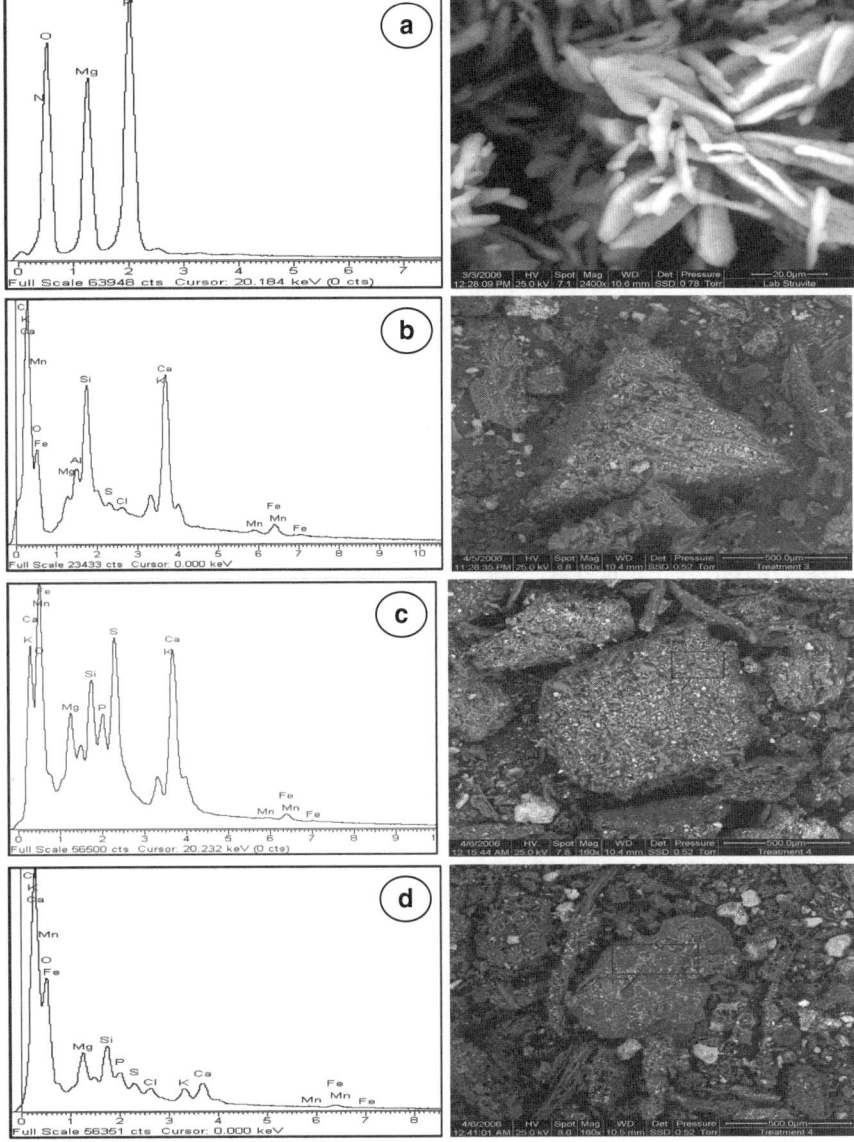

substances (Haung et al., 2004). The observed initial lower degree of degradation in treatments 3 and 4 could be attributed to the applied Mg and P salts. This observation is consistent with the report of Jeong and Kim (2001), which indicated that the application of these salts in excess of 20% (molar basis) of the initial N content may lead to considerable reduction in degradation (Jeong and Hwang, 2005). Thus the use of human urine for multi-supply of N, P and Mg appears to be a more desirable alternative since it has no depressive effect on the degree of degradation of the compost materials.

The dynamics of mineral N in this study showed that treatment 1 had gradual NH_4-N reduction across the sampling period which is consistent with observations in most composting operations. Higher loss of N has been reported to be associated with increase in temperature during the thermophilic stage as a result of mineralization of organic N compounds (Bernal et al., 1997). Since all the composting treatments had high initial N content (C:N = 30), relative loss of N is envisaged. The observed reduction of NH_4-N in treatments 2, 3 and 4 between composting days 14 and 56 could be attributed to the prevalence of the mesophilic phase

Table 2 Qualitative composition of struvite crystals identified from the three different composting mixtures with EDS analysis

Elements	Laboratory struvite	Compost 1	Compost 2	Compost 3
O	59.58	76.13	61.88	68.29
Mg	17.65	0	2.41	2.97
P	21.02	0	2.75	0.50
Ca	0	1.40	5.30	3.15
K	0	2.76	1.57	2.16
Mn	0	0.19	0.53	0.05
Fe	0	1.68	0	1.32
S	0	0	3.63	0.61

Note: Compost 1: compost treated with human urine only; compost 2: compost treated with human urine + MgO + SSP; compost 3: compost treated with human urine + MgO + RP

which is associated with nitrification and conversion of NH_4-N to NO_3-N (Pagan et al., 2006). The increase in the NO_3-N concentration was as a result of nitrification activities which occur during the mesophilic stage of composting. This is because temperatures greater than 40°C will inhibit the activities of nitrifiers (Morisaki et al., 1989). However, appreciable amount of NO_3-N was observed in all compost treatments at the maturation phase. The NH_4^+/NO_3^- ratio observed in this experiment for the five composting treatments was far below the recommended 0.16 required to describe mature compost (Bernal et al., 1997).

The potentials for struvite precipitation from human urine lie in its high ammonium content, which is a product of urea breakdown (Hanaeus et al., 1996). Subsequently, the combination of the ammonium with phosphorus and magnesium in the urine could result in struvite crystals under a wide range of alkaline pH (Uludag-Demirer et al., 2005). Direct recovery of N from human urine as struvite is low without

the addition of supplementary Mg and P, since their concentration is small in urine. The observed struvite in treatment 3 must have been influenced by the use of urine as supplementary source of N, since Mg and P salts were not added. This suggests that sole use of urine without mineral supplementation of Mg and P salt has the potentials to crystallize observable struvite crystals in composting. However, struvite formation has been observed in other studied where Mg and P salts were applied Jeong and Kim (2001). The observed nonsignificant difference in the final compost N content suggests that the observed struvite in treatments 2, 3 and 4 did not result in higher N yield compared with other treatments in this study.

The qualities of the four composts derived from our study are modified by the initial materials used. The five composts possess qualities that made them suitable for use as growing medium.

Conclusion

Co-composting human urine with pine bark and law clippings resulted in early maturity of the compost, while stability was reached at the end of the study. Consequently, human urine could adequately substitute for urea fertilizer as supplementary source of N in composting of organic waste. Struvite crystals were observed in compost treatment with human urine and in treatment where human urine was fortified with Mg and P salt. However, the quantity of the struvite and volume of NH_4^+-N conserved are not ascertained in this study. The chemical characteristics of the final compost showed that they are all suitable for use as horticultural growing medium. Further research is

Table 3 Chemical characteristics of the final compost

Treat	C:N	Polyphenol (%)	pH	EC	NH_4^+/NO_3^-	NH_4^+ (mg kg^{-1})	NO_3^- (mg kg^{-1})	Inorganic N (mg kg^{-1})	Total N (g kg^{-1})	P (g kg^{-1})	K (%)	Ca (%)	Mg (%)	Na (%)
1	27	0.46	4.96	4.63	0.014	1.18	764.0	775.4	11.5	1.363	0.839	0.310	0.553	0.063
2	29	0.89	4.66	5.75	0.004	3.5	859.0	861.9	10.9	1.463	1.260	0.103	0.240	0.285
3	27	0.76	5.20	10.9	0.009	7.8	783.0	790.1	10.1	5.891	0.862	0.660	0.453	0.178
4	27	0.43	6.24	5.48	0.003	2.0	591.0	592.7	10.9	5.583	0.959	0.433	0.470	0.163
Sig	Ns	Ns	**	**		**	Ns	Ns	Ns	**	Ns	**	**	**
LSD			0.35	1.61		0.013				0.877		0.211	0.078	0.073

Note: Treatment 1(pine bark + lawn clippings); treatment 2: (pine bark+ lawn clippings+ urea); treatment 3 (pine bark + lawn clippings + human urine); treatment 4 (pine bark + lawn clippings + human urine + MgO + single super phosphate); treatment 5 (pine bark + lawn clippings + human urine + MgO + rock phosphate)
**Significant at $P>0.01$

needed to determine the optimum quantity of other ions in a composting mixture that will permit N conservation through struvite precipitation.

Acknowledgements The authors wish to acknowledge Mr. Erol Kelly of Electron Microscope Unit, University of Fort Hare, for assistance with EDS/SEM analysis and Mr. Brian Whiting of Damien and Chrysler PYT, East London, South Africa, for assistance with EDS/SEM analysis. The financial support for this work was obtained from National Research Foundation, South Africa, and Govan Mbeki Research and Development Centre, University of Fort Hare, Alice.

References

Anderson JM, Ingram JSI (1993) Tropical soil biology and fertility: a handbook of methods. CAB International, Wallingford, CT, 221pp

Bernal MP, Lopez-Real JM, Scot KM (1993) Application of natural zeolite for the reduction of ammonia emission during the composting of organic waste in a laboratory composting simulator. Biores Technol 43:35–39

Bernal MP, Paredes C, Sanchez-Monedero MA, Cegara J (1997) Maturity and stability parameters of composts prepared with a wide range of organic wastes. Biores Technol 63:91–99

Bo-Bertil L, Zsofia B, Stefan B (2000) Nutrient recovery from human urine by struvite crystallization with ammonia adsorption on zeolite and wollastonite. Biores Technol 73:169–174

Brink N (1995) Composting of food waste with straw and other carbon sources for nitrogen catching. Acta Agic Scand B Soil Plant Sci 45:118–123

Daughton CG, Ternes TA (1999) Pharmaceuticals and personal care products in the environment: agent of subtle change? Environ Health Perspect 107(6):907–938

Day M, Krzymien M, Shaw K, Zaremba L, Wilson WR, Botden C, Thomas B (1998) An investigation of the chemical and physical changes occurring during commercial composting. Compost Sci Util 6(2):44–66

Hanaeus A, Hellstrom D, Johansson E (1996) Conversion of urea during storage of human urine. Vatten 52:263–270

Haung GF, Wong JWC, Wu QT, Nagar BB (2004) Effects of C/N on composting of pig manure with saw dust. Waste Manage 24(8):805–813

Jeong Y-K, Hwang S-J (2005) Optimum doses of Mg and P salts for precipitating ammonia into struvite crystals in aerobic composting. Biores Technol 96:1–6

Jeong Y, Kim J (2001) A new method for conservation of nitrogen in aerobic composting process. Biores Technol 79:129–133

Kirchmann H (1985) Losses, plant uptake and utilization of manure nitrogen during a production cycle. Acta Agric Scand Suppl 24. 175pp

Kristell S, Corre Le, Valsami-Jones E, Hobbs P, Parsons SA (2005) Impact of calcium on struvite crystal size, shape and purity. J Crystal Growth 283(3–4):514–522

Larsen TA, Peters I, Alder A, Eggen R, Maurer M, Muncke J (2001) Re-engineering the toilet for sustainable wastewater management. Environ Sci Technol 35(9):133–197

Liao PH, Jones L, Lau AK, Walkemeyer S, Egan B, Holbek N (1997) Composting of fish waste in a full scale in-vessel system. Biores Technol 59:163–169

Mahimairaja S, Bolan NS, Hedley MJ, Macgregor AN (1994) Losses and transformation of nitrogen during composting of poultry manure with different amendments: an incubation experiment. Biores Technol 47:265–273

Martins O, Dewes T (1992) Loss of nitrogenous compounds during composting of animal wastes. Biores Technol 42: 103–111

Mnkeni PSN, Adediran JA, Van Ranst E, Verplancke H (2006) The recycling of organic waste for soil fertility improvement and growing medium production in Eastern cape province, Chapter 9. In: Van Averbeke W et al (eds) Small holders farming and management of soil fertility in Eastern Cape, South Africa. UNISA Press, Cape Town, South Africa

Mnkeni PSN, Austin A, Kutu FR (2005) Preliminary studies on the evaluation of human urine as source of nutrient for vegetables in the Eastern Cape Province, South Africa. In ecological sanitation: a sustainable, integrated solution. Conference documentation of the 3rd international ecological sanitation conference, Durban, South Africa, pp 418–426

Morisaki N, Phae CG, Nakasaki K, Shoda M, Kobuta H (1989) Nitrogen transformation during thermophilic composting. J Ferment Bioeng 67:57–61

Ohlinger KN, Young TM, Schroeder ED (1999) Kinetic effect on preferential struvite accumulation in wastewaters. J Environ Eng ASCE 125(8):730–737

Okalebo JR, Gathua KW, Woomer PL (2002) Laboratory methods of soil and plant analysis: a working manual. TSBF program UNESCO – ROSTA soil science society of East Africa Technical Publication No. 1. Marvel EPZ Ltd; Nairobi, Kenya

Pagan E, Barena R, Font X, Sanchez A (2006) Ammonia emission from the composting of different organic wastes: dependency on process temperature. Chemosphere 62:1534–1542

Riddech M, Klammer S, Insam H (2002) Characterization of microbial communities during composting of organic wastes. In: Insam H, Riddech N, Klammer S (eds) Microbiology of composting. Springer, Heidelberg, pp 43–52

Ryckeboer J, Mergaert J, Coosemans J, Deprins K, Swings J (2003) Microbiological aspects of biowaste during composting in monitored compost bin. J Appl Microbiol 94:127–137

Sanchez-Monedero MA, Roig A, Paredes C, Bernal MP (2001) Nitrogen transformation during organic waste composting by the Rutgers system and its effects on pH, EC and maturity of the composting mixtures. Biores Technol 78:301–308

SAS (1999) Statistical analysis system Inc. SAS users guide, statistical analysis institute. Carry, NC, 112pp

Schulze-Rettmer R (1991) The simultaneous chemical precipitation of ammonium and phosphate in the form of magnesium ammonium phosphate. Water Sci Technol 23:659–667

Shin HS, Jeong YK (1996) The degradation of cellulosic fraction in composting of source separated wood waste and paper mixture with change of C:N ratio. Environ Technol 17:433–438

Sibbesen E, Lind AM (1993) Loss of nitrous oxide from animal manure in dung-heaps. Acta Agric Scand B Soil Plant Sci 43:16–20

Smith IE (1992) Pine bark as a seedling growing medium. Acta Hort (ISHS) 319:395–400

Sridhar MKC, Coker AO, Adeoye GO, Akinjogbin IO (2005) Urine harvesting and utilization for cultivation of selected crops: from Ibadan, South West Nigeria. In ecological sanitation: a sustainable, integrated solution. Conference documentation of the 3rd international ecological sanitation conference, Durban, South Africa, pp 331–336

Udert KM (2003) The fate of Nitrogen and phosphorus in source-separated urine. Schriftenreihe des Instituts fur Hydromechanics and Water resources management. Swiss federal Institute of Technology, Zurich

Uludag-Demirer S, Demirer GN, Chen S (2005) Ammonia removal from anaerobically digested dairy manure by struvite precipitation. Process Biochem 40(12):3667–3674

Witter E, Kirchmann H (1989) Peat, zeolite and basalt as adsorbents of ammoniacal nitrogen during manure decomposition. Plant Soil 115:43–52

Witter E, Lopez-Real J (1987) The potential of sewage sludge and composting in a nitrogen recycling strategy for agriculture. Biol Agric Hortic 5:1–23

Witter E, Lopez-Real J (1988) Nitrogen losses during the composting of sewage sludge, and the effectiveness of clay soil, zeolite, and compost in adsorbing the volatilized ammonia. Biol Wastes 23:279–294

Wolgast M (1993) Rena watten. Om tanker I Kretslopp. Creamon HB Uppsala, pp 1–186

Nutr Cycl Agroecosyst (2010) 88:79–90
DOI 10.1007/s10705-009-9332-1

RESEARCH ARTICLE

Extractable Bray-1 phosphorus and crop yields as influenced by addition of phosphatic fertilizers of various solubilities integrated with manure in an acid soil

E. W. Gikonyo · A. R. Zaharah · M. M. Hanafi ·
A. R. Anuar

Received: 5 March 2008 / Accepted: 25 November 2009 / Published online: 11 December 2009
© Springer Science+Business Media B.V. 2009

Abstract Soil extractable Bray-1 P (B1P) and response to phosphate (P) of *Setaria anceps* cv. *Kazungula* (Setaria grass) were monitored in a field trial bimonthly for 14 months in an acid soil fertilized with triple super phosphate (TSP), Gafsa phosphate rock (GPR) or Christmas Island phosphate rock (CIPR) integrated with or without manure. Extractable B1P from the same soil incubated with the same fertilizers in wet and dry 3-day cycles for 91 days was determined. Field experimental design was randomized complete block (RCB) with three replications. Results indicated that B1P magnitude for field and incubation trial were; TSP > GPR > CIPR, consistent with their solubility. An integration of manure and fertilizers resulted in much higher extractable B1P than sole fertilizers or manure. Over time, P availability decreased at a fast rate for the first 6 months and later was relatively constant. The dry matter yields (DMYs) exhibited quadratic relationships with P rates. Maximum DMYs (6–11 t ha^{-1}) were attained between 100 and 200 kg P ha^{-1}, above which they declined. Average DMYs were not significantly different for TSP, GPR and CIPR (6.1–6.6 t ha^{-1}). Maximum individual DMY were attained at 2–6 months and then declined to a minimum (2–4 t ha^{-1}) after 1 year. Cumulative yields (20–55 t ha^{-1}) also were not significantly different for the three fertilizers. Manure-CIPR integration increased DMY whilst in GPR and TSP/manure combinations DMYs were depressed. The PRs could supplement the expensive TSP without loss of yields but the non-reactive PR should be integrated with manure.

Keywords Manure · Phosphatic fertilizers · Phosphorus availability

E. W. Gikonyo (✉)
KARI-NARL, P.O. Box 14733-00800, Nairobi, Kenya
e-mail: estgikonyo@yahoo.com

Present Address:
E. W. Gikonyo · M. M. Hanafi
Institute of Tropical Agriculture, Universiti Putra
Malaysia, 43400 Serdang, Selangor, Malaysia
mmhanafi@Agri.upm.edu.my

A. R. Zaharah · A. R. Anuar
Faculty of Agriculture, Department of Land Management,
Universiti Putra Malaysia, 43400 Serdang, Selangor,
Malaysia
Zaharah@Agri.upm.edu.my
anuar@Agri.upm.edu.my

Introduction

Many acid soils in the world, especially in the tropics, present soil fertility problems that have constrained the development of successful sustainable agriculture. Malaysian soils are known to be highly weathered and are generally acidic and inherently low in phosphorus (P) and high in P fixing capacities (Owen 1947). They are characteristically high in iron (Fe) and

This article has been previously published in the journal "Nutrient Cycling in Agroecosystems" Volume 88 Issue 1.
A. Bationo et al. (eds.), Innovations as Key to the Green Revolution in Africa: Exploring the Scientific Facts. © 2010 Springer.

aluminium (Al) in their clay fractions and thus result in substantial P-fixation (Zaharah and Sharifuddin 1979) that can reduce the recovery rate of applied conventional water-soluble phosphate fertilizers. The P sources for Malaysian agricultural production have normally been obtained either from water soluble- or from water insoluble-P sources. It is widely accepted that the former includes triple superphosphate (TSP), single superphosphate (SSP), and diammoniam phosphate (DAP), while the latter is mainly phosphate rock (PR) materials. Because there are no phosphate mineral deposits found in Malaysia, the P sources are imported from various countries. The most significant P sources in decreasing order of imported amounts are Australia/Christmas Island (41.6%), Tunisia/Gafsa phosphate rock (18.2%), Algeria (14.1%) and China (12.8%) as shown by Yusda and Hanafi (2003).

However, the PRs vary widely in their effectiveness due to the differences in physico-chemical characteristics that influence their reactivity. Some PRs are considered reactive while, others are non-reactive. Unlike the reactive PRs, non-reactive PRs are less suitable for direct application and need to be acidified chemically or biologically (Rajan 1981) for sufficient P to become available at a rate matching plant demand. Chemical acidulation is not suitable for PRs containing high levels of Fe or Al because the metals reduce the solubility of P (Hammond et al. 1989). Some of the practical techniques of PR acidulation include integrating the PRs with organic materials such as farmyard manure (Ikerra et al. 1994; Oenema 1980), crop residues, and green manures (Zaharah and Bah 1997). However, to date there has been conflicting reports on the influence of organic manures on PR dissolution. Some workers have reported enhanced PR dissolution such as Ikerra et al. (1994), while Oenema (1980) observed reduction on PR solubility when PRs were combined with manure. In view of this the objectives of this study were:

1. to determine extractable Bray-1 P (B1P) from soil samples treated and incubated in the laboratory with a reactive or a non-reactive PR with or without farmyard manure (FYM) or TSP/FYM combination relative to TSP (the standard P fertilizer) and

2. to determine effects of applying reactive or a non-reactive PR and TSP with or without manure

combination in a field experiment on soil extractable B1P and Setaria dry matter yields (DMYs) sequentially sampled bimonthly for 14 months.

Materials and methods

Phosphate sources and soil

A laboratory incubation experiment was conducted to assess the dissolution of reactive and non-reactive PRs applied to an acid soil at different rates in the presence and absence of cattle manure relative to TSP. The PR dissolution was assessed through B1P from soils incubated with the two selected PR treatments compared to B1P from soils incubated with TSP. Gafsa phosphate rock (GPR) a naturally occurring phosphate ore originating from sedimentary [Francolite (Carbonate-Fluorapatite)] deposits in Tunisia was selected as the reactive PR. Gafsa PR producer is the 5th largest PR producer in the world exporting GPR to over 20 countries in the world including Malaysia (Elhaji 2003). On the other hand, Christmas Island PR (CIPR) grade A dust characterized with low chemical activity and low agronomic value when directly applied to crops was selected as the non reactive PR since it is import in the highest volume into Malaysia (41% of all PRs). Dissolution assessment of the PRs in the laboratory incubation trial (controlled environment) was followed by a field experiment to evaluate the same PRs relative to TSP with and without FYM addition. In the field trial PRs dissolution in the different treatments were measured by Bray-1 P, one of the most widely used P soil tests (Bray and Kutz 1945) and crop response in terms of DMYs.

Soil used for incubation experiment was sampled from the same field experimental site prior to the field experiment commencement. The trial site was Puchong experimental farm, Universiti Putra Malaysia (UPM) in Serdang, Selangor. The soil is classified as Bungor soil series, an Ultisol (Typic kandiudult, clayey, kaolinitic, isohyperthermic).

Soil, fertilizers and manure characterization

Top (0–15 cm) air dried soil sieved to pass through a 2 mm mesh was characterized by standard methods

(Page 1982). The soil pH_{water} and pH_{KCl} (1:2.5 soil to solution ratio) was 4.5 and 3.9, respectively; total carbon 30 g kg^{-1}; total P 280 mg P kg^{-1}; P sorption maximum determined from the Lamgmuir adsorption isotherm 769 mg P kg^{-1}; Bray-1 P 4.9 mg P kg^{-1} soil; citrate dithionite extractable Al_2O_3 and Fe_2O_3 15.2 and 30.8 g kg^{-1}, respectively; cation exchange capacity 5.4 cmol (+) kg^{-1}.

Characterization of the fertilizers and manure revealed that: TSP had total P 204 g kg^{-1}; CaO 192 g kg^{-1}; Al_2O_3 6 g kg^{-1}; Fe_2O_3 16 g kg^{-1}; solubility in water and in 2% citric acid and 2% formic acid were 380, 430 and 407 g kg^{-1}, respectively.

The GPR had total P 125 g kg^{-1}; CaO 460 g kg^{-1}; Al_2O_3 9 g kg^{-1}; Fe_2O_3 4 g kg^{-1}; solubility in water, 2% citric acid and 2% formic acid were 2, 100 and 110 g kg^{-1}, respectively. The CIPR had total P 140 g kg^{-1}; CaO 420 g kg^{-1}; Al_2O_3 31 g kg^{-1}; Fe_2O_3 107 g kg^{-1}; solubility in water, 2% citric acid and 2% formic acid were 1, 87 and 66 g kg^{-1}. Manure had total P of 10 g kg^{-1}, CaO 50 g kg^{-1}; Fe_2O_3 3 g kg^{-1}; Al_2O_3 0.4 g kg^{-1} and Mg 11 g kg^{-1}. Dissolved organic carbon (DOC) was 21 g kg^{-1}; C/N ratio 27; humic acid content 10.2 g kg^{-1} and fulvic acid content was only traces.

Closed incubation study

Soil samples (100 g air-dried sieved through 2 mm) in triplicates were treated with TSP, GPR and CIPR at four rates: 0, 100, 200 and 400 mg P kg^{-1}. Soils were wetted to approximately field capacity for 4 days at room temperature (28 ± 3°C), and then air-dried for another 3 days. The wetting and drying cycles were repeated 13 times (13 × 7) thus, adding to 91 days. The wetting and drying cycles were to simulate field environmental conditions. After the first cycle, soils were mixed to ensure uniform distribution of P. At the end of equilibration period soils were air dried, sieved through 2 mm and analyzed for extractable P by Bray-1 method to determine the effect of incubation and the different treatments on P availability.

Field experimental design and layout

Setaria anceps cv. *Kazungula* (Setaria grass) was used as the test crop because it has widespread occurrence, can survive low fertility but also respond well to nitrogen and phosphorus application. Additionally, it is persistent under frequent cutting or grazing hence was suitable for sequential harvesting required in the experiment. Experimental treatments for assessment of Setaria grass dry matter response to P from different sources and rates were factorial combinations consisting of 20 treatments ([nil + (3 P fertilizer sources × 3 P rates)] × 2 levels of manure) with three replications arranged in a randomized complete block design (RCBD). The three P sources were; TSP, GPR and CIPR applied at the P rates; 100, 200 and 300 kg P ha^{-1} applied alone (sole) or combined with cattle manure at zero (0) or 20 t ha^{-1}. Fertilizer rates were chosen from previous rates tested in 'P recapitalization' studies in tropical acid soils (Mutert and Fairhurst 2003). The experimental plots were 1.5 × 1.5 m^2. The P fertilizers and manure were applied once at the beginning of the experiment and treatments were incorporated into the top 10 cm depth of the soil 2 weeks prior to planting. At planting and after every harvest, all plots received a basal application of N and K at 60 kg N ha^{-1} as urea and 100 kg K ha^{-1} as murriate of potash, respectively. Setaria grass was established from seeds by broadcasting 0.3 g seeds (>80% germination) along five rows per plot at 30 cm inter-row width in all the plots. The seeds were only lightly covered with soil. Two weeks after germination, re-seeding was done where germination had failed. The plots were maintained weeds free by hand weeding at monthly intervals for the first 2 months but later, weeding was done after every harvest. Grass was harvested by cutting with a sickle at 3–5 cm above the soil surface. Harvesting was done from the three inner rows, discarding harvest from 0.25 m distance from each plot boundary (harvest area = 0.9 m^2). Harvesting was done after every 2 months for a period of 14 months.

The harvested grass was dried in a draught blown oven at 70°C until no further change in weight and DMY was recorded. A sub-sample of the dried grass was ground and analyzed for nutrient uptake by ashing in the muffle furnace at 300°C for 4 h and then 550°C for another 4 h. After cooling 2 ml concentrated HCl was added to the crucible and heated to dryness. Then, 10 ml of 20% HNO_3 was added and heated on a water bath for 1 h. The solution was then filtered and P, K, Ca, Mg and Zn in solution were measured using Inductively Coupled Argon Plasma

Spectrophotometry (Model: Thermal Elemental IRIS Advantage).

Results and discussion

Soil, fertilizers and manure characterization

The soil was strongly acid (pH_{KCl} 3.9) with low extractable P and substantial Fe and Al oxide, hence soil characterized by moderately high P sorption capacity (P sorption maximum = 765 mg kg^{-1}). Characterization of the fertilizers indicated that TSP had the highest P content, whilst GPR had the lowest amongst the three fertilizers. Additionally, TSP was the most soluble followed by GPR and hence CIPR was the least soluble in the three solvents used. Consequently, GPR was more reactive than CIPR based on the solubility (Heng 2003). Manure had 1% P content, which was considered high in manures perhaps associated with a high phosphate feed offered to the cattle. On the other hand, decomposition parameters revealed that dissolved organic carbon (DOC) and C/N ratio were both high (Zamora-Nahum et al. 2005). Moreover, the humic and fulvic acids were also low indicating manure used was not well decomposed.

Effect of treatments on extractable Bray-1 P from incubation study

Bray-1 P (B1P) levels were significantly influenced by the P source, P rate and manure addition ($P < 0.001$). Interactions were also significant ($P < 0.01$) with an exception of P source × manure. Generally, the magnitude of extractable P for the different P sources was in the order: TSP > GPR > CIPR consistent with their reactivity and hence dissolution rates (Zaharah and Sharifuddin 2002). Average extractable B1P from TSP, GPR and CIPR were 56.2, 39.0 and 20.4 mg P kg^{-1} of soil, respectively. Extractable B1P also increased with increasing P application rates. When P application rates were raised from 100 to 400 mg P kg^{-1} soil, the resulting B1P levels ranged from ≈27.0 to 72.7 mg P kg^{-1} soil. The increase of B1P with P application rates is related to increase in saturation of adsorption sites and hence reduced P fixation per added P (Zhang et al. 1995; Yusda 2003). Application of manure and P fertilizer combinations

led to significantly higher B1P levels ranging from ≈33.0 (sole P sources) to 44.0 mg P kg^{-1} soil (plus manure).

However, the overall magnitude of B1P was determined by the P source × P rate × manure interaction (Fig. 1). Extractable P increased with increasing P application rates in more or less a linear fashion in TSP treated soils. For instance, when TSP was applied at 100, 200 and 400 mg P kg^{-1}, the resulting B1P levels ranged from 29.6 mg P kg^{-1} (sole TSP) to 120.0 mg P kg^{-1} (TSP + manure). Manure + TSP combinations gave higher B1P levels at all the P application rates though with varying magnitudes. The additional amount of B1P extracted from TSP + manure and TSP at 100 kg P ha^{-1} rate was approximately equal to the difference in amount of B1P extracted at zero P, with and without manure [14.1 (plus manure treatment) − 3.8 (minus manure treatment)]. The latter was assumed to be equal to the amount of B1P originating from manure mineralization. Nonetheless, at 200 kg P ha^{-1}, the difference between plus and minus manure B1P was much higher but on the contrary almost negligible at 400 kg P ha^{-1} (Fig. 1). Extractable B1P from manure/P fertilizer combinations is possibly a product of several mechanisms. Decomposing manure may add P to the available P pool by mineralization, and concomitantly also releases net negative charges and organic anions that compete with P for sorption sites thus decreasing P sorption. In addition manure mineralization also releases Ca ions, Fe and Al oxides thus increasing P sorption depending on the quality and quantity of manure. Organic anions and

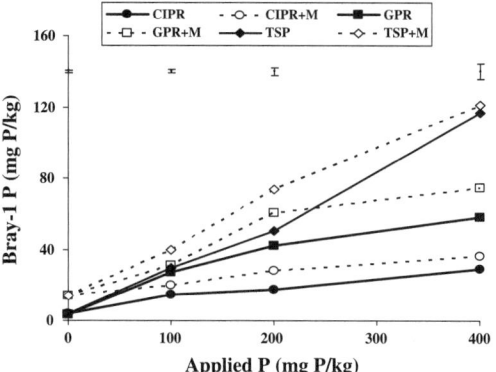

Fig. 1 Effects of P source × prate × manure on extractable Bray-1 P. Note the *bars* denote standard error

Nutr Cycl Agroecosyst (2010) 88:79–90

increased P sorption may also increase PR dissolution (Hanafi et al. 1992).

In the current study, manure addition increased B1P through P addition (from mineralization) and concurrently increased PSI (data not shown). Increased PSI in manure treated soil was due to significant amounts of oxalate and dithionite extractable iron oxide plus small amounts of aluminium oxide (Gikonyo 2006). Despite the sesquioxide's content of manure being low, the high rate of manure application (20 mg ha^{-1}) led to significant amounts of total sesquioxides in the soil (data not shown). Phosphate sorption index was inversely related to B1P. On the other hand, increasing P rates decreased PSI because as adsorption of P continues there is a shift of the soil particle surface to more negative values (Celi et al. 2000) consequently decreasing the bonding strength as increasing number of sites become occupied by P. Thus, an interaction of the various mechanisms described above determines the final amount of B1P.

For example, the significant difference between TSP and TSP + manure observed at 200 mg P ha^{-1} and the sudden increase of B1P for sole TSP at 400 mg P ha^{-1} could be explained to a great extent by the changes in PSI. Firstly, the PSI difference between sole TSP and TSP + manure at 100 kg P ha^{-1} was 34 mg P kg^{-1} and considering the low P rate, there was no significant difference between the two. On the other hand, at 200 mg P ha^{-1} the PSI difference between the manure and non-manure TSP combination was 57 mg P kg^{-1} and considering the higher P rate level the B1P difference was higher than at the previous rate. Similarly, the sudden increase in B1P at 400 kg P ha^{-1} for sole TSP was associated with a PSI decrease of 51 mg P kg^{-1} (minus manure) relative to 28 mg P kg^{-1} (manure combined) in addition to the high P level. Thus there was a sharp increase in the sole TSP and not the TSP + manure. Contrary, to the common knowledge, manure did not reduce the phosphate sorption capacity of the soils because of the presence of significant amounts of sesquioxides, particularly iron oxide. Additionally, the manure used exhibited low levels of organic acids, particularly fulvic acid (only traces present) which is one the most chemically and biologically active fraction of soil organic matter (SOM) fraction (Van Heels et al. 2005) thus explaining why P sorption was not reduced but increased by manure addition.

In PRs, B1P increased with increasing P rates up to 200 mg P kg^{-1}, but did not increase further at the highest P application rate (400 mg P kg^{-1}) thus exhibiting a linear plateau relationship. The relationship indicates a decline in PR dissolution rate above 200 kg P ha^{-1} that could be ascribed to the attainment of the specific PR's solubility constant. According to Chien and Black (1976), the solution P of a PR is fixed at a maximum value for each PR regardless of how much PR is added to the soil.

However, as indicated above B1P was influenced by P source × P rate × manure interaction. Gafsa PR treated soils gave higher B1P levels than CIPR at all application rates probably due to the higher reactivity of GPR relative to CIPR as shown by the solubility. Manure addition increased B1P in both PRs but the increases in GPR + manure were always higher than those in CIPR + manure. A similar scenario, to that observed in TSP at 200 mg kg^{-1} soil was observed in sole GPR and GPR + manure. The most significant difference between manure and minus manure treatment was observed at this level. Although, GPR + manure had the lower PSI, the difference was only ≈ 12 mg kg^{-1} but the difference in B1P was significant (18.7 mg kg^{-1}). Noteworthy, was the scenario in GPR at 400 mg P ha^{-1} where, unlike in TSP, GPR + manure gave higher B1P (by 16 mg kg^{-1}) than sole GPR although the PSI differed by a substantially higher magnitude (101.7 mg kg^{-1}). It was evident that the lower PSI value always resulted to a higher B1P though not proportionally.

Field experiment

Bray 1 extractable P

Extractable Bray-1 P was significantly influenced by P source (fertilizer type) ($P = 0.0001$) and fertilizer rate ($P = 0.0001$) but unlike the incubation trial, B1P was not significantly influenced by manure addition ($P = 0.064$). On average, the lowest B1P was extracted from CIPR treated soil (7.8 ± 1.20 mg kg^{-1}), while the highest was from TSP (22.1 ± 1.20 mg kg^{-1}). Moreover, the amount of B1P also increased with increasing fertilizer rates. For instance, increasing fertilizer rates from 100 to 300 kg P ha^{-1} increased B1P from an average P level of 10.5–26.8 mg kg^{-1}. Addition of manure increased B1P from an average of 12.4–15.1 mg kg^{-1} which was not statistically significant.

Statistical analysis of soil samples collected bimonthly at every grass harvest indicated varying significant effects of P sources, manure addition, P rates and their interactions. For instance, P sources, P rates and their interactions were significant for all the sampling times except the first one (prior to treatment application). In the treated soil samples, TSP gave the highest B1P at all sampling times (2nd–14th month) shown in Table 1. However, from the month 2–6 sampling TSP exhibited significant difference from the PRs but the latter were not significantly different. Nevertheless, from 8th to 14th month samplings the two PRs exhibited significant differences. Thus B1P levels were in the order TSP > GPR > CIPR in accordance to their solubility as shown in the incubation trial.

The different P application rates gave significantly different B1P values in increasing order from zero (0), 100 and 300 kg P ha^{-1}. The maximum B1P was attained at month 4 sampling (28.1 mg P kg^{-1} soil) and seemed to decline with time to a minimum at 14 months. Manure addition resulted in significant differences in the 2nd sampling by about 8 mg P kg^{-1}soil and month 4 sampling marginally by about 2.3 mg P kg^{-1} soil (Table 2). However, all through the experiment manure increased B1P levels although not significantly. At 2nd month sampling when manure was combined with TSP, GPR and CIPR at 100 kg P ha^{-1} B1P increased by 6.2, 6.1 and 6.5 mg P kg^{-1}, respectively. On the other hand, as P rate was increased to 200 kg P ha^{-1}, combinations with manure resulted in B1P increases of 30.7, 6.3

Table 1 Summary of statistical analysis of Bray-1 P data of soils sampled from 2nd to 14th months after experiment commencement

Treatments	Extractable Bray-1 P (mg kg^{-1})						
	Month 2	Month 4	Month 6	Month 8	Month 10	Month 12	Month 14
(a) P sources							
TSP	25.5 A	25.1 A	20.8 A	16.4 A	24.9 A	26.2 A	15.8 A
GPR	14.4 B	12.1 B	8.6 B	8.7 B	13.6 B	12.9 B	9.2 B
CIPR	11.1 B	9.9 B	6.9 B	5.8 C	8.0 C	8.1 C	5.0 C
(b) P rates (kg P ha^{-1})							
300	27.7 A	28.3 A	23.5 A	20.6 A	33.2 A	33.1 A	21.1 A
100	16.1 B	13.8 B	9.2 B	7.8 B	9.8 B	10.4 B	6.7 B
0 (Control)	7.1 C	4.9 C	3.7 B	2.7 C	3.5 C	3.9 C	2.3 C
(c) Manure addition							
Plus manure	21.0 A	17.0 A	13.2 A	11.5 A	17.1 A	16.7 A	11.1 A
Minus manure	13.0 B	14.3 A	11.0 A	9.2 B	13.9 A	14.8 A	8.9 A

Note: values followed by the same letters in one column within (a), (b) or (c) are not significantly different at 5% level DMRT

Table 2 Probability level of significance or non-significance of treatment effects on individual dry matter yield harvests

Treatment effects	Number of months at harvest from beginning of experiment						
	2	4	6	8	10	12	14
Model	0.0001	0.0011	0.0095	0.0001	0.0001	0.0011	0.0002
P source (S)	NS	NS	NS	NS	NS	NS	NS
P rate (R)	0.0001	0.0001	0.0001	0.0001	0.0002	0.00046	NS
Manure (M)	0.0003	0.025	NS	0.008	NS	NS	NS
S × R	0.0054	NS	NS	NS	NS	NS	NS
S × M	0.016	NS	NS	NS	NS	NS	NS
S × M × R	0.0001	NS	0.0033	0.003	NS	NS	NS

NS not significant

 Springer

and 0.8 mg P kg^{-1} for TSP, GPR and CIPR, respectively. Mineralization of manure gave about 2.5 mg P kg^{-1} estimated from B1P change in the control when manure was added. Consequently, it is evident that besides addition of P from manure mineralization, mechanisms influencing P release and P sorption were involved thus leading to quite a high increase in B1P from TSP and only marginal in CIPR. Addition of CIPR was found from the incubation trial to cause the highest increase in PSI hence explaining the low B1P increase. On the other hand, TSP had the highest increase although it had higher sesquioxides than GPR because the former is readily soluble hence satisfying the P sorption sites more than the slow releasing GPR finally resulting in higher B1P increases. In a field situation the enhancement of PR dissolution as the products of dissolution P and Ca are removed by plants uptake coupled with leaching of Ca (from both PRs and manure) by rainfall is possibly an additional mechanism determining final B1P (Hanafi and Syers 1994). The fact that the most significant manure effect was in the beginning (2nd month sampling) and not in the subsequent samplings implies that results of long term experiments where manure is added once at the beginning of the experiment may conflict with results obtained from long term experiments where manure is added every season/annually or one time experiments such as glasshouse trials. Long term experiments will report no significant manure effect while short-term experiments may report significant manure effects. This could partly explain some of the conflicting results on manure effects reported in literature. Nevertheless, the implication to the farming systems is that for sustainable manure effects, modest rates with repeated applications may be the most appropriate strategy. Additionally, there is need to conduct manure/PR substitution levels to achieve economical manure/PR combination rates for optimal soil B1P levels.

Combined analysis of the soil data for the eight sampling times revealed B1P was significantly influenced by the interaction of P source × fertilizer rate × manure over time. The B1P levels at the lower P application rate (100 kg P ha^{-1}) declined linearly up to the 6th–8th months after which, the levels were fairly constant (Fig. 2). At the higher P rate, the decline was very gradual throughout the 14 months

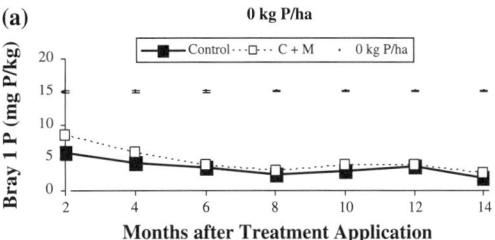

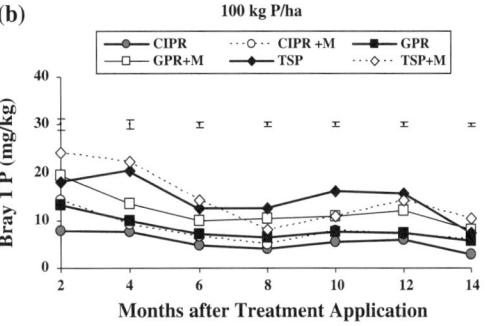

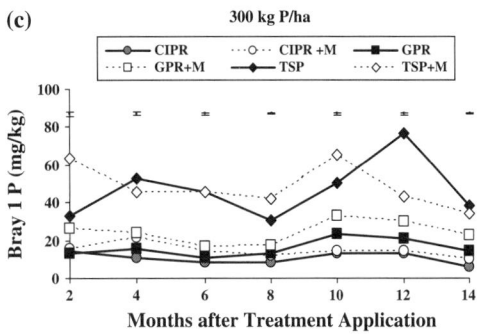

Fig. 2 Effects of different P sources with and without manure on extractable Bray-1 P over time. *Nb*: the *bars* denote SE

(Fig. 2). It should be noted that the data points of TSP at 12 months and TSP + manure at 10 months are considered as outliers. Throughout the experimental period, the P sources affected the magnitude of B1P in the order: TSP > GPR = CIPR. However, the B1P levels were not high despite the high P application rates particularly for the PRs. The highest B1P level in the PRs was <30 mg P kg^{-1} attained with GPR (300 kg P ha^{-1}) at the 10th month. On the other hand, TSP (300 kg P ha^{-1}) gave the highest B1P about 65 mg P kg^{-1} (Fig. 2).

Setaria dry matter yields

The yield data exhibited quadratic relationships in each harvest (Fig. 3). Statistical analysis of the individual DMY harvests indicated that treatment (P source, P rate, Manure) effects and their interactions exhibited variable significance levels among the seven harvests as shown in Table 1. The P sources were not significant in all the seven harvests while, on the other hand P rates were significant in all except the final harvest (14th month). Manure addition gave significant effects in 3 out of 7 harvests and the interaction; P source × manure and P rate × manure were only significant in the first harvest but not in all the others. However, interaction of P source × P rate × manure were significant in 3 out of 7 harvests.

The most significant P source × manure interaction was in harvest one (2nd month) and effects varied from one P rate to the other. At 100 kg P ha^{-1} sole TSP gave the highest DMYs (7.6 t ha^{-1}), whilst sole CIPR gave the lowest (6.8 t ha^{-1}) and was significantly similar to GPR (Fig. 3). This observation was similar to what was observed in the B1P levels (Table 1) implying B1P in this first harvest could substantially describe DMYs. However, combining manure with the three P sources at 100 kg P ha^{-1} increased DMY in the three P sources ranging from ≈7.0 t ha^{-1} (GPR + manure) to 10.4 t ha^{-1} (TSP + manure).

As P rate was increased from 100 to 200 kg P ha^{-1} DMY in the sole P sources increased to a maximum (11.0 t ha^{-1}) attained in TSP, whilst remaining

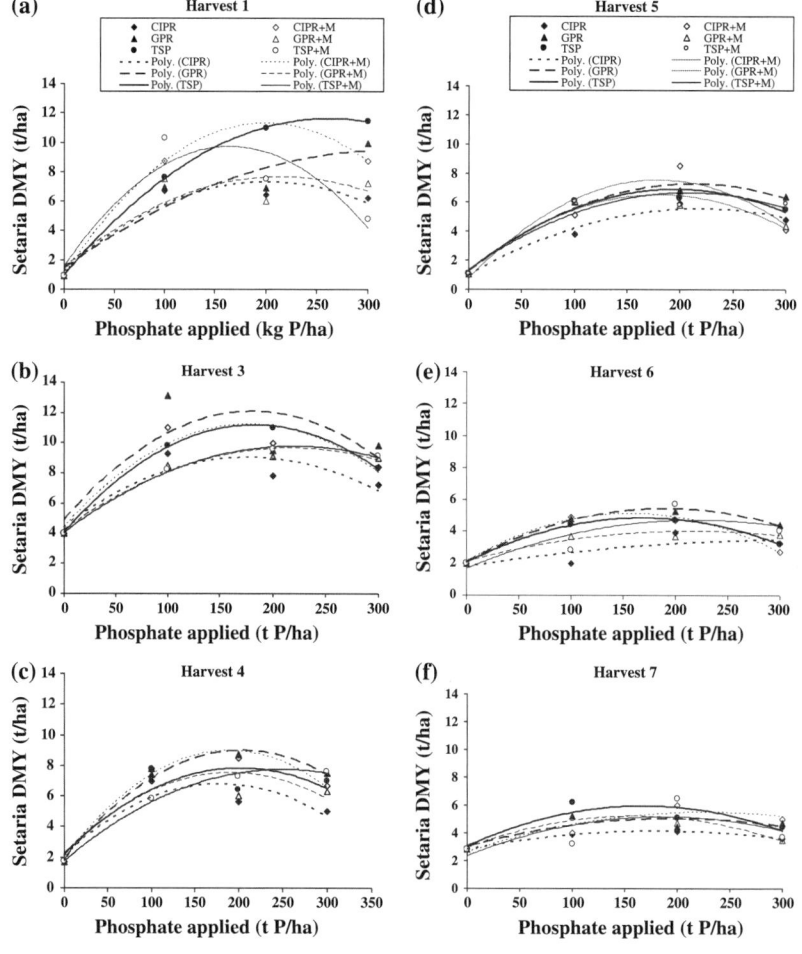

Fig. 3 Individual harvest dry matter yield curves for harvests of 2nd to 14th month

Nutr Cycl Agroecosyst (2010) 88:79–90

relatively constant and lower in the PRs (≈ 7.0 t ha^{-1}). The PRs had relatively lower maximum DMYs probably because available P from the PRs was lower probably below the critical P level required for the higher DMY (11 t ha^{-1}) as achieved in TSP since they were limited in the level of soil P they could rise to due to the low solubility product (Chien and Black 1976). This is supported by the fact that the B1P levels for PRs were about 50% that of TSP (Table 2). Interestingly, manure addition to GPR or TSP at this rate (200 kg P ha^{-1}) resulted in DMY depression to 6.0 and 7.5 t ha^{-1}, respectively (discussed later). Conversely, CIPR/manure combination resulted in increased DMY (11.0 t ha^{-1}).

As P application was increased to 300 kg P ha^{-1}, there was hardly any yield increase in sole TSP and CIPR (Fig. 3). On the other hand in GPR, there was an increase of DMY from 7.0 t ha^{-1} at 200 kg P ha^{-1} to DMY 9.9 t ha^{-1} (~ 3 t ha^{-1} increments). Addition of manure at this P rate led to DMY decline ranging from 2.5 (CIPR) to 6.0 t ha^{-1} (TSP). Some facts gathered from the above observations in harvest 1 (2nd Month) consists of; (a) observed DMYs changes were associated with changes in available P levels as shown by the different responses at the different P application rates and also in responses to P sources/manure combinations which also increased B1P levels (Fig. 2). (b) At 100 kg P ha^{-1} rate, the TSP being a highly soluble P source dissolved rapidly thus giving the highest B1P (Fig. 2) and hence the highest DMY (Fig. 3), whilst the most non reactive P source (CIPR) had the lowest B1P and hence the lowest DMY. This was in accordance to their reactivity (TSP > GPR > CIPR) as also shown by Zaharah and Sharifuddin (2002). (c) Manure/P sources combinations increased B1P (Fig. 2) and hence DMYs in the three P sources at 100 kg P ha^{-1} P rate (Fig. 3). (c) DMY response was a quadratic relationship with a defined maximum value above which, additional increase of the independent variable (P rate) led to decline of the DMY (dependent variable). Mead and Pike (1975) pointed out that for biological processes in general the most commonly used relationship is the quadratic.

In our study, the sole P sources attained maximum DMYs at P rates of 200–300 kg P ha^{-1} (Fig. 3). It was noted that the different P sources attained the maximum DMY at different P application rates (TSP < GPR < CIPR) according to their solubility constants (Chien and Black 1976). On the other hand,

CIPR attained the highest DMY (≈ 7.5 t ha^{-1}) at a P rate of 200 kg P ha^{-1} while, TSP at the same rate (200 kg P ha^{-1}) also yielded the maximum DMY (≈ 11.0 t ha^{-1}) (Fig. 3). The maximum yield attained by the different P sources at the same level of application were in the order TSP > GPR > CIPR implying the maximum DMYs were a function of dissolved P. Non-reactive PR sources such as CIPR may not give high yields as the water soluble P sources even if they are applied at high rates because the solution concentration P of the PRs is controlled by the solubility constant. Consequently, in CIPR unlike TSP/GPR, the maximum yield attained was mostly constrained by the limited dissolution of CIPR in the prevailing conditions. However, an addition of protons through manure mineralization most probably enhanced the PR dissolution (Van Straaten 2002) increasing available P and hence DMY. (v) It follows that, CIPR/manure combination at 200 kg P ha^{-1} increased DMY whilst depressing yields in GPR and TSP. In the latter two sources their combination with manure resulted to higher B1P levels than in CIPR hence the critical P level required for maximum DMY was probably exceeded in TSP and GPR and not CIPP and hence the latter was still in the positive side of the response curve (left hand side), while the other sources were in the negative response part (right hand side). For this reason, TSP and GPR when combined with manure resulted to depressed DMYs at both 200 and 300 kg P ha^{-1}, while a combination of manure and CIPR gave a positive effect at 200 kg P ha^{-1} but a negative effect at 300 kg P ha^{-1}. In the absence of the B1P data and the fact that high P rates were used in this experiment, it would have been difficult to explain the positive and negative effects of manure on the two PRs which has been the case in most literature.

Albeit, the cause of depression of DMY at high P rates in this work was not established but most probably it is associated with antagonistic effects between P and other plant nutrients such as Zn. Some workers reported P-induced Zn deficiency at high P application rates (Nyirongo et al. 1999). In the current work, evidence of antagonistic effects of P on Zn were shown by slightly higher Zn uptake at 100 kg P ha^{-1} than at 300 kg P ha^{-1} application rate (data not shown). Statistical analysis for harvest 1 indicated no significant difference between GPR and TSP but the former was different from CIPR and in turn CIPR and TSP were significantly different.

Nutr Cycl Agroecosyst (2010) 88:79–90

Similar findings that GPR was equally as good as or better than the water soluble, single superphosphate (SSP) were reported by Elhaji (2003).

In harvest 2 (not shown on Fig. 3), the control gave very high DMYs and only P rates and manure addition exhibited significant treatment effects. All the P rates were only significantly different from control. Interestingly manure addition depressed yields compared to control when applied alone and in combination with the other P sources at all rates with an exception of CIPR in which manure increased yields at all rates. The sharp increase in DMYs compared to harvest one and lack of response and negative effects of manure GPR and TSP is probably due to rhizosphere priming effect probably caused by ploughing in of the organic material in the field which had been fallow for a number of years (Kuzyakov 2002). The only evidence of this is the decline of the sodium bicarbonate extractable organic fraction which is associated with rapid mineralization to replenish P in the labile pool (Gikonyo et al. 2006). The fact that manure alone and addition of manure to GPR and TSP had negative effects even at 100 kg P ha^{-1} was evidence that there was additional available P from another source thus raising it above the P requirement for maximum yields. Most likely, even manure alone gave P concentration that was above the critical P requirement hence causing a negative effect.

In subsequent harvests (3–6), differences between the different P sources continued to decline and GPR

gave the highest DMYs whilst CIPR gave the lowest. In addition, maximum DMYs were attained between 100 and 200 kg P ha^{-1} for all the treatments and above that rate there were either no yield increases or the yields were influenced negatively. However, the negative effects particularly of manure integration continued to decline with time. GPR probably gave higher DMYs than TSP with time because TSP when applied quickly dissolved in the soil solution and was rapidly fixed by the soil constituents into less available P compounds whose availability to crops continually declines with time. On the other hand, P from GPR was released slowly as GPR gradually dissolved as evidenced the increase of hydrochloric extractable and residual P from GPR compared to TSP (Gikonyo et al. 2006). However, even the dissolved P from GPR was also converted to less available forms such as the sodium hydroxide fraction which acts as a sink. In fact from harvest 5 to 6, all the fertilizer rates gave maximum yields at 200 kg P ha^{-1} and yields from 100 and 300 kg P ha^{-1} were not significantly different from control. In the final harvest (7), there was no significant difference between all the P rates with control. However, GPR was doing better than TSP though all were not significantly different from one another. Nevertheless, these observations could not be explained by B1P which always exhibited TSP as the source giving the highest available P levels. It has been reported that B1P was better related to yields in the harvest following P application but not subsequent harvests (Gikonyo (2006). That's why B1P was better related to yield in harvest 1 but not the rest. The maximum yields achieved increased up to month 6 harvest and then declined with time (Fig. 3).

Cumulative yields showed similar trends to the individual harvests (Fig. 4). The fertilizers were not significantly different but yields were influenced mainly by an interaction of fertilizer type × P rate × manure ($P = 0.029$). Effects of P sources and manure combination on cumulative yields exhibited a similar trend to the individual harvests (GPR and TSP integration with manure depressing yields).

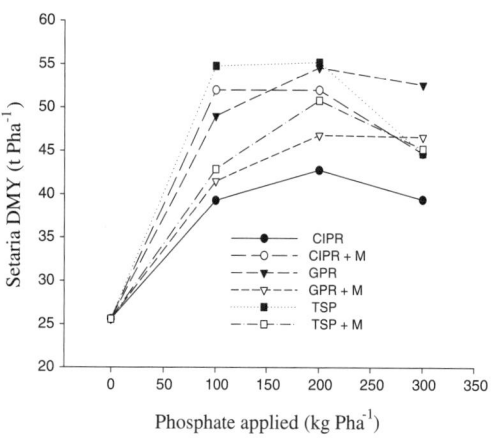

Fig. 4 Cumulative dry matter yields changes with fertilizer type, and aplication rate with or without manure

Conclusions

Phosphate rocks could supplement the expensive TSP without loss of yield. Effectiveness of both reactive

Nutr Cycl Agroecosyst (2010) 88:79–90

and non-reactive PRs was improved by combining them with manure. However, while the potential DMY of non-reactive CIPR was greatly enhanced by combining it with manure, depression of DMYs resulted from TSP/GPR manure combinations due to the high levels of soil solution P leading to nutrient imbalances. Nevertheless, the results of this work will be of great benefit because it has shown that higher yields comparable to ones achieved by TSP could be achieved by not only the reactive PRs but the non-reactive PRs sources. This will be particularly beneficial to organic farming where PRs and manures/composts are the only acceptable P sources. Nevertheless, further research is recommended to determine optimal rates of PRs/manure integration of different PRs/manures of varying qualities for maximum economic benefits. Furthermore, the negative effects of high P application rates and the fast decline of DMYs imply that the P 'recapitalization' strategy is not suitable for the tested soil/crop. The current study indicates that 100–200 kg P ha^{-1} is the optimum rate of P application for several crop harvests. However, there is need for further work in a variety of soils, crops and environments.

Acknowledgments We are grateful to The Institute of Tropical Agriculture for offering a post-doctoral fellowship and to the Third World Organization of Women in Sciences (TWOWS for the Ph.D. fellowship offer to the first author, Esther W. Gikonyo. The research was conducted at the Department of Land Management, Faculty of Agriculture, Universiti Putra Malaysia as part of a doctorate degree.

References

Bray RM, Kutz LT (1945) Determination of total, organic and available forms of phosphorus in soils. Soil Sci 59:39–45

Celi L, Lamacchia S, Barbeis E (2000) Interaction of inositol phosphate with calcite. Nutr Cycl Agroecosys 57:271–277

Chien SH, Black CA (1976) Free energy of formation of carbonate apatites in some phosphate rocks. Soil Sci Soc Am J 40:234–239

Elhaji E (2003) Researh on Gafsa (Tunisia) phosphate rock and its potential for direct application. In: Rajan SSS, Chien SH (eds) Direct application of phosphate rock and related appropriate technology: latest developments and practical experiences. Proc Int Meeting, Kuala Lumpur, 16–20 July 2001. Musle Shoals, USA, IFDC, pp 200–213

Gikonyo EW (2006) Dynamics of current and residual phosphorus in acid tropical soils. PhD Thesis. University Putra Malaysia, Serdang, Selangor, Malaysia

Gikonyo EW, Zaharah AR, Hanafi MM, Anuar AR (2006) Residual values evaluation of different fertilizers at

various rates used in phosphorus recapitalization of an acid tropical soil. J Sci Food Agric 86:2302–2310

Hammond LL, Chien SH, Roy AH, Mokwunye AU (1989) Solubility and agronomic effectiveness of partially acidulated phosphate rocks as influenced by their iron and aluminium oxide content. Fertil Res 19:93–98

Hanafi MM, Syers JK (1994) Agronomic and economic effectiveness of two phosphate rock materials in acid Malaysian soils. Trop Agric (Trinidad) 71(4):254–259

Hanafi MM, Syers JK, Bolan NS (1992) Leaching effect on the dissolution of two phosphate rocks in acid soils. Soil Sci Soc Am J 56:1325–1330

Heng LK (2003) Towards developing a decision support system for phosphate direct application in agriculture. In: Rajan SSS, Chien SH (eds) Direct application of phosphate rock and related appropriate technology: latest developments and practical experiences. Proc Int Meeting, Kuala Lumpur, 16–20 July 2001. Musle Shoals, USA, IFDC, pp 200–213

Ikerra TWD, Mnkeni PNS, Singh BR (1994) Effects of added compost and farmyard manure on P release from Minjingu phosphate rock and its uptake by maize Norwegian. J Agric Sci 8:13–23

Kuzyakov Y (2002) Review: factors affecting rhizosphere priming effects. J Plant Nutr Soil Sci 165(4):382–396

Mead R, Pike DJ (1975) A review of response surface methodology from biometric viewpoint. Biometrics 31: 803–851

Mutert E, Fairhurst T (2003) The use of phosphate rock in tropical upland improvement in Southeast Asia—past experience and future needs. In: Rajan SSS, Chien SH (eds) Direct application of phosphate rock and related appropriate technology: latest developments and practical experiences. Proc Int Meeting, Kuala Lumpur, 16–20 July 2001. Musle Shoals, USA, IFDC, pp 150–162

Nyirongo JCVB, Mughogho SK, Kumwenda JDT (1999) Soil fertility studies with compost and igneous phosphate rock amendments in Malawi. Afr Crop Sci J 7(4):415–422

Oenema O (1980) Combined application of organic manure and phosphate fertilizers, a literature review, Wageninngen university research centre, Wageningen, 50 pp

Owen G (1947) Retention of phosphate by Malayan soil types. J Rubber Res Inst Malaya 12:1

Page AL (1982) Methods of soil analysis, part 2, chemical and micro-biological properties, 2nd edition. Agron Monogr 9 ASA and SSSA, Madison, WI

Rajan SSS (1981) Use of low grade phosphate rocks as a biosuper fertilizer. Fertil Res 2:199–210

Van Heels PAW, Jones D, Finlay R, Godbold DL, Lundstrom US (2005) The carbon we do not see—the impact of low molecular weight compounds on carbon dynamics and respiration in forest soils. A review. Soil Biol Biochem 37:p1–p13

Van Straaten P (2002) Rocks for crops: agrominerals of Sub-Saharan Africa. International Centre for Agroforestry, Nairobi, p 338

Yusda H (2003) Efficiency of phosphorus fertilizer for cucumber (*Cucumis sativa* L.) grown on acid soils. PhD Thesis. Universiti Putra Malaysia

Yusda H, Hanafi MM (2003) Use of phosphate rock for perennial and annual crops cultivation in Malaysia. A

Nutr Cycl Agroecosyst (2010) 88:79–90

review. In: Rajan SSS, Chien SH (eds) Direct application of phosphate rock and related appropriate technology: latest developments and practical experiences. Proc Int Meeting, Kuala Lumpur, 16–20 July 2001. Musle Shoals, USA, IFDC, pp 63–67

Zaharah AR, Bah AR (1997) Effect of green manures on solubilization and uptake from phosphate rocks. Nutr Cycl Agroecosys 48:247–255

Zaharah AR, Sharifuddin HAH (1979) Phosphorus forms, fixation and availability in Malaysian soils. In: Yaacob O, Sharifuddin HAH (eds) Proceedings of a seminar on chemistry and fertility of malaysian soils. Malaysian soil Science Society and UPM, Kuala Lumpur

Zaharah AR, Sharifuddin HAH (2002) Phosphorus availability in an acid tropical soil amended with phosphate rocks. In: IAEA (ed) Assessment of phosphorus status and management of phosphatic fertilizers to optimize crop production. TECDOC-1272, pp 294–303

Zamora-Nahum S, Markovitch O, Tarchitzky J, Chen Y (2005) Dissolved organic carbon (DOC) as a parameter of compost maturity. Soil Biol Biochem 37:2109–2116

Zhang TQ, Mackenzie AF, Liang BC (1995) Long-term changes in Mehlich–3 extractable P and K in a sandy clay loam soil under continuous corn (*Zea mays* L.). Can J Soil Sci 75:361–367

Seedbed Types and Integrated Nutrient Management Options for Cowpea Production in the Southern Rangelands of Semi-arid Eastern Kenya

C.M. Githunguri, A.O. Esilaba, L.M. Kimotho, and L.M. Mutuku

Abstract The southern rangelands of Kenya are difficult environments prone to frequent droughts. The effect of the flat (farmer practice), tied-ridging and contour furrows water harvesting technologies (seedbed types) and five integrated nutrient management practices on the performance of rainfed cowpeas was studied on-farm at the southern rangelands of Kenya during the 2006 short rains season. Cowpea grain yield responded positively under manure at 10 t ha^{-1} irrespective of the water harvesting technology (seedbed) type. There was a negative effect on cowpea grain yields on addition of inorganic fertilizers probably due to the high water-holding property and capacity of manure. The highest cowpea grain yield was recorded in the treatments without inorganic fertilizers. Inorganic fertilizers tend to absorb water from their surroundings as opposed to manure that tend to hold water and release it to the surrounding plants. Tied-ridged seedbed type gave higher cowpea yield than contour furrows and flat seedbeds irrespective of the soil fertility option applied. Cowpea plots under the tied-ridging and contour furrows water harvesting technologies (seedbed types) applied together with manure at 10 t ha^{-1} produced significantly higher yields than those under the farmers practice. As such, if farmers are willing to apply manure at 10 t ha^{-1}, it is upon them to make a choice between the two seedbed types depending on the economic implications of their preparation. Keeping of small ruminants and poultry is one of the main economic activities in this area and as such, limited amounts of manure are readily available and affordable. In the absence of any soil fertility application option, the tied-ridging water harvesting technology should be recommended to farmers in the southern rangelands of Kenya.

Keywords Cowpeas · Drought · Integrated nutrient management · Seedbed types · Soil fertility

Introduction

The southern rangelands, which form part of Kenya's drylands, are fragile environments prone to vagaries of nature (Muga et al., 2007). Livelihood options are limiting and crop production is risky making food insecurity and poverty rampant. Estimates of average annual farm income for a household with an average family size of 5 in the drylands of southeast Kenya, for example, is Ksh. 69,820 per annum (Muga et al., 2007). This gives an annual income level of US $200 per person. Such income level is below the absolute poverty line estimated in 1997 at Ksh. 14,868 per person per annum.

In trying to achieve their economic ends, communities in the drylands engage in different farm activities (GoK, 2004). These include the growing of crops and trees to supplement returns from the main rangeland activities like livestock rearing. The crops and tree types grown are wide and varied depending on the need and preference of the farmer. However, the major impediment to crop and tree production is inadequate soil moisture to allow them reach physiological maturity. The drylands are characterized with low and erratic rainfall and high transpiration rates. In order

C.M. Githunguri (✉)
Katumani Research Centre, Kenya Agricultural Research Institute, Machakos, Kenya
e-mail: cgithunguri@kari.org; cyrusgithunguri@yahoo.com

A. Bationo et al. (eds.), *Innovations as Key to the Green Revolution in Africa*, DOI 10.1007/978-90-481-2543-2_43, © Springer Science+Business Media B.V. 2011

to improve crop and tree production in these areas, sustainable drought and dry spell mitigation farming methods through better on-farm rainwater management are required. The cowpea crop is one of the promising crop technologies which is being upscaled in this region. Intercropping and crop rotation studies showed cowpea had beneficial effects on maize and other crops (Mureithi et al., 1994, 1996; Rao and Mathuva, 2000; Vanlauwe et al., 2005). Under subsistence agriculture, average cowpea grain yields range from 100 to 300 kg ha^{-1} and 1000 to 4000 kg ha^{-1} under good management (CAB International, 2005).

For resource-poor subsistence farmer in the semi-arid areas of eastern Kenya, the principle source of nutrients for their crop is farmyard manure produced on their holdings. Its use provides a means of recycling nutrients and where animals have access to forage outside the croplands, a means of collecting nutrients from surrounding areas. Use of manure or inorganic fertilizer has been found to increase the yield of maize and beans. Higher yields can be achieved when both are combined. In both cases, water harvesting results in substantially more increase in grain yields of the crops (Itabari et al., 2004). In a recent study in the eastern arid and semi-arid zone, an analysis of grain yield data showed that a combination of farmyard manure and run-off harvesting significantly ($P \leq 0.05$) increased grain yield (Itabari et al., 2004). However, furrowing depressed yields probably due to occasional water logging observed while application of farmyard manure alone had no significant effect on grain yield. Application of farmyard manure without run-off harvesting gave fewer net benefits than the traditional practice of growing legume without application of farmyard manure indicating that unless a farmer has farmyard manure, buying manure to produce green grams without water harvesting was unprofitable (Itabari et al., 2004). Most soils in the arid and semi-arid areas of Kenya are fragile and prone to dramatic decline in fertility. Good soil fertility management must therefore accompany the use of soil and water conservation practices to give the farmer good crop yields. Highly fertile soils give high crop yields, good plant cover, and hence minimize the erosive effect of raindrops, runoff and wind. Fertile soils have a stable structure, which does not breakdown under cultivation and a high infiltration capacity. Practical ways of maintaining soil fertility must therefore be developed to enable farmers get the benefits of their

soil and water conservation efforts. The challenge for research is to achieve sufficient understanding of linkages between human actions, arid and semi-arid lands ecosystem processes and human well-being against a backdrop of diversity, vulnerability and transition to produce sustainable management of arid and semi-arid lands. In order to address some of these challenges researcher-farmer managed on-farm trials, and demonstrations were conducted at the southern rangelands (Kathonzweni) in Makueni District in Kenya to study the effect of different seedbed types (water harvesting technologies) and different integrated nutrient management practices on the performance of rainfed cowpeas (var. K80) (*Vigna unguiculata* L.).

Materials and Methods

Trials to study the effect of three water harvesting technologies (seedbed types) and five integrated nutrient management practices on the performance of rainfed cowpeas were conducted on-farm at the southern rangelands of Kenya during the 2006 short rains season. The trial treatments included three seedbed (water harvesting techniques) types: contour furrows, farmer practice (flat seedbed) and tied-ridging; five integrated nutrient management practices: control (without fertilizer), farmyard manure at 10 t ha^{-1}, farmyard manure at 10 t ha^{-1} + 20 kg N ha^{-1} + 20 kg P ha^{-1}, farmyard manure at 5 t ha^{-1} and farmyard at manure 5 t ha^{-1} + 20 kg N ha^{-1} + 20 kg P ha^{-1}. Each treatment was replicated four times in a randomized complete block design (RCBD). The trial was planted at the onset of the 2006 short rains and harvested 4 months later. During harvesting, total fresh weight was taken and the yields sub-sampled where necessary and whole fresh weight taken where the yields were very low, the sub-sample fresh weight was taken. The samples (both top and grain) were oven dried and dry weights taken. The weights taken were used to extrapolate the yields per hectare in each treatment. This was used for water harvesting and integrated nutrient management (INM) response comparison. The data collected were subjected to analysis of variance using the method described by Gomez and Gomez (1984) and the treatments' means were separated using the least significant difference (LSD) test using the SAS statistical package (SAS, 1990).

Results and Discussion

Table 1 shows initial chemical characterization of soils that had been carried out at Kampi ya Mawe, a KARI Sub-Centre situated near the trial site, by Miriti et al. (2007). Soil chemical properties indicated that initial soil N and P at the site were very low. This implies there is need to supplement the nutrients in order to achieve reasonable crop yields (Okalebo et al., 2002). For cowpea yields, water harvesting technologies and INM applications showed significant differences ($P<0.05$) (Table 2). However, interactions between water harvesting technologies and INM applications did not show significant differences between them ($P<0.05$) (Table 3). Tied-ridging gave the highest cowpea yield (1766 kg ha^{-1}) (Table 4) but was not significantly different from contour furrows. Farmers' practice (flat) had the lowest cowpea grain yield (1177 kg ha^{-1}) (Table 4) and was significantly different ($P<0.05$) from the other treatments. For INM

applications, the highest cowpea grain yield (1939 kg ha^{-1}) (Table 5) was recorded under manure at 10 t ha^{-1}, which was significantly different ($P<0.05$) from the other treatments (Tables 4 and 5). This implies that the two water harvesting technologies (tied-ridging and contour furrows) combine well with manure at 10 t ha^{-1} at the southern rangelands. This also gives farmers a choice between the two water harvesting techniques or seedbed types depending on the economic implications of their preparation. Therefore there is a need to carryout an economic analysis of the preparation of the various seedbed types.

When tied-ridging technology was applied, the highest grain yield (2129.6 kg ha^{-1}) was recorded where no INM was applied but was not significantly different from the other treatments (Table 6). Under contour furrows, manure at 10 t ha^{-1} recorded the highest grain yield (2219.5 kg ha^{-1}), which was significantly higher ($P<0.05$) than the other treatments (Table 6). The highest grain yield (1434.6 kg ha^{-1}) under the farmers practice was observed under manure

Table 1 Initial soil chemical properties at Kampi ya Mawe near Kathonzweni in the southern rangelands of Kenya during the 2006 short rains

Soil depth (cm)	pH	N (%)	P (ppm)	K (ppm)	Ca (ppm)	Mg (ppm)	Na (ppm)	CEC (cmol kg^{-1})	Org C (%)
0–20	5.87	0.09	8.34	381.93	765.14	234.45	502.57	6.05	1.16
20–40	5.65	0.08	5.34	273.53	808.06	258.44	451.59	5.76	0.97
40–60	5.70	0.06	2.59	158.75	564.32	275.06	431.99	4.84	0.97

Source: Miriti et al., 2007

Table 2 Analysis of variance of responses of cowpea yields (kg) under different integrated nutrient management and water harvesting technologies at Kathonzweni in the southern rangelands of Kenya during the 2006 short rains

Source of error	DF	SS	MS	F value	Pr > F
Model	9	9,929,951.06	1,103,327.90	2.45	0.01
Block	3	592,639.81	197,546.61	0.44	0.72
Water harvesting technology	2	4,847,647.40	2,423,823.70	5.38	0.01
Integrated nutrient management	4	4,489,663.85	1,122,415.96	2.49	0.05
Error	65	29,284,870.61	450,536.47		
Total	74	39,214,821.67			

$R^2 = 0.25$; CV $= 43.86$; root MSE $= 671.22$; grain mean $= 1530.37$

Table 3 Analysis of variance of responses of cowpea yields (kg) showing interactions between different integrated nutrient management and water harvesting technologies at Kathonzweni in the southern rangelands of Kenya during the 2006 short rains

Source of error	DF	Type III SS	MS	F value	Pr > F
Block	3	592,640	197,547	0.41	0.744
Water harvesting technology	2	4,847,647	2,423,824	5.07	0.009
Integrated nutrient management	4	4,489,664	1,122,416	2.35	0.065
Water harvesting technology × Integrated nutrient management	8	2,058,482	257,310	0.54	0.822

Table 4 Responses of cowpea grain yields (kg) under different water harvesting technologies at Kathonzweni in the southern rangelands of Kenya during the 2006 short rains

Water harvesting technology	N	Means	[a]t grouping
Tied-ridging	25	1765.9	A
Contour furrows	25	1647.9	A
Farmers practice (flat)	25	1177.3	B
LSD$_{(0.05)}$		379.16	

[a]Values with the same letter are not significantly different ($P<0.05$)

Table 5 Responses of cowpea yield (kg) under different integrated nutrient management technologies at Kathonzweni in the southern rangelands of Kenya during the 2006 short rains

Treatment	Means	N	[a]t grouping
Farmers practice (without fertilizer)	1602.9	15	BA
Manure 5 t ha^{-1}	1434.9	15	B
Manure 5 t ha^{-1} + 20 kg N + 20 kg P$_2$O$_5$ ha^{-1}	1190.2	15	B
Manure 10 t ha^{-1}	1939.2	15	A
Manure 10 t ha^{-1} +20 kg N+20 kg P$_2$O$_5$ ha^{-1}	1484.7	15	BA

[a]Values with the same letter are not significantly different ($P<0.05$)

10 t ha^{-1} but was not significantly different ($P<0.05$) from the other treatments (Table 6). Under contour furrows and manure at 5 t ha^{-1} the cowpea yields were significantly lower than the other treatments. On the other hand, the farmers practice (flat) with either levels of manure performed better than those treatments with inorganic fertilizers. The results suggest that the addition of inorganic fertilizers does not have beneficial effects on cowpea yields. It is possible that the addition of inorganic fertilizers could have led to increased vegetative growth at the expense of grain yield, which is important among communities that consume cowpea leaves as vegetables. However, this is subject to further investigations. Biological nitrogen fixation capability

of cowpeas is also another factor that needs to be considered in future trials. It is a well-established fact that manure improves soil texture, aeration, microbial activity that leads to the release of soil nutrients, water-holding capacity, cation exchange capacity; moderates soil temperature; and reduces soil erosion. Probably it is all these beneficial effects from manure or various combinations of them that lead to the improved performance of cowpeas.

Conclusions

Cowpea grain yield at Kathonzweni responded positively under manure at 10 t ha^{-1} irrespective of the water harvesting technology (seedbed) type. Even though the soils were inherently low in N, there was a negative effect on cowpea grain yields on addition of inorganic fertilizers and as such only manure should be applied on cowpeas. Inorganic N could have promoted growth in foliage at the expense of grain filling. The two water harvesting technologies (tied-ridging and contour furrows) combined well with manure at 10 t ha^{-1} at Kathonzweni and as such it is upon farmers to make a choice between the two seedbed types depending on the economic implications of their preparation. In the absence of any fertilizer applications the tied-ridging water harvesting technology should be recommended to farmers at Kathonzweni. There is need to carryout an economic analysis of the preparation of the various seedbed types in future trials.

Acknowledgements The authors would like to acknowledge the director, KARI, and the Desert Margins Programme for Africa for providing funds and support and for granting permission to publish this work. We acknowledge support of Kenya Forestry Research Institute, National Environment Management Authority and Ministry of Agriculture among

Table 6 Responses of cowpea yield (kg) under different integrated nutrient management and water harvesting practices at Kathonzweni in the southern rangelands of Kenya during the 2006 short rains

Treatment	Tied-ridging	Contour furrows	Farmers' practice
Manure 10 t ha^{-1}	2119.48	2219.45	1434.58
Manure 5 t ha^{-1}	1448.56	1554.90	1254.19
Manure 10 t ha^{-1} + 20 kg N + 20 kg P$_2$O$_5$ ha^{-1}	1663.48	1815.63	927.95
Manure 5 t ha^{-1} + 20 kg N + 20 kg P$_2$O$_5$ ha^{-1}	1343.66	1225.99	953.90
Farmers practice (without fertilizer)	2129.57	1436.66	1195.51
Mean	1740.95	1650.53	1153.23
LSD	1113.20	834.18	854.44

others for participating in experimental design, selection of trial farms and providing field technical assistance.

References

CAB International (2005) Crops protection compendium, 2005. CAB International, Wallingford, CT. www.cabicompendium. org.cpc

GoK (2004) National policy for the sustainable development of the arid and semi arid lands of Kenya, 3rd draft, 52 pp

Gomez KA, Gomez AA (1984) Statistical procedures for agricultural research, 2nd edn. Wiley, New York, NY, 680 pp

Itabari JK, Nguluu SN, Gichangi EM, Karuku AM, Njiru E, Wambua JM, Maina JN, Gachimbi LN (2004) Managing land and water resources for sustainable crop production in dry areas: A case study of small-scale farms in semi-arid areas of eastern, central and rift valley provinces of Kenya. In: Linda C (ed) Proceedings of agricultural research and development for sustainable resource management and food security in Kenya. Agriculture/livestock research support programme, phase II. End of programme conference, KARI, Nairobi, 11–12 Nov 2003

Miriti JM, Kironchi JO, Gachene CCK, Esilaba AO, Wakaba P (2007) Effects of water conservation tillage on water and nitrogen use efficiency in Maize-Cowpea cropping systems in semi-arid Eastern Kenya. A paper presented at the KARI mini-conference, Kenya Agricultural Research Institute, Nairobi, Kenya, 5–7 Nov 2007

Muga M, Wekesa L, Mutunga C, Muchiri D, Ng'ethe R (2007) Desert margins programme-acacia operations project demonstration sites on mechanised water harvesting in the southern rangelands of Kenya, 2006 Annual report, 13pp

Mureithi JG, Tayler RS, Thorpe W (1994) The effects of alley cropping with *Leucaena leucocephala* and of different management practices on the productivity of maize and soil chemical properties in lowland coastal Kenya. Agroforest Syst 27(1):31–51

Mureithi JG, Tayler RS, Thorpe W (1996) Effect of dairy cattle slurry and intercropping with cowpea on the performance of maize in coastal lowland Kenya. Afr Crop Sci J 4(3):315–324

Okalebo JR, Gathua KW, Woomer PL (2002) Laboratory methods of soil and plant analysis: a working manual, 2nd edn. TSBF-CIAT and Sacred Africa, Nairobi, p 128

Rao MR, Mathuva MN (2000) Legumes for improving maize yields and income in semi-arid Kenya. Agric Ecosyst Environ 78(2):123–137

SAS (1990) Statistical analysis system users guide, vol. 1, ACECLUS-FREQ version 6, 4th edn. SAS Institute, Cary, NC, 890pp

Vanlauwe B, Diels J, Sanginga N, Merckx R (2005) Long-term integrated soil fertility management in South-western Nigeria: crop performance and impact on the soil fertility status. Plant Soil 273(1/2):337–354

Land and Water Management Research and Development in Arid and Semi-arid Lands of Kenya

J.K. Itabari, K. Kwena, A.O. Esilaba, A.N. Kathuku, L. Muhammad, N. Mangale, and P. Kathuli

Abstract Increasing demographic pressure in the arid and semi-arid lands (ASALs) in Kenya has resulted in the use of non-sustainable farming practices and subsequent environmental degradation, characterized by declining soil fertility, widespread land degradation and loss of biomass and biodiversity. Crop production is low and insufficient to meet the food demands of the increasing population. Thus, food insecurity is a major threat to the livelihoods of the resource-poor smallholder subsistence farmers in these areas. Due to the limited possibilities of increasing the area under cultivation, the solution to meeting the demand for food and surplus for sale by the rapidly increasing population lies in increasing productivity through development of simple, effective and sustainable integrated soil fertility and water management technologies. Consequently, a lot of research work aimed at developing appropriate soil and water management technologies for these areas has been undertaken by research and development organizations in Kenya. This chapter reviews, summarizes and highlights the major findings of research on options for improving soil fertility and for conserving soil and water in the ASALs. Socioeconomic factors are considered as well as their interaction with intervention measures such as tillage, water harvesting, soil conservation, integrated nutrient management, soil and water

and crop management. The knowledge gaps in the various technological options tested are identified, and future research needs are suggested.

Keywords Runoff · Soil fertility · Soil water · Tillage · Water harvesting

Introduction

The arid and semi-arid lands (ASALs) of Kenya occupy agroclimatic zones IV–VII (Fig. 1) and account for 83% of the land area containing over 25% of the human population and over 50% of the livestock population. The semi-arid region occupies agroclimatic zones IV and V, with annual rainfall ranging from 550 to 900 mm, while the arid region occupies agroclimatic zones VI and VII, with annual rainfall rarely exceeding 300 mm and evapotranspiration regularly exceeding 3,000 mm (Sombroek et al., 1982; Jaetzold and Schmidt, 1983).

The major soil types are Luvisols, Acrisols and Vertisols. The semi-arid lands are characterized by mixed farming systems while the arid lands are characterized by intensive and semi-intensive grazing systems. The major crops grown include maize, sorghum, millet, beans, pigeon peas, cowpeas, citrus and mangoes. Livestock kept include cattle, camels, goats, sheep and donkeys.

A dominant feature of agricultural production systems in these regions is land degradation and low crop yields caused by intensification of land use without the use of appropriate land and water management practices (Nandwa et al., 2000). McCown and Jonnes (1992) have described the situation currently evident

J.K. Itabari (✉)
Katumani Research Centre, Kenya Agricultural Research Institute, Machakos, Kenya
e-mail: itabarijustus@yahoo.co.uk

A. Bationo et al. (eds.), *Innovations as Key to the Green Revolution in Africa*,
DOI 10.1007/978-90-481-2543-2_44, © Springer Science+Business Media B.V. 2011

Fig. 1 Agroclimatic zones of
Kenya (adapted from
Sombroek et al. 1982)

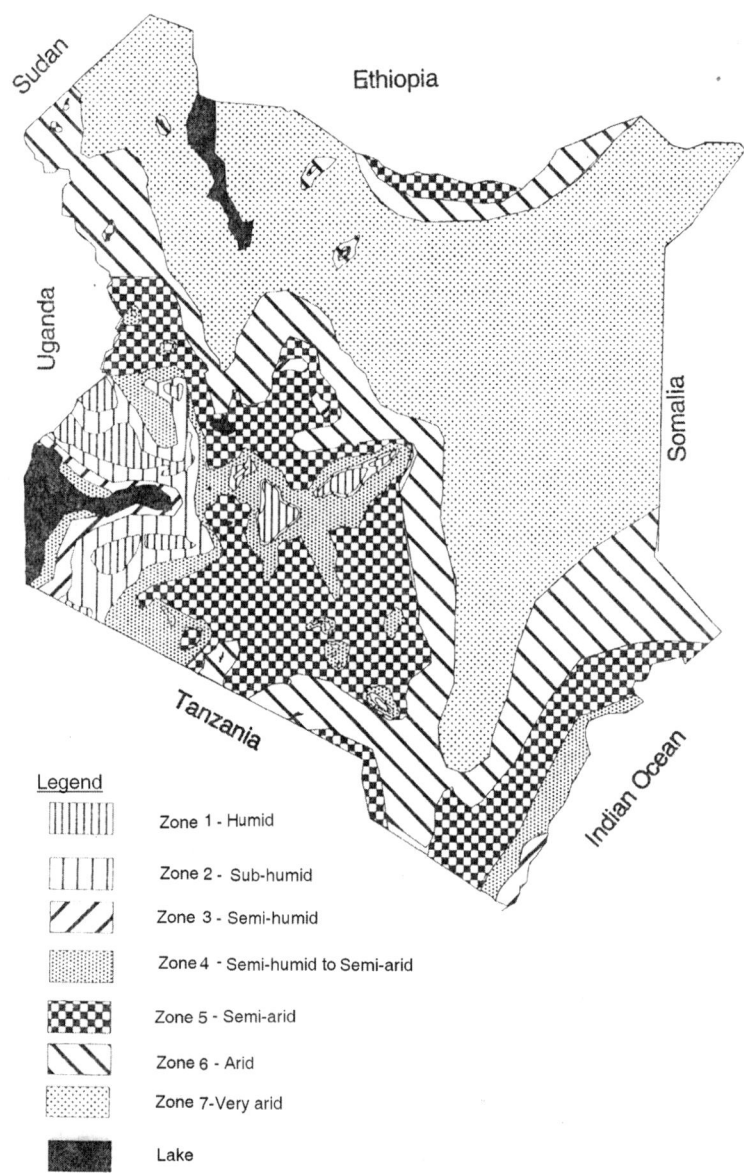

Legend

Zone 1 - Humid

Zone 2 - Sub-humid

Zone 3 - Semi-humid

Zone 4 - Semi-humid to Semi-arid

Zone 5 - Semi-arid

Zone 6 - Arid

Zone 7-Very arid

Lake

in most farms in these regions as a "poverty trap", in which the highly subsistent population living on degraded soils receives low income, affords low or no farm inputs and consequently gets low crop yields.

Numerous land and water management interventions have been suggested to alleviate these problems, but the level of adoption has remained very low due to a variety of reasons (Simpson et al., 1996). Further, studies have shown that those who attempt to adopt these measures do so in piecemeal and hence do not realize their full benefits. For instance, while it is true that farmers in these regions often adopt new crop varieties, a majority of them consistently ignore recommendations on improved land and water management technologies. Consequently, many farmers only achieve a small proportion of the potential productivity gains possible from adoption of new crop varieties. The net effect has been unabated decline in soil productivity and crop yields (Simpson et al., 1996).

This chapter highlights land and water management technologies so far developed for the ASALs. The main objective of this chapter is to expose and summarize a range of soil fertility and soil water management options which will empower farmers to adopt a specific technology. It is envisaged that the review will guide researchers and extension agents in their efforts to find the way forward towards improvement in soil fertility and soil water management and enhanced food security in Kenya's ASALs.

Soil Fertility Management

Most of the diagnostic surveys conducted in the ASALs of Kenya rank low soil fertility as the second key constraint to crop production after low soil moisture (Itabari et al., 2004). Soils in these regions are low in organic matter and deficient in the essential plant nutrients, notably nitrogen (N) and phosphorus (P) (Nadar and Faught, 1984; Okalebo et al., 1992). In addition, other studies have indicated that with intensive cropping, especially in sandy soils, there is a possibility that sulphur may become a limiting nutrient (Okalebo et al., 1992). Consequently, crop yields are very low even when rainfall is non-limiting. Declining soil fertility has been attributed to continuous cultivation without adequate replenishment of nutrients and loss of nutrients through erosion and leaching (Gachimbi et al., 2005). The outflow of nutrients in most farms in these regions far exceeds input flows (Gachene et al., 2000; Gachimbi et al., 2005). To address this problem, several options have been tried. Results of the various options are reviewed under the relevant sections.

Organic Fertilizers

Organic fertilizers that have been investigated in the ASALs include farmyard manure, compost and green manure cover crops. Farmyard manure (FYM) is the most widely used soil fertility input in semi-arid Kenya. Available estimates indicate that about 88% of the resource-poor subsistence farmers in semi-arid eastern Kenya use it as their principal soil fertility input (Omiti et al., 1999). However, due to diminishing herd sizes and poor manure management practices, the quantities produced are insufficient and of poor quality. Most of it contains only one-third of the N and P expected from fresh animal manure (Probert et al.,

1992). As a result, most farmers apply sub-optimal quantities. Nevertheless, studies in these regions have shown that FYM, when used judiciously, increases crop yields tremendously (Ikombo, 1984). They are, however, not unanimous on the optimal rate of application. For instance, in studies at Ithookwe and Kampi ya Mawe in eastern Kenya, Ikombo (1984) got yields similar to those obtained from the standard rate of mineral fertilizer (40 kg/ha of N and 17 kg/ha of P) with FYM at a rate of 8 t/ha. According to Ikombo (1984), 8 t/ha was capable of supplying 130 kg N and 17 kg P, which is adequate for a maize crop, despite the fact that not all these amounts would become available to the crop in the first season. Working in the same region, but at a different location, Kilewe (1987) got a crop as good as that reported by Ikombo (1984) with 8 t/ha from the highest input of fertilizer (capable of supplying 120 kg N/ha and 17 kg P/ha) with FYM at a rate of 40 t/ha. But, in a related study conducted at Machanga in eastern Kenya, Ngoroi et al. (1994) recommended a rate of 10 t/ha/year. This disparity in application rates may be attributed to variation in the quality of manures used, soil characteristics and the relative availability of this resource in the respective study sites. Studies on the residual effects of manure in these regions have found residual manure to give significant but less yields than fresh application (Probert et al., 1992; Watiki et al., 1999). This effect may last three to five seasons after the last application. However, to achieve maximum benefits from FYM, farmers are advised to spread and incorporate it into the soil at least 2 weeks before planting to allow the process of nutrient release to be initiated to benefit the early stages of crop growth and to reduce nutrient losses through volatilization.

Studies have shown that compost is even more suited to ASALs than a combination of farmyard manure and diammonium phosphate (DAP). In a study on maize in Machakos, for instance, Diop et al. (1997) found compost to outperform a combination of farmyard manure and DAP. It gave a higher maize yield, greater net cash benefits and a better return on labour. The maize yielded 2,018 kg/ha with the combination of 17 t/ha FYM plus 57 kg/ha DAP and 2,449 kg/ha when treated with 16.2 t/ha compost. However, its widespread adoption is constrained by scarcity of labour, shortage of organic materials, insufficient cattle, lack of transport, limited water supply and lack of technical know-how on large-scale production of high-quality compost (Hamilton, 1997; Onduru et al., 1999).

Gachene et al. (2000) found that incorporating green manure cover crops biomass increased maize yield substantially followed by mulching. The lowest yield was obtained when biomass was removed from the field. Thus, the average yields were in the following order: incorporation > mulching > removal. However, they noted that in seasons of low rainfall, the trend especially for *Mucuna pruriens* changed, the mulching treatment gave the highest grain yield (4.01 t/ha), followed by the incorporation treatment (1.90 t/ha), and the removal treatment gave the lowest yield (1.51 t/ha). The increase in yields in the mulched plots was attributed to soil moisture conservation.

Inorganic Fertilizers

Application of N and P fertilizers in the ASALs has shown very positive responses, particularly in cereals. For instance, working in eastern Kenya, Okalebo et al. (1980) found that application of N and P fertilizers at rates of 60 and 40 kg/ha, respectively, significantly increased maize grain yield. Subsequent studies at Katumani Dryland Research Station found 60 kg N/ha and 9 kg P/ha to be the optimal N and P fertilizer rates (Nadar and Faught, 1984). However, working in the same locality, Okalebo and Gathua (1987) found application of N fertilizer at rates of 60–120 kg N/ha to significantly increase maize grain yields while P fertilizer addition did not. Further, they found combined N and P applications from 30 to 106 kg N/ha and from 6 to 52 kg P/ha to significantly increase dry matter and P uptake by maize and sorghum crops during the vegetative stages of growth. In a survey conducted in semi-arid eastern Kenya, Okalebo and Simpson (1990) found available P levels to be very site specific, and 60% of the farms surveyed had available P levels of below 10 mg/kg. The survey also revealed that in 5 out of 12 test sites characterized by low P levels, application of fertilizer at rates equivalent to 60 kg N/ha and 17 kg P/ha gave maize grain yield increases of up to 283% above the control. However, working in Machakos District, Australian Centre for International Agricultural Research (ACIAR) researchers recommended a fertilizer application rate of 30–40 kg N/ha and 4 kg P/ha for maize (Lee, 1993). This variability

in fertilizer application rates may be attributed to differences in soil types and their characteristics, cropping history and the level of soil degradation in the respective farms or test sites.

The high cost of phosphate fertilizers has led researchers to undertake research into cheaper sources of P. For instance, Probert et al. (1992) compared the response of maize to three different P sources, prepared from Minjingu rock phosphate: beneficiated rock, single superphosphate and partially acidulated rock phosphate (PARP) in a semi-arid area of eastern Kenya, where the soil pH was 6.1. They found that the performance of the PARP was virtually identical to that of single superphosphate and that the rock phosphate source was less effective; it was unable to supply more than the equivalent of 12.5 kg/ha of P applied as superphosphate no matter how high a rate of application of the material was used.

Adoption of inorganic fertilizers is very low, mainly due to cash flow problems and lack of appropriate incentives. For instance, estimates available indicate that in Machakos District about 60% of farmers still do not use fertilizer, and among the users, 50% apply less than the recommended rate of the input (Omiti et al., 1999).

Biological Nitrogen Fixation (BNF)

Biological nitrogen fixation provides a cheap option for improving soil nitrogen status among most resource-constrained farmers, in both Kenya and the entire sub-Saharan Africa. Utilization of legume-fixed N by cereal crops is best accomplished in inter- and rotation cropping systems. For instance, in a study to investigate the effect of beans, cowpea, tepery beans and pigeon pea on the yield of associated and subsequent maize in intercropping and rotation systems, Nadar and Faught (1984) recommended the use of a maize–bean sole crop rotation to improve maize production as well as total crop production under the marginal rainfall conditions of semi-arid eastern Kenya. Simpson et al. (1992) showed that rotating a *Lablab*/pigeon pea or a cowpea/pigeon pea combination with maize or sorghum on an infertile soil gave an extra grain production of 300–400 kg/ha. The legumes contributed about 40 kg N/ha.

In another study, Nadar (1984) established that maize benefited more from fixed N when intercropped with beans, cowpeas or pigeon peas. Such systems were capable of increasing maize yields from 80 to 480 kg/ha (Lee, 1993). Most farmers in ASALs practise intercropping. However, not all soils are endowed with bean rhizobia (*Rhizobium phaseoli*). Inoculating beans with the right rhizobium strain has been shown to enhance germination (Kibunja et al., 1999) and to increase yield (Nadar and Chui, 1984). Further, studies have shown that inoculation and fertilization with 5–10 kg S/ha enhance N_2 fixation, especially in groundnuts (Karanja, 1985). But in spite of its apparent viability, BNF uptake by resource-poor smallholder farmers has remained very low, presumably due to lack of awareness, scarcity of land to practise crop rotation and lack of access to the inoculum.

Combining Organic and Inorganic Fertilizers

Combinations of FYM and chemical fertilizers appear to give better financial returns than either component alone, especially when combined with water harvesting. This view is supported by Itabari et al. (2004) who in a study in semi-arid Kenya on enhancing fertilizer use efficiency through water harvesting obtained much higher maize yields when manure was supplemented with N or with compound fertilizer (Table 1). Similar trends were observed in the same region by Ngoroi et al. (1994). Miriti et al. (2007), also working in the same region, reported a significant interaction between manure and nitrogen fertilizer, which gave higher maize stover yields. However, farmers are yet to fully embrace this concept. Most of them consider these inputs as substitutes rather than complementary and hence do not fully exploit the positive synergies that would optimize nutrient supply to crops.

Soil and Water Management

Soil and runoff losses cause nutrient depletion and decreased soil water availability, leading to decreased crop and pasture production. Measures aimed at appropriate soil and water conservation for increasing agricultural productivity on a sustainable basis are therefore imperative. To this end, a number of technologies have been developed. These technologies are reviewed under the relevant sections.

Rainwater Capture and Soil Conservation

Tillage

A number of investigators have studied the effects of various tillage methods on crop yields in the dry areas of Kenya. In his pioneering tillage studies in semi-arid eastern Kenya, Dagg (1969) reported that tillage experiments at Katumani showed little or no variation in soil moisture status under flat or tied ridge cultivation. Gicheru (1990), working on a clay soil (ferric Acrisols) in a semi-arid highlands area of Kenya, reported that tied ridging conserved the lowest amount of water and produced the lowest yields. He attributed this to no runoff to impound and high soil evaporation losses due to increased soil surface area. However, M'Arimi (1978) reported from a tillage study on an Alfisol (chromic Luvisol) in semi-arid eastern Kenya that soil water content, at the end of the rainy season, was highest under tied ridging and that significantly higher dry matter and grain yields of maize were obtained from the tied ridged plots (Table 2).

Table 1 The effect of soil fertility improvement on net benefits from maize (averaged over four clusters)

Fertility levels/ha	Net benefits (KShs/ha)/(US$/ha)		
	+Water harvesting	−Water harvesting	Difference
0 FYM	4,880 (74)	3,864 (59)	1,016 (15)
10 t FYM	8,245 (125)	4,354 (66)	3,891 (59)
20 t FYM	10,675 (162)	6,372 (97)	4,303 (65)
20 kg N	8,988 (136)	6,956 (105)	2,032 (31)
20 kg N + 9 kg P10 t FYM + 20 kg N	13,340 (202)	10,244 (155)	3,096 (47)
10 t FYM + 20 kg N + 20 kg	15,604 (236)	9,942 (150)	5,662 (86)
P_2O_5	19,166 (290)	12,710 (192)	6,456 (98)

Source: Itabari et al. (2004); values in brackets are equivalents in USD

Table 2 Effect of tillage methods on crop yield

Period	Crop	Crop yield (kg/ha)			
		Minimum tillage	Conventional tillage	Tied ridging	SE[a]
Long rains	Maize (dry matter)	1,068	1,047	1,105	±63
Short rains	Maize (dry matter)	2,040	1,920	1,760	±0.09
	Maize (grain yield)	337	221	513	±51

Source: M'Arimi (1978)

[a]Standard error

Itabari et al. (1998) reported that tied ridging under improved soil fertility conditions significantly increased grain yield of sorghum. Gichangi et al. (2003) also reported that tied ridging increased maize yield in a semi-arid area of eastern Kenya and bean yield in a semi-arid highlands area of central Kenya. Njihia (1979), working on an Alfisol in semi-arid eastern Kenya, investigated the effects of tied ridges, conventional tillage, crop residue mulch and farmyard manure on soil and water conservation. The results showed that maize stover effectively controlled runoff through increased surface storage, which in turn increased infiltration opportunity. Maize stover also minimized evaporation, surface sealing and crusting. Tied ridges effectively controlled runoff even from a large storm of 70 mm/day. Conventional tillage with or without farmyard manure resulted in about 40% runoff loss. A crop of maize was realized from tied ridged and maize stover mulch plots for a season of 171 mm whereas no yield was obtained from the conventional tillage plots with or without farmyard manure. Gicheru et al. (2004), working on a ferric Lixisol, in semi-arid eastern Kenya, reported that manure and mulching with minimum tillage had a greater effect on the water balance of crusted soils, maize emergence and yield. Application of manure and mulch increased rainfall infiltration rates and the amount of soil water stored in the 0–25 cm depth.

In a study to evaluate different tillage equipment on an Alfisol (chromic Luvisol) in semi-arid eastern Kenya, Muchiri and Gichuki (1983) reported that contour furrows were effective in controlling surface runoff, thereby conserving soil moisture. Kilewe and Ulsaker (1984), also working on the same Alfisol and in the same region, showed that conventional contour furrows, wide furrows and mini benches retained all the runoff and resulted in significantly higher water storage capacity than the flat tillage, which in turn resulted in increased yields of maize and water use efficiency. The wide furrow tillage produced significantly greater yields than all the other treatments (Table 3).

Kilewe and Mbuvi (1988) studied the effects of crop cover and residue management on runoff and soil loss. They reported that maize with minimum tillage reduced runoff by 0.8 and 39.2%, while application of 3 t/ha of maize residue reduced runoff by 58.7 and 78.6% during the 1983 long and short rains, respectively, as compared with that produced on bare fallow. During the two seasons, maize with minimum tillage reduced soil loss by 53 and 58.7%, while application of 3 t/ha of maize residue reduced soil loss by 94.4 and 64.1%, respectively. In the 1984 short rains and the 1985 long rains, maize with conventional tillage reduced runoff by 20.1 and 25.1%, beans alone reduced runoff by 41.1 and 29.7% while maize intercropped with beans on alternate rows reduced runoff by 42.0 and 29.2%, respectively, as compared to that produced on bare fallow. During the two seasons, reductions in soil losses were 57.9 and 51.8% by beans alone, 47.5 and 22.3% by maize intercropped

Table 3 Effects of tillage methods on water use, grain yield and water use efficiency of maize

Treatment	Total water use (mm)		Maize yield (kg/ha)		Water use efficiency (kg/ha/mm)	
	Short rains	Long rains	Short rains	Long rains	Short rains	Long rains
Flat	521.2	359.3	3,722	256	7.1	0.7
Conventional furrows	506.2	368.8	5,242	725	10.4	2.0
Wide furrows	509.2	351.4	5,458	844	10.7	2.4
Mini bench	524.2	370.1	4,680	643	8.9	1.7

Source: Kilewe and Ulsaker (1984)

with beans in alternate rows and 39.5 and 8.1% by maize with conventional tillage, respectively.

In a study to evaluate the effects of three tillage methods (conventional, strip and minimum) on the performance and yield of maize in a medium rainfall area of Kenya (1,081 mm per annum), where the soils are brown clays (eutric Nitisols) with a low organic matter content, Ngugi and Michieka (1986) showed that conventional tillage gave the highest yields during the two seasons of experimentation. Okwach and Simiyu (1999) studied sustainability of traditional and improved land management practices in a semi-arid area of Machakos District, eastern Kenya, for 19 seasons. They reported that both runoff and soil loss decreased as cover increased and that mulch cover was more effective than the growing crops' canopy. Mulch was more effective in soil loss reduction than in runoff reduction, and with additional nutrients, crop yield increased with cover. Conventional tillage was superior to minimum tillage in terms of runoff and soil erosion reduction. In on-farm studies carried out over a period of four seasons in semi-arid eastern Kenya, Biamah and Nhlabathi (2003) reported that subsoiling/ridging increased maize yields by, on average, 23% and biomass yield by, on average, 11% than conventional tillage. Mutua (2005) showed that direct planting on previously subsoiled plot had the highest maize grain yield at much lower production costs.

Agroforestry

The effectiveness of agroforestry in reducing runoff, soil erosion and soil evaporation has been investigated by a number of workers. In a study to assess the possibilities of improving rainwater use through various conservation measures (traditional or local, ridging, mulching and agroforestry), Liniger (1989) reported that runoff loss in the traditional practice was over 50% during heavy storms and that ridging did not fully control runoff. No runoff occurred under mulching and agroforestry (with mulching). Water loss through soil evaporation was higher under ridging and lowest under mulching. Mulching increased maize yield by 340% compared to that of the local method.

Kinama et al. (1999) investigated the effectiveness of contour hedgerows, grass strips and mulches in controlling soil erosion and runoff in a maize/senna agroforestry system in semi-arid eastern Kenya. They reported that mulch combined with hedgerow was most effective in controlling soil erosion and runoff compared to the sole crop (control), mulch, hedgerow and grass strip. Maize grain yields were, however, higher in the control plots than in the hedged, hedge and mulch, and grass strip plots. This was attributed to competition by the senna trees and grass with maize for water, light and nutrients.

Fallowing

Research to determine the effectiveness of fallowing in crop production in dryland areas has received very little attention. Studies conducted at Katumani (Whiteman, 1984), using a bare fallow, showed that it was possible to obtain a fivefold increase in sorghum grain yield after a 7-month fallow (land fallowed during the long rains cropping season – October/November to March) (1,720 kg/ha) compared with a virtually failed crop (320 kg/ha) after cereals and intermediate yield (881 kg/ha) after beans in a season that provided only 201 mm of rainfall. Njihia (1979), working in semi-humid and semi-arid Kenya, reported that the efficiency of moisture conservation by maintaining a clean fallow decreased from semi-humid to semi-arid areas. He concluded that the amount of water lost by evaporation from the clean fallow is sufficient to carry a short-maturity low-water demanding crop to maturity and increase a farmer's earnings. He recommended that a cropped fallow would be an attractive alternative to a clean fallow and that the crop should ideally be shallow rooted to exploit the moisture in the top 40–60 cm that is lost through evaporation in a clean fallow.

Water Harvesting

A number of investigators have attempted to determine the effectiveness of various water harvesting techniques on crop production in the semi-arid areas of Kenya. Their findings are reviewed under the various water harvesting techniques.

Modified (Enlarged) Fanya Juu Terraces

Studies have shown that these structures are very effective in controlling soil erosion but not appropriate for water conservation as runoff is impounded on only a small portion of the terrace, benefiting only a few rows of the crop (Kiome, 1992). However, crops grown on the trench have been shown to benefit greatly from the water retained there. The technique has been reported to be marginally economical in the long run if combined with soil fertility improvement (Kiome and Stocking, 1995). It has been widely adopted in the semi-arid areas despite its high labour requirement.

Semi-circular Bunds (Hoops)

Semi-circular bunds have been tested at various spacings for the rehabilitation of degraded grazing lands (Smith and Critchley, 1983; Kitheka et al., 1995) in the semi-arid areas of Kenya. Experimental results showed that restoration of the productivity of the degraded grazing lands could be achieved within three seasons. Adoption of the technique has been hampered by its high labour requirements.

Micro-catchment (Runoff–Runon)

Gibberd (1995), working in a semi-arid area of eastern Kenya, reported that runoff harvesting using a catchment to cultivated area ratio (C:CA) of 1:1 increased yields of most dryland crops by 30–90% (Table 4).

Similar beneficial effects of the technique have been reported elsewhere (Itabari et al., 2004; Critchley,

Table 4 Grain yields of various dryland crops using a catchment to cultivated area (C:CA) ratio of 1:1 runoff harvesting system at Machanga, Mbeere District

Crop	Control (no runoff harvesting)	1:1 C:CA runoff harvesting system
Pearl millet	665	998
Green gram	450	568
Sorghum	1,232	1,562
Cowpea	1,064	1,464
Maize	1,647	2,099
Sunflower	272	512
Cotton	757	1,128

Source: Gibberd (1995)

1989; Imbira, 1989). Despite the obvious benefits, its adoption has been hampered by its non-conformity with the traditional practice of planting the whole cultivated area, i.e. farmers are not willing to leave a portion of their land unplanted.

Zai Pits

The technique has been tested in semi-arid eastern Kenya (Mellis et al., 1997; Itabari et al., 1998) and has shown improved yields. However, adoption of the technique has been limited to a few farmers growing vegetables, e.g. tomatoes and kale, usually in various modifications, such as round and rectangular pits.

Weed Control

Weeds compete with crops for water, resulting in lower water use by the crops and hence lower yields. Makatiani (1970a), working with maize in semi-arid eastern Kenya, reported yield losses of 79 and 99% at Kampi ya Mawe and Katumani, respectively, due to weeds during a season with below average rainfall. Yield losses at the two sites during a season with above average rainfall were 31.7% for Kampi ya Mawe and 72% for Katumani (Table 5).

Plant Spacing and Population

Under the unreliable rainfall conditions of semi-arid eastern Kenya, Nadar (1984) found that planting maize at 75 cm row spacing would optimize maize yields under almost all rainfall conditions tested. The optimum population to be planted under favourable rainfall conditions was found to be around 70,000 plants per hectare. Under less than favourable rainfall conditions, 20,000 plants per hectare or less would produce the highest yields. In a "response farming" study using a crop model, CMKEN, Wafula et al. (1992) concluded that the additional benefit gained using this strategy even with ideal forecast was small compared with large benefits to be gained from using a standard plant population of three to four plants per square metre with a matching fertilizer rate of 30–40 kg N/ha.

Table 5 Effects of weeds on maize yields (kg/ha) at two locations in the short rains (SR) 1969/70 and the long rains (LR) 1970

Treatment	Kampi ya Mawe		Katumani	
	SR 1969/1970	LR 1970	SR 1969/1970	LR 1970
Control (no weeding)	130	1,960	20	1,170
Clean weeding for the first 3 weeks	740	2,960	1,990	3,870
Clean weeding after the first 3 weeks	560	2,900	1,600	3,400
Clean weeded throughout	619	2,870	2,330	4,240

Source: Makatiani (1970a)

Table 6 Reduction of maize yields (kg/ha) due to late planting at Katumani Research Station in the short rains 1969/1970 and long rains 1970

Season	Dry planted	Planted 3 weeks after the onset of the rains
Short rains 1969/1970	1,900	450
Long rains 1970	3,740	2,940

Source: Makatiani (1970b)

Time of Planting

In dryland crop production, time of planting has a significant effect on the optimization of soil water use. In a study conducted in semi-arid eastern Kenya, Makatiani (1970b) showed that delaying planting for 3 weeks after the onset of the rains reduced maize yields by 76% during a season in which rainfall was below average and by 21% during a season in which rainfall was above average (Table 6).

Conclusions

Due to the high cost of inorganic fertilizers and their unavailability at farm level, farmyard/boma manure produced on small holdings is the principal source of nutrients for crop production. Enhancing integration of crop–livestock systems, therefore, has the potential to increase productivity of the farms in a sustainable manner. To enhance this integration, future research should focus on identifying feed resources that will provide additional benefits to the farmers and strategies that will mitigate nutrient losses during handling and storage of manure.

Faced with declining livestock numbers, smallholder farmers are increasingly taking up composting as an option for augmenting their dwindling quantities of farmyard manure. However, the compost produced is of low quality, i.e. contains low levels of N and P. To enhance the role of compost in soil fertility management, future research should focus on identifying suitable composting materials, establishing the fertilizer equivalents of composts from various sources and strategies for improving the quality of composts.

Past research on integrated soil fertility management (ISFM) strategies has focused mostly on combinations of organic and inorganic fertilizers. Future research on ISFM should aim at combinations of many strategies to optimize all aspects of nutrient cycling.

Though research in the ASALs has shown that incorporating green manure cover crop (GMCC) residues increases crop yields, there has so far been no interest in this technology, possibly because the GMCCs that were evaluated were non-food legumes. To enhance adoption of this technology, future research should focus on identifying green manure legumes that can be used as food and green manure legumes that are less competitive with food crops for light, water and nutrients.

Tillage practices in the ASALs should take cognizance of the fragile soils and low rainfall in this environment. Future research on tillage should, therefore, focus on practices that improve management of soil and water resources to increase their productive capacity.

Agroforestry requires extra labour to establish tree/shrub seedlings, to prune them and to spread the prunings on the soil. The study in semi-arid eastern Kenya also showed a reduction in maize yield. There is, therefore, a need to find out if the reduction in crop yields and the extra labour required are worth the long-term benefits. The prunings, which are used as mulch, can also be used as fodder for livestock to increase milk, meat and manure production. The economic profitability of using the prunings as fodder or mulch needs to be investigated. There is also a need to identify trees/shrubs that are less competitive with crops for water and nutrients and to quantify the effect of row orientation on competition for light.

Water harvesting is an important strategy for increasing water availability to crops. However, harvesting of runoff of rainwater depends on rainfall; hence crops experience water stress during long dry spells, which are common during the rainy seasons. One way of overcoming this problem is to harvest and store runoff for supplementary irrigation. Future research on rainwater/runoff harvesting should, therefore, emphasize development of improved rainwater harvesting techniques and storage structures for supplementary irrigation during dry spells or for intensive forms of production such as kitchen vegetable gardens.

Survey reports indicate that adoption of improved soil and water management technologies in the ASALs is very low. This is attributed mainly to lack of farmer involvement in technology development and dissemination. A farmer-first or bottom-up approach enables problem identification and efficient allocation and use of scarce resources for research to deal with the real priorities of farmers. Future research should, therefore, lay more emphasis on farmer participation in all the stages of technology development and dissemination.

Another important constraint to adoption of improved land and water management technologies in the ASALs is land tenure. Most farmers do not have title deeds to the parcels of land they occupy. This militates against adoption of permanent soil and water conservation measures. No farmer is willing to make improvements on his/her land for fear of having to vacate it later. On the other hand, individuals who have title deeds are willing to invest in long-term land improvements with confidence of reaping benefits in future. The government of Kenya should, therefore, hasten land adjudication and issuance of title deeds to avert further land degradation in the ASALs.

Acknowledgements This review would not have been possible without financial support from the Desert Margins Programme (DMP) to whom the authors are immensely grateful. We are also grateful to the centre director, KARI Katumani Research Centre, for logistical support and the director, KARI, for permission to publish this chapter.

References

Biamah EK, Nhlabathi N (2003) Conservation tillage practices for dry land crop production in semi-arid Kenya: promotion of conservation tillage techniques for improving household food security in Iiyuni, Machakos, Kenya. In: Beukes D, Mkhize S, Sally H, Rensburg van L (eds) Proceedings of the symposium and workshop on water conservation technologies for sustainable dryland agriculture in sub-Saharan Africa (WCT), held at Bloem Spa Lodge and conference center, Bloemfontein, South Africa, 8–11 Apr 2003

Critchley WRS (1989) Runoff harvesting for crop production: experience in Kitui District, 1984–1986. In: Thomas DB, Biamah EK, Kilewe AM, Lundgren L, Mochoge BO (eds) Soil and water conservation in Kenya. Proceedings of the 3rd national workshop held at Kabete, Nairobi, Kenya, 16–19 Sep 1986, pp 396–405

Dagg M (1969) Report on the work of the physics division EAAFRO, on the water use efficiency of crops. Paper presented to the 11th specialist meeting on soil fertility and crop nutrition, Kampala, 25–27 Mar 1969

Diop JM, Onduru DD, Werf van der E, Kariuki J (1997) On-farm agro-economic study of organic and conventional techniques in medium and high potential areas of Kenya. KIOF/ETC, Nairobi

Gachene CKK, Mureithi JG, Anyika F, Makau M (2000) Incorporation of green manure cover crops in maize-based cropping systems in semi-arid and sub-humid environments of Kenya. In: Mureithi J, Gachene CKK, Muyekho FN, Onyango M, Mose L, Magenya O (eds) Participatory technology development for soil management by smallholders in Kenya. Proceedings of the 2nd scientific conference of the soil management and legume research network projects, Mombasa, Kenya, June 2000

Gachimbi LN, Keulen van H, Thuranira EG, Karuku AM, Jager de A, Nguluu S, Ikombo BM, Kinama JM, Itabari JK, Nandwa SM (2005) Nutrient balances at farm level in Machakos (Kenya), using a participatory nutrient monitoring (NUTMON) approach. Land Use Policy 22:13–22

Gibberd V (1995) Final report EMI Dry land farming and dry land applied research projects, 1988–1993

Gichangi EM, Njiru EN, Itabari JK, Wambua JM, Maina JN (2003) Promotion of improved soil fertility and water harvesting technologies through community-based on-farm trials in the ASALS of Kenya. Paper presented at the 1st adaptive research conference held at KARI Headquarters, Nairobi, 16–19 Jun 2003

Gicheru PT (1990) The effects of tillage and residue mulching on soil moisture conservation in Laikipia, Kenya. MSc. thesis, University of Nairobi, Nairobi, Kenya

Gicheru P, Gachene C, Mbuvi J, Mare E (2004) Effects of soil management practices and tillage systems on surface soil water conservation and crust formation on a sandy loam in semi-arid Kenya. Soil Tillage Res 75:173–184

Hamilton P (1997) Goodbye to hunger: the adoption and diffusion and impacts of conservation farming practices in rural Kenya. Association for Better Land Husbandry (ABLH), Nairobi

Ikombo BM (1984) Effects of farmyard manure and fertilizers on maize in semi-arid areas of Eastern Kenya. E Afr Agric For J 44:266–274

Imbira J (1989) Runoff harvesting for crop production in semi-arid areas of Baringo. In: Thomas DB, Biamah EK, Kilewe AM, Lundgren L, Mochoge BO (eds) Soil and water conservation in Kenya. Proceedings of the 3rd national workshop held at Kabete, Nairobi, 16–19 Sept 1986, pp 407–431

Itabari JK, Nguluu SN, Gichangi EM, Karuku AM, Njiru EN, Wambua JM, Maina JN, Gachimbi LN (2004) Managing land and water resources for sustainable crop production in dry areas: a case study of small-scale farms in semi-arid areas of Eastern, Central and Rift Valley Provinces of Kenya. In: Crissman L (ed) Agricultural research and development for sustainable resource management and food security in Kenya. Proceedings of end of programme conference, KARI, 11–12 Nov 2003, pp 31–42

Itabari JK, Wambua JM, Kitheka SK, Maina JN, Gichangi EM (1998) Effects of water harvesting and fertilizer on yield and water use efficiency of sorghum in semi-arid Eastern Kenya. Paper presented at the soil science society of East Africa conference held in Tanga, Tanzania, 14–18 Dec 1998

Jaetzold R, Schmidt H (1983) Farm management handbook of Kenya. Vol. II/C: Natural conditions and farm management information. Ministry of agriculture/GAT, Nairobi, Kenya

Karanja NK (1985) Effect of sulphur addition and inoculation on nodulation, dry matter yield and nitrogen content in groundnuts. E Afr Agric For J 50(3):61–68

Kibunja CN, Lijoh BO, Kitonyo EM, Muturi H, Salema MP (1999) Transfer of bio-fertilizer technologies to small-scale farming communities: opportunities and challenges for Kenya. A report submitted to FAO and IAEA

Kilewe AM (1987) Prediction of erosion rates and the effects of topsoil thickness on soil productivity. PhD. thesis, Department of Soil Science, University of Nairobi, Nairobi, Kenya

Kilewe AM, Mbuvi JP (1988) The effects of crop cover and residue management on runoff and soil loss. E Afr Agric For J 53:193–203

Kilewe AM, Ulsaker LG (1984) Topographic modification of land to concentrate and redistribute runoff for crop production. E Afr Agric For J 44:254–265

Kinama JM, Ong C, Ng'ang' JK, Stigter CJ, Gichuki F (1999) Erosion and runoff control in a maize/Senna agroforestry system in the semi-arid areas of Eastern Kenya. Proceedings of the 17th conference of soil science society of East Africa held in Kampala, Uganda, 6–10 Sept 1999

Kiome RM (1992) Soil and water conservation for improved moisture and crop production: an empirical and modelling study in semi-arid Kenya. PhD. thesis, University of East Anglia

Kiome RM, Stocking M (1995) Rationality of farmer perception of soil erosion: the effectiveness of soil conservation in semi-arid Kenya. Global Environ Change 5:281–295

Kitheka SK, Wambua JRM, Nixon DJ, Itabari JK, Watiki JM, Simiyu SC, Kihumba JN (1995) The use of pitting for the rehabilitation of denuded grazing lands in semi-arid Eastern Kenya. Proceedings of the 4th Biennial KARI scientific conference, Kenya, pp 231–245

Lee B (1993) Escaping from hunger: research to help farmers in semi-arid Kenya to grow enough food. ACIAR monograph no. 23. ACIAR, Canberra, Australia, p 52

Liniger H (1989) Water conservation in the semi-arid highlands of Laikipia. Paper presented at the workshop on soil and water conservation programme development in low potential areas (SIDA), Embu, 17–22 Sept 1989

Makatiani JBS (1970a) The effect of weeds on maize yields. Annual report. Ministry of agriculture, Kitale, Kenya, pp 58–60

Makatiani JBS (1970b) The reduction of maize yields due to late planting at Katumani research station. Annual report. Ministry of agriculture, Kitale, Kenya, pp 63–65

M'Arimi AM (1978) The effects of tillage methods and cropping systems on rainfall conservation in a semi-arid area of eastern Kenya. In: Soil and water conservation in Kenya. IDS Occasional paper No. 27, Nairobi, Kenya, pp 74–86

McCown RL, Jonnes RK (1992) Agriculture of semi-arid Kenya: problems and possibilities. In: Probert ME (ed) A search for strategies for sustainable dryland cropping in semi-arid Eastern Kenya. Proceedings of a symposium held in Nairobi, Kenya, 10–11 Dec 1990. ACIAR proceedings No. 41. Canberra, ACT, pp 8–14

Mellis D, Matsaert H, Micheni A (1997) Rainwater harvesting for crops in semi-arid areas. In: Kang'ara JN, Sutherland AJ, Gethi M (eds) Participatory dry land agricultural research east of Mount Kenya. Proceedings of a conference held at Izaak Walton Inn, Embu, Kenya, 21–24 Jan 1997

Miriti JM, Esilaba AO, Bationo A, Cheruiyot H, Kihumba J, Thuranira EG (2007) Tied ridging and integrated nutrient management options for sustainable crop production in semi-arid eastern Kenya. In: Bationo A, Waswa BS, Kihara J, Kimetu J (eds) Advances in integrated soil fertility management in sub-Saharan Africa: challenges and opportunities. Springer, Dordrecht, pp 435–441

Muchiri G, Gichuki FN (1983) Conservation tillage in semi-arid areas of Kenya. In: Thomas DB, Senga WM (eds) Soil and water conservation in semi-arid areas of Kenya. Proceedings of the 2nd workshop. University of Nairobi, Kenya

Mutua J (2005) Conservation agriculture pilot trials in Kenya – impacts and options for scaling up. In: Omanya G, Pasternak D (eds) Sustainable agriculture for the dry lands. Proceedings of the international symposium for sustainable dry land agriculture systems held in Niamey, Niger, 2–5 Dec 2003

Nadar HM (1984) Intercropping and intercrop component interaction under varying rainfall conditions in eastern Kenya. E Afr Agric For J 44(Special Issue):166–188

Nadar HM, Chui JN (1984) Evaluation of the effects of Rhizobium phaseoli strains on nodulation, dry-matter and grain yield of two bean (Phaseolus vulgaris) varieties. E Afr Agric For J 44(Special Issue):109–112

Nadar HM, Faught WA (1984) Maize yield response to different levels of nitrogen and phosphorus fertilizer application: a seven-season study. E Afr Agric For J 44(Special Issue): 147–156

Nandwa SM, Onduru DD, Gachimbi LN (2000) Soil fertility regeneration in Kenya. In: Hilhorst T, Muchena FM (eds) Nutrients on the move. Soil fertility dynamics in African farming systems. IIED, London, pp 119–132

Ngoroi EH, Gitari J, Njiru EN, Njakuthi N, Mwangi I, Muriithi C (1994) Effect of moisture conservation methods and fertilizer/FYM combinations on grain yield of maize in low rainfall areas. Annual report, KARI Regional Research Centre, Embu

Ngugi MN, Michieka RW (1986) Current findings on conservation tillage in a medium potential area of Kenya. In: Thomas DB, Biamah EK, Kilewe AM, Lundgren L, Mochoge BO (eds) Soil and water conservation in Kenya. Proceedings of the 3rd national workshop, Kabete, Nairobi, 16–19 Sept 1986, pp 155–162

Njihia CM (1979) The effect of tied ridges, stover mulch and farmyard manure on water conservation in a medium potential area, Katumani, Kenya. In: Lal R (ed) Soil tillage and crop production. IITA, Ibadan, pp 295–302

Okalebo JR, Gathua KW (1987) Nitrogen and phosphate fertilizers under marginal rainfall farming. NARC/KARI annual report 1987, KARI Muguga, Kenya

Okalebo JR, Mwaura TDN, Michobo WG (1980) Fertilizers and manures under marginal rainfall farming. ARD/KARI record of research annual report 1977–1980 (1980) KARI Muguga Kenya

Okalebo JR, Simpson JR (1990) The contribution of fertilizers and manure to crop production in semi-arid lands of Kenya. Transactions of the XIV congress of the international soil science society, Kyoto, Japan, 12–18 Aug 1990, pp 369–370

Okalebo JR, Simpson JR, Probert ME (1992) Phosphorus status in cropland soils in the semiarid areas of Machakos and Kitui Districts, Kenya. In: Probert ME (ed) A search for strategies for sustainable dry land cropping in semi-arid Eastern Kenya. Proceedings of a symposium held in Nairobi, Kenya, 10–11 Dec 1990. ACIAR proceedings No.41, Canberra, Australia

Okwach GE, Simiyu CS (1999) Effects of land management on runoff, erosion and crop production in a semi-arid area of Kenya. E Afr Agric For J 65:125–142

Omiti JM, Freeman HA, Kaguongo W, Bett C (1999) Soil fertility maintenance in eastern Kenya: current practices, constraints, and opportunities. CARMASAK working paper number 1. KARI-Katumani Research Centre, Kenya

Onduru DD, Gachini GN, de Jager A, Diop JM (1999) Participatory assessment of compost and liquid manure technologies in low potential areas of Kenya. Managing Africa's soils. Drylands programme IIED, London

Probert ME, Okalebo JR, Simpson JR, Jones RK (1992) The role of boma manure for improving soil fertility. In: Probert ME (ed) A search for strategies for sustainable dry land cropping in semi-arid eastern Kenya. Proceedings of a symposium held in Nairobi, Kenya, 10–11 Dec 1990. ACIAR proceedings No. 41, Canberra, Australia

Simpson JR, Karanja DR, Ikombo BM, Keating BA (1992) Effects of legumes in a cropping rotation on an infertile soil in Machakos District, Kenya. In: Probert ME (ed) A search for strategies for sustainable dryland cropping in semi-arid Eastern Kenya. Proceedings of a symposium held in Nairobi, Kenya, 10–11 Dec 1990. ACIAR proceedings No.41, Canberra, Australia, pp 44–49

Simpson JR, Okalebo JR, Lubulwa G (1996) The problem of maintaining soil fertility in eastern Kenya: a review of relevant research. ACIAR Monograph No. 41, 60p

Smith PD, Critchley WRS (1983) The potential of runoff harvesting for crop production and range rehabilitation in semiarid Baringo. In: Thomas DB, Senga WM (eds) Soil and water conservation in Kenya. Proceedings of the 2nd national workshop held in Nairobi, 10–13 Mar 1982. IDS Occasional Paper No. 42, pp 305–323

Sombroek WG, Braun HMH, van der Pouw BJA (1982) The exploratory soil map and agro-climatic zone map of Kenya (1980), scale 1: 1,000,000. Kenya Soil Survey E 1, Nairobi

Wafula BM, McCown RL, Keating BA (1992) Prospects for improving maize productivity through response farming. In: Probert ME (ed) A search for strategies for sustainable dry land cropping in semi-arid eastern Kenya. Proceedings of a symposium held in Nairobi, Kenya, 10–11 Dec 1990. ACIAR proceedings No. 41, Canberra, Australia, pp 101–107

Watiki JM, Gichangi EM, Itabari JK, Karuku AM, Nguluu SN (1999) The effects of rate and placement of boma manure on maize yield in semi-arid eastern Kenya. In: Proceedings of the 6th Biennial KARI scientific conference, Nairobi, Kenya, 9–13 Nov 1998

Whiteman P (1984) The importance of residual moisture reserves for increasing the reliability of cereal yields in East Africa. E Afr Agric For J 44:57

Evaluation of Establishment, Biomass Productivity and Quality of Improved Fallow Species in a Ferralic Arenosol at Coastal Region in Kenya

C. Walela, J.K. Ndufa, K. Balozi, O.V. Oeba, and M. Welimo

Abstract Inherent low soil fertility, continuous cultivation without nutrient inputs, nutrient losses through leaching and unsustainable farming methods of slash and burn leading to nutrient losses are some of the major causes of low crop yields at the coastal region. Short-term improved fallow technology has been successfully associated with soil fertility improvement and increase in crop yields in several parts of Kenya but has not been adequately evaluated at the coastal Kenya to give firm recommendations. An on-farm experiment was established in Malindi District as a randomized complete block design with eight treatments replicated thrice in 2006. Six species were evaluated at two intervals: 6 and 12 months after planting (MAP) for their site adaptability, biomass productivity, nutrient accumulation and leaf quality. The objective of the study was to determine the quantity of biomass produced, nutrient accumulation and leaf quality of improved fallow species. The data were analyzed using ANOVA procedures, and mean comparisons were done using orthogonal contrasts. *Mucuna pruriens* gave significantly ($p < 0.01$) higher foliage biomass yield of 2.3 and 22.9 t/ha at 6 and 12 MAP, respectively. *Sesbania sesban* had performed least with 0.03 and 0.2 t/ha of biomass foliage at 6 and 12 MAP, respectively. All species tested except *S. sesban* contained nitrogen >2.5%, with phosphorus and potassium contents being low in all species. Lignin contents were <15% except in *Tephrosia* species, but all species had low polyphenol contents of <4% at 6 MAP. *Tephrosia candida*

and *M. pruriens* had N contents of over 100 kg N/ha in the leaf biomass. These were substantially large amounts for recycled N by improved fallow species. In conclusion, *M. pruriens*, *T. candida* and *Tephrosia vogelii* were potential species for the coastal region, and their production can be enhanced in order to have recyclable amounts of nutrients that can be used to ameliorate the inherent low-fertility soils at the coast and subsequently enhance crop yields.

Keywords Biomass quantity and quality · Coastal region · Improved fallows

Introduction

Like most sub-Saharan African countries, the major constraint to smallholder farming in Kenya is declining soil fertility (Smaling et al., 1997). Smallholder farms, of about 2 ha on average, are usually cultivated continuously without adequate replenishment of plant nutrients resulting in removal of nutrients from soils mainly through crop harvests. An average of low crop (maize grain) yield <500 kg/ha has been reported in western Kenya (Nekesa et al., 1999) and in most nutrient-depleted areas in Kenya.

Food production at the coastal Kenya region has reportedly been very low compared to most parts of the country and has led to overreliance of external food aid. Inherent low soil fertility especially in soils surrounding Arabuko Sokoke Forest has led to low unsustainable crop yields. These soils are weakly developed arenosols characterized by a sandy texture with less than 15% clay (MOA, 1982). The soils also have low organic matter content, low cation

C. Walela (✉)
Kenya Forestry Research Institute, Gede Regional Research Centre, Malindi, Kenya
e-mail: cwalela@une.edu.au

A. Bationo et al. (eds.), *Innovations as Key to the Green Revolution in Africa*,
DOI 10.1007/978-90-481-2543-2_45, © Springer Science+Business Media B.V. 2011

exchange capacity and a low moisture storage capacity (MOA, 1982). Majority of farmers practise a slash and burn method of land preparation which leads to nutrient losses through leaching, denitrification and volatilization. Continuous cultivation without nutrient inputs and lack of crop diversification (planting mainly cassava and maize) have additionally led to crop failures in seasons of minimum rainfall in this region.

The most common method of maintaining soil fertility is the application of farmyard manure, but its quality is usually low because of poor handling and low-quality feeds for livestock (Lekasi et al., 1998). On the other hand, the high cost of inorganic fertilizers coupled with low returns and unreliable markets for agricultural produce have limited the use of fertilizers by the majority of smallholders in Kenya (Hassan et al., 1998). Traditionally, farmers left land fallow for several seasons as a means of restoring/rejuvenating soil fertility, but this method is no longer feasible because of diminishing size of land holdings in most parts of the country.

A major intervention for the coastal sandy soils is enhancement and maintenance of soil productivity for sustained crop production. This involves building the soil organic matter, an aspect that plays a key role in influencing soil fertility status by maintaining good soil physical and biological conditions including water-holding capacity, balanced supply of nutrients protected against leaching until released by mineralization, improving cation exchange capacity and greater recycling and supply of micronutrients or improved soil health.

The potential of improved fallow technology (targeted use of leguminous shrubs to attain one or more aims of the natural fallow within a short time in a smaller land area) has been widely demonstrated and has been recommended for scaling up in other regions. The benefits include soil fertility build up through nitrogen fixation (Gathumbi et al., 2002a; Giller and Wilson, 1991), deep N capture from the subsoil (Gathumbi et al., 2003; Jama et al., 1998a) and to increase crop yields. Hence deliberate planting of short-duration pure legume species fallows has shown great potential in improving soil fertility (Buresh and Tian, 1997) and is being increasingly adopted by smallholder farmers (Niang et al., 2002). Recognizing that agricultural technology is one aspect of a complex socio-economic–ecological system, there is a need to take into account the ecological variability, social and cultural institutions before such technologies can be extended to farmers in various regions of the country. It is against this background that an on-farm screening trial was established in Gede, Malindi District, Kenya, with an overall objective of identifying potential leguminous fallow species for possible adoption at the coastal region based on biomass productivity and leaf quality. The specific objective of this chapter was to determine the quantity of biomass produced, nutrient accumulation and leaf quality of selected improved fallow species within an interval of 6 and 12 months.

Materials and Methods

Description of Site and Soil Type

A field experiment was conducted on farmers' field in Gede, Malindi District. The soils at the site are classified in the coastal uplands class which consists of soils developed on unconsolidated sandy deposits. The soils fall into UCE1-sandy to loamy unit of very deep ferralic arenosols (Ministry of Agriculture, 1988). Baseline soil site characterization was carried out before treatment application using laboratory methods described in Okalebo et al. (2002).

Treatments and Experimental Design

The treatments were arranged in a randomized complete block design replicated three times during 2006 long rains. A summary of treatments and their sources are given in Table 1. Actual establishment was done during long rains (May/June 2006). Three maize seeds (Pwani Hybrid 4 (PH4) per hole) were planted at a spacing of 75 cm by 30 cm, and young plants were later thinned to one plant per hole. No prior treatments, e.g. organic/inorganic, were applied to the soils before maize planting. Fallows were introduced into a growing maize crop (1 month after planting) after the second weeding. These were introduced within the maize rows alternating at a spacing of 75 cm by 40 cm.

Table 1 Summary of treatments and their sources

Treatments	Source
1. Sole maize (continuous)	
2. Natural weed fallow	
3. Maize + *Crotalaria grahamiana*	Ex-Madagascar
4. Maize + *Sesbania sesban*	Siaya
5. Maize + *Tephrosia vogelii*	Ex-Yaounde Cameroon
6. Maize + *Mucuna pruriens*	Siaya
7. Maize + *Crotalaria paulina*	Ex-GBK Kenya
8. Maize + *Tephrosia candida*	Siaya

Measurements

Maize yields were determined at harvest while foliage, woody biomass yields and leaf quality (nutrient concentration) of the leguminous shrubs were assessed at two intervals, i.e. 6 months after planting (6 MAP) and 12 months after planting (12 MAP). Biomass at 6 MAP was determined by destructive harvesting of two internal rows on one side of the plot after discarding the border row. At 12 MAP, final harvest was done on all treatments per plot for biomass and nutrient accumulation measurements. Further, the subsamples were oven-dried at 70°C for 48 h and ground (20 mesh) for plant tissue analysis. Fallow biomass quality, lignin and active polyphenol concentrations were tested with lignin determined following the acid detergent fibre (ADF) method of Van Soest (1963). Total extractable polyphenols were analyzed from air-dried material by extraction using 50% aqueous methanol. The plant to extractant ratio was 0.1 g/50 ml, and polyphenols were analyzed colorimetrically using the Folin-CioCalteu method (Anderson and Ingram, 1993). Nutrient accumulation in fallows, nitrogen, phosphorus and potassium, was determined using methods described in Okalebo et al. (2002).

Statistical Analysis of Data

Analysis of variance (ANOVA) and orthogonal contrast procedures were conducted to determine treatment differences in biomass and nutrient yields of improved fallow species tested. Nitrogen uptake was transformed using square root procedures. One and five percent were used to declare significant differences among treatments unless otherwise stated. Data were managed in Microsoft Excel spreadsheet 2003 and analyzed using Genstat version 9 package.

Results and Discussion

Results

Soil Characterization

The results showed that soils sampled from the study site were slightly acidic (pH, H_2O, 5.92) and had very low percentage soil carbon (0.4%) and soil organic matter (SOM) content (0.63%) according to Okalebo et al. (2002). The samples further revealed that the nitrogen (0.02%) was very low and phosphorus was below the critical level (4 ppm). Total potassium and calcium levels were equally low at 91 and 562 ppm, respectively. The exchangeable cations Na and Mg were relatively high (Table 2).

Maize Grain Yields (Long Rains 2006)

Mean maize grain yields within the treatments showed a similar trend with an average yield of 2,605 kg/ha (Fig. 1). This is expected during this trial period, as the plots received no prior treatments, e.g. organic or inorganic fertilizers. Differences can only be expected during the second maize crop, as organic residues from different improved fallows will be incorporated into

Table 2 Mean values of soil chemical properties sampled 0–15 cm depth

Soil fertility parameters	Mean value
Soil pH (water)	5.92
EC (mS/cm)	0.03
Organic carbon (%)	0.40
Total N (%)	0.02
Soil organic matter (%)	0.63
P (Olsen method) ppm	4.00
Total Na (ppm)	112.0
Total K (ppm)	91.00
Total Ca (ppm)	562.00
Total Mg (ppm)	153.00
Sand (%)	89.20
Clay (%)	5.00
Silt (%)	5.00

Fig. 1 Mean foliage biomass yield for various improved fallow species

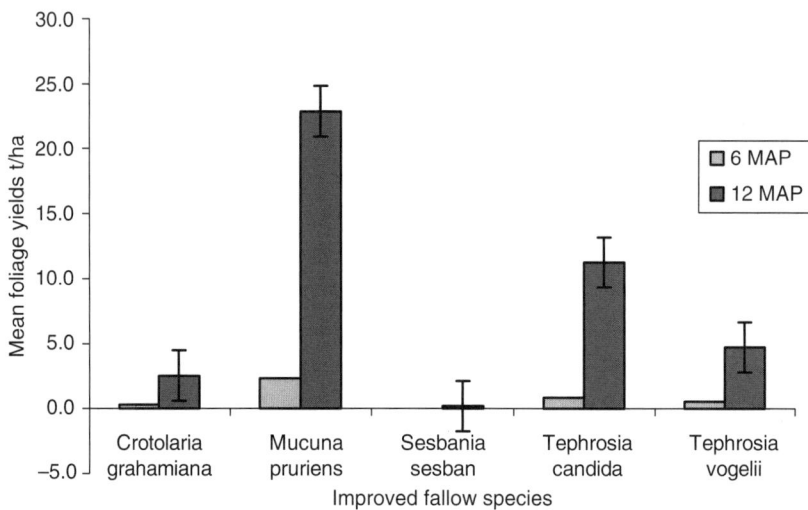

Table 3 Mean maize grain yields of two seasons in Malindi

Treatments	Maize yields (kg/ha) Long rains 2006 (baseline data)
Mucuna pruriens	2,455
Tephrosia candida	2,511
Tephrosia vogelii	2,739
Crotalaria grahamiana	2,893
Continuous plot	2,488
Weed fallow	2,232
Sesbania sesban	3,055

each plot. These yields therefore act as a baseline data for comparisons on yield improvements due to treatments in the subsequent seasons (Table 3).

Biomass Yield, Nutrient Concentration and Uptake

The results showed that at the age of 6 and 12 months after planting (MAP), the established improved fallow species performed differently in biomass yield (foliage and wood) where *M. pruriens* had the highest mean foliage biomass yield of 2.3 and 22.9 t/ha followed by *Tephrosia candida* at 0.8 and 11.3 t/ha, respectively (Table 4).

The above-mentioned fallow species also had the highest mean foliage nutrient uptake at 6 and 12 MAP. On the other hand, *M. pruriens* and *Tephrosia vogelii* had the highest mean nutrient N concentration at 6 MAP. In addition, there was a decrease in nutrient

concentration for all improved fallow species at the age of 12 MAP (Table 4).

Further analysis showed that there were highly significant differences ($p < 0.01$) among the improved fallow species in foliage and wood biomass yield at 6 and 12 MAP where the species, time and interaction between time and species contributed 39.2, 27.6 and 26.7% of total variability for foliage biomass, respectively. Similarly, species, time and their interaction effect contributed 31.3, 37 and 26.6% of total variability for wood biomass, respectively. These results showed that foliage and wood biomass from each improved fallow species significantly increased over time.

Consequently, comparing the foliage biomass yield using orthogonal contrasts among the improved fallow species showed that at 6 and 12 MAP, *Crotalaria grahamiana* and *M. pruriens* and *T. vogelii* and *M. pruriens* were significantly different ($p < 0.01$). Further, *C. grahamiana* and *T. candida* at 12 MAP were significantly different ($p = 0.015$) while *C. grahamiana* and *S. sesban* were not significantly different ($p = 0.432$) (Table 4).

Overall, the yield performance of foliage and biomass increased significantly across time for all species other than *S. sesban* (Figs. 1 and 2).

Foliage Nutrient Concentration

M. pruriens had the highest amount of nitrogen content (3.47%) while the lowest content was attained by

Table 4 Mean biomass yield, percentage foliage nutrient concentration and uptake of various improved fallow species

Time	Improved fallow species	Mean biomass yield (t/ha)		Foliage mean % nutrient concentration			Foliage mean nutrient uptake (kg/ha)		
		Foliage	Wood	N	P	K	N	P	K
6 months	*Crotalaria grahamiana*	0.3	0.3	2.6	0.2	0.8	7.2	0.7	2.0
	Mucuna pruriens	2.3	0.0	3.5	0.2	0.5	77.1	4.7	11.7
	Sesbania sesban	0.03	0.04	2.2	0.2	0.7	0.5	0.1	0.2
	Tephrosia candida	0.8	0.9	2.7	0.2	0.6	21.8	1.7	5.3
	Tephrosia vogelii	0.6	0.5	3.1	0.2	0.6	17.3	1.0	3.6
	s.e.d.	0.335	0.235	0.248	0.046	0.105	0.695	0.604	1.462
12 months	*Crotalaria grahamiana*	2.5	10.6	1.6	0.1	0.2	38.4	2.6	5.0
	M. pruriens	22.9	0.0	1.3	0.1	0.2	270.9	22.6	46.4
	S. sesban	0.2	2.6	1.1	0.1	0.1	1.1	0.1	0.2
	T. candida	11.3	22.3	1.4	0.1	0.2	152.3	10.0	19.8
	T. vogelii	4.7	10.1	1.6	0.1	0.2	71.4	4.7	9.1
	s.e.d.	2.831	2.519	0.338	0.027	0.020	2.238	5.69	6.47

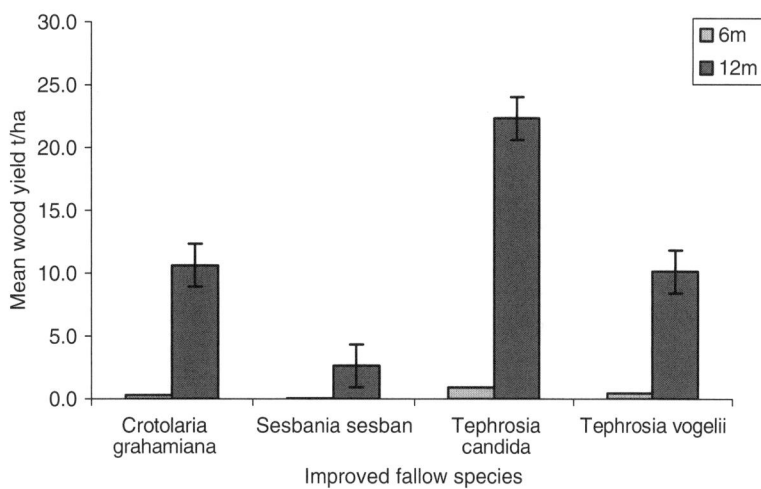

Fig. 2 Mean wood biomass yield for various improved fallow species

S. sesban (2.17%) at 6 MAP. ANOVA results showed that there was a significant difference ($p = 0.008$) in N concentration among the species at 6 MAP. This was evidenced mainly between *C. grahamiana* and *M. pruriens* ($p = 0.009$). On the other hand, there was no reason to believe that there was a difference in foliage N concentration between *C. grahamiana* and *S. sesban* ($p = 0.105$), *C. grahamiana* and *T. candida* ($p = 0.687$) and *M. pruriens* and *T. vogelii* ($p = 0.139$).

In addition, there was no significant difference ($p > 0.05$) in P and K concentrations at 6 MAP and N and P at 12 MAP in all improved fallow species. However, there was a significant difference ($p = 0.006$) in K concentrations at 12 MAP among the species. This was mainly evidenced between *C. grahamiana* and *S. sesban* ($p = 0.001$).

Foliage Nutrient Uptake

Analysis of nutrient uptake data showed that there was a significant difference ($p < 0.01$) in N uptake among different improved fallow species, which contributed the highest percentage (59.9%) of the total variability followed by time (21.4%) which was also highly significant ($p < 0.01$) and the interaction effect between species and time of growth (7.4%, $p = 0.027$). This implied that N uptake among various improved fallow species was considerably different at 6 and 12 MAP, where, for instance, *M. pruriens* had 77,271 kg/ha N and *T. candida* had 22,152 kg/ha N at 6 and 12 MAP, respectively.

Orthogonal comparisons showed that at 6 MAP, there were significant differences in N uptake between

C. grahamiana versus *M. pruriens* ($p < 0.01$), *C. grahamiana* versus *S. sesban* ($p = 0.021$), *M. pruriens* versus *T. candida* ($p < 0.01$) and *T. candida* versus *S. sesban* ($p < 0.01$) (Table 4). However, at 12 MAP, there were no significant differences between *C. grahamiana* versus *T. vogelii* ($p = 0.344$) and *M. pruriens* versus *T. candida* ($p = 0.103$).

Individual improved fallow species had significant improvement of N uptake at 6 and 12 MAP where *M. pruriens* ranked the highest and *S. sesban* performed the least (Fig. 3).

On P uptake, there were also significant differences ($p < 0.01$) among improved fallow species at 6 MAP where *M. pruriens* had the highest P uptake (4.7 kg/ha P) with *S. sesban* performed least (0.08 kg/ha P). At 12 MAP, there was sufficient evidence ($p = 0.027$) in differences in P uptake among improved fallows, where *M. pruriens* still performed the best (22.6 kg/ha P) and *S. sesban* least performed (0.1 kg/ha P).

Orthogonal contrasts showed that there were differences ($p < 0.01$) in P uptake between *C. grahamiana* versus *M. pruriens* and *T. vogelii* versus *M. pruriens*; however, there were no differences between *C. grahamiana* versus *S. sesban* ($p = 0.305$), *T. candida* and *C. grahamiana* ($p = 0.134$) at 6 MAP (Table 4). The differences in P accumulation were also similar at 12 MAP.

On the other hand, there was an increase in time of P uptake at 6 and 12 MAP among different improved fallows (Fig. 4).

Similarly, there were significant differences ($p < 0.01$) on K uptake among the different improved fallow species at 6 and 12 MAP where *M. pruriens* again had the highest uptake of 11.7 and 46.4 kg/ha K, respectively. *S. sesban* performed least among the other improved fallow species with 0.2 kg/ha K at 6 and 12 MAP. Mean comparison of K uptake among improved fallow species showed that there were differences between *C. grahamiana* versus *M. pruriens* and *T. candida* versus *M. pruriens* ($p < 0.01$) over the two sampling intervals (Table 4).

Equally, there was an increase in K uptake among different improved fallows except *S. sesban* (Fig. 5).

Biomass Quality and Characterization at 6 Months After Planting

The results showed that all species had polyphenol content below the threshold values (Delve et al., 2000) of less than 4% with *C. grahamiana* had the lowest value (1.9%) and *T. candida* had the highest value (3.5%). Lignin contents were below critical values <15% except *T. vogelii* and *T. candida* (Table 5).

Discussion

The results indicated that soils were of relatively low potential for subsistence farming. Decline in soil organic matter could be as a result of loss of carbon through oxidation to carbon dioxide during the prevalent slash and burn during land preparation, which

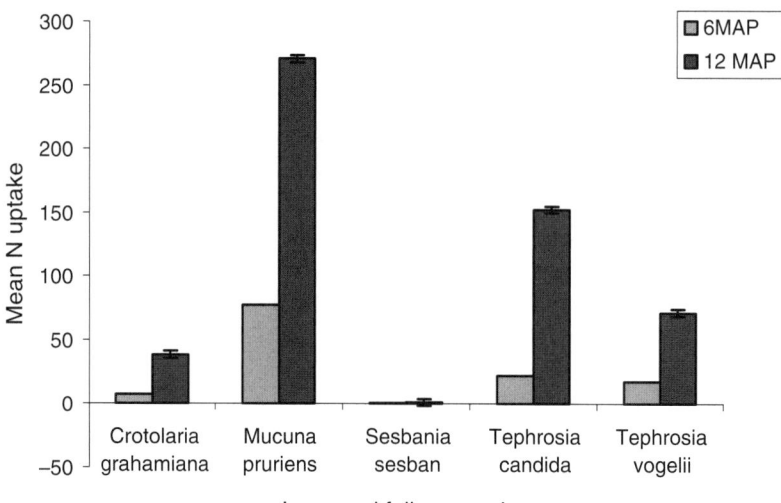

Fig. 3 Mean N uptake among different improved fallow species

Fig. 4 Mean P uptake among different improved fallow species

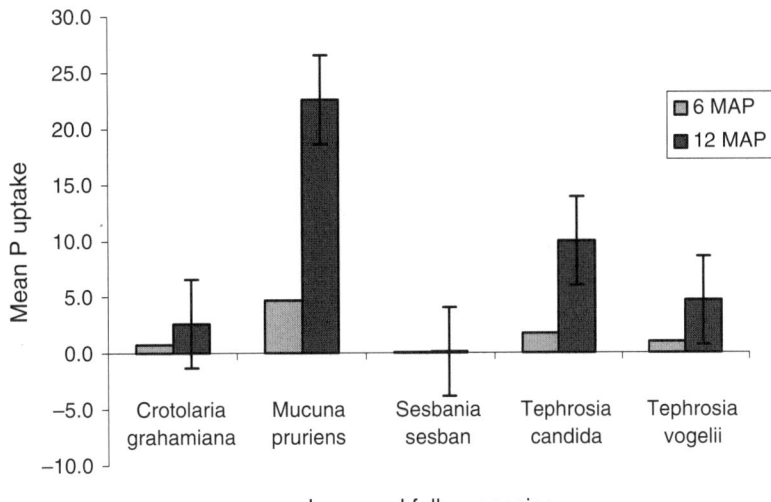

Improved fallow species

Fig. 5 Mean K uptake among different fallow species

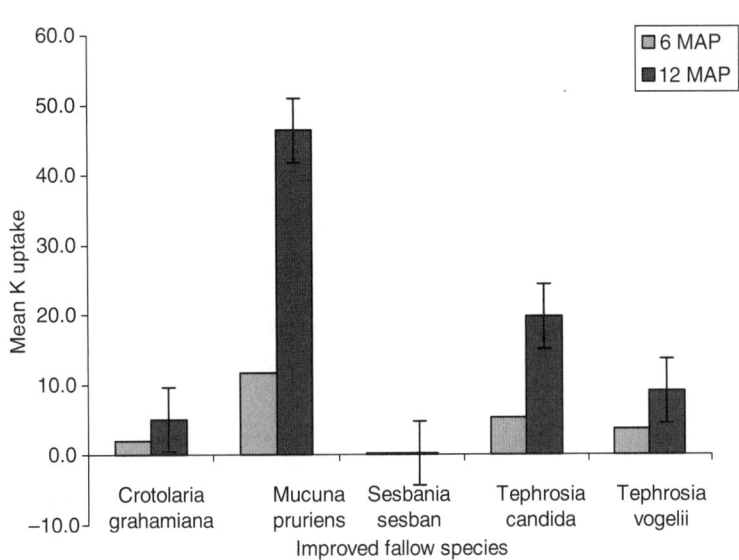

Improved fallow species

Table 5 Leaf quality of improved fallow species at 6 MAP

Improved fallow species	% Polyphenol	% Lignin	% ADF
Crotalaria grahamiana	1.9	7.7	28.9
Sesbania sesban	2.5	7.3	18.2
Tephrosia vogelii	2.9	19.3	36.8
Mucuna pruriens	2.4	15.7	31.8
Tephrosia candida	3.5	20.4	39.6

is the common practice in the region. Further loss is exacerbated by the high rate of mineralization from high coastal environmental temperatures. The low organic matter in these soils means that there are very little stable aggregates and hence high risk of soil erosion. Low levels of nitrogen in these soils are probably due to leaching of nitrogen, mainly in sandy soils through nitrate ions from surface layers to subsurface horizons. Further, the results show that in this site, the soils are below the critical P level which has been reported as 10 ppm P (Okalebo et al., 2002) and are thus deficient of phosphorus. A major intervention for the coastal sandy soils is enhancement and maintenance of soil productivity for sustained crop production. This involves building the soil organic matter, an aspect that plays a key role in influencing soil fertility status by maintaining good soil physical conditions, including water-holding capacity, balanced supply of nutrients protected against leaching until released by mineralization, improving cation exchange capacity and greater recycling and supply of micronutrients.

Biomass production of the improved fallows species tested in this study could be considered as fairly low except for *M. pruriens* and *T. candida* and may be due to low soil fertility on the site. *Crotalaria paulina* did not germinate/scanty germinate in all the replicates. *S. sesban* and *C. grahamiana* had poor growth performance with *S. sesban* established through seedlings, as establishment through direct seeding was poor. At 2 months of transplanting *S. sesban*, poor nodulation was noted. *T. vogelii* and *T. candida* fallows performed fairly well in the field with foliage biomass yield of 4 and 11 t/ha at 12 MAP, respectively. *T. candida* and *M. pruriens* had dry matter yield of >5 t/ha. It has been reported as a general rule that many organic materials when applied in modest amounts, i.e. >5 t dry matter/ha, contain sufficient N to match that of a 2 t crop of maize, but they cannot meet P requirements and must be supplemented by inorganic P in areas where P is deficient (Palm, 1995). Additions of organic residues in turn increase microbial pool sizes and activity, C and N mineralization rates and enzyme activities (Smith et al., 1993), all factors that affect nutrient cycling. *C. grahamiana* (2 t/ha) performance was very low at 12 MAP as compared to biomass yields of 20.5 t/ha reported by Ndufa et al. (2000) in a trial in western Kenya. Highest woody component yield was obtained from *T. candida* (22 t/ha at 12 MAP). This tentatively shows the potential of *T. candida* fallow for fuel wood production and soil fertility amelioration purposes at the coastal Kenya.

All the species tested, except *S. sesban*, had their nitrogen contents above the suggested critical level of 2.5% (Delve et al., 2000) at 6 MAP and are therefore a good source of nitrogen for N management at this age. The phosphorus contents in the foliage biomass were less than the critical concentration of P needed for remineralization of the microbially immobilized P and therefore cannot eliminate the need for

P. Notably, the process of remineralization of P has been reported to begin when the P content of a plant material reaches a critical level of 0.2% (Palm, 1995). However, most of the green manures have less than the critical concentration of P and cannot therefore eliminate the need for P (Palm, 1995). Considering the low P levels of these soils (4 mg P/kg soil), external inputs will have to be included for P replenishment in these soils as the improved fallows cannot meet this acute deficiency. Such remedies could be in form of rock phosphates (e.g. the reactive Minjingu rock phosphate) and combining organics with inorganic fertilizers. The low potassium levels in the soils could be as a result of leaching of soluble organic residues lost through unsustainable farming methods of slash and burn prevalent at the coastal region. This loss has a negative effect on availability of K for plant uptake. Remedy for potassium requirement in plants is therefore combining organic residues with inorganic potassium sources. Farmers should also abandon unsustainable farming methods such as slash and burn.

Low nutrient contents in the foliage biomass 12 MAP can be attributed to translocation of nutrients to storage organs as the plants were seeding at the time of sampling and only leaves were sampled. This therefore indicates that foliage biomass could be harvested slightly before flowering/seeding period so as to have recyclable biomass for the subsequent crop and benefit from the total nutrient uptake by the shrubs before nutrient translocation.

T. candida and *M. pruriens* had N uptakes of over 100 kg N/ha which were substantially large amounts for recycled N by improved fallows and are comparable to the current recommended rates of about 100 kg N/ha (FURP, 1994) for maize growth in Kenya in nutrient-depleted areas. Similar amounts of recyclable N by improved fallows have been reported by Niang et al. (1996) for western Kenya. However, phosphorus contents would be insufficient as the recommended rates are about 25 kg/ha (FURP, 1994) for most soils. Palm (1995) and Palm et al. (1997) reported that organic inputs are very low suppliers of P because of their low P concentrations. Low P concentrations in the soil may further have negative effects on N_2 fixation by legumes. In this case, there is a need for supplementing P from external sources as mentioned earlier for a better performance by improved fallows.

The results indicated that all species had their polyphenol contents less than the critical value at

6 MAP. Threshold values for lignin and polyphenols have been reported as less than 15 and 4%, respectively, by Palm and Sanchez (1991). Lignin and polyphenols are important intrinsic factors in plant materials for leaves of tropical legumes for net N mineralization to occur immediately. The effect of polyphenols in plant materials applied to the soil has been that of reduced N release through their protein-binding capacity. They also bind enzymes, which catalyse mineralization, in the nitrification process in particular. However, under aerobic conditions, most polyphenols are easily degraded. This is an important quality; *C. grahamiana* had the lowest polyphenol content of 1.85% while *T. candida* had the highest content of 3.47%. The highest lignin contents were also reported in *T. candida* and were 20.5%, followed by *T. vogelii* with 19.3%. Lowest lignin contents were reported in *S. sesban* (7.32%), followed by *C. grahamiana* (7.74%) and *M. pruriens* (15.7%).

Conclusion

Results of this study indicated that a 12-month fallow was important at the coast, as this would yield as much recyclable nutrients for the subsequent crop. The best performing improved fallow species based on high recyclable N and high biomass production were ranked as follows: *M. pruriens* > *T. candida* > *T. vogelii* > *C. grahamiana* > *S. sesban* > *C. paulina*. According to the decision tree for selecting organic inputs for nitrogen management, *Mucuna* spp., *Crotalaria* spp. and *Sesbania* species can be incorporated directly with annual crops, while *T. candida* and *T. vogelii* species will need to be mixed with inorganic fertilizers or high-quality organics. Results further indicate that *S. sesban* and *C. paulina* performed poorly in the study area. Overall results showed that *M. pruriens*, *T. candida* and *T. vogelii* are potential species for the coastal region, and their production can be enhanced in order to have recyclable amounts of nutrients that can be used to ameliorate the inherent low-fertility soils at the coast and subsequently enhance crop yields.

It is worth noting that although organic inputs from planted fallows could supply recyclable N for crop growth, they cannot supply enough phosphorus required by subsequent crops. Phosphorus recycling by fallow species through deep uptake is likely to be negligible owing to the very low concentrations of available P at the study site. The low phosphorus in the soils is further suspected to have highly negatively influenced N_2 fixation by the leguminous shrubs, and hence P additions from inorganic cheap sources such as Minjingu rock phosphate are recommended in order to alleviate P limitations and optimize crop yield benefits after the fallows. Evaluation of the economics of using various P sources is recommended. Studies on assessing the populations and effectiveness of the host rhizobia species are recommended where if found inadequate rhizobia inoculation is recommended for increasing the potential of the shrubs to fix N_2. Further investigation should also be carried out to establish reasons why *S. sesban* performed poorly and for the lack of germination in *C. paulina*.

Acknowledgements The authors wish to thank the KEFRI Management for providing the funds and research facilities to undertake this project through GoK funding. Special thanks to National Programme Coordinator, Oballa Phanuel, and Centre Director, Gede, Doris Mutta, for guidance through the project period. We appreciate efforts by technicians at Gede and KEFRI for their assistance and especially Florence Mwanziu and Mukirae Kamau for assisting in laboratory analysis and data collection. The efforts of Wim Buysse, ICRAF, and Vincent Oeba, KEFRI biometrician, in backstopping in data management aspects are highly acknowledged. KEFRI and KARI headquarters laboratory technicians are equally acknowledged for their endless effort in assisting chemical analysis of soils and fallow biomass. Lastly but not least many thanks to Prof. J.R. Okalebo, Moi University, for peer reviewing this work.

References

Anderson JM, Ingram JSI (1993) Tropical soil biology and fertility: a handbook of methods. CAB International, Wallingford, CT

Buresh RJ, Tian G (1997) Soil improvement by trees in sub-Saharan Africa. Agroforest Syst 38:51–76

Delve R, Gachengo C, Adams E, Palm C, Cadisch G, Giller KE (2000) The organic resource database. In: The biology and fertility of tropical soils: a TSBF report 1997–1998, pp 20–22

FURP (1994) Fertilizer use recommendations. Interim Phase. Siaya and Vihiga Districts. KARI-NARL, Nairobi

Gathumbi SM, Cadisch G, Buresh RJ, Giller KE (2002) ^{15}N natural abundance as a tool for assessing N_2 fixation of herbaceous shrubs and tree legumes in improved fallows. Soil Biol Biochem 34:1059–1071

Gathumbi SM, Cadisch G, Buresh RJ, Giller KE (2003) Subsoil nitrogen capture in mixed legume stands as assessed by deep ^{15}N placement. Soil Sci Soc Am J 67: 573–582

Giller KE, Wilson KJ (1991) Nitrogen fixation in tropical cropping systems. CAB International, Wallingford, CT

Hassan RM, Muriithi FM, Kama G (1998) Determinants of fertilizer use and gap between farmers' maize yields and potential yields in Kenya. In: Hassan RM (ed) Kenya in maize technology development and transfer. CAB international, Wallingford, UK, pp 137–161

Jama B, Buresh RJ, Ndufa JK, Shephered KD (1998a) Vertical distribution of roots and soil nitrate: tree species and phosphorus effects. Soil Sci Soc Am J 62:280–286

Lekasi JK, Tanner JC, Kimani SK, Harris PJC (1998) Manure management in the Kenyan highlands: practices and potential. High potential production system portfolio for the natural resources systems program renewable natural resources knowledge strategy. Department for International Development. A publication of the Henry Doubleday Research Association, UK, p 35

Ministry of Agriculture (1982) An outline of the major soils in Kenya

Ministry of Agriculture (1988) Fertilizer use recommendation project (Phase 1), final report Annex III. Kilifi district (Volume 29). NARL, Nairobi, Kenya

Ndufa JK, Gathumbi SM, Cadisch G (2000) Legume biomass production and maize yields in pure and mixed fallow systems: KEFRI conference proceedings, 2004, pp 72–85

Nekesa P, Maritim HK, Okalebo JR, Woomer PL (1999) Economic analysis of maize-bean production using a soil fertility replenishment product (PREP-PAC) in Western Kenya. Afr Crop Sci J 17:157–163

Niang AI, Amadalo BA, de Wolf J, Gathumbi SM (2002) Species screening for short-term planted fallows in the highlands of western Kenya. Agroforest Syst 56:145–154

Niang AL, Amadalo BA, Gathumbi SM, Otieno JHO, Obonyo CO, Obonyo E (1996) Maseno project progress report Sept 1995–1996. AFRENA report No.110. ICRAF, Nairobi, 71p

Okalebo JR, Gathua KW, Woomer PL (2002) Laboratory methods of plant and soil analysis: a working manual, 2nd edn. Tropical Soil Biology and Fertility Programme, Nairobi

Palm CA (1995) Contribution of agroforestry trees to nutrient requirements of intercropped plants. Agroforest Syst 30:105–124

Palm CA, Myers RJK, Nandwa SM (1997) Combined use of organic and inorganic nutrient sources for soil fertility maintenance and replenishment. In: Buresh RJ, Sanchez PA (eds) Replenishing soil fertility in Africa. SSSA special publication 51. SSSA, Madison, WI, pp 193–217

Palm CA, Sanchez PA (1991) Nitrogen release from the leaves of some tropical legumes as affected by their lignin and polyphenolic contents. Soil Biol Biochem 23: 83–88

Smaling EMA, Nandwa SM, Jansen BH (1997) Soil fertility in Africa is at stake. In: Buresh RJ et al (ed) Replenishing soil fertility in Africa. SSSA Spec.Publ.51.SSSA, Madison, WI, pp 47–61

Smith JL, Papendick RI, Bezdicek DF, Lynch JM (1993) Soil organic matter dynamics and crop residue management. In: Metting Jr FB (ed) Soil microbial ecology. Marcel Dekker, New York, NY, pp 65–94

Van Soest PJ (1963) Use of detergents in the analysis of fibrous feeds. II. A rapid method for the determination of fiber and lignin. J Assoc Off Anal Chem 46:829–835

Assessment of Potato Bacterial Wilt Disease Status in North Rift Valley of Kenya: A Survey

T.K. Kwambai, M.E. Omunyin, J.R. Okalebo, Z.M. Kinyua, and P. Gildemacher

Abstract A survey on bacterial wilt (*Ralstonia solanacearum*) prevalence was carried out in the major potato-growing areas of the North Rift region of Kenya in 2006. Limited information is available on bacterial wilt status in the region. The survey was conducted in Trans Nzoia, Uasin Gishu, Keiyo and Marakwet districts. A questionnaire was administered to 256 potato growers and field observations made in two major potato-growing divisions in each district. The study areas and potato fields were selected based on potato cropping intensity, current potato field size and crop stage. Characteristic plant and tuber bacterial wilt symptoms were the main criteria used to assess the disease. The survey established that potato was grown mainly in pure stand (63%) or in mixed (26%) cropping systems with other crops such as peas, beans, spring onions and maize depending on the area and/or community. Bacterial wilt prevalence varied significantly ($p \leq 0.01$) among districts, with the lowest (35%) and highest (99%) in Marakwet and Keiyo districts, respectively. The disease was absent in parts of Marakwet district that are at 2800 m above sea level or higher. Bacterial wilt incidence ranged from 0 to 33% and was significantly different ($p \leq 0.05$) among districts. Improper potato and bacterial wilt management practices by most respondents suggested that bacterial wilt has continued unabated due to inadequate farmer knowledge. Factors found to aggravate potato bacterial wilt included indiscriminate use and sourcing of seed, retention of volunteer potato plants, poor field hygiene and lack of management skills on bacterial wilt.

Keywords Bacterial wilt · Incidence · Kenya North Rift Valley · Potato · Prevalence

Introduction

In the North Rift Valley region of Kenya, potato is the third most important food crop after maize and beans and contributes 13% of total potato production in Kenya (Rees et al., 1997). However, its production and yields in the region are generally low, with a yield average of 12 t ha^{-1} compared to a potential yield of 40 t ha^{-1} (Kabira et al., 2006). The low yields are largely attributed to the unavailability of certified potato seed, low soil fertility, high incidence of late blight and bacterial wilt diseases (KARI, 2003).

Bacterial wilt caused by *Ralstonia solanacearum* is a serious problem in potato production in the world (Adipala et al., 2001; Lemay et al., 2003) including Kenya (Kinyua et al., 2001; Muriithi et al., 2001). Bacterial wilt has been reported to occur in the North Rift region of Kenya (Anon, 2004; Kinyua et al., 2004). Kinyua et al. (2004) reported bacterial wilt incidence of 34.3 and 32.3%, with higher limits of the disease latently in tubers in Mt. Elgon and Timboroa in the North Rift region. In spite of the above, there is insufficient information available on the disease status (prevalence, incidence, yield losses and management practices by farmers) in the region. The disease causes the greatest damage to small-scale farmers who have no access to good-quality seed and are unable to practise crop rotation because of limited land (Kidanemariam et al., 1998). This study was therefore carried out to assess the prevalence and incidence of bacterial wilt disease in major potato production areas.

T.K. Kwambai (✉)
National Agricultural Research Centre, Kenya Agricultural Research Institute, Kitale, Kenya
e-mail: tkkwambai2003@yahoo.com

A. Bationo et al. (eds.), *Innovations as Key to the Green Revolution in Africa*,
DOI 10.1007/978-90-481-2543-2_46, © Springer Science+Business Media B.V. 2011

Materials and Methods

The survey was carried out during the short rainy (SR) season (October to November 2006) in the highlands of the North Rift Valley of Kenya to assess the incidence and prevalence of bacterial wilt. A total of 256 farmers were interviewed in the major potato-growing areas of Trans Nzoia, Uasin Gishu, Keiyo and Marakwet districts. Potato plants with bacterial wilt symptoms in farmers' fields in Gitwamba, Trans Nzoia district, were observed, described and collected for field and laboratory diagnosis and isolation of *R. solanacearum* according to the procedure described by Englebrecht (1994).

Areas covered during the assessment fell within altitudes 2153–3102 m asl, latitude 0°01–1°16′N and longitude 34°75–35°60′E. The study areas were selected through the guidance of Ministry of Agriculture extension staff and secondary information from reports and previous research work on Irish potato. Two leading potato producing divisions and two to three locations were selected in each district. The study sites were selected based on the potato cropping intensity, current potato field size and crop growth stage. At least 30 potato fields in two divisions in each district with potatoes at various flowering stages were randomly selected at intervals of 1–10 km apart. In each selected potato field, a sampling area of at least 0.05 ha was considered sufficient. In each sampled field site, the number of potato rows, plants in six to eight representative rows and plants showing bacterial wilt symptoms were recorded. Bacterial wilt incidence in each farm was determined by the number of plants showing wilt symptoms expressed as a percentage of the total number of plants assessed (Zadoks and Schein, 1979). Disease prevalence was recorded as number of potato fields with bacterial wilt expressed as a percentage of the total number of fields assessed. Whenever there was doubt on the wilt symptoms, about two to three tubers were dug up from suspicious plants and a cross-section cut made using a surgical blade or knife for observation of brown ring or bacterial ooze in clear water. During the survey, a structured questionnaire was administered by enumerators to solicit information from farmers on general potato growing with emphasis on cropping systems, practices and bacterial wilt management. A geographical positioning system (GPS) unit was used to take altitude, latitude and longitude

of representative potato fields selected at random in the different areas. Data were analyzed using SPSS computer package.

Results and Discussions

Bacterial Wilt Identification

The survey revealed that most farmers could distinctly identify bacterial wilt through plant wilting and tuber oozing symptoms. However, they did not clearly understand the causal agent or the modes of spread other than through seed and soil. This suggested that farmers could manage the disease if they were equipped with the appropriate skills and knowledge on bacterial wilt epidemiology and control measures. There were positive field (wilting plants, oozing tubers, bacterial streaming) and laboratory (colony morphology) observations and/or tests for the presence of *R. solanacearum* in potato plants and tubers. Some colonies exhibited crescent shape on SMSA as shown in Fig. 1. Smith et al. (1996), while characterizing

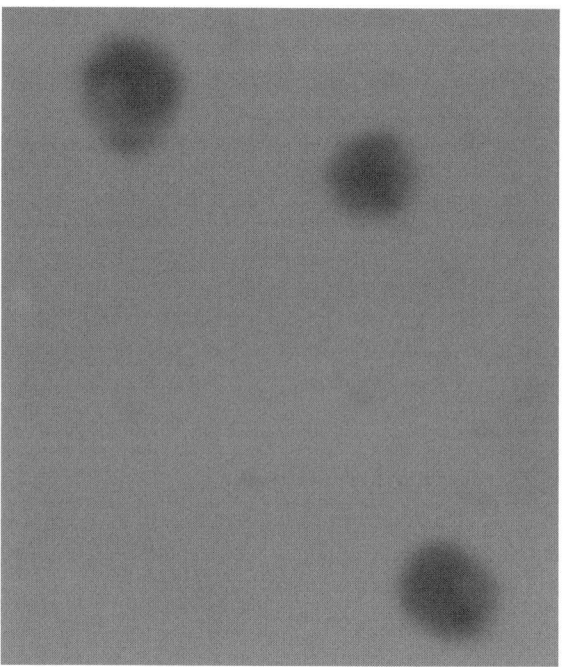

Fig. 1 Crescent-shaped colonies of *Ralstonia solanacearum* as observed on semi-selective media

R. solanacearum, found that race 3 biovar II is prevalent in the major potato-growing regions (highlands) of Kenya, where races 1 and 2 were absent. Priou and Aley (1999) also reported that *R. solanacearum* race 3 (biovar II) strains are the most common in higher elevations of the tropics up to 3400 m asl. Thus, the bacterial wilt pathogen identified in North Rift region during this study was considered to be race 3 biovar II.

Bacterial Wilt Prevalence

Bacterial wilt occurred in 79% of all farms visited. Prevalence of bacterial wilt was highest in Keiyo district (99%) followed by Uasin Gishu (95%) and Trans Nzoia (89%) and then lowest in Marakwet district (35%). This indicated that the disease was widespread in the North Rift region of Kenya. The high bacterial wilt prevalence in Keiyo, Uasin Gishu and Trans Nzoia districts was attributed to the time the disease found its way into these districts between early 1980s and 1990s, compared to Marakwet district, where it was purportedly introduced in the late 1990s to mid-2000s. This was suspected to have resulted from uncertified seed of new varieties such as Tigoni, which were acquired from other districts by some farmers. The occurrence of bacterial wilt in Marakwet district was recorded only in Kapcherop division, whereas in areas of Koisungur (<2815 m asl) in the same division, farmers indicated that bacterial wilt was new, having been introduced in 2004 through infected seed of the variety Tigoni acquired from other potato growers in neighbouring West Pokot district. These findings suggested that in the absence of knowledge and appropriate control, diseases could spread rapidly into new areas. The use of infected potato seed and accumulation of

R. solanacearum in the soil over time may contribute to the spread of bacterial wilt (Ateka et al., 2001). There was no bacterial wilt in Lelan and Kipyego areas in Kipyego division of the same district. The large distance separating these areas from other potato-growing districts, low temperatures as influenced by high altitude (above 2800 m asl) and the restricted access by road, which made these areas somehow "quarantined", could have contributed immensely to the absence of the disease. The occurrence of bacterial wilt in Marakwet district was low or none at altitudes above 2800 m asl, whereas all farms assessed in Uasin Gishu district within the same altitude range had the disease. This implied that the cool high altitudes were not a sufficient condition to eliminate bacterial wilt, but other bacterial wilt preventive measures were also required.

Bacterial Wilt Incidence and Spread

There was a significant difference ($p \leq 0.05$) in bacterial wilt incidence among the four districts, with Trans Nzoia (7.2%) and Marakwet (1.5%) districts giving the highest and lowest disease incidences, respectively (Fig. 2). This was lower than that reported by Kinyua et al. (2004) in two other areas of the North Rift region: Timboroa (32.3%) and Mt. Elgon (34.3%). Bacterial wilt incidence was significantly different ($p \leq 0.03$) and had an inverse correlation ($r = -0.35$) with altitude (Fig. 3). This was in agreement with the findings of Ateka et al. (2001), who while studying in eastern and central Kenyan districts found a decline of 51% in bacterial wilt incidence with rise in altitude from 1800 to 2700 m asl. The disease develops slowly when the soil temperature is lower than 20°C (Martin and French, 1985; Muriithi et al., 2001). The survey indicated that

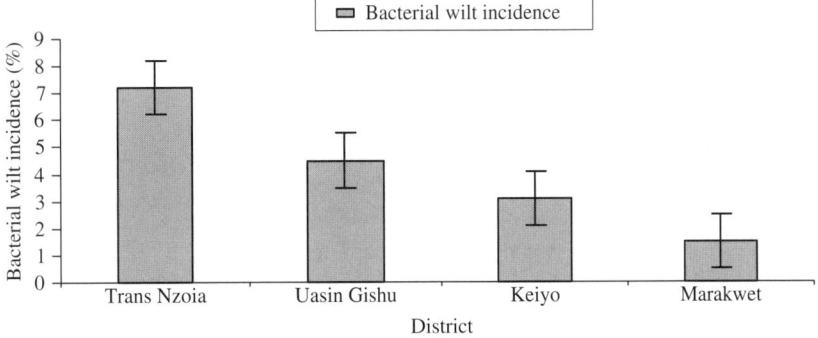

Fig. 2 Mean bacterial wilt incidence in four districts of the North Rift region of Kenya, October to December 2006. *Bars* represent standard error

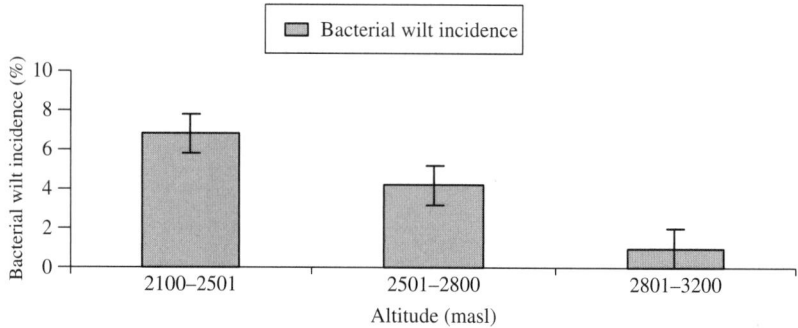

Fig. 3 Influence of altitude on bacterial wilt incidence in the North Rift region of Kenya during the short rainy season, October to December 2006. *Bars* represent standard error

bacterial wilt could be severe at altitudes below 2500 m asl, and farmers could experience higher crop loss with use of infected seed or fields at such altitudes than at mid- or higher altitudes. Average annual temperatures at the altitudes above 2800 m asl in the North Rift Valley range from 2 to 10°C (Jaetzold and Schmidt, 1983). Bacterial wilt incidence can be high in warm areas, moderate in mild areas and low in cool regions (French et al., 1996). Divisions in Keiyo and Trans Nzoia districts were not different in bacterial wilt incidence, possibly since they had near similar conditions. The significant difference in bacterial wilt incidence among districts was attributed to temperatures as influenced by altitude, cropping systems, farming practices and location in relation to other potato-growing areas. The high bacterial wilt incidence in Trans Nzoia, Uasin Gishu and Keiyo districts was attributed to contaminated soils due to continuous potato cultivation on same land over time, inadequate crop rotation intervals, use of infected seed due to lack of certified seed and poor bacterial wilt management practices.

Majority of the farmers (66%) interviewed indicated that bacterial wilt incidence was high during the short rainy season, while 34% said that it was high in the long rainy season. They further indicated that the disease initially entered their farms or spread into other farms mainly through seed potato and soils but it was also caused through black ants or rainwater flow (Table 1). Contaminated planting materials were the major source of inoculum in Kenya (Smith et al., 1996). The indication that black ants spread bacterial wilt could be attributed to their association by possibly feeding on the exudates from infected tubers, which may or may not spread the disease. The role of fertilizer in bacterial wilt spread was a perception that farmers had which could not be clearly understood, but may require further investigation. Generally the farmers indicated that they experienced bacterial wilt

Table 1 Farmers' views on initial source of bacterial wilt in their farms and modes of spread to other farms in the North Rift region, Kenya

Source/mode of spread of bacterial wilt	Farmer's response (%)	
	Initial source ($n = 213$)	Mode of spread ($n = 201$)
Seed tubers	75	86
Soil	16	12
Black ants	5	1.5
Through fertilizers	2	0
Rainwater flows	0	0.5
Not sure/don't know	1	0

problem yearly (67%), occasionally (17%) and sometimes depending on the seed (13%) or field (3%). The yearly bacterial wilt incidence experienced by farmers implied that most soils were contaminated with *R. solanacearum* or most of the seed available to farmers was infected. The high incidence in the short rainy season could be attributed to inoculum build-up in soil from infected potato crop in the long rainy season since farmers planted at least two potato crops in 1 year and rarely removed infected and/or crop residues and tubers from the field. Warmer temperatures in the short rainy season also could have led to increased bacterial wilt incidence, which subsequently lead to yield loss. Potato yield losses due to bacterial wilt were <10, 10–25, 25–50 and >50% as perceived by 49, 30, 14 and 7% of the respondents, respectively.

Potato and Bacterial Wilt Management Practices

Potato was indicated as the second most important crop enterprise after maize, but was the leading crop at altitudes above 2800 m asl, where temperatures were too low for maize production. The majority of the farmers (63%) practised monocropping compared to 26% who practised mixed cropping. A low but positive correlation ($r = 0.16$) between bacterial wilt incidence and monocropping system as practised by farmers was an indication of perpetuation of the *R. solanacearum* in field soils, which increased disease incidence on succeeding potato crops. However, a negative correlation ($r = -0.25$) between mixed cropping and bacterial wilt incidence suggested that mixed cropping system reduced bacterial wilt incidence and could be a useful practice for incorporation in integrated disease management. The crops used in the mixed cropping systems included peas, beans, spring onions and maize depending on the area and/community, with peas being the dominant companion crop. In Burundi, intercropping potatoes with beans resulted in less spread of *R. solanacearum* than with maize intercrop (French, 1994). Within-row intercropping of potato with maize or cowpea served to contain the spread of bacterial wilt from diseased to healthy plants when initial infection was caused by either latent infection of tubers or soil infestation in Mindanao, Philippines (French, 1994). The results suggested that farmers should be encouraged to have mixed cropping as this will not only ameliorate bacterial disease but also diversify diets and income. Farmers indicated that peas at low populations was preferred to other crops as a companion in mixed cropping system because the yields of either crop were not affected negatively, possibly due to the light foliage and height advantage of peas over such crops as beans which could be smothered by potatoes.

Generally farmers indicated that they practised crop rotation (78.4%) in potato growing, but of these 55% used it as a bacterial wilt control measure (Table 2). Farmers who did not practise crop rotation either had small land holdings (41%) or lacked knowledge on its importance in bacterial wilt control (59%). Crops used in rotation with potatoes included maize, cabbage, kales, beans, oats, peas, wheat and spring onions. Other farmers used crop rotation for improvement of soil fertility.

The majority of the farmers practised crop rotation programme at intervals of less than 1 (44%) to 2 (37%) years, which was insufficient to get rid of the *R. solanacearum* in the soil before returning a potato crop in the same field. Lemaga et al. (2001), working in Uganda, reported that planting different crops in two consecutive seasons reduced bacterial wilt compared to

Table 2 Bacterial wilt control methods practised by potato growers in the North Rift districts of Kenya, 2006

Control method	Farmers practising ($n = 202$) (%)
Crop rotation	55
Seed selection	18
Uprooting	12
Tolerant-resistant varieties	2
Mixed cropping	1
None	12

planting the same crop and that millet and sweet potato in the rotation gave the best results. They also reported that rotation of potatoes with maize, beans, onions carrots, millet, peas and sweet potato gave significantly higher tuber yields than potato monoculture.

Retention of volunteer potato plants was a common practice according to 77.5% of the respondents. The retention of volunteers suggested that such plants carried over the bacterial wilt pathogen from one season to another through infected plants and soils. Volunteers are unsuitable in potato cropping systems because they harbour the disease and make rotation ineffective, and efforts to use clean seed are negated if volunteers are present (Tusiime and Adipala, 2000). Survival of *R. solanacearum* is fostered by the presence of volunteer plants from tubers of the previous crops or voluntary carriers (French et al., 1996).

Farmers who uprooted infected plants (65%) indicated that they leave such plant materials in the field (Fig. 4). The uprooting of bacterial wilt-infected plants indicated positive efforts in the management of the disease. However, poor disposal of the same and possibly timing of the operation also aggravated the spread of the disease, since 58% of the farmers indicated

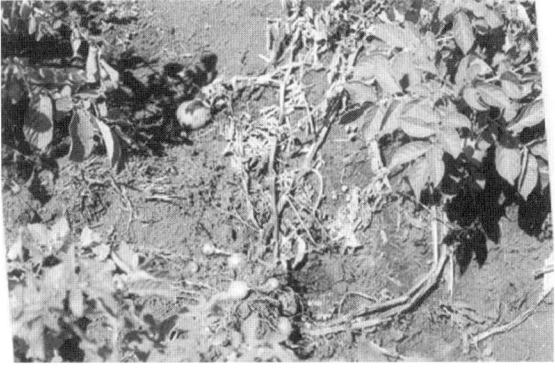

Fig. 4 Infected potato plants left in the field after uprooting

Fig. 5 Uncollected infected potato tubers in a (**a**) recently harvested field and (**b**) field planted with cabbage as a rotation crop after potatoes

that they left uprooted plants and tubers in the field, 29% left along the hedges of field, 7% sometimes fed to livestock and 6% buried or burned. It was also evident through field observations that farmers left infected tubers and plant debris in the field after harvest and even when a rotational crop has been planted (Fig. 5a, b). These practices could allow bacterial wilt pathogen to spread through runoff, human and livestock movement. Potato crop residues should be collected at harvest and given to animals, and rotten (bacterial wilt-infected) tubers buried away from waterways (Priou and Aley, 1999). The farmers' practices suggested a vicious cycle in the management of bacterial wilt, thus suggesting that education of farmers was of paramount importance in the reduction and eradication of the pathogen. Only 10% of the farmers had received some form of training on bacterial wilt management in the four districts.

Among the respondents, 44% took more than four cropping seasons before changing or replacing their seed potato for planting, while others took one (24%), two (23%) and three (9%) cropping seasons. Replacement was shorter areas with high bacterial wilt prevalence. Most farmers (72%) selected their own saved seed potato after harvest, and 26% selected both before and after harvest. This implied that the *R. solanacearum* inoculum from infected tubers in such farms persisted and inevitably spread to neighbouring farms through seed exchange, soil on tools and runoff water during rainy seasons.

Selection of seed mainly after harvest suggested that farmers unknowingly mixed tubers from healthy

Table 3 Various sources of seed potato used by farmers in the North Rift Valley districts of Kenya, 2006

Seed potato source	Farmers (253) (%)
Neighbours	43
Own saved seed	47
Local market	8
KARI	1
Others	1

and diseased plants, which may not be differentiated due to latent infection. This could be avoided through observation of healthy status of individual plants while actively growing. Selecting the best looking plants (positive selection) in a potato field as source of seed for the next season reduces the incidence and spread of bacterial wilt (CIP; 2007). None of the farmers interviewed had planted certified seed in the current season or in the previous 3 years. Farmers obtained their seed potato from various sources, but the most common were their own farms (47%) and neighbours (43%) (Table 3). The practice encouraged the spread of bacterial wilt within and across farms. These results supported the findings of Ateka et al. (2001) from survey studies in eastern and central Kenyan districts.

Conclusions

While there is widespread occurrence of bacterial wilt in potato-growing areas of the North Rift Valley districts of Kenya, there are some areas that are safe from the disease in Marakwet district. These are at altitudes above 2800 m asl, such as Lelan and Kipyego. The spread and incidence of the disease were greatly influenced by low altitude, indiscriminate use of seed tubers by and among farmers due to lack of certified seed and lack of technical knowledge on bacterial wilt management and general crop practices. There exist opportunities for bacterial wilt management, which have to be utilized. Bacterial wilt in general was spreading fast particularly through infected seed, threatening the currently safe potato production areas if immediate interventions are not taken both through control and preventive or legislative measures. The findings of the survey suggest that education of farmers was of paramount importance in curbing the bacterial wilt menace in the North Rift region of Kenya security in the North Rift Valley region of Kenya. Therefore, allocation of resources towards bacterial wilt management and prevention is critical to ensure continued potato production for food. Also, the crop losses due to bacterial wilt need to be quantified across altitudes and cropping systems so as to ascertain the magnitude of the problem. This could be done alongside studies on cumulative incidence of the disease. Crop rotation studies with crops such as maize, wheat, oats, cabbage and spring onions common in specific areas among others need to be investigated. The studies will need to include a series of crop rotation intervals in those farms whose sizes can allow the practice. Where farms are too small for crop rotation to be practised, alternative bacterial wilt management options will need to be investigated.

Acknowledgements We thank the director of Kenya Agricultural Research Institute for providing funding through the KARI/World Bank Scholarship. The contributions of Drs. L. Wasilwa and J.N. Kabira in the research plan development are highly appreciated. We are indebted to thank the centre director of National Agricultural Research Centre (NARC), Kitale, for providing facilities and enabling environment for the conduct of the research. It is our great pleasure to acknowledge the members of staff of NARC, Kitale, who participated during surveys and all farmers who offered their time and valuable information.

References

Adipala E, Tusiime B, Patel BK, Bega-Lemaga D, Olanya D (2001) Bacterial wilt management: a dilemma for sub-Saharan Africa. Afr Crop Sci Conf Proc 5:433–437

Anon (2004) Crops development 2003 annual report. Ministry of Agriculture, Trans Nzoia district, Kenya

Ateka EM, Mwang'ombe AW, Kimenju JW (2001) Reaction of potato cultivars to *Ralstonia solanacearum* in Kenya. Afr Crop Sci J 9(1):251–256

Englebrecht MC (1994) Modification of a semi-selective medium for the isolation and quantification of Pseudomonas solanacearum. In bacterial Wilt Newsletter Hayward AC 10:3–5. Australian Centre for International Agricultural Research, Canberra, Australia

French ER (1994) Integrated control of bacterial wilt of potatoes. Int Potato Center (CIP) Circular 20(2):8–11

French ER, Gutara L, Aley P (1996) Important diseases in potato seed-tuber production. Bacterial wilt manual. International Potato Center, Lima, pp 1–11

International Potato Center (2007) Select the best: positive selection to improve farm saved seed potatoes. Training manual. International Potato Center, Lima, pp 9–14

Jaetzold R, Schmidt H (1983) Farm management handbook of Kenya, vol II: natural conditions and farm management information, part B (Central and Rift Valley Provinces). Ministry of Agriculture, Kenya

Kabira JN, Wakahiu M, Wagoire W, Gildemacher P, Lemaga B (2006) Guidelines for production of healthy seed potatoes in East and Central Africa. Kenya Agricultural Research Institute, Nairobi, Kenya, pp 1–26

Kenya Agricultural Research Institute (2003) Kenya Agricultural Research Institute. Annual report 2003. National Potato Research Center, Tigoni, pp 7–8

Kidanemariam HM, Michieka A, Ajanga S, Njenga ad French E (1998) The role of plant resistance in the integrated management of potato bacterial wilt. Crop protection for resource-poor farmers in Kenya. In: Farrell G, Kibata GN (eds) Proceedings of the 2nd KARI Biennal Crop Protection Conference, 16–17 September 1998. KARI/DFID National Agriculture Project II. KARI: Nairobi, Kenya

Kinyua ZM, Kihara SN, Kinoti J (2004) Latently infected tubers and seed-flow channels contribute to the incidence of potato bacterial wilt in Kenya. KARI – National Agricultural Research Laboratories, Plant Pathology Section Annual Report 2004, pp 17–23

Kinyua ZM, Smith JJ, Lunga'aho C et al (2001) On farm success and challenges of producing bacterial wilt-free tubers in seed plots in Kenya. Afr Crop Sci J 9(1):271–285

Lemaga B, Siriri D, Ebanyat P (2001) Effect of Soil amendment on bacterial Wilt incidence and yield of potatoes in South Western Uganda. Afr Crop Sci Crop Sci J 9(1):1–8

Lemay A, Redlin S, Fowler G, Dirani M (2003) Pest data sheet: Ralstonia solanacearum race 3 biovar 2. USDA/APHIS/PPQ. Center for Plant Health Science and Technology, Plant Epidemiology and Risk Analysis Laboratory, Raleigh, NC

Martin C, French ER (1985) Bacterial wilt of potato Ralstonia solanacearum. International Potato Center 1996. Bacterial Wilt Manual. CIP, Lima, Peru, Sec 2-97-1–8

Muriithi LM, Wanyoike TK, Kamau K, Kariuki LK (2001) Current status of potato seed tubers and soil as major sources of bacterial wilt spread around the slopes of Mt. Kenya, pp 277–283

Priou S, Aley P (1999) Integrated management of potato bacterial wilt. International Potato Center (CIP), Lima, Peru, http://www.cipotato.org/potato/pest_diseases/bacterial_wilt/symptoms.asp

Rees DJ, Nkonge C, Wandera JL (eds) (1997) A review of agricultural practices and constraints in the North Rift Valley Province, Kenya. Kenya Agricultural Research Institute, Kitale, Kenya, pp 94–106. ISBN: 9966-879-36-6

Smith J, Simons S, Triagale A, Saddler G (1996) The development of a biological control agent for *Pseudomonas solanacearum* race 3 on potato in Kenya. Crop protection for resource-poor farmers in Kenya. In Proceedings of the 1st Biennial crop protection conference KARI/DFID NARP II, 27–28 Mar 1996, pp 14–17

Tusiime G, Adipala E (2000) Management of bacterial wilt of potato: approaches, limitations and role of stakeholders. Afr Potato Assoc Conf Proc 5:347–352

Zadoks JC, Schein RD (1979) Epidemiology and plant disease management. Oxford University Press, New York, NY, pp 100–115

Soil Fertility Variability in Relation to the Yields of Maize and Soybean Under Intensifying Cropping Systems in the Tropical Savannas of Northeastern Nigeria

J.D. Kwari, A.Y. Kamara, F. Ekeleme, and L. Omoigui

Abstract In northeast Nigeria, there are compound and bush fields which are variable in soil fertility. This variability influences the efficiency of resource use to increase crop yields on the fields. To address this variability, some soil fertility parameters in 0–15 cm depth soil samples from farmer's compound (64) and bush (73) fields in southern Guinea savanna (SGS), northern Guinea savanna (NGS), and Sudan savanna (SS) zones were related to soybean and maize yields. Sand and silt contents significantly influenced soybean yields in compound fields in SGS and maize yields in compound fields in NGS. Clay content had significant effect on soybean yield in compound fields in SGS and bush fields in NGS. Soil pH (range 5.23–8.03) was a significant parameter of soybean in bush fields in SGS and total nitrogen (range 0.14–4.90 g kg^{-1}) in bush fields in NGS. Organic C (range 2.2–23.6 g kg^{-1}) significantly influenced soybean yields in bush fields in SGS and maize yields in bush fields in SS. Available P (range 0.30–11.30 mg kg^{-1}) and exchangeable K (range 0.15–1.79 cmol kg^{-1}) were important variables for soybean yields in compound fields in SGS and for maize yields in bush fields in SS. Available phosphorus was also related to soybean yield in bush fields in NGS. Organic carbon and available P were deficient in some of the fields and were three times determinants of soybean or maize yields. The addition of organic inputs and phosphorus and the rotation of maize and soybean, but targeted to the specific fields, are recommended.

Keywords Available phosphorus · Bush fields · Compound fields · Organic carbon · Soil fertility

Introduction

Low organic matter and nitrogen and phosphorus are the main constraints to crop production in northeast Nigeria (Rayar, 1988; Kwari et al., 1999) and elsewhere in West Africa (Schlecht et al., 2006). Both mineral fertilizers and organic inputs are required to improve soil fertility (Vanlauwe et al., 2002). This view may, however, be contested because wide variations exist in soil fertility status among fields within a farm, as has been widely reported across Africa, eastern (Woomer et al., 1998; Murage et al., 2000; Vanlauwe et al., 2006), southern (Roberts et al., 2003), and western (Prudencio, 1993; Scoones and Toulmin, 1999). This variability is induced either by management practices or by the result of differences in texture.

In eastern Africa, Woomer et al. (1998) reported three types of fields in smallholder farming systems and all showed a soil fertility gradient from home sites to outfields. In South Africa, Roberts et al. (2003) showed higher fertility status and better management of home fields compared with outfields. Scoones and Toulmin (1999) reported on home fields and outfields in northeast Nigeria in which through the application of manures and fertilizers, as well as the management of organic materials, farmers maintain home fields that are more productive than outfields which generally are not manured or fertilized.

In the southern and northern Guinea and Sudan savannas (SGS, NGS, and SS) of northeast Nigeria, there are bush and compound fields which are variable

J.D. Kwari (✉)
Department of Soil Science, University of Maiduguri, Maiduguri, Nigeria
e-mail: jdkwari@yahoo.co.uk

A. Bationo et al. (eds.), *Innovations as Key to the Green Revolution in Africa*,
DOI 10.1007/978-90-481-2543-2_47, © Springer Science+Business Media B.V. 2011

in soil fertility. These fields differ in terms of distance from the villages and in management practices. Compound fields are situated within the village boundaries at a distance less than 0.5 km. Bush or distant fields are situated at more than 0.5 km from the villages. The compound fields are under continuous cropping of mainly maize and vegetables. The bush fields are cropped with cereal (maize or sorghum) and legume (groundnut or cowpea) intercrops with the occasional practice of bush fallow system where cropping is alternated with fallow periods of at least 2 years (PROSAB, 2004). The compound fields receive some small applications of animal manure and household wastes; small quantities of inorganic fertilizer are applied to bush fields when available. Maize has been replacing sorghum as a staple food crop and as a cash crop during the same period in the area because of its higher yield per unit area (Kassam et al., 1975).

Intensification of land use systems arising from increased population pressure combined with low fertilizer use has resulted in soil fertility depletion in northeast Nigeria (Rayar, 1988). This poses serious threats to maize production which has a high N requirement. Deficiency of N may therefore be a major cause of low maize yields. Therefore, it is necessary to identify promising nutrient management practices for intensive production. However, because of the differences in land use, fallowing, and use of manure and inorganic fertilizers between the compound and bush fields, soil fertility is likely to be variable in the two fields. Consequently, field heterogeneity or variability in smallholder farming system and the need for disaggregating and targeting management practices within this heterogeneity are becoming increasingly important in sub-Saharan Africa (Vanlauwe et al., 2006). In particular, resource-poor farmers are averse to taking risk with respect to soil fertility amelioration and are thus more likely to concentrate on managing better fields than improving poor fields within a farm.

The study assesses the initial soil nutrient status of farmers' compound and bush fields and relates them to the yields of maize and soybean. Maize is the major crop in the area, and soybean is being introduced to be grown in rotation with maize because of its capacity for improving soil fertility. This information will help to identify the soil parameters which are the major determinants of the yields of these crops on each type of field. This would guide in the prioritizing and targeting of sustainable soil management systems to deal with the heterogeneity within the fields.

Materials and Methods

Description of Study Area

The study area covers 30 communities in the southern part of Borno State, Nigeria. The study area lies between latitudes 10 and 11°N and longitudes 12°15′ and 14°E and covers three agro-ecological zones: southern Guinea savanna (SGS), northern Guinea savanna (NGS), and Sudan savanna (SS). It has an area of 16,100 km² and a population of about 700,000 (PROSAB, 2004). Average annual rainfall is 1100–1300 mm in SGS, 800–1000 mm in NGS, and 600–800 mm in SS (PROSAB, 2004). Temperatures are high throughout the study area, reaching a maximum of 40°C in April and a minimum of 18°C in December and January.

The soil types in the study area are related to parent materials from which they were developed. In SGS, basalt underlying the Biu plateau gives rise to heavy clay soils (vertisols) under conditions of poor internal and external drainage. Sandy clay loam/sandy clay/silt clay loam soils are developed on basalt and pyroclastic rocks on the Biu Plains on better drained sites in the remaining parts of the SGS and extending to NGS. In Kwaya Kusar/Wandali communities of NGS, and the whole of SS, sandstones give rise to sandy/sandy loam soils (Carroll and Klinkenberg, 1972).

Selection of Farmers' Fields, Soil Sampling, and Analysis

Six to ten farmers who were willing to provide their fields for participatory demonstration trials were identified in each of the 30 communities. Some of their fields were within the village boundaries at a distance less than 0.5 km (compound fields); other fields were situated more than 0.5 km from the villages (bush fields). A total of 137 composite topsoil (0–15 cm) samples were collected from compound (64) and bush (73) fields before planting began in May to early June 2004. Each topsoil sample contained 15 subsamples that were mixed together. For compound fields, there were 25 soil samples in SGS, 21 in NGS, and 18 in SS. For bush fields, there were 25 soil samples in SGS, 33 in NGS, and 15 in SS.

The soil samples were air-dried, ground to pass a 2-mm sieve, and stored in tightly sealed polythene bags. The soil samples were analyzed for texture, pH in water (1:2.5 soil–water ratio), organic carbon, total N, exchangeable K, and available P through methods outlined by Van Reeuwijk (1992). Potassium was determined by flame photometry and phosphorus was determined by spectrophotometry (Van Reeuwijk, 1992).

Field Trial Establishment and Cultural Practices

Each farmer's field was subdivided into three 20 m × 20 m plots. Each plot was cropped with soybean (either TGx 1448-2E or TGx 1904-6F in SGS and NGS and TGx 1830-2E in SS), improved maize (either 97 TZL COMP-1-W or TZE COMP-3 DT in SGS, TZE COMP-3 DT and 94 TZE COMP-5-W in NGS, and 95 TZEE-W and 94 TZE COMP-5-W in SS), or farmers' choice of maize varieties. It was intended to rotate the three crops in 2005 to improve the declining soil fertility status and demonstrate farmers the benefit of crop rotation.

Fertilizer was applied to maize at the rate of 100 kg N, 22 kg P, and 21 kg K ha^{-1} as NPK 15-15-15 and urea; 15 kg P ha^{-1} as single superphosphate was applied to soybean. Weeds were controlled manually by hoe weeding. At harvest, ears of maize and pods of soybean were removed and shelled, and grain yield was adjusted to 12% moisture. Simple statistics was used to compute mean grain yields of maize and soybean. Multiple regression analysis was carried out to relate maize and soybean grain yields with soil parameters using the PROC REG in SAS (SAS Institute, 1990).

Results and Discussions

Texture and Some Chemical Properties of the Soils

The crop yields, texture, and selected soil chemical properties are presented in Table 1. Except for available P and exchangeable K in SS, there are no significant differences in soil parameters between the fields in SGS and SS. However, there are significant differences in sand, silt, total N, available P, and exchangeable K contents between the fields in NGS. The sand content is higher in compound fields than bush fields in NGS, but silt, total N, available P, and exchangeable K contents are higher in bush fields than compound fields. In SS, available P and exchangeable P are higher in compound fields than bush fields.

The soil parameters were grouped into various ratings, and the proportions of the field types within each rating are presented in Table 2. Soil texture varied from sandy loam to clay in SGS and NGS and from loamy sand to clay loam in SS. The soils were grouped using clay contents into coarse (<180 g kg^{-1}), medium (180–350 g kg^{-1}), and fine (>350 g kg^{-1}) textures (FAO, 1974).

In NGS, most compound fields than bush fields coarse textured as settlements were sited on coarse- and medium-textured soils and bush fields were sited on medium- to fine-textured soils (Bulama Tsahuyam,

Table 1 Crop yields, texture, and macro-nutrient variability in farmers' fields in SGS, NGS, and SS in northeast Nigeria

Crop yields/soil properties	SGS		Probability level	NGS		Probability level	SS		Probability level
	Compound	Bush		Compound	Bush		Compound	Bush	
Maize (kg ha^{-1})	2442.2	2753.5	NS	2513.6	2549.6	NS	2240.8	1764.7	NS
Soybean (kg ha^{-1})	1798.8	2527.0	$P < 0.05$	1701.2	1906.4	NS	1816.3	1553.0	NS
Sand (g kg^{-1})	271.5	287.1	NS	463.8	330.4	$P < 0.01$	465.6	455.7	NS
Silt (g kg^{-1})	380.0	379.8	NS	264.3	381.3	$P < 0.01$	387.5	375.0	NS
Clay (g kg^{-1})	344.5	333.5	NS	272.0	289.4	NS	146.9	169.3	NS
pH	6.40	6.33	NS	6.48	6.42	NS	6.65	7.41	NS
Organic C (g kg^{-1})	12.1	10.9	NS	9.2	12.6	NS	8.9	8.2	NS
Total N (g kg^{-1})	1.6	1.5	NS	1.5	1.9	$P < 0.05$	1.5	1.5	NS
Available P (mg kg^{-1})	4.24	4.65	NS	4.70	3.68	$P < 0.01$	2.51	1.49	$P < 0.05$
Exchangeable K (cmol kg^{-1})	0.60	0.58	NS	0.54	0.68	$P < 0.05$	0.61	0.44	$P < 0.05$

NS, not significant

Table 2 Proportions (%) of field types and their physical and chemical status

Soil properties	SGS		NGS		SS	
	Compound (25)	Bush (25)	Compound (21)	Bush (33)	Compound (18)	Bush (15)
Coarse texture (<180 g kg^{-1} clay)	4.0	8.0	23.8	3.0	66.7	33.3
Medium texture (180–350 g kg^{-1} clay)	48.0	40.0	47.6	81.8	33.0	60.0
Fine texture (>350 g kg^{-1} clay)	48.0	52.0	28.6	15.2	0.00	6.7
pH						
Strongly/moderately acidic (5.1–6.0)	16.0	20.0	33.3	30.3	22.2	40.0
Slightly acidic/neutral (6.1–7.3)	80.0	80.0	57.1	63.6	66.7	60.0
Slightly alkaline (7.4–7.8)	4.0	0.00	9.5	6.1	11.1	0.00
Organic carbon (g kg^{-1})						
Very low/low (<10.5)	44.0	52.0	61.9	30.3	72.2	80.0
Moderate (10.5–15.0)	40.0	32.0	28.6	45.5	16.7	20.0
High/very high (>15.0)	16.0	16.0	9.5	24.2	11.1	0.00
Total N (g kg^{-1})						
Very low/low(<1.0)	12.0	28.0	28.6	18.2	38.9	6.7
Moderate–medium (1.01–2.00)	60.0	52.0	47.6	42.4	33.3	73.3
Moderate–high (>2.01)	20.0	12.0	19.0	36.4	27.8	20.0
Available P (mg kg^{-1})						
Low (<7.00)	80.0	68.0	90.5	100	94.4	100
Medium (7–20)	20.0	32.0	9.5	0.00	5.6	0.00
Exchangeable K (cmol kg^{-1})						
Very low/low (<0.30)	8.0	4.0	9.5	3.0	16.7	6.7
Moderate (0.30–0.60)	52.0	56.0	66.7	51.5	44.4	86.6
High/very high (>0.60)	40.0	40.0	23.8	45.5	38.9	6.7

Personal Communication). The texture of fields in the SS was coarser than that of those in the SGS and NGS because of the nature of the parent materials from which the soils were formed (Carroll and Klinkenberg, 1972). Texture is a stable soil parameter that cannot be easily changed through agronomic practice. However, in central Kenya, Murage et al. (2000) observed and attributed differences in the chemical and biological soil properties of productive and non-productive fields within a farm to past soil management since sand and clay contents did not vary between soil categories.

Soil Reaction

The pH range of both fields was variable in SGS and SS but was strongly acidic to slightly alkaline in NGS (overall range, 5.23–8.03). Across all zones, over 57.1% of the fields were slightly acidic to neutral in reaction. Most fields in SGS were in this category than were those in NGS and SS. This was also attributed to differences in parent materials (Carroll and Klinkenberg, 1972). Most of the fields that were

strongly/moderately acidic were in NGS and SS. In SS, most bush fields were strongly/moderately acidic than compound fields because the use of ashes from the household might have reduced acidity on the compound fields. The moderately acidic to strongly acidic soil reaction in some of the fields indicates fragility.

Organic Carbon

Organic carbon content range of both fields was variable in SGS and NGS but was very low to high in SS (overall range 2.2–23.6 g kg^{-1}). Organic C was very low to low in over 52% of the fields except compound fields in SGS and bush fields in NGS. The compound fields in NGS and fields in SS were more depleted in organic carbon partly because less rainfall and reduced vegetation cover resulted in a lower biomass production and a lower rate of addition of organic residues (Kwari et al., 1999).

Also, crop residues are burned or removed from the farmland for other uses (Kwari and Batey, 1991). Most fields in SS contained low organic carbon than did those in NGS because organic matter decayed faster as the soils in SS were mostly medium and coarse textured. The variation in organic carbon between the fields was not as clear as reported by Prudencio (1993), who observed that soil C varied at farm level from 2.0 g kg^{-1} in bush fields to 22.0 g kg^{-1} in homestead fields in two agro-climatic zones of the Mossi plateau in Burkina Faso.

Total Nitrogen

The range of total N in both fields (0.14–4.90 g kg^{-1}) was variable in all zones. Total N was moderate to medium in over 47.6% of the fields, except for bush fields in NGS and compound fields in SS. In western Kenya, Vanlauwe et al. (2006) observed different levels of soil fertility status for different fields within a farm, and total soil N decreased with the distance from the homestead (from 1.3 to 1.06 g kg^{-1}).

Most compound fields contained low total N than did bush fields in NGS and SS but the reverse in SGS possibly because continuous cropping leads to removal of N by crops. Bush fields in SGS are cultivated relatively more intensively than in NGS and SS because there is limited cultivable land due to the hilly/rocky nature of the terrain in the zone. This has serious implications for sustainable crop production since N is taken up by crops in greater quantities and limits crop production more than any other nutrient. To sustain crop growth, N needs to be applied yearly (Giller et al., 1997) and N sources are too expensive for the resource-poor farmers in the study area.

Available Phosphorus

Available P (range 0.30–11.30 mg kg^{-1}) was deficient in all fields. Vanlauwe et al. (2006) in western Kenya observed that Olsen-P decreased from 10.5 to 2.3 mg kg^{-1} with the distance from the homestead showing different soil fertility status for fields within a farm. The poor soil P status in the fields in SGS and NGS could be attributed to high P sorption due to high levels of Fe and Mn (Kwari et al., 1999). The low geological and organic P resources due to low organic matter and the sandy nature of the parent material from which the soils were derived could account for the deficient P status in fields in SS. During a pilot study in the area, a lack of response to P by crops was observed. Kwari and Batey (1991) and Smithson and Giller (2002) reported that P deficiency and/or severe P fixation are likely in acidic soils dominated by 1:1 clay minerals.

Since some legumes are capable of dissolving less available soil P sources, careful selection of crops that tolerate low soil P could increase P efficiency. Placement close to the seeds or crops to reduce the contact between fertilizer and soil could also improve P efficiency in such soils since P has a high affinity for soil constituents, is less mobile in the soil, and is required for early plant growth.

Exchangeable Potassium

Exchangeable K (range 0.15–1.79 cmol kg^{-1}) was moderate in over 51.5% of the fields except for compound fields in SS, suggesting adequate soil K. This agrees with Rayar (1988), who reported that non-intensive crop production in northern Nigeria may not be limited by K as farmers apply NPK fertilizers and root/tuber crops are not the major crops grown in the area.

Crop Yields

The yields of maize and soybean on farmers' fields are summarized (Table 3). Except for soybean yields in SGS, there are no significant differences in crop yields between the fields in all the zones. When disaggregated according to types of fields, the maize yields were 11% higher in bush fields than in compound fields in SGS and equal in NGS and 21% higher in compound fields than in bush fields in SS. The soybean yields were 29% higher in bush fields than in compound fields in SGS and 11% higher in NGS, but soybean yields were 14.5% higher in compound fields than in bush fields in SS. Across all fields, crop yields were higher in SGS and NGS than in SS.

The grain yields of maize and soybean from the bush fields in SGS and NGS were higher than those from compound fields, though there were no significant differences in soil parameters between the fields in SGS, and only small differences in NGS (Table 1). Other factors probably related to soil physical condition may explain the differences in yields between the

Table 3 Multiple regression of selected soil parameters with soybean and maize grain yields

Soil variables	Parameter estimates	Standard error	T value	Pr > t
SGS/soybean bush fields				
pH	5248.26	1377.49	3.81	0.0005
Organic C	3619.66	1285.67	2.82	0.0079
SGS/soybean compound fields				
Sand	101.35	24.27	4.18	0.0002
Silt	82.64	25.75	3.21	0.0030
Clay	108.28	24.93	4.34	0.0001
Available P	61.86	27.81	2.22	0.0333
Exchangeable K	1756.26	248.09	7.08	<0.0001
NGS/soybean bush fields				
Clay	469.50	227.00	2.07	0.0438
Total N	3505.08	1112.68	3.15	0.0028
Available P	162.07	62.20	2.61	0.0120
NGS/maize compound fields				
Sand	81.72	30.56	2.67	0.0123
Silt	126.30	48.13	2.62	0.0137
SS/maize bush fields				
Organic C	3594.69	1032.78	3.48	0.0029
Available P	633.04	174.76	3.62	0.0021
Exchangeable K	11341	2598.21	4.36	0.0004
SS/maize compound fields				
Organic C	3461.56	1025.41	3.38	0.0032

fields. The lower crop yields in SS than in SGS and NGS could be due to low productivity because of lower rainfall and water retention, and low fertilizer use efficiency since the soils were mainly coarse textured and contained low levels of organic matter.

In SS, mean crop yields were higher in compound fields than in bush fields because of widespread use of organic manure on the compound fields. Soil acidity (40.0% of the bush fields in SS were strongly/moderately acidic) could also explain this observation. Carsky et al. (1998) recorded the highest yields of maize (approximately 4 t ha^{-1}) on fields near the homestead (compound fields) where organic carbon values were 2.0–2.5% in the NGS of Nigeria and Vanlauwe et al. (2006) reported that maize grain yields decreased from 2.59 t ha^{-1} in homestead fields to 1.59 t ha^{-1} in distant fields in western Kenya, reflecting variable soil fertility status for fields within a farm.

Relationship Between Crop Yields and Soil Parameters

The multiple regression analysis relating soybean and maize yields with soil parameters is presented in Table 3. Soybean yield in SGS was a function of sand ($P = 0.0002$) and silt ($P = 0.0030$) contents in compound fields. Clay content significantly influenced soybean grain yield in compound fields in SGS ($P = 0.0001$) and bush fields in NGS ($P = 0.0438$). Soil pH significantly influenced soybean grain yield only in bush fields in SGS ($P = 0.0005$). Soybean yield was a function of organic carbon in bush fields ($P = 0.0079$) in SGS. Exchangeable K was related to soybean yield in compound fields in SGS ($P \leq 0.0001$). Total N had a significant effect on soybean grain yield in bush fields in NGS ($P = 0.0028$). The yield of soybean was related to available P in compound fields in SGS ($P = 0.0333$) and bush fields in NGS ($P = 0.0120$).

Maize yield in NGS was a function of sand ($P = 0.0123$) and silt ($P = 0.0137$) contents in compound fields. Maize yield in SS was a function of organic carbon in bush fields ($P = 0.0029$) and compound fields ($P = 0.0032$). The yield of maize was related to available P in bush fields in SS ($P = 0.0021$) despite its deficient status. Exchangeable K was an important variable for maize grain yields in bush fields in SS ($P = 0.0004$). The yields of maize and soybean were related to sand and silt contents (Table 3), maybe because the crops grow well on different soil types, provided these are deep and well drained; soils used for maize production are suitable for soybean (Javaheri and Baudoin, 2001). Clay content significantly influenced soybean grain yield in compound fields in SGS ($P = 0.0001$) and bush fields in NGS ($P = 0.0438$), probably because soybean, as compared with other legumes, is relatively tolerant to temporary water logging. Soil pH was related to soybean yields in bush fields in SGS ($P = 0.0005$), probably because the availability of nutrients such as Ca, Mg, K, and Mo to soybean depends on soil pH (Javaheri and Baudoin, 2001). Also N_2 fixation by soybean is affected by soil pH.

Organic carbon was related to soybean yields in bush fields in SGS and maize yields in fields in SS, probably because organic matter is a reservoir of soil nutrients and soil nutrients are better utilized for higher grain yields of crops. Organic matter also improves the nutrient use efficiency of applied fertilizer and increases crop yields through the enhanced availability of soil water (Ramamurthy and Shivashankar, 1996; Vanlauwe et al., 2001). Maize requires good nutrient and water availability (Carsky et al., 1998). Maize is known to be susceptible to drought during flowering and the first weeks of grain filling (Heisey and Edmeades, 1999). However, Carsky et al. (1998) reported that soil organic carbon was not related to maize yield in NGS of Nigeria because farmers applied variable amounts of fertilizers and manure and the maize crop was managed in diverse ways. The significant effect of total N on soybean yields in bush fields in NGS suggests that though soybean can obtain up to 85% of its N requirement through biological N_2 fixation, a starter amount may be essential if the N status of the soil is low (Javaheri and Baudoin, 2001).

The yield of soybean was related to available P in compound fields in SGS ($P = 0.0333$) and bush fields in NGS ($P = 0.0120$), while the yield of maize was related to available P in bush fields in SS ($P = 0.0021$). This suggests that P is an important nutrient for crop production in northern Nigeria (Vanlauwe et al., 2002). Phosphorus is required by soybean for root development and N_2 fixation (Danso, 1992). It was reported that four soybean genotypes responded to P application at 20 kg ha^{-1} (PROSAB, 2006).

Despite the moderate K status, it was an important variable for soybean grain yield in compound fields in SGS ($P \leq 0.0001$) and for maize grain yields in bush fields in SS ($P = 0.0004$) because soybean, in particular, removes much more K from the soil than did many other crops. Javaheri and Baudoin (2001) stated that 2.5 t ha^{-1} soybean consumes 45 kg K_2O compared with 5 t ha^{-1} of maize that consumes 35 kg K_2O.

Conclusions

The soil test results showed some variability in soil fertility parameters in the compound and bush fields in the zones. Variability in field soil fertility status in a farm influences the efficiency of resource use to increase crop yields and needs to be taken into account when prioritizing and targeting recommendations for farmers. The preferential use of resources where they are most profitable could improve productivity. The multiple regression results showed some differences between soybean and maize. Organic C was a determinant of soybean yield in bush fields in SGS and of maize yield in compound and bush fields in SS. Available P was a determinant of soybean yield in compound fields in SGS and in bush fields in NGS and of maize yields in bush fields in SS. Sand, silt, clay, and exchangeable K were two-time determinants and pH and total N were one-time determinants of soybean or maize yields. The addition of organic inputs (such as animal manure, compost, household waste/municipal refuse, and mulching with crop residues) to raise organic C, the application of inorganic fertilizer, especially P, and crop rotation using improved varieties of maize and soybean to improve P availability but targeted to the specific fields are recommended.

Acknowledgments This study was funded by the Canadian International Development Agency (CIDA) through the International Institute of Tropical Agriculture (IITA), Ibadan, Nigeria. Soil sampling and analysis was carried out by the staff of the Soil Science Department, University of Maiduguri, Nigeria.

References

Carroll DM, Klinkenberg K (1972) Soils. In: Tuley P (ed) The land resources of north-east Nigeria, vol I. The environment. Land resources study No. 9 and map 3, Surbiton, UK, pp 85–120

Carsky RJ, Nokoe S, Lagoke STO, Kim SK (1998) Maize yield determinants in farmer managed trials in the Nigerian Northern Guinea Savanna. Expl Agric 34:407–422

Danso SKA (1992) Biological nitrogen fixation in tropical agroecosystems: twenty years of biological nitrogen fixation research in Africa. In: Munlongoy K, Gueye M, Spencer DSC (eds) Biological nitrogen fixation and sustainability of tropical agriculture. Wiley, Chichester, pp 3–13

FAO (1974) Irrigation suitability classification. FAO Soils Bulletin 22. FAO, Rome

Giller KE, Cadisch G, Ehaliotis C, Adams E, Sakala WD, Mafongoya PL (1997) Building soil nitrogen capital in Africa. In: Buresh RJ, Sanchez PA, Calhoun F (eds)

Replenishing soil fertility in Africa. Soil Science Society of America Special Publication 51, Soil Science Society of America, Madison, WI, pp 151–192

Heisey PW, Edmeades GO (1999) Maize production in drought stressed environments: technical options and research resource allocation. In: World maize facts and trends. International Maize and Wheat Improvement Centre, Mexico D.F., pp 1–36, 1997/1998

Javaheri F, Baudoin JP (2001) Soybean (*Glycine max* (L.) Merril). In: Raemaekers RH (ed) Crop production in tropical Africa. Directorate General for International Cooperation (DGIC), Karmelietenstraat 15-Rue des Petitis Carmes 15, B-1000, Brussels, pp 809–828

Kassam AH, Kowal J, Dagg M, Harrison MN (1975) Maize in West Africa and its potential in the savanna. World Crops 27:73–78

Kwari JD, Batey T (1991) Effect of heating on phosphorus sorption and availability in some northeast Nigerian soils. J Soil Sci 42:381–388

Kwari JD, Nwaka GIC, Mordi RI (1999) Studies on selected soil fertility parameters in soils of northeastern Nigeria. I. Phosphate sorption. J Arid Agric 9:61–70

Murage EW, Karanja NK, Smithson PC, Woomer PL (2000) Diagnostic indicators of soil quality in productive and non-productive smallholders fields of Kenya's central highlands. Agric Ecosyst Environ 79:1–8

PROSAB (2004) Promoting sustainable agriculture in Borno state. Synthesis of livelihood analysis in three contrasting agroecological zones. PROSAB, Borno State, 45pp

PROSAB (2006) Annual progress report (April 2005–March 2006). International Institute of Tropical Agriculture (IITA) Ibadan, Nigeria, 51pp

Prudencio CY (1993) Ring management of soils and crops in the west African semi-arid tropics: the case of the Mossi farming systems in Burkina Faso. Agric Ecosyst Environ 47:237–264

Ramamurthy V, Shivashankar K (1996) Residual effect of organic matter and phosphorus on growth, yield and quality of maize (*Zea mays*). Ind J Agron 41:247–251

Rayar AJ (1988) Decline in fertility of a semi-arid savanna soil of north-eastern Nigeria under continuous cropping. J Arid Agric 1:227–241

Roberts VG, Adey S, Manson AD (2003) An investigation into soil fertility in two resource-poor farming communities in KwaZulu-Natal (South Africa). S Afr J Plant Soil 20:146–151

SAS (1990) Statistical analysis systems, SAS/STAT users guide, Version 6, 4th edn. SAS Institute, Cary, NC

Schlecht E, Buerkert A, Tielkes E, Bationo A (2006) A critical analysis of challenges and opportunities for soil fertility restoration in Sudano-Sahelian West Africa. Nutr Cycling Agroecosyst 76:109–136

Scoones I, Toulmin C (1999) Policies for soil fertility management in Africa: a report prepared for the department for international development. International Institute for Environment and Development, Russell Press, Nottingham, 128pp

Smithson PC, Giller KE (2002) Appropriate farm management practices for alleviating N and P deficiencies in low nutrient soils of the tropics. In: Adu-Gyamfi JJ (ed) Food security in nutrient stressed environments: exploiting plants genetic capabilities. Kluwer, The Netherlands, pp 277–288

Van Reeuwijk LP (1992) Procedures for soil analysis, 3rd edn. International Soil Reference and Information Centre, Wageningen

Vanlauwe B, Aihou K, Aman S, Iwuafor ENO, Tossah BK, Diels J, Sanginga N, Lyasse O, Merckx R, Deckers J (2001) Maize yield as affected by organic inputs and urea in the West African Savanna. Agron J 93:1191–1199

Vanlauwe B, Diels J, Lyassee O, Aihou K, Iwuafor ENO, Sanginga N, Merckx R, Deckers J (2002) Fertility status of soils of the derived savanna and northern Guinea savanna benchmarks and responses to major plant nutrients as influenced by soil type and land use management. Nutr Cycling Agroecosyst 62:139–150

Vanlauwe B, Tittonell P, Mukalama J (2006) Within-farm soil fertility gradients affect response of maize to fertilizer application in western Kenya. Nutr Cycling Agroecosyst 76: 171–182

Woomer PL, Bekunda MA, Karanja NK, Moorehouse T, Okalebo JR (1998) Agricultural resource management by smallholder farmers in East Africa. Nat Res 34:2

An Evaluation of Lucerne Varieties Suitable for Different Agro-ecological Zones in Kenya

B.A. Lukuyu, J.N. Methu, D. Mwangi, J. Kirui, S.W. Mwendia, J. Wamalwa, A. Kavatha, G.N. Ngae, and G.N. Mbure

Abstract In order to choose suitable varieties with high yield and good quality for cultivation, eight lucerne varieties, including seven foreign ones (WL 625 HQ, KKS 9595, WL 414, Robusta, KKS 3864, SA Standard, WL 525 HQ), and a local check Hunter River were studied on farm in seven different agro-ecological zones (AEZ) in the long and short rain seasons in 2006 using a randomized block design with two replications. Each plot was cut two times in both seasons to evaluate herbage production of the tested varieties. Results showed that dry matter yield (DM) from varieties was significantly different between sites in both short ($P <0.05$) and long ($P < 0.001$) rains seasons. Robusta and WL 525 HQ yielded significantly ($P < 0.05$) more DM compared to the local check Hunter River in both wet mid- and highland zones while SA Standard yielded significantly ($P < 0.001$) more DM in high dry land zones across all seasons. There was a significant AEZ × variety interaction on leaf:stem ratio in both short ($P < 0.001$) and long ($P < 0.05$) rains seasons. In both rain seasons, WL 414, WL 625 HQ and KKS 3864 had significantly ($P < 0.05$) higher leaf:stem ratio in that order to the local check Hunter River across all zones. Similarly, in both rain seasons, age at harvested significantly affected DM yield ($P < 0.05$) and leaf:stem ratio ($P < 0.005$) of all varieties across all zones.

Keywords DM yield · Leaf:stem ratio · Lucerne · Harvesting age · Varieties

Introduction

Despite the opportunities that dairy farming offers to smallholders, they often do not have the necessary feed resources to achieve their animals' full genetic potential for milk production. The poor quality (forages with low acid detergent fibre (ADF), neutral detergent fibre (NDF), lignin, stem, dry matter (DM) and low crude protein (CP), leaf:stem ratio, ash, digestibility) and quantity of feed resources, especially during the dry season, are the greatest constraints to increasing livestock production (of both milk and meat) in Kenya (Orodho, 2006). As dairy enterprises increase in number, forage deficits are predicted also to increase, particularly in densely populated areas with intensive farming systems and small (and diminishing) land holdings (Nicholson et al., 1999).

Lucerne or alfalfa (*Medicago sativa*) has the potential to have a substantial impact on the livelihoods of smallholder dairy farmers in East Africa (Harris et al., 2006). Without lucerne, the only feed available on the farm during dry season is dry grass and crop residues such as maize or sorghum stover, most of which are deficient in protein, energy and certain minerals, such as phosphorus and sulphur (Methu et al., 1996). Some farmers alleviate this problem by buying dairy concentrate (dairy meal), but this is of variable quality and too expensive to be used by the poorer farmers (Staal et al., 1998). By growing their own high-protein lucerne on farm, instead of buying dairy meal, farmers can immediately save money, while those who could not previously afford supplements can achieve substantial increases in milk production for income and/or family consumption by improving the nutrition of their animals. Promotion of lucerne to smallholder dairy

B.A. Lukuyu (✉)
Kenya Agricultural Research Institute, Nairobi, Kenya
e-mail: b.lukuyu@cgiar.org

A. Bationo et al. (eds.), *Innovations as Key to the Green Revolution in Africa*,
DOI 10.1007/978-90-481-2543-2_48, © Springer Science+Business Media B.V. 2011

farmers over the last 10–15 years has been hindered by its high specificity in its *Rhizobium* requirements, soil acidity levels preferring neutral soils, lack of specific management recommendations for different agro-ecological zones, lack of seed/planting material and lack of suitably adapted legume varieties (Foy, 1984). Lucerne has been identified as an appropriate legume with high digestibility and N content, but varieties suited to different agro-ecological zones are not available. Therefore, this study aimed at testing suitable lucerne varieties for different agro-ecological zones in Kenya.

Materials and Methods

Study Sites

The study was carried out at seven sites selected from different agro-ecological zones of Kenya. Details of experimental sites are shown in Table 1. The study was carried during the long and short rain seasons in 2006. Trials in the dry (LH5) and wet (LH2) highlands were set up with Wenyitie Dairy Club and Subukia FFS, respectively, while in the dry and wet midlands (UM4) they were set up with Rongai Livestock FFS and Amboni FFS, respectively. Trials in the frost-prone areas (UH2), lowland (LM1) and the irrigation system (UM4) were set up with 25 Comrades FFSs, Tosha farm and Nehemiah International Kisumu, respectively. However, due to various reasons, results for UH2, LM1 and the irrigation system are not reported here.

Experimental Designs

The design of trials comprised treatments of eight varieties × four harvesting stages replicated two times in a randomized completely block design in each of the sites. Seven new varieties, namely WL 625 HQ, KKS 9595, WL 414, Robusta, KKS 3864, SA Standard and WL 525 HQ, and a local check Hunter River were planted. Qualities of varieties were selected on the basis of suitability to different agro-ecological conditions, i.e. drought and cold tolerance, improved quality, disease resistance, persistence and early maturing. The varieties were imported from CROPLAN Genetics of United States, a subsidiary company of Land O' Lakes through their South African agent. Plots were designed to be harvested at four stages. The earliest harvest was planned at the onset of budding (up to 5% of the buds visible) while the second, third and fourth harvesting were planned at the interval of every 7 days thereafter. However, harvesting 1–2 and 3–4 were combined, respectively, due to field logistics and constraints in budget. A summary of the actual harvesting schedule is shown in Table 2.

Most experimental areas were sloping gently, and for this reason blocks and plots within blocks were drilled across the gradient. Plot sizes of 2 × 2 m were measured. Seedbeds were prepared finely. Seeds were broadcast at the start of rains. Seeds were sown at a rate of 5 kg per acre and covered to a depth of approximately 0.6 cm and made firm. About 100 kg/ha of triple superphosphate fertilizer was applied before planting. Weeds were uprooted until full ground cover was achieved.

To determine forage herbage production, plots were harvested approximately 3–5 cm above the ground from the entire plot. Total fresh weights of lucerne from the final harvest area of each plot were measured in the field using a 1000 g resolution spring balance. The whole amount of herbage was taken as a sample for analysis after weights had been taken. Samples were transported to the laboratory and fractionated into leaf and stem. A sub-sample of fresh material of each partition was collected and dried in a large ventilated

Table 1 The characteristics of trial sites

Agro-ecological zone	District zone	Elevation (m above sea level)	Soil pH	Average rainfall (mm) per annum
Wet highland (frost-prone area) (UH2)	Nyandarua	2400–3000	6.32	1200–1400
Dry highland (LH5)	Nyandarua	2190–2280	5.77	800–900
Wet highland (LH2)	Nakuru	2070–2400	5.46	1000–1200
Wet midland (UM4)	Nyeri	1830–2100	5.84	1000–1200
Dry midland (UM4)	Nakuru	1840–2010	5.82	800–900
Irrigation system (UM4)	Thika	1360–1520	6.80	700–900
Lowland (LM1) wet, hot and humid zone	Miwani, Kisumu	1200–1500	6.35	1200–1400

Table 2 The harvesting schedule

	Planting date	Stage 1 harvest date	Stage 2 harvest date	Stage 1 harvest (DPP)	Stage 2 harvest (DPP)
Dry highland (LH5)	11/04/2006	10/07/2006	27/07/2006	90	107
Wet midland (UM4)	07/04/2006	11/07/2006	27/07/2006	116	140
Dry midland (UM4)	06/04/2006	31/07/2006	24/08/2006	106	134
Wet highland (LH2)	13/04/2006	28/07/2006	25/08/2006	95	111

oven at 60°C to constant weight to determine DM. All the data were analysed statistically using the Genstat statistical package (Lawes Agricultural Trust, 2000). Data were also subjected to generalized regressions to determine whether differences between new varieties and the local check were significant. F ratios were considered significant at the $P \leq 0.05$ level.

Results

The main effect means for variety, agro-ecological zones and harvesting stage on DM yields and leaf:stem ratio are shown in Tables 3 and 4, respectively. All varieties behaved the same with respect to DM yields in different agro-ecological zones (Table 3). The highest increases in DM yields ($P < 0.05$) of 34, 20, 17 and 17% relative to the local check Hunter River were observed with SA Standard, WL 525 HQ, WL 625 HQ and KKS 9595, respectively (Table 3). A 16–23%

more ($P < 0.001$) biomass was recorded in the highlands compared to the midlands (Table 3). However, leaf:stem ratio increased ($P < 0.001$) in the midlands by 36–50% compared to the highlands (Table 4). The reduced leaf:stem ratio in the highlands may be attributed to differences in solar radiation and mean temperatures between mid- and highlands that affect partitioning between shoots and stems (Hamish et al., 2006). The highest increases in ($P < 0.001$) leaf:stem ratio of 81, 36 and 24% relative to the local check Hunter River were observed with WL 414, WL 625 HQ and KKS 3864, respectively (Table 4). Overall, harvesting lucerne at 2 weeks post-blooming rather than 5% blooming increased ($P < 0.001$) yields by 25% (Table 3); however, the leaf:stem ratio reduced ($P < 0.001$) by 20% during the initial harvest but increased ($P < 0.001$) by 18% during the first ratoon harvest (Table 4).

The effects of agro-ecological zone (AEZ) and variety on DM yields of lucerne are shown in Fig. 1.

Table 3 Main effect means for variety, agro-ecological zones and harvest stage on lucerne dry matter yields

Treatments		Initial harvest (t DM/ha)	1st Ratoon (t DM/ha)	Total yield (t DM/ha)
Variety	Hunter River	1.2	1.2	2.4
	KKS 3864	1.5	1.1	2.6
	KKS 9595	1.5	1.3	2.8
	Robusta	1.5	1.4	2.9
	SA Standard	1.9	1.4	3.2
	WL 414	1.2	1.3	2.5
	WL 525 HQ	1.5	1.3	2.9
	WL 625 HQ	1.4	1.4	2.8
AEZ	Dry highland (LH5)	1.6	1.4	3.1
	Wet midland (UM4)	0.9	1.3	2.3
	Dry midland (UM4)	1.6	1.0	2.6
	Wet highland (LH2)	1.6	1.4	3.0
Harvest stage	Onset of blooming	1.1	1.3	2.4
	2 weeks post-blooming	1.8	1.3	3.2
Sed and sig.	Variety	0.17**	0.13[ns]	0.26*
	AEZ	0.13***	0.10***	0.20***
	Harvest age	0.09***	0.06[ns]	0.13***
	AEZ × variety interaction	0.37[ns]	0.27[ns]	0.55[ns]
	AEZ × harvest stage	0.19***	0.18***	0.28***

***Significant at 0.001 level, **significant at 0.01 level, *significant at 0.05 level
ns not significant, *sed* standard error of difference

Table 4 Main effect means for variety, agro-ecological zones and harvest stage on lucerne leaf:stem ratios

Treatments		Initial harvest (leaf:stem ratio)	1st ratoon (leaf:stem ratio)	Mean (leaf:stem ratio)
Variety	Hunter River	1.5	1.3	1.4
	KKS 3864	1.8	1.7	1.7
	KKS 9595	1.8	1.4	1.6
	Robusta	1.6	1.3	1.4
	SA Standard	1.6	1.3	1.5
	WL 414	2.7	2.3	2.5
	WL 525 HQ	1.7	1.3	1.4
	WL 625 HQ	2.0	1.8	1.9
AEZ	Dry highland (LH5)	1.7	1.1	1.4
	Wet midland (UM4)	2.4	2.3	2.4
	Dry midland (UM4)	1.8	2.5	2.2
	Wet highland (LH2)	1.6	0.9	1.2
Harvest stage	Onset of blooming	2.0	1.4	1.7
	2 weeks post-blooming	1.6	1.7	1.7
Sed and sig.	Variety	0.2***	0.13***	0.13***
	AEZ	0.1***	0.10***	0.10***
	Harvest age	0.1***	0.05***	0.07^{ns}
	AEZ × variety interaction	0.37***	0.27***	0.28***
	AEZ × harvest stage	0.05*	0.18***	0.19***

*** Significant at 0.001 level, * significant at 0.05 level
ns not significant, *sed* standard error of difference

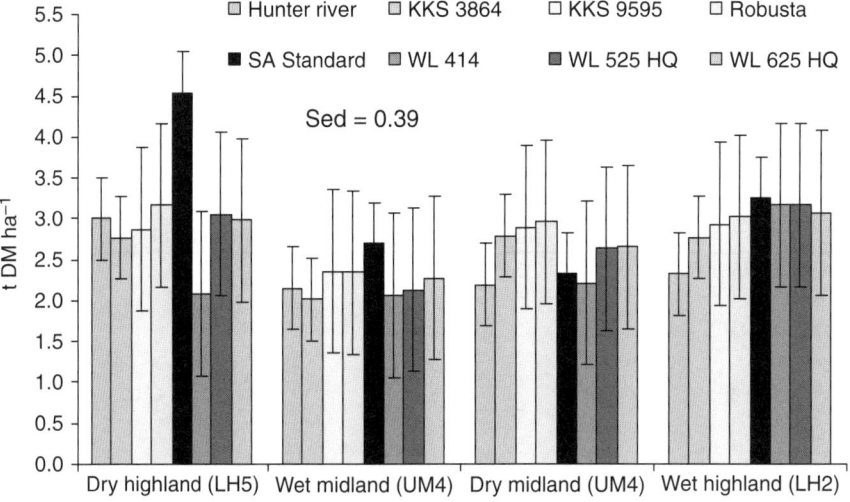

Fig. 1 Effects of agro-ecological zone and variety on yields of lucerne

Although there was no significant difference in lucerne yields between varieties within sites, the results indicate that SA Standard gave more DM yield by 26, 40 and 51% relative to the local check Hunter River in the wet midland (wet-UM4) and wet highland (LH2) and dry highland (LH5) zones, respectively. Robusta, KKS 9595, KKS 3864, WL 625 HQ and WL 525 HQ gave 35, 32, 27, 21 and 20% more DM yield relative to the local check in the dry midland (dry-UM4) zone.

In the LH2 all varieties recorded DM yield increases of between 19 and 40% relative to the local check. In addition to SA Standard, WL 414 and WL 625 HQ both gave 37% more DM yield relative to the local check (Fig. 1).

There was a significant ($P < 0.001$) interaction of AEZ × variety on lucerne leaf:stem ratio as shown in Fig. 2. WL 414 was the most outstanding variety in terms of leaf:stem ratio giving 50, 109, 115 and

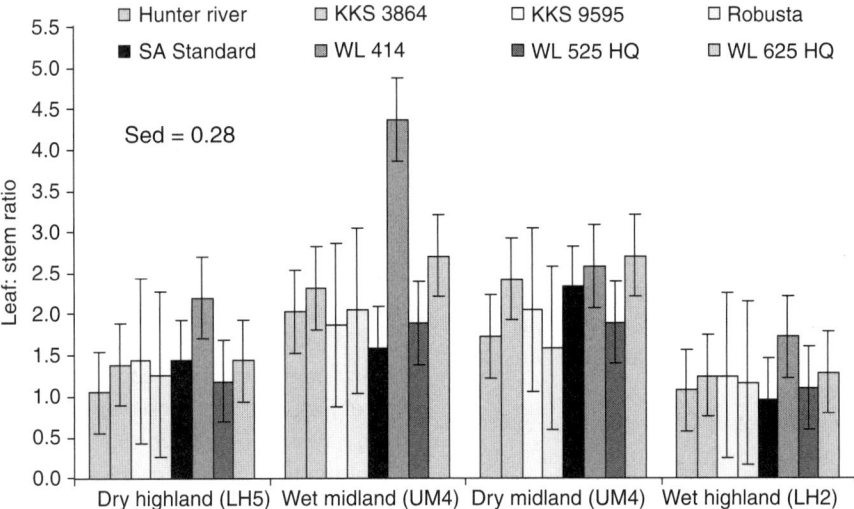

Fig. 2 Effects of agro-ecological zone and variety on leaf:stem ratio of lucerne

61% increase in leaf:stem ratio in dry-UM4, LH5, wet-UM4 and LH2, respectively, relative to the local check Hunter River. WL 625 HQ recorded an increase of 57, 37, 33 and 21% in leaf:stem ratio in dry-UM4, LH5, wet-UM4 and LH2, respectively, relative to the local check. SA Standard gave an increase of 35 and 37% in leaf:stem ratio in dry mid- and highlands, respectively, relative to the local check while KKS 9595 gave an increase of 36% in dry highland relative to the local check (Fig. 2).

The effects of agro-ecological zone (AEZ) and variety on DM yields and leaf:stem ratio of lucerne are shown in Table 5. The initial harvesting of lucerne at 2 weeks post-blooming rather than 5% blooming gave higher yields ($P < 0.001$) in all agro-ecological zones. Similar observation was made at ratoon harvest in wet-UM4 and LH2 only while yields decreased ($P < 0.001$) in dry-UM4 and LH5. However, leaf:stem ratio decreased ($P < 0.001$) when lucerne was cut at 2 weeks post-blooming rather than 5% blooming during the initial harvesting in all zones. On the other hand, leaf:stem ratio of the ratoon did not change in the LH5, reduced in the wet-UM4 and increased in dry-UM4 and LH2 (Table 5).

Table 5 The effects of agro-ecological zone and variety on dry matter yields and leaf:stem ratio of lucerne

Parameter	Harvest stage	Dry highland (LH5)	Wet midland (UM4)	Dry midland (UM4)	Wet highland (LH2)
Initial harvest yield (t DM/ha)	At 5% blooming	1.3	0.8	1.2	1.0
	2 weeks post-blooming	2.0	1.0	2.0	2.1
1st ratoon yield (t DM/ha)	At 5% blooming	1.5	1.1	1.4	1.2
	2 weeks post-blooming	1.4	1.6	0.6	1.5
Initial harvest (leaf:stem ratio)	At 5% blooming	1.8	3.0	2.0	1.7
	2 weeks post-blooming	1.7	1.7	1.7	1.4
1st ratoon (leaf:stem ratio)	At 5% blooming	1.1	3.0	1.1	0.8
	2 weeks post-blooming	1.1	1.7	3.9	1.0
Sed and sig.	Initial harvest yield	0.19***			
	1st ratoon yield	0.14***			
	Initial harvest (leaf:stem ratio)	0.16***			
	1st ratoon (leaf:stem ratio)	0.18***			

***Significant at 0.001 level
sed standard error of difference

Discussion

The current studies covered a wide range of environmental variables in soil type, rainfall, temperatures, pest and weed infestations and production systems. It provided a good representation of success probabilities in 1 year that may not be encountered for a number of years if working at one or two sites in one region. In addition, the trial was done in collaboration with farmers and provides farmers' perceptions of the new varieties (not reported here) that will be crucial to upscaling the 'best bets'.

The general trend in lower DM yields than expected across all sites may be ascribed to the fact that soil pH in all sites was on the acidic borderline ranging between 5.5 and 6.8. Soil pH of between 4.8 and 5.5 has been shown to reduce lucerne production (Foy, 1984). Field observation revealed that in most sites plants showed signs of yellowing and wilting mid-season following rainfall. These are signs of adverse effects of acidity on lucerne. Low yields may also be attributed to the fact that the data included results form the first harvesting. Results by Davis (1960) and Radcliffe and Judd (1970) have shown that to the first harvest will usually have low yields compared to subsequent yields. The tendency for higher DM yields from highlands than midlands irrespective of variety may be as a result of well above average rain during that period in the highlands.

WL 414 was outstanding in leaf:stem ratio. This may be because it has multifoliate (MF) leaves compared to other varieties that have trifoliate leaves (TF). Marinova et al. (2004) showed that multifoliate expression, based on the dry weight of MF leaves, was not correlated with dry matter (DM) yield but

was positively related with the percentage of leaves (leaf:stem ratio) in the total biomass. Marinova et al. (2004) further demonstrated that there is a positive relationship between DM yield and stem number per m^2. This may explain why there was a tendency for those varieties with high DM yield to have high stem:leaf ratio. The implication of this is that farmers may have to consider tradeoffs when selecting which varieties to plant between those that are high yielding and those with high quality (leaf:stem ratio). Table 6 summarizes possible 'best bets' for various agro-ecological zones deepening on whether farmers' objective is emphasis on yield or quality or a combination of both factors.

Cutting lucerne at a specific stage makes it possible to take into account the variation due to various environments and growth tempos. It is generally known that the quality of lucerne drops as the plant matures (Gonzalez et al., 2001). Results from this study showed that cutting lucerne at 5% blooming compared to 2 weeks post-blooming reduced yields by 20–52% but increases quality (leaf:stem ratio) by 15–43% in all agro-ecological zones. This confirms observations by Martiniello et al. (1997) that fodder yield and nutritive value are related to the phenological stage of the plants at harvest. They also agree with the results of Slake and Mason (1987) who found that cutting lucerne at the pre-flower bud stage, compared with cutting at the 10% bloom stage, reduced DM yield by 18% (16.4 vs. 13.5 t/ha) but increased crude protein content of the lucerne from 19.3 to 24%. In practical terms, changing harvest time would be more for influencing fodder quality. The choice of harvesting begins with the decision as what quality feed is required. Farmers who want high-quality lucerne will prefer a shorter stand and lower yield and vice versa. Therefore, the number of cuts,

Table 6 Potential 'best bets' based on yield, quality or a combination of both

AEZ	Varieties with a relatively balanced DM yield and leaf:stem ratio	Varieties with high DM yield but low leaf:stem ratio	Varieties with high leaf:stem ratio but low DM yield
Dry highland (LH5)	SA Standard	Robusta	KKS 3864, KKS 9595, WL 414 and WL 625 HQ
Wet midland (wet-UM4)	WL 414 and WL 625 HQ	KKS 3864, KKS 9595, Robusta, SA Standard and WL 525 HQ	KKS 3864
Dry midland (dry-UM4)	WL 625 HQ	SA Standard, KKS 9595 and Robusta	WL 414 and WL 625 HQ
Wet highland (LH2)	KKS 3864, KKS 9595 and WL 625 HQ	Robusta	WL 414 and SA Standard

AEZ agro-ecological zone

Table 7 Amount of extra fodder, supplement days and savings as a result of supplementing with lucerne for different agro-ecological zones

Parameter	Dry highland (LH5)	Wet midland (UM4)	Dry midland (UM4)	Wet highland (LH2)
Amount of extra fodder (kg DM)	29–262	32–94	102–131	119–158
Feeding days as supplement[a]	7–66	8–24	26–33	30–40
Savings by supplementing with lucerne (Ksh)[b]	348–3144	384–1128	1224–1572	1428–1896
Savings by supplementing with lucerne (US $)	5–49	6–18	19–25	22–30

[a]Supplementation at the rate of 4 kg DM/day
[b]Cost of dairy meal is Ksh 12 per kg (1 US $ = Ksh 64)

cutting date, stage of maturity, interval between cuts and cutting height are some of the factors that farmers may take into account in a harvest schedule. As has been demonstrated in this study, these different criteria will lead to differences in yield and quality. In addition to influencing yield and quality, it should be noted that regrowth after cutting lucerne may have already begun when the lucerne begins flowering. If harvesting is delayed, the growing points needed to produce the next harvest are cut off and regrowth retarded. On the other hand, cutting stage influences carbohydrate reserves which are stored in the form of non-structural carbohydrate in the roots and crown or in the remaining leaves and stems (Hamish et al., 2006). Frequent harvesting of immature lucerne (vegetative or bud stage) prevents sufficient vegetative regrowth to replace reserves of non-structural carbohydrates and leads to their depletion in the roots (Tabacco et al., 2000). It is therefore critical that farmers take into account these factors when developing a profitable harvest management programme.

In the present trial given the maturity age of between 106–161 and 90–154 days in the midlands and highlands, it is possible to have up to three and four harvests a year, respectively. However, it must be noted that lucerne must be allowed to grow out to full flowering once a year. This enables continued root development that ensures that lucerne is able to maintain a drainage buffer to at least a depth of 1.5 m and therefore contributes to a reduced drainage even on the most sodic and saline soils (Asseng et al., 2005). This helps in maintaining soil fertility.

The implication of these results shows that with DM yield increases of 17–34% from new varieties obtained in this study, farmers with 0.17 ha could produce between 102 and 262 kg DM of extra high-quality fodder. Staal et al. (1998) reported that 0.17 ha

is the mean land size on smallholder farm committed to growing napier grass. This amount of extra high-quality fodder could supplement (at the rate of 4 kg DM/day) a dairy cow weighing 350 kg for between 26 and 66 days. This could translate into a saving to a farmer who supplements his cow with commercial dairy concentrate at the rate of 4 kg/day of Ksh 1224–3144 (US $19–49; 1 US $: Ksh 64). Currently a kilogram of commercial dairy concentrate costs Ksh 12 (US $0.2). Amount of extra fodder, supplement days and savings as a result of supplementing with lucerne for different agro-ecological zones are shown in Table 7.

Conclusions

This study provides the justification for the recommendation of new lucerne variety packages for different agro-ecological zones. It shows that new lucerne varieties can be successfully established and grown in different agro-ecological zones, hence giving farmers' options. However, farmers have to make choices depending on whether their production objectives emphasize quantity or quality. SA Standard in LH5, WL 414 and WL 625 HQ in wet-UM4, WL 625 HQ in dry UM4, and KKS 3864, KKS 9595 and WL 625 HQ in LH2 were found to be potential 'best bets' that could meet the dual objective of quantity and quality. WL 414 was outstanding in leaf:stem ratio. If savings measured in this study can be repeated on smallholder farms it gives a major impetus to reducing milk production cost, hence improving farmers' profits. It reduces the cost of buying commercial dairy concentrates which are outreach to many smallholder farmers by up to 3144 (US $49).

Acknowledgements We are grateful to the extension staff from the Ministries of Agriculture and Livestock & Fisheries Development for assisting with data collection. We are grateful to farmers in the Farmer Field Schools for their cooperation in maintaining trials and data collection. This publication is an output from a collaborative research project between KARI and Land O' Lakes on lucerne funded by the United States Agency for International Development (USAID) through the SO7 project. The views expressed are not necessarily those of USAID.

References

Asseng S, Ward PR, Robertson MJ, Latta RA, Dolling PJ (2005) Soil water extraction and biomass production by lucerne in the south of Western Australia. Aust J Agric Res 56(4): 389–404

Davies WE (1960) The relative effect of frequency and time of cutting lucerne. Grass Forage Sci 15(3):262–269

Foy CD (1984) Physiological effects of hydrogen, aluminium and manganese toxicities in acid soils. In: Adams F (ed) Soil acidity and liming. Agronomy monograph 12, 2nd edn. American Social Agronomy, Madison, WI, pp 57–97

González J, Faría-Mármol J, Rodríguez CA, Alvir MR (2001) Effects of stage of harvest on the protein value of fresh lucerne for ruminants. Reprod Nutr Dev 41(5):381–392

Hamish E, Brown D, Moot J, Edmar IT (2006) Radiation use efficiency and biomass partitioning of lucerne (*Medicago sativa*) in a temperate climate. Eur J Agron 25(4):319–327

Harris R, Clune T, Peoples M, Swan T, Bellotti W, Chen W (2006) The role of nitrogen and in-crop lucerne suppression for increasing cereal performance in companion cropping systems. In: proceedings of the 13th Australian agronomy conference, Perth, Western Australia, Sept 2006

Lawes Agricultural Trust (2000) Genstat 6.1, Statistics Department, Rothamsted Experimental Station, AFRC Institute of Arable Crops Research, Harpenden, Hertfordshire, UK

Marinova D, Petkova D, Yancheva C (2004) Influence of expression of the multifoliolate trait on quantity and quality of lucerne (*M. sativa* L.) forage. Land use systems in grassland dominated regions. In: Proceedings of the 20th general meeting of the European grassland federation, Luzern, Switzerland, 21–24 June 2004

Martiniello P, Paoletti R, Berardo N (1997) Effect of phenological stages on dry matter and quality components in lucerne. Aust J Exp Agric 27(1):55–58

Methu JN, Owen E, Abate A, Scarr M (1996) Effect of level of offer of maize stover on the performance of lactating dairy cows. In: Proceedings of the 23rd Tanzania society of animal production conference, Arusha, 21–23 Aug 1996

Nicholson CF, Thornton PK, Mohammed L, Muinga RW, Mwamachi DM, Elbasha EH, Staal SJ, Thorpe W (1999) Smallholder dairy technology in coastal Kenya. An adoption and impact study. ILRI impact assessment series, No. 5, 59p

Orodho AB (2006) Intensive forage production for smallholder dairying in East Africa. In: Reynolds S, Frame J (eds) Grasslands: developments opportunities perspectives. FAO, Rome, Italy, 350p

Radcliffe P, Judd JC (1970) The influence of cutting frequency on the yield, composition and persistence of irrigated Lucerne. Aust J Exp Agric Anim Husb 10(42):48–52

Slarke RH, Mason WK (1987) Effect of growth stage at cutting on yield and quality of Lucerne cultivars from different dormancy groups in northern Victoria. Aust J Exp Agric 27(1):55–58

Staal SJ, Chege L, Kinyanjui M, Kimani A, Lukuyu B, Njubi D, Owango M, Tanner J, Thorpe W, Wambugu M (1998) Characterization of dairy systems supplying the Nairobi milk market. A pilot survey in Kiambu District for the identification of target groups of producers. Smallholder Dairy (R&D) Project. KARI, ILRI and Livestock Production Department (Ministry of Agriculture)

Tabacco E, Borreani G, Odoardi M, Reyneri A (2000) Effect of cutting frequency on dry matter yield and quality of lucerne (*Medicago sativa* L.) in the Po valley. Ital J Agron 6(1):27–33

Water Harvesting and Integrated Nutrient Management Options for Maize–Cowpea Production in Semi-arid Eastern Kenya

J.M. Miriti, A.O. Esilaba, A. Bationo, H.K. Cheruiyot, A.N. Kathuku, and P. Wakaba

Abstract Field experiments were conducted for 4 years at Emali, Makueni District, in Kenya to compare the effect of tied ridging and integrated nutrient management practices on the yield of rain-fed maize (*Zea mays* L.) and cowpeas (*Vigna unguiculata* L.). The main treatments were tied ridging and flat bed (traditional farmers' practice) as main plots. Farmyard manure (FYM) at 0 and 5 t ha^{-1} in a factorial combination with nitrogen (N) fertilizer at 0, 40, 80 and 120 kg N ha^{-1}, phosphorus (P) fertilizer at 0 and 40 kg P_2O_5 ha^{-1} and crop management were the sub-plots in a split plot in a randomized complete block design (RCBD). The results show that tied ridging significantly ($P < 0.05$) increased maize grain yields by 12% when compared to flat tillage. Maize grain and stover yields were significantly increased by 27 and 37%, respectively, when manure was applied during long rains seasons. Cowpea grain yields in tied ridging were 25% more than in flat tillage treatments, and the highest cowpea grain yield was 1354 kg ha^{-1}. Intercropping maize and cowpea lowered maize grain yield by more than 50 and 11% without and with nitrogen at 40 kg N ha^{-1}, respectively, and also reduced cowpea grain yields. However, crop rotation increased the yields of both maize and cowpea. Combining tied ridges with manure and inorganic fertilizers increased crop yields compared to when either of them was used separately. Thus, integration of in situ water management with integrated nutrient management has a potential in increasing food production in arid and semi-arid areas of Kenya.

Keywords Integrated nutrient management · Kenya · Semi-arid lands · Soil fertility · Water harvesting

Introduction

Agriculture is the mainstay of the Kenyan economy contributing approximately 55% of GDP. The sector further provides 80% employment and accounts for 60% export and 45% of the government revenue (Ragwa et al., 1998). However, more than 80% of Kenya is classified as arid and semi-arid lands (ASALs), characterized by low and erratic rainfall (between 100 and 700 mm per annum), high transpiration rates and fragile ecosystems that are unsuitable for permanent rain-fed agriculture. The low crop production in ASALs is often associated with inadequate rainfall, low soil plant nutrients and lack of appropriate farming strategies that are suited to the fragile ASAL ecosystems (Mbogoh, 2000). Farmers often use inappropriate farming practices that cause land degradation like soil erosion and soil fertility decline in the cultivated areas leading to crop yield reduction (Kilewe and Thomas, 1992). The combined effects of land degradation, low and erratic rainfall and low nutrient availability have resulted in food production lagging behind population growth to the extent that majority of smallholder farmers cannot adequately provide for their livelihood in these areas. For example, maize (*Zea mays* L.) yields in smallholder farm fields are as low as 1 t ha^{-1} which is much lower than the potential yields of 4 t ha^{-1} (Mbogoh, 2000).

J.M. Miriti (✉)
Desert Margins Programme, Kenya Agricultural Research Institute, Nairobi, Kenya
e-mail: jmmiriti@yahoo.co.uk

A. Bationo et al. (eds.), *Innovations as Key to the Green Revolution in Africa*,
DOI 10.1007/978-90-481-2543-2_49, © Springer Science+Business Media B.V. 2011

While the potential crop yields can be achieved by irrigation and judicial application of fertilizers, the high cost of acquiring these inputs remains beyond the reach of majority of smallholder farmers who form the bulk of agricultural producers. Efforts in improving land productivity in ASALs should therefore be aimed at introducing suitable on-farm rainwater harvesting and management systems that enable maximum utilization of rainwater for crop production.

The potential of improving crop production through use of rainwater harvesting and soil fertility management is widely cited (Biamah et al., 2000; Mellis, 1996; Njihia, 1977; Liniger, 1990). Different water harvesting techniques (i.e. level basin, tied ridges and conventional tillage) have been compared for their water retention, availability and suitability to crop production. Results from tied ridge techniques have given superior yields for different crops (Miriti et al., 2005; Itabari et al., 2003; Kipserem, 1996). However, some studies have shown that low crop yield levels may persist even with increased soil moisture if plant nutrients in the soil are inadequate due to the synergistic effects of combined water and nutrient management (Lal 1995; Fox and Rockström, 2003). Jensen et al. (2003) have reported enhanced crop response to rainfall and fertilizer, the soil supply of available N and crop yields as a result of ridging. Thus combined use of rainwater harvesting and nutrient management holds the key to ensuring higher and sustainable agricultural productivity in these areas.

Although integration of water harvesting and nutrient management is important in increasing and sustaining crop production, and also the maximization of the return from inputs such as fertilizers, there is limited knowledge on their interaction and crop response in the drylands of Kenya. The objective of this study was therefore to assess the effects of tied ridging and integrated nutrient management on maize and cowpea production in semi-arid eastern Kenya.

Materials and Methods

Site Description

The study was conducted for four seasons at Emali in the semi-arid areas of Makueni District in eastern Kenya for 4 years from 2003 to 2006 (Jaetzold

et al., 2006). The site has bimodal distribution of rainfall which is low and erratic. The short rains occur in October to January and the long rains in March to June. Temperatures are high giving rise to high evapotranspiration. The methods used for soil analysis were the following: pH (H_2O) in a 1:1 soil/water suspension, organic C by wet oxidation, available P by Bray2 and exchangeable bases (Ca^{2+}, Mg^{2+}, K^+ and Na^+) by ammonium acetate extraction (Okalebo et al., 2002). The soils are Ferralsols, and their chemical properties at 0–20 cm indicate that they are acidic (pH 4.8), while % organic carbon, % total N, available P and exchangeable K were low, moderate, high and very high, respectively (Okalebo et al., 2002).

Experimental Layout and Treatments

The experimental treatment arrangement was split–split–split plot with water harvesting versus conventional tillage as the main plots and manure versus no manure application as the sub-plots. The sub-plots were split into three cropping systems, i.e. (1) legume–cereal rotation, (2) legume–cereal intercrop and (3) continuous cereal. Each treatment was replicated four times in a randomized complete block design (RCBD).

Tillage

Tied ridges were used as the water harvesting method. Ridges (30 cm high) and ties (cross ridges, 20 cm high) were constructed to create a series of basins for storing water. The spacing of the ridges was 90 cm, and the cross ridges were made at 2.5 m interval using a hand hoe to prevent flow of runoff and ensure an even spread of captured water.

Fertilizer Application

In each cropping system different fertilizer treatments were applied. Goat manure was applied at 5 t ha^{-1} (Kihanda et al., 2004). The applied manure contained 1.7% N, 0.72% P, 3.66% K, 2.23% Ca and 0.59% Mg; org C 35% and ash 31.4%. Phosphorus (P) fertilizer was applied at 0 and 40 kg P_2O_5 ha^{-1} and nitrogen (N) at 0, 40, 80 and 120 kg N ha^{-1} depending on the treatment requirements. Triple superphosphate (TSP) and calcium ammonium nitrate (CAN) fertilizers were used as sources of P and N, respectively.

Planting

Maize (*Zea mays* L., dryland composite hybrid variety) and cowpea (*Vigna unguiculata* L., K80 variety) were used as the test crops. Maize and cowpeas were planted during the long rains (LR) and short rains (SR) from 2003 to 2006 in 25 m^2 plots. Maize and cowpea spacing was 90 × 30 cm in pure stands, while in the intercropping system, maize and cowpea were planted in the same row but in alternating hills at the same spacing. The crops were dry planted before the onset of the rains in each season. Thinning was done to a single plant per hill 1 month after planting. The thinned cowpea and maize plants were sampled for dry matter yield determination. The final maize and cowpea seed and dry matter yields were determined at harvesting time by sampling the inner 3 × 1.8 m area of each plot. All the data were subjected to statistical analyses using SAS package, and the treatment means were separated using the least significant difference (LSD) test. Each season was calculated separately, and later a combined comparison for all the seasons done.

Results

During the study period there were only five successful seasons (LR 2003, LR 2004, SR 2004, LR 2005 and LR 2006) where the crop yields were recorded. Only maize stover was harvested in 2 years (2003 and 2004) during the SR seasons. The other seasons there were crop failure due to inadequate rainfall. During the LR, cowpea grain yield was recorded only in 2004–2006. During LR 2003 season, tied ridges (water harvesting) did not have significant ($P < 0.05$) effects on maize and cowpea yields. Manure application (5 t ha^{-1}) had positive effects on both grain and stover yields. Water harvesting and manure application had a positive effect on both maize and cowpea yield ($P < 0.05$) level during the SR 2003 season.

During the LR 2004, water harvesting and manure application had a positive effect on both maize grain and stover yields. There were no maize yields recorded in SR 2004. Water harvesting and manure application had a positive effect on cowpea yields during LR 2004 and SR 2004 seasons.

During LR 2005 season, water harvesting and manure application had positive effects on both maize stover and grain yields. Manure application significantly reduced cowpea yields in LR 2005 season ($P < 0.05$ level). Both maize and cowpea crops failed during the SR 2005 season due to drought.

During LR 2006 season, tied ridging increased maize grain yield and lowered stover yields when compared with flat tillage. Manure treatments had higher maize grain yield than no manure treatments, and similar trend was observed with biomass. The cowpea grain yield under tied ridging recorded significantly higher grain and biomass yields than flat tillage, and a similar trend was recorded under biomass yield. Manure application had significantly higher cowpea yields than treatments without manure.

Maize Yields

Long Rains Season

The general crop response to manure application indicates that maize grain and stover yields increased by 27 and 37%, respectively. Rotating maize and cowpea under flat tillage did not result in better maize yields when compared to the continuous maize cropping treatments without inorganic fertilizers (control). However, when 40 kg P ha^{-1} + 40 kg N ha^{-1} fertilizer was applied to crop rotation, the grain yields improved from 942 to 1526 kg ha^{-1} (Table 1). This indicates that application of N is necessary in rotation of maize and cowpea in order to achieve reasonable maize yields. Intercropping maize and cowpea reduced maize yields except with the application of 40 kg P ha^{-1} + 80 kg N ha^{-1}, probably due to shading and competition for water and nutrients with cowpea.

Continuous planting of maize with application of 40 kg N ha^{-1} had the highest grain under continuous cropping, which was significantly different ($P < 0.05$) from the other continuous cropping systems. Application of N at 120 kg N ha^{-1} gave the

Table 1 Effects of tied ridges and INM on maize grain and stover yield (kg ha^{-1}) during the long rains seasons

| | Flat tillage | | | | Tied ridges | | | |
| | No manure | | Manure | | No manure | | Manure | |
Treatment	Grain	Stover	Grain	Stover	Grain	Stover	Grain	Stover
0 P + 0 N (*control*)	1368	2080	1095	2404	942	2301	1639	3046
Maize after cowpea								
40 P	892	1407	1269	2676	1109	2280	1342	2817
40 P + 40 N	1526	2431	1403	2694	1244	2494	1595	3339
Maize/cowpea intercrop								
40 P	1398	17,338	1401	2977	1025	1624	911	1801
40 P + 40 N	523	1293	1132	1750	965	2270	733	1684
Continuous maize cropping								
40 P	1016	1920	986	2589	1121	1923	1990	3597
40 P + 40 N	1106	2011	1361	3129	1534	2552	1630	3431
40 P + 80 N	1058	2223	1639	3068	1071	2269	1920	3848
40 P + 120 N	897	1944	1264	2431	862	1613	1591	2513
Mean	*1087*	*1894*	*1283*	*2635*	*1097*	*2147*	*1483*	*2897*
LSD	576	679	440	790	579	809	658	1346

lowest grain yield (Table 1). This indicates that application of 40 kg N ha^{-1} is adequate under flat tillage and no manure application. Similar trend in maize stover yields was also observed (Table 1).

In continuous maize cropping, the highest grain yield (1639 kg ha^{-1}) was recorded in 80 kg N ha^{-1} treatments when manure was applied in flat tillage. Although treatment differences were not significant, application of 120 kg N ha^{-1} resulted in yields lower than in 40 kg N ha^{-1}. Application of 40 kg N ha^{-1} to maize/cowpea intercrop reduced yields by 19% but increased the grain yields by 10% when compared to intercrop and rotations without N application. Stover yields had a similar trend as the grain yields, but the highest yield (3129 kg ha^{-1}) was recorded under continuous maize on application of 40 kg P ha^{-1} + 40 kg N ha^{-1} but was not significantly different ($P < 0.05$) from the other treatments.

When water harvesting technology (tied ridging) was used and no manure was applied, continuous maize cropping with 40 kg N ha^{-1} had the highest grain yield (1534 kg ha^{-1}). This was followed by maize/cowpea rotation with 40 kg N ha^{-1} (1244 kg ha^{-1}). The control, intercrops with 40 kg N ha^{-1} and continuous maize cropping with 120 kg N ha^{-1} treatments had the lowest grain yields. Maize stover yields had similar trends.

Maize under tied ridging in combination with manure gave the highest yields than undertied ridging without manure. In continuous maize cropping the highest grain yield of 1990 kg ha^{-1} was observed in treatments with 40 kg P ha^{-1}. Maize/cowpea intercrops gave the lowest yields compared to other treatments. For example, the intercrops with and without N application had yields of 911 and 733 kg ha^{-1}, respectively, while other treatments had over 1342 kg ha^{-1}. The stover yields had a similar trend except that the highest yield (3848 kg ha^{-1}) was recorded under continuous maize with application of 40 kg P ha^{-1} + 80 kg N ha^{-1} in the LR seasons. Maize/cowpea intercrops had the lowest stover yields.

Short Rains Season

In flat tillage without manure treatments, the highest maize stover yield (1511 kg ha^{-1}) was in continuous maize with 40 kg N ha^{-1} but was not significantly different ($P < 0.05$) from the other treatments (Table 2). Maize/cowpea rotation with 40 kg N ha^{-1} had the second highest stover yield (1112 kg ha^{-1}). The intercrops, continuous cropping with 120 kg N ha^{-1} and treatments without N application gave the lowest stover yields. Higher stover yields were recorded in flat tillage when manure was applied. When manure was applied to flat tillage, the highest yield of 2401 kg ha^{-1} was recorded in maize/cowpea rotation with 40 kg N ha^{-1} followed by continuous maize cropping (2231 kg ha^{-1}) with 80 kg N ha^{-1}. In the tied ridges without manure treatments, the highest stover yield

Table 2 Effects of tied ridges and INM on maize grain and stover yield (kg ha^{-1}) during the short rains seasons

| | Flat tillage | | | | Tied ridges | | | |
| | No manure | | Manure | | No manure | | Manure | |
Treatment	Grain[a]	Stover[b]	Grain[a]	Stover[b]	Grain[a]	Stover[b]	Grain[a]	Stover[b]
0 P + 0 N (control)	–	702	–	1064	–	1000	–	1523
Maize after cowpea								
40 P	–	590	–	1089	–	1701	–	2519
40 P + 40 N	–	1112	–	2401	–	1363	–	2101
Maize/cowpea intercrop								
40 P	–	880	–	1058	–	797	–	1453
40 P+40 N	–	770	–	1746	–	1222	–	1487
Continuous maize cropping								
40 P	–	770	–	916	–	1071	–	1621
40 P + 40 N	–	1511	–	1801	–	1657	–	2278
40 P + 80 N	–	1042	–	2231	–	2510	–	2940
40 P + 120	–	760	–	1826	–	551	–	2396
Mean	–	*904*	–	*1570*	–	*1319*	–	*2036*
LSD	–	631	–	1003	–	680	–	826

[a]There were crop failures during all the short rains (SR) season, and thus only maize stover was harvested in all the cropping years
[b]The stover recorded was for only the SR 2003 and SR 2004 seasons

(2510 kg ha^{-1}) was recorded in continuous maize cropping with 80 kg N ha^{-1} while the highest stover yield (2940 kg ha^{-1}) was recorded in continuous maize with manure and 40 kg N ha^{-1}. These yields were significantly ($P < 0.05$) higher than in the control. Although the yield differences were not significant the control had higher yields than applying 120 kg N ha^{-1} in the absence of manure application.

Cowpea Yields

Long Rains Season

During the long rains (LR) season, cowpea grain yield was recorded only in 2004, 2005 and 2006. The highest grain yield without manure application under flat tillage (1374 kg ha^{-1}) was observed under the maize/cowpea intercrops with 40 kg P ha^{-1} followed by maize/cowpea rotation with 40 kg P ha^{-1} (1120 kg ha^{-1}), but these yields were not significantly different ($P < 0.05$) from each other (Table 3). Application of N in both intercrop and rotation treatments did not result in significantly higher cowpea grain yields when compared with the control (with no fertilizer applied). The improved cowpea yields recorded when P was applied is an indication that the soils cannot supply adequate P required by cowpea. Cowpea did not respond to N,

which could be due to adequate supply of this nutrient by the soil, and therefore a starter N is not necessary or the cowpea is able to fix its own N. When manure was applied in flat tillage, the highest cowpea yields were recorded in maize–cowpea rotation. Addition of N in rotation and intercrop did not result in significant ($P < 0.05$) cowpea yield increases. Continuous cowpea cropping without inorganic fertilizers had the lowest yield (460 kg ha^{-1}).

During the LR, the highest cowpea grain yield under tied ridging without manure (975 kg ha^{-1}) was recorded in maize/cowpea rotation with 40 kg N ha^{-1} but was not significantly different ($P < 0.05$) from the other treatments. All treatments without 40 kg N ha^{-1} had low grain yields with the lowest (531 kg ha^{-1}) recorded in the control treatment (continuous cowpea without any inorganic fertilizers). The haulm yield had a similar trend as grain, but the highest yield (2870 kg ha^{-1}) was under maize/cowpea rotation on application of 40 kg P ha^{-1} + 40 kg N ha^{-1} and had significant differences from the other treatments.

The highest cowpea grain yield (990 kg ha^{-1}) under tied ridging with manure was recorded in maize/cowpea intercrop with 40 kg P ha^{-1} during the long rains. The yields from continuous cowpea cropping without inorganic fertilizer, intercrop with N and rotation without N application had the lowest and similar grain yields. The haulm yield trends were similar to the grain yields.

Table 3 Effects of tied ridges and INM on cowpea grain (kg ha^{-1}) and haulm yield (kg ha^{-1}) during the long rains seasons

| | Flat tillage | | | | Tied ridges | | | |
| | No manure | | Manure | | No manure | | Manure | |
Treatment	Grain	Haulms	Grain	Haulms	Grain	Haulms	Grain	Haulms
0 P + 0 N (*control*)	732	1768	460	1532	531	1638	618	2048
Cowpea after maize								
40 P	1120	2317	1526	3314	772	2399	640	2141
40 P + 40 N	658	1758	742	2169	975	2870	924	2407
Maize/cowpea intercrop								
40 P	1374	3084	770	2069	753	1899	990	2427
40 P + 40 N	934	2217	896	2260	804	2115	644	2241
Mean	*964*	*2229*	*879*	*2269*	*767*	*2184*	*763*	*2253*
LSD	360	599	582	761	248	362	450	621

Short Rains Season

Average cowpea grain yields under flat tillage without manure during the SR seasons were very low and ranged between 23 and 107 kg ha^{-1} in flat tillage without manure, and the lowest yields were obtained from continuous cowpea cropping without inorganic fertilizers (Table 4). When manure was applied in flat tillage, the highest cowpea yields of 377 kg ha^{-1} were recorded in maize–cowpea rotation. Addition of N in rotation and intercrop did not result in significant ($P < 0.05$) cowpea yield increases. There was a general yield increase when manure was added to flat tillage, and the highest grain yields were recorded in maize/cowpea rotation without N application. When tied ridges were used, the highest grain yields of 146 and 343 kg ha^{-1} were recorded in tied ridges without and with manure application, respectively. The intercrops in tied ridge treatments had lower yields than the rotations. Higher cowpea haulm yield was observed in treatments with manure than in treatment with 40 kg N ha^{-1}.

Discussions

There was a general increase in maize and cowpea yields when tie-ridging and manure were used. The improved crop yield was attributed to improved soil moisture conservation due to tie-ridging and manure application (Jensen et al., 2003). The effect of inorganic fertilizers on crop response was enhanced by tied ridges. The increased maize yields when nitrogen is supplemented with water harvesting (and manure) than either of them applied separately suggest enhanced water and fertilizer use efficiency because of the

Table 4 Effects of tied ridges and INM on cowpea grain and haulm yield (kg ha^{-1}) during the short rains (SR) seasons

| | Flat tillage | | | | Tied ridges | | | |
| | No manure | | Manure | | No manure | | Manure | |
Treatment	Grain[a]	Haulms[b]	Grain[a]	Haulms[b]	Grain[a]	Haulms[b]	Grain[a]	Haulms[b]
0 P + 0 N (*control*)	46	318	296	1350	82	494	161	623
Cowpea after maize								
40 P	76	343	298	1394	144	710	288	1612
40 P + 40 N	76	377	178	1094	146	784	343	1353
Maize/cowpea intercrop								
40 P	23	320	61	516	71	365	47	272
40 P + 40 N	107	665	58	512	103	759	59	388
Mean	*66*	*404*	*178*	*973*	*109*	*622*	*179*	*850*
LSD	97	239		544	110	171	281	520

[a]Grain yield was recorded for only the SR 2003

[b]Haulm yields were recorded for SR 2003 and SR 2004 seasons

combined response (Jensen et al., 2003). Jensen et al. (2003) have reported enhanced maize response to rainfall and fertilizer on maize grain yields where tied ridges have been used compared to flat cultivation. In their study in Tanzania, they found that 98% of the variance in yields could be attributed to the combined response to annual rainfall and N and P fertilizer under tied-ridged conditions.

The highest maize yields were observed with 80 kg N ha^{-1} when combined with manure. In the absence of manure, 40 kg N ha^{-1} gave the highest yields. The yield differences between the two N rates were, however, not statistically different at $P < 0.05$ probability level. This suggests that addition of 40 kg N ha^{-1} is adequate to meet maize N requirements at the site. Studies by Kamoni et al. (2003) in the semi-arid areas of Machakos District in Kenya have also indicated that N fertilizer application above 50 kg ha^{-1} did not increase maize yields.

Addition of phosphorus (P) fertilizer to cowpea in rotation and intercrop resulted in higher yields than in the control (cowpea without inorganic fertilizer). This indicates that there is need to supplement P in the soil. Addition of N to cowpea increased biomass and reduced grain yields. The increased cowpea biomass resulted in reduced maize yields in the intercrop which was attributed to moisture competition between the two crops.

Conclusion

Application of 40 kg N ha^{-1} under continuous cropping was found to be adequate for maize production since any higher level of N (80 and 120 kg N ha^{-1}) decreased grain yield. Rotation of cowpea/maize did not give enough N for sustained maize production, and hence addition of N (40 kg ha^{-1}) was necessary for increased maize yields. There is potential in combining water harvesting with manure and fertilizer application as shown by increased maize yields. Cowpea as a legume did not need external N application as it could fix adequate N. When maize/cowpea was grown in rotation, high cowpea yields were attained.

Acknowledgements The authors acknowledge Kenya Agricultural Research Institute (KARI) and Tropical Soil Biology and Fertility Institute of CIAT (TSBF-CIAT) through the Desert Margins Programme for funding this study. This chapter has been published with permission from the director of Kenya Agricultural Research Institute.

References

Biamah EK, Rockstrom J, Okwach GE (2000) Conservation tillage for dryland farming. Technical options and experiences in eastern and Southern Africa. Workshop report N0. 3. RELMA, SIDA, Kenya. 151p

Fox P, Rockström J (2003) Supplemental irrigation for dry-spell mitigation of rainfed agriculture in the Sahel. Agric Water Manage 1817:1–22

Itabari JK, Wambua JM, Kitheka SK, Maina JN, Gichangi EM (2003) Effects of water harvesting and fertilizer on yield and water use efficiency of sorghum in semi-arid Eastern Kenya. E Afr Agric For J 69(4):291–295

Jaetzold R, Smidt H, Hornetz B, Shisanya C (2006) Farm management handbook of Kenya. Vol. II/C: Natural conditions and farm management information East Kenya, 2nd edn. Eastern province subpart C1. Ministry of Agriculture/GAT, Nairobi, Kenya

Jensen JR, Bernhard RH, Hansen S, McDonagh J, MØberg JP, Nielsen NE, Nordbo E (2003) Productivity in maize based cropping systems under various soil-water-nutrient management strategies in semi-arid alfisol environment in East Africa. Agric Water Manage 59:217–237

Kamoni PT, Mburu MW, Gachene CKK(2003) Influence of irrigation and nitrogen fertiliser on maize growth, nitrogen uptake and yield in semi-arid Kenyan environment. E Afr Agric For J 69(2):99–18

Kihanda FM, Warren GP, Atwal SS (2004) The influence of goat manure application on crop yield and soil nitrates variations in semi-arid eastern Kenya. In: Bationo A (eds) Managing nutrient cycles to sustain soil fertility in sub-Saharan Africa. Academy of Sciences Publishers in association with the Tropical Soil Biology and Fertility Institute of CIAT, Nairobi, Kenya, pp 173–186

Kilewe AM, Thomas DB (1992) Land degradation in Kenya. A framework for policy and planning. Food and Rural Development Division, Commonwealth Secretariat, Marlborough House, Pall Mall London. Swiy SHX, March 1992, pp 148.

Kipserem LK (1996) Water harvesting on entric Fluvisols of Njemps Flats of Baringo district. A paper presented at the Dryland applied Research and Extension Project (DAREP) workshop at Embu from 4–6th June 1996

Lal R (1995). Tillage systems in the tropics. Management options and sustainability implications. FAO Soils Bulletin No. 71. FAO, Rome

Liniger HP (1990) Agroecology and water conservation for rainfed farming in the semi-arid footzone west and northwest of Mount Kenya: consequences on water resources and soil productivity. Geographica-Bernensia. African Studies Series Switzerland. Monographs v. A8, pp 95–105

Mbogoh S (2000) Makueni district profile: crop production and marketing 1988–1999. Dryland Research Working Paper 7. Crewkerne, UK, Dryland Research.

Mellis D (1996) Water harvesting and conservation tillage: Exploring and adapting with farmers in semi-arid Kenya. A Paper presented at the DAREP soil and water harvesting/conservation tillage workshop 4–6 June 1996. Embu, Kenya

Miriti JM, Odera MM, Kimani SK, Kihumba JN, Esilaba AO, Ngae GN (2005) On-farm demonstrations to determine farmers' subjective preference for technology-specific attributes in tied-ridges. In: Mugendi DN, Kironchi G, Gicheru PT, Gachene CKK, Macharia PN, Mburu M, Mureithi JG, Maina F (eds) Capacity building for land resource Management to meet the challenges of food security in Africa. 21st Annual conference of SSSEA, 1st–5th Dec, 2003, Eldoret, Kenya

Njihia CM (1977) The effect of tied ridges, stover mulch and farmyard manure on rainfall abstraction in a medium potential area, Katumani, Kenya. Paper presented at the international conference on role of soil physical properties,

International Institute of Tropical Agriculture (IITA). Ibadan, Nigeria, Dec 1977

Okalebo JR, Gathua KW, Woomer PL (2002) Laboratory methods of soil and plant analysis: a working manual, 2nd edn. TSBF-CIAT and Sacred Africa, Nairobi, Kenya, p 128

Ragwa PK, Kamau NR, Mbatia ED (1998) Irrigation and drainage branch position paper. Paper presented during the workshop and promotion of sustainable smallholder irrigation development in Kenya, 16–20 November. 1998. Embu. Kenya

The Potential of *Ipomoea stenosiphon* as a Soil Fertility Ameliorant in the Semi-arid Tropics

T. Mombeyarara, H.K. Murwira, and P. Mapfumo

Abstract There is potential for smallholder farmers in Zimbabwe to use locally available plant resources to improve soil fertility and crop production. The challenge is to identify species with high nutrient concentrations and large amounts of above-ground biomass. Such a potential species is *Ipomoea stenosiphon* (Hall) A. Meeuse, a plant species indigenous to Zimbabwe, Malawi, Mozambique and Zambia, and it is currently being used by farmers in Zimbabwe. The aim of this study was to determine its productivity and variation in shoot nutrient concentrations on different soil types, its N mineralisation potential and effect on maize productivity. *Ipomoea stenosiphon* grown on clayey soils had significantly higher shoot macronutrient concentrations than when grown on sandy soils (N 43 vs. 11 g kg^{-1} soil, P 4.6 vs. 2.5 g kg^{-1} soil, K 4 vs. 1.5%). Laboratory incubation of *I. stenosiphon* shoot biomass showed greater ($P < 0.05$) N mineralisation than did many other agroforestry species, although it was inferior to *Leucaena leucocephala* and *Acacia angustissima* because of their higher N content of 3.2 and 3.03%, respectively, compared to 2.3% for *I. stenosiphon*. Field evaluation as an organic nutrient source in semi-arid areas of Zimbabwe showed an increase in maize grain yields of 111 and 161% at two study sites after applying *I. stenosiphon* biomass at 75 kg N ha^{-1}. These yields showed an average N fertiliser (ammonium nitrate) equivalency of 83%. The study showed the high capacity for acquisition of soil nutrients in *I. stenosiphon* biomass, with corresponding high levels of mineralisation and crop production.

Keywords *Ipomoea stenosiphon* · Biomass transfer · Fertiliser equivalence · Nitrogen mineralisation

Introduction

Smallholder farmers in Zimbabwe and other parts of sub-Saharan Africa continue to seek alternative nutrient sources to complement the limited amounts of mineral fertilisers they can afford or access (Chikowo et al., 2004, Mtambanengwe et al., 2005). Availability of alternative sources of nutrients such as animal manure, green manures, agroforestry tree prunings and crop residues differs within agro-ecological regions, especially in the semi-arid areas of the tropics, where biomass production is lower (Snapp et al., 1998). Severe soil fertility constraints which form part of the "poverty complex" hamper crop production; hence immediate food security needs often take precedence over requirements for long-term investments in soil fertility (Scoones et al., 1996). Consequently, adaptive mechanisms to sustain agricultural production by smallholder farmers in semi-arid areas have been more out of desperation than by choice (Mapfumo and Giller, 2001). However, these adaptive mechanisms provide a starting point for tapping indigenous knowledge to develop locally viable soil fertility-improving methods. Knowledge of characteristics of locally available organic resources in terms of biomass production, mineralisation patterns and influence on the following crop productivity is important for planning their

T. Mombeyarara (✉)
Tropical Soil Biology and Fertility Institute of the International Centre for Tropical Agriculture (TSBF-CIAT), Harare, Zimbabwe
e-mail: t.mombeyarara@cgiar.org

A. Bationo et al. (eds.), *Innovations as Key to the Green Revolution in Africa*,
DOI 10.1007/978-90-481-2543-2_50, © Springer Science+Business Media B.V. 2011

use in soil fertility management (Palm et al., 1999). Decomposition and subsequent N mineralisation from organic resources depends on the quality and quantity of the organic inputs, soil mineralogy, acidity and biological activities (Myers et al., 1994). In biomass transfer systems, this helps in predicting the amount of biomass to be transferred to cropping fields for desired crop yields (Snapp et al., 1998). In this study, *Ipomoea stenosiphon* was evaluated to determine the variability in its biomass productivity, nutrient concentration and mineralisation, and N and P fertiliser equivalency when used as an organic resource.

Materials and Methods

Laboratory Experiment

A laboratory experiment was set up to compare N mineralisation from *I. stenosiphon* with other organic amendments. The organic amendments used in the incubation experiment were *Acacia angustissima* (Mill.) Kuntze, *Cajanus cajan* (L.) Millsp. Syn, *Leucaena leucocephala* (Lam.) de Wit, *Calliandra calothyrsus* Meissner, *Leucaena diversifolia* (Schldl.) Benth, *Leucaena esculenta* (DC.) Benth, *Leucaena pallida* Britton and Rose, *Macroptilium atropurpureum* (DC.) Urb. and *Lablab purpureus* (L.) Sweet. This wide range of species represents both high- and low-quality species which are already under use in many smallholder farming areas of Zimbabwe. All were collected from ICRAF sites at Domboshawa and Henderson Research Stations in Zimbabwe's Natural Region 2. The green prunings were air-dried under shade and ground to pass through a 0.15-mm mesh. The different organic amendments were analysed for total C, N, lignin and polyphenols using methods described below.

Soil samples were collected from the 0- to 20-cm layer of experimental plots at Domboshawa ICRAF Research Station, Zimbabwe, air-dried and ground to pass through a 2-mm sieve. The soil was a lixisol, with pH ($CaCl_2$) 4.8, total C (Walkley–Black) 1.6 g kg^{-1}, sand 92%, silt 3%, clay 5%, Ca 0.8 cmol kg^{-1}, Mg 0.14 cmol kg^{-1} and K 0.12 cmol kg^{-1} (extracted in NH_4OAc). N mineralisation was measured using leaching tube incubations following a modified method of Stanford and Smith (1972). Air-dried soil (50 g)

was mixed with acid-washed sand (1:2 w/w), and plant samples were added to give equal additions of nitrogen (100 kg N ha^{-1}). No plant sample was added to the control tubes. A layer of sand was added on top of the soil–sand mixture to avoid soil disturbance on addition of the leaching solution. To glass tubes of 30 mm internal diameter and 278 mm in length, different species were applied at a rate equivalent to 100 kg N ha^{-1} with three replications. Vials with 0.25 M NaOH were suspended in the leaching tube to trap carbon dioxide. The top of the leaching tube was covered with plastic paper fastened with rubber bands to retain the CO_2 released. Leaching was done on days 0, 3, 7, 14, 21, 28, 42 and 56 using a leaching solution of 1 mM $CaCl_2$, 1 mM $MgSO_4$, 0.1 mM KH_2PO_4 and 0.9 mM KCl (Cassman and Munns, 1980), which simulates soil cation composition. The leachate was then analysed for NO^{3-} and NH^{4+} using methods described by Anderson and Ingram (1993).

Field Experiments

The field experiments were conducted in Ngundu communal area (20°03′S, 31°03′E, 660 masl) and in Shurugwi communal area (19°41′S, 29°57′E 1050 masl). Both areas have a unimodal rainfall pattern (November to April). The annual average rainfall for Ngundu is 610 mm; but in the 2003/2004 season, 680 mm was recorded, while Shurugwi recorded 590 mm. Soils at both sites are predominantly infertile sandy soils (Haplic Lixisols—WRB) derived from granite, but dolerite intrusions give rise to smaller areas under more fertile red clay soils (luvisols).

Biomass Productivity and Nutrient Variability in Ipomoea stenosiphon and Soil

Biomass production for *I. stenosiphon* was determined in natural stands and farmers' fields by measuring four randomly selected quadrats (1 m × 1 m) on each site, and all the biomass enclosed in each quadrant were cut at ground level and measured. A sub-sample was oven dried to constant weight and used to calculate overall sample dry weight. Nutrient variation in *I. stenosiphon* was measured from the following environments: loamy soils with rock outcrops, clayey soils on termitaria mounts, contour ridges and abandoned agricultural fields with soils ranging from sandy (about 8% clay) to clay (about 60% clay content). Green biomass was collected from both the flowering and the senescing plants. Nutrient variation from *I. stenosiphon* habitats was determined by sampling the soil underneath the stands. Samples (0–20 cm) were collected from a radius of 50 cm from the base of *I. stenosiphon* stands. The samples were collected at 10 cm intervals from the stem of the plant and bulked to make one composite sample. All soil samples were air-dried, sieved (<2 mm) and analysed for total C using the Walkley–Black method and for N using the micro-Kjeldahl method (Brenner and Mulvaney, 1982), and pH was determined using 0.01 M $CaCl_2$. Soil-exchangeable K was determined by flame photometry, while Ca and Mg were determined by atomic absorption spectrophotometry after leaching with ammonium acetate. All leaf biomass was air-dried and analysed for N, P, K, Ca and Mg after digestion using methods described above.

Determination of N and P Fertiliser Equivalence Values

The experiments to determine the N fertiliser equivalence in *I. stenosiphon* and its effects on maize grain yields were established in two semi-arid areas of Zimbabwe, Ngundu and Shurugwi. The following treatments were evaluated in a completely randomised block design with four replicates.

1. Zero control
2. *Ipomoea stenosiphon* leaves at 75 kg N ha^{-1}
3. *Ipomoea stenosiphon* leaves at 50 kg N ha^{-1} + 25 kg N ha^{-1} applied as ammonium nitrate (AN)
4. 25 kg N ha^{-1} as AN
5. 50 kg N ha^{-1} as AN
6. 75 kg N ha^{-1} as AN
7. 100 kg N ha^{-1} as AN

With the exception of the control, basal fertiliser (20 kg P ha^{-1} and 25 kg K ha^{-1}) was added to all treatments. Nitrogen response curves were plotted using results from the control and mineral fertiliser treatments.

To determine the P fertiliser equivalence values the following set of treatments were used:

1. Zero control
2. *Ipomoea stenosiphon* leaves at 15 kg P ha^{-1}
3. *Ipomoea stenosiphon* leaves at 9 kg P ha^{-1} and 6 kg P ha^{-1} applied as single superphosphate (SSP)
4. 15 kg P ha^{-1} as SSP
5. 30 kg P ha^{-1} as SSP
6. 45 kg P ha^{-1} as SSP

Ammonium nitrate (120 kg N ha^{-1}) was added to all treatments except the control and *I. stenosiphon* treatments (1–3 above). Addition of *I. stenosiphon* leaves at 15 kg P ha^{-1} in treatment 2 gave an equivalent N application rate of 120 kg ha^{-1}. To achieve a target N application rate of 120 kg ha^{-1}, treatment 3 received additional mineral fertiliser at 45 kg N ha^{-1}.

The N and P fertiliser equivalencies of *I. stenosiphon* were determined by comparing yields from *I. stenosiphon* treatment to the N and P response curves from sole inorganic fertiliser treatments (Mutuo et al., 1999). All the fertiliser response curves followed a quadratic response represented by the equation $Y = ax^2 + bx + c$. The fertiliser equivalence (FE) value was therefore calculated as

$$X = \frac{-b \pm \sqrt{b^2 - 4ac}}{2a}$$

where $X = $ FE and a, b and c are constants. To compare the FE, FE% was calculated by dividing FE by the actual amount of N applied in crop residues and expressed as a percentage (Mutuo et al., 1999):

$$\% \, FE = \frac{FE \times 100}{N \, applied}$$

Apparent N Recovery

Total N uptake by the maize crop was obtained by multiplying total above-ground maize biomass by shoot nitrogen concentration. Apparent N recovery was then calculated as follows:

$$
\begin{aligned}
&Apparent \, N \, recovery \, (\%) \\
&= \frac{N \, uptake_{treatment} - N \, uptake_{control}}{N \, applied}
\end{aligned}
$$

Statistical Analysis

Analysis of variance was done for N release data for each sampling time and maize yield data, and treatment means were considered significant at $P < 0.05$ using the standard error of differences (SEDs) between means (Genstat 5 Committee, 1987). Simple linear regression analyses were used to determine the relationship between lignin + polyphenol and initial N content and between leaf nutrient variation and soil types. A regression analysis was used to analyse N mineralisation using the exponential decay model:

$$N_t = N_o \times \exp(-k \times t)$$

where N_t is the initial nitrogen at a particular rate k at time t.

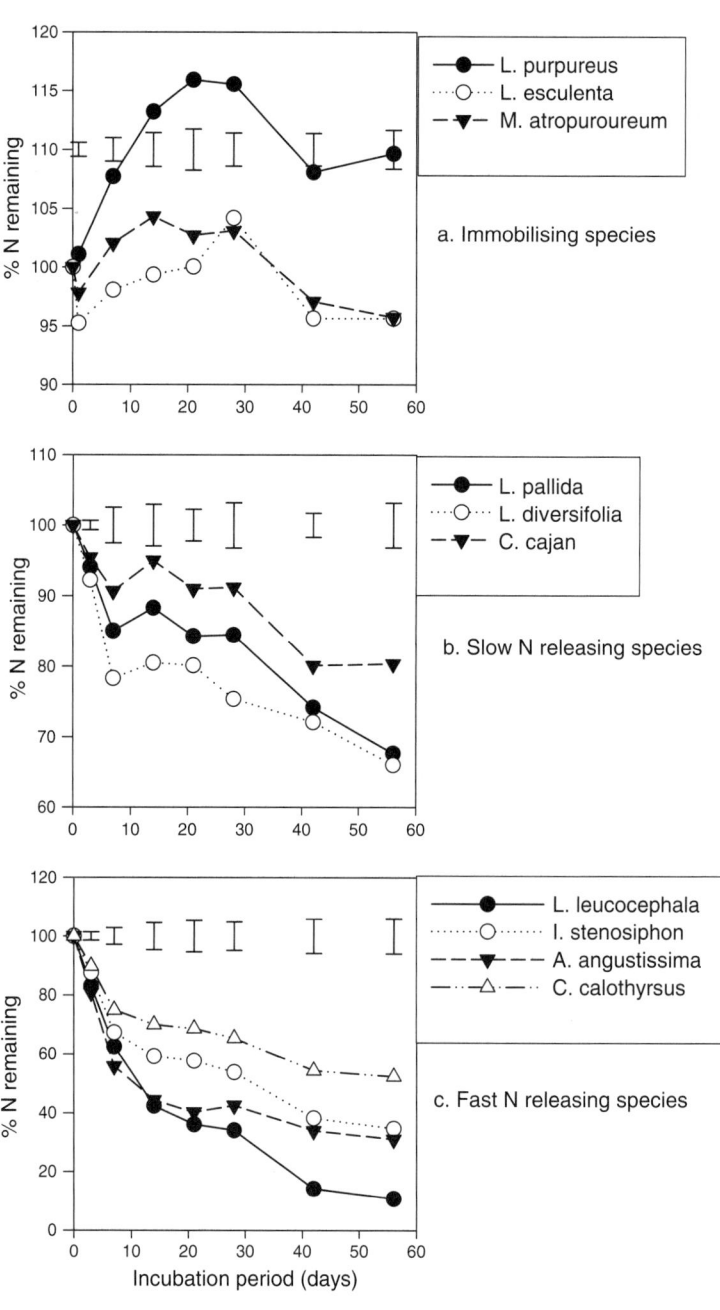

Fig. 1 N mineralisation from *I. stenosiphon* and other agroforestry species: (**a**) immobilising species (**b**); slow N-releasing species; (**c**) fast N-releasing species

Results and Discussion

Nitrogen Mineralisation

Most species released the bulk of their N within the first 2 weeks; *L. leucocephala*, which had the highest N content, mineralised about 90% N, while *I. stenosiphon* mineralised 65% over the 8-week incubation period (Fig. 1). The fast N-releasing species (Fig. 1c) were best fitted to the exponential decay model $N_t = N_0 \times \exp(-k \times t)$, with r^2 values ranging from 0.8 to 0.99. *Leucaena leucocephala* and *I. stenosiphon* had a low lignin + polyphenols:N ratio (Table 1) and therefore showed N release characteristics of a high-quality organic resource. Polyphenols have the capacity to bind proteins, thus reducing the degradation of protein N by microbes (Swift et al., 1979). *Ipomoea stenosiphon* is of superior quality compared to most organic resources found on farm. Crop residues (Snapp et al., 1998), cattle manure (Delve, 2004) and maize stover (Murwira and Kirchmann, 1993) are all of low quality, falling below the critical range of 1.8–2% N needed for N mineralisation.

Variation of Nutrient Concentration in Ipomoea stenosiphon *Shoot Biomass* with Soil Type

The N concentration of *I. stenosiphon* shoots ranged from 11 to 43 g N kg^{-1}, while the soils were generally low in N, ranging from 0.1 g N kg^{-1} in sandy soils to 2.2 g total N kg^{-1} in clay soils (Fig. 2, $P < 0.01$, Fig. 1a). Plants growing on clayey soils had significantly higher ($P < 0.05$) tissue N concentration compared to those on sandy soils. The same pattern was observed with P (Fig. 2b) as well as K, Ca and Mg (Fig. 3) and plants growing on clayey soils having shoot P concentration of 4.6 g P kg^{-1} compared with 2.5 g P kg^{-1} from sandy soils. The positive correlation between soil and shoot nutrient concentration suggested direct dependence of *I. stenosiphon* on the inherent fertility status of the soils. Soils in Ngundu are derived from granite and inherently low in CEC and organic matter (Nyamapfene, 1991). This resulted in low amounts of N in *I. stenosiphon* leaf biomass. However, despite the inherently low P status of the soils, *I. stenosiphon* still attained tissue P of 2.7 g P kg^{-1}, which is above the known threshold of 2.5 g P kg^{-1} required for net P mineralisation to occur (Palm et al., 1999). *Ipomoea stenosiphon* therefore has the capacity to accumulate P in its biomass even in poor soils and can be used to transfer large amounts of nutrients from non-accessible niches into crop lands. In Ngundu, farmers have access to hillsides where *I. stenosiphon* grows naturally and use the biomass as their N and P source. However, this study brings into question the sustainability of non-legume biomass transfer systems which have been cautioned in other studies to avoid fast degradation of the soil (George et al., 2001). *Ipomoea stenosiphon* is, however, deep rooted (>1.8 m deep) (Mombeyarara, 2005), which suggests deep nutrient capture which could account for the high nutrient concentrations in its biomass. Deep capture of nutrients that might otherwise be lost through leaching is a way of recycling them back into the farming system

Table 1 Chemical characteristics of plant prunings used in this study

Plant material	%N	%P	%K	%p/p	%Lignin	p/p:N	Lignin:N	Lignin+p/p:N
L. leucocephala	3.23	0.18	1.5	1.9	11.5	0.6	3.5	4.1
A. angustissima	3.03	0.10	1.2	8.7	7.9	2.9	2.6	5.5
C. calothyrsus	2.48	0.18	2.3	12.4	4.6	5.0	1.8	6.8
I. stenosiphon	2.27	0.29	3.5	4.7	6.8	2.1	3.0	5.1
L. diversifolia	2.12	0.14	1.3	8.6	7.8	4.1	3.7	7.1
L. pallida	2.08	0.11	0.6	10.9	9.9	5.3	4.7	10.0
C. cajan	1.80	0.13	0.5	1.9	10.5	1.1	5.9	7.0
M. atropurpureum	1.74	0.15	1.2	6.6	11.7	3.8	6.8	10.6
L. esculenta	1.37	0.08	0.5	9.7	8.3	7.1	6.1	13.2
L. purpureus	1.23	0.18	1.3	1.9	8.9	1.6	7.2	8.8

p/p is polyphenols

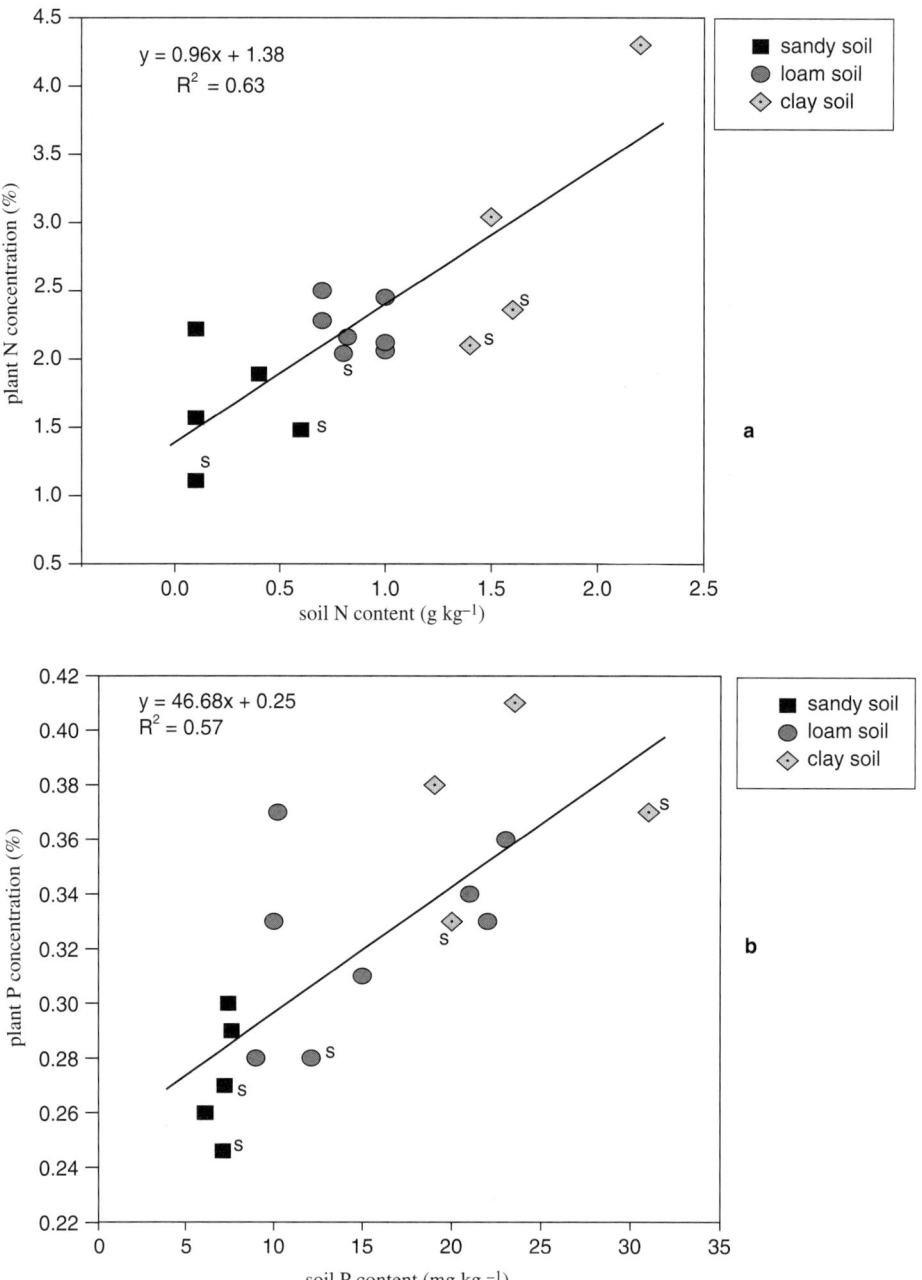

Fig. 2 Dependence of N and P contents in *I. stenosiphon* on (**a**) soil N and (**b**) soil P status in Ngundu communal area, Zimbabwe. Symbol "s" denotes plant undergoing senescence

Fig. 3 Plant variation in potassium (**a**), calcium (**b**) and magnesium (**c**) concentration with soil base status in Ngundu communal area during the 2003–2004 rainy season (symbol "s" denotes prunings from a senescing plant)

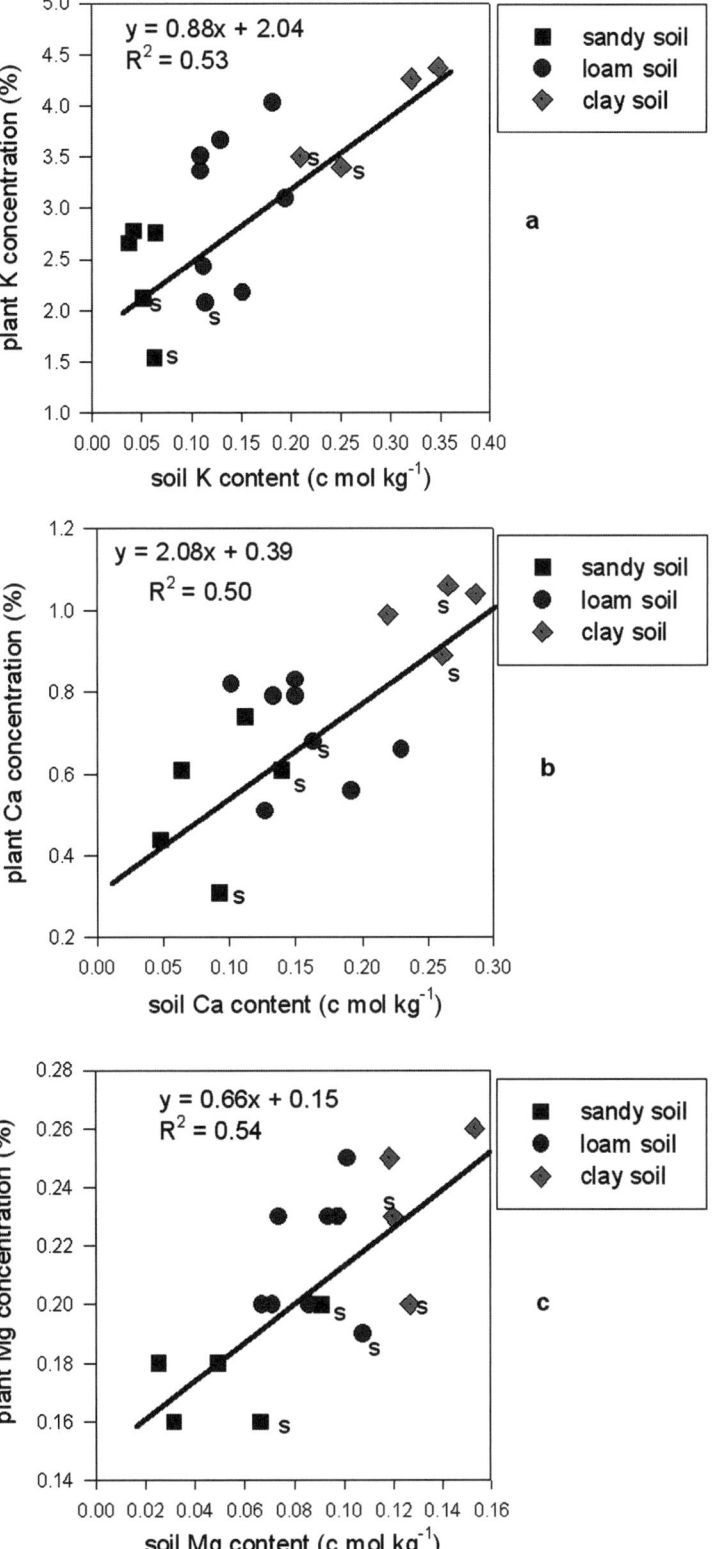

(van Noordvijk et al., 2001) and can be considered as a net nutrient input (Scroth, 1995), while lateral uptake constitutes nutrient redistribution (Buresh and Tian 1998).

Some plants have mechanisms to acquire nutrients which may be unavailable to other plant species, especially P (George et al., 2002). Although it is not clear why *I. stenosiphon* and *Tithonia diversifolia* share the same nutrient acquisition properties, studies in western Kenya suggest that *T. diversifolia* releases organic acids that help in solubilising P, thus improving its uptake (ICRAF, 1997). Pigeon pea was reported to be able to solubilise iron-bound phosphorus by releasing piscidic, malonic and oxalic acids which solubilise P from its unavailable Fe and Al compounds (Ae et al., 1996). Although no research has been done on nutrient uptake mechanisms of *I. stenosiphon*, nutrient accumulation in biomass suggests more efficient nutrient uptake from the soil.

Nitrogen and P Fertiliser Equivalence Values

Application of *I. stenosiphon* biomass at 75 kg N ha^{-1} gave maize yields of 3.15 and 4.26 t ha^{-1} in Ngundu and Shurugwi, respectively, compared to an average of 1.55 t ha^{-1} from the control treatment. There were significant differences in maize N uptake among different N sources in Ngundu ($P < 0.05$; Table 2). The highest N uptake of 97 kg N ha^{-1} was obtained

from the mineral fertiliser treatment which also had the highest N recovery of 81% (Table 2). Sole application of *I. stenosiphon* gave the lowest N recovery value of 58%. The yields for *I. stenosiphon* treatments in Ngundu and Shurugwi translated to fertiliser equivalence values of 81 and 85%, respectively. Maize grain yields also increased significantly with an increase in the rate of P application. Application of *I. stenosiphon* at 15 kg P ha^{-1} in Shurugwi increased maize grain yields by 116% over the control, while 9 kg P ha^{-1} of *I. stenosiphon* + 6 kg P ha^{-1} as SSP gave maize yields that were 122% higher than the control. The P fertiliser equivalence values for Ngundu and Shurugwi were 85 and 107%, respectively. Such high fertiliser equivalence values have also been reported elsewhere. *Tithonia diversifolia*, *Senna spectabilis* and *C. calothyrsus* had 119, 72 and 68% fertiliser equivalencies, respectively (Kimetu et al., 2003), while Delve (2004) reported FE values of about 100% for the same species. *Senna spectabilis* have also been reported to have an FE of more than 139% in a study by Murwira et al. (2002). However, exotic species such as *Acacia* spp. and *Calliandra* spp. are difficult to establish in the different semi-arid regions of Zimbabwe, in comparison to *I. stenosiphon* which is an indigenous species adapted to these agro-ecological regions. The P fertiliser equivalence was more than 100% in Shurugwi, an indication that it performed as well as mineral fertiliser. This can also be attributed to its high P content of above 0.25%, the critical value for P mineralisation (Palm et al., 1999). Farmers in these areas use very little mineral fertilisers (less than 40 kg N ha^{-1}) and in some cases, none at all because of financial

Table 2 Nitrogen and P uptake and recovery by maize in Ngundu and Shurugwi

Treatment (Ngundu)	N/P applied (kg N ha^{-1})	N/P uptake (kg N ha^{-1})		Apparent NIR	
		Ngundu	Shurugwi	Ngundu	Shurugwi
Control	0	36.2	45.9		
I. stenosiphon	75	79.8	100.6	58.0	73.0
I. stenosiphon + AN	75	86.8	116.4	73.0	94.0
AN	75	96.9	115.5	81.0	93.0
SED		*13.3*	*16.6*	*9.5*	*9.2*
Control	0	5.0	5.4	–	–
I. stenosiphon	15	11.6	13.9	44.9	56.4
I. stenosiphon + SSP	15	14.9	17.0	65.9	77.4
SSP	15	13.5	16.8	56.6	75.6
SED		*3.1*	*3.8*	*8.6*	*9.5*

constraints. The current yields attainable with these low nutrient application rates can be easily surpassed by applying 75 kg N ha^{-1} (3 t ha^{-1}) of *I. stenosiphon*. When established on good fertility areas on farm, *I. stenosiphon* can easily yield above-ground biomass of at least 8 t ha^{-1}, supplying more than 150 kg N. Use of *I. stenosiphon* has additional effects of other nutrients like Ca, Mg and K which are not present in mineral fertilisers (Mombeyarara, 2005).

High apparent N recovery values across sites, especially from the sole ammonium nitrate treatment (87%) and the combination treatment of ammonium nitrate and *I. stenosiphon* (83%), show that the bulk of the N applied was taken up by maize plants. This variation of N recovery values is due to different N release characteristics of resources used due to differing quality (Mafongoya et al., 1997). *Ipomoea stenosiphon* has high nutrient concentrations of Ca, Mg and K which could have contributed to higher yields. Significantly higher uptake of Ca, Mg and K was recorded in treatments amended with *I. stenosiphon* biomass compared to the control treatment (Mombeyarara, 2005), and therefore high yields are not solely because of nitrogen uptake.

Conclusion

Ipomoea stenosiphon is a species that can be used in smallholder farming systems to contribute significant amounts of N and P as well as other nutrients for increasing crop production. The higher maize yields attained with a combination of mineral and organic sources compared to sole mineral nutrient application make it possible for the resource-constrained farmers in the semi-arid tropics to achieve high yields through targeting small amounts of mineral fertilisers with additions of *I. stenosiphon*.

Acknowledgement The authors would like to acknowledge financial support from IFAD for this work and the University of Zimbabwe for laboratory space.

References

Ae N, Arihara J, Okada K, Yoshihara T, Johansen C (1996) Phosphate uptake by pigeon pea and its role in cropping systems of the Indian subcontinent. Science 248:477–480

Anderson JM, Ingram JIS (eds) (1993) Tropical soil biology and fertility. A handbook of methods, 2nd edn. CAB International, Wallingford, UK

Bremner JM, Mulvaney CS (1982) N-total. In: Page AL, Miller RH, Keeney DR (eds) Methods of soil analysis. Part 2. Agronomy 9. America Society of Agronomy, Madison, WI, pp 592–624

Buresh RJ, Tian G (1998) Soil improvement by trees in Sub-Saharan Africa. Agroforest Syst 38:51–76

Cassman KG, Munns DN (1980) Nitrogen mineralization as affected by soil moisture, temperature, and depth. Soil Sci Soc Am J 44:1233–1237

Chikowo R, Mapfumo P, Nyamugafata P, Giller KE (2004) Woody legume fallow productivity, biological N2-fixation and residual benefits to two successive maize crops in Zimbabwe. Plant Soil 262:303–315

Delve RJ(2004) Combating nutrient depletion in East Africa – the work of the SWNM program. In: Bationo A (ed) Managing nutrient cycles to sustain soil fertility in Sub-Saharan Africa. Afnet-CIAT, Nairobi, pp 127–136

Genstat 5 Committee (1987) Genstat reference manual. Clarendon Press, Oxford

George TS, Gregory PJ, Robinson JH, Buresh RJ, Jama BA (2001) *Tithonia diversifolia*: variations in leaf nutrient concentrations and implications for biomass transfer. Agroforest Syst 52:199–205

George TS, Gregory PJ, Robinson JS, Buresh RJ (2002) Changes in phosphorus concentrations and pH in the rhizosphere of some agroforestry and crop species. Plant Soil 246:65–73

ICRAF (1997) Using the wild sunflower, *Tithonia*, in Kenya for soil fertility and crop yield improvement. International Center for Research in Agroforestry (ICRAF), Nairobi, 12pp

Kimetu JM, Mugendi DN, Palm CA, Mutuo PK, Gachengo CN, Nandwa S, Kung'u JB (2003) Nitrogen fertiliser equivalence values for different organic materials based on maize performance at Kabete, Kenya. In: Bationo A (ed) Managing nutrient cycles to sustain soil fertility in sub-Saharan Africa. AfNet-CIAT, Nairobi

Mafongoya PL, Nair PKR, Dzowela BH (1997) Multipurpose tree prunings as a source of nitrogen to maize under semiarid conditions in Zimbabwe. 1. Nitrogen recovery rates in relation to pruning quality and method of application. Agroforest Syst 35:31–46

Mapfumo P, Giller KE (2001) Soil fertility management strategies and practices by smallholder farmers in semi-arid areas of Zimbabwe. International Crops Research Institute for the Semi-Arid Tropics (ICRISAT), P.O. Box 776, Bulawayo, Zimbabwe

Mombeyarara T (2005) *Ipomoea stenosiphon* as a multiple nutrient source for maize under field conditions in Zimbabwe. MSc thesis, University of Zimbabwe

Mtambanengwe F, Giller KE, Mpepereki S (2005) Tapping indigenous herbaceous legumes for soil fertility management by resource-poor farmers in Zimbabwe. Agric Ecosyst Environ 109:221–233

Murwira HK, Kirchmann H (1993) Comparison of C and N mineralisation of cattle subjected to different treatments, in Zimbabwean and Swedish soils. In: Merckx R, Mulongoy J (eds) The dynamics of soil organic matter in relation to sustainability of tropical agriculture. Wiley, Leuven, pp 189–198

Murwira HK, Mutuo P, Nhamo N, Marandu AE, Rabeson R, Mwale M, Palm CA (2002) Fertiliser equivalence values for organic materials of differing quality. In: Vanlauwe B, Diels J, Sanginga N, Merckx R (eds) Integrated plant nutrient management in sub-Saharan Africa. CAB International, Wallingford, CT

Mutuo PK, Marandu AE, Rabeson R, Mwale M, Snapp S, Palm CA (1999) Nitrogen fertilizer equivalencies based on organic input quality and optimum combinations of organic and inorganic N sources. Network trial results from East and Southern Africa. In SWNM Report on the Combating nutrient depletion – East Africa Highlands Consortium

Myers RJK, Palm CA, Cuevas E, Gunakilleke IUN, Brossard M (1994) The synchronization of nutrient mineralisation and plant nutrient demand. In: Woomer PL, Swift MJ (eds) Biological management of tropical soil fertility. Wiley–Sayce Co publication, New York, USA/Exeter, pp 81–116

Nyamapfene K (1991) Soils of Zimbabwe. Nehanda Publishers, Harare, Zimbabwe

Palm CA, Nziguheba G, Gachengo C, Gacheru E, Rao MR (1999) Organic materials as sources of phosphorus. Agroforest For 9(4):30–33

Scoones IC, Chibudu C, Chikura S, Jeranyama P (1996) Hazards and opportunities: farming livelihoods in Dryland Africa: lessons from Zimbabwe. Zed Books, London, UK

Scroth G (1995) Tree root characteristics as criteria for species selection and systems design in agroforestry. Agroforestry Syst 30:125–143

Snapp SS, Mafongoya P, Waddington S (1998) Organic matter technologies for integrated nutrient management in smallholder cropping systems of southern Africa. Agric Ecosyst Environ 71:185–200

Stanford G, Smith SJ (1972) Nitrogen mineralization potential of soils. Soil Sci Soc Am Proc 36:465–472

Swift MJ, Heal OW, Anderson JM (1979) Decomposition in terrestrial ecosystems. Studies in ecology, vol 5. Blackwell, Oxford, p 372

Van Noordwijk M, Toomich MTP, Verbist B (2001) Negotiation support models for integrated natural resource management in tropical forest margins. Conservation Ecol 5:21

Effect of Al Concentration and Liming Acid Soils on the Growth of Selected Maize Cultivars Grown on Sandy Soils in Southern Africa

C. Musharo and J. Nyamangara

Abstract A study was conducted from August 2002 to June 2004 to evaluate the effects of soil acidity on the growth of selected maize cultivars commonly grown in southern Africa under greenhouse and field conditions. Field experiments were conducted at Domboshawa Training Centre and Chendambuya smallholder area in central Zimbabwe. There were significant ($P < 0.001$) varietal growth differences to increasing Al concentration in terms of total root length (TRL), relative root length (RRE) and shoot and root dry matter weight. In terms of Al tolerance indices (ATI), DK 8031 and SC 517 were the most and least tolerant varieties, respectively. Liming significantly increased grain and stover yield at Mudzengerere site (relatively low Ca-to-Mg ratio) in all cultivars (grain, $P < 0.005$ and $P = 0.01$ for seasons 1 and 2, respectively; stover, $P = 0.02$ for season 2) except PAN 413. However, at Chisuko and Domboshawa sites (relatively high Ca-to-Mg ratio), liming depressed grain and stover yield in both seasons. It was concluded that root growth can be used to successfully screen maize cultivars commonly grown in southern Africa for tolerance to acidity. Response of maize to liming seems to depend on exchangeable soil Ca-to-Mg ratio.

Keywords Al tolerance · Al toxicity · Liming · Maize · Soil acidity

C. Musharo (✉)
Department of Soil Science and Agricultural Engineering, University of Zimbabwe, Mount Pleasant, Harare, Zimbabwe
e-mail: cmusharo@yahoo.com

J. Nyamangara (✉)
Department of Soil Science and Agricultural Engineering, University of Zimbabwe, Mount Pleasant, Harare, Zimbabwe
e-mail: jnyamangara@yahoo.co.uk

Introduction

Soil acidity has long been identified as a major constraint to sustainable crop production in tropical and sub-tropical areas and also in cultivated temperate areas. Of the world's tropical land area, 43% has been classified as acid comprising 27% of tropical Africa (Pandey et al., 1994). Aluminium and manganese toxicity and low P availability affect plant growth under acid soil conditions. The majority of soils in the smallholder areas of Zimbabwe are coarse-grained sands deficient in N, P and S and are often acidic in high rainfall areas. The use of agricultural lime in smallholder areas of Zimbabwe is limited because it is not easily accessible, and farmers rank it lower than other agricultural inputs such as seed and inorganic fertilisers (Dhliwayo et al., 1998).

Aluminium toxicity is an important factor under acid soils. Although there are benefits of low levels of Al on plant growth and mineral uptake (Foy, 1984), excess Al interferes with nutrient uptake, transport and use (Pandey et al., 1994). The primary expression of Al toxicity is the drastic inhibition of root growth presumed to be a result of Al binding to nuclear DNA, thereby reducing its replication (Moustakos et al., 1992). Root growth retardation results in reduced exploitation of nutrient and water supply in the soil.

The use of tolerant cultivars can allow for crop production without liming in soils where production would be impossible or uneconomic. Al-tolerant cultivars can be classified as either having symplasmic tolerance mechanism where Al is accumulated in the symplasm and chelated or sequestrated within internal compartments, e.g. vacuoles, or having exclusion tolerance mechanism whereby they release chelating

A. Bationo et al. (eds.), *Innovations as Key to the Green Revolution in Africa*,
DOI 10.1007/978-90-481-2543-2_51, © Springer Science+Business Media B.V. 2011

ligands in the rhizosphere which induce a higher pH so as to make Al less available (Schaffert et al., 2003). Tolerant cultivars are more effective in P and Ca uptake and utilisation in the presence of Al compared to non-tolerant varieties (Rout et al., 2001).

Maize cultivars grown in southern Africa are not screened for tolerance to acidity. Thus farmers are unable to select tolerant cultivars where soil is acidic and lime not available. The aim of this study was to screen some maize cultivars commonly grown in southern Africa for tolerance to Al and to determine the effect of liming on grain yield of some of the cultivars under field conditions.

Methods and Materials

Greenhouse Experiment

Maize cultivars locally grown in the southern African region were selected. The experiment was laid out in a completely randomised block design (CRBD). Pregerminated seeds of SC 403 (Seed-Co Zimbabwe), SC 517 (Seed-Co Zimbabwe), DK 8031 (MONSATO) and CZH 00013 and CZH 00017 (MZ00B-1269-3/4 Mozambique) maize cultivars were transplanted into 12 L 1/5 Steinberg nutrient solution (Foy et al., 1967) with 0, 4, 8 and 16 mg L^{-1} Al added as AlK(SO$_4$)$_2$.12H$_2$O. At planting the solutions were adjusted to pH 4.8 ± 0.2 and were not adjusted thereafter. Each treatment was replicated four times, and two seedlings were planted in each pot. The plants were grown in the greenhouse for 25 days (Mugwira et al., 1981). At harvest the final solution pH was determined.

The Effect of Al Concentration on Root Growth

Root lengths were measured during the course of the experiment. Relative root elongation (RRE) was calculated using root lengths at 4 days after transplanting and at harvest (Bennet, 1998). Total root length (L) was determined by the line intercept method, based on Buffon's needle problem (Anderson and Ingram, 1993). Aluminium tolerance index (ATI) was calculated using total root length measurements (Sapra et al., 1978). The ATI is given on a scale of 1–5 where 1 is highly sensitive and 5 highly tolerant.

Effect of Al Concentration on Root and Shoot Dry Matter Yield

After 25 days, roots and shoots were harvested, oven dried (60°C) and weighed. Relative dry weights (RDW), dry weight in relation to dry weight when no Al is added, were calculated in order to determine the relative decline in shoot and root dry matter yields at the different Al concentrations.

Field Experiments

Site Selection, Soil Sampling and Analysis

The field trials were conducted in Chendambuya smallholder area (18°10′S, 32°22′E, 1575 m altitude), north-eastern Zimbabwe, and at Domboshawa Training Centre (19°35′S, 31°14′E, 1474 m altitude), central. Both Chendambuya smallholder area and Domboshawa Training Centre are located in a sub-humid area (750–1000 mm a^{-1}), and the soils are predominantly sandy derived from granite and classified as Alfisols (USDA) or Lixisols (FAO). Experimental sites were selected based on soil pH status (pH (CaCl$_2$) < 5.0). Prior to planting, soil samples were collected using an auger from the top 0.15 m at each site to determine soil nutrient status and other properties. Ten sub-samples were randomly collected from each site. The samples were thoroughly mixed, air-dried and sieved through a 2-mm sieve before analysis.

Soil texture was determined by the hydrometer method (Gee and Bauder, 1986), soil pH by 0.01 M CaCl$_2$ and water method and exchangeable bases (Ca, Mg and K) by atomic absorption (emission for K) spectrophotometry (Anderson and Ingram, 1993). Available N and P were determined colorimetrically after extraction using 0.5 M KCl and 0.5 M sodium bicarbonate (pH 8.5), respectively (Anderson and Ingram, 1993).

Experimental Layout

The treatments for the experiment were two lime levels (limed and unlimed) and five maize cultivars (PAN 413, PHB 30G97, SC 403, SC 513 and SC 517). A split-plot design, with lime as the main plot and maize cultivar as the sub-plot, was used with four replications. The first (2002/2003) and second (2003/2004) season trials were established in the second week of November 2002 and first week of December 2003, respectively. Lime was broadcast on half of each field on the soil surface at 600 kg ha^{-1} on sandy soils and 1000 kg ha^{-1} on red soils in the first year just before planting and incorporated into the top 0.15 m of the soil using hand hoes. Lime was not applied in the second season.

Planting was done with 90 cm between rows and 30 cm between stations on 6 m × 5 m plots. Two pips were placed at each station and later thinned to one plant per station at 2 weeks after germination resulting in a plant population of 37,000 plants ha^{-1}. A basal application of compound D fertiliser (8% N, 7% P, 7% K and 8% S) was spot applied at 350 kg ha^{-1}. Ammonium nitrate (AN) was spot applied at 150 kg ha^{-1} split at 6 and 8 weeks after planting (WAP). Weed control was done by hand hoeing throughout the cropping season.

Grain and Dry Matter Yield

Maize grain yield was determined at physiological maturity (ca. 20 WAP) using 40 plants taken from the net plot area. Unshelled net plot cobs were weighed in the field and a sub-sample of four cobs taken for the determination of moisture content, shelling percentage and grain yield (adjusted to 12% moisture content).

For stover yield, three maize stalks were sampled from the net plot after cobs had been removed. Total above-ground dry matter yield was determined as weight of stalks and shelled cobs.

Results

Greenhouse Experiment

The Effect of Al on Maize Root Growth

There were significant ($P < 0.001$) varietal differences in response to Al toxicity in terms of total root length (TRL) (Fig. 1). Addition of small quantities of Al (4 mg L^{-1}) enhanced root growth for CZH 00013, CZH 00017 and DK 8031 but had a negative effect on SC 517 as shown by a rise in TRL (Fig. 1). Total root length for SC 403 was not significantly affected by Al (0–16 mg L^{-1}) addition. Aluminium tolerance indices (Table 1) showed that DK 8031 was the most tolerant variety while SC 517 was the least tolerant to Al.

Cultivars SC 403, CZH 00013 and CZH 00017 showed significant taproot elongation as concentration of Al was increased (Fig. 2). The response for DK 8031 peaked at 8 mg L^{-1} and then decreased, while that for SC 517 was variable. Relative root elongation for DK 8031 showed a general increase with increase in Al concentration while RRE for SC 517 was decreased (Table 2).

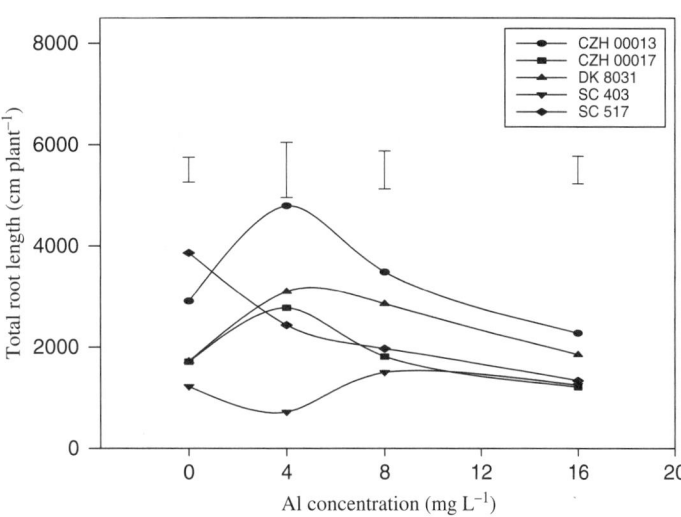

Fig. 1 The effect of Al concentration on total root length for five maize cultivars grown under greenhouse conditions (*error bars* represent standard error difference at 95% confidence)

Table 1 Aluminium
tolerance indices for five
maize cultivars (Indices were
calculated using total root
length data at 0 and 16 mg L^{-1}
Al concentration)

Cultivar	Root length (cm)		ATI (Al tolerance index)	
	0 mg L^{-1} Al	16 mg L^{-1} Al		
CZH 00013	2906	2275	3.39	
CZH 00017	1713	1215	3.09	
DK 8031	1724	1854	5.00	Most tolerant
SC 403	1223	1253	4.72	
SC 517	3863	1339	1	Least tolerant

Fig. 2 The effect of Al
concentration on final taproot
length of selected maize
cultivars after 25 days of
growth in a nutrient solution
(*error bars* represent standard
error difference of means at
95% confidence)

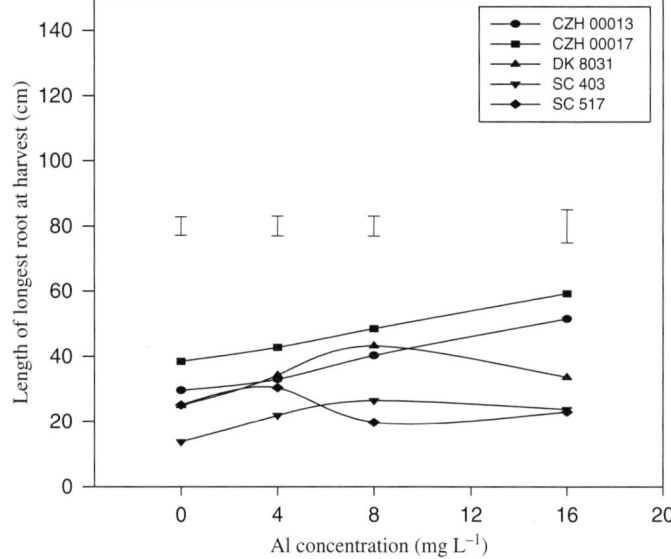

Table 2 Relative root elongations for DK 8031 and SC 517

Al concentration (mg L^{-1})	RRE (%)	
	SC 517	DK 8031
0	100	100
4	49	128
8	26	173
16	34	179

Relative root elongations (RRE) were expressed as a percentage
of root elongations at 0 mg L^{-1} Al

Effect of Al on Root and Shoot Dry Matter Yield

Root dry matter yield (RDMY) and shoot dry matter
yield (SDMY) varied inversely with Al concentration in all cultivars. Increasing Al (0–16 mg L^{-1})
significantly reduced SDMY for all cultivars ($P <$
0.001) (Fig. 3). Cultivar SC 517 had a higher SDMY
at 0 mg L^{-1} compared to the other three cultivars
(LSD 0.2623), but as Al concentration was increased
to >8 mg L^{-1} no varietal differences were observed.
Root dry matter yield for SC 403 was not significantly
reduced by Al concentration up to 16 mg L^{-1} while that

for DK 8031 was not significantly reduced up to 8 mg
L^{-1} Al (Fig. 3). Generally the decrease in RDMY was
lower compared to SDMY (Fig. 3).

Relative shoot dry matter weight (RSDW) was
reduced in all cultivars as Al concentration was
increased to 16 mg L^{-1} (Fig. 4). At 4 mg L^{-1} Al,
RSDW for SC 517 was reduced by 60% while that
for DK 8031 was reduced by only 11% of the weight
at 0 mg L^{-1} (Fig. 4). At 8 mg L^{-1} SDW for SC 403
was reduced by 25%, while that of other cultivars was
reduced by at least 60%. Shoot dry weights for all cultivars were reduced to less than 40% of the SDW in the
control treatment (no Al added) at 16 mg L^{-1}.

Field Experiment

All sites were very strongly acidic according to the
Zimbabwean classification (Nyamangara et al., 2000),
except the Chitsike site, which was in the strongly acid
range (Table 3). Cation exchange capacity was low on

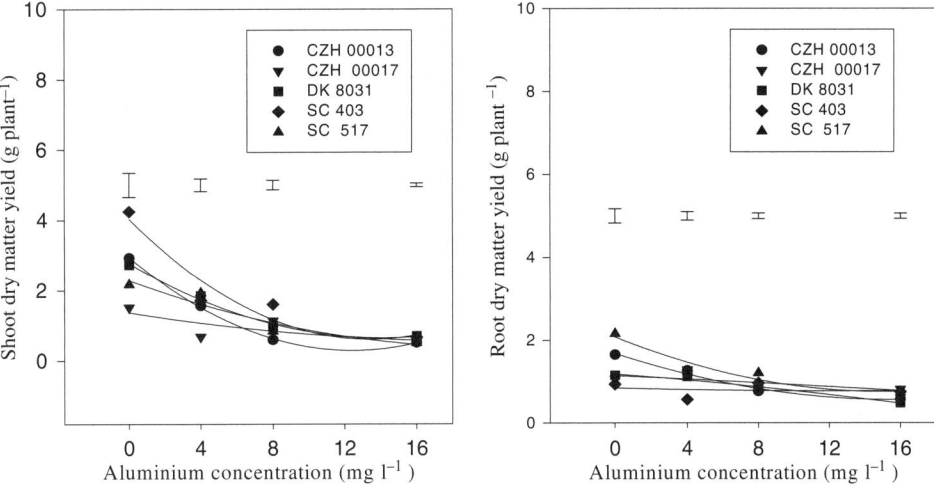

Fig. 3 The effects of Al concentration on shoot and root dry matter yield for five maize cultivars (*error bars* represent standard error difference of means at 95% confidence)

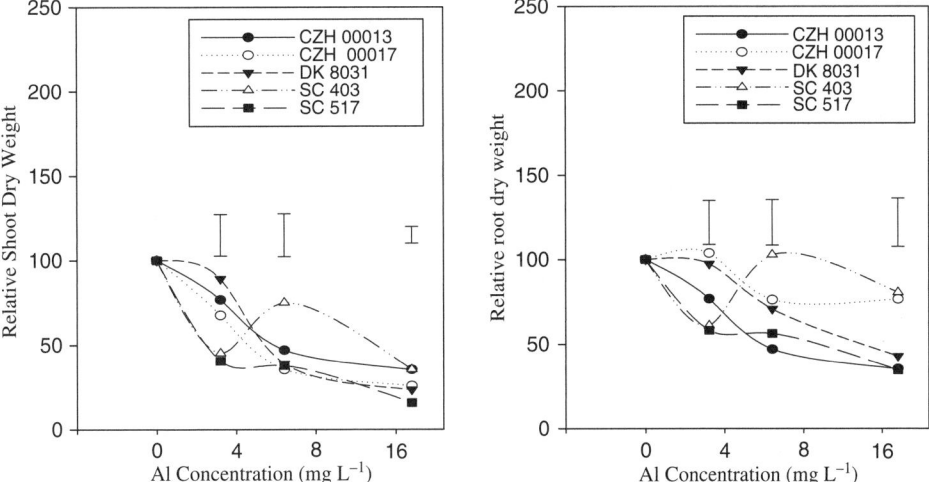

Fig. 4 The effect of Al concentration on relative shoot and root dry weights for five maize cultivars (*error bars* represent standard error difference of means at 95% confidence)

Table 3 Selected properties of soils taken from Chendambuya smallholder farming area and Domboshawa Training Centre trial sites

	Mudzengerere	Chisuko	Domboshawa sand
Clay (%)	5	6	2
pH (0.01 M CaCl$_2$/H$_2$O)	3.9/4.5	3.7/4.4	4.5/3.9
CEC (cmol$_c$ kg^{-1})	5.1	4.6	1.8
Exchangeable Ca (cmol$_c$ kg^{-1})	0.70	1.90	1.0
Exchangeable Mg (cmol$_c$ kg^{-1})	0.50	0.50	0.2
Exchangeable Na (cmol$_c$ kg^{-1})	0.06	0.06	0.1
Exchangeable K (cmol$_c$ kg^{-1})	0.09	0.17	0.1
Base saturation %	27	57	79
Exchangeable Al (me (100 g)$^{-1}$)	0.002	0.002	0.002
Available P (mg kg^{-1})	2.21	1.76	4.59

all sandy soils with the on-station site (Domboshawa) having the lowest CEC of 1.8 $cmol_c$ kg^{-1}. Base saturation (BS) was relatively low, e.g. 27% for Mudzengerere site.

Effect of Lime on Grain Yield

The Mudzengerere site showed significant increases in grain yield in all cultivars in both seasons ($P < 0.005$ and $P = 0.01$ for seasons 1 and 2, respectively) except PAN 413 where no significant response was observed (Fig. 5). Varietal differences at this site were significant in the second season ($P < 0.001$) but not in the first season ($P = 0.422$). Significant lime $\times$ variety interaction was significant with all varieties responding positively to lime except PAN 413 where no significant response was observed. PHB 30G97 yielded highest on both limed and unlimed plots, and the highest response to lime (92%) was observed with SC 517 site in the second season (Table 4).

At the Chisuko site, overall lime effect on grain yield was not significant ($P = 0.191$ and $P = 0.263$ for seasons 1 and 2, respectively) (Fig. 5). In the first year, varietal differences and lime $\times$ variety interaction were significant ($P < 0.001$). In the second season, varietal differences in yield were significant with PHB 30G97 yielding significantly lower than SC 513 (the highest yielding variety) on unlimed plots and PHB 30G97 and SC 517 yielding lower than SC 513 on limed plots (LSD = 0.63).

At Domboshawa, overall grain yield effects were also not influenced by lime in both seasons ($P = 0.062$ and $P = 0.179$, respectively). In the first season, varietal differences in grain yield were significant with SC 517 yielding lower grain on both unlimed and limed plots compared to highest yielding variety PHB 30G97 on unlimed plots and SC 513 on limed plots. Significant reductions in grain yield were observed with PHB 30G97 and SC 403 due to lime in the first season. In the second season, varietal differences in yield were significant and also lime $\times$ variety interaction effects ($P < 0.001$ and $P = 0.016$, respectively). Lime resulted in a decrease in yield for PAN 413 and SC 403 while yields for the other three varieties were not influenced by lime. PAN 413 and SC 403 yielded higher than the others on the unlimed plots while on limed plots there was no significant difference in yield between varieties.

Effect of Lime on Stover Yield

The Mudzengerere site was not sampled in the first season. In the second season, lime generally increased stover yield at harvest ($P = 0.020$). A 43% increase in stover yield was observed with PAN 413 (Fig. 6). Varietal differences in stover yield were also significant ($P < 0.001$). At the Chisuko and Domboshawa sites, stover yield was not influenced by lime in the first season ($P = 0.158$ and $P = 0.553$, respectively) but was significantly reduced by liming in the second season ($P = 0.032$ and $P = 0.050$, respectively). Effects of lime observed in the second season depended on cultivar ($P < 0.001$) (Fig. 6). There was significant lime $\times$ cultivar interaction effect on stover yield at harvest ($P < 0.001$).

Discussion

Greenhouse Experiment

Total root length and Al tolerance index have been documented as the most reliable growth indicators of Al tolerance (Sapra et al., 1978). Cultivars CZH 00013 and CZH 00017 exhibited a stress avoidance mechanism towards Al toxicity (Bennet, 1998). Their roots grew longer with increasing Al concentration in an attempt to move away from the source of stress (Fig. 2). These cultivars were also classified as tolerant based on the ATI (Table 1). Relative root elongation results show that in relatively tolerant cultivars (CZH 00013, CZH 00017, DK 8031 and SC 403) increase in Al concentration to 4 and 8 mg L^{-1} can enhance root growth while Al in non-tolerant cultivars (e.g. SC 517) can inhibit root elongation. Aluminium toxicity has been shown to damage root systems, particularly of susceptible cultivars thereby inhibiting root growth (Bennet et al., 1987).

Shoots were more sensitive to Al as SDMY was reduced by a larger percent than RDMY (Fig. 3), but varietal differences were not significant. This means that RDMY is more appropriate in screening maize for Al tolerance than shoots since varietal differences were significant (Fig. 4). A total decrease in leaf number and size and a decrease in shoot biomass have been documented as common responses by shoots to Al (Pietraszewska, 2001). The use of relative figures

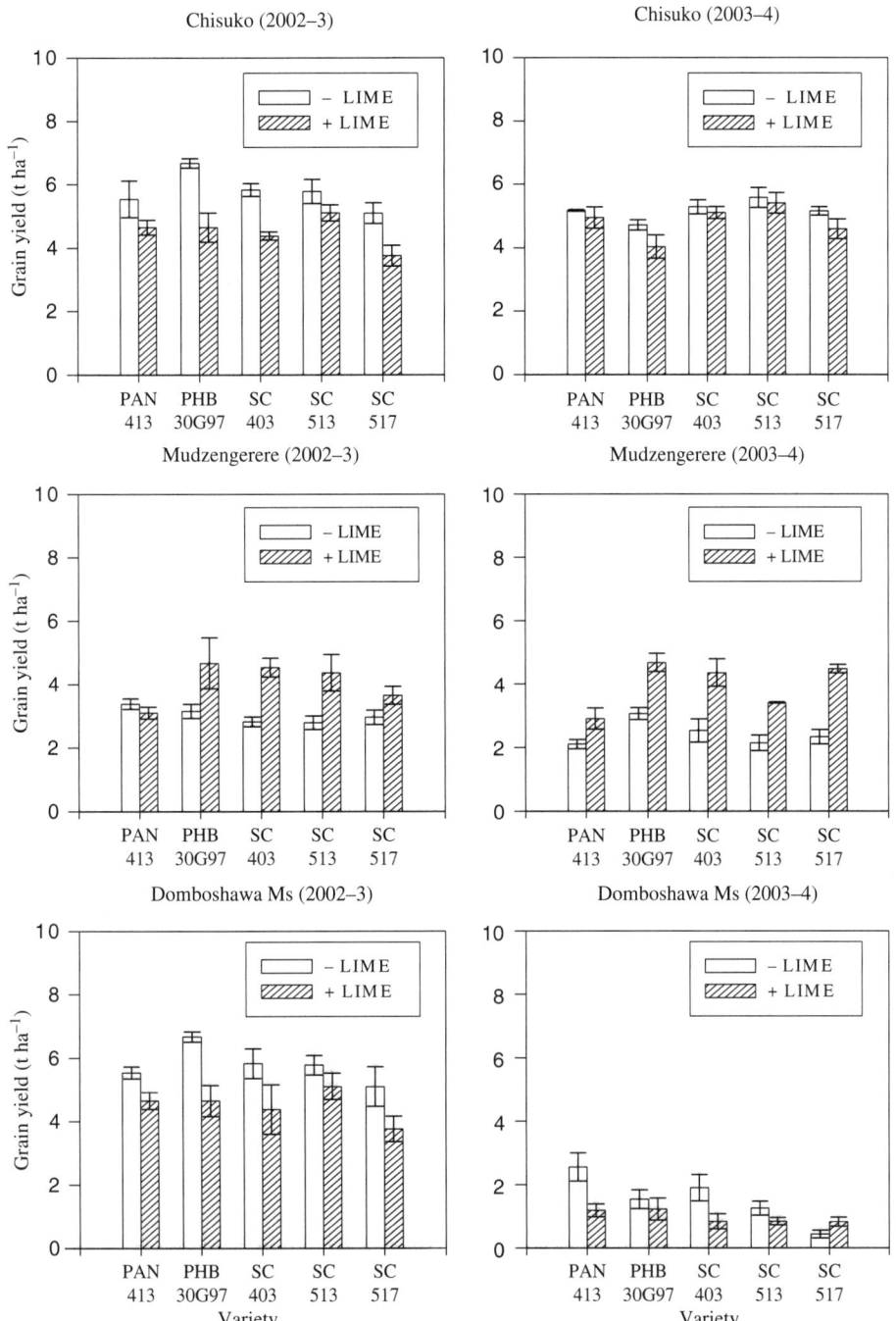

Fig. 5 Effect of liming sandy soils on grain yield for five maize cultivars (*error bars* represent standard error of means at 95% confidence)

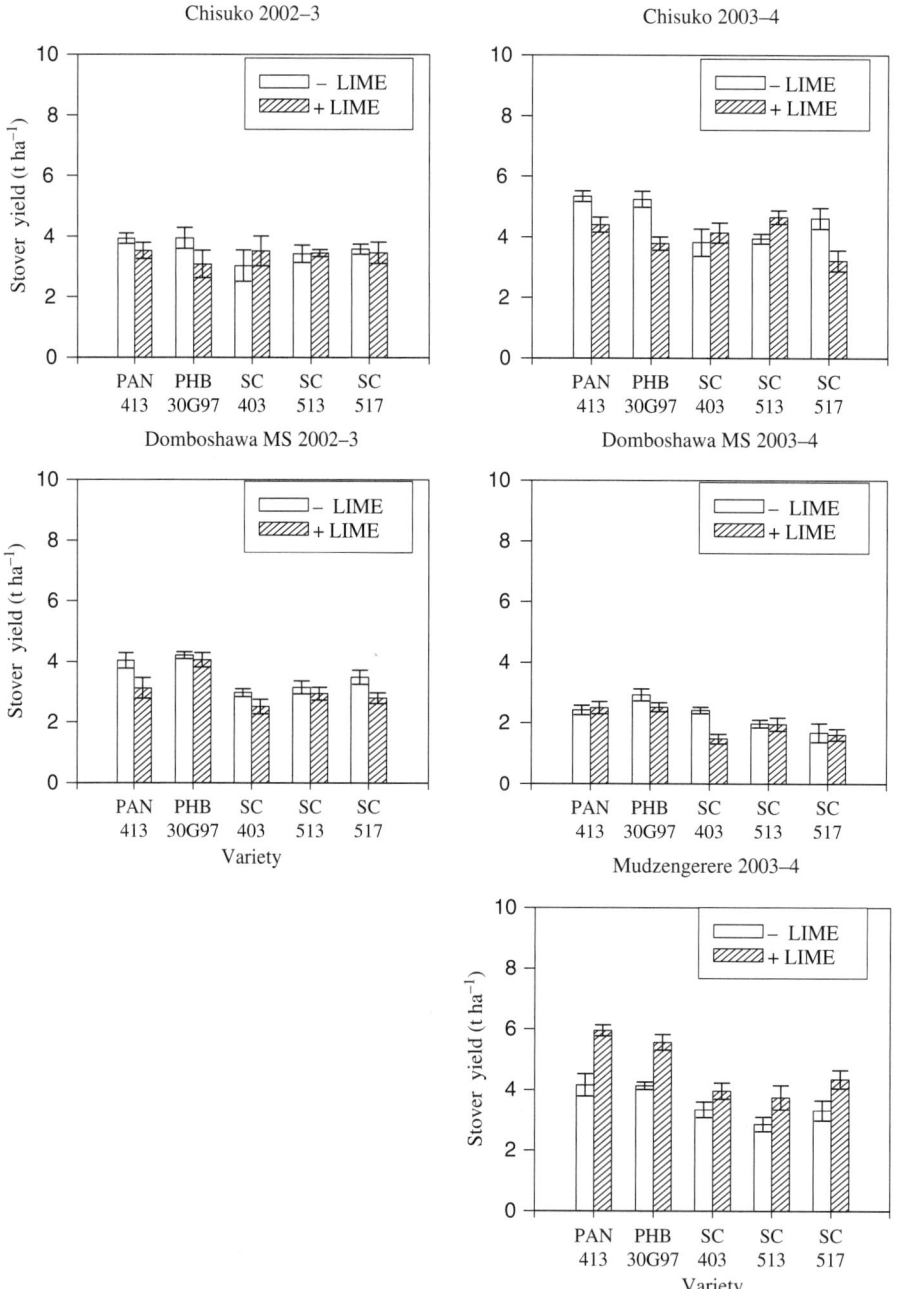

Fig. 6 Above-ground dry matter yield response to lime at harvest on sandy soils for five maize cultivars (*error bars* represent standard error of means at 95% confidence)

Table 4 Grain yield response (%) by five maize cultivars on sandy soils for seasons 1 and 2

Cultivar	Season 1 (2002–2003)			Season 2 (2003–2004)		
	Mudzengerere	Chisuko	Domboshawa	Mudzengerere	Chisuko	Domboshawa
PAN 413	–8	–23	–16	38	–4	–53
PHB 30G97	48	–8	–29	53	–15	–20
SC 403	61	14	–25	72	–3	–56
SC 513	57	–51	–12	59	–3	–33
SC 517	23	–34	–26	92	–11	91

Negative sign shows a reduction in stover yield

(relative root lengths and stover yields) reduces possible effects of differences in seed size and nutrient reserves.

Field Experiment

At two out of the three field sites, liming suppressed grain yield, implying that liming was not necessary despite the low pH. However, at Mudzengerere, with similar pH, liming significantly increased grain yield in all varieties except PAN 413 (Fig. 5). Comparing the chemical properties of the sites, Mudzengerere had lower levels of Ca (Table 5) and Ca-to-Mg ratio (1.4 compared to 3.8 and 5.0 for Chisuko and Domboshawa, respectively; Table 3), compared to the other sites. Thus lime could have acted primarily as a source of Ca and hence a positive response at Mudzengerere where Ca-to-Mg ratio was low. However, at sites with relatively high Ca-to-Mg ratio, addition of the calcitic lime further increased Ca levels and this may have reduced Mg uptake resulting in lower yields (Figs. 5 and 6).

The significant differences in the response of the maize cultivars to liming at Mudzengerere in the second year implied that farmers who can afford lime should select the more responsive varieties, e.g. grain yield for SC 517, an acid-sensitive cultivar (Table 1), increased by 92% compared to 38% for PAN 413 (Table 4). However, cultivars such as SC 403 and PHB 30G97, which give high yields with or without lime on acid soils, would be more suitable for smallholder areas where availability of lime is erratic.

Conclusions

Our results indicate that root growth (ATI, TRL and RRE) can be used to screen maize cultivars commonly grown in southern Africa for tolerance to Al toxicity. Maize growth response to liming depends on the Ca-to-Mg ratio with positive responses accruing when the ratio is low. It is therefore necessary to determine the Ca-to-Mg ratio, besides pH, before liming can be recommended.

Acknowledgements The authors would like to acknowledge The Rockefeller Foundation for financing the project through grant no. 2001 FS 140 and the farmers of Chendambuya for their cooperation.

References

Anderson JM, Ingram JSI (1993) Tropical soil biology and soil fertility. A handbook of methods. CAB International, Wallingford, CT

Bennet RJ (1998) The Al response network in wheat (*Triticum aestivum* L.). The root growth reactions. S Afr J Plant Soil 15:38–45

Bennet RJ, Breen CM, Fey MV (1987) The effects of Al on root cap function and root development in *Zea Mays* L. Environ Exp Bot 27:91–104

Dhliwayo DKC, Sithole T, Nemasasi H (1998) Soil acidity – is it a problem in maize-based production systems of the communal areas of Zimbabwe? In: Waddington SR, Murwira HK, Kumweda JDT, Hikwa D, Tagwira F (eds) Soil fertility research for maize-based farming systems in Zimbabwe. Soil Fert Net/CIMMYT, Harare, pp 217–221

Foy CD (1984) Physiological effects of hydrogen, Al, Mn toxicities in acid soils. In: Adams F (ed) Soil acidity and liming. Agronomy monograph 12, 2nd edn. American Society of Agronomy, Madison, WI, pp 57–97

Foy CD, Fleming AL, Burns GR, Armiger WH (1967) Characterisation of differential aluminium tolerance among varieties of wheat and barley. Soil Sci Soc Am J 31:513–521

Gee GW, Bauder JW (1986) Particle size analysis. In: Klute AL (ed) Methods of soil analysis. American Society of Agronomy, Madison, WI, pp 383–411

Moustakos M, Yupsanis T, Symeonids L, Karataglis S (1992) Al toxicity effects on durum wheat cultivars. J Plant Nutr 15:627–638

Mugwira LM, Floyd M, Patel SU (1981) Tolerance of triticale lines to Mn in soil and nutrient solution. Agron J 73:319–322

Nyamangara J, Mugwira LM, Mpofu SE (2000) Soil fertility status in communal areas of Zimbabwe in relation to sustainable crop production. J Sust Agric 16:15–29

Pandey S, Ceballos H, Magnavaca R, Bahla-Filha AFC, Duque-Vargas J, Vinasco LE (1994) Genetics of tolerance to soil acidity in tropical maize. J Crop Sci 34:1511–1514

Pietraszewska TM (2001) Effects of aluminium on plant growth and metabolism. Acta Biochim Pol 48:673–686

Rout GR, Samantaray S, Das P (2001) Aluminium toxicity in plants: a review. Agronomie 21:3–21

Sapra VT, Mugwira LM, Choudry A, Hughes JR (1978) Screening of triticale, wheat and rye germplasm for Al tolerance in soil and nutrient solution. Proceedings of the international wheat genetics symposium, New Delhi

Schaffert RE, Alves VMC, Parentoni SN, Raghothama KG (2003) Genetic control of phosphorus uptake and utilisation efficiency in maize and sorghum under marginal conditions. CIMMYT report on molecular approaches for genetic improvement of cereals for stable production in water limited environments, Mexico City, Mexico

The Role of Biological Technologies in Land Quality Management: Drivers for Farmer's Adoption in the Central Highlands of Kenya

J.K. Mutegi, D.N. Mugendi, L.V. Verchot, and J.B. Kung'u

Abstract We established hedges of calliandra, leucaena, napier, and their combinations along the contours on slopes of between 5 and 40% as options for soil and nutrient management on steep arable landscapes. Hedge biomass was harvested after every 2 months following proper hedge establishment and incorporated into the plots that were served by specific hedges. After 2.5 years farmers–research group interactions were terminated and farmers were left to continue independently. Three years later the region was surveyed for adoption rates, adoption drivers, and technology adaptation. We consistently observed significantly higher soil pH, exchangeable bases (Ca and Mg), and C in both sole leguminous hedge treatments and combination hedges at time 22 months in comparison to time 0 months ($P < 0.0001$). Consistent significant erosion differences between hedges were observed during the fifth season on slopes exceeding 10% ($P < 0.05$). Farmers' adaptations of hedges ranged from changes in type of trees used, contour hedge tree arrangement patterns, and frequency of pruning of hedge trees. The Logit model was significant at 10% level and predicted 72% of both adopters and non-adopters. The variables farmers' contact with extension agents, education level, farm income, livestock numbers, land size, membership to group or cooperative, sex, and age were significant in explaining contour hedge adoption. We conclude that contour hedges are capable of reducing soil losses and improving crop production and that households that have more educated heads with more livestock and higher farm income are more likely to adopt contour hedge technologies.

Keywords Adoption · Calliandra · Leucaena · Napier · Soil conservation

Introduction

Soil degradation is widely recognized as one of the most significant problems impacting the sustainability of agricultural productivity in many parts of the world (Veloz et al., 1985; Lutz et al., 1994; Barrett et al., 2002). In sub-Saharan Africa one of the regions where soil and nutrient degradation is significant is the central highlands of Kenya (Mugendi et al., 2003; Angima et al., 2003). This degradation is associated with, among others, the rapid population growth, cultivation on fragile ecosystems, and continous cropping. For example, Angima et al. (2002) estimated soil loss in central highlands to be in the range of 150–200 t ha^{-1} yr^{-1}. At a modest soil loss of 10 t ha^{-1} yr^{-1}, it is estimated that soils lose on average 28 kg N, 10 kg P, and 33 kg K ha^{-1} yr^{-1} (Mantel and van Engelen, 1999). In addition to causing serious monetary losses to farmers, soil loss pollutes rivers and other water bodies potentially causing eutrophication, bottom water hypoxia, and health hazards to both humans and animals (CAST, 1985; Justic et al., 1995).

Construction of physical soil conservation structures is expensive, laborious, and time consuming. In such an environment where resources like labor and capital are limited and farmers' strength

J.K. Mutegi (✉)
World Agroforestry Centre (ICRAF), Nairobi, Kenya
e-mail: mutegijames@yahoo.com

A. Bationo et al. (eds.), *Innovations as Key to the Green Revolution in Africa*,
DOI 10.1007/978-90-481-2543-2_52, © Springer Science+Business Media B.V. 2011

is diminishing due to high disease prevalence and hunger, adoption of such energy- and labor-intensive technologies is difficult. This has ultimately led to low adoption of conventional soil conservation technologies resulting in heavy soil and nutrient losses.

The usefulness of contour hedges as alternatives to physical soil conservation structures has been demonstrated in Kenya (Raintree and Torres, 1986; Angima et al., 2002) and Nigeria (Lal, 1989). Most of these trials have, however, been executed on a few slope categories and therefore do not capture the whole array of landscape differences in the farmers' fields. Additionally, most of the tests for contour hedge efficiency trials have mainly been carried out on-station using controlled conditions and have tended to focus on one attribute at a time such as soil conservation while leaving out the effects of these hedges on soil fertility and crop production (see Angima et al., 2002, 2003). This limits applicability of these earlier results across varying landscape types and farmers' challenges. Basically, contour hedgerows control soil erosion by two mechanisms: (1) the hedgerows act as permeable barriers for slowing the flow of runoff and (2) the pruned biomass which is deposited as green manure between the hedges provides a protective cover from raindrop impact (Young, 1997). Experience has shown that appropriate technologies are not always adopted, even where the need is obvious (Guerin, 1999). Farmers may reject or abandon many technologies that have been useful and adopt others in their place since they consider a variety of factors in deciding whether or not to adopt particular conservation practices (Brouwers, 1993; McDonald and Brown, 2000; Soule et al., 2000). This highlights the need to develop a better understanding of the conditions that encourage sustained adoption of conservation practices.

This work was therefore set out to explore and provide implicit data on the effect of contour hedges on soil erosion, soil fertility, and crop production on the steep landscapes of the central highlands of Kenya. We do understand that the purpose of agricultural technology is not yet fulfilled prior to farmers' uptake of the technology. It is due to this recognition that this study went further to investigate farmers' adaptation and determinants of farmers' uptake of contour hedge technologies.

Materials and Methods

Description of the Study Area

This study was conducted in Chuka division, a predominantly maize growing zone in the central highlands of Kenya. The area is on the eastern slopes of Mt Kenya at an altitude of approximately 1500 m above sea level. Mean annual rainfall is 1200 mm, distributed in two distinct seasons: the long rains (mid-March to June) with an average precipitation of 650 mm of rainfall and the short rains (mid-October to December) with an average of 550 mm of rainfall. The average monthly maximum temperature is 25°C and the minimum is 14°C. The long-term monthly average temperature is 19.5°C. The soils of this area are humic Nitisols (FAO, 1990) with an average soil reformation rate of 2.2–4.5 t ha^{-1} yr^{-1} for the top 0–25 cm soil depth and 4.5–10 t ha^{-1} yr^{-1} for the 25–50 cm soil depth (McCormack and Young, 1981; Kilewe, 1987). They are deep, well weathered with friable clay texture and moderate to high inherent fertility.

Experimental Design and Methodology

We selected 10 farms each with 5–10, 10–20, 20–30, and 30–40% slope categories. Within each farm and slope we then established and evaluated monospecific double hedges of calliandra, leucaena, napier, and combination hedges of calliandra + napier and leucaena + napier. The controls were continuous maize plots without vegetative hedges but with similar agronomic management. In each farm, all the treatments and the control were randomized across every slope category and replicated three times. Each hedge was made up of two rows of the above species arranged in interlocking/zigzag manner with inter-row spacing of 0.25 m and intra-row spacing of 0.5 m. The plots were 10 m long with variable inter-hedge spacing estimated according to Young's (1997) formula for biological hedge efficiency as influenced by degrees of arable land steepness stated as follows:

$$W = 200/S\% \tag{1}$$

where W = inter-hedge spacing in meters and $S\%$ = the percent slope.

Where there was a napier + either calliandra or leucaena, the tree row preceded the napier grass row upslope. Each farm represented a block. The aim of blocking was to minimize the effects of site variation so that the treatment effects could be more accurately quantified using statistical tests. Care was taken to ensure that none of the plots fell on obvious convex zones of higher than average net erosion or deposition zones of net sedimentation. We also trenched the plots on the upper lateral borders to prevent eroded sediments from upslope areas from entering into the test plots.

After planting, the hedges were left for 1 year to establish after which they were regularly pruned every 2 months to a height of 50 cm for trees and 10 cm for napier. This was meant to ensure that they did not overgrow the crop to pose significant aboveground competition. The resulting biomass from any one hedge was cut into fine pieces and incorporated into the test plot it served by use of hand hoe.

Soil Sampling and Analysis

Initial sampling for soil characterization was done on each farm before commencing the trials. The second set of soil samples was collected 22 months after establishment of the trials. For each collection date, at least six samples from each plot were collected. The six samples were mixed thoroughly and sub-sampled to form one composite sample for analysis. Field-moist sub-samples were refrigerated at 4°C immediately after collection. Twenty grams of this field-moist soil was extracted using 5 mL of 2 N KCl within 3 days of collection (ICRAF, 1995) by shaking for 1 h at 150 revolutions min^{-1}. The solution was filtered using a pre-washed Whatman No. 5 filter paper. Soil water content was determined gravimetrically from stored field-moist soil at the time of extraction and used for expression of inorganic N on dry weight basis. Nitrate plus NO_2^- were determined by Cd reduction method (Dorich and Nelson, 1984). Ammonium (NH_4^+) was determined by the salicylate hypochlorite colorimetric method (Anderson and Ingram, 1993). Soil bulk density was determined with the undisturbed core method

(Anderson and Ingram, 1993) and used for conversion of NO_3 values from milligrams per kilogram to kilograms per hectare.

For other analyses, soils were air dried and then crushed to pass through a 2 mm sieve. Soil pH was determined in H_2O (1:2.5 wt/vol) (McLean, 1982), exchangeable Ca and Mg by 1 M KCl extraction, and exchangeable K by 0.5 M $NaHCO_3$ + 0.01 M ethylene-diamine-tetra-acetic acid (EDTA) extraction. Soil texture/particle size was determined by use of Bouyoucos hydrometer method (Gee and Bauder, 1986). Total organic C was determined by digesting the soil at 130°C for 30 min with concentrated H_2SO_4 and $K_2Cr_2O_7$, after which C was determined colorimetrically (Anderson and Ingram, 1993).

Rainfall and Soil Loss Assessment

Rainfall was measured throughout the study period by use of two centrally placed rain gauges. Soil loss from contour hedge and control plots was estimated during the third and fifth seasons of hedge growth. It was assessed by use of plastic erosion pins (FAO, 1993) fixed at a spacing of 2×2 m on each plot. The measurements were taken to the nearest millimeter to allow any seasonal change in soil level to be clearly recognized. The resulting soil loss measurements were converted to tons per hectare by first calculating the volume of topsoil washed per plot by use of an equation:

$$\text{Plot volume} = (\text{average depth of washed soil}) \times \\ (\text{plot length}) \times (\text{alley width})$$

(2)

Using plot bulk density values the resulting volume values were converted to tons of soil lost per hectare.

Maize Yield Assessment

All the plots under evaluation were planted with maize (hybrid 513 variety) at a spacing of 0.75×0.25 m (53,000 plants ha^{-1}) which is the local agricultural extension recommendation. Maize was harvested from a net plot after removing the outer row to avoid the edge effect by cutting at root collar. It was weighed immediately to determine the total fresh weight (stover + unshelled cobs). The unshelled cobs were separated from the stover after which the total stover fresh weight was determined and a sub-sample taken for dry weight determination. To obtain grain yields, grains were separated from the core by hand shelling, weighed, and a sub-sample taken for dry weight determination. Similarly, empty cobs (without grains) were weighed and a sub-sample taken for dry weight determination. Dry weight was determined by drying the above sub-samples (cobs, stover, and grain) at 60°C for 3 days to a constant weight and then applying the resulting weight in calculation of dry weight of yield per hectare. In addition to the yield parameters, we measured maize crop height at the maize tasseling stage. The proportion of dry matter was calculated by use of the following formula:

$$\text{Yield (t/ha)} = (10 \times \text{TFW} \times \text{SSDW}) / (\text{HA} \times \text{SSFW}) \tag{3}$$

where 10 is a constant for conversion of yields in kg/m^2 to t/ha, TFW is total fresh weight (kg), SSDW is sub-sample dry weight (g), HA is harvest area (m^2), and SSFW is sub-sample fresh weight (g).

Field Survey

Stratified random sampling was used to identify 120 farmers (contour hedge adopters and non-adopters). Adopting households were defined as those households that had planted at least 50 m of contour hedge trees (from the time of research group/farmer contact) and maintained them for at least 2 years. Non-adopting farmers were defined as those with sloping farm but who had not planted trees in a hedge pattern on their arable farms since the time of farmer research/research group contact. Structured questionnaires were used as survey instruments. The questionnaires were pre-tested on 12 randomly selected adopters and non-adopters, analyzed, and then revised to incorporate farmers' suggestions on various observations and practices related to contour hedges on their farms and villages. Village-level data were collected from focused group interviews in the villages.

Analytical Model

To evaluate farmers' adoption decisions on contour hedges a Logit model (Maddala, 1983) was used. Logit analysis is used when the dependent variable takes on discrete categorical (0, 1) values rather than continuous numerical values. It was employed in this study because adoption can be considered a discrete, categorical variable equal to 1 if the farmer adopts contour hedges and 0 if the farmer does not. Logit model has widely been applied in adoption studies (Bagi, 1983; Polson and Spencer, 1991; Adesina and Sirajo, 1995). To define this model in a simple way, let Y be the decision to adopt contour hedge technology and $\mathbf{X}$ a vector of explanatory variables related to adoption. The model can be stated as $\mathbf{X} = F$ (social, institutional, physical, economic factors), where social factors include age, family size, and education; institutional factors include land tenure, membership to groups and cooperatives, and contacts with extension agents; physical factors include land size and slope; and economic factors include farm income, non-farm income, risks, and livestock numbers. The adoption decision of farmers is specified as $Y = f(\mathbf{X}, e)$, where e is an error term with logistic distribution. The conceptual model is stated as follows:

$$Y_{ik} = F(1_{ik}) = \frac{e^{Z_{ik}}}{1+e^{Z_{ik}}} \text{ for } Z_{ik}$$
$$= \mathbf{X}_{ik}\beta_{ik} \text{ and } -\infty < Z_{ik} < +\infty \tag{4}$$

where Y_{ik} is the dependent variable that takes on the value of 1 for the ith farmer who adopted contour hedge and its variants in zone k and 0 if no adoption occurred. $\mathbf{X}_{ik}$ is a matrix of explanatory variables related to adoption of contour hedges by the ith farmer in zone k, and β_{ik} are the vectors of parameters to be estimated. 1_{ik} is the implicit variable that indexes adoption. The Logit model was estimated using maximum likelihood techniques.

Definition of Variables Used in Empirical Model

The definition of all the variables in the empirical model was as shown in Table 1. Relevant variables were selected after thorough review of literature

Table 1 Definition of variables used in empirical/econometric models

Variable	Description
SEX	Dummy variable for gender of the plot owner; 1 if the owner is a man and 0 if the owner is a woman
AGE	Age of the farmer (years)
FSIZE	Family size
EDUC	Number of years spent in school
LVST	Livestock (TLU)
TENURE	Dummy variable for tenure status of the farmer; 1 if the farmer is the farm owner and 0 otherwise
CONTACT	Dummy variable for extension agent contact; 1 if a farmer has contacts with change agents and 0 otherwise
FINC	Variable for total annual farm income
NFINC	Variable for total annual non-farm income
GOCOOP	Dummy variable for membership to group or cooperative; 1 if a farmer is a group/cooperative member and 0 otherwise
AREA	Total farm area owned by the farmer (ha)
SLOPE	Average slope of the farm (degrees) determined by use of a clinometer
PERCEPTION	Dummy variable for farmers' perception of erosion occurrence in his farm; 1 if farmer perceived occurrence and 0 otherwise
RISK	Risk index of respondent

(see Ervin and Ervin, 1982; Atta-Krah and Francis, 1987; Adesina and Sirajo, 1995; Lapar and Pandey, 1999; Garcia, 1997) and from constant interaction with farmers in this region. The personal characteristics of farmers like level of education, age, and family size were expected to affect farmers' perception and adoption of contour hedges. Age (AGE) was expected to negatively affect adoption since younger farmers are likely to perceive a longer time horizon than older farmers. Education (EDUC), on the other hand, was expected to have a positive influence on adoption. Educated farmers have been found to have a great likelihood of adopting soil conservation technologies (Ervin and Ervin, 1982). Family size (FSIZE) is a measure of the size of the household. Large family sizes may indicate more labor availability to establish and manage contour hedges. Research on alley cropping, one of the variants for contour hedges, revealed that tree-based systems for soil management are labor intensive (see Atta-Krah and Francis, 1987) and therefore inappropriate for labor-stretched households.

CONTACT is a dummy variable that takes the value 1 if a farmer had contact with agroforestry extension agent within the last 5 years preceding interview and 0 if otherwise. Often extension agents expose and encourage farmers to take up technologies that have been shown to work for similar conditions elsewhere. It was hypothesized that CONTACT is positively related to adoption of contour hedges. We hypothesized that

farmers group, organizations/cooperative (GOCOOP) was positively correlated to adoption due to information flow among involved members.

FINC measures the income a farmer derives from his farming activities. Depending on FINC level, it might imply that either the farmer has more resources to engage additional labor to establish and manage hedges or he does not have. NFINC measures income associated with non-agricultural activities like off-farm employment. Studies have indicated mixed response of farmers' adoption behavior to availability of non-farm income. While a number of these studies have shown a positive correlation between non-farm income and adoption of agroforestry technologies, a number of others have shown an inverse or no relationship between non-farm income and adoption of such technologies (see Adesina et al., 2000). In this case the connection between non-adoption and non-farm income has been associated with availability of non-farm income to meet household requirements while adoption has been associated with ability to take agricultural risks and availability of supplemental income for financing conservation expenditure (Garcia, 1997).

TENURE is a dummy variable for the tenure status of land. Farmers with insecure tenure may not adopt soil conservation technologies due to uncertainty of capturing long-term benefits and vice versa for those with secure land tenure. The farmers' perception of risk (RISK), e.g., risks associated with new technologies like pests and diseases and loss of short-run income, affects technology adoption. Farmers who avoid risk may be reluctant to take up uncertain technologies. Farmers with higher risk index are therefore more likely to take up soil conservation technologies (Garcia, 1997).

Data Analysis

We analyzed data by use of GenStat for windows software (version 6.1, Rothamsted Experimental Station) (GenStat, 2002). We used analysis of variance (ANOVA) to test the hypothesis that leguminous contour hedges reduce losses of soil and enhance crop performance. Social data and models were run using Statistical Package for Social Sciences (SPSS).

Results

Rainfall Characteristics

We recorded an average annual rainfall of 1032 mm split into 467 mm during the long rains and 565 mm during the short rains. This annual rainfall was 14% lower than the long-term average for this area (i.e., 1200 mm). Rainfall peaks coincided with the months of April and November while the lowest precipitation was recorded in the months of February, June, and

September. This monthly rainfall distribution was in agreement with the expected rainfall pattern for this region.

Effects of Hedges on Soil Characteristics

We observed significantly higher soil pH ($P = 0.013$), Ca ($P = 0.001$), Mg ($P = 0.042$), and C ($P = 0.032$) after 22 months of trial, relative to the initial conditions. Inorganic N was higher at time 0 relative to 22 months later ($P < 0.0001$). We consistently observed significantly higher soil pH, exchangeable bases (Ca and Mg), and C in both sole leguminous hedge treatments and combination hedges at time 22 months in comparison to time 0 months ($P < 0.0001$) (Table 2). Soil exchangeable K increased significantly in the sole leguminous hedge plots after 22 months of experimentation ($P = 0.006$). We did not observe any significant differences in inorganic N concentration between treatments at time 0 ($P = 0.68$), but we did observe significantly higher inorganic N in the sole leguminous hedges relative to the control and napier after 22 months of trial ($P = 0.027$).

Soil Erosion and Maize Crop Performance

The third season (represents the effect of the two seasons old hedges) on average registered higher soil

Table 2 Properties of top 0–30 cm depth of soil at the start and after 22 months of experimentation in Chuka division, central highlands of Kenya

Treatment	pH H$_2$O[a]	Exchangeable Ca (cmol$_c$ kg^{-1})	Exchangeable Mg (cmol$_c$ kg^{-1})	Exchangeable K (cmol$_c$ kg^{-1})	Total organic C (g kg^{-1})	Inorganic N (NH$_4^-$ + NO$_3^-$) (kg ha^{-1})
Before establishment of trials (time 0 months)						
Control	4.8^c ± 0.09	4.2^c ± 0.18	2.1^a ± 0.07	0.5^{b,c} ± 0.03	17.4^{c,d} ± 0.28	62.7^a ± 5.88
Calliandra	4.7^d ± 0.11	3.8^c ± 0.18	1.4^{c,d} ± 0.04	0.4^c ± 0.04	17.0^d ± 0.13	60.9^a ± 4.31
Leucaena	4.9^c ± 0.15	3.9^c ± 0.14	1.1^e ± 0.06	0.5^b ± 0.02	17.1^{c,d} ± 0.22	66.1^a ± 3.39
Napier	4.6^d ± 0.08	4.2^c ± 0.15	1.7^b ± 0.02	0.5^{b,c} ± 0.02	17.1^{c,d} ± 0.25	70.3^a ± 3.11
Calliandra + napier	4.6^d ± 0.02	3.9^c ± 0.20	1.3^{d,e} ± 0.09	0.4^c ± 0.05	16.8^d ± 0.16	71.8^a ± 7.46
Leucaena + napier	4.8^c ± 0.03	4.0^c ± 0.08	1.6^{b,c} ± 0.03	0.5^{b,c} ± 0.02	17.6^{b,c,d} ± 0.27	70.3^a ± 4.07
After 22 months of experimentation (time 22 months)						
Control	4.9^c ± 0.09	3.8^c ± 0.10	1.0^e ± 0.16	0.2^d ± 0.03	18.3^{b,c} ± 0.26	26.6^d ± 3.40
Calliandra	5.2^b ± 0.06	4.7^b ± 0.12	1.5^{b,c} ± 0.04	0.6^b ± 0.05	21.7^a ± 0.82	38.7^{b,c} ± 0.85
Leucaena	5.1^b ± 0.03	4.6^{b,c} ± 0.21	1.5^{b,c,d} ± 0.04	0.7^a ± 0.03	20.8^a ± 0.76	43.7^b ± 0.26
Napier	4.7^d ± 0.13	3.3^d ± 0.07	1.3^{d,e} ± 0.13	0.4^c ± 0.06	18.2^{b,c} ± 0.15	27.4^d ± 0.22
Calliandra + napier	5.3^a ± 0.04	5.5^a ± 0.22	1.7^b ± 0.11	0.4^c ± 0.04	20.7^a ± 0.29	31.6^{c,d} ± 3.17
Leucaena + napier	5.2^{a,b} ± 0.05	5.1^a ± 0.10	1.8^a ± 0.07	0.6^b ± 0.09	18.6^b ± 0.22	30.0^{c,d} ± 0.27

Means within a column followed by different letters indicate significant difference based on Fisher's protected LSD test ($P = 0.05$); values are means ± SE

Table 3 Effect of vegetative hedges on soil losses on 5–40% slopes

Slope category (%)				
Treatment	5–10	10–20	20–30	30–40
Soil loss from two seasons old hedge plots (t ha^{-1})				
Control	16.80^a	79.50^a	77.40^a	67.53^a
Calliandra	15.52^a	37.53^b	34.87^b	26.47^b
Leucaena	14.70^a	46.63^b	37.50^b	29.60^b
Napier	12.64^a	20.87^c	22.90^c	20.62^b
Calliandra + napier	13.57^a	30.57^b	26.50^{b,c}	26.58^b
Leucaena + napier	14.17^a	35.18^b	33.73^b	21.78^b
Soil loss from four seasons old hedge plots (t ha^{-1})				
Control	16.48^a	79.61^a	79.25^a	78.90^a
Calliandra	11.00^a	26.14^b	28.92^b	22.18^b
Leucaena	12.31^a	29.68^b	28.62^b	23.50^b
Napier	10.10^a	10.21^c	11.90^c	9.67^c
Calliandra + napier	12.83^a	17.70^{b,c}	14.15^c	11.55^c
Leucaena + napier	10.66^a	17.67^{b,c}	13.38^c	12.98^c

For each slope category and season, means within a column followed by different letters indicate significant difference based on Fisher's protected LSD test ($P = 0.05$)

losses ($P = 0.004$) than the fifth season (represents the effect of the four seasons old hedges) for treatments with hedges and vice versa for the control (Table 3). Soil losses from plots on 5–10% slope had a narrow range (10–17 t ha^{-1} yr^{-1}) for different treatments and seasons in comparison to other slopes, and there were no significant differences between treatments ($P < 0.05$).

During the third season, we observed significantly lower ($P < 0.001$) soil losses in plots with hedges relative to the control on slopes exceeding 10% but with the exception of napier, no significant differences among different types of hedges. Consistent significant erosion differences between hedges were observed during the fifth season on slopes exceeding 10% ($P < 0.05$). Napier hedges were the most effective at reducing erosion losses in both seasons (Table 3). We observed significantly lower soil losses from the four seasons old combination hedges than individual tree hedges ($P = 0.012$). Soil loss on 10–20% slope category was higher than soil loss on any other slope category ($P = 0.043$). In an attempt to understand this seemingly unusual phenomenon we analyzed our particle size data while treating slope categories as factors and particle size/soil texture as variates. The soil textural characteristics for different slope categories were characterized by significantly lower ($P < 0.001$) clay content on the 10–20% slope relative to 20–30 and 30–40% slope (Table 4).

The presence of hedges had very little impact on maize crop yields during the first season (Table 5). Sole napier hedges suppressed yields during this season, but not during the fourth. The effects of the hedges

Table 4 Soil textural characteristics at different slope categories in Kirege

Slope category (%)	Particle size (g kg^{-1})		
	Sand	Silt	Clay
5–10	318.6^a	284.9^{a,b}	396.5^a
10–20	304.3^a	310.1^a	385.6^a
20–30	299.6^a	279.8^b	420.5^b
30–40	299.5^a	290.5^{a,b}	411.0^b

For each column, means followed by different letters indicate significant difference based on Fisher's protected LSD test ($P = 0.05$)

were more apparent during the fourth season. A number of treatments resulted in significant increases in yields, particularly when N-fixing trees were part of the system. Maize yield was higher during the fourth season than the first season for all the treatments with vegetative hedges, but lower on the control.

We observed significantly lower grain yield when cumulative soil loss exceeded 150 t ha^{-1} yr^{-1} ($P = 0.01$) and significantly lower plant height and stover weight when cumulative soil loss exceeded 150 t ha^{-1} yr^{-1} ($P < 0.0001$) (Table 6).

Household Characteristics of Adopters and Non-adopters of Contour Hedges

A higher percentage of male-headed households than female-headed households had adopted contour hedge technologies on their steep arable land (Table 7). Adoption rose with the level of formal education from 0–1 year of education category to 8–12 years of

Table 5 Maize yield at Chuka farms in plots served by various vegetative hedges during the first and fourth seasons of the trial

Treatment	First season of trial	Fourth season of trial	Treatment mean
	..................Maize grain (t ha^{-1} $\pm$ 1 SE)..................		
Control	2.2^a $\pm$ 0.5	2.0^a $\pm$ 0.3	2.1^a
Calliandra	1.9^a $\pm$ 0.4	2.9^b $\pm$ 0.4	2.4^{a,b}
Leucaena	2.1^a $\pm$ 0.6	3.1^b $\pm$ 0.5	2.6^{a,b}
Napier	0.9^b $\pm$ 0.1	2.1^a $\pm$ 0.4	1.5^c
Calliandra + napier	2.2^a $\pm$ 0.7	3.4^b $\pm$ 0.8	2.8^b
Leucaena + napier	2.3^a $\pm$ 0.8	3.6^b $\pm$ 0.6	2.9^b

For each column, means followed by different letters indicate significant difference based on Fisher's protected LSD test ($P = 0.05$); values are mean yield $\pm$ SE

Table 6 Relationship between observed soil erosion classes with selected maize growth parameters in Chuka farms

Soil loss (t ha^{-1} yr^{-1})	Grain weight (t ha^{-1})	Plant height (cm)	Stover weight (t ha^{-1})	TAGB (t ha^{-1})
40–100	1.9^a $\pm$ 0.2	247.3^a $\pm$ 5.0	7.0^a $\pm$ 0.2	10.2^a $\pm$ 0.5
100–150	1.5^a $\pm$ 0.2	259.0^a $\pm$ 8.5	7.3^a $\pm$ 0.2	4.1^b $\pm$ 0.3
150–200	1.5^a $\pm$ 0.2	226.2^b $\pm$ 8.6	5.6^b $\pm$ 0.5	3.2^{b,c} $\pm$ 0.1
>200	0.9^b $\pm$ 0.3	190.1^c $\pm$ 2.8	3.3^c $\pm$ 0.2	1.6^c $\pm$ 0.1

For each column, means followed by different letters indicate significant difference based on Fisher's protected LSD test ($P = 0.05$); values are means $\pm$ SE

TAGB – total aboveground biomass

Table 7 Household characteristics of adopters and non-adopters of contour hedge technology in Chuka division, Kenya (used $N = 120$)

Variable	Parameter	Non-adopters ($n = 60$)	Adopters ($n = 60$)
House head sex	Male (%)	45	55
	Female (%)	64	36
Education (%)	0–1 year	64	36
	1–4 years	51	49
	5–8 years	45	55
	8–12 years	40	60
	>12 years	78	22
Livestock – cattle	Cows	1	3
	Goats	2	2
	Sheep	2	3
Land tenure	Rented (%)	96	4
	Inherited (%)	68	32
	Bought (%)	56	44

Farmers' Adaptation of Contour Hedges

A number of adopters (55%) modified the originally demonstrated contour hedges (double interlocking hedges of calliandra, leucaena, and napier) with adaptations of their own (Table 8). Such adaptations ranged from changes in the type of trees used in contour hedges and contour hedge patterns to frequency of pruning of contour hedge trees. Majority of contour hedge modifiers cut the hedge trees at a height that was higher than the one that was demonstrated. High in the list of common modifications also were reduction in inter-row spacing and introduction of *Tithonia diversifolia* into the hedges (Table 8). Approximately 3% of the adopters left hedges to grow to trees. We found that such farmers were keen on using hedge species for fuelwood and as seed orchards.

education category and then declined sharply beyond that point. The adopters had on average more livestock than non-adopters. Adoption was highest among the farmers who had bought their land, low among the farmers who had inherited land, and lowest among those farmers who were on land rent arrangements.

Determinants of Adoption of Contour Hedges

The Logit model was significant at the 10% level. The model correctly predicted 72% of both adopters and non-adopters. Eight variables were significant in

Table 8 Percentage of adopters making modifications in their management of contour hedges in Chuka division, Kenya ($N = 120$)

Farmers' modification	Responses	Percentage
Inter-row spacing expanded	10	8
Inter-row spacing reduced	20	17
Introduced *Tithonia diversifolia* species into the double rows	18	15
Planted hedge as single row instead of double rows	9	8
Planted more than two rows in the same hedge	6	5
Height of tree cut back higher than demonstrated	25	21
Often allowed goats to graze on the hedges directly	4	3
Left hedges to grow into trees before cutting	4	3

Table 9 Econometric model results of factors affecting farmers' adoption of contour hedges in Chuka division, central highlands of Kenya

Variable	Estimate	Standard error	t-Statistic	P value
SEX	1.062	0.512	1.09	0.05
AGE	−0.03	0.031	1.31	0.06
FSIZE	0.152	0.127	1.71	NS
RISK	−0.302	0.533	0.59	NS
EDUC	3.39	1.12	2.91	0.005
LVST	1.59	0.821	1.93	0.02
FINC	1.61	0.512	1.95	0.01
NFINC	−0.191	0.622	0.46	NS
CONTACT	1.83	0.523	3.12	0.001
GOCOOP	2.43	0.921	2.753	0.05
SLOPE	0.66	0.473	1.25	NS
PERCEPTION	0.53	0.330	0.27	NS
AREA	−1.53	0.723	1.84	0.02
Intercept	−7.43	2.12	−3.56	0.005
Percent correct predictions	72.3			
Log of likelihood function	−54.21			

Source: Ruxton and Neuhauser (2010)

explaining the adoption of contour hedge technology at 5–10% level (Table 9).

They were farmers' contact with extension agents which was significant at 0.1%, level of education at 0.5% level, farm income at 1% level, livestock and land size at 2% level, membership to group or cooperative and sex at 5% level, and age at 6% level. The other variables did not significantly influence adoption of contour hedges. The coefficients for land size and age were negative and significant at 2 and 6% levels, implying that these two variables were inversely related to contour hedge adoption. Other variables like perception of soil erosion occurrence, slope, family size, and risk perceptions were not important in explaining contour hedge adoption behavior in this region.

Discussion

Soil Analytical Characteristics

The increase in soil pH on plots with tree hedges can be attributed to an increment in exchangeable bases (Ca^{2+} and Mg^{2+}). Increased calcium and magnesium react with acid soils replacing hydrogen and aluminum on the colloidal complex (Cahn et al., 1993). This adsorption of calcium and magnesium ions raises the percentage base saturation of the colloidal complex leading to corresponding increment in pH (Loomis and Connor, 1992). The high total soil organic carbon in the hedge plots after 22 months was most likely a result of transfer of hedge pruning into the plots.

Low inorganic N concentration at 22 months relative to initial levels can be attributed to weather and sampling time differences. The first sampling was done toward the end of September after a long dry spell and during land preparation for planting, while the second sampling (time 22 months) was done in July after the March to May rains (long rains) and July drizzles and at maize tasseling stage. So probably a lot of nitrate had been immobilized, leached, denitrified, or even taken up by the growing crop at the time of sampling. Maize has the highest demand for N at the tasseling stage (Karlen et al., 1988), so nitrogen would be locked in maize plant tissues at this time.

Soil Conservation and Maize Crop Performance

Lower soil losses during the fifth season on the contour hedge plots in comparison to the third season can be attributed to hedge species differences in stage of growth and natural terrace formation. During the fifth season, hedges were more mature and therefore formed a more intact barrier to sufficiently obstruct runoff and enhance deposition of the sediment load carried downslope by the runoff. Natural terraces form along contour hedges, advance, and become more effective in obstruction of soil movement with time due to entrapment of washed off soil on the upslope side of the hedge (Lal, 1989).

Napier hedge was overall the best vegetative hedge in soil conservation, possibly due to its rhizomatous rooting characteristics. These rhizomatous roots spread out superficially over a large area reinforcing soil around them and bringing about an increase in cohesion and hence in shear strength (Dissemeyer and Foster, 1985). It also sprouts many tillers within a short time, forming an intact hedge. Lower soil loss values on combination hedge plots as compared to single tree species hedge could partially be attributed to presence of napier component and the positive interaction between napier and leguminous tree species which recycle and fix N (NRC, 1983; Young, 1997).

The lower soil loss in the 20–30 and 30–40% slope categories relative to 10–20% slope confirms Angima et al.'s (2003) observation of lower soil losses on 40% slope than on 20% slope. It is probable that the low soil clay content we observed in the 10–20%

slope relative to higher slopes explains this observation. High soil clay content leads to surface sealing resulting in low soil particle detachment (Morgan and Rickson, 1995). High percentage of silt and fine sand decreases the raindrop energy required to break down soil clods increasing the susceptibility of soil particles to detachment and hence erosion (Morgan, 1986). This means that on steeper slopes, the ability of soil to resist detachment by runoff flow energy was probably higher than on the 10–20% slope category.

The inverse relationship between maize crop growth parameters and soil loss can be attributed to loss of topsoil, which is the most favorable soil for crop growth. The loss of topsoil inevitably reduces soil productivity, which in turn deters crop growth because the topsoil is usually the most fertile, containing natural plant nutrients, humus, and any fertilizers that farmers have applied (Lal, 1989). The fact that soil loss negatively affects crop growth parameters and contour hedges reduce soil loss implies that contour hedges can enhance crop production on sloping landscapes.

Farmers' Behavior and Adoption of Contour Hedges

Farmers' adaptation of contour hedges can be attributed to farmers' attempts to make contour hedge technologies more applicable to their local/individual conditions. Though we would have expected a strong and positive relationship between perception of soil erosion and slope with farmers' uptake of contour hedges, the two variables though positive did not significantly influence adoption of contour hedges. It is probable that steep landscapes increased farmers' risk index, hence negatively affecting farmers' willingness to invest their time and other resources in such land. Farmers' age was significantly related to likelihood of adoption at the 10% level of significance. The negative sign on AGE suggests that contour hedges are more likely to be adopted by younger farmers. This is probably because, as shown by a number of studies elsewhere, younger people are often better disposed to trying new innovations and have longer planning horizons to justify investments in tree-based technologies (Adesina et al., 2000; Adesina and Sirajo, 1995; Ervin and Ervin, 1982). The positive sign on land tenure suggests that the possession of rights over trees has positive influence on likelihood of contour hedge technology adoption. While the coefficient for farm income (FINC) was positive and significant, the coefficient of non-farm income (NFINC) though not significant was negative. This implies that farmers who made more resources from agricultural activities were more likely to adopt biological contour hedges than those that made less. On the other hand, farmers with higher non-farm income were more unlikely to take up contour hedges relative to those who did not have non-farm income or those with lesser non-farm income. Though risk perception was not a significant determinant of contour hedge adoption, the coefficient for risk was negative, indicating that farmers with a higher risk index were more unlikely to adopt contour hedges relative to those with lower risk index.

Conclusions

We conclude that contour hedges are capable of reducing soil losses and improving crop production. The farmers' level of education, age, land size, risk perception, farm income, and number of livestock are important variables in as far as adoption of contour hedgerows is concerned in the central highlands of Kenya.

Acknowledgments We gratefully acknowledge the support of National Agroforestry Project (NAFRP) in experimental setup. At the ICRAF laboratory, Mercy Nyambura, Vincent Mainga, and Robin Chacha were very instrumental in soil analysis. We are grateful to Chuka farmers for allowing us to use their farms for trials and responding to questions. This research was funded by Rockefeller Foundation.

References

Adesina AA, Mbila D, Nkamleu GB, Endamana D (2000) Econometric analysis of the determinants of adoption of alley farming by farmers in the forest zone of southwest Cameroon. Agric Ecosyst Environ 80:255–265

Adesina AA, Sirajo S (1995) Farmers' perceptions and adoption of new agricultural technology: analysis of modern mangrove rice varieties in Guinea-Bissau. Q J Int Agric 34(4):358–371

Anderson JM, Ingram JS (1993) Tropical soil biology and fertility: a handbook of methods. CAB International, Wallingford, UK

Angima SD, Stott DE, O'Neill MK, Ong CK, Weesies GA (2002) Use of calliandra-Napier grass contour hedges to control erosion in central Kenya. Agric Ecosyst Environ 91:15–23

Angima SD, Stott DE, O'Neill MK, Ong CK, Weesies GA (2003) Soil erosion prediction using RUSLE for central Kenyan highland conditions. Agric Ecosyst Environ 97: 295–308

Atta-Krah AN, Francis PA (1987) The role of on-farm trials in the evaluation of composite technologies: the case of alley farming in southern Nigeria. Agric Syst 23:133–152

Bagi FS (1983) A Logit model of farmers' adoption decisions about credit. Southern J Agric Econ 15:13–19

Barrett B, Place F, Aboud A, Brown DR (2002) The challenge of stimulating adoption of improved natural resource management practices in African agriculture. In: Barrett CB, Place F, Aboud AA (eds) Natural resources management in African agriculture: understanding and improving current practices. CAB International, Oxon, UK, pp 1–21

Brouwers JH (1993) Rural people's response to soil fertility decline: Adja case (Benin). Published PhD dissertation, Wageningen Agricultural University

Cahn M, Bouldin DR, Carro MS, Bowen WT (1993) Cation and nitrate leaching in an oxisol of Brazilian Amazon. Agron J 85:334–340

Council for Agricultural Science and Technology (CAST) (1985) Agriculture and ground water quality. Report number 103, May 1985, 62 pp

Dissemeyer G, Foster GR (1985) Modifying the universal soil loss equation for forestland. In: Swaify EL, Moldenhauer WC, Lo A (eds) Soil erosion and conservation S.A. Soil Science Society of America, Ankeny, IA, pp 480–95

Dorich RA, Nelson DW (1984) Evaluation of manual cadmium reduction methods for determination of nitrate in potassium chloride extracts of soil. Soil Sci Soc Am 48:72–75

Ervin CA, Ervin DE (1982) Factors affecting uses of soil conservation practices: hypotheses, evidence and policy implications. Land Econ 58(3):277–292

FAO (1990) Soil map of the world. Revised legend. World Resources Report 60. FAO, Rome

FAO (1993) Field measurement of soil erosion and runoff. Soils Bulletin No. 68. FAO, Rome, Italy

Garcia YT (1997) Analysis of decision models for upland soil conservation in Argao, Cebu. PhD dissertation, University of Philippines, Los Banos

Gee GW, Bauder JW(1986) Particle size analysis. In: Klute A (ed) Methods of soil analysis: physical and mineralogical methods. Soil Science Society of America, Madison, WI, pp 383–411

GenStat (2002) Version 6.1, Lawes Agricultural Trust, Rothamsted Experimental Station, UK

Guerin T (1999) An Australian perspective on the constraints to the transfer and adoption of innovations in land management. Environ Conserv 24(4):289–304

ICRAF (1995) Laboratory methods for soil and plant analysis. International Centre for Research in Agroforestry, Nairobi, Kenya

Justic D, Rabailis NN, Turner RE, Dortch Q (1995) Changes in nutrient structure of river dominated coastal waters: stoichiometric nutrient balance and its consequences. Estuar Coast Shelf Sci J 40:339–356

Karlen DL, Flannery RL, Sadler EJ (1988) Aerial accumulation and partitioning of nutrients by corn. Agron J 80:232–242

Kilewe AM (1987) Prediction of erosion rates and effects of topsoil thickness on soil productivity. PhD thesis, University of Nairobi, Kenya

Lal R (1989) Agroforestry systems and soil surface management of a tropical Alfisol. Water runoff, soil erosion and nutrient loss. Agroforest Syst 8:97–111

Lapar MLA, Pandey S (1999) Adoption of soil conservation: the case of the Philippine uplands. Agric Econ 21(3): 241–256

Loomis RS, Connor DJ (1992) Crop ecology: productivity and management in agricultural systems. Cambridge University Press, Cambridge

Lutz E, Pagiola S, Reiche C (1994) The costs and benefits of soil conservation: the farmer's viewpoint. World Bank Res Obs 9(2):273–295

Maddala GS (1983) Limited dependent variables and qualitative variables in econometrics. Econometric society monographs 3. Cambridge University Press, Cambridge

Mantel SD, van Engelen VM (1999) Assessment of the impact of water erosion on productivity of maize in Kenya: an integrated modeling approach. Land Degrad Dev 10: 577–592

McCormack DE, Young KK(1981) Technical and societal implications of soil loss tolerance. In: Morgan RPC (ed) Soil conservation problems and prospects. Wiley, Chichester, UK, pp 365–376

McDonald M, Brown K (2000) Soil and water conservation projects and rural livelihoods: options for design and research to enhance adoption and adaptation. Land Degrad Dev 11:343–361

McLean EO (1982) Soil and lime requirement. In: Page AL (ed) Methods of soil analysis. Part 2, 2nd edn. Agronomy Monograph 9. ASA and SSSA, Madison, WI, pp 199–224

Morgan RPC (1986) Soil erosion and conservation. Longman Group Ltd, Hong Kong

Morgan RPC, Rickson RJ (1995) Slope stabilization and erosion control: a bioengineering approach. Chapman & Hall, London

Mugendi DN, Kanyi MK, Kung'u JB, Wamicha W, Mugwe JN (2003) Mineral-N movement and management in an agroforestry system in central highlands of Kenya. In: Proceedings for soil science society of East Africa conference, Mombasa, 4–8 December 2000, pp 287–297

National Research Council (NRC) (1983) Calliandra, a versatile small tree for humid and tropics. Jakarta, Indonesia. National Academy Press, Washington, DC

Polson R, Spencer DSC (1991) The technology adoption process in subsistence agriculture: the case of cassava in southwestern Nigeria. Agric Syst 36:65–77

Raintree J, Torres F (1986) The agroforestry in farming systems perspectives: the ICRAF approach. IARC'S workshop on FSR, ICRISAT, Hyderabad, India, 17–21 Feb 1986

Ruxton D, Neuhauser M (2010) When should we use one-tailed hypothesis testing? Methods Ecol Evol 1:114–117

Soule JM, Tegene A, Wiebe DK (2000) Land tenure and the adoption of soil conservation practices. Am J Agric Econ 82(4):993–1005

Veloz A, Southgate D, Hitzhusen F, Macgregor R (1985) The economics of erosion control in a subtropical watershed: a Dominican case. Land Econ 61(2):145–155

Young A (1997) Agroforestry for soil management, 2nd edn. CAB International, Wallingford, UK

Biophysical Characterization of Oasis Soils for Efficient Use of External Inputs in Marsabit District: Their Potentials and Limitations

E.M. Muya, J.K. Lelon, M.G. Shibia, A.O. Esilaba, M. Okoti, G.N. Gachini, and A.L. Chek

Abstract The biophysical site characterization was carried out for irrigated oasis farming in Kalacha, where production has declined to an extent that farmers abandoned their farms and opted for forage production. The objective of carrying out the characterization work was to identify technologies and locally available resources not only for efficient utilization of limited water and nutrient but also to reverse the declining soil quality for long-term production. The land attributes used in differentiating the soils into mapping units were geomorphology, nutrient availability, salinity and sodicity hazards, infiltration rate and drainage conditions for various crops. Productivity index for each soil mapping unit was determined by the regressed relationships between crop production and the soil quality indicators. Unit PIL1 was classified as well-drained, Calcic Solonetz with productivity index of 22%. The most limiting factors were found to be nitrogen and high sodium concentrations. Unit PIL2 comprised well-drained, stratified Salic Fluvisols with productivity index of 26%. The most limiting factors were low nitrogen level and high sodium and salt concentrations. Unit PIL3 had excessively calcareous, moderately drained and strongly stratified soils, classified as Calcaric Fluvisols with productivity index of 21%. The major limitation was nitrogen. Unit PIL4 comprised firm soils, overlying cemented sub-soils with high carbonate concentration, classified as Calcic Solonchaks with imperfect to poor drainage conditions and having productivity index of 12%. The major limitations were poor drainage, high sodium concentration and low fertility status. The soil productivity can be improved by prescribing the required inputs to address the specified limitations.

Keywords Land quality · Potentials · Limitations

Introduction

Given the increasing population in arid regions with erratic rainfall and fragile environment, agricultural priority is to increase biological and economic yield per unit area, while conserving environment from further degradation. In irrigated agriculture around oasis, optimum utilization of resources involves not only efficient use of water but also improvement of land qualities that determine the productivity (Muya et al., 2005). Optimum utilization of resources starts with the matching of rated crop requirements with rated land quality indicators in order to establish the suitability (potential) and limitations of land for the envisaged crops. According to FAO (1986), the maximum yield level of a crop is determined by how well it is adapted to the prevailing soil and environmental conditions. Therefore, assessment of the potentials and limitations of soil for different crops provides the basis of formulating the appropriate management strategies which target specific management problem to improve the production of the specific crop (FAO, 1976). Environmental requirements of climate, soil and water for optimum growth and yield vary with crops and crop varieties. A careful selection of the crop and variety most suited to a given environment, coupled

E.M. Muya (✉)
National Agricultural Research Laboratories, Kenya
Agricultural Research Institute (KARI), Nairobi, Kenya
e-mail: edwardmuya@yahoo.com; kss@iconnect.co.ke

A. Bationo et al. (eds.), *Innovations as Key to the Green Revolution in Africa*,
DOI 10.1007/978-90-481-2543-2_53, © Springer Science+Business Media B.V. 2011

with management inputs prescribed on the basis of crop and soil requirements, is of paramount importance for obtaining high and efficient production. To achieve this, the relationships between different crops, climate, soil and biophysical processes must be examined as a basis of evaluating the potentials and limitations of soil and environmental resources to the envisaged farm enterprises. This practice corresponds to the precision agriculture recommended by Bouma (1997). In principle, the precision agriculture tries to fine-tune land management, with the objective of maximizing agricultural production and its quality while minimizing adverse environmental effects. For irrigated agriculture, management should be fine-tuned to minimize nutrient and water losses through leaching of excess water or deep percolation (Wesseling, 1998). This can be avoided by planning irrigation interval and scheduling, based on the understanding of variations of soil water requirements with different stages of growth as well as total water requirements over the entire growth period (Ekirapa, 1996). Therefore, information on the potentials and limitations of different soil types for different crops as well as their extent and management requirements is required for efficient use of nutrient and water resources in fragile environments and is required for planning sustainable land use system (Muya et al., 2006). This chapter provided that information by carrying out detailed biophysical characterization of an oasis area, which has been irrigated for over 20 years. In this area, agricultural production has declined to an extent that farmers abandoned crop farming and replaced it with forage production. This is because irrigated agriculture started through community initiative with no baseline information, where crop selection, management practices and irrigation methods applied were not based on biophysical evaluation of the area. To come up with sustainable fertility management strategies to improve the productivity of the production system, Kusewe and Guiragossian (1991) suggest detailed evaluation of all the agricultural elements and their relationships with physical, biological and socio-economic environments. The first step is to examine the overview of biophysical constraints of the area in order to come up with integrated analysis of biophysical conditions and socio-economic issues of the area (Mavua, 1985).

The main biophysical constraints of the research area are low organic matter content and poor soil structure. Therefore, a labile pool of organic matter,

soil structure stability and soil nutrients are important indices of soil productivity (Stewart et al., 2004). Consequently, loss of organic matter and soil structure stability due to unsustainable irrigation practices may lead to severe land degradation with adverse consequences on physical, biological and nutrient status of the soil. Biological degradation is aggravated by the increasing accumulation of salts and sodium within the soil profile. This creates the toxic environments that inhibit the microbial processes that sustain the ecosystem functions (Muya et al., 2005). In this context, the relevant land quality indicators selected were salinity, sodicity, nutrient availability and soil structure (determined by oxygen availability). Therefore biophysical site characterization was carried out with the following objectives:

(1) To assess the water quality and its suitability for irrigation.
(2) To assess the suitability of land for different crops to provide the basis of formulating appropriate management technologies to reverse the declining land quality and crop production.
(3) To provide agronomic, soil and water requirements for the most promising crops.

Materials and Methods

For the determination of irrigation water quality, a guideline provided by FAO (1977) was applied. According to this guideline, the quality of irrigation water is determined by the type and quantity of dissolved salts. The salts originate from rock weathering and soil, including dissolution of lime, gypsum and slowly dissolved minerals. These salts are applied with irrigation water, and they remain behind in the soil when water evaporates or is used by the crops. The suitability of water for irrigation was determined not only by the total amount of salts present but also by the type of salt. The soil characteristics used as a basis of evaluating water quality were those related to salinity, water infiltration rate, toxicity and a group of other miscellaneous problems (Table 1).

Systematic augering at a spacing of 100 by 100 m was done in all the sites, and at each observation point, detailed description of soils and topographical characteristics were recorded.

Table 1 Guideline for evaluating the quality of water for irrigation

Potential problems	Units	Degree of restriction on the use		
		None	Slight	Severe
Salinity				
Electrical conductivity of water (ECw)	dS/m	<0.7	0.7–3.0	>3.0
Total dissolved solids (TDS)	mg/l	<450	450–2000	>2000
Infiltration rate				
Sodium absorption ratio (SAR)		3–6	6–12	>12
Specific ion toxicity				
Sodium	SAR	<3	3–9	>9
Chloride	me/l	<4	4–10	>10
Boron	mg/l	0.7	0.7–3.0	>3.0
Miscellaneous				
Nitrogen	mg/l	<5	5–30	>30
Bicarbonate	me/l	<1.5	1.5–8.5	>8.5
pH	Normal pH range 6.5–8.4			

Source: FAO (1977)

The coding of soil mapping units was based on guidelines provided by Kenya Soil Survey Staff (1987). These units were described on the basis of soil depth, texture, calcareousness, cation exchange capacity, soil pH, salinity, sodicity, exchangeable cations and drainage conditions. The first entry in soil coding was physiography, followed by geology. The numbers 1–5 denoted differences in soil characteristics that were used to differentiate soil mapping units. The soil quality indicators used for evaluating land productivity and potentials for different crops were drainage (oxygen availability), depth, texture, structure, calcareousness, salt sodium levels, cation exchange capacity, exchangeable bases and phosphorous. Topographical survey was carried out using theodolite to provide the map plan to be used as the base maps for soil mapping.

Based on soil and topographical characteristics, representative soil profiles were identified, described, sampled and classified according to FAO-UNESCO (1997).

The productivity index (PI) was calculated based on the regressed relationships between maize yield and the soil quality attributes given in Table 2.

Having derived the sufficiency values of each of the soil quality attributes from the equation given in Table 2, the productivity index (PI) was calculated, based on the following equation as provided by Driessen and Konijn (1992)

$$PI = A \times B \times C \times D \times E$$

where *A*, *B*, *C*, *D* and *E* are the soil quality attributes with known relationships with the crop desired.

Evaluation of the potentials and limitations of soils for different crops was based on FAO framework for land evaluation (1976), while water quality was assessed, based on the guidelines given by FAO (1977). Land quality indicators used for evaluation were oxygen availability, salinity, sodicity and nutrient availability.

A land quality is an attribute of land, which acts in a distinct manner in its influence on land performance for specific land use. For each soil mapping unit, the following land qualities were assessed: available nutrients or fertility (FERT), oxygen availability (OXAV), salinity hazard (SAL) and sodicity hazard (SOD).

Table 2 Regressed relationships between maize yield and soil quality attributes

Soil quality attributes	Equation	R^2	Critical limit
Soil organic carbon (%SOC)	$Y = 0.31 + 0.56–0.11SOC$ Y = maize yield (kg/ha)	0.37	1.08
Phosphorous (ppm)	$Y = 1–0.95e^{-(0.204P)}$	0.88	7.6
Potassium (m.e.%)	$Y = 1–0.79e^{-(1.66 K)}$	0.43	0.83
Soil pH	$Y = 6.41+2.42 pH–0.97^{(pH)2}$	0.68	5.1

Source: Aune and Lal (1997)

Each of these soil mapping units has been characterized in terms of the soil attributes that constitute the land quality used for soil suitability evaluation (Table 4). These attributes are cation exchange capacity (CEC), phosphorous (P), calcium (Ca), magnesium (Mg) and potassium (K), which are compounded as one land quality, namely available nutrients or fertility (FERT). The other land qualities are oxygen availability (OXAV), salinity (SAL), measured by electrical conductivity (EC), and sodicity, measured by exchangeable sodium percentage (ESP).

Results and Discussion

Suitability of Water for Irrigation

The source of water for Kalacha irrigation scheme is the spring next to artesian well. The water is used for both irrigation and livestock. Therefore, for this water to be used sustainably for these purposes, appropriate irrigation practices must be identified, based on soil, water quality (Table 3). This will ensure efficient use of this limited resource for sustainable production. The irrigation water quality has been analysed and evaluated on the basis of the attributes shown in Table 4.

The degree of restriction of the use of the available water resources for irrigation is severe only with respect to chloride. Other attributes such as electrical conductivity, pH and sodium adsorption ratio are not limiting. The high chloride level recorded for the spring water may not limit agricultural production if appropriate irrigation practices are identified based on the hydraulic properties of soil. These include appropriate irrigation scheduling, based on the crop water and leaching requirements. Good management includes improving the drainage conditions of soil so that adequate excess chlorides can be leached out through deep percolation of irrigation water. Addition of manure or organic materials will improve soil structure and facilitate the leaching processes.

Suitability of Soils for Irrigation

All the soils are developed on lacustrine plain and limestone sediments. The physiography and geology were symbolized by the letter PI (lacustrine plain) and L (limestone), respectively.

Five soil mapping units were identified as PIL1, PIL2, PIL3, PIL4 and PIL5. It is expected that each unit differs from any other adjoining unit to an extent that it would respond differently to management. The land characteristics that influence the soil quality and productivity of each soil mapping unit as well as its extent are given in Table 5. Units PIL1, PIL2 and PIL3 are all deep. But they differ in terms of stratification, texture and calcareousness. These differences (resulting from pedogenetic processes) could be an explanation of the differences in chemical characteristics (Table 6). Unit PIL4 is different from the other three units because it is shallow. This shows that different soil units require different types and quantity of

Table 3 The quality of irrigation water in Kalacha irrigation scheme

Quality attributes	Spring next to artesian well	Artesian well
pH	8.5	8.25
Electrical conductivity (dS/m)	1.30	1.20
Sodium (me/l)	9.13	8.04
Potassium (me/l)	0.51	0.47
Calcium (me/l)	0.61	0.56
Magnesium (me/l)	4.58	4.17
Carbonates (me/l)	1.45	1.24
Bicarbonates (me/l)	6.14	5.70
Chlorides (me/l)	11.9	9.38
Sulphates (me/l)	47.4	41.0
Sodium adsorption ratio (SAR)	5.67	5.23

Table 4 Evaluation of water quality in Kalacha irrigation scheme

Water source	Degree on restriction on the use of water, based on the following attributes					
	ECw	pH	SAR	Chloride	Boron	Bicarbonate
Spring next to artesian well	Slight	Slight	None	Severe	Not determined	Slight
Artesian well	Slight	None	None	Slight	Not determined	Slight

Table 5 Characteristics of different soil mapping units

Soil mapping units (MU)	Extent (ha)	Characteristics	Soil classification
PIL1	36.9	The soils are well drained (WD), deep to very deep and calcareous clay	Calcic Solonetz
PIL2	22.9	The soils are well drained (WD), deep and calcareous and stratified loamy sand to sandy clay loam	Salic Fluvisol
PIL3	12.3	These soils are imperfectly drained (ID) to moderately drained (MD), deep, stratified and highly calcareous sandy clay loam to clay loam	Calcaric Fluvisol
PIL4	3.6	The soils are shallow, imperfectly drained (ID) to poorly drained and excessively calcareous clay	Calcic Solonchaks

fertilizer inputs, since variations in chemical attributes from one soil unit to the other indicate that different units have different soil fertility status, thus requiring different management approaches for sustainable production. Ignoring these differences through blanket recommendations will not only be unsustainable in the long run but also inefficient.

The productivity indices of the entire four soil mapping units (Table 7) were found to be much lower than 50%, indicating that all the soil quality attributes superseded their threshold values (the critical limits) due to increased land degradation caused by inappropriate irrigation practices. The results of investigation of soil-related constraints by Muya et al. (2005) in the same area indicated that agricultural production had decreased (Table 8) to a point where all the farmers stopped growing crops and reverted to forage production.

The variations in percentage decrease, shown in Table 8, for different crops indicate that different crops respond differently to change in soil quality attributes. This is because each crop has a different tolerance level to the adverse conditions resulting from land degradation. This has an important bearing on the development of the strategies intended to restore the degraded ecosystems. In this context, the development of the appropriate strategies should involve crop- and site-specific analysis of the cause of the decline in crop production, including all the physical factors that influence the dynamics, storage and availability of nutrient within each soil mapping unit so that the potentials and limitations of all the important parameters are identified for each crop per soil mapping unit.

Before applying any inputs, a strategy to address physical limitations must be put in place to ensure

efficient use of external inputs. This is because the physical conditions of soil determine the nutrient flows, storage and availability to plants. Since the soil mapping units differ in physical characteristics, each strategy must address the specific problem. For different soil mapping units, the differences in soil physical characteristics such as texture, degree of stratification and depth are reflected in the variations of the hydraulic properties of soils such as water infiltration, hydraulic conductivity and soil water holding capacity (Table 9). This has an important implication in deciding on the appropriate irrigation method and time for the application of the desired irrigation water depth (Ekirapa, 1996). According to Landon (1984), infiltration rate and hydraulic conductivity are moderate in units PIL1 and PIL2, while in units PIL3 and PIL4, they are low. This explains the variations in drainage conditions and oxygen availability between these units. The available soil moisture holding capacity is moderately high in units PIL1 and PIL2 but low in units PIL3 and PIL4.

The suitability and limitations for different crops are given in Tables 10 and 11, respectively. This is to provide the basis of crop selection and identifying the limiting factors for each crop. This information is necessary for deciding on the type, quantity and quality of the external inputs for optimal production, thus ensuring efficient use of these limiting resources. The major limitation across all the soil mapping units is extremely high exchangeable sodium percentage (ESP). This is far much above the environmental threshold value of 6, given by Landon (1984). It is followed by high salinity, which is only a problem in unit PIL2. A combination of the sodium and salt problem requires adequate leaching of excess salts. However, low infiltration rate

Table 6 Land quality per soil mapping unit

| | Nutrient availability (FERT) | | | | | | | | |
MU	P ppm	CEC me%	Ca	Mg	K	OXAV	EC	ESP	pH
PIL1	10.30	20.98	7.17	0.30	3.28	WD	3.40	48.8	8.8
PIL2	16.20	122.37	8.87	0.26	5.38	WD	35.5	88.1	11.8
PIL3	7.70	20.97	12.67	0.28	1.27	MD	0.16	6.8	8.0
PIL4	1.00	20.97	8.06	0.36	2.49	ID	0.70	50.0	9.3

Table 7 Productivity index of different soil mapping units

Mapping units	Productivity index (%)	Most limiting factors
PIL1	22	High sodium concentration in most soil mapping units disrupts the soil structure, causing the impeded drainage, reduced aeration, hence impaired ecosystem functions
PIL2	26	
PIL3	21	
PIL4	12	

Table 8 Production decline in Kalacha irrigation scheme in kg/ha

| | Year | | Percentage decrease in crop production |
Crops	1984	2004	
Maize	2000	480	76
Cabbage	2500	400	84
Kale	3500	350	86
Tomatoes	3700	250	93

Table 9 Hydraulic characteristics of soils for different mapping units

Mapping unit	Infiltration rate (cm/h)	Hydraulic conductivity (cm/h)	Available soil moisture holding capacity (mm/m)
PIL1	2.5	0.8	120
PIL2	2.0	0.5	110
PIL3	1.5	0.2	70
PIL4	0.5	0.1	55

Table 10 Suitability of soil mapping units for different crops

Unit	Maize	Tomato	Kale	Bean	Cowpea	Water melon	Sorghum	Cotton	Onion
PIL1	NS	S2	S1	S1	S2	S1	S1	S1	S1
PIL2	NS	NS	NS	NS	NS	NS	NS	NS	NS
PIL3	S3	NS	S3	S3	S3	S2	S2	S1	S1
PIL4	NS	NS	NS	NS	NS	S2	S2	S1	S1
PIL5	NS	NS	S3	NS	NS	S3	S1	S1	S3

Note: S1, highly suitable; S2, moderately suitable; S3, marginally suitable; NS, non-suitable

Table 11 The limitations per soil mapping unit for different crops

Unit	Maize	Tomatoes	Kale, cabbage, spinach	Bean	Cowpeas	Water melon	Sorghum	Cotton	Onions
PIL1	1	3	5	5	1, 2	5	5	5	5
PIL2	1, 2, 3	1, 2, 3	1, 2, 3	1, 2, 3	1, 2, 3	1, 2, 3	1, 2, 3	1, 2, 3	1, 2, 3
PIL3	3	1, 2, 3	3	1, 2	4	4	4	5	5
PIL4	1, 2, 3	1, 2, 3	4	1, 4	1, 4	1	1	5	1
PIL5	1, 3	3	3	1	3	2	5	5	1

Key for limiting factors: 1, sodicity; 2, salinity, 3, fertility, 4, oxygen availability

Table 12 Productivity and the most limiting factors of different soil mapping units

Mapping units	Soil structural conditions	Most limiting factors
PIL1	Poor	Low N and high Na concentration
PIL2	Poor	High Na concentration and low N
PIL3	Very poor	Low N and high Na concentration
PIL4	Extremely poor	Poor drainage conditions, high Na concentration and low fertility

and hydraulic conductivity in units PIL3 and PIL4 would impede the leaching processes unless appropriate strategy is employed to improve soil structure, and hence water uptake, transmission and retention characteristics. Application of agricultural inputs to improve the production may be rendered ineffective unless appropriate measures are taken to improve these soil physical conditions. This is because both salt and sodium are very much above the threshold value of 6 and 4.0 mmhos/cm, respectively, as provided by Mass and Hoffman (1977). P is low in all the soil mapping units, because the level is lower than the critical limit (20 ppm) provided by Gachimbi et al. (2005). The degrees of these limitations vary with different soil mapping units as is expressed by their productivity index. Although most of the nutrient bases are adequate in the soil mapping unit, there is bound to be nutrient deficiencies caused by nutrient imbalances, which are usually associated with very high pH (values above 7.5) (Landon, 1984).

The variations in suitability and limitations of soils in each soil mapping unit for different crops indicate that different management practices are required for different crops in different soil mapping units. This

is because different crops have different tolerance to adverse soil conditions; hence a management strategy that targets a specific management problem for a specific crop would be efficient. To facilitate the implementation of viable management practices, the major limitations and productivity of each mapping unit are summarized in Table 12. High sodium concentration is associated with poor soil structural conditions because of dispersive influence of sodium on clay, leading to structural collapse and poor pore size distribution.

These limitations can be addressed by the use of locally available organic resources in combination with inorganic fertilizers where necessary. However, at the implementation stage, it is important to take into consideration the requirements of the specific crop in relation to the specific limitations. Bouma (1997) defines this as precision agriculture, which tries to fine-tune land management practices according to the requirements for the specific soil and land use systems, with an objective of maximizing agricultural production and its quality while minimizing the adverse environmental side effects.

Based on suitability evaluation of the area, the most promising crops are sufflower, onions kale and

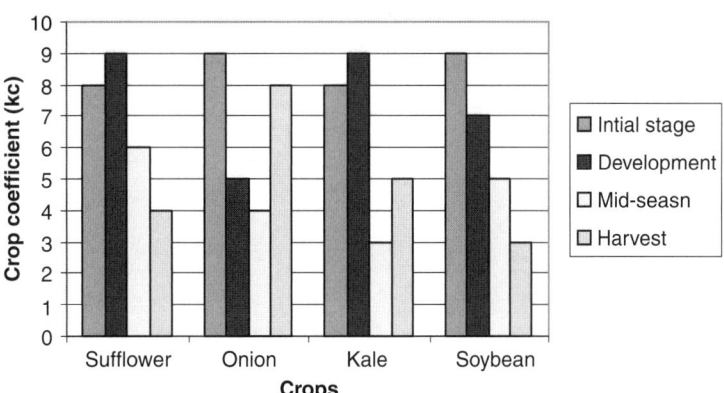

Fig. 1 The crop water requirements. *Source*: FAO (1986)

Table 13 The soil, environmental and water requirements of the most promising crops

Crop	Length of growing period in days	Temperature (°C)	Soil	Tolerance to salinity	Water requirements (mm)	Management requirements	
						Agronomic practices	Fertilizer application
Sunflower and soybean	200–300	20–30	Fertile, deep and well drained. For irrigation, medium textured soils are preferable	Tolerate up to ECe of 9.9 mmhos/cm, with 50% yield	Varies between 600 and 1200 mm, depending on the climate and length of growing period	Row spacing varies between 0.5 and 0.8 m with 15–35 plants per row	Under irrigation – 60–110 kg/ha N, 15–30 kg/ha P and 25–40 kg/ha K
Onion	130–175	15–20 for germination, the optimum being 15–25	Medium textured soil, with pH range of 6–7	ECe of 4.3 mmhos/cm with 50% yield	350–550	Sown in nurseries and transplanted after 30–35 days. Direct seedling is also practised. Usually planted in rows on raised beds, with two or more rows on a bed, with spacing of 0.3–0.5 × 0.05–0.1 m	60–100 kg/ha N, 25–45 kg/ha P and 45–80 kg/ha K
Kale	65–120	20–30	Well drained, pH range of 5.5–6.5	ECe 3.6 mmhos/cm with 50% yield	350–500	Spacing of 0.6–0.9 × 0.05–0.1 m. Depth of sowing is 2–5 cm	20–40 kg/ha N, 40–60 kg/ha P and 80–160 kg/ha K

Source: FAO (1986)

soybean. These crops are not only economically viable but also socially acceptable. The crop water requirements are different, and they vary with different stages of growth for all the crops as indicated by the crop water coefficient (kc) provided by FAO (1986) (Fig. 1). This has to be considered in planning irrigation scheduling and interval (Muchangi et al., 2005). The total water requirements for total growth period are given in Table 13. The agronomic requirements are also provided. Most of the soil and environmental requirements are met within the research area. However, for areas with severe limitations identified, application of the recommended agronomic practices may sustain the production at optimum level.

Conclusions

The water currently used for irrigation is suitable for the purpose, except for the chloride level, which is severe. However, improved drainage conditions and adequate leaching will reduce the level of the chlorides to an acceptable limit. Application of organic inputs will improve the soil structure and drainage conditions that will facilitate the leaching process.

Five soil mapping units were identified as PIL1, PIL2, PIL3, PIL4 and PIL5. The soils in unit PIL1 was found to be well drained and classified as Solonetz. The infiltration rate, hydraulic conductivity and available soil moisture holding capacity in this unit is 2.5 cm/h, 0.8 cm/h and 120 mm/m, respectively. Unit PLIL2 comprised well-drained, calcareous and stratified soils, classified as Fluvisols, with infiltration rate, hydraulic conductivity and available soil moisture holding capacity being 2.0 cm/h, 0.5 cm/h and 110 mm/m, respectively. Unit PIL3 had also stratified soils, classified as Fluvisols, but it differs from PIL2 since it was imperfectly drained and strongly saline. It has the lower infiltration rate (1.5 cm/h), hydraulic conductivity (0.5 cm/h) and soil moisture holding capacity (70 mm/m) than units PIL1 and PIL2. The soils in unit PIL4 were found to be imperfectly drained to poorly drained, excessively calcareous and overlying petrocalcic material, hence classified as Solonchaks. It has the lowest infiltration rate, hydraulic conductivity and soil moisture holding capacity, with values of 0.5 cm/h, 0.1 cm/h and 55 mm/m, respectively.

These soils have different potentials and limitations for different crops as follows:

- Unit PIL1: moderately to highly suitable for all the envisaged crops. Its productivity index is 22% and the major limiting factors are low N and high sodium concentration.
- Unit PIL2: non-suitable to all the crops. Its productivity index is 12% and the major limiting factors are high sodium concentration and high salinity.
- Units PIL3 and PIL4: non-suitable to highly suitable. Its productivity index is 21% and the major limiting factor is N and high Na concentration.
- Unit PIL4: non-suitable to marginally suitable for most crops. Its productivity index is 17%. The major limiting factor is high sodium concentration.

There are plenty organic resources within the area that should be well documented in terms of types, quantity and quality and applied to address the identified limitations in combination with external inputs.

Recommendations

Use of soil map is highly recommended to ensure efficient utilization of land resources for improved agricultural production. This is because it indicates geographical location, spatial distribution and extent of different soil types as well as their potentials, limitations for different crops and their management requirements. Before applying conventional agricultural practices such as growing high-yielding crop varieties supported by high level of agricultural inputs, the physical conditions of the soil must be improved using appropriate type and quality of organic inputs. This must be coupled with appropriate irrigation method and scheduling. Application of both organic and inorganic inputs must be done in consideration of the established potentials and limitations of soils for specific crops in order to ensure improved efficiency of using these limited resources in degraded and nutrient-deficient environment such as the research area. Assessment and evaluation of locally available agricultural inputs and their quality are highly recommended. Integrating external inputs with locally available resources is expected to be highly sustainable for these fragile environments. The capacity building in assessing land resources quality and productivity for various agricultural purposes is highly recommended. This is because it promotes precision agriculture, in which management inputs target well-defined management problems, with positive impacts on environment.

Acknowledgement We acknowledge the financial support by Desert Margin Programme for making this project a success story.

References

Aune JB, Lal R (1997) Agricultural productivity in the tropics and critical Limits of Properties of Oxisols, Ultisols and Alfisols. Trop Agric (Trinidad) 74(2):97–100

Bouma J (1997) The land use systems approach to planning sustainable land management at several scales. ITC J 3:237–241

Driessen PM, Konijn NT (1992) Land use systems analysis. Wageningen Agricultural University Publication, Wageningen, The Netherlands, pp 28–30

Ekirapa A (1996) Modeling of water and solute transport for irrigation and drainage in Taveta irrigation scheme. MSc thesis, Wageningen Agricultural University, pp 23–30

FAO (1976) Principles of land evaluation. FAO, Rome, pp 70–120

FAO (1977) Crop water requirements. Irrigation and drainage paper No. 24. FAO, Rome, pp 68–100

FAO (1986) Yield response to water. Irrigation and drainage paper No. 33. FAO, Rome, pp 15–75

FAO-UNESCO (1997) Soil map of the world. FAO, Rome

Gachimbi LN, van Keulen H, Thuranira EG, Karuku AM, Jager A, Nguluu S, Ikombo BM, Kinama JM, Itabari JK, Nandwa SM (2005) Nutrient Balances at Farm Level in Machakos (Kenya), Using a Participatory Nutrient Monitoring (NUTMON) Approach. Land Use Policy 22(1):13–22

Kenya Soil Survey Staff (1987) Manual for soil survey and land evaluation. Volume 1. Miscellaneous paper No. 24, Vol. 1. Ministry of Agriculture, Nairobi, Kenya, pp 29–101

Kusewe PK, Guiragossian V (1991) Research priority for enhancing crop productivity in marginal areas of Kenya. Workshop proceedings on agricultural research in Kenya: achievements, challenges and prospects held in Nairobi, 1991. KARI, Nairobi, Kenya, pp 21–36

Landon JR (1984) Bookers tropical soil manual. A hand book for soil survey and agricultural land evaluation in the tropics and subtropics. Longman, London and New York, pp 450

Mass EV, Hoffman GJ (1997) Crop salt tolerance current assessment. In: Tanji KK (ed) (1990) Agricultural salinity assessment and management. ASCE manuals and reports on engineering practices No. 71, US

Mavua JK (1985) Research-Extension-Farmer linkages. The approach and experience of Katumani. KARI, Nairobi, Kenya

Muchangi P, Sijali IV, Wendot H, Lemperiere P (2005) Improving the performance of irrigation in Africa: Irrigation systems, their operation and maintenance. Training source Book 1, March 2005. International Water Management Institute (IWMI) and Government of Kenya: Nairobi, Kenya

Muya EM, Lelon JK, Muga M, Maingi PM (2006) Characterization of Acacia Operation Project (AOP) sites for gums and rescind production: Integrating socio-economic issues with biophysical information for rehabilitation of the degraded arid and semi-arid lands of Kenya. DMP-KARI Publication, KARI, Nairobi, Kenya pp 12–40

Muya EM, Lelon JK, Sikunyi SM, Kimani M (2005) Biophysical site characterization of Kalacha irrigation scheme. Site evaluation report No. P78. DMP-KARI Publication

Stewart BW, Caesar-Ton That TC, Sara FW, Williams JD (2004) Organic matter addition, N, and residue burning effects on infiltration, biological and physical properties of an intensively tilled silt-loam soil. Soil Tillage Res 84: 154–167

Wesseling JG (1998) Poseidon: a tool to process the surface water output. Winand staring centre for integrated land, soil and water research output

Multi-functional Properties of Mycorrhizal Fungi for Crop Production: The Case Study of Banana Development and Drought Tolerance

D. Nwaga, A. Tenkouano, K. Tomekpe, R. Fogain, D.M. Kinfack, G. Tsané, and O. Yombo

Abstract This work discusses the potential of arbuscular mycorrhizal fungi (AM fungi) in improving nutrient uptake and drought tolerance of banana/plantain. Mycorrhizal fungi can also stimulate plant-beneficial properties such as flowering, development and crop yield in African soils. Inoculation of soils with selected strains could be used for restoration of soils with low activities of the fungus. Banana/plantain needs a lot of water and is very sensitive to drought stress and diseases. The potential benefit of symbiosis of banana–AM fungi was tested under nursery and farm conditions. For drought tolerance, test was done after 40 days of varied soil water field capacity: 90, 60 and 30%. Plant growth, leaf surface and dry weight are significantly stimulated by mycorrhiza in Oxisol and Andosol soils under nursery conditions. The correlation between water use efficiency and drought tolerance showed that AM fungal inoculation provided 30% more water to banana after 40 days of stress and AM fungi-inoculated micro-propagated banana provided bigger bunch than did non-inoculated ones under farm conditions in an Oxisol. These results indicate that inoculation with AM fungi could be a useful and practical approach towards sustainable banana production system. The cost/benefit ratio of inoculation could be determined to envisage its incorporation into nursery.

Keywords Arbuscular mycorrhizal fungi · Banana/plantain · Drought · Water use efficiency

D. Nwaga (✉)
Faculty of Science and Biotechnology Centre, University of Yaoundé I, 812, Yaoundé, Cameroon
e-mail: dnwaga@yahoo.fr

Introduction

Sub-Saharan Africa is the only region in the world where food production per capita is unstable since 40 years and about 180 million Africans do not have access to enough food for a more safe and productive life (Sanchez, 2002). Nine million people die of hunger every year; hunger kills more than do major diseases. This situation further increases their sensitivity to malaria, tuberculosis, AIDS and other diseases. Major biophysical causes of food insecurity in sub-Saharan Africa are reduction in soil fertility, pest and diseases, environmental constraints such as drought, soil erosion, high P fixation, high acidity with aluminium toxicity and low soil biodiversity. Nutrient deficiency is caused by lack of restoration of soil nutrients by farmers. Losses of –22 N, –2.5 P and –15 K kg/ha in 37 African countries in these last 30 years have been noticed according to Sanchez (2002). In the traditional slash-and-burn agriculture of humid forest zones, nutrient balance is more strongly negative in southern Cameroon.

Banana/plantain production is limited by decreasing soil fertility, root pests and leaf diseases. Improving banana from breeding strategies cannot by itself solve all theses constraints (Tomekpe et al., 2004). Since banana needs 1440–1920 mm of rain per year, a strong reduction in yield is noticed during the dry season when there is less than 30 mm for 3–4 months (Nkendah and Akyeampong, 2003). Banana/plantain needs adequate N, K and Mg fertilizer but not P and can defend itself against a short drought not exceeding 1 month (Anonymous, 2002). Water and K nutrition are probably the most limiting factor for banana/plantain yield in East

A. Bationo et al. (eds.), *Innovations as Key to the Green Revolution in Africa*,
DOI 10.1007/978-90-481-2543-2_54, © Springer Science+Business Media B.V. 2011

Africa (Okech et al., 2004). *Radopholus similis* and *Pratylenchus goodeyi* are the major nematodes damaging banana and plantain and thus facilitating the penetration of other root diseases and reducing nutrient uptake by roots (Sikora and Pocasangre, 2004). Soil-beneficial micro-organisms could be used as alternatives to chemical treatments since they contribute to agricultural production in Africa (Nwaga et al., 2000).

Mycorrhizae are multi-functional symbiotic associations between fungi and the roots of plants. In most cases, as a result of the symbiosis, the fungi obtain the sugars required to enable their growth and reproduction. The plants, in turn, may gain increased supplies of nutrients, which are captured from soil by the fungi, as well as enhanced resistance to disease and toxicity (Smith and Read, 1997; Read, 2001). They enhance uptake of nutrients, in particular phosphorus, and micro-nutrients such as Zn, Cu and Mo. Many crops can grow without AM fungi, but may be stunted, weak and low yielding. Some cultural practices such as tillage, long fallow, crop rotation and cropping systems and slash-and-burn agriculture are likely to modify their activity in the soil; but managing AM fungi for a sustainable agriculture using less fungicides and pesticides inputs must be the way forwards (Plenchette et al., 2005). According to Finlay (2004), mycorrhizal fungi must also be considered in their multi-functional perspectives including mobilisation of nutrients, mediation of plant response to drought, soil acidification, toxic metals, plant pathogens and interactions with other beneficial soil organisms. AM fungi and plant-beneficial relations such as tolerance to drought have been reported by many authors (Mosse and Hayman, 1971; Khalil and Loynachan, 1994; Augé, 2001; 2004), in contrary to Guissou et al. (2001). Although phosphorus is associated with drought tolerance, increase in water transport through fungal hypha, leaf and root water status and biomass could be independent of phosphorus increase (Smith and Read, 1997; Augé, 2001). Some studies have shown that AM fungi are able to increase growth and nutrient uptake in banana (Declerck et al., 1995; Geonaga and Irizarry, 2000; Fogain et al., 2001) or provide tolerance to pathogens such as nematodes, *Meloidogyne incognita* (Jaizme-Vega et al., 1997), *Radopholus similis, Pratylenchus coffeae* (Elsen et al., 2003) and fungi, *Fusarium oxysporum* (Jaizme-Vega et al., 1998). Work

on the application of AM fungi and beneficial microbes to micro-propagated banana has demonstrated benefits in terms of plant development and nutrient uptake (Jaizme-Vega et al., 2003; Rodríguez-Romero et al., 2005). Many authors have shown the multi-functional role of AM fungi in increasing crop production, disease tolerance, water use efficiency and also seed quality (McGoonigle, 1988; Johnson and Pfleger, 1992; Pfleger and Linderman, 1994; Newsham et al., 1995; Ngonkeu, 2003; Quilambo, 2003; Nwaga et al., 2004; Atayese, 2007; Nwaga et al., 2007). Despite the number of publications on AM fungi in sub-Saharan Africa, few of them have addressed environmental constraints on banana development and drought tolerance from nursery to the field. The aim of this study was to assess if inoculation of selected AM fungi such as *Glomus*, *Gigaspora* and *Scutellospora* was able to stimulate development, nutrient uptake and drought tolerance of micro-propagated banana plantlets during the nursery phase and bunch yield in the farm.

Materials and Methods

Experiment 1: Effect of AM Fungal Inoculation on the Development of Banana 'Grande Naine' Variety Growth Parameters and Yield

Biological Material, Inoculation and Experimental Design

A complete block design system with three treatments (control, AM fungus treatments CAM009 and Myco 5) using four replicates under screen house and micro-propagated banana variety provided by CARBAP 'Grande Naine' (*Musa acuminata* Colla AAA) and two types of AM fungal inocula: *Glomus* sp. (CAM009) or *Glomus clarum* + *Gigaspora margarita* (Myco 5). Inoculation was made by AM fungal propagules (roots, 500 spores per plant). CAM009 inoculum was produced on cowpea (*Vigna unguiculata*) by CARBAP and Myco 5 on millet (*Pennisetum americanum*) by the laboratory of soil microbiology (Nwaga et al., 2004). The control treatment received 2 ml per pot of bacterial extract obtained from the filtration of the AM fungal inoculum (25 g/l). For each treatment, 12 × 4 micro-propagated plants were used in 2.5 l of polypropylene sachets containing sterilised soil. The substrate used was sterilised sand for AM fungal inoculation during the early phase ('sevrage') and sterilised arable Oxisol soil from a *Chromolaena* fallow during the phase of acclimatisation.

Assessment of Variables

The soil characteristics were done. A ferraltic Oxisol from Yaoundé and a volcanic Andosol from Mbouroukou were used. Leaf area was estimated by a leaf area metre Li-3100 and root area and root length were evaluated by scanning using 'Data scan' software. For variable assessment, 12 plants were used per treatment three times during 3 months after inoculation. Root colonisation was made using trypan blue staining and the frequency of mycorrhizal according to Nwaga et al. (2004). Before plant inoculation, AM fungi were quantified and the number of spores and infective propagules (MPN) was determined according to the methods of Sieverding (1991) and Brundrett et al. (1996), respectively. Millet (*P. americanum*) was used as test plant for MPN determination. The frequency of mycorrhizal colonisation of plant roots was evaluated after staining and stereomicroscopic examination (Nwaga et al., 2004). Five-month-old micro-propagated banana previously treated was used for the evaluation of the yield under farm conditions. The farm was a sandy-loam Oxisol low in fertility. The experimental design was a non-randomised complete block design with a control and 2 mycorrhizal treatments and 10 replicates. The banana yield was assessed 16 months after planting with natural watering and no added fertilizers.

Statistical Analysis

Data were subjected to an analysis of variance using STAT ITCF software. When the main effect was significant ($P < 0.05$), differences between means were evaluated for significance by using Newman–Keuls test for comparing the different means.

Experiment 2: Effect of AM Fungal Inoculation and Water Stress on Leaf Area and Biomass of Two Contrasting Varieties of Banana After 40 Days of Drought

Biological Material, Inoculation, Drought Stress Application Method and Experimental Design

A factorial design system with 18 treatments was used (two banana varieties and three AM fungus treatments): a non-mycorrhizal control, two species of AM fungus (*G. clarum* and *Scutellospora gregaria*) and three levels of water stress regimes: 90, 60 and 30% field capacity using three replicates per treatment under screen house. AM fungi were selected isolates from a microbiology laboratory (Nwaga et al., 2004). The method used to apply drought stress is described by Tobar et al. (1994).

Micro-propagated banana/plantain, a hybrid variety 'PITA 21' and a local variety 'ELAT' (French moyen) from IITA Cameroon were used. This planting material was safe from pathogens and was homogenous. The substrate used was sterilised sand for AM fungal inoculation during the weaning phase and sterilised arable Oxisol soil from a *Chromolaena* fallow during acclimatising phase. This study was conducted in order to determine the effect of two arbuscular mycorrhizal fungal inoculation on the physiological behaviour of two tissue-cultured varieties of plantain (ELAT and PITA21) subjected to water stress in a 10-l container. Tissue-cultured plants were inoculated at the weaning phase using AM fungal propagules (roots and spores) of 140–180 spores per plant and were acclimated during 2 months. Selected AM fungi were a mixture of roots and spores from millet (*P. americanum*) coming from the resource bank of Soil Microbiology Laboratory of the University. The control treatment received 10 ml per pot of bacterial extract obtained from the filtration of the AM fungal inoculum (25 g/l).

Assessment of Variables

The soil characteristics were analysed. Plants were subjected to watering and nitrogen fertilizer (2 g per pot) after 40 days of water stress during which soil moisture was maintained at 90, 60 and 30% of the field capacity by weighing pots daily. The following parameters were measured: leaf surface, water use, water use efficiency and leaf proline content. For variable assessment, 12 plants per treatment and also 6 others for water consumption evaluation were used. Three indexes of plant tolerance to water stress (TI) were calculated on the basis of dry weight obtained, water consumption (WC) and proline accumulation as follows: the quantity produced by stressed plants divided by the quantity produced by non-stressed ones.

Statistical Analysis

Data were analysed using ANOVA with SAS software (SAS Institute, 1989). A two-way analysis of variance with randomised complete blocks was performed by using the following parameters as sources of variation for the second experiment: fungus, water, fungus–water interaction and error. When the main effect was significant ($P < 0.05$), differences between means were evaluated for significance by using Duncan's multiple range test in an orthogonal design.

Results and Discussion

The soil physicochemical characteristics of the two experiments were evaluated (Table 1). These Andosols and Oxisols were acidic (pH 4.60–5.97) and low in fertility (available P 5–6 ppm).

Experiment 1

Effect of AM Fungal Inoculation on the Development and Yield of Banana 'Grande Naine' Variety Under Controlled Conditions

From the results obtained, AM fungal inoculation of banana showed a significant increase in the parameters of the development in both Oxisol and Andosol after 3 months (Table 2). But the variation of the response of AM fungal inoculation over the control is more important in the Oxisol (RMD 181–185%) compared to the Andosol (RMD 113–146%). AM fungal inoculation increased leaf area by 105–125% in Oxisol and 112–124% in Andosol. This increase is more important in total root length (190–279%), followed by dry mass (175–183) and leaf area in Oxisol. In Oxisol of

Table 1 Analysis of the soils used for evaluating the effects of AM fungal inoculation on banana/plantain performances in humid forest zone of Cameroon

Soil type	Texture (C/L/S)	OM (g/kg)	C/N	Avail P (mg/kg)	Al (cmol/kg)	Al + H (cmol/kg)	Ca (cmol/kg)	Mg (cmol/kg)	K (cmol/kg)	Na (cmol/kg)	S (cmol/kg)	CEC (cmol/kg)	pH
Oxisol (Yaoundé)	41/9/50	2.34	9.0	5.0	0.0	0.08	1.02	0.38	0.19	0.09	1.68	7.42	5.90
Andosol (Mbouroukou)	53/09/38	4.58	11.5	6.0	2.36	3.56	1.17	0.40	0.22	1.10	1.89	8.39	4.60
Oxisol (Mbalmayo)	33/13/54	2.04	10.8	5.5	0.0	–	5.47	1.14	0.17	–	–	–	5,97

Table 2 Effect of AM fungal inoculation on the development of banana 'Grande Naine' variety under controlled conditions

Soil type	Treatments	Fresh mass (g)	Dry mass (g)	Leaf area (cm^2)	Total root length (m)	Root area (cm^2)	Myc col (%)	RMD (%)
Oxisol	Control	126[b]	12[b]	139[b]	6.1[c]	45.5[c]	0[b]	–
	Glomus sp. (CAM009)	355[a]	34[a]	313[a]	23.1[a]	154.3[a]	87[a]	185
	G. clarum+ G. margarita (Myco 5)	336[a]	33[a]	285[a]	17.7[b]	126.0[b]	80[a]	181
Andosol	Control	316[c]	28[c]	231[b]	13.2[c]	92.1[c]	0[b]	–
	Glomus sp. (CAM009)	583[a]	62[a]	517[a]	29.2[a]	225.5[a]	73[a]	146
	G. clarum+ G. margarita (Myco 5)	518[b]	50[b]	489[a]	23.9[a]	166.8[b]	70[a]	113

Myc col, root colonisation by AM fungi. Data taken 3 months after planting; data followed by the same letter (a, b, c) in each column do not significantly differ after Duncan's *t*-test ($P < 0.05$)

Mbouroukou the two inocula provide equivalent plant dry mass, while in Andosol, isolate CAM009 gave better results for most parameters. AM fungal inoculation has an important effect on total root length and also the total area occupied by the root system of banana in both soils. As for the dry mass and total root length, root colonisation frequency of AM fungi is higher in Oxisol compared to the Andosol. From these results one could say that AM fungi are very useful for micro-propagated banana development and nutrient uptake.

Effect of AM Fungal Inoculation on the Yield of Banana 'Grande Naine' Variety Under Farm Conditions

The results obtained with 16-month-old plants showed that nursery-inoculated banana gave a better yield than did non-inoculated control (Table 3). The mycorrhizal banana yield was 3–4 times higher than the control one in Oxisol. Under farm conditions, the inocula, CAM009 and Myco 5, were equivalent for the yield provided. Analysis of sporulation and infectivity (MPN) of soil from this farm during harvesting indicated that the two inocula were able to compete and provide a better propagation than did the native fungi since their activity is more important.

Experiment 2

Effect of AM Fungal Inoculation and Water Stress on Banana/Plantain Development and Tolerance to Drought After 40 Days of Stress Under Greenhouse Conditions

Plant Development

The two types of AM fungi had consistent effect on plant development parameters such as leaf area and plant dry weight for the three levels of stresses from 90 to 30% field capacity (Table 4). In non-mycorrhizal banana, the reduction of leaf area due to a severe drought was 39% for the tolerant variety ELAT and 50% for the sensitive one PITA21. In mycorrhizal banana, the reduction of leaf area was 27–29% for ELAT variety and 30–35% for PITA 21 variety. For the non-stressed plants, the effect was more pronounced on dry mass production with *Glomus* for PITA 21 variety, 54% response compared to 29% with *Scutellospora* inoculation. But when a more pronounced stress is applied (30% field capacity), for PITA 21, *Scutellospora* seems to be more efficient in biomass production than is *Glomus* since the increase in dry mass over the control is 73 and 54%, respectively. Applying a constraint of 30% field capacity severely reduced the plant growth and water

Table 3 Effect of AM fungal inoculation on the yield of banana 'Grande Naine' variety in Oxisol under farm conditions at Yaoundé

Treatments	Spore number/g	MPN/g sol	Fingers per bunch	Yield (t/ha)
Control	1.20[b]	1.44[b]	54[b]	2.0[b]
Glomus sp. (CAM009)	2.50[a]	4.36[a]	65[a]	7.9[a]
G. clarum + *G. margarita* (Myco 5)	3.48[a]	4.36[a]	75[a]	7.5[a]

Data taken after 16 months of planting; data followed by the same letter (a, b, c) in each column do not significantly differ after Duncan's *t*-test ($P < 0.05$)

Table 4 Effect of AM fungal inoculation and water stress on green leaf area (cm^2) and biomass (g dry weight) of banana/ plantain after 40 days of drought stress under controlled conditions in Oxisol at Mbalmayo site

Varieties	Treatments AM fungal species	Field capacity (%) 90	60	30	Field capacity (%) 90	60	30
ELAT	Control	3365[b]	2646[b]	1957[b]	63.07[c]	52.13[b]	40.18[c]
	Glomus sp.	4220[a]	3556[a]	2989[a]	82.74[b]	72.68[a]	56.36[a]
	Scutellospora sp.	4371[a]	3819[a]	3167[a]	85.38[b]	73.51[a]	57.77[a]
PITA21	Control	3008[b]	2478[b]	1547[c]	62.65[c]	42.25[b]	32.79[d]
	Glomus sp.	4060[a]	3588[a]	2780[a]	97.01[a]	69.62[a]	50.62[b]
	Scutellospora sp.	4189[a]	3532[a]	2815[a]	80.53[b]	70.10[a]	56.98[a]

Data followed by the same letter (a, b, c) in each column do not significantly differ after Duncan's *t*-test ($P < 0.05$)

consumption in non-mycorrhized plants. Both types of AMF allowed the plant to maintain its water content at high levels compared to non-inoculated plants.

Results using indexes of water stress tolerance show a reduction in banana aerial mass for both varieties and an increase in proline production (Table 5). One of the best criteria for characterising the two varieties is their response to water consumption index (WCI) after AM fungal inoculation. The ELAT variety is very stable (0.48–0.54), while PITA 21 is less stable (0.39–0.60) with banana–*Glomus* consuming less water when stressed compared to banana–*Scutellospora* consuming more water. When the water use efficiency (WUE) for mass production is used to compare the different treatments, the local variety ELAT is more efficient than PITA21. *Glomus*–banana is performing well since its WUE is close to twice that of the control one and that of *Scutellospora*–banana is close to 60% more than that of the control one.

Water Use Efficiency and Drought Tolerance Indexes

Water use efficiency (WUE) is expressed as the quantity of dry matter produced (in kg) for each cubic metre of water consumed (used and transpired) by the plant. WUE is significantly influenced by mycorrhiza and water stress. No varietal effect is noticed. *Glomus clarum* symbiotic banana gave the higher WUE value (1.40 kg m^3) and the non-symbiotic ones gave the lowest value (1.03 kg m^3). WUE increases from 90 to 60% field capacity in mycorrhizal plants, while it decreases in the non-mycorrhizal ones (Table 5). The ability of

the mycorrhizal fungal isolates to effectively maintain plant growth under water stress was related to lower levels of proline, higher levels of leaf water content and water use efficiency. Differences in proline accumulation in leaves among the treatments suggested that the AM fungi were able to induce different degrees of osmotic adjustment. The ELAT banana variety displayed the best tolerance to water stress, while *S. gregaria* induced the highest leaf water retention and lowest leaf proline content and seemed to be the most efficient fungus in improving dehydration tolerance of banana.

Mycorrhizal symbiosis may induce improvement in banana tolerance to drought according to AM fungal species and plant variety. At 60% field capacity, the two banana varieties showed a higher water consumption than did the non-mycorrhizal control. Mycorrhizae have maintained high levels of water content of plant. AM fungi are said to differ in their ability to improve drought tolerance, and *G. clarum* was shown to be more or less efficient under water stress (Augé, 2001). Because of the extension of their hyphae far from the rhizosphere zone, AM fungi are able to explore an important volume of soil, thus providing much more nutrients such as potassium and water to the plant. At 90% field capacity, AM fungi provide a biomass depending on the quantity of water consumed. At 60% field capacity, WUE increases in mycorrhizal plants; the quantity of water is enough to produce the same quantity of biomass like the one of non-mycorrhizal control receiving 90% field capacity. WUE can be improved by *Glomus macrocarpum*

Table 5 Effect of AM fungal inoculation of banana/plantain on water stress tolerance index under greenhouse conditions in Oxisol at Mbalmayo site

Variety	AM treatment	Biomass index	Proline index	WC index	WUE (kg DM m^3)
ELAT	Control	0.55^b	2.73^a	0.54^{ab}	0.91^c
	G. clarum	0.72^a	1.69^c	0.48^{bc}	1.76^a
	S. gregaria	0.72^a	1.77^c	0.53^{ab}	1.44^{ab}
PITA 21	Control	0.52^b	2.98^a	0.50^b	0.81^c
	G. clarum	0.61^{ab}	2.22^b	0.39^c	1.61^a
	S. gregaria	0.71^a	1.97^{bc}	0.60^a	1.29^b
ANOVA					
Variety		**	*	NS	NS
Mycorrhiza		**	**	**	**
Variety × mycorrhiza		*	NS	*	*
CV		8.40	9.18	9.61	

WCI, water consumption index; WUE, water use efficiency at 60% field capacity in kilogram dry matter per cubic metre of water consumed. Data followed by the same letter in each column do not differ after Duncan's test ($P < 0.05$); * Significant at 5%, ** significant at 1%

for *Eupatorium odoratum* and *Glomus monosporum* for wheat (Al-Karaki et al., 2004). Based on plant biomass, Guissou et al. (2001) have found that fruit trees inoculated by *Glomus aggregatum* were not able to improve stress tolerance index, while Ruiz-Lozano et al. (1995) have shown improvement of this index. AM fungi are more easy to apply in horticulture and micro-propagated crops (Azcón-Aguilar and Barea, 1997). External soil hyphae may play an important role in mycorrhizal influence on the high clay content soil structure and water relations of plants (Neergaard Bearden, 2001; Augé et al., 2003). Soil–plant–water relations depend on glomalin, a soil 'super glue'. Glomalin is a glycoprotein produced by AM fungal hyphae, with C storage ability and with benefits such as the following: acting as water and nutrient carrier, increased aggregate stability in soil and more beneficial microbial activities (Wright et al., 1999). Higher level of glomalin gives a better soil structure, a better root development and resistance to erosion which in turn lead to better plant production (Wright et al., 1996; Rillig et al., 2002).

The quality of many tropical soils may limit food production in annual cropping systems. Biological potentials of these soils may fulfil food production problems in low-fertility African soils. A holistic approach of soil fertility is now recommended; it is based on a sustainable management which integrates biological, chemical and socio-economic considerations (Woomer and Swift, 1994; Bationo et al., 2004; Nwaga et al., 2010).

Conclusions

AM fungi could stimulate the production of planting material of good quality with a high nutrient content and more adapted to environmental factors such as fertility and drought. Our results confirm the multifunctional properties of AM fungi to improve banana growth, nutrition and tolerance to drought during the nursery phase. They also showed that observations under nursery could be extended to the field by increasing banana bunch production. The management of this symbiosis represents a possible strategy for the micropropagation process in the banana-producing regions: micro-propagated plantlets would be planted in the field with an established mycorrhizosphere. This could

contribute to alleviate environmental constraints under field conditions. However, due to differences in the functioning of diverse AM fungal strains and preferences to banana varieties, a previous screening to select the best symbiotic combination should be done in order to optimise the results. Banana needs a lot of water and nutrients for a good yield, while AM fungi could provide not only water and inorganic nutrients but also tolerance to some root pathogens and nematodes. Developing some multi-functional properties of beneficial organisms could contribute to increasing productivity in sub-Saharan Africa and specifically water and nutrient use efficiency of mineral fertilizers. African continent needs agricultural biotechnology, since the technology could have a direct impact on food safety, poverty reduction and environmental protection. This is of importance specifically in the tropics because of low-input farming, mostly in developing countries, where little or no P fertiliser is used because of prohibitive costs and availability. How can we develop this approach in banana production system? The cost/benefit ratio of inoculation should be determined to envisage its incorporation into nursery, but more important is the incorporation of these beneficial properties at farmers' level.

Acknowledgements Thanks to the Ministry of Higher Education, CARBAP and IITA for partially funding these experiments; also special thanks to M. Hauser S. (IITA) for providing logistics of this work.

References

Al-Karaki GN, McMichael B, Zak J (2004) Field response of wheat to arbuscular mycorrhizal fungi and drought stress. Mycorrhiza 14:263–269

Anonymous (2002) Memento de l'agronome. Ministère des Affaires étrangères/CIRAD/GRET, 1691 pp

Atayese MO (2007) Field response of groundnut (*Arachis hypogea* L) cultivars to mycorrhizal inoculation and phosphorus fertiliser in Abeokuta, South west Nigeria. Am Eur J Agric Environ Sci 2(1):16–23

Augé R-M (2001) Water relations, drought and vesicular–arbuscular mycorrhizal symbiosis. Mycorrhiza 11:3–42

Augé RM (2004) Arbuscular mycorrhizae and soil/plant water relations. Can J Soil Sci 84:373–381

Augé RM, Moore JL, Cho K et al (2003) Relating foliar dehydration tolerance of mycorrhizal *Phaseolus vulgaris* to soil and root colonization by hyphae. J Plant Physiol 160:1147–1156

Azcón-Aguilar C, Barea JM (1997) Applying mycorrhiza biotechnology to horticulture: significance and potentials. Sci Hort 68(1–4):1–24

Bationo A, Kimetu J, Ikerra S et al (2004) The African network for soil biology and fertility: new challenges and opportunities. In: Bationo A (ed) Managing nutrient cycles to sustain soil fertility in sub-Saharan Africa. Academy Science Publishers (ASP)-TSBF CIAT, Nairobi, pp 1–23

Brundrett MC, Ashwath N, Jasper DA (1996) Mycorrhizas in the Kakadu region of tropical Australia. II. Propagules of mycorrhizal fungi in disturbed habitats. Plant Soil 184:173–184

Declerck S, Plenchette C, Strullu DG (1995) Mycorrhizal dependency of banana (*Musa acuminata* AAA group) cultivar. Plant Soil 176:183–187

Elsen A, Baimey H, Swennen R, De Waele D (2003) Relative mycorrhizal dependency and mycorrhizal-nematode interaction in banana cultivars (*Musa* spp.) differing in nematode susceptibility. Plant Soil 256:303–313

Finlay RD (2004) Mycorrhizal fungi and their multifunctional roles. Mycologist 18(2):91–96

Fogain R, Njifenjou S, Kwa M et al (2001) Mycorhization précoce et croissance de deux types de matériel végétal de plantain (*Musa* AAB). Cah Agric 10:195–197

Geonaga R, Irizarry H (2000) Irrigated banana: yield and quality of banana irrigated with fraction of class A Pan evaporation on an oxisol. Am Soc Agr 92:1008–1012

Guissou T, Ba AM, Plenchette C et al (2001) Effets des mycorhizes à arbuscules sur la tolérance à un stress hydrique de quatre arbres fruitiers: *Balanites zyphus aegyptiaca* (L) Del., *Parkia biglobosa* (Jacq) Benth, *Tamarindus indica* L et *Zimauritiana* Lam. Sècheresses 12(2):28–31

Jaizme-Vega MC, Rodriguez-Romero AS, Hermoso CM et al (2003) Growth of micropropagated banana colonized by root-organ culture produced arbuscular mycorrhizal fungi entrapped in Ca-alginate beads. Plant Soil 254:329–335

Jaizme-Vega MC, Sosa-Hernández B, Hernández-Hernández JM (1998) Interaction of arbuscular mycorrhizal fungi and the soil pathogen *Fusarium oxysporum* f.sp *cubense* on the first stages of micropropagated Grande Naine banana. Acta Horticulturae 490:285–295

Jaizme-Vega MC, Tenoury P, Pinochet J et al (1997) Interaction between the root knot nematode *Meloidogyne incognita* and the mycorrhizal association of *Glomus mossea* and Grande Naine banana. Plant Soil 196:27–35

Johnson NC, Pfleger FL (1992) Vesicular–arbuscular mycorrhizae and cultural stresses. In: Bethlenfalvay RG, Linderman GJ (eds) Mycorrhizae in sustainable agriculture. ASA Special Publ 54, American Society of Agronomy, Madison, WI, pp 71–99

Khalil S, Loynachan TE (1994) Soil drainage and distribution of VAM fungi in two toposequences. Soil Biol Biochem 26(8):929–934

McGoonigle TP (1988) A numerical analysis of published field trials with vesicular–arbuscular mycorrhizal fungi. Funct Ecol 2:473–478

Mosse B, Hayman DS (1971) Plant growth responses to vesicular arbuscular mycorrhiza. II. In unsterilized field soils. New Phytol 70:29–34

Neergaard Bearden B (2001) Influence of arbuscular mycorrhizal fungi on soil structure and soil water characteristics of vertisols. Plant Soil 229:245–258

Newsham K, Fitter AH, Watkinson AR (1995) Multifunctionality and biodiversity in arbuscular mycorrhizas. Tree 10:407–411

Ngonkeu MEL (2003) Biodiversité et potentiel des champignons mycorhiziens à arbuscules de quelques zones agroécologiques du Cameroun. Doctoral thesis, University of Yaoundé, Yaoundé

Nkendah R, Akyeampong E (2003) Données socio économiques sur la filière plantain en Afrique Centrale et de l'Ouest. InfoMusa 12(1):8–13

Nwaga D, Fankem H, Essono OG et al (2007) Pseudomonads and symbiotic micro-organisms as biocontrol agents against fungal disease caused by *Pythium aphanidermatum*. Afr J Biotechnol 6(3):190–197

Nwaga D, Jansa J, Abossolo Angue M, Frossard E (2010) The potential of soil beneficial micro-organisms for slash-and-burn agriculture in the humid forest zone of sub-Saharan Africa. Chap 5. In: Dion P (ed) Soil biology and agriculture in the tropics, Soil biology 21, Springer, Berlin, pp 81–107

Nwaga D, Ngonkeu MEL, Oyong MM et al (2000) Soil beneficial micro-organisms and sustainable agricultural production in Cameroon: current research and perspectives. In: Swift MJ (ed) The biology and fertility of tropical soils. TSBF Report 1998, UNESCO-TSBF, Nairobi, Kenya, pp 62–65

Nwaga D, The C, Ambassa-Kiki R et al (2004) Selection of arbuscular mycorrhizal fungi for inoculating maize and sorghum, grown in oxisol/ultisol and vertisol in Cameroon. In: Bationo A (ed) Managing nutrient cycles to sustain soil fertility in sub-Saharan Africa. ASP/TSBF Institute of CIAT, Nairobi, pp 467–486

Okech SH, van Asten PJA, Gold CS et al (2004) Effects of potassium deficiency, drought and weevils on banana yield and economic performance in Mbarara, Uganda. Uganda J Agric Sci 9:511–519

Pfleger FL, Linderman RG (1994) Mycorrhizae and plant health. APS Press, St Paul, MN, 344p

Plenchette C, Clermont-Dauphin C, Meynard JM et al (2005) Managing arbuscular mycorrhizal fungi in cropping systems. Can J Plant Sci 85:31–40

Quilambo OA (2003) The vesicular–arbuscular mycorrhizal symbiosis. Afr J Biotechnol 2(12):539–546

Read DJ (2001) Mycorrhiza. In: Wiley (ed) Encyclopedia of life sciences. New York, NY, Standard Article, available via http://www.mrw.interscience.wiley.com. Posting Date: 19 Apr 2001

Rillig MC, Wright SF, Eviner VT (2002) The role of arbuscular mycorrhizal fungi and glomalin in soil aggregation: comparing effects of five plant species. Plant Soil 238:325–333

Rodríguez-Romero AS, Guerra MSP, Jaizme-Vega MDC (2005) Effect of arbuscular mycorrhizal fungi and rhizobacteria on banana growth and nutrition. Agron Sustain Dev 25:395–399

Ruiz-Lozano JM, Azcon R, Gomez M (1995) Effects of arbuscular–mycorrhizal *Glomus* species on drought tolerance: physiological and nutritional plant responses. Appl Environ Microbiol 61(2):456–460

Sanchez PA (2002) Soil fertility and hunger in Africa. Science 295:2019–2020

SAS Institute, Inc. (1989) SAS user's guide: statistics, 5th edn. SAS Institute, Inc., Cary, NC, 846p

Sieverding E (1991) Vesicular–arbuscular mycorrhiza management in tropical ecosystems. GTZ, Eschborn, Germany, 371p

Sikora RA, Pocasangre LE (2004) Nouvelles technologies pour améliorer la santé des racines et augmenter la production. InfoMusa 13(2):25–29

Smith SE, Read DJ (1997) Mycorrhizal symbiosis. Academic, London, 223p

Tobar R, Azcon R, Barea JM (1994) The improvement of plants N nutrition from an ammonium-treated, drought stressed soil by the fungal symbiont in arbuscular mycorrhizae. Mycorrhiza 4:105–108

Tomekpe T, Jenny C, Escalant J-V (2004) Revue des stratégies d'amélioration conventionnelle de *Musa*. InfoMusa 13(2): 2–5

Woomer P, Swift MJ (1994) The biological management of tropical soil fertility. Wiley-Sayce, Chichester, 243p

Wright SF, Franke-Snyder M, Morton JB, Upadhyaya A (1996) Time-course study and partial characterization of a protein on hyphae of arbuscular mycorrhizal fungi during active colonization of roots. Plant Soil 181:193–203

Wright SF, Starr JL, Paltineanu IC (1999) Changes in aggregate stability and concentration of glomalin during tillage management transition. Soil Sci Soc Am J 63:1825–1829

Effect of Phosphorus Sources and Rates on Sugarcane Yield and Quality in Kibos, Nyando Sugar Zone

J.O. Omollo and G.O. Abayo

Abstract One of the causes for declined sugarcane yields is low soil nutrient levels, especially the most limiting ones, nitrogen (N) and phosphorus (P). This is further aggravated by inadequate or lack of nutrient replenishment. The study conducted at KESREF, Kibos, experimental field on a Eutric Vertisol soil evaluated the effect of four P sources, single superphosphate (SSP), triple superphosphate (TSP), diammonium phosphate (DAP), and rock phosphate (RP), and four P levels (0, 17, 34, and 52 kg P ha^{-1}) on yield and quality of sugarcane varieties KEN 82-808 and CO 421. P sources significantly influenced yield of variety CO 421 at second ratoon harvest, the trend being DAP > RP > TSP > SSP. Effect of P sources on quality was not significant. Application of P increased population of millable stalk and yield compared with the control (0 kg P ha^{-1}) in both KEN 82-808 and CO 421. Highest yield was recorded when P applied was 34 kg P ha^{-1} and lowest in control (0 kg P ha^{-1}), the trend being 34 > 17 > 52 > 0 kg P ha^{-1}. Effect of P rates on quality was not significant. It is concluded that fertilizer P sources can be applied to supply P. P plays a significant role on yield parameters than on quality parameters. The level 34 kg P ha^{-1} (80 kg P$_2$O$_5$ ha^{-1}) is appropriate to maintain the crop to second ratoon harvest for increasing the yield.

Keywords Phosphorus · Rates · Sources · Sugarcane · Yield

J.O. Omollo (✉)
Agronomy Programme, Crop Development Department, Kenya Sugar Research Foundation (KESREF), Kisumu, Kenya
e-mail: jac.omollo@gmail.com

Introduction

Sugarcane is a crop of great agro-economic importance in Kenya. Its cultivation is concentrated in parts of Western, Nyanza, and Rift Valley Provinces suitable for sugarcane production. The crop is grown mostly by small-scale farmers, who contribute up to 90%, while the remaining 10% is by large-scale farmers and the factory nucleus estates (KSB, 2003; Wawire et al., 2006). The current national mean yield of 69.2 t ha^{-1} is observed to be low since potential has been recorded to be over 100 t ha^{-1} nationally. KESREF (2004) observed cane yields as high as 180–200 t ha^{-1} under regular moisture supply throughout the crop growth.

Many causal factors to the declined yield at farm level have been documented, and one such cause could be attributed to declining level of soil nutrients, especially the most limiting ones. These are N and P which are the major nutrients required for higher and sustained sugarcane growth, yield, and quality. While most soils contain substantial reserves of total P, most of it remains relatively inert and less than 10% of the soil P enters the plant–animal cycle. Coupled with continuous sugarcane monocropping with limited replenishment, the P in soil solution becomes inadequate for sugarcane establishment (Malavolta, 1994). Phosphorus role in sugarcane is to stimulate early root formation and development. P deficiency leads to reduced metabolic rate and photosynthesis which then leads to reduced cane yield and quality (Blackburn, 1984; Malavolta, 1994).

One of the remedial measures is the application of P fertilizers to supplement P nutrients which the plant can obtain/withdraw from the soil. The P fertilizers are costly, and in developing countries like

A. Bationo et al. (eds.), *Innovations as Key to the Green Revolution in Africa*,
DOI 10.1007/978-90-481-2543-2_55, © Springer Science+Business Media B.V. 2011

Kenya, they are either imported or manufactured using imported raw materials. There are different, common commercially available P fertilizers varying in amount of available P nutrient and also their solubility in soil upon application. The methods which exist for evaluating P requirements are soil analysis, foliar diagnosis, and field trials (Black, 1993; Tisdale et al., 1993).

The Kenyan sugar industry has, as a matter of policy, applied fertilizer phosphorus (P) besides other agricultural inputs to sustain the productivity of the industry. The rate and source of P over the years has been largely blanket with the sugarcane zones under humid conditions receiving a blanket application ranging from 34 to 39 kg P ha^{-1}, while those in subhumid conditions, where Nyando sugar zone is situated, receiving 17–26 kg P ha^{-1} (KESREF, 2002).

The basis for fertilizer applications has been on the recommendation from studies that were conducted in the 1960s and 1970s. Moreover, these studies were conducted when the industry was dominated by only a few introduced varieties such as CO 331, CO 421, CO 617, and CO 945. To date, more locally released sugarcane varieties such as KEN 82-808, KEN 73-335 with high yields and sugar quality are available for commercial production (KESREF, 2002).

Considering the importance of P nutrition in sugarcane performance, the present study was undertaken to determine the effect of different P sources and rates on sugarcane yield and quality in Kibos, Nyando sugar zone.

Materials and Methods

The field experiment was conducted at Kenya Sugar Research Foundation, Kibos experiment farm (35°13'E, 0°06'S), Kisumu, during the period from January 2002 to November 2006. The experiment soil type was Eutric Vertisol (FAO, 1988), having soil pH 5.7, 32 ppm available phosphorus, 4.2 meq/100 g magnesium, 8.5 meq/100 g calcium, 1% organic carbon, and 1.7% organic matter.

The treatments comprised two factors, where factor 1 was four sources of phosphorus fertilizers, diammonium phosphate (DAP), triple superphosphate (TSP), single superphosphate (SSP), and rock phosphate (RP), while factor 2 was levels of P at 0, 17, 34, and 52 kg P ha^{-1}.

The treatments were applied to two sugarcane varieties KEN 82-808 (locally released) and CO 421 (introduced). The experiment was laid out in randomized complete block design with three replications. The gross plots measured 10 m × 1.2 m × 6 rows or 72 m^2, while the net plots measured 10 m × 1.2 m × 4 rows or 48 m^2 excluding the two outer rows (guard rows). Certified seed cane was planted on May 2002 where P treatments were applied at the time of planting. The crop was maintained for plant cane, first ratoon, and second ratoon harvests.

Nitrogen fertilizer was applied as urea at a uniform rate of 100 kg N ha^{-1} in all experimental units. At planting, plots with DAP received N at 18 kg N ha^{-1}; to equal N in DAP, the remaining N was applied as a side dress when the experiment was 3 and 6 months old in two equal splits. Weed control and other agronomic management practices were undertaken as recommended by KESREF (KESREF, 2002). Plant cane was harvested at 19 months after planting (MAP) in January 2004. First and second ratoon harvests were at 18 MAP in June 2005 and November 2006, respectively.

Data collected at harvest were quantitative and qualitative traits of yield. The quantitative traits were plant height, plant girth, stalk population, and stalk weight, while qualitative traits were pol % cane and fiber % cane. The data were analyzed statistically by Fisher's analysis of variance technique, and treatment means were compared using the least significance difference test at 0.05 P level as described by Gomez and Gomez (1984).

Results and Discussion

Effect of different P sources on population of millable stalk was not significant in both varieties and also in each crop year. Although statistically similar, difference in population of millable stalk was recorded between KEN 82-808 and CO 421 and also varied among the crop year harvest. KEN 82-808 was superior to CO 421 in this quantitative trait. Superiority of KEN 82-808 over CO 421 could be attributed to its intrinsic qualities of high vigor suitable for subhumid regions such as Nyando zone. CO 421 is an introduced variety in the 1970s and could be decreasing in vigor (KESREF, 2002).

There was significant influence of P sources on yield of variety CO 421 in second ratoon harvest. Application of DAP resulted in the highest yield of 70.7 t ha^{-1}, while application of SSP resulted in lowest yield, the trend in yield being DAP > RP > TSP > SSP.

P source did not significantly influence yield of KEN 82-808 in plant cane, first ratoon, and second ratoon harvests, likewise in CO 421 at plant cane and first ratoon harvest. This implies that any P source can be applied as source of P. However, for long-term P residual effect, P fertilizers such as DAP which has high P content and RP which slowly releases P should be applied at planting plant cane crop cycle. This will consequently satisfy the P requirement for subsequent ratoons (Black, 1993; Malavolta, 1994).

Data regarding the effect of different P sources on qualitative traits (Table 1) showed no significant influence of P sources on pol % cane and fiber % cane in KEN 82-808 and CO 421 and also in each crop year harvest. Effect of different P rates on quantitative traits, population of millable stalk, and yield of KEN 82-808 and CO 421 at plant cane, first ratoon, and second ratoon harvests is presented in Table 2. P rates significantly influenced population of millable stalk of CO 421 but not of KEN 82-808 (Table 3). The highest population was recorded when P rate applied was 52 kg P ha^{-1}, while the lowest was recorded in control plots with no P (0 kg P ha^{-1}), the trend in population being 57 > 34 > 17 > 0 kg P ha^{-1}.

Although statistically similar among the treatments on population of millable stalk of KEN 82-808 and CO 421 in first and second ratoon harvests, it was observed that population was lowest in control (0 kg P$_2$O$_5$ ha^{-1}) than where P was applied. This implies that P application had an effect in increasing the stalk number. Similar observations of enhanced stalk number upon P application were reported by Rahman et al. (1992), Perez and Melgar (1998), and Sreewarome et al. (2005).

P rates significantly influenced yield of KEN 82-808 in first ratoon and CO 421 in second ratoon harvest (Table 3). In both varieties, the highest yield was recorded when P applied was 34 kg P ha^{-1} and the lowest was recorded in control (0 kg P$_2$O$_5$ ha^{-1}), the trend in yield being 34 > 17 > 52 > 0 kg P ha^{-1}.

Also where statistical similarity was observed, the yields in control (0 kg P ha^{-1}) were lowest and highest among treatments. These results support those reported by Clements (1980), Rahman et al. (1992), and Sreewarome et al. (2005), who recorded heavier stalks with increased P application.

Data regarding the effect of P rates on qualitative traits showed no significant influence of P rates on pol % cane and fiber % cane in KEN 82-808 and CO 421 and also in each crop year harvest. Although statistically similar among treatment on quality parameters, it was generally noted that the lowest pol % cane and fiber % cane was recorded in control plots

Table 1 Effect of phosphorus sources on qualitative traits of KEN 82-808 and CO 421

P sources	Pol % cane			Fiber % cane		
	Plant cane	1st ratoon	2nd ratoon	Plant cane	1st ratoon	2nd ratoon
KEN 82-808						
SSP	14.1	12.3	14.3	16.0	16.1	14.9
TSP	14.0	11.7	14.0	16.4	16.6	15.1
RP	14.2	11.6	14.0	16.1	16.9	15.3
DAP	14.1	11.9	14.3	16.0	16.2	15.8
LSD $_{(P=0.05)}$	NS	NS	NS	NS	NS	NS
Mean	14.1	11.9	14.1	16.1	16.4	15.3
CO 421						
SSP	12.1	13.2	14.3	15.0	13.2	13.0
TSP	11.9	13.1	13.8	15.1	13.1	12.4
RP	12.2	12.9	14.5	15.4	13.6	12.5
DAP	12.0	13.0	14.2	15.1	12.9	12.2
LSD $_{(P=0.05)}$	NS	NS	NS	NS	NS	NS
Mean	12.1	13.1	14.2	15.1	13.2	12.5

NS, not significant; TCH, tonnes cane per hectare

Table 2 Effect of different P rates on quantitative traits of KEN 82-808 and CO 421

P sources	Population of millable stalks (stalks ha^{-1})			TCH (tonnes cane ha^{-1})		
	Plant cane	1st ratoon	2nd ratoon	Plant cane	1st ratoon	2nd ratoon
KEN 82-808						
0	132,123	113,542	113,331	139.5	101.5[b]	64
40	141,873	119,791	126,872	153.6	114.6[ab]	75.2
80	140,597	123,958	115,414	142.3	124.7[a]	68.6
120	140,783	115,104	121,873	142.9	118.7[ab]	68.5
LSD $(P = 0.05)$	NS	NS	NS	NS	21.2	NS
CV	23.44	12.55	17.69	9.73	17.89	21.7
Mean	138,836	118,098	119,373	144.6	114.9	69.08
CO 421						
0	98,306[c]	89,583	87,290	128.6	100.2	57.1[b]
40	109,764[bc]	91,146	86,873	142.4	101.4	67.5[a]
80	118,488[ab]	93,229	88,540	140.6	106.9	70.6[a]
120	133,253[a]	92,969	90,415	138.1	106.9	66.0[ab]
LSD$_{(0.05)}$	18,298	NS	NS	NS	NS	10.2
CV	115.4	16.5	16.30	14.1	18.9	14.6
Mean	114,952	91,731	88,280	137.4	103.9	65.3

Any two means not sharing a common letter differ significantly at 0.05 P level (LSD test)
NS, not significant; TCH, tonnes cane per hectare

Table 3 Effect of phosphorus rates on qualitative traits of KEN 82-808 and CO 421

P rates (P$_2$O$_5$ ha^{-1})	Pol % cane			Fiber % cane		
	Plant cane	1st ratoon	2nd ratoon	Plant cane	1st ratoon	2nd ratoon
KEN 82-808						
0	13.9	11.6	14.3	16.0	16	15.2
40	14.1	11.9	14.0	15.9	16.7	15.5
80	14.0	11.9	14.0	16.1	16.4	15.0
120	14.1	12	14.3	16.3	16.7	15.5
LSD$_{(P = 0.05)}$	NS	NS	NS	NS	NS	NS
Mean	14.0	11.9	14.1	16.1	16.5	15.3
CO 421						
0	12.0	12.8	14.0	14.9	13	12.6
40	12.1	12.8	14.0	15.2	12.8	12.6
80	12.4	13.1	14.5	15.1	13.5	12.3
120	12.3	13.5	14.2	15.3	13.5	12.7
LSD$_{(P = 0.05)}$	NS	NS	NS	NS	NS	NS
Mean	12.2	13.1	14.2	15.1	13.2	12.6

NS, not significant; TCH, tonnes cane per hectare

(0 kg P ha^{-1}), while the highest was when rate of P applied was 52 kg P ha^{-1}.

Conclusions

The source of phosphorus influenced yield of KEN 82-808 and CO 421 but not the number (population) of millable stalks. This could be attributed to the amount of nutrient P in fertilizer, the nature of fertilizer, and also soil properties where the fertilizer is applied. Single superphosphate (SSP) was superior in increasing yield at plant cane harvest, while diammonium phosphate (DAP) was superior at ratoon harvest. The fertilizer DAP has high P (19.8%) compared to SSP, which has 8.2% P; therefore the former leads to increased labile P where the P is subsequently slowly released for crop uptake.

Application of P positively influenced yield of KEN 82-808 and CO 421 as higher yields were recorded in treated soils, while the lowest yield was recorded in control (0 kg P ha^{-1}). Application rate of 34 kg P ha^{-1} (80 kg P$_2$O$_5$ ha^{-1}) recorded highest yield and population of millable stalks in all the varieties and in each crop year plant cane, first ratoon, and second ratoon harvests. Yields recorded at 34 kg P ha^{-1} were statistically similar to those recorded at 57 kg P ha^{-1} (120 kg P$_2$O$_5$ ha^{-1}).

The source of fertilizer P and their rates did not influence qualitative traits of KEN 82-808 and CO 421 harvested in plant cane and ratoon. In view of the results, it is recommended that studies on the interaction effect of the essential plant nutrients nitrogen, phosphorus, and potassium on sugarcane performance should be undertaken.

Acknowledgments The authors thank Dr. G. Okwach, Director, Kenya Sugar Research Foundation, Kisumu, for providing the authority and finance to undertake the study. Special mention must be made of the staff in KESREF, Department of Crop Development, Laboratory Services Section, and farm office, who ensured the success of the trial.

References

Black CA (1993) Soil fertility evaluation and control. Lewis Publishers, Boca Raton, FL

Blackburn F (1984) Sugarcane. Longman, Harlow

Clements HF (1980) Sugarcane logging and crop control: principles and practices. University of Hawaii, Honolulu

FAO – UNESCO – ISRC (1988) FAO/UNESCO soil map of the world revised legend. World soil resources report 60. FAO, Rome

Gomez KA, Gomez AA (1984) Statistical procedures for agricultural research. Wiley, New York, NY

Kenya Sugar Board (2003) Year book of sugar statistics. Nairobi

Kenya Sugar Research Foundation (2002) Sugarcane grower's guide. Kisumu

Kenya Sugar Research Foundation (2004) Annual Report. Kisumu

Malavolta E (1994) Fertilizing for high yield sugarcane. International Potash Institute, Basel

Perez O, Melgar M (1998) Sugarcane response to nitrogen, phosphorus and potassium application in Andisol soils. Better Crops Int 12(2):20–24

Rahman MH, Pal SK, Allan F (1992) Effect of nitrogen, phosphorus, potassium, sulphur, zinc, manganese nutrients on yield and sucrose content of sugarcane in flood-plain soil of Bangladesh. Indian J Agric Sci 62(7):450–455

Sreewarome A, Toomsan B, Limpinuntana V et al (2005) Effect of phosphorus on physiological and agronomic parameters of sugarcane cultivars in Thailand. Proc Int Soc Sugarcane Technol 25:126–131

Tisdale SL, Nelson WL, Beaton JD, Halvin JL (1993) Soil fertility and fertilizers, 5th edn. MacMillan, New York, NY

Wawire NW, Kahora FW, Wachira PM, Kipruto KB (2006) Technology adoption study in the Kenya sugar industry. Kenya Sugar Res Found Tech Bull 1:51–77

Natural and Entropic Determinants of Soil Carbon Stocks in Two Agro-Ecosystems in Burkina Faso

S. Youl, E. Hien, R.J. Manlay, D. Masse, V. Hien, and C. Feller

Abstract Impacts of two land-management systems on soil C content and stocks and their dynamic were assessed in two agro-ecosystems of South-West Burkina Faso. The study was carried out on 26 pedological profiles and 112 farmers' plots for a precise study compared to four plots in a close forest. The study goal was to assess differences between two cropping systems in terms of organic resources management. The itinerant one, based on yam production, developed by the native population and the permanent one based on cotton production, developed by the migrant population. The results show similarities as well as differences: (1) C content and stocks in soil are highly related to textural factors (clay content + tiny silt, %) in both systems; (2) for all uses the average stocks for all plots for the top 20 cm (0–10 and 10–20 cm) of soils are 14 and 12 t C ha^{-1} and (3) significant differences in C stocks occurred mainly in the 0–10-cm soil layer with higher stocks in the permanent cropping system than in the itinerant system or forest. No significant difference was noticed between cultivated and noncultivated plots. These are the main points of our study. This implies that management factor, as Cultural Intensity (CI) or the position of the plot in the cultural succession (POSISSUC) does not significantly improve C stocks prediction. Similarly texture does not improve C stocks prediction but participates in the variability of C stocks. Lastly, in order to simulate C stocks at the village territory level, taking into account different scenarios including biophysical or socio-economical parameters, the main stock data and equations are to be considered herein. The equations were established for the horizons 0–10 and 10–20 cm for cultivated and noncultivated plots, in permanent and itinerant cropping systems.

Keywords Burkina Faso · Carbon stocks · Cropping system · Farming system · Soil carbon · Savannah

Introduction

From an agro-ecological point of view, organic matter (OM) and its main component carbon (C) play an essential role in the functioning of agro-ecosystems. In fact they favorably affect physical, chemical and biological properties of soils (Batjes, 2001; Feller et al., 2001). OM represents a tank of available nutrients for plants after mineralization (Smith et al., 2000) (Zech et al., 1997). OM improves soil structure, water circulation and stock, aeration and root respiration (Detwiler, 1986). Finally, OM stimulates soil biological activities. OM content in soil results from a balance between: (1) photosynthetic production, (2) the part returned to the soil (exudation, above-ground litter and root restitution, dead vegetation) and (3) respiration of soil fauna and flora and micro-organisms.

From an environmental point of view, soil is an element of the global carbon cycle even if it seems that a dilemma still exists in relation to the desire to increase OM for the global change issue and agronomic needs as nutrient sinks for plants (Janzen, 2006). A change in management practices leading to an increase in C stocks (sink) represents a means to reduce CO_2 concentration in the atmosphere which is responsible for the increase of the Earth's surface temperature

S. Youl (✉)
IFDC-Ouaga 11 BP 82 CMS, Ouagadougou 11, Burkina Faso
e-mail: syoul@ifdc.org

A. Bationo et al. (eds.), *Innovations as Key to the Green Revolution in Africa*,
DOI 10.1007/978-90-481-2543-2_56, © Springer Science+Business Media B.V. 2011

(GIECC, 1997). Soil organic matter (SOM) constitutes an important resource to be well managed for agronomic and environmental challenges.

In the West African savannah, traditional agricultural practices were tightly bound to an alternation of short periods of cultivation and long periods of fallow (Jouve, 2000; Serpantié and Ouattara, 2000). Many factors such as population growth and others have led to rapid changes in farming systems. This last decade, present farming-system transformations with the introduction of cash crops have led to permanent cultivation of agricultural land (Serpantié, 2003). It's important to assess the influence of these land use changes on differences in organic C stocks in the soil–plant system at the village territory level (Augusseau et al., 2000). In the South-Western part of Burkina Faso, there is an area which receives migrants. It's called the "zone de front pionnier". In this area there is a cohabitation (in the same space) of two farming systems. The first one, practiced by natives is an itinerant system based on clearing after a long fallow period followed by yam cultivation for a year and then by a few years

of cereal before starting a new fallow period. The second one practiced by migrants, is based on a permanent cropping system, consisting of rotation of cotton and cereal.

The objective of this work is: (1) to identify and quantify natural and entropic determinants of C stocks in the two farming systems (Itinerant and Permanent); (2) to build a locally validated ecosystem model to predict C stocks in soil for further complex computer modeling.

Material and Methods

Study Area and Sampling

Torokoro village territory ($4°20'$–$4°30'$ latitude and $9°59'$–$10°05'$ longitude) is situated in the department of Mangodara (Western region of Burkina Faso). It covers an area of 15,000 ha and is located in the sudannian zone (Fig. 1). Its average temperature varies

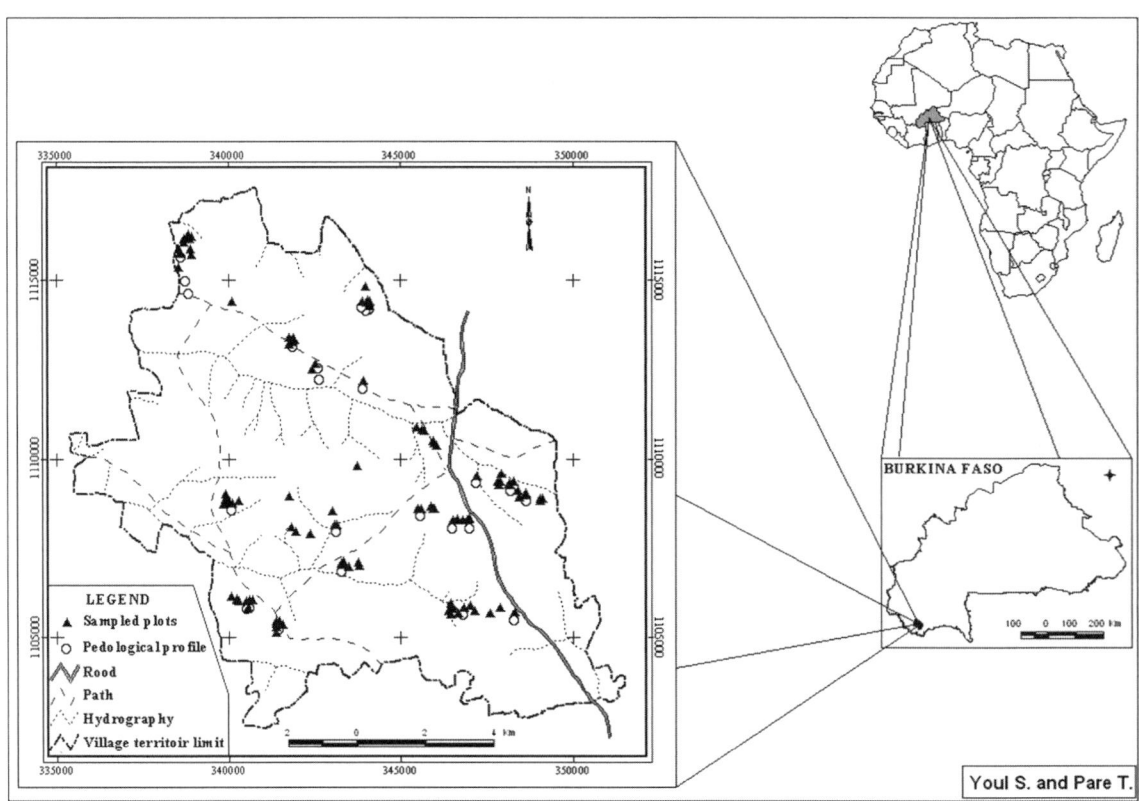

Fig. 1 Location of the village territory and sampled plots

between 27 and 28°C. Its climate consists of two contrasted seasons: a rainy one (from May to October) and a dry one (from November to April). The rainfall varies between 900 and 1,200 mm a year. Its natural vegetation is tree savannah in which the main species are *Vitelaria paradoxa*, *Terminalia laxiflora*, *Parkia biglobosa*, *Isoberlinia doka* which is associated in light forest with *Isoberlinia dalzielli*. The herbaceous tapi is mainly made of graminae in which the most important species are *Andropogon ascinodis*, *Schizachyrium sanguineum*, and *Hyparrhenia spp*. The village territory is situated in the western part of Burkina Faso and of note is its location on a granitic and schistic platform leading to ferruginous soils consisting of sandy material, sand and clay.

Soils

Soils are tropical leached indurated ferruginous, moderately deep and not deep and ferruginous with stains and concretions (CPCS, 1967) or lixisol epiplinthique and luvisol gleyique (BRM) (Annex 1). The first group represents the essential constitution of existing soils, while the second group is located in the lower part of the landscape. It is mainly the first group which is the purpose of the present study. Tropical ferruginous leached indurated soils can be divided into two categories: deep soils (>60 cm) and slightly deep soils (<60 cm). The stony layers are important in the first 20 cm for slightly deep and not deep soils which present an important gravely layer in layer B.

General Characteristics of Tropical Ferruginous Leached Indurated Soils (FLI) (Luvisol Endo-Plinthique, Plinthosol Epiplintique and Plinthosol Epipetrique)

In general, the profile color is brown, pale brown or light brown (7, 5YR6/4) when wet and brown (7, 5YR4/4) when humid, even dark brown (10YR4/3) when wet and dark grayish brown when humid. Texturally it is sandy-silty in the surface and silty-sandy in deeper horizons. Coarse elements increase from the surface to deeper horizons. The deeper horizons accumulate tiny elements like clay. The Bmcs horizon, which is composed of a carapace of indurations and cementation occurs between 20 and 80 cm.

General Characteristics of Tropical Ferruginous Leached Indurated Soils with Stains and Concretions (FLTTC) (Luvisol Gleyique)

These soils are generally light brownish gray (10YR6/2) when dry and very dark grayish brown (10YR3/2) when moist, and yellowish brown (10YR5/4) in the deeper horizons; coarse elements vary from 5 to 20% and stains and concretions appear in the deeper horizons. FLTTC presents sometimes a small water fissure in the last horizon.

Cropping Systems

Two cropping systems dominate:

- An itinerant cropping system (ICS) based on yam is the norm for native farmers. This cropping system involves successive cultivation of yam, maize, sorghum on an old fallow for 3–5 years, before renewed fallow or planting of *Anacardium occidentale*.
- A permanent cropping system with a triennial rotation (cereal/cotton/cereal) is implemented by migrant farmers; this system is developed on already cultivated soils, degraded and lying in fallow, which have been given to the migrants by the native population (Augusseau et al., 2000). A livestock system which is mainly an extensive one, practiced by natives, migrants and some nomadic groups, is being developed (Botoni et al., 2003).

These two cropping systems can be analyzed more precisely by two indicators

- Cultural Intensity (CI) is the number of years of cultivation for a total period including fallow length. The general formula of CI is the number of years of cultivation/(number year cultivation + fallow length) (Ruthenberg, 1971).
- The temporal position of a plot in the cropping succession (POSISSUC). According to convention, the forest position ranks 0 in the cropping succession, the following positions go from 1 to 5 corresponding to the number of years of cultivation and from 6 to 15 corresponding to the fallow length after a few years of cultivation (Fig. 2).

Annex 1: Summary of Pedological Profiles and Soil Types in the Village Territory

Profi_num	Plot_num	Classification CPCS	Classification BRM	Occupation	Longitude	Latitude	Depth (cm)	Depth class
1	9	FTLI	Luvisol plinthique	Natural vegetation	0°59,975'	04°23,042'	50–110	Moderately deep
2	138	FTLI	Luvisol plinthique	Anacardium	10°00,059'	04°23,845'	50–104	Moderately deep
3	93	FTLI TC	Luvisol gleyique	Fallow 1 year	10°00,117'	04°24,038'	>150	Deep
4	60	FTLI	Luvisol gleyique	Maize 2 years	10°04,111'	04°26,577'	50–100	Moderately deep
5	106	FTLI	Plinthosol epipétrique	Anacardium 8 years	10°03,791'	04°26,173'	>23	Slightly deep
6	42	FTLI TC	Luvisol gleyique	Yam	10°03,617'	04°26,149'	>160	Deep
7	310	FTLI	Luvisol plinthique	Natural vegetation	10°03,492'	04°25,456'	>53	Moderately deep
8	326	FTLI	Plinthosol epiplinthique	Fallow	10°04,915'	04°28,247'	46–146	Moderately deep
9	325	FTLI	Luvisol plinthique	Fallow	10°05,114'	04°28,308'	>66	Deep
10	324	FTLI	Plinthosol epipétrique	Anacardium 10 years	10°05,467'	04°28,380'	>47	Moderately deep
11	37	FTLI	Luvisol plinthique	Yam	10°04,686'	04°25,365'	>94	Deep
12	67	FTLI	Luvisol plinthique	Maize 2 years	10°04,679'	04°25,413'	>90	Deep
13	12	FTLI	Luvisol plinthique	Fallow old	10°04,730'	04°25,477'	>145	Deep
14	33	FTLI	Plinthosol epiplinthique	Yam	10°00,715'	04°25,793'	>34	Slightly deep
15	193	FTLI	Plinthosol epipétrique	Maize	10°02,067'	04°23,644'	>48	Moderately deep
16	155	FTLI	Luvisol plinthique	Fallow	10°01,945'	04°23,110'	>50	Moderately deep
17	203	FTLI	Luvisol plinthique	Cotton	10°01,791'	04°22,860'	>112	Deep
18	104	FTLI	Luvisol plinthique	Fallow	10°01,310'	04°25,885'	>55	Moderately deep
19	166	FTLI	Luvisol plinthique	Maize 1 year	10°01,559'	04°24,528'	52–130	Moderately deep
20	214	FTLI	Plinthosol epiplinthique	Maize 10 years	10°01,381'	04°23,744'	>44	Moderately deep
21	161	FTLI	Plinthosol epiplinthique	Fallow old	10°01,370'	04°24,016'	>44	Moderately deep
22[a]	307	FTLI	Plinthosol epiplinthique	Forest	30°0341867	UTM 113 1084	>22	Slightly deep
23	317	FTLI	Luvisol plinthique	Anacardium	10°01,624'	04°27,553'	>89	Moderately deep
24	23	FTLI	Luvisol plinthique	Fallow	09°59,829'	04°26,778'	>125	Deep
25	48	FTLI	Plinthosol epiplinthique	Yam	10°00,157'	04°27,258'	>41	Moderately deep
26	312	FTLI	Plinthosol epiplinthique	Sorghum-maize	10°00,135'	04°27,308'	>41	Moderately deep

[a]Profile 22: Forest: UTM coordinate

Plot Descriptions

Through survey, 112 farmers' plots were identified over the whole village territory: 78 from the itinerant system, 33 from the migrant system and four plots in a close forest. Within each system, we split plots into cultivated and noncultivated ones. At the beginning of the study noncultivated plots are under natural vegetation, have different fallow lengths, and different plantation ages.

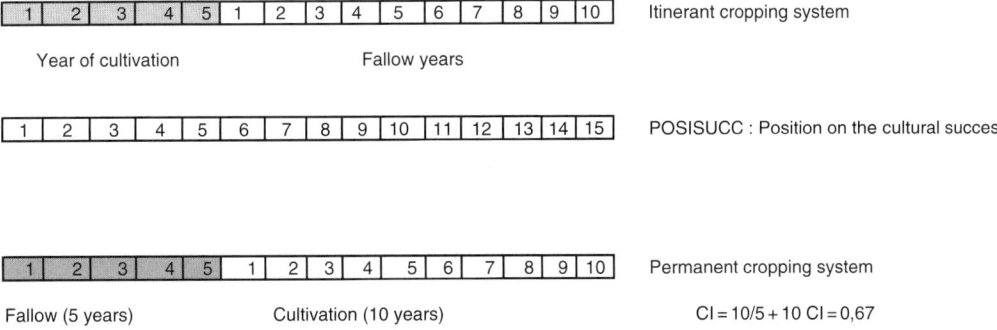

Fig. 2 Explanatory figure for the position of the cropping succession and cultural intensity

Soil Sampling

The sampling design for all plots was adapted from Hairiah et al. (2001); the soil was sampled every 4 m along a transect of 40 m, over the 0–10 and 10–20-cm intervals of a small profile, starting by sampling the deeper 10–20-cm layer. For yam, which is cultivated on mounds, we considered the mound to be the first 0–10-cm layer and we took the 10–20-cm layer under the mound after removal. These ten samples from across the transect were mixed to obtain a composite sample for this horizon.

In addition, 12 representative plots of the cropping system were systematically sampled along pedological profiles at the following depth horizons: 0–10, 10–20, 20–30, 30–50, 50–100, 100–150 cm. The final depth of each profile depended on the specific profile sampling restriction.

All soil samples were air dried, passed through 2 mm sieve and kept at room temperature before analysis.

Analytical Determinations

Coarse elements (CE), essentially ferruginous grit, larger than 2 mm were separated from fine soil (FS) particles (0–20 mm) by dry sieving followed by weighing.

Mechanical analysis was carried out on fine soil (FS) particles of the 0–10 and 10–20 cm samples according to the Robinson Pipette method. This resulted in the fractionation of the soil into five granular fractions: 2,000–200, 200–50, 50–20, 20–2, and 0–2 μm.

Bulk density (BD) was determined from an undisturbed soil sample of 100 cm^3. The density of soil was determined after drying at 105°C in an incubator for 24 h. A BD formula is linked to FS (BDFS) by taking into account the mass of the Coarse Element (CE). For the cylinder of 100 cm^3, the formula is:

$$\text{BDFS} = (\text{dry soil mass} - \text{CE})/100.$$

C content was determined by dry combustion with an auto analyzer CHN type NA 2000 N-Protein (FRISON Instruments).

C stocks were determined on the FS proportion after correction for bulk density (BD) as shown in the formula below:

$$\text{C stocks (Mg ha}^{-1}) = \text{BDFS (g cm}^{-3})$$
$$\times \text{C content (mg g}^{-1})$$
$$\times e \text{ (cm)} \times 0.1$$

C stocks in the whole profile are the sum (Σ_e) of pedological horizons or of soil layers.

$$\text{C stocks (Mg ha}^{-1}) = \Sigma_e\text{BDFS (g cm}^{-3})$$
$$\times \text{C content (mg g}^{-1})$$
$$\times e \text{ (cm)} \times 0.1$$

Predictions were made to assess the carbons tocks based on the land management options and the temporal C dynamics. Multiple linear correlations were performed to relate to the soil fine elements (clay + tiny silt) and the two soil occupation indicators (CI and POSISSUC) described above.

For every cropping system, the following models were tested:

$$Y = a \text{ (clay + tiny silt)} + b \text{ (CI)} + c$$
$$Y = a \text{ (clay + tiny silt)} + b \text{ (POSISSUC)} + c$$

Calculations were performed using the SPSS GLM module (SPSS, 1999).

Results

C Content and Stocks for the 0–10 and 10–20-cm Layers

The average content of tiny elements (clay + tiny silt) varies from 14 to 17% respectively for the 0–10 and 10–20-cm horizons. It is not statistically different between cultivated and noncultivated plots. The 0–10-cm horizon has a significantly higher clay + tiny silt content than the 10–20-cm horizon ($F = 11.01$; $p = 0.001$). The content of clay + tiny elements does not vary significantly between the two cropping systems and the forest. In the 0–10-cm horizon, tiny element and C content are strongly correlated but this is not the case in the 10–20-cm horizon (Fig. 3).

Bulk Density

Results in this section concern both pedological profiles and the 0–10 and 10–20-cm horizons of soils.

Average BDFS of the 0–10 and 10–20-cm horizons are respectively 1.14 and 1.24 for the itinerant cropping system and 1.19 and 1.07 for the permanent cropping system. In both cases BD is roughly higher in the upper horizon (0–10 cm) ($F = 7.101$; $p = 0.009$) than in the horizon below (10–20 cm). There is no significant difference between the cropping systems and (between) cultivated and noncultivated plots. Generally BDFS varies a lot in the profile due to variation of quantities of coarse elements in the different horizons (Fig. 4; Table 1).

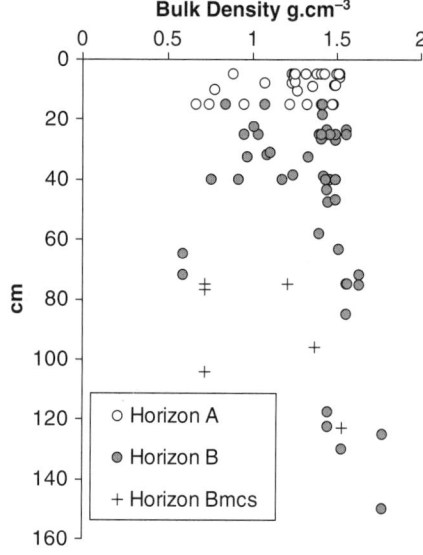

Fig. 4 Bulk density in pedological profiles

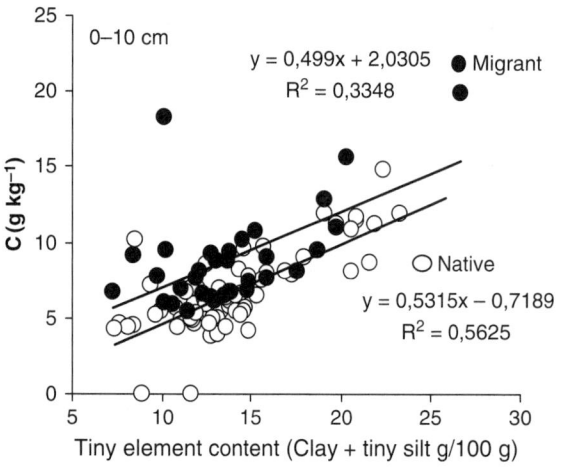

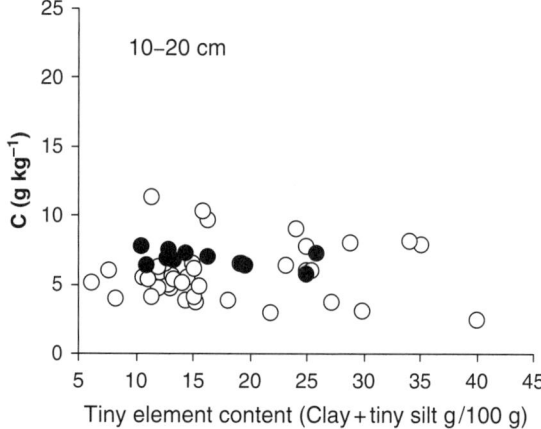

Fig. 3 C content according to tiny element content in the 0–10 and 10–20-cm layers

Table 1 Regression parameters of soil C content related to the depth according to the equation: C = a*Exp.(−b*p)

Compartment	n	a	b	r
All soils[a]	91	8.39[c]	0.16[c]	75
Deeper soils[a]	50	7.62[c]	0.13[b]	72
Slightly and moderately deep soils[a]	41	8.65[c]	0.21[c]	75

[a]Equation $C = a*Exp(−b*p)$
C content g kg^{-1}
p = horizon deep (cm)
n = number of measure a, b = estimated parameters
r = regression coefficient
[b]significant $P < 0.05$; [c]significant $P < 0.001$

C Content Inside Pedological Profiles

Details of C content and stocks are summed up in Annex 2; averages C content and stocks are presented per soil depth classes like: 1 = 0–10 cm; 2 = 10–20 cm; 3 = 20–30 cm; 4 = 30–50 cm; 5 = 50–100; 6 = 100–150 cm and more. On Table 2 C content for the first two horizons of plots are summarized. Inside pedological profile C content (Fig. 5) decrease significantly from the upper horizon to the deeper one (F = 33.42) for two classes of soils deep (Table 1). These contents do not vary significantly between cultivated and noncultivated plots or between the itinerant and permanent cropping system, except in the 0–10 cm horizon.

Annex 2: Content and Stocks of C Inside Pedological Horizons in the Two Cropping Systems

System	Class*	Depth_class	Cultivated	n	C (g kg^{-1})	C stocks (t ha^{-1})
Itinerant	1	Moderately deep	No	8	8.35	12.76
	1	Slightly deep	No	2	7.41	9.66
	1	Deep	No	6	6.20	10.78
	1	Deep	Yes	4	9.98	16.92
	2	Moderately deep	No	4	5.92	5.36
	2	Slightly deep	Yes	1	5.16	5.55
	2	Deep	No	3	5.47	7.46
	2	Deep	Yes	3	7.04	10.48
	3	Moderately deep	No	5	4.99	7.58
	3	Slightly deep	Yes	2	5.78	9.87
	3	Deep	No	5	4.24	9.21
	3	Deep	Yes	3	4.52	8.28
	4	Moderately deep	No	7	4.76	9.58
	4	Slightly deep	Yes	1	3.46	1.51
	4	Deep	No	5	3.34	12.71
	4	Deep	Yes	4	5.39	17.48
	5	Moderately deep	No	5	3.22	8.07
	5	Deep	No	5	3.15	20.15
	5	Deep	Yes	6	2.71	6.96
	6	Moderately deep	No	4	2.40	*
	6	Deep	No	5	3.06	17.84
	6	Deep	Yes	1	1.63	*
Permanent	1	Moderately deep	No	1	11.03	15.76
	1	Moderately deep	Yes	2	8.37	16.17
	2	Moderately deep	No	1	5.12	4.32
	2	Moderately deep	Yes	1	8.36	12.29
	3	Moderately deep	No	1	4.15	5.80
	3	Moderately deep	Yes	2	5.89	10.46
	4	Moderately deep	No	1	7.51	*
	4	Moderately deep	Yes	2	4.49	6.08

*Classes: 1 = 0–10; 2 = 10–20; 3 = 20–30; 4 = 30–50; 5 = 50–100; 6 = 100–150 cm

Table 2 C content and stocks for layers 0–10 and 10–20 cm in plots in the two systems

System	Cultivated	Horizon (cm)	n	C (g kg^{-1})	C (t ha^{-1})
Itinerant	Yes	0–10	33	6.65±0.37	8.29±0.40
	Yes	10–20	21	5.72±0.47	6.13±0.46
	No	0–10	33	7.09±0.47	8.62±0.37
	No	10–20	26	5.71±0.49	5.88±0.40
Permanent	Yes	0–10	28	8.94±0.65	10.84±0.82
	Yes	10–20	11	6.21±0.34	6.83±0.18
	No	0–10	5	9.04±1.91	11.61±1.97
	No	10–20	1	5.12	5.84

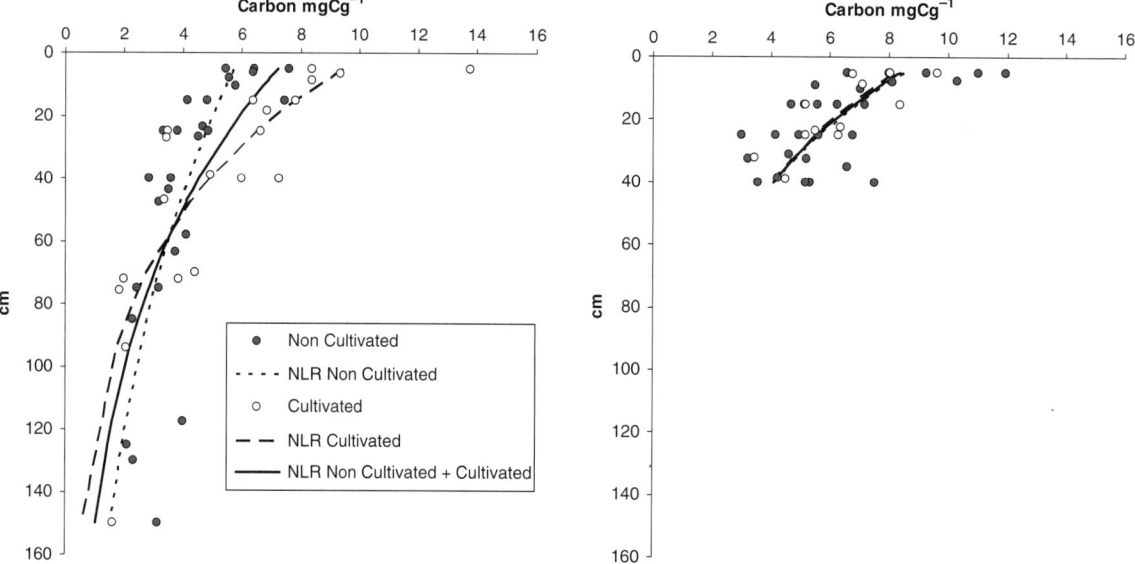

Fig. 5 Distribution of C content according to soil type. (**a**) Deep soils. (**b**) Moderately and slightly deep soils

In all cases, land occupation effect or the cropping system are not significantly related to the C content of horizons that are deeper than 20 cm.

Taking into account (1) the limited number of analyzed profiles, and (2) the only significant differences being observed for the upper horizon, a more detailed analysis was carried out for the 0–10 and 10–20-cm layers.

C Content and Stocks for the Upper Layers of Soil

C content and stocks by plot and per layer (0–10 and 10–20 cm) are summed up in Table 2.

We noted a significantly higher C content in the 0–10-cm layer compared to the 10–20-cm layer

(Table 2). Although the differences for each layer are not significant according to whether the plot is cultivated or noncultivated, C content in the upper horizon of the permanent cropping system is significantly higher than that in the upper horizon of the itinerant cropping system.

For C stocks, we have checked (results not shown herein) that the calculation of C stocks at "equivalent soil mass" (Ellert and Bettany, 1995), does not modify significantly the results in absolute values nor the trend between the different tested factors compared to the C stocks at "equivalent soil volume" as presented above.

C stocks are significantly higher in the 0–10-cm layer compared to the 10–20-cm layer. For each layer there is no significant difference between soil type occupation (cultivated vs. noncultivated) but significant differences exist among cropping systems (itinerant vs. permanent).

For the 0–10-cm layer the equation is:

Without any intercept: C stock (t ha^{-1})
$= 0.477 \times$ content of tiny elements (clay
$+$ silt) $+ 4.34 \times$ (Permanent) $+ 1.77$
$\times$(Itinerant) $+ 2.5 \times$ (Forest) ($r^2 = 0.94$)

With an intercept: C stock (t ha^{-1})
$= 1.77 + 0.477 \times$ (clay $+$ silt;%) $+ 2.57$
$\times$(Permanent) $+ 0.73 \times$ (Forest) ($r^2 = 0.44$)

For the 10–20-cm layer the equation is:

C stock (t ha^{-1}) $= 6.79 \times$ (Permanent) $+ 5.77$

$\times$ (Itinerant) ($r^2 = 0.92$)

The relationship between C content for a given layer and its tiny element content (clay + silt %) can be described by the equation presented in Table 4. It appears that for the 0–10-cm horizon, which shows a high content in the permanent cropping system, differences expressed by the intercept are respectively +2.18 g kg^{-1} for the content and 4.34 t C ha^{-1} for the stocks (Table 3).

If we take into consideration CI and POSISUCC in modeling C stocks in the upper soil layers (Table 4), we can see that introducing these indices IC and POSISSUC, if this is original, its does not improve significantly linear model establish related C content and stocks other textural parameters.

C stock $= 0.250$(POSISSUC) $+0.216$ (tiny element (clay + tiny silt) content) ($r^2 = 0.81$).

Therefore, it is CI which can be used for deeper investigation, both for the permanent and itinerant cropping systems due to its possible adaptability and also to its contribution to the regression. The tiny element content (clay + tiny silt %), according to the results below is the best indicator for modeling of content and stocks of soil C.

For the 0–10-cm horizon, two stock models were determined:

- C stock $= 4.265 + 0.34 \times$ tiny element (clay + silt) content ($r^2 = 0.44$)
- C stock $= (0.34 \times$ tiny element) $+ (3.29 \times$ natural vegetation) $+ (3.72 \times$ yam) $+ (3 \times$ maize) $+ (2.4$ Anacardium $+$ cereal) $+ (3.96 \times$ cereal) $+ (4.9 \times$ fallow 5 years) $+ (4.4 \times$ fallow 1year) $+ (4.26 \times$ forest) ($r^2 = 0.97$)

For the horizon 10–20 cm, two models

were also described for C storage:

C stock $= (4.47 \times$ natural vegetation) $+ (6.18 \times$ yam) $+ (5.3 \times$ Maize 3 years) $+ (5.7 \times$ Anacardium $+$ cereal 3 years) $+ (7.3 \times$ cereal 3 years) $+ (6.9 \times$ fallow 5 year) $+ (7.98 \times$ fallow 1 year) ($r^2=0.88$).

POSISSUC is adapted for the itinerant cropping system for which it has been defined while CI is used for the permanent system but the latter can also be used for the itinerant system. Neither have a significant effect.

Table 3 Equation of C content and stocks in the two horizons 0–10 cm and 10–20 cm related to content of tiny elements (percentage of clay + tiny silt)

| System | Horizon (cm) | C content (g kg^{-1}) | | | C stocks (t ha^{-1}) | | |
| | | Covariables | | | Covariables | | |
		Tiny element content	cte	r^2	Tiny element content	cte	r^2
Itinerant	0–10	0.488		0.94	0.477		0.94
	10–20	0.154	2.936	0.91	0.003	5.77	0.92
Permanent	0–10	0.488	2.184	0.94	0.477	4.343	0.94
	10–20	0.154	3.676	0.91	0.003	6.79	0.92

Table 4 Correlations between content and stocks of C and tiny element content

| System | Horizon (cm) | Covariables | | | |
		Tiny element content	POSISUCC	CI	r^2
Itinerant	0–10	0.574			0.96
	10–20	0.216	0.25		0.81
Permanent	0–10	0.617		2	0.93
	10–20	0.237		4.367	0.80

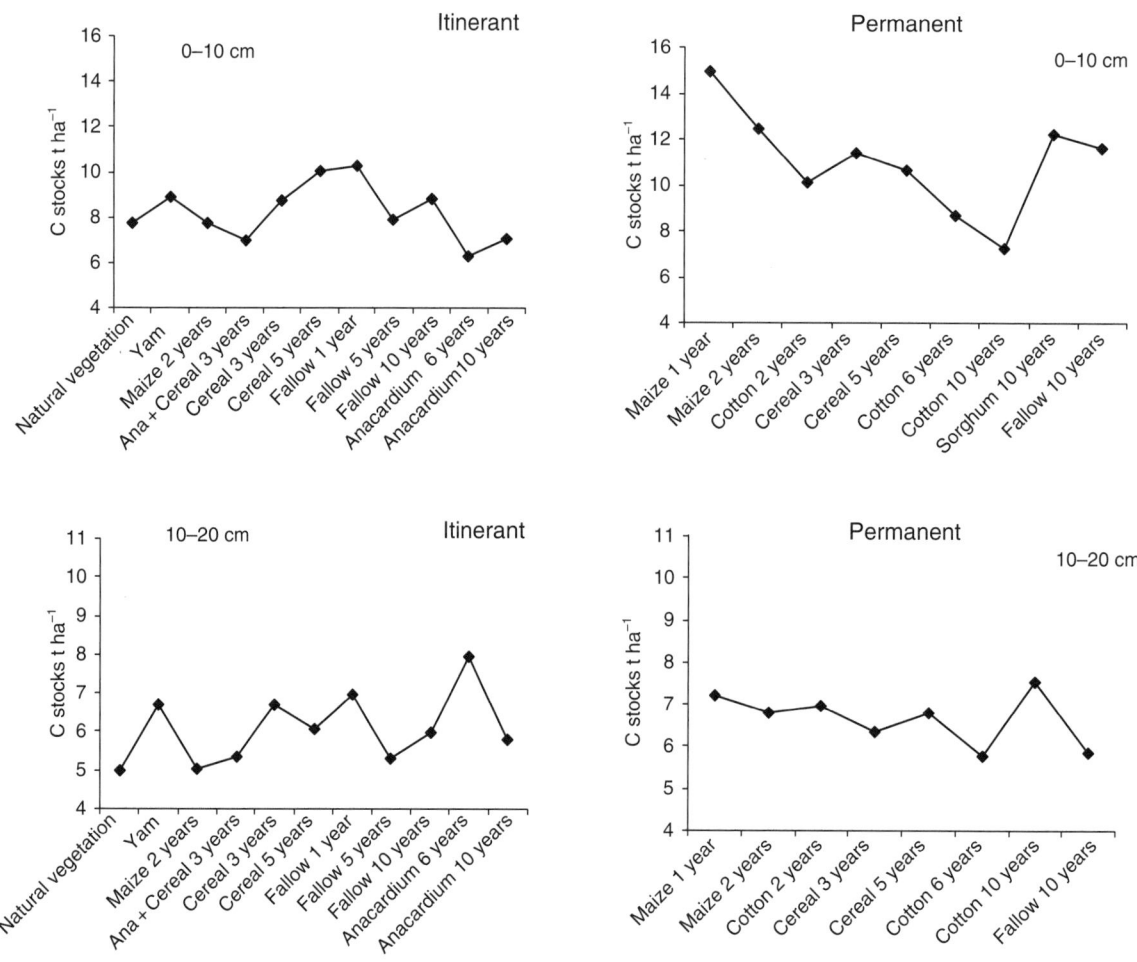

Fig. 6 C stocks for each land use cropping system (itinerant and permanent)

Effect of Land Use Change on C Stocks in the Upper Soil Layers

The effect is represented in Fig. 6. For the 0–10-cm horizon, we observe low variations for itinerant cropping systems (6–8 t ha^{-1}), whereas for the permanent cropping system we observe a big variation from 6 to 16 t ha^{-1}. For the 10–20-cm horizon, variations are about 6 t ha^{-1} in the two cropping systems.

Globally, in comparable situations for example maize 2 years, cereal 3 and 5 years, we find higher C stocks for permanent cropping than for the itinerant cropping system, particularly for the 0–10-cm layer. Notice also that in the itinerant cropping system, the lowest C stocks are the system under *Anacardium occidentale*, but in the permanent cropping system, plots

under cotton at 6 years old or more have the higher C content.

Discussion

Measured C stocks are 8±0.8 t C ha^{-1} for the 0–10-cm horizon and 6±0.9 t C ha^{-1} for the 10–20-cm horizon. Stocks for the first 0–20-cm horizon are 14 t C ha^{-1}. The average stocks of carbon measured in the village territory are close to values generally observed in these savannah agro-ecosystems (Manlay et al., 2002), (Serpantié et al., 2002), (Tschakert, 2004). These stocks are limited by the potential storage which is related to climate (Ingram and Fernandes 2001). In Nigeria stocks varying from 6 to 12 t ha^{-1} have also been established (Farage et al., 2003).

Annex 3: Equations of C Content in Two Upper Horizons Related to Tiny Element Content

Horizon (cm)	System	Cultivated	Parameters			Linear model
			n	a	b	
0–10	Itinerant	Cultivated	32	0.505^b	−0.248	C=0.505*Tiny element content−0.248
	Itinerant	Noncultivated	32	0.505^b	−0.166	C=0.505*Tiny element content−0.166
	Permanent	Cultivated	27	0.505^b	1.799^a	C=0.505*Tiny element content+1.799
	Permanent	Noncultivated	3	0.505^b	1.881^a	C=0.505*Tiny element content+1.881
	Forest	Noncultivated	4	0.505^b	0.26	C=0.505*Tiny element content+0.26
10–20	Itinerant	Cultivated	21	0.159^a	3.063	C=0.159*Tiny element content+3.063
	Itinerant	Noncultivated	13	0.159^a	3.011	C=0.159*Tiny element content+3.011
	Permanent	Cultivated	3	0.159^a	3.672	C=0.159*Tiny element content+3.672
	Permanent	Noncultivated		0.159^a	3.62	C=0.159*Tiny element content+3.62
	Forest	Noncultivated		0.159^a	−0.052	C=0.159*Tiny element content−0.052

Tiny element content = Content of clay + fine silt %
[a]Fprob<0.05; [b]Fprob<0.001

Relationship Between C and Texture (Natural Determinant of Storage)

Soil texture can be a major determining factor for the organic matter content, and then carbon. The regression in the case of the 0–10 cm horizon is C (t ha^{-1}) = 0.47*(clay+tiny silt %) + 2.7; $r^2 = 0.31$, $n = 103$). These interval values are the same as those of the equation of Feller (1995). Those relationships are similar to the results established for such an agro-ecological zone by Jones (1973).

Relationship Between C and Land-Management Options (Entropic Determinant)

The quick study of 26 pedological profiles shows evidence of a significant trend in organic matter stock levels according to the cropping system in the village territory. It also shows evidence that there is no further effect of land management under a depth of 20 cm. More intensive studies are however needed to confirm these findings.

The analyses of the two superficial layers show two curious behaviors:

- C in plots under natural vegetation or long length fallow is not statistically higher than C in cultivated land. The strong use of fallow and natural vegetation might be responsible for this (Botoni et al.,

2003). The state of degradation of natural vegetation due to grazing, frequent bush fires, cutting etc., affect the restitution to soil through vegetal biomass.

- C in the itinerant system is significantly higher than in the permanent system. The permanent system seems to improve the storage of carbon particularly in the surface horizon. As there is no textural difference between the soil of the two systems, this might be due to soil management. The differences in technical itinerary (mineral fertilization of cotton and the application of crop residues) might be the reason for these differences.

Management may affect mainly stocks in the 0–10-cm layers and less so the 10–20-cm layer due to the limited technical itinerary. The potential of these soils might be limited due to the intrinsic limit imposed by climatic factors (Ingram and Fernandes, 2001). Below the 20-cm depth, there is no evidence of an effect of management on C stocks. The distinction of deep, moderately deep, and slightly deep soils allowed a separate estimation of potential C stocks in these deeper horizons. For C stocks in the surface layer, the cropping system is the main factor affecting those stocks.

System, Soil Occupation and Sequestration

On the basis of these results, referring to an itinerant cropping system, the permanent cropping system seems globally to be a storage system for the 0–10 and 10–20-cm horizons. The net storage observed in Nigeria varies according to management, so in a system with recycling organic residues and using rotation of fallow and culture, it is +0.1 to 0.3 t ha^{-1} an^{-1} (Farage et al., 2003). The storage value is around 0.12 and 0.08 t ha^{-1} when mineral fertilization only is used. Agricultural practices might explain the differences between the two cropping systems, because soil textural parameters are not statistically different between the two.

In addition to the demographic pressure and the migrants' pressure in this pioneers' zone new stakeholders are appearing (Ouédraogo, 2000). This modifies agricultural systems. In this village territory, the study of farms evolution path shows the existence of many trajectories. The study indicates that many of them are possible, and then due to land constraint, yam surface might reduce (Augusseau et al., 2000). Are we going to expect a cotton production increase? The current development of animal traction and the existence of cash crops like maize can play a great role in that transformation.

Plots in the itinerant cropping system seem to have an impoverished C dynamic while plots in the permanent cropping system improve C status due to the management in addition to the supply of mineral fertilizer. *Anacardium occidentale* will create disturbance of nutrient cycling when digging out these trees (Aweto and Ishola, 1994) and yet *Anacardium occidentale* plantations practices in some regions follow the same strategies (Bertrand, 1993). The survey results have shown that the digging out of trees can be possible if land is scarce and if cereal production is risky.

Biennial rotation of cereal/cotton might be preferred first for its financial and alimentary benefits but also for its addition to soil nutrient stocks. Stocks of C are measured in situ on the plot. Fluxes are related to material flux, water erosion, animal transfers, farming, forestry, and other human activities. Our hypothesis is that at village territory level, these fluxes are a balance between input and output of C to the soil related to management using a global approach in the village territory.

Conclusions

The change of OM because of natural factors seems slow. The entropic factors seem to be the most important ones, particularly population growth. A demographic scenario with a dynamic simulation model clearly shows the collapse of the farming system when reaching high population growth rates. The results obtained in this study show clearly that in addition to population growth, farming system is one of the most important factors affecting organic C in the agroecosystem. We conclude that it is not sustainable to continue shifting cultivation when the population density increases. But as these results show organic stocks and then organic resources might be maintained in the system should an integrated management system be adopted (supply of mineral, recycling residue, etc.).

Acknowledgments I am grateful to friends and colleagues who have helped me with this topic, especially Ms. Paré Tahibou and Traore Parna. All other colleagues and reviewers are also gratefully acknowledged.

References

Augusseau X, Liehoun E, Kara A (2000) Evolution de l'organisation agraire dans deux terroirs d'accueil des migrants du sud ouest du Burkina Faso: UN même processus dans l'actuel front pionnier? In: FRSIT (ed) Actes du Forum sur la Recherche Scientifique et Technologique, 2000, CNRST, Burkina Faso. FRSIT, Ouagadougou, p 17

Aweto AO, Ishola MA (1994) The impact of cashew (Anacardium occidentale) on forest soil. Exp Agric 30(3):337–341

Batjes NH (2001) Options for increasing carbon sequestration in West African soils: an exploratory study with special focus on Senegal. Land Degradation Dev 12(2):131–142

Bertrand A (1993) La sécurisation foncière, condition de la gestion viable de ressources naturelles renouvelables? In: Ganry F, Cambell B (eds) Sustainable land management in African semi-arid and subhumid regions. Dakar, Sénégal, CIRAD, pp 313–327

Botoni E, Kara A, Augusseau X, Cornelius M, Saidi S, Daget P (2003) Evolutions agraires et construction des paysages végétaux: l'exemple du village de Torokoro en zone Sud soudanienne du Burkina Faso. In: Communication au colloque SAGERT, Montpellier, France, p 16

CPCS (1967) Commission de pédologie et couverture du sol

Detwiler RP (1986) Land use change and the global carbon cycle: the role of tropical soils. Biogeochemistry 2:67–93

Ellert BH, Bettany JR (1995) Calculation of organic matter and nutrients stored in soils under contrasting management regimes. Canadian Journal of Soil Science 75:529–538

Farage P, Pretty J, Ball A (2003) Biophysical aspects of carbon sequestration in Drylands. University of Essex, UK, 25pp

Feller C (1995) La matière organique du sol et la recherche d'indicateurs de la durabilité des systèmes de culture dans les régions tropicales semi-arides et subhumides d'Afrique de l'Ouest. In: Ganry F, Campbell B (eds) Sustainable land management in African semi-arid and sub humid regions. Proceedings of the SCOPE workshop, Dakar, Senegal, 15–19 Nov 1993. CIRAD, Montpellier, France, pp 123–130

Feller C, Albrecht A, Blanchart E, Cabidoche YM, Chevallier T, Hartmann C, Eschenbrenner V, Larre LMC, Ndandou JF (2001) Soil organic carbon sequestration in tropical areas. General considerations and analysis of some edaphic determinants for Lesser Antilles soils. Nutr Cycling Agroecosyst 61(1–2):19–31

GIECC (1997) Protocole de Kyoto à la Convention cadre des NATIONS UNIES sur les changements climatiques. UNO, Kyoto, Japan, p 24

Hairiah K, Sitompul SM, van Noordwijk M, Palm CA (2001) Methods for sampling carbon stocks above and below ground. International Centre for Research in Agroforestry, Bogor

Ingram JSI, Fernandes ECM (2001) Managing carbon sequestration in soils: Concepts and terminology. Agric Ecosyst Environ 87(1):111–117

Janzen HH (2006) The soil carbon dilemma: Shall we hoard it or use it. Soil Biol Biochem 36(2006):419–424

Jones MJ (1973) The organic matter content of the savannah soils of West Africa. J Soil Sci 24(1):42–53

Jouve P (2000) Jachères et systèmes agraires en Afrique sub-saharienne. In: Floret C, Pontanier R (eds) La Jachère en Afrique tropicale -vol II. De la Jachère Naturelle à la Jachère Améliorée: Le Point des Connaissances, Dakar, Sénégal, 13–16 Apr 1999. John Libbey, Paris, pp 1–20

Manlay RJ, Masse D, Chotte J-L, Feller C, Kaïré M, Fardoux J, Pontanier R (2002) Carbon, nitrogen and phosphorus allocation in agro-ecosystems of a West African savannah – II. The soil component under semi-permanent cultivation. Agric Ecosyst Environ 88(3):233–248

Ouédraogo M (2000) Nouveaux acteurs au BF. PNGT, Ouagadougou

Ruthenberg H (1971) Semi-permanent cultivation systems. In: Farming systems in the tropics. Clarendon Press, Oxford, pp 55–82

Serpantié G (2003) Persistance de la culture temporaire dans les savannes cotonnières d'Afrique de l'Ouest: Etude de cas au Burkina Faso. Doctorat de l'INA-PG-Agronomie, INA-PG, Paris, 321p

Serpantié G, Ouattara B (2000) Fertilité et jachère en Afrique de l'Ouest. In: Floret C, Pontanier R (eds) La Jachère en Afrique tropicale -Vol II. De la Jachère Naturelle à la Jachère Améliorée: le Point des Connaissances, Dakar, Sénégal, 13–16/04/1999. John Libbey, Paris, pp 21–83

Serpantié G, Yoni M, Hien V, Abbadie L, Bilgo A, Ouattara B (2002) Le carbone du sol dans les terroirs des savanes soudaniennes "cotonnières". Facteurs et dynamiques. In: Roose E (ed) Land-use management, erosion and carbon sequestration, Montpellier, France, 23–28 Sept 2002. Réseau Erosion – IRD – CIRAD

Smith P, Falloon P, Coleman K, Smith J, Piccolo MC, Cerri C, Bernoux M, Jenkinson D, Ingram J, Szabo J, Pasztor L (2000) Modelling soil carbon dynamics in tropical ecosystems. In: Lal R, Kimble JM, Stewart BA (eds) Global climate change and tropical ecosystems. CRC Press, Boca Raton, FL, pp 341–364

Spss (1999) SPSS Base 9.0: user's guide. Marketing Department, SPSS Inc., Chicago, IL, 740p

Tschakert P (2004) The costs of soil carbon sequestration: An economic analysis for small-scale farming systems in Senegal. Agric Syst 81:227–253

Zech W, Senesi N, Guggenberger G, Kaiser K, Lehmann J, Miano TM, Miltner A, Schroth G (1997) Factors controlling humification and mineralization of soil organic matter in the tropics. Geoderma 79(1–4):117–161

Nutr Cycl Agroecosyst (2010) 88:29–38
DOI 10.1007/s10705-008-9185-z

RESEARCH ARTICLE

Integrated soil fertility management involving promiscuous dual-purpose soybean and upland NERICA enhanced rice productivity in the savannas

S. O. Oikeh · P. Houngnandan · R. C. Abaidoo ·
I. Rahimou · A. Touré · A. Niang · I. Akintayo

Received: 5 March 2008 / Accepted: 22 May 2008 / Published online: 12 June 2008
© Springer Science+Business Media B.V. 2008

Abstract Integrated soil fertility management (ISFM) involving a nitrogen-fixing grain legume, limited chemical fertilizer, and a resilient rice variety may reduce the rate of soil fertility loss and enhance rice productivity in fragile upland rice ecosystems. A 2-year, on-farm study was carried out at Eglimé in the southern Guinea savanna (SGS) and Ouake in the northern Guinea savanna (NGS) of the Republic of Benin to evaluate the contribution of dual-purpose soybean cultivars (*Glycine max*) to grain yield of upland NERICA® rice receiving low fertilizer N. In 2005, four dual-purpose, promiscuous soybean varieties (cv. TGX 1440-IE, TGX 1448-2E, TGX 1019-2EB, and TGX 1844-18E), a popular soybean variety (cv. *Jupiter*), and a popular rice (control) were sown in ten farmers' fields. In 2006, resilient upland interspecific rice (NERICA1) and popular rice (IRAT-136) were sown in all plots with only 15 kg

N ha^{-1}. Soybean cv. TGX 1440-1E (late-maturing) ranked highest in nodulation, dry matter, shoot- and grain-N accumulation, and N-balance (21 kg ha^{-1}) in NGS, while TGX 1448-2E (medium-maturing) surpassed other varieties in the SGS. Nitrogen fertilizer replacement value for growing cv. TGX 1440-1E in NGS prior to rice ranged from 17 to 45 kg N ha^{-1} depending on the reference rice. Grain yield of NERICA1 following 1-year rotation with soybean cv. TGX 1440-1E or TGX 1019-2EB was 1.5 Mg ha^{-1} greater than the yield obtained from farmers' control of 2-year continuous IRAT 136 rice cropping. Results indicate that integrating appropriate dual-purpose soybean in an ISFM package can enhance rice productivity in resource-limited smallholder production systems.

Keywords Crop rotation · Degraded savannas · ISFM · NERICA rice · Promiscuous soybean

S. O. Oikeh (✉) · A. Touré · A. Niang · I. Akintayo
Africa Rice Center (WARDA), 01 BP 2031, Cotonou,
Benin
e-mail: S.Oikeh@aatf-africa.org

P. Houngnandan · I. Rahimou
Faculté des Sciences Agronomiques (FSA), Université
d'Abomey-Calavi (UAC), 01 BP 526 Recette Principale,
Cotonou, Benin

R. C. Abaidoo
International Institute of Tropical Agriculture (IITA),
Ibadan, Nigeria, c/o LW Lambourn & Co., Carolyn
House, 26 Dingwall Road, Croydon CR93 3EE, UK

Introduction

Rice currently sustains the livelihoods of about 100 million people in sub-Saharan Africa (SSA). It is an important crop in attaining food security and poverty reduction in many low-income, food-deficit African countries. However, the demand for rice far outstrips its production in Africa, which in the last 30 years has increased mainly due to land expansion,

This article has been previously published in the journal "Nutrient Cycling in Agroecosystems" Volume 88 Issue 1.
A. Bationo et al. (eds.), Innovations as Key to the Green Revolution in Africa: Exploring the Scientific Facts. © 2010 Springer.

Ⓐ Springer

with only 30% being attributable to an increase in productivity (Fagade 2000). To meet the shortfall in production, West Africa alone imports more than 6 million tonnes per annum into the subregion, costing about $1 billion in scarce foreign exchange annually.

The recent breakthrough in rice science by the Africa Rice Center (WARDA) in developing inter-specific rice from crosses between high-yielding *Oryza sativa* (Asian rice) and low-yielding resilient *Oryza glaberrima* (African rice)—named the New Rice for Africa (NERICA®)—is enhancing upland rice production in many parts of African (FAO 2007). NERICA varieties have been developed as low-management rice plant types for resource-limited, smallholder production systems (Dingkuhn et al. 1998), characterized by short duration, high yields, resistance to major local stresses, and good taste. Studies using participatory varietal selection (PVS) carried out in southwestern Nigeria on a wide range of upland cultivars (*O. sativa*, *O. glaberrima*, and NERICA) showed that farmers preferred varieties of NERICA because of their good tillering ability and high tolerance to major biotic and abiotic stresses (Okeleye et al. 2006). Some of the NERICA varieties have also been reported to be more weed competitive compared with their parents in the savannas of northeastern Nigeria (Ekeleme et al. 2007).

However, the upland rice production systems that account for almost one-half of the rice area and contribute to 29% of total rice production in West Africa are fragile and prone to soil degradation, nutrient depletion, and limited productivity due partly to limited use of nutrient inputs and land-use intensification (Buresh et al. 1997; Becker and Johnson 1999). Even though these production systems have the potential for 2–4 Mg ha^{-1}, rice yields are seldom above 1 Mg ha^{-1} in most smallholder farmers' fields because of constraints such as soil acidification, inherently low soil fertility, and limited use of fertilizers. Therefore, to optimize NERICA rice productivity and sustain farmers' interest in growing the crop, innovative strategies are needed to intensify production and stabilize yields in the upland production systems.

We hypothesize that integrated soil fertility management (ISFM) involving the use of appropriate dual-purpose, nitrogen-fixing grain legumes, limited chemical fertilizer, and resilient rice varieties may restore soil fertility and have a synergistic effect in enhancing rice productivity in the fragile upland rice ecosystems of West Africa. A number of studies have been carried out on biological management of soil fertility for upland rice production. The use of N$_2$-fixing legumes grown as preceding fallow cover crops was reported to enhance upland rice productivity and even suppress weed growth under intensified land-use systems in West Africa (Becker and Johnson 1998, 1999). Furthermore, the combined use of legume cover crops and sparingly soluble indigenous phosphate rock has also been recommended as a relatively inexpensive means to increase the supply of both N and P and enhance upland rice productivity in acid soils (Somado et al. 2003; Oikeh et al. 2008a). However, farmers in West Africa are often reluctant to adopt legume cover crops that are not for human consumption or without a direct economic benefit, in spite of the positive impact on restoring soil fertility (Oikeh et al. 1998; Vanlauwe et al. 2001a).

The integration of grain legumes such as dual-purpose promiscuous soybean (*Glycine max* [L.]) combined with a moderate level of inorganic N fertilizer into maize-based systems has been reported to greatly enhance maize productivity and the sustainability of the production systems in the West African savannas (Sanginga et al. 2003). The promiscuously nodulating soybean contributes residual soil-N to maize following in rotation, in addition to providing the farmers with seeds and fodder for food and feed, and income from the marketing of these products (Vanlauwe et al. 2001b; Sanginga et al. 2002, 2003). Some of the soybean varieties also have additional benefits of controlling nematodes and noxious parasitic weed (*Striga hermonthica*) in the production systems (Weber et al. 1995a, b; Carsky et al. 2000). Upland NERICA rice is grown in similar production systems to maize, but limited studies are available on integrating grain legumes such as promiscuously nodulating soybean into upland rice-based systems in the savannas. The identification of appropriate dual-purpose soybean varieties for the development of an ISFM package can enhance rice productivity and sustain NERICA rice production in resource-limited smallholder production systems.

The objective of this study was to identify the dual-purpose, promiscuous soybean variety most suitable for developing an ISFM package to enhance upland NERICA rice yield in the savannas.

 Springer

Materials and methods

Experimental plan

The experiment was conducted for 2 years during the 2005 and 2006 cropping seasons in ten farmers' fields, with five fields selected at Eglimé (7 33′ N, 2 75′ E; annual rainfall of 1,400 mm [bimodal]) in the southern (SGS) and another five at Ouake (9 46′ N, 1 35′ E; annual rainfall of 1,108 mm [monomodal]) in the northern Guinea savannas (NGS) of the Republic of Benin. Selected physical and chemical soil properties for the locations are presented in Table 1.

In 2005, soybean and upland rice crops were established in each of the farmers' fields in a randomized complete block design. Each farmer's field served as a replication. The soybean cultivars were four promiscuous dual-purpose cultivars: TGX 1440-1E and TGX 1844-18E (late maturing, >120 days), and TGX 1448-2E and TGX 1019-2EB (medium maturing, 110–120 days), together with NERICA1 and a popularly grown cultivar Jupiter ("local"). Promiscuous soybean seeds were collected from the International Institute of Tropical Agriculture, Ibadan, Nigeria. The upland rice cultivars were NERICA1 (interspecific; extra-early maturing, >90 days) obtained from the Africa Rice Center (WARDA) and IRAT-136 (a popular *O. sativa* in Benin; medium maturing, 110 days) obtained from the Benin National Agricultural Research Institute (INRAB). Farmers' practice of minimum tillage with a hoe was used. Soybean seeds were not inoculated with rhizobia. The seeds were sown by drilling at a spacing of 0.6 m between and 0.05 m within rows on the flat (266,667 plants ha^{-1}). Rice seeds were dibble-seeded (50 kg ha^{-1} seed rate) at a spacing of 0.20 m × 0.20 m. Plot size was 5 m wide and 6 m long (30 m^2). Seedlings were thinned at 18–21 days after seeding (DAS) to four per stand to give a final population of 10^6 plants ha^{-1}.

All plots received basal application of 13 kg P ha^{-1} as triple superphosphate and 25 kg K ha^{-1} as KCl just before the seeds were sown. The soybean plots were supplied with a starter dose of 20 kg N ha^{-1} (urea) at sowing while rice plots received the recommended topdressing of 60 kg N ha^{-1} in the form of urea (Oikeh et al. 2008b). Plots were kept weed-free throughout the experiment, and managed by the farmers with guidance from the research team.

Soybean grain was harvested and the aboveground dry biomass, except litter fallen before harvest, was removed from the field as practised by the farmers. Unfortunately, the rice crop, both stover and grain in almost all the fields, failed due to severe drought.

The experiment in 2006 aimed to assess the effects of the different previous cropping on performance of upland NERICA rice in systems where most of the aboveground residues had been removed from the field at harvest. Each of the preceding plots of soybean and NERICA1 rice was sown with NERICA1 rice and the previous plots of IRAT-136 rice were followed with IRAT-136 cultivar. NERICA1 had been reported to be a resilient cultivar that is tolerant to mild drought and low N (Rodenburg et al. 2006; Oikeh et al. 2008b). The seeding rate, spacing, P and K fertilizer application, and crop management were the same as in 2005. However, all plots were topdressed with only

Table 1 Physicochemical properties of experimental fields at Eglimé (SGS) and Ouake (NGS), 2005

Properties	Eglimé (Southern Guinea savanna)	Ouake (Northern Guinea savanna)
pH (water)	5.9 (5.5–6.5)[a]	5.5 (5–6.1)
pH (KCl)	5.5 (5–6.1)	6.3 (6–6.7)
Total N (g kg^{-1})	0.63 (0.53–0.84)	0.82 (0.7–1.4)
Total C (g kg^{-1})	13.8 (8.9–19.9)	13.2 (7.0–18.6)
C/N	22.2 (14.8–30.9)	17.2 (10.1–21.6)
Sand (mg kg^{-1})	511.9 (410–566.9)	764.1 (652.0–890.0)
Clay (mg kg^{-1})	85.0 (62.5–105.0)	64.8 (60.0–79.1)
Exchangeable (cmol$^+$ kg^{-1})		
K$^+$	0.25 (0.07–0.47)	0.22 (0.06–0.34)
Ca^{2+}	1.95 (0.84–2.78)	1.45 (0.52–2.06)
Mg^{2+}	1.66 (1.10–3.08)	1.56 (1.02–2.96)
Effective CEC	12.1 (7.0–26.0)	10.0 (7.0–13.0)

[a] Values in parenthesis are a range of parameters between farmers' fields

15 kg N ha^{-1} as urea in two splits of one-third at 21 DAS and two-thirds at 40–45 DAS.

Yield data were collected at physiological maturity from a net plot size of 6 m^2. Grain yield was corrected to a 140 g kg^{-1} moisture basis. Yield components, including number of panicles and tillers, and harvest index for dry matter (HI) defined as paddy yield per unit total dry matter, were also collected.

Soil sampling and measurement

Prior to establishment of the experiment in 2005, soil samples were taken to a depth of 0.20 m in each field for general characterization of the fields. To assess the contribution of the soybean cultivar to soil mineral N before sowing of the rice seeds in 2006, the soil was sampled at the Ouake location only to a depth of 0.20 m for mineral N content in each of the previous plots of soybean, rice, and the fallow plots used for estimating N fertilizer replacement value (NFRV). Samples were extracted with soil/1 N KCl solution (1:2 w/w) and analyzed for nitrate and ammonium using a Technicon AAII autoanalyzer (IITA 1989). Mineral N was a summation of nitrate and ammonium concentration.

Estimation of N fertilizer replacement value (NFRV) of soybean

In 2006 at Ouake only, an N response trial was carried out adjacent to each soybean field in previously fallow plots using the two rice cultivars (NERICA1 and IRAT-136) to establish the NFRV of the various soybean cultivars. The experiment was a split-plot design with nitrogen levels (0, 15, 30, and 45 kg ha^{-1}) in the main plots and the two rice cultivars in the subplots. Nitrogen was applied as urea in two splits, one-third at 21 days after sowing (DAS) and two-thirds at 40–45 DAS. The seeding rate, spacing, P and K fertilizer application, and crop management were as explained in the previous section. Yield data were collected and grain yield was corrected to a 140 g kg^{-1} moisture basis.

Estimation of biological nitrogen fixation (BNF) and surface area covered by soybean

At early podding stage (65–75 DAS) of soybean, xylem sap samples were taken from four plants of soybean cut at the first node above soil level in the same quadrant. A sterile syringe was used to collect the sap. Extracts were stored in vials with an equal volume of ethanol and kept at 4°C until they were ready for analysis at the Soil Microbiology Laboratory of IITA, Ibadan, Nigeria. The percentage N derived from N$_2$ fixation (%Ndfa) was estimated by determining the concentrations of ureides in the xylem sap as described by Peoples et al. (1989). The amount of N fixed (N$_f$) was estimated from %Ndfa and the amount of N taken up by the crop. The net contribution of N$_2$ fixation to soil N balance was estimated using the equation proposed by Peoples and Craswell (1992): net N balance = N$_f$ − N$_g$, where N$_g$ is the total N in the grain.

Whole roots of the same four plants used for sap collection were excavated and the nodules were carefully extracted and washed. Precautionary measures were taken during the excavation to avoid collecting roots from adjacent plants. The number of all the visible nodules extracted were counted and expressed per plant.

Also, at midflowering of soybean, the percentage ground surface covered by soybean plants was evaluated using a 1 m × 1 m quadrat. The evaluation was done in two positions in each plot.

Statistical analyses

Statistical analyses were conducted using the mixed model procedure with the restricted maximum-likelihood method (REML) for variance estimates (SAS Institute 2001). Data were analyzed by location and across locations. Locations and soybean cultivars were considered as fixed effects while the five farmers' fields in each location were considered as replications and random effects. Least-square means and standard errors are presented.

Results

Dry matter and N accumulation in soybean in farmers' fields

Soybean grain, grain-N, and shoot yields were similar in both savannas; but shoot-N was higher by 16%, and total amount of N fixed, N accumulated in the plant, and net N balance were 1.4–4 times greater in the

Table 2 Dry matter, N accumulation, N-fixation, N balance, and surface area cover at midflowering among promiscuous soybean cultivars grown at Eglimé (SGS) and Ouake (NGS), 2005

Variety	Dry matter (Mg ha^{-1})		N uptake (kg ha^{-1})		N fixation		Net N balance (kg ha^{-1})	Surface area cover (%)
	Shoot	Grain	Shoot	Grain	Ndfa[a] (%)	Amount (kg ha^{-1})		
Location: Eglimé (SGS)								
TGX 1440-1E	1.30	1.26	16.21	28.57	61.53	27.69	0.03	56.52
TGX 1448-2E	1.44	1.40	22.89	32.99	61.93	34.68	3.46	61.86
TGX 1019-2EB	1.34	1.18	20.15	28.37	64.35	30.69	4.12	55.05
TGX 1844-18E	0.69	0.60	9.45	14.86	65.29	16.32	1.46	66.15
Jupiter	1.09	1.30	20.27	31.89	60.49	31.46	1.23	39.13
Mean	1.17	1.15	17.79	27.34	62.72	28.21	3.52	55.74
SE ±	0.13*	0.14*	2.34*	3.25*	0.90^{ns}	1.20**	1.33^{ns}	4.60*
Location: Ouake (NGS)								
TGX 1440-1E	2.03	1.26	39.45	31.35	72.67	52.10	20.75	68.21
TGX 1448-2E	1.68	1.34	25.67	25.11	76.56	38.56	13.46	69.56
TGX 1019-2EB	1.93	1.14	31.56	24.80	73.48	41.58	16.78	73.91
TGX 1844-18E	1.68	1.18	27.08	24.40	69.85	36.17	11.77	78.08
Jupiter	1.14	0.75	15.80	14.26	72.13	21.69	7.42	56.52
Mean	1.69	1.13	27.91	23.98	72.94	38.02	14.03	69.26
SE ±	0.15*	0.10*	3.87*	2.75*	1.09^{ns}	1.30**	1.30**	3.63*

[a] Ndfa = percentage N derived from atmospheric N_2-fixation

* significant at $p < 0.05$; ** significant at $p < 0.01$; ns, not significant at $p < 0.05$

NGS than in the SGS (Table 2). Soybean cv. TGX 1448-2E (medium-maturing) ranked the highest in dry matter and N accumulation in the SGS. Soybean cv. TGX 1440-1E (late-maturing) gave the greatest dry matter yield, the highest N accumulation in both the grain and the shoot, and fixed the greatest amount of N with the largest net N balance in the NGS (Table 2). Averaging across locations, cv. TGX 1448-2E produced the highest grain yield of 1.37 Mg ha^{-1}.

Nodulation, BNF, and surface area coverage at flowering by soybean in farmers' fields

Nodules evaluated at podding (65–75 DAS) differed significantly between locations, with more than twice the nodules produced in the SGS than in the NGS. The significant interactions of location and soybean cultivar on nodule production (Fig. 1) showed that the greatest number of nodules was produced by cv. TGX 1440-1E at Eglimé (SGS) and the least by cv. TGX 1448-2E at Ouake (NGS), but the popularly grown cv. Jupiter produced the highest nodules in the NGS.

There was no significant effect of location on atmospheric N_2-fixation by the soybean cultivars, even though slightly more N (13%) was fixed in the NGS than in the SGS. Nitrogen fixation was similar among the soybean cultivars, ranging from 61% (cv. Jupiter) at Eglimé in the SGS to 77% (cv. TGX 1448-2E) at Ouake in the NGS (Table 2). Percentage area covered by the plants at midflowering was the lowest in cv. Jupiter (control) at both locations (Table 2), but among the improved cultivars at both locations, TGX 1844-18E covered the greatest surface area of more than two-thirds of the plot.

Rice yield and components of yield as influenced by preceding soybean cultivars

Mean NERICA1 rice paddy (grain) yields differences across the locations were highly significant ($P < 0.01$). Grain yield was more than twice as high in the NGS location than in the SGS (Table 3). Mean grain yield of NERICA1 supplied with only 15 kg N ha^{-1} following previous soybean cv. TGX 1019-2EB or TGX 1440-1E was 75–146% higher than 2-year

Fig. 1 Interaction of environment and genotype on nodule production in promiscuous soybean cultivars, 2005. The error bar indicates the standard error on the mean; significant at $P < 0.05$

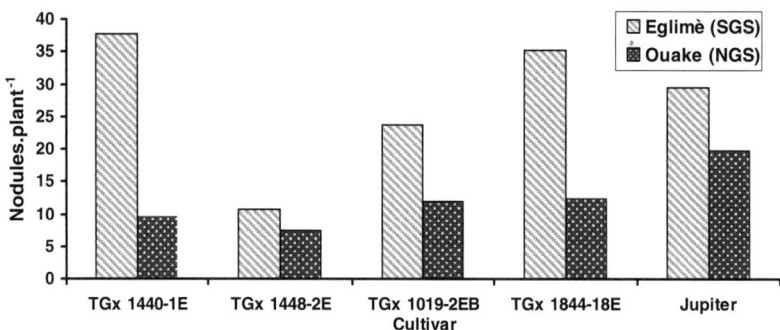

Table 3 Influence of soybean rotation and fallow on NERICA rice yield and some components of yields in farmers' fields in the SGS and NGS, 2006

Source of variation	Tillers (no. m^{-2})	Panicles (no. m^{-2})	Grain yield (Mg ha^{-1})	Straw yield (Mg ha^{-1})	HI
Location					
Eglimé	51	22	0.62	4.17	0.16
Ouake	117	94	1.40	4.56	0.22
Mean	84	58	1.01	4.36	0.19
SE ±	13	12	0.15	0.64	0.02
Rotation[a]					
TGX 1440-1E/NERICA1—15 N	90	65	1.57	5.22	0.23
TGX 1448-2E/NERICA1—15 N	100	68	1.02	5.12	0.16
TGX 1019-2 EB/NERICA1—15 N	87	63	1.74	3.99	0.36
TGX 1844-18E/NERICA1—15 N	95	64	1.37	4.41	0.27
Jupiter/NERICA1—15 N	86	63	1.05	4.41	0.21
NERICA1/NERICA1—15 N	82	56	0.99	3.87	0.19
Fallow/NERICA1—0 N	82	49	0.71	3.65	0.15
IRAT 136/IRAT 136—15 N	75	55	0.45	4.10	0.09
Fallow/IRAT 136—0 N (control)	58	36	0.19	4.53	0.04
Mean	84	58	1.01	4.36	0.19
SE ±	14	14	0.22	0.69	0.03

Source of variation	NDF	DDF	Probability level of F[b]				
			Tillers (no. m^{-2})	Panicles (no. m^{-2})	Grain yield (kg ha^{-1})	Straw yield (kg ha^{-1})	HI
LOC	1	9	0.005	0.002	0.004	0.667	0.013
Rotation	8	62	0.361	0.549	<0.001	0.476	<0.001
LOC × rotation	8	62	0.375	0.407	0.370	0.443	0.431
−2 Res log likelihood			753	708	1019	1154	−86

[a] Rotation systems of soybean-rice or fallow-rice or rice-rice supplied with either 15 kg N ha^{-1} (15 N) or without application of N fertilizer (0 N)

[b] Probability levels are fixed effects. NDF = numerator degree of freedom; DDF = denominator degree of freedom of covariance parameter for replication, its interaction with location, and residuals

continuous cropping of NERICA1 with or without low N (15 kg N ha^{-1}) application. A rotation of either of the two soybean cultivars with NERICA1 supplied with 15 kg N ha^{-1} gave over ninefold the yield obtained with the farmers' practice of growing their popular rice, cv. IRAT 136, without N fertilizer application (Table 3). When analyzed by location, previous medium-maturing soybean, cv. TGX 1019-2EB gave the highest rice grain yield (1.4 Mg ha^{-1}) at Eglimé (SGS), while the highest rice grain yield (2.0 Mg ha^{-1}) was obtained at Ouake (NGS) from rotation with the late-maturing soybean, cv. TGX 1440-1E.

Rice straw yields were similar at the two locations and were not influenced by the previous soybean cultivars. But the previous soybean cultivars influenced ($P < 0.001$) the partitioning of dry matter to the grain as measured by the harvest index (Table 3). The harvest index of NERICA1 in rotation with soybean, cv. TGX 1019-2EB, was the highest. The HI was 30% lower in the SGS than in the NGS (Table 3).

The number of tillers and panicles produced by NERICA1 was not influenced by the previous soybean cultivars (Table 2). Both traits were two- to three-fold higher in the NGS than in the SGS.

Mineral N as influenced by previous soybean and fallow in farmers' fields in the NGS

Mineral N measured at the beginning of 2006, just before the establishment of the rice crop, was 30–74% higher in the previous soybean plots than in the fallow plots in the NGS (Fig. 2). However, there were no significant differences in mineral N among the previous soybean cultivars. Unexpectedly, there was a large amount of mineral N from the previous NERICA1 plots, possibly because of the residual N from the application of the recommended 60 kg N ha^{-1} (urea) which was not taken up by the crop. The rice crop failed in 2005 due to severe drought.

NERICA response to N and NFRV of promiscuous soybean in farmers' fields in the NGS

The N response curves of NERICA1 and IRAT 136 at Ouake in the NGS (Fig. 3) showed that, when N was not applied, grain yield was three times higher for NERICA1 than for IRAT 136. NERICA1 gave a yield advantage of over 1.0 Mg ha^{-1} with application of 45 kg N ha^{-1} compared with IRAT 136. This difference may be attributed to the greater potential of NERICA1 to partition dry matter to the grain than the IRAT 136 cultivar, as indicated by a significantly higher HI in Table 3.

On the basis of the N response curves and the corresponding equations for the two rice cultivars, the 2.0 Mg ha^{-1} paddy yield obtained from rotation plots with soybean cv. TGX 1440-1E was equivalent to the application of 33 kg N ha^{-1} to NERICA1 or an application of over 60 kg N ha^{-1} to the farmers' variety, cv. IRAT 136 (Fig. 2). This gives an N contribution to the rice crop in the range of 33–60 kg N ha^{-1} less the 15 kg N ha^{-1} applied to the rice crop. Therefore, the NFRV of growing soybean cv. TGX 1440-1E in the NGS prior to a rice crop ranged from

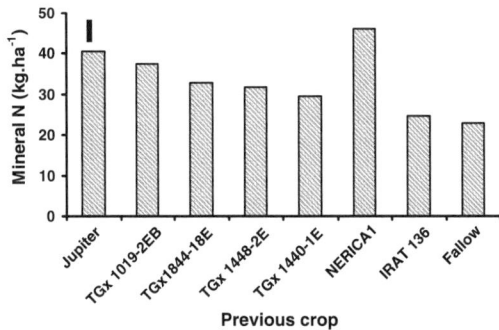

Fig. 2 Influence of previous crops and fallow on soil mineral N before rice planting in 2006 at Ouake (NGS). The error bar indicates standard errors on the mean; significant at $P = 0.09$

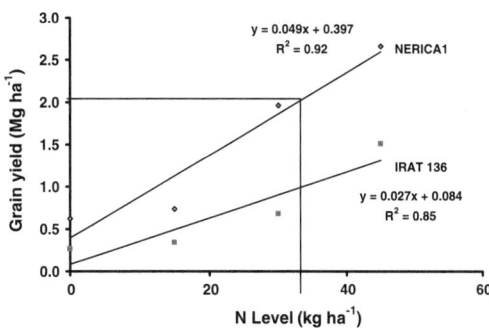

Fig. 3 Response of rice cvs. NERICA1 and IRAT 136 to N application and estimation of nitrogen fertilizer replacement value (NFRV) of soybean cv. TGX 1440-1E using NERICA1 and IRAT 136 as the reference crops in Ouake, NGS, 2006

17 to 45 kg N ha^{-1} depending on the reference rice crop.

Discussion

This study tried to develop an integrated soil fertility management (ISFM) package involving a nitrogen-fixing dual-purpose soybean, limited use of chemical fertilizer, and a resilient rice variety to restore soil fertility to highly degraded savanna soils and enhance rice productivity in fragile upland rice ecosystems of the Republic of Benin. Our results showed significant differences in dry matter and N accumulation among the promiscuous soybean cultivars on these degraded soils. Cultivar TGX 1448-2E, earlier reported to be high-yielding, resistant to frogeye leaf spot (*Cercospora sojina*), and well adopted in northern Nigeria (Sanginga et al. 2003) gave the best yield in this study.

The grain yields and proportions of N fixed by the promiscuous soybean cultivars used in this study were similar to the values reported for promiscuous soybean grown without inoculation in the Guinea savanna of Nigeria (Sanginga et al. 1997). However, the total N accumulation reported by Sanginga et al. (1997) was two- to threefold the values obtained in the present study. The reason may be due to the greater adaptation and enhanced colonization of the roots of the promiscuous soybean cultivars by the indigenous *Bradyrhizobium* sp. (TGX) populations in the on-station, savanna experimental site used for their study. Moreover, the cultivars were bred under similar environmental conditions to those used in the Sanginga et al. (1997) study compared with our study environments.

The high potential of cv. TGX 1844-18E to cover more than two-thirds of the soil surface by midflowering showed that this cultivar could play a significant role in smothering weeds in legume-rice systems where significant investment is made in the weeding of upland rice by hand. However, the weed-suppressive potential of the different soybean cultivars and its impact on the succeeding rice crop was not part of this evaluation.

Farmers have often been reluctant to devote their land solely to legume cover crops that would not provide food value for human consumption in spite of the positive benefits in weed suppression (Becker and

Johnson 1998, 1999) and impact in restoring soil fertility (Oikeh et al. 1998; Vanlauwe et al. 2001a). Therefore, the use of dual-purpose grain soybean varieties with potential for good surface cover to suppress weeds and for soil conservation particularly in the NGS and with grains for immediate economic benefits may be attractive to smallholder farmers in the Republic of Benin.

The previous soybean cultivars used in this study contributed N resource in the range of 30–40 kg N ha^{-1} to the succeeding rice crop, mostly from fallen leaf litter and root residues. A similar contribution of 33–40 kg mineral N ha^{-1} to lowland rice from previous legumes was reported in the Philippines by Ladha et al. (1996). However, legume residues containing 81–160 kg N ha^{-1} were returned to the soil prior to the rice crop in their study. Our values are, however, lower than the 75 kg N ha^{-1} earlier reported for a similar system in which the promiscuous soybean was harvested and exported from the field but the leaf litter and roots contributed N to the soil in soybean-maize rotation systems in the NGS of Nigeria (Oikeh et al. 1998).

The highest amount of soil N from previous plots of soybean cv. Jupiter (control) may have been due to an N-sparing effect. Even though all the soybean cultivars derived similar amounts of N from the atmosphere in the NGS, the total amount of N fixed and the N taken up by cv. Jupiter was one-half of the N absorbed by the promiscuous soybean cultivars, possibly due to limited N-use efficiency in cv. Jupiter, which merits further investigation.

The observed general performance of both soybean and rice in the NGS compared with that in the SGS was as expected. An earlier study had reported that maize had greater potential in the NGS than in the SGS because of higher levels of intercepted solar radiation, lower night temperatures, and reduced incidence of pests and diseases (Kowal and Kassam 1978). However, these parameters were not evaluated in the present study.

The mean grain yields obtained in this study for NERICA1 in rotation with the promiscuous soybean were higher than the values (0.3–1.1 Mg ha^{-1}) reported from rotation of upland rice after a 1-year fallow using a wide range of legumes in the NGS of Côte d'Ivoire (Becker and Johnson 1998). In their studies, the legumes were ploughed into the soil prior to the establishment of the rice crop, in contrast to the present study in which the soybean plants were harvested and exported from the field. Only the leaf litter and the roots contributed to the N economy of the rice in our study. Although the total plant N was similar in both studies, the percentage of atmospheric N_2-fixation estimated among the soybean cultivars used in our study was slightly higher (61–77%) than the values (32–72%) from the different fallow legumes used in the study of Becker and Johnson (1998). The difference in rice yields between both studies, therefore, could be attributed to the difference in the test crops used. Whereas the popular *O. sativa* rice (WAB-56-50) was used in the study of Becker and Johnson (1998), the present study used resilient, higher-yielding interspecific rice (NERICA1) bred specifically for low-management production systems with limited use of purchased inputs.

The N response curves of NERICA1 and IRAT 136 (*O. sativa*) in the NGS further confirmed the yield superiority of NERICA1. Although in the current study both NERICA1 and IRAT 136 produced similar number of tillers and panicles, NERICA1 gave higher grain yield at zero N and at 45 kg N ha^{-1} because of its greater potential to partition dry matter to the grain as indicated by a significantly higher HI than the IRAT 136 cultivar. A recent study also reported NERICA1 to be tolerant to mild drought and low N in the forest agroecosystems of Nigeria (Oikeh et al. 2008b).

The present study showed that the integrated use of promiscuous soybean cv. TGX 1019-2EB or TGX 1440-1E with resilient NERICA1 supplied with low N (15 kg ha^{-1}) gave a yield advantage of 1.5 Mg ha^{-1} compared with farmers' practice of growing their popular rice, cv. IRAT 136, without N fertilizer application. The positive effect of previous soybean to the NERICA rice could be attributed to N contribution by the soybean cultivars used in the present study as reported earlier (Carsky et al. 1997; Oikeh et al. 1998; Vanlauwe et al. 2001b) and partially due to the nitrate-sparing effect (Peoples et al. 1995) resulting from the previous soybean absorbing less soil N than the previous rice crop. In addition, soybean has a large belowground biomass including nodules which release N to subsequent crop through biomass decomposition (Sanginga et al. 1997; Abaidoo et al. 1999, 2007). Furthermore, other positive non-N benefits from soybean have been reported in northern Nigeria, including reduction in nematodes and *Striga hermonthica* (witchweed)

parasitism in cereals (Carsky et al. 2000; Weber et al. 1995a, b).

The benefit of integrating dual-purpose promiscuous soybean into an ISFM package to improve the N economy of the succeeding upland rice was further amplified in the NFRV, which showed that growing soybean cv. TGX 1440-1E in the NGS prior to a rice crop can save smallholder farmers as much as 17–45 kg N ha^{-1}. Although an economic analysis of this rice ISFM technology was not carried out in this study, a similar ISFM technology for maize-based systems in the derived savanna of Nigeria was reported to be highly profitable, with a net benefit as high as US $1233 ha^{-1} (Oyewole et al. 2001).

In conclusion, in production systems where soybean plants are harvested and exported from the field by the farmers, the integrated use of promiscuous soybean cv. TGX 1019-2EB or TGX 1440-1E with NERICA1 supplied with low N (15 kg ha^{-1}) can provide mineral N and enhance rice yields by as much as 1.5 Mg ha^{-1} compared with farmers' practice of growing their popular rice, cv. IRAT 136, without N fertilizer application.

Acknowledgements Financial support from UNDP-IHP Phase II Project and the African Development Bank through the African Rice Initiative is acknowledged. WARDA Manuscript No.: 070707.

References

Abaidoo RC, Dashiell KE, Sanginga N, Keyser HH, Singleton PW, Dashiell KE et al (1999) Time course of N$_2$-fixation of promiscuous soybean genotypes measured by the isotope-dilution method. Biol Fertil Soils 30(3):187–192. doi:10.1007/s003740050607

Abaidoo RC, Sanginga N, Okogun JA, Kolawole GO, Diels J, Tossah BK (2007) Evaluation of soybean genotypes for variations in phosphorus use efficiency and their contribution of N and P to subsequent maize crop in three agroecological zones of West Africa. In: Demand driven technologies for sustainable maize production in West and Central Africa, Proceedings of the 5th Biennial West and Central Africa Regional Maize Workshop, IITA-Cotonou, Benin, 3–6 May, 2005, WECAMAN/IITA, Ibadan, Nigeria

Becker M, Johnson DE (1998) Legumes as dry season fallow in upland rice-based systems of West Africa. Biol Fertil Soils 27:358–367. doi:10.1007/s003740050444

Becker M, Johnson DE (1999) The role of legume fallows in intensified upland rice-based systems in West Africa. Nutr Cycle Agroecosyst 53:71–81. doi:10.1023/A:10097675 30024

Buresh RJ, Smithson PC, Hellums DT (1997) Building soil phosphorus capital in Africa. In: Buresh RJ, Sanchez PA, Calhoun F (eds) Replenishing soil fertility in Africa. SSSA Special Publication 51. Soil Sci Soc Am, Madison, WI, USA, pp 111–149

Carsky RJ, Abaidoo R, Dashiel K, Sanginga N (1997) Effect of soybean on subsequent maize yield in the Guinea savanna zone of West Africa. Afr Crop Sci J 5:31–38

Carsky RJ, Berner DK, Oyewole BD, Dashiell K, Schulz S (2000) Reduction of *Striga hermonthica* parasitism on maize using soybean rotation. Int J Pest Manag 46: 115–120. doi:10.1080/096708700227471

Dingkuhn M, Jones MP, Johnson DE, Sow A (1998) Growth and yield potential of *O. sativa* and *O. glaberrima* upland rice cultivars and their interspecific progenies. Field Crops Res 57:57–69. doi:10.1016/S0378-4290(97)00115-9

Ekeleme F, Kamara AY, Oikeh SO, Chikoye D, Omoigui LO (2007) Effect of weed competition on upland rice production in northeastern, Nigeria. Afr Crop Sci Proc 8: 61–65. African Crop Science Society, El-Minia, Egypt

Fagade SO (2000) Yield gaps and productivity decline in rice production in Nigeria. International Rice Commission, FAO, pp 15

FAO (2007) FAO rice market monitor, vol X, no. 1. Available via http://www.fao.org/es/esc/en/index.html (click on rice)

International Institute of Tropical Agriculture (IITA) (1989) Automated and semi-automated methods for soil and plant analysis. Manual series, no. 7. IITA, Ibadan, Nigeria

Kowal JM, Kassam AH (1978) Agricultural ecology of savanna: a study of West Africa. Oxford University Press, Oxford

Ladha JK, Kundu DK, Angelo-Van Coppenolle MG, Peoples MB, Carangal VR (1996) Legume productivity and soil nitrogen dynamics in lowland rice-based cropping systems. Soil Sci Soc Am J 60:183–192

Oikeh SO, Chude VO, Carsky RS, Weber GK, Horst WJ (1998) Legume rotation in the moist tropical savanna: managing soil N dynamics and cereal yield in farmers' fields. Exp Agric 34:73–83. doi:10.1017/S0014479798001021

Oikeh SO, Somado EA, Sahrawat KL, Toure A, Diatta S (2008a) Rice yields enhanced through integrated management of cover crops and phosphate rock in P-deficient Ultisols in West Africa. Commun Soil Plant Anal 39 (in press)

Oikeh SO, Nwilene F, Diatta S, Osiname O, Touré A, Okeleye KA (2008b) Agro-physiological responses of upland NERICA rice to nitrogen and phosphorus fertilization in the forest agroecosystem of West Africa. Agron J 100(3):735–741. doi:10.2134/agronj2007.0212

Okeleye KA, Adeoti AYA, Tayo TO (2006) Farmers' participatory rice variety selection trials at Ibogun Olaogun village, Ogun State, Nigeria. Int J Trop Agric 24:643–649

Oyewole B, Schulz S, Tanko R (2001) Economic assessment of cereal-legume rotations in two agroecozones. Improvement of high intensity food and forage crop systems. IITA

Annual Report 2001, International Institute of Tropical Agriculture, Ibadan, Nigeria

Peoples MB, Crasswell ET (1992) Biological nitrogen fixation: investments, expectations and actual contributions to agriculture. Plant Soil 141:13–39. doi:10.1007/BF00011308

Peoples MB, Faizah AW, Rerkasem B, Herridge DF (1989) Methods for evaluating nitrogen fixation by nodulated legumes in the field. ACIAR Monograph No 11, ACIAR, Canberra, Australia

Peoples MB, Herridge DF, Ladha JK (1995) Biological nitrogen fixation: an efficient source of nitrogen for sustainable agricultural production? Plant Soil 174:3–28. doi: 10.1007/BF00032239

Rodenburg J, Diagne A, Oikeh S, Futakuchi K, Kormawa PM, Semon M et al (2006) Achievements and impact of NERICA on sustainable rice production in sub-Saharan Africa. Int Rice Comm Newsl 55:45–58

Sanginga N, Dashiell KE, Okogun JA, Thottapilly G (1997) Nitrogen fixation and N contribution by promiscuous nodulating soybeans in the southern Guinea savanna of Nigeria. Plant Soil 195:257–266. doi:10.1023/A:1004207530131

Sanginga N, Okogun JA, Vanlauwe B, Dashiell KE (2002) The contribution of nitrogen by promiscuous soybeans to maize-based systems in the moist savanna of Nigeria. Plant Soil 241:223–231. doi:10.1023/A:1016192514568

Sanginga N, Dashiell KE, Diels J, Vanlauwe B et al (2003) Sustainable resource management coupled to resilient germplasm to provide new intensive cereal-grain-legume-livestock systems in the dry savanna. Agric Ecosyst Environ 100:305–314. doi:10.1016/S0167-8809(03)00188-9

SAS Institute (2001) SAS technical report. SAS/STAT software: changes and enhancements. Release 8.02. SAS Inst., Cary, NC, USA

Somado EA, Becker M, Kuehne RF, Sahrawat KL, Vlek PLG (2003) Combined effects of legumes with rock phosphorous on rice in West Africa. Agron J 95:1172–1178

Vanlauwe B, Aihou K, Houngnandan P, Diels J, Sanginga N, Merckx R (2001a) Nitrogen management in adequate 'input' maize-based agriculture in the derived savanna benchmark zone of Benin Republic. Plant Soil 228:61–71. doi:10.1023/A:1004847623249

Vanlauwe B, Aihou K, Aman S, Iwuafor ENO, Tossah BK, Diels J et al (2001b) Maize yields as affected by organic inputs and urea in the West African moist savanna. Agron J 93:1191–1199

Weber G, Elemo K, Lagoke STO, Awad A, Oikeh S (1995a) Population dynamics and determinants of *Striga hermonthica* on maize and sorghum in savanna farming systems. Crop Prot 14:283–290. doi:10.1016/0261-2194(94)00004-R

Weber GK, Chindo PS, Elemo KA, Oikeh S (1995b) Nematodes as production constraints in intensifying cereal-based cropping systems of the northern Guinea savanna. Resource and Crop Management Research Monograph No.17. IITA, Ibadan, Nigeria

Nitrogen Use in Maize (*Zea mays*)–Pigeonpea (*Cajanus cajans*) Intercrop in Semi-arid Conditions of Kenya

S.W. Wanderi, M.W.K. Mburu, S.N. Silim, and F.M. Kihanda

Abstract   A field experiment was conducted at Jomo Kenyatta University of Agriculture and Technology between 2001 and 2002 to determine nitrogen use in maize–pigeonpea intercrop system. The experiment was laid out as a randomized complete block design replicated four times. Treatments included two pigeonpea maturity types: two long-duration types (erect and semi-erect) and one medium-duration type intercropped with maize (Katumani Composite) or sole crop. Data on plant total N uptake, litter fall, N fixed and soil mineral N at key phenological stages were determined. Results showed that intercropping maize and pigeonpea increased maize grain N concentration compared to sole maize, an indication of nutritional quality improvement. Long-duration cultivars had the highest plant N uptake and contributed high amount of N through litter fall and biological fixation compared to medium duration. Soil mineral N increased over time, probably due to soil N mineralization or pigeonpea N contribution through litter fall decomposition which ranged from 3.9 to 7.6 t/ha. Maize yield and N uptake in subsequent season after pigeonpea were higher in plots previously planted with pigeonpea than those planted continuously with maize. In conclusion, this study showed that long-duration pigeonpeas may play an important role in low-input maize production systems primarily through N cycling (probably through capture of deep soil N pool and litter) and through biological nitrogen fixation and this improves maize yield and quality.

Keywords Nitrogen uptake · N fixation · Maize–pigeonpea intercrop · Residual effect · Soil mineral N

Introduction

Nitrogen is one of the major nutrients that limit crop production in Kenya because of the high demand by crops and off-farm export through crop harvests (Giller et al., 1997; Sanchez et al., 1997). Farmers rarely use chemical fertilizers in cereal-based production systems because of high cost, lack of credit, poor transport and marketing infrastructure. Organic sources of N are a practical soil fertility remedial option for small-scale farmers but are not available in large quantities (Kapkayai et al., 1998). Legume intercrops are a source of plant N through atmospheric fixation that can offer a practical complement to inorganic fertilizers (Jerenyama et al., 2000) and reduce competition for N from cereal component (Fujita et al., 1992). Legumes also contribute to the economy of intercropping systems either by transferring N to the cereal crop during the growing period (Ofori et al., 1987) or as residual N that is available to the subsequent crop (Papastylianou, 1988). One such leguminous crop is pigeonpea widely grown in semi-arid areas as an intercrop with cereals such as maize, sorghum and millet. Pigeonpea has a great potential to replenish soil N through atmospheric nitrogen fixation and litter decomposition (Giller et al., 1997) and avails extra N in the low-input production system. However, N utilization in the intercrop system depends on several factors such as soil type, crop varieties, crop management and competitive ability of the component crop (Kumar Rao et al., 1996).

S.W. Wanderi (✉)
Kenya Agricultural Research Institute (KARI), Embu, Kenya
e-mail: wanderi_susan@yahoo.com

A. Bationo et al. (eds.), *Innovations as Key to the Green Revolution in Africa*,
DOI 10.1007/978-90-481-2543-2_58, © Springer Science+Business Media B.V. 2011

In addition, the positive responses of cereals following legumes have been attributed largely to enhanced availability of N to the cereal crop (Sanginga, 2003). Therefore, in order to develop management strategies that would improve the productivity and sustainability of the intercrop system, information on N utilization in maize intercropped with pigeonpeas is needed. This research was therefore designed to determine nitrogen use in maize–pigeonpea intercrop system and the residual effect of pigeonpeas on maize yield.

Materials and Methods

Study Area

This study was conducted at Jomo Kenyatta University of Agriculture and Technology, Thika. The area is located at 1549 m above sea level and it receives an average annual rainfall of about 768 mm p.a. with an average maximum temperature of 24.1°C and minimum temperature of 13.5°C and evaporation of 105 mm p.a. The soils are described as Eutric Cambisol with dark brown colour on all the horizons with sandy clay texture.

Experimental Design

The experiment was laid out as a randomized complete block design with seven treatments replicated four times. The treatments were the following: three pigeonpea varieties (one medium – ICEAP 00557 and two long duration [semi-erect – ICEAP 00040 and erect – ICEAP 00053]) either sole or intercropped with maize (Katumani Composite). The intercrop planting arrangement was three rows of maize interspersed with two rows of pigeonpea (75 cm by 20 cm). Farmers accept this planting arrangement because it does not reduce maize yields, which is considered to be the main crop (ICRISAT, 1995). The intercrop plots were 10 m long by 4 m wide, and the monocrop plots were 10 m long by 3 m wide. The plant population of maize was 44,289 and 26,700 plants/ha while pigeonpea was 66,633 and 26,500 plants/ha in the sole and intercrop plots, respectively. The first experiment was sown during the short rains (30 October 2001) while the second maize season was sown in the same plots on 28 February 2002. The residual effect experiment was sown on 18 October 2002 to check the residual nutrient effects of the pigeonpeas on maize yield and N uptake.

Plant Sampling and Analysis

Four plants of pigeonpeas and maize were sampled from each plot at the middle of the season, i.e. 115 days after planting (DAP), and at the end of the season (220 DAP), ensuring that the samples were taken far from the gap. The samples were taken to the laboratory where they were dried in an oven at 70°C and their dry mass determined. The oven-dried plant samples were later finely ground and weighed into 0.3 g digestion tubes and mixed with salicylic acid, hydrogen peroxide, concentrated sulphuric acid and selenium powder. The tubes were put into a digestion block and heated to 110°C for 1 h and cooled by adding hydrogen peroxide. Temperature was later raised to 330°C for the solution to turn colourless. The tubes were removed from the digestion block and allowed to cool to room temperature. Twenty-five millilitres of distilled water was added and mixed well until no more sediment dissolved. The tubes were allowed to cool and the content made up to 50 ml with distilled water. The tubes were allowed to settle, and a clear solution was taken from the top of the tube for analysis. Total N was later determined using steam distillation as described by Okalebo et al. (2002).

Litter Fall Collection

Litter fall was determined in the entire season by putting 1 m^2 wire mesh below the pigeonpea canopy in both the sole and the intercrop to collect the falling leaves. Litter fall was collected after every 14 days and later oven dried at 70°C (to constant weight) to determine dry mass and ground for total N determination as described by Okalebo et al. (2002).

Estimation of Biological Nitrogen Fixed

N fixation was determined using the N difference method (Giller and Wilson, 2001) using cotton as the reference crop. The N derived from atmospheric fixation (Ndfa) was calculated as the difference between total N in the pigeonpea crop and total N in the non-fixing crops. Cotton (Hart 89 M and Uka 59/146) was used as the reference crop. Preliminary identification of an appropriate reference crop in a greenhouse study was done as by Wanderi et al. (2002).

Soil Sampling and Analysis

Soil samples were taken 1 week before planting, at the middle and at the end of the crop-growing season. The soil samples per plot were collected at 0–20, 20–50 and 50–100 cm, and two samples per plot were bulked. The soil samples for total N were air dried, ground manually with a pestle and mortar and passed through a 2 mm sieve. A soil sample (0–20 g) was taken and digested and analysed for total N using micro-Kjeldahl method (Okalebo et al., 2002). Soil samples for soil mineral N determination were stored in iceboxes for transportation and later in a deep freezer for mineral N determination. Thirty grams of the thawed soil was weighed into a 200 ml plastic container and 100 ml 2 M KCl added to the soil and put in a shaker for 1 h.

The contents were filtered. NO_3^--N and NH_4^+-N were determined by steam distillation of the KCl extract using 0.4 g Devarda's alloy and 0.2 g MgO as catalysts. The distillate was collected in H_3BO_3 solution and titrated with dilute H_2SO_4 for the N determination.

Data Analysis

Data were analysed using GENSTAT (1995), and means were separated using LSD test at a significance level of 5%.

Results and Discussion

Total Nitrogen Uptake

Maize N uptake was higher in the second season than in the first season (Table 1). It is likely that the higher soil moisture content permitted maize to absorb more N which is reflected by the higher total rainfall received in the second season (618 mm) compared to the first season (148 mm) (Fig. 1). Intercropping maize with long-duration erect pigeonpea in the second season significantly influenced N uptake, but this was not observed in maize intercropped with medium-duration pigeonpea. This could be attributed to larger

Table 1 Maize N partitioning (kg/ha) and nitrogen concentration (% N) of cobs, grains and stems in the intercrop and sole crop at 115 and 220 DAP (JKUAT, Thika)

Cropping system	Crop	N concentration (%)			Amount of N (kg/ha)			
		Stem	Cob	Grain	Stem	Cob	Grain	Total
Season 1								
Sole crop	MZ	0.55	0.43	0.66	17.1	4.81	21.8	43.7
Intercrop	MMD	0.53	0.39	1.06	12.6	2.97	25.2	40.8
	MLDSE	0.48	0.26	0.99	7.7	1.69	24.7	34.1
	MLDE	0.52	0.27	1.17	8.8	1.49	18.7	29
	SED	0.06	0.11	0.21	5.58	1.22	5.96	5.58
Season 2								
Sole crop	MZ	1.61	1.15	1.05	42.9	3.11	29.5	75.5
Intercrop	MMD	1.38	1.10	1.66	34.7	1.91	32.9	69.5
	MLDSE	1.17	1.33	1.37	38.1	2.31	18.4	58.8
	MLDE	0.74	1.26	1.49	22.1	1.81	23.0	46.9
	SED	0.79	0.73	0.73	16.88	1.89	12.4	17.43

MZ is maize sole whereas MLDSE, MLDE and MMD represent maize intercropped with long-duration semi-erect, long-duration erect and medium-duration pigeonpea, respectively
SED standard error of difference

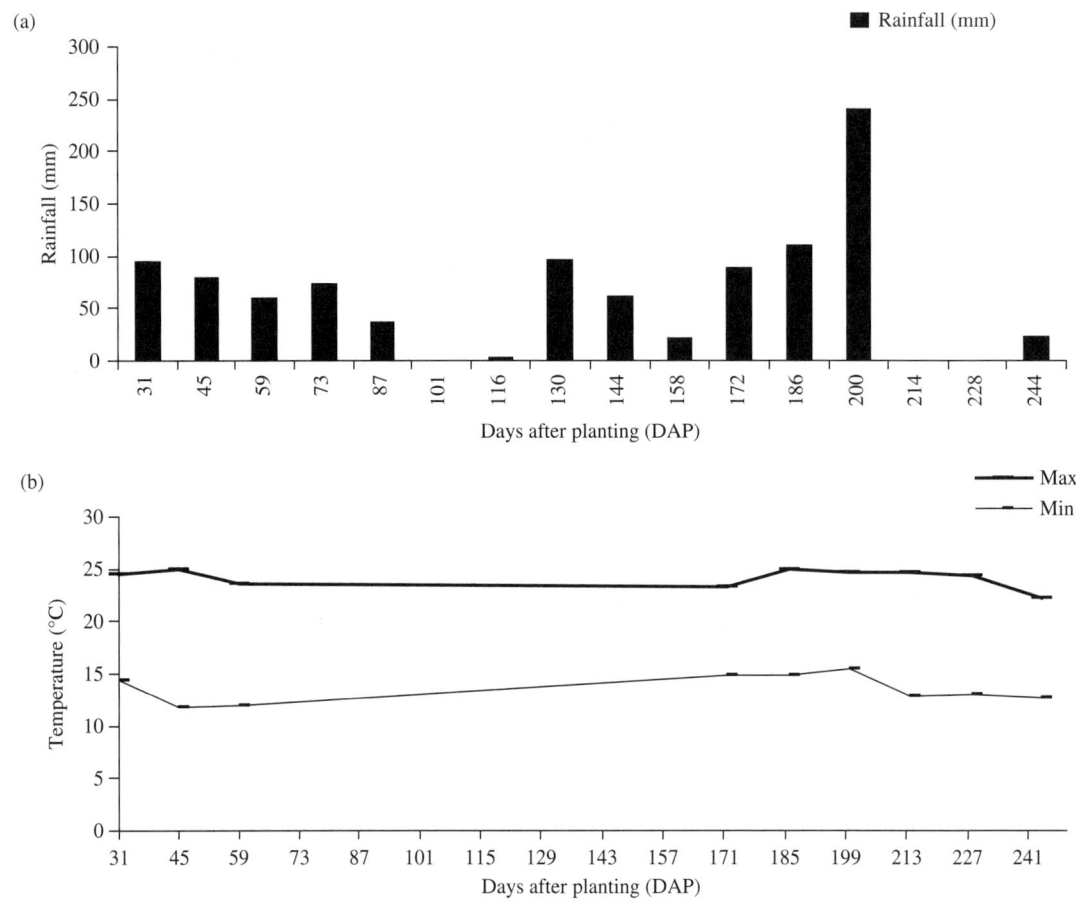

Fig. 1 (**a**) Biweekly rainfall (mm) and (**b**) diurnal minimum and maximum temperature (°C) amount from October 2001 to June 2002 at JKUAT, Thika

canopy of the former than the latter reducing dry matter accumulation for maize resulting in low N uptake. This implies that the medium-duration pigeonpea can be intercropped with maize without substantial maize N reduction. Intercropped maize had higher grain N concentration (1.6%) compared to sole maize (1.1%), an indication of improved grain nutritional quality. Higher N uptake of pigeonpeas than maize was probably due to capture of soil N deep in the profile and self-fertilization from the litter fall and biological N fixation. Long-duration pigeonpeas' total N uptake (1266 kg N/ha) was significantly higher ($P = 0.01$) than for medium-duration pigeonpea (345 kg N/ha) at the end of the season (Table 2). This is because total N of legumes is related to maturity other than the location or the season (Taylor et al., 1982), and the early duration of the medium duration may have been responsible

for the lower N. Long-duration pigeonpea had low N concentration in stem (2.2%) and husk (1.4%) but higher grain N concentration of 4.5% compared to 3.6% in the medium duration, an indication of efficient N remobilization to the grains. Though pigeonpea grains had higher N concentrations (3–4%) than stems (1.65–2.99%), stems had the highest total N yield. This is because N uptake depended more on dry matter production rather than to changes in N concentration, indicating that dry matter yield is the overriding factor influencing total N uptake rather than N concentration (Holderbaum et al., 1990). Pigeonpea allocation of total N uptake was 82, 2 and 16% to the stems/leaves, husks and grains, respectively, excluding litter mass. This therefore implies that allowing pigeonpea stem to decompose in the field rather than using them as fuel would play a vital role in farm N economy.

Table 2 Pigeonpea N partitioning (kg/ha) and nitrogen concentration (% N) of husks, grains and stems in the intercrop and sole crop at 220 DAP at JKUAT, Thika

220 DAP		N concentration (%)			Amount of N (kg/ha)			
		Stem + leaves	Husk	Grain	Stem + leaves	Husk	Grain	TDM
Sole crop	MD	2.43	1.59	3.67	249	10.49	95.6	345
	LDSE	2.38	1.38	3.67	992	13.07	214.7	1220
	LDE	2.24	1.38	4.53	1046	19.74	200.6	1266
Intercrop	MD	2.3	1.35	3.61	268	7.99	65.5	341
	LDSE	1.68	1.39	2.33	623	7.45	52.9	683
	LDE	2.11	1.26	4.20	859	16.71	101.2	977
	SED	0.38	0.15	0.37	129.3	1.1	21.4	133.9

MD medium duration, LDSE long-duration semi-erect, LDE long-duration erect pigeonpea, TDM total dry matter, SED standard error of difference

Litter Fall

Long-duration pigeonpeas litter mass ranged from 3.9 to 7.6 t/ha while for medium duration it was 2.7 t/ha (Table 3). Intercropping reduced litter fall because of lower plant population density in the intercrop compared to sole crop. The sole and the intercrop of long-duration erect pigeonpea had higher leaf fall than either the long-duration semi-erect or the medium-duration pigeonpea. This is an indication of excessive late season litter fall as from 186 days after planting. Total N (kg/ha) supply expected from litter fall based on litter biomass and nitrogen concentration % N ranged from 56 to 132 kg N/ha with the long-duration erect having the highest N (kg/ha), therefore having a substantial contribution to soil fertility through litter decomposition. Other values have been reported (64–86 kg N/ha) in the leaves that fall during the growth of pigeonpeas (Kumar Rao et al., 1996).

Nitrogen Derived from Atmospheric Fixation (Ndfa)

The amount of nitrogen fixed derived from the atmosphere (Ndfa) by pigeonpea was calculated using two cotton varieties (Hart 89 M and Uka 59/146) as the reference crops. N derived from fixation followed the same pattern as total biomass accumulation, high in cultivars with high dry matter (long-duration pigeonpea 39–74 kg/ha) and low in cultivars with low dry matter (medium-duration pigeonpea 4.8–7.3 kg/ha) (Table 4). These results corroborate findings of Kumar Rao (1990) that long-duration varieties fix more N than early-maturity groups. The Ndfa in medium-duration pigeonpea was unreliable because the reference crop (cotton) grew too large, hence the underestimated amount of N fixed. In addition, the small amount of N fixed by medium-duration pigeonpea cultivars was attributed to their low biomass productivity and early

Table 3 Seasonal litter fall (kg/ha) and total N (kg/ha) of pigeonpea at JKUAT, Thika

Cropping system	Pigeonpea variety	N %	Litter (kg/ha)	Total N (kg/ha)
Sole crop	MD	2.30	2650	61
	LDSE	2.45	3870	94
	LDE	1.68	7600	132
Intercrop	MD	2.81	1760	50
	LDSE	2.82	3370	96
	LDE	1.8	5180	93
SED		0.3	698	21.1

N % represents nitrogen concentration
MD medium, LDSE long-duration semi-erect, LDE long-duration erect pigeonpea, SED standard error of difference

Table 4 Estimated amount of N fixed by different pigeonpea varieties using two cotton varieties (Hart 89 M and Uka 59/146) as the reference crops at JKUAT, Thika

| | Amount of N fixed (kg/ha) | | | |
| | 115 DAP | | 220 DAP | |
Variety	Hart	Uka	Hart	Uka
LDSE	39.1	48	39.2	38.4
LDE	−12.3	−9.8	74.2	73.4
MD	4.8	7.3	−12.2	−13
SED	15.9	44.0	14.8	14.8

MD medium, LDSE long-duration semi-erect, LDE long-duration erect pigeonpea, SED standard error of difference

maturity. In the present study long-duration erect produced the highest total dry matter, which probably means that it made more photosynthates available to the nodules and hence fixed more atmospheric nitrogen. This also suggests that as with grain yield, high total dry matter ensures higher atmospheric N fixation. In the intercrop system, no significant difference was observed in biological N fixed.

Soil Mineral N

Soil mineral N increased over time and down the profile throughout the experimental period (Fig. 2), probably due to mineralization that could be related to increased rainfall as from 130 days after planting (DAP) to 200 DAP (Fig. 1). Soil mineral N has been reported to increase with increasing soil water content (Ma et al., 1999). Most of the increase was found in NH_4^+-N fraction compared to NO_3^--N fraction; probably NO_3^--N was the preferential uptake of the crops or it was leached down the profile. Nitrate ions move freely to the plant root either by diffusion or by mass flow, and most of the plants absorbed nitrogen in this form (Tisdale et al., 1990). However, NO_3^--N continued to decline down the profile in the long-duration pigeonpea plots compared to maize plots (Fig. 1), maybe the former took up more N from the deeper layers of the soil than the latter, thereby minimizing the competition for nutrients. The highest levels of N as both NO_3^--N and NH_4^+-N at 0–20 cm depth were observed in the medium-duration plots, which reflects lower demand of N or less efficient in N uptake as compared to the long-duration types. Soil mineral N was higher in pigeonpea plots than maize plots, possibly due to contribution through litter fall decomposition.

Residual Effect of Pigeonpea on Subsequent Maize Growth

Maize yield increased by about 15% from plots previously intercropped with long-duration semi-erect than after maize. Similarly, total N uptake and total biomass increased by about 39% from plots previously intercropped with long-duration semi-erect than after plots with continuous maize crop except in the intercropped plots with the medium-duration pigeonpea (Table 5). The increase may be related to the N added in the soil through decomposition of the litter fall, and from this study the amount of N from litter fall ranged from 50 to 132 kg N/ha. Kumar and Goh (2000) reported that the magnitude of the yield increase of the subsequent crop is related to the amount of material returned to the soil.

Conclusions and Recommendations

Cotton reasonably estimated N fixed by long-duration pigeonpea, but it was difficult to estimate N fixed by the medium-duration pigeonpea because cotton accumulated more biomass, hence the underestimated amount of N fixed. Therefore, a more reliable method is required to estimate the amount of N fixed, especially for medium-duration pigeonpea.

Intercropping maize and long-duration types of pigeonpeas in the first season followed by sole crop of pigeonpea in the second season would be more suitable to reduce N competition effect because of large pigeonpea canopy in semi-arid conditions. However, the medium-duration pigeonpea can be intercropped with maize without substantial maize N reduction. Intercropping also increased maize grain N concentration compared to sole maize, indication of nutritional quality improvement.

Soil mineral N increased over time, due to mineralization of soil N and possibly pigeonpea contributing N through litter fall decomposition and/or recycling from deep soil horizons. N contribution through the litter fall

Fig. 2 Soil mineral nitrogen (kg/ha) in the soil profile at the beginning (before planting), middle (115 DAP) and end of the season (220 DAP) at depths 0–20, 20–50 and 50–100 cm, respectively. **a, b** and **c** represent maize, medium-duration and long-duration erect sole and intercropped plots, respectively, at JKUAT, Thika. *Bars* represent SED values ($P = 0.05$)

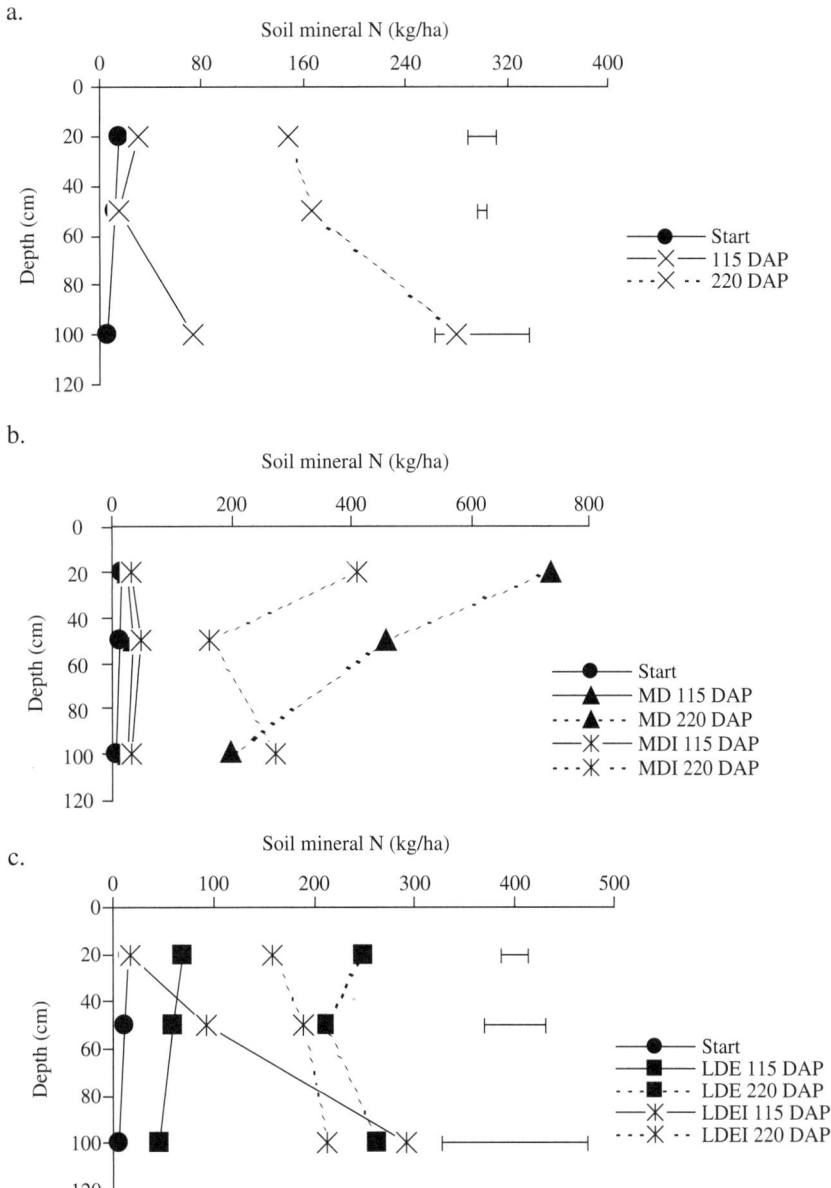

was beneficial to subsequent maize crop in plots previously intercropped with pigeonpea compared with continuous maize cropping. This would make a maize–pigeonpea intercrop in the first season followed by sole pigeonpea in the second season and by sole maize in the third season attractive to farmers.

Long-duration pigeonpea may play an important role in low-input maize production systems primarily through N cycling (probably through capture of deep soil N pool) and to a lesser extent through biological nitrogen fixation, and this improves maize yield and quality. In addition, pigeonpea stems would play an important part in improving soil fertility if incorporated back to the soil or used as animal beddings. However, decomposition rates of the stems need to be determined.

Table 5 Maize dry matter accumulation and N uptake grown after harvesting maize and pigeonpea at JKUAT, Thika

Cropping system	Crop	Dry matter (kg/ha)				N uptake (kg/ha)			
		Stem	Cob	Grain	TDM	Stem	Cob	Grain	TN
Continuous plot	MZ	1187	450	2161	3380	19.1	5.98	35.9	54.2
Sole plots	MD	1302	412	1583	3297	21.0	5.49	26.3	52.7
	LDSE	863	322	1524	2200	13.9	4.17	25.3	35.4
	LDE	1744	555	1134	3576	28.1	6.98	18.9	57.1
Intercrop plots	MMD	1125	368	1917	3410	18.1	6.05	31.8	53.4
	MLDSE	1674	529	2506	4709	27.0	7.03	41.6	75.6
	MLDE	1476	575	2241	3870	23.8	7.64	37.2	61.8
	SED	296.3	88.2	403.6	696.7	88.2	1.00	6.70	11.54

MZ is sole maize whereas MLDSE, MLDE and MMD represent maize intercropped with long-duration semi-erect, long-duration erect and medium-duration pigeonpea, respectively

TN total nitrogen uptake, *TDM* total dry matter, *SED* standard error of difference

Acknowledgement The authors are grateful to Rockefeller Foundation for funding the research project and to the University of Nairobi and ICRISAT for technical support during implementation.

References

Fujita K, Ofusu-Budu KG, Ogata S (1992) Biological nitrogen fixation in mixed legume-cereal cropping systems. Plant Soil 141:155–175

Genstat (1995) Genstat 5 release 3.2 for windows 95. Lawes Agricultural Trust, Rothamsted Experimental Station, Hertfordshire

Giller KE, Cadisch G, Ehaliotis C, Sakala WD, Mafongoya PL (1997) Building soil nitrogen capital in Africa. In: Buresh RJ, Sanchez PA, Calhoun FJ (eds) Replenishing soil fertility in Africa. Soil science society of America special publication 51, Madison, WI, pp 155–192

Giller KE, Wilson KJ (2001) Nitrogen fixation in tropical cropping systems, 2nd edn. CAB International, Wallingford, UK

Holderbaum JT, Decker AM, Meisinger JJ, Ulford FR, Vough LR (1990) Fall-seeded legume cover crops for no tillage corn in the humid east. Agron J 82:117–124

ICRISAT annual report (1995) ICRISAT southern and eastern Africa region, pp 13–14

Jerenyama P, Hestermann OB, Waddington SR, Hardwood RR (2000) Relay intercropping of sunnhemp and cowpea into a smallholder maize system in Zimbambwe. Agron J 92: 239–244

Kapkayai JJ, Karanja NK, Woomer PL, Qureshi JN (1998) Soil organic carbon fractions in a long-term experiment and the potential for their use as a diagnostic assay in the highland farming systems of central Kenya. Afr Crop Sci J 6:19–28

Kumar K, Goh KM (2000) Crop residues and management practices: effects on soil quality, soil nitrogen dynamics, crop yield and nitrogen recovery. Agron J 688:199–319

Kumar Rao JVDK (1990) Pigeonpea: nitrogen fixation. In: Nene YL, Hall SD, Sheila VK (eds) The pigeonpea. CAB International, Wallingford, UK, pp 233–256

Kumar Rao JVDK, Johansen C, Tobita TS, Ito O (1996) Estimation of nitrogen fixation by the natural ^{15}N abundance technique and nitrogen uptake by pigeonpea genotypes of different maturity groups grown in an inceptisol. Agron J 177:129–138

Ma BL, Dwyer LM, Gregorich EG (1999) Soil nitrogen amendment effects on seasonal nitrogen mineralization and nitrogen cycling in maize production. Agron J 91:1003–1009

Ofori F, Pate JS, Stern WR (1987) Evaluation of N_2 fixation and nitrogen economy of a maize/cowpea intercrop system using ^{15}N dilution methods. Plant Soil 102:149–160

Okalebo JR, Gathua KW, Woomer PL (2002) Laboratory methods of soil and plant analysis: a working manual, 2nd edn. Tropical Soil Biology and Fertility Programme, Nairobi, pp 29–32

Papastylianou I (1988) The ^{15}N methodology in estimating N_2 fixation by vetch and pea grown in pure stand or in mixtures with oats. Plant Soil 107:183–188

Sanchez PA, Shephered KD, Soule MJ, Place FM, Buresh RJ, Izac AMN (1997) Soil fertility replenishment in Africa. An investment in natural resource capital. In: Buresh RJ et al (eds) Replenishing soil fertility in Africa. SSSA Special Publication No. 51 SSSA, Madison, WI, pp 1–46

Sanginga N (2003) Role of biological nitrogen fixation in legume based cropping systems: a case study of West Africa farming systems. Plant soil 252:25–39

Taylor RW, Griffin JL, Meche GA (1982) Yield and quality characteristics of vetch species for forage and cover crops, University of Wisconsin–Madison, Department of Agronomy. Prog Rep 15:53–56

Tisdale SL, Nelson WL, James DB (1990) Soil fertility and fertilizers. Macmillan Publishing Company, New York, NY

Wanderi SW, Mburu MWK, Silim SN, Nkonge IG (2002) Identification of a suitable reference crop to determine amount of N fixed by pigeonpea. In: Proceedings of 5th regional meeting of the forum for agricultural resources husbandry, 12–16 Aug 2002, Entebbe

Nutr Cycl Agroecosyst (2010) 88:59–77
DOI 10.1007/s10705-009-9303-6

ORIGINAL ARTICLE

Nitrogen and phosphorus capture and recovery efficiencies, and crop responses to a range of soil fertility management strategies in sub-Saharan Africa

R. Chikowo · M. Corbeels · P. Mapfumo ·
P. Tittonell · B. Vanlauwe · K. E. Giller

Received: 5 March 2008 / Accepted: 11 July 2009 / Published online: 1 August 2009
© Springer Science+Business Media B.V. 2009

Abstract This paper examines a number of agronomic field experiments in different regions of sub-Saharan Africa to assess the associated variability in the efficiencies with which applied and available nutrients are taken up by crops under a wide range of management and environmental conditions. We consider N and P capture efficiencies (NCE and PCE, kg uptake kg^{-1} nutrient availability), and N and P recovery efficiencies (NRE and PRE, kg uptake kg^{-1} nutrient added). The analyzed cropping systems employed different soil fertility management practices that included (1) N and P mineral fertilizers (as sole or their combinations) (2) cattle manure composted then applied or applied directly to fields through animal corralling, and legume based systems separated into (3) improved fallows/cover crops-cereal sequences, and (4) grain legume-cereal rotations. Crop responses to added nutrients varied widely, which is a logical consequence of the wide diversity in the balance of production resources across regions from arid through wet tropics, coupled with an equally large array of management practices and inter-season variability. The NCE ranged from 0.05 to 0.98 kg kg^{-1} for the different systems (NP fertilizers, 0.16–0.98; fallow/cover crops, 0.05–0.75; animal manure, 0.10–0.74 kg kg^{-1}), while PCE ranged from 0.09 to 0.71 kg kg^{-1}, depending on soil conditions. The respective NREs averaged 0.38, 0.23 and 0.25 kg kg^{-1}. Cases were found where NREs were >1 for mineral fertilizers or negative when poor

R. Chikowo (✉) · P. Mapfumo
Soil Science & Agricultural Engineering Department,
University of Zimbabwe, Box MP 167, Mt Pleasant,
Harare, Zimbabwe
e-mail: rchikowo@agric.uz.ac.zw;
regischikowo@yahoo.co.uk

M. Corbeels · P. Tittonell
Département Persyst, Centre de Coopération
Internationale en Recherche Agronomique pour le
Développement (CIRAD), Av Agropolis TA B-102/02,
34398 Montpellier Cedex 5, France
e-mail: corbeels@cirad.fr

P. Mapfumo
The Soil Fertility Consortium for Southern Africa
(SOFECSA), CIMMYT- Zimbabwe, Box MP 163, Mount
Pleasant, Harare, Zimbabwe

B. Vanlauwe
Tropical Soil Biology and Fertility Institute of CIAT
(TSBF-CIAT), Nairobi, Kenya

K. E. Giller
Plant Production Systems, Department of Plant Sciences,
Wageningen University, P.O. Box 430, 6700 AK
Wageningen, The Netherlands

This article has been previously published in the journal "Nutrient Cycling in Agroecosystems" Volume 88 Issue 1.
A. Bationo et al. (eds.), Innovations as Key to the Green Revolution in Africa: Exploring the Scientific Facts. © 2010 Springer.

quality manure immobilized soil N, while response to P was in many cases poor due to P fixation by soils. Other than good agronomy, it was apparent that flexible systems of fertilization that vary N input according to the current seasonal rainfall pattern offer opportunities for high resource capture and recovery efficiencies in semi-arid areas. We suggest the use of cropping systems modeling approaches to hasten the understanding of Africa's complex cropping systems.

Keywords Nutrient use efficiency ·
Sub-Saharan Africa · Nutrient mining ·
Fertilizers · Manure · Legumes ·
Cropping systems modeling

Introduction

Soils with poor nutrient contents, particularly of N and P, are widespread in sub-Saharan Africa (SSA), and this has been widely recognized as one of the pivotal causes of poor agricultural productivity. Compared to other parts of the world were agricultural green revolutions have been stimulated by mechanization and high fertilizer use, SSA soil nutrient balances remain largely negative (Smaling et al. 1997). Where nutrients are applied, albeit often in small doses, their capture and utilization by crops has been poor largely due to nutrient imbalances (Kho 2000). It is well established that efficient nutrient recovery by crops is a function of a multitude of factors that should be in a balanced state (Janssen 1998). Recovery efficiencies of added nutrients depend on soil and plant characteristics, crop management, fertilizer dosage and timing, and season quality. For example, while crop rooting density requirements to remove nitrate from soil is small in relation to that required for less mobile nutrients such as P, nitrate not taken up may be quickly lost through leaching in sandy soils during periods of high rainfall when residence time is short (Cadisch et al. 2004; Chikowo et al. 2003). On the other hand, P availability is often heavily restricted by the iron and aluminum oxides which are common in highly weathered tropical soils (e.g. Vanlauwe et al. 2002; Sanchez et al. 1997). These are the some of the many difficult scenarios that resource-constrained smallholder farmers in Africa have to grapple with in their production systems.

Short-range spatial variability in soils commonly exist within and between farms due to localized differences in parent material and/or management (Tittonell et al. 2005; Mtambanengwe and Mapfumo 2005; Samaké et al. 2006), with major implications on water and nutrient use efficiency. In most cases fields that are poor in N and/or P will give poor returns even when these nutrients are amply supplied through fertilizers, as there would be other nutrients limiting production, beyond N and P (e.g. Vanlauwe et al. 2005; Wopereis et al. 2006; Zingore et al. 2007). Therefore, any fertilization strategy that seeks to optimize resource use efficiency by crops has to recognize the important role of the inherent and distinct capacity of different soils to supply nutrients to the crops.

In the face of limited external resources, the question of how to efficiently target the available nutrients on the farms in a continuum of circumstances becomes critical (Giller et al. 2006). Therefore, a key objective of this study was to analyze nutrient use management options in SSA agriculture and obtain insights on the associated nutrient use efficiencies, a vital step for magnifying cropping systems or system components that offer opportunities for intensification. We illustrate the performance of different cropping systems in the different regions of SSA using data from key publications based on field experiments spanning over the past two decades. The various data in the publications were re-analyzed, taking into account the indigenous nutrient supplying (INS) capacity of the soils, and the fertilizers and organic amendments added to estimate N and P availability and associated nutrient capture efficiencies.

Description of database and data computation

Literature searches were done on various electronic library platforms using key words such as nitrogen, phosphorus or nutrient use efficiency. Relevant articles published from 1990 to date were reviewed and those based on field experiments in SSA were identified. This involved a large array of cropping systems that managed soil fertility in equally varied approaches. The principal cereal crop is maize, but there is a significant component of the small grains (millet and sorghum) in the semi-arid parts of southern Africa and the Sahelian region, and upland and lowland rice in West Africa.

 Springer

Throughout this study we make reference to two slightly different nutrient use efficiency terms: (1) N and P capture efficiency (NCE or PCE, respectively, for N and P) as the amount of nutrient captured per unit of nutrient availability, and (2) N and P recovery efficiency (NRE or PRE) as the amount of nutrient recovered per unit added through the different fertilization strategies. To compute NCE for the different experiments, N availability was taken as the sum of the external N supply through mineral or organic fertilizers and the indigenous soil N supply.

i.e. NCE = Crop N uptake/(externally supplied N + indigenous soil N supply).

Indigenous soil supply of N was estimated from data on N uptake from a treatment in which all other nutrients were amply supplied except for N. Similarly, the indigenous soil P supply was estimated from P uptake in plots where other nutrients had been amply supplied except P. To be included in the database, it was desirable, though not strictly necessary, that the trials contained treatments that resemble this description. Where the treatments were such that this information could not be extracted easily, the soil organic carbon (SOC) content was used to estimate potential N or P availability through mineralization using the transfer functions used in the model QUE-FTS, that were derived from experimental data from East Africa (Janssen et al. 1990). In cases where the authors did not provide information on nutrient uptake we assumed an internal N conversion efficiency of 55 kg grain (kg N uptake)$^{-1}$, a value slightly higher than the intermediate between the physiologically possible maximal dilution and maximal concentrations for maize (Janssen et al. 1990). Estimations for other crops were done using their respective average internal N efficiencies, 35 kg kg^{-1} for millet and sorghum, and 55 kg kg^{-1} for rice, given from an extensive review by van Duivenbooden et al. (1996).

Results and discussion

The database

Restricting our scope to SSA excluded a large volume of information on N and P use efficiencies available worldwide. A lot of set backs were encountered during literature retrieval, as many authors only

provided information just enough to meet their immediate objectives. As a result many potentially useful articles could not be included in our database. An overview of the literature data grouped into crop responses to N and P fertilizers, legume cover crop/fallow and manure presented as summary statistics reveal the existence of broad ranges in nutrient availability and use efficiencies for the different cropping systems, with the indigenous soil N supply ranging from 10 to 91 kg ha^{-1} (Table 1). The number of experiments testing N fertilizers was considerably larger than for P or manure. When all data from experiments that involved N and P fertilizers with maize were pooled, it was clear that other than N availability, there were other important explanatory variables that explained N uptake (Fig. 1). The description of the cropping systems and of some of the key experiments that constitute the database is given in the following sections.

Capture and recovery efficiency of N and P from mineral sources for maize

A wide array of experiments with N and P fertilizers have been carried out both on-station and on-farm with equally varied responses (Tables 2, 3, 4, 5 and 6). In an experiment that was carried out over three seasons in Togo, Wopereis et al. (2006) reported responses of maize to N and P on farmers' fields that had received organic inputs for at least 10 years (infields) and those that did not (outfields). Being on the same soil type, the main difference between infields and outfields on an individual farm was SOC content. Averaged over three seasons at sole 100 kg ha^{-1} N application, NCE was significantly higher on infields compared to outfields (0.52 vs. 0.38 kg kg^{-1}). In a related experiment on degraded and non-degraded soils in Togo, Fofana et al. (2005) also demonstrated that NCE was always superior on a non-degraded soil. Significant improvements in the NRE of applied N and overall NCE on the degraded soil was only realized when N and P were simultaneously applied. Upon application of 40 kg P ha^{-1} on the degraded soil, NCE increased from 0.29 to 0.46 kg kg^{-1} at N rate of 50 kg ha^{-1}, and from 0.29 to 0.38 kg kg^{-1} at 100 kg N ha^{-1} N. As expected, NCE was lower at higher N application rate as N availability increased and its shortage relative to other production resources decreased. The data shows

Table 1 Soil N availability, maize yields, NCE, NRE and summary statistics of the variables for experiments that involved (a) NP fertilizers (b) improved fallows/cover crops and (c) animal manure, in sub-Saharan Africa

Variable/fertility management practice	n	Minimum	Maximum	Mean	Median	Standard deviation	Coefficient of variation
(a) NP fertilizers							
Indigenous soil N (kg ha^{-1})	41	10	91	34	30	18	54
Total N availability (kg ha^{-1})	86	13	191	94	100	49	52
Maize yields (Mg ha^{-1})	84	0.29	7.70	2.61	2.1	1.68	64
N uptake (kg ha^{-1})	85	6	136	53	44	32	61
NCE (kg kg^{-1})	64	0.16	0.98	0.53	0.52	0.19	35
NRE (kg kg^{-1})	58	0.05	1.00	0.38	0.35	0.22	56
(b) Improved fallow/cover crops							
Indigenous soil N (kg ha^{-1})	26	12	82	33	30	17	51
N availability (kg ha^{-1})	64	13	400	146	147	86	59
Maize yields (Mg ha^{-1})	69	0.30	8.20	2.14	1.92	1.47	68
N uptake (kg ha^{-1})	68	8	149	48	46	29	61
NCE (kg kg^{-1})	51	0.05	0.75	0.34	0.32	0.19	56
NRE (kg kg^{-1})	51	0	0.66	0.23	0.20	0.17	71
(c) Animal manure							
Indigenous soil N (kg ha^{-1})	10	15	70	36	31	19	54
N availability (kg ha^{-1})	21	25	433	132	115	103	78
Maize yields (Mg ha^{-1})	23	0.40	5.90	2.42	2.1	1.27	52
N uptake (kg ha^{-1})	23	8	107	46	37	26	58
NCE (kg kg^{-1})	15	0.10	0.74	0.37	0.35	0.20	53
NRE (kg kg^{-1})	15	0.0	0.65	0.25	0.19	0.19	77
(d) P relations							
Indigenous soil P (kg ha^{-1})	10	5	18	9	8.5	4.2	47
P availability (kg ha^{-1})	28	5.0	113	49	44	31	62
Maize yields (Mg ha^{-1})	32	0.50	8.4	3.06	2.0	2.3	77
P uptake (kg ha^{-1})	27	2.90	34.0	12.4	11.0	7.7	62
PCE (kg kg^{-1})	22	0.09	0.71	0.25	0.20	0.17	69
PRE (kg kg^{-1})	23	0	0.29	0.16	0.17	0.07	45

Summary statistics for P relations involving experiments with P fertilizers are shown in (d)

n number of publications on the data set

that N fertilization alone significantly increased P uptake by maize and that a moderate P rate of 20 kg ha^{-1} was sufficient for maximum P uptake.

Experiments with N fertilizers and rock phosphate in Mali resulted in NCE ranging between 0.33 and 0.50 kg kg^{-1} for maize over a 4-year period (Bationo et al. 1997; Table 4). In southern and eastern Tanzania, application of a large amount of P fertilizer (80 kg ha^{-1}) across sites with acid P-fixing soils resulted in NCE range of 0.16–0.42 kg kg^{-1} (Msolla et al. 2005). Across four sites, NCE ranged from 0.28 to 0.48 kg kg^{-1} with P fertilizer, and from 0.25 to

0.35 kg kg^{-1} when an equivalent amount of rock P was used. The highly P fixing soils responded poorly to N application, marginally increasing NCE from 0.10 kg kg^{-1} without P to 0.16 kg kg^{-1} when P was applied.

In an experiment that spanned over a 6-year period on two contrasting soils in Zimbabwe, NCE varied between 0.24 and 0.50 kg kg^{-1} on poor farmers' fields, compared with NCE ranging between 0.52 and 0.77 kg kg^{-1} at an on-station site (Waddington and Karingwindi 2001). Despite the annual addition of 18 kg ha^{-1} P, apparent recovery of applied N at on-farms sites was in some

 Springer

Nutr Cycl Agroecosyst (2010) 88:59–77

Fig. 1 The relationships between N availability and **a** maize grain yield **b** N uptake **c** N capture efficiency (NCE) and **d** N recovery efficiency (NRE) for experiments that involved N and P fertilizers

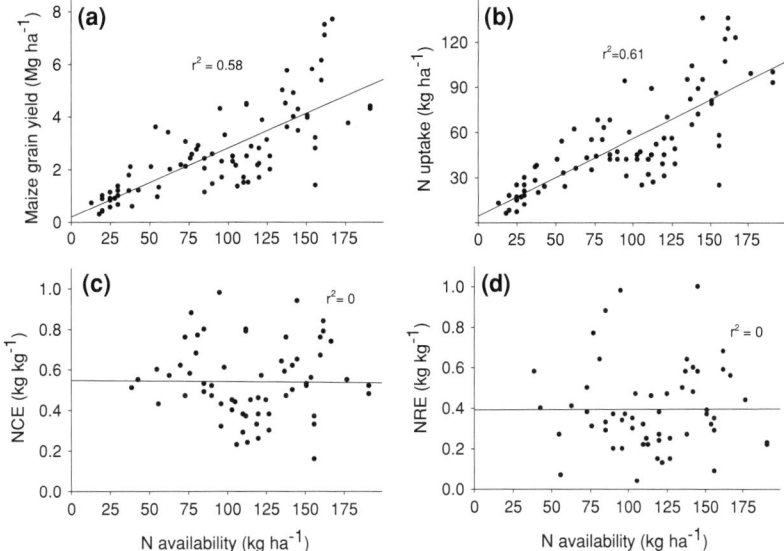

cases as low as 0.1 kg kg^{-1}, a scenario that has forced some croplands to be abandoned. This is considered as a classical example of little mileage gained when N and P fertilizers are added to a soil with multiple constraints that may include acute micronutrient deficiencies and soil acidity.

On an Alfisol and Oxisol in Nigeria, average NCE was 0.57 kg kg^{-1} when 45 kg ha^{-1} N and 12 kg ha^{-1} P were annually applied to maize fields over a 10-year period (Kang et al. 1999). Another long-term experiment on a Ferric Lixisol in the same region under ample P supply had average NRE of 0.32 kg kg^{-1} when 120 kg ha^{-1} N was applied, which increased to 0.68 kg kg^{-1} at a reduced N application rate of 60 kg ha^{-1} (Vanlauwe et al. 2005). Application of adequate N and P to five maize varieties in the moist savanna of Nigeria resulted in average NCE of 0.50 kg kg^{-1} and NRE of 0.30 kg kg^{-1} (Oikeh et al. 2003). In Cameroon, application of P fertilizer on two basaltic soils led to variable yield responses by maize (Osiname et al. 2000). Response to P was significant at both sites with grain yield increasing with P rates up to 88 kg ha^{-1} at one site, compared with no additional yield gains beyond an application rate of 22 kg ha^{-1} P at another site, in spite of the low soil P test. Large responses at low rates are encouraging, as it is possible for resource poor farmers to benefit from small amounts of fertilizer P, at the same time avoiding the

degradation of soil P status through small maintenance fertilization rates.

In Kenya, Probert and Okalebo (1992) showed that under non-N limiting conditions and when extractable P was 8 µg g^{-1}, maize responded to P application but there was no significant difference between three P application rates (20, 40 and 60 kg ha^{-1}) or any tendency for the higher rates to give increased yields of maize. At application rates of 20 kg P ha^{-1}, PRE by maize averaged 0.14 kg kg^{-1} (Table 2). However, in a separate experiment where the same authors employed surface soil management through mulching and tied-ridging, there was a significant response to P inputs as high as 60 kg ha^{-1}. A plausible explanation for this could be that conservation measures resulted in more water retention and thus better root growth and exploitation of P. This is another example on resource interactions and the importance of balanced resource availability for increased resource use efficiency. In southern Malawi, farms usually stretch through three landscape positions, from steep eroded slopes through *dambo* margins to *dambo* valleys, with increasing fertility towards the valley. Phiri et al. (1999) studied the effect of landscape position on the utilization of fertilizer N, and got significantly higher NRE on the *dambo* and *dambo* margin positions (0.46 kg kg^{-1} N) compared with poor recovery (0.22 kg kg^{-1}) on the steep slopes.

Table 2 Indigenous soil P supply (IPS), total P availability, and estimated P capture and apparent recovery efficiencies for experiments that involved N and P fertilizers with maize and millet and rice

Treatment description (ha⁻¹)	Soil conditions	IPS (kg ha⁻¹)	P availability (kg ha⁻¹)	Grain yields (kg ha⁻¹)	P uptake (kg ha⁻¹)	PCE (kg kg⁻¹ P)	PRE (kg kg⁻¹ P)	Country and region / climatic conditions	Reference
Maize									
40 kg P	Degraded, 0.4% C	10	50	0.50	3.60			Togo, coastal savana	Fofana et al. (2005)
50 kg Urea-N + 40 kg P			50	1.30	10.0	0.20	0.17		
100 kg Urea-N + 40 kg P			50	1.70	11.9	0.24	0.22		
60 kg N, 43 kg P	Clay, kaolinitic	7	50	1.01	10.1	0.21	0.12	Benin, west Africa	Saïdou et al. (2003)
100 kg P	Vertisols	13	113	3.60	13			Kenya	Sigunga et al. (2002)
100 kg NP			113	5.80	22	0.19	0.09		
100 kg NP + drainage			113	7.70	34	0.30	0.21		
Control	Sandy loam, 0.43% C	11	31	0.59	2.9			Kenya, semi arid	Probert and Okalebo (1992)
90 kg N				3.86	11				
90 kg N + 20 kg P				4.54	13.8	0.19	0.14		
Control (25 kg P)	Kaolinitic, P fixing	5	30	0.80	4	0.13	0	Kenya, humid tropics	Gachengo et al. (1999)
Senna spectabilis			15	1.50	6	0.4	0.2		
Tithonia diversifolia			22	2.00	9	0.41	0.29		
Senna + 25 kg P			40	2.00	8	0.2	0.11		
Tithonia + 25 kg P			47	3.20	12	0.26	0.19		
Millet									
60 kg P	Loamy sand, 0.26% C	7	67	0.90		0.1		Niger, semi arid	Kho (2000)
180 kg N, 60 kg P			67	1.34		0.12	0.19		
90 kg N, 30 kg P			37	1.20		0.2	0.23		
Rice									
100 kg N	Ultisol	5	5	0.75	5			Ivory coast, humid	Sahrawat et al. (1997)
100 kg N + 45 kg P	(Cultivar 1)		55	2.05	9	0.09	0.09		
100 kg N + 90 kg P			95	2.35	12	0.09	0.08		

Nutr Cycl Agroecosyst (2010) 88:59–77

Table 2 continued

Treatment description (ha^{-1})	Soil conditions	IPS (kg ha^{-1})	P availability (kg ha^{-1})	Grain yields (kg ha^{-1})	P uptake (kg ha^{-1})	PCE (kg kg^{-1} P)	PRE (kg kg^{-1} P)	Country and region / climatic conditions	Reference
100 kg N	(Cultivar 2)	5	5	1.07	5.5				
100 kg N + 45 kg P cv2			55	1.69	10	0.10	0.11		
100 kg N + 90 kg P			95	1.61	11	0.11	0.07	Senegal, West Africa	Haefele and Wopereis (2005)
20 kg P, 50 kg K	Alluvial Vertisols	10–38	44	4.0					
151 kg N 20 kg P			44	7.8			0.11		
151 N, 20 P, 50 kg K			44	8.4			0.18		
Control	Loamy, 0.4% C	18	18	1.80	6			Mauritania, Sahelian	van Asten et al. (2005)
175 kg N			18	5.00	18				
175 kg N, 13 kg P			31	5.60	22	0.71	0.29		
175 kg N, 26 kg P			44	6.10	24	0.55	0.21		
175 kg N, 26 kg P + straw			44	6.20	28	0.64	0.25		

WA west Africa, *EA* east Africa, *SA* southern Africa, *CA* central Africa

Table 3 Indigenous soil N supply (INS), total N availability, and estimated N capture and apparent recovery efficiency for experiments that involved manures, short term fallows, and NP fertilizers on rice in sub-Saharan Africa

Treatment description	General conditions	INS (kg ha⁻¹)	N availability (kg ha⁻¹)	Grain yields (Mg ha⁻¹)	N uptake (kg ha⁻¹)	NCE (kg kg⁻¹)	NRE (kg kg⁻¹)	Country	Reference
100 kg N	Ultisol	15	115	0.75	12	0.11		Ivory coast	Sahrawat et al. (1997)
100 kg N + 45 kg P	(WAB 56–125)		115	2.05	34	0.29	0.22		
100 kg N + 90 kg P			115	2.35	39	0.34	0.27		
100 kg N	(Local CV)		115	1.07	17	0.15			
100 kg N + 45 kg P cv2			115	1.69	28	0.24	0.11		
100 kg N + 90 kg P			115	1.61	26	0.22	0.10		
20 kg P, 50 kg K	Alluvial Vertisols	18–78	44	4.0				Niger	Haefele and Wopereis (2005)
151 kg N, 20 kg P			195	7.8	130	0.66	0.34		
151 N, 20 P, 50 kg K			195	8.4	140	0.71	0.41		
Weed fallow	Alfisol	10	10	0.32	6			Ivory coast	Becker and Johnson (1999)
81 kg Legume fallow[a]			19	1.01	17	0.16	0.12		
Control	Loamy, 0.4% C	39	39	1.80	21			Mauritania	van Asten et al. (2005)
175 kg N			214	5.00	78	0.36	0.32		
175 kg N, 13 kg P			214	5.60	94	0.43	0.41		
175 kg N, 26 kg P			214	6.10	94	0.43	0.41		
175 kg N, 26 kg P + straw			214	6.20	130	0.60	0.52		
Control	0.5% C	25	25	0.97	21			Sierra Leone	Bar et al. (2000)
60 kg N			85	1.33	29	0.35	0.14		
Sesbania rostrata fallow			119	1.36	30	0.25	0.10		
Sesbania + 30 kg N			150	2.14	47	0.31	0.21		

[a] Mean of five legumes (*Calopogonium, Canavalia, Centrosema, Mucuna, Pueraria*)

Table 4 Indigenous soil N supply (INS), total N availability, and estimated N capture and apparent recovery efficiencies for experiments that involved manures, short term fallows, and NP fertilizers on sorghum in sub-Saharan Africa

Treatment description	General conditions	INS (kg ha⁻¹)	N availability (kg ha⁻¹)	Grain yields (Mg ha⁻¹)	N uptake (kg ha⁻¹)	NCE (kg kg⁻¹)	NRE (kg kg⁻¹)	Country/region	Reference
Control	Sandy loam	45	45	1.54	43			Niger	Zaongo et al. (1997)
Mulch			45	2.10	60				
50 kg N ha⁻¹			95	2.37	67	0.71	0.45		
50 kg N ha⁻¹ + mulching			95	2.49	71	0.75	0.52		
Control		48	48	1.68	48	1.00	0.04	Uganda	Hagedorn et al. (1997)
112 kg N (Tephrosia)			160	1.87	53	0.33	0.59		
52 kg manure N			100	2.76	79	0.79			
Control (site 1)	Lixisol	33	33	1.16	33			Burkina Faso	Ouédraogo et al. (2001)
65 kg compost N (site 1)			98	1.68	48	0.49	0.23		
Control (site 2)	Ferric Lixisol	15	15	0.40	11				
130 kg compost N (site 2)			142	1.38	39	0.27	0.21		
Control	Average 0.7% C	28	35	0.86–1.04				Mali	Bationo et al. (1997)
7 kg N + 11 kg P				1.24–1.92	35–54	>1	>1		
12 kg N + 27 kg P rock phosphate				1.16–2.21	33–63	0.64–1.0	>0.5		
Control	Ferric Lixisol	35	31	1.10				Burkina Faso	Zougmoré et al. (2004)
50 kg Urea-N			81	2.10	60	0.74	0.58		
50 kg compost-N			81	2.27	65	0.80	0.68		
Control (year 2)		35	31	1.16	33	1.07			
50 kg Urea-N (year 2)			81	1.40	40	0.49	0.14		
50 kg compost-N (year 2)			81	2.38	68	0.84	0.70		
65 kg N + 10 kg P only	Ferric Lixisol (year 1)	30	95	1.12	32	0.33		Burkina Faso	Mando et al. (2005)
10 t cattle manure + 65 kg N + 10 kg P			205	2.53	72	0.35			
65 kg N + 10 kg P only	Ferric Lixisol (year 2)		95	0.62	18	0.19			

Table 4 continued

Treatment description	General conditions	INS (kg ha⁻¹)	N availability (kg ha⁻¹)	Grain yields (Mg ha⁻¹)	N uptake (kg ha⁻¹)	NCE (kg kg⁻¹)	NRE (kg kg⁻¹)	Country/region	Reference
10 t cattle manure + 65 kg N + 10 kg P			205	2.35	67	0.32			
60 kg P	Loamy sand, 0.26% C	25	25	0.90	25			Niger	Kho (2000)
180 kg N, 60 kg P			205	1.34	43	0.32	0.19		
90 kg N, 30 kg P			115	1.20	30	0.31	0.23		
Control	Sandy	12	12	0.37	10	0.18	0.07	Niger	Rockström and de Rouw (1997)
5 t Manure			82	0.52	15				
30 kg N, 13 P kg P fertilizer			42	0.65	18	0.44	0.28		
13 kg N + 22 kg P fertilizer	Sandy (0.2–0.6% C)	20	33	0.41–0.97	11–27	0.48–0.78	0.33–0.47	Mali	Bationo et al. (1997)
23 kg N +27 kg P rock phosphate			43	0.35–0.96	10–27	0.28–0.64	0.05–0.41		
Control	Sandy, 0.13% C, low P	25	25	0.58	25			Niger	Sangaré et al. (2002)
3 t Low quality manure			83	0.88	37	0.45	0.21		
3 t High quality manure			98	0.92	39	0.39	0.19		
3 t Low quality manure + mulch			92	1.15	49	0.53	0.36		
3 t High quality manure + mulch			106	1.41	58.5	0.55	0.41		
Control (crest)	0.2% C, 85% sand	15	15	0.36	12			Niger	Brouwer and Powell (1998)
1.5 t manure (crest)			46	0.81	27	0.59	0.48		
8.5 t manure (crest)			183	0.97	32	0.18	0.12		
Control (concave)			15	0.24	8.1				
2.9 t manure (concave)			74	0.35	12	0.16	0.07		
Control	>90% sand (nitisols)	20	20	0.48	16			Niger	Gandah et al. (2003)
3.53 t manure			74	1.10	36	0.49	0.38		

A range indicates data is derived from multiple seasons

Table 5 Indigenous soil N supply (INS), total N availability, and estimated N capture and apparent recovery efficiencies for experiments that involved manures, short term fallows, and NP fertilizers on millet in sub-Saharan Africa

Treatment description	General conditions	INS (kg ha⁻¹)	N availability (kg ha⁻¹)	Grain yields (Mg ha⁻¹)	N uptake (kg ha⁻¹)	NCE (kg kg⁻¹)	NRE (kg kg⁻¹)	Country/ region	Reference
60 kg P	Loamy sand, 0.26% C	25	25	0.90	25			Niger	Kho (2000)
180 kg N, 60 kg P			205	1.34	43	0.32	0.19		
90 kg N, 30 kg P			115	1.20	30	0.31	0.23		
Control	Sandy	12	12	0.37	10	0.18	0.07	Niger	Rockström and de Rouw (1997)
5 t Manure			82	0.52	15	0.44	0.28		
30 kg N, 13 P kg P fertilizer			42	0.65	18				
13 kg N + 22 kg P fertilizer	Sandy (0.2–0.6% C)	20	33	0.41–0.97	11–27	0.48–0.78	0.33–0.47	Mali	Bationo et al. (1997)
23 kg N +27 kg P rock phosphate			43	0.35–0.96	10–27	0.28–0.64	0.05–0.41		
Control	Sandy, 0.13% C, low P	25	25	0.58	25			Niger	Sangaré et al. (2002)
3 t Low quality manure			83	0.88	37	0.45	0.21		
3 t High quality manure			98	0.92	39	0.39	0.19		
3 t Low quality manure + mulch			92	1.15	49	0.53	0.36		
3 t High quality manure + mulch			106	1.41	58.5	0.55	0.41		
Control (crest)	0.2% C, 85% sand	15	15	0.36	12			Niger	Brouwer and Powell (1998)
1.5 t manure (crest)			46	0.81	27	0.59	0.48		
8.5 t manure (crest)			183	0.97	32	0.18	0.12		
Control (concave)			15	0.24	8.1				
2.9 t manure (concave)			74	0.35	12	0.16	0.07		
Control	>90% sand (nitisols)	20	20	0.48	16			Niger	Gandah et al. (2003)
3.53 t manure			74	1.10	36	0.49	0.38		

A range indicates data is derived from multiple seasons

 Springer

Table 6 Indigenous soil N supply (INS), total N availability, and estimated N capture and apparent recovery efficiencies by maize in experiments that involved manure application (with or without NP fertilizers) in sub Saharan Africa

Treatment description	General conditions	INS (kg ha⁻¹)	N availability (kg ha⁻¹)	Grain yields (Mg ha⁻¹)	N uptake (kg ha⁻¹)	NCE (kg kg⁻¹)	NRE (kg kg⁻¹)	Country	Reference
Control	Loamy sand	25	25	1.15	25			Zimbabwe	Nyamangara et al. (2003)
12 t manure			142	2.20	49	0.35	0.20		
12 t manure + 60 kg N			202	3.51	78	0.38	0.29		
37 t manure			373	3.19	71	0.19	0.13		
37 t manure + 60 kg N			433	4.05	90	0.20	0.15		
17 t manure	Poor sandy	15	148	0.40	15	0.10	0.08	Zimbabwe	Chikowo et al. (2004)
17 t manure + 40 kg N			188	1.60	25	0.13	0.12		
2 t manure	Degraded soil	18	58	1.40	34	0.58	0.40	Tanzania	Baijukya et al. (2006)
2 t manure + 50 kg N			108	2.30	63	0.58	0.50		
Control	Sandy loam	20		0.40	8			Mali	Kaya and Nair (2001)
10 t manure				1.12	20				
Control	Alfisol	30	30	1.58	25			Ethiopia	Lupwayi et al. (1999)
3 t manure			117	2.00	32	0.28	0.08		
40 kg N, 30 kg P			127	2.15	33	0.26	0.19		
Control (site 1)	Humic nitisols, acidic	69	69	2.30	42			Kenya	Smaling et al. (1992)
5 t manure			144	4.50	82	0.56	0.53		
Control (site 2)	Clayey, moderate fertility	70	70	3.20	58				
5 t manure			145	5.90	107	0.74	0.65		
Control (site 3)	Sandy loam, low NPK	32	32	1.60	29				
5 t manure			107	1.20	21	0.20	0		
Control	(Semi-arid)		37	1.78	37			Tanzania	Jensen et al. (2003)
7.5 t manure (site 1)	Sandy loam	37	104	2.17	44	0.42	0.10		
7.5 t manure (site 2)	Sandy	45	115	3.46	69	0.60	0.32		

Low-lying areas with Vertisols can constitute locally productive soils in otherwise largely unproductive areas. However, these soils are often not fully exploited due to excess water problems during periods of high rainfall, as their high content of expansive clays prevents the rapid drainage of excess water. Sigunga et al. (2002) investigated the effect of improved drainage on N and P utilization efficiencies on such soils using 0.4–0.6 m deep furrows. At 100 kg ha^{-1} N and P fertilizer application, drainage increased NCE from 0.56 to 0.74 kg kg^{-1} N and PRE from 0.09 to 0.21 kg kg^{-1} P (Table 2).

Capture and recovery efficiency of N and P
with rice, millet and sorghum

Rice is an important crop in West Africa, with various alternatives currently being proposed towards its intensified production. Becker and Johnson (1999) investigated the role of several legume accessions on rice yields when grown for 6 months during the dry season, under a range of hydrological and soil conditions. Overall, legumes increased rice yields by 0.23 Mg ha^{-1}, while some five selected legumes raised rice yields from 0.32 to 1 Mg ha^{-1} (Table 3). In Mauritania, application of N fertilizer increased rice yields, and addition of straw had a positive effect, independent of fertilizer dose or soil type (van Asten et al. 2005). On neutral soils, NRE ranged from 0.32 kg kg^{-1} in the absence of P to 0.41 kg kg^{-1} when P was added, and further increased to 0.52 kg kg^{-1} in the presence of rice straw. On alkaline soils the NRE was lower and the range was 0.20–0.42 kg kg^{-1}. Recovery of P ranged between 0.1 and 0.35 on both alkaline and neutral soils, but was high and confined between 0.21 and 0.29 kg kg^{-1} on neutral soils (Table 2). Haefele and Wopereis (2005) demonstrated the significance of localized soil variability to nutrient use efficiency in rice on a 3 ha experimental farm. They reported NRE ranging from 0.34 to 0.41 kg kg^{-1} N, and PRE ranging from 0.11 to 0.19 kg kg^{-1} P, depending on K addition and the indigenous soil N or P supply. The suitability of *Sesbania rostrata* as green manure in combination with N fertilizer for lowland rice production was evaluated in Sierra Leone (Bar et al. 2000). Rice recovered 0.14 kg kg^{-1} N added as urea, while NCE was 0.35 kg kg^{-1}. Recovery of *Sesbania* N alone was 0.10 kg kg^{-1}, and overall

recovery efficiency doubled when 30 kg urea-N was added as top dressing.

A summary of calculated capture and recovery efficiencies of N by millet and sorghum for various systems is presented in Tables 4 and 5. Among the important findings was that mulching alone significantly increased yields in dry environments. Millet and sorghum experiments with small doses of N fertilizers and rock phosphate in Mali resulted in NCE ranging between 0.05 and 0.41 kg kg^{-1} for millet and at least 0.50 kg kg^{-1} for sorghum over a 4-year period (Bationo et al. 1997). On a ferric Lixisol in Burkina Faso, Zougmoré et al. (2004) reported large sorghum yield increase from either urea or compost application in 1 year, with urea NRE of 0.58 kg kg^{-1}, but poor NRE of only 0.14 kg kg^{-1} for urea during the following season that was linked to in-season dry spells.

Capture and recovery efficiency of N
from animal manure

Animal manure is an important resource on smallholder farms, as nutrients are concentrated from common rangelands. Animal ownership is therefore a strong determinant for farm SOC and N management. The potential and pitfalls for efficient utilization of N through crop-livestock systems have been recently reviewed (Rufino et al. 2006). We summarize results of experiments with manure in Table 6. Some early experiments with cattle manure in several agroecological zones in Kenya showed that maize response to manure application was different across sites (Smaling et al. 1992). At a site with P-fixing soils, manure was shown to be particularly effective in increasing maize yields, with application of 5 Mg ha^{-1} increasing yields from 2.3 to 4.5 Mg ha^{-1}. Contrasting results across sites were also found in Tanzania, where at one site NRE was only 0.1 kg kg^{-1} compared with 0.32 kg kg^{-1} at the other site, when 7.5 Mg manure of similar quality were applied to maize (Jensen et al. 2003). In Zimbabwe, NRE ranged between 0.15 and 0.29 kg kg^{-1} when manure was applied alone or in combination with N fertilizer (Nyamangara et al. 2003), while on a degraded sandy soil application of 17 Mg ha^{-1} poor quality manure alone could not supply sufficient nutrients to a maize crop (Chikowo et al. 2004). Mando et al. (2005) reported the long-term effects of tillage and manure application on sorghum in the Sudano-Sahelian conditions. The increase in yields was associated with increased N availability due to manure and greater water availability that improved the efficiency of use of the applied fertilizer.

Capture and recovery efficiency of N and P in fallow/cover crop-cereal crop sequences

Inputs of N from N_2-fixation in tropical cropping systems are limited by both the small proportion of legumes actually grown and by the restrictions placed on the fixation rate by drought and nutrient deficiencies (Giller 2001). Nitrogen cycling through leguminous shrubs has had mixed fortunes on many smallholder farms, with N recovery from organic materials on light textured soils found to be pitifully poor in some cases and promising in others (e.g. Chikowo et al. 2004; Mafongoya and Dzowela 1999; Mtambanengwe and Mapfumo 2006). Carsky et al. (1999) showed that many legumes accumulated large amounts of N, but there also were large N losses during the long dry season resulting in poor translation into improved rotational maize yields. In contrast to these results, Sanginga et al. (1996) estimated *Mucuna pruriens* N fertilizer replacement value of 120 kg N ha^{-1} in a derived savanna where the dry season is only 3 months long. So far, the general experience with the cover crops is that they have to be targeted on fields that are not yet extremely depleted and acidic, to be able to produce acceptable biomass of at least 2 Mg ha^{-1}.

In Kenya, Stahl et al. (2002) reported the contribution of 22-month fallows of *Sesbania* sesban and *Calliandra* on two subsequent maize crops. The immediate post-fallow maize crop suffered from drought, resulting in poor N recovery. During the more favorable second season, *Sesbania* more than doubled maize yields, with similar effects as 60 kg ha^{-1} fertilizer N. Nitrogen recovery efficiency with *Calliandra* was comparatively poor, in line with its low quality. In eastern Zambia, Kwesiga and Coe (1994) demonstrated that maize yields following 2-year *Sesbania* fallows were equivalent to application of 112 kg N ha^{-1}. Mafongoya and Dzowela (1999) also showed *Sesbania* as a promising improved fallow species on Alfisols in Zimbabwe, though research on sandy soils (Chikowo et al. 2004)

indicated that *Sesbania* produced very little biomass and was therefore unsuitable.

In western Kenya Gachengo et al. (1999) reported increased N uptake and NRE when the *Senna spectablis* or *Tithonia* where applied in combination with P fertilizer. Highest yields were obtained with *Tithonia* plus P fertilizer treatment. Overall, PRE ranged from 0.11 to 0.29 kg kg^{-1}. The range of plant materials with critical total P concentrations of 2.4 g kg^{-1} with the propensity to cause net P release is narrow and a soil fertility strategy that involves replenishing soil P with plant materials alone seems to be bound to failure in many African cropping systems (Kwabiah et al. 2003; Palm et al. 2001). When all data from experiments that involved improved fallows and cover crops were pooled, only a weak relationship between N availability and N uptake could be established, while a general tendency for reduced NRE with increased N availability existed (Fig. 2).

Resource capture and recovery efficiencies in grain legume-cereal rotations

Grain legumes fortify food security in many rural communities in SSA through strengthening sustainable production of cereals grown in sequence. For example, a review by Mpepereki et al. (2000) indicates that promiscuous soybean varieties with low N harvest indices have been successfully grown in rotation with maize over years by smallholder farmers in southern Africa. The benefits to the rotational cereal crop have also been demonstrated, e.g. Osunde et al. (2003) and Sanginga et al. (2002) in West Africa (Table 7). However, the reasons for the increased rotational maize yields are often not straight forwardly related to the N balances as a result of the legume N fixation. For example, in the Guinea savanna zone of Nigeria, Osunde et al. (2003) showed that despite a cumulative negative N balance of 147 kg N ha^{-1} after two successive cropping of soybean with stover exported, maize yields were at least 2 Mg ha^{-1} (45 kg ha^{-1} N uptake) compared with 0.5 Mg ha^{-1} (11 kg ha^{-1} N uptake) for the fallow plots (Table 7). Also, in the moist savanna, Sanginga et al. (1996, 2002) used five soyabean lines and investigated their residual effects on maize. Soybean net N input from fixation ranged from −8 to 43 kg ha^{-1} N, and the rotational effects on maize were all positive with soyabean N contribution

Nutr Cycl Agroecosyst (2010) 88:59–77

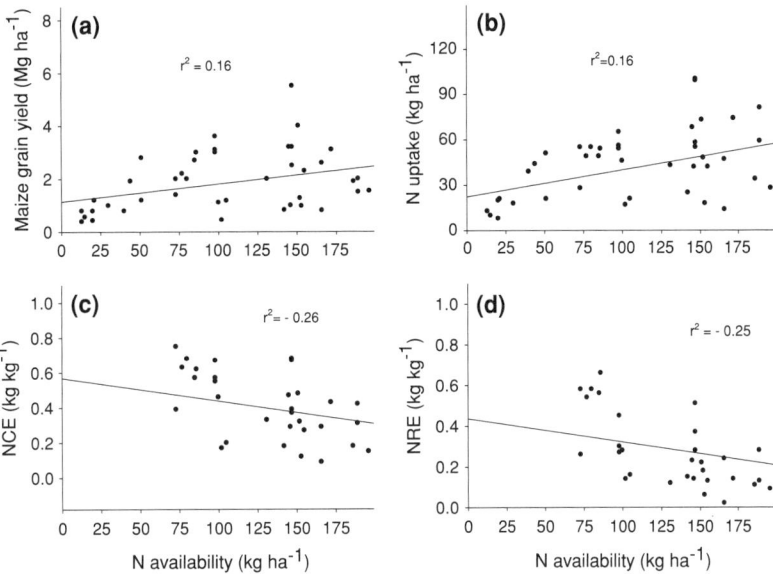

Fig. 2 The relationships between N availability and **a** maize grain yield **b** N uptake **c** N capture efficiency (NCE) and **d** N recovery efficiency (NRE) for experiments that involved improved fallows/cover crops

to maize N uptake ranging from 16 to 27 kg ha^{-1} N, again in spite of net N depletion by one of the soybean lines. However, in Benin, Ogoke et al. (2003) showed that only modest positive N contributions of up to 10 kg ha^{-1} N are attainable in a soyabean-cereal cropping system, this only possible when the soybean variety is late maturing, some P is applied and all the soybean residues are retained in the field. In principle, N balances are often used to estimate potential impact on the in-coming crop, but there seems to be no direct link when grain legumes are involved. Therefore, an analysis of N capture efficiency by maize following grain legumes, based on estimated net-N contribution and N balance approach alone, is not sufficient as it fails to capture the other positive 'rotational effects' that result in increased maize yields in cases were N balances are negative. Other than open statements, we are yet to come across publications where the 'sparing effects' of grain legumes have been quantitatively given for different environments. An attempt towards this objective will be useful for the development of useful algorithms for modeling grain legume-cereal sequences.

Nutrient capture and soil water relations

Nitrogen and P utilization in experiments that span across several cropping seasons were found to be

variable, and arguably one of the factors responsible for this variability is rainfall. A fragmented approach in which the focus is on single elements of the farming systems such as nutrient supply or soil and water conservation will likely fail to generate substantial impact in semi arid areas, where crop production is equally limited by soil water availability and nutrients. High water losses are associated with poorly managed soil surfaces in hot environments with high evaporative demand. For example, in rain-fed millet in West Africa, Wallace (2000) reported that soil evaporation losses constituted 30–45% of rainfall, runoff and drainage constituted 40–50%, while only 15–30% of the rainfall was available for transpiration. In fields where farmers establish sparse crop stands as a management strategy under poor fertility, evaporation losses are likely to be higher. Generally, the high runoff losses are a result of infrequent but intensive rainfall, and the tendency of sandy soils to form crusts with low infiltration rates. To manage variable rainfall environments, Piha (1993) devised and successfully tested a flexible system of fertilization, in which theoretically optimum rates of phosphorus, potassium and sulfur fertilizers are applied based on yield potential in an average rainfall season, while N is applied as a series of split applications, which are adjusted during the season according to the degree of water stress

Table 7 Estimated net N input from N₂-fixation, continuous or rotational cereal grain yields, total N uptake, and the legume contribution to maize N uptake for grain legume-cereal sequences in sub-Saharan Africa

Treatment description/crop sequences	General conditions	Net N input from N$_2$-fixation (kg ha^{-1})	Grain yields (Mg ha^{-1})	N uptake (kg ha^{-1})	Legume contribution to cereal N uptake (kg ha^{-1})	Country	Reference
Fallow	Guinea savanna		0.50	10		Nigeria	Osunde et al. (2003)
Soyabean (promiscuous)—maize		−65	2.10	38	28		
Soyabean (specific variety)—maize		−47	2.60	47	37		
Continuous maize	Alfisol, savanna soil		1.22	41		Nigeria	Sanginga et al. (2002)
Soybean variety 1- maize		−8	1.54	57	16		
Soybean variety 2- maize		11	2.42	68	27		
Soybean variety 3- maize		15	3.02	67	26		
Soybean variety 4- maize		30	1.45	58	17		
Soybean variety 5- maize		43	1.98	64	23		
Continuous maize	Loamy sand (site 1)		0.38	8		Zimbabwe	Kasasa et al. (1999)
Soybean (promiscuous)—maize		13	1.59	30	22		
Soybean (specific variety)—maize		7	1.11	23	15		
Continuous maize	Loamy sand (site 2)		0.36	7			
Soybean (promiscuous)—maize		26	1.62	36	30		
Soybean (specific variety)—maize		−7	1.17	25	18		
Maize–maize	Sandy soil		0.20	6		Zimbabwe	Chikowo et al. (2004)
Soybean–maize		8	0.50	15	7		
Continuous millet	0.16–0.5% C	Not given	0.94	26		Niger	Bagayoko et al. (2000)
Cowpea –millet		Not given	1.26	36	10		
Continuous sorghum			0.40	12			
Cowpea—sorghum		Not given	0.56	16	4		
Continuous maize	Loamy sand		2.46	44		Zimbabwe	Waddington and Karingwindi (2001)
Groundnut-maize		Not given	4.61	84	40		

 Springer

observed. This system optimized resource use efficiency during good rainfall seasons, while ensuring minimum losses in case of drought due to the reduced fertilizer inputs.

Conclusions and future perspectives

This study is an exhibit of research carried out on soils that have been degraded and run-down over years due to lack of soil fertility investments and therefore decades of nutrient mining. Cases of naturally fragile soils and tropical ecosystems e.g. extremely sandy soils and P-fixing acid soils, also presented challenges to increased nutrient use efficiency. The complexity of systems across Africa calls for complementary exploration with modeling tools. Recently, the NUANCES modeling framework, which recognizes the heterogeneity between farmers and within farming systems, allowing the exploration of trade-offs between different options, has been developed and tested (Tittonell et al. 2008, 2009).

The study has been an ambitious project to define N and P use efficiencies in cropping systems across SSA as we endeavor to have an in depth understanding of the systems. This work indicated that N and P use efficiencies in SSA cropping systems are diverse, being a logical consequence of poor correlation between yields and N or P availability in environments with other multiple constrains. For example, NCE ranged from as little as 0.05 to >0.70 kg kg^{-1}. Numerous examples were found in which response to nutrients applied were meager when other resources were limiting. Flexible systems of fertilization that vary N input according to the current seasonal rainfall pattern offer opportunities for high resource capture and recovery efficiencies in semi-arid areas. In much of our work, we employ integrated soil fertility management (ISFM) as a gateway to increased resource use efficiencies, in the process strongly subscribing to the need to balance nutrient inputs for efficient use as discussed in 'Efficient use of nutrients—an art of balancing' (Janssen 1998).

This study will probably direct some readers into a 'so what' mode. In the years ahead, scientists working across SSA will continue to re-design and execute their experimental research programs that will produce extra scientific information. Those as optimistic as us will continue to hope that strides are being made towards the coveted Green revolution for Africa in the light of the many challenges we have highlighted in this paper. Undoubtedly, there is another constituency of scientists who are getting weary and frustrated by what they perceive as an extremely slow sub-continent.

Acknowledgments We are grateful to the European Commission for funding of the Africa-NUANCES project under the INCO program. Funding from the Soil Fertility Consortium for Southern Africa (SOFECSA), through its activities under the Forum for Agricultural Research in Africa (FARA)'s sub-Saharan Africa Challenge Program (SSA-CP), enabled the completion of this study.

References

Bagayoko M, Buerkert A, Lung G, Bationo A, Romheld V (2000) Cereal/legume rotation effects on cereal growth in Sudano-Sahelian West Africa: soil mineral nitrogen, mycorrhizae and nematodes. Plant Soil 218:103–116

Baijukya FP, de Ridder N, Giller KE (2006) Nitrogen release from decomposing residues of leguminous cover crops and their effect on maize yield on depleted soils of Bukoba district, Tanzania. Plant Soil 279:77–93

Bar AR, Baggie I, Sanginga N (2000) The use of Sesbania (*Sesbania rostrata*) and urea in lowland rice production in Sierra Leone. Agrofor Syst 48:111–118

Bationo A, Ayuk E, Ballo D, Koné M (1997) Agronomic and economic evaluation of Tilemsi phosphate rock in different agroecological zones of Mali. Nutr Cycl Agroecosyst 48:179–189

Becker M, Johnson DE (1999) The role of legume fallows in upland rice-based systems of West Africa. Nutr Cycl Agroecosyst 53:71–81

Brouwer J, Powell JM (1998) Increasing nutrient use efficiency in West-African agriculture: the impact of micro-topography on nutrient leaching from cattle and sheep manure. Agric Ecosyst Environ 71:229–239

Cadisch G, de Willigen P, Suprayogo D, Mobbs DC, van Noordwijk M, Rowe EC (2004) Catching and competing for mobile nutrients in soils. In: van Noordwijk M, Cadisch G, Ong CK (eds) Below-ground interactions in tropical agroecosystems: concepts and models with multiple plant components. CAB International, Wallingford, pp 171–191

Carsky RJ, Oyewole B, Tian G (1999) Integrated soil management for the savanna zone of W. Africa: legume rotation and fertilizer N. Nutr Cycl Agroecosyst 55:95–105

Chikowo R, Mapfumo P, Nyamugafata P, Nyamadzawo G, Giller KE (2003) Nitrate-N dynamics following improved fallows and spatial maize root development in a Zimbabwean sandy clay loam. Agrofor Syst 59:187–195

Chikowo R, Mapfumo P, Nyamugafata P, Giller KE (2004) Maize productivity and mineral N dynamics following different soil fertility management practices on a depleted sandy soil in Zimbabwe. Agric Ecosyst Environ 102:119–131

Fofana B, Tamelokpo A, Wopereis MCS, Breman H, Dzotsi K, Carsky RJ (2005) Nitrogen use efficiency by maize as affected by a mucuna short fallow and P application in the coastal savanna of West Africa. Nutr Cycl Agroecosyst 71:227–237

Gachengo CN, Palm CA, Jama B, Othieno C (1999) Tithonia and senna green manure and inorganic fertilizers as phosphorus sources for maize in Western Kenya. Agrofor Syst 44:2–36

Gandah M, Bouma J, Brouwer J, Hiernaux P, van Duivenbooden N (2003) Strategies to optimize allocation of limited nutrients to sandy soils of the Sahel: a case study from Niger, West Africa. Agric Ecosyst Environ 94:311–319

Giller KE (2001) Nitrogen fixation in tropical cropping systems, 2nd edn. CAB International, Wallingford

Giller KE, Rowe EC, de Ridder N, van Keulen H (2006) Resource use dynamics and interactions in the tropics: scaling up in space and time. Agric Syst 88:8–27

Haefele SM, Wopereis MCS (2005) Spatial variability of indigenous supplies of N, P and K and its impact on fertilizer strategies for irrigated rice in West Africa. Plant Soil 270:57–72

Hagedorn F, Steiner KG, Sekayange L, Zech W (1997) Effect of rainfall pattern on nitrogen mineralization and leaching in a green manure experiment in south Rwanda. Plant Soil 195:365–375

Janssen BH (1998) Efficient use of nutrients: an art of balancing. Field Crops Res 56:197–201

Janssen BH, Guiking FCT, van der Eijk D, Smaling EMA, Wolf J, van Reuler H (1990) A system for quantitative evaluation of tropical soils (QUEFTS). Geoderma 46:299–318

Jensen JR, Bernhard RH, Hansen S, McDonagh J, Moberg JP, Nielsen NE, Nordbo E (2003) Productivity in maize based cropping systems under various soil-water-nutrient management strategies in a semi-arid, Alfisol environment in East Africa. Agric Water Man 59:217–237

Kang BT, Caveness FE, Tian G, Kolawole GO (1999) Long-term alley cropping with hedgerow species on an Alfisol in southwestern Nigeria—effect on crop performance, soil chemical properties and nematode populations. Nutr Cycl Agroecosyst 54:145–155

Kasasa P, Mpepereki S, Musiiwa K, Makonse F, Giller KE (1999) Residual nitrogen fixation benefits of promiscuous soybeans to maize under smallholder field conditions. Afr Crop Sci 7:375–382

Kaya B, Nair PKR (2001) Soil fertility and crop yields under improved fallow systems in southern Mali. Agrofor Syst 52:1–11

Kho RM (2000) On crop production and the balance of available resources. Agric Ecosyst Environ 80:71–85

Kwabiah AB, Stoskopf NC, Palm CA, Voroney RP (2003) Soil P availability as affected by the chemical composition of plant materials: implications for P-limited agriculture in tropical Africa. Agric Ecosyst Environ 100:53–61

Kwesiga FR, Coe R (1994) The effect of short rotation Sesbania sesban planted fallows on maize yield. For Ecol Manag 64:199–208

Lupwayi NZ, Haque I, Saka AR, Siaw DEKA (1999) Leucaena hedgerow intercropping and cattle manure application in

the Ethiopian highlands II. Maize yields and nutrient uptake. Biol Fertil Soil 28:196–203

Mafongoya PL, Dzowela BH (1999) Biomass production of tree fallows and their residual effects on maize in Zimbabwe. Agrofor Syst 47:139–151

Mando A, Ouattara B, Somado AE, Wopereis MCS, Stroosnijder L, Breman H (2005) Long-term effects of fallow, tillage and manure application on soil organic matter and nitrogen fractions and on sorghum yield under Sudano-Sahelian conditions. Soil Use Manag 21:25–31

Mpepereki S, Javaheri F, Davis P, Giller KE (2000) Soyabeans and sustainable agriculture: promiscuous soyabeans in southern Africa. Field Crop Res 65:137–149

Msolla MM, Semoka JMR, Borggaard OK (2005) Hard Minjingu phosphate rock: an alternative P source for maize production on acid soils in Tanzania. Nutr Cycl Agroecosyst 72:299–308

Mtambanengwe F, Mapfumo P (2005) Organic matter management as an underlying cause for soil fertility gradients on smallholder farms in Zimbabwe. Nutr Cycl Agroecosyst 73:227–243

Mtambanengwe F, Mapfumo P (2006) Effects of organic resource quality on soil profile N dynamics and maize yields on sandy soils in Zimbabwe. Plant Soil 281:173–191

Nyamangara J, Bergström LF, Piha MI, Giller KE (2003) Fertilizer use efficiency and nitrate leaching in a tropical sandy soil. J Environ Qual 32:599–606

Ogoke IJ, Carsky RJ, Togun AO, Dashiell K (2003) Effect of P fertilizer application on N balance of soybean crop in the guinea savanna of Nigeria. Agric Ecosyst Environ 100:153–159

Oikeh SO, Carsky RJ, Kling JG, Chude VO, Horst WJ (2003) Differential N uptake by maize cultivars and soil nitrate dynamics under N fertilization in West Africa. Agric Ecosyst Environ 100:181–191

Osiname OA, Meppe F, Everett L (2000) Response of maize (Zea mays) to phosphorus application on basaltic soils in Northwestern Cameroon. Nutr Cycl Agroecosyst 56:209–217

Osunde AO, Bala A, Gwam MS, Tsado PA, Sanginga N, Okogun JA (2003) Residual benefits of promiscuous soybean to maize grown on farmers' fieds around Minna in the southern Guinea savanna zone of Nigeria. Agric Ecosyst Environ 100:209–220

Ouédraogo E, Mando A, Zombré NP (2001) Use of compost to improve soil properties and crop productivity under low input agricultural system in West Africa. Agric Ecosyst Environ 84:259–266

Palm CA, Gachengo CN, Delve RJ, Cadisch G, Giller KE (2001) Organic inputs for soil fertility management in tropical agroecosystems: application of an organic resource database. Agric Ecosyst Environ 83:27–42

Phiri ADK, Kanyama-Phiri GY, Snapp S (1999) Maize and Sesbania production in relay cropping at three landscape positions in Malawi. Agrofor Syst 47:153–162

Piha MI (1993) Optimizing fertilizer use and practical rainfall capture in a semi-arid environment with variable rainfall. Exp Agric 29:405–415

Probert ME, Okalebo JR (1992) Effects of phosphorus on the growth and development of maize In: Probert ME (ed) A

Nutr Cycl Agroecosyst (2010) 88:59–77

search for strategies for sustainable dryland cropping in semi-arid Kenya. Proceedings of a symposium held in Nairobi, Kenya, 10–11 December 1990. Canberra, Australia, ACIAR proceedings No. 41

Rockström J, de Rouw A (1997) Water, nutrients and slope position in on-farm pearl millet cultivation in the Sahel. Plant Soil 195:311–327

Rufino MC, Rowe EC, Delve RJ, Giller KE (2006) Nitrogen cycling efficiencies through resource-poor African crop-livestock systems. Agric Ecosyst Environ 112:261–282

Sahrawat KL, Jones MP, Diatta S (1997) Direct and residual fertilizer phosphorus effects on yield and phosphorus efficiency of upland rice in an Ultisol. Nutr Cycl Agroecosyst 48:209–215

Saïdou A, Janssen BH, Temminghoff EJM (2003) Effects of soil properties, mulch and NPK fertilizer on the maize yields and nutrient budgets on ferralitic soils in southern Benin. Agric Ecosyst Environ 100:265–273

Samaké O, Smaling EMA, Kropff MJ, Stomph TJ, Kodio A (2006) Effects of cultivation practices on spatial variation of soil fertility and millet yields in the Sahel of Mali. Agric Ecosyst Environ 109:335–345

Sanchez PA, Sherperd KD, Soule MJ, Place FM, Buresh RJ, Izac AM, Mokunywe AU, Kwesiga FR, Ndiritu CG, Woomer PL (1997) Soil fertility replenishment in Africa: an investment in natural resource capital. In: Buresh RJ, Sanchez PA, Calhoun F (eds) Replenishing soil fertility in Africa. SSSA Spec. Publ. 51 ASA and SSSA, Madison

Sangaré M, Rernández_Rivera S, Hiernaux P, Bationo A, Pandey V (2002) Influence of dry season supplementation for cattle on soil fertility and millet (*Pennisetum glaucum* L.) yield in a mixed crop/livestock production system of the Sahel. Nutr Cycl Agroecosyst 62:209–217

Sanginga N, Ibewiro B, Houngnandan P, Vanlauwe B, Okogun JA, Okobundu IO, Versteeg M (1996) Evaluation of symbiotic properties and the nitrogen contribution of mucuna to maize grown in the derived savanna of West Africa. Plant Soil 179:119–129

Sanginga N, Okogun J, Vanlauwe B, Dashiell K (2002) The contribution of nitrogen by promiscuous soybeans to maize based cropping in the moist savanna of Nigeria. Plant Soil 241:223–231

Sigunga DO, Janssen BH, Oenema O (2002) Effects of improved drainage and nitrogen source on yields nutrient, uptake and utilization efficiencies by maize (*Zea mays* L.) on Vertisols in sub-humid environments. Nutr Cycl Agroecosyst 62:263–275

Smaling EMA, Nandwa SM, Prestele H, Roetter R, Muchena FN (1992) Yield response of maize to fertilizers and manure under different agro-ecological conditions in Kenya. Agric Ecosyst Environ 41:241–252

Smaling EMA, Nandwa SM, Janssen BH (1997) Soil fertility in Africa is at stake. In: Buresh RJ, Sanchez PA, Calhoun F (eds) Replenishing soil fertility in Africa. SSSA Publication 51 SSSA and ASA, Madison, pp 47–61

Stahl L, Nyberg G, Högberg P, Buresh RJ (2002) Effects of planted tree fallows on oil nitrogen dynamics, above ground and root biomass, N2-fixation and subsequent

maize crop productivity in Kenya. Plant Soil 243:103–117

Tittonell P, Vanlauwe B, Leffelaar PA, Shepherd KD, Giller KE (2005) Exploring diversity in soil fertility management of smallholder farms in western Kenya. II. Within-farm variability in resource allocation, nutrient flows and soil fertility status. Agric Ecosyst Environ 110:166–184

Tittonell P, Corbeels M, van Wijk MT, Vanlauwe B, Giller KE (2008) Targeting nutrient resources for integrated soil fertility management in smallholder farming systems of Kenya. Agron J 100:1511–1526

Tittonell P, van Wijk MT, Herrero M, Rufino MC, de Ridder N, Giller KE (2009) Beyond resource constraints— Exploring the biophysical feasibility of options for the intensification of smallholder crop-livestock systems in Vihiga district, Kenya. Agric Syst 101:1–19

van Asten PJA, van Bodegom PM, Mulder LM, Kropff MJ (2005) Effect of straw application on rice yields and nutrient availability on an alkaline and a pH-neutral soil in a Sahelian irrigation scheme. Nutr Cycl Agroecosyst 72:255–266

Van Duivenbooden N, de Wit CT, van Keulen H (1996) Nitrogen, phosphors and potassium relations in five major cereals reviewed in respect to fertilizer recommendations using simulation modelling. Fertil Res 44:37–49

Vanlauwe B, Diels J, Lyasse O, Aihou K, Iwuafor ENO, Sanginga N, Merckx R, Deckers J (2002) Fertility status of soils of the derived savanna and northern guinea savanna and the response to plant nutrients, as influenced by soil type and land use management. Nutr Cycl Agroecosyst 62:139–150

Vanlauwe B, Diels J, Sanginga N, Merckx R (2005) Long-term integrated fertility management in South-western Nigeria: crop performance and impact on the soil fertility status. Plant Soil 273:337–354

Waddington SR, Karingwindi J (2001) Productivity and profitability of maize + groundnut rotations compared with continuous maize on smallholder farms in Zimbabwe. Exp Agric 37:83–98

Wallace JS (2000) Increasing agricultural water use efficiency to meet future food production. Agric Ecosyst Environ 82:105–119

Wopereis MCS, Tamélokpo A, Ezui K, Gnakpénou D, Fofana B, Breman H (2006) Mineral fertilizer management of maize on farmer fields differing in organic inputs in the West African savanna. Field Crop Res 96:355–362

Zaongo CGL, Wendt CW, Lascano RJ, Juo ASR (1997) Interactions of water, mulch and nitrogen on sorghum in Niger. Plant Soil 197:119–126

Zingore S, Murwira HK, Delve RJ, Giller KE (2007) Influence of nutrient management strategies on variability of soil fertility, crop yields and nutrient balances on smallholder farms in Zimbabwe. Agric Ecosyst Environ 119:112–126

Zougmoré R, Mando A, Stroosnijder L, Ouédraogo E (2004) Economic benefits of combining soil and water conservation measures with nutrient management in semiarid Burkina Faso. Nutr Cycl Agroecosyst 70:261–269

Greenhouse Evaluation of Agronomic Effectiveness of Unacidulated and Partially Acidulated Phosphate Rock from Kodjari and the Effect of Mixed Crop on Plant P Nutrition

E. Compaore, J.-C. Fardeau, and J.-L. Morel

Abstract Greenhouse study was conducted to determine the agronomic effectiveness of unacidulated (KPR) and partially acidulated (KAPR) Kodjari phosphate rock relative to triple superphosphate (TSP). The available phosphorus of a ferruginous soil (pH 6.2) was labelled with a $^{32}PO_4^{3-}$ solution, and the soil was amended with 50 mg P kg^{-1} as ground KPR, KAPR and TSP. Two plants, maize and cowpea, were grown alone or in association. After 60 days of growth period, shoot yield, P uptake and real coefficient of P utilization (RCU%) were determined. Dry matter yields and P in dried tops increased by KAPR and TSP application; however, with KPR, no significant difference with the control was recorded. Also in KPR treatment, the dry matter yield and P uptake for maize and cowpea, grown in association, decreased in all treatments. The agronomic effectiveness of KPR, KAPR and TSP in terms of RCU% was classed in this order: KPR < KAPR < TSP. The average values of RCU% in KPR treatments of pure or mixed culture were very low (0.3%), indicating that KPR was not dissolved in the soil. In contrast, they were high in KAPR and TSP treatments. In KPR treatments planted with maize alone or in association, the RCU% values were higher than those measured with cowpea. This indicated that maize seemed to use KPR more efficiently than did cowpea due to organic acid anions excreted by its roots and higher root densities. At last, the association did not significantly increase the use of KPR by plants.

Keywords Phosphate rock · Agronomic effectiveness · *Zea mays* L. · *Vigna unguiculata* L. Walp · Mixed cropping · Dissolution

Introduction

In West Africa, the available soil phosphorus (P) is one of the most limiting factors of crop production. But it is possible to increase the level of soil-available P by adding P fertilizers (Fardeau and Frossard, 1991; Compaoré et al., 2001) as water-soluble P, phosphate rock (PR) or manure. Due to the low income of West African farmers, there is an increasing interest in the use of cheaper alternative P fertilizers as PR, for direct application or after partial acidulation. Burkina Faso, following the example of several West African countries, has large deposits of low-grade rock phosphate, the Kodjari phosphate rock (KPR). The use of this indigenous phosphate rock may improve the phosphate status of soils at low cost if the rock phosphate amended is dissolved in soil and effectively used by crops. The Kodjari phosphate rock is generally regarded as being a rock of low chemical reactivity (Paul, 1988). Thus, its slow release characteristics make it poorly suited for direct application. Legumes are known to be efficient users of PR because of the acidifying effect of N_2 fixation on the rhizosphere. A decrease in the rhizosphere pH can enhance PR dissolution (Khasawneh and Doll, 1978) and therefore increase uptake of rock-derived P by these plants, which acidify the rhizosphere (Bekele et al., 1983; Haynes, 1983, 1992). If legume can enhance PR dissolution, then there is an opportunity to use them for

E. Compaore (✉)
LSE/ENSAIA, Vandoeuvre-lès-Nancy Cedex, France; Station de Recherches Agricoles de Farako-Bâ, Institute of the Environment and Agricultural Research Institute (INERA), Bobo-Dioulasso 01, Burkina Faso
e-mail: ecompaorez@hotmail.com

increasing the agronomic effectiveness of PR. Cowpea is the most used legume plant for feeding in Africa, particularly in the Sahelian zone (north Benin, Burkina Faso, Mali, Niger, north Nigeria, etc.). It grows in West Africa mainly in association with sorghum, millet, maize, cassava yam and sesame (Stoop, 1979). The main arguments advanced to justify this cultural system are the judicious use of space and the nutrient because the legume plant may actively fix N_2. But there are no data on the capacity of cowpea to acidify its rhizosphere and favour the use of PR.

This chapter reports on the performance of unacidulated and acidulated Kodjari phosphate rock relative to TSP with two crops species, a legume plant (cowpea) and a cereal plant (maize) grown alone or in association on a ferruginous soil labelled with ^{32}P using the method described by Fardeau (1996). The proposition that legumes may enhance PR agronomic effectiveness was examined.

Materials and Methods

Soil

The 0–20-cm layer of an unfertilized ferruginous soil from a field cultivated since 16 years was used. The soil collected in dry season was ground to 2 mm size before use. Some properties of the soils are listed in Table 1. The available P and the P fixing capacity were determined by isotopic method (Frossard et al., 1994; Fardeau, 1996).

Table 1 Some properties of the soil studied

pH (H_2O)	6.2
pH (KCl)	5.2
% Clay	8.9
% Silt	21.9
% Sand	69.2
% Organic matter	0.5
CEC (meq 100 g^{-1})	3.5
% N	0.041
% C	0.35
Al_d (mg g^{-1})	0.4
Fe_d (mg g^{-1})	4
Si_d (mg g^{-1})	0.8

Phosphate Fertilizers

Three P fertilizers, unacidulated (KPR), partially acidulated (KAPR) Kodjari phosphate rocks and triple superphosphate (TSP) were used. The mineralogical composition of KPR determined by X-ray diffraction techniques gives 63.7% francolites, 27% quartz and 9.1% amorphous. It contains 11.2% P with water solubility of 0.03%.

The PAPR was prepared in the laboratory by treating Kodjari phosphate rock at 25% with sulphuric acid and monoammonium phosphate. The KAPR obtained had a total P content of 9.4% of which 12.9% was water soluble. The TSP used in this experiment as the water-soluble fertilizer contained 20.6% total P.

Experimental Procedure

The experiment was carried out in a greenhouse, where the temperature ranged from 20 to 30°C in the day and from 15 to 20°C in the night. The 2 mm fraction of soil was divided into four sets of 15 kg corresponding to the four treatments (control, KPR, KAPR and TSP).

Each set of 15 kg of soil was placed in a heavy plastic recipient and 450 mL of a solution containing 0.82 M Bq L^{-1} of $^{32}PO_4^{3-}$ was added and homogeneously mixed with the soil. Then powdered P fertilizers applied as KPR, KAPR and TSP were thoroughly mixed with the soil at rate equivalent to 50 mg P kg^{-1}. After P addition, the equivalent of 1 kg dry soil labelled was placed into plastic bags. Five replicates of each treatment were then prepared giving a total of 60 pots. The pots were then sown with pre-germinated maize (*Zea mays* L. var. KPB) and cowpea (*Vigna unguiculata* L. Walp. var. KN-1) seeds, one maize seed or one cowpea seed per pot and one maize seed associated with one cowpea seed per pot. Basal nutrient solutions were added to each pot to ensure that P was the only element-limiting yield. The following elements were added per pot: 164 mg K as K_2SO_4, 147 mg N as NO_3NH_4, 34 mg Mg as $MgSO_4 \cdot 7H_2O$, 3.9 mg Mn as $MnSO_4 \cdot H_2O$, 2.7 mg Cu as $CuSO_4 \cdot 5H_2O$ and 0.5 mg Zn as $ZnSO_4 \cdot 7H_2O$. The soils in the pots were watered daily at 70–80% of field capacity by weight using deionized water.

Plants were grown for 2 months (60 days) and then whole tops were cut at ground level and were oven dried at 70°C for 72 h. Dry matter (DM) obtained was weighed and then was mineralized in an oven at 550°C. The resulted ashes were dissolved in dilute nitric acid (HNO_3). The total P in all extracts was measured colorimetrically by molybdenum blue method (John, 1970). In the same extracts the radioactivity was measured by Cerenkov radiation counting using a liquid scintillation.

Data Analysis

To compare the agronomic performance of the different P fertilizers, the real coefficient of P utilization (RCU%) of the fertilizers was calculated. This parameter was determined using isotopic dilution concepts:

$$RCU\% = 100 \times [1/E \times (P_e - r_e/r_t \times P_t)]$$

where E is the quantity of P fertilizer applied, P_e is the P content in the plant harvested on the fertilized soil, P_t is the P content in the plant harvested on the unfertilized soil, r_e is the radioactivity in the plant harvested on the fertilized soil and r_t is the radioactivity in the plant harvested on the unfertilized soil.

Statistical analysis of data was performed by analysis of variance and Turkey procedure was used to make comparison among means.

Results and Discussion

Dry Matter Yield and P Uptake by Plants

Figure 1 shows the dry matter (DM) yield of maize and cowpea obtained from different P treatments on the ferruginous soil. Without P fertilizers, the average DM yield was very low, about 0.50 g DM per pot. It increased by the addition of KPR (0.90 g DM per pot), but this response to KPR was not significant. In contrast, significant response to P was obtained with KAPR (5.0 g DM per pot) and TSP (4.2 g DM per pot). These results indicated that KAPR was effective as TSP in increasing DM yield. The DM yield of maize or cowpea grown alone in KPR treatments was similar to the DM yield of maize or cowpea in the association. While the DM yield of sole maize or sole cowpea was significantly higher than that of the maize or the cowpea mixed cropped in KAPR and TSP treatments. The competition between plants may explain this result.

The P uptake by plants increased in KPR treatments but was not significantly different with P uptake by unfertilized plants (Fig. 2). But with KAPR, the quantity of P taken up by plants was highly superior to control, but significantly inferior to P taken up by plants in TSP treatments. The trends in P uptake data were not consistent with those of DM yield. According to the DM recorded in KAPR and TSP treatments, the P uptake in these treatments would be similar. This suggested that the production of DM in KAPR treatments was not only the result of P supply. The presence of others nutrients like calcium derived from the phosphate rock used for preparing KAPR can favour the vegetative development observed.

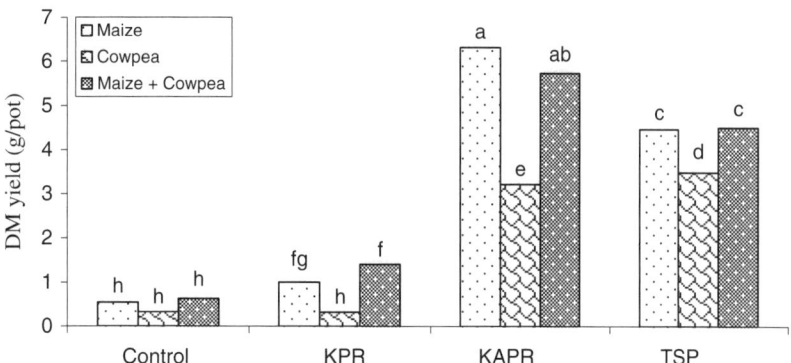

Fig. 1 Dry matter response to P application as KPR, KAPR and TSP

Fig. 2 P uptake from each source

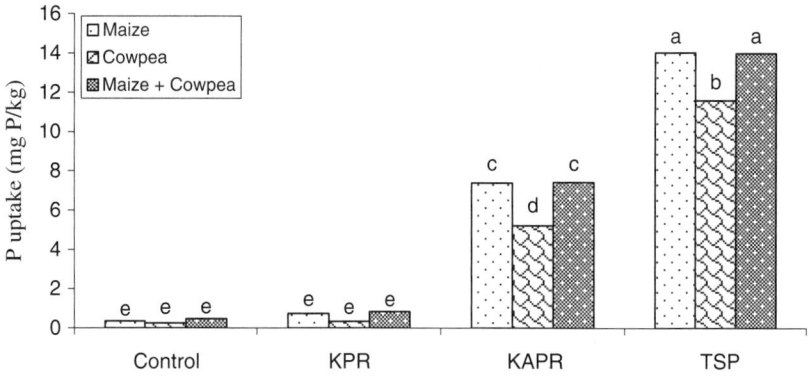

The P uptake into the plant tops increased with increasing water-soluble P in the fertilizer. The average P uptake by plant in KAPR and TSP treatments over control was 18- and 37-folds, respectively. In each case, plants grown alone absorbed more P than plants grown in association. The P uptake in KPR treatments by cowpea was inferior to P uptake in KPR treatments by maize. This result suggested that cowpea did not enhance the solubilization of KPR. As the buffering capacity of the soil used is low (Compaoré et al., 2003), the P derived from eventually dissolved KPR could increase the P uptake by plants, but this did not. Therefore the N_2 fixing, which must cause the rhizosphere acidification and consequently the KPR dissolution, did not occur. This is probably due to the bad development of cowpea in this P-deficient soil or relatively high pH of the soil. Rhizosphere acidification occurs also when plants are using ammonium nitrate (NH_4NO_3) as their major N source. The presence of ammonium N creates an acid environment around the root when NH_4^+ ions are absorbed. In the present pot experiment, N fertility was supplied as ammonium nitrate (NH_4NO_3); consequently this N fertilizer must favour the KPR dissolution (Bolan and Hedley, 1991). But the KPR dissolution did not occur. Therefore lack of KPR dissolution was then mostly due to its mineral composition.

Effectiveness of P Fertilizers

The average values of RCU% were 0.3, 12.0 and 25.0 for KPR, KAPR and TSP, respectively (Fig. 3). The relatively high RCU% obtained with TSP was probably due to the low available P content and the low P adsorption capacity of the soil studied. The poor performance of KPR was due to the chemical nature and the low water-soluble P content of KPR (Khasawneh and Doll, 1978; Paul, 1988), and probably to the soil pH, which was relatively high, i.e., pH 6.2. This value seems to be the upper soil limit for significant utilization of rock phosphate (Fardeau et al., 1988).

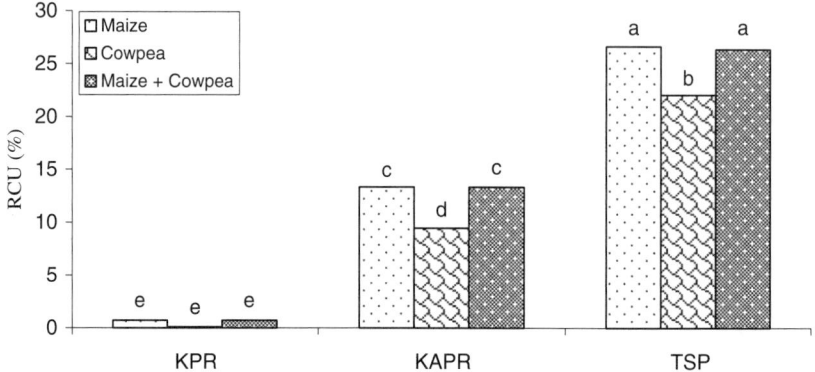

Fig. 3 RCU% of the different P fertilizers

The coefficient of soil P utilization (CU%) is always lower for cowpea than for maize in limiting conditions (KPR and control treatments). This means that cowpea requires more P than does maize. In the KPR treatments, both cowpea and maize plants used more soil P than did those of KPR. The supply of TSP increased the CU% values by 30-fold for maize and 72-fold for cowpea. This confirms that the cowpea plant did not use efficiently the unacidulated Kodjari phosphate rock.

For the sole crop, the RCU% values of maize were higher than those of cowpea in all treatments. The higher RCU% obtained with maize in KPR treatment can be attributed to specific plant abilities to absorb P from KPR such as root exudate production. The maize roots excrete some organic acids (Petersen and Bottger, 1991) that could increase the dissolution of the phosphate rock. As Kpomblekou and Tabatabai (1994) have shown organic acids can dissolve KPR. The maize plant could also use KPR better compared to cowpea plant because of well-developed roots (Hammond et al., 1986). Plant with higher root densities in the surface layers of the soil in which the PR is located would be expected to acquire dissolved P from the PR. The average dry root weight greater for maize (1.16 $\pm$ 0.38 g per pot) than for cowpea (0.41 $\pm$ 0.09 g per pot) recorded in this study supports this hypothesis.

Evidently the cowpea growing in this study was not relying significantly on symbiotically fixed N_2 since its growth was not lowered rhizosphere pH and then it could not use KPR. So that the quantity of P derived from KPR used by cowpea is very low.

Effect of Association on Maize and Cowpea P Nutrition

The quantities of P uptake by maize and cowpea in the case of association (maize + cowpea) did not increase significantly in comparison with those by maize and cowpea grown alone; in contrast, its trends seemed to decrease. This result was in contradiction to the common view that the association cereal + legume can enhance the mineral nutrition of the cereal. In the present study, the association did not absorb more P than did sole maize crop. The maximum decrease of P uptake in the association observed was 32% for maize and 63% for cowpea in TSP and KPR treatments, respectively. This result was probably the consequence of competition between plant species for P and light. The competition is less marked in the KPR treatments owing to their low available P content and the low growth of the plants. The maize also has exerted on the cowpea a competition for light, which contributed to reduction in the dry matter yield production. The competition in the rhizosphere for P led to decrease of P uptake by maize and cowpea. However, the cowpea plant was more affected by the interspecific competition than was the maize plant. Thus the average level of maize and cowpea P nutrition was not enhanced in the case of association. In the present pot experiment, the mixed cropping system had no benefit effect on the plant P nutrition.

Conclusions

The agronomic effectiveness of KPR was near zero. This result confirms that KPR is not suited for direct application. The use of cowpea, a legume susceptible to acidifying its rhizosphere and lead to KPR dissolution did not occur. The association (maize + cowpea) never increased the plant P nutrition because some constraints appeared. The severe deficiency of soil in available P and probable competition for P and light made more damage to cowpea, consequently its dry matter yield and its P uptake were drastically reduced.

The partial acidulation of KPR to KAPR increased its solubilization and enhanced its agronomic effectiveness significantly. TSP was the most effective fertilizer.

The maize plant seemed to use KPR more efficiently than did cowpea plant because maize root excretes some organic acids which can enhance KPR dissolution or because of high root densities.

In this pot experiment the small volume of 1 kg soil per pot used might be a limiting factor, because the soil was rapidly explored by plant roots before they achieved their maximum development. Therefore it would be useful to conduct again this kind of experiment with 5 kg soil per pot that might favour a better growth of plants.

References

Bekele T, Cino BA, Ehlert PAI et al (1983) An evaluation of plant-borne factors promoting the solubilization of alkaline rock phosphates. Plant Soil 75:361–378

Bolan NS, Hedley MJ (1991) Processes of soil acidification during nitrogen cycling with emphasis on legume based pastures. In: Wright RJ, Baligar VC, Murmann RP (eds) Plant soil interactions at low pH. Kluwer, Dordrecht, pp 169–179

Compaoré E, Fardeau JC, Morel JL et al (2001) Le phosphore biodisponible des sols: une des clés de l'agriculture durable en Afrique de l'Ouest. Cah Agric 10(2):81–85

Compaoré E, Frossard E, Fardeau JC et al (2003) Influence of land-use management on isotopically exchangeable phosphate in soils from Burkina Faso. Commun Soil Sci Plant Anal 34(1&2):201–223

Fardeau JC (1996) Dynamics of phosphorus in soils. An isotopic outlook. Fertil Res 45:91–100

Fardeau JC, Frossard E (1991) Processus de transformation du phosphore dans les sols de l'Afrique de l'Ouest semi-arides: Application au phosphore assimilable. In: Tiessen H, Frossard E (eds) Phosphorus cycles in terrestrial and aquatic ecosystems: regional workshop 4: Africa. S.C.O.P.E./UNEP, Nairobi, 18–22 Mar 1991

Fardeau JC, Morel C, Jahiel M (1988) Does long contact with the soil improve the efficiency of rock phosphate? Results of isotopic studies. Fertil Res 17:3–19

Frossard E, Fardeau JC, Brossard M et al (1994) Soil isotopically exchangeable phosphorus: a comparison between E and L values. Soil Sci Soc Am J 58:846–851

Hammond LL, Chien SH, Mokwunye AU (1986) Agronomic value of unacidulated and partially acidulated phosphate rocks indigenous to the tropics. Adv Agron 40:89–140

Haynes RJ (1983) Soil acidification induced by leguminous crops. Grass Forage Sci 38:1–11

Haynes RJ (1992) Relative ability of a range of crop species to use phosphate rock and monocalcium phosphate as P sources when grown in soil. J Sci Agric 60:205–211

John MK (1970) Colorimetric determination of phosphorus in soils and plant materials with ascorbic acid. Soil Sci 109:214–220

Khasawneh FE, Doll EC (1978) The use of phosphate rock for direct application to soils. Adv Agron 30:159–206

Kpomblekou K, Tabatabai MA (1994) Effect of organic acids on release of phosphorus from phosphate rocks. Soil Sci 158(6):442–452

Paul I (1988) Caractérisation physico-chimique et évaluation de l'efficacité de phosphates bruts ou partiellement acidifies provenant d'Afrique de l'Ouest. Thèse INPL, Nancy, France, 296p

Petersen W, Bottger M (1991) Contribution of organic acids to the acidification of the rhizosphere of maize seedlings. Plant Soil 132:159–163

Stoop WA (1979) Cereal-based intercropping systems for the West African semi-arid tropics, particularly Upper Volta. In: Proceedings of the international workshop on intercropping, Hyderabad, 10–13 Jan 1979

Effect of Continuous Mineral and Organic Fertilizer Inputs and Plowing on Groundnut Yield and Soil Fertility in a Groundnut–Sorghum Rotation in Central Burkina Faso

E. Compaore, P. Cattan, and J.-B.S. Taonda

Abstract Two field groundnut–sorghum experiments were conducted at Saria in the center of Burkina in order to assess the effect of chemical and organic fertilizers on groundnut yield and its components and soil fertility on two ferruginous tropical soils of different texture. Groundnut haulm yield, pod yield, number of pods, % pod two-seeded, % pod rot, seed yield, and 100-seed weight as well as sorghum dry shoot and grain yields were measured for 8 years. Most of the different variables were affected by continuous cropping without fertilizer application on the two soils, in particular on the coarser one. The effect of a likely deficiency in some nutrients (P, K, and Ca) was observed. The mineral fertilizers allowed maintaining yield, but their supply was not able to replenish the nutrient uptake by plants. The effect of compost on the crop production was weak but enhanced during the years 7 and 8. Tillage had also an effect on yield, but this effect was limited and varied with soil type. In the control, the initial status of organic matter reduced in 5 years and did not increase with the application of fertilizers and compost. In contrast the addition of fertilizers increased the content of total N and Bray-I P. Nitrogen, P, K, and Ca balance was negative in almost all treatments without mineral fertilizers. However, the surplus of production did not provide significant profits. The cultural techniques improved sorghum growth, while groundnut responded better than sorghum on the soil eroded.

Keywords Groundnut–sorghum rotation · Mineral and organic fertilizers · Soil fertility · Yields

Introduction

Groundnut is a major cash crop in both small and large farms in West Africa and aboveground forage (haulm) constitutes an important livestock feed in the West African savannah. It is a traditional smallholder low-input crop in Burkina Faso and it grows mainly rotated with cereals (sorghum, millet, maize). Soils in Burkina Faso are typically deficient in N and P (Stoorvogel et al., 1993; Compaoré et al., 2003), and phosphate-containing fertilizers are commonly used for groundnut (*Arachis hypogea* L.) production. Farmers are most likely to apply a NPK fertilizer for groundnut production while single superphosphate and triple superphosphate are little used due to higher cost. The use of fertilizers poses two main challenges: (i) their effectiveness in different cultural systems and climatic conditions which sometimes is unfavorable and (ii) long-term maintenance of their efficacies, for maintaining crop yield and soil fertility in continuously cropped soil. The steadily declining productivity and soil fertility decrease have been noticed in continuous cultivation in Burkina Faso (Piéri, 1989; Cattan and Schilling, 1992; Taonda et al., 1995). Other studies showed that crop yields have been observed to decline with continuous cropping, even when fertilizers are used, while nutrient supply and soil organic

E. Compaore (✉)
Station de Recherches Agricoles de Farako-Bâ, Institute of the Environment and Agricultural Research Institute (INERA), Bobo-Dioulasso, Burkina Faso
e-mail: ecompaorez@hotmail.com

matter decreased and soil became more acidic (Aguilar et al., 1988; Bell et al., 1995; Taonda et al., 1995; Manna et al., 2005). Studies by Bhandari et al. (2002) and Regmi et al. (2002) attributed the reduced productivity to declining soil organic matter, decreased soil fertility, occurrence of nutrient imbalances, and inappropriate fertilizer practices. But application of 100% NPK along with manure is suggested to stop deterioration in crop productivity and soil fertility and can lead to considerable buildup in the availability of P, K, and S (Nand, 2000). Also a continuous significant yield increase with increasing NPK level has been reported (Singh et al., 2001). Assessment of the effect of long-term cultivation and fertilizer use on productivity and soil fertility is necessary to suggest soil management practices for the sustainability of Burkina Faso agrosystems. The objective of this research was to evaluate the effects of organic and inorganic fertilizers and of tillage on the productivity and soil fertility of a continuously cropped groundnut–sorghum rotation on a ferruginous tropical soil in Central Burkina Faso.

Materials and Methods

Location, Climate, and Soils

Two long-term field experiments were conducted: one on a coarse sand soil and the other on a sandy loam soil at INERA Experimental Station, Saria, Burkina Faso (latitude 12°16′N and longitude 2°9′W) in the Soudano-sahelian zone of Burkina Faso. The typical annual rainfall varies between 700 and 800 mm frequently with erratic distribution. The soils of the experimental site are leached ferruginous tropical soils found in most of Burkina Faso. Initially, the sandy loam soil had higher content of organic matter, available P (Bray-I P), and total N than the coarse sand one (Table 1).

Field Experiments

The field trials were set up in 1988. The following treatments were applied in a split-plot factorial design with six replications. The main plot consisted of three

Table 1 Chemical and physical characteristics of the two soils studied

Texture	Soil	
	Coarse sand	Sandy loam
pH (H_2O)	6.1	5.8
Organic matter	0.6	0.65
% clay	9.6	13.5
% fine silt	6.1	7.0
% coarse silt	18.7	18.0
% fine sand	22.0	32.0
% coarse sand	43.6	29.5
CEC (meq 100 g^{-1})	1.84	2.86
Bases saturation (meq 100 g^{-1})	0.81	1.97
N total (mg N kg^{-1})	460	575
Bray-I P (mg P kg^{-1})	4.9	6.3

cultural techniques (HO = hoeing, PLO = plowing, COMP = plowing + compost) combined with two levels of fertilization (F0 = without fertilization and F1 = with fertilization). Also the main plots were divided into sub-plots, which corresponded to the years of planting (88, 89, 90, 91, etc.). The sub-plots were planted with groundnut–sorghum rotation. Groundnut (cv. CN94C, 90 days to maturity) was sown at 40 × 15 cm spacing and sorghum (*Sorghum bicolor*) (cv. ICSV 1049, 120 days to maturity) at a spacing of 80 × 40 cm; the sub-plot size was 12 × 3.2 m. Compost was applied at a rate of 3 t ha^{-1} to all plots once every two years, because only the sorghum straw produced was composted. Averages of some chemical constituents of the compost are given in Table 2. A rate of 150 kg ha^{-1} of NPK (12 N–24 P–12 K) was supplied to the plots that received mineral fertilizers. Also a rate of 50 kg ha^{-1} of urea was added to all plots cropped with

Table 2 Chemical constituents of the compost used

Constituents	
pH	7.6
MO %	33
C %	19.18
N %	1.37
C/N	14
P %	0.13
K %	0.29
Ca %	1.0
Mg %	0.28
Na %	0.026

sorghum at 6 weeks after planting. The rates of fertilizers applied correspond to rates recommended on low-input agricultural system in Burkina Faso.

The effect of continuous cultivation has been more often assessed by the chemical and physical parameters of the soils as well as by the whole plant yield. Consideration of intermediate representative variables of different stages for final grain production and relying on yield decomposition in components elaborated at successive times during the plant cycle may allow a best evaluation of continuous cultivation. Thus, groundnut haulm yield, pod yield, number of pods, percentage (%) pod two-seeded, % pod rot, seed yield, and 100-seed weight were measured. For sorghum, dry matter and grain yield were determined.

Soil Sampling and Analysis

Soil samples were collected periodically (every 2 years) after sorghum harvest from the topsoil (Ap) horizon (0–20 cm) of all the six replications of the six treatments. The soil samples were mixed thoroughly and passed through a 2-mm sieve before use. Soil organic matter, total N, K, and Ca were determined

by routine analysis. Available P was determined by the most commonly used Bray-I extractant (0.05 N HCl + 0.03 N NH$_4$F).

Statistical Analysis

The data obtained were subjected to analysis of variance using SAS (1996) statistical procedures. The data analyzed here concern those collected from 1988 to 1996.

Results and Discussion

Effects of Cultivation and Cultural Techniques on the Groundnut Yield

The effects of cultivation and mineral and organic fertilizers on groundnut yield and its components are given in Table 3.

On the gravely sandy soil, the continuous cultivation without fertilization (check plots) affected mainly the reproductive parts of the plant. Thus, the pod

Table 3 Effect of cultivation and management techniques on groundnut yield and its components

	Haulm	Pods	Seed	Number of pod	% pod two-seeded	% pod rot	100-seed weight	Seed number/haulm weight	Pod weight/haulm weight
Coarse sand soil									
Control, year 0	993	836	584	113	63.5	13.1	32.1	1.95	0.93
Effect 8 years	−25	−186*	−161*	−7.8	−17.2*	+8.8*	−3.6*	−0.45*	−0.21*
Mean HO F0	780	525	359	80	60.2	14.6	29.9	1.65	0.75
Effect of compost	+101	+158*	+116*	+19*	+1.7*	−2.0*	+1.2	+0.11	+0.07
Effect of labor	+101	+116	+84	+15*	+0.4	−1.6*	+0.8	+0.11	0.06
Effect of fertilizers	+496*	+314*	+218*	+43*	+1.2*	−0.3	+0.5	−0.12	−0.05
Sandy loam soil									
Control, year 0	1541	1557	1123	188	70.8	11.4	37.6	2.06	1.08
Effect 8 years	−316*	−428	−342	−49.3*	−8.6*	+2.1	−4.1*	−0.06	−0.08*
Mean HO F0	1290	995	696	139	66.1	12.4	32.7	1.69	0.8
Effect of compost	+243*	+203*	+159*	+23.0*	+3.5*	−2.0*	+0.6	+0.05	+0.02
Effect of labor	+111*	+91	+74	+10.3	+1.5	−1.5*	+0.3	+0.03	+0.01
Effect of fertilizers	+512*	+337*	+236*	+42.1	+3.4	−1.0*	+0.1	−0.04	−0.01

Control, year 0 = control at first year of cultivation; HO F0 = hoeing without fertilization; * = Significant at $p < 0.05$

Table 4 Some chemical constituents of the soil before cultivation and after 3 and 5 years of cultivation without fertilization

	pH (H$_2$O)	MO, %	Nt, mg kg^{-1}	P Bray-1, mg kg^{-1}	CEC meq 100 g^{-1}
Coarse sand soil					
Year 0	6.10	0.58	457	5.50	1.86
	6.11	0.62	463	4.26	1.83
Year 3	5.98	0.52	523	4.89	2.24
	5.83	0.56	550	6.31	2.95
Year 5	5.91	0.44	625	3.86	2.65
	5.84	0.45	591	10.76	1.92
Sandy loam soil					
Year 0	5.86	0.62	525	7.89	3.04
	5.79	0.68	625	4.73	2.68
Year 3	5.94	0.67	716	6.17	2.78
	5.53	0.70	814	12.35	2.98
Year 5	5.73	0.56	685	8.21	2.34
	5.61	0.62	724	11.93	2.60

weight, % pod two-seeded, 100-seed weight, number of seeds/haulm weight ratio or growth efficacy, and % pod rot decreased by continuous cropping (Table 3), except for the pod number ($p > 0.05$). The haulm yield was not also significantly decreased, therefore the pod weight by haulm weight ratio decreased. On the sandy loam soil, with no fertilization, haulm yield, pod number, % two-seeded pods, and 100-seed weight were reduced with continuous cropping but pod and seed yields were not decreased. The pod yield and yield components were more affected on the coarse sand soil than on the sandy loam one (Table 3). This was probably due to the difference in physical and chemical fertility between the two soils. The decrease

Table 5 Mineral balance (kg ha^{-1}) over 8 years of cultivation for the different treatments

	N	P	K	Ca
Coarse sand soil				
HO + F0	−60	−34	−93	−54
PLO + F0	−79	−42	−114	−63
COMP + F0	54	−19	−110	98
HO + F1	26	191	−124	−118
PLO + F1	4	178	−153	−131
COMP + F1	123	199	−141	33
Sandy loam soil				
HO + F0	−173	−84	−219	−106
PLO + F0	−183	−90	−245	−118
COMP + F0	−76	−78	−270	32
HO + F1	−70	146	−224	−165
PLO + F1	−119	25	−268	−182
COMP + F1	4	144	−270	−27

of haulm yield and yield variables caused by the continuous cropping without fertilizer addition was due to soil organic matter degradation and to nutrient decline resulting from depletion of soil available nutrients (P, K, N, Ca) (Tables 4 and 5). Then, the decrease of 100-seed weight observed in this study was explained by soil Ca deficiency (Table 5). This was in accordance with the finding of Adams and Hartzog (1980), which showed that unfilled pods were caused by Ca deficiency. A Ca-depleted soil may produce high % empty pods. We obtained similar result on a Ca-deficient soil at Farako-Bâ research station, Burkina Faso, in 2003 and 2004 (unpublished). However, the significant increase of % pod rot measured in the coarse sand soil indicated that pod rot was not only due to Ca deficiency because the sandy loam soil was more deficient in Ca than the coarse sand soil but also caused by *Pythium myriotylum* Drechs. and *Rhizoctonia solani* Kuehn (Hallock and Garren, 1968; Walker and Csisnos, 1980). Therefore this result may be caused by these fungi. The % pod two-seeded reported as above significantly decreased in the two soils. It was shown that % pod two-seeded is negatively related to soil available K (Bockélé Morvan, 1964; Gillier and Gautreau, 1971). Thus, the result obtained here was probably caused by a K deficiency of the soils studied (Table 5). Soil P levels required for groundnut are often lower than those required for other crops (Cope et al., 1984). But the initial soil test P levels of the two soils are low (Table 1), therefore P may be the possible reason associated with the haulm and pod yield decline noticed in the unfertilized plots.

The addition of NPK increased significantly the yield over the control. The mineral fertilization acted mainly on the % pod two-seeded and therefore on the seed number concomitant to its effect on the vegetative growth. The addition of fertilizers in the rotation system allowed to obtain best crop yield and to reduce year-to-year yield variation. The fertilizers' efficacy seemed to be maintained over time according to the yield trends. However, the fertilization did not allow keeping a pod weight by haulm weight ratio similar to initial conditions. This suggested that long-term recommended mineral fertilization even in a rotation system on these soils does not maintain the yield production. These results are consistent with the current literature. Thus, Nambiar and Abrol (1989) found a declining trend even with NPK in many fertility experiments in India. Also, Bhandari et al. (1992) and Watanable and Liu (1992) have reported similar results. But Singh et al. (2001) showed that yields of rice and wheat increased significantly with increasing rate of NPK alone up to 100% of the recommended dose.

The compost application significantly increased the pod yield (Table 3, $p < 0.05$). But this increase might be higher if the rates of compost applied were higher than 3 t ha^{-1} $year^{-1}$, because in soils where the % sand is high, the application of 5 t ha^{-1} $year^{-1}$ of organic materials was not sufficient to maintain the stock of soil organic matter in continuous cultivation (Piéri, 1989). This conclusion supports the slight increase of pod yield with the application of 3 t ha^{-1} of compost, one time by 2 years in the present field experiments. The average effect of compost on the pod weight by haulm weight ratio was not significant for 8 years ($p > 0.05$). However, it was observed that this ratio increased markedly during the last 2 years of cropping, 1996 and 1997, in opposition to the effect of mineral fertilization. This ratio increased by 0.15 and 0.08 (ratio of 0.48 and 0.6), respectively, on the coarse sand and sandy loam soils over the control. The application of organic compound to soil was generally accepted as the primary method to maintain soil organic matter; therefore, the addition of compost and the probable progressive accumulation of organic matter originating from root biomass have favored the building of a stock of soil organic matter during the last years (1996 and 1997).

The plowing significantly increased groundnut dry matter yield on the sandy loam soil but not on the coarse textured soil. It had no effect on groundnut pod yield on the two soils. The tillage improved also the plants' early development (data not shown). Soil mechanical resistance due to high bulk densities limits root growth; the plowing reduced bulk density and thereby increased the soil porosity and water infiltration and facilitates root proliferation, which results in better use of soil nutrients by roots. This could explain the small but significant increase in the number of pods and the decrease of the % pod rot on the coarse sand soil, the increase of haulm yield and the number of pods, and the decrease of the % pod rot on the sandy loam soil (Table 3). At last the results obtained by using tillage and compost varied with soil type. On the sandy loam soil, the lack of tillage and compost supply led to a significant decrease of pod number. On the coarse textured soil, the decrease of the pod weight by haulm weight ratio was more marked than on the sandy loam one. Also, a high % pod rot characterized the gravely soil. These results indicated that the coarse sand soil seemed to be more affected by the lack of tillage and compost addition. This may be due to soil initial N and P content, soil texture (Table 1), and increase of soil erosion because soil aggregates deteriorate under cultivation (Elliot, 1986). These results suggest that whatever the soil type, the continuous cultivation deceases the potential production of the soil.

Soil Fertility Changes

The changes in pH, organic matter, available P, and total N as a result of continuous application of mineral fertilizers and compost after 8 years of the groundnut–sorghum cropping are presented in Table 4.

Soil pH

Soil pH declined slightly in the two soils after 5 years of cropping without fertilizer addition (Table 4). Soil pH decreases from an initial 6.1 to 5.9 and 5.8 to 5.7 in the coarse sand soil and sandy loam soil, respectively. The continuous application of fertilizers (NPK and urea) led to similar variations in soil pH in the two soil types. Significant decreases in pH have been reported with continuous cropping systems (Taonda et al., 1995; Bell et al., 1995).

Soil Organic Matter

Initially soil organic matter was about 0.6% in the two soils. With continuous cropping of groundnut–sorghum for 5 years, a declining trend in soil organic matter was observed in the coarse sand soil (Table 4). After 5 years, maximum loss of 27% in the initial level was noted in the control plot of this soil. The observed decline in organic matter with duration of cultivation is in good agreement with the relationship established by Nand (2000). In the sandy loam soil, there was a decrease of organic matter from the initial period until the third cropping followed by an increase while a decrease after the fifth cropping was obtained. It is, however, not clear which factors led to these variations of soil organic matter content in this soil with cropping. With the use of NPK and urea the loss was similar to the unfertilized plots. The supply of compost did not allow maintaining the rate of initial soil organic matter especially on the gravely sand soil. The compost added was not effective in maintaining levels of soil organic matter because the quantities applied were very low and it degrades quickly under tropical conditions.

Soil Nutrient Status

The patterns of soil N and P status were affected by continuous cultivation and fertilizer use. Initially total extractable N was 460 and 575 mg N kg^{-1} on the coarse sand soil and the sandy loam soil, respectively. Over the years, an increasing trend in N content was observed in the two soil types and in the control and fertilized treatments. This increase was higher in NPK + urea plots. The increase in total N contrasted with the global decline in soil organic matter level. The total N was probably supplied by the use of water-soluble fertilizers, NPK, and urea and relatively rapid mineralization of native organic matter and root biomass. The initial level of Bray-I P was 4.9 and 6.3 mg P kg^{-1} on the coarse sand soil and the sandy loam soil, respectively. The extractable Bray-I P content was in the range of low fertility class for P availability. Continuous cropping of groundnut–sorghum without adding P for 5 years decreased the soil available P in the coarse sand soil. In the sandy loam soil there was a decrease of Bray-I P from the initial period until the third cropping followed by an increase while a decrease after the fifth cropping was obtained

(Table 4). Addition of NPK after 5 years increased the values of extractable Bray-I P up to 10 P kg^{-1} considered as a critical value; above it the plant P nutrition will not limit crop production in tropical soils.

Apparent N, P, K, and Ca balances were inputs and outputs measured during the 8 years of the study (Table 5).

The apparent N balance was positive in compost treatments and NPK treatments in the coarse sand soil and it was negative in all treatments except for those of NPK + compost on the sandy loam soil. In the sandy loam soil the production of yield was more important, hence uptake of N was also high, which is why the apparent N balance was markedly negative. The apparent P balance was positive in NPK treatments and negative in other treatments on the coarse sand and sandy loam soils. The positive P balance in mineral fertilized plots suggests that the current fertilizer P recommendations are adequate to maintain short-term soil-supplying capacity. The apparent K balance was negative in all treatments and in both the coarse sand soil and the sandy loam soil. The K balance was still negative, despite the addition of NPK. The large negative K balance suggests that the system will not be able to sustain the K supply in the medium run. The apparent Ca balance on the two soils was negative in all treatments except for those of compost. On the sandy loam soil the Ca balance was positive in the compost treatment alone and slightly negative in the compost + mineral fertilizer treatment. This suggests that the rate of compost applied is not adequate to maintain short-term soil-supplying capacity of Ca. Thus, the results clearly show that neither recommended dose of NPK nor urea or compost at present level of supply could sustain the initial level of productivity.

Conclusions

Our study demonstrates that the continuous cultivation of a groundnut–sorghum rotation affected groundnut yield and its components and also sorghum yield. The lack of fertilization (mineral and organic) and tillage increased the effect of continuous cropping and seemed to strongly degrade the soil. The soil is becoming more and more deficient in nutrients (N, P, K, and Ca) with low organic matter content. The two soils studied did not resist the degradation although the differences of texture and fertility between these soils were notable. Moreover, the use of fertilization and plowing permitted to limit the effect of continuous cultivation. These results clearly reveal that current fertilizer recommendations are inadequate in the short run. The total input of N, P, K, and compost should be optimal to ensure a sufficient nutrient supply for acceptable yields.

The use of intensive agricultural systems is a necessity not only in the viewpoint of food production increase but also for maintaining soil fertility. A common soil management strategy among farmers relies actually on the crop characteristics, which leads to growing sorghum or pearl millet after the first years of land clearance and to continue with groundnut crop that responds better to degraded and nutrient-deficient soils. This cultural system is then a mining one looking for maximum enhanced value of soils until the abandonment of the plots cropped. The successful changes from the now dominant mining and shifting system to an enhanced cropping system require widespread use and best management of mineral and organic fertilizers.

References

Adams F, Hartzog DL (1980) The nature of yield of Florunner peanuts to lime. Peanut Sci 7:120–123

Aguilar R, Kelly EF, Heil RD (1988) Effect of cultivation on soils in northern Great Plains rangeland. Soil Sci Soc Am J 52:1081–1085

Bell MJ, Harch GR, Bridge BJ (1995) Effects of continuous cultivation on ferrosols in subtropical Southeast Queensland. I. Site characterization, crop yields and soil chemical status. Austr J Agric Res 46:237–253

Bhandari A, Ladha JK, Pathak H et al (2002) Yield and soil nutrient changes in a long-term rice–wheat rotation in India. Soil Sci Soc Am J 58:185–193

Bhandari AL, Sood A, Sharma KN et al (1992) Integrated nutrient management in rice–wheat system. J Indian Soc Soil Sci 40:742–747

Bockélé Morvan A (1964) Etude sur la carence potassique de l'arachide au Sénégal. Oléagineux 19(10):603–609

Cattan P, Schilling R (1992) Evaluation expérimentale de différents systèmes de culture incluant l'arachide en Afrique de l'Ouest. Oléagineux 47(11):635–644

Compaoré E, Frossard E, Fardeau JC et al (2003) Influence of land-use management on isotopically exchangeable phosphate in soils from Burkina Faso. Commun Soil Sci Plant Anal 34(1&2):201–223

Cope JT, Starling JG, Ivey HW et al (1984) Response of peanut and other crops to fertilizers and lime in two long-term experiments. Peanut Sci 11:91–94

Elliot ET (1986) Aggregates structure and carbon, nitrogen and phosphorus in native and cultivated soils. Soil Sci Soc Am J 50:627–633

Gillier P, Gautreau J (1971) Dix ans d'expérimentation dans la zone à carence potassique de patar au Sénégal. Oléagineux 26(1):33–38

Hallock DL, Garren KH (1968) Pod breakdown, yield and grade of Virginia type peanut as affected by Ca, Mg, and K sulphates. Agron J 60:253–257

Manna MC, Swarup A, Wanjari RH, Ravankar HN, Mishra B, Saha MN, Singh YV, Sahi DK, Sarap PA (2005) Long-term effect of fertilizer and manure application on soil organic carbon storage, soil quality and yield sustainability under sub-humid and semi-arid tropical India. Field Crops Res 93:264–280

Nambiar KKM, Abrol IP (1989) Long-term fertilizer experiments in India: an overview. Fertil News 34:11–20

Nand R (2000) Long-term effects of fertilizers on rice-wheat-cowpea productivity and soil properties in a mollisols. In: Abrol IP, Bronson KF, Duxbury JM, Gupta RK (eds) Long-term soil fertility experiments in rice-wheat cropping systems. Rice-wheat consortium paper series 6. New Delhi, India, pp 50–55

Piéri C (1989) Fertilité des terres de savanes. Bilan de trente ans de recherche et de développement agricoles au Sud du Sahara. Montpellier. Ministère de la coopération et du développement et CIRAD/IRAT, Paris, 444p

Regmi AP, Ladha JK, Pathak H et al (2002) Yield and soil fertility trends in a 20-year rice-rice-wheat experiment in Nepal. Soil Sci Soc Am J 66:857–867

SAS Institute (1996) SAS user's guide: statistics. SAS Institute, Cary, NC

Singh KN, Prasad B, Sinha SK (2001) Effect of integrated nutrient management on a typic haplaquant on yield and nutrient availability in a rice–wheat cropping system. Austr J Agric Res 52:855–858

Stoorvogel JJ, Smaling EMA, Jansen BH (1993) Calculating soil nutrient balances at different scale I. Supra-national scale. Fertil Res 35:227–235

Taonda SJ-B, Bertrand R, Dickey J et al (1995) Dégradation des sols en agriculture minière au Burkina Faso. Cah Agric 4:363–369

Walker ME, Csisnos AS (1980) Effect of gypsum on the yield grade and incidence of pod rot in five peanut cultivars. Peanut Sci 7:109–113

Watanable I, Liu CC (1992) Improving nitrogen fixing systems and integrating them to sustainable rice farming. Plant Soil 141:57–67

Soil Inorganic N and N Uptake by Maize Following Application of Legume Biomass, Tithonia, Manure and Mineral Fertilizer in Central Kenya

J. Mugwe, D.N. Mugendi, M. Mucheru-Muna, and J.B. Kung'u

Abstract In the smallholder farms of central Kenya soils suffer from nitrogen (N) deficiency due to inability to replenish it through application of chemical fertilizers and/or manure. This study evaluated the effect of some organic materials such as *Mucuna pruriens*, *Crotalaria ochroleuca*, *Calliandra calothyrsus*, *Leucaena trichandra*, cattle manure and *Tithonia diversifolia* applied solely or combined with inorganic fertilizer on soil mineral N dynamics and N uptake by maize. Soils and maize samples were taken at 0, 2, 4, 6, 8, 12, 16 and 20 weeks after planting maize (WAP) during 2002 long rain (LR) and 2004 LR seasons and analysed. The study showed that amounts of soil inorganic N and uptake of N by maize varied among the different sampling dates, treatments and between seasons. There was a general increase of mineral N after the start of the season followed by a drastic reduction during 6 and 4 WAP during 2002 and 2004 LR, respectively. This trend was attributed to the decomposition of organic materials at the beginning of the season followed by leaching due to intense rainfall during this period. Treatments that had tithonia, leucaena and calliandra applied recorded the highest amounts of soil inorganic N and also the highest N uptake by maize. Poor rainfall in 2004 LR restricted N uptake and was responsible for lower N uptake by maize in 2002 LR than in 2004 LR. At the end of the growing season, there were high amounts of mineral N at 100–150 cm soil depth that was probably due to leaching. This mineral N is below the rooting zone of most maize plants, consequently not available to maize crop and is therefore of concern.

Keywords Organic materials · Inorganic fertilizer · Soil mineral N · Uptake of N by maize

Introduction

One of the greatest biophysical constraints to increasing agricultural productivity in Africa is low soil fertility (Sanchez et al., 1997). The need to improve soil fertility management has therefore become a very important issue in the development policy agenda because of the strong linkage between soil fertility and food security on the one hand and the implications on the economic well-being of the population on the other. Declining soil fertility in central highlands of Kenya due to intensive cultivation without adequate soil nutrient replenishment has resulted in low returns to agricultural investment, decreased food security and general high food prices (Odera et al., 2000). Though the area has high potential for food production because of favourable seasonal precipitation, many of the soils are deficient in nutrients, particularly nitrogen (N) and phosphorus (P).

Nitrogen is one of the major plant nutrients and in the soil is broadly subdivided into organic and inorganic forms. Inorganic N is available for plant uptake while organic forms slowly become available for plant uptake through microbial decomposition and mineralization. Mineralization involves the microbial conversion of organic N into soil inorganic N or mineral N. The principal forms of inorganic N in soils are ammonium (NH_4^+) and nitrate (NO_3^-) and any nitrogen in

J. Mugwe (✉)
Department of Agricultural Resource management, School of Agriculture and Enterprise Development (SAED), Kenyatta University, Nairobi, Kenya
e-mail: jaynemugwe@yahoo.com

A. Bationo et al. (eds.), *Innovations as Key to the Green Revolution in Africa*,
DOI 10.1007/978-90-481-2543-2_62, © Springer Science+Business Media B.V. 2011

the soil that is available to the crop is almost always in one of the two forms (Barrios et al., 1998). Both nitrate-N and ammonium-N may be recycled within the soil biota, taken up by the crop, retained within the soil matrix or lost through leaching, volatilization, nitrification and denitrification processes. In central highlands of Kenya the rate of N loss through soil erosion, leaching and crop harvests is higher than the rate of replenishment resulting in negative balances and severe N deficiencies in most of the soils. Farmers usually lack the financial resources to replenish N through mineral fertilizers (Mugwe et al., 2004). The few farmers who use inorganic N fertilizers apply them at very low rates of 15–25 kg N ha^{-1} (Kihanda, 1996).

Options for replenishing N are therefore receiving a lot of attention from scientists and include the use of organic materials such as agroforestry trees, leguminous cover crops and animal manures. The quantity of organic materials is, however, inadequate at the farm level and in the recent past there has been increased interest in devising ways of optimizing nutrient availability by combining the use of organic and inorganic N resources. The use of organic materials and their combinations with fertilizer to optimize nutrient availability to annual crops presents a challenge. This is because mineralization, which is dependent upon the quality of organic materials, environmental factors and proportions (Palm et al., 2001), will have to occur in order to release inorganic N into the soil before plant can utilize. Research in western Kenya showed that organic materials increased soil inorganic N and N mineralization in the plow layer (0–15 cm depth) compared with continuous unfertilized maize (Barrios et al., 1997). These measures of soil N availability were significantly lower following tree legumes with fast decomposing litter than with slow decomposing litter as assessed from the (lignin + polyphenol)/N ratio in their leaves (Barrios et al., 1997). Nitrogen uptake by the crop is also influenced by the quality of organic materials applied to the soil. In the central highlands of Kenya, organic materials being introduced include herbaceous legumes, biomass of leguminous trees and livestock manure, all of which have varying qualities in terms of C:N ratio and the content of polyphenols. An understanding of how these different organic materials will influence soil inorganic N availability and subsequent uptake by the crops would help provide strategic management guidelines for optimizing N utilization in farming systems, in the region and elsewhere, that use organic materials to replenish soil fertility. Such information is scanty, especially in the central highlands of Kenya where only limited studies have been undertaken. This study sought to monitor amounts of soil inorganic N within a growing season and associated N uptake by maize following application of selected organic materials applied solely or combined with inorganic fertilizer.

Research Methodology

The Study Area

The study was conducted in Chuka division of Meru South District of Kenya. Meru South District lies between latitudes 00°03′47•N and 0°27′28•S and longitudes 37°18′24•E and 28°19′12•E. Meru South District covers an area of 1032.9 km^2 and Chuka division covers an area of 169.6 km^2. According to agro-ecological conditions (based on temperature and moisture supply), the area lies in the Upper Midland Zone (UM2-UM3) (Jaetzold et al., 2006) on the eastern slopes of Mt. Kenya at an altitude of 1500 m above sea level with an annual mean temperature of 20°C and a total bimodal rainfall of 1200–1400 mm. The rainfall is in two seasons: the long rains (LR) lasting from March through June and short rains (SR) from October through December. The soils are mainly humic Nitisols (Jaetzold et al., 2006), which are deep, well weathered with moderate to high inherent fertility. The district is a predominantly maize growing zone with small land sizes ranging from 0.1 to 2 ha with an average of 1.2 ha per household.

The area is characterized by rapid population growth, low agricultural productivity, increasing demands on agricultural resources and low soil fertility (GOK, 2001). The main cash crops are coffee (*Coffea arabica* L.) and tea (*Camelina sinensi* (L.) O. Kuntze) while the main staple food crop is maize (*Zea mays* L.), which is cultivated from season to season mostly intercropped with beans (*Phaseolus vulgaris* L.). Other food crops include potatoes (*Ipomea batatas* (L.) Lam), bananas (*Musa* spp. L.) and vegetables that are mainly grown for subsistence consumption. Livestock production is a major enterprise, especially dairy cattle that is of improved breeds. Other livestock in the area include sheep, goats and poultry.

Before planting, soil characterization was carried out. The soil was sampled in March 2000 at 0–15 cm depth and analysed for pH, exchangeable magnesium (Mg), calcium (Ca), potassium (K), available phosphorus (P), total organic carbon (C) and total nitrogen (N). All the analyses were carried out at the International Centre for Research in Agroforestry (ICRAF) laboratories using procedures outlined in the ICRAF laboratory manual (ICRAF, 1995; Anderson and Ingram, 1993). The results showed that pH of the soil was 5.2 while total N and C was 0.21 and 1.8%, respectively. Available P was 7.1 Cmol kg^{-1}, K was 0.3 Cmol kg^{-1}, Ca was 3.4 Cmol kg^{-1} and Mg was Cmol kg^{-1}.

Experiment Establishment and Management

This study was carried out in an experiment that had been established in March 2000 in Chuka division, Meru South District. The experiment had 14 treatments comprising 6 organic resources applied solely or combined with inorganic fertilizer, inorganic fertilizer and a control. The organic resources were two herbaceous legumes *Mucuna pruriens* and *Crotalaria ochroleuca* (intercropped with maize); two leguminous shrubs *Calliandra calothyrsus*, *Leucaena trichandra* (biomass transfer); cattle manure; and *Tithonia diversifolia* (biomass transfer) (Table 1). The experiment was a completely randomized block design with three replications. Plot sizes measured 6 m × 4.5 m and maize was planted at a spacing of 0.75 and 0.5 m inter- and intra-row spacing, respectively. Compound fertilizer (23:23:0) was the source of inorganic N and was applied at sowing during the four seasons. A uniform P application was done in all the plots at the recommended rate (60 kg P ha^{-1}) as triple superphosphate (TSP). Other agronomic procedures for maize production were appropriately followed after planting.

The herbaceous legumes (*Mucuna* and *Crotalaria*) were intercropped between two maize rows 1 week after planting maize. After maize was harvested, legumes were left to grow in the field till land preparation for the subsequent season when they were harvested, weighed, chopped and incorporated into the soil to a depth of 15 cm. The weight of the herbaceous legume biomass applied during the study period varied

Table 1 Treatments in the experiment at Kirege School, Chuka, Kenya

Treatment	Amount of N supplied (kg ha^{-1})	
	Organic	Inorganic
Mucuna pruriens alone	a	0
Mucuna + 30 kg N ha^{-1}	a	30
Crotalaria ochroleuca alone	a	0
Crotalaria + 30 kg N ha^{-1}	a	30
Cattle manure alone	60	0
Cattle manure + 30 kg N ha^{-1}	30	30
Tithonia diversifolia	60	0
Tithonia + 30 kg N ha^{-1}	30	30
Calliandra calothyrsus	60	0
Calliandra + 30 kg N ha^{-1}	30	30
Leucaena trichandra	60	0
Leucaena + 30 kg N ha^{-1}	30	30
Recommended rate of fertilizer	0	60
Control (no inputs)	0	0

[a]Total N applied varied among seasons and depended on the amount of biomass produced during the previous season

across the seasons (Table 2). The amount of N contributed into the soil via the incorporated biomass was calculated by multiplying the amount of biomass (kg) with N concentration in the biomass (%).

The other organic materials (calliandra, leucaena, tithonia and cattle manure) were incorporated into the soil to a depth of 15 cm during land preparation. Nitrophosphate fertilizer (23:23:0) was the source of inorganic N and was applied at sowing during the four seasons. A blanket application of P at 60 kg N ha^{-1} was applied to all the plots to prevent P deficiency, which was observed at the start of this experiment. Other agronomic procedures for maize production were appropriately followed after planting. Subsamples of all organic materials were collected uniformly at the beginning of each season and analysed. The samples were first washed with distilled water and oven dried at 65°C for 48 h. Samples were ground, packed in polythene bags and stored under dry conditions before N was determined at ICRAF laboratories (ICRAF, 1995; Anderson and Ingram, 1993). The dry plant samples were analysed for total N, P, K, Ca and Mg by Kjeldahl digestion with concentrated sulphuric acid (Anderson and Ingram, 1993). Nitrogen and phosphorus were determined colorimetrically while potassium was by flame photometry (Okalebo et al., 2002). Magnesium and calcium was by atomic absorption spectrophotometer (Anderson

Table 2 Amount of herbaceous legumes produced and their N contribution in the soil during 2002 LR to 2003 SR at Chuka, Kenya

Treatment	2002 LR	2002 SR	2003 LR	2003 SR	2004 LR	Average	Mean N kg ha^{-1} season^{-1}
	................Biomass in t ha^{-1} season^{-1}................						
Mucuna	1.7	2.8	0.8	0.2	0.3	1.16	28.6
Mucuna + 30 kg N ha^{-1}	1.9	3.2	0.9	0.3	0.9	1.44	37.7
Crotalaria	2.3	2.3	0.6	0.2	0.6	1.2	33.8
Crotalaria + 30 kg N ha^{-1}	2.8	2.5	0.8	0.3	0.7	1.42	39.1
SED	0.19	0.17	0.11	0.11		0.08	

Table 3 Average nutrient composition (%) of organic materials applied in the soil during the study period

Treatment	N	P	Ca	Mg	K	Ash
Cattle manure	1.3	0.2	1.0	0.4	1.8	45.9
Tithonia	3.2	0.2	2.1	0.6	3.0	13.0
Calliandra	3.3	0.2	1.0	0.4	1.2	5.9
Leucaena	3.6	0.2	1.4	0.4	1.8	8.5
SED	0.4	0.004	0.04	0.01	0.05	0.27

and Ingram, 1993). Table 3 shows the mean nutrient composition of organic materials used during the four seasons under study.

Soil Sampling and Determination of Soil Inorganic Nitrogen

Soil samples were taken in March 2002 and 2004 at the beginning of the season (before planting) at 0–15 cm depth. Subsequent samples were taken at 4, 6, 8, 12, 16 and 20 weeks after planting maize (WAP) at the same depth during the two seasons but in 2004 LR sampling was also done at 2 WAP. Due to the variability of inorganic N in the field 10 soil cores per plot were taken and bulked to give one composite sample. The soil samples were taken using a metal sampling tube, which was driven into the soil and removed. After sampling the soil samples were packed in cooled boxes and delivered to ICRAF laboratory within 24 h. The samples were stored in the fridge at 5°C to restrict mineralization prior to extraction.

During 2002 LR, at harvesting (20 WAP), soil samples were taken at four depths (0–20, 20–50, 50–100 and 100–150 cm) in all plots and samples stored in the fridge at 5°C to restrict mineralization before delivery to ICRAF laboratories. In the laboratory, soil extraction was done using 2 N potassium chloride (ICRAF, 1995; Anderson and Ingram, 1993). The filtrates were then analyzed for extractable nitrate (NO_3^-) by cadmium (Cd) reduction column method (ICRAF, 1995;

Anderson and Ingram, 1993) and extractable ammonium determined using colorimetric method (ICRAF, 1995; Anderson and Ingram, 1993).

Determination of Uptake of N by Maize

Destructive random sampling of maize was carried out at 4, 6, 8, 12, 16 and 20 (WAP) for determination of dry matter and N concentration in the plant tissues during 2002 and 2004 LR. At 20 WAP the maize grains and stover were analysed separately. Nitrogen concentration was determined by Kjeldahl acid digestion followed by colorimetry (Okalebo et al., 2002). Nitrogen uptake by maize crop was determined by multiplying the dry matter yields (kg) with the nitrogen concentration (%).

Statistical Analysis

Statistical analysis was performed using Genstat 5 for windows (Release 8.1) computer package (Genstat, 2005). After testing for normality the data were subjected to analyses of variance (ANOVA) that was used to compare treatment means. Significant differences were declared significant at $p \leq 0.05$ and treatment means found to be significantly different were separated by least significant differences (LSD) at $p \leq 0.05$.

Results

Soil Inorganic N at the Plow Layer at Different Sampling Periods

The bulk (almost 90%) of inorganic N found in the soil (0–15 cm) at all sampling periods during 2002

Table 4 Soil nitrate-N and ammonium-N at 0–15 cm soil depth sampled at different periods during 2002 LR at Chuka, Meru South District

Treatment	. . . . .Weeks after planting.						
	0	4	6	8	12	16	20
	. . . . .Nitrate-N (NO_3^--N) in kg ha^{-1}.						
Mucuna alone	2.3	11.8	9.2	5.6	16.1	3.4	11.0
Crotalaria alone	6.9	21.8	10.3	6.6	22.3	4.2	10.8
Mucuna + 30 kg N ha^{-1}	10.3	21.2	10.1	5.0	20.6	4.5	11.3
Crotalaria + 30 kg N ha^{-1}	20.6	31.7	11.9	7.0	26.8	5.0	15.0
Manure	3.4	19.3	10.9	8.0	22.6	2.5	19.6
Manure + 30 kg N ha^{-1}	22.8	33.8	10.6	6.9	17.4	4.0	14.8
Tithonia alone	30.7	40.5	13.9	10.1	29.7	7.2	15.5
Calliandra alone	16.0	24.4	11.6	8.0	26.2	9.0	15.0
Leucaena alone	19.1	30.5	11.5	6.7	23.6	4.3	19.4
Tithonia + 30 kg N ha^{-1}	27.5	40.2	13.3	8.5	17.0	2.4	13.4
Calliandra + 30 kg N ha^{-1}	25.3	35.4	12.2	5.9	30.3	6.8	12.6
Leucaena + 30 kg N ha^{-1}	19.4	30.5	10.5	5.4	21.1	3.9	12.5
Fertilizer (60 kg N ha^{-1})	19.1	47.5	6.4	4.2	23.6	4.2	8.8
Control	9.4	20.4	6.3	3.1	12.9	5.3	9.8
p-value	0.001	0.001	<0.001	<0.001	0.005	0.001	<0.001
SED	6.3	6.9	1.4	1.0	4.0	1.3	2.0
	. . . . .Ammonium-N (NH_4^+-N) in kg ha^{-1}.						
Mucuna	0.7	1.4	2.4	2.4	2.1	3.7	0.6
Crotalaria	1.2	1.8	4.7	3.3	3.2	6.0	1.3
Mucuna + 30 kg N ha^{-1}	1.3	2.0	3.1	3.1	2.5	5.3	3.7
Crotalaria + 30 kg N ha^{-1}	2.1	2.8	2.8	2.8	1.4	5.7	2.0
Manure	0.3	1.0	2.4	2.9	2.1	6.9	2.2
Manure + 30 kg N ha^{-1}	0.4	1.1	2.5	2.5	2.3	5.2	1.3
Tithonia	1.0	1.7	3.7	3.7	2.6	8.3	3.2
Calliandra	0.8	1.3	3.5	4.0	2.0	7.6	3.1
Leucaena	1.3	1.9	3.0	3.0	3.0	7.4	2.2
Tithonia + 30 kg N ha^{-1}	1.2	1.8	3.4	3.4	2.3	6.9	2.5
Calliandra + 30 kg N ha^{-1}	1.6	2.3	3.4	3.4	3.2	6.5	3.0
Leucaena + 30 kg N ha^{-1}	1.3	1.9	3.9	3.9	3.0	6.4	2.3
Fertilizer (60 kg N ha^{-1})	0.9	1.6	3.4	3.4	2.5	3.5	2.1
Control	1.3	1.9	3.2	3.2	2.8	4.3	2.1
p-value	0.23	0.22	0.53	0.84	0.79	0.012	0.17
SED	0.6	0.6	0.9	0.8	0.9	1.2	0.9

LR was in the form of nitrate-N (NO_3^--N) with ammonium-N (NH_4^+-N) contributing less than 10% (Table 4). Nitrate-N was significantly different among the treatments at the different sampling periods while NH_4^+-N was rather uniform across the different treatments at all sampling periods in 2002 LR (Table 3). During this season, there was also a general increase of NO_3^--N from 0 to 4 WAP, followed by a decrease at 6 and 8 WAP and then an increase at 12 WAP followed by a decrease at 16 WAP. In addition, NO_3^--N content was highest at 4 WAP (ranging from 11.8 to 47.5 kg N ha^{-1}) among all the sampling periods with

inorganic fertilizer, tithonia, tithonia +30 kg N ha^{-1}, manure + 30 kg N ha^{-1} and calliandra + 30 kg N ha^{-1} recording among the highest amount of inorganic N. The inorganic fertilizer treatment had notably the highest amount of NO_3^--N at 4 WAP. The control treatment and herbaceous legumes alone treatment consistently maintained the lowest amount of NO_3^--N across the season.

During 2004 LR NO_3^--N was also higher than NH_4^+-N at all sampling periods but NH_4^+-N tended to accumulate from 8 WAP (Table 5). During this season NO_3^--N was significantly different among the

Table 5 Soil nitrate-N and ammonium-N nitrogen at 0–15 cm soil depth sampled at different periods during 2004 in Chuka, Meru South District

	........Weeks after planting........							
	0	2	4	6	8	12	16	20
Treatment	........Nitrate-N (NO_3^--N) in kg ha^{-1}........							
Mucuna	16.2	27.3	5.0	9.5	14.5	22.1	19.7	19.7
Crotalaria	5.6	15.5	4.3	14.7	13.5	19.7	18.3	23.4
Mucuna + 30 kg N ha^{-1}	9.8	20.6	3.9	5.7	15.1	21.5	21.9	21.9
Crotalaria + 30 kg N ha^{-1}	5.4	15.1	3.5	8.5	13.5	16.9	14.1	14.1
Manure	9.5	18.6	3.5	11.6	21.7	19.4	21.1	21.1
Manure + 30 kg N ha^{-1}	7.9	17.0	3.8	9.8	18.1	19.4	20.4	20.4
Tithonia	28.3	44.8	7.2	17.1	16.7	26.6	36.4	36.4
Calliandra	39.8	56.9	5.2	22.7	17.3	28.3	31.8	31.8
Leucaena	31.4	53.8	6.4	16.6	12.7	32.6	38.8	38.8
Tithonia + 30 kg N ha^{-1}	21.4	32.6	4.2	8.3	30.3	16.6	24.5	24.5
Calliandra + 30 kg N ha^{-1}	23.2	32.9	6.7	21.0	21.0	25.5	20.6	20.6
Leucaena + 30 kg N ha^{-1}	17.2	39.2	4.6	18.3	9.9	27.7	29.9	29.9
Fertilizer (60 kg N ha^{-1})	3.8	10.5	2.4	9.6	9.4	14.9	18.0	18.0
Control	3.2	10.9	2.7	11.4	9.0	15.2	17.2	17.2
p-value	0.001	<0.001	0.003	0.013	0.003	0.005	<0.001	0.006
SED	4.5	9.5	0.7	4.4	3.8	4.2	4.4	5.8
	........Ammonium-N (NH_4^+-N) in kg ha^{-1}........							
Mucuna	2.5	1.4	0.5	2.0	7.3	1.2	8.5	14.5
Crotalaria	0.4	0.5	0.6	1.9	10.4	6.8	10.6	16.6
Mucuna + 30 kg N ha^{-1}	2.4	1.6	0.4	1.7	13.8	3.2	8.1	14.1
Crotalaria + 30 kg N ha^{-1}	0.4	0.6	0.4	7.0	8.5	6.2	7	13.0
Manure alone	0.4	1.2	0.4	3.1	19.1	4.0	9.6	15.6
Manure + 30 kg N ha^{-1}	0.8	1.7	0.4	1.4	15.1	7.8	10.1	16.1
Tithonia	9.8	3.3	0.9	7.3	13.6	9.8	19.0	25.0
Calliandra	12.8	5.7	0.4	6.3	17.7	5.1	10.8	16.8
Leucaena	16	3.6	0.5	4.9	10.3	7.4	18.0	24.0
Tithonia + 30 kg N ha^{-1}	2.9	1.7	1.2	2.5	14.1	4.1	9.3	15.3
Calliandra + 30 kg N ha^{-1}	2.1	2.4	0.6	3.3	10.1	8.3	10.4	20.2
Leucaena + 30 kg N ha^{-1}	14.5	2.5	0.5	3.5	11.7	8.1	11.2	17.2
Fertilizer	0.2	1.5	0.3	2.6	8.0	6.3	7.4	13.4
Control	0.4	1.3	0.4	2.6	8.2	4.2	10.2	16.2
p-value	0.002	0.007	0.64	0.041	0.083	0.41	0.113	0.115
SED	2.3	1.1	0.4	2.0	3.8	3.3	3.8	3.9

treatments at the different sampling periods while NH_4^+-N was different among the treatments at 2, 6 and 8 WAP (Table 4). Nitrate-N in soils sampled before the season (at 0 weeks) differed significantly among the various treatments with treatments that had sole application of calliandra, leucaena and tithonia recording the highest NO_3^--N with 39.8, 31.4 and 28.3 kg N ha^{-1}, respectively (Table 4). This was followed closely by leucaena + 30 kg N ha^{-1}, calliandra + 30 kg ha^{-1} and tithonia + 30 kg ha^{-1}. During this season NO_3^--N increased in the first 2 WAP, dropped at 4 WAP and then increased gradually to the 16 WAP and remained constant between the 16 and 20 WAP. Overall the highest amount of NO_3^--N among

the sampling periods was recorded at 2 WAP, which ranged from 10.5 (control treatment) to 56.9 kg N ha^{-1} (calliandra alone treatment). The lowest amount was recorded at 4 WAP, which ranged from 2.4 (fertilizer treatment) to 7.2 kg N ha^{-1} (tithonia alone treatment). At 20 WAP NO_3^--N was highest in leucaena alone, tithonia alone and calliandra alone treatments with 38.8, 36.4 and 31.8 kg N ha^{-1}, respectively (Table 5).

A difference was observed between the two seasons in relation to the amounts of NH_4^+-N across the seasons. Whereas NH_4^+-N remained almost constant throughout the growing period, during 2002 LR, NH_4^+-N tended to accumulate from the 8 WAP. The accumulation can be explained by the rainfall distribution. Termination of the rains early in the season resulted in NH_4^+-N increasing slowly and to become a significant component of the residual mineral N at 16–20 WAP during 2004 LR (Table 4).

Total soil inorganic N (sum of nitrate-N and ammonium-N) at 0–15 cm was generally lower during 2002 LR than in 2004 LR at all sampling periods (Tables 4 and 5). For example, at 20 WAP, total soil inorganic N in 2002 LR ranged from 10.9 (fertilizer treatment) to 21.8 kg N ha^{-1} (manure alone treatment) while it ranged from 27.1 (crotalaria + 30 kg N ha^{-1}) to 62.8 g N ha^{-1} (leucaena alone treatment) in 2004 LR.

Residual Soil Inorganic N at Different Depths at the End of 2002 LR

Generally, amount of soil inorganic N in all treatments decreased with increase in soil depth from 0–20 to 20–50 cm, after which it increased to 100–150 cm depth (Table 6). At the 0–20 cm soil depth, inorganic N ranged from 27.7 (control treatment) to 63.2 kg N ha^{-1} (tithonia alone treatment). There were significant differences among treatments at this soil depth. At the 20–50 cm soil depth, calliandra + 30 kg N ha^{-1} and crotalaria alone treatments had the highest inorganic N of 57 and 41.7 kg N ha^{-1}, respectively. The lowest inorganic N at this depth was recorded in manure + 30 with 16.2 kg N ha^{-1}, and tithonia + 30 with 16.5 kg N ha^{-1}.

There were significant differences ($p = 0.05$) in soil inorganic N among the treatments at the 50–100 cm depth with tithonia alone and leucaena + 30 kg N ha^{-1} having the highest inorganic N of 75 and 69.4 N ha^{-1}, respectively. The lowest amount of inorganic N at this depth was recorded in crotalaria + 30 kg N ha^{-1} (17 kg N ha^{-1}) followed by mucuna alone (20.6 kg N ha^{-1}) and tithonia + 30 kg N ha^{-1} (20.6 kg N ha^{-1}).

At the 100–150 cm depth the highest mineral N was recorded in treatments that had tithonia, calliandra, leucaena applied either alone or in combination with fertilizer. Also, all treatments generally recorded higher amount of mineral N than the control with the exception of the recommended level of inorganic fertilizer.

Nitrogen Uptake by Maize at Different Sampling Periods

Nitrogen uptake by maize was lower during 2004 LR than 2002 LR due to lower rainfall amounts received during 2004 LR than 2002 LR. However, during both seasons, N uptake by maize was slow at 0–4 WAP after which increased up to 12 WAP and decreased from 16 to 20 WAP (Table 7). During 2002 LR, N uptake by maize was highest during 6–8 and 8–12 WAP sampling periods in all treatments. However, during these periods, there were significant differences in

Table 6 Amount of soil inorganic N (kg N ha^{-1}) in soils sampled at different depths at the end of 2002 LR at Chuka, Meru South District, Kenya

Treatment	. . . . . . Soil depth in cm.			
	0–20	20–50	50–100	100–150
Mucuna	32.9	17.9	20.6	20.4
Crotalaria	55.5	41.7	27.9	52.9
Mucuna + 30 kg N ha^{-1}	32.0	18.7	39.2	40.7
Crotalaria + 30 kg N ha^{-1}	51.8	19.7	17.0	42.2
Manure alone	49.1	17.5	20.5	24.9
Manure + 30 kg N ha^{-1}	39.6	16.2	25.8	43.4
Tithonia	63.2	26.5	75.0	90.2
Calliandra	42.9	19.9	28.1	79.5
Leucaena	50.7	25.3	39.2	85.3
Tithonia + 30 kg N ha^{-1}	39.9	16.5	20.6	95.6
Calliandra + 30 kg N ha^{-1}	43.2	57.0	54.5	197.6
Leucaena + 30 kg N ha^{-1}	40.2	20.9	69.4	166.2
Fertilizer + 60 kg N ha^{-1}	35.9	17.2	24.2	30.1
Control	27.7	21.2	33.7	56.9
p-value	0.076	0.011	0.017	<0.001
LSD$_{(0.05)}$	20.8	25.4	32.8	56.2

Table 7 Nitrogen uptake (kg ha^{-1}) by maize at different sampling periods during 2002 and 2004 LR at Chuka, Kenya

Treatment	0–4	4–6	6–8	8–12	12–16	16–20	Total N uptake
			.....Time in weeks......				
			......2002 LR......				
Mucuna	3	10	14	12	9	3	51
Crotalaria	3	18	16	25	8	5	75
Mucuna + 30 kg N ha^{-1}	4	16	17	28	12	7	84
Crotalaria + 30 kg N ha^{-1}	5	19	26	37	10	9	106
Manure	5	12	14	19	13	6	69
Manure + 30 kg N ha^{-1}	4	32	38	58	11	6	149
Tithonia	5	29	34	48	14	8	138
Calliandra	5	23	25	38	10	6	107
Leucaena	5	18	22	39	14	5	103
Tithonia + 30 kg N ha^{-1}	6	29	37	58	15	6	151
Calliandra + 30 kg N ha^{-1}	6	20	28	42	14	6	116
Leucaena + 30 kg N ha^{-1}	5	22	32	46	12	7	124
Fertilizer + 60 kg N ha^{-1}	5	21	23	35	9	5	98
Control	3	12	10	15	12	2	54
p-value	0.05	0.02	0.004	0.001	0.02	0.01	
SED	1.5	4.5	5.9	6.2	3.2	1.6	
			......2004 LR......				
Mucuna	4	5	12	26	4	0.5	52
Crotalaria	3	6	9	10	5	0.5	34
Mucuna + 30 kg N ha^{-1}	4	8	15	25	5	1	58
Crotalaria + 30 kg N ha^{-1}	4	5	14	29	5	1	58
Manure	5	4	6	12	4	0.5	32
Manure + 30 kg N ha^{-1}	4	9	19	30	6	1	69
Tithonia	5	10	23	35	7	2	82
Calliandra	6	6	14	18	8	2	54
Leucaena	5	12	12	17	7	2	55
Tithonia + 30 kg N ha^{-1}	6	14	26	37	6	1	90
Calliandra + 30 kg N ha^{-1}	5	9	13	20	6	1	54
Leucaena + 30 kg N ha^{-1}	3	7	18	22	7	1	58
Fertilizer + 60 kg N ha^{-1}	5	7	22	27	7	0.5	69
Control	3	4	9	12	4	0.2	32
p-value	0.03	0.001	0.02	0.05	0.02	0.03	
SED	1.5	3	3.5	5.6	1.2	0.3	

N uptake among the different treatments. The highest N uptake during these sampling periods was recorded in manure + 30 kg N ha^{-1}, followed by tithonia + 30 kg N ha^{-1} and leucaena + 30 kg N ha^{-1}. At harvest, cumulative N ranged from 51 to 149 for 2002 LR and 24–88 kg N ha^{-1} for 2004 LR. Nitrogen uptake was highest in tithonia + 30 kg N ha^{-1} (151 kg N ha^{-1}), followed closely by manure + 30 kg N ha^{-1} (149 kg N ha^{-1}), calliandra alone (138 kg N ha^{-1}) and leucaena alone (136 kg N ha^{-1}) treatments during 2002 LR. The lowest cumulative N uptake during this season was recorded in mucuna alone treatment (51 kg N ha^{-1}), control treatment (54 kg N ha^{-1}), manure alone (69 kg

N ha^{-1}) and crotalaria alone treatment (75 kg N ha^{-1}). During 2004 LR, the highest cumulative N uptake was recorded in tithonia + fertilizer and tithonia treatments.

Discussions

The different pattern observed in the amounts of soil inorganic N among the different sampling periods in the two seasons under study (2002–2004 LR) was attributed to differences in rainfall patterns between the two seasons. During 2002 LR, rainfall was fairly distributed during the growing season, with almost a normal season for the area (Fig. 1) with a peak at around 5 and 6 WAP. On the other hand, rainfall during 2004 LR was very poor as it only rained for the first 3 weeks. The rains started on the 20th of April and rained for the next 3 weeks at 130.5, 84 mm and 131.8 during the first, second and third weeks, respectively and dry conditions prevailed for the rest of the season (Fig. 1). The dry conditions meant that there was low soil moisture content and this could have been the major cause of low N taken up by maize at the different sampling periods, and also the low cumulative amounts of N taken up by maize.

Ammonium-N content was low at all sampling periods because it probably nitrified as fast as mineralization occurred. Nitrification takes place in soils that are well drained, moist and with a pH of 5.5 to about 10 (Tisdale et al., 1990). The soils under this study have these characteristics and thus the bulk of inorganic N as expected was in the nitrate form. Similar low quantities of soil NH_4^+-N compared to NO_3^--N were reported in a kandiudalfic Eutrodox in Western Kenya (Mekonnen et al., 1997; Maroko et al., 1998) and at a humic Nitisol at Embu, central Kenya (Mugendi et al., 2003).

The accumulation of soil NH_4^+-N during 2004 LR from 8 WAP could be attributed to the dry nature of the soil that prevailed during this period. Chikowo et al. (2004) reported a similar accumulation of soil NH_4^+-N towards the end of a growing season during the ninth week when soil moisture was low. Similarly, Barrios et al. (1998) working in an Ustic Rhodustalf in Zambia reported NH_4^+-N content of between 1.1 and 12.3 mg

kg^{-1} while NO_3^--N was between 2.1 and 20.1 mg kg^{-1}. This relatively low NH_4^+-N was attributed to rapid nitrification since this process is increased by soil wetting and the soil was sampled 2–3 weeks after the onset of rains. In the following year other soil samples collected on the same place before the onset of the rains (low soil moisture) had higher NH_4^+-N content ranging between 4.9 and 25.2 mg kg^{-1} than NO_3^--N that ranged from 0.7 to 10.2 mg kg^{-1}.

The increase in mineral N from 0 to 4 WAP in 2002 LR and 0–2 WAP in 2004 could be explained by mineralization of organic materials and the flush of nitrogen following rapid mineralization after a dry spell known as the 'Birch effect' (Birch, 1964). These corroborate the findings of Mugendi et al. (1999), Chikowo et al. (2004) and Ayuke et al. (2004) who reported nitrogen flush at the start of the rainy season at Embu, Kenya, Zambia and Western Kenya, respectively. The relatively higher amount of NO_3^--N in inorganic fertilizer, tithonia and tithonia + 30 kg N ha^{-1} at 4 WAP compared to all other treatments indicate a higher risk of leaching in these three treatments. While on the other hand it could imply that more N was available in these treatments for crop uptake, but considering the fact that maize roots are not yet developed at 4 WAP, this observation means that high amounts of NO_3^--N might be availed to the crop at a stage it does not require such large amounts of N, implying lack of synchrony. Nitrate-N is vulnerable and is susceptible to losses through leaching (Giller, 2002; Pypers et al., 2005) and the reduction of NO_3^--N at 6–8 WAP during 2002 LR and 4 WAP during 2004 LR was probably

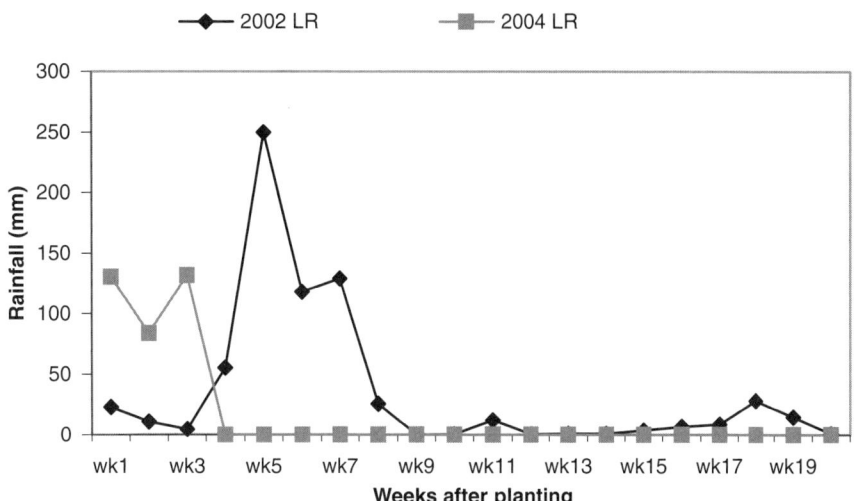

Fig. 1 Comparison of rainfall distribution during 2002 and 2004 LR at Chuka, Meru South District, Kenya

mainly due to plant uptake coupled with leaching. The loss through leaching was more likely at this period due to the high amounts of rainfall received during this period. Such a phenomenon was reported by Thornton et al. (1995) who estimated 40–50% of the mineralized N being lost under high rainfall environments through leaching and Myers et al. (1997) who noted that leaching of mineral N in the soil increased as the rainfall amount increased. The generally higher total inorganic N in the treatments that had organic materials incorporated throughout the cropping season could be explained by the long-term nature of nutrient availability from organic materials. Ayuke et al. (2004) reported similar results in their study in Western Kenya.

The decline in nitrate in the top 20–50 cm depth would suggest that maize roots were able to utilize this nutrient at this depth. Mugendi et al. (2003) reported more than 80% of maize roots to be concentrated in the top 0–60 cm depth and thus could absorb mineral N at 20–50 cm depth. The high mineral N at 100–150 cm soil depth could be attributed to greater N mineralization compared to plant uptake of topsoil immediately after the onset of the rainy season and subsequent nitrate leaching, an observation also reported by Phiri et al. (1998) in Malawi, and Njinue and Wagger (2003) at the Kenyan coast, respectively. This explanation is exemplified by the data on mineral N at different periods, which showed substantial amounts of mineral N at 4 WAP followed by a reduction at 6 WAP. In this region, there is nitrate buildup in the soil early in the season with N (from mineralization) exceeding maize-crop demand during the first 6 weeks of each cropping season (Mugendi et al., 1999; Nyamangara, 2004).

Nitrate leaching followed by adsorption of nitrate on positively charged soil surfaces at the 0.3–1.5 m depth has been reported in Western Kenya (Hartemink et al., 1996; Mekonnen et al., 1997). This mineral N observed in the 100–150 cm depth is below the rooting zone of most maize plants and may not be available to the maize crop (Mugendi et al., 2000; Kibuja et al., 2002). It may also not be readily transformed (denitrified or assimilated) because of the limited microbial population and available C at this depth (Paramasivam et al., 2001).

This high concentration of mineral N at this soil depth is therefore of concern as it is prone to leaching or may percolate to the water table or it could find its way to streams and rivers. Measures to curb downward movement of N like timely application of inputs

and split fertilizer application that have been reported to reduce leaching should be encouraged (Randall and Mulla, 2001). In addition trees if well planted and integrated in the farming systems may assist in recovering some of this N that is lost to annual crops as observed by Mugendi et al. (2003) who reported that treatments with tree hedges recorded a lower amount of mineral N in the 100–300 cm depth than treatments without tree hedges. This is mainly because trees root deep into sub-soil layers and way beyond the rooting depth of most annual crop-inaccessible nutrients (Mekonnen et al., 1997; Shepherd et al., 2000) and therefore are capable of taking up nutrients from zones where crop roots do not reach.

During 1–6 WAP, N uptake by maize is low and this is the time when there is substantial amount of mineral N recorded within the plow layer resulting from nitrogen flush at the start of the season and decomposition of the organic materials. There was thus a likelihood of N supply being higher than the demand thus creating asynchrony.

The highest N uptake recorded at 6–8 WAP and 8–12 WAP during 2002 and 2004 LR, respectively, was due to maize growing very fast at these stages therefore taking up a lot of nutrients. In this region, maize grows rapidly from 4 to 12 WAP (silking/grain-filling stage) and this is the phase when maize has the highest demand for N (Karlen et al., 1988; Kihanda, 2003). Increased N uptake in manure, tithonia and leucaena combined with inorganic fertilizer treatments in the 4–6 and 8–12 WAP is an indication of more availability of plant nutrients in these treatments than all the other treatments. This agrees with the findings of Makumba et al. (2005) and Hawara et al. (2006) who reported increased N uptake by maize in treatments that had tree prunings combined with inorganic fertilizers applied. However, this supply and uptake is highly dependent on rainfall, where lack of adequate rainfall restricts maize N uptake consequently causing accumulation of mineral N mainly in form of ammonium-N in the plow layer as observed in this study.

Conclusions and Recommendations

This study showed variation in amounts of soil inorganic N among the treatments, at the different sampling periods (0, 2, 4, 6, 8, 12, 16 and 20 weeks after planting) and between the two seasons under study (2002 and 2004 LR). Generally treatments that had tithonia, leucaena and calliandra incorporated into the soil recorded higher total soil inorganic N than the control, herbaceous legumes and fertilizer treatments in most sampling periods, especially during 2004 LR. The same treatments recorded highest amounts of N taken up by the maize crop that ranged from 120 to 158 kg N ha^{-1}. This is an indication that these organic materials (tithonia, leucaena and calliandra) can be used as sources of N for growing crops.

During 2002 LR, when rainfall was fairly distributed, soil inorganic N increased from 0 to 4 weeks after planting maize and generally reduced from 6 up to 20 WAP. On the other hand, during 2004 LR when the rainfall was poor and rained intensely for the first 3 weeks, inorganic N dropped drastically at 4 weeks after planting maize after which it increased gradually up to 20 WAP. The reduction in soil inorganic was attributed to leaching coupled with plant uptake of N. Indeed, the cumulative N uptake curves showed that the highest uptake of N by maize occurred from 6 to 12 WAP. High rainfall amounts at the start of the season, when the ground is bare, consequently causing leaching is a big challenge in the region and a solution needs to be sought.

During 2004 LR accumulation of both $NO_3^- $-N and $NH_4^- $-N from the 6 WAP was associated with dry conditions that also restricted uptake of N by the maize crop. The dry conditions were also associated with accumulation of NH_4^+-N that occurred from around 8 WAP up to the end of the season. It was evident that the amount of soil inorganic N and N uptake by maize was dependent on organic materials and soil moisture availability where lack of rainfall restricted N uptake by maize and consequently caused accumulation of soil inorganic N.

These results confirmed the nitrogen flush (Birch effect) that occurs at the beginning of the rainy season. Unfortunately most of this N seems to get lost and does not benefit the crop. More investigations are therefore needed on synchrony between N release from decomposing organic materials and plant demand for N to be able to develop a system that maximizes N use.

At the end of the growing season, there was high amounts of mineral N at 100–150 cm soil depth that was probably due to greater N mineralization compared to plant uptake of topsoil immediately after the onset of the rainy season and subsequent nitrate leaching. This mineral N observed in the 100–150 cm depth is below the rooting zone of most maize plants, consequently not available to maize crop and is therefore of concern as it is prone to leaching, or percolation to the water table or many streams and rivers. Measures to curb downward movement of N, such as timely application of inputs and split fertilizer application, need to be investigated. In addition if trees are well integrated in the farming systems, they may assist in recovering some of this N that is lost to annual crops as earlier reported by other studies. This is because trees root deep into sub-soil layers and way beyond the rooting depth of most annual crop-inaccessible nutrients.

Acknowledgements This work was supported by the Bentley Fellowship and the Rockefeller Foundation. We would also like to acknowledge the financial assistance provided for analysing the soil by the African Network (AfNet) of Tropical Soil Biology and Fertility Programme of CIAT. We also appreciate the contributions of collaborators from the Kenya Agricultural Research Institute, Kenya Forestry Research Institute and Kenyatta University. The staff at World Agroforestry Centre Laboratory in Kenya are thanked for assisting in the analysis. Pre-review of the paper by Job Kihara is also appreciated.

References

Anderson JM, Ingram JS (1993) A handbook of methods. Tropical Soil Biology and Fertility (TSBF) and CAB International, Wallingford, UK

Ayuke FO, Rao MR, Swift MJ, Opondo-Mbai ML (2004) Effect of organic and inorganic nutrient sources in soil mineral N and maize yields in Western Kenya. In: Bationo A (ed) New challenges and opportunities. Managing nutrient cycles to sustain soil fertility in sub-Saharan Africa. Science Academy Publishers and TSBF-CIAT, Nairobi, Kenya, pp 66–76

Barrios E, Kwesiga F, Buresh RJ, Sprent JI (1997) Light fraction soil organic matter fraction and available nitrogen following trees and maize. Soil Sci Soc Am J 61:826–831

Barrios E, Kwesiga F, Buresh RJ, Sprent JI, Coe R (1998) Relating preseason soil nitrogen to maize yield in tree legume-maize rotation. Soil Sci Soc Am J 62:164–169

Birch HF (1964) Mineralisation of plant nitrogen following alternative wet and dry conditions. Plant Soil 20:43–49

Chikowo R, Mapfumo P, Yamgafata P, Giller KE (2004) Maize productivity and mineral dynamics following different soil fertility management practices on a depleted sand soil in Zimbabwe. Agric Ecosyst Environ 102:109–131

Genstat (2005) Genstat Release 8.1 for Windows. Lawes Agricultural Trust, Rothamsted Experimental Station, UK

Giller KE (2002) Targeting management of organic resources and mineral fertilizers: can we match scientists fantasies with farmers' realities Africa. In: Vanlauwe B, Diels J, Sanginga N, Merckx R (eds) Integrated plant nutrients management in sub Saharan Africa: from concept to practice. CABI, Wallingford, CT, pp 155–172

Government of Kenya (GOK) (2001) The 1999 Kenya National Census results. Ministry of Home Affairs, Nairobi, Kenya

Hartemink AE, Buresh RJ, Jama B, Janssen BH (1996) Soil nitrate and water dynamics in Sesbania fallow, weed fallow, and maize. Soil Sci Soc Am J 60:568–574

Hawara R, Lehmann J, Akkinifes iF, Fernandes E, Kanyama-Phiri G (2006) Nitrogen dynamics in maize based agroforestry systems as affected by landscape position in Malawi. Nutr Cycling Agroforest Syst 75:271–284

ICRAF (1995) International Centre for Research in Agroforestry: Laboratory methods of soil and plant analysis. ICRAF: Nairobi, Kenya

Jaetzold R, Schmidt H, Hornet ZB, Shisanya, CA (2006) Farm management handbook of Kenya. Natural conditions and farm information, 2nd edn, vol 11/C. Eastern Province. Ministry of agriculture/GTZ, Nairobi, Kenya

Karlen DL, Kramer LA, Logsdon SD (1998) Field-scale nitrogen balance associated with long-term continuous corn production. Agron J 90:644–650

Kibunja CN, Mugendi DN, Mwaura F, Salema MP (2002) Nitrogen transformations in a long-term maize-bean cropping systems amended with repeated applications of organic and inorganic nutrient sources. In: Mwariri M, Ogutu S, Kimani IW, Ayemba JA (eds) Demand driven agricultural research for sustainable natural resource base, food security, food security and incomes; proceedings of the 8th KARI biennial scientific conference. Kenya Agricultural Research Institute (KARI), Nairobi, pp 276–279

Kihanda FM (1996) The role of farmyard manure in improving maize production in the sub-humid highlands of central Kenya. PhD thesis, The University of Reading, UK

Kihanda FM (2003) Effect of composted manures with high quality organic residues on dry matter accumulation, nitrogen uptake and maize grain yield. East Afr Forest Agric J 69(1):63–68

Makumba W, Janssen B, Oenema O, Akkinifesi FK (2005) Influence of time of application on the performance of Gliricidia sepium prunings as a source of N fro maize. Exp Agric 42:51–63

Maroko JB, Buresh RJ, Smithson PC (1998) Soil nitrogen and availability as affected by maize fallow systems on two soils in Kenya. Biol Fertil soils 26:229–234

Mekonnen K, Buresh RJ, Jama B (1997) Root and inorganic distributions of Sesbania fallows, natural fallow and maize. Plant soil 188:319–327

Mugendi DN, Kanyi M, Kung'u JB, Wamicha W, Mugwe JN (2003) The role of agroforestry trees in intercepting leached nitrogen in the agricultural systems of the central highlands of Kenya. East Afr Agric Forest J 69:69–79

Mugendi DN, Nair PKR, Graetz DA, Mugwe JN, O'Neill MK (2000) Nitrogen recovery by alley-cropped maize and trees from 15 N-labelled tree biomass in the sub-humid highlands of Kenya. Biol Fertil Soils 31:97–101

Mugendi DN, Nair PKR, Mugwe JN, O'Neill MK, Woomer P (1999) Alley cropping of maize with calliandra and leucaena in the subhumid highlands of Kenya. Part 1. Soil fertility changes and maize yield. Agroforest Syst 46: 39–50

Mugwe J, Mugendi DN, Okoba BO, Tuwei P, O'Neill M (2004) Soil conservation and fertility improvement using leguminous shrubs in central highlands in central highlands of Kenya. In: Bationo A (ed) Managing nutrient cycles

to sustain soil fertility in Sub-Saharan Africa. Academy Science, Nairobi, pp 277–297

Myers RJK, Van Noordwijk M, Vitayakon P (1997) Synchrony of nutrient release and demand: plant litter quality, soil environment and farmer management options. In: Cadisch G, Giller KE (eds) Driven by nature: plant litter quality and decomposition. CAB International, Wallingford, CT, pp 215–229

Njunie MN, Wagger MG (2003) Use of herbaceous legumes for improving soil fertility. East Afr Forest Agric J 69(1):49–61

Nyamangara J (2004) Mineral N distribution in the soil profile of a maize field as affected by cattle manure and mineral N additions under humid sub-tropical conditions. In: Bationo A, Kimetu J, Kihara J (eds) Improving human welfare and environmental conservation by empowering farmers to combat soil fertility degradation. Book of abstracts of the African network (AfNet) meeting. Yaounde, Cameroon, p 13

Odera MM, Kimani SK, Musembi F (2000) Factors influencing adoption of integrated use of manure and inorganic fertilizer in central highlands of Kenya. In: Mukisira EA, Kiriro FH, Wamuongo JW, Wamae LW, Muriithi FM, Wasike W (eds) Collaborative and participatory research for sustainably improved livelihoods. Proceedings of the 7th Biennial Scientific conference. Kenya Agricultural Research Institute (KARI), Nairobi, pp 58–64

Okalebo JR, Gathua KW, Woomer PL (2002) Laboratory methods of soil and plant analysis: a working manual, 2nd edn. TSBF-CIAT and SACRED Africa, Nairobi, Kenya

Paramasivam S, Alva AK, Fares A, Sanjwan KS (2001) Estimation of nitrate leaching in an entisol under optimum citrus production. Soil Sci Soc Am J 65:914–921

Phiri RH, Snapp S, Kanyama-Phiri Y (1998) Undersowing maize with Sesbania sesaban in southern Malawi: 2. Nitrate dynamics in relation to N source at three landscape positions. In: Waddington SR, Murwira HK, Kumwenda JD, Hikwa D, Tagwira F (eds) Soil fertility research for maize-based farming systems in Malawi and Zimbabwe. Proceedings of the soil Fertility Network Workshop. African University, Mutare, pp 63–69

Pypers P, Verstraete S, Cong PT, Merckx R (2005) Changes in mineral nitrogen, phosphorus availability and salt-extractable aluminium following the application of green manure residues in two weathered soils of South Vietnam. Soil Biol Biochem 37:163–172

Randall GW, Mulla DJ (2001) Nitrate-nitrogen in surface waters as influenced by climatic conditions and agricultural practices. J Environ Q 30:337–344

Sanchez PA, Shepherd JD, Soule MJ, Place FM, Buresh RJ, Izac AMV, Mukwonye AU, Kwesiga FR, Ndiritu CG, Woomer PL (1997) Soil fertility replenishment in Africa: an investment in natural resource capital. In: Buresh RJ, Sanchez PA,, Calhoun F (eds) Replenishing soil fertility in Africa. SSSA Special publication No. SSSA, Madison, WI, pp 1–46

Shepherd G, Buresh RJ, Gregory PJ (2000) Land use effects on the distribution of inorganic soil nitrogen in smallholder production systems in Kenya. Biol Fertil Soil 31:348–335

Thornton PK, Saka AR, Singh U, Kumwenda JD, Brink JE, Dent J (1995) Application of a maize crop simulation model in the central region of Malawi. Exp Agric 31:213–226

Tisdale SL, Nelson WL, Beaton JD (1990) Soil fertility and fertilizers. Macmillan, New York, NY

Changes in δ^{15}N and N Nutrition in Nodulated Cowpea (*Vigna unguiculata* L. Walp.) and Maize (*Zea mays* L.) Grown in Mixed Culture with Exogenous P Supply

P.A. Ndakidemi and F.D. Dakora

Abstract A two-factorial experiment, involving three levels of phosphorus (0, 40, and 80 kg P ha^{-1}) and four cropping systems (mono, inter-row, intra-row, and intra-hole maize/cowpea culture), was conducted in the field for 2 consecutive years in 2003 and 2004 to assess the effects of P supply and cropping system on plant growth and symbiotic performance. Adding P to cowpea plants significantly increased the growth of all organs (shoots, pods, roots, and nodules) and whole plants in year 2, more than in year 1. Mixed culture however depressed plant growth. Isotopic analysis of both cowpea and maize plants revealed significantly decreased δ^{15}N values in shoots, roots, pods, and whole plants in year 2, but less so in year 1, with P supply. This resulted in significantly increased N derived from fixation in organs and at whole-plant level in the legume. Relative to monoculture, mixed culture decreased the δ^{15}N in all cowpea organs in year 2, and in shoots and pods in year 1. As a result, significantly more N was derived from fixation in intercropped cowpea plants compared to monoculture, with intra-hole cowpea showing the highest dependency on symbiotic fixation for its N nutrition. However, the growth suppression in cowpea caused by the mixed culture resulted in significantly lower actual amounts of N fixed in intercropped cowpea relative to monoculture.

Keywords Intercropping · Phosphorous application · Nitrogen fixation

Introduction

Symbiotic N-fixing systems contribute significant amounts of N to cropping systems. The rhizobia–legume symbioses can fix N$_2$ at rates of 50–300 kg N ha^{-1} year^{-1} (Dakora and Keya, 1997). As a result, the component species of mixed cultures involving symbiotic legumes and cereals can increase the availability of nutrients to the other species. For example, some studies have shown that the symbiotically fixed N from the legume may be transferred to the non-legume partner (Giller et al., 1991; Frey and Schüepp, 1993; Elgersma et al., 2000; Høgh-Jensen and Schjoerring, 2000; Chu et al., 2004). Conversely, the cereal can also make other nutrients more available to the legume through the activity of root exudate molecules such as phytosiderophores (Dakora and Phillips, 2002).

Although available evidence suggests that some factors such as crop species, planting patterns, and crop densities can affect biological N$_2$ fixation (Tothill, 1985; Fujita et al., 1990), few data exist to confirm this view. In farmers' field, planting systems such as intra-row and intra-hole cropping are widely used, yet few detailed studies exist which have addressed the effects of these cropping systems on mineral nutrition in N$_2$ fixation of symbiotic legumes. Although there is evidence that intercropping can reduce the growth of the symbiotic legume via the overshadowing effects of the cereal partner (Willey and Osiru, 1972; Willey, 1979; Mead and Willey, 1980; Horwith, 1985), it is still unclear whether this is genetically pre-determined for the species or only a phenotypic trait. Although a number of studies have been done on cowpea/maize intercropping, these have largely involved the use of inbred lines, whose response is likely to differ from that of landraces.

P.A. Ndakidemi (✉)
Research & Technology Promotion, Cape Peninsula University of Technology, Keizersgracht, Cape Town 8000, South Africa
e-mail: ndakidemip@cput.ac.za

A. Bationo et al. (eds.), *Innovations as Key to the Green Revolution in Africa*,
DOI 10.1007/978-90-481-2543-2_63, © Springer Science+Business Media B.V. 2011

This study assesses the effects of exogenous P supply and cropping system on plant growth and N_2 fixation in cowpea intercropped with maize in the Western Cape Province of South Africa.

Materials and Methods

Experimental Site

Field experiments were conducted at the Agricultural Research Council Nietvoorbij site (33°54′S, 18°14′E) in Stellenbosch, South Africa, during the 2003 and 2004 summer seasons. The site lies in the winter rainfall region of South Africa at an elevation of 146 m above sea level. The mean annual rainfall on the farm is 713.4 mm and mean annual temperatures range from 22.6°C at day to 11.6°C at night.

The experimental site in 2003 had a previous history of grape cultivation, whereas in 2004 it was under grass fallow. The soil type was sandy loam (Glenrosa, Hutton form). Following land preparation, but prior to planting, soil samples were collected and analysed for nutrients.

Experimental Design and Treatments

The experimental treatments consisted of three P levels (0, 40, and 80 kg P ha^{-1}) and four cropping systems (namely mono crop, maize/cowpea inter-row, maize/cowpea intra-row, and maize/cowpea intra-hole cropping). The experimental layout followed a split-plot design with P levels as the main plots, and cropping system, the subplots. There were four replicates per treatment, and the plots measured 4.5 m × 3.2 m. All maize plots had inter-row spacing of 90 cm and intra-row spacing of 40 cm, giving a density of 55,555 plants per hectare. Mono cowpea was sown with inter-row spacing of 60 cm and intra-row spacing of 20 cm to produce plant density of 166,666 per hectare. The within-row spacing of cowpea in the maize/cowpea inter-row cropping system was 20 cm, resulting in cowpea density of 111,111 plants per hectare. The maize/cowpea intra-row planting distance was also 20 cm, giving a density of 55,555 plants per hectare

(identical to that of maize). The intra-hole planting produced a plant density of 55,555 per hectare. The intra-row and intra-hole planting mimicked the practice of traditional smallholder farmers in Africa (Fawusi et al., 1982). Planting was done after ploughing, harrowing, and P application to the respective plots. A local maize variety (Kamapilaa) and farmer-selected cowpea variety (Bengpilaa) were used. Three seeds were planted per hole for each species, and later thinned to two at 2 weeks after planting. The rhizobial inoculant used in this study was peat-based *Bradyrhizobium* strain CB756, which was applied at the rate of 10^9 cells g^{-1} of inoculant. Weeding was done manually with a hoe at 3 and 8 weeks after planting.

Measurement of Soil pH and Organic Matter

The pH of soil was measured in 0.01 M CaCl$_2$ solution using a 1:2.5 soil-to-solution ratio. Organic carbon in soil was determined using the wet digestion method of Walkley and Black (Jackson, 1967). Soil organic matter (SOM) was determined in air-dried soil samples as loss on ignition at 450°C for 24 h after drying at 105°C for 12 h.

Characterization of Selected Soil Chemical Properties

The determination of S in soil was done by adding 20 g of soil in 0.01 M Ca(H$_2$PO$_4$)$_2$ H$_2$O extracting solution (FSSA, 1974), followed by filtering, and S determined by direct aspiration on a calibrated simultaneously inductively coupled plasma (ICP) spectrophotometer (IRIS/AP HR DUO Thermo Electron Corporation, Franklin, Massachusetts, USA).

The extractable P, K, Na, Ca, and Mg were determined by citric acid method as developed by Dyer (1894) and modified by the Division of Chemical Services (DCS, 1956) and Du Plessis and Burger (1964). A 20 g air-dried soil sample was extracted in 200 mL of 1% (w/v) citric acid, heated to 80°C, shaken for 2 min at 10-min intervals over a total period of 1 h and filtered. A 50 mL aliquot was heated to dryness on a water bath, digested with 5 mL of concentrated HCl and HNO$_3$, evaporated to dryness on a water bath, and 5 mL of concentrated HNO$_3$ and 20 mL of de-ionized water added. The mixture was heated to dissolve the dry residue, and the sample filtered. Measurements of P, K, Na, Ca, and Mg were then done directly by direct aspiration on the calibrated simultaneous ICP.

The micronutrients Cu, Zn, Mn, Fe, and Al were extracted from soil using di-ammonium ethylenedi-aminetetraacetic (EDTA) acid solution (Trierweiler and Lindsay (1969), as modified by Beyers and Coetzer (1971)). The extractants were analysed for Cu, Zn, Mn, Fe, and Al using the calibrated simulta-neous ICP spectrophotometer. Boron in the soil was determined following the method of FSSA (1974) and values measured using the ICP spectrophotometer.

Measurement of Soil N

Soil samples were analysed for total N concentra-tions by a commercial laboratory (BemLab, De Beers RD, Somerset West, South Africa), using a LECO-nitrogen analyser (LECO Corporation, St Joseph, MI, USA) with Spectrascan standards (Drobak, Norway) as described by McGeehan and Naylor (1988).

Plant Harvest and Sample Preparation

At 60 days after planting, cowpea and maize plants were sampled for dry matter weight and nitrogen anal-ysis. About 16 and 8 plants were sampled for cowpea and maize, respectively, from the middle rows of each plot. The border plants within each row were excluded. The plants were carefully dug out with their entire root system, washed, and cowpea plants separated into nod-ules, roots, shoot, and pods, while maize plants were divided into roots and shoots. The plant organs were oven-dried at 60°C for 48 h weighed and ground into a fine powder.

Analysis of δ^{15}N and Estimation of Plant Dependence on N$_2$ Fixation

The ratio of ^{15}N/^{14}N and the concentrations of N in plant organs were measured using a Carlo Erba NA 1500 elemental analyser (Fisons Instruments SpA, Strada Rivoltana, Italy) coupled to a Finnigan MAT 252 mass spectrometer (Finnigan MAT Gmbh, Bremen, German) via a Conflo II open-split device.

The ^{15}N natural abundance technique was used to estimate the legume dependence on N$_2$ fixation as follows:

$$\% \text{ N derived from fixation} = \frac{\delta^{15}\text{N reference plant} - \delta^{15}\text{N legume}) \times 100}{\delta^{15}\text{N (reference plant)} - B} \tag{1}$$

where B is the δ^{15}N value of the legume organ rely-ing entirely on N$_2$ fixation for its N nutrition. B values used for cowpea in this study were 1.759‰ for shoot, 0.94‰ for roots, and 1.4713‰ for pods. Maize was used as a reference plant. The amount of N per organ was estimated as a product of % N and dry mass of the organ.

Statistical Analysis

A two-factorial design (two-way ANOVA) was used to analyse for plant growth and symbiotic perfor-mance. The analysis was done using the software of STATISTICA program 1997. Fisher's least significant difference was used to compare significant treatment means at $P \leq 0.05$ (Steel and Torrie, 1980).

Results

Soil Chemical Properties at Planting

The soil chemical properties measured before estab-lishing the experiments in years 1 and 2 are shown in Table 7. The total N concentration in the soil was 9.0 and 13.7% g kg^{-1} for years 1 and 2, respectively.

Effects of P Supply and Mixed Plant Culture on Growth of Cowpea and Maize

Exogenous supply of P to cowpea plants numerically, but not significantly, increased growth of all organs

Table 1 Effect of P supply and cropping system on dry matter yield of organs and whole plants of maize and cowpea planted in 2003 (year 1) and 2004 (year 2)

Plant dry matter (g plant⁻¹)

	Total shoot mass	Total root mass	Pods	Cowpea nodules	Whole plant	Total shoot mass	Total root mass	Pods	Cowpea nodules	Whole plant
	Year 1					Year 2				
A: Cowpea										
P levels (kg P ha⁻¹)	**Main treatments**									
P0	12.1 ± 1.03a	1.7 ± 0.15a	4.2 ± 0.34a	0.36 ± 0.06a	13.8 ± 1.01a	**13.2 ± 2.30b**	**1.0 ± 0.10b**	**8.1 ± 1.06b**	**0.25 ± 0.04b**	**14.3 ± 2.40b**
P40	13.4 ± 1.28a	1.8 ± 0.20a	4.7 ± 0.45a	0.40 ± 0.04a	15.2 ± 1.31a	**20.9 ± 2.50a**	**1.5 ± 0.14a**	**12.5 ± 1.47a**	**0.43 ± 0.07a**	**22.4 ± 2.60a**
P80	14.3 ± 1.25a	2.1 ± 0.27a	5.0 ± 0.44a	0.43 ± 0.05a	16.3 ± 1.26a	**20.4 ± 3.60a**	**1.4 ± 0.14a**	**11.7 ± 2.12a**	**0.43 ± 0.06a**	**21.8 ± 3.70a**
Cropping system	**Sub-treatments**									
Mono cowpea	14.6 ± 1.54a	1.8 ± 0.15a	5.1 ± 0.53a	0.43 ± 0.05a	16.5 ± 1.55a	**31.2 ± 4.00a**	**1.7 ± 0.14a**	**18.2 ± 1.91a**	**0.52 ± 0.06a**	**33.0 ± 4.10a**
Inter-row cowpea	13.9 ± 1.14a	1.7 ± 0.18a	4.7 ± 0.39a	0.36 ± 0.04a	15.6 ± 1.27a	**18.2 ± 1.60b**	**1.4 ± 0.16b**	**10.9 ± 1.08b**	**0.47 ± 0.08b**	**19.6 ± 1.80b**
Intra-row cowpea	12.9 ± 1.22a	2.1 ± 0.35a	4.5 ± 0.43a	0.43 ± 0.08a	15.0 ± 1.32a	**12.5 ± 2.10c**	**1.1 ± 0.11c**	**7.6 ± 1.42c**	**0.32 ± 0.05c**	**13.6 ± 2.20c**
Intra-hole cowpea	11.6 ± 1.55a	1.8 ± 0.27a	4.1 ± 0.54a	0.37 ± 0.06a	13.4 ± 1.70a	**10.8 ± 1.70c**	**0.9 ± 0.10c**	**6.3 ± 1.01c**	**0.16 ± 0.04d**	**11.7 ± 1.80c**
Significance of F-values										
P levels (df = 2)	0.97	0.67	0.97	0.51	0.86	4.98*	4.62*	5.61**	4.35*	5.15*
Cropping systems (df = 3)	0.79	0.47	0.78	0.39	0.71	17.24**	8.48***	20.94***	8.32***	17.31**
P × cropping systems (df = 11)	0.26	1.74	0.25	0.37	0.35	1.81	0.75	2.82	0.94	1.79
B: Maize										
P levels (kg P ha⁻¹)	**Main treatments**									
P0	**64.6 ± 5.62b**	16.6 ± 1.66a	–	–	**81.1 ± 5.53b**	**54.1 ± 6.90b**	11.1 ± 1.77a	–	–	**65.2 ± 8.60b**
P40	**65.4 ± 4.86b**	16.7 ± 1.04a	–	–	**82.1 ± 4.70b**	**72.4 ± 7.90a**	13.7 ± 1.16a	–	–	**86.1 ± 8.20a**
P80	**81.8 ± 5.37a**	17.1 ± 2.73a	–	–	**98.9 ± 6.17a**	**70.2 ± 9.40a**	13.8 ± 0.85a	–	–	**84.0 ± 9.00a**
Cropping system	**Sub-treatments**									
Mono maize	70.2 ± 4.62a	18.9 ± 3.25a	–	–	89.1 ± 6.29a	82.0 ± 12.40a	13.5 ± 1.60a	–	–	95.5 ± 12.20a
Inter-row maize	74.3 ± 7.19a	15.5 ± 0.94a	–	–	89.8 ± 7.72a	57.3 ± 9.60a	12.1 ± 1.43a	–	–	69.4 ± 10.50a
Intra-row maize	67.9 ± 6.30a	15.3 ± 1.48a	–	–	83.2 ± 7.08a	60.6 ± 7.80a	13.2 ± 1.67a	–	–	73.8 ± 8.60a
Intra-hole maize	69.9 ± 7.79a	17.6 ± 2.47a	–	–	87.4 ± 9.31a	62.4 ± 6.90a	12.6 ± 1.59a	–	–	75.0 ± 8.20a
Significance of F-values										
P levels (df = 2)	**3.17***	0.03	–	–	2.22	**1.47***	1.30	–	–	1.75
Cropping systems (df = 3)	0.18	0.63	–	–	0.16	1.37	0.18	–	–	1.34
P × cropping systems (df = 11)	0.99	1.57	–	–	1.11	0.65	0.19	–	–	0.69

Values followed by dissimilar letters in the same column (bold type) differ significantly at $P \leq 0.05$. NB: Total shoot mass in cowpea = shoot weight + pod weight; total root mass in cowpea = root weight + nodule weight

Values presented are means ±SE, $n = 4$. *; **; *** = significant at $P \leq 0.05$, $P \leq 0.01$, $P \leq 0.001$ respectively, SE = standard error. Means followed by dissimilar letters in a column are significantly different from each other at $P = 0.05$ according to Fischer least significance difference

and whole plants in year 1 (Table 1). However, in year 2, shoots, pods, roots, nodules, and whole plants of cowpea were significantly increased with exogenous P application at both 40 and 80 kg P ha^{-1} relative to zero-P control (Table 1). Applying mineral P to maize also significantly increased the growth of shoots and whole plants in both years 1 and 2 relative to zero-P control (Table 1). The dry matter yield of organs and whole plants of cowpea was numerically, but not significantly, decreased in mixed culture relative to monoculture in year 1. However, in year 2, biomass of shoots, pods, roots, nodules, and whole plants was markedly decreased in mixed culture relative to monoculture (Table 1). However, with maize, plant growth was not affected by planting pattern in both years (Table 1).

Effect of P Supply and Cropping System on N Concentration in Organs of Nodulated Cowpea and Maize

The application of P to cowpea had no effect on N concentrations in organs except for nodules where supplying P significantly increased % N of cowpea nodules in year 1 (Table 2). External supply of P also had no effect on N concentration in maize organs.

The cropping system showed a significant effect on the N concentration of only cowpea roots and maize shoots in year 2. Intra-hole cowpea roots showed significantly higher N concentrations relative to monoculture and inter-row planted cowpea (Table 2). Shoot N concentration in maize was significantly more increased in intra-hole plants relative to monoculture (Table 2).

Effect of P Supply and Cropping System on δ^{15}N Values in Organs of Nodulated Cowpea and Maize

As shown in Table 3, supplying P to cowpea plants significantly decreased the δ^{15}N of shoots, roots, pods, and whole plants in year 2 and to a lesser extent in year 1. The application of P also decreased the δ^{15}N values of maize roots in 2 years of experimentation,

leading to a significantly lowered δ^{15}N at the whole-plant level (Table 3).

The cropping system also affected the δ^{15}N of maize and cowpea organs. Relative to monoculture, intercropping decreased the δ^{15}N of shoots, roots, pods, nodules, and whole plants of cowpea in year 2 and only in shoots and pods in year 1 (Table 3). Similarly, shoots, roots, and whole plants of maize showed significantly decreased δ^{15}N when grown in mixed culture compared with monoculture.

Effect of P Supply and Cropping System on Percent of Nitrogen Derived from Atmosphere (Ndfa) in Organs of Nodulated Cowpea

Applying P to symbiotic cowpea significantly increased the % N derived from fixation in virtually all organs and at whole-plant level in that of legume in 2 years of experimentation (Table 4). Cropping system similarly altered the % Ndfa in various organs of cowpea. In both years 1 and 2, significantly more N was derived from fixation in intercropped cowpea plants as compared to monocultures, with intra-hole cowpea showing the highest dependency on symbiotic fixation for its N nutrition (Table 4).

Effect of P Supply and Cropping System on Nitrogen Content of Organs of Cowpea and Maize

Plant total N in cowpea increased significantly with P supply as a result of higher N levels in nodules in year 1, as well as elevated N content of shoots, roots, and nodules in year 2 (Table 5). Whole-plant N also increased in maize with P supply due to greater shoot N accumulation. Although cropping system showed insignificant effect in year 1, N levels in shoots, roots, pods, and nodules were significantly decreased by intercropping relative to monoculture (Table 5). As a result, whole-plant N content was markedly reduced in cowpea plants grown in mixed cultures (Table 5). With maize, however, there was no effect of intercropping on total N of organs or whole plants (Table 5).

Table 2 Effect of P supply and cropping system on N concentration in organs and whole plants of nodulated cowpea and maize sown in 2003 (year 1) and 2004 (year 2)

	Cowpea					Maize		
	Shoots	Roots	Pods	Nodules	Whole plant	Shoots	Roots	Whole plant
Year 1: Effect of P on % N in cowpea and maize								
P0	2.6 ± 0.05a	0.77 ± 0.06a	3.5 ± 0.06a	**4.8 ± 0.18b**	2.9 ± 0.05a	0.93 ± 0.05a	0.09 ± 0.03a	0.50 ± 0.03a
P40	2.5 ± 0.07a	0.77 ± 0.06a	3.6 ± 0.06a	**5.2 ± 0.20ab**	3.0 ± 0.06a	0.93 ± 0.05a	0.09 ± 0.02a	0.50 ± 0.03a
P80	2.6 ± 0.05a	0.75 ± 0.07a	3.4 ± 0.08a	**5.4 ± 0.14a**	3.1 ± 0.06a	1.04 ± 0.07a	0.09 ± 0.04a	0.55 ± 0.04a
Year 1: Effect of cropping system on % N in cowpea and maize								
Sole cowpea	2.6 ± 0.07a	0.79 ± 0.06a	3.4 ± 0.07a	5.2 ± 0.22a	3.0 ± 0.07a	0.88 ± 0.05a	0.07 ± 0.03a	**0.48 ± 0.03b**
Inter-row cowpea	2.5 ± 0.07a	0.70 ± 0.09a	3.6 ± 0.11a	5.1 ± 0.18a	3.0 ± 0.07a	0.89 ± 0.05a	0.09 ± 0.02a	**0.49 ± 0.03ab**
Intra-row cowpea	2.5 ± 0.07a	0.73 ± 0.06a	3.4 ± 0.04a	5.4 ± 0.23a	3.0 ± 0.07a	1.06 ± 0.07a	0.13 ± 0.06a	**0.61 ± 0.05a**
Intra-hole cowpea	2.6 ± 0.05a	0.83 ± 0.07a	3.5 ± 0.08a	4.9 ± 0.21a	3.0 ± 0.07a	1.03 ± 0.06a	0.07 ± 0.02a	**0.56 ± 0.03a**
Significance of F-values								
P levels (df = 2)	0.82	0.04	1.80	3.00*	1.00	1.38	0.001	1.61
Cropping systems (df = 3)	0.44	0.62	1.87	1.17	0.37	2.66	0.59	3.63*
P × cropping systems (df = 11)	0.71	0.44	0.38	0.34	0.45	0.49	0.31	0.74
Year 2: Effect of P on % N in cowpea and maize								
P0	1.9 ± 0.08a	0.96 ± 0.04a	3.3 ± 0.22a	3.4 ± 0.22a	2.4 ± 0.07a	1.04 ± 0.04a	0.57 ± 0.03a	0.80 ± 0.03a
P40	2.0 ± 0.07a	0.97 ± 0.02a	3.0 ± 0.13a	3.4 ± 0.34a	2.4 ± 0.10a	1.04 ± 0.04a	0.62 ± 0.05a	0.85 ± 0.04a
P80	2.0 ± 0.07a	0.97 ± 0.03a	3.0 ± 0.11a	3.5 ± 0.32a	2.4 ± 0.08a	1.06 ± 0.03a	0.57 ± 0.03a	0.80 ± 0.03a
Year 2: Effect of cropping system on % N in cowpea and maize								
Sole cowpea	1.9 ± 0.09a	**0.93 ± 0.03b**	3.1 ± 0.15a	**3.0 ± 0.28a**	**2.3 ± 0.08b**	**0.90 ± 0.04b**	0.54 ± 0.03a	**0.70 ± 0.02b**
Inter-row cowpea	1.9 ± 0.09a	**0.91 ± 0.03b**	3.0 ± 0.14a	3.3 ± 0.36a	**2.3 ± 0.09b**	**1.04 ± 0.04ab**	0.54 ± 0.03a	**0.80 ± 0.03ab**
Intra-row cowpea	2.0 ± 0.08a	**0.99 ± 0.03ab**	3.2 ± 0.12a	3.3 ± 0.36a	**2.4 ± 0.10ab**	**1.10 ± 0.04a**	0.61 ± 0.06a	**0.85 ± 0.04a**
Intra-hole cowpea	2.1 ± 0.08a	**1.05 ± 0.03a**	3.2 ± 0.18a	4.1 ± 0.31a	**2.6 ± 0.09a**	**1.15 ± 0.02a**	0.65 ± 0.05a	**0.90 ± 0.03a**
Significance of F-values								
P levels (df = 2)	0.50	0.11	2.07	0.06	0.14	0.15	0.59	0.33
Cropping systems (df = 3)	1.37	4.00*	0.22	1.75	3.13*	10.40***	1.65	8.23***
P × cropping systems (df = 11)	0.29	0.23	0.23	0.53	0.62	1.35	0.52	0.83

Values followed by dissimilar letters in the same column (bold type) are significant at $P \leq 0.05$

Values presented are means ±SE, $n = 4$. *, **, *** = significant at $P \leq 0.05$, $P \leq 0.001$ respectively, SE = standard error. Means followed by dissimilar letters in a column are significantly different from each other at $P = 0.05$ according to Fischer least significance difference

Table 3 Effect of P supply and cropping system on δ^{15}N values (‰) of organs and whole plants of nodulated cowpea and maize sown in 2003 (year 1) and 2004 (year 2)

Treatment	Cowpea					Maize		
	Shoots	Roots	Pods	Nodules	Whole plant	Shoots	Roots	Whole plant
Year 1: Effect of P supply on δ^{15}N of nodulated cowpea and maize								
P0	1.7 ± 0.14a	6.4 ± 0.23a	2.6 ± 0.12a	12.0 ± 0.28a	3.6 ± 0.09a	2.9 ± 0.32a	8.0 ± 0.18a	5.5 ± 0.20a
P40	1.4 ± 0.27a	6.4 ± 0.30a	2.4 ± 0.12a	11.8 ± 0.30a	3.4 ± 0.18a	2.8 ± 0.17a	7.5 ± 0.12b	5.2 ± 0.09a
P80	0.75 ± 0.26b	6.3 ± 0.24a	2.3 ± 0.17a	11.7 ± 0.34a	3.1 ± 0.15a	2.6 ± 0.18a	7.6 ± 0.12b	5.1 ± 0.12a
Year 1: Effect of cropping system on δ^{15}N of nodulated cowpea and maize								
Sole cowpea	1.9 ± 0.20a	6.4 ± 0.27a	2.8 ± 0.16a	11.8 ± 0.31a	3.7 ± 0.17a	3.3 ± 0.31a	7.7 ± 0.13a	5.5 ± 0.17a
Inter-row cowpea	1.5 ± 0.22ab	6.4 ± 0.34a	2.2 ± 0.17b	11.8 ± 0.37a	3.4 ± 0.16a	2.2 ± 0.25b	7.8 ± 0.24a	5.0 ± 0.19a
Intra-row cowpea	1.1 ± 0.29b	6.4 ± 0.27a	2.3 ± 0.12b	12.1 ± 0.26a	3.3 ± 0.16a	2.7 ± 0.21ab	7.6 ± 0.19a	5.2 ± 0.17a
Intra-hole cowpea	0.60 ± 0.30c	6.2 ± 0.32a	2.5 ± 0.15ab	11.6 ± 0.45a	3.1 ± 0.16a	2.9 ± 0.22ab	7.8 ± 0.15a	5.3 ± 0.16a
Significance of F-values								
P levels (df = 2)	5.29**	0.05	1.83	0.20	2.47	0.61	3.95*	2.05
Cropping systems (df = 3)	5.31**	0.16	2.90*	0.28	2.45	2.81	0.54	1.22
P × cropping systems (df = 11)	0.61	0.03	0.48	0.43	0.27	0.52	1.49	0.42
Year 2: Effect of P supply on δ^{15}N of nodulated cowpea and maize								
P0	1.1 ± 0.12a	4.4 ± 0.035a	1.5 ± 0.10a	10.1 ± 0.35a	2.3 ± 0.076a	4.1 ± 0.25a	6.6 ± 0.30a	5.4 ± 0.24a
P40	1.0 ± 0.14a	4.2 ± 0.016b	1.4 ± 0.13a	10.2 ± 0.34a	2.2 ± 0.088a	3.8 ± 0.34a	5.4 ± 0.30b	4.6 ± 0.24b
P80	0.79 ± 0.17b	4.1 ± 0.019c	1.2 ± 0.15b	9.7 ± 0.31a	2.0 ± 0.100b	3.3 ± 0.43a	4.5 ± 0.12c	3.9 ± 0.23c
Year 2: Effect of cropping system on δ^{15}N of nodulated cowpea and maize								
Sole cowpea	1.7 ± 0.13a	4.3 ± 0.047a	1.9 ± 0.11a	10.5 ± 0.33a	2.6 ± 0.069a	4.8 ± 0.29a	5.9 ± 0.41a	5.3 ± 0.31a
Inter-row cowpea	0.97 ± 0.13b	4.3 ± 0.047a	1.5 ± 0.14b	10.5 ± 0.41a	2.2 ± 0.086b	3.6 ± 0.43b	5.8 ± 0.39a	4.7 ± 0.31ab
Intra-row cowpea	0.68 ± 0.10bc	4.3 ± 0.043a	1.1 ± 0.10bc	9.6 ± 0.34ab	2.0 ± 0.070c	3.8 ± 0.38ab	5.1 ± 0.34b	4.5 ± 0.28b
Intra-hole cowpea	0.53 ± 0.08c	4.2 ± 0.039b	0.97 ± 0.09c	9.3 ± 0.36b	1.9 ± 0.052c	2.9 ± 0.32b	5.1 ± 0.38b	4.0 ± 0.29b
Significance of F-values								
P levels (df = 2)	2.20*	55.0***	2.38*	0.62	7.16**	1.61	16.67***	11.02***
Cropping systems (df = 3)	20.21***	8.3***	14.79***	2.60*	26.64***	4.65**	1.90*	4.88**
P × cropping systems (df = 11)	0.47	0.7	1.22	0.95	0.89	0.60	0.42	0.26

Values followed by dissimilar letters in the same column (bold type) are significant at $P \leq 0.05$

Values presented are means ±SE, $n = 4$. *, **; *** = significant at $P \leq 0.05$, $P \leq 0.01$, $P \leq 0.001$ respectively, SE = standard error. Means followed by dissimilar letters in a column are significantly different from each other at $P = 0.05$ according to Fischer least significance difference

Table 4 Effect of P supply and cropping system on % N derived from fixation (Ndfa) in organs and whole plants of nodulated cowpea sown in 2003 (year 1) and 2004 (year 2)

	Shoots	Roots	Pods	Whole plant
Year 1: Effect of P supply on % Ndfa				
P0	$72 \pm 2.3c$	$18 \pm 2.9a$	$56 \pm 1.9b$	$48 \pm 1.3b$
P40	$74 \pm 3.5bc$	$19 \pm 3.8a$	$60 \pm 2.0a$	$51 \pm 2.5ab$
P80	$81 \pm 2.4a$	$19 \pm 3.1a$	$61 \pm 2.8a$	$54 \pm 2.2a$
Year 1: Effect of cropping system on % Ndfa				
Sole cowpea	$68 \pm 3.3c$	$18 \pm 3.5a$	$53 \pm 2.7b$	$46 \pm 2.5b$
Inter-row cowpea	$73 \pm 3.2bc$	$18 \pm 4.3a$	$62 \pm 2.8a$	$51 \pm 2.4ab$
Intra-row cowpea	$78 \pm 3.4ab$	$18 \pm 3.4a$	$62 \pm 2.0a$	$53 \pm 2.1ab$
Intra-hole cowpea	$82 \pm 2.4a$	$21 \pm 4.1a$	$59 \pm 2.5a$	$54 \pm 2.3a$
Significance of F-values				
P levels (df = 2)	3.13^*	0.04	1.82^*	1.58^*
Cropping systems (df = 3)	3.69^*	0.16	2.90^*	1.82^*
P × cropping systems (df = 11)	0.30	0.03	0.48	0.15
Year 2: Effect of P supply on % Ndfa				
P0	$83 \pm 1.9b$	$38 \pm 0.49b$	$77 \pm 1.5b$	$66 \pm 1.2b$
P40	$84 \pm 2.3b$	$41 \pm 0.22a$	$78 \pm 2.1b$	$68 \pm 1.4b$
P80	$88 \pm 2.6a$	$43 \pm 0.27a$	$81 \pm 2.3a$	$70 \pm 1.6a$
Year 2: Effect of cropping system on % Ndfa				
Sole cowpea	$74 \pm 2.1b$	$39 \pm 0.66b$	$71 \pm 1.7b$	$61 \pm 1.1b$
Inter-row cowpea	$85 \pm 2.1ab$	$40 \pm 0.65ab$	$77 \pm 2.2ab$	$67 \pm 1.3ab$
Intra-row cowpea	$89 \pm 1.5a$	$40 \pm 0.60ab$	$82 \pm 1.5a$	$71 \pm 1.1a$
Intra-hole cowpea	$92 \pm 1.2a$	$42 \pm 0.54a$	$85 \pm 1.4a$	$73 \pm 0.8a$
Significance of F-values				
P levels (df = 2)	2.39^*	55.00^{***}	2.38^*	6.68^{**}
Cropping systems (df = 3)	20.21^{***}	8.30^{***}	14.79^{***}	26.49^{***}
P × cropping systems (df = 11)	0.47	0.67	1.22	0.90

Values followed by dissimilar letters in the same column (bold type) are significant at $P \leq 0.05$
Values presented are means $\pm$SE, $n = 4$. *; **; *** = significant at $P \leq 0.05$, $P \leq 0.01$, $P \leq 0.001$ respectively, SE = standard error. Means followed by dissimilar letters in a column are significantly different from each other at $P = 0.05$ according to Fischer least significance difference

Effect of P Supply and Cropping System on Amounts of N₂ Fixation in Cowpea

Providing external P to cowpea increased N_2 fixation and accumulation of fixed N in this species. P application increased the amount of fixed N in shoots, pods, and/or roots of cowpea, leading to significantly increased amount at the whole-plant level (Table 6). Measurements of N_2 fixation, expressed on per-hectare basis, were also markedly greater with external P supply relative to control, irrespective of the plant density (Table 6).

Although there was no effect of cropping system in year 1, fixed N levels were significantly lower in the intercropping treatments relative to monoculture

for all cowpea organs (Table 6). As a result, the measured values of N_2 fixation were similarly decreased by intercropping, irrespective of whether the cowpea densities were different or equalized (Table 6).

Discussion

Effects of P and Cropping System on Plant Growth

In Africa, the most limiting mineral nutrients to increased crop production are N and P. Although the former can be obtained cheaply from the legume

Table 5 Effect of P supply and cropping system on total N (mg plant^{-1}) in organs and whole plants of cowpea and maize sown in 2003 (year 1) and 2004 (year 2)

Treatment	Cowpea					Maize		
	Shoots	Roots	Pods	Nodules	Whole plant	Shoot	Root	Whole plant
Year 1: Effect of P supply on N content of cowpea and maize								
P0	197.2 ± 18.0a	10.3 ± 1.2a	153.9 ± 13.4a	16.2 ± 2.4b	377.6 ± 30.1b	593.0 ± 44.5b	13.5 ± 3.4a	606.5 ± 44.4b
P40	221.4 ± 21.6a	12.1 ± 1.9a	158.9 ± 15.7a	19.7 ± 2.1ab	412.1 ± 36.6b	748.1 ± 57.2a	15.8 ± 3.8a	763.9 ± 57.1a
P80	246.9 ± 24.4a	9.3 ± 0.9a	167.6 ± 18.1a	23.6 ± 3.0a	447.4 ± 39.3a	657.6 ± 59.8ab	15.9 ± 5.2a	673.4 ± 59.9ab
Year 1: Effect of cropping system on N content of cowpea and maize								
Sole cowpea	236.5 ± 30.2a	10.7 ± 0.8a	186.0 ± 14.0a	22.3 ± 3.1a	455.5 ± 41.8a	602.2 ± 41.2a	10.9 ± 4.1a	613.1 ± 40.9a
Inter-row cowpea	232.1 ± 22.9a	8.7 ± 0.8a	170.4 ± 13.2a	17.7 ± 2.3a	429.0 ± 34.0a	655.8 ± 74.1a	13.5 ± 3.0a	669.3 ± 72.8a
Intra-row cowpea	222.9 ± 21.4a	11.5 ± 2.5a	139.1 ± 15.8a	22.3 ± 3.5a	395.7 ± 37.0a	691.9 ± 48.2a	22.5 ± 7.3a	714.4 ± 47.6a
Intra-hole cowpea	195.9 ± 26.1a	11.2 ± 1.7a	145.1 ± 25.0a	17.0 ± 2.9a	369.2 ± 50.3a	715.0 ± 85.7a	13.2 ± 3.3a	728.3 ± 87.0a
Significance of F-values								
P levels (df = 2)	1.21	1.24	0.18	1.86*	1.87*	2.27*	0.09	2.38*
Cropping systems (df = 3)	0.49	0.75	1.31	0.86	0.76	0.68	1.01	0.77
P × cropping systems (df = 11)	0.54	2.44	0.16	0.29	0.36	1.91	0.31	2.02
Year 2: Effect of P supply on N content of cowpea and maize								
P0	91.2 ± 19.3b	7.5 ± 0.7b	278.6 ± 44.0a	8.9 ± 1.9b	386.1 ± 64.0b	498.3 ± 70.6b	9.8 ± 3.5a	508.1 ± 70.4b
P40	162.6 ± 21.2a	9.9 ± 0.8a	390.3 ± 56.8a	13.2 ± 2.2a	576.1 ± 74.3a	640.6 ± 64.4ab	10.9 ± 1.8a	651.4 ± 64.3ab
P80	180.8 ± 34.3a	9.2 ± 0.7ab	340.8 ± 63.7a	14.1 ± 2.4a	544.8 ± 97.3a	699.8 ± 84.6a	14.7 ± 5.8a	714.4 ± 83.3a
Year 2: Effect of cropping system on N content of cowpea and maize								
Sole cowpea	249.2 ± 43.5a	11.3 ± 1.1a	567.9 ± 70.1a	15.7 ± 2.6a	844.1 ± 112.2a	722.2 ± 123.0a	9.8 ± 4.1a	731.9 ± 122.1a
Inter-row cowpea	140.8 ± 17.6b	8.3 ± 0.7b	324.0 ± 27.1b	14.7 ± 2.7a	487.8 ± 41.4b	477.4 ± 61.7a	10.2 ± 2.3a	487.6 ± 62.4a
Intra-row cowpea	93.7 ± 12.4b	7.5 ± 0.6b	256.6 ± 61.1b	10.8 ± 2.2ab	368.7 ± 73.4b	619.7 ± 79.7a	18.8 ± 7.8a	638.4 ± 78.8a
Intra-hole cowpea	95.7 ± 19.5b	8.2 ± 0.8b	197.7 ± 30.7b	7.1 ± 2.0b	308.7 ± 43.2b	632.3 ± 64.3a	8.4 ± 1.7a	640.7 ± 63.6a
Significance of F-values								
P levels (df = 2)	5.54**	3.33*	1.74	1.79*	2.97*	1.85*	0.38	1.96*
Cropping systems (df = 3)	9.83***	4.47**	10.99***	2.74*	12.37***	1.33	0.97	1.35
P × cropping systems (df = 11)	1.42	0.49	1.28	0.82	1.51	0.38	0.68	0.36

Values followed by dissimilar letters in the same column (bold type) are significant at $P \leq 0.05$

Values presented are means ±SE, $n = 4$. *; **; *** = significant at $P \leq 0.05$, $P \leq 0.01$, $P \leq 0.001$ respectively, SE = standard error. Means followed by dissimilar letters in a column are significantly different from each other at $P = 0.05$ according to Fischer least significance difference

Table 6 Effect of P supply and cropping system on fixed N in organs and whole plants of cowpea sown in 2003 (year 1) and 2004 (year 2)

Treatment	Shoots	Roots	Pods	Whole plant	N fixed (different plant densities)	N fixed (equal plant densities)
	(mg plant^{-1})				(kg N ha^{-1})	
Year 1: Effect of P supply on fixed N						
P0	**143.2 ± 14.7b**	1.7 ± 0.28a	86.8 ± 9.0a	**231.7 ± 21.3b**	**22.4 ± 3.4b**	**38.6 ± 3.6b**
P40	**162.6 ± 18.4ab**	2.2 ± 0.43a	94.6 ± 9.2a	**259.3 ± 25.7ab**	**25.6 ± 4.2ab**	**43.2 ± 4.3ab**
P80	**200.2 ± 20.9a**	2.0 ± 0.47a	99.9 ± 9.6a	**302.1 ± 27.6a**	**29.6 ± 4.5a**	**50.3 ± 4.6a**
Year 1: Effect of cropping system on fixed N						
Sole cowpea	158.4 ± 19.4a	2.0 ± 0.52a	99.9 ± 10.4a	260.3 ± 27.0a	43.4 ± 4.5a	43.4 ± 4.5a
Inter-row cowpea	175.4 ± 23.4a	1.7 ± 0.51a	106.8 ± 9.3a	283.9 ± 29.6a	31.5 ± 3.3b	47.3 ± 4.9a
Intra-row cowpea	176.3 ± 20.9a	2.0 ± 0.44a	85.2 ± 9.5a	263.4 ± 27.7a	14.6 ± 1.5c	43.9 ± 4.6a
Intra-hole cowpea	164.6 ± 25.0a	2.1 ± 0.41a	83.2 ± 12.6a	249.9 ± 35.8a	13.9 ± 2.0c	41.6 ± 6.0a
Significance of F-values						
P levels (df = 2)	**2.17***	0.26	0.45	**1.71***	**1.80***	**1.71***
Cropping systems (df = 3)	0.15	0.11	1.00	0.21	21.28***	0.21
P × cropping systems (df = 11)	0.35	0.34	0.10	0.27	0.58	0.27
Year 2: Effect of P supply on fixed N						
P0	**74.9 ± 15.1c**	**2.9 ± 0.29b**	**215.4 ± 35.7b**	**293.2 50.2b**	**32.9 ± 8.9b**	**48.9 ± 8.4b**
P40	**132.9 ± 15.7b**	**4.1 ± 0.35a**	**298.6 ± 43.1a**	**435.6 54.7a**	**46.0 ± 9.8a**	**72.6 ± 9.1a**
P80	**150.7 ± 25.7a**	**3.9 ± 0.29a**	**261.8 ± 43.9a**	**416.3 67.2a**	**49.9 ± 13.4a**	**69.4 ± 11.2a**
Year 2: Effect of cropping system on fixed N						
Sole cowpea	**185.0 ± 34.1a**	**4.6 ± 0.47a**	**406.1 ± 55.1a**	**595.7 87.0a**	**99.3 ± 14.5a**	**99.3 ± 14.5a**
Inter-row cowpea	**118.8 ± 14.6b**	**3.4 ± 0.31b**	**249.2 ± 21.7b**	**371.4 31.3b**	**41.3 ± 3.5b**	**61.9 ± 5.2b**
Intra-row cowpea	**84.9 ± 11.7b**	**3.1 ± 0.28c**	**210.8 ± 49.6b**	**298.7 59.5bc**	**16.6 ± 3.3c**	**49.8 ± 9.9b**
Intra-hole cowpea	**89.2 ± 18.7b**	**3.5 ± 0.32b**	**168.4 ± 26.8c**	**261.0 38.6c**	**14.5 ± 2.1c**	**43.5 ± 6.4b**
Significance of F-values						
P levels (df = 2)	**5.38****	**5.08***	**1.36***	**2.48***	**1.94***	**2.48***
Cropping systems (df = 3)	**5.49****	**3.80***	**6.34****	**6.99****	**6.99****	**6.99****
P × cropping systems (df = 11)	0.98	0.54	0.78	0.92	1.34	0.92

Values followed by dissimilar letters in the same column (bold type) are significant at $P \leq 0.05$

Values presented are means ±SE, $n = 4$. *; **; *** = significant at $P \leq 0.05$, $P \leq 0.05$, $P \leq 0.01$, $P \leq 0.001$ respectively, SE = standard error. Means followed by dissimilar letters in a column are significantly different from each other at $P = 0.05$ according to Fischer least significance difference

Table 7 Concentration of extractable nutrients in bulk soil sampled prior to P application and planting

Year	pH (CaCl₂)	SOM	C	P	K	Ca	Mg	S	Na	Cu	Zn	Mn	B	Fe	Al
						Concentration of mineral nutrients (mg kg⁻¹)									
2003	6.3 ± 0.1	3200 ± 100	900 ± 100	40.0 ± 3.5	103.5 ± 10.3	842.1 ± 80.2	121.5 ± 12.2	3.1 ± 0.2	0.10 ± 0.001	8.4 ± 0.3	2.47 ± 0.3	20.5 ± 1.9	0.7 ± 0.04	122.8 ± 11.5	–
2004	6.3 ± 0.1	3100 ± 200	1900 ± 100	8.8 ± 0.8	141.8 ± 5.9	521.3 ± 40.1	121.5 ± 3.7	3.7 ± 0.2	0.067 ± 0.002	0.8 ± 0.04	1.79 ± 0.2	9.7 ± 0.1	0.3 ± 0.02	131.5 ± 5.02	0.7 ± 0.04

Each value represents an average of 15 soil samples collected from different points within each replicate plot

symbioses with root-nodule bacteria, the latter is exhaustible (as rock phosphate), and its deficiency negatively affects N yield from the legume–rhizobia symbiosis. Understanding P use efficiency in cropping systems involving legumes and cereals has prospects for overcoming production constraints posed by N and P. Assessing the response of maize and cowpea to exogenous P supply showed no changes to cowpea growth in year 1, although with maize there was a significant increase in shoot and whole-plant growth. Because the endogenous soil P at the site of year 2 experiment was 4.5-fold lower than that of year 1 (Table 7), whole-plant growth and that of all organs except roots was significantly increased by external P application in both cowpea and maize (Table 1).

The response obtained here for legume and cereal growth in soils with low endogenous P, but not at higher P concentration, is consistent with reports by Dakora (1984), Israel (1987), Chang and Shibles (1985), Ssali and Keya (1986), Pereira and Bliss (1987), Giller et al. (1998), Ndakidemi et al. (1998), Buerkert et al. (2001), Tang et al. (2001), Carsky (2003), and Jensen et al. (2003) who showed that increasing external supply increased organ development and overall plant growth.

Effects of P on Cowpea Nodule Function and Metabolism in Maize

The external supply of P to nodulated legumes is known to improve symbiotic performance and N_2 fixation in these species (Robson et al., 1981; Jacobsen, 1985; Singleton et al., 1985; Israel, 1987). In this study, the application of P to maize and cowpea in the sole and mixed culture had no effect on the N concentration of organs in either species, except for % nodule-N which rose with P supply in year 1. However, the $\delta^{15}N$ values of cowpea organs and whole plants were significantly decreased with exogenous P supply in years 1 and 2 (Table 3), clearly indicating P-related enhancement of N_2 fixation. Interestingly, the $\delta^{15}N$ values of maize roots and whole plants were also significantly reduced by P application, in a manner similar to the legume. Such a decrease in $\delta^{15}N$ of tissues could either stem from mycorrhizal transfer of fixed N from the root zone to the cereal or direct rhizosphere imports of secreted fixed N by vigorous growing maize roots.

In general, the lower the $\delta^{15}N$ values, the greater the N derived from fixation (Shearer and Kohl, 1986). As expected, the lowering of $\delta^{15}N$ in shoots, pods, roots, and whole plants of cowpea resulted in significantly greater proportion of % N derived from fixation in those organs in years 1 and 2 (Table 4). This increase in Ndfa with P supply was reflected as a rise in total plant N, especially in year 2 where mineral P provision to a soil low in endogenous P (Table 7) resulted in markedly greater total N of shoots, roots, nodules, and whole plants (Table 5). Shoot and whole-plant N was similarly increased in years 1 and 2 with P supply, possibly as a result of improved acquisition by mycorrhizal activity or a well-developed root system from enhanced P nutrition (Bianciotto and Bonfante, 2002; Rengel and Marschner, 2005).

The $\delta^{15}N$ Ndfa and total N values truly reflect nodule function as fixed N levels in organs were found to closely mirror those of the various symbiotic traits. As reported by Israel (1987), N_2 fixation in nodules increased with provision of exogenous P to cowpea plants (Table 6). Whether measured on the basis of the different cowpea densities used in the field or on an equalized density basis, N_2 fixation was markedly increased by P supply to cowpea plants (Table 6).

Effects of Cropping System on Cowpea Nodule Function and Metabolism in Maize

The $\delta^{15}N$ values of cowpea shoots, pods, roots, nodules, and whole plants were also decreased by intercropping when compared to those of monoculture, clearly suggesting greater nodule activity (Shearer and Kohl, 1986). As a result, the % Ndfa in these cowpea organs was significantly higher with intercropping relative to sole culture (Table 4). Interestingly, the magnitude of this increase in % Ndfa from intercropping was also greater in year 2 relative to year 1, possibly due to differences in endogenous soil P levels. However, total plant N was found to decrease in cowpea with intercropping relative to monocropping (Table 5), a true reflection of the decrease in whole-plant biomass with intercropping (Table 1).

As a result of the observed decrease in total legume N with intercropping, fixed N per organ or whole plant was also significantly reduced by intercropping compared to sole culture (Table 6). Consequently, the contribution of cowpea to N economy of the cropping system was also decreased by mixed culture relative to monoculture (Table 6). Irrespective of whether the fixed N measurements were adjusted, or not, the differences in cowpea density used in each cropping system, there was still a marked variation in the levels of N_2 fixation. This clearly suggests that the smaller amount of N fixed with intercropping was not merely due to differences in plant numbers, but rather to the effect of intercropping on plant function.

Although the levels of N fixed may be low with intercropping, the values obtained in this study are comparable to those of other studies where legumes were intercropped with cereals (Willey and Osiru, 1972; Willey, 1979; Mead and Willey, 1980; Horwith, 1985). The unadjusted data shown in Table 6 for fixed N in intercropped cowpea are probably close to the amounts obtained in farmers' fields, where cowpea is sparsely cropped with maize or sorghum.

Acknowledgements This study was supported with a competitive grant awarded by the Department of Science and Technology, Pretoria, under the auspices of the Southern African Development Community Science and Technology Research Fund, as well as a grant from the National Research Foundation and financial assistance from the Cape Peninsula University of Technology.

References

Beyers CPDL, Coetzer FJ (1971) Effect of concentration, pH and time on the properties of di-ammonium EDTA as a multiple soil extractant. Agrochemophysica 3:49–54

Bianciotto V, Bonfante P (2002) Arbuscular mycorrhizal fungi: a specialised niche for rhizospheric and endocellular bacteria. Antonie Leeuwenhoek 81:365–371

Buerkert A, Bationo A, Piepho HP (2001) Efficient phosphorus application strategies for increased crop production in sub-Saharan West Africa. Field Crops Res 72:1–15

Carsky RJ (2003) Response of cowpea and soybean to P and K on *terre de barre* soils in southern Benin. Agric Ecosyst Environ 100:241–249

Chang JF, Shibles RM (1985) An analysis of competition between intercropped cowpea and maize II. The effect of fertilization and population density. Field Crops Res 12: 145–152

Chu GX, Shen QR, Cao JL (2004) Nitrogen fixation and transfer from peanut to rice cultivated in aerobic soil in an intercropping system and its effect on soil N fertility. Plant Soil 263:17–27

Dakora FD (1984) Nodulation and nitrogen fixation by groundnut in amended and unammended field soils in Ghana. In: Ssali H, Keya SO (eds) Proceedings of the First Conference of the African Association for biological Nitrogen Fixation (AABNF) held in Nairobi, Kenya, 23–27 Jul 1984. *Rhizobium* Microbiological Resource Centre, The Nairobi, pp 324–339

Dakora FD, Keya SO (1997) Contribution of legume nitrogen fixation to sustainable agriculture in Sub-Saharan Africa. Soil Biol Biochem 29:809–817

Dakora FD, Phillips D (2002) Root exudates as mediators of mineral acquisition in low-nutrient environments. Plant Soil 245:35–47

Division of Chemical Services (1956) Analytical methods. Division of Chemical Services, Department of Agriculture, Pretoria

Du Plessis SF, Burger RDT (1964) A comparison of chemical extraction methods for the evaluation of phosphate availability of top soils. South Afr J Agric Sci 8:1113

Dyer B (1894) On the analytical determinations of probably available "mineral plant-food in soil". J Chem Soc 65:115

Elgersma A, Schlepers H, Nassiri M (2000) Interactions between perennial ryegrass (*Lolium perenne* L.) and white clover (*Trifolium repens* L.) under contrasting nitrogen availability: productivity, seasonal patterns of species composition, N2 fixation, N transfer and N recovery. Plant Soil 221:281–299

Fawusi MOA, Wanki SBC, Nangju D (1982) Plant density effects on growth, yield, leaf area index, and light transmission in intercropped maize and *Vigna unguiculata* (L.) Walp. in Nigeria. J Agric Sci 99:19–23

Fertilizer Society of South Africa (1974) Manual of soil analysis methods. FSSA Publication no 37

Frey B, Schüepp H (1993) A role of vesicular-arbuscular (VA) mycorrhizal fungi in facilitating interplant nitrogen transfer. Soil Biol Biochem 25:651–658

Fujita K, Ogata S, Matsumoto K, Masuda T, Godfred K, Ofosu-Budu KG, Kuwata K (1990) Nitrogen transfer and dry matter production in soybean and sorghum mixed cropping system at different population densities. Soil Sci Plant Nutr 36: 233–241

Giller KE, Amijee F, Brodrick SJ, Edje OT (1998) Environmental constraints to nodulation and nitrogen fixation of *Phaseolus vulgaris* L in Tanzania II. Response to N and P fertilizers and inoculation with *Rhizobium*. Afr Crop Sci J 6:171–178

Giller KE, Ormsher J, Awah F (1991) Nitrogen transfer from Phaseolus bean to intercropped maize measured using 15N-enrichment and 15N-isotope dilution methods. In: Dommergues YR, Krupa SV (eds) Soil microorganisms and plants. Elsevier, Amsterdam, pp 163–203

Hogh-Jensen H, Schjoerring JK (2000) Below-ground nitrogen transfer between different grassland species: direct quantification by ^{15}N. Plant Soil 227:171–183

Horwith B (1985) A role for intercropping in modern agriculture. BioScience 35:286–291

Israel DW (1987) Investigation of the role of phosphorus in symbiotic dinitrogen fixation. Plant Physiol 84:835–840

Jackson ML (1967) Soil chemical analysis. Prentice Hall of India Private Limited, New Delhi, 498 pp

Jacobsen I (1985) The role of phosphorus in nitrogen fixation by young pea plants (*Pisum sativum*). Physiol Plantarum 64:190–196

Jensen JR, Bernhard RH, Hansen S, McDonagh J, Møberg JP, Nielsen NE, Nordbo E (2003) Productivity in maize based cropping systems under various soil–water–nutrient management strategies in a semi-arid, alfisol environment in East Africa. Agric Water Manage 59:217–237

McGeehan SL, Naylor VV (1988) Automated instrumental analyses of carbon and N in plant and soil samples. Commun Soil Sci Plant Anal 19:493–505

Mead R, Willey RW (1980) The concept of a "Land Equivalent Ratio" and advantages in yields from intercropping. Exp Agric 16:217–228

Ndakidemi PA, Nyaky AS, Mkuchu M, Woomer PL (1998) Fertilization and inoculation of *Phaseolus vulgaris* in Arusha, Tanzania. In: Dakora FD (ed) The proceedings of eighth congress of the African Association for Biological Nitrogen Fixation, 23–27 Nov 1998. University of Cape Town, South Africa, pp 166–167

Pereira PAA, Bliss FA (1987) Nitrogen fixation and plant growth of common beans (*Phaseolus vulgaris* L) at different levels of phosphorus availability. Plant Soil 104:79–84

Rengel Z, Marschner P (2005) Nutrient availability and management in the rhizosphere: exploiting genotypic differences. New Phytol 168:305–312

Robson AD, O'Hara GW, Abbott LK (1981) Involvement of phosphorus in nitrogen fixation by subterranean clover (*Trifolium subterraneum* L). Austr J Plant Physiol 8:427–436

Shearer G, Kohl DH (1986) N$_2$ fixation in field settings: estimates based on natural ^{15}N abundance. Austr J Plant Physiol 13:699–756

Singleton PW, Abdel-Magid HM, Tavares JW (1985) Effect of phosphorus on effectiveness of strains of *Rhizobium japonicum*. Soil Sci Soc Am J 49:613–616

Ssali S, Keya SO (1986) The effects of phosphorus and nitrogen fertilizer level on nodulation, growth and dinitrogen fixation of three bean cultivars. Trop Agric (Trinidad) 63: 105–109

Steel RGD, Torrie JH (1980) Principles and procedures of statistics: a biometrical approach, 2nd ed. McGraw-Hill, New York, NY

Tang C, Hinsinger P, Drevon JJ, Jaillard B (2001) Phosphorus deficiency impairs early nodule functioning and enhances proton release in roots of *Medicago truncatula* L. Ann Bot 88:131–138

Tothill JC (1985) The role of legumes in farming systems of sub-Saharan Africa. In: Haque I, Jutsi S, Neate PJH (eds) Potentials of forage legumes in farming systems of sub-Saharan Africa. ILCA, Addis Ababa, Ethiopia, pp 162–185

Trierweiler JF, Lindsay WL (1969) EDTA-Ammonium carbonate soil test for zinc. Soil Sci Soc Am Proc 33:49–54

Willey RW (1979) Intercropping – its importance and research needs. Part I competition and yield advantages. Field Crops Abstr 32:1–10

Willey RW, Osiru DSO (1972) Studies on mixtures of maize and beans (*Phaseolus vulgaris*) with particular reference to plant population. J Agric Sci 79:519–529

Cation Flux in Incubated Plant Residues and Its Effect on pH and Plant Residue Alkalinity

G.M. Sakala, D.L. Rowell, and C.J. Pilbeam

Abstract Plant residues offer a viable alternative to the costly and non-readily available commercial lime in addressing the constraint of soil acidity among the rural smallholders. The liming potential of the residues, attributable to excess cations over inorganic anions, exists in either available or non-available forms. This study investigates cation flux and its effect on pH and plant residue alkalinity of four plant residues, maize (*Zea mays*), soya beans (*Glycine max*), Leucaena (*Leucaena leucocephala*) and Gliricidia (*Gliricidia sepium*), upon incubation for 100 days with and without application of lime in an acidic Zambian Ferralsol. Initial characterization of base cation content ranged from 239 to 879 (Ca^{2+}), 188 to 458 (Mg^{2+}) and 298 to 477 $mmol_c$ kg^{-1} plant material (K^+). Of these, 26–60, 62–92 and 76–96% in that order were water soluble. On incubation, up to 70% Ca^{2+} and at least 80% Mg^{2+} and K^+ added in the residues were initially present eventually increasing to 84 and 95%, respectively. Potential alkalinity values were 373 (maize), 1264 (soya beans), 794 (*Leucaena*) and 1024 (*Gliricidia*) $mmol_c$ kg^{-1}. Of these, between 42 (*Gliricidia*) and 52% (*Leucaena*) constituted available alkalinity. Exchangeable aluminium was absent or appeared in insignificantly very low amounts towards the end of the incubation, while base cations were fixed. There was initial dependence of pH on both total cation concentration and residue alkalinity, but this relationship was later lost, suggesting incomplete activation of the non-available fraction of the potential alkalinity. Nitrogen mineralization affected both cation flux and residue alkalinity. This study highlights the importance of residues on amelioration of major cation deficiencies and aluminium phytotoxicity.

Keywords Cation flux · Nitrogen mineralization · Plant alkalinity · Soil acidity · Zambia

Introduction

Many soils throughout the world may be termed 'highly weathered', resulting from dissolution and leaching processes over extremely long periods or when precipitation is high, from rapid breakdown of easily weathered parent materials over shorter period time spans (van Wambeke, 1992). The clay mineralogy of highly weathered soils is dominated by kaolinite and oxides of iron and aluminium. These clay minerals have low cation exchange capacity (CEC), complicated further by the fact that CEC is dependent upon pH and ionic strength of the soil solution (van Raij and Peech, 1972; Uehara and Gillman, 1981; Gillman et al., 2001). Soil acidity increases the concentrations of phytotoxic monomeric inorganic aluminium species in the solution of mineral soils. This leads to increased soil exchangeable aluminium content (Wong and Swift, 1995). Aluminium phytotoxicity together with deficiencies of nutrients such as calcium, magnesium, potassium and phosphorus is the major cause of acid-induced soil infertility. Manganese phytotoxicity and nitrogen deficiency are also common in these soils. Liming alone does not always alleviate the problem as addition of nutrients may also be required and the risk

G.M. Sakala (✉)
Zambia Agriculture Research Institute, Mount Makulu Research Station, Chilanga, Zambia
e-mail: godfreysakala@yahoo.co.uk

A. Bationo et al. (eds.), *Innovations as Key to the Green Revolution in Africa*,
DOI 10.1007/978-90-481-2543-2_64, © Springer Science+Business Media B.V. 2011

of over liming can cause other nutrient deficiencies. An option is to use organic matter additions as part of an integrated approach to ameliorate acid soils (Noble et al., 1996; Wong et al., 2000).

Plant materials differ in their ability to ameliorate soil acidity (Bessho and Bell, 1992; Wong et al., 1995; Sakala et al., 2004). Cation amelioration of soil acidity is possible because of the mineralized cations (Yan et al., 1996; Pocknee and Sumner, 1997). However, with the low CEC and the erosivity of rainfall in the tropics, soluble components from plant residues are liable to loss through erosion if not properly managed. This is more likely to happen in surface-applied residues practised in minimum tillage systems. The general low contents of base cations in these highly weathered soils even in pristine condition, and little capacity to retain those nutrients when applied in water soluble form, call for appropriate management options of the sources of these nutrient ions. Development of the use of these materials in an integrated approach to the management of soil acidity therefore requires a better understanding of the cation flux. This study investigates cation flux with time of incubated plant residues and its effect on pH and plant residue alkalinity.

Materials and Methods

Soil

The soil used in the experiment was an acid soil (Rhodic Ferralsol) from northern Zambia. The mean annual rainfall and temperature of the area were 1247 mm and 19.7°C, respectively. The soil was sampled from the top 20 cm from cleared land that had been fallowed for at least 5 years with no known fertilizer history. Particle size analysis was determined by the hydrometer method (Carter, 1993). Soil pH was determined in 10 mM CaCl$_2$ at a soil solution ratio of 1:2.5. Exchangeable base cations were determined as described by Thomas (1982). Exchangeable aluminium and hydrogen were determined by the NaOH/NaF method (Rowell, 1994).

Plant Material

The residues used were Leucaena (*Leucaena leucocephala*), Gliricidia (*Gliricidia sepium*), soya beans

(*Glycine max*) and maize (*Zea mays*). The plant materials were characterized for total composition, and forms of alkalinity were determined as described by Sakala et al. (2004).

The Incubation Experiment

To 300 g of soil (four replicates arranged in a randomized complete block design) was added 9 g of residue and incubated with or without application of lime (0.291 g of AR grade Ca(OH)$_2$) followed by thorough mixing and the addition of 45 ml of H$_2$O. A control without residue was included. They were incubated at 30°C in polythene bags loosely folded and occasionally opened and shaken to ensure adequate aeration. Water was added to maintain the moisture content.

Samples were extracted from all treatments before the first addition of water to give data at zero incubation time and then at 14-day intervals. pH was measured in 10 mM CaCl$_2$ at a 1:2.5 soil/solution ratio. Twenty grams of soil was extracted with 100 ml of 1 M KCl for 2 h, filtered and analysed for Ca^{2+}, Mg^{2+}, Al^{3+}, H$^+$, NH$_4^+$ and NO$_3^-$. Potassium and Na$^+$ were extracted from 6 g of soil shaken with 30 ml of 1 M ammonium acetate buffered at pH 7.

Analysis of variance was performed at each sampling interval to determine statistical differences between treatments and cross sections of individual response curves to ascertain changes with time using GenStat program. Any two treatment means for any variable could be compared by the standard error of differences (SED) given. Any two means in the body are significantly different at 5% level if they differ by more than 1.96 × SED.

Results and Discussion

Table 1 presents the properties of the soil. For the pH range 4.2–6.5 the buffer curve is linear.

Plant Residues

Chemical characteristics of the residues are shown in Table 2. Calcium, Mg^{2+} and K$^+$ are major cations,

Table 1 Soil properties

Soil property	
pH	4.24
Exchangeable cations (cmol_c kg^{-1})	
Ca^{2+}	0.39
Mg^{2+}	0.16
K^+	0.27
Na^+	0.06
NH_4^+	0.04
Al^{3+}	0.55
H^+	0.14
ECEC	1.60
Al saturation	33%
Base saturation	56%
Total nitrogen	0.07%
Organic C	0.86%
Sand:silt:clay	51:8:41
Buffer capacity	12.9 mmol OH$^-$ kg^{-1} pH^{-1}

while SO_4^{2-} is the major inorganic anion. Maize had the highest C:N ratio. Soya beans had the highest potential alkalinity, while maize had the lowest. A similar trend is seen for available alkalinity.

Amounts of cations and nitrogen added through the residues on incubation are shown in Table 3. The values reflect the constituent concentrations of the parameters in the residues as shown in Table 2, with soya beans adding the largest amounts of cations to the soil.

Table 4 shows amounts of base cations present at the beginning and at the end of incubation after 100 days with and without application of lime. Soya beans released the largest amounts of Ca^{2+} at the beginning and at the end in the absence of lime, followed by

Gliricidia. A similar trend was exhibited for Mg^{2+} at both lime levels. With application of lime, *Gliricidia* released more Ca^{2+} than did the other treatments receiving residues, but similar amounts were released as soya beans at the end. *Gliricidia* released most K in the absence of lime both at the beginning and at the end of the incubation compared to the other amended treatments. With the application of lime, similar amounts were released at zero incubation time between *Gliricidia* and soya beans, while *Gliricidia* released most K^+ at the end. Cation release from *Leucaena* was intermediate between amounts released by *Gliricidia* and soya beans, while maize released least amounts of cations among the amended treatments. This cation release pattern was exhibited at both lime levels. Sodium was least among the cations with and without application of lime, reflective of the small amounts of the cation content in the residue (Table 2). The control released least cations compared to treatments receiving residues at both lime levels.

Figure 1 shows changes in pH, inorganic N pools and acidic cations of H^+ and Al^{3+} during incubation. pH values both without (Fig. 1a) and with the application of lime (Fig. 1b) increase in the first 14–28 days and then fall back again at both lime levels. Treatments receiving lime were buffered at higher pH values than those that had not received lime. The greatest pH shift was exhibited by *Gliricidia* where the pH rose to over 7 at both lime levels between 14 and 28 days. Figure 1c–f shows changes in the inorganic N pools both without and with the application of lime. *Gliricidia*-amended treatments mineralized the largest amounts of NH4$^+$-N which were maintained throughout the incubation period (Fig. 1c, d). This was

Table 2 Chemical characteristics of the plant residues

	Ca^{2+}, mmol_c kg^{-1} (%)	Mg^{2+}, mmol_c kg^{-1} (%)	K^+, mmol_c kg^{-1} (%)	Na^+, mmol_c kg^{-1} (%)	Cl^-, mmol_c kg^{-1} (%)	$H_2PO_4^-$(P)	SO_4^{2-}(S)	Potential alkalinity	Available alkalinity	C (%)	N (%)
Maize	239 (0.48)	189 (0.23)	298 (1.17)	1.1 (<0.01)	74 (0.26)	32 (0.10)	259 (0.41)	373	181	47	1.1
Soya beans	879 (1.76)	444 (0.54)	360 (1.41)	2.9 (<0.01)	18 (0.07)	102 (0.32)	345 (0.55)	1264	577	46	1.66
Leucaena	547 (1.15)	238 (0.29)	374 (1.46)	4.7 (0.01)	10 (0.04)	53 (0.16)	499 (0.80)	794	415	51	4.26
Gliricidia	790 (1.58)	278 (0.34)	456 (1.78)	0.2 (<0.01)	97 (0.34)	66 (0.20)	350 (0.56)	1024	427	49	3.2

All values are means of duplicates

Table 3 Amounts of cations and nitrogen added at the beginning of the incubation

Residue	Ca^{2+} (mmol_c kg^{-1} soil)	Mg^{2+} (mmol_c kg^{-1} soil)	K^+ (mmol_c kg^{-1} soil)	Na^+ (mmol_c kg^{-1} soil)	N (μg g^{-1} soil)
Maize	7.17	5.67	8.94	0.03	330
Soya beans	26.37	13.32	10.8	0.09	498
Leucaena	16.41	7.14	11.22	0.14	1278
Gliricidia	23.7	8.34	13.68	0.01	960

Table 4 Base cations present at the beginning (initial) and at the end (final) of the incubation with and without application of lime

| | Without lime application | | | | | | | | With application of lime | | | | | | | |
| | Initial | | | | Final | | | | Initial | | | | Final | | | |
Residue	Ca^{2+} (cmol$_c$ kg^{-1} soil)	Mg^{2+} (cmol$_c$ kg^{-1} soil)	K$^+$ (cmol$_c$ kg^{-1} soil)	Na$^+$ (cmol$_c$ kg^{-1} soil)	Ca^{2+} (cmol$_c$ kg^{-1} soil)	Mg^{2+} (cmol$_c$ kg^{-1} soil)	K$^+$ (cmol$_c$ kg^{-1} soil)	Na$^+$ (cmol$_c$ kg^{-1} soil)	Ca^{2+} (cmol$_c$ kg^{-1} soil)	Mg^{2+} (cmol$_c$ kg^{-1} soil)	K$^+$ (cmol$_c$ kg^{-1} soil)	Na$^+$ (cmol$_c$ kg^{-1} soil)	Ca^{2+} (cmol$_c$ kg^{-1} soil)	Mg^{2+v}	K$^+$ (cmol$_c$ kg^{-1} soil)	Na$^+$ (cmol$_c$ kg^{-1} soil)
Maize	0.76	0.72	1.28	0.024	0.86	0.76	1.20	0.029	3.08	0.72	1.18	0.021	3.00	0.66	1.16	0.027
Soya beans	1.91	1.28	1.10	0.022	2.50	1.38	1.24	0.029	3.85	1.31	1.33	0.026	4.11	1.08	1.18	0.019
Leucaena	1.15	0.80	1.27	0.026	1.63	0.81	1.29	0.035	3.27	0.76	1.30	0.030	3.99	0.76	1.29	0.031
Gliricidia	1.24	0.92	1.46	0.019	2.27	0.90	1.55	0.029	4.32	0.80	1.34	0.021	4.19	0.86	1.57	0.024
Control	0.27	0.14	0.29	0.012	0.26	0.12	0.24	0.017	2.66	0.17	0.25	0.020	2.73	0.14	0.26	0.013

SED: Ca, 0.05; Mg, 0.02; K, 0.02; Na, 0.003

Any two means in the body are significantly different at 5% level if they differ by more than $1.96 \times$ SED

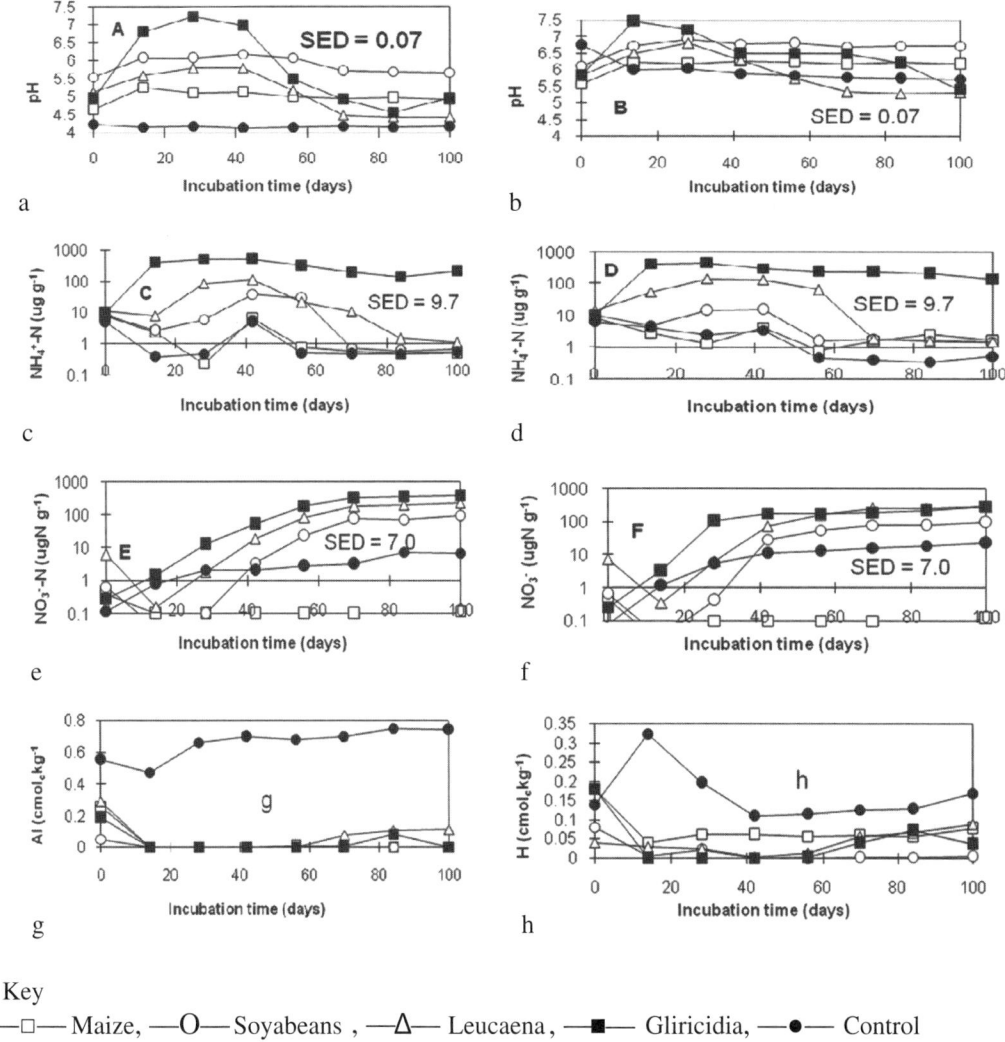

Key

—□— Maize, —O— Soyabeans , —△— Leucaena , —■— Gliricidia, —●— Control

Fig. 1 Changes in pH in the absence (**a**) and presence (**b**) of lime; and inorganic N pools: NH_4^+-N without lime (**c**) and NH_4^+-N with lime (**d**); NO_3^--N without lime (**e**) and NO_3^--N with lime (**f**); exchangeable Al^{3+} (**g**) and H^+ (**h**), both in the absence of lime. Note the logarithmic scale on the Y-axis for inorganic N (**c–f**); the missing data represent zero values which cannot be plotted on a log scale

contrary to the observation made for the other amendments where $NH4^+$-N reached a peak between 28 and 42 days after which the levels declined progressively, almost reaching background levels at the end of the incubation period. Apart from *Gliricidia* and the control treatments, NO_3^--N declined between 0 and 14 days after which the amounts also increased reaching a peak at the end of the incubation (Fig. 1e, f). The increases in pH over the first 40 days were to some extent in line with increases in NH_4^+ concentrations

with increases in NO_3^- aligning with subsequent decreases in pH. Levels of exchangeable Al^{3+} (g) and H^+ (h) appeared only in treatments that had not received lime and did so only in very minimal amounts.

Figure 2 shows cations released relative to the totals added through the residues. Proportionally, up to 70% of Ca^{2+} that was added in the residue was released at zero incubation time, increasing to 84% in almost all the residues in the absence of lime at the end of the

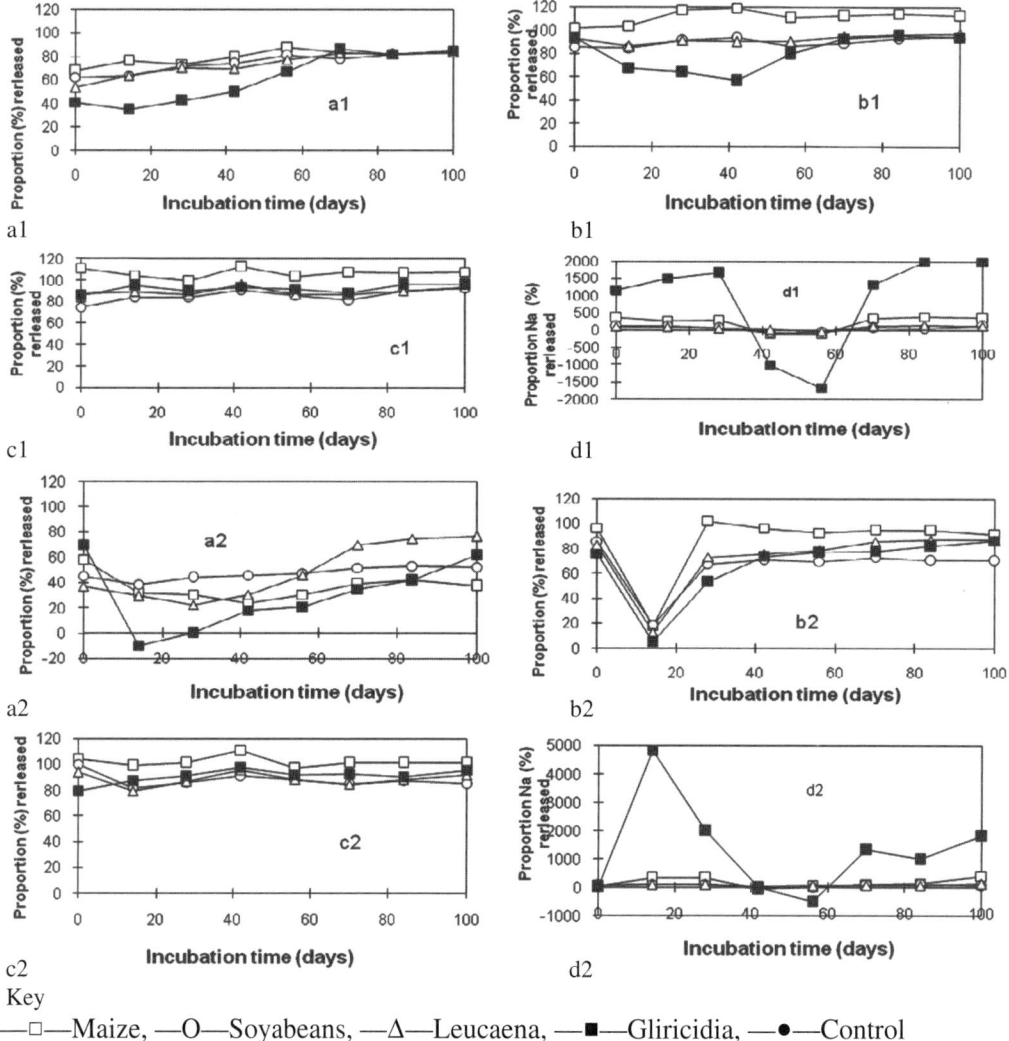

Key

—□—Maize, —O—Soyabeans, —△—Leucaena, —■—Gliricidia, —●—Control

Fig. 2 Proportion of cations added in the residue released over 100 days in the absence of lime (*top* three charts: **a1** = Ca^{2+}; **b1** = Mg^{2+}, **c1** = K^+ and **d1** = Na^+) and in the presence of lime (*bottom* three charts: **a2** = Ca^{2+}; **b2** = Mg^{2+}, **c2** = K^+ and **d2** = Na^+)

incubation (Fig. 2a1). Addition of lime affected the amounts of Ca^{2+} that was measured, with *Gliricidia* being the most affected where the amount of the cation fell below background levels (Fig. 2a2). In general, maize released the largest proportion of cations compared to the other amended treatments at both lime levels, and for K^+ there was an almost 100% release of the cation from the residue at the very beginning at both lime levels (Fig. 2c1, c2). In addition, more K^+ was released from the maize residue than what was added in the residue, indicating release of K^+ from the non-exchangeable pool in the course of the incubation,

more so for Na^+ with *Gliricidia*-amended treatments (Fig. 2d1, d2) and similarly for Mg^{2+} in the absence of lime (Fig. 2b1). Larger proportions of Mg^{2+} and K^+ that were added in the residues became solubilized at zero incubation time than those for Ca^{2+} (Fig. 2b1, c1). At least 80% of Mg^{2+} and K^+ became solubilized at the very beginning of the incubation at both lime levels. There was a huge reduction in the amount of exchangeable Mg^{2+} between 0 and 14 days for all the amended treatments receiving lime, reducing to less than 20% of the amounts present at zero incubation time, which ranged between 80 and 100%. However, most of the

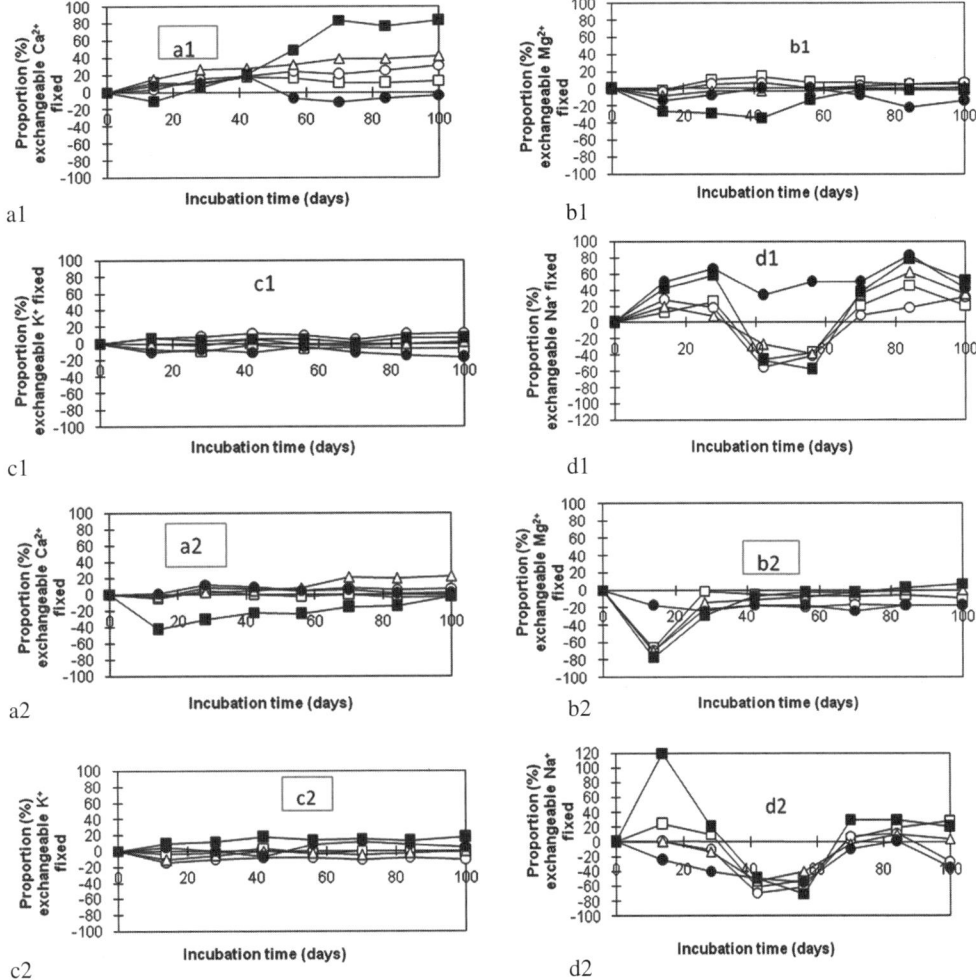

a1

b1

c1

d1

a2

b2

c2

d2

Fig. 3 Cation flux during incubation (i) in the absence of lime: (**a1**) Ca^{2+}, (**b1**) Mg^{2+}, (**c1**) K^+ and (**d1**) Na^+ and (ii) in the presence of lime: (**a2**) Ca^{2+}, (**b2**) Mg^{2+}, (**c2**) K^+ and (**d2**) Na^+. Values above the X-axis represent proportions in excess of the amounts at 0 time and the values below the X-axis represent proportion by which the amounts at zero incubation time are reduced

cations rapidly became available again, rising to almost their original levels at 28 days.

Figure 3 shows proportions of exchangeable base cations relative to the amounts present at the beginning of the incubation. Taking the amount of base cations at zero incubation time to consider changes in cation flux (F) with time, proportion cation flux was determined by $F = 100(X_1 - X_0)/X_0$, where X_0 is the amount of cations present at zero incubation time and X_1 is the amount of cations present at each sampling interval. Positive values indicate amounts in excess of that present at zero incubation time, while negative values indicate a magnitude by which the initial amounts are reduced.

There was a general increase in the amount of Ca^{2+} for all the amended treatments during incubation in the absence of lime, with *Gliricidia* exhibiting the largest increase (Fig. 3a1). There was little or no increase in exchangeable K^+ (Fig. 3c1) for all the amended treatments with time. The same flux pattern is seen for Mg^{2+} apart from treatments receiving *Gliricidia* that showed an initial reduction of up to 40% between 0 and 56 days (Fig. 3b1). Proportionally, sodium fluxes both without (Fig. 3d1) and with the application of lime (Fig. 3d2) exhibit a reduction between 28 and 70 days at both lime levels. However, the amount of exchangeable Na^+ was low both in the residues and in the soil, generally less than 0.04 $cmol_c$ kg^{-1} soil.

Application of lime affected cation flux. Using the control treatment as a basis where 2.62 $cmol_c$ kg^{-1} soil of Ca^{2+} was added through lime, about 90% was available at zero incubation time (Table 4). While other treatments including the control showed little or no increase in Ca^{2+}, *Gliricidia*-amended treatment had a reduction of up to 40% of the cation initial concentration (Fig. 3a2). There was a large reduction in Mg^{2+} flux at 14 days, where there was up to 80% reduction in the amount of the cation present at zero incubation time, but there was an immediate increase of an almost equal magnitude (Fig. 3b2). Beyond 14 days, all the treatments including the control continued exhibiting reduced amounts of Mg^{2+} of an almost constant magnitude. Sodium flux was similar to that exhibited in the absence of lime.

There was a positive relationship between amounts of cations fixed and nitrogen mineralization (Fig. 4a). Up to 70% of the variability in cation fluxes is accounted for by net NH_4^+ present at any one time during incubation in the presence of lime. On the other hand, only 32% of the variability in cation fluxes could

be accounted for in the absence of lime (results not shown). Net NH_4^+ was calculated by finding the difference between the two forms of inorganic N pools at any one time: NH_4^+-N–NO_3^--N. The relationship between pH and total cation concentration was good at zero incubation time but poor at the end (Fig. 4). This is contrary to findings of Pocknee and Sumner (1997), who reported a very strong dependence ($r^2 = 92$) of pH on total basic cations at the end of a 574-day incubation of nine residues, suggesting inadequate incubation period in the current study resulting in incomplete decomposition of basic cation containing organic compounds.

Cation flux exhibited in this study could be accounted for on four basic principles: (i) water-soluble cations, (ii) dissociation, (iii) decomposition and (iv) transition between exchangeable and non-exchangeable forms. Passive dissolution of cations could be limited to zero incubation time. This pool would either increase or decrease depending on soil conditions and extent of decomposition of the residues. Microbial decomposition of base cation containing

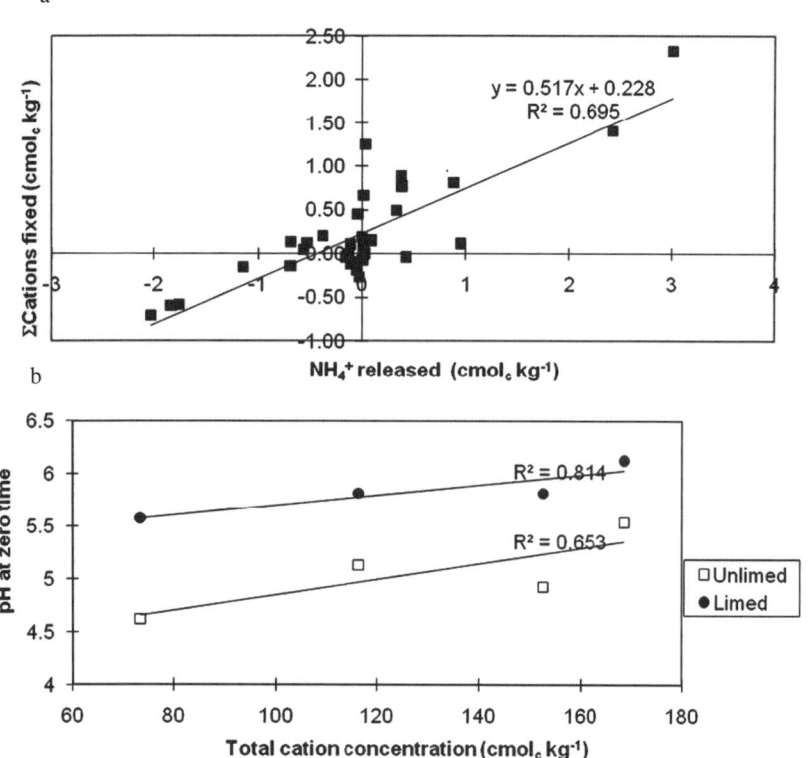

a

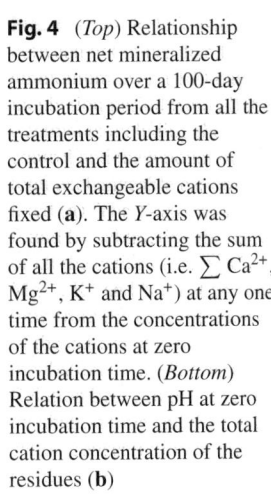

Fig. 4 (*Top*) Relationship between net mineralized ammonium over a 100-day incubation period from all the treatments including the control and the amount of total exchangeable cations fixed (**a**). The *Y*-axis was found by subtracting the sum of all the cations (i.e. $\sum Ca^{2+}$, Mg^{2+}, K^+ and Na^+) at any one time from the concentrations of the cations at zero incubation time. (*Bottom*) Relation between pH at zero incubation time and the total cation concentration of the residues (**b**)

Fig. 5 Maize grain yield response to increased levels of spot-applied lime in the planting basins in Monze, southern Zambia (from CIMMYT trials, 2005/2006 cropping season)

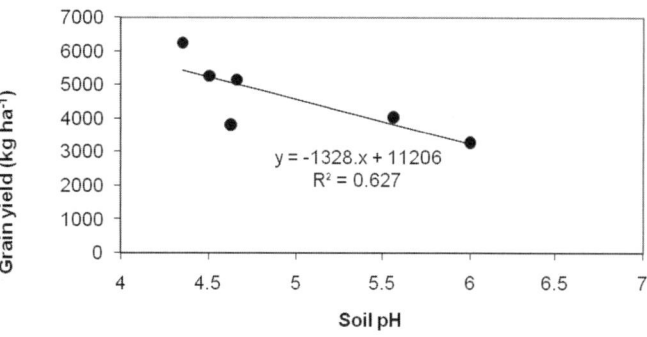

organic compounds would release cations, thereby increasing base cation flux. The influence of nitrogen mineralization on pH would have a bearing on the exchangeability of cations, both in soil solution and on exchange sites. Hydroxyls from nitrogen mineralization would increase the pH of the soil solution. This would result in the association of base cations with organic ligands. This process would reduce the amount of available cations. Decomposition would further contribute to the ECEC. Cations associated with the structural make-up of the plant would be made available through the action of the soil microbial biomass.

The conversion of exchangeable cations to non-exchangeable form in the presence of lime has been reported by other workers (Adams, 1984, 1986; Rowell, 1994). Rowell (1994) reported up to 71% fixation of exchangeable Mg^{2+} in a Nigerian Ultisol treated with 6.20 $cmol_c$ kg^{-1} $Ca(OH)_2$, while K^+ was not affected. This is in agreement with the observation made in this study. Gouveia and Eudoxie (2002) demonstrated the influence of competing cations ($Na^+ > Ca^{2+} > K^+$) on release of fixed $NH4^+$ in a range of Trinidad soils. Cation binding by the applied residues could have reduced further available cations because of the rise in pH. Carboxylic groups could have reacted with the cations, binding them firmly by complexation reactions resulting in the formation of chelates which is favoured by high pH (Rowell, 1994).

The selective binding of Ca^{2+} and Mg^{2+} over K^+ is not unusual since multivalent cations are preferred to monovalent cations (Talibudeen, 1991). Mg^{2+} was more affected than Ca^{2+} because of its position in the Irvin William series for the stability of metal complexes, with trivalent Fe and Al being most strongly held followed by divalent ions (Wild, 1988).

Implication of the Results for Minimum Tillage Systems

The findings in this study have implication for spot-applied lime in planting basins as well as surface-applied residues practised in minimum tillage systems. The fact that most of the cations are readily solubilizable in water is of significance for surface-applied residues as the base cations will solubilize very early in the growing season. This is at a time when crops are still young and their root system is still developing. The nutrients, therefore, risk being lost as they will not be contained by the low CEC of the highly weathered soils. With the erosivity of rainfall that is received in the tropics, surface-applied residues will suffer loss in quality as the soluble fraction will easily be washed away during a heavy rainfall event. The soluble fraction constitutes a larger and more effective component of plant residue alkalinity in ameliorating soil acidity (Sakala et al., 2004). Moreover, the high concentration of spot-applied nutrients in the planting basins will have competing cations in soil solution. Addition of lime into the basins meant to correct soil acidity further compounds the problem. Significant loss in yield may result. This observation was made in CIMMYT-supported minimum tillage trials on liming carried out in southern Zambia, where maize yields were depressed with increasing levels of lime in planting basins (Fig. 5).

Conclusion

There was conversion of cations from exchangeable to non-exchangeable forms at both lime levels. Treatments receiving lime were more affected than treatments that had not received lime, with divalent cations Ca^{2+} and Mg^{2+} being more affected than the monovalent cations (K^+). Calcium in *Gliricidia*-amended treatments became non-exchangeable in the early days of the incubation. This was more pronounced in treatments receiving lime. However, this situation was only temporary as most of the cations became exchangeable again, almost rising to original levels by the end of the incubation. Exchangeable Al levels were absent or insignificantly very low at the end of the incubation, highlighting the tenacity with which aluminium is held by organic matter. Nitrogen mineralization affected cation flux. The presence of copious amounts of NH_4^+ from readily mineralizing residues affected the availability of other basic cations in the soil. This study highlights the importance of residues on amelioration of major cation deficiencies and aluminium phytotoxicity. It also highlights the potential loss of these cations through erosion considering their readily solubilizable nature, especially in minimum tillage systems.

Acknowledgements This work was partly to fulfil the requirements for a degree of doctor of philosophy at Reading University and I would like to thank my supervisors, Drs D.L. Rowell and C.J. Pilbeam for their guidance, to colleagues in the department who were also involved in soil acidity studies and last but not least to the Government of Zambia for the financial support.

References

Adams F (1984) Crop response to liming in the United States. In: Adam F (ed) Soil acidity and liming, 2nd edn. American Society Agronomy Monograph 12, Madison, WI, pp 211–266

Adams SN (1986) Interaction between liming and form of nitrogen fertilizer on established grassland. J Agric Sci Camb 106:509–513

Bessho T, Bell LC (1992) Soil solid and solution phase changes and mung bean response during amelioration of aluminium toxicity with organic matter. Plant Soil 140:183–206

Carter MR (ed) (1993) Soil sampling and methods of analysis. Canadian Society of Soil Science. Lewis Publishers, Boca Raton, FL, pp 499–511

Gillman GP, Burkett DC, Coventry RJ (2001) A laboratory study of application of basalt dust to highly weathered soils: effect on soil cation chemistry. Aust J Soil Res 39: 799–811

Gouveia G, Eudoxie G (2002) Relationship between ammonium fixation and some soil properties and effect of cation treatment on fixed ammonium release in a range of Trinidad soils. Commun Soil Sci Plant Anal 33(11 and 12): 1751–1765

Noble AD, Zenneck I, Randall PJ (1996) Leaf litter ash alkalinity and neutralization of soil acidity. Plant Soil 179: 293–302

Pocknee S, Sumner ME (1997) Cation and nitrogen contents of organic matter determine lime potential. Soil Sci Soc Am J 61:86–92

Rowell DL (1994) Soil science: methods and applications. Longman, Harlow

Sakala GM, Rowell DL, Pilbeam CJ (2004) Acid–base reactions between an acidic soil and plant residues. Geoderma 123:219–232

Talibudieen O (1991) Cation exchange in soils. In: Greenland DJ, Hayes MHB (eds) The Chemistry of Soil Processes. Wiley, Chichester, pp 115–173

Thomas GW (1982) Exchangeable cations. In: Page AL, Miller RH, Keeney DR (eds) Methods of soil analysis. Part 2. Chemical and microbiological properties, 2nd edn. ASA, SSSA, Madison, WI, pp 159–165

Uehara G, Gillman GP (1981) The mineralogy, chemistry, and physics of tropical soils with variable charge clays. Westview Press, Boulder, CO

van Raij B, Peech M (1972) Electrochemical properties of some Oxisols and Alfisols of the tropics. Proc Soil Sci Soc Am 36:587–593

Van Wambeke A (1992) Soils of the tropics: properties and appraisal. McGraw-Hill, New York, NY

Wild A (ed) (1988) Russel's soil conditions and plant growth, 11th edn. Longman Group UK Ltd, Harlow

Wong MTF, Akyeampong E, Nortcliff S, Rao MR, Swift RS (1995) Initial response of maize and beans to decreased concentrations of monomeric aluminium with application of manure and tree prunings to an Oxisol in Burundi. Plant Soil 171:175–182

Wong MTF, Gibbs P, Nortcliff S, Swift RS (2000) Measurement of the acid neutralizing capacity of agroforestry tree prunings added to tropical soils. J Agric Sci 134:269–276

Wong MF, Swift RS (1995) Amelioration of aluminum phytotoxicity with organic matter. In: Date RA et al (eds) Plant soil interactions at low pH: principles and management. Kluwer, Dordrecht, The Netherlands, pp 41–45

Yan F, Schubert S, Mengel K (1996) Soil pH increase due to biological decarboxylation of organic anions. Soil Biol Biochem 28(4/5):617–624

A Study of the Agronomic Efficiency of Human Stool and Urine on Production of Maize and Egg Plant in Burkina Faso

M. Bonzi, F. Lompo, I.D. Kiba, A. Kone, N. Ouandaogo, and P.M. Sédogo

Abstract   Poor rural populations are exposed to sanitary risks arising from improper management of the environment. Indeed the majority of the illnesses contracted from water are related to the management of human excreta in this zone, where sometimes ca. 80% of households do not have a latrine. The hypothesis under discussion is that human stool and urine collected from adapted and less expensive latrines are usable in agriculture to increase production and to improve the income of small-scale farmers. This work was carried out on the peri-urban site in Burkina Faso. Urine and stool, collected separately from specific latrines, then stocked in cans (urine) for 4 weeks and in a closed pit (stool) for 6 months, have been tested respectively on eggplant (*Solanum melongena*) and on maize (*Zea mays*). Three doses determined according to the chemical composition of excreta were compared to mineral fertiliser on clay-sandy soil for eggplant and, on ferric oxisoil for maize in the farmer's fields, over 2 years. The optimal doses for urine and stool are respectively 1.2 L per plant for eggplant and 980 kg ha^{-1} for maize. They increased yields respectively for eggplant and maize by +84 and +90% compared to the control, and also improved the farmer's income. Extension material facilitating the agricultural use of human manure was prepared. In conclusion the use of collected excreta aids in purifying the environment while at the same time improving agricultural production and income.

Keywords Stool · Urine · Eggplant · Maize · Income

M. Bonzi (✉)
Institute of the Environment and Agricultural Research
(INERA), Ouagadougou, Burkina Faso
e-mail: bouabonzi@yahoo.fr

Introduction

In Burkina Faso, 80% of households do not have toilets (CREPA, 2004). People used to urinate and defecate directly in open areas, thus contributing to environmental pollution of surface water, air and even underground water through infiltration. Where toilets exist, they are the type where urine and faeces are mixed in an entirely buried pit, relatively very deep and generally not waterproof. This established fact indicates that these kinds of toilets are sources of pollution through infiltration of wastewater. This situation shows that the population which has a food self-sufficiency priority doesn't perceive sanitary risks from biodegradable solid waste and wastewater. There is a lack of general knowledge about elemental sanitary risks of human waste. All things considered, the population continues to be poor and strongly exposed to diseases which originate from their own behaviour. In addition, other problems affecting the population include nutrient depletion in their agricultural soils, and lack of adequate sustainable soil fertility management practices. Indeed, the low availability of manure remains a problem to be solved; the cost of chemical fertilisers is increasing and the use of these especially nitrogen and phosphorus fertilisers often poses risks of pollution and eutrophication of water (Bado, 1994; Bonzi et al., 2004).

In view of the above, research needs to determine socially and economically adapted methods that increase agricultural production in both the poor urban and rural areas and in the whole environment. The study of excreta use in agriculture could provide an alternative method of improving agricultural production. Herein we give a description of a 2-year

A. Bationo et al. (eds.), *Innovations as Key to the Green Revolution in Africa*,
DOI 10.1007/978-90-481-2543-2_65, © Springer Science+Business Media B.V. 2011

agronomic survey of the use of excreta in market (vegetable) gardening (eggplant) and for cereal (maize). The hypothesis is that urine and human stool constitute a not very expensive alternative to the use of mineral manures to increase productivity of the soils and to increase to a significant degree the income of the farmers.

Material and Methods

Site Descriptions

The experiment was carried out in Saaba (near Ouagadougou) where the following was observed: (1) the existence of real problems of sanitation and water-borne diseases, (2) high population density with 68 persons per km^2 against an average of 38 for the country, (3) 85% of the population are concentrated in sectors of agricultural production and breeding on agricultural soils with nutrient depletion. The soils, the majority of which are oxisols, have low nitrogen levels (<1%) and organic matter (<2%), are strongly deprived of phosphorus, vulnerable to hydrous and wind erosion and, subject to intensive agricultural utilization.

The market-gardening site is located at the edge of Saaba's dam on clay-silt hydromorphic soils. These soils which are indeed suitable for market (vegetable) gardening production have been exploited for this purpose for tomato production over the last 3 years. The fertilisation was always through addition of manure of various types and sometimes a low dose of mineral mixed fertiliser (NPK). These soils are temporarily flooded from August to mid-November. The type of irrigation used for the experimentation is gravitational starting from the dam water with aspiration via a motorbike pumping unit. Water is supplied according to the needs of the crops.

The cereal site is located on a farm between Ouagadougou and the town of Saaba. The climate is of the soudano-sahelian type (Guinko, 1984) with only one rainy season from mid-May to mid-October. This farm has been used by the same family for over 10 years for the production of cereal (sorghum especially) and occasionally a mixture of sorghum and cowpea. Temperatures at this site varied between 17 and 40°C. The soils are oxisols. Rainfall generally varies from 600 to 800 mm per year with important variations in space.

Materials for Collection and Transport of the Excreta

Human excreta was collected using EcoSan toilets of the type "TECPAN" (CREPA, 2004) which have two pits separated by a wall. The two pits are used alternatively. When a pit is full, it is closed for hygienisation while the second is brought into service. The collection of solar energy is done through a removable lid intended for access, reversal and draining of the products in the pit. The lid is metallic and tilted at 45° and is situated at the back of the latrine and painted black (sheet punt or undulated). The urine was collected using plastic tanks of 20-L capacity, which are connected to the latrine by plastic piping. A bicycle was used to collect and transport tanks of urine and stool bags to the field.

Plant Varieties

The variety of aubergine (eggplant) used was *Violette longue hâtive*. It is a fast and very vigorous growing variety. It produces elongated fruits from 20 to 25 cm. Its flesh is firm, tasty and its period of production is long. It is usually used by the farmers of the area and is adapted to local ecological conditions. The variety of maize used was *Kamboinsé Extra Précoce Blanc* (*KEB*; synonym: *TZEEW*). It is a variety developed by SAFGRAD and INERA for areas with annual rainfall of less than 900 mm.

Fertilisers and Manures

Mineral fertilisers used were: urea 46% *N*; super triple phosphates (TSP), 20.06% *P*; potassium sulphate (49.8% *K*) in market gardening and potassium chloride (49.8% *K*) in cereal. The urine and human stools were used as nutrient sources in comparison with the mineral fertiliser. The Zaï system was used for treatment with faeces before sowing. The urine was diluted to 100% with water.

Experimental Design

The experimental design was a Completely Randomised Block with four (4) replications. Treatments were as follows:

In eggplant: N was the factor studied because of the high N content in urine.

1. Control without fertiliser;
2. Vulgarised fertiliser (research recommended): 92 *N* (*N* from urea) – 80.22 *P* (*P* from TSP) – 174.3 *K* (*K* from KCl);
3. Urine + PK: 92 *N* (*N* from urine) – 80.22 *P*–174.3 *K*;
4. Control PK: 80.22 *P*–174.3 *K*.

In maize: P was the factor studied because of P deficiency in soil and as a major factor in maize production.

1. Control without fertiliser;
2. Control NK: 67 N ha^{-1} + 17.43 K ha^{-1} (recommended amount);
3. Faeces (Q) + NK (Q = 15.04 P ha^{-1} from faeces) = 980 kg faeces ha^{-1};
4. Faeces (Q/2) + NK (Q/2 = 7.54 P ha^{-1} from faeces) = 490 kg faeces ha^{-1};
5. Faeces (Q + Q/2) + NK (Q + Q/2 = 22.67 ha^{-1} from faeces) = 1,470 kg faeces ha^{-1};
6. Vulgarised fertiliser (research recommended): 150 kg ha^{-1} (NPK 14-23-14) + 100 kg ha^{-1} (urea 46% *N*) in two fractions of 2/3 at sowing and 1/3 at flowering;
7. Mixed: ½ faeces + ½ urine.

The quantity of faeces in treatment 3 is calculated according to the amount of P vulgarised fertiliser. Treatments 4 and 5 are calculated respectively from the half of P recommended and P recommended and a half. The amount of N and K of treatments 3, 4, 5, 6 and 7 took into account the amount of N and K in the urine and faeces from their composition.

Indicators of Follow-Up

Soil samples were collected during two periods: (1) before experimentation and (2) after experimentation at 0–20 and 20–40 cm depth. Soil pH, N, P, K

and organic matter were determined in the laboratory. Indicators measured for eggplant were germination rate, total number of fruits per hectare, fresh fruits biomass, the average weight of a fruit. Similarly, indicators measured for maize included germination rate, growth rate at sixtieth day, grain, dry straw and the weight of 1,000 grains.

Methods of Chemical Analysis

Walkley and Black's (1934) method for soil and the ashes method for faeces according to Okalebo et al. (2002) were used. The pH was measured by the electrometric method with a pH-meter with a soil/water ratio of 1/2.5 according to AFNOR (1981). Soil N and P were obtained using Skalar determination after mineralisation using the Kjeldahl method reprised by Novozansky et al. (1983). Total potassium was determined by a Jencons flame photometer according to the method proposed by Walinga et al. (1989).

Statistical Analysis

The ANOVA (ANalysis Of VAriance) of the data was carried out with the software SPSS 11.5 using the general linear model (GLM). The averages were compared with the tests of Student-Newman–Keuls and the test of Bonferonni.

Results and Discussion

Fertilising Values of the Human Excreta

The results are presented in Table 1. They are the average values of several samples of hygienised excreta taken in latrines from the various households. The urine is rich in nitrogen, relatively weak in phosphorus and low in potassium and their pH is basic. Stool is on the other hand very rich in major nutrients (N-P-K). It is basic and the C/N ratio is lower than 20. In comparison with the contents of cow manure (organic matter of reference) determined by Sedogo (1981) and Bonzi (1989), stool is 8-times richer in

Table 1 Chemical characteristics of human excreta

Excreta	Total N	Total P	Total K	Total C	C/N	pH water
	(g kg^{-1} faeces and g L^{-1} of urine)			(%)		
Faeces	33.7	15.36	22.32	53.1	16.4	8.2
Urine	2.702	0.370	0.314	–	–	8.9

phosphorus (2.2 g P kg^{-1} in cow manure), 2-times richer in nitrogen (17.5 g kg^{-1} in cow manure), and has a similar potassium content to cow manure which is on average 21 g kg^{-1}.

Effects of the Hygienised Excreta on Soils and Crop Production

Effects of the Urine on Aubergine Production

The results show that the urine had a harmful effect on the recovery of the seedlings compared to the control without fertiliser (Table 2). However, there are no significant differences in the rate of recovery between the urine treatments and the recommended mineral fertiliser. The urine-generated fruits and biomass yields (respectively 17.6 and 1.3 t ha^{-1}) are very similar to those of the mineral fertiliser (17.8 and 1.2 t ha^{-1}). The yields obtained with the urine and those obtained with urea are not statistically different. It is noted that the absence of nitrogen in the treatment with only P and K resulted in a very high loss of fruit yields of almost –75% compared to the complete mineral fertiliser (4.5 t ha^{-1} against 17.8 t ha^{-1}). This shows well that the nitrogen is the factor determining the production of aubergines.

Effects of Faeces on Maize Production

The study revealed that there was no treatment effect on the lifting of maize (Table 3). One observes a tendency towards growth improvement by increasing the amount of faeces. Measurements of height on the sixtieth day show that only the mineral fertiliser approximates the values obtained by the faeces group, whereas the control and NK treatment are homogeneous and give the lowest yields. The three faeces treatment amounts increased grains and straw production statistically to the same level as the mineral fertiliser. The best production is obtained with an amount of 980 kg faeces ha^{-1}.

Effects of Mixed Urine and Faeces on Maize Production

The results are presented in Table 4.

Effects of the Urine on Soil After Aubergine Production

Table 5 presents the effects of the urine on soils after aubergine production. The data does not reveal a significant difference for the lifting of maize between

Table 2 Effects of urine on the survival and the yields of aubergines

Treatment	Recovery rate (%)	Fruits (number of fruits ha^{-1})	Average weight of a fruit (g)	Yield fruits (t ha^{-1})	Biomass (t ha^{-1})
Control (without fertiliser)	99[a]	56,661[a]	49.5[a]	2.8[a]	0.33[a]
Vulgarised fertiliser	94[ab]	185,185[b]	96.9[b]	17.8[b]	1.18[b]
Urines + PK	86[b,c]	195,332[b]	87.1[b]	17.7[b]	1.28[b]
Control PK	96[a,c]	74,364[a]	57.4[a]	4.5[a]	0.45[b]
Significance	*S*	*THS*	*THS*	*THS*	*THS*
Probability	*0.045*	*<0.001*	*<0.001*	*<0.001*	*<0001*

The averages affected by the same letter in the same column are not significantly different at the level of 5% by the method of Student-Newman–Keuls. *THS* = Very Highly Significant ($P < 0.001$); S = Significant; a, b, c = Numbers followed by the same letter are not significantly different

Table 3 Effects of faeces on the lifting, growth and maize yields

Treatment	Rate of lifting (%)	Weight of 1,000 grains (g)	Height to the 60th day after sowing (cm)	Grains yield (t ha^{-1})	Straw yield (t ha^{-1})
Control without fertiliser	96	110.78^a	44.1^a	0.13^a	0.25^a
Control NK	92	124.43^a	61.9^a	0.22^a	0.47^a
Faeces (Q) + NK	99	155.32^b	112.1^b	1.35^b	3.32^b
Faeces (Q/2) + NK	99	140.57^a	111.5^b	1.16^b	2.64^b
Faeces (Q + Q/2) + NK	99	163.70^c	117.7^b	1.09^b	3.17^b
Vulgarised fertiliser	97	122.86^a	109.2^b	1.05^b	2.54^b
Significance	*NS*	*THS*	*THS*	*THS*	*THS*
Probability	*0.057*	*<0.001*	*<0.001*	*<0.001*	*<0.001*

The averages affected by the same letter in the same column are not significantly different at the level of 5% by the method of Student-Newman–Keuls. *THS* = Very Highly Significant ($P < 0.001$); *NS* = Non-significant ($P > 0.05$); a, b, c = Numbers followed by the same letter are not significantly different

Table 4 Effects of mixed urine and faeces on the lifting, growth and yield of maize

Treatment	Rate of lifting (%)	Weight of 1,000 grains (g)	Height to the 60th day after sowing (cm)	Grains yield (t ha^{-1})	Straw yield (t ha^{-1})
Control without fertiliser	96	110.78^a	44.1^a	0.13^a	0.25^a
Vulgarised fertiliser	97	122.86^a	109.2^b	1.05^b	2.54^{b,c}
Urine Q/2 + PK	99	125.89^a	96.7^b	0.79^b	1.67^b
Faeces Q/2 + NK	99	140.57^b	111.5^b	1.16^b	2.64^{b,c}
Faeces Q/2 + Urine Q/2	99	149.85^b	137.5^c	2.15^c	3.6^c
Significance	*NS*	*S*	*THS*	*THS*	*THS*
Probability	*0.057*	*0.001*	*<0.001*	*<0.001*	*<0.001*

The averages affected by the same letter in the same column are not significantly different at the level of 5% by the method of Student-Newman–Keuls. *THS* = Very Highly Significant ($P < 0.001$); *NS* = Non-significant ($P > 0.05$); *S* = Significant; a, b, c = Numbers followed by the same letter are not significantly different

Table 5 Effects of urine on the chemical characteristics of the soil after aubergine production

Treatment	Horizon (cm)	ΔN total (g kg^{-1} dry soil)	Δ P_2O_5 total (mg kg^{-1} dry soil)	Δ K_2O total (mg kg^{-1} dry soil)	ΔpHeau*
Vulgarised fertiliser	0–20	–0.02	+68.16	+1.47	–0.22
	20–40	–0.01	+22.5	–151.09	–0.10
Urine	0–20	+0.14	+45.01	+181.19	–0.15
	20–40	+0.18	+22.19	–9.59	–0.33
Control PK	0–20	+0.21	+136.14	+208.43	+0.25
	20–40	+0.13	+21.87	–8.04	+ 0.32

*Difference between initial and terminal rate.

treatments. The combination of urine with faeces significantly improves the growth of the maize. These results also show that for the grains and straw yields the mixed treatment is better. It permits a grains yield of 2.15 t ha^{-1}, significantly higher than the other treatments which all give less than 1.5 t ha^{-1}.

In the top soil, the mineral fertiliser does not improve the stock of nitrogen while it improves potassium and phosphorus. On the other hand, the urine improves stocks of all three elements. At lower depths, at depth (20–40 cm), one notes an improvement of the phosphorus and nitrogen stock with the urine and only an improvement of the phosphorus stock with the mineral fertiliser. In addition, it is noted that the urine does not improve the acidity of the soil.

Effects of Faeces on the Soil After the Production of Maize

It was noticed that all the amounts of faeces improve the stock of nitrogen in the top soil and only the Q/2 amount improves both horizons (Table 6). The improvement is marked with low dose Q/2 (490 kg ha^{-1}) at depth and the strong dose (1,470 kg ha^{-1}) in the top soil. The stock of phosphorus shows an important improvement with the low dose and the strong respectively at depth and on the surface. The combination of the urine with faeces has a positive effect on the stock of nitrogen, organic matter and on the acidity of the soil. However, this combination does not improve the stock of phosphorus in the two horizons. The amount Q improves the content of organic matter in the two horizons, but this improvement is much more marked in the top soil. The acidity of the soil shows a light improvement with faeces. Indeed, the amount Q increased the pH in the two horizons and the Q/2 and (Q+Q/2) amounts have the same effect only in the top soil.

Discussions

Agronomic Value of Human Excreta

The urine is rich in nutrients and constitutes an important source of nitrogen which could supplement or replace nitrogen fertilisers. In effect these fertilisers are not only expensive but also their chemical quality has often not been well controlled since the liberalisation of the fertilisers market in Burkina Faso. The results are consistent with those of Esray et al. (2001), where the capacity of production of urine for one adult person is also indicated as approximately 400 L of urine per year containing 4 kg of nitrogen, 0.4 kg of phosphorus and 0.9 kg of potassium. Moreover, these authors showed that there is a greater quantity of nutrients in the urine than in the mineral fertilisers used in agriculture and also the heavy metal concentrations in human urine are much lower than those found in mineral fertilisers.

The C/N ratio of faeces is interesting at the agronomic level (<20). This ratio indicates a relative facility towards the mineralisation of the organic matter (Godefroy, 1979; Sedogo, 1981; Guiraud, 1984). The P content of faeces is of interest in areas where this element is less prevalent in Burkina Faso and in the soils of other major west African countries and even in manure which according to Lompo (1993) is a reflection of the deficiency of phosphorus in the soils from which its materials are derived.

The nutrient content of the human excreta is lower than that found by Esray et al. (2001) for Swedish excreta (10 g of nitrogen per litre of urine). These differences are probably due to the food. In this connection, Franceys et al. (1995) advanced that adults who have a fibre-rich diet and live in the rural zone produce more abundant faeces than the children or

Table 6 Effects of faeces on the chemical characteristics of the soil after maize

Treatment	Horizon (cm)	ΔN total (g kg^{-1} dry soil)	Δ P$_2$O$_5$ total (mg kg^{-1} dry soil)	Δ OM (%)	ΔpHeau*
Control NK	0–20	+0.06	–0.23	–0.02	–0.56
	20–40	+0.11	–74.20	–0.01	–0.35
Faeces (Q)	0–20	+0.06	–75.11	+0.06	+0.06
	20–40	0	–50.84	+0.01	+0.14
Faeces (Q/2)	0–20	+0.06	–26.34	+0.07	+0.19
	20–40	+0.37	+242.51	–0.01	–0.15
Faeces (Q+Q/2)	0–20	+0.43	+266.56	–0.02	+0.18
	20–40	0	–51.30	+0.01	–0.03
Vulgarised fertiliser	0–20	0	–26.79	–0.12	–0.49
	20–40	0	–76.03	–0.08	–0.57
Faeces Q/2 + Urine Q/2	0–20	+0.06	–100.53	+0.02	+0.23
	20–40	+0.11	–100.53	0	+0.16

OM = Organic matter
*Difference between initial and terminal rate.

adults of a certain age who live in the urban zone and have a fibre-poor diet. The consumption of proteins and certain strongly nitrogenised food also influences the fertilising values of the excreta. Indeed in our own analyses the samples had variable contents from one household to another.

Effects of Human Excreta on Agricultural Production

The weak survival of aubergines with urine use is not very worrying, because there is no significant difference to mineral fertilisation which is the normal country practice. However, to avoid any risk, it is necessary to apply urine only after the seedlings have fully emerged (at least 2 weeks after planting). Also, split application can help to mitigate possible mortalities. The weak survival of the seedlings is consistent with the results of Niang (2004). This author, through an experiment carried out in Senegal, showed that early application of urine before the emergence of seedlings caused significant plant death, sometimes higher than 50%, whereas no problem was observed if its application was done after the emergence of seedlings.

The results show that one can produce aubergines with urine as well as with the mineral fertiliser. According to our observations on soil, the urine has three advantages which are: (1) the external aspect of the fruits which seemed improved with the urine (more brilliant fruits); (2) the improvement in biomass with the urine which lets one think that if one continued to harvest the seedlings which received the urine may produce more fruits considering the abundance of foliage and their very green aspect; (3) prolongation of harvest with the urine, which allows a more beneficial offer to the market. The weak production of aubergines obtained with treatment PK reflects the importance of nitrogen in the production of aubergines, confirming the remarks of Bélem (1990), Bélanger et al. (1994), and Lebot et al. (1997). Indeed according to these authors, the nitrogenised nutrition is involved in an increase in photosynthesis, enhancing assimilation for the formation of the fruits. However, the poor yields may also be ascribable to a chemical imbalance between the various nutrients caused by the incomplete formula PK.

The statistical analyses show that there is no treatment effect on the lifting of maize when one sows in the pit which received faeces. The faeces amount (Q) of 980 kg ha^{-1} resulted in a better production of grains and straw compared to the Q/2 amounts and Q+Q/2. The amount Q seems more interesting as it makes it possible to achieve a double production goal (grains and straw) and also the increase in the amount from 980 to 1,470 kg deposit ha^{-1} is not economic because it does not involve a significant increase in yield.

Of all the treatments, the mixed treatment faeces + urine gave the best grains and straw yields of maize. One could qualify this as a "synergistic effect" between the two types of fertilisers. All results obtained look like a combination of mineral fertiliser with organic matter. Indeed the beneficial effects of such a combination were shown by Pichot et al. (1981), Ganry (1990), Hien (1990), Bationo and Mokwunye (1991), and Bado et al. (1997). These authors advance that the concomitant use of organic matter and mineral fertiliser reduced losses and increased mineral and hydrous plant nutrition leading to an increase in their productivity. The mixed faeces + urine product provides a better answer to the agronomic objective of this survey of human excreta because the supplier is the same (Man).

Effects of Human Excreta on the Soils

Under aubergine, the incomplete formula PK seems to unbalance the system, and prevent nutrient uptake by the plant, which probably explains the improvement of stock N-P-K by this treatment. It can be a question for example of a deficiency induced by the lack of N, from where P and K become excessive and are not used by the plant.

The losses of the organic matter stock observed with the mineral fertiliser and treatment NK could be related to the mineralisation supported by the nitrogen fertiliser, which caused an acidification of the soil. The urine behaves like the mineral fertiliser towards soil acidity and the availability of N, P, K. One can use them like maintenance nitrogen fertilisers for plants.

Faeces, contrary to the mineral fertilisers, imply an organic contribution which improves the organic matter content of the in-depth soil or of the top soil and probably allows an improvement in the acidity of the soil. For this reason, they can be used as amendments. The beneficial effects of faeces with respect to the chemical characteristics of the soil are related to those of the organic substrates of the moment (cow manure, compost) which were highlighted by several researchers such as Bonzi (1989), and Kambiré (1994). Because of the low improvement of the organic matter rate, it seems to be indicated that to really improve this factor, one would need much higher quantities of faeces.

Conclusion

The human excreta are rich in nutrients and like the mineral fertiliser, improve market (vegetable) gardening and also cereal production. They have good agronomic value. Indeed the urine is rich in N, at least 3 g L^{-1} of N. Urine also contains other major elements such as P and K, which can improve soil fertility. For faeces analyses showed that they are very rich in major elements with 34 g N kg^{-1}, 15 g P kg^{-1} and 22 g total K kg^{-1}. In comparison with the average fertilising value of manure, the local amendment reference for soil, faeces are 2-times richer in N, 8-times richer in P and equivalent in K.

The results obtained for eggplant cropping show that urine can be an alternative source of nitrogen fertiliser. Indeed it has no harmful effect on the recovery rate of seedlings. The fruits collected under urine fertilisation do not present any undesirable physical difference; to the contrary they are sometimes as beautiful, if not more, than fruits treated with the mineral fertiliser. Fruit yields obtained with urine are higher than those obtained without fertilisation or with PK only (the difference is very highly significant). While considering the economic aspects, it is clear that production using urine as fertiliser is very profitable because of the low costs compared to mineral fertiliser. The cropping calendar using urine as fertiliser is prolonged in time compared to the mineral fertiliser which gives a grouped production. This observation is important because production spread out over a longer period permits management of stocks and stabilisation of prices in the market by a good balance of offer/request. The urine does not improve soil nutrient stock and neither does it improve soil pH in spite of its alkalinity. In maize cropping there is no risk with

seeding after immediate faeces application. Faeces can be used as fertilisation to improve soil organic matter content and productivity. A combination of urine with faeces is beneficial.

The agronomic assessment of human excreta not only makes it possible to improve soil productivity, but also can reduce sanitary risks and improve the framework of life of local populations. These results obtained after two cropping years are interesting and one would gain by scaling up this technology. Urine is efficient at the amount of 1.2 L per seed hole for truck crop (eggplants). Urine must be applied in three equal fractions of 0.4 L each (2 weeks after transplanting, 3 weeks after the 1st application and 4 weeks after the 2nd application). The mode of application of urine according to the order hoeing-urine-water simultaneously either in the seed hole or in the irrigation line is to be scrupulously respected in order to avoid mortality risks to young seedlings less resistant to urine corrosion in the case of our study. Faeces must be applied at the amount of 980 kg ha^{-1} for grain crop. Faeces must be applied before seedling crop and incorporated well in the soil. One can also split the faeces treatment into two equal applications, with complementary nitrogen fertiliser to be closer to the combination faeces + urine. Two agronomic charts are available for better use of the excreta in agriculture. Latrines of the double-pit type (TECPAN), or the EcoSan latrine (CREPA, 2004) must be popularised so that populations can impregnate mode of use, collection and process of hygienisation of human excreta. Once this is done, the techniques involved in the use of urine and hygienised faeces in agriculture must be transferred to farmers, through charts which we prepared following the results obtained and the observations carried out during the two programs of experiments.

References

Afnor (1981) Détermination du pH. (association française de normalisation) NF ISO 103 90. In: AFNOR (ed) Qualité des sols. Paris, France, pp 339–348

Bado BV (1994) Modification chimique d'un sol ferralitique sous l'effet de fertilisants minéraux et organiques: conséquences sur les rendements d'une culture continue de maïs. INERA Doc. Burkina Faso, 57p

Bado BV, Sedogo MP, Cescas MP, Lompo F, Bationo A (1997) Effets à long terme des fumures sur le sol et les rendements du maïs au Burkina Faso. Agricultures 6(6):571–575

Bationo A, Mokwunye AU (1991) Role of manures and crop residue in alleviating soil fertility constraints to crop production with special reference to the Sahelian and Sudanian zones of West Africa. Kluwer, Netherlands, pp 217–225

Bélanger G, Gastral F, Warembourg F (1994) Carbon balance of tall fescue (Festuca arundinacea Schreb): effects of nitrogen fertilisation and the growing season. Ann Bot 74: 653–659

Bélem J (1990) Effets de l'enrichissement carboné et du type de plateau multicellulaire sur la croissance et la productivité de transplants de légumes de champs. Grade Maître ès Sciences. Université de Laval, Canada, 67p

Bonzi M (1989) Etudes des techniques de compostage et évaluation de la qualité des compost: effets des matières organiques sur les cultures et la fertilité des sols. Mémoire de fin d'études IDR, université de Ouagadougou, Ouagadougou, 66p

Bonzi M, Lompo F, Sedogo MP (2004) Effet de la fertilisation minérale et organo-minérale du maïs et du sorgho en sol ferrugineux tropical lessivé sur la pollution en nitrates des eaux. Communication à la 6è édition du FRSIT, Ouagadougou, Burkina Faso; 18 pp; (Lauréat du Prix du Groupe ETSHER/EIER)

CREPA (2004) Programme régional Assainissement Ecologique. CREPA doc. Burkina Faso, 18p

Esray SA, Jean G, Dare R, Ron S, Mayling SH, Jeorje V (2001) Assainissement ecologique éd Winblad. Winblad ed., England, 91p

Francey R, Pickard J, Reed R (1995) Guide de l'assainissement individuel. OMS, Genève, 221p

Ganry F (1990) Application de la méthode isotopique à l'étude des bilans azotés en zone tropicale sèche. Thèse Sciences naturelles, université de Nancy I. 354p

Godefroy J (1979) Composition de divers résidus organiques utilisés comme amendement organo-minéral. Fruits 34(10):579–584

Guinko S (1984) Végétation de la Haute Volta. Thèse de Doctorat d'Etat Sciences Naturelles. Université de Bordeaux III, 318p

Guiraud G (1984) Contribution du marquage isotopique à l'évaluation des transferts d'azote entre les compartiments organiques et minéraux dans les systèmes sol-plante. Thèse de doctorat ès sciences naturelles, Université P. et M. Curie, Paris VI, 335p

Hien V (1990) Pratiques culturales et évolution de la teneur en azote organique utilisable par les cultures dans un sol ferralitique du Burkina Faso, Thèse docteur, INPL, 149p

Kambiré SH (1994) Systèmes de culture paysan et productivité des sols ferrugineux lessivés du plateau central (Burkina Faso): effets des restitutions organiques. Thèse doctorat troisième cycle, université de Dakar, 188p

Lebot J, Andriolo JL, Adamowicz G, Robin P (1997) Dynamics of N accumulation and growth of tomato plants in hydroponics: an analysis of vegetative and finit compoartiments. Colloques INRA, France, pp 121–139

Lompo F (1993) Contribution à la valorisation des phosphates naturels du Burkina Faso: études des effets de l'interaction phosphates naturels-matières organiques. Thèse Docteur Ingénieur. Université nationale de Côte d'Ivoire, 249p

Niang Y (2004) Amélioration du rendement de la tomate par l'utilisation des urines comme source de fertilisation. Communication au premier forum du réseau CREPA 35:36

Novozansky I, Houba VJG, Van eck R, van vark w (1983) A novel digestion technique for multi-element analysis. Commun Soil Sci Plant Anal 14:239–249

Okalebo JR, Gathua KW, Woomer PL (2002) Laboratory methods of soil and plant analysis: a working manual, 2nd edn. Uganda, 127p

Pichot J, Sedogo MP, Poulain JF (1981) Evolution de la fertilité d'un sol ferrugineux tropical sous l'influence des fumures minérales et organiques. Agron Trop 36:122–133

Sedogo PM (1981) Contribution à l'étude de la valorisation des résidus culturaux en sol ferrugineux et sous climat tropical semi-aride. Matière organique du sol, nutrition azotée des cultures. Thèse Docteur Ingénieur, INPL NANCY, 135p

Walinga I, Van Vark W, Houba VJG, Van der Lee JJ (1989) Plant analysis procedures. Department of Soil Science, Plant nutrition, Wageningen Agricultural University, The Netherlands, Syllabus, Part 7, pp 197–200

Walkley A, Black JA (1934) An examination of the Detjareff method for determining soil organic matter and a proposed modification of the chromatic acid titration method. Soil Sci 37:29–38

Potential for Reuse of Human Urine in Peri-urban Farming

O. Semalulu, M. Azuba, P. Makhosi, and S. Lwasa

Abstract The possibility of recycling human urine for maize and vegetable growing was assessed on-farm in a peri-urban Kyanja parish, Kampala District, Uganda. The objectives were to demonstrate to farmers and other stakeholders, the potential for using urine and develop guidelines for use of urine in farming. Field plots measuring 1.5 × 6 m were established on 20 farmers' fields and planted with maize (*Zea mays* L.), nakati (*Solanum aethiopicum*), kale (*Brassica oleracea* L.) and spinach (*Spinacia oleracea* L.). Urine was applied at 10% (0.5:5 urine to water), 20 and 30%. Each concentration was applied weekly, bi-weekly and monthly. Urine:water mixtures (20 L) were applied to each bed while the control received water only. Urine application significantly increased maize height and fresh yield; 30% urine weekly application gave highest benefits. Weekly application of 10% urine increased nakati yield from 8.3 to 22.2 kg/plot, within 2 months and yielded nearly the same biomass as that treated to 20% urine weekly. Weekly application of 20% urine increased the yield of kale from 2.4 to 5.5 kg/plot, and spinach from 6.6 to 17.1 kg/plot within 2 months. These represented the highest, most economical biomass yields compared to other treatments. Weekly application outperformed bi-weekly and least on once a month. From these findings, we propose the following guidelines: maize, apply 30% urine weekly for 8 weeks; nakati, apply 10% urine weekly for 8 weeks; kale, apply 20% urine weekly; spinach, apply 20% urine weekly. Since kale and spinach can grow for about 1 year, prolong urine application for continued harvesting.

Keywords Ecological sanitation · Closing the loop · Nutrient recycling · Urban agriculture · Food security

Introduction

The World Summit on Sustainable Development (2000) adopted the Millennium Development Goals (MDGs) which emphasized among other things, provision of adequate sanitation and food security by 2015. About 2.4 billion persons, mostly in developing countries, lack adequate sanitation (WEHAB, 2002); majority of farmers in rural and urban areas experience shortage of water and nutrients in agriculture. Increasing population growth and high rate of urbanization puts more pressure on land for increased food production, requiring a threefold increase in fertilizer use in the twenty-first century to sustain food production. This is in contrast with the decreasing level of nutrient reserves worldwide, especially P and K. Alternative fertilizer sources are urgently required to meet nutrient deficits, one of which is to reuse nutrients in human excreta. By providing affordable sanitation emphasizing recycling of by-products for agriculture, the potential contribution of ecological sanitation to the realization of the MDGs is enormous.

Ecological sanitation (EcoSan) is based on the principle that urine, faeces and waste water are resources in the ecological loop. It uses dry toilets to separate urine from faeces and through recycling, uses the two to support crop production. Important features in EcoSan are separation at source, destruction

O. Semalulu (✉)
National Agricultural Research Laboratories (NARL),
Kampala, Uganda
e-mail: o.semalulu@gmail.com

of pathogens and recycling of urine, faeces and grey water, thereby closing the loop. Nutrients and organic matter (OM) in human excreta can be recycled to benefit soil organisms and plant life and to produce human food. Consumed plant nutrients leave the human body through excreta. In adults, almost all nutrients in the food eaten are excreted. Jonsson and Vinneras (2004) developed descriptive equations for predicting nutrient contents of human excreta based on the available food supply in different countries as per FAO statistics:

$$N = 0.13 \times (\text{total food protein}) \qquad (1)$$

$$P = 0.011 \times (\text{total food protein} + \text{vegetal food protein}) \qquad (2)$$

where N and P refer to the nitrogen and phosphorus content in human excreta.

Based on equations (1) and (2) it is estimated that in Uganda, 2.2, 0.3 and 1.0 kg of N, P and K, respectively, per person per year are excreted in urine (Jonsson and Vinneras, 2004). Reuse of urine in farming would greatly enable recovery of these nutrients.

Human excreta contain pathogens which if not carefully handled can affect health. Excreta recycling thus calls for safe handling and management guidelines to exploit its crop nutritional value safely. According to Hoglund (2001), urine from a single household is safe for all types of crops if the crop is for household's own consumption, and 1 month allowed between fertilizing and consumption. Urine used on cereals presents negligible risk of transmission of infection. Schonning and Stenstrom (2004) advised that the amount of urine used should be based on the N recommended when fertilizing with urea-based fertilizers. Where no recommendations are available, a rule of thumb is to apply urine collected from 1 person during 1 day to 1 m^2 of land per growing season. According to Hoglund (2001), urine can be assumed to have pH of 8.8 and N content of at least 1 g L^{-1}. Where no data are available, urine N content of 3–7 g L^{-1} can be used (Vinneras, 2002; Jonsson and Vinneras (2004)). For urine collected from different households, the risk of infection decreases during 6 months of storage (WHO, 2006). Urine should be applied close to the ground to avoid aerosol formation and incorporated into soil. Urine can be applied diluted or without dilution. Dilution varies from 1 part water:1 part urine to 10:1, but 3:1 is most common (Jonsson et al., 2004). In areas with heavy rainfall, repeated application of urine during the growing season rather than single doze may be an insurance against leaching.

Examples worldwide (India, South Africa, Botswana, Australia, Sweden, Mexico, among others) show high potential for EcoSan and nutrients reuse. In China, over 90% of garbage and human excreta are recycled as night soils (Edwards, 1992). Simons and Clemens (2004) observed that the N effect of urine on barley corresponded to 90% or higher than that of an equal amount of mineral fertilizer. Urine-fertilized Swiss chard yielded four times higher than the unfertilized in Ethiopia (Sundin, 1999). In Zimbabwe, lettuce, spinach and tomatoes grown on 3:1 water/urine mixture applied three times per week yielded two to seven times higher than unfertilized (Morgan, 2003).

In Lusaka and Dar es Salaam cities, urban agriculture contributes immensely to food consumption and livelihoods (Drangert, 1997) and will no doubt continue to expand in cities like Kampala, hence increased realization of the need to use urine in farming. While information exists from elsewhere on urine reuse, differences in diets lead to differences in nutrient (NPK) content of excreta. Thus, guidelines developed elsewhere need to be adapted to local conditions to take into consideration the differences in diets, handling, storage-application conditions and local crops to be grown. Further, many farmers would need advice on urine dilution rates and application frequently to prior to field use. Rarely do they bother about local fertilizer recommendations, let alone nutrient content in urine. For leafy vegetables, studies recommend 65–75 kg N/ha for Soroti area, eastern Uganda (sandy soils, lower fertility than in central Uganda) (National Horticulture Programme, personal communication). No fertilizer recommendations were found for vegetables in central Uganda, hence a further focus of the study.

This work was part of Kampala City Council (KCC)'s efforts towards improving sanitation and living standards of low-income communities living in lowland areas with poor sanitation through achieving a "closed-loop" approach. The objectives were to raise awareness of farmers, technical staff, local leaders and other stakeholders on the value and potential of using human urine in farming and to develop safe re-reuse and application guidelines for urine in farming.

Materials and Methods

Field Procedures

The study was conducted in Kyanja parish, Nakawa division, Kampala District, central Uganda. The area is 1250 m a.s.l. and receives 1390 mm of rainfall annually, bi-modally distributed between two seasons. The first season runs from March to June, the second season from August to November. The mean annual temperature is 21.4°C (Mugisha, 1998). Twenty farmers were randomly selected from seven zones of Kyanja parish to participate in the study. They were first sensitized on proper handling and the benefits of recycling urine in farming. Farmers selected the test crops as maize (*Zea mays*, L.) and nakati (*Solanum aethiopicum*, a popular local vegetable) in first season and kale (*Brassica oleracea* L.) and spinach (*Spinacia Oleracea* L.) in the second season. Farmers prepared the land and demarcated four, 1.5 m by 6 m plots for establishing the demonstration plots. Two plots were planted to maize and two to nakati during the first season, then kale and spinach during the second season of 2005.

Farmers collected urine from their individual households and stored it for at least 2 weeks in closed plastic jerricans. During the first season, urine was diluted with water in ratios of 0.5 urine:5 water, 1.0 urine:5 water and 1.5 urine:5 water, representing 10, 20 and 30% urine, respectively. Twenty litres of each urine–water mixture was prepared and applied once a week, once every 2 weeks and once a month (20 L of mixture per treated bed and 20 L water for the control). In the second season, urine was mixed with water in the ratios 10, 20, 30 and 40% urine and applied weekly or bi-weekly. Treatments were allocated to farmers' plots taking into account the soil nutrient status, based on the soil test and urine analysis results.

Soil and Urine Sampling and Analyses

Soil (500 g) was sampled from 0 to 30 cm of each bed before planting to make a representative composite sample for each farmer. Air-dried soil samples were analysed using standard methods as follows: soil pH on the 1:2.5 soil to water mixture (Okalebo et al., 1993). Organic matter (OM) by the Walkley and Black (1934) wet oxidation procedure modified for colorimetric determination of OM using sucrose, total N by the micro-Kjeldahl wet ashing procedure (Bremner, 1965), available P, extractable K, Ca and Mg by the ammonium lactate (pH 3.8) procedure (Foster, 1972) and soil particle size by the hydrometer method (Okalebo et al., 1993). Freshly collected urine (300 ml) was sampled from each of the 20 households and kept refrigerated at 4°C for at least 2 weeks prior to analysis. Urine nutrient content was determined on the wet ash digest following the micro-Kjeldahl wet oxidation procedure (Bremner, 1965). The N and P were determined colorimetrically, and K on the flame photometer.

Field Data Collection and Analysis

Data were collected on yield characteristics of test crops. For maize, data included number and size of cobs per plot; for nakati, number of plants and weight of leaf vegetable at each harvest and for kale and spinach, number of plants, weight and number of leaves at each harvest. Using Genstat (3.2), data were analysed as a completely randomized design to determine the effect of urine treatments (concentrations, frequency of application) with different farmers representing replicates. Statistically significant differences among means were determined at 5% level.

Results and Discussion

Soil Characteristics

Table 1 presents soil characteristics from the participating farmers' plots. Mean soil pH and organic matter (OM) were above the critical values; only 7 out of 20 samples had OM below the critical value. Mean total N was 0.12% with values below the critical value in 19 out of 20 farmers' fields. Soil P varied widely with 8 out of 20 farmers' fields having values below the critical value of 5.0 mg kg^{-1}. Overall, soil N was most limiting followed by P.

Urine Characteristics

Urine mean pH was within the range reported (Table 2). Urine N and P contents were lower

Table 1 Selected soil characteristics

Soil property	Mean[a]	Range	Critical values[b]
pH	6.95	6.10–8.40	5.2
OM (%)	3.7	2.3–7.1	3.0
Total N (%)	0.12	0.06–0.31	0.2
Extractable P (mg kg^{-1})	51.9	0.8–301.5	5.0
Extractable K (cmol$_c$ kg^{-1})	1.5	0.12–8.18	0.4
Extractable Ca (cmol$_c$ kg^{-1})	6.7	1.8–24.6	0.9
Sand (%)	58.7	48.4–69.1	Na
Silt (%)	29.5	19.6–41.6	Na
Clay (%)	11.8	7.3–17.6	Na
Textural class	Sandy loam		

[a]Number of observations = 20
[b]Below these values, soils are deficient or poor (Foster, 1972)
na = not applicable

Table 2 Selected mean chemical characteristics of urine used in the study

Property	Mean[a]	Range	Reported values[b]
pH	9.1	8.5–9.4	8.8
N, g L^{-1}	2.26	0.98–4.4	8.0
P, g L^{-1}	0.06	0.04–0.12	0.8
K, g L^{-1}	1.63	0.8–2.4	1.8

[a]No. of observations = 20
[b]SEPA (1995)

than those reported, reflecting a much lower protein and P intake in local diets (Jonsson et al., 2004). Nevertheless, considering the low soil N levels, addition of urine could supply the much needed N, enhancing crop growth. Potassium was within the range reported for urine.

Effect of Urine Application on Maize Fresh Yield and Income

Being peri-urban, farmers preferred to harvest maize fresh. The number of cobs from each bed was counted, and from the market price of an average cob size from each bed, expected income from each bed was computed. Plots receiving 10% urine weekly produced more and bigger cobs fetching a higher price than the control (Table 3). This translated into higher income from the treated plots than control. Applying 20% urine increased the number and size of cobs, resulting in higher income than the control. Applying 30% urine weekly resulted in even greater number of cobs (and therefore income) over the control. Thus, at the same frequency (weekly), higher urine concentrations resulted in increasing number and size of cobs, translating into higher income.

Effect of Urine Application on Nakati Growth

Overall, nakati field stand was better on urine-treated beds than the control. For many farmers, urine-treated nakati beds were harvested earlier, more frequently, and yielded higher than the control, although they were planted on the same day. Figure 1 presents nakati weight harvested over time for different urine treatments. First harvested leaves were from 20% urine weekly followed by 10% urine weekly, harvested at 58 and 61 days after planting (DAP), respectively. The 30% urine bi-weekly was first harvested at 66 DAP

Table 3 Effect of urine application on maize fresh yield and income

Urine concentration	Application frequency	Number of cobs	Av. market price per cob (Ug. Shs)	Income from the 9 m^2 bed (Ug. Shs)
Control		36	50	1,817
10% urine	Once a week	43	75	3,225
20% urine	Once a week	58	100	5,800
30% urine	Once a week	72	100	7,200
SED		6.1	10	957

Fig. 1 Cumulative nakati biomass following different urine treatments

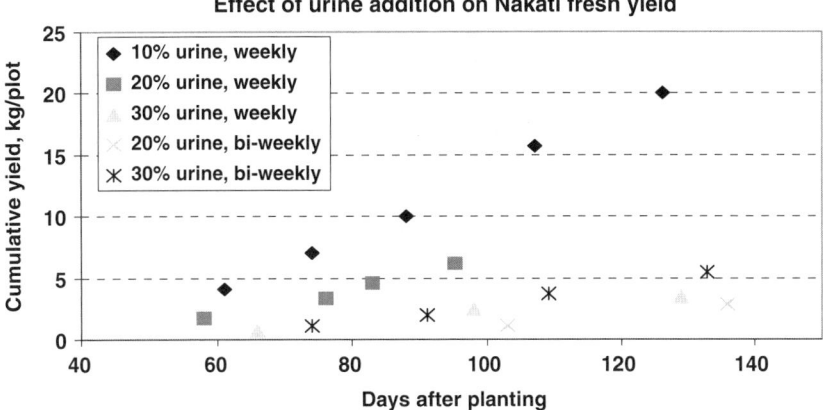

Effect of urine addition on Nakati fresh yield

◆ 10% urine, weekly
■ 20% urine, weekly
▲ 30% urine, weekly
× 20% urine, bi-weekly
✳ 30% urine, bi-weekly

while those from 30 to 20% bi-weekly were realized 74 and 103 DAP, respectively. The 10% urine weekly generated more biomass and maintained it throughout the monitoring period. The 20 and 30% weekly and 20 and 30% bi-weekly maintained a linear but lower leaf biomass compared to 10% urine weekly.

Effect of Urine Concentration on Total Nakati Leaf Yield

Total nakati yields reported are ones obtained within 2 months (July to August 2005) monitoring. Overall yields were likely higher since nakati can be harvested up to 4 months if soil moisture is not limiting. Results showed a significant interaction between urine concentration and frequency of application affecting

nakati yields ($P<0.01$). Weekly application of 10 or 20% urine significantly increased nakati yield over untreated plots (Table 4). The 20% urine treatment gave higher fresh yield than 10% (26.2 kg vs 22.2 kg) ($P<0.05$).

Effect of Urine Application on the Different Yield Parameters of Nakati

Average nakati yield per plant was higher for urine-treated plots than control ($P<0.05$) (Table 4). Yield per plant was similar for plots receiving 10 or 20% urine weekly and for plots receiving 20 or 30% urine bi-weekly. At 20% urine, yield per plant was higher for bi-weekly than weekly urine, suggesting increased lateral branching at bi-weekly compared to 20% urine

Table 4 Effect of urine application on nakati fresh yield plus associated economic gains

Urine concentration	Application frequency	Fresh yield (kg/plot)	Yield per plant (g/plant)	No. of plants to yield 700 g	[a]Gross income (Ug. Shs)/plot	[b]Total N applied (kg N/ha)	Yield (kg) per kg of N applied	[c]Nitrogen 'cost' (Shs/ha)	Returns (Shs)/ Sh 'spent' on N
Control		4.9	14.7	49.8	4,495	0	–	–	–
10% urine	Once a week	22.2	23.8	29.4	20,195	24	927	41,749	485.1
20% urine	Once a week	26.2	23.1	30.3	23,845	82	356	142,005	179.3
20% urine	Once in 2 weeks	2.8	35.0	20.0	2545	42	75	72,806	36.5
30% urine	Once in 2 weeks	5.4	34.6	20.2	4,926	31	196	53,194	99.3
SED		3.4	3.2	5.8	2,628	23	15	25,344	112

[a]Based on the prevailing market price: A 550 g bundle of fresh nakati leaves costs Ug. Shs. 500
[b]Total N (kg N/ha) applied over 2 months was calculated as follows:
$N = (N_{urine} \times V \times n \times 10{,}000 \text{ m}^2 \text{ ha}^{-1})/(D.F \times A \times 1000 \text{ g kg}^{-1})$
where N_{urine} = N concentration in urine, g L^{-1}; V = volume of urine–water mixture applied per plot (20 L); n = number of times the urine–water mixture was applied over a period of 2 months; D.F = dilution factor (e.g. 0.5 L diluted to 5 L gives a D.F of 10); A = area of demonstration plot (1.5 m by 6 m)
[c]Based on the prevailing market price for urea (50 kg bag urea costs Ug. Shs. 40,000)

weekly. The FAO recommends a daily consumption of 700 g of leafy vegetables for a family of six members (father, mother and four children aged 2, 5, 12 and 14 years) (FAO, 2006). The number of nakati plants to be harvested to produce 700 g of fresh vegetable was smaller on all urine-treated plots than control ones, more so on 20 or 30% urine bi-weekly treated plots. Thus, urine use in farming can contribute to food security, better nutrition and income, thus improved people's livelihoods.

Economic Assessment of Urine Reuse for Nakati Production

The monetary value of nakati obtained from farmers' plots was computed based on prices in the nearby *Kalerwe* market, where most of Kyanja farmers sell their produce. Weekly application of 10 and 20% urine yielded higher than the control resulting in huge financial gains compared to control (Table 4). Income from 10 to 20% urine weekly treated beds was, respectively, four and five times higher than that of the control; bi-weekly application of 20 or 30% did not significantly increase the income from nakati over the control. Data also show that applying 20 or 30% urine bi-weekly gave lower nakati yields (hence lower income) than did 10% urine concentration applied weekly. Consistent with Jonsson et al. (2004), these results stress the need for frequent application of soluble fertilizer materials to minimize nitrogen leaching.

Total N applied was computed based on urine N content from each farmer, dilution factor and application frequency and expressed on a hectare basis (Table 4). The monetary value of N from urine was computed based on urea market price. Based on expected income from nakati sales, expected returns to 'investment' were computed. Clearly, 20% urine weekly corresponded to the highest N rate while 10% urine weekly gave the lowest. Applying 10% urine weekly resulted in the highest nakati yields per kilogram of urine N applied, thus representing the highest N recovery, hence a more profitable management practice. The N recovery was lower for 20% weekly than 10% suggesting luxury consumption or possible N leaching. Bi-weekly application of 20 or 30% urine gave lower nakati yield per unit of urine N applied, and correspondingly lower returns to 'investment' compared to weekly and more so, 10% urine weekly.

Since urine is free and available, recycling of urine would save huge sums of money otherwise be spent on N purchase.

Relationship Between Nakati Yield and Soil Parameters

A detailed look at nakati yields from participating farmers showed that sites with deficient to low N (0.0–0.15%) and medium P (20–65 mg kg^{-1}) gave the highest response to urine. These sites had soil pH 6.7–7.2 and low OM levels, suggesting a fairly good soil medium for plant growth, except for low N. Added urine supplied N, thus increasing crop yields. Soils with high P (>150 mg kg^{-1}) and medium to high N (>0.16–0.31%) gave no response to urine. Apparently, these farmers had used cow dung, chicken droppings in their backyard gardens (also reflected in high OM levels). Despite high soil P and fairly good N levels, however, nakati yields observed on these farmers' fields were lower compared to other farmers, suggesting other limiting factors (possibly micronutrient(s)) (Marschner, 1993). This possibility is supported by high soil pH (7.8–8.4) and Ca (10–25 cmol$_c$ kg^{-1}). However, soil micronutrients were not tested in this study.

Effect of Urine Application on Kale Leaf Biomass Accumulation

The first harvested kale leaves were from 40% urine weekly followed by 20% urine weekly, harvested at 35 and 37 DAP, respectively (Fig. 2). First harvest from 30% urine bi-weekly was realized at 46 DAP while those from 40% urine bi-weekly and 10% weekly were realized at 51 and 53 DAP, respectively. Compared to other treatments, 40% urine weekly maintained more biomass throughout. Leaf biomass from 10 to 20% urine weekly plus 30 and 40% bi-weekly treatments increased linearly with time but remained lower than that of 40% urine weekly.

Total Kale Fresh Yields

Table 5 presents the effect of urine application on the total kale fresh yield as observed within 2 months of monitoring. Data are a summation of the observed

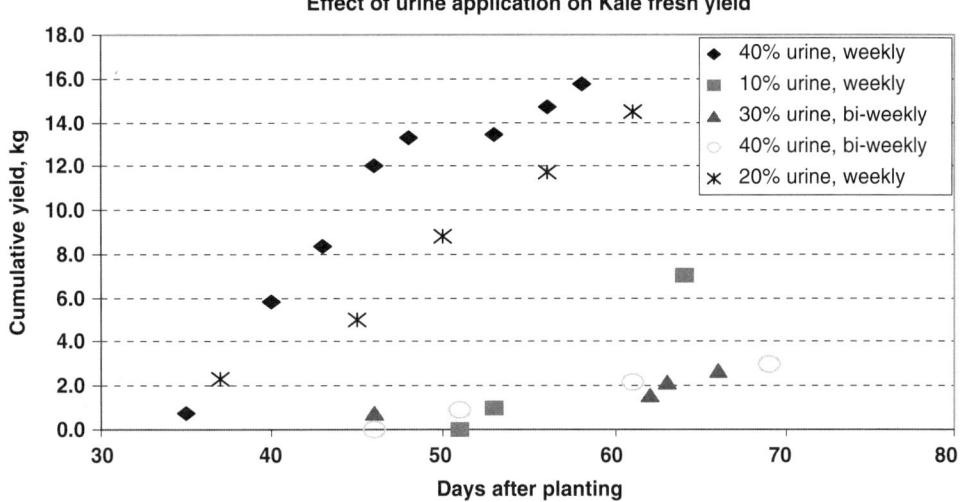

Fig. 2 Kale leaf biomass accumulation following fertilization with urine

Table 5 Effect of urine application on the kale fresh yield and associated economic gains

Urine concentration	Application frequency	Fresh yield (kg/plot)	Average leaf wt (g/leaf)	No. of leaves to yield 700 g	Gross income (Ug. Shs)/plot	Total N applied (kg N/ha)	Yield (kg) per kg of N applied	Nitrogen 'cost' (Shs/ha)	Returns (Shs)/(Sh) 'spent' on N
Control		6.8	16.9	42.9	5,632				
10% urine	Once a week	7.1	19.8	35.4	5,872	27.2	287.9	48,120	135.9
20% urine	Once a week	14.5	23.6	29.6	12,079	55.8	288.6	98,757	135.9
40% urine	Once a week	15.7	20.4	34.4	13,115	260.3	67.2	460,446	31.6
20% urine	Once in 2 weeks	1.9	20.8	33.6	1,561	33.2	62.6	58,814	29.5
30% urine	Once in 2 weeks	2.7	20.2	34.6	2,240	29.6	100.8	52,419	47.5
40% urine	Once in 2 weeks	3.0	20.0	35.1	2,511	99.9	33.5	176,756	15.8
SED		1.9	1.1	2.7	1,569	45.2	58.4	79,979	27.5

yield for different piecemeal harvests. Applying 10% urine weekly had no significant effect on kale fresh yield. Applying 20 or 40% urine weekly more than doubled the kale yields ($P<0.05$). Bi-weekly application of 20, 30 or 40% urine actually depressed kale yield ($P<0.05$). This may be due to moisture limitations.

Average Leaf Weight

Average leaf weight was determined by dividing total weight of kale by the total number of leaves harvested. Weekly urine treatment increased average leaf weight up to 20% urine, beyond which it decreased. Table 5 shows the number of kale leaves harvested to produce FAO daily recommended 700 g of fresh leaf vegetable was smaller on urine-treated plots than the control, more so on plots receiving 20% urine applied weekly

($P<0.05$). This is due to increased biomass per leaf following urine addition. Terry et al. (1981) reported increased kale leaf biomass due to N fertilization. The results of this study show 20% urine weekly as the optimum rate for kale.

Economic Assessment of Urine Use for Kale Production

Income from kale sale was derived as described for nakati. On average, a bundle of kale leaves sold for Shs. 500 weighed 600 g and contained 62 leaves. Based on this, applying 10% urine weekly did not significantly increase the income (Table 5). Applying 20 and 40% urine weekly resulted in higher vegetable yield (hence income) ($P<0.05$). Income from 20 to 40% urine weekly was not significantly different. Kale yield per kilogram of N applied was highest at 20%

> 10% urine weekly > 30% urine bi-weekly > 40% weekly > 20% bi-weekly > 40% bi-weekly. This suggests that 10 and 20% urine weekly were the most efficient treatments at recovering the applied N and transforming it into leaf biomass. If the urine-derived N was to be purchased from the market, the cost would be highest for 40% urine weekly and lowest for 10% urine weekly. The returns from such an investment would be highest for 10 and 20% urine weekly, and lowest for 40% urine bi-weekly. These findings suggest that 10 and 20% urine applied weekly are viable management options for urine on kale compared to other treatments investigated in this study.

Considering leaf biomass results above, although economic benefits could favour 10 and 20% urine weekly treatments, total yield (and leaf weight) gain at 10% urine over the control, although significant, was smaller compared to 20% urine weekly (Table 5). Based on this and other reasons discussed above, we propose 20% urine weekly as the most viable for kale growing. For leafy vegetables, studies recommend 65–75 kg N/ha for Soroti area, Eastern Uganda (sandy soils, lower fertility than in central Uganda) (National Horticulture Programme, personal communication). The present study suggests the optimum urine rate of 20% applied weekly for kale in central Uganda, corresponding to N rate to be 56 kg N/ha.

Effect of Urine Application on Total Spinach Fresh Leaf Biomass

Weekly application of 10, 20 and 30% urine increased spinach fresh yield from 3.8 to 4.2, 17.1 and 18.0 kgplot^{-1}, respectively (Table 6). These responses were significant except at 10% urine. Bi-weekly application

of 30 or 40% urine had no significant effect on spinach yield compared to the control. Spinach yields were lower for bi-weekly urine-applied plots than the weekly ones. Biemond (1996) reported increased biomass accumulation with increasing N application on spinach. Splitting N application had smaller effect on biomass accumulation.

Average Leaf Size

Average leaf weight was determined as described for kale. There was a significant increase in leaf weight following weekly urine addition up to 20% urine, beyond which it declined ($P<0.05$) (Table 6). Biemond (1995) observed that the size of mature spinach leaves increased with increasing N levels up to a certain level. Our results also show that bi-weekly application of 30 and 40% urine significantly depressed average leaf weight ($P<0.05$), possibly due to moisture stress. The number of spinach leaves harvested to produce the FAO daily recommended 700 g of fresh leafy vegetables was overall smaller on urine-treated plots than the control, more so on plots receiving 20% urine applied weekly (Table 6). Thus, urine application can directly contribute to food security, better nutrition and improved people's livelihoods.

Economic Assessment of Urine Use for Spinach Production

Income from spinach sale was derived as described for nakati. On average, a bundle of spinach leaves sold for Shs. 500 weighed 450 g and contained 18 leaves. Based on this, 10% urine weekly had no significant

Table 6 Effect of urine application on spinach fresh yield and associated economic benefits

Urine concentration	Application frequency	Fresh yield (kg/plot)	Average leaf weight (g/leaf)	No. of leaves to yield 700 g	Gross income (Ug. Shs)/plot	Total N applied (kg N/ha)	Yield (kg) per kilogram of N applied	Nitrogen 'cost' (Shs/ha)	Returns (Shs)/(Sh) spent' on N
Control		3.8	10.8	107.3	4189.6				
10% urine	Once a week	4.19	21.8	32.1	4658	27.2	171.2	48.1	107.6
20% urine	Once a week	17.12	34.5	20.3	19022	75.7	251.2	134	157.8
40% urine	Once a week	17.98	24.2	29	19982	260.3	76.8	460.4	48.2
30% urine	Once in 2 weeks	1.36	2.5	280	1511	29.6	51	52.4	32
40% urine	Once in 2 weeks	1.17	2.5	278	1300	99.9	13	176.8	8.2
SED		2.2	3.8	40.2	2467.2	47.8	48.5	84.5	30.5

effect on income over the control (Table 6). Applying 20% urine weekly resulted in highly significant increase in income over both the control and 10% urine ($P<0.01$). Increasing urine concentration further to 40% did not increase income over 20% urine weekly.

Table 6 shows that spinach yield per kg N applied was highest at 20% urine weekly, least at 40% bi-weekly. This suggests 20% urine weekly as the most efficient in transforming applied N into biomass. If the urine-derived N were to be purchased from the market, the cost would be highest for 40% urine weekly and lowest for 10% urine weekly (Table 6). Returns from such an investment would be highest for 20% urine weekly and lowest for 40% urine bi-weekly. For leafy vegetables, studies recommend 65–75 kg N/ha for Soroti area, Eastern Uganda (sandy soils, lower fertility than in central Uganda) (National Horticulture Programme, personal communication). Results of this study suggest 76 kg Nha^{-1} corresponding to 20% urine weekly.

Conclusions and Recommendations

This study has demonstrated the benefits of using human urine in maize and vegetable (nakati, kale and spinach) production. Application of urine significantly increased maize fresh yield, number and size of cobs. Higher urine concentrations (at least up to 30% urine) favoured better maize growth. Urine application also increased nakati yields with 10% urine weekly being economically more feasible than 20 or 30% urine. Kale and spinach production was enhanced by urine application with the 20% urine weekly being the best management practice. For all crops, weekly application was more effective than bi-weekly or one a month. Response to urine was most pronounced on soils of low N and P. Urine adds nutrients to the generally low fertility soils, favouring maize and vegetable (nakati, kale and spinach) growth and yield. Economic assessment shows huge economic benefits arising from urine reuse, particularly on vegetables.

Recommendations

Use of urine in farming is agronomically feasible and economically viable. Being readily available, urine reuse should be promoted among farmers, especially smallholders who may not readily invest resources to purchase external inputs for improving soil fertility. Further to the guidelines developed elsewhere, findings from this study suggest the following local guidelines:

Maize (*Z. mays* L.): Apply 30% urine (3:10 urine to water) weekly for 8 weeks.

Nakati (*S. aethiopicum* L): Apply 10% urine (1:10 urine to water) weekly for 8 weeks.

Kale (*B. oleracea* L): Apply 20% urine (2:10 urine to water) weekly for 8 weeks

Spinach (*S. oleracea* L): Apply 20% urine (2:10 urine to water) weekly for 8 weeks.

Acknowledgements Our appreciation go to Sida and KCC for funding, Kyanja farmers and extension workers for participating in the study, the Kampala District Urban Agriculture Officer, Mrs. Azuba and Ms. Anna Richert Stintzing of the SwedEnviron Cooperation for the Environment Consulting Group for their technical advice and all EcoSan project staff for their support.

References

Biemond H (1995) Effects of nitrogen on development and growth of the leaves of vegetables. 3. Appearance and expansion of leaves of Spinach. Neth J Agric Sci 1995(2):247–260

Biemond H (1996) Effects of nitrogen on accumulation and partitioning of dry matter and nitrogen of vegetables. 3. Spinach. Neth J Agric Sci 1996(3):227–239

Bremner JM (1965) Total nitrogen. In: Black CA et al (eds) Methods of soil analysis, part 2. Agronomy, vol 9. American Society of Agronomy, Madison, WI, pp 1149–1178

Drangert JO (1997) Fighting the urine blindness to provide more sanitation options. Institute of Water and Environmental Studies, Linkoping University, Linkoping

Edwards P (1992) Reuse of human waste in aquaculture. A technical Review. UNDP-World Bank, Water and Sanitation Program, 350pp

FAO (2006) Improving nutrition through home gardening. Practical nutrition for field workers. FAO corporate document repository. http://www.fao.org

Foster HL (1972) Rapid routine soil and plant analysis without automatic equipment. I. Routine soil analysis. East Afr Agric Forest J 37:160–170

Hoglund C (2001). Evaluation of microbial health risks associated with the reuse of source separated human urine. PhD thesis, Department of Biotechnology, Royal Institute of Technology, Stockholm, Sweden. ISBN 91-7283-039-5. http://www.lib.kth.se/sammanfathnigar/hoglund 010223. pdf

Jonsson H, Stintzing AR, Vinneras B, Salomon. E (2004) Guidelines on the use of urine and feaces in crop production. Report 2004-2. EcoSanRes. SEI

Jonsson H, Vinneras B (2004) Adapting the nutrient content of urine and faeces in different countries using FAO and Swedish data. In: EcoSan – closing the loop. Proceedings of the 2nd international symposium on ecological sanitation, incorporating the 1st IWA specialist group conference on sustainable sanitation. Lubeck, 7–11 Apr 2003, pp 623–626

Marschner H (1993) Mineral nutrition of higher plants. Academic, London

Morgan P (2003) Experiments using urine and humus derived from ecological toilets as a source of nutrients for growing crops. Paper presented at the 3rd world water forum 16–23 Mar 2003. http://aquamor.tripod.com/KYOTO.htm

Mugisha OR (1998) Uganda districts information handbook. Fountain Publishers. http://www.naro.go.ug/UJAS/Papers/vol.12,no1-7.pdf

Okalebo JR, Gathua KW, Woomer PL (1993) Soil pH and Electroconductivity. In: Laboratory methods of soil and plant analysis: a working manual. Tropical soil biology and fertility programme, Nairobi, 88p

Schonning C, Stenstrom. TA (2004) Guidelines for the safe use of urine and faeces in ecological sanitation systems. Report 2004-1. EcoSanRes. Programme. SEI

SEPA (1995) Vad innehaller avlopp fran hushall? (Content of wastewater from households). Report 4425, Swedish Environmental Protection Agency, Stockholm

Simons J, Clemens J (2004) The use of separated human urine as mineral fertilizer. In: EcoSan – closing the loop. Proceedings of the 2nd international symposium on ecological sanitation, incorporating the 1st IWA specialist group conference on sustainable sanitation. Lubeck, pp 595–600, 7–11 Apr 2003

Sundin A (1999) Human urine improves the growth of Swiss chard and soil fertility in Ethiopian urban Agriculture. Thesis and seminar projects No. 112. Department of Soil Science, Swedish University of Agricultural Sciences

Terry N, Waldron LJ, Taylor SE (1981) Environmental influences on leaf expansion. In: Dale JE, Milthorpe FL (eds) The growth and functioning of leaves. Cambridge University Press, Cambridge, pp 179–193

Vinneras B (2002) Possibilities for sustainable nutrient recycling by fecal separation combined with urine diversion. PhD thesis, Swedish University of Agricultural Sciences, Uppsala

Walkley A, Black IA (1934) An examination of the method for determining soil organic matter and a proposed chromic acid titration method. Soil Sci 37:29–38

WEHAB (2002) A framework for action on water and sanitation. World Summit on Sustainable Development, Johannesburg, 40 pp, Aug 2002

WHO (World Health Organization) (2006) Guidelines for the safe use of wastewater, excreta and greywater. Excreta and greywater use in agriculture, vol 4. http://www.who.int/water_sanitation_health/wastewater/gsuweg4/en/index.html

Towards Sustainable Land Use in Vertisols in Kenya: Challenges and Opportunities

E.C. Ikitoo, J.R. Okalebo, and C.O. Othieno

Abstract Vertisols are heavy clay soils with 30 to over 70% clay dominated by smectite mineralogy. They are the most widely distributed soils worldwide located mainly in tropical and sub-tropical regions. In Africa, they occupy about 3.5% or 99 million hectares of the landmass located mainly in eastern Africa. In Kenya, they occupy about 5% or 2.8 million hectares of the landmass and occur mainly in arid and semi-arid areas (ASALs). They are potentially very productive soils because of their high cation exchange capacity and water retention, moderate soil fertility and low salinity/alkalinity problem. However, their poor infiltration and internal drainage result in severe management problems. Traditionally, the Vertisols in Kenya were and still are used largely for extensive livestock grazing, however, due to increased population pressure and emigration of people from humid high potential areas to the ASALs; they are being converted to arable cropping. Vertisols cropping in Kenya is, however, a new utility system being embraced enthusiastically in spite of the serious management constraints observed. The major constraints include water logging and tillage difficulties, which pose serious challenges to their utilization for cropping in the country. In this chapter, problems and challenges associated with the ecosystem, population pressure and traditional and current land use including livestock grazing and arable cropping in the Vertisols are considered in relation to sustainable land use and management. In conclusion, views on the way forward towards sustainable land use and management of the Vertisols including proposals on stocking rates and cropping systems for maintaining the soil fertility and productivity while minimizing soil erosion and land degradation are given. Use of the drainage landforms (DLs) increased maize and sunflower grain yields by 5–57% and above-ground total biomass significantly.

Keywords Vertisols · Land use/utilization · Cropping · Sustainable · Management

Introduction

Vertisols are heavy clay soils dominated by smectite mineralogy, principally the montmorillonite. Referred to in Kenya as *black cotton soils*, they are the most widely dispersed soils in the world spreading over five continents from 45°S to 45°N, mainly in tropical and sub-tropical regions, and occupy about 1.8% of the world's landmass or approx. 257 million hectares in 76 countries (Dudal, 1963; FAO, 1965; Donahue et al., 1977; Probert et al., 1987). In Africa, the soils occupy about 3.5% of the land mass or about 99 million hectares, with the vast majority occurring in eastern Africa (Santana, 1989). Sudan has about 500,000 km^2 of Vertisols, which constitute about 20% of the country's landmass, located in the western, central and eastern clay plains, where precipitation ranges from 200 to 800 mm per annum. In Ethiopia, they occupy about 9 million hectares with about 70% being located in the highland plateau and 30% in the lowlands (Beneye and Regassa, 1989).

In Kenya, Vertisols occupy about 5% of the landmass or approx. 2.8 million hectares scattered all over the country with about 80% of them being located in

E.C. Ikitoo (✉)
Department of Soil Science, Moi University, Eldoret, Kenya
e-mail: ikitoo.caleb@gmail.com

A. Bationo et al. (eds.), *Innovations as Key to the Green Revolution in Africa*,
DOI 10.1007/978-90-481-2543-2_67, © Springer Science+Business Media B.V. 2011

arid and semi-arid lands (ASALs) and the remaining 20% located in more humid areas (Van de Weg, 1987). Developed on a variety of parent materials including pre-Cambrian basement system rocks (ferromagnesian gnesses, etc.), volcanic rocks (basalt, etc.) to alluvial and colluvial deposits derived from different rocks, the soils occupy gently undulating pene-plains, rolling uplands, alluvial and volcanic plains and flood plains, valley bottoms and bottomlands (ILACO, 1974; Van de Weg and Mbuvi, 1975; De Costa, 1973; Muchena and Gachene, 1985).

Vertisols are potentially very productive soils especially in the semi-arid region of world, where most of them occur, because of their high water retention, high cation exchange capacity (CEC) and moderate soil fertility. However, due to their unique morphological and physical–chemical properties including high clay content, swell–shrink characteristics, poor infiltration rate, low internal drainage and fine to massive soil structure, they pose serious tillage and management problems when used for arable cropping. Major among the problems include water logging during wet seasons, tillage difficulties due to narrow soil moisture range for effective workability, high erodibility index and accelerated soil erosion on exposed sites with slope greater than 2%, all culminating in poor yield (Hubble, 1984; Probert et al., 1987; Ahmad, 1989; Ikitoo, 1989). Lack of suitable land forming implement for smallholder farmers is a major problem in Kenya (Muckle, 2002).

where most of the Vertisols occur (Van de Weg and Mbuvi, 1975; Siderius and Njeru, 1976; Van de Weg, 1987; Herlocker, 1999). The practice is prevalent in the ASALs' agro-ecological zones V-1, V-2, VI-1 and VII-1 in Marsabit, Moyale, Garissa, Narok, Kajiado, Taita-Taveta, Kwale and Tana River districts. However, some Vertisol ASALs with rapidly increasing populations are being converted to arable cropping (Van de Weg and Mbuvi, 1975; Sombroek et al., 1982; Van de Weg, 1987; Ikitoo, 1989).

(b) Rain-fed arable cropping:

Rain-fed cropping of the Vertisols is currently practiced in western Kenya's Kano plains and Kitale/Endebess area on the slope of Mt Elgon, where sugarcane, maize, cotton, sorghum, beans and horticultural crops are grown. Further rain-fed cropping of the Vertisols is practiced in the central Kenyan plateau areas of Mwea, Masinga, Matuu and Athi-River where maize, beans, green grams, chickpeas and pigeon peas are grown (Van de Weg and Mbuvi, 1975; Van de Weg, 1987; Ikitoo, 1989).

(c) Intensive irrigated cropping

Vertisols are successfully used for irrigated crop production of rice, cotton, maize and horticultural crops in Mwea, Ahero, West Kano, Bunyala, Hola and Bura irrigation schemes (ILACO, 1974; Van Gassel, 1982; Njihia, 1984; RoK Statistical Abstracts, 2004).

Challenges of Changing Land Use Scenario in Vertisols in Kenya

Land Use

(a) Nomadic pastoralists system

The Vertisols in Kenya occur mainly in the ASALs generally referred to as rangelands. The rangelands traditionally supported a pastoral subsistence economy aimed at continued food supply with minimum risks to both man and livestock (Herlocker, 1999). Thus, the Vertisols in the rangelands provide mainly pasture for extensive livestock grazing, in particular, during dry seasons or drought periods when the grazing is extended to valley bottoms and depressions

Social–Ecological Impact

(a) Population pressure

The current trend towards intensive cropping of Vertisols in some ASAL areas in Kenya is a function of increased population pressure on land resulting in expansion of land under cultivation, which is associated with emigration of people from the humid and semi-humid high to medium potential areas to the ASALs. The population in Kenya rose sharply from 15 million in 1979 to 22, 28 and 35 million in 1989, 1995 and 2003, respectively (RoK Economic Survey, 1996; RoK Statistical Abstract, 2004). With annual growth rate of about 3%, the population was estimated at 37 million for 2007, with an average density of about 50 persons km^{-2}. However, disparity in the population density is significant with some areas having over 500 persons km^{-2} while others with as low as 5 persons km^{-2}. Densely populated areas include high potential central Kenya highlands and Lake Victoria basin, which have significant emigration of people to sparsely populated adjacent ASALs (RoK Economic Survey, 1996; RoK Statistical Abstract, 2004).

(b) Ecological balance

To increase land productivity and quality of life of people utilizing Vertisols in the tropics, an understanding of how ecosystems in the tropics function is essential as it provides basic information required in choosing suitable cropping systems at farm, catchments and regional levels (Scholes et al., 1994). The productivity of an ecosystem is a function of its genetic resources, climate and soil characteristics interacting over time. Thus, natural ecosystems in the tropics have evolved over time and are well adapted to their environments with nutrients in their soils being resilient and in balance. It should not, however, be assumed that all natural ecosystems that produce large amounts of biomass in the tropics are suitable for continuous arable cropping and will remain productive under continuous cropping. Experience has shown that under continuous cropping, the nutrient status in most soils in the tropics drop to new much lower steady-state levels, with drastic reduction in land productivity (Brown et al., 1994; Woomer et al., 1994; Alegre and Cassels, 1994; Juo and Manu, 1994).

(c) Problems associated with land use change

In traditional pastoral systems practiced in the rangelands, various strategies were used to achieve delicate ecological balance between man, livestock and pasture. Such strategies included livestock mobility, dry season grazing reserves, splitting of herds, maintaining high proportion of breeding females, retaining certain herd groups beyond prime age, maximizing the livestock to human number ratio and practicing a social-security system of stock loans (Herlocker, 1999). Thus, under traditional livestock management regimes, natural ecosystems in the Vertisol areas remained stable with minimum net loss as long as stocking rates were optimized (Baumer, 1987; Brown et al., 1994; Herlocker, 1999). However, conversion of land use in some Vertisol areas from livestock grazing to arable cropping is a new utility system adopted albeit with

serious management problems that require development of suitable technologies for sustainable land use.

The soil morphological and physical properties that distinguish Vertisols from other soils (ISSS/ISRIC/FAO, 1994; FAO, 2002) confer them with unique characteristics that have profound influence on their management requirements for successful arable cropping (Probert et al., 1987; FAO, 1984; Ahmed, 1989). Vertisols in Kenya have high clay content, i.e. about 40–75%, dominated by the smectite mineralogy, in particular montmorillonite, hence, their pronounced shrink–swell characteristics and poor infiltration rate and internal drainage (ILACO, 1974; Van de Weg, Mbuvi, 1975; Muchena and Gachene, 1985; Van de Weg, 1987; Ikitoo, 1989). In Table 1, high clay content, i.e. 50–70%, low to moderate % organic C, i.e. 0.8–2.8%, and very low % total N, i.e. 0.07–0.18%, were indicated for some cultivated Vertisols. The CEC was generally high to very high, i.e. 41–80 C mol$_c$ kg^{-1}. However, effective CECs in the Mwea irrigation scheme and Ahero rice research station Vertisols were moderate and similar to those in the more humid areas of Kitale, South Kano and Mohoroni. On the other hand, the effective CECs of the rain-fed Mwea area, i.e. Gategi and Kakindu sites, and Amboseli/Kibwezi, which had lower rainfall, were relatively high. Similarly, the pH of the Vertisols in the humid, semi-humid and irrigated areas was moderately acid to neutral, i.e. 5.5–7.0, while that of the Vertisols in lower rainfall areas including rain-fed Mwea areas, Amboseli/Kibwezi and Kindaruma was neutral to alkaline, i.e. 7.0–8.0.

Based on the Vertisol properties indicated in Table 1, the major problems associated with cropping of the soils in Kenya are considered as follows:

Tillage

Tillage in Vertisols is necessary to loosen the soil for land forming to allow surface water removal, planting and soil conservation (Ahmad, 1989). However, problems of water logging, extreme stickiness when wet and excessive hardness when dry, make timely tillage for early planting in the Vertisols difficult (Probet et al., 1987; Coughlan et al., 1987). In Kenya, tillage is a major problem in Vertisols of the sugar growing zone in the Kano plains and irrigated Mwea rice fields, due to soil compaction resulting from heavy machinery

Table 1 Chemical characteristics[a] of topsoils (0–30 cm) of some Vertisols used for arable cropping in Kenya

Location (site)	% Clay	% Org C	% N	CEC (C mol_c kg^{-1})	% Base saturation	pH	Available P (mg kg^{-1})	Available cations (C mol_c kg^{-1})				References
								Ca	Mg	K	Na	
Selected farms in Mwea irrigation settlement scheme; no. = 40 farms	61	1.3	NA	NA	NA	7.2	5.7	19.4	12.5	0.1	1.3	Njihia (1984)
Mwea, Gategi (P.M. Nzau's farm)	60	0.67	0.07	60	NA	8.1	3.9	32.4	25.5	0.4	1.2	Ikitoo and Muchena (1992)
Mwea, Kakindu (S.M. Kithuku's farm)	47	1.6	0.09	80	NA	7.5	12.1	56.1	5.8	0.1	16.7	Ikitoo et al. (2005)
Kitale (ADC Namandala farm, western Kenya)	49	2.8	0.28	29	NA	6.1	22.5	20.4	4.2	0.8	3.3	Ikitoo et al. (2005)
Kano Plains (Mohoroni expt site – Egerton University)	61	1.9	0.18	42	65	6.1	N/A	18.1	8.0	0.7	0.7	Sigunga (1997)
South Kano, (western Kenya)	69	2.2	NA	41	77	5.7	68.5	18.2	6.40	1.8	2.8	Ikitoo (1989)
Amboseli/Kibwezi area	53	0.8	NA	52	100	8.2	130	51.0	13.5	1.9	4.2	Van de Pour and Van Engelen (1983)
NIB, rice research station, Ahero (Kisumu, western Kenya)	NA	1.5	0.16	NA	NA	6.3	84	16.5	6.0	1.1	1.4	ILACO (1974)
Kindaruma area (eastern Kenya)	55	0.8	NA	NA	100	8.0	NA	NA	NA	NA	NA	Van de Weg and Mbuvi (1975)

[a]Mean value of samples of variable sizes ranging from 5 to 40; NA: not available

operated on wet soils, hence, high bulk densities of 1.40–1.87 (Ikitoo, 1989 and Kondo et al., 2001).

Available Water and Germination

Vertisols retain large amounts of water compared to most other soils while their wilting point (–15 bar) water content is equally high, typically 0.25–0.50 cm^3 cm^{-3} (Probert et al., 1987). Table 2 shows pF/pressure values and percentage of water retention for major soil types in Kindaruma and Mwea areas of the central Kenya plateau, while Table 3 shows the same values for Vertisols of Mwea (Kakindu) and Kitale (ADC Namandala farm) in western Kenya highlands. Compared to the other soils, Vertisols not only retained the highest soil MC, i.e. 20–38%, but also had the highest wilting point (pF 4.2 or atm 15) soil MC, i.e. 25–36%. This demonstrates their high water retention

capacity, which could lead to seed germination problems, if initial rainfall is low. Thus, depending on the length of the dry season, air-dry soil moisture content in Vertisols can drop to levels well below the wilting point, occasionally to less than 5% v/v water content and germination problems would occur when initial rainfall at planting is light, because to wet the upper 10 cm of air-dry Vertisols required 20–40 mm of rainfall and to wet the entire profile to 50 cm required 75–175 mm of rainfall (Probert et al., 1987).

Continuous Cropping and Decline in Soil Fertility

Soil fertility depletion in smallholder farms is the main cause of decline in per capita food production in the sub-Saharan Africa due to continuous land cultivation with little or no fertilizer application, thus, nutrient depletion resulting mainly from crop removal, leaching and soil erosion (Sanchez et al., 1997). In Zimbabwe,

Table 2 Water retention: volume % water held in the soil at different pF values or pressures for topsoils of major soil classes in the Kindaruma area.

Soil group	pF: 2.0 atm: 1/10	2.5 1/3	3.0 1	3.7 5	4.0 10	4.2 15	% Water retained
Arenosols	12.6	9.1	6.4	3.5	3.3	2.8	9.8
Ferralsols BU	32.0	25.4	22.0	18.4	17.3	16.8	15.2
Ferralsols VK	41.9	36.3	30.7	25.5	24.6	23.7	18.2
Luvisols	18.8	15.1	12.5	10.8	9.4	9.2	9.6
Vertisols	56.1	41.0	47.6	43.9	37.9	35.7	20.4

Source: Van de Weg and Mbuvi (1975)

Table 3 Water retention: Percentage moisture (g/g) held at different pF values or pressures (atm) for three soil depths at Kakindu, Mwea and ADC Namandala, Kitale

Site/soil depth	pF: 0 atm: 0	1.5 1/13	2.0 1/10	2.3 10/33	2.5 1/3	4.2 15	% Water retained
Mwea							
0–15 cm	48.2	43.2	41.9	40.9	40.4	25.0	23.2
16–30 cm	45.2	42.7	40.7	40.8	40.3	24.6	20.6
31–45 cm	46.2	43.5	42.3	41.3	40.9	25.1	21.7
Mean	46.6	43.1	41.6	41.0	40.5	24.9	21.8
SE ±	3.7	1.8	0.9	2.4	1.5	0.7	N/A
SE of a mean soil depth ±	2.1	1.1	0.5	0.9	0.9	0.4	N/A
Kitale							
0–15 cm	62.7	61.3	59.5	56.1	52.3	25.5	37.2
16–30 cm	63.7	60.1	54.9	54.4	48.6	25.5	38.2
31–45 cm	63.8	58.9	54.6	53.4	50.5	26.6	37.2
Mean	63.4	60.1	56.3	54.6	50.5	25.9	37.5
SE ±	7.3	3.9	3.6	3.3	4.3	0.6	N/A
SE of a mean soil depth ±	5.9	2.3	2.1	1.9	2.5	0.4	N/A

Source: Ikitoo (2008). PhD thesis in soil science, Moi University

studies on the Chisumbanje Vertisols cultivated from 0 to 50 years indicated steady decline in organic C, KCL-extractable N and folia concentration of total K and Ca in maize with increasing cultivation period (Mugabe et al., 2002). Similarly, long-term studies on Vertisols under irrigation in the Gezira, Sudan, indicated that continuous cultivation affected the SOM quality (Elias et al., 2002). Thus, there is urgent need to replenish soil fertility in the sub-Saharan Africa region and where new lands are opened for cultivation; technologies for sustainable land management should be adopted (Woomer et al., 1994; Alegre and Cassels, 1994; Juo and Manu, 1994).

Soil Erosion and Land Degradation

Development in general is driven by increasing human population and in agricultural systems where both livestock and arable cropping are practiced, there is a tendency for farmers to overstock, which predisposes the land to soil erosion. The situation is made worse by the impact of grazing animals on the soil's physical properties (Devendra and Thomas, 2002; Taddese et al., 2002; Lodge et al., 2003; Wanyoike and Wahome, 2004). Vertisols are highly pre-disposed to rill and gully erosion in areas with slope $\geq 2\%$ due to poor infiltration (Probert et al., 1987). Thus, suitable soil conservation measures should be incorporated in the tillage approaches and cropping systems used.

Water Management for Increased Productivity

(a) Paddy rice production

Water use efficiency is a major problem in irrigated paddy rice production in Kenya. Studies on water use efficiency in paddy rice grown in the Vertisols of the Mwea irrigation scheme indicated significant losses resulting from deep percolation, drainage flows and evapo-transpiration in uncropped land. The water losses varied across units and months and the use efficiency was estimated at 40–45% before transplanting and 65–75% at the peak period of the season when management was best (Van Gassel, 1982). Variation in yield ranged from 1.5 to 9.3 t ha^{-1}, due to various factors including farmers ability to

apply recommended practices, drainage conditions, slope of the land, soil permeability and exchangeable sodium percentage (ESP) (Njihia, 1984). Ploughing paddy fields is a major problem; however, studies on soil physical characteristics indicated that ploughing during dry seasons minimized water percolation into deep layers along cracks, thus improving traffic ability (Kondo et al., 2001). Studies in Bhopal, Madhya Pradesh, India, showed that paddling intensity, i.e. four or eight passes, affected soil physical properties by increasing depth of puddle layer, amount of easily dispersible clay plus silt particles and total porosity, in paddy rice (Mohanty and Painuli, 2003).

Table 4 presents data for irrigated paddy rice production in Kenya for the period 1997/98 to 2003/04. Average annual production and area under cultivation varied considerably from 18,000 to 49,000 metric tons (MT) and 7,000 to 30,000 ha, respectively. However, it is notable that maximum yields of up to 5.3, 4.4, 3.9 and 3.2 MT ha^{-1} were achieved in Mwea, Ahero, Bunyala and West Kano irrigation schemes, respectively, during the 7-year period under review, which indicated that with improved crop management, average yields could be increased up to 5.0 t ha^{-1}.

However, the large fluctuations in production and area under cultivation suggested serious management problems, which could have resulted from some or all of the factors indicated for low yields in the Mwea irrigation scheme. Thus, an in-depth study on the factors responsible for the low and highly variable rice yields/production and possible remedial measures is required.

(b) Surface water removal for increased productivity

The ICRISAT Centre has developed technologies and strategies for managing the large Vertisols in the central India region at farm and watershed levels (Virmani et al., 1981; Ryan and Sarin, 1981; Kampen, 1982; Virmani and Tandon, 1984). In Kenya, studies on drainage landforms (DLs) for surface water removal including ridge and furrow (RF), broad bed and furrow (BBF), BBF tied (BBFt), cambered bed (CB) and bed with furrow after 5 m (5-MCH) were initiated in Mwea during the 1990s (Ikitoo and Muchena, 1992; Muchena and Ikitoo, 1993).

Results of the initial research work done in Mwea are summarized in Tables 5, 6 and 7. In the research two control treatments, i.e. FB(a) and FB(b), were

Table 4 Statistics on recent paddy rice production on Vertisols in Kenya

	Year						
Location (site)	1997/1998	1998/1999	1990/2000	2000/2001	2001/2002	2002/2003	2003/2004
Mwea irrigation scheme							
Hectares cropped	6000	6052	8617	10,590	6054	15,800	10,000
No. of plot holders	3392	3381	3500	3381	3835	3200	3400
Paddy rice (MT)	21,352	31,876	44,830	45,810	14,802	35,550	46,875
Calculated yield (kg ha^{-1})	3559	5267	5203	4326	2445	2250	4688
Ahero irrigation scheme							
Hectares cropped	268	600	764	690	202	N/A	N/A
No. of plot holders	155	426	449	531	125	N/A	N/A
Paddy rice (MT)	968	1836	1497	1222	880	225	750
Calculated yield (kg ha^{-1})	3612	3060	1959	1771	4356	N/A	N/A
West Kano irrigation scheme							
Hectares cropped	900	663	569	552	N/A	N/A	N/A
No. of plot holders	600	536	703	693	693	N/A	N/A
Paddy rice (MT)	1606	1976	1580	1742	N/A	N/A	N/A
Calculated yield (kg ha^{-1})	1784	2980	2777	3156	N/A	N/A	N/A
Bunyala							
Hectares cropped	213	216	200	206	N/A	N/A	N/A
No. of plot holders	132	132	125	134	134	N/A	N/A
Paddy rice (MT)	728	837	500	491	N/A	N/A	N/A
Calculated yield (kg ha^{-1})	3418	3875	2500	2384	N/A	N/A	N/A
Total annual cultivated area (ha)	7381	7531	10,150	12,038	7136	29,662	10,580
Total no of plot holders	4279	4475	4777	4739	4787	3713	5014
Total annual production (MT)	24,654	36,525	48,407	49,265	18,047	35,775	47,625

Sources: Government of Kenya, Statistical Abstracts (2003), Government of Kenya, Economic survey (2005)
N/A = not available

Table 5 Effect of DLs on the maize grain yield (kg ha^{-1}) in the Mwea Vertisols at P.M. Nzau's farm during the experimental period 1989–1994

Drainage landform	1989[a] LR	1989[a] SR	1990[a] LR	1990[b] SR	1991[b] LR	1991[b] SR	1992[c] LR	1992[c] SR	1993[c] LR	1993[c] SR	1994[c] LR	Mean
Flat bed (FB) (a)	645	1968	459	5136	4664	2565	3517	1674	1564	1397	1480	2279
Flat bed (FB) (b)	–	–	–	–	2210	1640	2919	2002	883	2627	723	1858
Ridge and furrow (RF)	305	2121	453	4206	3545	2287	2258	2641	1265	1745	1932	2069
Broad bed and furrow (BBF)	367	2121	389	4726	4128	2294	3385	2419	1785	1814	1633	2278
Cambered bed (CB)	733	2066	414	4837	2912	2703	3190	2697	1411	1877	1654	2227
Furrow after 5 m wide bed (5-MCH)	497	2082	415	5359	4907	2440	3586	2780	1812	2273	1487	2513
Mean	*509*	*2072*	*426*	*4853*	*3728*	*2322*	*3143*	*2369*	*1453*	*1956*	*1485*	NA
SE ±	110	182	92	573	551	385	212	177	93	148	145	N/A
CV%	26	12	30	17	21	4	14	8	7	8	10	N/A
F-test ($\alpha = 0.05$)	**	NS	NS	NS	**	*	*	*	NS	*	**	N/A

Sources: [a]Ikitoo and Muchena (1992); [b]Muchena and Ikitoo (1993); [c]Ikitoo (Unpublished) PhD. thesis in soil science, Moi University
*symbolizes $\alpha = 0.05$ (5% significance level); **symbolizes $\alpha = 0.01$ (1% significance level)

used. The FB(a), which was located in the main experimental block, tended to behave like the 5-MCH because of significant drainage effect on two of its perimeter sides necessitating inclusion of an isolated control plot, i.e. FB(b) with effect from 1991 LR, for more realistic control treatment effect. Results of the later research done in Mwea and Kitale are summarized in Table 8. In the initial research work,

Table 6 Effect of DLs on the sunflower grain yield (kg ha^{-1}) in the Mwea Vertisols at P.M. Nzau's farm during the experimental period 1989–1994

Drainage landform	1989[a] LR	1989[a] SR	1990[a] LR	1990[b] SR	1991[b] LR	1991[b] SR	1992[c] LR	1992[c] SR	1993[c] LR	1993[c] SR	1994[c] LR	Mean
Flat bed (FB) (a)	1055	1223	500	2152	2316	1668	1482	1133	243	341	758	1170
Flat bed (FB) (b)	–	–	–	–	2071	1654	695	1223	70	180	660	933
Ridge and furrow (RF)	1229	1779	765	2093	2349	1411	1209	1974	320	445	973	1323
Broad bed and furrow (BBF)	906	1540	695	2719	2719	1501	1446	1668	403	563	917	1380
Cambered bed (CB)	1061	1751	681	2460	2766	1626	1788	1877	438	598	1091	1467
Furrow after 5 m wide bed (5-MCH)	1589	1557	689	2152	2427	1529	1487	1758	445	549	980	1378
Mean	*1164*	*1590*	*666*	*2330*	*2441*	*1565*	*1351*	*1606*	*320*	*443*	*897*	*NA*
SE ±	267	201	94	204	220	169	98	80	108	111	54	NA
CV%	32	18	20	12	14	1	7	5	34	25	6	NA
F-test ($\alpha = 0.05$)	NS	NS	NS	NS	*	NS	NS	NS	NS	*	NS	NA

Sources: [a]Ikitoo and Muchena (1992); [b]Muchena and Ikitoo (1993); [c]Ikitoo (Unpublished) PhD thesis in soil science, Moi University
*symbolizes $\alpha = 0.05$ (5% significance level)

Table 7 Effect of drainage landforms on the grain yield of maize and sunflower grown in the Vertisols in Mwea

	Mean grain yield for 11 seasons (1989–1994)			
	Maize grain yield		Sunflower grain yield	
Drainage landform	Yield (kg ha^{-1})	% yield increase above FB	Yield (kg ha^{-1})	% yield increase above FB
Flat bed (FB)	1858	N/A	933	N/A
Ridge and furrow (RF)	2069	11.4	1323	41.8
Broad bed and furrow (BBF)	2278	22.6	1380	47.9
Cambered bed (CB)	2227	19.9	1467	57.2
Beds with furrows 5 m apart (5-MCH)	2493	34.2	1378	47.7
Mean	2185	N/A	1296	N/A
Range	635	N/A	534	N/A
SE	106.2	N/A	93.7	N/A
CV %	10.9	N/A	16.2	N/A

Sources: Ikitoo and Muchena (1992), Muchena and Ikitoo (1993) and Ikitoo (unpublished)
N/A = not applicable

grain yield increases of 11–34% for maize and 42–48% for sunflower due to the application of DLs were observed with the 5-MCH and CB having the best treatment effects in maize and sunflower, respectively. In the later work presented in Table 8, maize grain yield increases of 7–16% and 19–46% due to DLs were observed in Mwea and Kitale, respectively, and the BBFt and 5-MCH were the best DLs in Mwea and Kitale, respectively. Further, DLs significantly increased the total above-ground biomass.

The result of the studies (Tables 5, 6, 7 and 8) indicated that management of soil MC using the DLs removed surface water, thereby improving soil aeration and increasing yield. In the later studies (Sigunga et al., 2002; Ikitoo et al., 2005), it was observed that tying the beds in mid-season for water harvesting in low rainfall seasons increased the yield and by combining the DLs with fertilizer N, further yield increases were realized, indicating synergies between the DLs and fertilizer N.

However, the challenge rests on possibility for scientific breakthrough in developing technological capacity for making right choices in determining when to apply DLs and when to tie the beds. The choices have to be made prior to the onset of the rainy season. This is so because during seasons with high rainfall the DLs have positive effect on plant growth and yield, while in seasons with low rainfall they have negative effect (Ikitoo and Mechena, 1992; Muchena and Ikitoo, 1993; Ikitoo et al., 2005).

Table 8 Effect of drainage landforms on the grain yield and total biomass of maize grown in the Vertisols in Mwea and Kitale

| | Mean grain yield for three seasons during 2001–2003 | | | | | | | |
| | Mwea site (S.M. Kithuku's farm) | | | | Kitale site (ADC Namandala farm) | | | |
Drainage landform	Grain yield (kg ha^{-1})	% yield increase above FB	Total biomass (kg ha^{-1})	% yield increase above FB	Grain yield (kg ha^{-1})	% yield increase above FB	Total biomass (kg ha^{-1})	% yield increase above FB
Flat bed (FB)	1332	N/A	5741	N/A	2278	N/A	7812	N/A
Broad bed and furrow (BBF)	1423	6.8	5515	3.9	2705	18.8	8601	10.1
Broad bed and furrow tied (BBFt)	1551	16.4	5802	1.1	2742	20.4	8770	12.3
Beds with furrows 5 m apart (5-MCH)	1455	9.2	5962	3.9	3332	46.3	9782	25.2
Mean	*1440*	*N/A*	*5755*	*N/A*	*2764*	*N/A*	*8741*	*N/A*
Range	219	N/A	447	N/A	1054	N/A	1970	N/A
SE	45.1	N/A	92.6	N/A	216.6	N/A	404.9	N/A
CV %	6.3	N/A	3.2	N/A	15.7	N/A	9264	N/A

Sources: Ikitoo et al. (2005); Ikitoo, E.C (2008) – PhD thesis in soil science, Moi University
N/A = not applicable

Opportunities for Sustainable Land Management in Vertisols: The Way Forward

Sustainable land productivity in Vertisols depends on how the soils are managed with respect to *soil quality*, which is defined as the capacity of a soil to function within its ecosystem boundaries and interact positively with the environment external to that system (Larson and Pierce, 1991). Based on this definition, *soil quality* is a key factor in four major objectives of sustainable land management, namely agronomic, ecological, micro-economic and macro-economic sustainability (Lourance, 1990). Larson and Pierce (1991) highlighted and amply discussed the elements and indicators of *soil quality*, on which premise, soil management systems are sustainable only if they maintain or enhance the *soil quality*. Practically, it is necessary to develop and adopt suitable technologies for sustainable land management to enhance productivity. Thus, some important aspects with respect to sustainable land management in Vertisols are considered below:

(a) Crop germination

Given the high wilting point soil *moisture content* (MC), i.e. 0.25–0.50 cm^3 cm^{-3}, in Vertisols (Probert et al., 1987) and highly erratic initial rainfall, i.e. ≤ 20 to ≥ 400 mm month^{-1}, in the ASALs of Kenya where most of the Vertisols in the country occur (Brown, 1975; Stewart and Wang'ati, 1981; Van de

Weg, 1987 and Stewart, 1988), seed germination in the Vertisols often is a major problem. Poor germination of early planted crop was observed in the Mwea Vertisols due to excessive soil MC (Ikitoo and Mechena, 1992; Muchena and Ikitoo, 1993) while in the Kitale Vertisols, low germination percentage resulted from soil MC deficit (Ikitoo et al., 2005; Ikitoo, 2008). Further, when rainfall was poorly distributed, cumulative precipitation of 200 mm that was sufficient to germinate and sustain maize growth on lighter textured soils, i.e. Andisols and Cambisols, in the Kitale/Endebess area was inadequate to germinate maize seed in the Vertisols at the ADC Namandala farm, which required over 300 mm of rainfall (Ikitoo, 2008). In India, mean monthly and annual rainfall coefficients of variation of 39–225 and 32%, respectively, were recorded in Vertisol areas (Kalyal et al., 1987). Thus, development of technologies that improve drainage at the time of planting and conserve soil MC under conditions of moisture deficit, while minimizing loses in the Vertisols, is essential to ensure high germination percentage.

(b) Soil fertility management

- *Tillage and soil quality:* Tillage is necessary in Vertisols; however, continuous tillage or cultivation results in soil organic matter (SOM) decline, deterioration of soil structure and loss of nutrients (Woomer et al., 1994; Brown et al., 1994; Sanchez et al., 1997; Mugabe et al., 2002; Elias et al., 2002; Jayasree and Rao, 2002; Hullugale et al., 2002 and

Knowles and Singh, 2003). In India, studies on thin sections of Vertisols indicated that continuous cultivation resulted in soil structural deterioration through changes in pore size distribution (Jayasree and Rao, 2002) while in Cameroon, similar studies on soil structure indicated that macro-aggregates from ploughed and cropped Vertisols collapsed in one step into semi-liquefied micro-aggregates and primary particles, whereas macro-aggregates from fallow and zero-tillage plots collapsed in a step-wise manner (Obale-Ebanga et al., 2003). In Queensland, Australia, zero tillage and no-traffic were observed to minimize environmental degradation effects including soil erosion while increasing grain yield. Further, the amount of rain required to initiate run-off in irrigated cotton was increased by retaining crop residues and other materials that increased soil cover while perennial pasture rotations improved soil fertility (Silburn and Glanville, 2002; Armstrong et al., 2003).

■ *Land forming for surface water drainage:* Poor drainage is a major problem in Vertisols due to water logging of early planted crops; thus, universally, Vertisols are cropped post-rain season with plant growth being sustained largely by residual soil MC. The use of DLs to remove surface water facilitates early planting and enhances water use efficiency while use of reservoirs to store the surplus water provides source of irrigation water. In India and Ethiopia, use of low-cost broad bed and furrow (BBF) DL systems increased crop yields significantly (Virmani et al., 1981; Ryan and Sarin, 1981; Kampen, 1982; (Probert et al., 1987; Gebre et al., 1988; Ahmad, 1989; Reeves, 1990; Astatke et al., 2002; Gupta, 2002; Yadav et al., 2003).

In Kenya, studies on application of DLs to facilitate early and timely planting indicated that DLs significantly increased yield (Njihia, 1984; Ikitoo, 1989; Ikitoo and Muchena, 1992; Muchena and Ikitoo, 1993; KARI-FURP, 1994; Sigunga et al., 2002; Tables 5, 6, 7 and 8). Thus, as more Vertisols are put to arable cropping in the country, priority should be given to use of management systems that increase productivity while minimizing loss of water, nutrients and soil (Virmani et al., 1981; Probert et al., 1987; Ahmad, 1989). Minimum tillage, crop rotation and other cropping systems that improve drainage while minimizing soil erosion should be applied concurrently.

However, lack of a suitable land forming implement for the smallholder farmers is one of the main limiting factors to adoption of the DL technology. An ox-drawn ridger has been developed for land forming in the Vertisols; however, the ridger has not been tested extensively on farm (Muckle, 2002). Thus, for successful adoption of the DL technology, the ridger and the DL management system have to be tested extensively at farm and catchments levels.

■ *Application of fertilizers N and P:* Vertisols are universally low in N (Katyal et al., 1987; Probert et al., 1987; Ahmad, 1989; FAO, 2002) and variable in P (Ahmad, 1989; Guo et al., 2000; Singh et al., 2003). Thus, significant responses to fertilizer N have been observed in many crops grown in Vertisols in India and Ethiopia including maize, wheat, cotton, sorghum, pigeon peas, soybeans and beans (Badanur and Deshpande, 1987; Burford, 1987; Katyal et al., 1987; Bansal et al., 1988; Erkossa et al., 2000; Tarekegne et al., 2000; Patil et al., 2001; Nalatwadmath et al., 2003). In Kenya, application of fertilizer N significantly increased crop yields of both rain-fed and irrigated crops (Njihia, 1984; KARI-FURP, 1994; Kondo et al., 2001). However, better yield responses were observed when the fertilizer N was combined with DLs (Sigunga, 1997; Sigunga et al., 2002; Ikitoo et al., 2005; Ikitoo, 2008). P availability in Vertisols depends on the soil sorption capacity, sorption/desorption rate and amount of P in the soil (Probert et al., 1987; Wild, 1988 and Chaudhary et al., 2003). Responses were observed in wheat, maize, soybeans and pigeon peas and were better when the P was combined with N, manure and P-solubilizing microorganisms (cooper et al., 1989; Ae et al., 1993; Dubey, 2000; Babhulkar et al., 2000; Rao and Rupa, 2000; Tiwari and Kulmi, 2005). There is need to undertake further studies on the use of fertilizer and manure in relation to cropping systems in Kenya.

(c) Cropping systems

Studies in the USA, Australia and Ethiopia indicated that by combining zero, minimum and conservation tillage with rotation cropping improved long-term soil quality and environmental protection including soil erosion control and reduction in land degradation (Lopez-Bellido et al., 1996; Chan et al., 1997; Castano et al., 2000; Nicolova et al., 2001; Morrison and

Sanabria, 2002; Guled et al., 2002; Salinas-Garcia et al., 2002; Gupta, 2002; Astatke et al., 2003; Armstrong et al., 2003). Studies on stocking rates in Vertisols in the Ethiopian highlands, conducted between 1996 and 2000, indicated that moderate grazing pressure (stocking rate of 1.8 animal unit months (AUM) ha^{-1}) was better than heavy grazing pressure (stocking rate of 3.0 AUM ha^{-1}) and removal of grass cover enhanced cracking of the soils. The effect of livestock trample was also higher in the heavily grazed plots (3.0 AUM ha^{-1}) than in the non-grazed plots with the infiltration rate and soil moisture content being low in the heavily grazed plots. Low to moderately grazed plots had minimum negative effects on the soil hydrological conditions (Taddese et al., 2002). In southern Kenya, studies on stocking rates showed that lighter density stocking (4 heifers ha^{-1}) maintained the highest output per head while heavy stocking density (16 heifers ha^{-1}) resulted in the highest soil loss, lowest infiltration rate and highest bulk density (Mworia et al., 1996). In Tasmania, Australia, studies of long-term management systems indicated that pasture paddocks developed stronger soil structures and smaller aggregates than cropped paddocks, which had larger clods (Cotching et al., 2002). Where livestock and arable crops co-exist, there is need to balance livestock numbers, pasture and cropped lands in order to maintain and/or enhance the soil physical properties (Mworia et al., 1996; Taddese et al., 2002). In a crop–animal system in Southeast Asia, in which agriculture is characterized by complex production systems in the humid and sub-humid climate with forests and grasslands, land degradation is a major threat to sustainability of the farming systems (Devendra and Thomas, 2002). In the system, as in many others, development of management and cropping systems that enhance and maintain land productivity is essential. Studies in Australia indicated that the best land quality could be achieved from low to medium stocking rates combined with field resting for 4 or 12 weeks, based on a *sustainability index* from data on animal production, pasture, soil health and soil water (Lodge et al., 2003).

Conclusion

In conclusion, we wish to state as follows with regard to the development of sustainable land management systems on Vertisols in Kenya. With rapidly growing population, there is increasing emigration of people from humid high potential areas to the low potential ASALs; associated with increasing use of fragile soils and unsuitable lands for agricultural production. There is urgent need for the government to formulate a comprehensive land use and tenure policy that encourages development and establishment of sustainable land management systems. This is essential for economical productivity of the land and conservation of the environment. In view of the various management constraints to arable cropping of the Vertisols in Kenya, there is need to undertake further studies on the soils in order to develop the required technologies for sustainable land use in the Vertisols in the country.

Acknowledgement I would like to acknowledge the Director, KARI, and the Agricultural Research Fund (ARF) for providing funds for the drainage research work. Further, I wish to acknowledge Moi University and my co-authors Prof. J.R. Okalebo and Prof. C.O. Othieno for providing research facilities for analytical work and for critical evaluation of materials for part of this chapter.

References

Ae N, Arihara J, Okada K, Yoshihara T, Otani T, Johansen C, Randall PJ, Delhaize E, Richards RA, Munns R (1993) The role of piscidic acid secreted by pigeon pea roots grown in an Alfisol with low P fertility: genetic aspects of plant mineral nutrition. The fourth international symposium on genetic aspects of plant mineral nutrition, Canberra, Australia, pp 279–288, 30th Sept–4th Oct 1991

Ahmad N (1989) Management of Tropical Vertisols. In: Ahn PM, Elliot CR (eds) Vertisol management in Africa. IBSRAM proceedings No.9, Bangkok, Thailand, pp 29–62

Alegre JC, Cassel DK (1994) Soil Physical Dynamics under Slash-and-Bun Systems. In: Sanchez PA, Van Houten H (eds) Alternatives to slash-and-burn agriculture. 15th international soil science Congress, Acapulco, Mexico, pp 47–61

Armstrong RD, Millar G, Halpin NV, Reid DJ, Standley J (2003) Using zero tillage, fertilizers and legume rotations to maintain productivity and soil fertility in opportunity cropping systems on a shallow Vertisol. Aust J Exp Agric 43(2):141–153

Astatke A, Mohammad J, Mohammad-Saleem MA, Erkossa T (2002) Development and testing of low cost animal drawn minimum tillage implements: experience on Vertisols in Ethiopia. AMA, Agric Mechan Asia Afr Latin Am 33(2): 9–14

Babhulkar PS, Wandile RM, Badole WP, Balpande SS (2000) Residual effect of long-term application of FYM and fertilizers on soil properties (Vertisols) and yield of soybean. J Indian Soc Soil Sci 48(1):89–92

Badanur VP, Deshpande PB (1987) Soil moisture and nutrient interaction effects on yield and nutrient uptake in rabi sorghum grown on a Vertisol. J Indian Soc Soil Sci 35(3):404–411

Bansal RK, Kshirsagar KG, Sangle RD (1988) Efficient utilization of energy with an improved farming system for selected semi-arid tropics. Agric Ecosyst Environ (Netherlands) 24(4):381–394

Baumer M (1987) Le role possible de l'agroforesterie dans la lutte contre la desertification at al degradation de l'environnement. Technical Centre for Agriculture and Rural Cooperation, ACP-EEC Lome Convention, Wageningen, The Netherlands, CTA: 16

Beyene D, Regassa H (1989) Research work on the management of vertisols in Ethiopia: experience of the Institute of Agricultural Research. In: Ahn PM, Elliot CR (eds) Vertisol management in Africa. IBSRAM proceedings No. 9, Bangkok, Thailand, pp 12–21

Brown KJ (1975)Rainfall reliability in cotton in Central Kenya. Cotton Grow Rev 52:38–45

Brown S, Anderson JM, Woomer PL, Swift MJ, Barrios E (1994) Soil biological processes in tropical ecosystems. In: Woomer PL, Swift MJ (eds) The biological management of tropical soil fertility. Wiley, New York, NY, pp 15–46

Burford JR (1987)Strategies for maintenance of soil fertility.. In: Latham M, Ahn P (eds) Management of Vertisols under Semi-Arid Conditions. Proceedings of the First regional seminar on management of Vertisols under semi-arid conditions, Nairobi, Kenya 1986. IBSRAM Proceedings No. 6, Thailand, pp 311–323

Castano CA, Herrera GO, Madero MEE (2000) Effect of tillage on some physical properties of an ustert and on corn yield. Acta Agronomica, Universidad Nacional de Colombia 50: 1–2, 48–58

Chan KY, Bowman AM, Friend JJ (1997) Restoration of soil fertility of degraded Vertisols using a pasture including a native grass (Astrebla lappacea). Trop Grasslands 31(2): 145–155

Chaudhary EH, Ranjha AM, Gill MA, Mehdi SM (2003) Phosphorus requirements of maize in relation to soil characteristics. Int J Agric Biol 5(4):625–629

Cooper PJM, Keatinge JDH, Kukula S (1989) Influence of environment on the management and productivity of cereals on a Vertisol at Jindiress, Syria. Management of Vertisols for improved agricultural production. In: Proceedings of IBSRAM workshop, India, pp 195–211, 18–22 Feb 1985

Cotching WE, Cooper J, Sparrow LA, McCorkell BE, Rowley W, Hawkins K (2002) Effect of agricultural management on vertisols in Tasmania. Austr J Soils Res 40(8):1267–1286

Coughlan KJ, Mc Garry D, Smith GD (1987) The physical and mechanical characterization of Vertisols. In: Latham M, Ahn P (eds) Management of Vertisols under semi-arid conditions' – proceedings of the first regional seminar on management of Vertisols under semi-arid conditions, Nairobi, Kenya 1986. IBSRAM proceedings No. 6, Thailand, pp 81–105

Da Costa VPFX (1973) Characterization and interpretation of the soils of the Kano plains for irrigation agriculture. Thesis submitted in partial fulfillment of the requirements for the degree of MSc (Agriculture) in the University of East Africa, 1973, Nairobi/Kampala

Devandra C, Thomas D (2002) Crop-animal systems in Asia: importance of livestock and characterization of agroecological zones. Agric Syst 71(1/2):5–15

Donahue RL, Miller RW, Schickluna JC (1977) Soils – an introduction to soils and plant growth. Prentice-Hall, New Jersey

Dubey SK (2000) Effectiveness of rock phosphate and super-phosphate amended with phosphate solubilizing microorganisms in soybean grown on Vertisols. J Indian Soc Soil Sci 48(1):71–75

Dudal R (1963) Dark clay soils of tropical and subtropical regions. Soil Sci 95:264

Elias AE, Alaily F, Siewert C (2002) Characteristics of organic matter in selected profiles from the Gezira Vertisols as determined by thermo-gravimetry. Int Agro-Phys 16(4):269–275

Erkossa T, Mamo T, Kidane S, Abebe M (2000) Response of some durum wheat landraces to nitrogen application on Ethiopian Vertisols. The eleventh regional wheat workshop for Eastern, Central and Southern Africa, Addis Ababa, Ethiopia, pp 229–238, 18–22 Sept 2000

FAO (1965) Dark clay soils of tropical and Sub-tropical regions. In: Dudal R (ed) FAO agricultural development paper No. 83. FAO, Rome, 161 pp

FAO (1984) Guidelines: land evaluation for rain-fed agriculture. FAO soils bulletin No 52. FAO and Agriculture Organization of the United Nations, Rome, pp 137–156

FAO (2002) Lecture notes on the major soils of the world, Vertisols. In: Driessen P, Deckers J, Spaargaren O, Nachtergaele F (eds) Major soils of the world. World Soil Resource report No. 94: FAO. FAO, Rome, pp 73–87, ISSN 0532-0488

Gebre H, Jutzi SC, Haque I, McIntire J, Stares JES (1988) Crop agronomy research on Vertisols in the central highlands of Ethiopia: IAR's experience. Management of Vertisols in Sub-Saharan Africa. In: Proceedings of a conference held at ILCA, Addis Ababa, Ethiopia, pp 321–334, 31 Aug–4 Sept 1987

Guled MB, Gundlur SS, Hiremath KA, Yarnal RS, Surakod VS (2002) Impact of different soil management practices on run-off and soil loss in Vertisols. Karnataka J Agric Sci 15(3):518–524

Guo F, Yost RS, Hue NV, Evensen CL, Silva JA (2000) Changes in phosphorus fractions in soils under intensive plant growth. Soil Sci Soc Am J 64(5):1681–1689

Gupta RK (2002) Natural resource conservation technologies for black clay soil region of peninsular India. J Indian Soc Soil Sci 50(4):438–447

Herlocker D (ed) (1999) Rangeland Resources in eastern Africa: their ecology and development. German Technical Cooperation (GTZ), Nairobi, Kenya

Hubble GD (1984) The cracking clay soils: definition, distribution, nature, genesis and use. In: McGarity W, Hoult EH, So HB (eds) The properties and utilization of cracking clay soils. Reviews in Rural Science, 5. University of New England, Armidale, pp 3–13

Hulugalle NR, Entwistle PC, Weaver TB, Finlay LA (2002) Cotton- based rotation systems on a sodic Vertisol under irrigation: effects on soil quality and profitability. Aust J Exp Agric 42(3):341–349

Ikitoo EC (1989) Some properties of Vertisols in Kenya and their current level of management for crop production.

In: Ahn PM, Elliot CR (eds) Vertisol management in Africa. IBSRAM Proceedings No.9, Bangkok, Thailand, pp 194–208

Ikitoo EC (2008) The influence of surface water management and fertilizer use on growth and yield of maize (*Zea mays* L.) in vertisols of Kenya. Theses submitted to the school of agriculture and biotechnology, Moi University in partial fulfillment of the requirements for the award of a Doctor of Philosophy (D. Phil) in Soil Science

Ikitoo EC, Muchena FN (1992) Management of Vertisols in semi-arid areas of Kenya: a case study in the Mwea area. Reports and papers on the management of Vertisols – ABRAM/AFRICALAND. International Board for Soils Research and Development, Network Document No. 1:97–115

Ikitoo EC, Othieno CO, Okalebo JR (2005) Management of vertisols in semi-arid areas of Kenya: project over view of results. In: Proceedings of the KARI/DIFID end of project workshop, Kenya Agricultural Research Institute, Nairobi, Aug 2003

ILACO BV (1974) Irrigation station Ahero, Kenya: general report. A report prepared for the Food and Agriculture Organization of the United Nations (acting as executing agency for the United Nations Development Program). ILACO B.V. International Land Development Consultants, Arnhem, The Netherlands

ISSS/ISRIC/FAO (1994) World reference base for soils resources. International Society of Soil Science, International Soils Reference and Information Centre and the Food and Agriculture Organization of the United Nations. Wageningen/Rome, pp 11–110

Jayasree G, Rao YN (2002) Effect of different long-term management practices on micro-structure of Vertisols and Alfisols. J Indian Soc Soil Sci 50(4):452–456

Juo ASR, Manu A (1994) Chemical Dynamics in Slash-and-Burn Agriculture. In: Sanchez PA, Van Houten H (eds) Alternatives to slash-and-burn agriculture. 15th international Soil Science Congress, Acapulco, Mexico, pp 62–76

Kampen J (1982) An approach to improved productivity on deep Vertisols. Information Bulletin No. 11.. International Crops Research Institute for the S-semi-Arid tropics, Patancheru, India

KARI-FURP (1994) Fertilizer use recommendations. Vol. 8-Kisii District, Vol. 12-Siaya District, Vol. 13-Kisumu District, Vol. 14-South Nyanza District, Vol. 15-Kericho District, Vol. 18-Northern Rift Valley districts (West Pokot, Marakwet, Baringo),vol. 22, Kirinyaga District. Kenya Agricultural Research Institute, NARL, Nairobi, Kenya

Katyal JC, Hong CW, Vlek PLG (1987)Fertilizer management in Vertisols.. In: Latham M, Ahn P (eds) Management of Vertisols under semi-arid conditions. Proceedings of the first regional seminar on management of Vertisols under semi-arid conditions, Nairobi, Kenya 1986. IBSRAM Proceedings No. 6, Bangkok, Thailand, pp 247–266

Knowles TA, Singh B (2003) Carbon storage in cotton soils of northern New South Wales. Austr J Soil Res 41(5): 889–903

Kondo M, Oto T, Wanjogu R (2001) Physical and chemical properties of Vertisols and soil nutrient management for intensive rice cultivation in the Mwea area in Kenya. Jpn J Trop Agric 45(2):126–132

Larson WE, Pierce FJ (1991) Conservation and enhancement of soil quality. In: Evaluation for sustainable land management in the developing world. Volume 2: Technical Papers. Int Board Soils Res Manage – IBSRAM Proc 12(2):175–203

Lodge GM, Murphy SR, Harden S (2003) Effects of grazing and management on herbage mass, persistence, animal production and soil water content of native pastures. 2. A mixed native pasture, Manilla, North West Slopes, New South Wales. Austr J Exp Agric 43(7/8):891–905

Lopez-Bellido L, Fentes M, Castillo JE, Lopez-Garrido FJM, Fernandez EJ (1996) Long-term tillage, crop rotation, and nitrogen fertilizer effects on wheat yield under rain-fed Mediterranean conditions. Agron J 88(5):783–791

Lourance R (1990) Research approaches for ecological sustainability. J Soil Water Conserv 45:51–54

Mohanty M, Painuli DK (2003) Land preparatory tillage effect on soil physical environment and growth and yield of rice in a Vertisol. J Indian Soc Soil Sci 51(3):217–222

Morrison JE, Sanabria J (2002) One-pass and two-pass spring tillage for conservation row-cropping in adhesive clay soils. Trans ASAE 45(5):1263–1270

Muchena FN, Gachene CKK (1985) Properties, management and classification of Vertisols in Kenya. Fifth meeting of the Eastern Africa Sub-committee for Soil Correlation and Land Evaluation, Wadi Medani, Sudan, 1983. World Soils Resources Reports No. 56, FAO, Rome, pp 23–30

Muchena FN, Ikitoo EC (1993) Management of Vertisols in semi-arid areas of Kenya: the effect of improved methods of surplus/surface water drainage on crop performance. Report of the 1992 Annual meeting on AFRICALAND Management of Vertisols in Africa, 11–13 June 1992, Accra, Ghana. Network Document No. 3. IBSRAM, Bangkok, Thailand

Muckle TB (2002) Selection and testing of potentially viable implements for tillage/drainage of Vertisols and early planting (Activity 2.1.7/8). Final Report of Consultancy Workplan 2001–02. Agriculture/Livestock Research Support Programme, Phase II (ARSP II), Land and Water Management Research Programme. KARI, Nairobi

Mugabe FT, Nyamangara J, Mushiri SM, Nyamudeza P, Kamba E (2002) Sustainability of the current crop production system on the Chisumbanje Vertisols in South-East Zimbabwe. J Sustain Agric 20(3):5–19

Mworia JK, Musimba NKR, Orodho AB (1996) Vegetation and hydrological responses to stocking density in a semi-arid rangeland in Kenya. Focus on agricultural development in a changing economic environment: proceedings of the 5th KARI scientific conference, KARI Headquarters, Nairobi, pp 231–242, 14th–16th Oct 1996

Nalatwadmath SK, Rao MSRM, Patil SL, Jayaram NS, Bhola SN, Arjun P (2003) Long-term effects of integrated nutrient management on crop yields and soil fertility status in Vertisols of Bellary. Indian J Agric Res 37(1): 64–67

Nikolova D, Borisova M, Mitova T, Dimitrov I (2001) Influence of the different soil, tillage systems and fertilization on productivity and extraction of nutrient elements in three field crop rotations. Pochvoznanie Agrokhimiya I Ekologiya 36(4–66):231–233

Njihia CM (1984) Causative factors of rice yield variation and decline at Mwea irrigation settlement, Kenya. Irrigation and Drainage Research Project, Ministry of Agriculture and Livestock Development, IDRP-NAL, Nairobi, Kenya

Obale-Ebanga F, Sevink J, de Groot W, Nolte C (2003) Myths of slash and burn on physical degradation of savannah soils: impact on Vertisols in northern Cameroon. Soils Use Manage 19(1):83–86

Patil SI, Rama-Mohan-Rao MS, Nalatwadmath SK, Manamohan S (2001) Moisture, plant population and nitrogen doses and their interactions on nutrient availability and production of rabi sorghum in the Vertisols of Deccan Plateau. Indian J Dryland Agric Res Dev 16(1):9–15

Probert ME, Fergus IF, Bridge BJ, McGarry D, Thompson CH, Russell JS (1987) The properties and management of Vertisols. CAB International, Wallingford, UK

Rao AS, Rupa TR (2000) Effects of continuous use of cattle manure and fertilizer phosphorus on crop yields and soil organic phosphorus in a Vertisol. Biosource Technol 75(2):113–118

Reeves J (1990) ILCA develops technology to improve Vertisol productivity. Entwicklung-Und landlicher-Raun (Germany) 24(5):28–29

RoK Economic Surveys (1996) Republic of Kenya economic survey report 1995. Central Bureau of Statistics, Office of the Vice President and Ministry of planning and National Development, Nairobi, pp 117–130

RoK Economic Surveys (2005) Republic of Kenya economic survey report 2004. Central Bureau of Statistics, Ministry of planning and National Development, Nairobi, Kenya, pp 142–156

RoK Statistical Abstracts (2004) Republic of Kenya statistical abstracts for 2003. Central Bureau of Statistics, Ministry of planning and National Development, Nairobi, pp 123–132

Ryan JG, Sarin R (1981) Economics of technology options for vertisols in the relatively dependable rainfall regions of the Indian semi-arid tropics. In: Improving the management of India's deep blacks soils. Proceedings of the seminar on management of deep black soils for increasing production of cereals, pulses, and oilseeds. ICRISAT, Patancheru, New Delhi, 21 May 1981

Salinas-Garcia JR, Velazquez-Garcia J, Gallardo-Valdez M de J, Diaz-Mederos P, Caballero-Hernandez F, Tapia-Vargas LM, Rosales-Robles E, de J, Velazquez-Garcia J, Franzluebbers A (2002) Tillage effects on microbial biomass and nutrient distribution in soils under rain-fed corn production in central-western Mexico: conservation tillage and stratification of soil properties. 15th Meeting of the International Soil Tillage Research Organization, Fort Worth, Texas, USA, July 2000. Soil Tillage Res 66(2):143–152

Sanchez PA, Shepherd KD, Soule MJ, Place FM, Buresh RJ, Izac AN, Mo'Kivunye AU, Kwesiga FR, Nderitu CG, Woomer PL (1997) Soil fertility replenish in Africa. An investment in natural resource capital: replenishing soil fertility in Africa. SSSA Special Publication No.51: Soil Science Society of America and American Society of Agronomy, Madison, WI, pp 1–46

Santana R (1989) African Vertisols: their characteristics, extent, and some specific examples of development. In: Ahn PM, Elliot CR (eds) Vertisol management in Africa. IBSRAM Proceedings No 9, Bangkok, Thailand, 19–28

Scholes MC, Swift RMJ, Heal OW, Sanchez POA, Ingram JSL, Dalal R (1994) Soil fertility research in response to the demand for sustainability. In: Woomer PL, Swift MJ (eds) The biological management of tropical soil fertility. Wiley, New York, NY, pp 1–14

Siderious W, Njeru EB (1976) Soils of Trans Nzoia District. Site Evaluation Report No 28, July 1976. KSS, National Agricultural Laboratories, Ministry of Agriculture, Nairobi, Kenya

Sigunga DO (1997) Fertilizer nitrogen use efficiency and nutrient uptake by maize (Zea Mays L.) in Vertisols in Kenya. Theses submitted to the University of Wageningen for the award of a PhD

Sigunga D, Janssen BH, Oenema O (2002) Effect of improved drainage and nitrogen source on yields, nutrient up-take and utilization efficiencies by maize (Zea mays, L.) on Vertisols in sub-humid environments. Nutr Cycling Agroecosyst 62(3):263–275

Silburn DM, Glanville SF (2002) Management practices for control of runoff losses from cotton furrows under storm rainfall. 1. Runoff and sediment on a black Vertisol. Austr J Soil Res 40(1):1–20

Singh SK, Baser BL, Shyampura RL, Narain P (2003) Phosphorus fractions and their relationship to weathering indices in Vertisols. J Indian Soc Soil Sci 51(3):247–251

Sombroek WG, Braun HMH, Van de Pour BJA (1982) Exploratory soil map and Agro-climatic zone map of Kenya, 1980, scale 1:1,000,000. Exploratory Soil Survey Report No E1, Kenya Soil Survey. NAL, Ministry of Agriculture, Kenya

Stewart JI (1988) Development of management strategies for minimizing the impact of seasonal rainfall variations. Drought Research Priorities for the Dryland Tropics. ACRISAT, Patancheru, Andhra Pradesh, India, pp 131–150

Stewart JI, Wang'ati FJ (1981) Research on crop water use and drought responses in East Africa. Agroclimatological Research Needs of the Semi-Arid Tropics. In: Proceedings of the international workshop on the agroclimatological research needs of the semi-arid tropics, 22–24 Nov 1978, Hyderabad, India. ACRISAT, Patancheru, India, pp 170–180

Taddese G, Mohamed Saleem MA, Ayalneh W (2002) Effect of livestock grazing on physical properties of a cracking and self mulching Vertisol. Austr J Exp Agric 42(2):129–133

Tarekegne A, Tanner DG, Tessema T, Mandefro C (2000) Agronomic and economic evaluation of on-farm N and P response of bread wheat grown on two contrasting soil types in central Ethiopia. The eleventh regional wheat workshop for eastern, Central and Southern Africa, Addis-Ababa, Ethiopia, 18–22 Sept 2000, pp 239–252

Tiwari PN, Kulmi GS (2005) Effect of biofertilizers, organic and inorganic nutrition on growth and yield of isabgol (Plantago ovvata Forsk. Res Crops 6(3):568–571

Van de Pour BJA, Van Engelen VWP (1983) Soils and vegetation of the Amboseli-Kibwezi area. Reconnaissance Soil Survey Report No R6, Kenya Soil Survey, Nairobi, Kenya

Van de Weg RF (1987) Vertisols in Eastern Africa. In: Latham M, Ahn P (eds) Management of Vertisols under semi-arid conditions'. Proceedings of the first regional seminar on management of Vertisols under semi-arid conditions, Nairobi, Kenya 1986. IBSRAM Proceedings No. 6, Thailand, pp 45–50

Van de Weg RF, Mbuvi JP (1975) Soils of the Kindaruma area (Quarter Degree Sheet 136). Reconnaissance Soil Survey Report No. R1. Republic of Kenya, Ministry of Agriculture, National Agricultural Laboratories (Kenya Soil Survey), Nairobi, Kenya

Van Gassel JM (1982) Irrigation and drainage research project: Mwea water use study (water management part). Republic of Kenya, Ministry of Agriculture, SRD and Kingdom of the Netherlands, Ministry of Foreign Affairs, Department of International Technical Assistance, Nairobi, Kenya

Virmani SM, Tandon HLS (1984) Watershed – based dryland farming in black and red soils of peninsular India. In: Proceedings of a Workshop held on 3–4 October 1983 at ICRISAT Centre, Patancheru, India, pp 27–47

Virmani SM, Willey RW, Reddy MS (1981) Problems, prospects and technology for increasing cereal and pulse production from deep black soils. Improving the Management of India's Deep Black Soils. ICRISAT, Patancheru, India, pp 21–36

Wanyoike MM, Wahome RG (2004) Research and development. Cattle production in Kenya. Workshop proceedings, KARI Headquarters, 15–16 December, 2003. KARI/EU, Nairobi, Kenya

Wild A (1988) Russell's soil conditions and plant growth, 11th edn. Longman Scientific and Technical, Longman Group, England, pp 608–625, 663

Woomer PL, Martin A, Albrecht A, Resch DVS, Scharpenseel HW (1994) The importance of management of soil organic matter in the tropics. In: Woomer PL, Swift MJ (eds) The biological management of tropical soil fertility. Wiley, New York, NY, pp 47–80

Yadav AL, Goswami B, Aggarwal P, Arya M (2003) Few important soil management technologies for sustainable crop production – a review. Agric Rev 24(3):211–216

Potential Nitrogen Contribution of Climbing Bean to Subsequent Maize Crop in Rotation in South Kivu Province of Democratic Republic of Congo

L. Lunze and M. Ngongo

Abstract Nitrogen has become the most limiting nutrient in the Eastern Highlands of DR Congo characterized by high-potential soils on volcanic deposits. However, because of high population density which imposes intensive and continuous cropping without external inputs, smallholder farmers are facing declining land productivity in the region. Among low-input technologies to increase production, climbing bean is promoted and accepted by farmers because of its high production potential. Besides, this bean type is perceived as a potential crop to contribute to soil fertility and sustainable cropping system through its high biomass production and N-fixing capacity. On-station and on-farm farmer participatory trials were conducted to assess the beneficial effects of the climbing bean on the subsequent maize crop in rotation compared to bush bean and continuous maize cropping systems. Maize response to mineral nitrogen applied at rates of 0, 33, and 66 kg N ha^{-1} was evaluated in three different rotations, climbing bean–maize, bush bean–maize, and continuous maize, over three cropping seasons. Maize grain yield was generally higher in climbing bean–maize rotation compared to bush bean–maize and continuous maize cropping system. Without applied nitrogen fertilizer, the average maize grain yield increase over three cropping seasons in response to the preceding climbing bean effect was 489 and 812 kg ha^{-1} compared to bush bean and maize as preceding crops, respectively, which is 17.5 and 33.8% increase. However, better yield advantage of climbing bean over continuous maize was obtained in the long-rain cropping season, 43.2% compared to 24.2% in short-rain season. The residual effect measured that the second season following climbing bean was even better, and the yield difference was 1,300 kg ha^{-1}, i.e., 81% yield increase. Nitrogen contribution from the climbing bean to the system estimated as N fertilizer replacement values varied from 15 to 42 kg N ha^{-1} in the first season of rotation. The potential of climbing bean to improve soil fertility was confirmed by farmers' evaluation in on-farm trials.

Keywords Climbing bean · Maize · Rotation · Nitrogen · Soil fertility

Introduction

Kivu province in eastern DR Congo is characterized by relatively high-fertility soils of recent volcanic origin. However, because of the high population density, farmers are faced with rapid soil fertility decline as a result of continuous cropping and inappropriate cropping systems with very little or no external input to replenish soil fertility. Farmers are generally poor without resources to purchase external inputs to replenish soil fertility. In such farming conditions, N has become the most limiting nutrient in food production. Among technologies to increase production, climbing bean was promoted in the Great Lakes Region with the objective to intensify productivity and has been well accepted by farmers and is readily adopted by most resource-poor farmers (Sperling et al., 1994).

L. Lunze (✉)
Centre de Recherche de Mulungu, INERA, D.S. Bukavu, D.R. Congo
e-mail: llunze@yahoo.fr

A. Bationo et al. (eds.), *Innovations as Key to the Green Revolution in Africa*, DOI 10.1007/978-90-481-2543-2_68, © Springer Science+Business Media B.V. 2011

Climbing bean is by far the bean type with high biomass production and probably high N fixation capacity; therefore, considerable green manure effect is expected, and its integration in the production system can improve productivity and contribute to sustain crop productivity. In fact, climbing bean develops extensive nodulation three times more than bush bean (Van Schoonhoven and Pastor-Corrales, 1994), which is an indication of higher capacity for N fixation. It has been suggested that the contribution from N-fixing legume in rotation was responsible for most of the beneficial rotation effect to succeeding corn in rotation (Sanginga, 2003; Bado, 2002; Baldock et al., 1981). The evidence of net positive soil N balance by climbing bean has been reported (Kumarasinghe et al., 1992). In this study, they reported that at the late pod-filling stage, the climbing bean had accumulated 119 kg N ha^{-1}, 84% being derived from fixation, 16% from soil, and only 0.2% from the ^{15}N fertilizer. Wortmann (2001) reports in a long-term experiment improved sorghum yield in rotation with climbing bean and an estimate of nitrogen derived from the atmosphere to be 40–57% of plant N depending on the estimation method used. It is clear that evidence of benefit of climbing bean cultivation exists, either as rotational effects or as improved nitrogen nutrition. There is necessity of rational use of this potential to develop farming practices that are economically viable.

Although climbing bean is being extensively promoted in the potential regions to intensify productivity, the benefit of other high-yield potential has been very little studied and exploited. The quantity of nitrogen fixed in climbing bean-based systems, green manure effects, and N nutrition effect in rotation and even soil conservation effect is not well known. It is, however, assumed that climbing bean promotion is the appropriate strategy for higher productivity and sustainability for the smallholder farmers.

The objective of the study was to determine N contribution by climbing bean to the subsequent crop in rotation, to assess the N fertilizer saved in this cropping system, and to develop a soil fertility management decision guide.

Materials and Methods

The effects of previous legumes cropping and maize treatments on grain yield and N response of a subsequent maize crop were evaluated in on-station and on-farm participatory trials. The trials conducted were on station at Mulungu Research Station with soil generally of good fertility potential, but with intensive cropping, where N has become a limiting factor (Lunze, 2000). On-farm trials were conducted in locations in vicinity of Mulungu Research Station on soils of varying fertility levels.

On-station, maize response to N applied at the rates of 0, 33, and 66 kg N ha^{-1} (which correspond to 0, $\frac{1}{2}$, and 1 recommended rate, applied as urea in two split applications) is evaluated following climbing bean, bush bean, and maize. The experimental design was a split plot, with preceding crop as main plot and N rates as subplot, with four replicates. Three rotations were studied: climbing bean–maize, bush bean–maize, and continuous maize monocropping.

In on-farm trials, the protocol is somewhat modified, the objective being to study the effects of the soil variability on the climbing bean effects on succeeding the maize crop in the rotation. The preceding crops were climbing bean, bush bean, and the maize or short-duration fallow, 6 months maximum which, according to Lunze (1988), is common in the region. The actual fields for experiments were selected based on native vegetation as indicators of soil fertility (Ngongo and Lunze, 2000) in order to have similar soil fertility in the same location. Four fields were selected in each location. We selected four locations, Kashusha 1, Kashusha 2, Tshirumbi, and Bushumba, representing high-, medium-, low- and least fertile soil, respectively. On the least fertile soil of Bushumba, compost was applied at the rate of 5 t ha^{-1} prior to experiment initiation to allow acceptable crop growth, including climbing bean which is known to require fertile or amended soils (Sperling et al., 1994).

Both bean types were harvested and biomass produced worked in prior to planting maize in the following season, while maize crop biomass of the preceding season was used as mulch.

Climbing bean variety was VCB 91012, bush bean variety Kirundo, and maize variety Kasai 2.

We used the fertilizer replacement value method to estimate the N contribution of climbing bean to maize in rotation (Bado, 2002), although this method is believed to overestimate the amount of N supplied by the legumes in the systems (Bullock, 1992).

The data were analyzed using GenStat Discovery Edition 2.

Results and Discussion

Maize crop yields are followed for three consecutive seasons, 2002A–2003A and residual effect evaluated in 2004. The maize grain yield as affected by N fertilizer in rotation with climbing bean and bush bean and in continuous maize monocropping from 2002 through 2003 is presented in Table 1. The results indicate that without added fertilizer nitrogen, maize yield was the best in rotation with climbing bean, followed by bush bean. The lowest maize grain yield was obtained in continuous maize cropping system. The statistical analysis indicates positive system and nitrogen effects, as well as positive interaction systems × nitrogen effects. In fact, maize yield under continuous maize cropping system was low, varying from 1,902 to 2,724 kg ha^{-1} without applied nitrogen. In rotation with climbing bean, the maize grain yields amounted from 2,847 to 3,800 kg ha^{-1}, which is a yield advantage of 15–46%. Positive effect on maize yield was obtained in rotation with bush bean as well, the maize yield advantage ranging from 4 to 31%, although no contribution from N would be anticipated from bush bean crop. Maize grain yield increase would therefore be attributable partly to rotational effect of climbing bean and bush bean, other than N contribution, as reported for other legume crops (Sanginga, 2003). These rotational effects could be, according to Wani et al. (1995), improved nutrient availability, improved soil structure, reduced pests and diseases and hormonal effects.

The yield advantage of climbing bean over continuous maize was obtained in the long-rain cropping season compared to short-rain season.

Maize response to nitrogen observed in all systems tended to be better following maize, and the least after climbing bean crop. These data indicate clearly a substantial N contributed in the systems by the climbing bean. The N fertilizer replacement values due to climbing bean varied from 15 to 42 kg N ha^{-1} in the first year of rotation. Similar values were reported by Bado (2002) for cowpeas in West Africa. Considering the higher potential for total biomass production and rotation benefit, intercropping is a viable alternative to conventional corn monoculture.

Residual Effects

Maize yield and yield response to applied nitrogen in three rotations and two seasons after application are presented in Table 2. The results indicate that maize yield remains greater in the rotation with climbing bean as preceding crop, followed by bush bean, and the lowest in continuous maize monocropping system. The statistical analysis reveals that the cropping system's effects were highly significant ($P<0.001$) while N effects were no longer significant. The yield advantage from climbing bean over continuous maize was 1,302 kg ha^{-1}. This indicates that the climbing bean effect on succeeding crop was apparent and significant over two seasons. While maize grain yield sharply declined the third season in continuous maize monocropping, climbing bean rotation allowed slightly constant maize yield over two seasons. This yield trend is an indication of improved and sustained productivity attributable to climbing bean in rotation.

Table 1 Effects of preceding crop and fertilizer N on maize grain yield over three seasons

Preceding crop	N rate (kg ha^{-1})	Maize grain yield (kg ha^{-1})		
		Season 2002A	Season 2002B	Season 2003A
Bush bean	0	2,911	2,735	2,625
	33	4,542	3,738	3,656
	66	5,297	4,206	4,003
Climbing bean	0	3,821	2,847	3,072
	33	4,869	4,115	4,000
	66	5,019	4,375	4,278
Maize	0	2,724	2,477	2,102
	33	4,141	3,556	2,872
	66	4,836	3,931	3,897
Mean yield		4,540	3,558	3,336
LSD (0.05)		334.9	707.1	379.2
CV (%)		6.5	9.8	10.0

Table 2 Residual effect of preceding crop and fertilizer N on maize yield two seasons later, 2004A

Preceding crop	Maize grain yield (kg ha^{-1})		
	0 kg N	33 kg N	66 kg N
Bush bean	2,396	2,630	2,878
Climbing bean	2,904	3,333	3,190
Maize	1,602	2,539	2,227
Mean	2,300	2,834	2,767
LSD (0.05)		230.8	
CV (%)		21.5	

The residual effects of climbing bean in the rotation climbing bean–maize–maize remain significant on maize yield two seasons later. On the other hand, there is very little carryover effect of mineral N.

Farmers' Participatory Evaluation

Maize grain yield response to rotation with climbing and bush bean and response to applied N were evaluated in on-farm nonreplicated trials on soils with varied characteristics. In total, 12 demonstration and evaluation trials were conducted. Beneficial rotation effects of climbing bean observed in on-farm trials are presented in Table 3. It appears that maize yield varied widely among sites. The best yields were obtained on high-fertility soils, where climbing bean effects were highest because of the largest quantity of biomass produced.

At Kashusha with high-potential soil, considerable yield improvement was obtained following climbing bean. On the other hand, at Bushumba, maize yield was generally low and no advantage of integration of climbing bean in the system could be demonstrated.

Table 3 Average maize grain yield in on-farm evaluation trials at four locations with varying soil fertility levels

Preceding crop	Maize grain yield (kg ha^{-1})			
	Kashusha 1	Kashusha 2	Tshirumbi	Bushumba
Control	3,385	2,604	1,672	781
Bush bean	3,645	3,210	1,671	1,346
Climbing bean	4,687	3,950	2,494	1,440

Conclusion

Maize grain yields were substantially enhanced following climbing bean compared with bush bean as preceding crop or continuous maize or at all N rates. Climbing bean, a commonly grown crop, appears to be a potential crop to be used when integrated in the cropping system to improve and sustain productivity in the highlands of high-potential soils. The maize yield advantage obtained from climbing bean as preceding crop in the rotation amounts up to 46%. Moreover, the economy of fertilizer N is evident in the rotation and is amounted up to 42 kg N ha^{-1}. Climbing bean is being used to develop sustainable cropping systems in the highlands of eastern DR Congo with high potential of production and in similar environments. We recommend, however, that bean breeders develop climbing bean genotypes capable of high biomass production.

Acknowledgments This work was supported by CIAT/ECABREN. We are grateful to all technicians who have been instrumental in the accomplishment of the field work on station and on farm, particularly Mr Chidorho Rwizibuka and farmers who have hosted the trials.

References

Bado BV (2002) Rôle des légumineuses sur la fertilité des sols ferrugineux tropicaux des zones guinéenne et soudanienne du Burkina Faso. Thèse de Philosophiae Doctor (Ph.D.) Département des Sols et de Génie Agroalimentaire, Faculté des Sciences de l'Agriculture et de l'Alimentation Université Laval, Québec, Decembre 2002

Baldock JO, Higgs RL, Paulson WH, Jackobs JA, Shrader WD (1981) Legume and mineral N effects on crop yields in several crop sequences in the Upper Mississippi Valley. Agron J 73:885–890

Bullock DG (1992) Crop rotation. Crit Rev Plant Sci 11(4): 309–326

Kumarasinghe KS, Danso SKA, Zapata F (1992) Field evaluation of N-2 fixation and N-partitioning in climbing bean (*Phaseolus vulgaris L.*) using N-15. Biol Fertil Soils 13: 142–146

Lunze L (1988) Effect of Traditional Cropping Systems on Soil Fertility in South Kivu, Zaïre. In: Wortmann CS (ed) Soil fertility research for bean cropping systems in Africa. Proceedings of a workshop. Addis Ababa, Ethiopia, 5–9 Sept 1988. CIAT African Workshop Series, n 3. CIAT, Ethiopia

Lunze L (2000) Fertilité des sols et possibilité de leur gestion au Sud-Kivu. ISDR-BUKAVU, Cahier de CERPRU, No.14

Ngongo M, Lunze L (2000) Espèce d'herbe dominante comme indice de productivité du sol et de la réponse du haricot commun à l'application du compost. Afr Crop Sci J 8:251–261

Sanginga N (2003) Role of biological nitrogen fixation in legume based cropping systems: a case study of West Africa farming systems. Plant Soil 252(1):25–39

Sperling L, Scheidegger U, Buruchara R, Nyabienda P, Munyanesa S (1994) Intensifying production among smallholder farmers: the impact of improved climbing bean in Rwanda. CIAT African Occasional Publication Series No. 12. CIAT/RESAPAC, Butare, Rwanda, p13

van Schoonhoven A, Pastor-Corrales MA (1994) Système standard pour l'évaluation du germoplasme du haricot. Publication CIAT, Cali, Colombia, p 207

Wani SP, Rupela OP, Lee KK (1995) Sustainable agriculture in the semi-arid tropics through biological nitrogen fixation in grain legumes. Plant Soil 174(1–2):29–49

Wortmann CS (2001) Nutrient dynamics in a climbing bean and sorghum crop rotation in the Central Africa Highlands. Nutr Cycling Agroecosyst 61(3):267–272

Investigation on the Germination of *Zanthoxylum gilletii* (African Satinwood) Seed

M.M. Okeyo, J.O. Ochoudho, R.M. Muasya, and W.O. Omondi

Abstract *Zanthoxylum gilletii* is an indigenous tropical tree species that is valued for its structural timber, agroforestry and medicinal properties. Seed of many *Zanthoxylum* species have been reported to have poor germination. The study investigated the germination of fresh *Z. gilletii* seed harvested at two maturity stages, green and red (ripe) follicles, and harvested from two provenances: Kakamega and Koiwa in November and December 2006, respectively. Follicles were dried at controlled temperature of 20°C and relative humidity of 20%. During harvesting, Satinwood seed had high MC of 26%, which decreased to 12% after processing and drying. Seed dried to 12.5% MC were divided into two: one seed lot remained unwashed while the other was washed with sodium hydroxide (NaOH) solution. Unwashed seed from Kakamega and Koiwa provenances sown on sand in the glass house gave a germination of 3 and 8%, compared to seed washed with NAOH solution which germinated up to 10 and 23%, respectively, by the 17th week. Washed seed sown on 1% agar in incubators at various constant temperatures (20, 25, 30, and 35°C±1) and alternating temperatures (20/30°C and 15/35°C±1) recorded poor germination of less than 3%, probably due to the presence of chemicals which inhibited germination. These findings suggest that the hard seed coat and oil on the testa influence seed germination and possibly contribute to dormancy exhibited by *Z. gilletii* seed. Germination of seed from mature green and ripe fruits was similar and therefore *Z. gilletii* fruits should be harvested when they start to ripen. African Satinwood seed should be washed with soap solution to remove the oil film and fruit appendages on the testa before incubation. These results are still low and there is need for further investigation to improve germination of *Z. gilletii* seed.

Keywords Germination of *Zanthoxylum gilletii* · Maturity stages · Provenances · Washing seed with 10% sodium hydroxide

Introduction

The African Satinwood [*Zanthoxylum gilletii* (De Wild.) Waterman; synonyms: *Fagara macrophylla* (Oliv.) Engl. and *F. gilletii* De Wild.] is an indigenous species of the African continent and is found growing in highlands to medium highlands of moist forests in West, Central, South and East Africa (Beentje and Adamson, 1994; Kokwaro, 1982). *Zanthoxylum* genus belongs to the Rutaceae family and is distributed worldwide from the tropics to temperate zones with more than 200 species, which range from small shrubs to large trees. The genus is characterized by sharp thorns/prickles on the stem and/or foliage; leaves are ash-like in appearance. In Kenya, *Z. gilletii* is found growing in Kakamega, South Mau, Nandi and in the slopes of Aberdares and Mount Kenya forests (Beentje and Adamson, 1994). It is typically a tropical rainforest species, distributed between altitude ranges of 900 and 2,300 m above sea level, with an annual rainfall of 1,200 mm and above (Kokwaro, 1982). Over the years, most of the indigenous forests have

M.M. Okeyo (✉)
Kenya Forestry Research Institute (KEFRI), Nairobi, Kenya;
Londiani Regional Research Centre, Londiani, Kenya
e-mail: mikemairura@yahoo.com

A. Bationo et al. (eds.), *Innovations as Key to the Green Revolution in Africa*,
DOI 10.1007/978-90-481-2543-2_69, © Springer Science+Business Media B.V. 2011

been under intense pressure from agriculture, human settlement, forest fires, poor regeneration and illegal forest poachers (Odhiambo et al., 2004; Njunge, 1996). *Z. gilletii* is an important timber species in the tropics and is popular with many local communities; most members of the *Zanthoxylum* genus are recognized for their medicinal qualities that include treating stomach ache, toothache, intestinal worms, rheumatism, scabies, snake bites, fever and cholera (Adesina, 2005). *Zanthoxylum* species are a source of essential oils and are good ornamental trees/shrubs (Setzer et al., 2005).

The inflorescence is terminal with axillary pyramidal panicles that are 20–34 cm long; flowers are cream/yellow. Most members of the *Zanthoxylum* genus are dioecious and this applies to *Z. gilletii*. Mature fruits are reddish in colour, with sub-globoid follicles 3.5–6 mm in diameter; each follicle contains one seed, which is 2.5–3.5 mm in diameter, and rarely two seed. The seed is black with a shiny testa (Omondi et al., 2004; Beentje and Adamson, 1994). *Z. gilletii* regenerates poorly in its natural habitat and this has been partly attributed to poor seed germination and destruction of young seedlings and wildings by defoliators and wild animals (Odhiambo et al., 2004; Njunge, 1996). Poor germination of *Z. gilletii* seed is attributed to its seed coat oiliness and unclear dormancy problems (WAC, 2005; Sanon et al., 2004; Rodriguez, 1995). Its seed biology is not clearly known as some authors classify it as recalcitrant (WAC, 2005; Owuor, 1999) while others classify it as orthodox (Sanon et al., 2004; Rodriguez, 1995). Seed of other *Zanthoxylum* species have been reported to germinate poorly and this is attributed to seed dormancy apparently imposed by the seed coat (Rodriguez, 1995; Bonner, 1974). Francis (1991) reported only 5% germination of untreated seed of *Zanthoxylum* species. Sanon et al. (2004) reported that fresh *Z. zanthoxyloides* seed gave an initial germination of 2% and retained the same when dried to 3% MC. The authors attributed poor germination in *Zanthoxylum* species to seed immaturity and dormancy and recommended the need to investigate stages of maturity and optimum germination conditions for the species.

For successful propagation and conservation of tropical indigenous trees and shrubs, there is need to generate scientific information on proper seed handling, processing and germination (Sacande et al., 2004; Baskin and Baskin, 1998). Despite such high species diversity in the tropics, only a handful of tree species are used in forest restoration, agroforestry systems and planting programmes (Sacande et al., 2004). Other fast regeneration methods such as tissue culture techniques for the species are not yet documented. Tree exploitation from the natural habitat leads to genetic erosion of important tropical species (Sacande et al., 2004; Njunge, 1996). As a contribution to species domestication and conservation, this study investigated germination of fresh *Z. gilletii* seed. Information generated in this study will assist various government institutions, NGOs, farmers and interested groups in raising *Z. gilletii* seedlings for domestication and habitat restoration programmes.

Materials and Methods

Fruit Harvesting, Processing, Moisture Content and Weight Determination

Zanthoxylum gilletii fruits at two maturity stages (green and red) were harvested from the crown in Kakamega in November and Koiwa in December 2007. Whole inflorescences were severed from the branches and kept in well-aerated rooms during the fruit harvesting/collection process to avoid seed deterioration. During collection, follicles were separated into two classes: mature green and ripe (red) follicles. Munsell colour chart (Munsell, 1972) was used as a guide for separating mature green and ripe (red) follicles. Fresh harvested follicles were handled according to recommendations by IPGRI protocol (Sacande et al., 2004) and the *Tree Seed Handbook of Kenya* (Omondi et al., 2004). After collection, fruits were transported to the National Gene Bank of Kenya (GBK); follicles were placed in small net bags and dried at controlled temperatures of 20°C and relative humidity of 20% until follicles dehisced.

Fresh seed weight and initial moisture content of each seed lot were determined and was repeated for extracted and dried seed. Fruits for fresh MC were transported in cold boxes to avoid desiccation. Moisture content of each seed lot was also determined before individual germination tests. *Z. gilletii* seed and pulp were manually separated from fresh fruits for initial MC. Follicles (with seed), seed and pulp for MC were randomly sampled; five replicates of each lot weighing 2 g or 50 seed/follicles were oven dried at 103°C for 17 h for moisture determination according to ISTA (2003).

Percentage seed moisture content
$$(\%MC) = \left[(W_i - W_f)/W_f \right] \times 100$$

where W_i = initial seed weight (g) and W_f = seed weight after oven drying (g).

Two replicates of 100 g extracted seed were sampled using the sleeve-type sampler and used for purity analysis according to ISTA (2004). Eight replicates of 100 pure seed from three random samples were used for the determination of the weight of 1,000 seed (ISTA, 2004).

Imbibition and Germination Tests

Seed weighing 3.0 g were placed on moist filter papers and were allowed to imbibe for 12-h intervals up to a maximum of 96 h; three replicates were used for every 12-h interval. At the end of each 12-h interval, seed were dried using blotter paper to remove wetting water. Dried seed were weighed and discarded while the others were allowed to continue imbibing (Turner et al., 2005).

Seed from mature green and ripe seed lots with 11.5% MC were subjected to various germination tests. Four replicates of 25, 50 and 100 whole seed were sown on 1% agar and blotter papers in Petri dishes, in germination boxes and on sand in germination trays, respectively, at various constant and alternating temperatures. Two replicates of 10 exercised embryos were sown on 1% agar and incubated at 30°C. Germination was scored weekly because *Z. gilletii* seed took a long time to germinate. Seed sown in incubators were considered to have germinated when the radicle protruded by 2 mm, while seed sown in the glass house were considered to have germinated when the hypocotyl emerged 2 mm (ISTA, 2004). Total germination percentage, standard deviation (SD) and standard error (SE) of total germination and coefficients of variation (CV%) of total germination between replicates were calculated. Germination tests continued for a period of 17 weeks after incubation (ISTA, 2004; Sacande et al., 2004). Germination morphology of *Z. gilletii* seed was observed and documented

during the germination process. Completely randomized design (CRD) was adopted for germination experiments.

- Seed were sown (25 per replicate) in incubators at constant temperatures of 20, 25, 30, and 35±1°C at 12/12 h day/night.
- Seed were sown (25 per replicate) in incubators at alternating temperatures of 30/20°C and 35/15±1°C at 8/16 h light/dark.
- Seed were sown in the KEFRI glass house at alternating temperatures of 35/18±4°C at 12/12 h day/night. Seed were sown on sand (100 seed per replicate) and covered with a thin layer of sand. Light watering was done once a day (at 9.00 am).

a. Germination capacity (GC) and the mean germination time (MGT) were computed for all germination tests. GC and MGT were calculated by the following formulae:

   i. GC = (number of germinated seed/total number of seed sown) × 100

   ii. MGT (weeks) = $\sum (t_i n_i)/\sum n_i$
where t_i is the number of days starting from the day of sowing and n_i is the number of seed completing germination on day t_i (Tigabu et al., 2007). Germination period continued up to 16 weeks from the time seed were sown.

b. Median length of germination (MLG) and mean length of germination (mean LG) of the germination period were calculated for all seed that germinated as a measure of dormancy. Standard deviation of mean germination time and that of total germination time until the last seed that germinated were calculated (Sautu et al., 2006).

Statistical Analyses

General statistical package (GenStat, 2003) was used to analyze germination capacity and mean germination time. Microsoft Excel (2003) was used for data management, exploratory data analyses and drawing of graphs and bar charts. Standard error of the mean used to form error bars was computed using Excel. Germination capacity (GC), mean germination time (MGT) and median length of germination period (MLG) were calculated. Total germination percentages were square root $\left[\sqrt{(0.5 + X)}\right]$ transformed (Gomez and Gomez, 1984) and subjected to analysis of variance to quantify the differences between the applied treatments (all tables contain non-transformed data). Treatment means were separated using the least significant differences (LSD) at $P \leq 0.05$ and $P \leq 0.001$.

Results reported and discussed in this chapter are those of seed sown on sand at the Kenya Forestry Research Institute glass house at 35/18±4°C because seed sown in most incubators and at various temperatures failed to germinate after 17 weeks incubation period.

Results

Fruit Harvesting, Processing, Moisture Content and Weight Determination

Mature green and red fruits were harvested from Kakamega in November and from Koiwa in December 2006. After drying fruits at controlled temperatures (20°C) and humidity (20%) at GBK, seed were extracted, dried and stored at 4°C. Seed from green fruits had significantly higher MC (29.03%) at $P \leq 0.001$ compared to seed from red fruits with MC of 27.12%. Fresh follicles and seed had high MC of 55–57% and 27–29%, respectively, but the fruit pulp had the highest MC of 63–64%; thus, most of the water is in the pulp (Table 1).

Mean weight of 1,000 seed of *Z. gilletii* from red fruits was significantly heavier ($P \leq 0.001$) than that of seed from green fruits in Kakamega provenance.

Table 1 Moisture content of fresh fruits, seed and pulp after harvesting (mean ± SE)

Fruit/seed/pulp	Moisture content (%) of fruits, seed and pulp	
	Green follicles	Red follicles
Fruit	57.15±0.06	55.47±0.06
Seed	29.03±0.04	27.15±0.17
Pulp	64.71±0.05	63.43±0.03

LSD was used to separate significantly different moisture contents of various seed lots at $P \leq 0.001$. LSD (seed) = 0.3833, LSD (fruit) = 0.1303, LSD (pulp) = 0.1378

Mean weight of 1,000 seed from green and red fruits was not significantly different ($P \leq 0.05$) in the Koiwa provenance. The number of *Z. gilletii* seed per kilogram ranged from 31,000 to 44,000 (Table 2). During the cutting tests (results not shown), *Z. gilletii* seed were observed to have hard and brittle seed coat.

Imbibition and Germination Tests

Results indicated that unwashed *Z. gilletii* seed initially imbibed faster than washed seed up to 48 h, after which washed seed imbibed higher (33.6%) up to 84 h compared to unwashed seed (Fig. 1). Imbibition of *Z. gilletii* seed continued up to 84 h; after that, there was no further imbibition. Within the first 24 h, average seed imbibition was 20.7% and thereafter imbibition rate slowed down (Fig. 1). Washed seed from green fruits imbibed significantly higher ($P \leq 0.05$) compared to seed from red fruits, but the differences were not significant ($P \leq 0.05$) for unwashed seed from green and red fruits.

Seed sown in most incubators at constant and alternating temperatures failed to germinate after 17 weeks incubation period; evaluation of sown seed after concluding germination tests revealed that 17.2% were firm, 11.8% were empty and 71% were rotten. Seed sown on 1% agar at 20±1°C constant and 35/15±1°C alternating temperatures germinated poorly (3.5%) after 17 weeks incubation period. Exercised embryos sown on 1% agar failed to germinate.

Seed from Koiwa provenance recorded significantly higher ($P \leq 0.001$) germination capacity and lower mean germination time ($P \leq 0.05$) compared to seed from Kakamega provenance. Seed washed with 10% NaOH showed significantly higher ($P \leq 0.001$) germination compared to unwashed seed. Similarly, washed seed from the two provenances gave high germination ($P \leq 0.001$) compared to unwashed seed (Fig. 2). There were no significant differences ($P \leq 0.05$) in the germination of seed from green and ripe fruits. Washed seed from Kakamega and Koiwa had a median length of germination of 7.8 and 6.8 weeks, respectively (Table 3).

Mean germination time of washed seed from Koiwa was significantly lower ($P \leq 0.05$) compared to washed seed from Kakamega provenance. Germination started on the fifth week and continued up to the 17th week

Table 2 Weight of *Z. gilletii* seed harvested from Kakamega and Koiwa provenances (mean ± SE)

Provenance and fruit maturity stages	Kakamega		Koiwa	
	Green	Red	Green	Red
Weight (g) of 1,000 seed	22.683±0.022a	25.543±0.010b	31.588±0.010c	32.148±0.008c
Number of seed per kilogram	44,087	39,150	31,658	31,106
LSD (0.05)	0.078		0.078	

LSD was used to separate significantly different treatments at $P \leq 0.05$. Letters a, b and c indicate significantly different parameters along the rows. Seed used were at 11.5% MC

Fig. 1 Imbibition percentage of *Z. gilletii* seed over 96-h period for unwashed and washed seed from green and red fruits. Washed seed from green fruits (△), unwashed seed from green fruits (▲), washed seed from red fruits (□), unwashed seed from red fruits (■) and average imbibition (○). Error bars represent SE during the hourly imbibition period, $n = 5$ (3 g of seed per replicate)

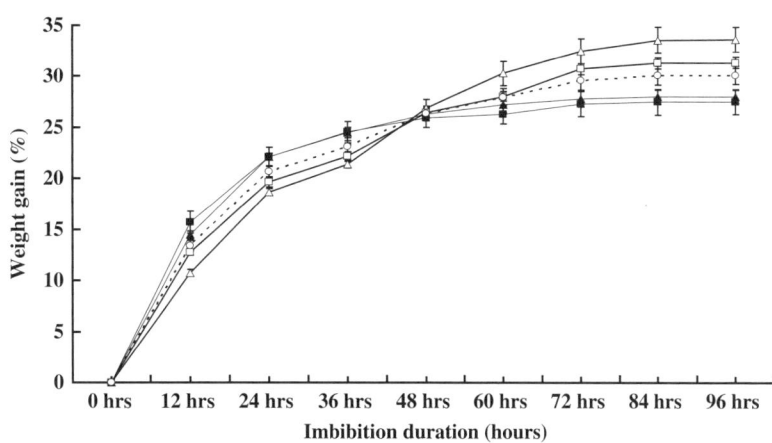

Fig. 2 Germination capacity (A) and mean germination time (MGT) (B) of *Z. gilletii* seed from Kakamega (1–4) and Koiwa (5–8) provenances after 17 weeks incubation period. Unwashed seed from green (1 and 5) and red (2 and 6) fruits; washed seed from green (3 and 7) and red (4 and 8) fruits. Seed were sown on sand in the glass house (35/18±4°C at 12/12 h day/night). Letters a, b and c indicate significant differences at $P \leq 0.05$. Error bars represent SE for different treatments; $n = 4$ (100 seed per replicate)

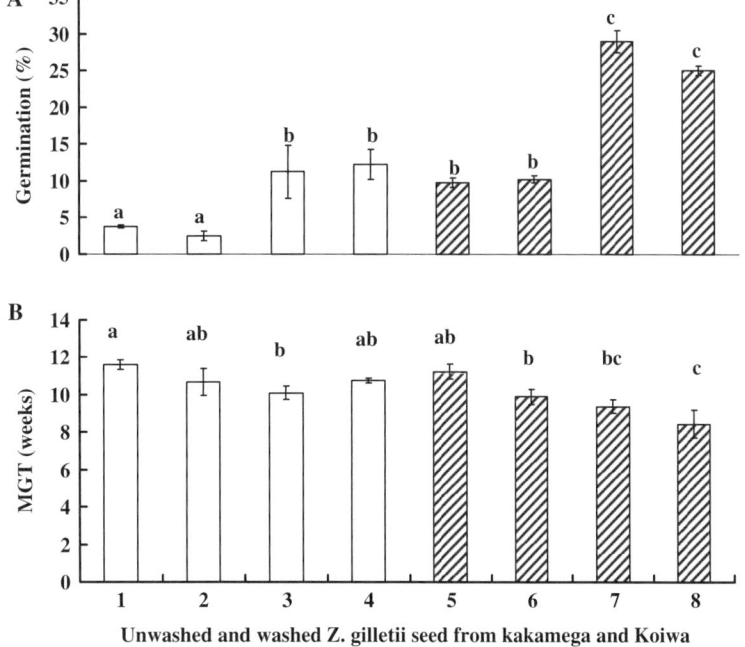

Table 3 Germination capacity (GC) and mean germination time (MGT) of interactions: (site × maturity) and (site × seed condition) and median length of germination (MLG) of *Z. gilletii* seed after 17 weeks incubation period. Seed were sown on sand in the glass house (35/18±4°C at 12/12 h day/night). The CV% in brackets is that of transformed germination data. *n* = 4 (100 seed per replicate)

GC and MGT	GC% ± SE		MGT ± SE	
Provenance/maturity/seed condition	Kakamega	Koiwa	Kakamega	Koiwa
Green	7.50±1.049bd	19.37±1.049be	10.84±0.352ad	10.30±0.352bd
Red	7.37±1.049bd	17.62±1.049be	10.71±0.352ae	9.18±0.352ad
Unwashed	3.12±1.049ad	10.00±1.049ae	11.14±0.352ad	10.56±0.352bd
Washed	11.75±1.049 cd	27.00±1.049ce	10.41±0.352ae	9.18±0.352ad
LSD (0.05)	3.085	3.085	1.034	1.034
CV%	22.9 (11.8)	22.9 (11.8)	9.7	9.7
			Kakamega (MGL)	Koiwa (MGL)
Median germination length (MGL), washed seed			7.79±0.12	6.75±0.95

Letters a, b and c indicate significant differences in GC and MGT along the column, while d and e indicate differences along the rows at *P* ≤ 0.001

after sowing (Fig. 3). Germination percentage of the interactions between the two provenances and washing seed with NaOH were significant ($P \leq 0.001$), but the other interactions between (sites × maturity stages), (maturity stages × washing) and (sites × maturity stages × washing) were not significant at $P \leq 0.05$ (Table 3).

The fruit of *Z. gilletii* is a dehiscent follicle with one round seed and rarely two flat black shiny seed. Seed are small with a diameter of 2.5–3.5 mm; each seed weighs $2.3–3.2 \times 10^{-2}$ g^{-1} (Table 3). The embryo is bean shaped with a radicle and two cotyledons. During germination, seed were observed to exhibit epigeal germination. *Z. gilletii* seed are non-endospermic with one-step testa rapture germination process. The only two visible signs in the germination of *Z. gilletii* seed are testa rapture and radicle protrusion.

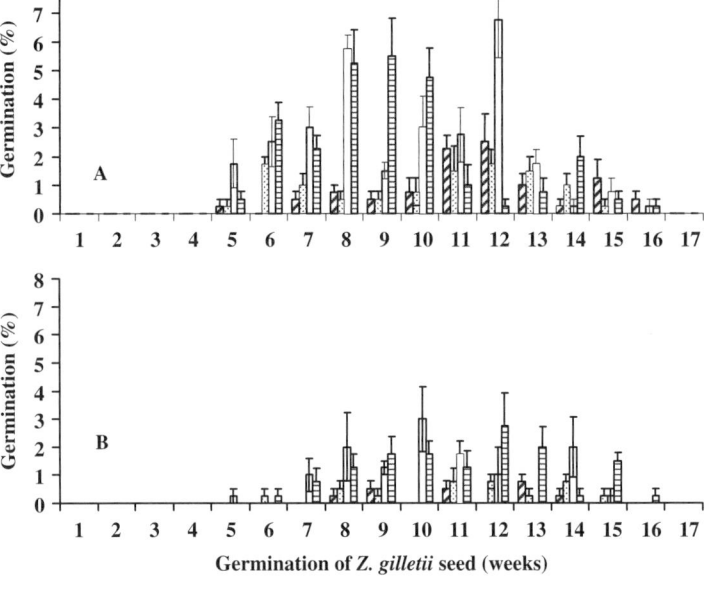

Fig. 3 The spread of germination of unwashed and washed *Z. gilletii* seed harvested from Koiwa (**a**) and Kakamega (**b**) provenances. Seed were germinated on sand in the glass house (35/18±4°C at 12/12 h day/night). Error bars represent SE during the incubation period; *n* = 4 (100 seed per replicate)

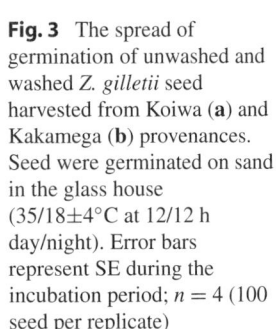

Discussion

Z. gilletii trees were recorded as having male and female flowers borne on separate plants. This concurs with work done by Kokwaro (1982) and Hedberg and Edwards (1989) on the *Rutaceae* family. Most members of the *Zanthoxylum* genus are dioecious (Hedberg and Edwards, 1989; Kokwaro, 1982) but some authors (Omondi et al., 2004; Beentje and Adamson, 1994) failed to point out clearly that male and female flowers are borne on separate plants. In some instances, where the female trees are solitary or when there is poor synchronization of male and female trees flowering, fruits form without pollination and the resultant seed are empty and/or infested with insect larvae. These observations agree with earlier research which attributed poor germination of *Zanthoxylum* species to empty and/or insect-infested seed (WAC, 2005; Sanon et al., 2004).

At natural dispersal stage, *Z. gilletii* seed had a high moisture content of 25.5% which decreased further during processing and drying to 9.4% (Table 1). This suggests that *Z. gilletii* seed are either intermediate or orthodox in their storage behaviour and not recalcitrant as implied by WAC (2005) and Owuor (1999). Most intermediate species can be dried to low moisture content without losing viability (Sacande et al., 2004; Hong and Ellis, 1996). Seed of wild species whose biology is not known or well understood should be collected at the natural dispersal stage or when fruits are ripe and sown immediately (Baskin and Baskin, 1998). The number of *Z. gilletii* seed per kilogram (39,000–44,000 for Kakamega and 31,000–32,000 for Koiwa provenances) compares closely to the 44,000–46,000 seed per kg reported by Omondi et al. (2004) (Table 2). The difference in the number of seed per kilogram could be due to provenance and year to year variations in seed weight.

Washing *Z. gilletii* seed with a soap solution improved overall imbibition compared to unwashed seed (Fig. 1). *Z. gilletii* seed imbibed to a maximum of 34% in 84 h and after that there was no further increase in imbibition. Unwashed seed had 29.1% oil while washed seed had 16.3% oil; thus, washing removed 12.8% oil and fruit appendages which were primarily on the seed coat. Fruit appendages on the seed coat could have contributed to higher initial imbibition

(up to 48 h) for unwashed seed compared to the seed washed with 10% NaOH solution.

Germination of seed from green and ripe fruits was similar showing that *Z. gilletii* fruits may be harvested either green or red (ripe) so long as they have reached physiological maturity (Fig. 2). Change of fruit colour from green to red and dehiscing of mature fruits are good indicators of seed maturity for *Z. gilletii*. Change of fruit colour from green to red or yellow has been shown to coincide with physiological maturity in many plant species (Samarah et al., 2004; Bewley and black, 1994). Washed *Z. gilletii* seed from both provenances gave higher germination compared to unwashed seed. Sodium hydroxide solution removed the oil and fruit appendages on the seed coat and possibly leached out inhibitors, thus improving imbibition and overall germination (Figs. 1 and 2). These results agree with Rodriquez (1995) who reported that washed seed of *Z. killermanii* (P. Wilson) and *Z. mayanum* improved germination from 47 and 5% (unwashed) to 90 and 100% (washed), respectively. Seed from Koiwa provenance recorded higher germination and were heavier in seed weight compared to those from Kakamega (Table 2). This could be attributed to provenance differences arising from altitude and rainfall variations during the flowering and fruiting period. Koiwa had higher rainfall during the flowering period compared to Kakamega provenance. Low germination of seed from Kakamega provenance could have been due to *Apion* species weevils that fed on the whole or part of the embryo. The *Apion* sp. weevils were found only on fruits from Kakamega provenance, suggesting that seed from Koiwa are of high quality as opposed to those from Kakamega. Odhiambo et al. (2004) attributed poor germination of *Z. gilletii* seed to phytophagous insects. These findings suggest that the oil on the seed coat retards germination of *Zanthoxylum* species and possibly contributes to their dormancy. Most of the essential oil in *Zanthoxylum* species is unsaturated (Adesina, 2005; Setzer et al., 2005); the oil becomes rancid during seed processing and storage and possibly contributes to dormancy and poor germination of *Zanthoxylum* species. Other results (not shown) revealed that a filtrate from *Z. gilletii* seed significantly inhibited the germination of fast germinating seed species such as *Acacia albida*, *Calliandra calothyrsus*, *Parkinsonia aculeata* and *Corchorus olitorius*. This explains why *Z. gilletii* seed sown in Petri dishes and closed germination boxes germinated poorly or failed to germinate. This is further supported by de Feo et al. (2002) who found that oil extract from *Ruta graveolens* seed (Rutaceae) inhibited germination and radicle elongation of radish seed.

Germination of *Z. gilletii* seed was spread from the 5th to the 17th week of incubation (Fig. 3). Seed of most *Zanthoxylum* species including those of *Z. killermanii* and *Z. zanthoxyloides* have been reported to start germinating after 5–6 weeks and continue to germinate up to 13 weeks after sowing (Rodriquez, 1995). The long duration the seed took to start and complete germination was attributed to some form of seed dormancy (Baskin and Baskin, 2004). Germination of most wild species and in particular tree species takes a long time and is spread out. *Z. gilletii* seed (unwashed and washed) from Koiwa provenance germinated faster (MGT of 9 weeks) compared to Kakamega seed with an MGT of 10 weeks (Fig. 2). These findings suggest that Koiwa seed are of high quality compared to those from Kakamega. Seed with a median length of germination of more than 30 days are considered to be dormant according to dormancy classification by Baskin and Baskin (2004). *Z. gilletii* seed recorded an MLG of 8 and 7 weeks for Kakamega and Koiwa, respectively (Table 3), suggesting that they exhibit dormancy. Pritchard et al. (2004) attributed poor germination of *Z. zanthoxyloides* from Burkina Faso to physiological dormancy. Dormancy is the inability of the embryo to germinate due to some inherent inadequacy even when given optimum germination conditions. Dormancy is useful in that it allows distribution of germination in time and space, therefore guaranteeing species survival and continuity (Bewley, 1997).

Germination morphology of *Z. gilletii* is non-endospermic with one-step testa rapture and is similar to that of *Brassica* and pea seed. Germination of *Z. gilletii* is complete with testa rapture followed by initial radicle protrusion (Finch-Savage et al., 2006). Germination starts with rapid initial water uptake by a quiescent seed (phase I, imbibition) followed by a plateau phase (phase II, enzymatic activity). Phase III is a further increase in the imbibition process that takes place after the germination process is completed (Bewley, 1997). Water uptake in phase III causes hydraulic growth of the embryo and the emerged seedling. In dormant seed, as was the case for *Z. gilletii*, abscisic acid (ABA) has been shown to inhibit phase III, thus preventing completion of the germination process (Finch-Savage et al., 2006).

Conclusion

Results obtained in this study confirm that *Z. gilletii* trees are dioecious with female and male flowers on separate plants. There were no significant differences in the germination of seed from green and red fruits, therefore *Z. gilletii* fruits should be harvested when fruit colour starts to change from green to red. These findings further suggest that *Z. gilletii* seed are desiccation tolerant and can be dried to low moisture content of 8% without losing viability. *Z. gilletii* seed started to germinate on the 5th week and continue up to the 17th week after incubation. Seed washed with 10% NaOH (soap solution) removed oil on the seed coat and improved germination capacity to 29% and mean germination time to 8.5 weeks. *Z. gilletii* seed should therefore be washed with a soap solution and sown on sand in germination trays and boxes which drain off excess water. Seed that were sown on 1% agar germinated poorly or failed to germinate probably due to the presence of inhibitors. Seed from Koiwa provenance recorded higher germination capacity of 29% with a lower mean germination time of 8.5 weeks and are therefore of high quality compared to seed from Kakamega provenance. Kakamega seed were infested with weevils (*Apion* sp.) and that probably contributed to their poor germination compared to Koiwa seed which did not have weevils and were also heavier in seed weight. Poor germination of *Z. gilletii* seed was probably due to multiple dormancy condition caused by seed coat oil and hard seed coat.

Acknowledgements This study was funded by Kenya Forestry Research Institute and Seed for Life Project and supervised by Moi University. I thank my supervisors Professor Reuben M. Muasya and Dr. Julius Ochuodho of Moi University and Mr. William O. Omondi of Kenya Forestry Research Institute. All those who assisted in seed collection, processing and monitoring experiments are acknowledged.

References

Adesina SK (2005) The Nigerian *Zanthoxylum* chemical and biological values. Afr J Tradit Complement Altern Med 2(3):282–301

Baskin CC, Baskin JM (1998) Seeds: ecology, biogeography, and evolution of dormancy and germination. Academic, San Diego, CA, 666pp

Baskin CC, Baskin JM (2004) A classification system for seed dormancy. Seed Sci Res 14:1–6

Beentje H, Adamson J (1994) The description of Rutaceae family: in Kenya trees, shrubs and lianas. 365–374

Bewley JD (1997) Seed germination and dormancy. Plant Cell 9:1055–1066

Bewley JD, Black M (1994) Seeds: physiology of development and germination. Plenum, New York, NY, 445pp

Bonner FT (1974) *Zanthoxylum* L., prickly-ash. In: Schopmeyer CS (ed) tech. cord. Seeds of woody plants in the USA. Agric. handbook. 450. USDA Forest Service, Washington, DC, pp 859–861

De Feo V, De Simone F, Senatore F (2002) Potential allelochemicals from the essential oil of *Ruta graveolens*. Phytochemistry 61(5):573–578

Finch-Savage and Metzger WE and Metzger GL (2006) Seed dormancy and the control of germination. New Physiol 171:501–523

Francis JK (1991) *Zanthoxylum martinecence* (Lam.) DC. espino rubial. SO-ITF-SM-42. USDA Forest Service, Southern Forest Experimental Station, New Orleans, LA, 5pp

GenStat (2003) GenStat for Windows (GenStat 7th Edition)

Gomez KA, Gomez AA (1984) Least significant differences and Duncan's multiple range test. Statistical procedures for agricultural research. Wiley, New York, NY, pp 188–215

Hedberg I, Edwards S (eds) (1989) The description of *Rutaceae* family. Flora of Ethiopia, 3: Addis Ababa and Asmara Ethiopia Uppsala Sweden, pp 419–432

Hong TD, Ellis RH (1996) A protocol to determine seed storage behaviour. In: Engels JMM, Toll J (eds) Technical bulletin 2. International Plant Genetic Resources Institute, Rome, Italy

ISTA (2003) Handbook on seedling evaluation. 3rd edn. The International Seed Testing Association, Bassersdorf

ISTA (2004) International rules for seed testing. Seed Sci Technol 32(Suppl)

Kokwaro JO (1982) Flora of tropical East Africa, Rutaceae family 1–45. Prepared at the Royal Botanic Gardens/Kew with the assistance from the East African Herbarium. A.A. Balkema, Rotterdam

Microsoft Excel (2003) Microsoft Excel computer package

Munsell Colour Division (1972) Munsell colour charts for plant tissues, 2nd edn. Munsell Colour Division Kollmorgen Corporation, Baltimore

Njunge JT (1996) Species composition and regeneration at South Nandi Forest, Kenya. PhD thesis, University of Wales, Bangor, UK

Odhiambo KO, Njunge JT, Opondo-Mbai ML (2004) Some factors influencing regeneration of selected trees in Kakamega forest, Kenya. Discov Innovat 16(2/3):117–124

Omondi W, Maua JO, Gachathi FN (2004) Tree seed handbook of Kenya second edition. Kenya Forestry Research Institute (KEFRI), Nairobi, Kenya, pp 257–284

Owuor BO (1999) An ethnobotanical and phytochemical study of the herbal remedies of Migori District, Kenya. MSc thesis, University of Nairobi, Kenya

Pritchard HW, Sacandé M, Berjerk P (2004) Biological aspects of tropical tree seed desiccation and storage responses. In: Sacande M, Joker D, Duloo ME, Thomas KA (eds.) Comparative storage biology of tropical tree seed. IPGRI, Rome, Italy, pp 319–338

Rodriquez L (1995) Tratamientos pregerminativos para algunas especies forestales nativas, de la Region Huetar Norte de Costa Rica. Simposio Latino Americano sobre semillas forestales. Managua, Nicaragua, 16–24 October, 1995

Sacande M, Joker D, Duloo ME, Thomas KA (2004) Comparative storage biology of tropical tree seeds. International Plant Genetic Resources Institute (IPGRI), Rome, Italy

Samarah NH, Allataifeh N, Turk MA, Tawaha AM (2004) Seed germination and dormancy of fresh and air-dried seed of common vetch (*Vicia sativa* L.) harvested at different stages of maturity. Seed Sci Technol 32:11–19

Sanon MD, Gamene CS, Sacande M, Neya O (2004) Desiccation and storage of *Kigelia africana, Lophira lanceolata, Parinari curatellifolia* and *Zanthoxylum zanthoxyloides* seeds from Burkina Faso. 16–23

Sautu A, Baskin JM, Baskin CC, Condit R (2006) Studies on the seed biology of 100 native species of trees in a moist tropical forest, Panama, Central America. Forest Ecol Manage 234:245–263

Setzer WN, Noletto JA, Lawton RO, Haber WA (2005) The leaf essential oil composition of *Zanthoxylum* species from Monteverde, Costa Rica. Mol Divers 9:3–13

Tigabu M, Fjellstrom J, Oden PC, Teketay D (2007) Germination of *Juniperus procera* seed in response to stratification and smoke treatments, and detection of insect-damaged seed with VIS + NIR spectroscopy. New Forests 33:155–169

Turner SR, Merritt DJ, Baskin CC, Dixon KW, Baskin JM (2005) Physical dormancy in seed of six genera of Australian Rhamnaceae. Seed Sci Res 15:51–58

World Agroforestry Centre (WAC) (2005) *Zanthoxylum gilletii*. From Agroforestry tree Database: a tree species reference and selection guide

Combining Ability for Grain Yield of Imidazolinone-Resistant Maize Inbred Lines Under Striga (*Striga hermonthica*) Infestation

I.H. Rwiza, M. Mwala, and A. Diallo

Abstract *Striga hermonthica* infests cereal crops, particularly maize, leading to severe yield reduction. In Tanzania, maize is grown on about 2 million hectares but the yield obtained is very low, estimated at 1.3 tons ha^{-1}; this is due to various factors including striga weeds. The control measures used have not been effective. One of the promising strategies in controlling striga is the use of imidazolinone-resistant (IR) maize seed coated with imazapyr herbicide. A research was conducted to investigate the inheritance of this trait in maize inbred lines; 93 testcrosses were evaluated under striga infestation conditions in the Lake zone of Tanzania and Kisumu-Kenya in the 2006 season using alpha (0,1) lattice design. Maize grain yield was used as a proxy for resistance to the herbicide; resistant materials were selected as suitable candidates in striga-infested areas with the use of herbicide. The results also showed differences in both general combining ability (GCA) and specific combining ability (SCA) effects for grain yield. GCA effects ranged from –0.57 to 0.78. SCA effects were different within each tester. The SCA effects with tester A ranged from –0.67 to 0.58, tester B from –0.70 to 0.32 and tester C from –0.62 to 0.80. The contribution of GCA and SCA to entry sums of squares for grain yield was relatively higher for GCA than for SCA at 38 and 32%, respectively. This suggested that the additive gene effects were the most important source of variation on herbicide resistance.

Keywords General combining ability · Imazapyr herbicide · Imidazolinone-resistant maize · Specific combining ability · *Striga hermonthica*

Introduction

Maize (*Zea mays* L.) is one of the most important cereal crops in the world. It is used as food and feed in developing and developed countries, respectively (FAO, 1980). Yields in industrialized countries are more than 8 tons ha^{-1}, while in the developing countries these are less than 3 tons ha^{-1} (Pingali, 2001). In Africa, maize production is mostly under rain-fed conditions, where 95% is produced by small- and medium-scale farmers, who own less than 10 ha (Heisey and Mwangi, 1997).

Moshi et al. (1990) reported that in East Africa the yields of 1.5 tons ha^{-1} for Kenya, 1.3 tons ha^{-1} for Tanzania and 1.2 tons ha^{-1} for Uganda were common. In the region maize production is facing various problems including biological, environmental and physical stresses.

In Tanzania, more than 80% of the population depends on maize as a major food crop (GoT, 2003). The annual per capita consumption of maize is estimated at 112.5 kg, while the national maize consumption is approximately 3 million tons per year (Kaliba et al., 2000). An average of 2 million hectares of maize are grown in the high potential areas of the country, such as Southern Highlands, Lake and Northern zones (Moshi et al., 1990). In the country maize production is limited by biotic and abiotic stresses. Striga weed (*Striga hermonthica*) is among the major contributing

I.H. Rwiza (✉)
Agriculture Research Institute Ukiriguru, Mwanza, Tanzania
e-mail: rwizaih@hotmail.com

A. Bationo et al. (eds.), *Innovations as Key to the Green Revolution in Africa*,
DOI 10.1007/978-90-481-2543-2_70, © Springer Science+Business Media B.V. 2011

factors for yield reduction. De Groote and Wangare (2002) reported that striga infestation can result in zero harvest.

The weed control measures that have been experimented by the international and national research centres are mostly based on agronomic strategies; most of them have not been effective and were not adopted by farmers. The most promising approach to suppressing striga parasitism tried so far is the application of herbicide as seed coating on herbicide-resistant maize varieties. A research conducted by CIMMYT scientists has shown that coating imidazolinone-resistant (IR) maize seeds with imazapyr herbicide gives effective striga control and reduces striga seed bank when continuously applied on the same field for more than three seasons (Kanampiu et al., 2001). They also reported that the use of herbicide-treated seed does not affect the sowing of herbicide-sensitive intercrops like beans and cowpea in intercropping; therefore this technology can be used by traditional small-scale farmers who practise intercropping systems. The development of IR maize varieties is therefore a relevant strategy for maize production in striga-infested areas.

Little information on combining abilities for grain yield of imidazolinone-resistant maize is available to enable strategic use of the trait in breeding for herbicide resistance. It is against this background that this study has been undertaken to determine the combining ability for grain yield of IR maize inbred lines under striga environment so as to identify the useful materials for developing IR maize varieties.

Estimation of Combining Ability

The combining ability for IR maize inbred lines and broad-based testers in line × tester mating design was estimated using a formula established by Singh and Chaudhary (1985). The line × tester analysis provides information about the general and specific combining abilities of the material evaluated for the trait of interest; at the same time it estimates various types of gene effects. The crossing plan involves '*l*' lines and '*t*' testers. All of these '*l*' lines are crossed to each of the '*t*' testers and therefore line × tester (*l* × *t*) full-sib progenies are produced. These progenies along

with or without parents, that is, lines and testers, are tested in a replicated trial using suitable field design (Comstock and Robinson, 1948; Singh and Chaudhary, 1985; Tyagi and Lal, 2005).

The formula presented by Singh and Chaudhary (1985) for estimation of the effects due to the general and specific combining abilities is as follows:

(a) Estimation of general combining ability effects:

  (i) Lines: $g_i = x_i.../tr - x.../ltr$
  (ii) Testers: $g_i = x_j.../lr - x.../ltr$
      where g is the general combining ability effects; x the number of crosses; l the number of lines; t the number of testers; and r the number of replications.

(b) Estimation of specific combining ability effects:

$$s_{ij} = (x_{ij}/r) - (x.../tr) - (x_i.../lr) - (x.../ltr)$$

where s_{ij} is the specific combining ability effects for cross l, x_{ij} the grand total for cross l, x_i the grand total of lines for cross l, x_j the grand total of testers for cross l, $x...$ the grand total crosses, l the number of lines, t the number of testers; and r the number of replications.

Materials and Methods

The research was conducted in the Lake zone of Tanzania and Kisumu-Kenya in the year 2006; the materials were planted in five different striga-infested sites representing striga-prone areas in the countries.

The experimental materials were generated by crossing 31 IR maize lines (Table 1) with three testers of different heterotic groups using line × tester mating design. Two testers were single cross hybrids including CML373-IR/CML393-IR (group A) and CML202-IR/CML395-IR (group B). The third tester was an open pollinated variety namely IR OPV (Synthesis 2000-IR) (group C). After crossing the inbred lines and testers, 93 testcrosses were developed and used in the study. Three commercial maize varieties which are less tolerant to striga were incorporated in the experiment as checks.

Table 1 List of inbred lines

Line no.	Pedigree	Line no.	Pedigree	Line no.	Pedigree
1	CML247-IR(BC0)-B-B-B-108	11	CML373-IR(BC0)-B-B-38-B	21	CML445-IR(BC0)-B-B-117-B
2	CML247-IR(BC0)-B-B-B-110	12	CML373-IR(BC0)-B-B-55-B	22	CML445-IR(BC0)-B-B-23-B
3	CML247-IR(BC0)-B-B-B-46	13	CML373-IR(BC0)-B-B-70-B	23	CML445-IR(BC0)-B-B-37-B
4	CML247-IR(BC0)-B-B-B-99	14	CML373-IR(BC0)-B-B-93-B	24	CML445-IR(BC0)-B-B-5-B
5	CML312-IR(BC0)-B-B-B-60	15	CML384-IR(BC0)-B-B-B-126	25	CML445-IR(BC0)-B-B-60-B
6	CML312-IR(BC0)-B-B-B-65	16	CML390-IR(BC0)-B-B-6-B	26	CML78-IR(BC0)-B-B-B-123
7	CML312-IR(BC0)-B-B-B-67	17	CML395-IR(BC0)-B-B-41-B	27	CML78-IR(BC0)-B-B-B-129
8	CML312-IR(BC0)-B-B-B-79	18	CML444-IR(BC0)-B-B-131-B	28	CML78-IR(BC0)-B-B-B-130
9	CML312-IR(BC0)-B-B-B-92	19	CML444-IR(BC0)-B-B-154-B	29	CML78-IR(BC0)-B-B-B-26
10	CML373-IR(BC0)-B-B-180-B	20	CML444-IR(BC0)-B-B-158-B	30	CML78-IR(BC0)-B-B-B-62
				31	CML78-IR(BC0)-B-B-B-7

Experimental Design and Data Collection

The crosses were planted at a spacing of 75×50 cm in an alpha (0, 1) lattice design with three replications. Each plot had 22 plants, and observations from 18 plants were recorded for anthesis date, anthesis silking interval, plant height, ear height, *Puccinia sorghi*, *Exserohilum turcicum*, grain yield and striga count at 6th, 8th, 10th and 12th weeks after planting.

Management

Cultural operations like weeding and fertilizer and insecticide application for stem borer control were done, leaving striga weeds undisturbed. Phosphate fertilizer was applied after land preparation at a rate of 40 kg P_2O_5 ha^{-1}, while nitrogen fertilizer was applied as side dressing at a rate of 100 kg N ha^{-1}.

Data Analysis

The agronomic data were analysed using statistical analysis system (SAS), while GCA and SCA effects were estimated using a formula presented by Singh and Chaudhary (1985). The analysis of variance was done assuming a randomized complete block design. Environments were considered as random effects and entries as fixed effects. The variances for GCA (σ^2_{gca}), SCA (σ^2_{sca}) and error were calculated from expected mean squares of the analysis of variance; the error variances (σ^2_e) were equal to mean square error.

Results and Discussion

The study looked into the combining ability for yield of IR maize inbred lines under striga (*S. hermonthica*) infestation so as to identify the useful lines for use in developing IR maize varieties. The grain yield was the key trait for assessing their performance.

Grain Yield and Plant Characteristics

Significant differences among testcrosses across all sites were observed for grain yield ranging from 0.8 to 3.1 tons ha^{-1} (Table 2). The checks produced lower grain yield compared to some of the testcrosses; an yield of 1.5–2.3 tons ha^{-1} was obtained. The differences were observed between testcrosses across testers and within testers suggesting that the inbred lines had different contributions to yielding ability in their hybrid combinations and indeed the testers were not equally suitable as parents in hybrid combinations.

Similar results were reported by Diallo (2005) when evaluating 320 IR maize testcrosses under striga in Kenya. He identified 16 testcrosses that gave yields above 9 tons ha^{-1} and yielded significantly higher than most of the checks. He noted that hybrids that gave higher performance were the ones that also showed

Table 2 Mean grain yield (GY) (tons ha^{-1}) and other important agronomic characters of IR maize testcrosses tested across five sites

Agronomic characteristics	Mean	Significance	Minimum	Maximum	Signif. sites
GY – across site	1.95	**	0.81	3.13	5
Anthesis date (days)	67.50	**	60.40	73.30	5
ASI (days)	5.70	**	3.10	9.40	3
Plant height (cm)	186.00	**	158.40	222.40	4
Ear height (cm)	79.10	**	62.30	99.90	5
P. sorghi (scores)	1.30	**	1.00	1.90	4
E. turcicum (scores)	2.40	*	1.60	3.00	2
Striga count (6th WAP)	0.10	ns	0.00	2.10	–
Striga count (8th WAP)	0.10	**	0.00	6.70	4
Striga count (10th WAP)	0.30	**	0.10	9.20	5
Striga count (12th WAP)	0.40	**	0.20	11.60	4

** ($P < 0.01$), *($P < 0.05$), ns: not significant

higher resistance to the prevailing stresses such as ear rot, grey leaf spot, *P. sorghi, E. turcicum* and striga. He concluded that IR maize is the best option for poor resource farmers who are faced with striga problem in their farms.

Disease Infection

Results from the current study showed that testcrosses were significantly different ($P < 0.01$) in their reaction to *P. sorghi* at four sites (Table 2). Crosses CML445-IR(BC0)-B-B-60-B/TESTER C, CML445-IR(BC0)-B-B-60-B/TESTER A and CML247-IR(BC0)-B-B-B-99/TESTER A scored 1.0; this was the lowest score. The highest scores of 1.9 and 1.8 were obtained in crosses CML373-IR(BC0)-B-B-93-B/TESTER C and CML373-IR(BC0)-B-B-38-B/TESTER C, respectively. The scores recorded from the checks ranged from 1.2 to 1.8. This suggests that most of the testcrosses were resistant to *P. sorghi*. As was noted by Diallo (2005), such testcrosses give better performance.

Similarly, significant differences ($P < 0.05$) in the incidence of *E. turcicum* among crosses assessed were observed (Table 2). The low scores of 1.6, 1.7 and 1.9 were obtained from crosses CML373-IR(BC0)-B-B-70-B/TESTER B, CML395-IR(BC0)-B-B-41-B/TESTER B and CML390-IR(BC0)-B-B-6-B/TESTER B, respectively. Some of the crosses like CML247-IR(BC0)B-B-B-99/TESTER C, CML444-IR(BC0)-B-B-131-B/TESTER A and CML444-IR(BC0)-B-B-158-B/TESTER A scored highly, 3.0, 3.0 and 2.9, respectively, similar to checks

with scores ranging from 2.4 to 3.0. These results strongly suggest differences in the reaction to *E. turcicum*, though the disease had no effect on grain yield.

Striga Count

Striga counts as an indication of resistance to imidazolinone herbicide varied among testcrosses. Significant difference was observed at the 8th, 10th and 12th weeks after planting, not at earlier stages (Table 2). At the 8th week testcross CML247-IR(BC0)-B-B-B-46/TESTER A was the only one with significant count of 0.1. By the 10th week, three testcrosses, CML78-IR(BC0)-B-B-B-130/TESTER B, CML312-IR(BC0)-B-B-B-60/TESTER C and CML445-IR(BC0)-B-B-37-B/TESTER A, were significantly infested with mean counts of 0.4, 0.3 and 0.3, respectively. The situation at the 12th week was almost similar with only two testcrosses, CML445-IR(BC0)-B-B-117-B/TESTER A and CML247-IR(BC0)-B-B-B-99/TESTER A, showing significant mean striga counts of 1.1 and 0.9, respectively. This indicates how the testcrosses had different levels of resistance to imidazolinone herbicide. The variation in herbicide resistance resulted in different levels of striga infestation.

Combining Ability

The study focused on GCA and SCA for grain yield of imidazolinone-resistant maize inbred lines under striga conditions; grain yield was used as the trait proxying

Table 3 Combined analysis of variance for grain yield (GY) (tons ha^{-1}) across five sites in Tanzania and Kenya

Source of variation	DF	MS	F-value	P-value
Location	4	80.98	357.64**	0.00
GCA (line)	30	1.81	2.40**	0.00
GCA (line) × location	120	0.28	1.24^{ns}	0.08
Tester	2	21.41	28.39**	0.00
Tester × location	8	0.85	3.75**	0.00
SCA (line × tester)	60	0.75	1.61**	0.00
SCA (line × tester) × location	240	0.23	0.48^{ns}	1.00
Pooled error	755	0.47		
Total	1219			
σ^2_{gca}		0.55		
σ^2_{sca}		0.14		

** ($P<0.01$); ns: not significant; DF: degrees of freedom; MS: mean square

the resistance. The analysis across sites showed significant effects ($P < 0.01$) for GCA, SCA and GCA × location (Table 3). Previous research done by Kaya (2004) showed that GCA and SCA effects were significant for all traits under the study. The significance of GCA and SCA effects for the traits showed the importance of both additive and non-additive gene effects. The significant SCA effects detected in the traits imply the contribution of non-additive gene effects to the phenotypic variation among the hybrids. Sakila et al. (2000) observed the same differences on the lines, testers and line × tester interactions.

Everett et al. (1995) evaluated the optimal combining ability patterns among promising populations for inbred lines development in tropical mid-altitude zones and detected highly significant differences for GCA, SCA and GCA × environment interaction. From the results, they suggested the need for selecting different parental lines for hybrid development for specific environment.

GCA Effects

Significant positive GCA effects for grain yield (GY) were observed in 12 inbred lines (Table 4). Lines 22 and 29 were the best general combiners expressing grain yield increase of +0.57 and +0.78 tons ha^{-1}, respectively. The poorest line was L14 with a yield reduction of –0.57 tons ha^{-1}. The positive GCA effects indicate that the additive gene effects contributed to the increase of grain yield. Similar results were reported by Nass et al. (2000), where a

line with large positive GCA and another with negative GCA were identified. In their study more than half of the lines had positive GCA effects, indicating that on average these lines contributed to grain yield increase. They also pointed out that, genetically, the GCA effects from one environment contain the average GCA effects and GCA × environment interaction. As a result, when estimates of GCA × environment interaction are near to zero, the average GCA effects will approach the GCA effects obtained for each environment. Therefore, the selection of the best parents based on the average GCA effects can be done if there is interest in single cross hybrids adapted to all environments.

The estimation of variance due to general combining ability (σ^2_{gca}) and specific combining ability (σ^2_{sca}) showed that the former was higher (0.55) than the latter (0.14) for grain yield (Table 3). This indicated a predominantly additive gene action for this trait. Variances of 0.55 and 0.14 were estimated for GCA and SCA, respectively (Table 3). The SCA variances are used to determine the homogeneity with which inbred lines transmit yielding abilities to the progenies. Relative magnitudes of GCA and SCA variances are useful in identifying superior lines. According to Griffing (1956), the ideal SCA variance is one. High SCA variance designates high variability in transmitting yielding ability to the progenies, whereas low SCA variance indicates less variability in transmitting yielding ability. The inbred lines assessed had lower SCA variance (0.14), which implies that the lines had less variation in transmitting yielding ability. On the basis of the above, comparing the magnitudes of GCA and SCA variances, lines 3, 4, 15, 21, 22, 23, 24, 25, 27, 29, 30 and 31 were identified as superior lines.

Table 4 GCA effects for grain yield (GY)

Line no.	Pedigree	GY (tons ha^{-1})	Line no.	Pedigree	GY (tons ha^{-1})
L1	CML247-IR(BC0)-B-B-B-108	−0.18	L17	CML395-IR(BC0)-B-B-41-B	−0.35
L2	CML247-IR(BC0)-B-B-B-110	−0.09	L18	CML444-IR(BC0)-B-B-131-B	−0.12
L3	CML247-IR(BC0)-B-B-B-46	0.04	L19	CML444-IR(BC0)-B-B-154-B	−0.21
L4	CML247-IR(BC0)-B-B-B-99	0.31	L20	CML444-IR(BC0)-B-B-158-B	−0.13
L5	CML312-IR(BC0)-B-B-B-60	−0.20	L21	CML445-IR(BC0)-B-B-117-B	0.46
L6	CML312-IR(BC0)-B-B-B-65	−0.05	L22	CML445-IR(BC0)-B-B-23-B	0.57
L7	CML312-IR(BC0)-B-B-B-67	−0.24	L23	CML445-IR(BC0)-B-B-37-B	0.22
L8	CML312-IR(BC0)-B-B-B-79	−0.24	L24	CML445-IR(BC0)-B-B-5-B	0.40
L9	CML312-IR(BC0)-B-B-B-92	−0.09	L25	CML445-IR(BC0)-B-B-60-B	0.30
L10	CML373-IR(BC0)-B-B-180-B	−0.22	L26	CML78-IR(BC0)-B-B-B-123	−0.31
L11	CML373-IR(BC0)-B-B-38-B	−0.43	L27	CML78-IR(BC0)-B-B-B-129	0.25
L12	CML373-IR(BC0)-B-B-55-B	−0.01	L28	CML78-IR(BC0)-B-B-B-130	−0.18
L13	CML373-IR(BC0)-B-B-70-B	−0.20	L29	CML78-IR(BC0)-B-B-B-26	0.78
L14	CML373-IR(BC0)-B-B-93-B	−0.57	L30	CML78-IR(BC0)-B-B-B-62	0.07
L15	CML384-IR(BC0)-B-B-B-126	0.05	L31	CML78-IR(BC0)-B-B-B-7	0.52
L16	CML390-IR(BC0)-B-B-6-B	−0.13	Mean		0.00
			LSD$_{(0.05)}$		0.02

Non-significant difference for the GCA × E interactions (Table 3) implies that the contribution of additive genes to expression of grain yield is the same in the different environments where the testcrosses were evaluated. However, testing lines at different sites is important because it ensures selection of correct inbred lines, which are stable in their performance under targeted stress across locations. Scott (1967) mentioned that, in the absence of significant GCA × E interaction, it is advisable to select stable genotypes.

The contribution of GCA sum of squares to entry total sums of squares for grain yield was relatively higher than from the SCA, 38% against 32%. This suggests that the additive gene effects were more important than the non-additive gene effects for this set of entries and were the main source of variation in inbred lines on striga resistance. Similar results were reported by Nass et al. (2000): for the hybrids which expressed higher grain yields than the hybrid checks, the contributions of the GCA and SCA were approximately 58 and 44%, respectively. These results showed that for the best single crosses, both GCA and SCA effects were important for the grain yield but GCA effects were more important than the SCA effects due to predominant additive gene effects. Beck et al. (1989) also mentioned that high proportion of GCA sums of squares compared to SCA sums of squares for grain yield indicates the importance of additive gene effects against non-additive gene effects for the trait of interest.

SCA Effects

The highly significant positive SCA effects were expressed in 14 lines with tester heterotic group A, 21 lines with tester heterotic group B and 16 lines with tester heterotic group C. The highest effects were observed in line L30 (heterotic group A), L23 (heterotic group B) and L11 (heterotic group C) with yield increase of 0.58, 0.32 and 0.80 tons ha^{-1}, respectively (Table 5). The positive SCA effects imply that the variation on yield among inbred lines is contributed by non-additive gene effects. Nass et al. (2000) also reported the highly significant positive and negative SCA effects. The largest positive and negative SCA effects for grain yield from their study were observed in L5 × L8 and L6 × L7 crosses, respectively. Therefore, the positive GCA and SCA effects expressed by the lines showed that the variation among inbred lines on grain yield was due to both additive and non-additive gene effects, but additive gene effects were more predominant than non-additive gene effects. These findings agree with a study conducted by Garay et al. (1996) who found additive and non-additive gene effects for grain yield. This signifies that the differences in grain yield among inbred lines were caused by both gene effects.

SCA (line × tester) mean squares for grain yield were significantly different ($P < 0.01$). The interactions among line, tester and locations (line ×

Table 5 SCA effects for grain yield (GY) (tons ha^{-1}) and heterotic groups for hybrid and open pollinated testers

		SCA effects		
Line no.	Pedigree	CML373-IR/CML393-IR (heterotic group A)	CML202-IR/CML395-IR (heterotic group B)	IR OPV (Synthesis 2000-IR) (heterotic group C)
L1	CML247-IR(BC0)-B-B-B-108	0.21	−0.18	−0.03
L2	CML247-IR(BC0)-B-B-B-110	−0.06	0.08	−0.03
L3	CML247-IR(BC0)-B-B-B-46	−0.12	0.29	−0.17
L4	CML247-IR(BC0)-B-B-B-99	−0.02	0.13	−0.11
L5	CML312-IR(BC0)-B-B-B-60	−0.13	−0.18	0.31
L6	CML312-IR(BC0)-B-B-B-65	0.12	−0.27	0.15
L7	CML312-IR(BC0)-B-B-B-67	−0.08	0.12	−0.03
L8	CML312-IR(BC0)-B-B-B-79	0.57	0.05	−0.62
L9	CML312-IR(BC0)-B-B-B-92	−0.18	0.23	−0.04
L10	CML373-IR(BC0)-B-B-180-B	−0.12	−0.21	0.33
L11	CML373-IR(BC0)-B-B-38-B	−0.67	−0.13	0.80
L12	CML373-IR(BC0)-B-B-55-B	−0.46	0.16	0.30
L13	CML373-IR(BC0)-B-B-70-B	−0.22	0.10	0.13
L14	CML373-IR(BC0)-B-B-93-B	0.20	0.14	−0.34
L15	CML384-IR(BC0)-B-B-B-126	−0.54	0.04	0.50
L16	CML390-IR(BC0)-B-B-6-B	−0.20	0.03	0.17
L17	CML395-IR(BC0)-B-B-41-B	0.12	−0.31	0.19
L18	CML444-IR(BC0)-B-B-131-B	0.16	−0.04	−0.12
L19	CML444-IR(BC0)-B-B-154-B	0.53	0.03	−0.56
L20	CML444-IR(BC0)-B-B-158-B	0.14	−0.13	0.00
L21	CML445-IR(BC0)-B-B-117-B	−0.28	0.14	0.14
L22	CML445-IR(BC0)-B-B-23-B	0.25	0.06	−0.32
L23	CML445-IR(BC0)-B-B-37-B	0.05	0.32	−0.37
L24	CML445-IR(BC0)-B-B-5-B	0.32	0.12	−0.44
L25	CML445-IR(BC0)-B-B-60-B	−0.13	0.04	0.09
L26	CML78-IR(BC0)-B-B-B-123	−0.20	0.24	−0.04
L27	CML78-IR(BC0)-B-B-B-129	0.26	0.14	−0.41
L28	CML78-IR(BC0)-B-B-B-130	−0.21	0.02	0.19
L29	CML78-IR(BC0)-B-B-B-26	−0.23	0.18	0.05
L30	CML78-IR(BC0)-B-B-B-62	0.58	−0.70	0.12
L31	CML78-IR(BC0)-B-B-B-7	0.35	−0.50	0.15
Mean		0.00	0.00	0.00
LSD$_{(0.05)}$		0.04	0.04	0.04

tester × environment) did not show significant difference for grain yield (Table 3). The significance of SCA mean squares for grain yield implies that there are variations in the trait that are controlled by non-additive genes, and these can enable identification of promising crosses based on SCA effects.

SCA × environment (E) interaction shows how the lines express themselves on their performance under different locations. The present study showed that the SCA × E interaction had no influence on grain yield change within lines. This implies that the difference in yield performance had not been influenced by non-additive gene effects and locations. Kaya

(2004) also reported that, from the study, the hybrid × environment interactions were not significantly different across locations. The non-significant genotype × environment (G × E) interactions imply the possibility of evaluating genotypes without diverse environmental conditions influencing the traits.

Conclusion

Selection of superior inbred lines that are resistant to imidazolinone herbicide should not be based on agronomic characters even if they are significant. The emphasis will be on grain yield expressed by the lines.

From the study 12 inbred lines 3, 4, 15, 21, 22, 23, 24, 25, 27, 29, 30 and 31 were identified as potential lines for grain yield; the crosses developed from these lines produced high grain yield. The lines can be used in developing imidazolinone-resistant maize varieties through application of appropriate selection methods, such as backcross selection and recurrent selection that can efficiently pyramid desirable alleles for the trait.

Acknowledgements We are grateful to Rockefeller Foundation for provision of funds and all members of staff in the Department of Crop Science at the University of Zambia for technical support. We also thank CIMMYT staff in Kenya for provision of IR maize crosses used in the study and their technical contributions. The Agricultural Research Institute Ukiriguru Mwanza Tanzania staff are also appreciated for their technical assistance and advices during the period of the study.

References

Beck DL, Vasal SK, Crossa J (1989) Heterosis and combining ability of CIMMYT'S tropical early and intermediate maturity maize (Zea mays L.) germplasm. Maydica 35:279–285

Comstock RE, Robinson HF (1948) The components of genetic variance in populations of biparental progenies and their uses in estimating the average degree of dominance. Biometrics 4:254–266

De Groote H, Wangare L (2002) Potential market for Striga resistant maize seed. In: Kanampiu F (ed) Herbicide resistant maize method for Striga control. Meeting to explore the commercial possibilities, 4–6th July, 2002. Proceedings of Kisumu. CIMMYT, Nairobi, Kenya, pp 18–23 (unpublished)

Diallo AO (2005) Summary of breeding activities. CIMMYT, Nairobi, Kenya, pp 2–6

Everett LA, Eta-ndu JT, Ndioro M, Walker P (1995) Combining ability among source populations for tropical mid-altitude maize inbreds. Maydica 40:165–171

FAO (1980) Improvement and production of maize, sorghum and millets vol 2 – Breeding, agronomy and seed production. Publication division, Food and Agriculture Organization of the United Nations, Rome, Italy, pp 1–14

Garay G, Igartua E, Aivarez A (1996) Genetic changes associated with S_1 recurrent selection in flint and dent synthetic populations of maize. Crop Sci 36:1129–1134

GoT (Government of Tanzania) (2003) Population and housing census 2002 General Report. Central Census Office, National Bureau of Statistics, President's Office, Planning and Privatization, Dar es Salaam, Tanzania

Griffing B (1956) Concept of general and specific combining ability in relation to diallel crossing systems. Austr J Biol Sci 9:463–493

Heisey PW, Mwangi W (1997) Fertilizer use and maize products. In: Byrlee D, Eicher CK (eds) Africa's emerging maize revolution. Lynne Rienner, Boulder, CO

Kaliba ARM, Verkuijl H, Mwangi. W (2000) Factors affecting adoption of improved maize seeds and use of inorganic fertilizer for maize production in the intermediate and lowland zones of Tanzania. Agric Appl Econ 32(1): 35–37

Kanampiu FK, Ransom JK, Gressel. J (2001) Imazapyr seed dressings for Striga control on acetolactate synthase target site resistant maize. Crop Prot 20:885–895

Kaya Y (2004) Determining combining ability in sunflower (Helianthus annuus L). Turkey Agric J 29:243–250

Moshi AJ, Mduruma JO, Lyimo NG, Marandu WF, Akoonay HB (1990) Maize breeding for target environment in Tanzania. In: Moshi AJ, Ramson JK (eds) Maize research in Tanzania. Proceedings of the First Tanzania National Maize Research Workshop, Dar es Salaam, Tanzania, pp 11–16

Nass LL, Lima M, Vencovsky R, Gallo B (2000) Combining ability of maize inbred lines evaluated in three environments in Brazil. Sci Agric 5:7

Pingali PL (2001) CIMMYT 1999–2000 world maize facts and trends meeting world maize needs: technological opportunities and priorities for the public sector. CIMMYT, Mexico, D.F

Sakila M, Ibrahim SM, Kalamani A, Backiyarani. S 2000. Evaluation of Sesame hybrids through line × tester analysis. Sesame and Safflower newsletter No. 15

Scott GE (1967) Selecting for Stability of yield in maize. Crop Sci 7:549–551

Singh RK, Chaudhary. BD (1985) Biometrical methods in quantitative genetic analysis. Kalyani Publishers, New Delhi, Ludhiana, pp 205–214

Tyagi AP, Lal P (2005) Line × tester analysis in sugarcane (Saccharum officinarum). Department of Biology, School of Pure and Applied Sciences, The University of the South Pacific, Suva, Fiji. S Pac J Nat Sci 23:30–36

Identification of Plant Genetic Resources with High Potential Contribution to Soil Fertility Enhancement in the Sahel, with Special Interest in Fallow Vegetation

S. Tobita, H. Shinjo, K. Hayashi, R. Matsunaga, R. Miura, U. Tanaka, T. Abdoulaye, and O. Ito

Abstract The sandy soil in the Sahel is characterized as low inherent fertility, that is, having nutrient deficiency (total N and available P), low organic matter and high risk of erosion. Under the concept of integrated soil fertility management (ISFM), possible contribution of natural inhabitant plants to the improvement of soil fertility in the Sahel was evaluated. A broad variation in $\delta^{15}N$ values was observed among the plant species commonly found in cropland and fallow land of the Sahelian zone. Annual leguminous herbs, *Cassia mimosoides* (Caesalpiniaceae) and *Alysicarpus ovalifolius* (Papilionaceae), had low $\delta^{15}N$ values, showing their higher dependency on biological nitrogen fixation. They will be efficiently utilized as an extensive means of soil fertility management, for example, through more encouraged incorporation into the fallow vegetation. *Ctenium elegans*, *Eragrostis tremula* and *Schizachyrium exile*, greatly dominating annual grass species in the fallow land, though their $\delta^{15}N$ values were high, would contribute to the soil fertility by supplying a significant amount of organic matter.

Keywords Biological nitrogen fixation · Fallow vegetation · ISFM · Leguminous plants · Soil organic matter

Introduction

As well as uncertain rainfall and locust outbreak, poor nutrient (carbon, nitrogen and phosphorus) status of the sandy soils is one of the causes for low and unstable productivity in agroecosystems in the Sahel. Therefore, the Sahel is very marginal for agricultural production in the world, where the success of green revolution is most wanted among the African continent.

In general, improvement of soil fertility through indigenous organic matter management is proposed with high possibility and practicality for smallholder farmers in sub-Saharan Africa (Lal, 2007). Concerning nitrogen, a most limiting factor in the soil, biological nitrogen fixation (BNF) by plant–microbe interactions (symbiotic and associative) can be efficiently utilized in croplands and fallow lands (Sanginga, 2003; Ito et al., 2005).

In croplands of the Sahel, cowpea (*Vigna unguiculata* [L.] Walp.) is the most important leguminous crop, not only for human food but also for animal fodder. Therefore, the identification of such dual-purpose cowpea cultivars which adapt well to the Sahel environment and have higher BNF ability could contribute to the improvement of soil fertility in this area. In collaboration with IITA (International Institute of Tropical Agriculture) and INRAN (Institute National de la Recherche Agronomique du Niger), a total of 140 cowpea lines were evaluated (Matsunaga et al., 2006, 2007) and best cultivars identified for dissemination to farmers of the Sahel (Matsunaga and Tobita, 2007).

Fallow systems have been locally practiced by farmers for soil fertility maintenance and restoration as an extensive means of management in the Sahel.

S. Tobita (✉)
JIRCAS, Tsukuba, Ibaraki, Japan
e-mail: bita1mon@jircas.affrc.go.jp

A. Bationo et al. (eds.), *Innovations as Key to the Green Revolution in Africa*,
DOI 10.1007/978-90-481-2543-2_71, © Springer Science+Business Media B.V. 2011

However, soil N supplying ability evaluated by N uptake by pearl millet plants (*Pennisetum glaucum* L.), staple crop in the Sahel, was low from fallow lands as compared with those from other soil fertility management, e.g. animal corralling and manure or household waste transporting (Suzuki et al., 2007). Therefore, it is strongly required to improve the fallow system for more effective soil fertility management especially in remote fields that are far away from the farmers' household.

Fallow vegetation has a role in field to retain and capture the wind-borne fertile materials from cropped area by quantitative aerodynamic observations (Herrmann, 1996; Ikazaki et al., 2007). It may lead to propose the physical improvement of fallow vegetation with temporal and spatial rearrangements.

Another aspect of the improvement of fallow vegetation could be biological, say, encouragement of plant species which contribute more to soil fertility. In this study, an attempt was made to evaluate the natural fallow plant species in the Sahel for biological nitrogen fixation as measured by natural abundance of ^{15}N, i.e. δ^{15}N (‰).

Materials and Methods

The site of this study is located in the Fakara area in Dantiandou District of Tillaberi Prefecture (80 km east of Niamey, the capital of Niger), a typical region of the Sahel with a good mixture of cultivators and pastoralists.

Plants of higher dominance in fallow lands were photographed and collected in the dry season of 2003 and rainy season of 2004 and 2005, from fallow lands of different ages and crop fields at the Fakara region. Each plant species was identified by a scientific name from local nomenclature (Zarma and Peul) with use of the "Lexique de noms vernacularies des plantes du Niger", published by INRAN (1977), and verified by ordinary ways of plant taxonomy. This work was provisionally summarized as a pictorial dictionary of "Fakara Plants" to be accessed through the web (Miura and Tobita, 2005).

The plant samples were dried and finely powdered with use of a multi-beads shocker (Yasui Kagaku Co. Ltd., Osaka, Japan) and a vibrating sample mill (CMT, Iwaki, Japan). An adequate quantity (around 3 mg) of the samples was then introduced to an element analyser (Flash EA-1112, Carlo Erba, Milan, Italy) connected with a continuous-flow isotope ratio mass spectrometry (Delta XP^{plus}, ThermoFinnigan, Hamburg, Germany) connected with an element analyser (Flash EA-1112) to determine the compositional deviation of N stable isotopes (^{15}N to ^{14}N) in total plant N, expressed as δ^{15}N (‰). δ^{15}N has been used as an indicator of their dependency on air nitrogen (N_2), because air and soil are exclusive sources of N and they generally have a sufficient difference in their δ^{15}N values (0‰ versus +5 to +10‰). Thus, lower δ^{15}N values (occasionally negative) in plants show their higher dependency on N_2 fixation. In this study, by means of the natural abundance method, the δ^{15}N values of fallow plant species were measured for the estimation of their dependency level on BNF and its contribution to the maintenance and improvement of N fertility of the sandy soils in the Sahel.

Results and Discussion

The principle of the δ^{15}N natural abundance method is based on two prerequisites as follows (Unkovich and Pate, 2001): (i) only two N pools (soil N and air dinitrogen) for plant growth and (ii) significant difference in natural δ^{15}N abundance between the two N pools. To satisfy the first requisite, all plant samples were taken from fields with no fertilizer-N application. The δ^{15}N value of non-N-fixing pearl millet plants from field with no N application was +5.1‰, which was thought to reflect the δ^{15}N value of the soil N. Therefore, the second requisite was also fulfilled because it deviated enough from that of air N, viz., theoretically 0‰. If the dependency of the plant on BNF is higher, δ^{15}N value becomes more negative because the ratio of N from soil to N from the air is decreased.

For the plants collected in the dry season of 2003, out of 28 species comprising 12 families, the lowest δ^{15}N (highest dependence on BNF) was recorded in *Cassia mimosoides* (Caesalpiniaceae), an annual leguminous herb, having −0.97‰. Young twigs of *Acacia albida* (Mimosaceae), a leguminous tree, had its δ^{15}N value of +1.02‰. It is interesting that some of annual grass species such as *Ctenium elegans*, *Eragrostis tremula* and *Schizachyrium exile* (all belonging to Gramineae) had lower δ^{15}N value around +2.0 to +2.5‰, especially from lands with a longer fallow duration. This may suggest a possibility of biological nitrogen fixation by associative microorganisms in non-leguminous plant species.

In rainy season, fallow vegetation was more abundant in biomass and number of plant species, as compared with dry season. Samples of a total of 45 species from 18 families were collected in August 2004. Some of them, such as *Alysicarpus ovalifolius* (Papilionaceae) and *Commelina forskalaei* (Commelinaceae), were revealed to have higher dependency on BNF, +0.57 and +1.85‰, respectively, as well as *C. mimosoides* having −1.76‰, the lowest.

In the 2005 rainy season, native plants were collected from fallow lands on which the period of fallow experiences was recognized. The vegetation of the 11-year-old fallow was dominated by annual Gramineae species, *C. elegans* and *S. exile*, and perennial shrub, *Guiera senegalensis* (Combretaceae). The $\delta^{15}N$ values in this fallow vegetation ranged from −2 to +8‰ for all plant species (Fig. 1a). Three leguminous herbs, *C. mimosoides*, *A. ovalifolius* and *Indigofera pilosa*, had $\delta^{15}N$ values at less than 0‰, showing higher dependency on BNF.

Samples of more number (22) of plant species were collected from the 3-year-old fallow land,

where non-leguminous broad-leaf annual herbs, *Mitracarpus scaber* (Rubiaceae), *Pergularia tomentosa* (Asclepiadaceae) and *Jacquemontia tamnifolia* (Convolvulaceae), were dominant. Averages and ranges in $\delta^{15}N$ values of the plants, shown in Fig. 1b, clearly demonstrate much higher dependency on BNF in leguminous plant species (*C. mimosoides*, etc.) than others. It is noteworthy that some non-leguminous *Hibiscus* sp. (a wild relative of *Hibiscus sabdarifa*), *G. senegalensis* and *S. exile* had relatively lower $\delta^{15}N$ values (+3 to +4‰). This may suggest possible N_2 fixation by non-legumes associated with diazotrophic microorganisms in the soil of the Sahel.

In the new fallow site, a total of 27 species were collected but the total biomass was not high, because of pearl millet cultivation and weeding activity by farmers during the last season. The range of $\delta^{15}N$ values shifted to more positive values, from +2 to +14‰ for all plants (Fig. 1c), as compared with those from the 11- and 3-year-old fallow lands. This could be attributed to much higher $\delta^{15}N$ in soil N after several years of crop cultivation, which might be partially supported by relatively lower $\delta^{15}N$ in perennial *G. senegalensis* and *Piliostigma reticulatum* (a leguminous tree). Moreover, $\delta^{15}N$ values in leguminous herbs such as

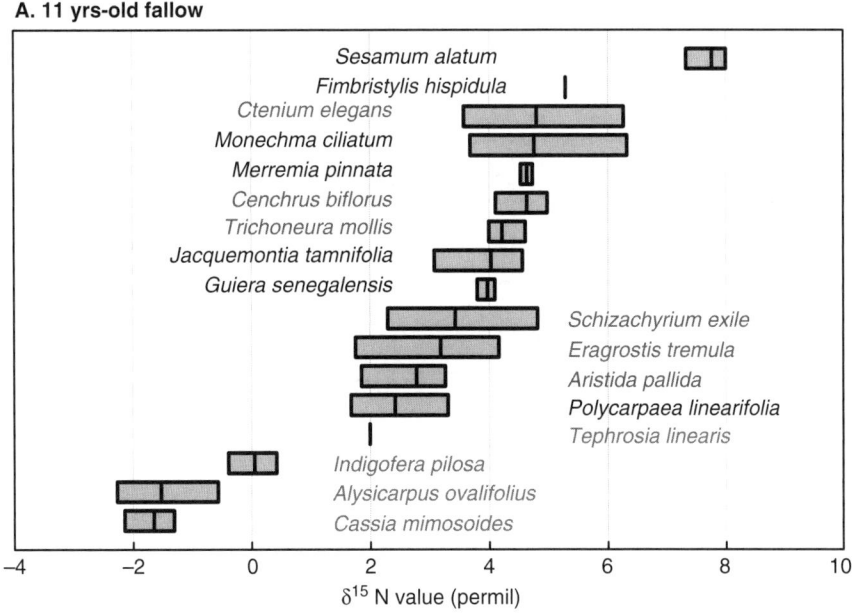

Fig. 1 Ranges (*extent of each horizontal bar*) and averages (*vertical line in the bar*) in $\delta^{15}N$ values of the native plants which were collected at the site undergoing 11-year (**a**) and 3-year (**b**) fallow periods and new fallow land (**c**). Scientific nomenclatures are written beside the bars

B. 3 yrs-old fallow

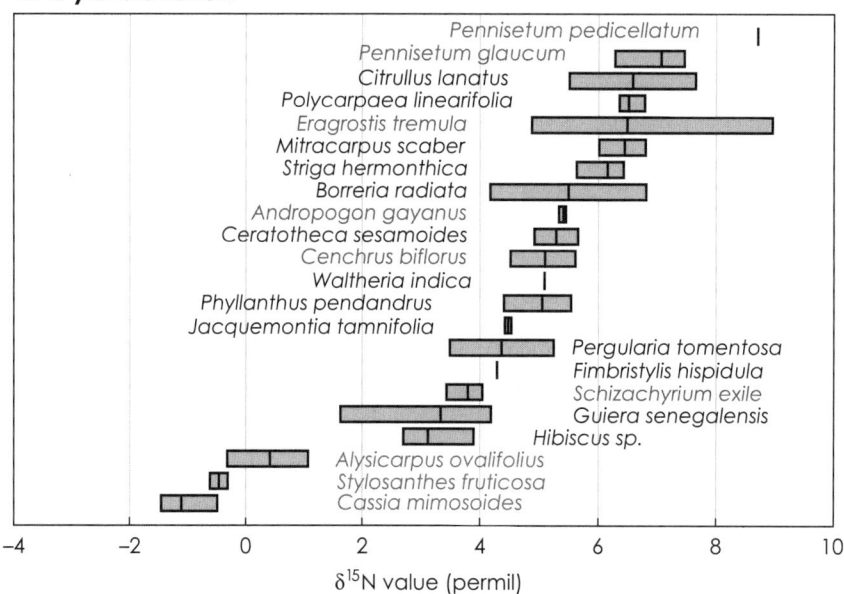

C. New fallow after one-year cultivation

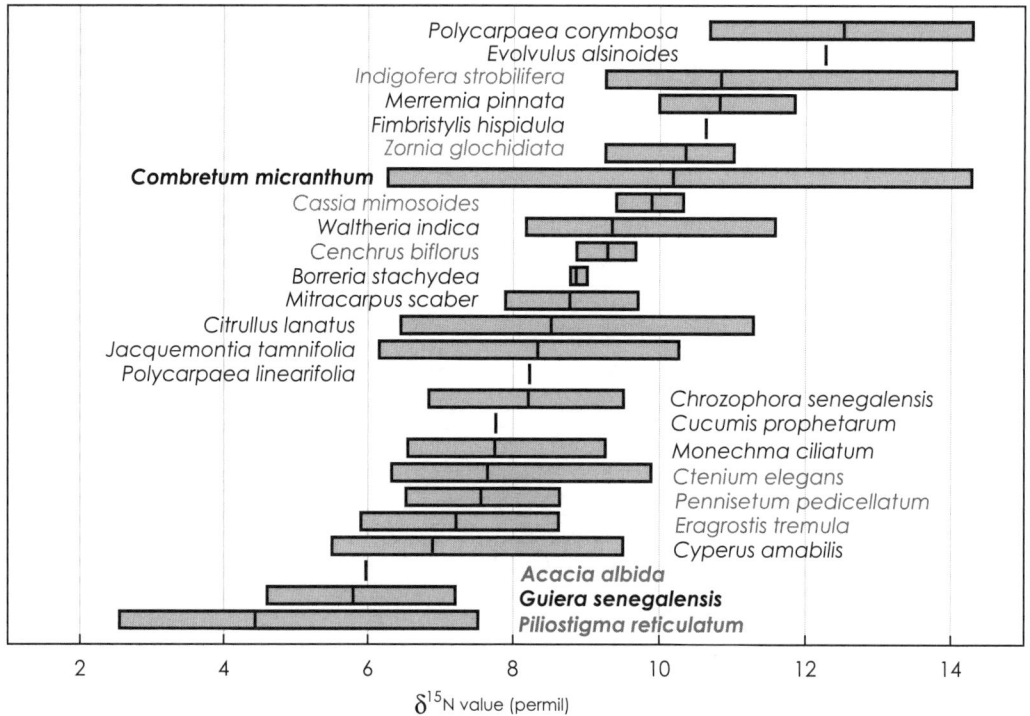

Fig. 1 (continued)

C. mimosoides were higher as other non-leguminous plants in this new fallow land. It is speculated that BNF by the legumes–rhizobia symbiosis may not be fully activated in lands just after being used for intensive cropping years.

Elucidation of the N cycle with precise quantification of each flow between the agro-ecosystem components is a prerequisite for the full deployment of BNF in potential plants of the Sahel. Studies on farming systems in the wet savannah of southern Mali (Ramesch, 1999) and southern Senegal (Manlay et al., 2002) are excellent precedents of the N balance sheet work, but the contribution of BNF was not estimated. It is shown in this study that plant $\delta^{15}N$ could change as affected not only by its dependency on BNF but also by the age of the fallow where it habitats. For quantitative estimation of N fixation by fallow vegetation, therefore, studies have been going on to compare $\delta^{15}N$ among plant species grown on homogenous soil which has a uniform N profile in content and isotopic ratio.

Information on the ability of nitrogen acquisition from the air will be utilized for identification of fallow plant species with high potential of the contribution to soil fertility improvement, coupled with annual biomass production data through the organic matter input to the soil. Table 1 is a summarized list of natural fallow plant species with possible high contribution to the soil fertility improvement of the Sahel. They will be necessarily examined with information on ecological and livestock-related issues, say, possible interspecific competition, fodder value and grazing tolerance (Hiernaux, 1998).

On efficient utilization of limited nutrients (mostly organic matters) in sandy soils of the Sahel, Gandah et al. (2003) pointed out that the traditional fertility management ways such as cattle manure application and pearl millet stover returning had shown substantial effects on soil nutrients status, but they did not mention clearly about the contribution of fallow plants. Our concurrent activities to quantify the N budget of the agro-ecosystem in the target site, Fakara area, are showing that technology options associated with fallow system should be developed for the improvement of soil fertility in the Sahel, especially in remote fields that are far away from farmers' household. This corresponds to a part of the integrated soil fertility management (ISFM) concept (CGIAR, 2002), focusing on the utilization of nitrogen-fixing plants and soil microorganisms.

Conclusions

Among the natural fallow plant species in the Sahel, an annual leguminous herb, *C. mimosoides* (=*Chamaecrista mimosoides*), Caesalpiniaceae, had the lowest $\delta^{15}N$ value, showing its higher dependency on biological nitrogen fixation. It will be efficiently utilized as an extensive means of the soil fertility management, for example, through more encouraged incorporation into the fallow vegetation. Annual grass species, *C. elegans*, *E. tremula* and *S. exile*, were greatly dominating in the fallow land as well as a shrub

Table 1 A list of natural fallow plant species with potential contribution to the soil fertility improvement in the Sahel, as evaluated by frequency, biomass production and dependency on BNF

Species	Local nomenclature	Family	Frequency	Biomass	Dependency on air N
Acacia albida	Gao	Mimosaceae	Very high	High	High
Alysicarpus ovalifolius	Gadagi	Papilionaceae	Low	Low	High
Andropogon gayanus	Andro	Gramineae	High	High	Low
Aristida longiflora	Subu Kware	Gramineae	Very high	Very high	Low
Cassia mimosoides	Ganda bani	Caesalpiniaceae	High	High	Very high
Cenchrus biflorus	Dani	Gramineae	High	Low	Low
Ctenium elegans	Bata	Gramineae	Very high	Very high	Low
Guiera senegalensis	Sabara	Combretaceae	Very high	Very high	Low
Jacquemontia tamnifolia	Hurkutu	Convolvulaceae	Very high	High	Low
Mitracarpus scaber	Hinkini a kangi	Rubiaceae	High	Low	Low
Pergularia tomentosa	Fattakka	Asclepiadaceae	High	High	Low
Schizachyrium exile	Subu kirey	Gramineae	Very high	Very high	Low
Sida cordifolia	Kongoria	Malvaceae	High	High	Low
Zornia glochidiata	Marbku	Papilionaceae	Very high	High	High

tree, *G. senegalensis*. Though their $\delta^{15}N$ values were high, they would contribute to the soil fertility by supplying a significant amount of organic matter into the soil.

Acknowledgements This study was conducted as part of the collaborative research project between JIRCAS and ICRISAT on the "Improvement of Fertility of Sandy Soils in the Semi-Arid Zone of West Africa through Organic Matter Management". The authors thank the staff of ICRISAT-WCA, especially Mr. Sodja Amadou for his assistance in Fakara area.

References

CGIAR (2002) Combating soil fertility degradation in Africa. Paper presented in the Annual General Meeting 2002, Washington, DC, USA, 30–31 Oct 2002

Gandah M, Bouma J, Brouwer J, Hiernaux P, van Duivenbooden N (2003) Strategies to optimize allocation of limited nutrients to sandy soils of the Sahel: a case study from Niger, west Africa. Agric Ecosyst Environ 94:311–319

Herrmann L (1996) Straubdeposition auf Böden West Afrikas. Eigenschaften und herkunfisgebiete der Stäube und ihr Einflub auf Boden und Standortseigenschaften. PhD thesis, Universität Hohenheim, Stuttgart, Germany, 239pp

Hiernaux P (1998) Effects of grazing on plant species composition and spatial distribution in rangelands of the Sahel. Plant Ecol 138:191–202

Ikazaki K, Shinjo H, Tanaka U, Kosaki T (2007) Development of a new sediment catcher to evaluate the effect of wind erosion on carbon dynamics in the Sahel, west Africa. Abstracts of the international symposium on organic matter dynamics in agro-ecosystems, Poitier, France, 16–19 July 2007

INRAN (1977) Lexique de Noms Vernaculaires de Plantes du Niger, 2nd edn, INRAN, Niamey, Niger

Ito O, Matsunaga R, Kumashiro T (2005) Challenge for improvement of soil fertility in West Africa. Paper presented at a J-FARD & JIRCAS symposium on perspectives of R & D for improving agricultural productivity in Africa, Tokyo, Japan, 14–15 July 2005

Lal R (2007) Promoting technology adoption in sub-Saharan Africa, South Asia. CSA News 52(7):10

Manlay RJ, Kaire M, Masse D, Chotte J-L, Ciornei G, Floret C (2002) Carbon, nitrogen and phosphorus allocation in agro-ecosystems of a West African savanna. I. The plant component under semi-permanent cultivation. Agric Ecosyst Environ 88:215–232

Matsunaga R, Singh BB, Adamou M, Tobita S, Hayashi K, Kamidohzono A (2006) Evaluation of cowpea germplasm in the infertile sandy soil of the Sahelian zone. II. Grain and fodder yield. Abstracts of the 99th annual meeting of the japanese society of tropical agriculture, Tsukuba, Japan, 31 Mar–1 Apr 2006

Matsunaga R, Singh BB, Adamou M, Tobita S, Hayashi K, Kamidohzono A (2007) Evaluation of cowpea germplasm in the infertile sandy soil of the Sahelian zone. III. Nitrogen fixation and phosphorus utilization. Abstracts of the 101st annual meeting of the Japanese society of tropical agriculture, Tokyo, Japan, 30–31 Mar 2007

Matsunaga R, Tobita S (2007) Identification of improved dual-purpose cowpea varieties for the Sahel. JIRCAS Annual Report 2006, pp 28–29

Miura R, Tobita S (2005) Fakara Plants – A photographic guide to common plants of Sahel – (Preliminary Version). http://ss. jircas.affrc.go.jp/project/africa_dojo/Fakara_plants/Fakara_Plants_home.html

Ramesch J (1999) In the balance? Evaluating soil nutrient budgets for an agro-pastoral villages of Southern Mali, Managing Africa's Soils no. 9. Russell Press, Nottingham

Sanginga N (2003) Role of biological nitrogen fixation in legume based cropping systems; a case study of West Africa farming systems. Plant Soil 252:25–39

Suzuki K, Matsunaga R, Hayashi K, Okada K (2007) The effect of the different managements continued by farmers on the nitrogen supplying capacity of the soils in the Fakara region, Niger. Abstracts of the 101st annual meeting of the Japanese society of tropical agriculture, Tokyo, Japan, 30–31 Mar 2007

Unkovich M, Pate JS (2001) Assessing N_2 fixation in annual legumes using ^{15}N natural abundance. In: Unkovich M, Pate J, McNeill A, Gibbs DJ (eds) Stable isotope techniques in the study of biological processes and functioning of ecosystems. Kluwer, Dordrecht, pp 103–118

Within-Farm Variability in Soil Fertility Management in Smallholder Farms of Kirege Location, Central Highlands of Kenya

J.M. Muthamia, D.N. Mugendi, and J.B. Kung'u

Abstract Smallholder farms in Central Highlands of Kenya exhibit a high degree of heterogeneity, determined by a complex set of socio-economic and biophysical factors. The farms consist of multiple plots managed differently in terms of allocation of crops, nutrient inputs and labour resources, making within-farm soil fertility gradients caused by management strategies a common feature. In most cases, nutrient inputs are preferentially allocated to home fields, whilst outfields are neglected. A monitoring study involving nutrient inputs, flows and balances was conducted in Kirege location, where nine case study farms were used. The study was to compare the intensity of soil fertility management between home fields, mid-fields and outfields. It also compared soil fertility management practices between three different resources endowment classes to reveal important differences in patterns of fertility management. The farms were visited to record movement of nutrient-containing materials using a monitoring protocol covering household, crops, livestock, soil and socio-economic aspects of the farm. Data obtained was analyzed using IMPACT program version 2.0 to obtain total nutrient inputs and balances at field and farm levels and statistical analysis done using GenStat Discovery edition 2. Results revealed that mean N inputs over all resource endowment classes decreased with distance to the homestead (from 94 to 22.9 kg ha^{-1}), as did P (from 54.6 to 15.6 kg ha^{-1}) and K (from 193 to 34 kg ha^{-1}). Due to this heterogeneity in smallholder farms, there is a need for a more targeted approach to soil fertility intervention that differentiates between farm fields, agro-ecological zone and resource endowment status.

Keywords Heterogeneity · Home fields · Nutrient inputs · Outfields · Soil fertility gradient

Introduction

Decline in soil fertility has been described as the fundamental constraint to productivity of smallholder farms in sub-Saharan Africa (Sanchez et al., 1997). Most farmers cultivate crops continuously on the same plots with little additions of nutrient resources, which has led to severe depletion of soil fertility. Traditionally, smallholder farmers relied on long fallow periods under shifting cultivation to replenish soil fertility. Shifting cultivation, however, has disappeared as increasing population density and pressures on land use led to intensive, sedentary agriculture on small-scale landholdings and expansion of agriculture into marginal areas. This intensification of agriculture in small landholdings has typically not been accompanied by sufficient inputs of nutrients through biological nitrogen fixation, organic materials and mineral fertilizers to match the outputs of nutrients through harvested products and losses.

The processes of nutrient depletion and soil degradation, however, are spatially heterogeneous, as determined by the underlying parent material and geomorphology and by (current and historical) management (Smaling et al., 1997). Causes of variability in soil fertility status at different scales (i.e. region, village, farm and field) are both biophysical and socio-economic.

J.M. Muthamia (✉)
Department of Environmental Sciences, Kenyatta University, Nairobi, Kenya
e-mail: muthamiajoses@yahoo.com

A. Bationo et al. (eds.), *Innovations as Key to the Green Revolution in Africa*, DOI 10.1007/978-90-481-2543-2_72, © Springer Science+Business Media B.V. 2011

Variability at regional scale is determined by climate and dominant soil types, presence of and access to factor and product markets and historical, socio-cultural and ethnic aspects defining land use. The variability between different farm types (resource endowment groups) is associated with differences in soil fertility management between poor and wealthy households (Crowley and Carter, 2000). For instance, Murage et al. (2000) in a study in central Kenya reported differences in chemical and biological soil properties of productive and non-productive fields within a farm. Since clay and sand contents did not vary between soil categories in their study, they suggested that these differences in chemical and biological soil properties are not inherent but result from past soil management. Their findings reveal that farmers are more likely to allocate their limited organic resources and fertilizers to higher value crops in more productive areas of the farm than to attempt amelioration in fertility-depleted fields.

Here, we describe a monitoring study that was undertaken to understand within-farm variability in soil fertility management in smallholder farms of Kirege location, Central Highlands of Kenya. This was seen as a necessary step in identifying spatial–temporal niches for targeting of soil fertility management strategies and technologies. The objectives were (i) to construct farm typologies that reflect potential access of households to resources for managing their soils, (ii) to determine the magnitude of the nutrient flows and balances at farm scale, (iii) to compare soil nutrient inputs between home fields and outfields and (iv) to assess the influence of resource endowment on soil fertility management and nutrient balances.

Materials and Methods

The Study Site

The study was conducted in Kirege location, Chuka Division, in Meru South District. This area is a predominantly maize-growing zone. It is in the upper midland zones two and three (UM2–UM3) (Jaetzold and Schimdt, 1983). The area lies on the eastern slopes of Mt. Kenya at an altitude of approximately 1,500 m above sea level within an annual mean temperature of 20°C. It has an annual rainfall ranging from 1,200 to 1,400 mm and is bimodal, falling in two distinct seasons. The long rains (LR) occur from March

to June and the short rains (SR) from October to December. The soils are deep, well-drained, weathered humic nitisols (commonly called red Kikuyu loams) with moderate to high inherent fertility (Jaetzold and Schimdt, 1983).

The area is highly populated with a population density of about 700 persons per km^2 (Mutegi, 2004). Land is owned individually under freehold system of land tenure. Smallholder mixed farming is the most predominant farming system in the area. A wide variety of cropping systems as well as species and breeds of crops and livestock are found within individual farm holdings. Coffee (*Coffea arabica*) and tea (*Camellia sinensis*) are the major cash crops, while maize (*Zea mays*) and beans (*Phaseolus vulgaris*) are the main food crops in the area. Other food crops include potatoes, cassava, bananas, sweet potatoes, and various fruits and vegetables. Cattle, sheep, goats and poultry are the most common livestock species in the area.

Development of Farm Topology and Selection of Case Study Farms

A community meeting was organized and focus group discussions conducted to identify farmer criteria to be used as a basis for grouping themselves into different wealth classes. Farmer-identified indicators of wealth status were ranked, and this formed a basis for grouping farmers into different wealth status. A rapid survey was conducted using a sample of 50 households randomly selected out of the list of households in Kirege obtained from the local chief's office to characterize and classify the farms into three different groups (rich, medium and poor). During the survey and farm walks, it was observed that there were no significant differences in the biophysical characteristics (climate and soil type) of the farms in the village. Having confirmed the resource endowments of the farms through farm walks, nine case study farms were randomly selected for detailed resource flow mapping. There were three from each of the three wealth categories (referred to as resource groups or farm types) that had been identified.

Development of Field Topologies

The farms selected for detailed study above were visited to sensitize the farmers on the nutrient monitoring exercise. During the visit, the researcher together with the farmers drew sketch farm maps to indicate location of farm plots under various activities/enterprises. An inventory was conducted to identify the important features of the farm to be studied, such as fields, crops, animals, compost pits, household composition, farm size, farm implements and facilities. Area of farm plots, their coordinates and distance from the homestead were obtained by use of a Global Positioning Unit (GPS). The fields within each farm visited were also classified using a field typology that described resource allocation patterns and internal (within farm) nutrient flows that affect soil fertility (Mapfumo and Giller, 2001). Land use and distance from the homestead were the main criteria used to classify field types (home fields, mid-fields and outfields) (Tittonell, 2003).

Resource Flow Mapping and Calculation of Partial Nutrient Balances

The farms were first visited in August 2006 and farmers were asked to draw schematic maps and indicate all production units and the flows of nutrients to and from the units identified. Additionally, the type and number of crops grown, use or destination of the outputs, type and amount of inputs used, timing of crop and soil management activities and sequential order within the farm, sources of labour, off-farm income, average yields and general crop and livestock husbandry practices adopted were recorded using datasheets of IMPACT. A seasonal time frame was used, considering the long rains (March to August) of 2006 and farmers were asked recall questions relating to the above-mentioned aspects of farm management. In the second season (2006/2007 short rains), a monitoring approach was adopted where farms were regularly visited from September when farmers were preparing their land to March when the crops were harvested to monitor the flow of nutrients in the farms. During the monthly visits, farmers were interviewed to provide information on crop and livestock husbandry practices between then and the previous visit.

During resource flow mapping, farmers indicated quantities of inputs and outputs to the different production units/fields in local units, such as *tins* (±2 kg of grains), *debes* (±16 kg of grains), bags (±90 kg

of grains), bunches of bananas (±40 kg) and head loads (±40 kg of Napier grass or maize stover), and these were converted into SI units. Many of the values in kilogram given to local units were taken from previous work in the region and farmers' own experiences with the products. Parameters such as dry matter and nutrient contents (N, P and K) of materials that were most frequently used and therefore core determinants of nutrient movements were taken from literature (Rotich et al., 1999, Palm et al., 2001, TSBF, 2001) The main groups of these products included crop products, crop residues, manure and compost. This data was entered into IMPACT version 2.0 model for analysis.

Data Analysis and Presentation

Data obtained during the study was analyzed by use of IMPACT program version 2.0. Nutrient inputs and balances were calculated both for the farms as units and for field(s), separating nutrient sources into off-farm and on-farm sources. Data obtained was subjected to ANOVA using GenStat Discovery edition program with farm types and field types used as factors and nutrient inputs and balances as variates. Comparisons were made between 'farm types' and 'field types' for both nutrient inputs and balances and their means separated using least square difference (LSD) at 5% level of significance.

Results and Discussion

General Description of Households in Kirege

In order to provide a contextual background to farmers' soil fertility management practices and to explain their choice of strategy, this section examines the empirical results of a household survey on socio-economic characteristics (Table 1).

About 79% of the sampled households were male headed as compared to 21% female-headed households. Seventy-two percent of the household heads were married with spouse present, 13% married with spouse absent, 11% widows or widowers and 4% single. In many communities, gender influences access

Table 1 Socio-economic characteristics of households in Kirege ($n = 50$)

Characteristics	Type	Frequency	Age (%)
Gender of household head (HH)	Male	37	79
	Female	10	21
Age of HH	25–40	25	53
	41–60	14	30
	>60	8	17
Marital status of HH	Single	2	4
	Widow/er	5	11
	Married spouse present	34	72
	Married spouse absent	6	13
Education level of HH	None	1	2
	Primary	22	47
	Secondary	20	43
	Tertiary	3	6
Average family size	0–4	23	48
	5–8	24	52

to resources which are vital in farm management in general and soil fertility management in specific.

Slightly over half of the households (53%) were headed by people of 25–40 years of age as compared to 30% of 41–60 years and 17% of above 60 years. Slightly over 50% of households consisted of ≤4 people, while 48% of households had 5–8 people. The average family size for households in Kirege was found to be about four people. Considering the age of the household head and the family structure is important because it introduces the concept of the 'farm developmental cycle' (Crowley and Carter, 2000). The attitudes towards risk (investments) and innovation are highly variable according to the phase of the farm developmental cycle in which the household is (land, capital and/or labour constraints are also related to this). Young people work hard to improve their status,

are receptive of new ideas and are therefore more likely to adopt new technologies for soil fertility replenishment.

Education level of the household head also influences the kind of decisions made regarding general farm management. At least 96% of the household heads had basic (primary) education, while about 50% had received at least secondary education and only 6% had tertiary education.

Results from participatory wealth ranking revealed that type of housing is an important indicator of wealth in the community. They said that rich households have permanent houses (concrete floor, stone wall and tiled roof), while medium households have semi-permanent houses with concrete floor, timber wall and iron sheet roofing and poor households have houses with earthen floor, timber and at times mud wall and iron sheet roofing. Other indicators identified by farmers were type of livestock housing, livestock ownership, intensity of use of mineral fertilizer and the households' frequency of hiring or selling labour (Table 2). Most farmers in the medium and poor resource groups use their family as the main source of farm labour and they rely on reciprocal arrangements with neighbours to provide extra hands for planting, weeding or harvesting. Hired labour is used by better-off farmers, mainly in exchange for cash or food.

Farmers' soil fertility management strategies are also shaped by the size of the farm. Owning more land allows a farmer to grow a wider range of crops and to use different niches, thereby increasing the household's food security. Poor farmers have relatively larger farms compared to rich farmers in Kirege (Table 3). Livestock are a key productive asset and a major component of the farming system. They not only influence soil fertility by providing manure but can also be sold to purchase fertilizer. Poor farmers with no

Table 2 Indicators of the wealth status of the farmers and the characteristics of the different groups at Kirege

Indicator of wealth status	Rich	Medium	Poor
Type of housing	Concrete floor, stone wall, tiled roofing	Concrete floor, timber wall, iron roofing	Earth floor, timber/mud wall, iron roofing
Livestock ownership	Own more than two cattle	Own one cattle	No cattle but own goats/sheep and chicken
Type of livestock housing	Roofed and with concrete floor	Roofed without concrete floor	No roofing and concrete floor
Production orientation	Produce surplus for sale	Produce mainly for subsistence	Produce mainly for subsistence
Mineral fertilizer use	Use regularly and in large amount	Use regularly but in small amounts	Not regularly
Hire or sell labour	Afford to hire regularly	Do not afford to hire regularly	Sell labour locally

Table 3 Average resource endowment on the farms in the location ($n = 50$) and in selected case study farms ($n = 9$) in the different farmer resource groups in Kirege

Level	Farm type	No. of farms	No. of plots	Farm size (ha)	No. of cattle	No. of shoats	No. of chicken
Village	Rich	10	7	0.54	3	5	20
	Medium	21	5	1.08	2	3	12
	Poor	16	7	0.9	1	2	5
Case farms	Rich	3	7	0.5	2	5	18
	Medium	3	5	0.5	2	3	4
	Poor	3	7	1.2	1	3	6

cattle may not gain access to manure because they are not likely to afford to purchase; instead it is them who sell the little amounts of manure they have to the rich. Rich farmers own significantly more cattle than does any other group. Poor farmers try to raise goats, sheep and poultry and use their dung to fertilize their land.

Most households have extra earning from non-agricultural activities. Only the relatively rich farmers generate any significant income from the sale of crop and livestock produce: farmers in the resource endowment classes earn very little in this way. Most households in the area reported that the largest share of total family monetary expenses goes to meeting basic household needs, followed by expenditure on school fees, medication for family members and agricultural inputs such as fertilizers. Further discussions with farmers revealed that poorer households spent relatively more on food and other basic household needs, while richer farmers spent more on manure, fertilizer and improved seeds. This suggests that poor farmers have limited financial resources available to purchase inputs for maintaining soil fertility.

Categorizing and Describing Field Types

Different field types were identified within a farm, varying in enterprises/production activities, resource allocation and management practices, as revealed by the farm transects (see example in Fig. 1). Crop diversification is one of the strategies that farmers have adopted to cope with declining land sizes and changes in livelihood in many parts of sub-Saharan Africa. In a previous study in Central Highlands, Njuki and Verdeaux (2001) found that farmers were growing between six and seven crops because of reduction in land size, loss of market for old crops and opening of new markets for new crops. Crop allocation

in smallholder farms of Kirege was most diversified in the home fields which had an average of eight crops compared to outfields with four crops (Table 4) in agreement with previous studies. The home fields were small fields around the homestead, with a variety of crops sharing small pieces of land or intercropped (grains and pulses are normally intercropped). High-value crops such as fruits and vegetables were allocated to most of the home fields, while low-value crops were allocated to the outfields (see example in Table 4). The home fields were normally managed by women and often the first fields to be planted and weeded, receiving kitchen wastes and the sweepings from the house. The home fields were also receiving spills of manure from animal shed or manure stored in heaps due to their proximity to these structures. In some cases, the cattle manure is collected in compost pits instead of heaped.

The mid-distance and outfields were those in which more extensive crops were grown. The diversity of crop types decreased with increasing distance from the homestead, hence outfields had the lowest diversity of crops (Table 4). In the mid-fields, an intermediate management situation was found, strongly influenced by the farm type. In wealthy farms, they were managed in a similar way to the home fields, though input use was less intense. In the mid-fields, most of the cash crops such as tea and coffee were planted. The outfields were distant and/or difficult to access, and the crop produce was more prone to theft, particularly in areas of steep slopes. In this type of field, associated with poor-quality land, farmers planted their woodlots or crops that are known to produce under conditions of poor soil fertility, such as sweet potatoes, cassava or Napier grass. In some farms, outfields were located in the flood plain (river banks) and in such cases, farmers planted vegetables such as arrowroots and kales (Fig. 1).

Fig. 1 Example of a farm transect drawn during farm walk in Kirege, Central Kenya (scanned from original field notes)

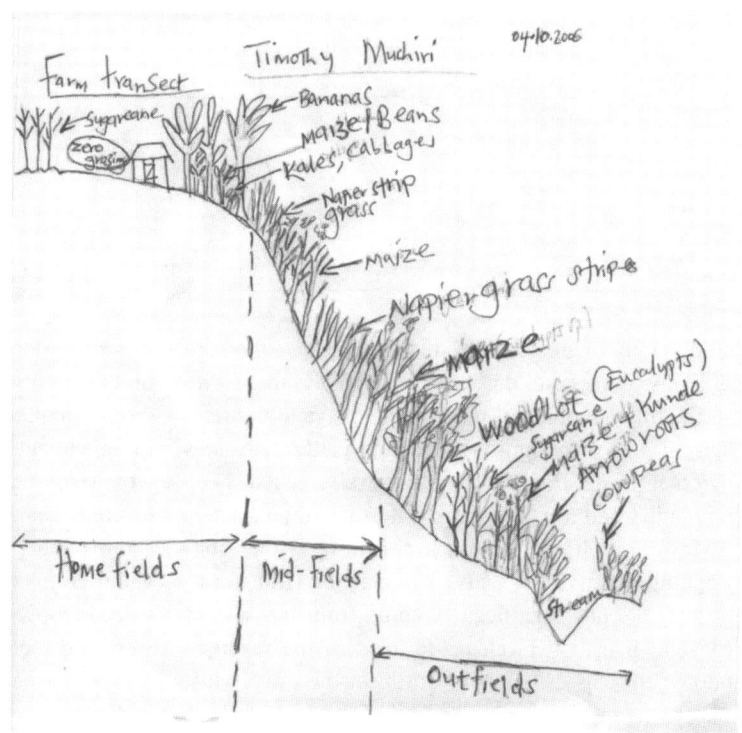

Variability in Resource Allocation Within Farms

As previously suggested (Brouwer et al., 1993; Carter and Murwira, 1995; Tittonel et al., 2005; Zingore et al., 2006), distance from the homestead tended to affect the allocation of production activities and resources (Tables 4, 5 and 6, Figs. 2 and 3a–d). Resource allocation in the different field types within a farm varied widely, as illustrated by the nutrient inputs calculated from the results of the resource flow mapping (Table 5). The use of organic resources varied clearly for different field types and was strongly influenced by distance from the homestead. Variations in resource allocation were also observed with regard to farmers' level of resource endowment. The wealthy farmers used large amounts of organic resources, which provided an average of 70 kg N ha^{-1}, 25 kg P ha^{-1} and 85 kg N ha^{-1} compared to the poor who used about 25 kg N ha^{-1}, 14 kg P ha^{-1} and 50 kg K ha^{-1}.

Vegetable crops grown in the home fields received most of the organic resources, followed by the cash and grain crops grown in the mid-distance fields. Very

little organic resources were applied to the outfields, due to the extra effort required to transport coarse materials to distant parts of the farm (Table 5). Crop residues were used as fodder, composted to make manure or incorporated in situ. Residues used as fodder were transported to the homestead and fed to animals restricted in stalls. In some instances, crop residues were taken from the field to a compost pile or compost pit, mixed with animal manure, ashes and kitchen wastes and used as organic fertilizers in planting holes, while in others, a small proportion of the residues were incorporated into the soil directly.

Mineral fertilizers were used with varying intensities in the different field types (Table 5). The wealthy farmers applied them in all field types, and relatively high rates (34 kg N ha^{-1} and 23 kg P ha^{-1}) were used in the outfields compared to poor farmers where no fertilizer was reportedly used in the outfields (Table 5) because the resources were not enough to be used in all the fields. With regard to farm type, there was a large gap in amounts of mineral fertilizers used by the wealthiest farmers (>21 kg N ha^{-1} and 14 kg P ha^{-1}) and the poorest farmers (<5 kg N

Table 4 Average area, distance from the homestead and most frequently grown crops for the different field types, averaged over all farm types at Kirege ($n = 9$ farms)

	Home fields ($n = 35$)	Mid-distance fields ($n = 15$)	Outfields ($n = 12$)
Average area (ha)	0.07	0.18	0.23
Average distance[a] (m)	25	84	158
Minimum (m)	5	45	85
Maximum (m)	40	80	180

Most frequently grown crops (frequency %)[b]

Maize/beans	35	83	0
Maize	25	7	18
Beans	12	0	7
Bananas	19	0	0
Coffee	2	20	0
Tea	0	7	0
Vegetables[c]	23	13	6
Potatoes	18	9	0
Napier grass	9	13	41

[a]Distance from the homestead
[b]Only one season data used
[c]Include kales, cabbages, tomatoes and onions

Table 5 Allocation of organic resources and mineral fertilizer to different field types by different resource endowment groups at Kirege

Resource group	Field type	Organic nutrient inputs (kg ha^{-1})			Mineral nutrient inputs (kg ha^{-1})		
		N	P	K	N	P	K
Rich	Home fields	75	61.3	379	12.5	7	0
	Mid-fields	60	40.7	298	23	11.6	0
	Outfields	19	9.1	55	34.1	23	0
	Mean/farm	70	25.8	85.7	21.4	13.9	0
	LSD	45.8	26.1	87.6	9.4	9.4	0
Medium	Home fields	73	42.7	145	28.6	24.9	0
	Mid-fields	57	36.2	119	1.8	6.3	0
	Outfields	15.5	16.1	33	5.6	2	0
	Mean/farm	49.2	39.4	244	13.8	11.1	0
	LSD	44.3	26.1	194.7	20.3	22.1	0
Poor	Home fields	54.2	30.5	106	2.6	2.6	0
	Mid-fields	15.5	8.5	30	2.3	2.3	0
	Outfields	8.1	4.6	16	0	0	0
	Mean/farm	25.9	14.5	50.7	1.6	1.6	0
	LSD	24.5	16.1	55.9	1.6	1.6	0

ha^{-1} and 5 kg P ha^{-1}). The wealthy farmers distributed mineral fertilizers evenly across their farms but preferentially targeted organic resources (read manure) to the plots closest to the homesteads, which received about 75 kg N ha^{-1} and 61 kg P ha^{-1} from manure compared with 60 kg N ha^{-1} and 40 kg P ha^{-1} on the mid-fields and 19 kg N ha^{-1} and 9 kg P ha^{-1} on the outfields.

Total Nutrient Inputs and Partial Nutrient Balances

As observed previously with regard to allocation of organic resources and mineral fertilizers, total nutrient inputs calculated for different field and farm types indicated that most inputs (N, P and K) were applied

Table 6 Total nutrient inputs and partial balances of different field types for the three different resource endowment groups in Kirege

Resource group	Field type	Total nutrient inputs (kg ha^{-1})			Partial nutrient balances (kg ha^{-1})		
		N	P	K	N	P	K
Rich	Home fields	109.1	72.9	379	41	46.8	29
	Mid-fields	72.5	47.7	298	8	33.8	−20
	Outfields	42	32.1	55	2	21.2	−15
	Mean/farm	74.5	50.9	244	17	67.8	−2
	LSD	38.6	23.6	194.6	24.2	14.4	30.6
Medium	Home fields	101.6	67.3	145	16.1	39	47
	Mid-fields	58.8	42.5	119	2.1	34	4
	Outfields	21.1	18.1	33	−43.2	11	−84
	Mean/farm	60.5	42.6	99	−8.3	28	−11
	LSD	46.6	28.4	67.6	35.6	19.2	77
Poor	Home fields	56.8	32.8	106	−29	27.9	−57
	Mid-fields	17.3	11.1	30	−77	6	−171
	Outfields	8.1	4.6	16	−169	2.6	−188
	Mean/farm	27.4	16.2	50.7	−91.7	12.2	−138.7
	LSD	29.8	16	55.8	81.2	14.8	28.2

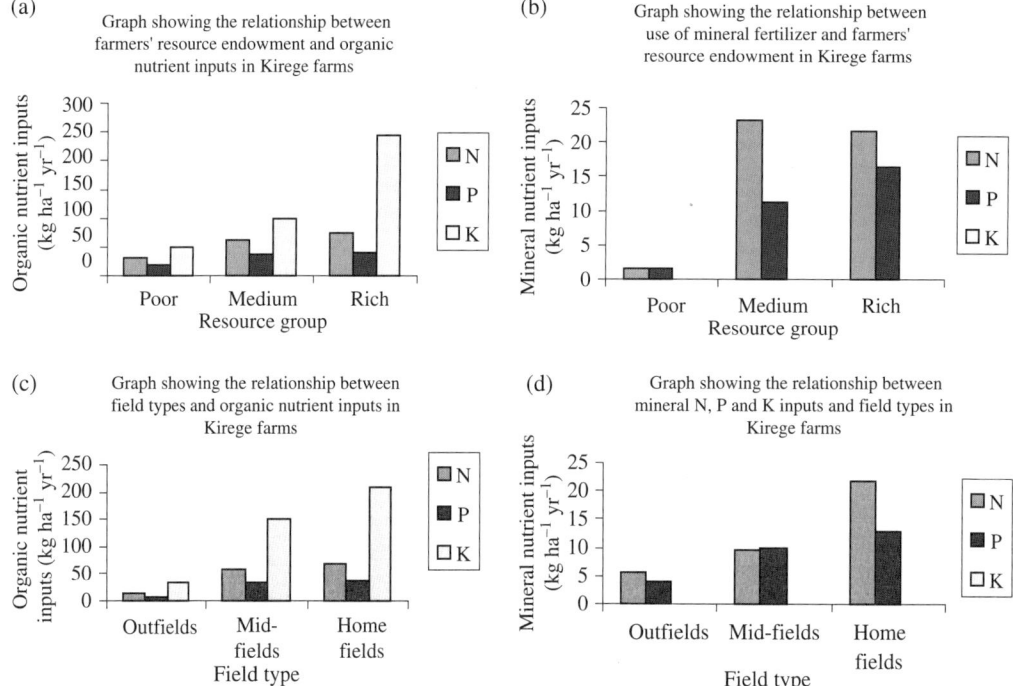

Fig. 2(a–d) Allocation of organic resources and mineral fertilizer to different fields and farm types in Kirege, Central Kenya

to the home fields. Although the average rates of nutrient inputs at farm level for wealthy and medium groups were close to the recommended (60 kg N ha^{-1} and 60 kg P ha^{-1}), calculations at field level revealed that little nutrients were applied to the outfields and especially in the medium and poor farms (Table 6). Total N and P inputs differed little in the home fields of rich and medium farms, i.e. 101 kg N ha^{-1}, 72 kg P ha^{-1} and 101 kg N ha^{-1}, 67 kg P ha^{-1} respectively. On average, large amounts of K (>50 kg ha^{-1}) were

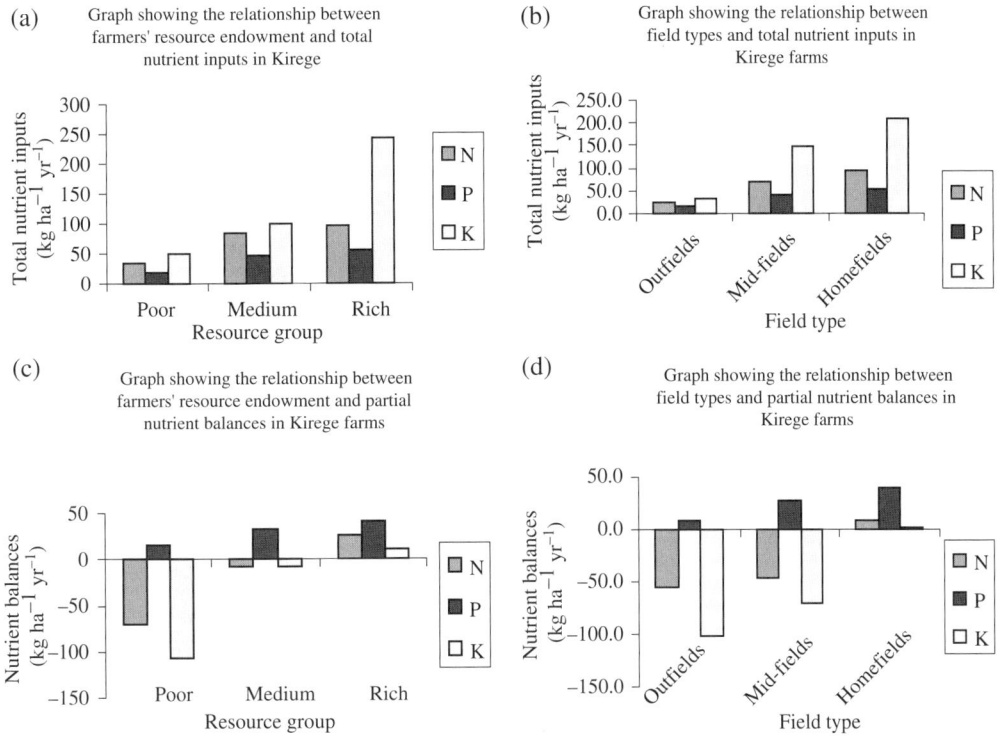

Fig. 3(a–d) Total nutrients inputs and partial balances for the different fields of the case study farm types (rich, medium and poor) at Kirege, Central Kenya

applied to all farm types and all this was obtained from organic resources as no K was available from mineral fertilizers because farmers used nitrogen- and phosphorus-based fertilizers whose potassium content, if any, is negligible.

Partial nutrient balances at field scale revealed the existence of N 'accumulation' areas within the wealthy farms and home fields of medium farms. N, P and K partial balances were largest on the wealthy farms, averaging 17 kg N ha^{-1}, 67 kg P ha^{-1} and –2 kg K ha^{-1}. The partial balances on the wealthy farms were largest on the home fields (41 kg N ha^{-1}, 46 kg P ha^{-1} and 29 kg K ha^{-1}) but decreased (8 kg N ha^{-1}, 33 kg P ha^{-1} and –20 kg K ha^{-1}) in the mid-fields and (2 kg N ha^{-1}, 21 kg P ha^{-1} and –15 kg K ha^{-1}) in the outfields. The partial N balances were negative for outfields in medium farms and all fields of the poor farms, illustrating that the amount of N added from both organic and mineral fertilizers was obviously less than the amount of N harvested with the biomass removed (Fig. 3a–d). P balances were found to be positive in all the farms and field types, although in some fields, the situation was almost in equilibrium (see Table 6, Fig. 3c and d).

In agreement with earlier observations (Mapfumo and Giller, 2001), the areas being depleted were much larger than the areas of 'accumulation', leading to an overall negative nutrient balance at farm scale for medium and poor farms. Household wastes and crop residues from other fields were brought to the home fields in the form of compost. Besides, nutrients accumulated in the home fields would not be efficiently used by grain and pulse crops often sparsely planted and shaded by banana plants and trees, affecting the magnitude of nutrient outflows as harvested crop parts. Typically, most inputs (e.g. fertilizers, manure, improved seeds) were applied in the home fields and farmers reported that their productivity was very high (80–90%).

Conclusions

The participatory monitoring approach adopted in this work helped to increase the understanding of the management aspects of smallholder farms that affect soil fertility. The heterogeneity in agricultural productivity, in terms of the intensity of nutrient depletion, and the allocation of resources and production activities to the different fields within the farm varied in magnitude between farm types. Distance from the homestead and level of resource endowment was found to influence allocation of crops and resources to different fields in the farms. Since scarce resources and investments are preferably allocated to less risky land units, such a pattern results in increased within-farm variability in soil fertility management. Management decisions at farm scale, which are affected by both biophysical and socio-economic factors, have an important impact on the resulting soil fertility. Due to this heterogeneity in smallholder farms, there is a need for a more targeted approach to soil fertility intervention that differentiates between farm fields, agro-ecological zone and resource endowment status.

Acknowledgements We acknowledge Vlaamse Interuniversitaire Raad (VLIR) for providing the necessary financial support to carry out this study in the framework of the project on 'Integrated Soil Fertility Improvement Technologies for Increased Food Production in Smallholder Farms of Chuka Division, Central Highlands of Kenya'. We also acknowledge Monicah Mucheru-Muna (KU) and Stanley Karanja (ILRI) for their skilled assistance in developing the research concept and data management using IMPACT program, respectively.

References

Brouwer J, Fusell LK, Herrmann L (1993) Soil and crop growth micro-variability in the West African semi-arid tropics: a possible risk-reducing factor for subsistence farmers. Agric Ecosyst Environ 45:229–238

Carter S, Murwira H (1995) Spatial variability in soil fertility management and crop response in Mutoko Communal Area, Zimbabwe. Ambio 24:77–84

Crowley EL, Carter SE (2000) Agrarian change and the changing relationships between toil and soil in Maragoli, western Kenya (1900–1994). Hum Ecol 28:383–414

Jaetzold R, Schimdt H (1983) Farm management handbook of Kenya. Natural conditions and farm information, vol II/C. East Kenya. Ministry of Agriculture, Nairobi

Mapfumo P, Giller KE (2001) Soil fertility management strategies and practices by smallholder farmers in semi-arid areas of Zimbabwe. International Crops Research Institute for the Semi-Arid Tropics (ICRISAT), Bulawayo

Murage EW, Karanja NK, Smithson PC, Woomer PL (2000) Diagnostic indicators of soil quality in productive and non-productive smallholders fields of Kenya's Central Highlands. Agric Ecosyst Environ 79:1–8

Mutegi JK (2004) Use of Calliandra calothyrsus and Leucaena tricandra tree species for soil nutrient enhancement in Chuka division, central highlands of Kenya. MSc thesis, Department of Environmental Foundations, Kenyatta University, Nairobi, Kenya

Njuki J, Verdeaux F (2001) Changes in land use and land management in the eastern highlands of Kenya: before land demarcation. International. Centre for Research in Agroforestry, Nairobi

Palm CA, Gachengo CN, Delve RJ, Cadisch G, Giller KE (2001) Organic inputs for soil fertility management in tropical agroecosystems: application of an organic resource database. Agric Ecosyst Environ 83(2001):27–42

Rotich DK, Ogaro VN, Wabuile E, Mulamula HHA, Wanjala CMM, Defoer T (1999) Participatory characterisation and on-farm testing in Mutsulio village, Kakamega district. Pilot project on soil fertility replenishment and recapitalisation in western Kenya, Report no. 11. Kenya Agricultural Research Institute, Kakamega, Kenya

Sanchez P, Shepherd K, Soule M, Place FM, Buresh R, Izac AMN (1997) Soil fertility replenishment in Africa: an investment in natural resource capital. In: Buresh RJ, Sanchez PA (eds) Replenishing soil fertility in Africa. ASA, CSSA, SSSA, Madison, WI, pp 1–46

Smaling EMA, Nandwa SM, Janssen BH (1997) Soil fertility is at stake. In: Buresh RJ, Sanchez PA, Calhoun F (eds) Replenishing soil fertility in Africa. Special publication No. 51. American Society of Agronomy and Soil Science Society of America, Madison, WI, pp 47–61

Tittonell P (2003) Soil fertility gradients in smallholder farms of Western Kenya. Their origin, magnitude and importance. Quantitative approaches in system analysis no. 25. The C.T. de Wit Graduate School for Production Ecology & Resource Conservation, in co-operation with the Tropical Soil Biology and Fertility Institute of the International Centre for Tropical Agriculture (TSBF-CIAT), Wageningen, 233pp

Tittonell P, Vanlauwe B, Leffelaar PA, Rowe E, Giller KE (2005) Exploring diversity in soil fertility management of smallholder farms in western Kenya. I. Heterogeneity at region and farm scale. Agric Ecosyst Environ 110: 149–165

TSBF (2001) Folk Ecology. Report on the findings of preliminary community interviews and group discussions undertaken on August 29, September 17, 18 and 19, 2001 in Emuhaya, Busia and Teso of western Kenya. Tropical Soil Biology and Fertility Programme, Nairobi, Kenya, 34pp

Zingore S, Murwira HK, Delve RJ, Giller KE (2006) Influence of nutrient management strategies on variability of soil fertility, crop yields and nutrient balances on smallholder farms in Zimbabwe. Agric Ecosyst Environ 119: 112–126

Residual Effects of Applied Phosphorus Fertilizer on Maize Grain Yield and Phosphorus Recovery from a Long-Term Trial in Western Kenya

W.M.H. Kamiri, P. Pypers, and B. Vanlauwe

Abstract Phosphorus (P) application is essential for crop production in the weathered, P-fixing soils of Western Kenya. It is hypothesized that a single large application of phosphorus fertilizer can shift soil-available P levels above a critical threshold, while further seasonal applications are requisite for sustaining yields. A field study was conducted in Siaya district to evaluate maize yield, P uptake, soil P balance and economic returns from P applied at different initial P rates and further seasonal P additions. In the first season, triple superphosphate (TSP) was added at rates of 0, 15, 30, 50, 100, 150 and 250 kg P ha^{-1}, and maize yield and P uptake were assessed during 10 consecutive seasons. Additional treatments were included where an initial application of 100 kg P ha^{-1} was supplemented with seasonal additions of 7 kg P ha^{-1}, supplied as TSP, manure or *Tithonia diversifolia*. Residual benefits of maize in terms of increased grain yields were high with cumulative yields ranging from 17.4 to 54.8 t ha^{-1} when P was applied at rates above 100 kg P ha^{-1}. Resin-extractable P increased significantly with initial P addition but decreased rapidly with time, particularly for treatments with one-time high dose of P application. Economic evaluation of these technologies revealed that application of initial P as 100 kg P ha^{-1} with seasonal additions of 7 kg P ha^{-1} as TSP would give the best marginal returns to investment.

Keywords Maize yield · P uptake · P recovery · P balance

Introduction

In recent years, the use of phosphorus (P) fertilizer has been associated with improvement in crop yields and soil fertility. In Western Kenya, P is often limiting in the majority of farmlands (about 80% of farms in Siaya and Vihiga districts) most of which have been estimated to have <5 mg P kg^{-1} soil of bicarbonate-extractable P in soils and exhibit high P sorption capacity (Jama et al., 1998; Sanchez, 1999; Shepherd et al., 1993). Soil tests have also shown a continuous decline in plant-available P, as a consequence of increased crop production without adequate replacement of removed P. Information regarding fertilizer type and use has increased over the years with different farmers using both organic and inorganic sources. However, the rates applied by farmers have always been lower than what is recommended. To restore phosphorus fertility, continued applications of P fertilizer are required to maintain a given level of available P and to ensure plant productivity. The buildup of soil P is predicted to be achieved using several ways, with one-time application of high rates of P fertilizer (Buresh et al., 1997) or using seasonal additions of inorganic phosphorus as sole or in combination with organic sources (Gachengo et al., 1999; Nziguheba et al., 1998). Several long-term fertilizer experiments have been conducted in Western Kenya on the use of combined organic and inorganic P sources on grain yield (Gachengo et al., 1999) and on soil phosphorus availability (Nziguheba et al., 2002). Data on long-term fertilizer experiments

W.M.H. Kamiri (✉)
Tropical Soil Biology and Fertility Institute of the International Centre for Tropical Agriculture (TSBF-CIAT), Nairobi, Kenya
e-mail: wangechikamiri@yahoo.com

A. Bationo et al. (eds.), *Innovations as Key to the Green Revolution in Africa*,
DOI 10.1007/978-90-481-2543-2_73, © Springer Science+Business Media B.V. 2011

involving large versus small phosphorus application rates in maize cropping systems are, however, unavailable in Western Kenya.

This study aims to compare the direct and residual effects of (i) a single application of P fertilizer of various rates and (ii) seasonal additional application of inorganic or organic P of small rates on the maize yield, maize P utilization and soil-available P. Economic benefits of P additions were addressed and compared.

Materials and Methods

Study Site

The study was carried out on a farmer's field located in Nyabeda, Siaya district ($0°08'$N, $34°25'$E; altitude $= 1420$ m.a.s.l), in the highlands of Western Kenya, characterized by a sub-humid climate. The rainfall (1800 mm year^{-1}) follows a distinct bimodal pattern with the highest precipitation during March to August ("long rains") and September to January ("short rains"). Previously, the site was under continuous maize cropping without manure or fertilizer inputs. The topsoil (0–15 cm) is characterized by a clay texture with 58% clay, 28% sand and 14% silt. Some chemical characteristics before the initiation of the trial are as follows: pH (1:2.5 soil:water suspension) $= 5.4$, total organic carbon $= 15.4$ g kg^{-1}, total nitrogen $= 0.18$ g kg^{-1}, total phosphorus 0.06 g kg^{-1}, exchangeable bases $= 0.13$ cmol$_c$ K$^+$ kg^{-1}, 4.9 cmol$_c$ Ca^{2+} kg^{-1} and 1.8 cmol$_c$ Mg^{2+} and exchangeable acidity $= 1.2$ cmol$_c$ kg^{-1}. Resin-extractable P (Sibbesen, 1978) of the topsoil prior to trial establishment equalled 2.4 mg kg^{-1}. The soil is classified as humic ferralsol (WRB, 1998) and is considered as moderately P fixing.

Experimental Design and Field Plot Management

The trial was established in March 1998 in a complete randomized block design with 11 treatments and 4 replicates, and measurements were taken during 10 consecutive seasons. Treatments included a control and triple superphosphate (TSP) application at six P rates

(15, 30, 50, 100, 150 and 250 kg P ha^{-1}), applied only in the first season. In another three treatments, TSP was applied at an initial rate of 100 kg P ha^{-1} and seasonally, 7 kg P ha^{-1} was supplied as TSP, *Tithonia diversifolia* (Tit) leaf residues or farmyard manure (FYM) (Table 1). The organic materials were applied at dry matter (DM) rates of 2 t ha^{-1}, supplying approximately 7 kg P ha^{-1} (Table 1). An additional treatment with an initial application of 250 kg P ha^{-1} and seasonal additions of TSP at 30 kg P ha^{-1} was included and assumed to represent conditions without P limitation. The experimental plots measured 5 m by 5.25 m. At the beginning of each season, a blanket of N and K was applied as urea and potassium chloride, respectively, at rates of 100 kg N ha^{-1} and 100 kg K ha^{-1}. Nitrogen fertilizer was split applied: one-third at planting (broadcasted and incorporated) and two-thirds at 4 weeks after planting (applied banded). Prior to planting, the organic materials and fertilizers used in the study were broadcasted uniformly onto the soil surface and incorporated into the top 10 cm soil using hand hoe. Maize (*Zea mays* L.), hybrid 513, was planted at a spacing of 75 cm by 25 cm (53,333 plants ha^{-1}). Plots were handweeded three times per season. Maize stock borer was controlled using Bulldock at 5 weeks after planting (WAP). Termites were controlled by spraying the base of the maize plants with Gradiator. After harvesting, the maize stover was removed from the trial.

Soil Sampling and Analysis

Soil sampling was done at the beginning of each cropping season (before ploughing) for resin-extractable P analysis (Sibbesen 1978), and the colorimetric P determination was based on the method of Murphy and Riley (1962).

Table 1 Chemical characteristics of *Tithonia diversifolia* leaf residues and farmyard manure used in the study. Average of 6 (nitrogen and potassium) and 10 (phosphorus) readings (range indicated in parenthesis)

Organic material	Total % N[a]	Total % P[a]	% C[b]	Total % K[a]
Tithonia	4.3 (3.9–4.5)	0.3 (0.18–0.45)	42.0	3.7 (3.63–3.90)
Manure	1.3 (0.7–2.1)	0.4 (0.26–0.56)	41.5	0.7 (0.62–0.76)

[a]Acid digestion using H_2SO_4 and H_2O_2 (Okalebo et al., 2002)
[b]Nelson and Sommers (1975)

Plant Analyses

Maize grain and stover yields and P uptake were determined during 10 consecutive seasons. Total P in the plant samples was determined using Kjeldahl digestion with concentrated H_2SO_4 (Anderson and Ingram, 1993) and colorimetric analysis (Parkinson and Allen, 1975). In order to compare the treatment effects over various seasons, yields were expressed relative to the control:

$$\text{Yield increase (\%)} = (\text{Yield}_{\text{treatment}} - \text{Yield}_{\text{control}})/ \text{Yield}_{\text{control}} \times 100.$$

Economic Analysis

Partial budget analysis was carried out following the methodology of CIMMYT (1998). Partial budgets were used to evaluate extra costs and benefits derived from the use of different rates of fertilizer P and organic resources, i.e. above the costs and benefits in the control. Benefits include the maize grain yield returns and residue yield returns (used for livestock feeding or as fuel wood). Costs include the purchase of inputs, transport, handling and application costs of inputs and the labour invested. Only yield-dependent labour operations were considered since additional costs and benefits from the control were budgeted. Cash opportunity costs for fertilizer purchase and opportunity costs (land allocation) for growing *Tithonia* were taken into account, as current niches cannot satisfy farmers' demands (Nziguheba et al., 2002). Dominance analysis was done by sorting the treatments on the basis of costs, listing them from the lowest to the highest together with their respective net benefits. In moving from the lowest to the highest, any technology that costs more than the previous one but yielded less net benefits was declared to be dominant and consequently eliminated from further analysis. The marginal rate of return (MRR) is an indicator of what farmers expect to gain on average in return for their investment when they decide to change from one practice to another. Marginal rate of return was calculated as follows:

$$\text{MRR \%} = \frac{(\text{Added benefits of technology } B - \text{Added benefits of technology } A) \times 100}{(\text{Added costs of technology } B - \text{Added costs of technology } A)}$$

where A is the technology with lowest costs and B the technology with highest costs. Other parameter values and assumptions are presented in Table 2.

Statistical Analysis

An analysis of variance was carried out to determine treatment effects on the parameters studied (maize yield, P uptake, resin-extractable P) using the mixed linear model procedure and the replicate number considered as a random-effect parameter (SAS Institute, 1995). Treatment comparisons were conducted by computing differences of least-square means and represented using standard errors of difference in means (SED).

Results and Discussions

Rainfall Distribution During the Maize Cropping Seasons

The annual rainfall during the cropping years was bimodally distributed. During this time, annual rainfall recorded was 1591, 1733, 1200 and 1650 mm and 1426 in the 5 years from 1998 to 2002 as compared with 1556 mm for long-term average. Monthly rainfall distribution during the cropping years is shown in Fig. 1. Maize crop responded positively to the seasonal changes in rainfall whereby in the short-rain seasons (SR), maize grain yields were lower than in the long-rain seasons (LR). This was evident in SR 1998 which had a crop failure due to below normal rainfall averaging to 427 mm.

Table 2 Parameters used for calculating costs and benefits of maize production with differing P application regimes

Parameter	Value	Units	Source
Returns			
Maize selling price	0.2	USD kg^{-1}	Nziguheba et al. (2002)
Maize stover selling price	0.014	USD kg^{-1}	
Cost of inputs			
TSP fertilizer price	0.41	USD ha^{-1}	Nziguheba et al. (2002)
Baseline fertilizer transport and application costs (150 kg ha^{-1})	1.5	USD ha^{-1}	Jama et al. (1997)
Additional fertilizer application costs	0.2	% of baseline cost kg^{-1}	Jama et al. (1997)
Cash opportunity costs	20	%	Jama et al. (1997)
Tithonia opportunity costs for growing (land allocation)	0.040	USD (kg DM)$^{-1}$	Nziguheba et al. (2002)
Tithonia costs for cutting, transportation and application	0.029	USD (kg DM)$^{-1}$	Nziguheba et al. (2002)
Manure costs for buying	0.013	USD (kg DM)$^{-1}$	Jama et al. (1997)
Manure labour costs for transportation and application	0.46	labour days (kg P ha^{-1})$^{-1}$	Jama et al. (1997)
Manure costs for transportation and application	0.56	USD (kg P ha^{-1})$^{-1}$	Jama et al. (1997)
Labour cost			
Labour cost	**1.21**	USD (labour day)$^{-1}$	Jama et al. (1997)
Labour for maize harvesting (yield-independent operations only)	0.016	labour days (kg grains)$^{-1}$	Jama et al. (1998)
Labour costs for maize harvesting (yield-dependent operations only)	0.01936	USD (kg grains)$^{-1}$	Jama et al. (1998)

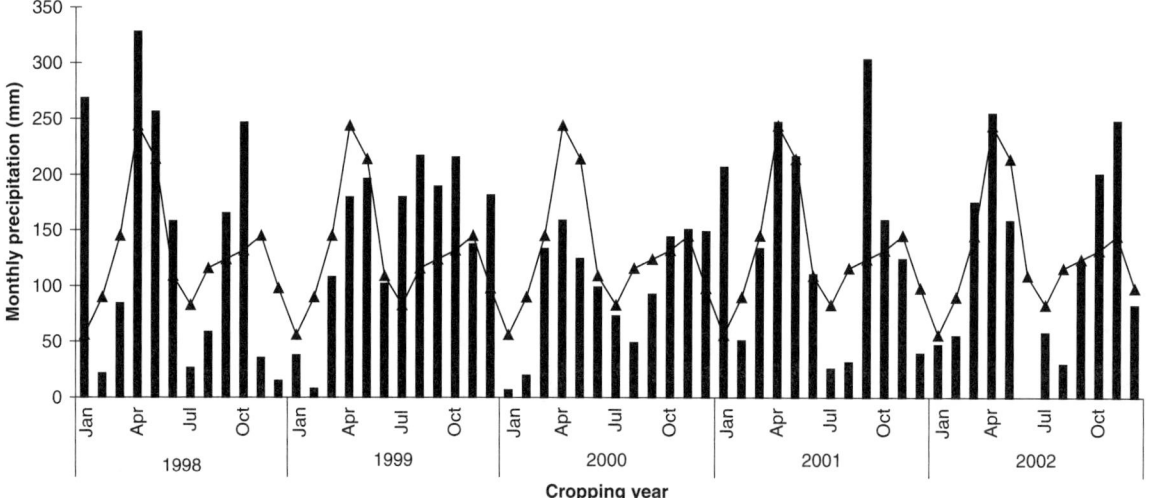

Fig. 1 Monthly rainfall and long-term averages in Siaya-western Kenya. The measured rainfall (■) was assessed by the summation of the rainfall for each month during the cropping years from 1998 to 2002 while long-term monthly averages (▲) were obtained from rainfall averages of each month for 15 years

Maize Grain Yield Over 10 Successive Seasons After Fertilizer P Application

The dry matter yield of grain during the 10 cropping seasons revealed that the P rates had statistically significant effect on grain yield at $P = 0.05$ except SR 1998 which was severely affected by low rainfall (Fig. 1). Phosphorus fertilization increased dry matter yield in the medium to high P rates (100–250 kg P ha^{-1}) (Fig. 2). However, minimal response to fertilization was observed in the low P application rates of 15–50 kg P ha^{-1}. There seemed to be no increasing or decreasing trend of grain yield during the experiment period in all treatments. Yields declined most rapidly with time when P was applied at low rates, while decline was below 10% after 2 years when P was applied at 250 kg P ha^{-1}.

The grain yield from 100P+7P(Tit), 100P+7P(FYM) and 100P+7P(TSP) treatments

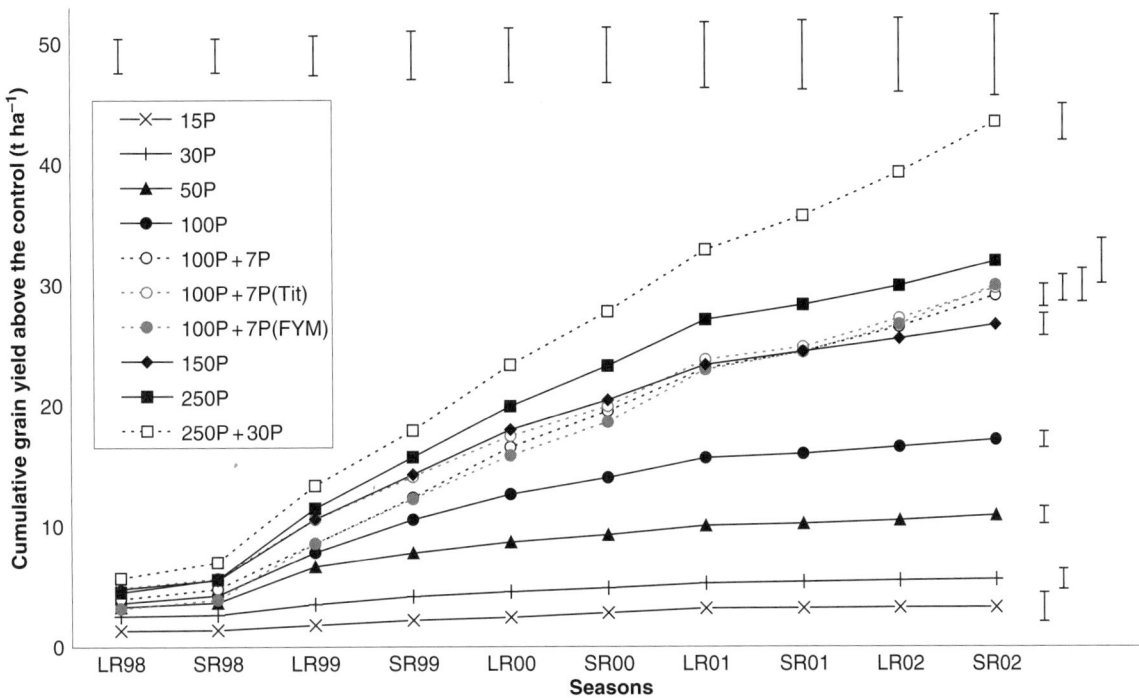

Fig. 2 Cumulative grain yields above the control grain yield as affected by P application for seasons SR98–SR01. Error bars on *top* represent LSD$_{0.05}$ between treatments for each season; error bars on the *left* represent LSD$_{0.05}$ between seasons for each treatment

was statistically not different in all seasons. However, seasonal variations in grain yield were observed during the long-rain season as compared to the short-rain season (Fig. 2). In LR 1998 season, maize grain increased from 0.32 t ha^{-1} (0 P) to 6.0 and 5.1 t ha^{-1} in 250P+30P and 100P+7P(FYM), respectively. Overall, the three treatments with moderate seasonal P additions 100P+7P(TSP), 100P+7P(FYM) and 100P+7P(Tit) gave similar grain yields while repeated application of phosphorus as 7 kg P ha^{-1} (TSP) increased grain yield above the single application of 100 kg P ha^{-1}. The highest grain output above the control was obtained from the 250P+30P treatment (43.6 t ha^{-1}) while the lowest was in the 15P input treatments (3.26 t ha^{-1}).

Addition of fertilizer P in a wide range of rates and sources has always been associated with an increase in crop yield in Western Kenya (Nziguheba et al., 2002; Jama et al., 2000). The rates of inorganic fertilizer used in this study provided adequate amounts of P to meet the needs of maize and to affect maize response. Phosphorus fertilizer nearly doubled maize yield in the crop to which it was applied in both the long- and short-rain seasons. The applied P fertilizer

was effective in maintaining average grain yields above 2.5 t ha^{-1} for four successive seasons for 100P, 150P and 250 kg P ha^{-1} (Fig. 2). This was as a result of declining available P in the soil as the crop continued to deplete the phosphorus pool (Fig. 3). Similar results were obtained by Linquist et al. (1996) where dry matter yield of rice declined by 54% with application of 150 kg P ha^{-1} as compared to 15% with application of 930 kg P ha^{-1} in four seasons, in highly P-fixing soils. On average, grain yields were highest from the 250P+30P treatment and significantly different from the other treatments of 100P+7P(TSP), 100P+7P(Tit) and 100P+7P(FYM) (Fig. 2). This study has revealed that repeated applications of fertilizer P in every single crop are not advisable since the P fertilizer clearly has significant residual value when applied at moderate rate, and higher rates are not likely to be adopted by resource-poor farmers in this region. Even though 100P+7P(FYM) treatment gave lower yield, the difference between 100P+7P(TSP) and 100P+7P(Tit) was not significant indicating either option, depending on the economic feasibility would be acceptable. The results also suggest that to maintain grain yields of about 2.5–3.0 t ha^{-1}, a rate of 100 kg P ha^{-1} could be

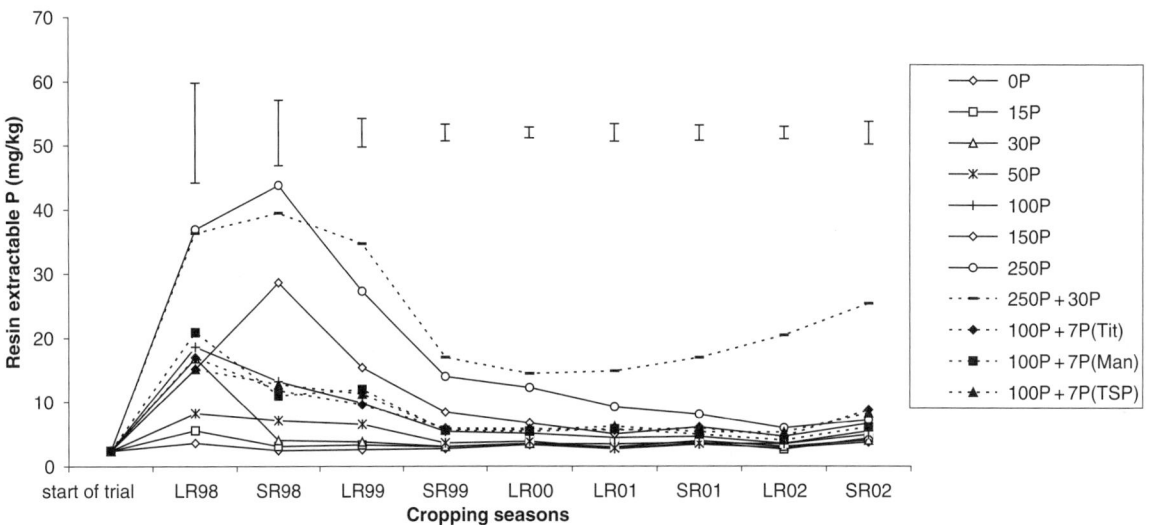

Fig. 3 Effect of single applications of P and initial-high rate of P with additional seasonal-small applications of phosphorus on resin-extractable P. *Vertical bars* represent standard error of the means for each season ($n = 4$)

applied at the beginning of the cropping period as this will be sufficient to support grain yield of up to four seasons. To obtain a grain yield of more than 3.5 t ha^{-1}, a seasonal application of P as modest as 7 kg P ha^{-1} could be applied as either inorganic fertilizer or organic residues of manure or *Tithonia*.

Effects of Phosphorus Application on Resin-Extractable Phosphorus

Initial amount of resin-extractable P measured at the beginning of the trial was equivalent to 2.4 mg P kg^{-1} soil and was little influenced by fertilizer application at the lower rates (15–50 kg P ha^{-1}) except in the first season LR 1998 (Fig. 3). The resin-extracted P declined rapidly with one-time application of high doses of phosphorus from a high of 18.6, 15.2 and 36.9 mg P kg^{-1} in the application season to a low of 5.5, 6.7 and 7.3 mg P kg^{-1} in the 100P, 150P and 250P treatments, respectively, at the end of the cropping period. In treatments receiving low rates of phosphorus, the decline was steady from 5.6 to 4.3 mg P kg^{-1} and 8.2 to 4.9 mg P kg^{-1} in the 15P and 50P treatments, respectively.

The resin-extractable available P with time shows that accumulation of P was highest in the first three seasons (LR 1998 to LR 1999) and was directly related to the rate of P application (Fig. 3). For example,

application of 250 kg P ha^{-1} with seasonal addition of 30 kg P ha^{-1} resulted in an accumulation of 221.9 mg kg^{-1} soil while application of 250 kg P ha^{-1} resulted in 176.3 mg kg^{-1} soil. Addition of P as *Tithonia*, manure or TSP had similar resin-extractable P accumulation of 77.6, 78.1 and 78.1 mg kg^{-1} soil, respectively, as compared to application of TSP at 100 kg as initial application which gave 72.7 mg kg^{-1} soil.

Resin-extractable P declined in each successive season possibly due to fixation and crop removal. On average the largest decline in resin P was realized in treatments that received low to moderate P addition (81, 71, 69 and 62% in 30P, 100P+7P(FYM), 100P and 100P+7P(Tit) treatments). Earlier studies by Chang (1975) and Medhi and De Tatta (1997) found that the residual effect of soluble P such as that from TSP on subsequent crops is considerable but decreases rapidly with cropping seasons whereas that of insoluble P sources is relatively low, but persists for a longer time. Similar findings were reported by Blake et al. (2006) in Rothamsted and Skierniewice where P availability in plots receiving phosphorus as superphosphate were considerably higher than plots receiving farmyard manure. Treatments receiving additional seasonal P as organic *Tithonia* and manure had similar resin-extractable phosphorus, which equally declined with seasons. The amount and form of plant material added to the soil (as *Tithonia* or manure) could have a significant bearing on the magnitude of the available and residual P pools and hence a significant effect on the residual value of the P fertilizer for maize. Thus, this could partly explain the increased level of available phosphorus in the treatments with organic residues (*Tithonia* and manure) above the control even though there was a decline in the available resin-extractable P during the cropping seasons. Besides improving phosphorus availability, organic residues have also been found to enhance the fertility status of the soil through their effects on soil structure, pH and nutrient retention capacity (Buresh et al., 1997).

The decline in extractable P over time could have possibly been due to increased P sorption by the soil, plant uptake and P microbial immobilization over time (Nguluu et al., 1996). The extractable resin P measured in the soil after crop harvest in all the seasons indicated a sharp decline, which could also be attributed to P transformation and transport involving the process of microbial immobilization and mineralization and/or soil sorption of P (Medhi and De Tatta, 1997). We are unable to determine the relative importance of these processes in the resin-extractable P since none of the variables was directly determined. Resin P measurements in this study were used as a means of ranking the treatments from their available P potential and effects on maize grain yield. No attempts were made to correct resin P for P sorption or buffering capacity of the soil, or P immobilization from the soil microbial biomass, or the stimulation effect on P availability of added fertilizer due to P fixation or desorption process.

Phosphorus Uptake, Recovery and Balance as Affected by Different P Rates

For 8 out of the 10 seasons phosphorus uptake by the maize crop was calculated from the crop yield per hectare, expressed as the dry weight of the harvested crop, and the percentage of P in the dry material. As expected grain P concentrations in the treatments with high P applications were greater than the treatments with low P application, there was no significant difference in P concentrations among the treatments except during the LR 2000 and LR 2001 seasons (Table 3). However, significant differences among seasons were observed which could be attributed to rainfall regimes between the long-rain and short-rain seasons.

Maize took up 20–41% of the P applied in one-time application (50–250P kg ha^{-1} treatments) and 15–28% in the treatments with seasonal P additions (Table 4). The highest P inputs generally resulted in the greatest P off-take with the most efficient uptake response to mineral P occurring in the 100P+7P(TSP) treatment. Cumulative recovery of P applied over the period of 10 seasons was lowest in the high P application treatments (250P+30P and 250P treatments) at 11 and 14% and highest in the medium application rates of 100P+7P(FYM) and 100P+7P(TSP) (Fig. 4). The results suggest that less than 60% of the applied P (rates 15, 30 and 50) was retained in the soils as residual P after 5 cropping years.

Table 3 Phosphorus concentration in maize grain as affected by P application for seasons SR98–SR02

	LR98	SR98	LR99	SR99	LR00	SR00	LR01	SR01	LR02	SR02	LSD$_{0.05}$[b]
0P	nd	0.10	0.14	0.10	0.16	0.12	0.16	0.16	nd	0.22	ns
15P	nd	0.06	0.18	0.08	0.11	0.11	0.13	0.16	nd	0.25	0.12
30P	nd	0.08	0.15	0.10	0.14	0.16	0.16	0.23	nd	0.23	ns
50P	nd	0.08	0.11	0.08	0.12	0.07	0.13	0.17	nd	0.16	0.08
100P	nd	0.11	0.17	0.06	0.14	0.08	0.20	0.17	nd	0.18	0.10
150P	nd	0.12	0.13	0.13	0.15	0.09	0.15	0.16	nd	0.15	ns
250P	nd	0.07	0.08	0.09	0.09	0.17	0.15	0.20	nd	0.22	0.09
250P+30P	nd	0.08	0.12	0.15	0.22	0.15	0.26	0.22	nd	0.25	0.15
100P+7P(Tit)	nd	0.06	0.16	0.07	0.16	0.12	0.13	0.16	nd	0.11	0.10
100P+7P(FYM)	nd	0.07	0.16	0.11	0.11	0.10	0.16	0.16	nd	0.18	ns
100P+7P(TSP)	nd	0.08	0.11	0.10	0.12	0.13	0.15	0.16	nd	0.16	0.08
LSD$_{0.05}$[a]		ns	ns	ns	0.11	ns	0.08	ns		ns	

nd: not determined; ns: not significantly different at $P < 0.05$; LSD$_{0.05}$
[a]Least significant difference at $P < 0.05$ between treatments; LSD$_{0.05}$
[b]Least significant difference at $P < 0.05$ between seasons

Table 4 Total fertilizer phosphorus applied, P off-take in maize grain and stover and P balance (–) released, (+) leached or fixed over 10 consecutive maize crops between 1998 and 2002

P rate	Total P applied (kg P ha^{-1})	Total P off-take (kg P ha^{-1})	Available P in soil (kg P ha^{-1})			P loss (kg P ha^{-1})[c]
			Initial[a]	After[b]	Change in P	
0P	0	14.4	2.4	29.9	27.5	–39.5
15P	15	14.1	2.4	35.1	32.7	–29.4
30P	30	19.7	2.4	47.7	45.3	–32.6
50P	50	20.3	2.4	46.6	44.2	–12.1
100P	100	29.4	2.4	72.7	70.3	2.7
150P	150	42.7	2.4	99.3	96.9	12.8
100P+7P(Tit)	163	39.4	2.4	77.6	75.2	50.8
100P+7P(FYM)	163	40	2.4	78.1	75.7	49.7
100P+7P(TSP)	163	44.9	2.4	78.1	75.7	44.8
250P	250	50.4	2.4	176.3	173.9	28.1
250P+30P	520	78	2.4	221.9	219.5	224.9

[a]Soil-available P at start of trial LR 1998
[b]Cumulative soil-available P at end of season 10 (SR 2002)
[c]By subtracting total off-take and change in P from total added and initial P

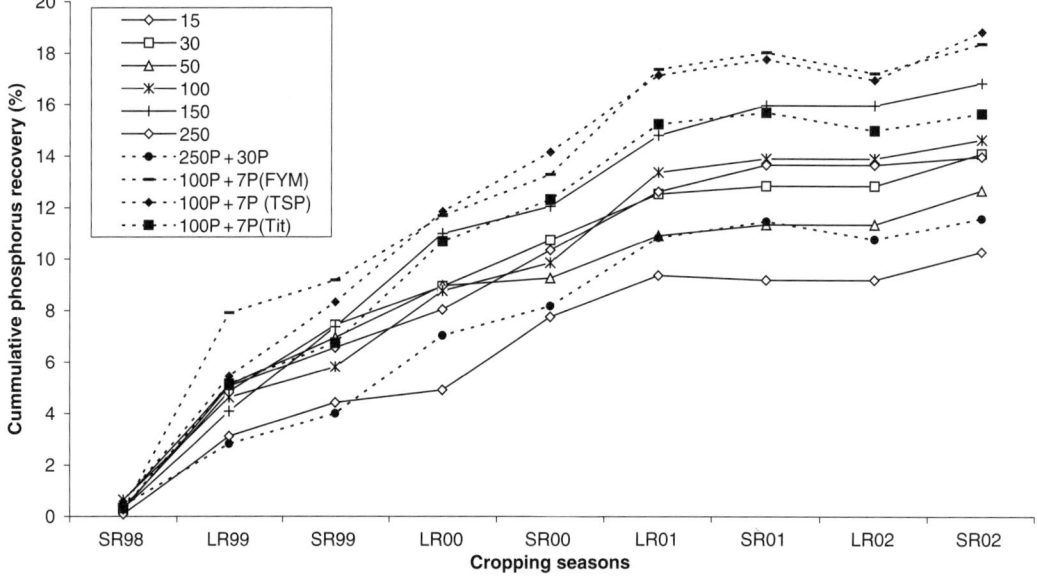

Fig. 4 Cumulative phosphorus recovery calculated as total uptake of P treatments (15P, 30P, 50P, 100P, 150P, 250P, 250P+30P, 100P+7P(TSP), 100P+7P(FYM) and 100P+Tit) minus cumulative recovery of the control (0P) during the maize cropping seasons

P recovery by maize depended on the rates of P application and was observed to decline as the number of cropping years progressed in all treatments. The most efficient utilization of P was attained in the 100P+7P(FYM) and 100P+7P(TSP) treatments (Fig. 4). In most trials percentage of phosphorus recovery is usually lower than 15% (Morel and Fardeau, 1990), thus in this study most of the P applied was not used by the plant and probably was leached, fixed or transformed to microbial biomass.

In the high P application treatments, despite a soil P balance of 224.9 kg P ha^{-1} in the 250P+30P and 28.1 kg P ha^{-1} in the 250P treatments after the 10th season, there was a 28 and 54% reduction in grain yield, respectively, in relation to season 1 (LR 1998). The highest decline was realized in the low P application rates (15P–50P) which had over 80% reduction while lowest decline was realized in the 100P+7P(Tit) treatment at a 3% with a P balance of 50.8 kg P ha^{-1}. The results from this study agree with those from the studies by Linquist et al. (1996) who observed declining trends in dry matter yield of corn but a high P balance from treatments receiving high doses of phosphorus. When P is added to soil, the phosphorus is subjected to precipitation and adsorption processes that reduce concentration of P in soil solution, P availability to plants and recovery of P fertilizer by crops. Therefore, continued application of P fertilizer is required to maintain a given level of available P and optimum crop growth. The fertilizer rate required to obtain a particular target uptake thus drastically decreases in course of time due to the residual effect of the previously applied fertilizer. Since initial soil P levels were low (2.4 mg kg^{-1}) and plant uptake is only one possible fate of the mineralized P, the contribution by mineralization to plant-available P is small. Given the high amount of P remaining in the soil and the massive decline in grain and stover yield (100<150<250 kg P ha^{-1}), respectively, a more efficient management strategy for this soil would be to apply small amounts of P to each crop as indicated in the seasonal application treatments.

Economic Analysis

Net benefits derived from application of fertilizer phosphorus at different rates were determined from the difference between maize grain value and added cost of production due to fertilizer use (Table 5). In this study labour for weeding and land preparation was assumed to be constant for all the P applications and therefore had no effect on the net benefit of the various treatments. This study did not take into account that collection, and application of *Tithonia* could be a constraint under farmers' condition when this activity coincides with the period of peak labour requirement (planting and land preparation). Input cost (stated in USD) for the 10 seasons was lowest where phosphorus was applied at the lowest rates as one-time application. The cost was USD 169 and 312 in the 30 and 50 kg P ha^{-1}, respectively. For moderate P additions, the cost increased due to extra cost of transport and application of fertilizers up to 3 to 10 times that of the lowest P rate. Added costs were highest at USD 1889 and 1863 for 250P+30P and 100P+7P(Tit) treatments, respectively (Table 5).

Among the six single (initial) applications of P, the highest net benefit was derived from plots which received from the 250P treatment (USD 5698 ha^{-1}) followed by those receiving 150P (USD 4890 ha^{-1}). The net benefits from the maize harvest taken from 100P+7P(Tit) plots were lowest among the treatments receiving small-seasonal additions of P. As expected the unfertilized control (0P) treatment yielded the lowest net benefits of USD 642 ha^{-1}. Added benefits were positive and highest in 250P+30P at USD 7415 while the benefits were similar in the 100P+7P(FYM) and 100P+7P(TSP) at USD 5340 and 5306, respectively.

Despite the fact that 100P+7P(Tit) had moderately higher grain yields than the 100P+7P(FYM) and 100P+7P(TSP) treatments, the costs associated with switching to this technology were sufficiently high enough not to warrant a switch from the 250P option. This could have been due to costs associated with cutting and transportation of *Tithonia* biomass and labour required for preparation and incorporation. Further dominance analysis revealed that 100P+7P(Tit) treatment was more expensive and thus was excluded from the analysis of marginal rate of return (Fig. 5). In this study, the lowest rate of return to investment was obtained from the shift from no P input (0P) to

Table 5 Effects of phosphorus additions on added benefits and costs for 10 maize cropping seasons on a long-term trial in Western Kenya

P rate	Total added costs (USD ha^{-1})[a]	Total added benefits (USD ha^{-1})[a]	Marginal rate of return (%MRR)
15P	93	642	690
30P	169	1056	551
50P	312	2065	705
100P	533	3164	496
150P	819	4890	605
100P+7P(TSP)	891	5306	571
100P+7P(FYM)	1064	5340	19
250P	1122	5698	617
250P+30P	1889	7415	224

[a]Sum of 10 season's data

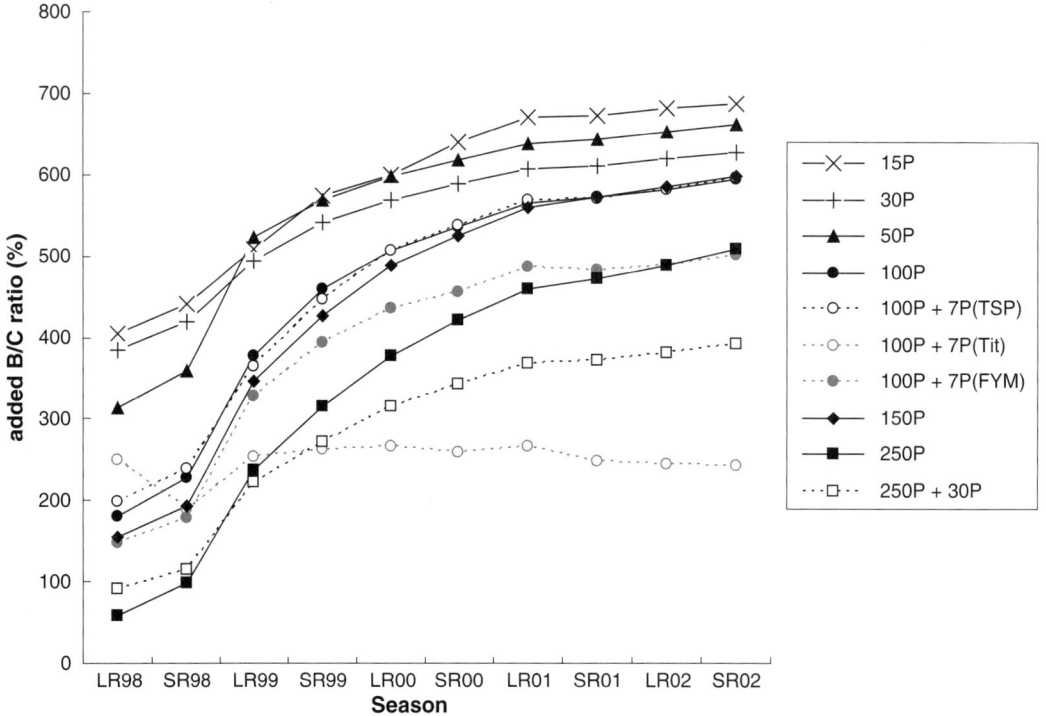

Fig. 5 Marginal rate of return of maize crop grown for 10 consecutive seasons when phosphorus was applied as a single initial application or as initial plus small seasonal additions. Seasons 1–10 represent LR98, SR98, LR99, SR99, LR00, SR00, LR01, SR01, LR02 and SR02, respectively

100P+7P(FYM) and 250P+30P which gave a return of 19 and 224%, respectively. The highest MRR was realized with the moderate P inputs of 100P+7P(TSP) and 150P which gave returns of over 571 and 605%, respectively, as shown in Table 5.

The low but positive net benefits obtained in this study with low P application were due to diminishing crop yield. By switching from 250P to 100P+7P(Tit) technology, the farmer would incur an additional cost of USD 714 but would realize a loss of net benefits of USD 1172. By switching from 100P+7P(TSP) to 100P+7P(FYM) the marginal rate of return is only 19% and while it is profitable in the sense that the added revenue generated by the technology covers the added expense, it would not be best to recommend this technology since the rate of return is below what the farmer considers to be acceptable. This would mean investing another USD 173 and obtaining an additional net return of only USD 34 or a return of 19% on the last USD 173 invested. If it assumed that a minimum acceptable marginal rate of return (MRR) of 100% is required for a technology to be attractive to

farmers, then by switching from 150P (initial single application) to 100P+7P(TSP) (small-seasonal addition) a farmer's net income would improve by spending less by USD 73 and benefiting more by USD 416. A switch from 100P+7P(TSP) to 250P and 250P+30P would result in 170 and 211% returns, respectively. However, a shift from 150P to 250P and 250P+30P would result in 266 and 236% MRR. This implies that for each dollar invested in the 250P and 250P+30P technologies, the farmer could expect to recover an additional return of USD 1.7 and 2.1 when shifting from 100P+7P(TSP) to 250P and 250P+30P, respectively, and 2.66 and 2.36 when he shifts from 150P to 250P and 250P+30P technologies, respectively. On the other hand, a shift from 150P to 100P+7P(TSP) would yield him USD 5.7 for each dollar invested. Hence the best technology to recommend to farmers in this area would probably be 100P +7P where an initial moderate rate of P is followed by minimal seasonal P additions. As illustrated above, it is not always the technology that returns the highest yield, net benefits and marginal returns that are best ones to recommend.

Conclusions

Soil P can be replenished with either soluble P fertilizers or in combination with organic residues. This can be applied either as large single application or gradually with moderate seasonal applications sufficient to increase availability of soil P. In this case, application of 100 kg P ha^{-1} with seasonal additions was found to be sufficient to supply soil P for at least three seasons and beyond and to give adequate grain yield. The added benefits of 100P+7P(TSP) treatment were considerable (USD 5306) with the highest returns if the farmer opted to shift to this technology. Thus this technology would be considered suitable for selection by smallholder farmers in this region. For soils similar to the highly weathered acid soils of Western Kenya, applying large amounts of phosphorus to quench the P fixation of the soil may not be the most cost-effective strategy despite the high benefits obtained.

Acknowledgements We gratefully acknowledge the BMZ-ISFM through TSBF-CIAT for providing financial support for this study. The assistance of John Mukalama and Wilson Ngului in field data collection and laboratory analysis is highly acknowledged.

References

Anderson JM, Ingram JSI (1993) Tropical soil biology and fertility. A hard book on methods. CAB International, Wallingford, UK, 221pp

Blake L, Mercik S, Koerschens M, Moskal S, Poulton PR, Goulding KWT, Weigel A, Powlson DS (2006) Phosphorus content in soil, uptake by plants and balance in three European long-term field experiments. Nutr Cycling Agroecosyst 56:263–275

Buresh RJ, Smithson PC, Hellums DT (1997) Building soil phosphorus capital in Africa. In: Buresh RJ, Sanchez PA, Calhoun F (eds) Replenishing Soil Fertility in Africa. Soil Sci. Soc. Am. J. Special Publication 51, 193–217. SSSA, Madison, WI, pp 111–149

Chang SC (1975) Utilization and maintenance of the natural fertility of paddy soils. Experimental Research Bulletin No. 61. ASPAC. Food and Fertilizer Technology Center, Taipei, Taiwan

CIMMYT (1998) From agronomic data to farmer recommendations: an economics training manual. CIMMYT_Mexico, D.F., Mexico

Gachengo CN, Palm CA, Jama B, Othieno C (1999) Combined use of trees, shrubs and inorganic fertilizers for soil fertility improvement. Agroforest Syst 44:21–36

Jama B, Buresh RJ, Place FM (1998) Sesbania tree fallows on phosphorus-deficient sites: maize yield and financial benefit. Agron J 90:717–726

Jama B, Palms CA, Buresh RJ, Niang AL, Gachengo C, Nziguheba G (2000) Tithonia as a green manure for soil fertility improvement in western Kenya. A review. Agroforest Syst 49:201–221

Jama B, Swinkles RA, Buresh RJ (1997) Agronomic and economic evaluation of organic and inorganic sources of phosphorus for maize. Agron J 89:597–604

Linquist B, Singleton PW, Cassman KG, Keane K (1996) Residual phosphorus and long-term management strategies for an Ultisol. Plant Soil 184:47–55

Medhi DN, Tatta De SK (1997) Residual effect of fertilizer phosphorus in lowland rice. Nutr Cycling Agroecosyst 46:195–203

Morel C, Fardeau JC (1990) Uptake of phosphate from soils and fertilizers as affected by soil P availability and solubility of phosphorus fertilizers. Plant Soil 121:217–224

Murphy J, Riley JP (1962) A modified single solution for determination of phosphate in natural waters. Anal Chem Acta 27:31–36

Nelson DW, Sommers LE (1975) A rapid and accurate method for estimating organic carbon in soil. Proc Indiana Acad Sci 84:456–462

Nguluu SS, Probert ME, Myers RJ, Waring SA (1996) Effect of tissue phosphorus concentration on the mineralization of nitrogen from stylo and cowpea residues. Plant Soil 191: 139–146

Nziguheba G, Merckx R, Palm CA, Mutuo P (2002) Combining Tithonia diversifolia and fertilizers for maize production in a phosphorus deficient soil in Kenya. Agroforest Syst 55: 165–174

Nziguheba G, Palm CA, Buresh RJ, Smithson PC (1998) Soil phosphorus fractions and adsorption as affected by organic and inorganic sources. Plant Soil 198:159–168

Okalebo JR, Gathua KW, Woomer PL (2002) Laboratory methods of soil and plant analysis: a working manual, 2nd edn. TSBF-CIAT and Sacred Africa, Nairobi, Kenya

Parkinson JA, Allen SE (1975) A Wet Oxidation procedure suitable for the determination of Nitrogen and mineral nutrients in biological materials. Commun Soil Sci Plant Nutr 6: 1–11

Sanchez PA (1999) Improved fallows come of age in the tropics. Agroforest Syst 47:3–12. Kluwer, Netherlands

SAS Institute (1995) SAS users guide, 6th edn. SAS Institute, Cary, NC

Shepherd KD, Olson E, Okalebo JR, Ndufa JK, David S (1993) A statistic model of nutrient flow in mixed farms in the highlands of western Kenya to explore the possible impact of improved management. Paper submitted at the international conference on livestock and sustainable nutrient cycling in mixed farming systems of Sub-Saharan Africa, Addis Ababa, Ethiopia, 22–26 Nov 1993

Sibbesen E (1978) An investigation of the anion exchange resin method for soil phosphate extraction. Plant Soil 50:305–321

WRB (1998) World reference base for soil resources. World soil resources report 84. FAO ISRIC ISSS, Rome, Italy

Combined Effect of Organic and Inorganic Fertilizers on Soil Chemical and Biological Properties and Maize Yield in Rubona, Southern Rwanda

A. Mukuralinda, J.S. Tenywa, L.V. Verchot, and J. Obua

Abstract Effects of *Calliandra calothyrsus* Meissner, *Tithonia diversifolia* Hensley A.Gray and *Tephrosia vogelii* Hook.f green manure applied independently or combined with triple superphosphate (TSP) on soil chemical and biological properties that influence maize yield were evaluated on an Ultisol of Rubona, Rwanda. Treatments compared in randomised complete block design were the control, limestone at 2.5 t ha^{-1}, TSP at 25 and 50 kg P ha^{-1}, leaf of *Calliandra*, *Tithonia* and *Tephrosia* applied independently at 25 and 50 kg P ha^{-1} and each combined with TSP at equivalent rates of 25 and 50 kg P ha^{-1}, respectively. Lime led to significant increases in soil pH followed by *Tithonia* combined with TSP at a rate of 50 kg P ha^{-1}. All treatments significantly reduced exchangeable acidity and aluminium compared to the control. A combination of organic materials with TSP at a rate of 50 kg P ha^{-1} improved soil organic carbon, microbial biomass carbon (MBC) and phosphorus (MBP). Compared to TSP applied alone, only the combination of *Tithonia* with TSP increased labile inorganic P fractions by 1.6–52.2%. In the fourth season, application of *Tithonia* green manure combined with TSP at a rate of 50 kg P ha^{-1} resulted in higher maize yield (25% increase) than TSP and *Tithonia* (9% increase) applied alone. Plant quality residues, labile inorganic P fractions, soil organic carbon, MBC and MBP values were correlated with maize yield, confirming the crucial role of plant residues quality and the soil properties in improving maize yield.

Keywords Biomass transfer · Maize yield · Soil chemical and biological properties

Introduction

In Rwanda, like elsewhere in sub-Saharan Africa (SSA), continuous cropping without adequate fertilizer application has led to soil fertility depletion, negative nutrient balance and low crop yield. For instance, NPK uptake from soils and other losses have led to negative nutrient balance in arable lands during 1981–1985, 1986–1990, 1991–1995 and 1996–1999 at –151.5, –136.3, –128.4 and –123.8 kg NPK ha^{-1}, respectively (Henao and Baanante, 1999).

Among these nutrients, phosphorus is the most limiting nutrient, particularly in acid soils. Mainly, P deficiency is due to low native soil P, removal in crop residues, high soil P – adsorption capacities and erosion (Ikerra and Kalumuna, 1992; Warren, 1992). Use of mineral fertilizers or organic amendments can reverse declining crop yield (Ikerra et al., 2006). However, use of mineral fertilizers is constrained by several socio-economic limitations such as high price, high transport cost, lack of credit facilities and low price of agricultural crop, which do not allow farmers to get surplus money to buy inorganic fertilizers needed (Heisey and Mwangi, 1996 and Kazombo-Phiri, 2005). Most organic amendments are low in P (Palm et al., 1997) although they can improve soil organic matter (SOM) and pH, as well as reduce exchangeable acidity (EXAC) and exchangeable aluminium (EXAl) that influence maize yield.

A combination of locally available organic materials like *Calliandra calothyrsus* (CC), *Tephrosia vogelii*

A. Mukuralinda (✉)
World Agroforestry Centre (ICRAF), Rwanda
e-mail: mukuratha@yahoo.com

(TV) and *Tithonia diversifolia* (TD), with moderate amounts of inorganic P, could be a cheaper and appropriate option for small-scale farmers to increase maize yield. Use of organic materials and inorganic P applied independently or combined has been documented in East Africa (Iyamulemye et al., 1996; Jama et al., 1997; Nziguheba et al., 2002; Ochwoh et al., 2005). However, these studies have been limited to the final effects of amendments on available P and maize yield. Little attention has been paid to investigate the mechanisms influencing maize yield increases. This study has attempted to determine the effects of P resources on soil chemical and biological properties, which in turn influence maize yield in Rubona, southern province of Rwanda.

Therefore, the objectives of this study were (i) to determine the effects of P resources and lime on soil chemical and biological properties, (ii) to determine the effects of P resources on maize grain yields and (iii) to establish the relationships between plant residues quality, soil chemical and biological properties and maize grain yields.

Materials and Methods

Study Site Description and Characteristics

The field experiment was conducted in Rubona, southern province of Rwanda ($2°35'$ South and $29°43'$ East). Rubona is located at an altitude of 1,734 m above sea level, with a mean annual temperature of $20.2°C$ and mean annual rainfall of 1,400 mm distributed over two cropping seasons. The soil at the experimental site was a typic hapludult (Soil Survey Staff, 1994). The soil physical and chemical properties of the experimental site are described in Table 1.

Experimental Design

The treatments compared were the following: (a) organic P sources such as *Calliandra*, *Tithonia* and *Tephrosia* applied independently each at rates of 25 and 50 kg P ha^{-1}, (b) inorganic P source applied alone at a rate of 25 and 50 kg P ha^{-1}, (c) the combination

Table 1 Selected initial physicochemical properties of the topsoil (0–20 cm depth) of the experimental site at Rubona, southern province of Rwanda

Soil properties	Value
Soil pH in water (1:2.5)	4.6
Exchangeable Ca (cmol kg^{-1})	2.1
Exchangeable Mg (cmol kg^{-1})	1.2
Exchangeable K (cmol kg^{-1})	0.1
Extractable P (mg P kg^{-1} soil)	6.1
Total soil carbon (%)	1.3
Clay (%)	25.5
Sand (%)	69
Silt (%)	5.5

of organic materials listed above with inorganic P at rates of 25 and 50 kg P ha^{-1}, (d) lime as $Ca(OH)_2$ applied at a rate of 2.5 t ha^{-1} and (e) the control without amendments.

This study was conducted in the field during the long and short rain seasons of 2001 and 2004. The experiment was laid out in a randomised complete block design with 16 treatments replicated 3 times. Tree leaf biomass was incorporated in the top 0–20 cm soil layer by hoeing once in each season and 1 day before planting maize. To avoid confounding effects of nitrogen and potassium, N was applied in the form of urea (46% N) at the rate of 100 kg N ha^{-1} while K was applied as muriate of potash (52% K) at the rate of 100 kg K ha^{-1}. Nitrogen was applied at the rate of 50 kg N ha^{-1} at planting and 50 kg N ha^{-1} at 5 weeks after planting while K was broadcasted at planting. Experimental plots measured 9 × 4 m and maize ZM 607 variety was planted at 0.70 m between rows and 0.25 m within rows.

Soil Sampling and Analysis

Each season, after maize harvest, soil samples were randomly collected from 10 locations in each individual experimental plot at 0–20 cm, put in bucket and mixed with hand, and two composite samples were taken. One composite sample was kept fresh in cold room at $4°C$ for soil biological properties determination. The other part of the soil composite sample was air-dried, grounded and sieved through 0.5 mm and subsequently analysed for soil chemical properties.

The soil biological proprieties were analysed for microbial biomass carbon (MBC) using the fumigation–extraction method described in Tate et al. (1988), while microbial biomass phosphorus (MBP) was determined according to the method of Kuono et al. (1995). Acid phosphatase activity (APA) was determined based on the method developed by Tabatabai and Bremner (1969) and modified by Sinsabaugh and Linkins (1990) and Carreiro et al. (2000), and adapted by Verchot and Borelli (2005) for degraded tropical soils.

For soil chemical properties, soil pH was determined in water (1:2.5 ratio). Exchangeable acidity and aluminium were analysed by the method of Anderson and Ingram (1993). Carbon and nitrogen in the soil were determined by the dry combustion method using (CN) Analyser LECON – 2000 described by Wright and Bailey (2001). Phosphorus fractions were determined sequentially using the method of Tiessen and Moir (1993). All biological properties were analysed only in season 4, while soil chemical parameters were analysed each season.

Plant Sampling and Analysis

In each season, fresh leaves of the agroforestry shrubs were picked, mixed and a composite sample of leaves from each species was randomly taken. The composite sample was air-dried, grounded and sieved through 0.5 mm. The total plant P in the samples was determined using the sulphuric acid Kjeldahl digestion method (Anderson and Ingram 1993). The soluble C and P were determined using the method described by TAPPI (1988). Lignin was determined according to the acid detergent fibre (ADF) method of Van Soest (1963) and total extractable polyphenol was analysed according to the method of Constantinides and Fownes (1994). Total C and N in the leaves were determined by the dry combustion method using (CN) Analyser as described earlier in the section of soil analysis.

Data Analysis

Analysis of variance (ANOVA) of the data was carried out using GenStat for Windows Discovery Edition 9 (GenStat, 2006). Mean comparisons were done using Duncan multiple range test (DMRT). Linear and non-linear regression analyses were carried out to determine the relationships between plant quality residue, soil chemical and biological properties and total maize grain yields. The following models were used:

Linear model:

$$Y = ax + b \tag{1}$$

where Y represents total maize grain yields, a the slope, x the total P, N and C, b the intercept

Exponential decay model:

$$Y = a \times \exp(-a \times x) \tag{2}$$

where Y represents total maize grain yields, x the plant quality parameters, a the fitted parameter

Hyperbola model:

$$Y = a \times x/(a + x) \tag{3}$$

where Y represents total maize yield, a the fitted parameter, x the soil chemical and biological properties.

Results and Discussions

Chemical Composition of the Leaf Agroforestry Species

The average values of nutrient content for four seasons are given in Table 2. All species were fairly rich in total N, being above the critical level of 2.5%. Palm and Rowland (1997) reported that above this critical level, decomposition and N mineralisation occurred fast. *Tithonia* had the highest value of N, while *Tephrosia* had the lowest. In terms of total and soluble P, *Tithonia* was superior to both *Calliandra* and *Tephrosia* with a difference of about 50% for total P content (Table 2). Both *Calliandra* and *Tephrosia* had similar values of total P concentration and were below the critical limit (<0.24%), which indicates that immobilisation would be expected (Palm and Rowland, 1997). On the other hand, *Tithonia* had the highest soluble C than *Calliandra* (7.4%) and *Tephrosia* (5.6%).

Table 2 Selected chemical characteristics of agroforestry inputs used in the study

Chemical constituents	Calliandra	Tithonia	Tephrosia	Critical value (%)	Reference
Total N	3	3.3	2.8	>2.5	Palm and Roland (1997)
Total C	39.7	30.9	38.6	–	–
Total P	0.2	0.4	0.2	>0.2.4	Palm and Roland (1997)
Soluble P	0.1	0.2	0.1	–	–
Soluble C	7.4	8.1	5.6	–	–
Polyphenol	10.5	3.1	3.9	<4	Palm and Roland (1997)
Lignin	16.2	7.8	8.2	<15	Palm and Roland (1997)
Soluble C/soluble P	123	36	48	–	–
Soluble C/ total P	45	25	30	<30	Nziguheba (2001)
Total C/P	247	93	206	<300	Iyamulemye and Dick (1996)
N/P	18	9.2	14.7	–	–
C/N	13	9.4	13.9	<30	–
Polyphenol/N	4	1	1	–	–
Lignin/N	6	2	3	–	–
Polyphenol + Lignin/N	9	3	4	–	–

The values are presented in % where applicable

Lignin and soluble polyphenol contents in *Tithonia* and *Tephrosia* leaf biomass were below the critical levels of 15 and 4%, respectively, and lignin and soluble polyphenol of *Calliandra* were above the critical levels of >15 and >4%, respectively, at which *Calliandra* would have been immobilised (Palm and Rowland, 1997). In terms of lignin:total N ratio, *Calliandra* had the highest ratio, while *Tithonia* had the lowest. The ratios of polyphenol:total N and lignin + polyphenol:total N had similar trends as that of lignin:total nitrogen.

All the species studied had C:N ratio < 19, indicating that net immobilisation was unlikely to occur. Frankenberger and Abdelmagid (1985) reported a critical ratio of <19 for net mineralisation to occur. However, the C:N ratio has been found to be a poor predictor of decomposition and N release in most agroforestry studies, with exception studies of Sandhu et al. (1990), Tian et al. (1992), Mtambanengwe and Kirchman (1995). In the literature, this index rarely relates to P mineralisation. In this study, C:N ratio seemed to be a good index for predicting P mineralisation.

Among the three species, *Calliandra* had the highest C:P ratio (247), followed by *Tephrosia* (206), but both *Calliandra* and *Tephrosia* had C:P ratio below the critical level (C:P ratio > 300). *Tithonia* had the lowest C:P ratio (92.9). However, C:P ratio for all species tested were lower than the critical level of >300 above which immobilisation was expected (Iyamuremye et al., 1996). Several workers (Dalal,

1979; Iyamuremye and Dick, 1996) suggested that whereas P mineralisation occurs in residues with C:P ratio < 200, P immobilisation occurs when the ratio is >300.

The soluble C:total P ratio was lowest for *Tithonia* and highest for *Calliandra* (Table 2). Only *Tithonia* was below the critical level (<30), and net mineralisation was evident in this study as expected (Nziguheba, 2001). The trend of decomposable organic materials was in the order of *Tithonia* > *Tephrosia* > *Calliandra*.

Effect of Organic and Inorganic Fertilizer Inputs on Soil Chemical Properties

Soil pH

The soil pH measurements ranged from 4.2 to 5.5 during the four seasons (Table 3). Lime increased pH consistently across the seasons compared to other treatments. This was due to the effect of $CaCO_3$ on exchangeable aluminium. Except lime, only the treatments including *Tithonia* increased soil pH from 0.4–0.9 units compared to the control. The increase in soil pH due to *Tithonia* could be attributed to self-liming (alkalinity) caused by the mineralisation of C and release of basic cations, and from the release of OH^- produced by NH_4^+ during the decomposition and ammonification of *Tithonia*. Haynes and Mokolobate

Table 3 Effects of organic and inorganic fertilizer inputs on soil chemical properties

Treatments	pH in water (1:2.5 ratio)				Exch. acidity (cmol kg⁻¹)				Exch. Al³⁺ (cmol kg⁻¹)				SOC (g ckg⁻¹)
	S1	S2	S3	S4	S1	S2	S3	S4	S1	S2	S3	S4	
Control	4.5	4.6	4.2	4.5	1.5^a	2.7^a	1.3^a	2.0^a	0.8^a	1.0^a	0.2^a	0.9^a	8.2^c
Lime	5.5	5.4	5.3	5.5	0.1^d	0.2^d	0.2^d	0.3^d	0.1^b	0.1^d	0.02^d	0.1^c	9.1^c
TSP25P	4.8	4.7	4.2	4.6	0.4^c	2.4^a	1.2^a	1.9^a	0.3^b	0.4^b	0.2^{ab}	0.6^b	8.9^c
TSP50P	4.6	4.5	4.4	4.6	0.8^b	2.2^a	1.2^a	1.6^a	0.2^b	0.5^b	0.2^{ab}	0.5^{bc}	9.3^c
CC25P	4.7	4.7	4.3	4.6	0.3^{cd}	1.3^b	1.1^{ab}	1.4^b	0.1^b	0.2^{bc}	0.1^{bc}	0.4^{bc}	9.7^{bc}
CC50P	4.6	4.7	4.5	4.7	0.1^d	0.5^{bcd}	0.9^b	1.1^{bc}	0.2^b	0.1^d	0.1^{bc}	0.3^c	10.5^{bc}
CC12.5P+12.5P	4.8	4.7	4.6	4.8	0.1^d	1.0^{bcd}	0.9^b	1.4^b	0.1^b	0.2^{bc}	0.1^{bc}	0.3^c	10.0^{bc}
CC25P+25P	4.8	4.6	4.2	4.7	0.2^d	1.1^{bcd}	1.0^b	1.4^b	0.1^b	0.2^{bc}	0.2^{ab}	0.4^{bc}	12.7^a
TD25P	4.9	5	4.5	5	0.2^d	0.6^{bcd}	0.6^c	0.9^c	0.1^b	0.1^d	0.1^{bc}	0.2^c	9.3^c
TD50P	4.9	5	4.7	5.3	0.2^d	0.4^{bcd}	0.6^c	0.6^{cd}	0.1^b	0.1^d	0.1^{bc}	0.1^c	10.3^{bc}
TD12.5P+12.5P	5	5	4.6	4.8	0.1^d	0.4^{bcd}	0.6^c	1.1^{bc}	0.1^b	0.1^d	0.04^d	0.2^c	9.9^{bc}
TD25P+25P	5	5.5	4.5	5.1	0.1^d	0.3^d	0.5^c	0.8^{cd}	0.1^b	0.1^d	0.1^{bc}	0.1^c	12.0^a
TV25P	4.7	4.6	4.3	4.7	0.1^d	1.2^{bcd}	1.0^b	1.4^b	0.2^b	0.3^{bc}	0.1^{bc}	0.3^c	9.2^c
TV50P	4.8	4.7	4.4	4.7	0.2^d	1.0^{bcd}	1.0^b	1.3^{bc}	0.1^b	0.2^{bc}	0.1^{bc}	0.3^c	10.6^b
TV12.5P+12.5P	4.7	4.8	4.3	4.7	0.2^d	1.2^{bcd}	0.9^b	1.3^{bc}	0.1^b	0.2^{bc}	0.2^{ab}	0.3^c	9.3^c
TV25P+25P	4.8	4.7	4.3	4.7	0.2^d	0.7^{bcd}	0.9^b	1.2^{bc}	0.2^b	0.3^{bc}	0.2^{ab}	0.6^b	12.6^a

In the columns, means with the same letter are not significantly different at $P < 0.05$ according to Duncan's multiple range test

TD, *Tithonia diversifolia*; CC, *Calliandra calothyrsus*; TV, *Tephrosia vogelii*; TSP, triple super phosphate; S1 to S4, season one to season four; Exch., exchangeable acidity; Al³⁺, aluminium ion; SOC, soil organic carbon; 12.5P = 12.5 kg P ha⁻¹; 25P = 25 kg P ha⁻¹; 50P = 50 kg P ha⁻¹

(2001) reported that the process of ammonification increases soil pH as OH$^-$ is produced by NH_4^+ that is mineralised from organic N. These results are also consistent with the work of Phan Thi Cong (2000), George et al. (2002) and Ikerra et al. (2006) who reported increase in soil pH on similar soils due to *Tithonia* application in Vietnam, Kenya and Tanzania.

Second, the increase in soil pH could be due to the ligand exchange or specific adsorption of organic acid molecules produced during plant residue decomposition.

Exchangeable Acidity and Aluminium

Lime and P resources tested reduced exchangeable acidity and aluminium significantly ($P \leq 0.05$) relative to the control. The decrease in exchangeable acidity and aluminium might have been due to chelation of exchangeable aluminium by organic matter produced during the decomposition of green manure. These results are in agreement with the work of Wong et al. (1995) and Haynes and Mokolobate (2001) who reported that organic matter produced during the decomposition of green manure reduces exchangeable aluminium in soil solution.

Soil Organic Carbon (SOC)

The cumulative effect of application of P resources showed that SOC varied from 8.2 to 12.7 g C kg^{-1} (Table 3). The combination of green manure with TSP at a rate of 50 kg P ha^{-1} increased SOC significantly ($P < 0.05$). The significant effects due to the combination of green manure and TSP at a rate of 50 kg P ha^{-1} could be attributed to the quantity of green manure and P added compared to other treatments. Supplying high amounts of C, N and P through high amount of green manure and TSP applied could have a positive effect on microorganism activities which decomposed fast organic residues to increase soil organic C. Goyal et al. (1999) observed a 44% increase in soil organic C two seasons after addition of a combination of wheat straw and fertilizers.

Effects of Organic and Inorganic P Fertilizer Inputs on Inorganic Labile P Fractions

Resin P

The concentration of resin P ranged from 0.5 to 18.2 mg P kg^{-1} (Table 4).

The combination of *Tithonia* green manure with TSP at a rate of 50 kg P ha^{-1} contributed to increase resin P fraction consistently and significantly more than *Calliandra* and *Tephrosia* for the same rate of applied P ($P < 0.05$). This was due to the quality of green manure from *Tithonia* that was rich in P (Table 2). Furthermore, *Tithonia* decomposes fast (Gachego et al. 1999) and releases high amount of P during decomposition, which must have contributed to high resin P.

In western Kenya, Nziguheba (2001) reported that seasonal applications of 25 kg P ha^{-1} gradually replenished resin P. The findings of this study did not agree with the result reported above because application of 25 kg P ha^{-1} did not differ significantly with the control. The soils of western Kenya have moderate soil P sorption capacity while that of Rubona has high P sorption capacity (Mutwewingabo, 1989), probably P released was reabsorbed by sesquioxides.

Inorganic Bicarbonate P Fraction (BPi)

The concentration of BPi varied from 3.7 to 20.5 mg P kg^{-1} (Table 4). The combination of organic materials with TSP at a rate of 50 kg P ha^{-1} and TSP at a rate of 50 kg P ha^{-1} applied independently showed significant ($P < 0.05$) higher BPi compared to the control. *Tithonia* residues combined with TSP at a rate of 50 kg P ha^{-1} consistently increased BPi across the seasons. This finding was in agreement with the results on *Tithonia* residues reported by Ikerra et al. (2006) in the experiment conducted in the Magadu area of the Sokoine University of Agriculture, Tanzania. The absence of a trend of BPi across the seasons implies that this fraction is a major source of P to maize crop.

Table 4 Effects of organic and inorganic fertilizer inputs on labile P fractions in Rubona, southern Rwanda

Treatments	Resin P (mg P kg^{-1})				BPi (mg P kg^{-1})			
	S1	S2	S3	S4	S1	S2	S3	S4
Control	0.5^d	2.2^c	0.5^d	1.3^f	5.9^C	4.5^e	6.2^d	3.7^d
Lime	1.3^{cd}	3.1^{bc}	0.7^{cd}	2.9^{cdef}	6.3^C	5.2^{de}	6.6^d	4.8^d
TSP25P	1.3^{cd}	4.5^{bc}	1.7^{bcd}	5.5^{bc}	7.7^C	7.7^{bcde}	9.3^{cd}	15.0^b
TSP50P	2.6^{bcd}	9.3^b	3.6^b	6.7^b	9.3^{bc}	10.6^{abcd}	13.1^{bc}	20.5^a
CC25P	1.1^{cd}	5.1^{bc}	1.4^{bcd}	2.2^{def}	7.1^c	6.7^{bcde}	8.1^d	6.1^{cd}
CC50P	1.2^{bcd}	6.6^{bc}	2.1^{bcd}	5.5^{bc}	7.6^c	7.9^{bcde}	9.9^{cd}	7.6^{cd}
CC12.5P+12.5P	2.5^{bcd}	5.7^{bc}	1.9^{bcd}	3.6^{bcdef}	8.5^{bc}	6.4^{cde}	8.7^d	8.1^{cd}
CC25P+25P	3.2^{abc}	9.15^b	2.7^{bcd}	5.7^{bc}	11.1^{ab}	11.4^{abc}	14.7^b	12.5^{bc}
TD25P	2.0^{bcd}	6.7^{bc}	2.4^{bcd}	4.2^{bcdef}	8.1^{bc}	6.9^{bcde}	8.6^d	7.0^{cd}
TD50	2.1^{bcd}	9.0^b	3.4^{bc}	5.8^{bc}	9.4^{bc}	11.2^{abc}	13.2^{bc}	9.5^{cd}
TD12.5P+12.5P	4.0^{ab}	8.9^b	2.2^{bcd}	4.7^{bcde}	11.0^{ab}	8.2^{bcde}	10.2^{cd}	8.2^{cd}
TD25+25P	5.0^a	18.2^a	5.6^a	10.2^a	13.2^a	13.7^a	18.7^a	19.9^a
TV25P	0.9^{cd}	4.7^{bc}	1.1^{bcd}	1.9^{ef}	6.7^c	6.8^{bcde}	8.0^d	5.2^d
TV50	1.2^{cd}	8.2^{bc}	2.1^{bcd}	4.3^{bcdef}	7.6^c	7.1^{bcde}	10.2^{cd}	6.3^{cd}
TV12.5+12.5P	1.9^{bcd}	5.8^{bc}	1.8^{bcd}	2.1^{def}	8.2^{bc}	8.0^{bcde}	8.6^d	7.4^{cd}
TV25+25P	3.3^{abc}	8.3^{bc}	3.4^{bc}	5.3^{bcd}	10.9^{ab}	12.3^{ab}	14.8^b	12.8^{bc}

In the columns, means with the same letter are not significantly different at $P < 0.05$ according to Duncan's multiple range test
TD, *Tithonia diversifolia*; CC, *Calliandra calothyrsus*; TV, *Tephrosia vogelii*; TSP, triple super phosphate; S1 to S4 season one to season four; Bpi, inorganic bicarbonate phosphorus; 12.5P = 12.5 kg P ha^{-1}; 25P = 25 kg P ha^{-1}; 50P = 50 kg P ha^{-1}

Effects of Organic and Inorganic Fertilizer Inputs on Soil Biological Properties

Acid Phosphatase Activity (APA)

Acid phosphatase activity (APA) ranged from 0.2 to 0.4 μmol pNP g soil^{-1} h^{-1} (Table 3). The green manure applied independently and combined with TSP at a low rate of P (25 kg P ha^{-1}) seemed to stimulate APA significantly ($P < 0.05$) than the control, lime and TSP. The results from this study showed that the treatments with high rate of P tended to suppress APA. However, in western Kenya, it was reported that application of 250 kg P ha^{-1} did not suppress acid phosphatase activity (Radersma and Grierson, 2004). The results of this study disagree with this result but agree with several previous studies (Tabatabai, 1982; Tarafdar and Marschner, 1994), which indicated that P deficiency in soil stimulates higher acid phosphatase activity.

Soil Microbial Biomass Carbon (MBC)

Microbial biomass C varied from 104.7 to 226.9 mg C kg^{-1} (Table 5). All treatments increased MBC significantly ($P < 0.05$) than the control except TSP and *Calliandra* applied independently at a rate of 25 kg P ha^{-1}. Green manure combined with TSP applied at a rate of 50 kg P ha^{-1} significantly increased MBC than all treatments ($P < 0.05$). This was due to the quantity of C and P added through green manure. In western Kenya, the results reported by Nziguheba (2001) support this finding.

Microbial Biomass Phosphorus (MBP)

Microbial biomass phosphorus ranged from 0.5 to 8.4 mg P kg^{-1} (Table 5). The combination of *Tithonia* with TSP at a rate of 50 kg P ha^{-1} increased MBP significantly ($P < 0.05$) compared to the remaining treatments. This may reflect effects of the highly soluble C and P provided by *Tithonia* and soluble P from TSP on microbial activity. Chauhan et al. (1981) and Sharpley (1985) stressed the necessity of both added C and P sources to build up soil P. The carbon source provides substrate for microbial growth and activity, and subsequent microbial turnover results (Nziguheba, 2001). The addition of P without a C source or addition of a highly soluble C source (*Tithonia*) with or

Table 5 Effects of P
resources and P levels on acid
phosphatase activity (APA),
microbial biomass carbon
(MBC) and microbial biomass
phosphorus (MBP)

Treatments	APA (μmol pNP g soil^{-1} h^{-1})	MBC (mg C kg^{-1})	MBP (mg P kg^{-1})
Control	0.3^b	104.7^d	0.5^c
Lime	0.2^b	144.4^c	0.8^{bc}
TSP25P	0.3^{ab}	118.2^{cd}	2.7^{bc}
TSP50P	0.3^{ab}	135.1^{cd}	4.0^b
CC25P	0.4^{ab}	134.3^{cd}	0.9^{bc}
CC50P	0.3^{ab}	174.0^{bc}	2.6^{bc}
CC12.5P+12.5P	0.4^a	166.1^{bc}	1.6^{bc}
CC25P+25P	0.3^{ab}	207.9^{ab}	3.2^{bc}
TD25P	0.4^a	144.8^c	4.4^b
TD50P	0.3^{ab}	178.1^b	4.7^b
TD12.5P+12.5P	0.3^{ab}	173.8^{bc}	3.6^{bc}
TD25P+25P	0.3^b	226.9^a	8.4^a
TV25P	0.4^a	141.1c	1.4^{bc}
TV50P	0.4^a	157.1^{bc}	1.6^{bc}
TV12.5P+12.5P	0.3^{ab}	171.8^{bc}	2.6^{bc}
TV25P+25P	0.3^{ab}	202.4^{ab}	4.8^b

In the columns, means with the same letter are not significantly different at $P < 0.05$ according to Duncan's multiple range test

APA, acid phosphatase activity; MBC, microbial biomass carbon; MBP, microbial biomass phosphorus; TD, *Tithonia diversifolia*; CC, *Calliandra calothyrsus*; TV, *Tephrosia vogelii*; 12.5P = 12.5 kg P ha^{-1}; 25P = 25 kg P ha^{-1}; 50P = 50 kg P ha^{-1}

without P resulted in differing microbial P and C dynamics. Thus, the coupling of C and P is critical in plant–soil–biota-cycling systems. The soil microbial biomass may be regarded as the driving force of the P cycling by which organic residues are taken up, converted into new products (nutrients), and subsequently released (Steward, 1992; Sharpley, 1996).

Maize Response to Organic and Inorganic Fertilizer Inputs

Maize grain yield varied across the cropping seasons from 0.9 and 7.1 t ha^{-1} (Table 6). Lowest maize yield were recorded in the third season due to the low amount of rainfall, which was also poorly distributed over this period. Maize yield harvested in the control ranged between 0.9 and 1.6 t ha^{-1} while lime increased maize yield from 1.1 to 3.3 t ha^{-1}. Only in the first season, application of lime resulted in significant higher maize yield than the control ($P < 0.05$). During the cropping seasons, organic residues showed higher significant ($P < 0.05$) maize grain yields compared to the control and lime. This increment was due to the recycling of nutrients from leaf biomass of

agroforestry shrubs through microbial activity, which added multiple nutrients at the same time (N, P and K) to the soil and benefit to the maize grain yields.

Application of inorganic fertilizers even at low rate (25 kg P ha^{-1}) also increased maize yield significantly ($P < 0.05$) compared to the control and lime throughout the cropping seasons (Table 6). In the fourth season, *Tithonia* combined with TSP applied at a rate of 50 kg P ha^{-1} produced similar maize yield as *Tithonia* applied independently at similar rate, but significantly higher than other remaining treatments ($P < 0.05$). Niang et al. (1996) found greater maize grain yields following incorporation of *Tithonia* biomass than residues of other common shrubs and trees in western Kenya. They associated this with the high biomass quality of *Tithonia* green manure.

Relationships Between Maize Yield and Plant Residues Quality Parameters

Variation in maize yield was highly influenced by the plant quality residues parameters such as total P, C and N contents, and C:P and C:N ratios. The high

Table 6 Influence of P resources, rates of P and seasons on maize grain yields (t ha^{-1}) at Rubona in Rwanda

Treatments	P level (kg ha^{-1})	S1	S2	S3	S4	Mean total maize yield
Control	0	1.6^e	1.2^e	0.9^c	1.1^f	4.8^e
Lime	0	3.3^d	1.2^e	1.1^c	1.2^f	6.8^e
TSP25P	25	4.3^{abc}	4.6^{dc}	3.0^{ab}	5.0^{cde}	16.9^{cd}
TSP50P	50	5.0^{ab}	5.3^{abc}	3.9^{ab}	5.7^{bc}	19.8^{bc}
CC25P	25	3.9^{bcd}	4.0^d	2.9^{ab}	4.5^{de}	15.3^d
CC50P	50	4.9^{ab}	4.8^{bcd}	3.6^{ab}	5.0^{cde}	18.2^{bc}
CC12.5+12.5P	25	4.7^{abc}	4.6^{dc}	3.0^{ab}	5.2^{cde}	17.5^{cd}
CC25+25P	50	5.1^a	5.1^{abc}	4.2^a	5.8^{bc}	20.2^b
TD25P	25	5.0^{ab}	5.1^{abc}	3.6^{ab}	5.8^{bc}	19.5^{bc}
TD50P	50	5.1^a	5.4^{abc}	3.7^{ab}	6.5^{ab}	20.7^{ab}
TD12.5+12.5P	25	5.3^a	5.5^{ab}	4.5^a	6.3^b	21.5^{ab}
TD25+25P	50	5.2^a	5.8^a	4.5^a	7.1^a	22.6^a
TV25P	25	3.7^{dc}	4.0^d	2.3^b	4.2^e	14.3^d
TV50P	50	4.5^{abc}	4.6^{dc}	3.3^{ab}	4.8^{cde}	17.1^{cd}
TV12.5+12.5P	25	4.8^{ab}	4.6^{dc}	3.2^{ab}	5.2^{cd}	17.8^{cd}
TV25+25P	50	4.4^{abc}	4.9^{bc}	4.0^{ab}	5.6^{bc}	18.8^{bc}

In the columns, means with the same letter are not significantly different at $P < 0.05$ according to Duncan's multiple range test
S1, season 1; S2, season 2; S3, season 3; S4, season 4; TSP, triple superphosphate; CC, *Calliandra calothyrsus*; TD, *Tithonia diversifolia*; TV, *Tephrosia vogelii*

regression coefficients to predict maize yield were obtained with total plant N (83%), C:N ratio (81%), total plant P (74%) and C:P ratio (61%) (Fig. 1). The relationships between total maize yield and total plant N, P and C were linear but negative for C. *Tithonia* green manure with low total plant C content was associated with high maize yield, while *Calliandra* and *Tephrosia* green manure with high total plant C content were associated with low maize yield (Fig. 1). This could be due to the higher soluble C in *Tithonia* residues than *Calliandra* and *Tephrosia* residues. *Tithonia* green manure, which was the richest in P, showed higher total maize yield than *Calliandra* and *Tephrosia* residues both of which had similar maize yield (Fig. 1). The same trend was observed for total plant N. On the other hand, the narrow ratios of C:P and C:N was associated with high maize yield, while the wide ratios were associated with low maize yield (Fig. 1). Frankenberger and Abdelmagid (1985) reported that a critical C:N ratio of <19 leads to net mineralisation and this implies that N release was available to maize crop, while a wide C:N ratio indicates that N was slowly released and less available to maize crop. Iyamuremye et al. (1996) reported that C:P < 300 leads to net mineralisation. *Tithonia* green manure had C:P of 92.9 (Table 2), which leads to net P mineralisation and high maize yield.

Relationships Between Total Maize Yield, Soil Chemical and Biological Properties

The relationship between maize grain yields and soil chemical (inorganic labile P fractions and soil organic carbon) and biological properties (MBC and MBP) were highly significant ($P < 0.0001$) (Table 7). For inorganic labile P fractions and SOC, the relationships between maize grain yields, P fractions and SOC were weak (less than 50%), while it was high for MBP. These weak relationships suggest other limitations such as N, K, Ca and Mg, which play an important role in maize production. Linquist et al. (1996) reported weak relationships between P pools and crop yields.

Relationships between total maize yield and soil chemical and biological properties were highly significant, which confirmed the important role of soil chemical and biological properties in maize production.

Conclusion

Substantial maize yield was obtained from *Tithonia* green manure combined with TSP applied at a rate

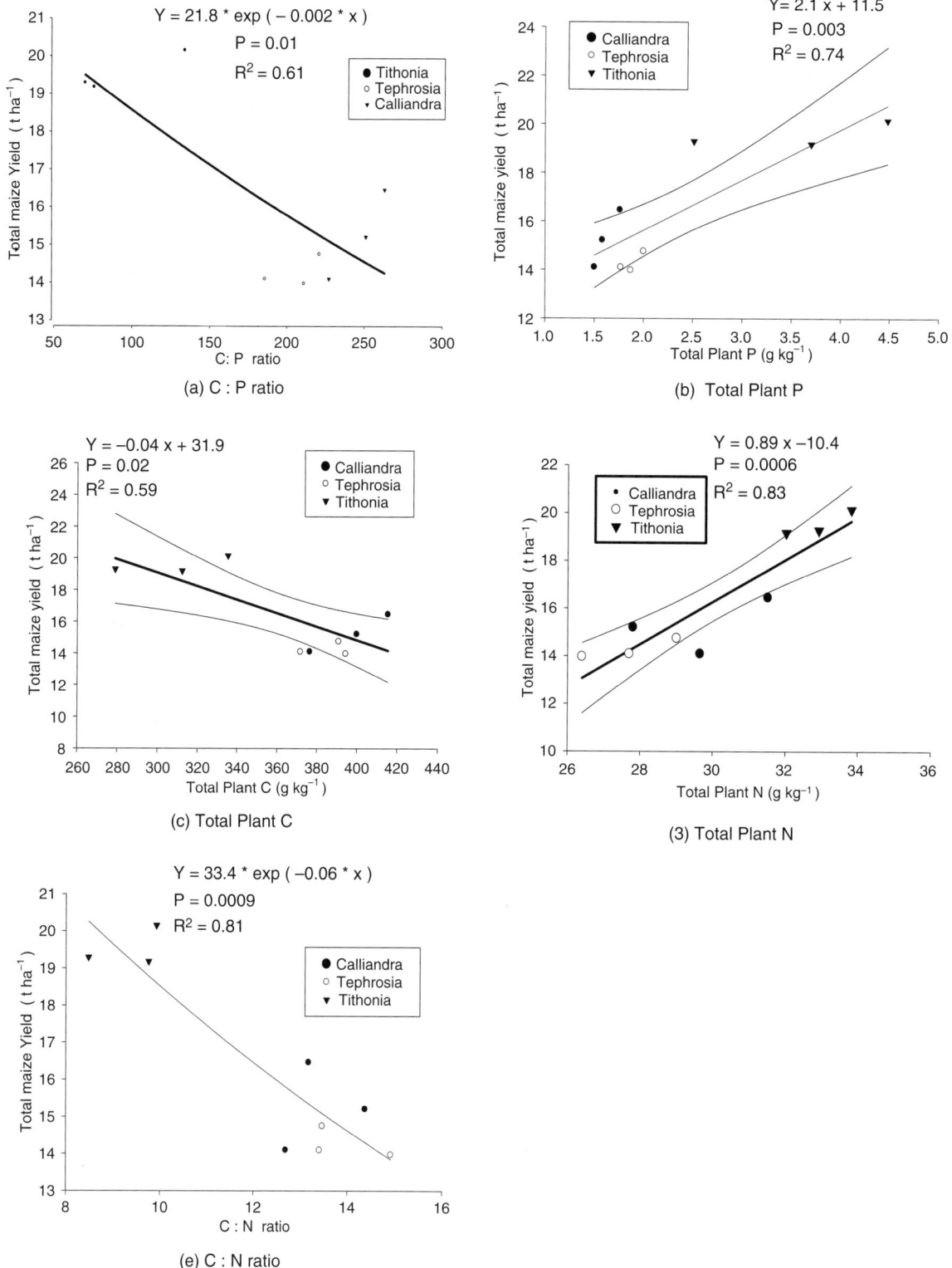

Fig. 1 Relationship between total maize grain yields and plant quality parameters

Table 7 Relationships between total maize grain yields and selected soil chemical and biological properties

Response variable	Explanatory variable	Model function	R^2 (%)	Probability
Total maize grain yields	Resin P	$Y = 8.1 \times x/(2.5 + x)$	46	$P < 0.0001$
	BPi	$Y = 8.6 \times x/(6.1 + x)$	48	$P < 0.0001$
	SOC	$Y = 73.3 \times x/(139.6 + x)$	41	$P < 0.0001$
	MBC	$Y = 20.7 \times x/(509.3 + x)$	28	$P < 0.0001$
	MBP	$Y = 7.1 \times x/(0.8 + x)$	58	$P < 0.0001$

Bpi, inorganic bicarbonate P fraction; SOC, soil organic carbon; MBC, microbial biomass carbon; MBP, microbial biomass phosphorus

of 50 kg P ha^{-1}. *Tithonia* was superior to *Calliandra* and *Tephrosia* serving as a P source for maize production in an acid-high P-fixing Ultisol in Rubona, Rwanda. The plant quality residue determined can be used to predict decomposition of tree leaf biomass of agroforestry shrub species, P release, screening agroforestry trees/shrubs as P sources and maize yield. Integrated use of organic residues with moderate dose of TSP at 25 kg P ha^{-1} is a recommendable option to the small-scale farmers as a way to improve maize production and gradually replenishing soil chemical and biological properties in an acid soil of Rubona. The soil chemical and biological properties can be used to predict total maize yield. The challenges for adoption and scaling up of green manure combined with TSP options to millions of small-scale farmers in Rwanda are the issues of quality of green manure, nutrient balancing, labour to collect and transport organic inputs and their management, which need to be optimised.

References

Anderson JM, Ingram JSI (1993) A handbook on methods of soil analysis, 2nd edn. Tropical Soil Biology and Fertility (TSBF), CABI, Wallingford, UK, p 140

Carreiro MM, Sinsanbaugh RL, Repert DA, Pankhurt DF (2000) Microbial enzyme shifts explain litter decay responses to simulated nitrogen deposition. Ecology 81:2359–2365

Chauhan BS, Stewart JWB, Paul EA (1981) Effect of labile inorganic phosphate status and organic carbon additions on the microbial uptake of phosphorus in soil. Can J Soil Sci 61:373–385

Constantinides M, Fownes JH (1994) Tissue-to-solvent ratio and other factors affecting determination of soluble phenolics in tropical leaves. Commun Soil Sci Plant Anal 25:3221–3227

Dalal RC (1979) Mineralisation of carbon and phosphorus from carbon-14 and phosphorus-32 labelled plant materials added to soil. Soil Sci Soc Am J 43:913–916

Frankenberger WT, Abdelmagid HM (1985) Kinetic parameters of nitrogen mineralisation rates of leguminous crops incorporated into soil. Plant Soil 87:257–271

Gachengo CN, Palm CA, Jama B, Othieno C (1999) Combined use of trees, shrubs and inorganic fertilizers for soil fertility management. Agroforest Syst 44:21–36

GenStat (2006) GenStat for windows, 9th edn. VSN International, Oxford, UK

George TS, Gregory PJ, Robinson JS, Buresh RJ, Jama B (2002) Utilisation of soil organic P by agroforestry and crop species in the field, Western Kenya. Plant Soil 246:53–63

Goyal S, Chander K, Mundra MC, Kapoor KK (1999) Influence of inorganic fertilizers and organic amendments on soil organic matter and soil microbial properties under tropical conditions. Biol Fertil Soils 29:196–200

Haynes RJ, Mokolobate MS (2001) Amelioration of Al toxicity and P deficiency in acid soils by additions of organic residues: a critical review of the phenomenon and the mechanisms involved. Nutr Cycling Agroecosyst 59:47–63

Heisey PW, Mwangi W (1996) Fertilizer use and maize production in sub-Saharan Africa. Economics Working Paper 96, D.F CIMMYT, Mexico, p 34

Henao J, Baanante CA (1999) Estimating rates of nutrients depletion in soils of agricultural lands of Africa. Int Centre Soil Fert Agricl Dev (IFDC) 31(1):126:36–45

Ikerra ST, Kalumuna MC (1992) Phosphorus adsorption characteristics of soils and their influence on maize yields to P application along Mlingano catena, Tanga, Tanzania. In: Zake JYK et al (eds) Proceedings of the Eleventh Annual General Meeting of the Soil Science of East Africa. Mukono, Kampala, Uganda, pp 85–94

Ikerra ST, Semu E, Mrema JP (2006) Combined *Tithonia diversifolia* and minjingu phosphate rock for improvement of P availability and maize grain yields on a chromic acrisol in Morogoro, Tanzania. Nutr Cycling Agroecosyst 76:249–260

Iyamulemye F, Dick RP (1996) Organic amendments and phosphorus sorption by soils. Adv Agron 56:139–185

Iyamulemye F, Dick RP, Baham J (1996) Organic amendments and phosphorus dynamics: II. Distribution of soil phosphorus fractions. Soil Sci 161:436–443

Jama B, Swinkels AR, Buresh JR (1997) Agronomic and economic evaluation of organic and inorganic sources of phosphorus in Western Kenya. Agron J 89:597–604

Kazombo-Phiri SFM (2005) Main causes of declining soil productivity in Malawi: effects on agriculture and management technologies employed to alleviate the problems. In: Omonya GO, Pasternak D (eds) Proceedings of the International Symposium for Sustainable Dryland Agriculture Systems. Niamey, Niger, pp 59–66

Kuono K, Tuchiya Y, Ando T (1995) Measurement of microbial biomass phosphorus by anion exchange membrane method. Soil Biol Biochem 27:1353–1357

Linquist BA, Singleton PW, Cassman KG, Keane K (1996) Residual phosphorus and long-term management strategies for an Ultisol. Plant Soil 184:47–55

Mtambanengwe F, Kirchman H (1995) Litter from tropical savanna woodland (miombo) chemical composition C and N mineralisation. Soil Biol Biochem 27:1639–1651

Mutwewingabo B (1989) Geneses, characteristics and management constraints of sombric acidic profound horizon in Rwanda high land (Genèse, caractéristiques et contraintes d'aménagement des sols acides a horizon sombre de profondeur de la région de haute altitude du Rwanda). Soltrop 89°: actes du premier séminaire franco-africain de Pédologie tropicale, Lomé, 6–12 Février 1989. ORSTOM, Paris, pp 353–385

Niang AI, Gathumbi SM, Amadalo B (1996) The potential of improved fallow for crop productivity enhancement in the highlands of western Kenya. East Afr Agric Forum J 62:103–124

Nziguheba G (2001) Improving phosphorus availability and maize production through organic and inorganic amendments in phosphorus deficient soils in western Kenya. PhD thesis of the Leuven Catholic University, Belgium, p 110

Nziguheba G, Merckx R, Palm CA (2002) Soil phosphorus dynamics and maize response different rates of phosphorus fertilizer applied to an acrisol in western Kenya. Plant Soil 243:1–10

Ochwoh VA, Claassens AS, de Jager PC (2005) Chemical changes of applied and native phosphorus during incubation and distribution into different soil phosphorus pools. Commun Soil Sci Plant Anal 36:535–556

Palm CA, Myers RJK, Nandwa SM (1997) Combined use of organic and inorganic nutrient sources for soil fertility maintenance and replenishment. In: Buresh RJ, Sanchez PA, Calhoum F (eds) Replenishment soil fertility in Africa. Soil Science Society of America Journal Specialisation, Publication 51, Madison, WI, pp 193–217

Palm CA, Rowland A (1997) Chemical characterisation of plant quality for decomposition. In: Cadish G, Giller KE (eds) Driven by nature: plant litter quality and decomposition. CAB International, Wallingford, CT, pp 379–392

Phan Thi Cong (2000) Improving phosphorus availability in selected soils from the uplands of South Vietnam by residue management. A case study: Tithonia diversifolia. PhD thesis Nr 439 Kartholieke Universtitet, Leuven, Belgium

Randersma S, Grierson PL (2004) Phosphorus mobilization in agroforestry: organic anions, phosphatase activity and phosphorus fractions in the rhizosphere. Plant Soil 259:209–219

Sandhu JM, Simha M, Ambasht RS (1990) Nitrogen release from decomposing litter of Leucaena leucocephala in the dry tropics. Soil Biol Biochem 7:171–177

Sharpley AN (1985) Phosphorus cycling in unfertilized and fertilized agricultural soil. Soil Sci Soc Am J 49:905–911

Sharpley AN (1996) Availability of residual phosphorus in manured soils. Soil Sci Soc Am J 60:1459–1466

Sinsabaugh RI, Linkins AE (1990) Enzymatic and chemical analysis of particulate organic matter from a boreal river. Freshwater Biol 23:301–309

Soil Survey Staff (1994) Keys to soil taxonomy, 6th edn. Soil Conservation Service, United States Department of Agriculture, Government Printer office, Washington, DC

Stewart T (1992) Land-use options to encourage forest conservation on a tribal reservation in the Philippines. Agroforest Syst 18(3):225–244

Tabatabai MA (1982) Soil enzyme. In: Page AL, Millar EM, Keehey DR (eds) Methods of soil analysis. ASA and Soil Science Society of America, Madison, WI, pp 501–538

Tabatabai MA, Bremner JM (1969) Use of p-nitrophenylphosphate for assay on soil phosphatase activity. Soil Biol Biochem 1:301–307

TAPPI (1988) Water solubility of wood and pulp T 207 OM-88. Technical Association of the Pulp and Paper Industry, Atlanta, GA

Tarafdar JC, Marschner H (1994) Phosphatase activity in the rhizosphere and hyposphere of VA mycorrhizal wheat supplied with inorganic and organic phosphorus. Soil Biol Biochem 26:395

Tate KR, Ross DJ, Feltam CM (1988) A direct extraction method to estimate soil microbial C: effects of experimental variables and some different calibration procedures. Soil Biol Biochem 20:329–335

Tian G, Kang BT, Brussaard L (1992) Effects of chemical composition on N, Ca, and Mg release during incubation of leaves from selected agroforestry and fallow plant species. Biochemistry 16:103–119

Tiessen H, Moir JO (1993) Characterisation of available P by sequential extraction. In: Caster MR (ed) Soil sampling and methods of analysis. Canadian Society of Soil Science, Levis Publishers, London

Van Soest PJ (1963) Use of detergents in analysis of fibrous feeds. II. A rapid method for the determination of fiber and lignin. Assoc Off Agric Chem J (AOACJ) 46:829–835

Verchot LV, Borelli T (2005) Application of para-nitrophenol (pNP) enzymes assays in degraded tropical soils. Soil Biol Biochem 37(4):625–633

Warren GP (1992) Fertilizer phosphorus sorption and residues values in tropical African soils. Natural Resource Institute, Chathan, England. NRI, Bulletin 37

Wong MTF, Akyeampong E, Nortcliff S, RaO MR, Smith RS (1995) Initial response of maize and beans to decreased concentrations of monomeric inorganic aluminium with application of manure or tree prunings to an Oxisol in Burundi. Plant Soil 171:275–1995

Wright AF, Bailey JS (2001) Organic carbon, total carbon, and total nitrogen determinations in soils of variables calcium carbonate contents using a LECOCN-2000 dry combustion analyser. Commun Soil Sci Plant Anal 32(19&20):3243–3258

Phenotypic Characterization of Local Maize Landraces for Drought Tolerance in Kenya

I.M. Tabu, S.W. Munyiri, and R.S. Pathak

Abstract Maize is one of the most important staple food crops in Kenya. Drought is a major constraint contributing to the low crop yield. Improved crop genotypes have been developed but farmers continue to grow the local maize landraces. An experiment was conducted to identify the local landraces suited for the dry areas and determine their phenotypic characteristics as a step towards yield improvement. The experiment was conducted at the Kenya Agricultural Research Institute (KARI), Masongaleni farm, Kibwezi, located at an elevation of 650 m above sea level, altitude 2°21.6′S and longitude 38°7.3′E, and agro-ecological zone VI with about 400 mm of rainfall per annum. Sixty-four landraces were evaluated in an alpha lattice design experiment under irrigation with moisture stress at flowering. The genotypes varied significantly in the days to tasselling, days to silking (ASI) and yield. Some local landraces had low ASI (1–5 days) compared to those from the highlands with ASI of up to 16 days. Stress significantly reduced maize yield by up to 82%. Genotypes with the lowest ASI had the lowest yield loss. The controls (Katumani composite B and dryland composite) had losses of 62 and 68% when subjected to moisture stress at flowering. The results led to the conclusion that some local landraces hold potential for improving maize for drought tolerance.

Keywords Maize landraces · Drought tolerance · Flowering yield

Introduction

Most of Kenya's landmass (82%) is arid and semi-arid land with water scarcity as the main constraint to crop production. The area receives low rainfall that is poorly distributed. According to the agricultural dry spell analysis, maize in East Africa is exposed to at least 1 day dry spell of 10 days or more in 74–80% of the seasons (Barron et al., 2003). In addition, even the medium- and high-potential agricultural lands also experience transient droughts that negatively affect crop production. The ongoing climatic changes attributable to global warming have also increased the water-limited environments and the need for more food. The challenge therefore is to increase food production in the face of the increasingly low and unreliable rain. The rapid increase in population estimated at about 35 million people by the year 2010 in Kenya implies that food production has to be increased if the millennium development goal of self-sufficiency is to be achieved.

The drought problem can be tackled by management, irrigation and/or development of drought-tolerant genotypes. Irrigation is however too expensive for the predominantly low-resource farmers and is becoming a lesser solution as global water demand increases (Boyer and Westgate, 2004). Improved genotypes, an important source of tolerance and yield stability, can be conveniently packaged and easily adopted. The first step towards maximizing yield in the drought-prone areas is matching phenology of cultivars to the

I.M. Tabu (✉)
Department of Crops, Horticulture and Soils, Egerton University, Egerton, Kenya
e-mail: immtabu@yahoo.com

pattern of rainfall. In cases where the rainy season is short, drought escape where the plant completes critical physiological processes before drought sets is recommended (Edmeandes et al., 1997). Drought tolerance enables one genotype to be more productive than another under similar conditions. Drought resistance is therefore a combination of drought escape and tolerance where physiological mechanisms allow the plant a reasonable level of production despite the presence of drought (Mugo et al., 1998). The ability to yield well in dry environments may depend on drought avoidance, drought tolerance or both. Short season (early maturing) cultivars escape drought that occurs at the end of a rainy season but normally they have limited yield potential. Escape through early maturity is rarely sufficient in Kenya since the occurrence of drought is erratic, often varying in intensity and timing as it occurs at any stage of growth (Mugo et al., 1998).

Maize (*Zea mays* L.) is one of the most important staple food crops in Kenya. In a bid to achieve self-sufficiency in food production, breeding has mainly aimed at increasing crop yield. High-yielding maize varieties such as Hybrid 626 have been developed for medium- and high-potential agricultural areas, while the open-pollinated ones are recommended for the dry areas. Maize-growing areas in Kenya have further been identified and matched to crop varieties and digitally mapped. Generally, farmers continue to realize suboptimal yields (less than 1 t/ha compared to a potential yield of 8 t/ha) because of inappropriate genotypes (Hassan et al., 1998; Sallah et al., 2002; Laffitte et al., 2007). For the past 30 years, CIMMYT has been attempting to develop a protocol for drought tolerance selection (Banziger et al., 1999; Bruce et al., 2002).

The risk of drought is highest at both the start and the end of the growing season. Since maize is a monoecious plant in which male and female flowers are separated by up to 1 m, it is particularly susceptible to drought during flowering. It is thought to be more susceptible at flowering than other rainfed crops because its female florets develop virtually at the same time and are usually borne on a single ear on a single stem (Banziger et al., 2000). When drought stress occurs just before or during the flowering period, a delay in silking is observed, resulting in an increase in the length of anthesis–silking interval (ASI) and a decrease in grain yield (Ribaut et al., 1996). Grain yield under stress still remains the primary and most important trait during selection. Heritability however reduces under drought because genetic variance for yield decreases more rapidly than the environment. Secondary traits whose genetic variance increases under stress can therefore increase selection efficiency. Development of tolerant genotypes also requires germplasm with variability to adaptive traits. The unpredictable nature of drought however implies that improved genotypes must perform well in both favourable and stressed environments. A combination of stressed and unstressed environments is therefore usually used in selection of genotypes for drought-stressed areas.

While impressive progress has been achieved through conventional breeding, the potential for genetic improvement of maize production under drought conditions remains large (Ceccarelli, 1994). In addition to the search for materials with high inherent yielding ability across the diverse maize-growing environments, dealing with agro-climatic diversity remains crucial for developing a sound maize-breeding strategy (Hassan et al., 1998). Unfortunately, while maize-growing regions in Kenya have been delineated and digitally mapped with respect to crop genotype, development has been slow especially in dry areas. In marginal areas, farmers use local landraces because unlike hybrids, landraces are perceived to perform better under low or no input use (Sammons, 1987; Bellon et al., 2006). Some of the local landraces also possess more desirable attributes such as resistance/tolerance to diseases and pests, tolerance to drought and productivity under low soil fertility.

While the existence of the local maize populations is not doubted, their characteristics have rarely been documented. For example, lack of adoption of improved varieties of maize in some *Striga*-prone areas was because the local 'Nyamula' landrace performed better in many aspects. The existence of the highly valuable landraces represents a potential wealth of genetic material already adapted to widely varying environments of Kenya in which maize is produced (Mirache, 1966; FAO, 2005). In an effort to understand these landraces, CIMMYT in conjunction with KARI collected and stored them. The evaluation of the genetic diversity existing in germplasm is essential if the full potential value of genotype is to be revealed. A study was carried out to identify and phenotypically characterize the local maize landraces for drought tolerance.

Materials and Methods

Site Description

The experiment was carried out at Kenya Agricultural Research Institute (KARI), Masongaleni farm, Kibwezi, Makueni District Eastern province. The farm is located in agro-ecological zone VI, latitude 2°21.6′S and longitude 38°7.3′E, at an elevation of about 650 m above sea level. The area receives about 400 mm of rainfall per annum with mean maximum temperature between 24 and 30°C. The soils are well drained, deep, dark reddish brown, friable, clay loam classified as *pentroplinthite ferralsols* (KARI, 2004).

Experimental Design

(i) Phenological characteristics of landraces

Sixty-four landraces collected from environmental niches in Western, Eastern, Coast and Nyanza provinces were evaluated and inbred lines from CIMMYT-Nairobi, KARI-Kitale and KARI-Katumani were used as checks. The 64 landraces were planted in an 8×8 lattice design replicated three times. The crop was grown under normal rain supplemented with irrigation. Diammonium phosphate at the rate of 40 kg N/ha and 102 kg P/ha was applied at planting. Nitrogen fertilizer was applied as a topdress a month after planting at the rate of 48 kg Na/ha. Dipterex® granules were applied at 30 cm height for protection against stalk borer attack. Each experimental plot consisted of 10 plants in a row at a spacing of 75 cm × 30 cm. Normal rainfall was supplemented by overhead sprinkler irrigation. The experiment was used to quantify and document the phenological characteristics of the landraces.

(ii) Response to drought

Landraces with different anthesis–silking interval (ASI), based on phenological characteristics, were planted in split plot randomized complete block design (RCBD), replicated three times. The main plot was moisture stress level, while the subplots were landraces. Landraces with similar ASI were grouped together for synchronized planting and easier management of stressed conditions during flowering. In the moisture-stressed plots, irrigation continued until 1 week to male flowering (anthesis) when it was stopped. Irrigation was resumed after 40 days when at least 80% male flowering had been achieved. This ensured at least there was grain formation.

Measurements

Days to anthesis (AD) and silking (SD) were calculated from the date in which 50% of the plants had begun shedding pollen or had silks emerging from the husk. Anthesis–silking interval (ASI) was obtained by subtracting AD from SD. Plant height was measured from the ground to the point of flag leaf insertion. Visual scores of leaf rolling were taken on two occasions during stress. At maturity, ears were harvested. An ear was considered fertile if it had one or more grains on the rachis. All harvested ears were shelled and grain weight determined. Grain moisture was determined using a Dickey Johns® moisture tester. The grain was then adjusted to the standard 14% moisture content.

Data Analysis

Data was first tested for normality, and ASI was normalized using $\log_e \sqrt{(ASI +10)}$. Analysis of variance was carried out using the general linear model (SAS, 1997) and means separated by Duncan's multiple range test.

Results and Discussions

The maize genotypes varied significantly in their phenotypic characteristics (Table 1). The yield of the genotypes varied from 0.69 to 4.25 t/ha. The variation in yield was mainly attributed to differences in growth rate (height), flowering, shelling percent and ears per plant. The maize genotypes varied significantly in their phenotypic characteristics. The days to silking varied from 51 to 78 days. The genotypes from the highlands took longer to silk compared to those from the lowlands. The anthesis-to-silking interval varied from 1 to 16 days. Some of the genotypes with the shortest ASI were local genotypes.

Table 1 Phenotypic characteristics (means) of maize landraces in Kenya

Genotype	Phenotypic characteristic						
	Plant height (m)	Days to tasselling	Days to silking	Anthesis-to-silking interval	Yield (t/ha)	Shelling percent	Ears per plant
Control (KCB)*	1.54 (0.21)**	43 (4.51)	53.67 (4.9)	10.67 (2.65)	1.99 (0.68)	75.56 (6.4)	0.69 (0.15)
Landrace	1.74 (0.04)	64.5 (0.92)	72.36 (0.99)	7.81 (0.54)	2.37 (0.14)	68 (1.3)	0.75 (0.03)

*Katumani composite B; **Standard error in brackets

This was expected because the genotypes were collected from different environments within Kenya. Diallo et al. (2001) similarly found large variations in maize genotype characteristics in the moist mid-altitudes of East Africa.

Water stress reduced the yield of maize by more than 50%. The influence however varies with genotype. The yield reduction was mainly via longer anthesis-to-silking interval (ASI) together with reduced growth rate (height), number of ears per plant and number of kernels per ear (Table 2). Moisture stress prolonged the ASI and ultimately reduced the grain yield (Figs. 1 and 2). The check genotype Katumani composite B (KCB) had longer ASI (16 days). This was expected because KCB is recommended for the dry areas because of the escape strategy

Table 2 Correlation (r) between yield and other plant components

	Stress	Yield (kg/ha)	Ears/plant	Plant height	Shelling (%)	ASI	Tassel size
Stress	1.00						
Yield (kg/ha)	0.661***	1.000					
Ears/plant	0.643***	0.677***	1.00				
Plant height	0.457***	0.455***	0.340***	1.00			
Shelling (%)	0.338***	0.536***	0.475***	0.184**	1.00		
ASI	−0.192**	−0.311***	−0.345***	−0.009 NS	−0.328***	1.00	
Tassel size	0.186**	0.079 NS	−0.030 NS	0.420***	−0.062 NS	0.267***	1.00

ASI stands for anthesis-to-silking interval; *, ** and *** significant at 10, 5 and 1%, respectively

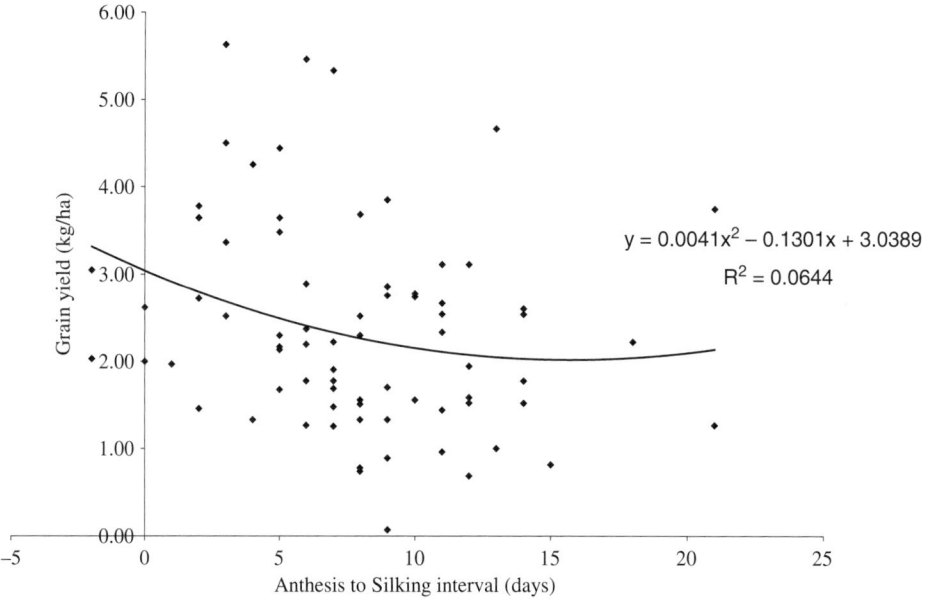

$$y = 0.0041x^2 - 0.1301x + 3.0389$$
$$R^2 = 0.0644$$

Fig. 1 Relationship between maize yield (kg/ha) and anthesis-to-silking interval under water stress conditions

Fig. 2 Relationship between yield (kg/ha) and anthesis to silking under normal water conditions

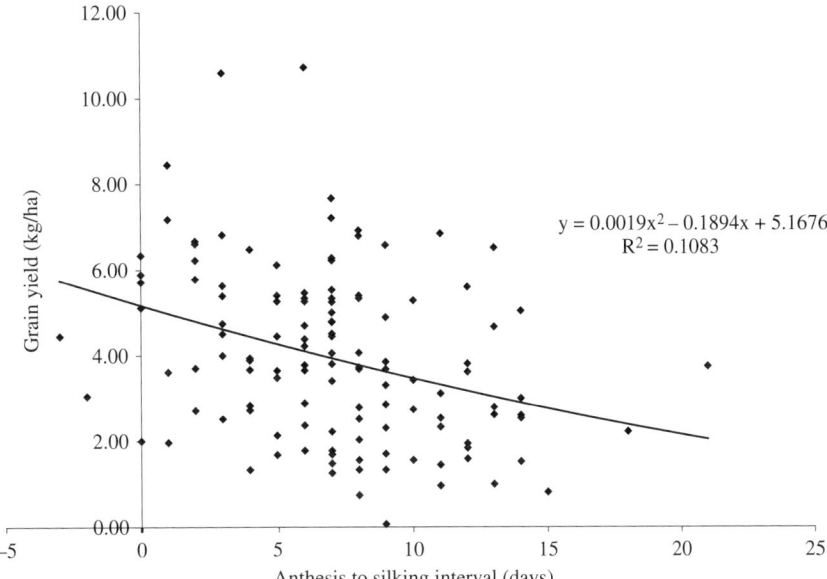

$$y = 0.0019x^2 - 0.1894x + 5.1676$$
$$R^2 = 0.1083$$

(Diallo et al., 2001). Other local genotypes had ASIs as low as 5 days compared to others that ranged up to 16 days. Gebre (2005) in Ethiopia observed comparable levels of yield reduction by water stress in some Kenyan maize germplasms. Tassel size was only correlated with ASI, water stress and plant height. While most of the correlations were positive, ASI was negatively correlated with water stress, grain yield, number of ears per plant, plant height and the shelling percent. The result was consistent with findings by Gebre (2005) that water stress increases the ASI. The effect of water stress has been attributed to the constrained partitioning of photosynthates to the ear (Boyer and Westgate, 2004; Richards, 2006; Campos et al., 2004).

When photosynthesis per plant at flowering is reduced by drought, silk growth is delayed leading to an easily measured increase in the anthesis–silking interval (Banziger et al., 2000). Drought at flowering also causes pollen sterility, abortion of embryos and premature end of grain filling, which reduce seed set and yield by up to 70% (Barron et al., 2003; Passioura, 2006).

Drought reduced the shelling by between 11 and 22% and the number of ears per plant by 4.5–33%. Anthesis-to-silking interval was however increased by about 26%. Water stress also increased leaf rolling by about between 80 and 180% depending on the genotype of the plant.

Conclusions

Maize genotypes varied significantly even under optimal moisture conditions. The variation was experienced in growth rate, flowering (ASI), yield and yield components. Drought significantly reduced maize yield. The reduction however varied with genotypes. The variation was associated with length of ASI, reduced number of ears per plant and shelling percent. Genotypes with longer ASI tended to have lower grain yield under water stress compared to those with shorter ASI. The variation showed that there is still a lot of maize germplasm potential that can be exploited for improvement of maize yield in dry areas.

Acknowledgement We are grateful to RUFORUM for providing funds for this work. Thanks also to the Kenya Agricultural Research Institute for providing the farm and Egerton University for facilitating the work.

References

Banziger M, Edmeades GO, Beak D, Bellon M (2000) Drought and nitrogen stress tolerance in maize: from theory to practice. CIMMYT, Mexico, DF

Banziger M, Edmeades GO, Edmeandes HR (1999) Selection for drought tolerance increases maize yield over a range of nitrogen levels. Crop Sci 39:1035–1040

Barron J, Rockstrom J, Gichuki F, Hatibu N (2003) Dry spell analysis and maize yields for two semi-arid locations in East Africa. Agric Forest Meteorol 117:23–37

Bellon MR, Adato M, Mindek D (2006) Poor farmers perceived benefits from different types of maize germplasm: the case of creolization in lowland tropical Mexico. World Dev 34(1):113–129

Boyer JS, Westgate ME (2004) Grain yield with limited water. J Exp Bot 55(407):2385–2394

Bruce WB, Edmeades GO, Barker TC (2002) Molecular and physiological approaches to maize improvement for drought tolerance. J Exp Bot 366(53):13–25

Campos H, Cooper M, Habben JE, Edmeades GO, Schussler JR (2004) Improving drought tolerance in maize: a view from industry. Field Crops Res 90:19–34

Ceccarelli S (1994) Specific adaptation and breeding for marginal conditions. Euphytica 77:205–219

Diallo AO, Kikafunda J, Wolde L, Odongo O, Mduruma ZO, Chivatsi WS, Friesian DK, Mugo S, Banziger M (2001). Drought and low nitrogen tolerant hybrids for the moist midaltitude ecology of east Africa. A paper presented during the seventh Eastern and Southern Africa regional maize conference held between 11th and 15th February 2001

Edmeades GO, Bolanos J, Chapman SC (1997) Value of secondary traits in selecting for drought tolerance in tropical maize. In: Edmeades GO, Banziger M, Mickelson HR, Pena-Valivia CB (eds) Developing drought and low nitrogen tolerance in tropical maize. Proceedings of a symposium held on March 25th–29th. CIMMYT El, Bata Mexico

FAO (2005) FAO.STAT data 2005. Faostat. Fao. org. Feb 2006

Gebre GB (2005) Genetic variability and inheritance of drought and plant density adaptive traits in maize. Philosophiae Doctor. University of the Free State, Department of Plant Sciences, Bloemfontein, South Africa

Hassan RM, Corbett JD, Njoroge K (1998) Combining geo-referenced data survey with agro-climatic attributes to characterize maize production systems in Kenya. In: Hassan RM (ed) Maize technology development and transfer – a GIS application for research planning in Kenya (pp 43–68). CAB International, Oxfordshire

Kenya Agricultural Research Institute (KARI) (2004) KARI commercialization plan. Agricultural Research Investments and Services, Nairobi, Kenya, pp 1–18

Laffitte HR, Yonsheng G, Yan S, Li Z-K (2007) Whole plant responses, key processes and adaptations to drought: the case of rice. J Exp Bot 58(2):169–175

Mirache MP (1966) Maize in tropical Africa. University of Wisconsin Press, Madison, WI

Mugo SN, Smith ME, Banziger M, Setter TL, Edemeades GO, Elings A (1998) Performance of early maturing Katumani and Kito maize composites under drought at the seedling and flowering stages. Afr Crop Sci J 6(4):329–324

Passioura J (2006) Increasing crop productivity when water is scarce-from breeding to field management. Agric Water Manage 80:176–196

Ribaut JM, Hoisington DA, Deutsch JA, Jiang C, Gonzalez de Leon (1996) Identification of quantitative trait loci under drought conditions in tropical maize. 1. Flowering parameters and the anthesis-silking interval. Theor Appl Genet 92(7):905–914

Richards RA (2006) Physiological traits used in breeding of new cultivars for water-scarce environments. Agric Water Manage 80:197–211

Sallah PY, Obeng-Antwi KK, Ewool MB (2002) Potential of elite maize composites for drought tolerance in stress and non-drought environments. Afr Crop Sci J 10(1):9–19

Sammons DJ (1987) Origin and early history of maize in East Africa with special reference to Kenya. Egerton University Research Paper Series, Njoro, Kenya, pp 2–11

SAS (1997) SAS Proprietary Software Release 6.12. SAS Institute, Inc, Cary, NC

Targeting Resources Within Diverse, Heterogeneous and Dynamic Farming Systems: Towards a 'Uniquely African Green Revolution'

P. Tittonell, B. Vanlauwe, M. Misiko, and K.E. Giller

Abstract Smallholder farms in sub-Saharan Africa (SSA) are highly diverse and heterogeneous, often operating in complex socio-ecological environments. Much of the heterogeneity within the farming systems is caused by spatial soil variability, which results in its turn from the interaction between inherent soil/landscape variability and human agency through the history of management of different fields. Technologies and resources designed to improve crop productivity often generate weak responses in the poorest fields of smallholder farms. Thus, options for soil fertility improvement must be targeted strategically within heterogeneous farming systems to ensure their effectiveness and propensity to enhance the efficiency of resource (e.g. land, labour and nutrients) use at farm scale. Key issues in design of approaches for strategic targeting of resources include (1) inherent soil variability across agroecological gradients; (2) social diversity, farmers' production orientations and livelihood strategies; (3) farmer-induced gradients of soil fertility, their causes and consequences of efficient allocation of scarce resources; (4) competing objectives and trade-offs that farmers face between immediate production goals and long-term sustainability and (5) the complexity of farmers' own indicators of success. We used an analytical framework in which systems analysis is aided by survey, experiments and simulation modelling to analyse farming futures in the highlands of East Africa. Our work contributes to the development of 'best-fit' or tailor-made technologies, using combinations of mineral fertilizers and organic matter management from N_2-fixing legumes and animal manures. Thus, we hope to contribute to the design of a 'uniquely African green revolution' called for by UN Director General Kofi Annan, which fits technology interventions to the diverse and heterogeneous smallholder farming systems of sub-Saharan Africa.

Keywords Sub-Saharan Africa · Farm typology · Markets · Soil fertility · Resource use efficiency · Agricultural inputs

Introduction

Smallholder farms in sub-Saharan Africa are highly diverse and heterogeneous, often operating in complex socio-ecological environments. Much of the heterogeneity within the farming systems is caused by spatial soil variability, which results in its turn from the interaction between inherent soil/landscape variability and human agency through the history of management of different fields (e.g. Prudencio, 1993; Tittonell et al., 2005c). Classical 'green revolution' technologies designed to improve crop productivity often generate weak responses in the poorest fields of smallholder farms, as evidenced by, e.g., the large variability in fertilizer use efficiencies within single farms observed in East, West and Southern Africa (Vanlauwe et al., 2006; Wopereis et al., 2006; Zingore et al., 2007). Thus, options to restore productivity must be targeted strategically within heterogeneous farming systems to ensure their effectiveness and propensity to enhance

P. Tittonell (✉)
Plant Production Systems, Department of Plant Sciences, Wageningen University, 6700 AK Wageningen,
The Netherlands; Tropical Soil Biology and Fertility Institute of the International Centre for Tropical Agriculture (TSBF-CIAT), Nairobi, Kenya
e-mail: Pablo.Tittonell@wur.nl

A. Bationo et al. (eds.), *Innovations as Key to the Green Revolution in Africa*,
DOI 10.1007/978-90-481-2543-2_76, © Springer Science+Business Media B.V. 2011

the efficiency of resource (e.g. land, labour and nutrients) use at farm scale. Such options should be evaluated not only in terms of immediate benefits (which can be crucial in determining the adoption of a certain technology by farmers) but also by assessing their contribution to livelihood strategies and system sustainability in the long term (Giller et al., 2006).

Systems analysis aided by simulation modelling constitutes a means to evaluate options for sustainable intensification of farming systems, considering (1) their diversity, spatial heterogeneity and variability in time; (2) the scaling-up in space and time of the effect of single interventions operating at field plot scale, to infer consequences at farm and village scales in medium- to-long-term horizons (i.e. strategies) and (3) the possibility to perform scenario analysis with prospective or explorative purposes, evaluating the potential impact of factors that are external to the farm system, such as climate change or market developments. A first step in systems analysis and scenario evaluation consists in defining representative prototypes of fields, cropping sequences, farms or localities that must capture the key managerial, socio-economic and agroecological aspects of the systems under study. Their heterogeneity and diversity at different scales should be categorised, relying on solid understanding of the key drivers of such variability and using methodologies that allow comparisons across systems. Such cross-scale categorisation may also serve to define recommendation domains or socio-ecological niches (e.g. Ojiem et al., 2006) to which resources/technologies can be targeted.

The drivers of diversity, heterogeneity and farming systems dynamics can be grouped, in decreasing order of spatio-temporal scale, as site-specific conditions (agroecology, markets, population, ethnicity, etc.), soil–landscape associations, farm resource endowment, land use (crop types and livestock system), and long- and short-term management decisions (current soil fertility status and operational resource and labour allocation, respectively). Rather than static entities, farming systems are dynamic, subject to changing socio-economic and environmental contexts and risks (through, e.g., climatic or market variability). This chapter illustrates our conceptualisation of complex farming systems and presents examples of application of an integrative analytical approach (combining farm typologies, participatory research, experiments and modelling within the NUANCES analytical framework – www.africanuances.nl) to identifying intervention opportunities and pathways towards the sustainable intensification of smallholder systems in sub-Saharan Africa. The impact of the above factors and their implication for the promotion of green revolution technologies are examined using examples from mid- to high-potential agricultural areas of Kenya and Uganda (Table 1). Examples from a number of studies conducted in these sites and published elsewhere are used here for illustration.

The Biophysical and Socio-economic Environments

In targeting interventions to improve livelihoods through agricultural policy, investment in infrastructure or technology promotion, two main dimensions determining opportunities and constraints across locations are often considered: agroecological potential and market opportunities. To illustrate this, the six sites in Kenya and Uganda described in Table 1 are placed within a schematic plane defined by these two dimensions (Fig. 1a). Market opportunities are defined by the size, development and accessibility of major markets (e.g. proximity to urban and export markets, vial infrastructure, market information and transaction costs). For example, Meru South and Mbeere vary widely in agricultural potential but both districts are located close to the city of Nairobi (with an international airport) and are surrounded by the relatively highly populated areas and mid-sized towns of central Kenya, well connected through major national roads. Soils are inherently more fertile in Meru South and Mbale, located on the foot slopes of Mt. Kenya and Mt. Meru, respectively, and receiving ample rainfall. Soil organic C is a good proxy for the inherent soil fertility and agricultural potential of different sites in this case: soils with proportionally more clay under cooler and wetter climates tend to accumulate more organic matter due to larger primary productivity (more water and nutrient availability for plant growth) and slower rates of organic matter decomposition (lower temperatures and physicochemical protection of C within the soil matrix).

However, biophysical potential and market opportunities, which are also frequently correlated, are not

Table 1 Main characteristics of the case-study sites in three sub-regions of East Africa (from Tittonell et al., 2010)

	Central Kenya		Western Kenya		Eastern Uganda	
Characteristics	Meru South	Mbeere	Vihiga	Siaya	Mbale	Tororo
Biophysical						
Altitude (masl)	1,500	1,100	1,600	1,200	1,600	1,100
Rainfall (mm)[a]	1,600	700	1,800	1,400	1,200	1,100
Dominant soil types (FAO)	Nitosols, ferralsols	Lixisols, arenosols	Nitosols, ferralsols	Ferralsols, acrisols	Ferralsols, acrisols	Acrisols, vertisols
Landscape	Strongly undulating, slopes up to 45%	Fairly flat to gently undulating, slopes <5%	Gently undulating, slopes 5–20%	Fairly flat, slopes <3%	Gently undulating, slopes 5–10%	Fairly flat, slopes <3%
Socio-economic						
Population density (km^{-2})	800	400	1,000	350	350	250
Farm sizes (ha)	0.5–3	1–10	0.3–2	0.5–5	0.5–5	1–8
Production activities						
Major food crops	Maize, beans	Sorghum, cowpeas	Maize, beans	Maize, cassava	Bananas, beans	Cassava, sorghum
Major cash crops	Coffee, tea	Miraa, groundnuts	Tea, coffee	Sugar cane, cotton	Coffee	Cotton, tobacco
Livestock system	Zero grazing dairy systems and cultivation of fodder crops; improved cattle	Free-ranging local zebu and goats; night corralling	Tethered cattle grazing in compound and communal fields, and zero grazing	Free grazing and tethered local cattle, used for traction (ox-ploughing)	Zero grazing of cattle and goats; donkey used for transport	Free grazing in communal grasslands; local zebu

[a]In all sites, rainfall takes places in a bimodal pattern (i.e. long and short rainy seasons)

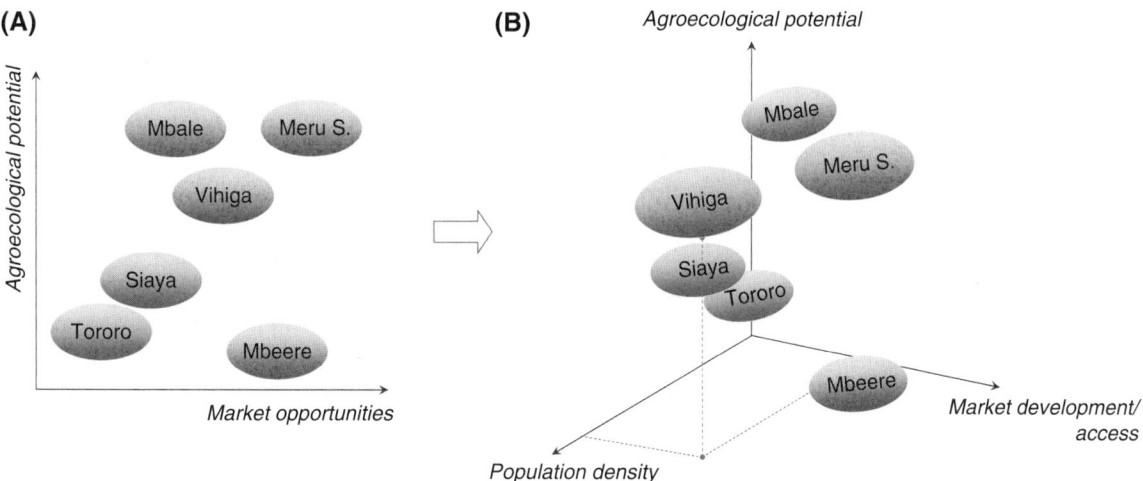

Fig. 1 Different sites in Kenya and Uganda ordered by their agricultural potential and market opportunities (**a**), and by these two factors plus population density (**b**). Details on the six sites are presented in Table 1. For guidance, the intersections with the market and population axes are indicated for Vihiga

enough to explain the observed diversity of livelihood strategies across and within locations. Historical, political and demographic processes in combination with local variability among households will ultimately determine the space of opportunities and constraints in which households develop. In Fig. 1b, the same locations are now placed within a space defined by agroecology and markets plus a third dimension representing population density. Intuitively, one may expect higher population densities in areas with the highest agroecological potential and best market opportunities. That is not the case in this example, with more than 1,000 inhabitants km^{-2} in many areas of Vihiga district, due to ethno-cultural and historical backgrounds (Crowley and Carter, 2000). Population densities beyond a certain (site-specific) threshold are often inversely proportional to the availability of resources per household, but more people living in a certain area may also create more local market and/or job opportunities in rural communities.

Household Diversity and Livelihood Strategies

The diversity of livelihood strategies, which determine to a large extent production orientation and household objectives, has important implications for the targeting of agricultural technologies. Considering the two dimensions discussed above, natural resources and local markets, Dorward et al. (2001) distinguish three main livelihood strategies of the poor in rural areas, briefly (1) 'hanging in', which takes place in situations of poor natural resource potential and market opportunities, and where households engage in activities to maintain their current livelihood level (subsistence farming); (2) 'stepping up', in situations of high agricultural potential and where investments in assets are made to expand current production activities (semi-commercial farming) and (3) 'stepping out', when activities are engaged to accumulate assets that may eventually allow moving into different activities, not necessarily farming (i.e. migration to cities and/or local engagement in non-farm activities). At local scale, these strategies and their determinants are nuanced by differences between households in terms of resource endowment and social capital. In areas of high resource potential and ample market opportunities such as Meru South (cf. Table 1), different households may hang in, step up or step out, or pursue mixed strategies, such as investing in lucrative cash crops and re-investing their income into higher education for their children (to eventually step out). By

contrast, areas of poorer natural and market potential will force most households to hang in.

Next to agroecology, markets and population density, rural–urban connectivity and off-farm opportunities contribute to shaping livelihood strategies. Access to non-farm income through remittances or employment in urban areas, or to off-farm income from selling labour locally in rural areas has been used in combination with indicators of production orientation and resource endowment to categorise household types in East Africa (Tittonell et al., 2005b). This constitutes a *functional* typology of households in which the position of the household in the farm developmental cycle is also considered (Fig. 2), expanding the more frequent approach of *structural* farm typologies used to categorise households (e.g. wealth rankings through indicators of resource endowment – Mettrix, 1993).

The various farm types thus defined engage in different income-generating activities, exhibit contrasting patterns of resource allocation and prioritisation of investments, and pursue different long-term livelihood strategies. For example, farms of type 1 and 5, relying largely on off-/non-farm income, have stepped out of agriculture as their main activity. In promoting technologies, farms of type 2 and 3 constitute the most promising target groups, since agricultural production represents their main source of income. In western Kenya, while type 2 includes wealthier households headed by respected aged farmers, type 3 includes mostly households headed by younger, enterprising farmers that show a high degree of participation in extension activities such as farmer field schools (Misiko, 2007).

Different sites across sub-Saharan Africa vary in their propensity to promote hanging-in, stepping-up or stepping-out livelihood strategies. Within a certain location, individual farms and decision makers differ in resource endowments, objectives, individual attitudes and ability to innovate. Although this variability must be recognised and categorised for better targeting of technologies, the socio-economic context should not be overlooked: most households in the study areas presented in Table 1 are below the poverty line, as indicated by the latest poverty mapping in the region (www.worldbank.org/research/povertymaps), and our categorisations basically distinguish between very poor, poor and less poor households. The ultimate beneficiaries of green revolution technologies in Africa – those that must be

(A)

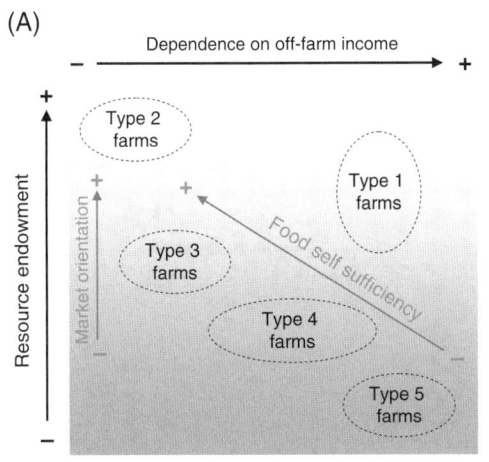

(B)

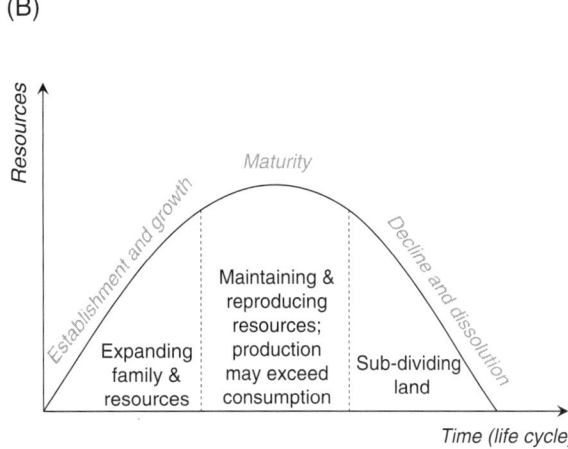

Fig. 2 (a) A typology of smallholder farms of East Africa, considering classical wealth indicators (resource endowment) plus source of income and production orientation to categorise households (adapted from Tittonell et al., 2010). (b) A schematic representation of the developmental cycle of farm households (Forbes, 1949) and its implications for resource allocation

targeted – are poor families, lacking cash and assets, and farming on small pieces of (frequently degraded) land.

Farm Heterogeneity and Resource Use Efficiency

Soil–landscape variability interacts with the regional and local factors discussed in the previous section to determine patterns of land use and resource allocation; such patterns lead in the long term to the creation of (human-induced) spatial heterogeneity within individual farms or soil fertility gradients. Resource use efficiency (units of output per unit of resource available) results from two components: resource capture (units of resource intercepted per unit of resource available) and resource conversion efficiency (units of output per unit of resource intercepted). Farm heterogeneity affects resource use efficiencies operating mostly on the efficiency of resource capture, leading to resource use efficiency gradients (Tittonell et al., 2007b) and to different patterns of responsiveness to technology interventions (e.g. non-responsive poor fields, responsive fields and non-responsive fertile fields – Tittonell et al., 2008). In cropping systems, poor resource use efficiencies may result from resource imbalances (Kho, 2000) or deficient agronomic management (Tittonell et al., 2007a).

Soil organic C was mentioned earlier as a good indicator of the agricultural potential of different sites, with greater contents in soils of finer texture (Fig. 3a). The fluctuation in average soil C levels for soils of similar texture (i.e. for a narrow range of clay+silt content) can be explained by climatic differences across sites, by local soil–landscape variability and, in particular, by management-induced farm heterogeneity. Farmers typically allocate more resources (labour, cash and nutrients) to fields that are perceived as more fertile or that are more convenient to manage (Fig. 3b), reinforcing this variability. In intensively cultivated farming systems, resource imbalances are also often closely associated with soil fertility gradients. For example, the pattern of variation in soil organic C and P availability illustrated in Fig. 3c can be commonly observed across farming systems; while soils with poor P availability may have relatively high or low soil C contents, high P availability is found only in soils with soil C contents above a certain threshold value. Such co-variability in C and P has been induced by farmers' management, with high concentrations of both C and P often found in gardens and fields closer to the homesteads. Another type of management-induced interaction is the use of improved cultivars in combination with nutrient inputs (Fig. 3d). In soils of similar organic C contents, farmers obtained larger yields when they planted hybrid maize compared with local varieties. Although improved cultivars may improve resource conversion efficiencies

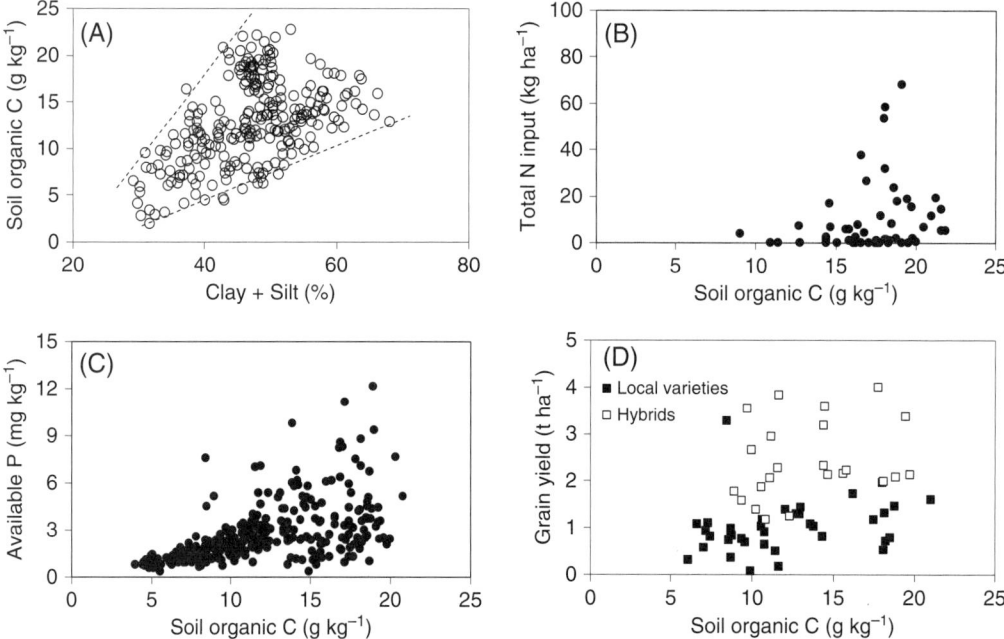

Fig. 3 (**a**) Weighted average soil C vs. average clay+silt contents at farm scale in 250 farms from Kenya and Uganda (Table 1). (**b**) N input as organic or inorganic fertilizers allocated to fields of different soil C contents in 15 farms of western Kenya. (**c**) Relationship between soil available (Olsen) P and soil organic C for 190 fields sampled in western Kenya. (**d**) Grain yield of local varieties and hybrids of maize vs. soil organic C content measured in 45 farms of western Kenya

(e.g. through larger harvest indexes), the yield differences in Fig. 3d were also partly due to the more frequent application of nutrient inputs to fields planted with hybrids.

Smallholder farmers are normally careful to allocate their limited production resources to the most profitable activities (in theory, to those that yield the highest marginal returns – Ellis, 1993). In systems where pressure on the land restricts the possibility of fallow or the availability of common grazing land, nutrient resources may enter the farming system chiefly via fertilizers and feedstuffs for livestock (if present). Let us consider the example of using animal manure in farms of western Kenya (Table 2). If all the manures potentially available for application to crops in these case-study farms were evenly spread over their total area, the average application rates would vary between ~10–40 kg ha^{-1} for N and 1–6 kg ha^{-1} for P. As these rates are unlikely to induce substantial crop responses, farmers tend to concentrate on the available nutrient resources creating zones of soil fertility within their farms.

Cattle densities are low in most of SSA (e.g. between 1 and 5 heads km^{-2}, with values

> 5 heads km^{-2} only in certain spots within the highlands of East Africa – cf. www.ilri.org/gis), and cropland-to-grassland ratios are ever increasing due to human population growth in high potential areas (www.earthtrends.wri.org). This reinforces the argument that the use of mineral fertilizers must be strengthened to sustain productivity of smallholder farms of sub-Saharan Africa. Given the wide diversity and heterogeneity of farming systems, however, the performance of fertilizers and their adoption may easily fluctuate from success to failure. Due to economic and environmental considerations, mineral fertilizers should be judiciously targeted to ensure high capture and utilisation efficiencies. Strategic targeting should consider not only the referred spatial heterogeneity but also the dynamics of the farming systems.

Long-Term System Dynamics and Interventions

The way in which green revolution technologies are (or have been, so far) promoted depends largely on

Table 2 Potential availability of manure and C, N and P for application to crops in farms from different wealth classes in western Kenya as derived from resource flow analysis (adapted from Tittonell, 2003)

Village[a]	Resource endowment	Land cropped (ha)	Livestock heads (TLUs)	Potential manure availability (t year^{-1})	Potential application rates (kg ha^{-1})[b]		
					C	N	P
Ebusiloli	High	2.1	4.0	8.4	960	38	6.1
	Medium	1.1	2.2	3.6	785	31	5.0
	Poor	0.5	0.8	1.1	528	21	3.3
Among'ura	High	2.3	2.3	3.5	212	8	1.3
	Medium	2.2	2.0	2.9	218	9	1.4
	Poor	1.0	1.7	2.0	408	16	2.6

[a]Ebusiloli (Vihiga district) is located in a highly populated area ($\sim$1,000 inhabitants km^{-2}), closer to urban centres with easier access to markets; intensive (zero grazing, Friesian) livestock production systems predominate. Among'ura (Teso district) area is less populated (200–300 inhabitants km^{-2}), land is available for fallow, markets are far and the local (zebu) livestock graze in communal land

[b]Calculated over the total area of cropped land, assuming optimum manure handling and an average dry matter content of 80%, C content of 30%, N content of 1.2% and P content of 0.19%

the way in which farming systems are conceptualised. Input-based intensification often rests on the assumption that input use efficiencies are independent of resource stocks or availability. To clarify this point, Fig. 4 presents two simplified models of resource utilisation. The term 'stock' is used generically to indicate the value of a state variable. In this simplest model, there is no formal distinction between stock and availability; a fraction of the inputs (I) added temporarily increases the stock (S), and output (O) is produced by transforming a fraction of the increased stock:

$$\text{Stock} = S_0 + \Delta\text{Stock} \times \text{Time} \quad (1)$$

$$\Delta\text{Stock} = \text{Input} \times R_i - \text{Output} \quad (2)$$

$$\text{Output} = (\text{Stock} + \text{Input} \times R_i) \times R_o \quad (3)$$

The three parameters of the model (cf. Fig. 4a) are the initial stock (S_0), the fraction of input added that is retained to increase the stock (R_i) and the fraction of the stock removed in the output (R_o). Note that, generically, R_i and R_o may represent resource capture and resource conversion efficiencies, respectively. The simplest model A was initialised with $S_0 = 100$, $I = 10$, $R_i = 0.4$ and $R_o = 0.2$; that is, inputs were added at a rate of one-tenth of the original stock of 100, 40% of the input added was captured and 20% of the new stock (increased by input addition) was removed in the output. Equilibrium was reached at $S_e = 16$ (when, Input $\times R_i =$ Output $= 4$), after the model was run for 34 iterations (Fig. 5a – cf. 'constant rates').

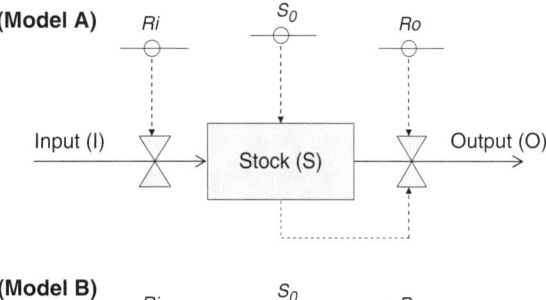

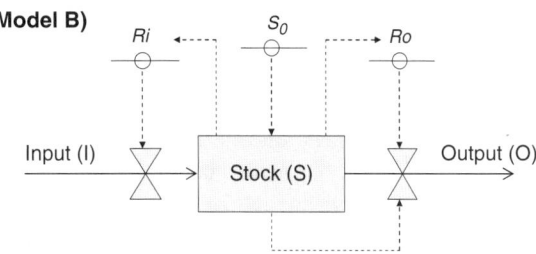

Fig. 4 Diagrams of two simplified models to illustrate resource use efficiency within farming systems. (**a**) The simplest case in which resource capture (R_i) and conversion (R_o) efficiencies are kept constant. (**b**) A case in which both R_i and R_o depend on resource availability (stock). See further explanation in the text

This simple model can be used to illustrate the effect of different types of interventions to restore productivity (Fig. 5b). Doubling the rate of inputs after equilibrium was reached will lead to doubling the value of S_e. Improving the efficiency of resource capture (R_i) by 50% will also increase S_e by 50%, often incurring less costs rather than doubling inputs (e.g. in cropping systems, R_i can be increased by planting on time, weeding frequently or using mulches). Improving the efficiency of resource conversion (R_o) by 50% will

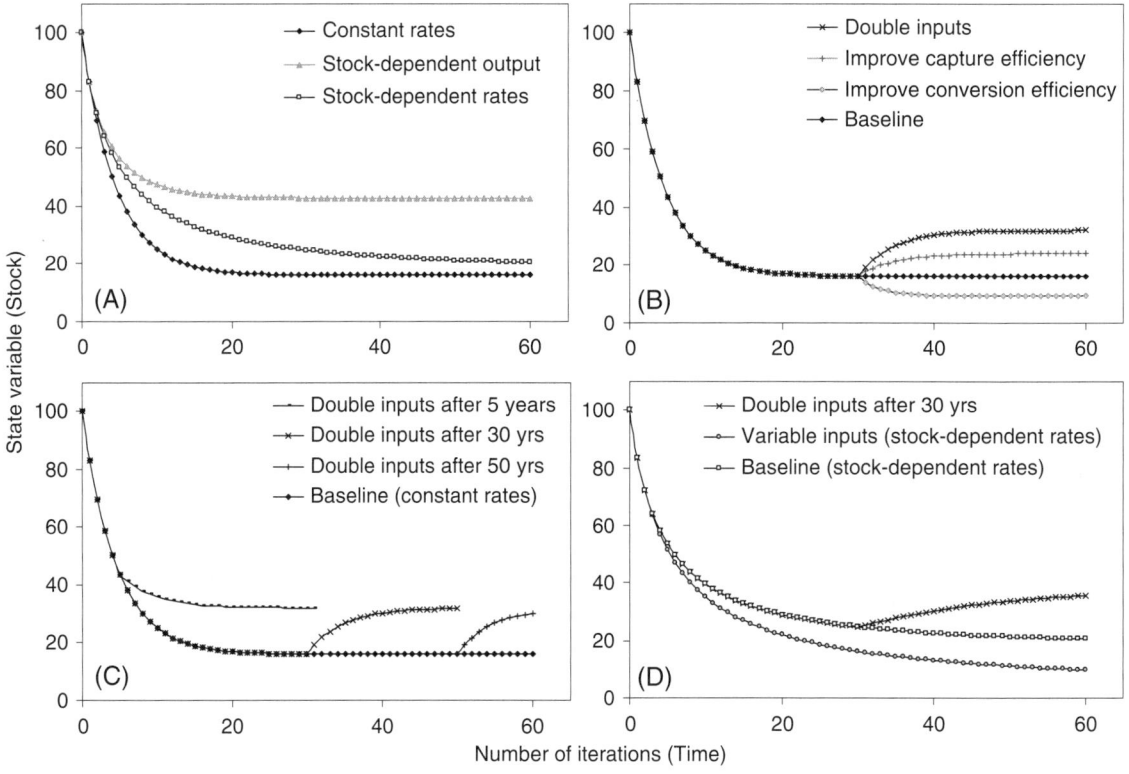

Fig. 5 Simulation of long-term system dynamics with models A and B depicted in Fig. 4. See explanation in text

reduce S_e by 50%. The latter is the case, in cropping systems, when farmers allocate less exigent crops in terms of soil fertility to fields that were cultivated for long periods without inputs; for instance, in western Kenya, maize growing is replaced with Napier grass in the outfields (Tittonell et al., 2005c), which has higher nutrient conversion efficiencies and further accelerates soil depletion.

A model slightly more complex is presented in Fig. 4b, in which both R_i and R_o are stock dependent, i.e. the efficiencies of resource capture and conversion decrease when the stock decreases. Within heterogeneous smallholder systems, resource capture efficiencies vary more broadly than the more conservative (e.g. crop type dependent) resource conversion efficiencies (Tittonell et al., 2008). In model B, R_i and R_o have been expressed simply as linear functions of S ($R_i = 0.004 \times S$ and $R_o = 0.002 \times S$). Due to such self-regulation mechanisms (or negative feedbacks), the rate of stock depletion is slower and equilibrium is reached only after 120 years at $S = 19$ (Fig. 5a – cf. 'stock-dependent rates'). This represents

a more realistic case, as observed when, e.g., poorer crop yields are obtained in the poor outfields of a farm and therefore comparatively less nutrients are removed in the harvest. An intermediate case is that in which R_i remains constant and only R_o is stock dependent (Fig. 5a – cf. 'stock-dependent output'). Equilibrium is reached at higher values of S ($= 43$), and this can be the case when more than one resource is considered simultaneously. For example, the availability of a second resource that depends also on the stock S (e.g. water) affects the efficiency of conversion of the resource represented by S (e.g. a certain nutrient) (cf. Kho, 2000).

The underlying assumption behind 'blanket recommendations' in agriculture is often in correspondence with a resource utilisation model closer to model A, largely ignoring the long-term dynamics of farming systems. The different fields of an individual farm or a village have a history of use and management that places them in a certain position along the curve describing the trajectory of S (years under cultivation/fallow), or on a totally different trajectory of S

when the values of S_0, I, R_i or R_o would have been different. In model A, doubling the amount of inputs added will eventually double the value of S_e irrespective of the time at which the intervention takes place (Fig. 5c). However, field experimentation has demonstrated variable patterns of responsiveness to applied nutrients to fields with different history of use (e.g. Vanlauwe et al., 2006). By contrast, the initial response to doubling inputs will be slower in a self-regulatory system close to equilibrium, often discouraging farmers to continue applying inputs due to the poor initial response in productivity (Fig. 5d).

In reality, not only the stock but also the output rate participates in a feedback mechanism via farmers' decision making, i.e. fewer inputs are added as output rates decrease (a positive feedback). This will lead to even faster decline and lower values of S_e, in spite of the self-regulatory mechanisms (Fig. 5d – cf. 'variable inputs'). Most fields in intensively cultivated smallholder systems of SSA are in this situation and represent the target of green revolution technologies. When the system under study is the farm, the village or the region, this process is often termed 'downward spiral' or 'poverty trap' (e.g. Walker et al., 2006). The dynamics of a system cannot be understood without taking into account the dynamics and cross-scale influences of the processes above and below it. The context in which the system operates may be important in regulating the rate of inputs, through, e.g., market incentives (which may induce input use) or risks (which may deter input use), or when households are facing different stages of the farm developmental cycle (cf. Fig. 2). Finally, not only the rate of addition but also the timing, opportunity and quality of the inputs added play a role in shaping the dynamics of the system. For example, N inputs can be brought into the farm via mineral or organic fertilizers, feed concentrates for livestock or atmospheric N_2 fixation by legumes, inducing different efficiencies of N capture within the system.

Farmers' Objectives, Indicators and Trade-offs

Farmers' objectives and aspirations can be translated into quantifiable indicators by understanding the system attributes that concern the achievement of such objectives (López-Ridaura, 2005). For example, economic profitability and crop yields are two different indicators pertaining to the same attribute of a farming system, productivity. Nutrient balances are often used as indicators of system stability in the long term, although they do not really comply with all the desirable properties of a good indicator, among others, being easily measurable, having established thresholds and being easy to communicate to stakeholders. It is easily intelligible that negative nutrient balances may lead to nutrient depletion in the long term. However, nutrient balances are often most negative for the best yielding fields, and this is a puzzling concept to discuss with farmers (Tittonell et al., 2005a). Although farmers tend to concentrate resources in these fields (cf. Fig. 3b), the input rates are yet insufficient. Partial nutrient balances calculated in western Kenya indicate alarming rates of nutrient depletion, with removal in crop harvests often more than doubling nutrient inputs (Fig. 6a). The two encircled points in Fig. 6a represent two fields belonging to different farms; in spite of being cultivated with different rates of N inputs, the partial N balance was –22 kg N ha^{-1} in both cases. This highlights the need to analyse nutrient balances in relation to nutrient stocks, expressing the results in relative terms.

Even when major limiting nutrients such as N and P are applied to compensate crop removals, balances are negative for other nutrients that are not applied (e.g. K, S and Mg). At farm scale, maximising production often implies cultivating all the fields of a farm, even those of marginal fertility or with permanent impediments (e.g. steep slopes). This may lead to a trade-off between productivity and efficiency (e.g. nutrient capture efficiency), of which farmers are not always aware (Fig. 6b). Another typical example, and one that threatens the dissemination of conservation agriculture in SSA, is the trade-off between retaining crop residues in the field and using them to feed cattle, as fuel, sell them locally or add them to the compost. The best option would depend on the characteristics of the system and objectives, but the removal of crop residues from fields that received N or P fertilizers contributes to farmers' generalised perception that fertilizers 'spoil the soil' (Misiko, 2007).

The best indicator to analyse the system or the performance of a certain green revolution technology would also be different for different stakeholders. For formal comparisons across systems, indicators that are

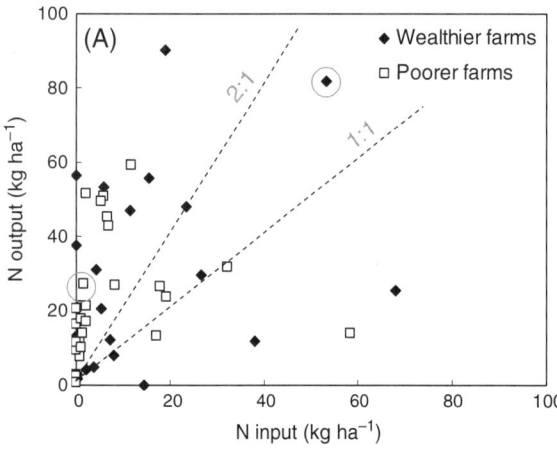
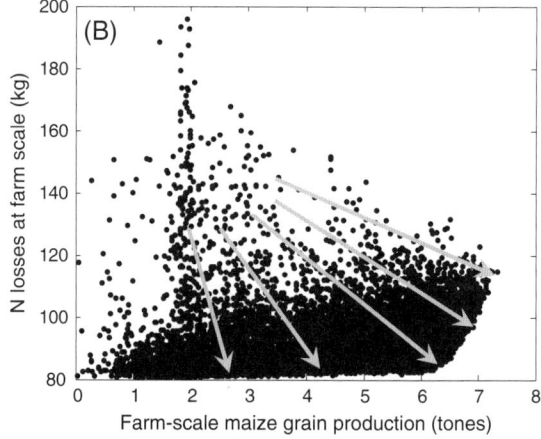

Fig. 6 (**a**) N outputs in crop harvests plotted against N inputs to the soil as organic and mineral fertilizers and crop residues in 15 farms of western Kenya. (**b**) Trade-off curve between maize productivity and N losses at farm scale obtained through optimisation using inverse modelling techniques (from Tittonell et al., 2007c)

often less obvious may yield important information. For instance, a system-level indicator of efficiency could be derived expressing soil fertility as kilogram of soil nutrients available per family member (e.g. for N, it could be calculated as *soil N content × soil depth × bulk density × area cropped/number of family members*). For instance, in Tororo (Table 1), soil P availability is often low (2–3 mg kg^{-1}) but average farm sizes are large (up to 8 ha). The kilograms of nutrients available per family member are a prerequisite to achieving food self-sufficiency; then, it is a question of how efficient is the production system (or what is the contribution of a certain technology) to capture and convert those nutrients into food.

'Green Evolution' – Promoting Inputs or Designing New Systems?

Promoting green revolution technologies under the same paradigm by which these technologies have been promoted in the past would most likely lead to new failure. A green revolution has to be 'uniquely African', as called for by Kofi Annan, due to the following particularities of smallholder systems in SSA:

1. Farms are heterogeneous and complex; variability within and between farms may yield promising

green revolution technologies useless in terms of boosting productivity and long-term sustainability. Truly integrated soil fertility management must consider the various components of complex systems; for example, recommendations on the use of manure plus fertilizer must be based on realistic rates of application (in line with manure availabilities at farm scale), nutrient contents (often very poor in reality) and labour availability on the farm.

2. Smallholder farms are not necessarily commercially oriented; rural livelihood strategies are diverse, conditioned by agroecology and markets, and determined by household objectives, resource endowment and individual preferences of the decision maker. While some families 'make a living' out of agriculture, most keep the family land for a number of other reasons (e.g. social insurance) and regard agriculture as a secondary (or complementary) activity.

3. Land tenure and demographic processes are closely linked to culture and vary broadly across sites. The fact that in many cases, farmers do not have property rights on their land has led economists to argue that farmers (i) may lack motivation to invest in improving their soils and (ii) are not able to access credits to purchase agricultural inputs or reproduce their assets.

4. Most rural families in Africa are below the poverty line and farming already degraded land; assuming that promoting the use of agricultural inputs

through price policies or subsidies will automatically boost productivity and improve livelihoods is too simplistic. This is particularly the case when rural families have diverse sources of income and/or the (short or long term) aspiration to step out of agriculture.

Green revolution technologies should be targeted to diverse, heterogeneous and dynamic farming systems. Having one specific recommendation for each individual farm would be ideal, but impracticable, and thus it is necessary to categorise patterns of variability and identify possible entry points. Ideally, such patterns and opportunities should be easily recognisable by farmers, whose capacity for decision making should be built on solid knowledge about the systems they manage and their context. Far from simply promoting the use of agricultural inputs, a uniquely African 'green evolution' should contribute to the design of new systems, promoting improved resource use efficiencies, organisational skills and extension systems that involve farmers (shared knowledge and learning) and the development of rural markets. This calls for a truly interdisciplinary research and development effort, which must surpass the boundaries of agricultural disciplines.

References

Crowley EL, Carter SE (2000) Agrarian change and the changing relationships between toil and soil in Maragoli, western Kenya (1900–1994). Hum Ecol 28:383–414

Dorward AR, Anderson S, Clark S, Keane B, Moguel J (2001). Asset functions and livelihood strategies: a framework for pro-poor analysis, policy and practice. EAAE seminar on livelihoods and rural poverty, Wye, Kent, Imperial College, London

Ellis F (1993) Peasant economics. Farm households and agrarian development. Cambridge University Press, Cambridge, UK, 312pp

Forbes M (1949) Time and social structure. An Ashanti case study. In: Fortes M (ed) Social structure: studies presented to A.R. Radcliffe-Brown. Clarendon, Oxford, pp 54–84

Giller KE, Rowe E, de Ridder N, van Keulen H (2006) Resource use dynamics and interactions in the tropics: scaling up in space and time. Agric Syst 88:8–27

Kho RM (2000) On crop production and the balance of available resources. Agric Ecosyst Environ 81:223–223

López-Ridaura S (2005) Multi-scale sustainability evaluation. A framework for the derivation and quantification of indicators for natural resource management systems. PhD thesis, Wageningen University, Wageningen, 202pp

Misiko M, 2007. Fertile ground? Soil fertility management and the African smallholder. PhD thesis, Wageningen University, 208pp

Ojiem JO, de Ridder N, Vanlauwe B, Giller KE (2006) Socio-ecological niche: a conceptual framework for integration of legumes in smallholder farming systems. Int J Agric Sust 4:79–93

Prudencio CF (1993) Ring management of soils and crops in the West African semi-arid tropics: the case of the Mossi farming system in Burkina Faso. Agric Ecosyst Environ 47:237–264

Tittonell P, 2003. Soil fertility gradients in smallholder farms of western Kenya. Their origin, magnitude and importance. Quantitative approaches in systems analysis No. 25, Wageningen, The Netherlands, 233pp

Tittonell P, Misiko M, Ekise I, Vanlauwe B (2005a) Feeding-back the result of soil research: the origin, magnitude and importance of farmer-induced soil fertility gradients in smallholder farm systems. Emanyonyi Farmer Field School, Vihiga, western Kenya, July 27, August 17 of 2005. TSBF-CIAT, Nairobi, 23pp

Tittonell P, Muriuki AW, Shepherd KD, Mugendi DN, Kaizzi KC, Okeyo J, Verchot L, Coe R, Vanlauwe B (2010) The diversity of rural livelihoods and their influence on soil fertility in agricultural systems of East Africa- a typology of smallholder farms. Agric Syst 103(2):83–97

Tittonell P, Shepherd KD, Vanlauwe B, Giller KE (2007a) Unravelling factors affecting crop productivity in smallholder agricultural systems of western Kenya using classification and regression tree analysis. Agric Ecosyst Environ 123:137–150

Tittonell P, Vanlauwe B, Corbeels M, Giller KE (2008) Yield gaps, nutrient use efficiencies and response to fertilizers by maize across heterogeneous smallholder farms of western Kenya. Plant Soil 313(1–2):19–37

Tittonell P, Vanlauwe B, Leffelaar P, Rowe E, Giller K (2005b) Exploring diversity in soil fertility management of smallholder farms of western Kenya I. Heterogeneity at region and farm scales. Agric Ecosyst Environ 110:149–165

Tittonell P, Vanlauwe B, Leffelaar PA, Shepherd KD, Giller KE (2005c) Exploring diversity in soil fertility management of smallholder farms in western Kenya. II. Within-farm variability in resource allocation, nutrient flows and soil fertility status. Agric Ecosyst Environ 110:166–184

Tittonell P, Vanlauwe B, de Ridder N, Giller KE (2007b) Heterogeneity of crop productivity and resource use efficiency within smallholder Kenyan farms: soil fertility gradients or management intensity gradients? Agric Syst 94:376–390

Tittonell P, van Wijk MT, Rufino MC, Vrugt JA, Giller KE (2007c) Analysing trade-offs in resource and labour allocation by smallholder farmers using inverse modelling techniques: a case-study from Kakamega district, western Kenya. Agric Syst 95(1–3):76–95

Vanlauwe B, Tittonell P, Mukalama J (2006) Within-farm soil fertility gradients affect response of maize to fertilizer application in western Kenya. Nutr Cycling Agroecosyst 76:171–182

Walker B, Gunderson L, Kinzig A, Folke C, Carpenter S, Schultz L (2006) A handful of heuristics and some propositions for understanding resilience in social–ecological systems. Ecol Soc 11(1):13. URL: www.ecologyandsociety.org

Wopereis MCS, Tamélokpo A, Ezui K, Gnakpénou D, Fofana B, Breman H (2006) Mineral fertilizer management of maize on farmer fields differing in organic inputs in the West African savanna. Field Crop Res 96:355–362

Zingore S, Murwira HK, Delve RJ, Giller KE (2007) Soil type, historical management and current resource allocation: three dimensions regulating variability of maize yields and nutrient use efficiencies on African smallholder farms. Field Crop Res 101:296–305

Exploring Crop Yield Benefits of Integrated Water and Nutrient Management Technologies in the Desert Margins of Africa: Experiences from Semi-arid Zimbabwe

I. Nyagumbo and A. Bationo

Abstract The benefits of integrating locally adaptable water and nutrient management technologies were explored in semi-arid Zimbabwe. On-farm maize-based experiments were set up on six farmers' fields in Ward 5, Shurugwi. Three tillage systems, namely post-emergence tied ridging (PETR), rip and potholing (RPH) and conventional mouldboard ploughing (CMP) were integrated to three nutrient management regimes, i.e. a control with no fertility amelioration, pit-stored cattle manure band applied at 10 t/ha and the latter with an additional top dressing of ammonium nitrate (34.5% N) at 100 kg/ha. On each site the treatments were set up as a completely randomized split-plot block design replicated three times with tillage (water management) as the main treatment and fertility as the sub-treatment. CMP mimicked the farmer's common land preparation practice, while PETR and RPH systems represented the improved water harvesting tillage techniques. Results revealed significant nutrient management effects right from the first season giving 3-year means of 1298, 1977 and 2490 kg/ha for the control, manure and manure plus fertilizer treatments, respectively. On the other hand, water harvesting tillage effects were insignificant initially (2003/2004) but had beneficial effects in subsequent seasons (2004/2005 and 2005/2006) with 3-year grain yield means of 1624, 2032 and 2108 kg/ha for CMP, PETR and RPH, respectively. Maximum yield benefits from integrating PETR and RPH with manure + AN fertility ameliorants amounted to 218 and 261%, respectively, compared to CMP with no fertility amendment. The results, therefore, showed increased benefits when in situ water harvesting tillage techniques are integrated with appropriate nutrient ameliorants giving realizable food security benefits to the farmer.

Keywords Crop yield · In situ water harvesting · Nutrient management · Tillage · Water

Introduction

Sub-Saharan Africa continues to suffer from stagnant or declining per capita food production, despite considerable advances in technology across the whole world. While the UN's Millennium Development Goals seek to half global poverty and hunger by 2015, for Africa this remains a serious challenge.

Among other factors, water and nutrient management are key factors to any meaningful progress towards increased productivity among smallholder farmers of the continent. About 60% of the southern African region is semi-arid or arid and suffers from periodic droughts. In Zimbabwe, for example, maize production patterns are distinctly determined by the quality of each rainfall season (Fig. 1). The consumptive water use of most crops grown in the region exceeds 600 mm, yet annual rainfall in the semi-arid regions ranges between 450 and 600 mm, thereby making rain-fed cropping in the region a risky undertaking. Reliability of rainfall also increases as one moves from the high potential sub-humid to the semi-arid and arid regions of the country. Most cropping seasons

I. Nyagumbo (✉)
Department of Soil Science and Agricultural Engineering, University of Zimbabwe, Mount Pleasant, Harare, Zimbabwe
e-mail: inyagumbo@agric.uz.ac.zw

A. Bationo et al. (eds.), *Innovations as Key to the Green Revolution in Africa*,
DOI 10.1007/978-90-481-2543-2_77, © Springer Science+Business Media B.V. 2011

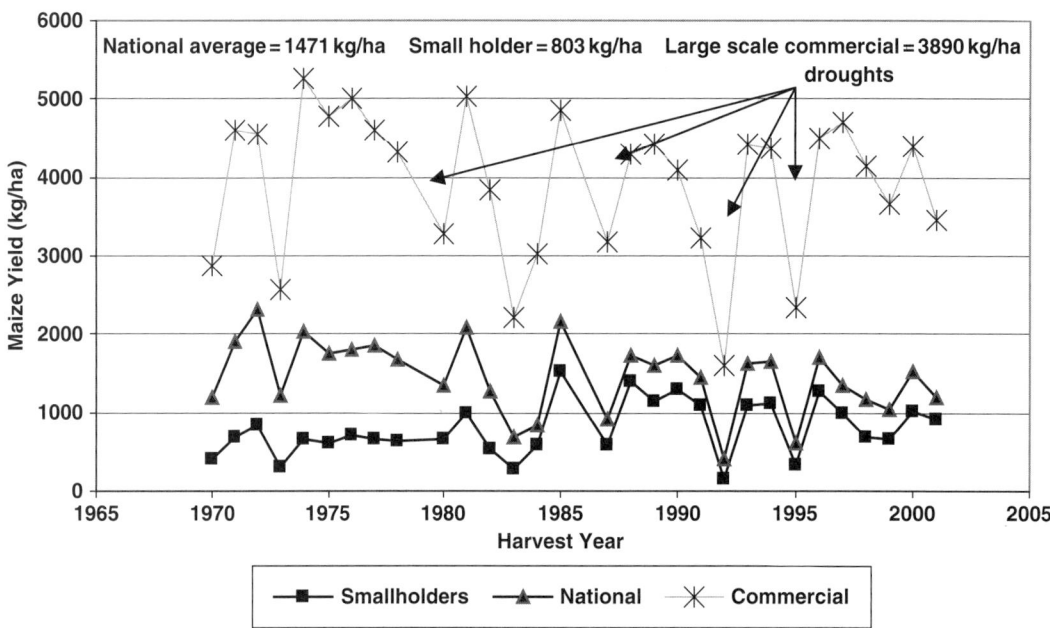

Fig. 1 Maize yield trends by agricultural sector in Zimbabwe since 1970. *Source*: Adapted from MLARR (2001)

are characterized by mid-season dry spells which seriously reduce yield potential, making water the greatest limitation to crop productivity in the region (Bratton, 1987). Furthermore annual evaporation surpasses rainfall, typically exceeding 1400 mm, meaning that most of the rain is lost through evaporation.

Despite its potential, less than 7% of the cropland in Zimbabwe is under irrigation due to the high costs and the lack of requisite infrastructure. It follows, therefore, that the scope for increased production lies in efficiently utilizing natural rainfall so as to get more 'crop per drop'. It is of no surprise, therefore, that improving land and water management has been given high priority by farmers and other stakeholders in SADC (Nhira and Mapiki, 2005). Furthermore, recommendations to policy makers for action from a regional eastern and southern Africa conference on agricultural water management emphasized *Exploitation of the potential of rain-fed agriculture* as one of the five critical factors towards the reduction of poverty and hunger in the region (IMAWESA, 2007).

It is therefore clear that unless rain-fed production systems become more efficient in utilizing available rainfall and nutrients, productivity and food security in Zimbabwe's smallholder farming sector will continue to remain below desired levels. It is estimated that the smallholder farmers in Zimbabwe use on average

18 kg/ha compared to 290 kg/ha in the large-scale sector, which is way below crop requirements that are usually more than 200 kg/ha (FAO, 1999) and results in net nutrient mining (Stoorvogel and Smaling, 1990).

Limited capacity of farmers to ameliorate soil fertility deficits through inorganic fertilizers and manures results in farmers failing to realize higher yields during good rainfall seasons.

The benefits of combining inorganic and organic fertility ameliorants have since been recognized in Zimbabwe (Grant, 1981), yet the cost of acquiring such fertilizers is increasingly becoming prohibitive while manure quantities and handling methods remain major constraints (Chuma et al., 2000, 2001; Mapfumo and Giller, 2001). The use of inorganic fertilizers by smallholders declined in the 1990s compared to the 1980s following sharp increases in fertilizer costs and farmers resorted to use of manure as a mitigatory measure (Chuma et al., 2000). According to the latter, the use of leaf litter, ash and anthill soil increased by at least 50%, between 1990 and 1996, in Chivi (Masvingo province). Anaerobically composted manures (pit stored) have also been found to result in significantly higher maize yields compared to heap storing, a standard farmer practice, particularly during the year of application (Nzuma and Murwira, 2000). Anaerobically composted pit-stored manures have also been found to significantly increase the nutrient content of cattle manure notably N, P, Ca and K compared to aerobically composted ones (Nzuma, 2002). Most cattle manures in communal areas of Zimbabwe are generally of low quality with dry matter N content ranging between 0.3 and 1.4% (Mugwira, 1984) compared to up to 2.2% N in anaerobically composted ones (Nzuma, 2002). Banded application of manure in planting furrows was also found in different studies to be superior to the widely used farmer practice of broadcasting method in terms of maize yields (Munguri, 1996; Nzuma, 2002). However, not much is known about the benefits of integrating such anaerobically pit-stored manures with improved water conserving tillage practices.

The use of alternative tillage techniques such as conservation tillage is increasingly seen worldwide as a means of redressing water deficit and other problems through maintaining soil organic carbon and increasing infiltration, crop water-use efficiency and groundwater recharge (Batjes, 1999; Beukes et al., 1999). For example at Hatcliffe (sub-humid north of Zimbabwe), significant benefits from conservation tillage systems were obtained in terms of enhanced potential for groundwater recharge and runoff reduction from mulch ripping and no-till tied ridging, averaging 3 and 9% of seasonal rainfall, respectively, compared to 22% for conventional mouldboard ploughing. Significant benefits were also found in terms of crop water-use efficiency from mulch ripping (Nyagumbo, 2002). At Makoholi (semi-arid southern Zimbabwe), higher yield increases of between 20 and 300% from deep ripping (300 mm) over hand planting were measured (Mashavira et al., 1997), thereby suggesting better moisture conservation from such ripping techniques through plough pan breaking.

Unfortunately few studies in Zimbabwe have attempted to explore the benefits of integrating such water management to fertility management techniques, driven mainly by institutional disciplines and their respective research mandates. Nevertheless, sorghum yields increased by between 50 and 200% when basal fertilizer (compound D) application increased from 100 to 200 kg/ha in Chiredzi under the tied furrow system compared to the flat, suggesting increased water-use efficiency when both water and nutrient status were enhanced (Nyamudeza and Nyakatawa, 1995). Also at Makoholi (semi-arid) under conventional ploughing, maize water-use efficiency increased

from 1.6 to 4.9 kg/ha/mm, while at Marondera (sub-humid) it increased from 2.4 to 5.4 kg/ha/mm in the season 1992/1993 when nitrogen fertilizer was applied (Shamudzarira, 1994). Elsewhere multi-locational water balance studies in Niger also showed that the use of fertilizer increases water-use efficiency (Bationo et al., 1998).

Therefore, to derive maximum benefit from water conserving tillage systems, it is also necessary to optimize soil nutrient status. Besides these issues draught power and labour constraints coupled to the prevalence of HIV/AIDS leading to poor health and women labour necessitate the use of labour saving technologies such as conservation tillage.

This study investigated the synergy of integrating soil tillage and nutrient management technologies in terms of crop yields and related parameters in a semi-arid region of Zimbabwe over three seasons from 2003/2004 to 2005/2006.

Materials and Methods

The study was carried out in Ward 5 of Shurugwi district of Zimbabwe. The area falls in agro-ecological region IV receiving 450–650 mm rainfall per annum (Vincent and Thomas, 1960). The soils are typically shallow and poorly developed granite-derived sands of low inherent fertility (Grant, 1981). Typical profile chemical analysis data from two of the six locations on which experiments were conducted are presented in Table 1. Average pH ($CaCl_2$) ranged from 4.6 to 6.1, while organic carbon is generally low ranging from 0.2 to 0.9% (Table 2).

Three tillage treatments namely (i) post-emergence tied ridging (*PETR*), (ii) rip and potholing (*RPH*) and (iii) conventional mouldboard ploughing (*CMP*) were compared in combination with three soil fertility management levels, i.e. (a) control with no additions (*residual*), (b) 10 t/ha pit-stored manure (*manure*) and (c) 10 t/ha pit-stored manure plus 100 kg/ha ammonium nitrate with 34.5% N (*manure + AN*).

RPH is a relatively new tillage technique developed by the authors as an alternative to the no-till tied ridging technique where crop emergence may be poor when seeds are planted in the ridge. It has potential to reduce soil compaction problems and thereby improve water infiltration. Ripping was carried out

Table 1 Soil chemical analysis data from two of the six study sites used in the study

Depth (cm)	Site 1						Site 2			
	0–16	16–31	31–44	44–73	73–110	110–125	0–13	13–23	23–34	34–120
Texture	cS	cLS	mS	cS	cS	cS	cS	cLS		
Clay %	4	5	4	3	2	2	4	9	–	–
Silt %	4	4	4	4	4	4	5	5	–	–
Fine sand %	2	26	46	21	29	26	35	34	–	–
Medium sand %	42	35	26	35	25	34	29	28	–	–
Coarse sand %	48	30	20	38	40	34	26	25	–	–
pH (CaCl$_2$)	5.8	6.8	6.9	6.9	6.8	6.9	4	4.9	–	–
Ex Ca (me %)	5.2	8.1	4.1	3.4	2.8	2.5	2	1.6	–	–
Ex Mg (me %)	0.8	0.9	0.5	0.7	0.6	0.6	0.9	0.8	–	–
Ex Na (me %)	0.14	0.19	0.06	0.06	0.18	0.04	0.06	0.06	–	–
Ex K (me %)	0.27	0.29	0.18	0.18	0.16	0.14	0.18	0.18	–	–
TEB (me %)	3.7	2.9	2.3	1.7	1.7	2.1	1.7	1.6	–	–
CEC (me %)	3.7	2.9	2.3	1.7	1.7	2.1	1.7	1.6	–	–
Base sat %	100	100	100	100	100	100	100	100	–	–
E/C	96.9	55.6	64.3	66	79.3	91	40.8	18.3	–	–
S/C	96.9	55.6	64.3	66	79.3	91	40.8	18.3	–	–
ESP	3.9	6.4	2.6	3.6	10.9	1.9	3.5	3.7	–	–
EKP	7.2	9.9	7.8	10.8	9.7	6.7	10.4	11.1	–	–
Organic carbon %	0.39	0.52	0.37	0.31	0.4	0.44	0.29	1.63	–	–

Table 2 Average topsoil pH and organic carbon (%) measured in 2006 on five sites in Shurugwi after three seasons

Site	pH (CaCl$_2$)	Organic carbon (%)
Mukandabvute	5.9	0.20
Gweru	5.8	0.29
Siziba	5.4	0.92
Mfiri	4.6	0.71
Shura	6.1	0.88

using an animal-drawn *Palabana* subsoiler to depths varying between 15 and 25 cm depending on field conditions at the time of planting. Crop inter-rows were potholed every 2–3 m using a locally developed donkey tool. This system has potential to reduce labour for land preparation and to improve soil and water conservation as no conventional ploughing is carried out. In the RPH treatment, ripping thus constituted the primary land preparation and was carried out each year prior to planting in the ripped furrows. Potholes were installed at or soon after planting to create micro-depressions that facilitated water harvesting and increased infiltration.

PETR involved ridging and cross-tying the crop when maize was about knee height. Primary land preparation in this case was the same as for conventional ploughing. Unlike the no-till tied ridging system where ridges are semi-permanent (Elwell and Norton,

1988), in post-emergence or post-planting tied ridges, planting was initially carried out on the flat in the conventional way with ridges (12–20 cm height) and cross-ties (half the height of ridges) being constructed 3–4 weeks after planting or when the maize crop was about knee height. Tied ridging is generally a well-established technique reducing soil loss and runoff to sustainable magnitudes (Nyagumbo, 2002), but suffers from poor crop emergence if planting within is carried out when the ridges are not fully moist. This problem arises from poor uniformity in the progression of the wetting front in ridge forms as compared to flat cultivation culminating in a drier ridge and wetter furrows at the start of the wet season (Twomlow and Breneau, 2000). Establishing ridges on a flat planted crop was therefore considered a viable alternative to planting on top of ridges as practiced in the no-till tied ridging system. Conventional ploughing was conducted using the animal-drawn mouldboard plough to mimic current farmer practice (i.e. shallow ploughing to 10–12 cm depth) just before planting.

Fertility management techniques included a control with no fertility amendments, 10 t/ha pit-stored cattle manure banded in crop rows and pit-stored cattle manure plus ammonium nitrate top dressing at 100 kg/ha spot applied in rows when maize was about knee height. Pit-stored manure was prepared by storing cattle kraal manure in a soil covered pit for approximately 3 months, excavating it and immediately band apply it and cover in planting furrows to avoid N loss by volatilization based on previous research findings (Nzuma, 2002). Top dressing ammonium nitrate was applied in the manure + AN plots as a side dressing to each plant at about 100 kg/ha at approximately 4 weeks after planting when rainfall conditions were conducive.

Experiments were set up on six farmers' fields following an equipment demonstration workshop prior to the start of season 2003/2004 after which six host farmers were selected from volunteers. The trials were researcher managed and farmer implemented. Each trial was laid out in a completely randomized block split-plot design with three replicates per main treatment. Main plots were 24 m long and 8 m wide. Fertility sub-plots were installed by sub-dividing each main plot into three sub-sections resulting in 8 × 8 m plots on which fertility sub-treatments were randomly allocated. Planting was carried out on the same day on each site following at least 20 mm of rain between November and mid-December each year.

Maize crop performance parameters measured included crop emergence, growth rates and crop yield. A short season hybrid maize variety (SC 513) selected by the farmers was used on all sites, planted and thinned after emergence to achieve 34–36,000 plants/ha in line with local extension recommendations. Emergence of maize seedlings was assessed between 2 and 3 weeks after planting during the first season by counting the number of emerged plants in each sub-plot over a distance of 6 m on three positions, i.e. third row from the top, middle row and the third row at the bottom of the sub-plot. The same plots and treatments were repeated over the three seasons in a similar fashion. Maize grain and biomass yields were assessed from check plots in each sub-plot measuring four rows wide by 5 m long. The moisture content of a sample of maize grain from each check plot was determined gravimetrically and used to calculate grain yield corrected to 12.5% moisture content (wet basis), used as the local standard for maize.

Rainfall utilization efficiency, a parameter used to provide crude estimates of water-use efficiency, was calculated as grain yield (kg/ha) divided by the total seasonal rainfall. The term *water-use efficiency* has been used in many studies to provide a measure of the efficiency of water utilization in various cropping systems based on actual crop evapotranspiration or ET

(Moyo and Hagmann, 1994; (van Duivenbooden et al., 1999).

A *combined analysis of variance across sites* derived from split-plot arrangements on each trial was used to test for tillage (main treatments) and fertility (sub-treatment) effects on yield parameters during each season. An overall analysis involving all farmers and seasons was carried out to test if treatment performances were consistent over consecutive seasons. Season was considered a random variable.

Results

Rainfall Seasons

Seasonal rainfall totals were 657, 645 and 845 mm for three seasons 2003/2004, 2004/2005 and 2005/2006, respectively (Fig. 2). Although the first two seasons had very similar rainfall totals, they differed a lot in its distribution. Season 2003/2004 had better rainfall distribution than 2004/2005. December and January of 2003/2004 were rather dry, while the wettest month was March. Most maize was planted between November and early December during that season (2003/2004). On the other hand, 2004/2005 was characterized by large and intermittent storms with long dry spells in between and in particular between March and April, periods which coincided with tasselling and grain filling stages of maize. In contrast the last season 2005/2006 was characterized by a rather late start resulting in most maize in the area being planted in December. Thus, despite the high total amount the season 2005/2006 tailed off too early in March resulting in a long dry spell right up to the end (Fig. 2).

Maize Seedling Emergence

The mean plant counts/6 m lengths at emergence was 40.1 (Table 3) giving an average spacing of 15 cm/plant during the first season. No significant differences in crop emergence were found between tillage and fertility management treatments, although there were significant differences between farmers ($p = 0.007$).

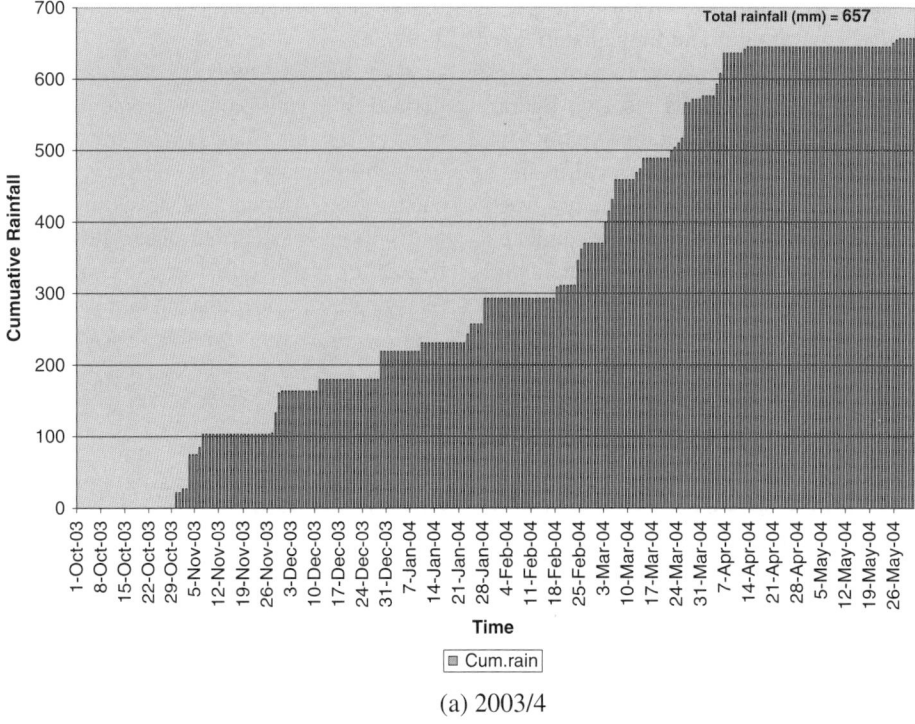

(a) 2003/4

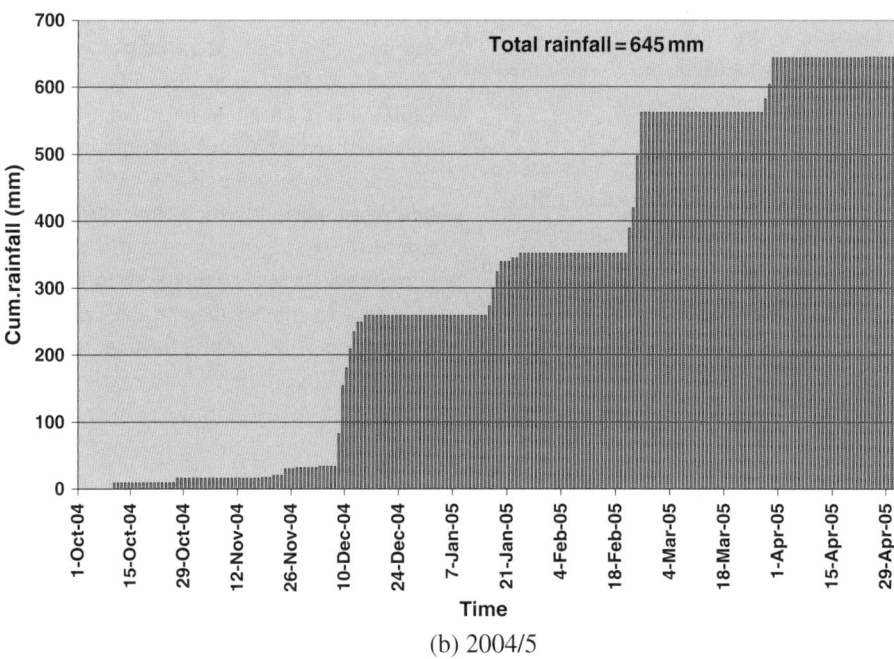

(b) 2004/5

Fig. 2 Cumulative daily rainfall amounts over three seasons in ward 5 of Shurugwi district of Zimbabwe (**a**) 2003/2004, (**b**) 2004/2005 and (**c**) 2005/2006

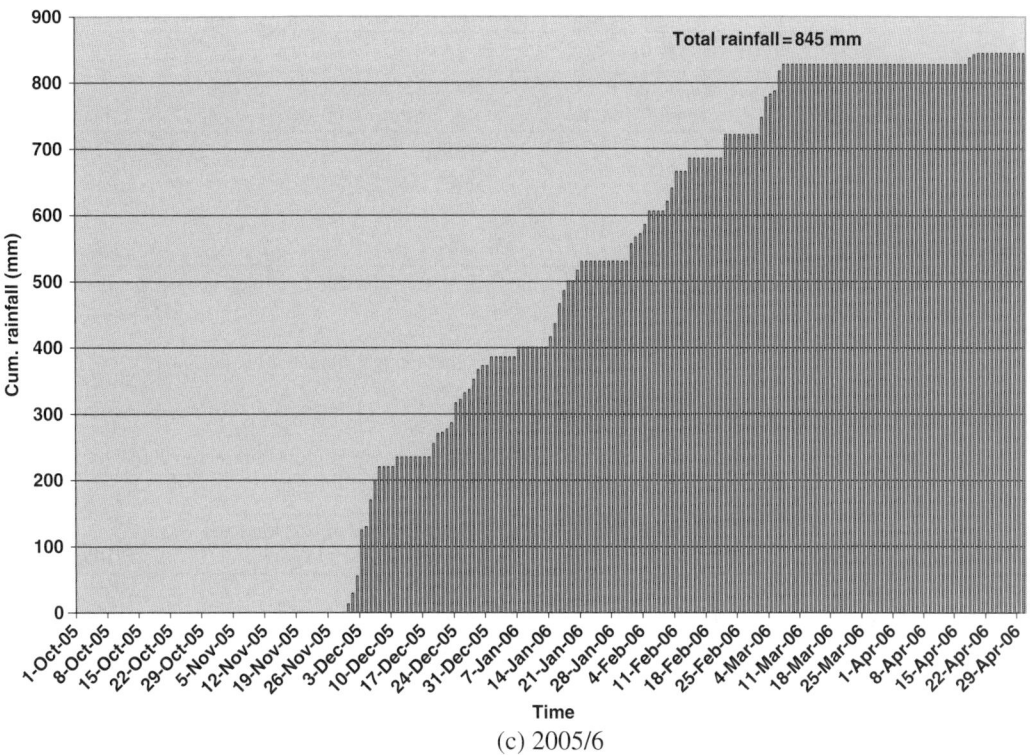

(c) 2005/6

Fig. 2 (continued)

Table 3 Effects of tillage and fertility management on maize seedling emergence in Shurugwi in season 2003–2004

| Tillage system | Fertility management system | | | |
	Control	Pit-stored manure (10 t/ha)	Pit-stored manure (10 t/ha) and AN fertilizer (100 kg/ha)	Mean plant counts/6 m row by tillage system
CMP	39.7	40.0	40.6	40.1[x]
PETR	39.4	40.1	38.5	39.3[x]
RPH	41.0	39.7	40.7	40.5[x]
Mean plant counts/6 m row by fertility management system	40.0[a]	39.9[a]	39.9[a]	40.1

Note: PETR = post-emergence tied ridging; RPH = rip and pothole; CMP = conventional mouldboard ploughing. l.s.d (fertility) = 1.36 plants; l.s.d (tillage) =1.74 plants; l.s.d (tillage × fertility) = 2.35 plants; CV = 8.1%; $N = 135$
Means in the same column or row followed by the same superscript letters are not significantly different at $p < 0.05$

Maize Yields

During the first season (2003/2004) there were no significant tillage effects on maize yields (Table 4). Overall average maize yields over the six sites were 2.6 t/ha. However, fertility management effects were significant with the highest yields coming from manure and top dressing combinations. As expected, differences between farmers were significant. Farmer × fertility management interactions were found to be significant due to differences between farmers and quality of manure used but there were no significant interaction effects between tillage and fertility.

In the second season both tillage ($p = 0.002$) and fertility ($p < 0.001$) yield effects became significant (Table 5) and there were no significant interaction between the two factors. Interaction between farmers and tillage were also insignificant. However, significant differences between farmers ($p = 0.002$) were

Table 4 Effects of tillage and fertility management on maize grain yields (kg/ha) in Shurugwi in season 2003–2004

| Tillage system | Fertility management system | | | Tillage mean (kg/ha) |
	No manure	Pit-stored manure (10 t/ha)	Pit-stored manure (10 t/ha) and AN fertilizer (100 kg/ha)	
CMP	2023	2725	3505	2751[a]
PETR	1884	2173	3937	2665[a]
RPH	1514	2624	3578	2572[a]
Fertility mean	1807[a]	2507[b]	3673[c]	2662

PETR = post-emergence tied ridging; RPH = rip and pothole; CMP = conventional mouldboard ploughing. l.s.d.$_{05, 72df}$ (fertility) = 334 kg/ha; l.s.d.$_{05,24df}$ (tillage) = 414 kg/ha; l.s.d$_{0.05, 28.03df}$ (tillage × fertility) = 618 kg/ha; CV = 32.7%; $N = 162$
Means in the same column or row followed by the same letter are not significantly different at $p < 0.05$

Table 5 Effects of tillage and fertility management on maize grain yields (kg/ha) in Shurugwi in season 2004–2005

| Tillage system | Fertility management system | | | Tillage mean (kg/ha) |
	No manure	Pit-stored manure (10 t/ha)	Pit-stored manure (10 t/ha) and AN fertilizer (100 kg/ha)	
CMP	690	918	704	770[x]
PETR	785	1373	1292	1150[y]
RPH	1025	1492	1284	1267[y]
Fertility mean	833[a]	1261[b]	1093[c]	1062

Note: PETR = post-emergence tied ridging; RPH = rip and pothole; CMP = conventional mouldboard ploughing. l.s.d$_{05, 72df}$ (fertility) = 141.0 kg/ha; l.s.d.$_{05,24df}$ (tillage) = 261 kg/ha; l.s.d$_{0.05, 56.16df}$ (tillage × fertility) = 323 kg/ha; CV = 34.6%; $N = 162$
Means in the same column or row followed by the same superscript letters are not significantly different at $p < 0.05$

Table 6 Effects of tillage and fertility management on maize grain yields (kg/ha) in Shurugwi in season 2005–2006

| Tillage system | Fertility management system | | | Tillage mean (kg/ha) |
	No manure	Pit-stored manure (10 t/ha)	Pit-stored manure (10 t/ha) and AN fertilizer (100 kg/ha)	
CMP	935	1353	1768	1352[x]
PETR	1232	2643	2970	2282[y]
RPH	1595	2492	3372	2486[y]
Fertility mean	1254[a]	2162[b]	2704[c]	2040

Note: PETR = post-emergence tied ridging; RPH = rip and pothole; CMP = conventional mouldboard ploughing. l.s.d$_{05, 72df}$ (fertility) = 237.1 kg/ha; l.s.d$_{05,24df}$ (tillage) = 380.6 kg/ha; l.s.d$_{0.05, 65.44df}$ (tillage × fertility) = 498.5 kg/ha; CV = 30.3%; $N = 162$
Means in the same column or row followed by the same superscript letters are not significantly different at $p < 0.05$

again observed. Due to persistent dry spells at the time of application top dressing AN was not applied to the manure plus AN plots in this second season. Observed effects were thus due to manure and perhaps some residual N effects from the previous season's top dressing. Plots receiving manure only gave the highest yields followed by the manure plus AN plots and lastly the control plots which did not receive any additional nutrients. With respect to tillage, PETR and RPH resulted in significantly higher yields ($p = 0.002$) compared to CMP, although the former two were not significantly different from each other (Table 3), with RPH giving the highest yields.

During the third season (2005/2006) there were no significant differences between farmers. However, performance of tillage depended on fertility status, i.e. tillage × fertility interaction was significant. Manure and manure and fertilizer plots enabled yields to almost double in tied ridges and rip and pothole treatments compared to conventional mouldboard ploughing plots with manure or manure plus fertilizer (Table 6). This interaction between tillage and fertility was further explored and is illustrated in Fig. 3.

Figure 3 shows that during the first season (2003/2004), the two tillage systems PETR and RPH resulted in negative relative yield benefits of –7 and –25%, respectively, compared to ploughing only (0) at low fertility levels, i.e. at control level where no fertility amendments were made. Positive benefits were, however, obtained from all three tillage systems when manure was added and even better when this manure was supplemented with 100 kg/ha of ammonium nitrate with up to 95% gain under PETR but as stated earlier significant differences only featured from the fertility treatments during the first year. In the next season no negative yield gains were observed even at

the control fertility level, but the margin between CMP and the other two tillage systems at higher fertility levels increased to as much as 116% under RPH integrated to the manure treatment. A somewhat drop in this margin was, however, observed in the manure + fertilizer plots that had previously received AN in the first year but did not receive it in the second season due to persistent mid-season dry spells, although the total rainfalls were similar for the two seasons (657 compared to 645 mm).

During the third season (2005/2006), the benefits of integrating both water conservation techniques and fertility management clearly pronounced themselves with margins ranging from 167 to 261% for both RPH and PETR at manure and manure plus AN levels with the highest yield returns being generated from RPH combined with manure + AN, hence the significant tillage × fertility interaction.

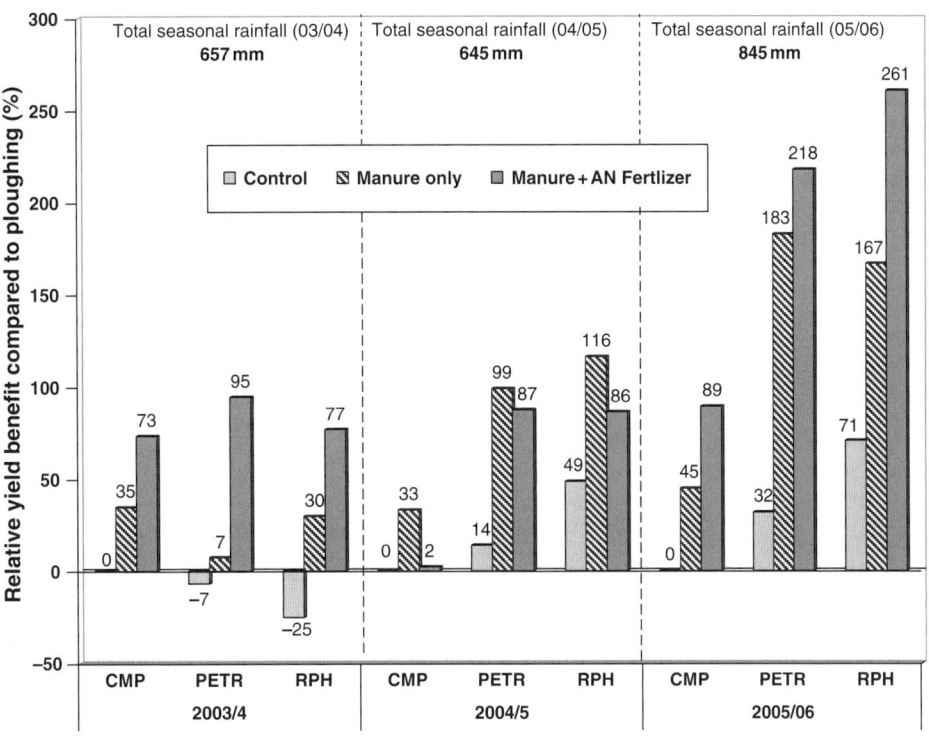

Fig. 3 Relative grain yield benefits (%) of *post-emergence tied ridging* (PETR) and *rip and pothole* (RPH) systems compared to *conventional mouldboard ploughing* (CMP) over three cropping seasons in Shurugwi district, Zimbabwe. *Note*: Control = no fertility amendments; manure only = 10 t/ha pit-stored manure banded in crop rows; manure + AN fertilizer = 10 t/ha pit-stored manure plus 100 kg/ha ammonium nitrate fertilizer

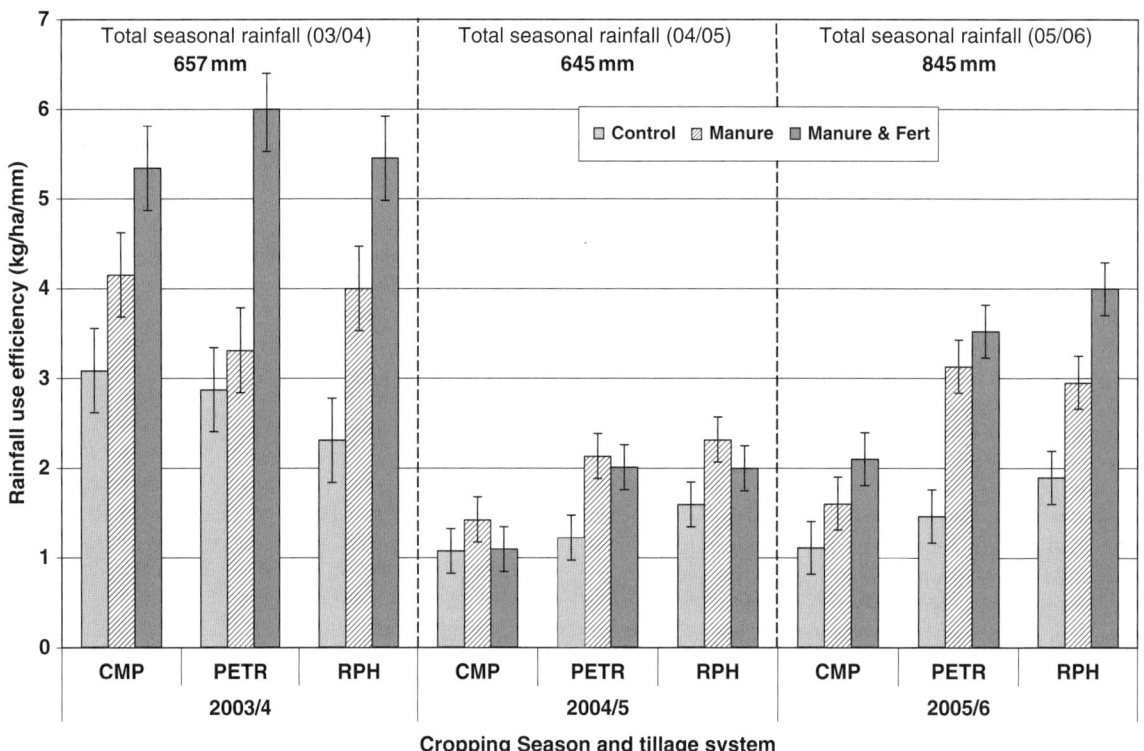

Fig. 4 Rainfall use efficiencies of three tillage systems *post-emergence tied ridging* (PETR), *rip and pothole* (RPH) and *conventional mouldboard ploughing* (CMP) integrated to three fertility management regimes over three cropping seasons in Shurugwi district, Zimbabwe. *Note*: Control= no fertility amendments; manure only = 10 t/ha pit-stored manure banded in crop rows; manure + fertilizer = 10 t/ha pit-stored manure plus 100 kg/ha ammonium nitrate fertilizer. *Error bars* denote *tillage × fertility* interaction l.s.d$_{(0.05)}$ for comparing means within each year

Rainfall Use Efficiencies

The highest rainfall use efficiencies (Fig. 4) were obtained during the first season (657 mm) followed by the last season with 845 mm and lastly the second season (645 mm). The results also show that the use of water conservation techniques (PETR and RPH) in combination with manure or manure and fertilizer generally resulted in superior rainfall utilization efficiency compared to the control treatments under CMP. The first season (2003/2004) as shown in Fig. 2a had well-distributed rainfall throughout, while the second season had long dry spells that stressed crops excessively and hence resulted in the poorest rainfall utilization efficiencies. In contrast the last season offered a very good start to crops from planting all the way to tasselling but tailed off too early resulting in crops being stressed again before maturing leading to surprisingly lower yields compared to the

other two seasons given the total rainfall received. Of the three seasons, the highest rainfall use efficiencies (6 kg/ha/mm) were obtained from PETR and manure + AN combination during the first season and the lowest (approx 1 kg/ha/mm) from CMP with no fertility amendment combination during the season 2004/2005.

Discussion

The negative yield gains at low fertility levels from the water conservation techniques in the first season were probably due to poor mineralization of the poor nutrient reserves from the soil from which farmers normally benefit under conventional ploughing. This negative effect was more pronounced under RPH where no ploughing was carried out before planting, a phenomenon often characteristic of conservation tillage systems. It is, however, not clear why this phenomenon was also experienced in the PETR system which was also ploughed just before planting. These negative yield gains generally contributed to lack of significant yield differences from the three tillage systems during the first year, despite the significant fertility effects. The results from the first year suggested that fertility level was the more limiting factor than the water conservation method and hence place more emphasis on the need for fertility amelioration when soils have very poor nutrient reserves for any meaningful responses to water management techniques.

Results from the second and third seasons suggest that once fertility limitations have been overcome, water conservation becomes an important factor. Various studies on conservation tillage systems in Zimbabwe have shown that yield benefits only accrue at high fertility levels in situations where residues are retained as surface mulches (Nyagumbo, 2008). It is worth noting that the RPH treatment was not ploughed.

In the second season the manure plus fertilizer treatment which was in fact not top dressed in that season actually yielded numerically lower than the manure-only treatment under both PETR and RPH. This effect was attributed to increased mineralization and more efficient extraction of nutrients particularly phosphorus and micronutrients from the applied manure in this treatment during the first season resulting in all the nutrient reserves from the manure being exhausted by the crop in the first season and little if any residual benefit to the subsequent crop. A look at Table 3 shows that the use of AN fertilizer resulted in an additional yield gain of 1166 kg/ha or 46.5% compared to the manure-only treatment in the first season. Such a yield gain therefore resulted in higher extraction of nutrients leaving the next crop to rely entirely on nutrients from the freshly applied manure that second season. Furthermore, availability of nutrients such as P enhances root growth and results in efficient water and nutrient absorption from the soil.

The practice of applying top dressing depending on rainfall conditions is common in Zimbabwe due to the risk of fertilizer burning experienced when applied under dry conditions. Research studies have also shown that conditional application of N depending on rainfall received resulted in 86, 59 and 60% yield increases in natural regions II, III and IV, respectively (Piha, 1993; Machikicho, 1999). In this case failure to apply the AN in the manure + AN treatment probably saved the crop from even higher yield depressions

due to fertilizer burning. Results from the third season, however, further confirm the additional benefit of using AN when conditions are conducive (25% increase compared to manure only). The increased margin of differences between the CMP control and the other treatments suggests the incremental benefits that farmers may accrue from repeated additions of organic or combinations with inorganic for sustained productivity in the long run.

Results from the second and third seasons also show that the integrated use of nutrient and water management technologies results in synergic effects on crop yields. For example, the use of PETR and RPH without any fertility amendments resulted in 14 and 49% yield increase in 2004/2005 and 32 and 71% in 2005/2006, respectively, compared to CMP control (Fig. 3). However, when fertility amendments were included, PETR and RPH resulted in yield increments ranging from 86 to 261% in the two seasons compared to CMP control. Neither could such benefits be derived from the use of these fertility amendments in the ploughed treatments where maximum of 33 and 89% yield increases were realized in 2004/2005 and 2005/2006, respectively (Fig. 3).

It is also apparent from this study not only that integration of water and nutrient management technologies enhances rainfall utilization efficiency (Fig. 4) but also that high rainfall if poorly distributed may not necessarily translate to higher efficiency. In general gain water-use efficiency based on ET, a more accurate measure than rainfall use efficiency, ranges between 3.5 and 10 kg/ha/mm for tropical grains in rain-fed environments (Rockström et al., 2001) but figures exceeding 16 kg/ha/mm for maize have been reported in South Africa (Beukes et al., 1999). The results obtained from this study were therefore within the range of those based on ET when fertility and water management technologies are integrated but way below this range when CMP was used without any amendments. In another study in the sub-humid north of Zimbabwe over 5 years, average water-use efficiencies of 9.25 and 11.06 kg/ha/mm for CMP and no-till tied ridging, respectively, were measured (Nyagumbo, 2002; Moyo, 2003). No-till tied ridging is a technique similar to PETR but involves the planting of crops on semi-permanent ridges compared to annually prepared ridges used in this study where the maximum rainfall use efficiency was about 6 kg/ha/mm. It is therefore apparent from here that investments by farmers in tillage techniques at low fertility levels may not yield benefits at least in the short term.

The results from this work therefore serve as further proof that improved water availability to the plant without the nutrient is not helpful just as nutrients availability without the moisture is futile and hence the importance of their integrated use (Zougmore et al., 2000). Studies elsewhere have also shown that without an efficient nutrient management system, poor water-use efficiency is inevitable. Studies by Fofana et al. (2003) concluded that addressing soil nutrient imbalances in the soil increases water-use efficiency by a factor of 3–5. Water-use efficiency was also increased by a factor of 2–3 when nutrient sources were added to a soil and water conservation technology compared to the one with no nutrient addition (Bationo and Mokwunye, 1991). Complementary studies following this work have gone further to show that any organic amendments used by farmers in this area other than pit-stored cattle manure positively impact on grain yields (Nyamangara and Nyagumbo, 2010). It is therefore clear that to optimize the synergy between water conservation and nutrient availability, there is need for an integrated approach whereby water and nutrient imbalances are addressed simultaneously.

Conclusions

The study showed that integration of fertility and water management technologies results in high returns to both investments compared to when not integrated. More than 200% yield gains above the CMP control with no fertility amendments were obtained from PETR and RPH treatments when integrated to manure plus fertilizer combinations. Over the three season studies, use of water or fertility management technologies alone without the other failed to match any of the yield benefits from integration with a maximum of 71% obtained from RPH in the third season above the CMP control. It is therefore clear that an integrated approach whereby water and nutrient imbalances are addressed simultaneously provides optimum conditions for synergy between water and nutrient management technologies. It is also clear that investments in tillage techniques by farmers (labour, equipment, etc.) at low fertility levels may not translate to yield benefits to farmers at least in the short term.

The study also showed not only that integration of water and nutrient management technologies enhances rainfall utilization efficiencies but also that high rainfall if poorly distributed may not necessarily translate to rainfall use efficiency thereby placing emphasis on technologies enabling farmers to fully capitalize on the entire rainfall season as well as ensuring maximum capturing of any rainfall received.

Recommendations

The study recommends further studies to evaluate economic benefits of the tested technologies and their viability among farmers in the district. The need to scale out this work is also emphasized and since 2005/2006 through other initiatives there has been overwhelming uptake of potholing and tied ridging techniques as well as fertility management technologies but main constraints include labour, access to equipment and other inputs. The need to link farmers to input suppliers, e.g. equipment, seed and fertilizer suppliers, is thus a prerequisite for success while increased awareness among both extension staff and farmers will prove worthwhile for innovating farmers.

Acknowledgements We thank TSBF-AFNET Desert Margins Program for funding this work and to the Soil Fertility Consortium of Southern Africa for supporting outscaling initiatives in the district from 2005.

References

Bationo A, Lompo F, Koala S (1998) Research on nutrient flows and balances in West Africa: state-of-the-art. Agric Ecosyst Environ 1350:1998

Bationo A, Mokwunye AU (1991) Alleviating soil, water and nutrients management constraints to increased crop production in West Africa. Fertil Res 29:95–115

Batjes NH (1999) Management options for reducing CO2-concentrations in the atmosphere by increasing carbon sequestration in the soil. Dutch national research programme on global air pollution and climate change & technical Paper 30. International Soil Reference and Information Centre (ISRIC), Wageningen, 114 pp

Beukes DJ, Bennie ATP, Hensley M (1999) Optimization of soil water use in the dry crop production areas of South Africa. In: Van Duivenbooden MP, Studer C, Bielders CL (eds) Efficient soil water use: the key to sustainable crop production in the dry areas of West Asia, and North and Sub-Saharan Africa. Proceedings of the 1998 (Niger) and

1999 (Jordan) workshops of the Optimizing Soil Water Use (OSWU) Consortium. ICARDA and ICRISAT, Aleppo, Syria Patancheru, India, pp 165–191

Bratton M (1987) Drought, food and the social organisation of small farmers in Zimbabwe. In: Glantz M (ed) Drought and hunger in Africa denying famine a future. Cambridge University Press, New York, NY, pp 31–35

Chuma E, Mombeshora BG, Murwira HK, Chikuvire J (2000) The dynamics of soil fertility management in communal areas of Zimbabwe. In: Hilhorst T, Muchena F (eds) Nutrients on the move: soil fertility dynamics in African farming systems. International Institute for Environment and Development, London, pp 45–64

Chuma E, Mvumi B, Nyagumbo I (2001) A review of sorghum and pearl millet-based production systems in the semi-arid regions of Zimbabwe. SADC/ICRISAT Sorghum and Millet Improvement Program (SMIP), PO Box 776, Bulawayo, Zimbabwe; International Crops Research Institute for the Semi-Arid Tropics (ICRISAT), Nairobi, (Limited distribution), 72 pp

Elwell HA, Norton AJ (1988) No-till tied ridging: a recommended sustainable crop production system. Institute of Agricultural Engineering, Harare, Sept 1988

FAO (1999) A fertilizer strategy for Zimbabwe, 1999. Food and Agriculture Organisation of the United Nations and African Centre for Fertilizer Development, Rome, Italy, 105pp. ftp://ftp.fao.org/agll/docs/fszimb.pdf

Fofana B, Wopereis M, Zougmore R, Breman H, Mando A (2003) Integrated soil fertility management, an effective water conservation technology for sustainable dryland agriculture in sub-saharan Africa. In: Buekes D, de Villiers M, Mkhize S, Sally H, van Rensburg L (eds) Proceedings of the symposium and workshop on water conservation technologies for sustainable Dryland agriculture in sub-saharan Africa. ARC-South Africa, Bloem Spa Lodge and Conference Centre, Bloemfontein, pp 109–117, 8–11 April 2003

Grant PM (1981) The fertilization of sandy soils in peasant agriculture. Zimbabwe Agric J 78:169–175

IMAWESA (2007) Agricultural water management, a critical factor in the reduction of poverty and hunger: principles and recommendations for action to guide policy in Eastern and Southern Africa, June 2007. The IMAWESA, Improving Management of Agricultural Water in Eastern and Southern Africa project, Nairobi, Kenya, 30 pp, www.asareca.org/swmnet

Machikicho JT (1999). Socio-economic and agronomic evaluation of a soil management package for increased crop production by communal area farmers. MPhil thesis, Department of Soil Science and Agricultural Engineering, University of Zimbabwe, Harare, 152 pp

Mapfumo P, Giller KE (2001) Soil fertility management strategies and practices by smallholder farmers in semi-arid areas of Zimbabwe. Limited Distribution, 2001. International Crops Research Institute for Semi-Arid Tropics and Food and Agriculture Organization of the United Nations, Bulawayo, Zimbabwe, 60 pp

Mashavira TT, Dhliwayo HH, Mazike PS, Twomlow SJ (1997) Agronomic consequences of reduced tillage. In: Ellis-Jones J, Pearson A, O'Neill D, Ndlovu L (eds) Improving the productivity of draught animals in sub-Saharan Africa.

25–27 February 1997, UZ, AGRITEX, DR&SS, CVTM, SRI, Institute of Agricultural Engineering, Hatcliffe, Harare, Zimbabwe, pp 85–92

MLARR (2001) The agricultural sector of Zimbabwe: statistical bulletin-2001. Ministry of Lands Agriculture and Rural Resettlement, Harare, p 60

Moyo A (2003) Assessment of the effect of soil erosion on nutrient loss from granite derived sandy soils under different tillage systems in Zimbabwe. DPhil thesis, Department of Soil Science and Agricultural Engineering, University of Zimbabwe, Harare, 228 pp

Moyo A, Hagmann J (1994) Growth effective rainfall in maize production under different tillage systems in semi-arid conditions and granitic sands of Southern Zimbabwe. In: Jensen BE, Schjonning P, Mikkelsen SA, Madsen KB (eds) Soil tillage for crop production and protection of the environment. Proceedings of the 13th International Conference, International Soil Tillage Research Organisation (ISTRO). ISTRO, Aalborg, pp 475–480

Mugwira LM (1984) Relative effectiveness of fertilizer and communal area manures as plant nutrient sources. Zimbabwe Agric J 81:85–90

Munguri MW (1996) Inorganic fertilizer and cattle manure management for dryland maize (Zea Mays L.) production under low input conditions. MPhil thesis, Department of Crop Science, University of Zimbabwe, Harare, 162 pp

Nhira C, Mapiki A (eds) (2005) Regional situational analysis land and water management in the SADC region. Vol. Workshop proceedings no.1 2004. Southern African Development Community (SADC) Secretariat, Gaborone, 164 pp

Nyagumbo I (2002) Effects of three tillage systems on seasonal water budgets and drainage of two Zimbabwean soils under maize. Dphil thesis, Department of Soil Science and Agricultural Engineering, University of Zimbabwe, Harare, 270 pp

Nyagumbo I (2008) A review of experiences and developments towards conservation agriculture and related systems in Zimbabwe. In: Goddard T, Zoebisch MA, Gan YT, Ellis W, Watson A, Sombatpanit S (eds) No-till farming systems. Special publication No. 3. World Association of Soil and Water Conservation, Bangkok, pp 345–372

Nyamangara J, Nyagumbo I (2010) Interactive effects of selected nutrient resources and tied-ridging on plant growth performance in a semi-arid smallholder farming environment in Central Zimbabwe. Nutr Cycling Agroecosyst 88(1):103–109

Nyamudeza P, Nyakatawa EZ (1995) The effect of sowing crops in furrows of tied ridges on soil water & crop yield in Natural Region V of Zimbabwe. In: Twomlow SJ, Ellis-Jones J, Hagmann J, Loos H(eds) Proceedings of a technical workshop, soil & water conservation for smallholder farmers in semi-arid Zimbabwe-transfers between research and extension. Integrated Rural Development Programme, Belmont Press, Masvingo, Silsoe Research Institute, Report OD/95/16, Masvingo, Zimbabwe, pp 32–40, 3–7 Apr 1995

Nzuma JK (2002) Manure management options for increasing crop production in the smallholder sector of Zimbabwe. DPhil thesis, Department of Soil Science and Agricultural Engineering, University of Zimbabwe, Harare, 173 pp

Nzuma JK, Murwira HK (April, 2000) Improving the management of manure in Zimbabwe. Manage Africa's Soil 15:1–20, Apr 2000

Piha MI (1993) Optimizing fertilizer use and practical rainfall capture in a semi-arid environment with variable rainfall. Exp Agric 29:405–415

Rockström J, Barron J, Fox P (2001) Rainwater management for increased productivity among small-holder farmers in drought prone environments. In: 2nd WARFSA/Waternet Symposium: Integrated Water Resources Management: Theory, Practice, Cases, Cape Town, pp 319–330, 30–31 Oct 2001

Shamudzarira Z (1994) Maize water use as affected by nitrogen fertilization. In: Craswell ET, Simpson J (eds) Proceedings of ACIAR/SACCAR workshop: soil fertility and climatic constraints in Dryland agriculture, 30 August to 1 September 1993. ACIAR Canberra Australia, Harare, Zimbabwe, pp 15–18

Stoorvogel JJ, Smaling EMA (1990) Assessment of soil nutrient depletion in sub Saharan Africa: 1983–2000. Rep. No. 28.

DLO Winand Staring Ctr. Integrated Land, Soil and Water Research, Wageningen, Netherlands, 258 pp

Twomlow SJ, Breneau PMC (2000) The influence of tillage on semi-arid soil-water regimes in Zimbabwe. Geoderma 95:33–51

van Duivenbooden MP, Studer C, Bielders CL (eds) (1999) Efficient soil water use: the key to sustainable crop production in the dry areas of West Asia, and North and Sub-Saharan Africa. In: Proceedings of the 1998 (Niger) and 1999 (Jordan) workshops of the Optimizing Soil Water Use (OSWU) consortium. ICARDA and ICRISAT, Aleppo, Syria Patancheru, India, 496 pp

Vincent V, Thomas RG (1960) An agricultural survey of Southern Rhodesia: part I: the agro-ecological survey. Government Printer, Salisbury, 76 pp

Zougmore R, Guillobez S, Kambou NF, Son G (2000) Runoff and sorghum performances affected by the spacing of stones lines in the semi arid Sahelian zone. Soil Tillage Res 56: 175–183

Nutr Cycl Agroecosyst (2010) 88:91–101
DOI 10.1007/s10705-009-9325-0

ORIGINAL ARTICLE

Population dynamics of mixed indigenous legume fallows and influence on subsequent maize following mineral P application in smallholder farming systems of Zimbabwe

T.P. Tauro · H. Nezomba ·
F. Mtambanengwe · P. Mapfumo

Received: 25 April 2008 / Accepted: 8 October 2009 / Published online: 27 October 2009
© Springer Science+Business Media B.V. 2009

Abstract Developing soil fertility management options for increasing productivity of staple food crops is a challenge in most parts of Sub-Saharan Africa, where soils are constrained by nitrogen (N) and phosphorus (P) deficiencies. A study was conducted to evaluate the response of indigenous legume populations to mineral P application, and subsequently their benefits to maize yield. Mineral P was applied at 26 kg P ha^{-1} before legume species were sown in mixtures at 120 seeds m^{-2} species^{-1} and left to grow over two rainy seasons (2 years). Application of P increased overall biomass productivity by 20–60% within 6 months, significantly influencing the composition of non-leguminous species. Dinitrogen fixation, as determined by the N-difference method, was increased by 43–140% although legume biomass productivity was apparently limited by nutrients other than P and N. *Crotalaria pallida* and *C. ochroleuca* accounted for most of the fixed N. Improved N supply increased the abundance of non-leguminous species, particularly *Conyza sumatrensis* and *Ageratum conyzoides*. However, abundance of common weed species, *Commelina benghalensis, Richardia scabra* and *Solanum aculeastrum,* declined by up to18%. Application of P did not significantly influence productivity of those legume species that reached maturity within 3 months. There was increased N$_2$-fixation and biomass productivity of indifallows as influenced by specific legume species responding to P application. Compared with natural (grass) fallows, indigenous legume fallows (indifallows) increased subsequent maize grain yields by $\sim$40%. Overall, 1- and 2-year indifallows gave maize grain yields of >2 and 3 t ha^{-1}, respectively, against <1 t ha^{-1} under corresponding natural fallows. Two-year indifallows with P notably increased maize yields, but the second year gave low yields regardless of P treatment. Because of their low P requirement, indigenous legume fallows have potential to stimulate maize productivity under some of the most nutrient depleted soils.

Keywords Indigenous legumes · Maize yield · N$_2$-fixation · Species composition · Species abundance · Nutrient depleted sandy soils

T. P. Tauro · H. Nezomba · F. Mtambanengwe ·
P. Mapfumo
Department of Soil Science and Agricultural Engineering,
University of Zimbabwe, P.O. Box MP167, Mount
Pleasant, Harare, Zimbabwe

T. P. Tauro
Department Research and Specialist Services (DR&SS),
Chemistry and Soil Research Institute, P.O. Box CY 550,
Causeway, Harare, Zimbabwe

P. Mapfumo (✉)
Soil Fertility Consortium for Southern Africa
(SOFECSA), CIMMYT, Southern Africa,
P.O. Box MP 163, Mount Pleasant, Harare, Zimbabwe
e-mail: p.mapfumo@cgiar.org

🖄 Springer

This article has been previously published in the journal "Nutrient Cycling in Agroecosystems" Volume 88 Issue 1.
A. Bationo et al. (eds.), Innovations as Key to the Green Revolution in Africa: Exploring the Scientific Facts. © 2010 Springer.

Introduction

Maize yields on most smallholder farms in Sub-Saharan Africa (SSA) have remained <1 t ha^{-1}, threatening household food security for over 70% of the rural populations primarily drawing their livelihoods from agriculture. Addressing food security challenges for smallholder farmers in SSA requires technical innovations that draw on both conventional and indigenous knowledge systems. Most locally available organic amendments used by smallholder farmers to improve soil fertility, such as cattle manure, crop residues, termitarium soil and leaf litter contain low concentrations of nitrogen (N) and phosphorus (P), and are often applied at rates far short of meeting requirements for most cereal food crops (Mapfumo and Giller 2001). Apart from mineral fertilizers, legumes have been known to make realistic N additions to cropping systems through biological nitrogen fixation (BNF). Increase in maize (*Zea mays L.*) yields by twofolds from 1.6 to 2.27 t ha^{-1} and by fourfolds from 1.2 to 5.59 t ha^{-1} have been reported following inclusion of legumes in cropping systems compared with unfertilized maize (Kwesiga and Coe 1994). However, efforts to improve the N economy of nutrient-depleted soils in SSA using legumes have often failed partly due to poor stand establishment (Chikowo et al. 2004) and low biomass productivity (Mapfumo et al. 1999). Most legumes used for soil fertility improvement have failed to produce >2 t ha^{-1} of biomass, resulting in insignificant N contributions to cropping systems.

Studies by Mapfumo et al. (2005) have revealed scope for manipulating indigenous legume species as an initial step for addressing soil fertility problems in maize based cropping systems. The indigenous legumes were found to grow on soils abandoned by farmers due to frequent failure of crops such as maize. The abandoned soils were characterized by low soil organic carbon ($\sim$4 g C kg^{-1} soil), P deficiency (<10 mg kg^{-1} soil) and low clay (<100 g kg^{-1} soil) (Mapfumo et al. 2005). Establishment of legumes is affected by a number of nutritional factors, which reduce the potential benefits associated with legume BNF in cropping systems (Giller 2001). Legume-rhizobium symbiosis is often limited by molybdenum (Mo) deficiency in soils, exacerbated by sulphur (S) and zinc (Zn) deficiencies that are associated with poor root development. Calcium is known to promote root infection and nodule formation, while cobalt is required by N$_2$-fixing bacteria for electron transport. However, P deficiency is commonly the most limiting nutrient to legume biomass productivity and N$_2$-fixation (Giller 2001). For example, application of 8 kg P ha^{-1} + 4 kg S ha^{-1} was shown to improve biomass of *Mucuna pruriens*, *Cajanus cajan* and *Tephrosia vogelii* in Malawi (Malwanda et al. 2002). Bationo et al. (1990) showed that cereal crops such as millet would not respond to N fertilizer application until P demands were met. In most tropical soils, there is limited scope to supply P through organic nutrient inputs. Phosphorus nutrition is therefore primarily improved through inorganic P fertilizer application.

Application of single super phosphate (SSP) to sunnhemp (*Crotalaria juncea*) and velvet bean (*M. pruriens)* increased subsequent maize grain yield by 150% compared to direct application of SSP to the maize crop (Muza 2002). Furthermore, it was shown that maize could not fully utilize the applied P, but would benefit from the increase in biomass of legumes following P application. Despite emerging evidence on the potential of the indigenous legume fallows (indifallows) in improving the N economy in maize-based cropping systems (Mapfumo et al. 2005), there is limited quantitative information on the influence of P application on plant population dynamics, N$_2$-fixation and biomass productivity. This study investigated the population dynamics and P response of indigenous legumes as an alternative nutrient resource in smallholder farming systems of Zimbabwe. Specific study objectives were to determine the influence of mineral P application on: (a) establishment and species composition in indifallows; (b) N$_2$-fixation of indigenous legumes; and (c) yields of maize following 1- and 2-year indifallows.

Materials and methods

Study site

The study was conducted over three rainfall seasons, from 2005/2006 to 2007/2008, under smallholder farming conditions in Chinyika Resettlement Area (18°13'S, 32°22'E) and Chikwaka Communal Area (17°44'S, 31°29'E). Parallel studies were also conducted on-station at Domboshawa Training Centre

Nutr Cycl Agroecosyst (2010) 88:91–101

(Domboshawa) (17°35′S, 31°14′E) and Makoholi Experimental Station (Makoholi) (19°47′S, 30°45′E). Domboshawa and Chikwaka, which are 30 and 80 km northeast of Harare respectively, are in agro-ecological region (NR) II of Zimbabwe which receives over 800 mm of rainfall annually. Chinyika is 250 km east of Harare, in NR III which receives rainfall of 650–750 mm year^{-1}. Makoholi is about 300 km south of Harare, under NR IV receiving between 450–650 mm year^{-1}. Rainfall at all sites is unimodal and is received between November and March while the soils are granite-derived sands to loamy sands, classified as Arenosols and Lixisols (World Reference Base 1998). Studies were conducted on fields abandoned by farmers due to poor soil fertility. Composite soil samples collected from the top 20 cm per field, were air dried and analysed for texture, pH, organic carbon (OC), bases, cation exchange capacity (CEC), available P (Olsen method) and plant available mineral N. The soils had <10 ppm of P at all sites except Domboshawa, and organic C ranged from 0.4 to 0.7% (Table 1).

Experimental treatments

Indigenous legume fallow (indifallow) and natural grass fallow (natural fallow) treatments were established during the 2005/2006 rainy season. The following treatments were imposed in plots measuring 6 m by 4.5 m:

(1) Indifallow
(2) Indifallow + single super phosphate (SSP)
(3) Natural fallow
(4) Natural fallow + SSP

Single super phosphate (SSP) was broadcast and incorporated to 15–20 cm depth at 26 kg P ha^{-1} following recommendations for promoting P build up in light-textured soils in Zimbabwe. A randomized complete block design (RCBD) with three replicates per treatment was used. The indifallows were established by broadcasting mixtures of *Crotalaria laburnifolia* (L.), *C. ochroleuca* G. Don, *C. pallida* (L.), *C. pisicarpa* Welw.ex Baker, *C. cylindrostachys* Welw.ex Baker, *Eriosema ellipticum* Welw.ex Baker, *Neonotonia wightii* (Wight & Arn.) J.A. Lackey, *Chamaecrista mimosoides* (L.), *C. absus* (L.), *Indigofera astragalina* DC and *I. arrecta* Hochst. Ex A. Rich, at a seeding rate of 120 seeds m^{-2} species^{-1}. The seed used was provided by the farming communities following earlier interventions by Mapfumo et al. (2005). The natural fallows were established by leaving the selected plots to natural vegetation (mainly grass) following land preparation.

Measurement of plant biomass and abundance

Each plot was divided into several quadrats measuring 0.5 m × 0.5 m. Three quadrats were then randomly selected for measuring plant biomass through destructive sampling. The biomass sampling was done at 3, 6, 15 and 20 months, spreading over two rainy seasons. All standing plants in the quadrats were cut just above the soil surface, and separated into individual species which were then oven-dried to constant weight at 60°C. The biomass yield for each species was then determined and calculated per hectare. Relative species abundance was estimated as:

Table 1 Chemical characteristics of soils at different study sites where indigenous legume species were established in Zimbabwe

Site	Organic C (%)	Total N (%)	Available P (mg kg^{-1})*	pH (CaCl$_2$)	Mineral N (mg kg^{-1})**	Ca (cmol$_{(c)}$ kg^{-1})	Mg (cmol$_{(c)}$ kg^{-1})	K (cmol$_{(c)}$ kg^{-1})
Domboshawa	0.7 (0.08)	0.06 (0.003)	15 (0.3)	4.8 (0.15)	35 (1.2)	0.8 (0.06)	0.4 (0.01)	0.1 (0.005)
Makoholi	0.4 (0.02)	0.03 (0.005)	3 (0.5)	4.6 (0.13)	17 (0.5)	0.3 (0.02)	0.3 (0.02)	0.1 (0.004)
Chikwaka	0.4 (0.06)	0.04 (0.006)	4 (0.1)	4.5 (0.11)	22 (1)	0.4 (0.01)	0.5 (0.01)	0.2 (0.001)
Chinyika 1	0.5 (0.05)	0.05 (0.005)	5 (0.4)	4.1 (0.05)	19 (0.8)	0.7 (0.01)	0.5 (0.02)	0.2 (0.005)
Chinyika 2	0.6 (0.04)	0.06 (0.002)	9 (0.2)	4.5 (0.12)	28 (1.3)	1.0 (0.02)	0.8 (0.04)	0.3 (0.004)
Chinyika 3	0.4 (0.05)	0.04 (0.007)	4 (0.6)	4.8 (0.13)	18 (1.1)	0.6 (0.03)	0.7 (0.03)	0.1 (0.003)

Chikwaka = Nyamayaro farm; Chinyika 1 = Mudange farm; Chinyika 2 = Chikodzo farm; Chinyika 3 = Masara farm

* Olsen P method

** Mineralizable N after 2 weeks of anaerobic incubation; figures in parentheses denote standard errors

Species abundance $(\%)$

$$= \frac{\text{Species biomass}\left(\text{kg ha}^{-1}\right)}{\text{Total system biomass productivity}\left(\text{kg ha}^{-1}\right)} \times 100$$

The dried legume and non-legume plant biomass were separately ground in a Wiley Mill to pass through a 1-mm sieve before analysis for tissue P concentration. The P was determined through the modified micro-Kjeldhal procedure (Anderson and Ingram 1993) and measured colorimetrically at 880 nm. Phosphorus uptake by legume and non-legume species was obtained by multiplying the tissue P concentration by the corresponding biomass.

Estimation of N_2-fixation

The N-difference (ND) method (Peoples et al. 1995) was used for estimating N_2-fixation by indigenous legumes at 3 and 6 months after seeding. Non-leguminous plants exhibiting corresponding growth habits under natural fallow plots were used as reference crops for the different legume species (Nezomba et al. 2008). The ND method measures N_2-fixation by comparing N accumulated in N-fixing plant with that taken up by a non-fixing reference plant of corresponding growth habit. The difference between N accumulated by the fixing legume and a non-fixing reference crop is assumed to be the N derived from fixation. Some of the reference plants selected include *Ageratum conyzoides*, *Bidens pilosa*, *Leucus martinicensis* and *Richardia scabra*. Identification of unknown non-legume species was done with the assistance of botanists from the National Herbarium and Botanic Gardens under the Department Research and Specialist Services of the Zimbabwe Ministry of Agriculture. The dried legume and reference plant biomass samples were ground in a Willey Mill to pass through a 1-mm sieve, followed by total N analysis using the micro-Kjeldhal digestion method (Anderson and Ingram 1993). The legume and reference plant N uptake were obtained by multiplying the tissue N concentration by the corresponding biomass, and N_2-fixation calculated as follows:

Amount of N fixed $\left(\text{kg ha}^{-1}\right)$
$= \text{Legume N uptake} - \text{Reference plant N uptake}$

Determining maize yield under indifallows

The biomass (leaf litter and standing stems) under 1- and 2-year indifallows was incorporated manually by hand-hoeing at planting. Maize cultivar SC 513, an early maturing variety, was planted during the 2006/2007 rainfall season with the first effective rains in November. The maize was planted at a spacing of 0.9 m inter-row and 0.3 m within rows with two seeds per planting station which were thinned to one at 2 weeks after emergence (WAE). No basal fertilizer was applied at planting but all plots received 45 kg N ha^{-1}, being 50% of the recommended N rate in smallholder maize production (Cooper and Fenner 1981). The N fertilizer was applied in three splits: 30% at 2 WAE; 40% at 6 WAE and the remaining 30% at 9 WAE. The experimental plots were kept weed-free through hand-hoeing and the maize crop received a single dose of Dipterex (1% Thiodan) at 6 WAE to control maize stock borer (*Busseola fusca* Fuller). At physiological maturity, maize grain yield was measured at 12.5% moisture content, following determination of total biomass yield. A second maize crop was planted on same plots in the 2007/2008 season to test the residual effects of the indifallows on maize performance. Agronomic management for maize was kept the same as in the previous year.

Data analyses

Data were analyzed by using GENSTAT statistical package (GENSTAT 2005). Analysis of variance was used to separate treatment effects on legume species biomass, maize grain yields, and P uptake. All biomass is reported on a dry matter basis. The Shannon-Wiener diversity index (H′) was used to test for changes in plant species diversity, with comparisons of system diversity done at a critical t value of 1.96 and 774 degrees of freedom. All mean comparisons were considered at $P < 0.05$ significance.

$$\% \text{ N from } N_2 \text{ - fixation} = \frac{\text{Total legume N uptake} - \text{Total reference plant N uptake}}{\text{Total legume N uptake}} \times 100$$

Nutr Cycl Agroecosyst (2010) 88:91–101

Results

Effect of P fertilization on species composition in indifallows

At Domboshawa, application of P increased the abundance of the grasses *Eleusine indica* and *Cynodon dactylon* in indifallow by 10 and 4%, respectively. On the other hand, there was decline in the abundance

of *Commelina benghalensis, Richardia scabra,* and *Solanum aculeastrum* by 8, 6 and 18%, respectively (Fig. 1). Application of P increased abundance of *Crotalaria pallida* by 20 and 30% at Domboshawa and Makoholi respectively (Fig. 2). The application of P significantly increased species diversity under natural fallow and this was particularly evident 6 months after establishment. The P added plots for both indifallows and natural fallows had the highest

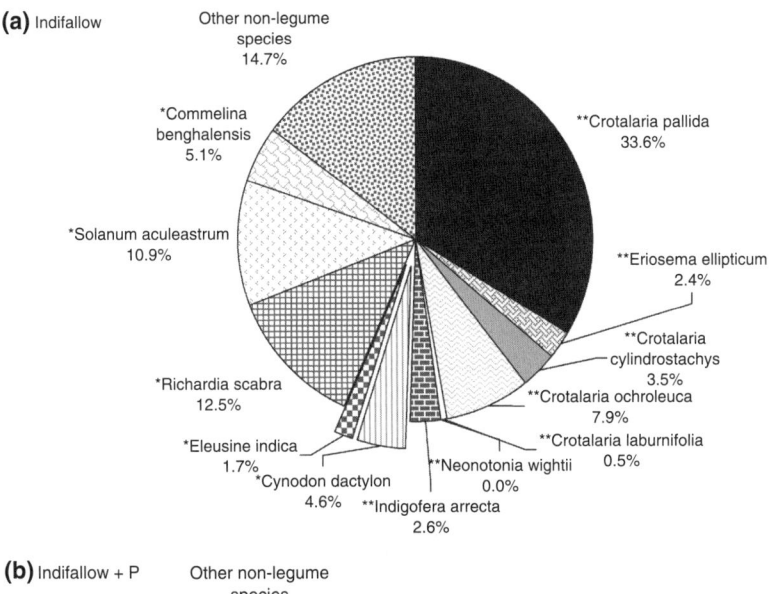

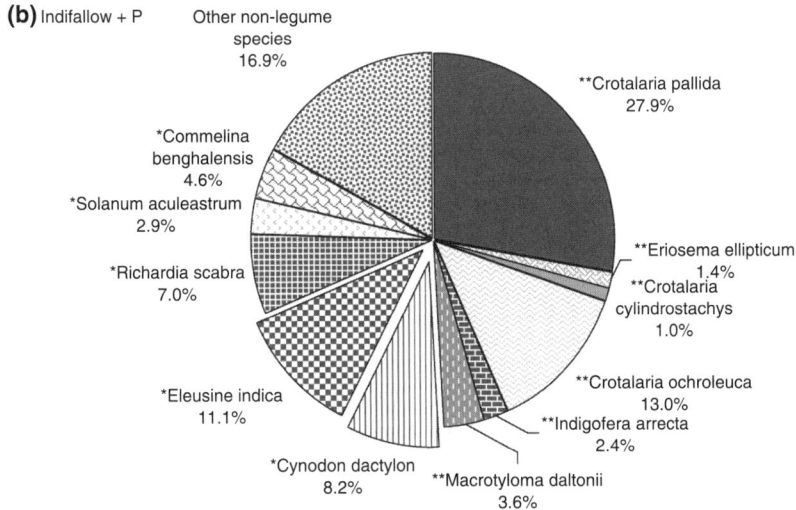

Fig. 1 The relative abundance of indigenous legume and non-legume species at 3 months after establishment under indigenous legume fallow (indifallows) without (**a**) and with (**b**) P at Domboshawa in Zimbabwe (*non-legume species, **indigenous legume species)

 Springer

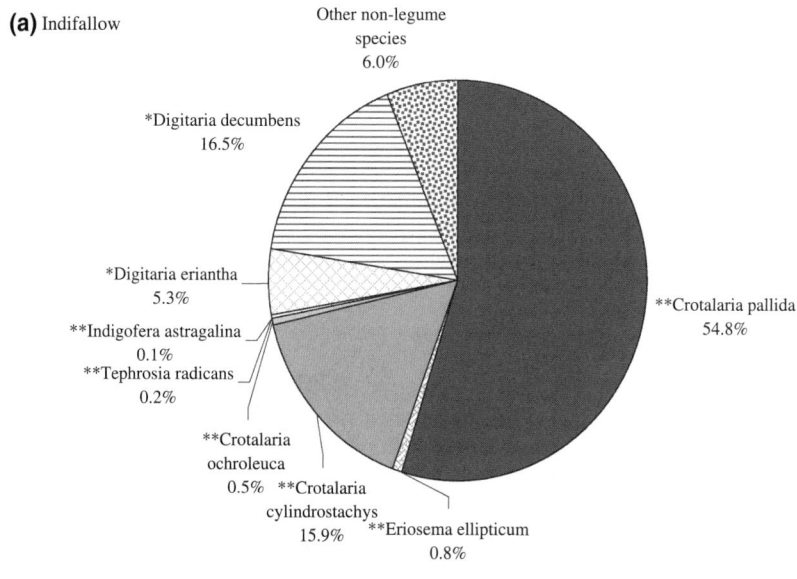

Fig. 2 The relative abundance of indigenous legume and non-legume species at 6 months after establishment under indigenous legume fallows (indifallows) with (**a**) and without (**b**) P application at Makoholi in Zimbabwe (*non-legume species, **indigenous legume species)

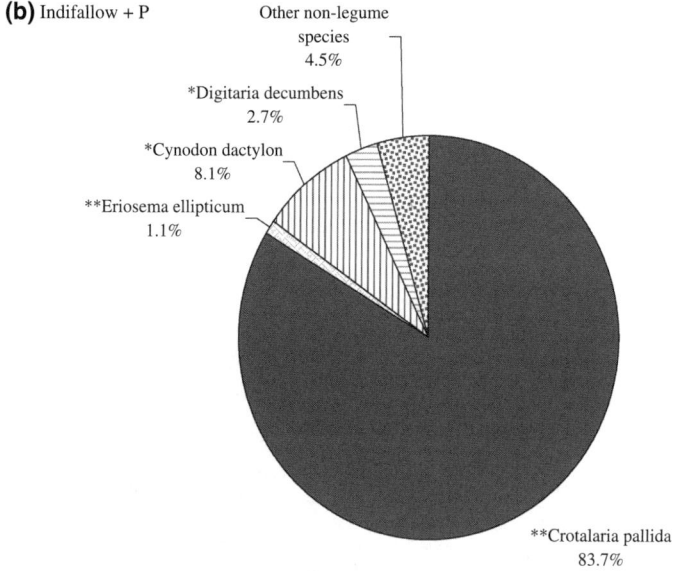

non-legumes species diversity and evenness. At Chikwaka, application of P increased the abundance of non-legume species under indifallow and apparently stimulated the appearance of new species such as *Conyza sumatrensis* and *Ageratum conyzoides*. The non-legume species reached peak biomass at 3 months after sowing, compared with 6 months for the legumes. During the second season, non-legume species out-competed indigenous legume species, most likely due to increased availability of N from the legumes during the preceding season (Fig. 3). The abundance of *Cyperus esculentus* and *R. scabra* in the indifallow was reduced during the second season, possibly due to competition from increased abundance of other N responsive species such as *C. dactylon*.

Nutr Cycl Agroecosyst (2010) 88:91–101

Fig. 3 Changes in total biomass of leguminous and non-leguminous species under P fertilized indigenous legume fallows (indifallow) over two seasons at Domboshawa, Zimbabwe

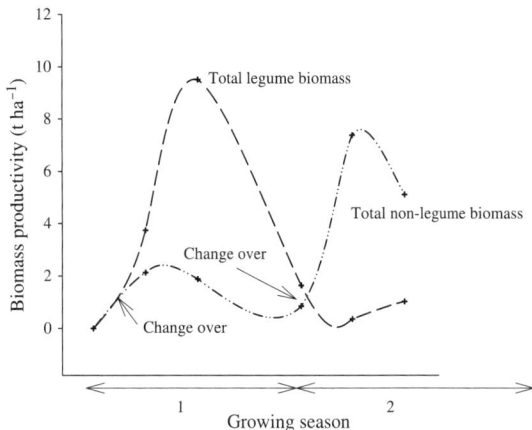

Influence of P application on indifallow biomass productivity

Overall, indifallow biomass productivity (legume and non-legume) increased with the addition of P at all sites, and this was significantly evident 6 months after sowing (data not shown). Phosphorus application significantly increased total biomass productivity under natural fallow as measured after 3 months of establishment, while there was no significant effect on total legume biomass under indifallows over the same period. Only three legume species (*Crotalaria cylindrostachys, C. glauca* and *C. pisicarpa*) had reached peak biomass after 3 months, but their contribution to overall productivity of the indifallow was not significant. Of the three legume species which had peak biomass at 3 month, only *Crotalaria cylindrostachys* responded to P application from 0.2 to 0.4 t ha^{-1}. At 6 months after sowing, indifallow biomass productivity had increased by between 20

and 60% in response to P application, with *Crotalaria pallida* accounting for most of the observed yield (Table 2).

Shoot P concentration of the legumes ranged from 0.05% for *C. cylindrostachys* to 0.09% for *E. ellipticum* and *C. pallida*. In contrast, non-legume species had a mean shoot P concentration of 0.12%. Application of P did not significantly influence total P uptake in total indifallow biomass, apparently due to poor P response by the legumes. The P accumulated in indifallow ranged from 8 to 9 kg ha^{-1} at Domboshawa and 2–7 kg ha^{-1} at Makoholi (Fig. 4).

Effects of P application on N$_2$-fixation

Addition of P increased N$_2$-fixation by indigenous legume species at all sites except at Makoholi which had severely depleted soils (Table 3). *Crotalaria pallida* and *C. ochroleuca* contributed much of the N$_2$ fixed in the indifallows due to their relatively high

Table 2 Effect of phosphorus application on species biomass productivity (t ha^{-1}) under indigenous legume fallows (indifallows) at 6 months after sowing at two contrasting sites in Zimbabwe

Indigenous legume species	Makoholi (low rainfall; <650 mm year^{-1})			Domboshawa (high rainfall; >750 mm year^{-1})		
	Plus P	Minus P	SED	Plus P	Minus P	SED
Crotalaria pallida	7.8	7.6	2.10	9.2	8.9	2.70
Crotalaria ochroleuca	ND	ND	ND	0.3	4.0	1.88
Crotalaria cylindrostachys	0.5	0.9	0.10	0.7	0.2	0.53
Eriosema ellipticum	0.1	0.1	0.04	0.04	0.5	0.27
Indigofera astragalina	0.08	0.1	0.03	ND	ND	ND

ND not determined as species was not used during establishment of indifallow

Fig. 4 Total phosphorus uptake in indifallows and natural fallows after 6 months of establishment under (**a**) high rainfall conditions at Domboshawa (> 750 mm year^{-1}) and (**b**) low rainfall conditions at Makoholi(450–650 year^{-1}) in the 2005–2006 season. Bars represent least significant difference (LSD) at $P < 0.05$

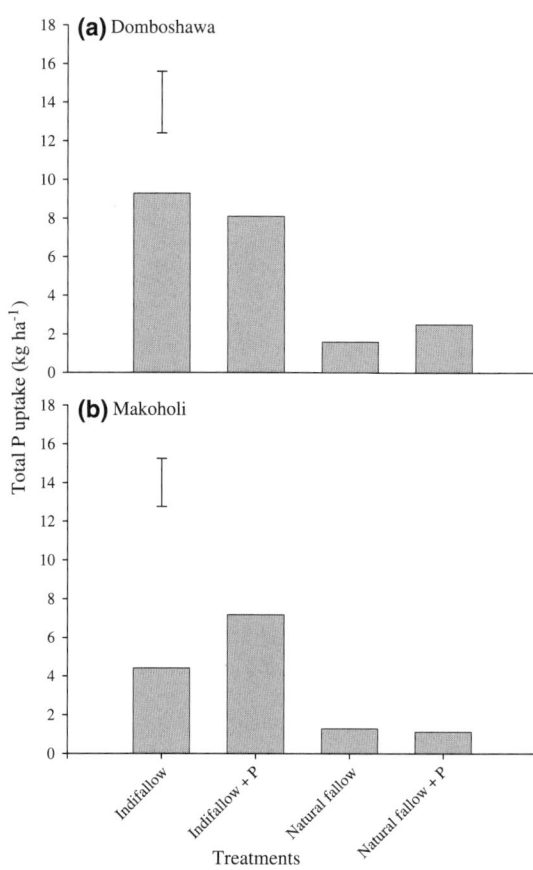

Table 3 Total amount of N$_2$-fixed (kg N ha^{-1}) (as determined by the difference method) under indigenous legumes fallows (indifallows) 3 months after establishment under different rainfall conditions in Zimbabwe

Site	Treatments		
	Indifallow	Indifallow + P	SED
Domboshawa (>750 mm year^{-1})	84	120	10
Makoholi (450–650 mm year^{-1})	91	63	17
Chikwaka (>750 mm year^{-1})**	6	13	3
Chinyika (650–750 mm year^{-1})*	15	26	6

** Nyamayaro farm

* Mudange farm

abundance. At Domboshawa, P addition increased the total amount of N$_2$ fixed by 43%, yielding up to 120 kg N ha^{-1}. On farmers' fields in Chikwaka and Chinyika, P application increased the amount of N$_2$ fixed by 141 and 81%, respectively (Table 3), but there was high variation in yields across plots.

Maize yields after indifallow termination

In the first maize cropping season (2006/2007 rainfall season), maize following 1-year indifallow produced the highest grain yield of ~2.2 t ha^{-1}. The yields were higher than maize after natural fallow which

 Springer

Table 4 Maize grain yield response following phosphorus (P) fertilized indigenous legume fallow (indifallow) and natural (grass) fallow systems at Domboshawa in Zimbabwe

Treatments	Maize yields (t ha^{-1})	
	2006/2007 rainfall season	2007/2008 rainfall season
1 Year indifallow	2.23[a]	0.97[a]
1 Year indifallow + P	1.93[a]	0.87[a]
1 Year natural fallow	1.21[b]	0.82[a]
1 Year natural fallow + P	1.39[ab]	0.87[a]
2 Year indifallow	NA	2.69[b]
2 Year indifallow + P	NA	3.22[b]
LSD	0.42	0.85

NA not applicable: the indifallows were left to grow for 2 years. Figures in the same column followed by the same letter are not significantly different

yielded ~1 t ha^{-1} with or without P applied (Table 4). Maize yields following indigenous legume fallows were about 40% more than those obtained after natural. While the first maize crop responded significantly to indifallow treatments, the yields for the second crop were <1 t ha^{-1} regardless of P treatment. However, the first maize crop following 2-year indifallow with added P produced the highest maize grain yield of 3.2 t ha^{-1}, 20% more than the corresponding treatment without P (Table 4).

Discussion

Species composition as affected by P application in indifallows

The increase in abundance of non-legume species such as *Cynodon dactylon* and *Eleusine indica* in response to P application can be attributed to the fibrous and dense rooting characteristics of grass species, which enabled their ready access to soil P (Cadisch et al. 1998). This could account for increased non-legume species diversity under both indifallow and natural fallow following P addition. However, the increase in non-legume species diversity under indifallow can also be partly due to increased availability of N as a result of contributions from N$_2$-fixation by the legumes. At Makoholi, lack of response by non-legume species to P application at a seemingly high rate of 26 kg P ha^{-1} suggested that nutrients other than P could have been limiting. The soils at Makoholi are highly prone to

nutrient leaching and exhibit multiple nutirent deficiencies (Mtambanengwe and Mapfumo 2006). Even though legumes have a higher P requirement than grasses (Cassman et al. 1993), they have a slow initial growth rate and therefore respond less rapidly to added nutrients than grasses.

Influence of P on biomass productivity under indifallow system

Indigenous legume species produced high biomass without P at Domboshawa and Chinyika where soil available P was between 9 and 15 mg kg^{-1}, suggesting that these P levels did not significantly limit their growth. This confirms earlier findings by Mapfumo et al. (2005) that indigenous legumes are adapted to P deficient soils from where maize could not yield >1 t ha^{-1} of grain. Furthermore, the particularly low P content (P < 1%) of these indigenous legume species and low total P uptake of indifallow systems confirm that these legumes utilize low amounts of P, and may not highly respond to P application. Among the species within the genera *Crotalaria*, only *C. pallida* and *C. ochroleuca* responded to P application at 6 months after establishment. Rao et al. (1999) reported that plants differ in their ability to utilize available forms of P in the soil and large differences exist among species even of the same genera. Slow growing biennial species such as *I. arrecta*, *N. wightii* and *E. ellipticum* did not respond to P application in contrast to the fast growing annuals *Crotalaria pallida* and *C. ochroleuca*.

Influence of P on N$_2$-fixation

The general increase in N$_2$-fixation under the indifallows following P application, as opposed to
biomass yield, suggests that P was a limiting nutrient
to N$_2$-fixation of indigenous legume species. According to Giller (2001), P application is a major
requirement for N$_2$-fixation in both cropping and
natural systems. The lack of response to N$_2$-fixation
at Makoholi following P application might be linked
to the inherent low soil fertility. Similar challenges
have been cited in efforts to enhance productivity of
soyabean and other grain legumes under severely
depleted soils in sub-Saharan Africa (Mapfumo et al.
1999; Snapp et al. 2002). N$_2$-fixation by legumes
under these soils has often been limited by nutrients
other than P (Grime and Curtis 1976). Working on
similar depleted soils, Zingore et al. (2007) reported a
yield of 44 kg N ha^{-1} for soyabean amended with
manure compared to 25 kg N ha^{-1} following P
application, and attributed this to additional nutrients
supplied in manure.

Influence of P application on subsequent
maize yields

The supply of N from indigenous legumes resulted in
high maize grain yields after indifallows, out-yielding
natural fallows which were predominantly comprised
of non-legume species. According to Nezomba et al.
(2008), biomass of these indigenous legume species
are of high quality and can contribute significantly to
the N demands of a subsequent maize crop in the
short-term. The enhancement of maize yields following P application under indifallows could be attributed to the improved N supply. Vanlauwe et al.
(2000) also observed increased yields of subsequent
maize following P application to *Mucuna pruriens*
and *Dolichos lablab*. The significant decrease in
maize yield in the second cropping season following
a 1-year indifallow system might be attributed to a
rapid depletion of nutrients accumulated in the fallow
phase, particularly N. Kwesiga and Coe (1994) also
showed a similar decline in maize yield in the second
year of cropping following a tree-based improved
fallow system. The relatively high maize yield from
2-year indifallow following P application could have
been due to the additional biomass and N inputs
accumulated over two growing seasons.

Conclusions

Phosphorus application to indigenous legume fallows
influenced non-legume species composition but did
not affect establishment and composition of the
legume species. Different legume species of the same
genera responded differently to P application, exhibiting the advantage of sowing them in mixtures.
Application of P significantly increased N$_2$-fixation of
indifallows, which translated into higher grain yields
of subsequent maize than was attainable from natural
fallows. Application of P increased N$_2$-fixation by
indigenous legume species as well as overall productivity of the fallows. There was evidence to suggest
that the indigenous legumes had a relatively low P
requirement in terms of biomass productivity, and
their biomass yields were apparently limited by
nutrients other than P and N. The study demonstrated
the applicability of indigenous legume fallows in
improving maize productivity on otherwise abandoned soils on smallholder farms, but also suggest that
a P fertilisation strategy for legume-maize rotations
demands attention to challenges associated with
multiple nutrient deficiencies of these predominantly
mono-cropped soils.

Acknowledgments The study was funded by The Rockefeller
Foundation through Grant 2004 FS 105. We are grateful to the
support from Soil Fertility Consortium for Southern Africa
(SOFECSA), Department Research and Specialist Services
(DR&SS) of the Government of Zimbabwe's Ministry of
Agriculture, and participating farmers in Chikwaka and
Chinyika smallholder areas.

References

Anderson JM, Ingram JSI (1993) Tropical soil biology and
  fertility: a handbook of methods, 2nd edn. CAB Interna-
  tional, Wallingford, p 221
Bationo A, Chien SH, Henso J, Christianson CB, Mokwunye
  AU (1990) Agronomic evaluation of two unacidulated and
  partially acidulated phosphate rocks indigenous to Niger.
  Soil Sci Soc Am J 54:1772–1777
Cadisch G, de Oliveira OC, Cantarutti R, Carvalho E, Urquiaga
  S (1998) The role of legume quality in soil carbon
  dynamics in Savannah ecosystems. In: Bergström L,
  Kirchmann H (eds) Carbon and nutrient dynamics in
  natural and agricultural tropical ecosystems. CAB Inter-
  national, Wallingford, pp 51–53
Cassman KG, Singleton PW, Linquist BA (1993) Input/output
  analysis of the cumulative soybean response to phosphorus on an Ultisol. Field Crops Res 34:23–26

Chikowo R, Mapfumo P, Nyamugafata P, Giller KE (2004) Maize productivity and mineral N dynamics following different soil fertility management practices on sandy soil in Zimbabwe. Agric Ecosyst Environ 102:119–131. doi:10.1007/s10705-005-2651-y

Cooper GRC, Fenner RJ (1981) General fertilizer recommendations. Zim Agric J 78:123–128

GENSTAT (2005) GENSTAT edition 2 reference manual. Laws agricultural trust (rothamsted experimental station). Clarendon Press, Oxford Science Publications, UK

Giller KE (2001) Nitrogen fixation in tropical cropping system, 2nd edn. CAB International, Wallingford

Grime JP, Curtis AV (1976) The interaction of drought and mineral nutrient stress in calcareous grassland. J Eco 64:976–998

Kwesiga F, Coe R (1994) Potential of short rotation sesbania fallows in eastern Zambia. For Eco Manage 64:161–170

Malwanda AB, Mughogho SK, Sakala W, Saka AR (2002) The effect of phosphorus and sulphur and the yield of subsequent maize in Northern Malawi. In: Waddington SR (ed) Grain legumes and green manures for soil fertility in southern Africa: taking stock of progress. Proceedings of a conference held October 8–11, 2002, Leopard Rock Hotel, Vumba, Zimbabwe. Soil fert net and CIMMYT-Zimbabwe, Harare, pp 197–204

Mapfumo P, Giller KE (2001) Soil fertility management strategies and practices by smallholder farmers in semi-arid areas of Zimbabwe. P.O. Box 776, Bulawayo, Zimbabwe: international crops research institute for the semi-arid tropics (ICRISAT) with permission from the food and agriculture organization of the United Nations (FAO). Bulawayo, Zimbabwe, pp 60

Mapfumo P, Giller KE, Mpepereki S, Mafongoya PL (1999) Dinitrogen fixation by pigeonpea of different maturity types on granitic sandy soils in Zimbabwe. Symbiosis 27:305–318

Mapfumo P, Mtambanengwe F, Giller KE, Mpepereki S (2005) Tapping indigenous herbaceous legumes for soil fertility management by resource-poor farmer in Zimbabwe. Agric

Ecosyst Environ 109:221–233. doi:10.1016/j.agree.2005.03.015

Mtambanengwe F, Mapfumo P (2006) Effects of organic resource quality on soil profile N dynamics and maize yields on sandy soil in Zimbabwe. Plant Soil 281:173–191. doi:10.1007/s11104-005-4182-3

Muza L (2002) Green manuring in Zimbabwe from 1900 to 2002. In: Waddington SR (ed) Grain legumes and green manures for soil fertility in southern Africa: taking stock of progress. Proceedings of a conference held October 8–11, 2002, Leopard Rock Hotel, Vumba, Zimbabwe. Soil fert net and CIMMYT-Zimbabwe, Harare, pp103–112

Nezomba H, Tauro TP, Mtambanengwe F, Mapfumo P (2008) Nitrogen fixation and biomass productivity of indigenous legumes for fertility restoration of abandoned soils in smallholder farming systems. S Afr J Plant Soil 25:161–171

Peoples MB, Herridge DF, Ladha JK (1995) Biological nitrogen fixation: an efficient source of nitrogen for sustainable agricultural production. Plant Soil 174:3–28

Rao IM, Friessen DK, Hoorst WJ (1999) Opportunities for germpalsm selection to influence phosphorus acquisition from low phosphorus soils. Agrofor For 9(4):13–16

Snapp SS, Rohrbach DD, Simtowe F, Freeman HA (2002) Sustainable soil management options for Malawi: can smallholder farmers grow more legumes? Agric Ecosyst Environ 91:159–174. doi:10.1016/s0167-8809(01)00238-9

Vanlauwe B, Diels J, Sanginga N, Carsky RJ, Deckers J, Merkx R (2000) Utilization of rock phosphate by crops on representative topsequence in the northern Guinea Savanna zone of Nigeria: response by *Mucuna pruriens*, Lablab and maize. Soil Biol Biochem 32:2063–2077

World Reference Base for Soils, FAO/ISRIC/ISSS (1998) World soil resources report no. 84. Food and Agriculture Organization, Rome

Zingore S, Murwira HK, Delve RJ, Giller KE (2007) Variable grain legume yields, responses to phosphorus and rotational effects on maize across soil fertility gradients on African smallholder farms. Nutr Cycl Agroecosyst 101:296–305. doi:10.10007/s10705-007-9117-3

Formulating Crop Management Options for Africa's Drought-Prone Regions: Taking Account of Rainfall Risk Using Modeling

J. Dimes

Abstract Few smallholder farmers in Africa's extensive semi-arid regions use fertilizer and virtually none use recommended high levels of application. Essentially, Africa's farmers have ignored the formal fertilizer recommendations of national research and extension systems. Because of this, productivity gains from fertilizer use remain grossly under-exploited. The existing fertilizer recommendations are one clear example of an information constraint that has proven intractable, despite more than 15 years of farmer participatory research in Africa. Due largely to training, researchers are generally preoccupied with identifying and reporting only the best option – the near-maximum yield result. While such optima may be correct from an agro-climatic perspective, in drought-prone regions, the risk associated with seasonal rainfall variations can determine whether or not farmers are likely to adopt a new technology and in what form. Yet, almost no research and extension recommendations given to farmers in Africa include any estimates of the variability in technology response that can be expected due to climatic risk. ICRISAT and partners have been pursuing a range of improved crop management options for the semi-arid tropics through crop systems simulation and farmer participatory research. This chapter presents some examples of how the application of crop modeling can provide a cost-effective pathway to formulation of crop management options under variable rainfall conditions and for farmers with a range of resource constraints. It includes examples of fertilizer recommendations, crop cultivar selection, and residue management in semi-arid regions.

J. Dimes (✉)
International Crops Research Institute for the Semi Arid Tropics (ICRISAT), Bulawayo, Zimbabwe
e-mail: j.dimes@cgiar.org

Keywords Crop systems simulation · Farmer participatory research · Fertilizer recommendation · Technology adoption · Rainfall risk · Drought

Introduction

It has long been recognized that blanket recommendations for fertilizer use are inadequate – although progress on addressing this issue has been painfully and unnecessarily slow.

Malcolm Blackie, Malawi, 1994.

Few smallholder farmers in Africa's extensive semi-arid regions use fertilizer and almost none use recommended high levels of application (see Chapter "Micro-dosing as a Pathway to Africa's Green Revolution: Evidence from Broad-Scale On-Farm Trials" this volume). Essentially, Africa's farmers in these regions have ignored the formal fertilizer recommendations of national research and extension systems. Partly because of this, productivity gains from fertilizer use remain grossly under-exploited in Africa. Inappropriate fertilizer recommendations for resource-poor farmers are one clear example of an information constraint that has proven intractable, as highlighted in the Blackie statement above. Alarmingly, this statement is still largely applicable more than a decade on, particularly in regard to semi-arid cropping regions. The problem obviously is not only inappropriate blanket recommendations. A more fundamental issue is the methods and process used by researchers and extension agents to pursue such outcomes in the first place, and, despite more than a decade of participatory research initiatives, to persist with them in the second.

Part of the problem is that on-farm participatory experiments tend to yield highly variable season-

A. Bationo et al. (eds.), *Innovations as Key to the Green Revolution in Africa*,
DOI 10.1007/978-90-481-2543-2_79, © Springer Science+Business Media B.V. 2011

and management-specific results that are difficult to interpret and draw conclusions from, while on-station research tends to give atypical results that reflect high levels of management and soil fertility. Overriding these technical constraints, and largely as a consequence of training, researchers are generally preoccupied with identifying and reporting only the best option – the near-maximum yield or economically "optimal" result. In the process, the smallholder farmer's reality of having limited resources with respect to the technology input, as well as competitive demands for these resources, is overlooked, as is the fact that the highest marginal returns are at the lower input levels on the response curve.

In developed world agriculture, it has been shown that the two risk-related factors with greatest impact on adoption decision are risk aversion and relative riskiness of a technology (Abadi Ghadim, 2000, described in Marra et al., 2003). Hence, it is reasonable to assume that resource-poor, strongly risk-averse farmers in Africa will be most interested in technologies that have limited risk and offer the highest payoff to input of limited resources. They will also tend to prefer technologies and management practices that constitute incremental changes in current farming practices and be willing to accept incremental rather than optimal benefits in productivity because of lower risk (Ahmed et al., 1997). Research and extension recommendations provide little advice on how to manage the necessary trade-offs associated with technology investment choices (e.g., Dimes et al., 2003) and say even less about associated risks – be it climatic risk, market risk, pest risk, or information risk, all leading to uncertainty. And uncertainty itself is a major factor in adoption decision (Marra et al., 2003). Lastly, the wide variations in household resource status of smallholder farmers imply that extension recommendations need to offer a range of options rather than the traditional optimal solution, which, even though correct from an agro-climatic perspective, are realistically only affordable by the wealthiest of farmers (Rohrbach, 1998).

A significant risk to technology adoption faced by smallholder farmers in semi-arid regions of Africa is the unreliable rainfall patterns of inter-seasonal as well as intra-seasonal distributions. One question then is how can research and extension better formulate technology options for a wide spectrum of farmers in this environment that includes indicators of associated rainfall risk and yield uncertainties to allow farmers to make more informed decisions about technology adoption? This chapter describes the application of crop simulation modeling as a tool to assist in the formulation of such options. First, it describes the application of modeling to the case of fertilizer recommendations for dry regions, by quantifying and comparing the seasonal risk of recommended and small dose fertilizer technology (see Chapter "Microdosing as a Pathway to Africa's Green Revolution: Evidence from Broad-Scale On-Farm Trials" this volume) across agro-ecological regions in Zimbabwe. It will extend this analysis to one of the most successful examples of technology adoption known in Africa, that of improved crop germplasm. Lastly, it will consider the issue of residue management central to the conservation agriculture (CA) concept currently been widely promoted in parts of Africa.

Materials and Methods

The Model

ICRISAT's applied simulation work in southern Africa uses the *Agricultural Production Systems Simulator* (APSIM). APSIM is a modeling environment that uses various component modules to simulate cropping systems (McCown et al., 1996; Keating et al., 2002). Modules can be biological, environmental, managerial, or economic. The modules are not directly linked with each other and can therefore be plugged in or pulled out of the modeled scenario depending on the specifications for the simulation task.

APSIM has the ability to simulate the growth of a range of crops (Table 1) in response to a variety of management practices, crop mixtures, and rotation sequences, including pastures and livestock. Importantly, this is accomplished in such a way that the soil accrues the effects of the different agricultural practices such as cropping and particular crops, fallowing, residue management, and tillage. In this way, APSIM can simulate long-term trends in soil productivity due to fertility depletion and erosion. APSIM contains modules that permit the simulation of soil organic matter rundown, nutrient leaching, soil erosion, soil structural decline, acidification, and soil phosphorus. There is however no current capability to deal with effects of salinization, insects, diseases, or biodiversity loss.

Table 1 APSIM crop, soil, and management modules

APSIM crop modules	Maize, sorghum, millet[a], wheat, sugarcane, chickpea, mung bean, soybean, barley, groundnut, canola, cotton[b], faba bean, lupin, pigeon pea[a], mucuna[a], hemp, sunflower, lucerne, annual medic, trees, weeds[a]
APSIM soil and related modules	Soil N, soil P, soil wat, SWIM[c], solutes, residue, manure[a], erosion, soil pH[d]
APSIM management modules	Manager, fertilize, irrigate, accumulate, operations, canopy

[a]Developed in association with ICRISAT and CIMMYT
[b]By arrangement with CSIRO Cotton Research, Australia
[c]By arrangement with CSIRO Land and Water, Australia
[d]Developed in association with CSIRO Land and Water

The suitability of APSIM to simulate crop productivity in smallholder farming systems in semi-arid tropical Africa has been tested over several years and in a number of regions. Building on the work of Keating et al. (1991) in Kenya, the APSIM model has been tested and used to simulate surface runoff and erosion (Okwach et al., 1999), N fertilizer response (Shamudzarira and Robertson, 2002), manure and P responses (Carberry et al., 2002), water use efficiencies (Dimes and Malherbe, 2005), legume rotational effects (Ncube et al., 2007), crop–weed interactions (Dimes et al., 2003), and extrapolation of research findings to other sites (Rose and Adiku, 2001).

Model Inputs

Long-term daily climate data for Harare (1951–2000), Masvingo (1951–1998), Beitbridge (1952–1997), and Bulawayo (1951–1999) were used to simulate maize yields across the agro-ecological regions of Zimbabwe. The cropping season (November–April) mean annual rainfall for the four sites are Harare, 780 mm, Masvingo, 580 mm, Beitbridge, 300 mm, and Bulawayo, 550 mm. Twomlow et al. (2007, these proceedings) report the main features of the smallholder farming system in Zimbabwe.

The technology options simulated are maize response to alternative N fertilizer investments and long (sc601) and short (sc401) duration cultivars. The baseline simulation for farmer practice is no N inputs (all other nutrients are assumed non-limiting and there are no pest and disease constraints). The simulated N fertilizer inputs are 1 or 3 bag(s) ammonium nitrate (AN) fertilizer (17.5 and 52 kg N/ha) at 35 days after sowing. The three bags of top-dress fertilizer is the extension recommendation that broadly applies across the agro-ecological regions of Zimbabwe.

Maize response is simulated for a shallow sand (PAWC = 60 mm, 1 m rooting depth) of low fertility (OC% = 0.6) or a deep sand (PAWC = 120 mm, 1.7 m rooting depth) of medium fertility (OC% = 1.0). In the simulations, a maize crop is planted each year of the climate record when a planting rain occurs between November 20 and January 10. Seasons were simulated independently by re-initialization of water and N (PAW = 0, mineral N = 10 kg N/ha, OC% = 1.0 or 0.6%) on 19th November each year. Plant population was 2 plants m^{-2} (approx. farmer's population in SAT regions) when comparing cultivar response and the extension recommendation of 3.7 plants m^{-2} when comparing N response. Re-setting PAW to zero assumes that pre-sowing rainfall is largely lost via soil evaporation and/or weed growth. Re-setting OC% each year ensures simulated yield outputs are not confounded by effects of soil fertility decline. All simulations assume no weed competition. For all scenarios, maize residues are removed at crop harvest.

Where value cost ratios for N investment with maize is presented, the price of maize grain is Z$7500 per ton and the price of AN fertilizer is Z$800 per bag. These prices last applied in Zimbabwe in 2001 when the N:maize price ratio was 6.3. This ratio is similar to the current ratio of 6.7 applicable in Republic of South Africa at the time of reporting (August 2007).

Results and Discussion

Regional Responses to N Top-Dress Fertilizer

Figure 1a shows simulated maize yields for Harare and the three N fertilizer treatments – no applied fertilizer, the recommended 3 bags/ha (52 kg N/ha), and a

Fig. 1 Simulated maize yield (cultivar SC401) on a deep sand soil at (**a**) Harare, (**b**) Masvingo, and (**c**) Beitbridge, Zimbabwe, for climate records starting in 1952 and N inputs of 0 (farmer practice), 17 (1 bag AN/ha), and 52 kgN/ha (Recommended, 3 bags/ha)

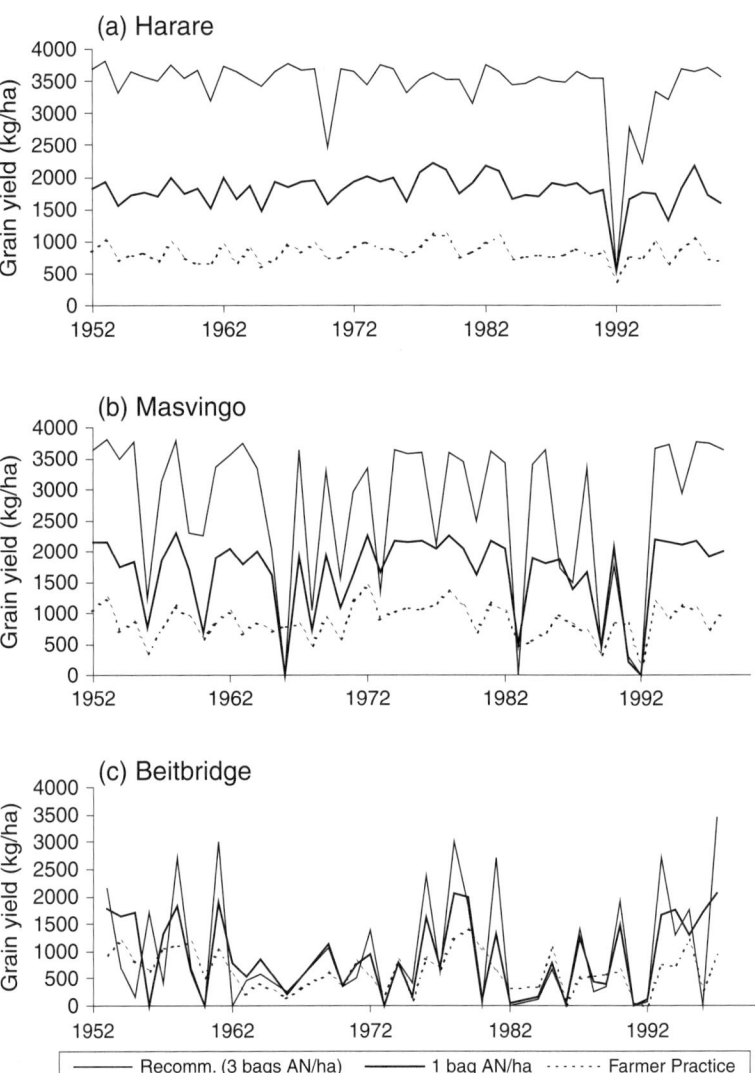

smaller investment of 1 bag/ha (17 kg N/ha). Simulated yields are very stable at Harare, except for the 1992 season. This result reflects the reliable rainfall in this region, and as a consequence, there is a consistent and clear response to the application of N fertilizer.

In Fig. 1b, c simulated maize yields for Masvingo and Beitbridge and the three N fertilizer treatments are shown. Simulated yields are highly variable at Masvingo, reflecting the variable rainfall in this region. For the recommended treatment, there are many years with good responses to N fertilizer but also years when there is no yield advantage. At lower N inputs, the response to N is more stable and mostly above that with no N fertilizer input.

In contrast, at Beitbridge, simulated yields are mostly low and highly variable, reflecting the extremely low and variable rainfall in this region. There are a few years with good responses to N fertilizer but most years there is no yield advantage to N application.

Figure 2 shows the Z$ return in maize grain production per Z$ invested in N fertilizer for simulated crops at Masvingo for 1 bag/ha (17 kg N/ha) compared with the recommended 3 bags/ha (52 kg N/ha). Returns on fertilizer investments are often high (>Z$7/Z$ invested) and at the lower level of investment, reasonably stable across the 45 years of simulation. Returns at the higher level are much more

Fig. 2 Z$ return in maize grain production per Z$ invested in N fertilizer for simulated crops at Masvingo for (**a**) 1 bag AN/ha (17 kg N/ha) and (**b**) the recommended 3 bags AN/ha (52 kg N/ha)

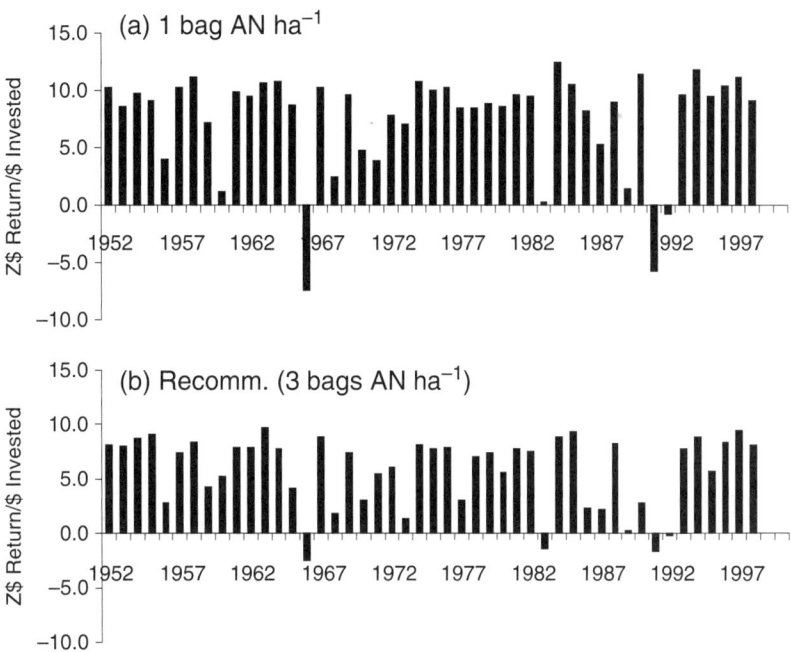

variable with many seasons having little or no return on investment.

Figure 3 graphs use the simulated returns on investment for each year and re-plot the data as cumulative probabilities of achieving a Z$ return per Z$ invested in fertilizer for Harare, Bulawayo, Masvingo, and Beitbridge. Cumulative probability plots quantify

the riskiness of different investment options. The above graphs indicate that a fertilizer investment of 1 bag/ha has only low probabilities of loss at Harare and Bulawayo, a slightly higher chance of loss at Masvingo, but very high loss probabilities at Beitbridge. A fertilizer investment of the recommended 3 bags/ha has low chance of loss at Harare,

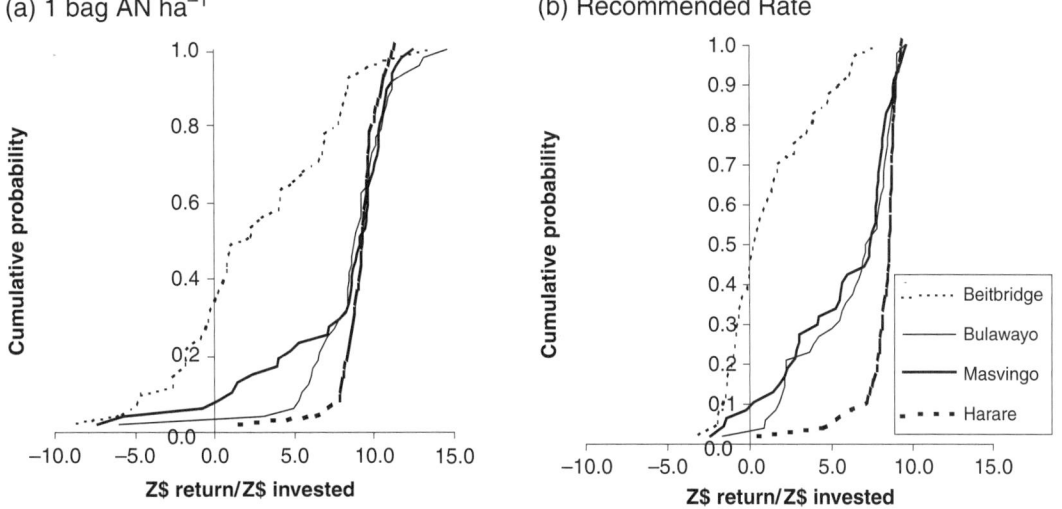

Fig. 3 Cumulative probability distributions for simulated returns on investment in fertilizer for Harare, Bulawayo, Masvingo, and Beitbridge – (**a**) 1 bag/ha (17 kg N/ha) and (**b**) recommended, 3 bags/ha (52 kg N/ha)

Table 2 The percentage of years that maize grain yield can be expected to attain various yield thresholds at Harare and Masvingo in response to N inputs

Yield (kg ha^{-1})	Harare			Masvingo		
	0N	Low N	Recomm. N	0N	Low N	Recomm. N[a]
<500	2	0	0	11	11	11
500–1000	85	2	2	51	6	0
1000–1500	13	4	0	38	4	9
1500–2000	0	79	0	0	40	6
>2000	0	15	98	0	38	74

[a]Recommended N

but is in the order of 10% of years at Bulawayo and Masvingo and 50% at Beitbridge.

The above results provide different approaches to using simulation output as a means to quantifying the climatic risk of fertilizer investments across rainfall gradients in Zimbabwe. However, they are not in a format readily understood by farmers or extension officers for that matter. Table 2 is an example of how the same data for two of the regions might be presented to smallholder farmers who are thinking about investing in top-dress fertilizer and who are restricted to 1 ha of cropland.

At Harare, a smallholder farmer who does not currently apply top-dress N and is interested in ensuring food security can learn from Table 2 that a small investment will shift the odds dramatically away from a food-insecure situation (<1000 kg) to one of food security with a higher chance of surplus grain than that of deficits. On the other hand, a farmer wanting to venture into the commercial grain market would see that one could make the necessary fertilizer investment in line with the extension recommendation with very little risk of crop failure.

At the drier Masvingo site, there are about 10% of years when drought will seriously limit crop yields irrespective of the N management (in the absence of any weed, pest, or disease constraints). However, a small investment in N will allow the farmers maize to make more efficient use of the rainfall in the majority of seasons, such that food deficits could reduce from over 60% of years to around 15%. For the more commercially orientated farmer at Masvingo, the recommended N rate should provide surplus grain for sale in about 75% of years. However, in approximately 20% of years, there will be insufficient grain for sale to re-coup the fertilizer investment, after allowing for household consumption. While some smallholder farmers in this region will have the resource status and

risk aversion profile to take up this option, the majority will not.

Cultivar Responses in a Semi-arid Rainfall Environment

Simulated maize yield for long- and short-season cultivars (representing traditional and improved, respectively) with no N inputs for shallow sand at Bulawayo is shown in Fig. 4. The output clearly shows that the short-duration cultivar provides fewer seasons of complete crop failure compared to the long duration (2 vs 8). This is consistent with the expected benefits and rationale of breeding programs targeting short-duration varieties for this environment. However, the simulated long-term average grain yield for both cultivars is low (long = 664 kg/ha, short = 680 kg/ha) and the year-to-year variability high, although substantially less for the short-season cultivar (stdev = 298 vs 436 kg ha^{-1})

In Fig. 5, results in Fig. 4 are converted into an annualized difference for the cultivar responses. The effect of applying N fertilizer is also included in Fig. 5. In Fig. 5, a positive value in any year represents the yield advantage in that season for the technology indicated above the x-axis, and a negative value represents the yield advantage of the alternative technology indicated below the x-axis.

With no N applied (Fig. 5a), the yield advantage of the short-season cultivar averages 300 kg/ha and is achieved in 48% of years. In comparison, the long-season type has an average yield advantage of 250 kg/ha and is achieved in 52% of years. If a small amount of N is applied (Fig. 5b), then there is a considerable shift in favor of the short-season cultivar – average yield advantage is 600 kg/ha, and an advantage is seen in 60% of years. But in 40% of years, the

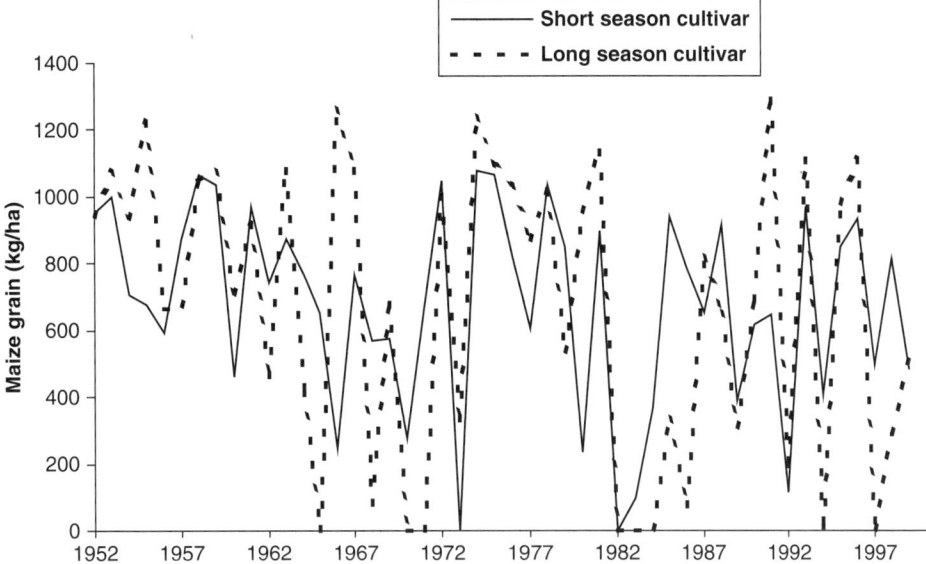

Fig. 4 Simulated maize grain yield for long- and short-season cultivars with no N inputs on shallow sand at Bulawayo for cropping seasons 1951–1998

long-season cultivar still outperforms the short-season cultivar with an average grain advantage of 390 kg/ha.

The cultivar analysis presented here shows that in these environments, rainfall distribution patterns can actually favor the long-season cultivar in a high proportion of seasons. This fact has largely been overlooked in breeding and extension programs for drier areas – which tend to concentrate on short-season varieties to avoid terminal moisture stress. The analysis also helped to highlight that water productivity increases in this environment only really come about with investment in fertility management (Figs. 4 and 5b).

Residue Management and Conservation Agriculture in Dry Areas

Conservation agriculture is promoted as a more sustainable approach to crop production with more efficient use of rainfall and protection of the soil resources. Currently, it is being widely promoted in smallholder agriculture in sub-Saharan Africa, including the semi-arid regions (Twomlow et al., 2006). One of the cornerstones of this technology is retention of crop residues as a surface mulch to reduce runoff and soil erosion. To this end, CA advocates a minimum of 30% ground cover and in Zimbabwe's maize cropping systems, it has been established that at least 2 t/ha of maize residues is required to comply with the 30% threshold. CA proponents generally acknowledge that in the mixed farming systems common in the semi-arid regions there will be competition for crop residues as a livestock feed. However, there is less recognition of the residue production potential of cropping in these environments and implications for achieving the desired ground cover threshold.

In Table 3, the stover yields associated with maize grain yields displayed in Fig. 1b, c have been analyzed to provide estimates of the percentage of years in which residue thresholds will be achieved with varying levels of N input. At Masvingo, model output suggests that 90% of years will produce sufficient stover to achieve the 2 t/ha threshold, even with no N input. With increasing N inputs, increasing amounts of residue could be fed to animals while retaining the desired mulch cover. However, only at the highest N input is sufficient excess residues produced to feed animals commensurate to the existing feeding regimes (i.e., 0N treatment, approx. 2 t/ha of crop residues) and then only in 65% of seasons. As suggested above, only the wealthiest of farmer in this environment will have the resources to pursue this level of N investment and associated climatic risk.

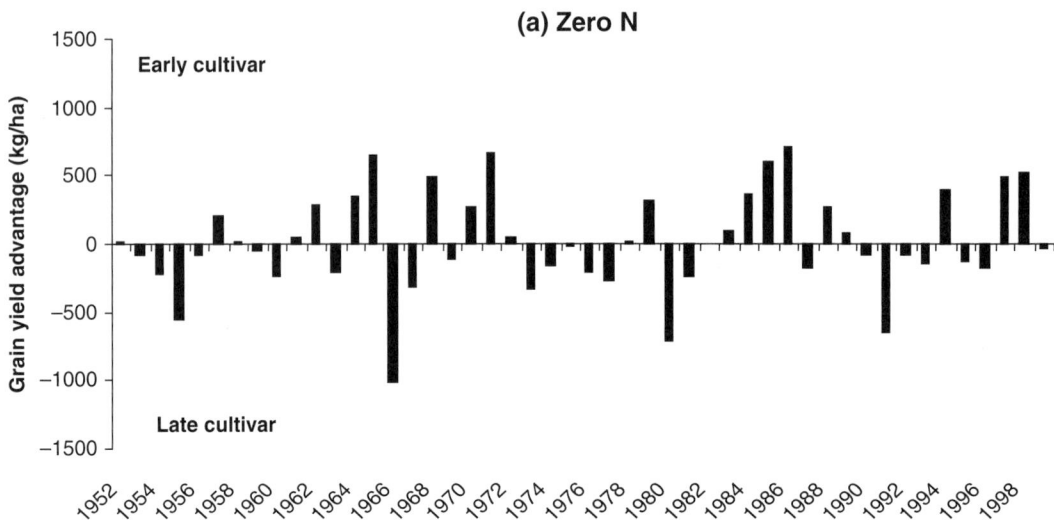

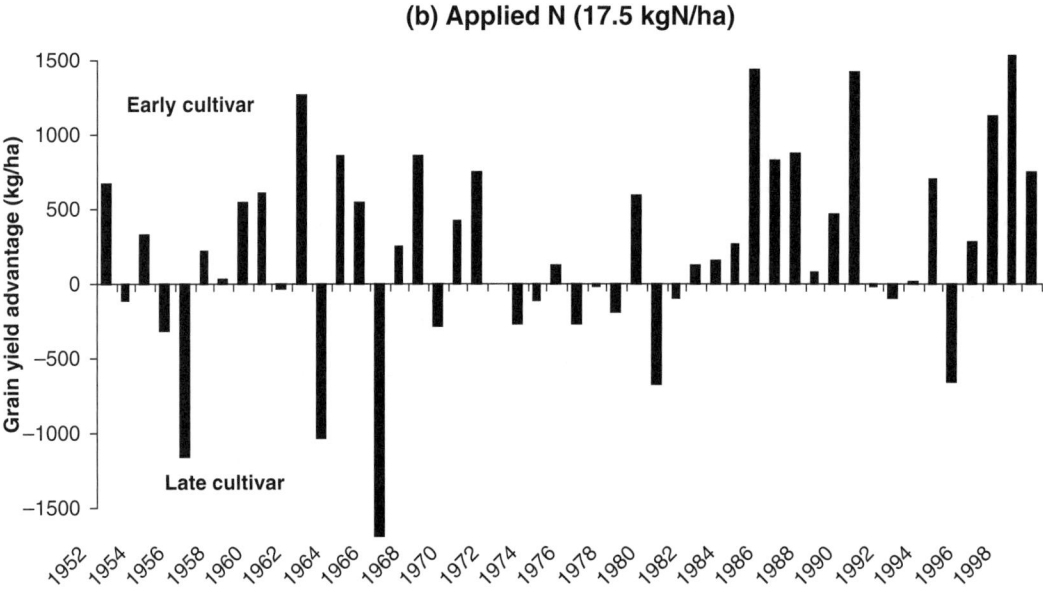

Fig. 5 Annual grain yield difference between short- and long-duration maize cultivars simulated for Bulawayo (**a**) without and (**b**) with N fertilizer applied

Table 3 The percentage of years that maize stover yield can be expected to attain various yield thresholds at Masvingo and Beitbridge in response to N inputs

	Masvingo			Beitbridge		
Stover (kg ha^{-1})	0N	Low N	Recomm. N	0N	Low N	Recomm. N[a]
<1000	2	6	8	15	17	23
1000–2000	8	2	0	60	33	25
2000–3000	90	33	2	25	48	29
3000–4000	0	58	25	0	2	23
>4000	0	0	65	0	0	0

[a]Recommended N

At Beitbridge, production of crop residues is much lower, and 75% of years do not reach the 2 t/ha threshold with 0N input. Even with N inputs, a deficit to the threshold will occur in approximately 50% of years. Of course, in this water-limited environment, it might be expected that the water conservation offered by a mulch would have a significant effect on maize yield and stover production and their responses to N inputs. This effect is not included in the simulation output used in this analysis.

Conclusions

The focus of this chapter has been crop improvement technologies for maize cropping systems under highly variable rainfall regimes. By definition resource-poor, smallholder farmers in such environments are strongly risk averse and seasonal rainfall variability will be a major risk factor in any technology adoption decision. Yet, almost no crop management recommendations given to farmers by research and extension in Africa include any estimates of the variation in technology response that can be expected due to climatic risk.

As an example, the area-specific fertilizer recommendations developed in Malawi and reported by Benson (1998) were undoubtedly a response to the Blackie statement of 1994. The new recommendations were based on over 1600 yield response trials to fertilizer inputs across all ecological cropping zones of Malawi. This research effectively resulted in the formulation of seven fertilizer recommendations to cover all areas of the country in place of the previous blanket recommendation. The new formulations are designed to take account of four farmer production objectives and two soil texture combinations. While this is a clear step in the right direction, nowhere do the new recommendations provide any information on expected yield variations due to seasonal rainfall conditions.

This chapter has hopefully demonstrated that crop simulation models provide a cost-effective pathway to assist formulation of crop management options that can take variable rainfall conditions into account. Only when research and extension are able to report both the positive and negative responses of a technology due to variable rainfall conditions will there be improved learning by both researchers and farmers. Such information is essential if risk-averse farmers are to be encouraged in their adoption of improved management technologies, especially in drier areas.

Of the 205 abstracts received for this symposium, only this chapter and one other included the word "risk" in its title. For a green revolution in sub-Saharan Africa to be realized, this suggests that there needs to be a dramatic turnaround by research and extension in its focus on climatic risk.

Acknowledgments The author gratefully acknowledges the funding support provided by ACIAR in conducting this research.

References

Abadi Ghadim AK (2000) Risk, uncertainty and learning in farmer adoption of a crop innovation. PhD thesis, University of Western Australia

Ahmed MM, Rohrbach DD, Gono LT, Mazhangara EP, Mugwira L, Masendeke DD, Alibaba S (1997) Soil fertility management in communal areas of Zimbabwe: current practices. constraints and opportunities for change. ICRISAT southern and Eastern Africa region working paper No. 6. PO Box 776. ICRISAT, Bulawayo, Zimbabwe

Benson T (1998) Developing flexible fertilizer recommendations for smallholder maize production in Malawi. In: Waddington SR, Murwira HK, Kumwenda JDT, Hikwa D, Tagwira F (eds) Soil fertility research for maize-based farming systems in Malawi and Zimbabwe. The soil fertility network for Maize based cropping systems in Malawi and Zimbabwe. CIMMYT, Zimbabwe, pp 37–244

Carberry PS, Probert ME, Dimes JP, Keating BA, McCown RL (2002) Role of modeling in improving nutrient efficiency in cropping systems. Plant Soil 245:193–303

Dimes JP, Malherbe J (2005) Climate variability and simulation modeling – challenges and opportunities. Proceedings of water and food project inception workshop – "increased food security and income in the Limpopo Basin – integrating crop water and soil fertility options and public-private partnerships", Polokwane, RSA, 25–27 Jan 2005

Dimes J, Muza L, Malunga G, Snapp S (2003) Trade-offs between investments in nitrogen and weeding: on-farm experimentation and simulation analysis in Malawi and Zimbabwe. In: Proceedings of the eighth Eastern and Southern African regional Maize conference, Nairobi, Kenya, 11–15 Feb 2002

Keating BA, Carberry PS, Hammer GL, Probert ME, Robertson MJ, Holzworth D, Huth NI, Hargreaves JNG, Meinke H, Hochman Z, McLean G, Verburg K, Snow V, Dimes JP, Silburn M, Wang E, Brown S, Bristow KL, Asseng S, Chapman S, McCown RL, Freebairn DM, Smith CJ (2002) The agricultural production systems simulator (APSIM): its history and current capability. Eur J Agron 18:267–288

Keating BA, Godwin DC, Watiki JM (1991) Optimising nitrogen inputs in response to climatic risk. In: Muchow RC, Bellamy JA(eds) Climatic risk in crop production: models and management in the semiarid tropics and subtropics. CAB International, Wallingford, CT, pp 329–358

Marra M, Pannell DJ, Abadi Ghadim A (2003) The economics of risk, uncertainty and learning in the adoption of new agricultural technologies: where are we on the learning curve? Agric Syst 75:215–234

McCown RL, Hammer GL, Hargreaves JNG, Holzworth DP, Freebairn DM (1996) APSIM: a novel software system for model development, model testing, and simulation in agricultural research. Agric Syst 50:255–271

Ncube B, Twomlow SJ, van Wijk MT, Dimes JP, Giller KE (2007) Farm characteristics and soil fertility management strategies across different years in smallholder farming systems under semi-arid environments of southwestern Zimbabwe. PhD. Thesis chapter in Ncube B. Understanding cropping systems in the semi-arid environments of Zimbabwe: options for soil fertility management. Wageningen University, The Netherlands

Okwach GE, Huth N, Simiyu CS (1999) Modeling surface runoff and soil erosion in semi-arid eastern Kenya. In: Okwach GE, Siambi MM (eds) Agricultural resource management for sustainable cropping in semi-arid Eastern Kenya. Proceedings of the first review workshop of the CARMASAK project, Machakos, Kenya, 27–29 May 1997, CARMASAK Proceedings No. 1: 104–140

Rohrbach D (1998) Developing more practical fertility management recommendations. In: Waddington SR, Murwira HK, Kumwenda JDT, Hikwa D, Tagwira F (eds) Soil fertility Research for maize-based farming systems in Malawi and Zimbabwe. The soil fertility network for Maize based cropping systems in Malawi and Zimbabwe. CIMMYT, Zimbabwe, pp 237–244

Rose CW, Adiku S (2001) Conceptual methodologies in agro-environmental systems. Soil Tillage Res 58:141–149

Shamudzarira Z, Robertson MJ (2002) Simulating response of maize to nitrogen fertiliser in semi-arid Zimbabwe. Exp Agric 38:79–96

Twomlow S, Rohrbach D, Hove L, Mupangwa W, Mashingaidze N, Moyo M, Chiroro C (2007) Conservation farming by basins breathes new life into smallholder farmers in Zimbabwe. In: Mapiki A, Nhira C (eds) Land and water management for sustainable agriculture. Proceedings of the EU/SADC land and water management applied research and training programmes inaugural scientific symposium, Malawi institute management, Lilongwe, Malawi, 14–16 Feb 2006. Paper 7.2

Twomlow SJ, Steyn JT, du Preez CC (2006) Dryland farming in southern Africa. In: Dryland agriculture. Agronomy monograph No. 23, 2nd edn. American Society of Agronomy. Madison, WI, pp 769–836

Nutr Cycl Agroecosyst (2010) 88:121–131
DOI 10.1007/s10705-008-9216-9

Residue quality and N fertilizer do not influence aggregate stabilization of C and N in two tropical soils with contrasting texture

R. Gentile · B. Vanlauwe · A. Kavoo ·
P. Chivenge · J. Six

Received: 5 March 2008 / Accepted: 23 September 2008 / Published online: 4 October 2008
© Springer Science+Business Media B.V. 2008

Abstract To address soil fertility decline, additions of organic resources and mineral fertilizers are often integrated in sub-Saharan African agroecosystems. Possible benefits to long-term C and N stabilization from this input management practice are, however, largely unknown. Our objectives were (1) to evaluate the effect of residue quality and mineral N on soil C and N stabilization, (2) to determine how input management and root growth interact to control this stabilization, and (3) to assess how these relationships vary with soil texture. We sampled two field trials in Kenya located at Embu, on a clayey soil, and at Machanga, on a loamy sand soil. The trials were initiated in 2002 with residue inputs of different quality (no input, high quality *Tithonia diversifolia*, medium quality *Calliandra calothyrsus*, and low quality *Zea mays* (maize) stover), incorporated at a rate of $4 \text{ Mg C ha}^{-1} \text{ year}^{-1}$ alone and in combination with $120 \text{ kg N ha}^{-1} \text{ season}^{-1}$ mineral fertilizer. Maize was grown in the plots each season, and a section of the plots was left uncropped. All aboveground maize residues were removed from the plots. Soil samples (0–15 cm) were collected in March 2005 to assess aggregation and C and N stabilization. The fine-textured soil at Embu was more responsive to inputs than the coarse-textured soil at Machanga. Residue additions increased macroaggregation at Embu, and cropping increased aggregation at Machanga. At Embu adding organic residue, regardless of the quality, and cropping significantly increased total soil C and N. This increase was also observed in the macroaggregate and microaggregate-within-macroaggregate fractions. Input treatments had little effect on C and N contents of the whole soil or specific fractions at Machanga. Nitrogen fertilizer additions did not significantly alter C or N content of the whole soil or specific fractions at either site. We conclude that residue quality does not affect the stabilization of soil organic C and N. Inputs of C and soil stabilization capacity are more important controls on stabilization of soil organic matter.

Keywords Aggregation · Fertilizer · Residue quality · Roots · Soil organic matter

R. Gentile (✉) · P. Chivenge · J. Six
Department of Plant Sciences, University of California,
One Shields Ave., Davis, CA 95616, USA
e-mail: rgentile@ucdavis.edu

B. Vanlauwe
TSBF-CIAT, c/o ICRAF, Gigiri, P.O. Box 30677-00100,
Nairobi, Kenya

A. Kavoo
Tropical Soil Biology and Fertility Institute of the International
Centre for Tropical Agriculture (TSBF-CIAT), Nairobi, Kenya,
agneskavoo@yahoo.com

Introduction

Soil fertility decline is often cited as the most important constraint to crop production in sub-Saharan Africa (SSA) (Sanchez and Jama 2002). As mineral and organic fertilizers are often limited in

This article has been previously published in the journal "Nutrient Cycling in Agroecosystems" Volume 88 Issue 1.

🌱 Springer

quantity and quality, soil fertility research in SSA has focused on developing integrated management strategies (Vanlauwe et al. 2002). Efficient short-term management of organic inputs is dictated by their quality, which influences the rate of decomposition and nutrient release. A synthesis of residue quality research on short-term decomposition and nutrient release led to the development of a Decision Support System (DSS) for organic residue management (Palm et al. 2001). This decision tree divides organic residues into four quality classes based on N, lignin and polyphenol contents, and recommends whether organic resources should be combined with mineral fertilizer or not. High quality residues (class I) have high N, low lignin and low polyphenol contents (>2.5% N; <15% lignin; <4% polyphenols). Medium quality residues have high N, high lignin and high polyphenol contents (>2.5% N; >15% lignin; >4% polyphenols) as class II, or low N and low lignin contents (<2.5% N; <15% lignin) as class III. Low quality residues (class IV) have low N and high lignin contents (<2.5% N; >15% lignin). The DSS advocates that medium quality residues be combined with mineral fertilizer to optimize residue decomposition and nutrient release rates, while high quality residues can be directly incorporated and low quality residues should be surface applied. The combined use of organic and mineral inputs recognizes possible benefits derived from a temporary immobilization of mineral N due to addition of a medium quality residue in improving the N synchrony and reducing N losses. When applied in combination, organic resources can also alleviate other constraints to crop production besides the nutrients added with the fertilizer and thus improve the use efficiency of the latter (Vanlauwe et al. 2002).

While there is much evidence that residue quality parameters such as N, lignin, and polyphenol contents control short-term C and N dynamics (e.g., Constantinides and Fownes 1994; Trinsoutrot et al. 2000; Vanlauwe et al. 2005), less is known about the medium- to long-term fate of C and N from different quality residues. Soil organic C and N can be stabilized through physical, chemical, and biochemical mechanisms (Six et al. 2002). Due to the recalcitrance of lignin and polyphenols, residues with high concentrations of these compounds may form an important component of soil C via biochemical stabilization. However, recent research characterizing

the chemical structure of soil organic matter has revealed a low stability of lignin in soils (Gleixner et al. 2002; Rasse et al. 2006).

As soil aggregates physically protect soil organic matter (Tisdall and Oades 1982) and influence microbial community structure (Hattori 1988), their structure and dynamics can play an important role in controlling C and N release and stabilization from inputs. Microaggregates formed within macroaggregates have been found to be the structures in which C is preferentially stabilized in both temperate and tropical soils (Six et al. 2000; Denef et al. 2007). The stabilization of C within microaggregates is controlled by the rate of macroaggregate turnover, which has to be slow enough to allow for the formation of microaggregates within macroaggregates, yet fast enough to allow for the incorporation of newly added residues (Six et al. 2000; Plante and McGill 2002). Since high quality residues and mineral N are postulated to increase the rate of macroaggregate turnover (Harris et al. 1963; Six et al. 2001), input quality may influence the physical stabilization of C and N within aggregates.

In addition to input management, root growth greatly affects soil aggregate formation and breakdown through a suite of physical, chemical and biological mechanisms (Degens 1997). The physical effects of roots on aggregation are evident through root penetration breaking apart soil aggregates (Materechera et al. 1994; Denef et al. 2002), localized drying and wetting cycles from root water uptake increasing aggregate formation (Reid and Goss 1981; Denef et al. 2002), and root enmeshment of soil particles as temporary aggregate binding agents (Tisdall and Oades 1982; Jastrow et al. 1998). In addition to these physical effects, root exudates can increase aggregate formation and stability (Morel et al. 1991), and stimulate microbial activity, which further increases aggregate formation. Lastly, roots are sources of C input and can promote aggregate formation as they decompose. Root-derived C may be more readily stabilized in soils than shoot-derived materials as roots provide a more intimate association with aggregates, contribute continuous inputs through exudates and fine root turnover, and generally have a lower residue quality (Gale et al. 2000; Puget and Drinkwater 2001). Root growth distributions are altered by environmental conditions and respond to nutrient inputs. Maize root growth is stimulated in

Nutr Cycl Agroecosyst (2010) 88:121–131

regions of high N availability (Durieux et al. 1994; Chassot et al. 2001). Therefore, root growth may interact with input management to control physical stabilization of organic matter.

The physical and chemical stabilization of C and N in soil aggregates will also be influenced by soil texture. The greater reactive surface area in finer textured soils will lead to a higher aggregation level and a decreased susceptibility to disruptive forces (Kemper and Koch 1966). Therefore, fine textured soils may predominantly form more aggregates with greater stability with organic matter additions leading to a greater protection of C and N than coarse textured soils. Kölbl and Kögel-Knabner (2004) found that soils with higher clay contents preserved more C in occluded organic matter fractions.

We conducted a study to examine the linkage between input management and root growth as they influence physical stabilization of soil organic matter. Our objectives were (1) to evaluate the effect of residue quality and mineral N on soil C and N stabilization, (2) to determine how input management and root growth interact to control this stabilization, and (3) to assess how these relationships vary with soil texture. We hypothesized that high quality residue and N fertilizer additions would decrease the stabilization of C and N. We expected root growth would increase C and N stabilization, but that root effects would be greater in high quality input treatments leading to an interaction of input management with cropping. Lastly, the low protection capacity of coarse textured soils was anticipated to result in low stabilization of C and N.

Materials and methods

Experiments were conducted at two sites in the central highlands of Kenya, representing different soil textures. Embu (0°30′ S, 37°27′ E) is characterized by a red clay Humic Nitisol, whereas Machanga (0°47′ S, 37°40′ E) is a loamy sand Ferric Alisol (FAO 1998). Soil characteristics for each site are summarized in Table 1. Embu has a mean annual temperature of 20°C and rainfall of 1,200 mm, while Machanga experiences a mean annual temperature of 26°C with 900 mm of precipitation. Rainfall occurs in two distinct seasons from March to June and October to December, which correspond with the two growing seasons per year.

The field trials were initiated at each site in March 2002 as long-term trials to evaluate the repeated application of different quality organic residues alone and in combination with N fertilizer on soil organic matter dynamics. The experimental design was a split-split plot with the organic residue input as the main plot, N fertilizer application as the subplot, and cropped versus bare soil as the sub-subplot. There were three replicate blocks at each site. The organic residue treatments consisted of a control plot with no inputs and inputs of high quality *Tithonia diversifolia*, medium quality *Calliandra calothyrsus*, and low quality *Zea mays* (maize) stover residues applied at rates of 4 Mg C ha^{-1} year^{-1}. Low quality (class IV) residues were not included in this study as they should be surface applied and thus tillage would be a confounding factor within the treatments. Nitrogen fertilizer was applied in the form of calcium ammonium nitrate at a rate of 120 kg N ha^{-1} season^{-1}. Maize was planted in the cropped plots each growing season while the bare plots were left unplanted. Subplot sizes were 6 × 5 m in Embu and 6 × 6 m in Machanga with areas of 3 × 2.5 m and 3 × 3 m, respectively, reserved for the bare plots.

At the onset of the March growing season of each year, the organic residues were analyzed for total C content (Anderson and Ingram 1993), which was used to determine the quantity of residue to apply. Organic residues were subsequently analyzed for total N, lignin, and soluble polyphenols (Anderson and Ingram 1993). The mean quality characteristics for the residues used during the first 3 years of the study are presented in Table 2. The organic residues were broadcasted and incorporated by hand hoe to 15 cm. All plots received a basal fertilizer application of 60 kg P ha^{-1} and 60 kg K ha^{-1} each growing season.

Soil samples (0–15 cm) were collected 3 years after trial establishment in March 2005 to assess treatment influence on aggregation and C and N stabilization. Five soil cores (15-cm diameter) per plot were taken at a depth of 0–15 cm. The cores for each plot were combined, sieved through an 8-mm sieve and air-dried. Soil samples were separated into three aggregate size fractions by wet sieving according to a modified method of Elliott (1986). Two sieves were used to separate soil into macroaggregates (>250 μm), microaggregates (53–250 μm), and silt and clay particles (<53 μm). A 80-g subsample

Table 1 Initial soil characteristics (0–15 cm) at Embu and Machanga

Soil property	Embu	Machanga
Organic C (g kg^{-1})	29.32 (0.51)[a]	5.27 (0.58)
Total N (g kg^{-1})	2.77 (0.06)	0.60 (0.07)
Extractable P (mg P kg^{-1})	2.25 (0.27)	5.18 (0.31)
Exchangeable Ca (cmol + kg^{-1})	5.48 (0.48)	1.77 (0.11)
Exchangeable Mg (cmol + kg^{-1})	3.52 (0.25)	0.82 (0.06)
Exchangeable K (cmol + kg^{-1})	0.77 (0.08)	0.55 (0.13)
pH (1:1 water)	5.81 (0.08)	6.13 (0.25)
Texture		
Sand (%)	17 (3)	66 (3)
Silt (%)	18 (1)	11 (1)
Clay (%)	65 (3)	22 (2)
Clay mineralogy	Kaolinite (dominant), gibbsite, goethite, hematite	Mica, kaolinite, halloysite, gibbsite

[a] Values are means with standard errors in parentheses ($n = 3$)

Table 2 Quality parameters of organic residues applied in Embu and Machanga during 2002–2004

Organic residue	C (%)	N (%)	C:N	Lignin (%)	Polyphenols (%)	Quality class[a]
Tithonia diversifolia	38.3 (1.3)[b]	3.2 (0.5)	12.5 (2.1)	8.9 (1.7)	1.7 (0.8)	I
Calliandra calothyrsus	44.2 (0.3)	3.3 (0.2)	13.3 (0.6)	13.0 (9.4)	9.4 (1.6)	II
Zea mays—Embu	40.3 (1.3)	0.7 (0.1)	59.4 (11.2)	5.4 (1.2)	1.2 (0.2)	III
Zea mays—Machanga	40.1 (1.2)	0.8 (0.2)	59.1 (12.9)	5.7 (0.4)	1.2 (0.1)	III

[a] Residue quality classifications according to Palm et al. (2001)

[b] Values are means with standard errors in parentheses ($n = 3$)

for Embu, or a 100-g subsample for Machanga, was evenly spread over the 250 μm sieve and submerged in 1 cm of water for 5 min. The soil was subsequently sieved for 2 min by moving the sieve up and down with 50 repetitions. Macroaggregates remaining on the sieve were transferred to a pre-weighed beaker for drying. Soil and water that passed through the sieve was transferred to the 53 μm sieve and the sieving procedure repeated. Microaggregates remaining on the 53-μm sieve were transferred to a second beaker and all water plus silt and clay particles passing through the sieve were transferred to a third beaker. The three aggregate fractions were oven dried at 105°C and weighed.

Macroaggregates were further separated according to the microaggregate isolation methods of Six et al. (2000). A subsample of macroaggregates was submersed in water over a 250 μm sieve and shaken on a reciprocal shaker with 50 metal beads (4-mm diameter) under continuous water flow. The method was adapted for each soil to allow for complete dispersion of the macroaggregates without disrupting the microaggregates within macroaggregates. For Embu, 5-g subsamples were soaked overnight in 50 ml of water and then shaken for 6 min at 250 rpm. With the Machanga macroaggregates, a 15-g subsample was soaked in water for 20 min and then shaken for 4 min at 150 rpm. Water-stable microaggregates passing through the 250 μm mesh were collected on a 53 μm sieve and wet-sieved as previously. The coarse particulate organic matter remaining on the 250 μm mesh, microaggregates (53–250 μm), and silt and clay (<53 μm) fractions isolated from the macroaggregates were oven dried at 105°C and weighed. A 1–2 g subsample of the whole soil and each aggregate and macroaggregate size fraction was ground and analyzed

Springer

for C and N with a PDZ Europa Integra C–N isotope ratio mass spectrometer (Cheshire, United Kingdom).

Analyses of variance were performed on all measured variables using Proc MIXED in the SAS statistical software (SAS Institute, Cary, NC). Data from each site were analyzed separately. The main effects of residue input, N fertilizer and cropping and their interactions were treated as fixed effects. Block, block × residue, and block × residue × N effects were considered random. Effects were deemed to be significant at $P < 0.05$, and treatment means were subsequently separated using the PDIFF option of the LSMEANS statement. As none of the main effect interactions were found to be significant, only the means for each main effect treatment are presented.

Results

Aggregate distributions

Adding organic residues increased macroaggregation at Embu, thereby increasing the proportion of macroaggregates by an average of 7.1 g 100 g^{-1} soil in the residue amended treatments compared to the control (Table 3). Residue quality did not influence the amount of macroaggregation as all residue inputs had the same magnitude of effect. Cropping also produced a slight statistically significant increase in aggregation at Embu by reducing the proportion of the silt and clay fraction. At Machanga, cropping was the only treatment to lead to a small increase in aggregation. Macroaggregates and microaggregates combined were 2.2 g 100 g^{-1} greater in the cropped versus bare treatment. Adding N fertilizer did not influence aggregation, nor were there any significant interactions between the main treatment effects at either site.

There were no treatment differences in the proportion of coarse particulate organic matter, microaggregates or silt and clay fractions isolated from the macroaggregates of each site (data not shown). At Embu, the mean fraction distributions were 2.2, 73.1, and 24.7 g 100 g^{-1} g macroaggregates for the coarse particulate organic matter, microaggregates, and silt and clay, respectively. At Machanga, the coarse particulate organic matter, microaggregates, and silt and clay fractions

accounted for 64.9, 30.3, and 5.1 g 100^{-1} g macroaggregates, respectively.

Soil C and N

At Embu adding residue increased C and N contents of the whole soil by an average of 5.82 g C kg^{-1} soil and 0.52 g N kg^{-1} soil (Table 4). As with the macroaggregation, there was no difference between residue types in the amount of C or N stabilized. The increase in C and N with the addition of residue was due to greater C and N in the macroaggregate and microaggregate within macroaggregate fractions. Cropping also significantly increased whole soil C by 1.30 g C kg^{-1} soil, as a result of greater C in the microaggregate within macroaggregate fraction. In contrast, the C content of the silt and clay fraction was higher for the bare plots due to the lower aggregation in this treatment. Nitrogen fertilizer had little influence on C and N contents. Additionally, there were no consistent significant interactions between the treatment effects.

There were minimal treatment effects on C and N concentrations of the soil fractions at Machanga and no change in whole soil C and N contents with any treatment (Table 5). The *Tithonia* residue had greater C, and the *Tithonia* and *Calliandra* residues greater N in the coarse particulate organic matter fraction than the other residue inputs. Adding nitrogen fertilizer increased the C and N contents of the silt and clay fraction. As at Embu, the treatment effects did not produce any consistent significant interactions.

Discussion

Between the two sites, the fine-textured soil at Embu was more responsive to the input treatments than the coarse-textured soil at Machanga. This observation supports our hypothesis that the coarse-textured soil would have lower C and N stabilization of residue additions than the fine-textured soil. However, the input treatments failed to have any significant effect on aggregation or C and N contents at Machanga, while residue additions stimulated aggregation and increased C and N contents of specific fractions at Embu. Silt and clay content is an important determinant of aggregate formation and soil organic matter levels (Feller and Beare 1997) as clay minerals have a

Table 3 Aggregate size distribution response to residue inputs, N fertilizer, and cropping at Embu and Machanga

Main effect	Embu (g 100 g⁻¹ soil)			Machanga (g 100 g⁻¹ soil)		
	M[a]	m	s + c	M	m	s + c
Residue						
Control	63.6 b[b]	30.8 a	5.6	24.5	66.2	9.3
Maize	71.0 a	24.3 b	4.6	25.4	65.8	8.8
Calliandra	70.4 a	24.7 b	4.9	23.9	67.2	8.9
Tithonia	70.5 a	24.7 b	4.8	27.0	63.6	9.4
SED	2.2	1.9	0.3	1.6	1.5	0.8
N fertilizer						
0 kg ha⁻¹	69.4	25.7	4.9	25.1	66.0	8.9
120 kg ha⁻¹	68.3	26.6	5.1	25.3	65.4	9.3
SED	1.3	1.2	0.2	0.7	0.7	0.3
Cropping						
Bare	68.8	25.8	5.5 a	24.7 b	65.1 b	10.2 a
Cropped	69.0	26.5	4.6 b	25.7 a	66.3 a	8.0 b
SED	0.9	0.7	0.2	0.4	0.5	0.2

[a] Soil fractions are referred to as M, macroaggregates; m, microaggregates; s + c, silt and clay

[b] Within each main effect and size fraction, means followed by a different letter are significantly different ($P < 0.05$)

greater reactive surface area with which to bind organic matter and interact with other particles. Higher clay contents would increase a soils capacity for stabilization of organic matter, both through physical protection in aggregates and by chemical stabilization to mineral surfaces (Six et al. 2002).

Contrary to what we hypothesized, residue quality did not influence C and N stabilization. All of the residues added, regardless of quality, increased whole soil C and N as compared to the control in the fine-textured soil at Embu. Therefore, the biochemical recalcitrance of residue inputs appears to have little measurable influence on the stabilization of soil C and N after 3 years. The lack of C preservation under inputs of residues with higher lignin and polyphenol concentrations, supports recent research showing low stability of litter-derived C compounds in soil organic matter (Gleixner et al. 2002; Rasse et al. 2006; von Lützow et al. 2006). A large portion of soil organic matter stems from microbial products (von Lützow et al. 2006); therefore, the capacity of a soil to protect a higher amount of C may be more important than the type of C compounds being added.

The increase in soil C and N with residue addition at Embu was accompanied by an increase in stable macroaggregate formation. In the residue addition treatments, the promotion of macroaggregation by residues increased the amount C and N stabilized within the macroaggregate and microaggregate within macroaggregate fractions, which in turn contributed to the increase in whole soil C and N. The observed pattern of C and N accumulation in the microaggregate within macroaggregate fraction supports the aggregate hierarchy model of soil organic matter stabilization for the red clay soil at Embu (Oades 1984; Six et al. 2000). Six et al. (2001) hypothesized that a decreased rate of aggregate turnover with low quality residue would increase soil C stabilization. However, their study showed only a difference in the particulate organic matter C fraction after 6 year, but there was no difference in total soil C seen after 6 years (Six et al. 2001) and 10 years (van Kessel et al. 2006) of different quality pasture residues. Likewise, residue quality did not change the accumulation of C and N within aggregate fractions and total soil C in the present study. Even in the microaggregate within macroaggregate fraction, which is the most sensitive fraction to C changes (Denef et al. 2004; Kong et al. 2005), residue quality did not affect C and N stabilization. Therefore, residue quality appears to have little influence on

 Springer

Table 4 Carbon and nitrogen content of aggregate size fractions response to residue inputs, N fertilizer, and cropping at Embu

Treatment effect	Whole soil	Whole soil fractions			Macroaggregate fractions		
		M[a]	m	s + c	cPOM	mM	s + cM
	Carbon (g C kg⁻¹ soil)						
Residue							
Control	26.00 b[b]	17.35 b	8.09	1.74	0.56	12.91 b	5.71
Maize	31.66 a	23.74 a	7.53	1.65	1.00	16.14 a	8.20
Calliandra	31.27 a	22.86 a	7.69	1.74	1.04	15.72 a	7.36
Tithonia	32.56 a	23.59 a	8.02	1.78	1.03	16.85 a	7.86
SED	1.57	1.65	0.62	0.08	0.16	0.85	0.93
N fertilizer							
0 kg ha⁻¹	29.87	22.04	7.68	1.68	0.89	15.40	7.35
120 kg ha⁻¹	30.87	21.73	7.98	1.78	0.93	15.41	7.21
SED	0.73	0.91	0.31	0.05	0.07	0.43	0.63
Cropping							
Bare	29.72 b	21.58	7.58	1.82 a	0.83	15.07 b	7.12
Cropped	31.02 a	22.19	8.08	1.64 b	0.98	15.74 a	7.44
SED	0.35	0.34	0.28	0.05	0.07	0.23	0.32
	Nitrogen (g N kg⁻¹ soil)						
Residue							
Control	2.27 b	1.45 b	0.70	0.16	0.03	1.24 b	0.43
Maize	2.76 a	2.04 a	0.66	0.16	0.06	1.50 a	0.69
Calliandra	2.76 a	1.99 a	0.68	0.18	0.08	1.49 a	0.62
Tithonia	2.84 a	2.04 a	0.71	0.17	0.06	1.56 a	0.61
SED	0.14	0.16	0.05	0.01	0.01	0.07	0.08
N fertilizer							
0 kg ha⁻¹	2.60	1.89	0.67	0.15 b	0.06	1.46	0.61
120 kg ha⁻¹	2.71	1.87	0.71	0.18 a	0.06	1.43	0.57
SED	0.08	0.09	0.03	0.00	0.00	0.05	0.04
Cropping							
Bare	2.66	1.86	0.69	0.18 a	0.05	1.41	0.57 b
Cropped	2.65	1.89	0.69	0.15 b	0.06	1.47	0.61 a
SED	0.03	0.03	0.00	0.00	0.00	0.03	0.02

[a] Soil fractions are referred to as M, macroaggregates; m, microaggregates; s + c, silt and clay; cPOM, coarse particulate organic matter; mM, microaggregates within macroaggregates; s + cM, silt and clay within macroaggregates

[b] Within each main effect and size fraction, means followed by a different letter are significantly different ($P < 0.05$)

medium- to long-term C and N dynamics. In terms of physical stabilization of C and N within aggregates, a stimulation of aggregate turnover rates with high quality residues may not significantly influence C and N stabilization on a long-term timescale. Additionally, an outside factor, such as tillage or wetting and drying cycles, may exert a greater control over macroaggregate turnover rates in the present study, thus overriding any effect of residue quality. Macroaggregate turnover rates have been found to increase with both tillage (Six et al. 1999; Denef et al. 2004) and drying and wetting cycles (Denef et al. 2001), and these forces may have

Table 5 Carbon and nitrogen content of aggregate size fractions response to residue inputs, N fertilizer, and cropping at Machanga

Treatment effect	Whole soil	Whole soil fractions			Macroaggregate fractions		
		M^a	m	s + c	cPOM	mM	s + cM
	Carbon (g C kg^{-1} soil)						
Residue							
Control	5.43	1.29	3.10	1.23	0.48 b^b	0.43	0.34
Maize	4.50	1.29	2.56	1.11	0.43 b	0.30	0.30
Calliandra	5.10	1.70	2.84	1.29	0.76 b	0.28	0.26
Tithonia	6.62	2.39	3.64	1.08	1.19 a	0.60	0.40
SED	1.18	0.48	0.59	0.18	0.17	0.16	0.10
N fertilizer							
0 kg ha^{-1}	5.23	1.60	2.98	1.11 b	0.63	0.39	0.30
120 kg ha^{-1}	5.60	1.73	3.09	1.25 a	0.80	0.41	0.35
SED	0.33	0.19	0.14	0.06	0.09	0.04	0.04
Cropping							
Bare	5.90	1.76	3.06	1.20	0.76	0.40	0.32
Cropped	4.92	1.57	3.01	1.16	0.67	0.41	0.33
SED	0.29	0.16	0.08	0.04	0.04	0.02	0.02
	Nitrogen (g N kg^{-1} soil)						
Residue							
Control	0.51	0.10	0.28	0.14	0.03 b	0.04	0.02
Maize	0.45	0.09	0.24	0.12	0.03 b	0.02	0.02
Calliandra	0.53	0.13	0.27	0.13	0.06 a	0.03	0.02
Tithonia	0.68	0.19	0.35	0.14	0.07 a	0.05	0.03
SED	0.12	0.04	0.06	0.02	0.01	0.01	0.01
N fertilizer							
0 kg ha^{-1}	0.53	0.12	0.28	0.12 b	0.04	0.03	0.03
120 kg ha^{-1}	0.56	0.14	0.28	0.14 a	0.05	0.04	0.03
SED	0.03	0.02	0.02	0.01	0.00	0.00	0.00
Cropping							
Bare	0.52	0.13	0.26 b	0.14 a	0.05	0.03	0.03
Cropped	0.56	0.12	0.31 a	0.13 b	0.04	0.04	0.03
SED	0.02	0.01	0.01	0.00	0.00	0.00	0.00

[a] Soil fractions are referred to as M, macroaggregates; m, microaggregates; s + c, silt and clay; cPOM, coarse particulate organic matter; mM, microaggregates within macroaggregates; s + cM, silt and clay within macroaggregates

[b] Within each main effect and size fraction, means followed by a different letter are significantly different ($P < 0.05$)

been more dominant in controlling aggregate dynamics in the present study than residue quality.

Similar to the effects of residue quality, additions of N fertilizer did not influence aggregation or soil C and N stabilization at either site. This lack of N fertilizer effect further supports our observations that

increased N availability, whether through high N content residues or mineral N fertilizer, does not affect soil organic matter stabilization. Previous studies have shown much variation in the degree and direction of response of organic matter decomposition to nitrogen additions (Fog 1988). In a meta-analysis of the effects of N on litter decomposition, Knorr et al. (2005) found that the stimulation or inhibition and amount of change in decomposition was controlled by fertilization rates, litter quality, and time. In this analysis, decomposition was stimulated by high rates of N fertilization >20 times ambient N deposition and high residue quality (<10% lignin). However, whereas studies of <24 months duration generally showed a stimulation, decay periods of >24 months showed an inhibition of decomposition. While the combined application of organic residue and mineral fertilizer inputs is advocated to improve the synchrony of short-term nutrient dynamics (Vanlauwe et al. 2002), we found that adding fertilizer has no impact on longer-term soil organic matter formation.

As with residue additions, adding C through root growth increased soil aggregation and stabilization of C at Embu. While the quantity of C increase with cropping was relatively low (1.30 g kg^{-1} soil), it was statistically significant and shows a trend for increased soil C with C inputs. However, contrary to what we hypothesized there were no significant interactions between residue and N fertilizer inputs and root growth on aggregation or organic matter stabilization at either site. The high quality *Tithonia* and medium quality *Calliandra* residues and N fertilizer stimulated maize growth in these field trials (Chivenge et al. unpublished). Therefore, we would anticipate a greater root contribution to aggregation and C and N stabilization in the treatments with higher quality and fertilizer inputs. A lack of interaction between cropping and inputs indicates that the contribution of roots to soil C was relatively minor in comparison to residue inputs. Residues were applied at a high rate of 4 Mg C ha^{-1} year^{-1} in order to observe treatment differences over a relatively short period of time, and this amount of C input likely masked the contributions of roots to soil C in these trials.

In contrast to Embu, adding organic residues did not promote stable macroaggregation at Machanga. Instead, cropping was the only treatment to slightly improve aggregation, which may indicate that forces of physical enmeshment are more important for aggregation in the coarse-textured soil than C additions (Degens 1997). The lack of stable aggregate formation in the coarse-textured soil of Machanga prevented the physical protection of the residue

inputs and stabilization of C and N. The increase in C and N content of the coarse particulate organic matter fraction with high quality residues indicates a direct input of residues into this fraction as they are broken down into smaller pieces. Vanlauwe et al. (2000) found that maize N uptake was related to particulate organic matter N concentrations. The higher C and N contents observed in the coarse particulate organic matter fraction in the *Tithonia* and *Calliandra* treatments implies improved short-term fertility with high quality residues, but this C and N failed to become further stabilized within protected aggregate fractions in the coarse-textured soil.

Conclusions

Residue quality or N fertilizer additions did not affect the stabilization of soil organic C and N. Soil texture greatly influenced soil organic matter stabilization as in the fine-textured soil C inputs led to increased soil C, whereas in the coarse-textured soil residue management had no effect on C stabilization. To manage inputs for the long-term maintenance of soil organic matter, organic residues should be applied regardless of their quality or their use with additional mineral fertilizer. However, residue applications will only have a beneficial effect on soil organic matter in soils with the capacity to physically protect these additions.

Acknowledgments   Helen Wangechi and the field technicians at Embu and Machanga are gratefully acknowledged for the maintenance and soil sampling of the trials. Gard Okello provided invaluable assistance in the lab. This research was supported by a grant from the National Science Foundation (DEB: 0344971).

References

Anderson JM, Ingram JSI (1993) Tropical soil biology and fertility: a handbook of methods. CAB International, Wallingford

Chassot A, Stamp P, Richner W (2001) Root distribution and morphology of maize seedlings as affected by tillage and fertilizer placement. Plant Soil 231:123–135. doi:10.1023/A:1010335229111

Constantinides M, Fownes JH (1994) Nitrogen mineralization from leaves and litter of tropical plants: relationship to nitrogen, lignin and soluble polyphenol concentrations. Soil Biol Biochem 26:49–55. doi:10.1016/0038-0717(94)90194-5

Degens BP (1997) Macro-aggregation of soils by biological bonding and binding mechanisms and the factors affecting these: a review. Aust J Soil Res 35:431–459. doi:10.1071/S96016

Denef K, Six J, Paustian K, Merckx R (2001) Importance of macroaggregate dynamics in controlling soil carbon stabilization: short-term effects of physical disturbance induced by dry-wet cycles. Soil Biol Biochem 33:2145–2153. doi:10.1016/S0038-0717(01)00153-5

Denef K, Six J, Merckx R, Paustian K (2002) Short-term effects of biological and physical forces on aggregate formation in soils with different clay mineralogy. Plant Soil 246:185–200. doi:10.1023/A:1020668013524

Denef K, Six J, Merckx R, Paustian K (2004) Carbon sequestration in microaggregates of no-tillage soils with different clay mineralogy. Soil Sci Soc Am J 68:1935–1944

Denef K, Zotarelli L, Boddey RM, Six J (2007) Microaggregate-associated carbon as a diagnostic fraction for management-induced changes in soil organic carbon in two oxisols. Soil Biol Biochem 39:1165–1172. doi:10.1016/j.soilbio.2006.12.024

Durieux RP, Kamprath EJ, Jackson WA, Moll RH (1994) Root distribution of corn—the effect of nitrogen fertilizer. Agron J 86:958–962

Elliott ET (1986) Aggregate structure and carbon, nitrogen and phosphorus in native and cultivated soils. Soil Sci Soc Am J 50:627–633

FAO (1998) World reference base for soil resources. FAO, Rome

Feller C, Beare MH (1997) Physical control of soil organic matter dynamics in the tropics. Geoderma 79:69–116. doi:10.1016/S0016-7061(97)00039-6

Fog K (1988) The effect of added nitrogen on the rate of decomposition of organic matter. Biol Rev Camb Philos Soc 63:433–462. doi:10.1111/j.1469-185X.1988.tb00725.x

Gale WJ, Cambardella CA, Bailey TB (2000) Surface residue- and root-derived carbon in stable and unstable aggregates. Soil Sci Soc Am J 64:196–201

Gleixner G, Poirier N, Bol R, Balesdent J (2002) Molecular dynamics of organic matter in a cultivated soil. Org Geochem 33:357–366. doi:10.1016/S0146-6380(01)00166-8

Harris RF, Allen ON, Chesters G, Attoe OJ (1963) Evaluation of microbial activity in soil aggregate stabilization and degradation by the use of artificial aggregates. Soil Sci Soc Am Proc 27:542–545

Hattori T (1988) Soil aggregates as habitats of microorganisms. Rep Inst Agric Res Tohoku Univ 37:23–36

Jastrow JD, Miller RM, Lussenhop J (1998) Contributions of interacting biological mechanisms to soil aggregate stabilization in restored prairie. Soil Biol Biochem 30:905–916. doi:10.1016/S0038-0717(97)00207-1

Kemper WD, Koch EJ (1966) Aggregate stability of soils from western United States and Canada. USDA-ARS Technical Bulletin 1355. U.S. Govt. Print. Office, Washington

Knorr M, Frey SD, Curtis PS (2005) Nitrogen additions and litter decomposition: a meta-analysis. Ecology 86:3252–3257. doi:10.1890/05-0150

Kölbl A, Kögel-Knabner I (2004) Content and composition of free and occluded particulate organic matter in a

differently textured arable Cambisol as revealed by solid-state ^{13}C NMR spectroscopy. J Plant Nutr Soil Sci 167:45–53. doi:10.1002/jpln.200321185

Kong AYY, Six J, Bryant DC, Denison RF, van Kessel C (2005) The relationship between carbon input, aggregation, and soil organic carbon stabilization in sustainable cropping systems. Soil Sci Soc Am J 69:1078–1085

Materechera SA, Kirby JM, Alston AM, Dexter AR (1994) Modification of soil aggregation by watering regime and roots growing through beds of large aggregates. Plant Soil 160:57–66. doi:10.1007/BF00150346

Morel JL, Habib L, Plantureux S, Guckert A (1991) Influence of maize root mucilage on soil aggregate stability. Plant Soil 136:111–119. doi:10.1007/BF02465226

Oades JM (1984) Soil organic matter and structural stability: mechanisms and implications for management. Plant Soil 76:319–337. doi:10.1007/BF02205590

Palm CA, Gachengo CN, Delve RJ, Cadisch G, Giller KE (2001) Organic inputs for soil fertility management in tropical agroecosystems: application of an organic resource database. Agric Ecosyst Environ 83:27–42. doi:10.1016/S0167-8809(00)00267-X

Plante AF, McGill WB (2002) Soil aggregate dynamics and the retention of organic matter in laboratory-incubated soil with differing simulated tillage frequencies. Soil Tillage Res 66:79–92. doi:10.1016/S0167-1987(02)00015-6

Puget P, Drinkwater LE (2001) Short-term dynamics of root- and shoot-derived carbon from a leguminous green manure. Soil Sci Soc Am J 65:771–779

Rasse DP, Dignac M-F, Bahri H, Rumpel C, Mariotti A, Chenu C (2006) Lignin turnover in an agricultural field: from plant residues to soil-protected fractions. Eur J Soil Sci 57:530–538. doi:10.1111/j.1365-2389.2006.00806.x

Reid JB, Goss MJ (1981) Effect of living roots of different plant species on the aggregate stability of two arable soils. J Soil Sci 32:521–541. doi:10.1111/j.1365-2389.1981.tb01727.x

Sanchez PA, Jama B (2002) Soil fertility replenishment takes off in East and Southern Africa. In: Vanlauwe B, Diels J, Sanginga N, Merckx R (eds) Integrated plant nutrient management in sub-Saharan Africa: from concept to practice. CABI, Wallingford, pp 23–46

Six J, Elliott ET, Paustian K (1999) Aggregate and soil organic matter dynamics under conventional and no-tillage systems. Soil Sci Soc Am J 63:1350–1358

Six J, Elliott ET, Paustian K (2000) Soil macroaggregate turnover and microaggregate formation: a mechanism for C sequestration under no-tillage agriculture. Soil Biol Biochem 32:2099–2103. doi:10.1016/S0038-0717(00)00179-6

Six J, Carpentier A, van Kessel C, Merckx R, Harris D, Horwath WR, Luscher A (2001) Impact of elevated CO_2 on soil organic matter dynamics as related to changes in aggregate turnover and residue quality. Plant Soil 234:27–36. doi:10.1023/A:1010504611456

Six J, Conant RT, Paul EA, Paustian K (2002) Stabilization mechanisms of soil organic matter: implications for C-saturation of soils. Plant Soil 241:155–176. doi:10.1023/A:1016125726789

Tisdall JM, Oades JM (1982) Organic matter and water-stable aggregates in soils. J Soil Sci 33:141–163. doi:10.1111/j.1365-2389.1982.tb01755.x

Nutr Cycl Agroecosyst (2010) 88:121–131

Trinsoutrot I, Recous S, Bentz B, Linères M, Chèneby D, Nicolardot B (2000) Biochemical quality of crop residues and carbon and nitrogen mineralization kinetics under nonlimiting nitrogen conditions. Soil Sci Soc Am J 64:918–926

van Kessel C, Boots B, de Graaff M-A, Harris D, Blum H, Six J (2006) Total soil C and N sequestration in a grassland following 10 years of free air CO_2 enrichment. Glob Chang Biol 12:2187–2199. doi:10.1111/j.1365-2486.2006.01172.x

Vanlauwe B, Aihou K, Aman S, Tossah BK, Diels J, Lyasse O, Hauser S, Sanginga N, Merckx R (2000) Nitrogen and phosphorus uptake by maize as affected by particulate organic matter quality, soil characteristics, and land-use history for soils from the West African moist savanna zone. Biol Fertil Soils 30:440–449. doi:10.1007/s003740050022

Vanlauwe B, Diels J, Aihou K, Iwuafor ENO, Lyasse O, Sanginga N, Merckx R (2002) Direct interactions between N fertilizer and organic matter: evidence from trials with ^{15}N-labelled fertilizer. In: Vanlauwe B, Diels J, Sanginga N, Merckx R (eds) Integrated plant nutrient management in Sub-Saharan Africa: from concept to practice. CAB International, New York, pp 173–184

Vanlauwe B, Gachengo C, Shepherd K, Barrios E, Cadisch G, Palm CA (2005) Laboratory validation of a resource quality-based conceptual framework for organic matter management. Soil Sci Soc Am J 69:1135–1145

von Lützow M, Kögel-Knabner I, Ekschmitt K, Matzner E, Guggenberger G, Marschner B, Flessa H (2006) Stabilization of organic matter in temperate soils: mechanisms and their relevance under different soil conditions—a review. Eur J Soil Sci 57:426–445. doi:10.1111/j.1365-2389.2006.00809.x

Interaction Between Resource Quality, Aggregate Turnover, Carbon and Nitrogen Cycling in the Central Highlands of Kenya

A. Kavoo, D.N. Mugendi, G. Muluvi, B. Vanlauwe, J. Six, R. Merckx, R. Gentile, and W.M.H. Kamiri

Abstract Combined use of organic resource (OR) and mineral resource (MR) of nutrients is accepted as one of the most appropriate ways to address the problems of declining soil fertility and poor crop yields facing small-scale farming in sub-Saharan Africa. A field study was conducted at Embu in Central Kenya to investigate the effect of OR and MR management on aggregate turnover, C sequestration and N stabilization. The study comprised of ORs of differing quality: *Tithonia diversifolia* (high quality), *Calliandra calothyrsus* (medium quality), *Zea mays* stover (medium quality), *Grevillea* robust sawdust (low quality) and farmyard manure applied at a rate of 4 ton C ha^{-1} with or without 120 kg N ha^{-1} mineral fertilizer. Soil organic matter (SOM) fractions from soils sampled from the top soil (0–15 cm depth) at the establishment of the field trial in 2002 and before the long rains in 2005 were analysed for C, N and C-13 signatures. In 2005, SOM fractions C and N quantity was higher for both sole and combined application of *Tithonia, Calliandra*, stover and manure compared to the initial (2002) total soil C and N. High-quality ORs had the highest SOM input compared to low-quality ORs while medium-quality ORs contributed most to the formation of stable macroaggregates and SOM accumulation. Therefore, both OR quality and MR should be considered when devising soil management options for soil fertility and crop production.

Keywords Aggregate turnover · Soil organic matter · Organic resource quality · Mineral N fertilizer · Crop production

Introduction

Soil fertility decline is a major problem facing small-scale farming in sub-Saharan Africa. Soil fertility can be maintained if the output of plant nutrients through harvested products and losses in the form of leaching and gaseous emissions is compensated by an equivalent input (Kirchmann and Thorvaldsson, 2000). The realization that "Green Revolution" technologies reliant purely on mineral fertilizers have failed to take hold in Africa, and equally that approaches based on organic inputs cannot provide the required increments in agricultural production, demands effective use of both types of resources (Giller, 2002). It is logical that efficient use of organic resources (ORs) supplemented with mineral fertilizers may be an optimal strategy for smallholder farmers (Giller, 2002).

In recent years, a growing consensus has emerged on the need for both organic matter and fertilizer to reverse the negative nutrient balances in cropping systems in agriculture in sub-Saharan Africa (SSA) as continuous application of either of these inputs tends to create soil-related constraints to crop production (Vanlauwe et al., 2002b). One important aspect of simultaneously applying OM and fertilizer is the potential for positive interactions between both inputs, leading to added benefits in the form of extra grain yield or improved soil fertility and reduced losses of nutrients (Giller, 2002).

A. Kavoo (✉)
Tropical Soil Biology and Fertility Institute of the International Centre for Tropical Agriculture (TSBF-CIAT), Nairobi, Kenya
e-mail: agneskavoo@yahoo.com

A. Bationo et al. (eds.), *Innovations as Key to the Green Revolution in Africa*,
DOI 10.1007/978-90-481-2543-2_81, © Springer Science+Business Media B.V. 2011

However, to fully exploit the potential benefits of combined applications and optimize N use efficiency requires a better understanding of underlying mechanisms and soil processes. Primary mechanisms were identified by Vanlauwe et al. (2001) in the form of a direct and an indirect hypothesis. The *direct hypothesis* stated that "temporary immobilization of applied mineral N improves the synchrony between the supply and demand of N by the plant and leads to an increase in system-wide N use efficiency". Nitrogen immobilization prior to maximum plant demand due to C availability (Sakala et al., 2000; Vanlauwe et al., 2002a) can be beneficial at the ecosystem level as it reduces N losses through leaching and/or denitrification (Robertson, 1997; Scow and Johnson, 1997). The *indirect hypothesis* stated that "any organic resource applied in addition to a mineral resource leads to an improved soil condition which improves plant growth and consequently enhances system-wide resource use efficiency". This growth-limiting factor can be present at the plant nutritional, the soil physico-chemical, or the soil (micro) biological level.

The beneficial interactions between OR and MR remain poorly understood. However, it is known that organic resource quality is an important modifier of these interactions (Vanlauwe et al., 2001). In particular, residue quality and mineral N level governs the rate of aggregate formation and breakdown (i.e. the rate of aggregate turnover) (Browning and Milam, 1944; Harris et al., 1963; Six et al., 2001). Therefore, we postulate that aggregate turnover plays a fundamental role in interactions between OR and MR. Aggregates physically protect soil organic matter (e.g. Tisdall and Oades, 1982; Six et al., 2000), influence microbial community structure (e.g. Hattori, 1988), limit O_2 diffusion (e.g. Sexstone et al., 1985), regulate water flow (e.g. Prove et al., 1990), determine nutrient adsorption and desorption (e.g. Linquist et al., 1997), and reduce run-off and erosion (e.g. Barthes and Roose, 2002).

The objectives of this study were (i) to determine the effect of organic resource quality on the rate of aggregate turnover, (ii) to understand the effect of integrating/combining organic resources of different qualities with mineral fertilizer on the rate of aggregate turnover, and (iii) to determine the effect of combining organic and inorganic nutrient sources on C sequestration and N stabilization within aggregates.

Materials and Methods

Experimental Sites

The experiment was carried out in Embu district which is located in the Central highlands of Kenya at 0°30′ S, 37°27′ E and at an altitude of 1,380 m above sea level. The average temperature is about 20°C. The rainfall is bimodal with the long rains (LR) from March to May and short rains (SR) from mid-October to December. The average annual rainfall is 1,200 mm. The physical and chemical characteristics of soils sampled at the establishment of the experiment in 2002 are presented in Table 1.

Experimental Design

The experiment was set up in March 2002 as part of long-term trials aimed at determining the influence of repeated application of organic resources (ORs) of different quality and quantities on soil organic matter dynamics. Five organic resources belonging to the four quality classes proposed by Palm et al. (1997) were applied at a C rate of 4 t C ha^{-1}. The OR treatments applied were class I: *Tithonia diversifolia*, class II: *Calliandra calothyrsus*, class III: *Zea mays* stover, class IV: *Eucalyptus saligna*, sawdust, and high-quality farmyard manure. The ORs were broadcast and hand-incorporated to a depth of 0.15 m

Table 1 Chemical and physical characteristics of soils sampled at 0–15 cm depth at Embu in March 2002

Soil property	Quantity
Organic C (g kg^{-1})	29.35
Total N (g kg^{-1})	2.66
Exchangeable Ca (cmol$_+$ kg^{-1})	5.45
Exchangeable Mg (cmol$_+$ kg^{-1})	2.98
Exchangeable K (cmol$_+$ kg^{-1})	0.82
Exchangeable Na (cmol$_+$ kg^{-1})	0.07
CEC (cmol$_+$ kg^{-1})	16.03
Base saturation (%)	58.52
pH	5.43
Texture	
Sand (%)	32
Silt (%)	30
Clay (%)	38
Clay mineralogy	Kaolinite (dominant), gibbsite, goethite, haematite

using a hand hoe once a year at planting during the long rain (LR) season. A control treatment where no residues were applied was also included. The plot sizes were 12 by 6 m laid out in a randomized complete block design in a factorial experiment with three replicates. Maize was planted as the test crop in both the long rain and short rain seasons. The plots were split in half where 120 kg N ha^{-1} N fertilizer was added in one-half of the main plot. One third of the N fertilizer was applied 3 weeks after planting and the remainder was applied 8 weeks after planting by broadcasting and incorporating in the soil. All plots received a blanket application of P at 60 kg P ha^{-1} and K at 60 kg K ha^{-1} at planting. Before incorporation, the five residue types were sub-sampled and analysed for total organic C, N, lignin and polyphenol (Table 2). These values were used to convert the C equivalent application rate to application rates of fresh biomass per hectare.

Soil Sampling

Initial soil sampling was carried out in April 2002. Subsequent soil sampling was carried out at the beginning of the long rains in March 2005 before land preparation and residue application. Soil samples were taken at a depth of 0–15 cm at the sub-sub-plot level following a systematic scheme, across both diagonals of the plot. Five disturbed cores were taken from the cropped plots. The soils from the cores were combined and the dry weight of the soil samples was obtained. Soils were sieved through an 8-mm sieve and sub-samples of approximately 2 kg were taken for SOM fractionation.

Wet Sieving

Air-dried macroaggregates (<8 mm) were separated into four size fractions by wet sieving (Elliott,

1986): large macroaggregates (>2,000 μm), small macroaggregates (250–2,000 μm), microaggregates (53–250 μm), and silt + clay-associated particles (<53 μm). Eighty (80) g of air-dried soil samples were immersed in water on sieves of 2 mm, 250 μm, and 53 μm (one by one) and manually moved up and down 3 cm and 50 times during a 2-min period. Soil aggregates that retained on each sieve were then backwashed into pre-weighed containers, oven-dried at 105°C overnight and weighed thereafter. To sub-sample silt and clay fraction, a bottle-based method was used and a representative sub-sample of 250 ml was taken, oven-dried at 105°C and weighed. All fractions were analysed for C, N and δ^{13}C.

Determination of the Isotope Carbon-13 (δ^{13}C)

Analysis of delta carbon-13 (δ^{13}C) for pulverized soil samples was done on an automated nitrogen and carbon (ANCA-analyzer) mass spectrometer (Diels et al., 2001). Carbon isotope composition was expressed in delta-13 (δ^{13}C) units using the international Pee De Belemnite (PDB) reference standard:

$$\delta^{13}C\%o = \frac{[^{13}R_{sample} - 1]}{^{13}R_{standard}} \times 1000$$

where

$$^{13}R = \frac{^{13}C}{^{12}C}.$$

Statistical Analyses

Data collected from this study was analysed as a split–split plot design. Each response variable for the field study was analysed with the PROC MIXED procedure

Table 2 Quality of organic resources used in the study area at the time of field incorporation at Embu

Organic material	% C	% N	% Lignin	% Polyphenol
Calliandra	44 (43–46)	3 (3–3)	13 (7–16)	10 (6–14)
Stover	40 (38–42)	1 (0.5–0.9)	5 (3–7)	1 (1–2)
Manure	27 (17–35)	2 (1–3)	13 (11–18)	1 (0.3–2.5)
Sawdust	44 (43–45)	0.2 (0.1–0.2)	18 (17–19)	0.5 (0.4–0.6)
Tithonia	38 (36–40)	3 (2–3)	11 (7–15)	2 (0.8–3.3)

Average of five readings (range indicated in parenthesis)

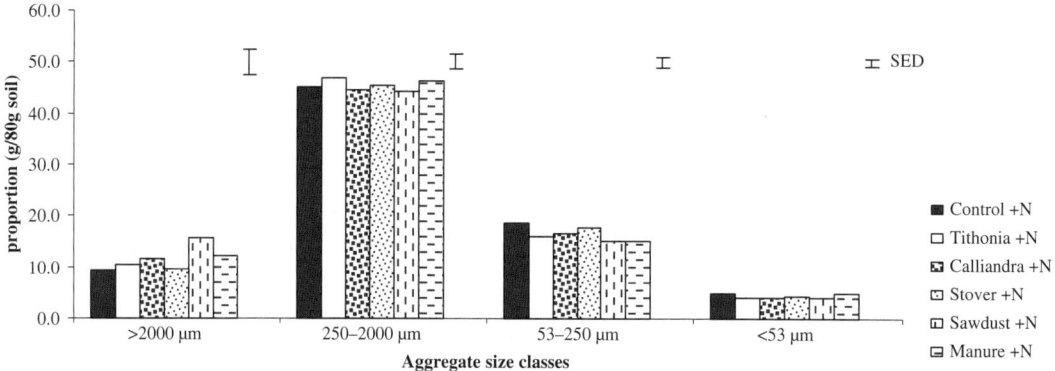

Fig. 1 Effect of combining organic resource quality and mineral fertilizer on the proportion of aggregate size classes at Embu in March 2005

of SAS, with the effect of replicate, organic resource (OR) class, and mineral resource (MR) addition as fixed effects. The means were separated at ($p < 0.05$) and the standard error of difference (SED) of the means was used to compare the responses between OR treatments with and/or without mineral fertilizer.

Results

Effect of Organic Resource Quality and Mineral Fertilizer on Aggregate Proportions

Organic resource (OR) quality did not significantly ($p \leq 0.05$) affect aggregate proportions across the aggregate size classes (Fig. 1). The combined application of sawdust and mineral fertilizer however led to a significant formation of large macroaggregates.

Effect of Resource Quality and Mineral Fertilizer on Carbon and Nitrogen Content of Whole Soil

C and N contents of whole soils sampled per plot before the long rains in March 2005 were compared to the soils sampled at the establishment of the Embu trials in 2002. In 2005, sole application of organic resources did not significantly affect the whole-soil C. However, significant differences in whole-soil N were observed between the control and sole *Tithonia*, *Calliandra*, maize stover and manure treatments. Soil amended with manure had the highest N concentration (2.8 g N kg^{-1} whole soil) while the control had the least N content (2.3 g N kg^{-1} whole soil) (Table 3).

However, this fast release and utilization of available N by the maize crop leads to depletion of C and N over time as observed with the whole-soil C and N in 2005 compared to 2002 at the establishment of the experiment.

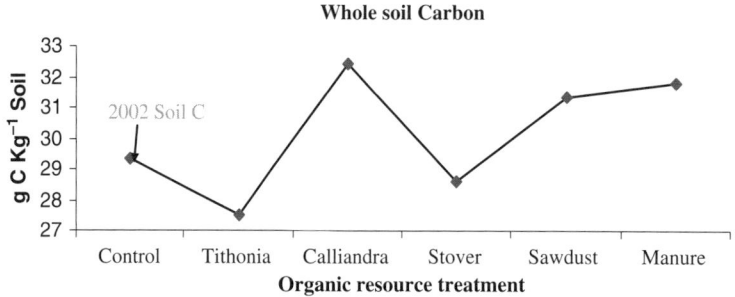

Table 3 Effect of resource quality and the combination of OR and MR on the quality of nutrients (C and N) of whole soil at Embu in March 2005

Treatment	C (g C kg^{-1} WS)		N (g N kg^{-1} WS)	
	−N	+N	−N	+N
Control	27.5	26.2	2.3	2.2
Tithonia	31.8	35.8	2.7	3.0
Calliandra	31.2	32.5	2.7	2.8
Stover	31.3	32.4	2.7	2.8
Sawdust	28.6	27.6	2.4	2.3
Manure	32.5	32.0	2.8	2.9
SED	1.8	1.95	0.16*	0.18

SED = standard error of difference of means; SED* = difference between OR treatments is statistically significant at *p* < 0.05 and/or there is an interaction between resource quality and mineral fertilizer

The effect of combined application of organic resources (ORs) and mineral fertilizer (MR) on C and N contents of whole soils in 2005 was only significant on the *Tithonia* plus N treatment (Table 3). The combined application of *Tithonia* and MR had a higher concentration of C (35.8 g C kg^{-1} whole soil) and N (3.0 g N kg^{-1} whole soil) compared to the sole applications (31.8 g C kg^{-1} whole soil and 2.7 g N kg^{-1} whole soil). Higher concentrations of C and N for the combined applications were also observed on the *Calliandra*, maize stover and manure (only for N) treatments compared to the sole application treatments.

Effect of Resource Quality and Mineral Fertilizer on Carbon and Nitrogen Content of Aggregate Size Fractions

The fraction under *Tithonia* amendment had the highest C content of 35.75 g C kg^{-1} and was the only fraction statistically different from the control within the large macroaggregate size class (>2,000 μm) (Fig. 2). The N content of the large macroaggregate size class fractions amended with maize stover and manure was significantly different from the control. The manure-amended fraction had the highest N content of 3.19 N kg^{-1}. Within the small macroaggregate size class (250–2,000 μm), only the C and N content of the fraction of soil amended with *Tithonia*, maize stover and manure were statistically different from the control. The highest C concentration (35.75 g C kg^{-1}) was observed with the fraction under *Tithonia*

treatment followed by the fractions under manure (34.0 g C kg^{-1}) and maize stover (32.46 g C kg^{-1}) treatments. The maize stover-amended aggregate fraction had the highest N content (2.85 N kg^{-1}). The same trend was followed for the microaggregate size class (53–250 μm) except that the aggregate fraction under manure treatment was also statistically different from the control. The *Tithonia*-amended fraction had the highest C and N content (32.11 g C kg^{-1} and 2.70 g N kg^{-1}, respectively) while the fractions under control and sawdust treatment had the least C (25.904 g C kg^{-1}) and N (2.23 N kg^{-1}) contents.

The combined application of ORs and MR led to a reduced level of C and N contents of small macroaggregate size class fractions. Within the large macroaggregate size class, fractions under combined application of manure, *Calliandra*, stover and sawdust with MR led to a reduction of C and N while that under *Tithonia* plus MR treatment had a positive N increment of 0.04 g N kg^{-1} fraction compared to the fractions under sole ORs application treatments (Fig. 3).

Higher C contents were noted within the microaggregate fractions under *Tithonia*, manure, *Calliandra* and maize stover plus N treatments compared to the sole applications. Combined application of *Tithonia* plus MR and *Calliandra* plus MR increased the N contents of the microaggregate fractions by 0.03 and 0.04 g N kg^{-1} fraction (for *Tithonia* and *Calliandra*, respectively) compared to the sole OR application treatments. Increased contents of N among combined treatments were observed within the silt and clay aggregate size fraction and these were in the order of *Tithonia*>*Calliandra*>maize stover>manure>control>sawdust (Fig. 3).

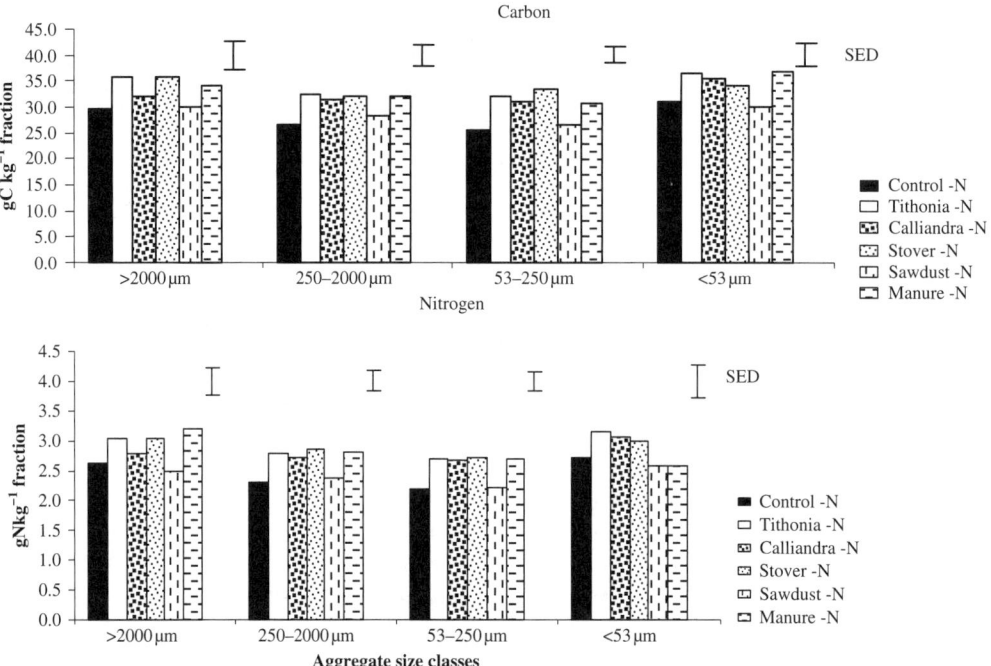

Fig. 2 Effect of resource quality on the C and N concentration of aggregate size classes at Embu in March 2005

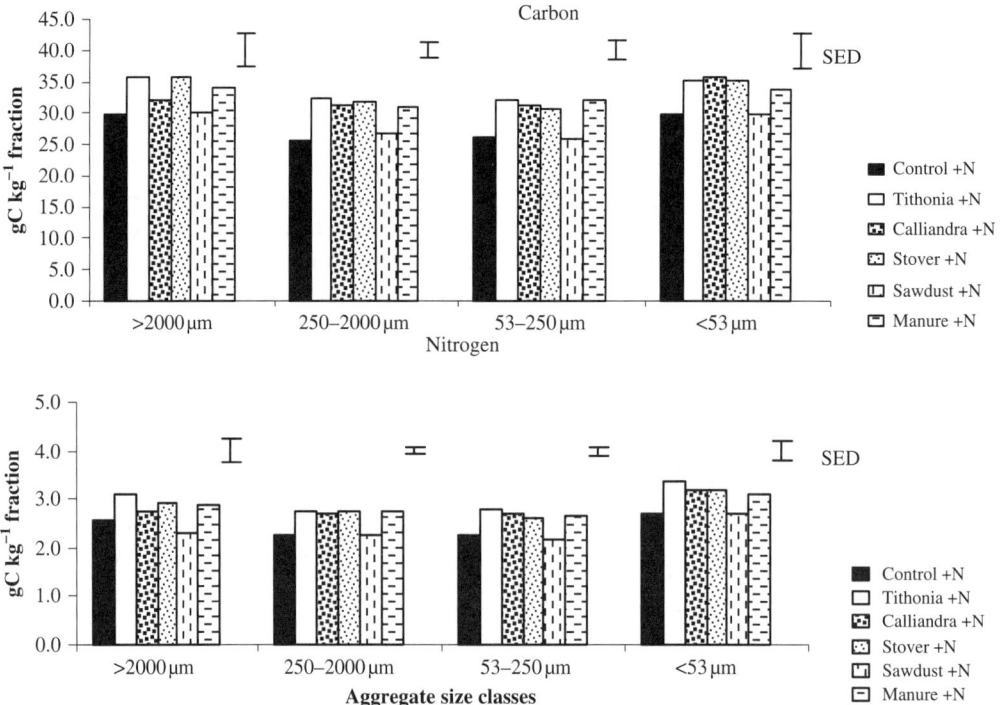

Fig. 3 Effect of combining organic resource quality and mineral fertilizer on the C and N concentration of aggregate size classes at Embu in March 2005

Effect of Resource Quality and Combining of Organic Resource Quality and Mineral Fertilizer on Carbon-13 Signatures

Whole-Soil Carbon-13 Signatures

Significant differences ($p \leq 0.05$) in $\delta^{13}C$ values were observed between the sole OR treatments on the whole-soil samples (Table 4). Manure minus N had the most negative $\delta^{13}C$ value (−19.7‰), followed by *Tithonia* (−19.3‰), *Calliandra* (−19.0‰), sawdust (−18.7‰), maize stover (−18.4‰) and control (18.1‰). Interaction between OR and mineral fertilizer on the whole soil was only significant for the manure treatment. Manure plus N had a $\delta^{13}C$ value of −19.7‰ while manure minus N had a $\delta^{13}C$ value of −19.2‰.

Table 4 Effect of resource quality and the combination of organic resources (ORs) and mineral resources (MRs) on whole-soil C-13 signatures at Embu in March 2005

| Treatment | Delta PDB (‰) | |
	−N	+N
Control	−18.1	−18.3
Tithonia	−19.3	−20.1
Calliandra	−19.0	−19.1
Stover	−18.4	−18.7
Sawdust	−18.7	−18.2
Manure	−19.7	−19.2
SED	0.45*	0.48*

SED = standard error of differences of means; SED* = difference between OR treatments is statistically significant at $p < 0.05$ and/or there is an interaction between resource quality and mineral fertilizer; −N = organic resource only; +N = organic resource plus fertilizer; PDB = Pee De Belemnite

Effect of Resource Quality and Combining of OR and MR on Aggregate Size Class Fractions Carbon-13 Signatures

Organic resource (OR) quality and combined organic resource quality and mineral fertilizer (MR) did not significantly affect the $\delta^{13}C$ values within the large macroaggregate size class (>2,000) (Table 5). However, significant OR treatment and MR effects were observed on the small macroaggregate size class (250–2,000) (Table 5). Among the sole OR treatments, only ^{13}C signatures of soils amended with *Tithonia*, *Calliandra* and manure were significantly different from the control. Interaction between MR and OR was only statistically significant in the soil under manure treatment. The sole application of manure had a more negative ^{13}C signature compared to the manure plus MR treatment. A similar trend was noted within the microaggregates (53–250) and s+c (<53) fractions. Overall, the ^{13}C signatures of all the aggregate classes were more negative for the soils under *Tithonia*, *Calliandra* and sawdust treatments compared to the control and maize stover treatment.

Discussion

Aggregate Proportions

The results indicated that the use of organic resources (ORs) solely does not significantly affect the formation and distribution of aggregates. However, the

Table 5 Effect of resource quality and combining of OR and MR on aggregate size class fractions carbon-13 (^{13}C) signatures at Embu in March 2005

| Treatment | Delta ^{13}C(‰) | | | | | | | |
| | >2,000 µm | | 250–2,000 µm | | 53–250 µm | | <53 µm | |
	−N	+N	−N	+N	−N	+N	−N	+N
Control	−18.46	−18.63	−18.12	−18.20	−18.15	−18.32	−18.93	−18.83
Tithonia	−19.68	−19.25	−19.44	−19.45	−19.63	−19.80	−19.93	−19.72
Calliandra	−18.54	−18.79	−19.17	−19.21	−19.27	−19.18	−19.78	−19.65
Stover	−18.73	−18.85	−18.62	−18.59	−18.72	−18.74	−19.04	−19.44
Sawdust	−18.54	−17.94	−18.82	−18.32	−18.65	−18.33	−18.76	−19.02
Manure	−20.12	−19.26	−19.91	−19.43	−19.86	−19.49	−20.34	−19.69
SED	0.51	0.72	0.34*	0.37	0.41*	0.43	0.33*	0.40

SED = standard error of differences of means; SED* = difference between OR treatments is statistically significant at $p < 0.05$ and/or there is an interaction between resource quality and mineral fertilizer

combination of sawdust and mineral N showed a positive interaction and resulted in an increase in large macroaggregate formation compared to the sole application of sawdust. Organic resource quality has been found to be an important modifier of the interaction between OR and MR (Vanlauwe et al., 2001). Therefore, the high quantity of macroaggregates observed with the combined application of MR and sawdust (low-quality OR) could be attributed to immobilization of applied mineral N by microbial activity induced by addition of sawdust. This in turn leads to the formation of aggregates around the applied OR.

Residue quality and mineral N level govern the rate of aggregate formation and breakdown (i.e. aggregate turnover) (Harris et al., 1996). Therefore, the high lignin and low N content of sawdust (low-quality OR) could also have contributed to N immobilization, slow decomposition of sawdust (Tian et al., 1992) and a consequent reduction in aggregate turnover. This is consistent with the findings of Harris et al., (1996) who found that N-limiting conditions reduce aggregate breakdown because microbes slowly decompose the small amounts of C-rich binding agents that are produced. Hence the addition of MR leads to a faster release of the C-rich binding agents due to increased microbial activity and a higher initial formation of macroaggregates as shown in Fig. 1.

Carbon and Nitrogen Content of Whole Soil

There was a general increase in C and N for the *Tithonia*, *Calliandra*, maize stover and manure for both combined and sole application treatments (Table 3) in 2005 compared to the total C and N contents of the soil at the beginning of the trials (29.9 g C kg^{-1} soil and 2.66 g N kg^{-1} soil for C and N, respectively) in 2002. This indicates that continued application of nutrient sources leads to a build-up of soil organic matter in the long term. This finding is consistent with other studies by Cadisch and Giller (2000), who showed that continuous inputs of ORs influence the levels of soil organic matter (SOM) and the quality of some or all its pools. In particular, organic resources serve as precursors to SOM and

influence nutrient availability through mineralization–immobilization patterns (Vanlauwe et al., 2002a). This finding also confirms the argument that the combined application of mineral inputs and ORs possibly generates added benefits either in terms of extra crop yield or extra C build-up (Palm et al., 1997).

Effect of Sole Organic Resources on C and N Contents of Aggregate Fractions

Sole application of organic resources (ORs) led to a higher accumulation of SOM, compared to the combined application of ORs and mineral resources and this is important for improving and maintaining soil fertility. In particular, the contribution of the *Tithonia* treatment to the SOM pool was significant within the macroaggregate and microaggregate size classes. This could be attributed to the high initial C and N content (N > 2.5%) associated with this class of ORs (high-quality OR) and is therefore consistent with the findings by Constantinides and Fownes (1994) who found that N accumulation was most strongly correlated with initial N concentration in the litter material whereas soluble polyphenols were the controlling factors for green material decomposition. Tisdall and Oades (1982) also reported a higher formation of macroaggregates as a result of increased C and N accumulation. This accumulation can only be termed as a short-term benefit because, as other studies have shown, high-quality ORs cause N mineralization (Palm and Sanchez, 1991) and hence C and N loss in the long term.

The high N concentration observed with the maize stover minus N (medium-quality organic resource) treatment within the small macroaggregates could be related to the high N concentration of maize stover observed within the whole-soil characteristics as well as limited N release associated with this class of ORs (class III ORs). Class III residues have been found to cause substantial N immobilization (Vanlauwe et al., 2005). The N immobilization creates a slow N-releasing SOM pool and this induces an intermediate aggregate turnover and optimizes N use.

Overall, the silt and clay fraction had higher C and N concentrations across all sole applied OR treatments compared to the other aggregate classes. This is consistent with the findings by Christensen (1996) who suggested that SOM can be stabilized in the soil through several mechanisms, such as chemical association with silt and clay particles and physical protection inside aggregate structures.

Effect of Combining Organic Resources and Mineral Fertilizer on C and N

The combined application of organic resources (ORs) and mineral resources (MRs) led to a reduced level of C and N contents of large and small macroaggregates compared to the sole application of ORs. An opposite trend was noted within the lower SOM pools; C and N tended to accumulate within the microaggregates and the mineral fraction (silt and clay fraction). The reduced C and N concentrations observed on the combined treatments within the large and small macroaggregate size classes could be attributed to a high aggregate turnover induced by the addition of MR to ORs. According to Harris et al., 1996, addition of mineral N without a C source induces microbial decomposition of the C-rich binding agents, thus leading to a faster aggregate turnover.

In terms of OR quality, *Tithonia* and *Calliandra* contributed significantly to the accumulation of C and N in the lower SOM pools (i.e. microaggregates and silt plus clay fractions). This could be attributed to the high N content (>2.5%) of *Tithonia* (high-quality OR) and *Calliandra* (medium-quality OR). In addition, the high N concentration within the silt and clay fraction could also be attributed to the stabilization of N in the lower SOM pools either as a result of N mineralization (for *Tithonia*) or due to the presence of leached phenolics which bind mineralized N compounds (for *Calliandra*). This confirms the hypothesis that combined application of MR and ORs enhances C sequestration and N stabilization and is consistent with the findings of Constantinides and Fownes (1994) who found that N is an important modifier of N release processes, i.e. N immobilization and mineralization. Since the lower SOM pools (i.e. microaggregates and silt and clay fractions) are important in the formation of macroaggregates, continued application of ORs with or without mineral N leads to long-term sequestration of C and N as well as a continued formation of stable macroaggregates.

Effect of Organic Resources and Mineral Fertilizer on Carbon-13 Signatures

Sole application of manure, *Tithonia*, *Calliandra* and sawdust contributed significantly to the incorporation of particulate organic matter as indicated by the carbon-13 signatures of whole-soil macroaggregates, microaggregates and silt plus clay fractions. This therefore indicates a more C3 input compared to the maize stover and control and this confirms the role played by C3 OR input to C sequestration. The ^{13}C natural abundance technique has been applied in plant/soil systems where the ^{13}C signal of the C input is different from that of the native SOM, for example, where C3 plants ($\delta^{13}C$-28‰) grow on soils derived from C4 (with $\delta^{13}C$-12‰) (Balesdent et al., 1987). Based on this change in ^{13}C-signature of SOM, Cerri et al. (1985) and Balesdent et al. (1987) concluded that the relative contribution of new versus old soil organic C can be quantified using the mass balance of stable isotope contents. Therefore, the finding from this study confirms the expectation that the $\delta^{13}C$ values of *Tithonia*, *Calliandra*, manure and sawdust should be more negative than that of the maize stover (maize residue), since maize crop had been grown in the Embu experiment. Overall, the ^{13}C values of the aggregate classes tended to be more negative than the whole-soil ^{13}C values indicating that recently incorporated organic material tended to accumulate in the various aggregate classes.

Conclusion

From the above results, it can be concluded that maximum short-term benefits in terms of quantity SOM can be achieved by applying high-quality ORs such as *Tithonia* and high-quality manure or by combining ORs of medium quality and mineral fertilizer. Consequently, long-term build-up of SOM can be achieved through sole application of medium-quality ORs such as maize stover and *Calliandra*. On the other hand, application of low-quality ORs such as sawdust, though important in the initial formation of macroaggregates, requires more experimental time to establish their role in SOM build-up in the long term.

Acknowledgements We acknowledge the funding by the National Science Foundation through TSBF-CIAT which has supported this study. We appreciate the hard work of Muriithi in the daily management of the field experiment and the valuable remarks of anonymous reviewers.

References

Balesdent J, Mariotti A, Gulliet B (1987) Natural ^{13}C abundance as a tracer for studies of soil organic matter dynamics. Soil Biol Biochem 19:25–30

Barthes B, Roose E (2002) Aggregate stability as an indicator of soil susceptibility to runoff and erosion; validation at several levels. Catena 47:133–149

Browning GM, Milam FM (1944) Effect of different types of organic materials and lime on soil aggregation. Soil Sci 57:91–106

Cadisch G, Giller KE (2000) Soil organic matter management: the role of residue quality in carbon sequestration and nitrogen supply. In: Rees RM et al (eds) Sustainable management of soil organic matter. CAB International, Wallingford, CT, pp 97–111

Cerri CC, Feller C, Balesdent J, Victoria R, Plenecassagne A (1985) Application du traçage isotopique naturel en 13C, à l'étude de la dynamique de la matiére organique dans lês sols. Comptes Rendus de l'Academie de Science de Paris 9:423–428

Christensen BT (1996) Straw incorporation and soil organic matter in macroaggregate and particle size separates. J Soil Sci 37:125–135

Constaninides M, Fownes JH (1994) Nitrogen mineralization from leaves and litter of tropical plants: relationship to nitrogen, lignin and soluble polyphenol concentrations. Soil Biol Biochem 26:49–55

Diels J, Vanlauwe B, Sanginga N, Coolen E, Merckx R (2001) Temporal variations in plant ^{13}C values and implications for using the ^{13}C technique in long-term soil organic matter studies. Soil Biol Biochem J 33(9):1245–1251

Elliot ET (1986) Aggregate structure and carbon, nitrogen and phosphorous in native and cultivated soils. Soil Sci Soc Am J 50:627–633

Giller KE (2002) Targeting management of organic resources and mineral fertilizers: can we match scientists' fantasies with farmers' realities? In: Vanlauwe B, Sanginga N, Diels J, Merckx R (eds) Balanced nutrient management systems for the moist savanna and humid forest zones of Africa. CAB International, Wallingford, pp 155–171

Harris RF, Allen ON, Chesters G, Attoe OJ (1963) Evaluation of microbial activity in soil aggregate stabilization and degradation by the use of artificial aggregates. Soil Sci Soc Am Proc 27:542–545

Harris RF, Chesters G, Allen ON (1996) Dynamics of soil aggregation. Adv Agron 18:107–169

Hattori T (1988) Soil aggregates as habitats of microorganisms. Rep Inst Agric Res Tohoku Univ 37:23–36

Kirchmann H, Thorvaldsson G (2000) Challenging targets for future agriculture. Eur J Agron 12:145–161

Linquist BA, Singleton PW, Yost RS, Cassman KG (1997) Aggregate size effects on the sorption an release of phosphorus in an ultisol. Soil Sci Soc Am J 61:160–166

Palm CA, Myers RJK, Nandwa SM (1997) Combined use of organic and inorganic nutrient sources for soil fertility maintenance and replenishment. In: Buresh RJ, Sanchez PA, Calhoun F (eds) Replenishing soil fertility in Africa. Soil Science Society of America special publication number 51. Soil Science Society of America, Madison, WI, USA, pp 193–217

Palm CA, Sanchez PA (1991) Nitrogen release from the leaves of some tropical legumes as affected by their lignin and polyphenolic contents. Soil Biol Biochem 23:83–88

Prove BG, Loch RJ, Foley JL, Anderson VJ, Younger DR (1990) Improvements in aggregation and infiltration characteristics of a krasnozem under maize with direct drill and stubble retention. Austr J Soil Res 28:577–590

Robertson GP (1997) Nitrogen use efficiency in row crop agriculture. Crop nitrogen use and soil nitrogen loss. In: Jackson LE (ed) Ecology in agriculture. Academic, New York, NY, pp 347–365

Sakala W, Cadisch G, Giller KE (2000) Interactions between residues of maize, pigeonpea, and mineral N fertilizers during decomposition and N mineralization. Soil Biol Biochem 32:699–706

Scow KM, Johnson CR (1997) Effect of sorption on biodegradation of soil pollutants. Adv Agron 58:1–56

Sexstone AJ, Revsbech NP, Parkin TB, Tiedje JM (1985) Direct measurement of oxygen profiles and denitrification rates in soil aggregates. Soil Sci Soc Am J 64: 2149–2155

Six J, Carpentier A, van Kessel C, Merckx R, Harris D, Horwath WR, Luscher A (2001) Impact of elevated CO_2 on soil organic matter dynamics as related to changes in aggregate turnover and residue quality. Plant Soil 234:27–36

Six J, Elliott ET, Paustian K (2000) Soil macroaggregate turnover and microaggregate formation: a mechanism for C sequestration under no-tillage agriculture. Soil Biol Biochem 32:2099–2103

Tian G, Kang BT, Brussaard L (1992) Biological effects of plant residues with contrasting chemical compositions under humid tropical conditions – decomposition and nutrient release. Soil Biol Biochem 24:1051–1060

Tisdall JM, Oades JM (1982) Organic matter and water-stable aggregates in soils. J Soil Sci 33:141–163

Vanlauwe B, Aihou K, Aman S, Iwuafor ENO, Tussah BO, Judies N, Sanginga N, Lyasse O, Merckx R, Dickers J (2001) Maize yield as affected by organic inputs and urea. Agron J 93:1191–1199

Vanlauwe B, Diels J, Sanginga N, Merckx R (2002a) Integrated plant nutrient management in Sub-Saharan Africa: from concept to practice. CAB International, Wallingford, CT, 352pp

Vanlauwe B, Diels J, Aihou K, Iwuafor ENO, Lyasse O, Sanginga N, Merckx R (2002b) Direct interactions between N fertilizer and organic matter: evidence from trials with 15N labelled fertilizer. In: Vanlauwe B, Diels J, Sanginga N, Merckx R (eds) Integrated plant nutrient management in sub-Saharan Africa: from concept to practice. CAB International, Wallingford, UK, pp 173–184

Vanlauwe B, Diels J, Sanginga N, Merckx R (2005) Long-term integrated soil fertility management in South-western Nigeria: crop performance and impact on the soil fertility status. Plant Soil 273:337–354

Performances of Cotton–Maize Rotation System as Affected by Ploughing Frequency and Soil Fertility Management in Burkina Faso

K. Ouattara, G. Nyberg, B. Ouattara, P.M. Sédogo, and A. Malmer

Abstract On-farm experiments were conducted on two soil types (Lixisol and Luvisol) in the western cotton area of Burkina Faso with the objective to develop sustainable water and soil fertility management techniques that improve cotton–maize productivity. The hypothesis that reducing ploughing frequency with addition of organic and mineral fertilizers may improve cotton (*Gossypium hirsutum*) and maize (*Zea mays L.*) productions was tested. The treatments were combination of two tillage regimes (annual oxen ploughing, AP and ploughing/manual scarifying, RT) with compost (Co) and without compost (nCo) application. The treatment annual ploughing with compost addition (AP + Co) had the highest soil water content (WC) on both the Lixisol and the Luvisol. The cotton yield increase was 46 and 36% on reduced tillage plot with compost additions (RT + Co) compared to the control in the Lixisol and the Luvisol, respectively. In the Lixisol the highest maize grain yield was recorded in the annual ploughing plot with the additional amount of nitrogen equivalent to the compost nitrogen content. Reduced tillage together with compost additions had the highest maize yield in the Luvisol. These results confirmed the hypothesis that reduced tillage with organic and mineral fertilization improved cotton and maize productions.

Keywords Ploughing frequency · Compost · Cotton–maize · Yield · Soil water content · Burkina Faso

Introduction

In developing countries of Africa, the smallholders in the tropical semi-arid areas practice subsistence agriculture with low input of mineral fertilizer and organic material (Bationo et al., 1998; Krogh, 1999; Vanlauwe et al., 2000). In most of cases, agriculture intensification is related to cash crop production because farmers earned financial incomes that make the market fertilizers affordable. For instance, Vanlauwe et al. (2006) reported that in Uganda, crop residues from the remote fields are harvested, fed to livestock and manures applied to crops intended for the market. Alternatively industrial companies support the cash crop production with equipments and chemical inputs. In this system the staple food crops were still cropped without input to compensate the soil nutrient exportation through harvests. One way to make food crops to benefit from some input, at least the remaining effect of fertilizers used for the cash crop, was to put them in a rotation system with cash crop production. Thus, cash crop can be used as the entry point for sustainable soil fertility management in the smallholders' farming systems. In Burkina Faso, the development of cotton–maize production has been done along with the mechanization (animal-drawn equipments and small tractors) and the use of mineral fertilizers, pesticides and herbicides (The World

K. Ouattara (✉)
Institute of Environment and Agricultural Research (INERA),
04 BP 8645 Ouagadougou 04, Burkina Faso
e-mail: korodjouma_ouattara@hotmail.com

Bank, 2004). In the short term, these practices gave satisfactory results. Annual ploughing with mineral fertilizer increased crop production (Nicou and Poulain, 1972; Mando et al., 2005). Chopart and Nicou (1976) demonstrated that root development in Ferric Lixisols was far better under mouldboard ploughing than under soil surface scarification. If performed correctly, the seedbed preparation using ploughing gives the best results with regard to yield and soil water stock (Le Moigne, 1981; Ellmer et al., 2000; Drury et al., 2003).

However, it has also induced advanced soil degradation processes (Lal, 1993; Fall and Faye, 1999; Alcazar et al., 2002; Terzudi et al., 2006). The advanced soil degradation currently observed in ploughing farming systems is mainly due to tillage technique in relation to the rainfall pattern and to soil erosion. This negative effect of ploughing is a concern globally and the concepts of minimum tillage, reduced tillage, conservation tillage, no-till or zero-tillage have developed (Hulugalle and Maurya, 1991; Ellmer et al., 2000). The idea in these concepts is to buffer or to suppress the negative consequences of ploughing.

With the increasing population and human pressure on soil, the need for sustainable farming system became a big concern. This necessity is crucial in the areas where the impact of soil degradation may be particularly severe, as people living on the land are resource-poor farmers. The sustainable cropping system in the context of economic limited conditions should increase not only the yields of the cash crop to build the farmer's financial capacity but also the yield of the food crop to meet the food requirement.

Few studies dealt with on-farm performances of tillage and reduced tillage in combination with organic matter in the cotton–maize production system in dry area context. The objective of this study was to contribute to the development of alternative and sustainable water and soil fertility management techniques for improved cotton–maize production in Burkina Faso. It was based on combination of varied tillage frequency, mineral fertilizers and compost application. The hypothesis was that reducing the frequency of annual oxen ploughing for an alternative shallow soil tillage system, with addition of organic material, could enhance soil fertility and improve cotton and maize yields.

Material and Methods

Site Description

This study was carried out on a farm at Bondoukuy (11°51′N, 3°46′W, 360 m.a.s.l.), located in the western cotton zone in Burkina Faso. There are two main characteristic morphopedological units in the area (Kissou, 1994; Ouattara et al., 2006): (i) soils of loamy texture classified as Ferric or Gleyic Luvisol formed on the "low glacis" at low topographic position (~300 m.a.s.l) and (ii) the "plateau" at high topographic position (~380 m.a.s.l) having sandy-loam soils classified as Ferric Lixisol (FAO, 1998).

Thirty years (1969–1998) mean annual rainfall of the area is 850 mm based on the national isohyets' map drawn using the data of the National Meteorology Service. The monthly mean rainfall is monomodally distributed between May and October (Son et al., 2003). The daily maximum temperatures vary between 31 and 39°C, and the average annual potential evapotranspiration amounts to 1900 mm (Somé, 1989). The physical and chemical characteristics of the soils are given in Tables 1 and 2, respectively, and the seasonal total rainfalls during the years of the experiment are given in Table 3.

Experimental Design

The experiments were started in 2003 on eight fields (each cropped for more than 10 years), four on each of the two soil types described above. The field plots did not contain any trees, which is an increasingly common feature with mechanically tilled fields. The treatments, in a split-plot design, were combinations of ox ploughing/hand hoeing, and organic and mineral fertilizers in a cotton–maize (*Gossypium hirsutum–Zea mays*) rotation. The main factor per field was the tillage regime and fertilization regime was investigated in sub-plots. The sub-plots measured 10 m × 8 m, and each field represented one replicate.

Two fertilization treatments were included in the design during the second year of the experimentation to investigate (i) the remaining effect, during the

Table 1 Initial physical properties at 0–140 cm depth in the Lixisol and in the Luvisol of the Bondoukuy area, Burkina Faso

Horizon	BD	Sand (%)	Silt (%)	Clay (%)	Field capacity (v v^{-1}, %)	Wilting point (v v^{-1}, %)
Lixisol						
0–10	1.53 ± 0.05	74.7 ± 2.2	18.9 ± 2.2	6.4 ± 1.6	10.6 ± 0.7	3.8 ± 1.4
10–20	1.52 ± 0.05	72.6 ± 4.3	18.5 ± 3.3	8.9 ± 2.2	12.3 ± 1.3	7.2 ± 1.5
20–30	1.52 ± 0.06	69.6 ± 4.5	18.4 ± 2.1	12.1 ± 3.2	12.7 ± 1.3	8.4 ± 1.4
30–60	1.53 ± 0.08	56.4 ± 4.8	18.6 ± 2.3	25.1 ± 4.1	19.8 ± 1.6	15.4 ± 5.0
60–90	1.53 ± 0.14	43.2 ± 6.8	19.0 ± 1.7	37.8 ± 5.9	23.3 ± 1.1	17.5 ± 2.3
90–100	1.50 ± 0.17	40.4 ± 8.2	19.7 ± 2.2	39.9 ± 6.9		
Luvisol						
0–10	1.51 ± 0.05	40.7 ± 9.2	47.1 ± 8.6	12.2 ± 0.7	20.8 ± 4.3	10.7 ± 3.7
10–20	1.51 ± 0.07	35.4 ± 12.1	42.3 ± 6.3	22.3 ± 5.8	20.5 ± 1.9	16.8 ± 3.4
20–30	1.51 ± 0.05	30.6 ± 12.2	38.9 ± 3.7	30.5 ± 8.5	21.7 ± 3.7	18.5 ± 5.8
30–60	1.51 ± 0.05	27.8 ± 10.0	30.7 ± 2.1	41.5 ± 9.5	26.0 ± 2.4	20.7 ± 1.7
60–90	1.53 ± 0.06	25.6 ± 6.8	30.1 ± 1.4	44.3 ± 5.4	26.4 ± 1.3	20.9 ± 2.2
90–100	1.55 ± 0.05	23.7 ± 7.8	31.7 ± 0.9	44.6 ± 31.7		

Table 2 Initial chemical properties at 0–60 cm depth in the Lixisol and in the Luvisol of the Bondoukuy area, Burkina Faso

Horizon	C (%)	N total (%)	P total (%)	P-Bray (mg kg^{-1})	pH$_{water}$	Base cations (mg g^{-1})
Lixisol						
0–10	0.36 ± 0.07	0.025 ± 0.005	0.0110 ± 0.0009	6.2 ± 0.5	6.3 ± 0.3	0.394 ± 0.091
10–20	0.34 ± 0.07	0.025 ± 0.005	0.0117 ± 0.0013	6.6 ± 0.7	6.2 ± 0.2	0.347 ± 0.065
20–40	0.24 ± 0.04	0.022 ± 0.004	0.0110 ± 0.0012	6.2 ± 0.6	6.2 ± 0.2	0.311 ± 0.042
40–50	0.24 ± 0.03	0.022 ± 0.002	0.0110 ± 0.0009	6.2 ± 0.5	6.0 ± 0.3	0.355 ± 0.135
Luvisol						
0–10	0.56 ± 0.04	0.041 ± 0.004	0.0124 ± 0.0010	7.0 ± 1.0	6.2 ± 0.5	0.528 ± 0.127
10–20	0.43 ± 0.03	0.035 ± 0.008	0.0126 ± 0.0025	7.1 ± 1.4	5.8 ± 0.3	0.549 ± 0.084
20–40	0.35 ± 0.07	0.034 ± 0.008	0.0120 ± 0.0030	6.8 ± 1.7	5.5 ± 0.5	0.531 ± 0.084
40–50	0.28 ± 0.02	0.029 ± 0.004	0.0114 ± 0.0015	6.4 ± 0.8	5.3 ± 0.3	0.561 ± 0.104

Table 3 Total annual rainfall (mm) during the years of the experiment in each soil type zone at Bondoukuy

	2003	2004	2005	2006
Lixisol zone	825	654	688	1088
Luvisol zone	705	550	794	1038

year when maize was cropped, of compost + NPK applied (rCo) in cotton growing year (T6 and T7) and (ii) the additional effect of urea N, equivalent to the amount of nitrogen in the compost (eqN) and to the mineral fertilizer on no compost plots (nCo; T5 and T8), in order to evaluate the N contribution to an eventual compost effect. At each farmer's field there were eight treatments as described in Table 4.

The mineral fertilizer (NPK) was applied at 100 kg ha^{-1} NPK (14 N, 23 P, 14 K) and 50 kg ha^{-1} urea (46% N) for cotton and 100 kg ha^{-1} urea for maize. The compost (15.6 C, 1.01 N, 0.19 P, 0.58 K), made with crop residues and cow dung in a pit, was spread and ploughed at 5 Mg ha^{-1} (dry weight) each other year. In the first year of the experiment, 400 kg ha^{-1} of Burkina natural rock phosphate (27.59 P and 0.53 K) was applied uniformly.

Table 4 Descriptions of treatments applied in the experiments conducted at Bondoukuy in 2003–2006

Treatments		Cotton (2003)	Maize (2004)	Cotton (2005)	Maize (2006)
T1	Annual ploughing (AP)	Ploughing	Ploughing	Ploughing	Ploughing
AP+nCo	NPK, no compost (nCo)	nCo	nCo	nCo	nCo
T2	Annual ploughing (AP)	Ploughing	Ploughing	Ploughing	Ploughing
AP+Co	NPK + compost (Co)	Co	nCo	Co	nCo
T3	Reduced tillage (RT)	Ploughing	Scarifying	Ploughing	Scarifying
RT+Co	NPK + compost (Co)	Co	nCo	Co	nCo
T4	Reduced tillage (RT)	Ploughing	Scarifying	Ploughing	Scarifying
RT+nCo	NPK, no compost (nCo)	nCo	nCo	nCo	nCo
T5	Annual ploughing (AP)	Ploughing	Ploughing	Ploughing	Ploughing
AP+eqN	NPK + equivalent N in compost (eqN)	eqN	nCo	eqN	nCo
T6	Annual ploughing (AP)	Ploughing	Ploughing	Ploughing	Ploughing
AP+rCo	Remaining (NPK + compost) (rCo)	Co	No fertilizer	Co	No fertilizer
T7	Reduced tillage (RT)	Ploughing	Scarifying	Ploughing	Scarifying
RT+rCo	Remaining (NPK + compost) (rCo)	Co	No fertilizer	Co	No fertilizer
T8	Reduced tillage (RT)	Ploughing	Scarifying	Ploughing	Scarifying
RT+eqN	NPK + equivalent N in compost (eqN)	eqN	nCo	eqN	nCo

During cotton cropping years (2003 and 2005), the mineral fertilizer (NPK) was spread at thinning, while the urea was applied at cotton flowering. In the maize cropping years (2004 and 2006), the mineral fertilizer was applied in two times, the first application (NPK + 50 kg ha^{-1} urea) was done at maize thinning and the second (50 kg ha^{-1} urea) at flowering. Fertilizer regimes were based on normal practices, i.e. research-based recommendations from Ministry of Agriculture and compost applications on what could be realistic to farmers.

Plant Material

The cotton variety used in the experiment was STAM-59 A (115 days, at first open boll) developed at the research station of Anié Mono (Togo). It has a potential of 2.6 Mg ha^{-1} on research station and 1.1 Mg ha^{-1} at farmer's conditions. The maize cultivar was SR-22 (105 days, at maturity) developed by IITA Ibadan (Nigeria) and has a potential yield ranging between 4.2 and 5.1 Mg ha^{-1} on station and between 2.6 and 3.7 at farmers' conditions.

Sampling and Measurements

Daily rainfall was recorded using a direct reading rainfall bucket placed on each soil type. Before the experiment, two soil composite samples (bulk of four sub-samples) were taken at the experiment site of each farmer in 0–10-, 10–20-, 20–40- and 40–50-cm soil layers. Soil samples were air-dried, sieved through 2 mm, stored at room temperature and analysed for soil C, N and P contents, exchangeable bases and soil pH. Soil particle size distribution was determined per plot on composite samples, taken using the method described above, from 0- to 10-cm, 10- to 20-cm, 20- to 30-cm, 30- to 60-cm, 60- to 90-cm and 90- to 100-cm layers.

During the rainy seasons of 2004 and 2005, soil moisture was monitored in situ using time domain reflectometer (TDR, IMKO Micromodultechnik, Ettlingen Germany). An IMCO TRIME-FM (Ettlingen, Germany) instrument with a Trime-T3 was used to measure the volumetric soil water contents (SWCs). One tube was installed into the soil for each of the treatments T1–T4 and allowed the soil moisture measurements from 0 to 160 cm soil depth with an increment of 20 cm. Two experiments per soil type were equipped with SWC measurement devices. The measurements were made weekly and after each rainfall event larger than 10 mm during the rainy season.

The biomass (kg ha^{-1}), after harvesting and drying, and grain yield (kg ha^{-1}) were used to assess and compare crop productivities per treatment.

Laboratory Analysis

Soil organic carbon was determined using the Walkley and Black method on composite samples from three sampling points per plot, total N by the Kjeldahl method, soil total P by acid extraction (mixture of H_2SO_4–Se) and soluble P by the Olsen–Dabin method (Baize, 1988; Walinga et al., 1989). The pH_{water} was determined by potentiometric methods in a suspension of soil to water at the ratio of 1:2.5.

Data Analysis

Because of the interaction effect between treatments and soil type, the data was analysed per soil type. ANOVA and comparison of treatments were made using Genstat package (Ver. 9.2, General Statistic, Rothamsted Experimental Station, UK) and least significant differences (LSD) at $P = 0.05$. Repeated measurement analysis was made over the 2 years for the productions of each crop.

Results

Treatment Effects on Soil Surface (0–20 cm) Water Contents over Time

During both 2004 and 2005, annual ploughing with compost addition (AP + Co) had the highest soil water content (WC) on both the Lixisol and the Luvisol from the beginning of the measurements to September (Fig. 1). The lowest soil WC was recorded on the annual ploughing plot without compost application

(AP + nCo) and the reduced tillage plot without compost addition (RT + Co) in the Lixisol, while the lowest soil WC was recorded on RT + nCo in the Luvisol. At the end of September 2004 (21–28), the soil water contents reached the wilting point in both soil types on all the treatments (Fig. 1a). During the cropping year 2005, soil water contents reached the wilting point in all the treatments in October (Fig. 1b).

Effects of Treatments on Soil Total C, N and P Contents

Effect of Tillage Regimes on Soil Nutrient Contents

After 3 years of experimentation, the tillage regimes did not modify significantly the soil C, N and P contents compared to the beginning of the experiment for the Lixisol (Fig. 2a, c, e). Soil management regimes tended to increase the soil N contents ($P = 0.09$) and the soil total P contents ($P = 0.04$) throughout the soil profile in the Luvisol (Fig. 2b, d, f).

Effect of Fertilization Regimes on Soil Nutrient Contents

The compost application plots (Co and rCo) and the application of the N equivalent to the compost's N content (eqN) had significantly higher soil N contents than the control and the initial N contents of the Lixisol (Fig. 3c). At the Luvisol site, the compost application (Co) had higher soil C content in the 0–10 cm soil depth and higher N and P contents throughout the soil profile than the initial soil C, N and P contents (Fig. 3b, d, f).

Main Effects of Soil Type, Year Conditions and Treatments on Cotton and Maize Yields

Over the 2 years of cultivation of each crop type, the soil type and the interaction of year conditions and tillage regime affected significantly cotton fibre and

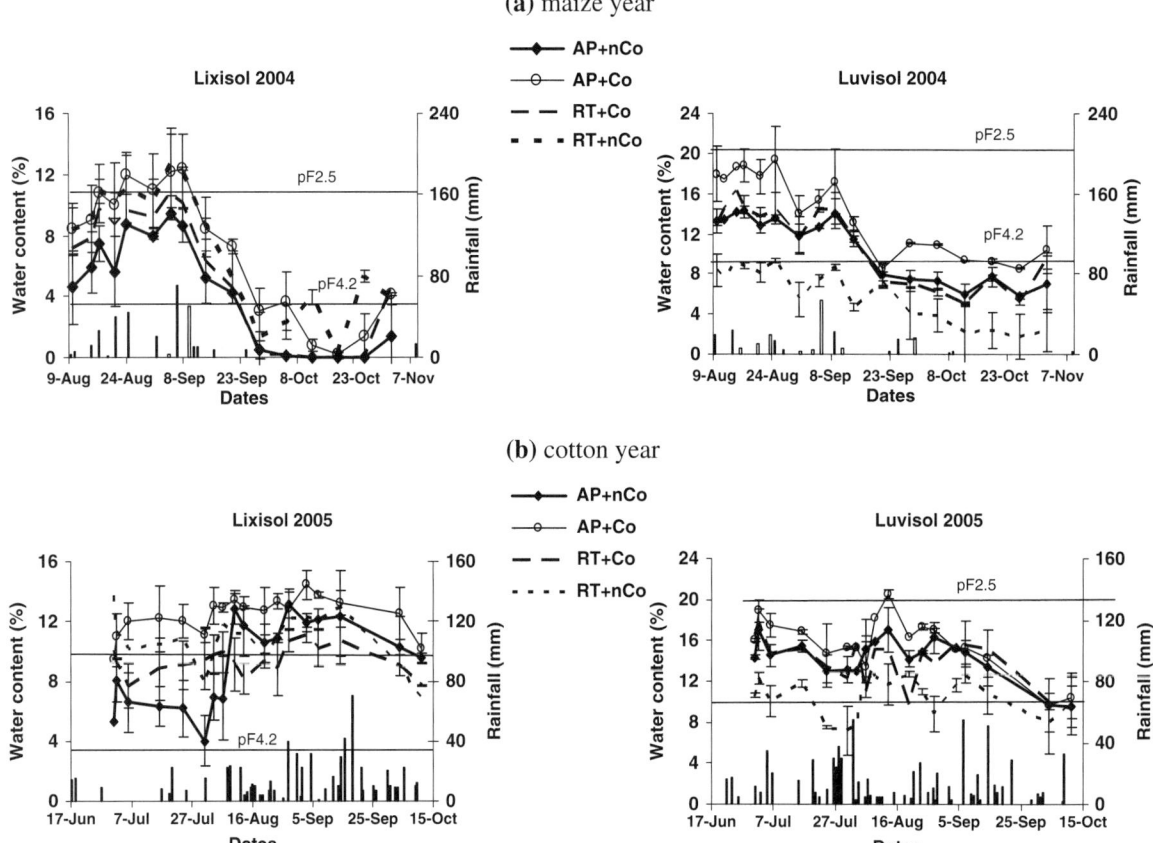

Fig. 1 Rainfall (*bars*, mm) and treatment effects on soil moistures (lines, v/v, %) in 0–20-cm soil layer during the wet season in 2004 and 2005 in the two soil types. *Errors bars* represent standard deviations (std dev.). AP, annual ploughing; RT, reduced tillage; Co, compost; nCo, no compost. (**a**) Maize year and (**b**) cotton year

maize stalk yields, while the fertilization and the combined effect of tillage and fertilization (treatments) affected cotton fibre, cotton and maize stalk yields. The tillage regime effect was significant on maize grain and stalk production. The interaction of year condition–tillage–fertilization significantly affected maize stalk yield (Table 5).

Treatment Effects on Cotton Production

During the first cotton cropping year (2003), all the plots were ploughed and only the fertilization applications were different. Compost application (Co) plot produced 31 and 40% (+687 kg ha^{-1}) more cotton fibre

in the Lixisol and the Luvisol, respectively, than did the plot where compost was not applied (nCo) (Fig. 4a).

In the second cotton cropping year (2005), the productions were not significantly different on the tillage system plots both in the Lixisol and in the Luvisol. The compost application plots in the Lixisol yielded 37% more cotton fibre than did those where compost was not applied. The combinations AP + rCo and AP + Co produced the highest cotton yield whatever be the soil type (Fig. 4b). Over the two cotton cropping years in both the Lixisol and the Luvisol, reduced tillage with addition of compost (RT + Co) had the highest cotton yield and produced +46 and 36% more cotton fibre in the Lixisol and the Luvisol, respectively, than did the control (AP + nCo).

In the Luvisol, cotton biomass production in 2003 was +946 kg ha^{-1} higher on compost application plot

Fig. 2 Changes in soil C, N and P content profiles under ploughing regimes in the Lixisol (**a, c, e**) and the Luvisol (**b, d, f**) in 2005. *Error bars* represent standard error of means (SE). AP, annual ploughing; RT, reduced tillage; Initial, the beginning of the experiment

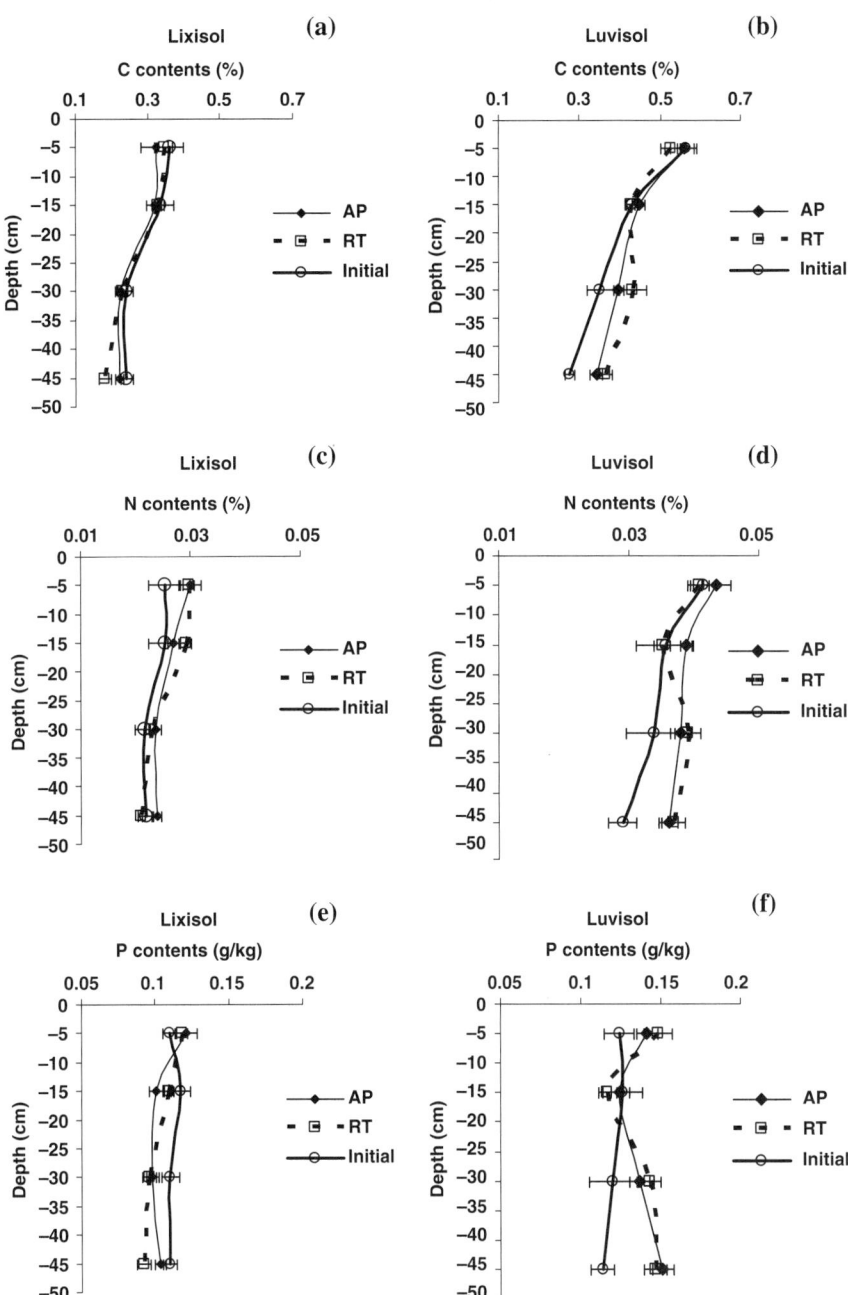

than the plot without compost input. In the Lixisol, there was no significant difference between compost and no compost application plots (Fig. 5a) although the tendency ($P = 0.08$, +623 kg ha^{-1}) was towards higher biomass in compost-treated plot compared to no compost application plot. In 2005, tillage treatment, RT, produced more cotton biomass than did AP plot in the Lixisol, while there was no difference between tillage treatments in the Luvisol. In the Lixisol, the compost-treated plots, Co and rCo, yielded +345 and +366 kg ha^{-1} more cotton biomass, respectively, than did the nCo treatment. In the Luvisol, there was no significant difference in cotton biomass production between fertilization treatments (Fig. 5b).

Fig. 3 Changes in soil C, N and P content profiles under fertilization regimes in the Lixisol (a, c, e) and the Luvisol (b, d, f) in 2005. *Error bars* represent standard error of means (SE). Co, compost; eqN, N equivalent to the compost's N content; nCo, no compost; rCo, remaining compost; Initial, the beginning of the experiment

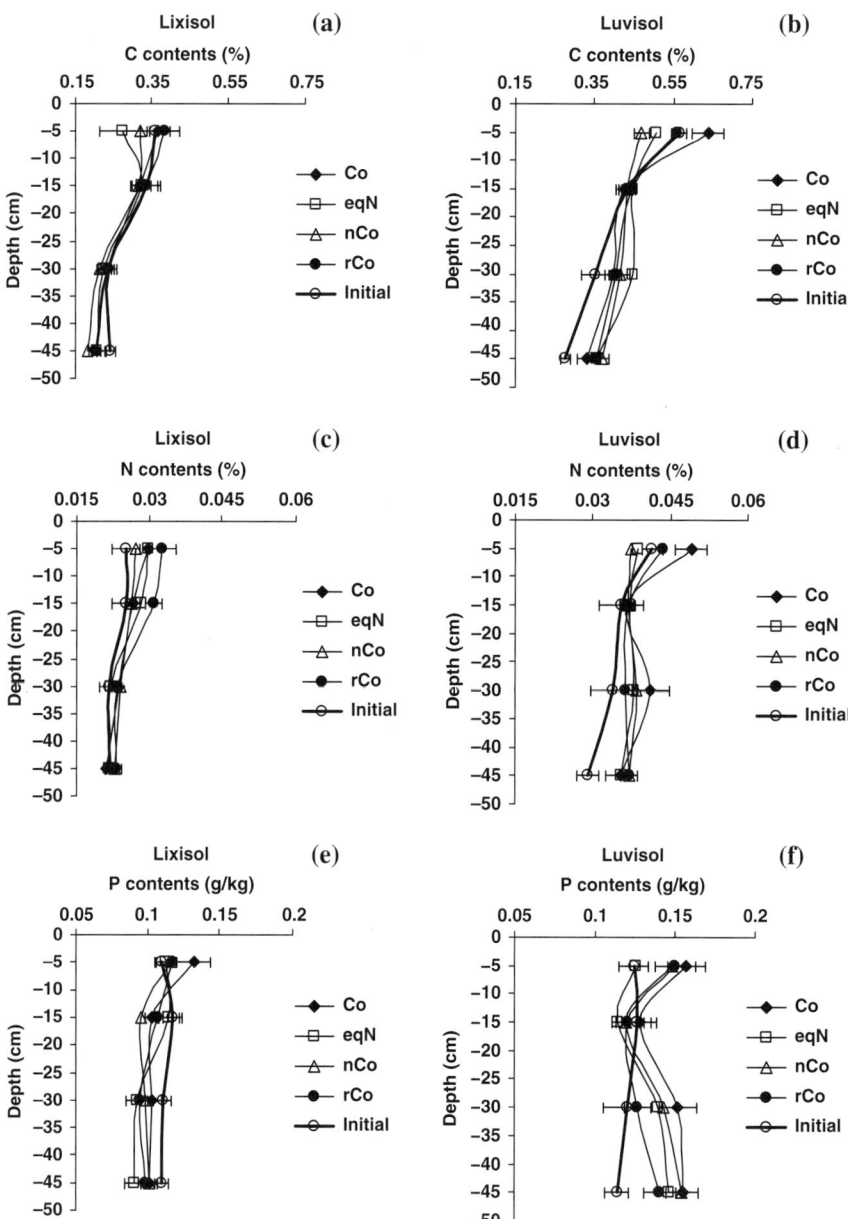

Treatment Effects on Maize Production

In the maize cropping year 2004, in the Lixisol, there were no significant differences in maize grain yields between tillage treatments, while in the Luvisol, annual ploughed plot (AP) yielded 45% (+337 kg ha^{-1}) more than did reduced tillage plot (RT). The compost-treated plots (Co) had the highest yields in both soil types and the lowest maize yields were

recorded on the remaining compost plots without mineral fertilizer application (rCo) in the two soil types (Fig. 6a). In the Lixisol, the combination treatment AP + Co had the highest maize yield and AP + rCo the lowest compared to other treatments, while in the Luvisol, the treatments AP + nCo and AP + Co had three times more maize yields than did the treatment RT + rCo (Fig. 6a). During the second year of maize production (2006), in the Lixisol, RT + rCo-treated plot had the highest maize grain yield, while in the

Table 5 Factors main and interaction effects on cotton and maize yields

| | P values | | | |
| | Grain yields | | Stalk yields | |
Factors	Cotton	Maize	Cotton	Maize
Soil type	**0.014**[*]	0.111	0.096	**0.010**[*]
Year	0.231	**<0.001**[***]	**<0.001**[***]	**<0.001**[***]
Tillage	0.431	**0.030**[*]	0.334	**<0.001**[***]
Fertilizer	**<0.001**[***]	0.320	**0.002**[**]	**0.006**[**]
Soil.year	**0.004**[**]	**0.033**[*]	**<0.001**[***]	0.167
Soil.tillage	0.859	0.162	0.770	0.077
Year.tillage	**0.045**[*]	0.238	0.463	**0.001**[**]
Soil.fertilizer	0.076	0.689	0.406	0.205
Year.fertilizer	0.422	0.080	0.418	**<0.001**[***]
Tillage.fertilizer	**<0.001**[***]	0.124	**0.009**[**]	**0.003**[**]
Soil.year.tillage	0.168	0.602	0.738	0.332
Soil.year.fertilizer	0.260	0.825	0.937	0.793
Soil.tillage.fertilizer	0.345	0.732	0.360	0.822
Year.tillage.fertilizer	0.850	0.458	0.634	**<0.001**[***]
Soil.year.tillage.fertilizer	0.911	0.978	0.887	0.758

[*]Significance at p = 0.05; [**]Significance at p = 0.01; [***]Significance at p < 0.01

Luvisol there were no significant differences between tillage, fertilization and the combination tillage × fertilizer regimes. Although the tendency was to a higher maize yield on RT + Co plot compared to other treatments in this soil type ($P = 0.06$) (Fig. 6b).

The results of maize biomass in 2004 showed higher production on annually ploughed plot (AP) than reduced tillage plot, +411 kg ha^{-1} and +769 kg ha^{-1}, respectively, in the Lixisol and the Luvisol. The compost application plot (Co) had the highest biomass production compared to the rest of fertilization regime in both soil types. In both soil types, the combination of AP + Co yielded the highest biomass (Fig. 7a). In 2006, in the Luvisol, there were no significant differences in maize biomass production between the different treated plots, while in the Lixisol the treatment RT+rCo had the highest biomass production (Fig. 7b). Over the 2 years of maize production, the treatments AP + eqN and RT + Co had the highest grain yields in the Lixisol and the Luvisol, respectively, while in both soil types, the maize biomass production was the highest in the treatment AP + Co.

Correlations Between Crop Production and Soil Water Content

There was a positive relationship between cotton and maize yields and soil water contents in 0–20-cm layer during the period of July to September (Fig. 8a, b).

Discussion

Soil C, N, and P contents did not change substantially in the two soil types compared to the initial soil nutrient contents. That can be ascribed to the short-term evaluation of the soil chemical characteristics and also the amount of fertilizer used. The soil nutrient content measurement has been done 3 years after the application of the different treatments and the compost application occurred in 2 years out of 3 years of experiment. In the Luvisol the compost application plots had higher carbon content compared to the mineral fertilized plots. Soil carbon content tended to decrease in the plots where only mineral fertilizer was applied, probably because of the gradual mineralization and loss of soil organic matter. In the Lixisol there were no significant differences between treatments for soil carbon contents, maybe because of the slow build-up of the organic matter in this sandy-texture soil. In agricultural lands, soil carbon content changes slowly with time and this change is difficult to detect until enough time has elapsed for the change to be larger than the spatial and analytical variabilities in the soil (Entry et al., 1996). Alvarez (2005) reported in a review paper that the accumulation of soil organic carbon under reduced tillage was an S-shaped, time-dependent process, which peaks around 5–10 years, and reached a steady state after 25–30 years.

Cotton and maize production in the two soil types have been differently affected by the applied

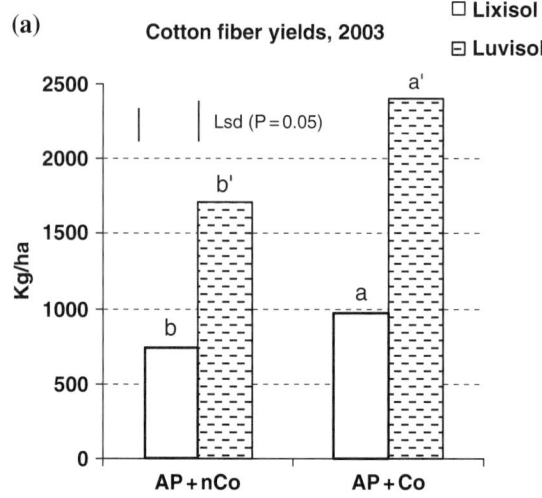

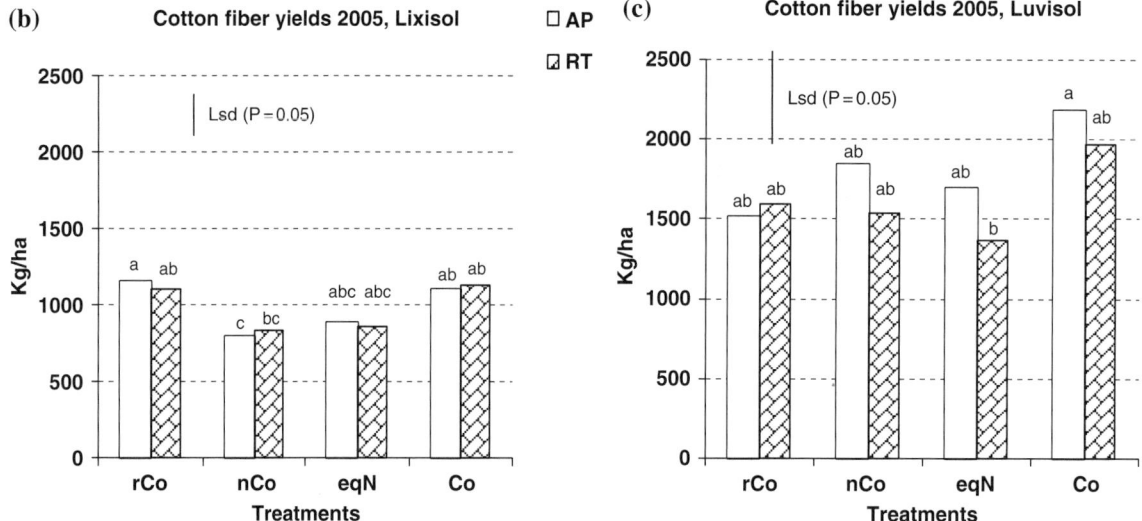

Fig. 4 Effects of treatments on cotton fibre yields (kg ha^{-1}) in the Lixisol and the Luvisol at Bondoukuy during 2003 (**a**) and 2005 (**b, c**). Columns with the same letter were not statistically different

treatments. Interacting with treatment effect, the seasonal rainfalls and their distributions over time also affected significantly the crop performances. Water supply from rainfalls, and the dynamics of water in the soil resulting from the effect of the different treatments, determined the level of crop yields. That was highlighted by the positive relationship between soil water contents and crop yields (Fig. 8). Soil water contents for both soil types were highest in the AP + Co treated plots in 2004 and 2005 because tillage modifies the soil surface, where the partitioning of rainfall into runoff, infiltration and evaporation occurs

(Mambani et al., 1990; Kaumbutho et al., 1999; Keller et al., 2007). Additional application of compost stabilizes pores of different size and therefore increases the persistence of these pores over time (Kay and VandenBygaart, 2002). With RT + Co, this additional benefit of organic material to tillage effect may persist for longer time because of the reduction of organic matter mineralization rate during scarification years. Tillage regime with addition of organic material modifies soil surface structure and total porosity, and therefore has great influence on soil moisture characteristics, water transmission, the depth of the wetting or

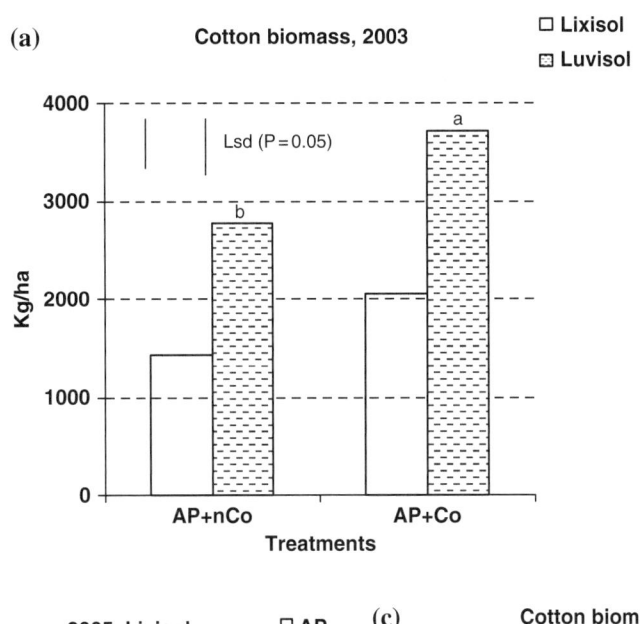

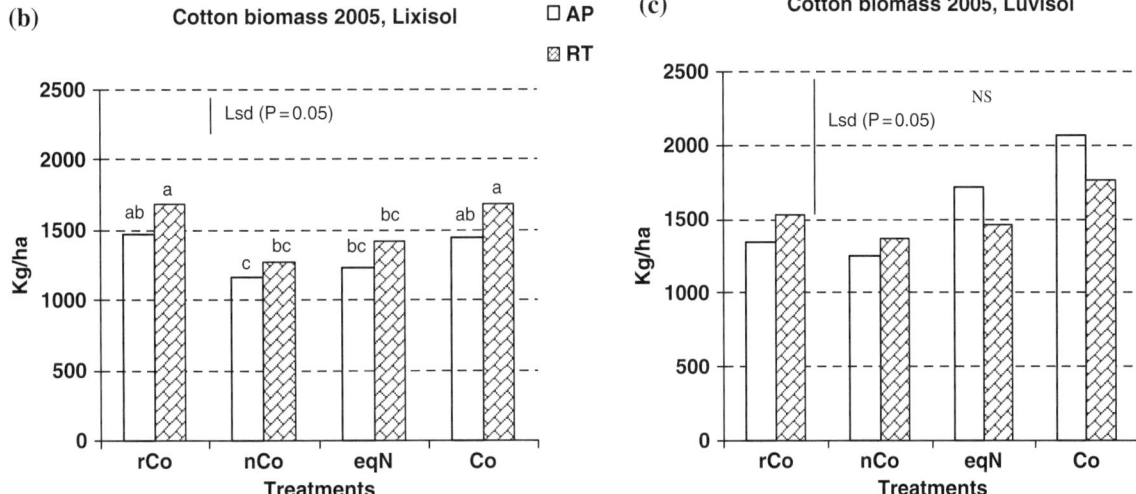

Fig. 5 Effects of treatments on biomass production (kg ha^{-1}) of cotton in the Lixisol and the Luvisol at Bondoukuy during 2003 (**a**) and 2005 (**b, c**). Columns with the same letter were not statistically different. NS, not significant

the drying front and water extraction patterns through the soil profile (Ghuman and Lal, 1984; Scopel et al., 2001; Ouattara et al., 2007).

In tropical environments, when rainfall during the cropping season is limited and/or irregularly distributed, rainfed agricultural productivity is strongly determined by crop water use (Scopel et al., 2001; Somé and Ouattara, 2005). In contrast, when rainfall during the crop season is evenly distributed over time, crop production depends mainly on nutrient availability and weed control (Jourdain et al., 2001). These

results are supported by ours on cotton and maize yields at Bondoukuy. The rainy seasons 2003 and 2005 can be considered as average rainfall years with the annual rainfall ranging between 700 and 800 mm in the areas of the two soil types. In contrast, the crop growing season 2004 was a dry year and 2006 an excess rainfall year. These differences in rainfall patterns interacted with tillage and fertilization regime to affect cotton and maize productivities. Over the 2 years (2003 and 2005) the production of cotton was the highest in reduced tillage with addition of compost

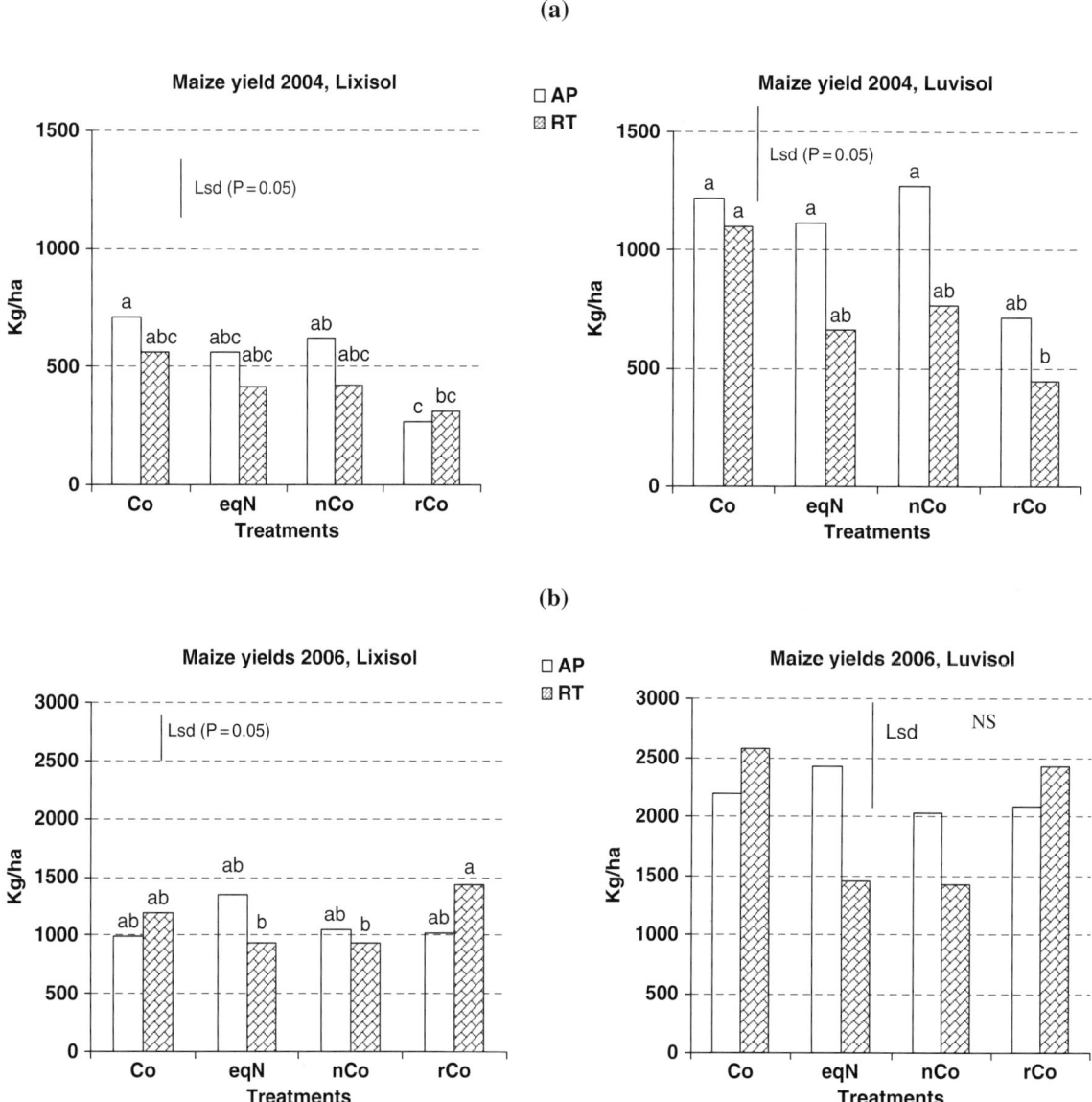

Fig. 6 Effects of treatments on maize grain yields (kg ha^{-1}) in the Lixisol and the Luvisol at Bondoukuy during 2004 (**a**) and 2006 (**b**). Columns with the same letter were not statistically different. NS, not significant

(RT + Co) in the Lixisol, while the highest production was recorded in the treatment of annual ploughing with compost addition (AP + Co) in the Luvisol (Figs. 4 and 5). Maize production was low in the dry year and better in the excess rainfall year, over which AP + eqN and AP + Co treatments had the highest production in the Lixisol and the Luvisol, respectively (Figs. 6 and 7). Reduced tillage had negative impact on maize yield during the dry year because maize crop suffered from

drought stress, and maize is very susceptible to drought during flowering and the first weeks of grain filling (Vanlauwe et al., 2001). Several authors have also shown that reduced tillage and no tillage have considerable potential for stabilizing production in semi-arid zones, but can have contrasting consequences on water regime and yields (Lal et al., 1978; Chopart and Koné, 1985). Furthermore, reduced tillage in some ecosystems and on-farm conditions lead to loss of yield

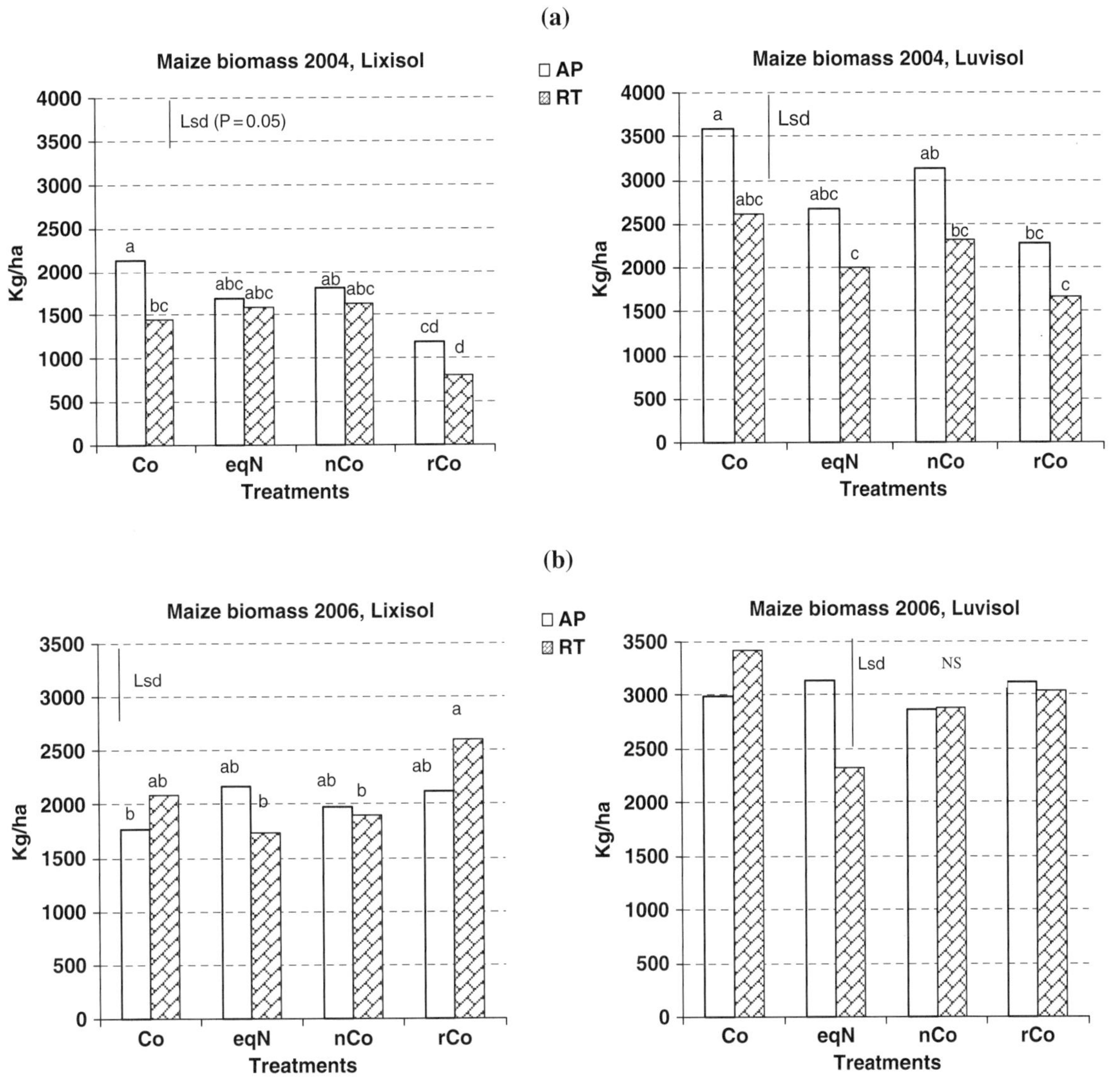

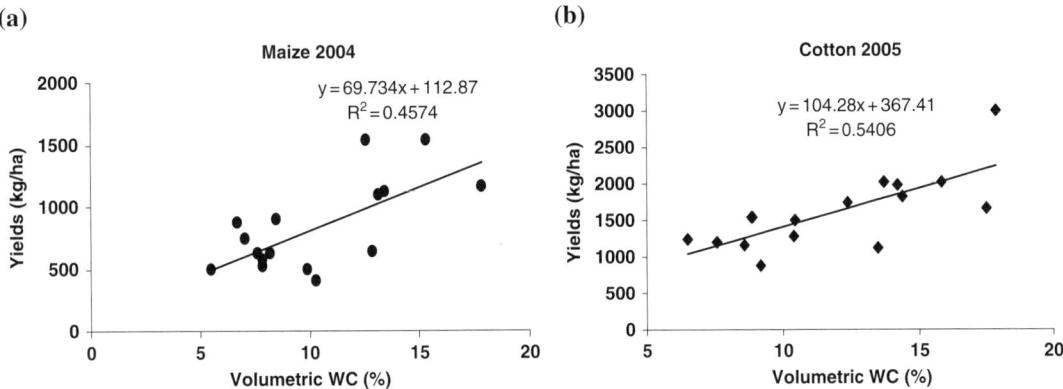

(a)

Maize biomass 2004, Lixisol

□ AP
▨ RT

Maize biomass 2004, Luvisol

(b)

Maize biomass 2006, Lixisol

□ AP
▨ RT

Maize biomass 2006, Luvisol

Fig. 7 Effects of treatments on maize biomass production (kg ha^{-1}) in the Lixisol and the Luvisol at Bondoukuy during 2004 (**a**) and 2006 (**b**). Columns with the same letter were not statistically different. NS, not significant

(a)

Maize 2004

$y = 69.734x + 112.87$
$R^2 = 0.4574$

(b)

Cotton 2005

$y = 104.28x + 367.41$
$R^2 = 0.5406$

Fig. 8 Relationship between soil water contents (0–20 cm depth) and maize (**a**) and cotton (**b**) yields

caused by increases in weed populations and topsoil compaction (Randy et al., 2000; Scopel et al., 2001). In our study, reduced tillage was a sequence of ploughing/scarification over 2 years with compost application or without compost. In the long term, reduced tillage with compost application may improve soil structure and minimize the risk of compaction of these soils that are prone to hardening and give favourable soil condition for crop growth and production. The positive effect of compost and mineral fertilizer addition on cotton and maize production confirmed the generally accepted idea that to increase crop production in West Africa, both inorganic and organic inputs are needed (Vanlauwe et al., 2001). Organic inputs are needed to maintain the physical and chemical health of soils, while fertilizers are needed to supply a readily available amount of nutrients to the crop.

Conclusion

The applied treatments, and also the climatic conditions, have affected cotton and maize performances during the experiment. After 2 years of cultivation of each crop species, cotton and maize, the main results were as follows:

(i) In both the Lixisol and the Luvisol, cotton yield was improved on reduced tillage regime with addition of compost compared to the recommended practice, annual ploughing with application of the complex fertilizer NPK.

(ii) In the Lixisol, the highest production of maize has been recorded on the treated plot of annual ploughing with addition of supplementary nitrogen equivalent to the compost N content. In the Luvisol, the best maize yield was harvested on the plot of annual ploughing with addition of compost.

(iii) During a deficit or an erratic rainy season when scarification was applied, the maize yield was less on reduced tillage than annual ploughing.

The decrease in crop performances on reduced tillage treatment in dry year is expected to be minimized in the long term with the additional effect of organic matter.

References

Alcazar J, Rothwell RL, Woodard PM (2002) Soil disturbance and the potential for erosion after mechanical site preparation. North J Appl For 19:5–13

Alvarez R (2005) A review of nitrogen fertilizer and conservation tillage effects on soil organic carbon storage. Soil Use Manag 21:38–52

Baize D (1988) Guide des analyses courantes en pédologie, Choix – Expressions – Présentation – Interprétation. INRA, Paris

Bationo A, Lompo F, Koala S (1998) Research on nutrient flows and balances in west Africa: state-of-the-art. Agric Ecosyst Environ 71:19–35

Chopart JL, Koné D (1985) Influence de différentes techniques de travail du sol sur l'alimentation hydrique du maïs et du cotonnier en Côte d'Ivoire. L' Agronomie tropicale 40:223–229

Chopart JL, Nicou R (1976) Influence du labour sur le développement radiculaire de differentes plantes cultivées au Sénégal, conséquences sur l'alimentation hydrique. Agron Trop 31:7–28

Drury CF, Tan CS, Reynolds WD, Welacky TW, Weaver SE, Hamill AS, Vyn TJ (2003) Impacts of zone tillage and red clover on corn performance and soil physical quality. Soil Sci Soc Am J 67:867–877

Ellmer F, Peschke H, Kohn W, Franl-M CF, Baumecker M (2000) Tillage and fertilizing effects on sandy soils. Review and selected results of long-term experiments at Humboldt—University Berlin. J Plant Nutr Soil Sci 163:267–272

Entry JA, Mitchell CC, Backman CB (1996) Influence of management practices on soil organic matter, microbial biomass and cotton yield in Alabama's Old Rotation. Biol Fertil Soils 23:353–358

Fall A, Faye A (1999) Minimum tillage for soil and water management with animal traction in the West-African region. In: Kaumbutho PG, Simalenga TE (eds) Conservation tillage with animal traction. A resource book of the ATNESA. Hararé, Zimbabwé, pp 141–147

FAO (1998) World reference base for soil resources. In World soil resources reports, FAO (ed), p 98. http://www.fao.org/docrep/W8594E/W8594E00.htm, Rome

Ghuman BS, Lal R (1984) Water percolation in a tropical Alfisol under conventional ploughing and no-till systems of management. Soil Tillage Res 4:263–276

Hulugalle NR, Maurya PR (1991) Tillage systems for the West African semi-arid tropics. Soil Tillage Res 20:187–199

Jourdain D, Scopel E, Affholder F (2001) The impact of conservation tillage on the productivity and stability of maize cropping systems: a case study in Western Mexico. CIMMYT economics working paper 01-02. CIMMYT, Mexico, DF, p 20

Kaumbutho PG, Gebresenbet G, Simalenga TE (1999) Overview of conservation tillage practices in East and Southern Africa. In: Kaumbutho PG, Simalenga TE (eds) Conservation tillage with animal traction. A resource book of the animal traction network for Eastern and Southern Africa (ATNESA), Harare, Zimbabwe, p 173

Kay BD, VandenBygaart AJ (2002) Conservation tillage and depth stratification of porosity and soil organic matter. Soil Tillage Res 66:107–118

Keller T, Arvidsson J, Dexter AR (2007) Soil structures produced by tillage as affected by soil water content and the physical quality of soil. Soil Tillage Res 92:45–52

Kissou R (1994) Les contraintes et potentialités des sols vis-à-vis des systèmes de culture paysans dans l'Ouest Burkinabé. Cas du Plateau de Bondoukui. In: Institut du Developpement Rural. Université de Ouagadougou, Ouagadougou, p 94

Krogh L (1999) Soil fertility variability and constraints on village scale transects in Northern Burkina Faso. Arid Soil Res Rehabil 13:17–38

Lal R (1993) Tillage effects on soil degradation, soil resilience, soil quality, and sustainability. Soil Tillage Res 27:1–8

Lal R, Wilson GF, Okigbo BN (1978) No-tillage farming after various grasses and leguminous cover crops in tropical Alfisols I. Crop performance. Field Crops Res 1:71–84

Le Moigne M (1981) Contraintes posées par l'insertion de la mécanisation dans les unités de production agricoles en zone sahélienne. Etude méthodologique. CEEMAT, Paris

Mambani B, Datta SK, Redulla CA (1990) Soil physical behaviour and crop responses to tillage in lowland rice soils of varying clay content. Plant Soil V 126:227–235

Mando A, Ouattara B, Sédogo M, Stroosnijder L, Ouattara K, Brussaard L, Vanlauwe B (2005) Long-term effect of tillage and manure application on soil organic fractions and crop performance under Sudano-Sahelian conditions. Soil Tillage Res 80:95–101

Nicou R, Poulain JF (1972) Les effets agronomiques du travail du sol en zone tropicale sèche. Mach Agric Trop 37:35–41

Ouattara K, Ouattara B, Nyberg G, Sédogo MP, Malmer A (2007) Ploughing frequency and compost application effects on soil infiltrability in a cotton–maize (*Gossypium hirsutum–Zea mays L.*) rotation system on a Ferric Luvisol and a Ferric Lixisol in Burkina Faso. Soil Tillage Res. doi:10.1016/j.still.2007.01.008

Ouattara B, Ouattara K, Serpantié G, Mando A, Sédogo MP, Bationo A (2006) Intensity cultivation induced effects on soil organic carbon dynamic in the western cotton area of Burkina Faso. Nutr Cycling Agroecosyst 76:331–339

Randy L, Raper D, Reeves W, Schwab EB, Burmester CH (2000) Reducing soil compaction of Tennessee Valley soils in conservation tillage systems. J Cotton Sci 4(2):84–90

Scopel E, Tardieu F, Edmeades GO, Sebillotte M (2001) Effects of conservation tillage on water supply and rainfed maize production in semi-arid zones of West–central Mexico, NRG paper. CIMMYT, Mexico, DF

Somé L (1989) Diagnostic agropédologique du risque de sécheresse au Burkina Faso. Etude de quelques techniques agronomiques améliorant la résistance sur les cultures de sorgho, de mil et de maïs. USTL Montpellier, Montpellier, p 268

Somé L, Ouattara K (2005) Irrigation de complément et culture du sorgho au Burkina Faso. Agron Afr XVII:163–253

Son G, Bourarach EH, Ashburner J (2003) The issue of crops establishment in Burkina Faso western area. CIGR J Sci Res Dev 5:1–10

Terzudi CB, Mitsios J, Pateras D, Gemtos TA (2006) Interrill soil erosion as affected by tillage methods under cotton in Greece. CIGR Ejournal VIII:1–21

The World Bank (2004) Cotton cultivation in Burkina Faso – A 30 year success story. In: Scaling up poverty reduction: a global learning process and conference Shanghai, May 25–27, 2004

Vanlauwe B, Aihou K, Aman S, Iwuafor ENO, Tossah BK, Diels J, Sanginga N, Lyasse O, Merckx R, Deckers J (2001) Maize yield as affected by organic inputs and urea in the west African moist savanna. Agron J 93:1191–1199

Vanlauwe B, Aihou K, Aman S, Tossah BK (2000) Nitrogen and phosphorus uptake by maize as affected by particulate organic matter quality, soil characteristics, and land-use history for soils from the West African moist savanna zone. Biol Fertil Soils 30:440–449

Vanlauwe B, Tittonell P, Mukalama J (2006) Within-farm soil fertility gradients affect response of maize to fertiliser application in western Kenya. Nutr Cycling Agroecosyst 76:171–182

Walinga I, Vork W, Houba VJG, Lee JJ (1989) Soil and plant analysis, Part 7. Department of Soil Science and Plant Nutrition, Wageningen Agricultural University, Wageningen, 263p

Developing Standard Protocols for Soil Quality Monitoring and Assessment

B.N. Moebius-Clune, O.J. Idowu, R.R. Schindelbeck, H.M. van Es, D.W. Wolfe,
G.S. Abawi, and B.K. Gugino

Abstract Africa's agricultural viability and food security depend heavily on its soil quality. However, while approaches to measuring air and water quality are widely established, standardized, publicly-available soil quality assessment protocols are largely non-existent. This chapter describes the process we have used in selecting and developing a set of inexpensive, agronomically meaningful, low-infrastructure-requiring indicators of soil quality (SQ), which make up the Cornell Soil Health Test (CSHT). In 2006, the CSHT was made available to the public in New York State (NYS), United States, similar to the widely available soil nutrient tests. Case studies show the CSHT's success at measuring constraints in agronomically essential soil processes and differences between management practices in NYS. It thus helps farmers to specifically target management practices to alleviate quantified constraints. Such indicators have the potential to be developed into standardized soil quality tests for use by African agricultural non-governmental and government organizations and larger commercial farmers to better understand agricultural problems related to soil constraints and to develop management solutions. Low cost and infrastructure requirements make these tests excellent tools for numerous low-budget extension and NGO-based experiments established in collaboration with local small farmers, as well as to quantify the status and trends of soil degradation at regional and national scales.

Keywords Soil health · Soil quality · Soil quality assessment · Soil quality indicators · Soil quality monitoring

Introduction

Soil Quality Degradation and the Need for Standards

Declining soil quality (SQ) is emerging as an environmental and economic issue of increasing global concern as degraded soils are becoming more prevalent due to intensive use and poor management, often the result of over-population (Eswaran et al., 2005). Pressing problems such as erosion, compaction, acidification, organic matter losses, nutrient losses, and desertification reduce agricultural production capacity. SQ decline severely impacts the environment and agricultural viability, and thus ecosystems and the population's health, food security, and livelihoods.

Tests to monitor air and water quality have been standardized and widely adopted internationally (Riley, 2001). However, although an estimated 65% of the land area worldwide is degraded (FAO, 2005), no standardized SQ tests exist currently, especially for use in the tropics (Winder, 2003). The World Soils Agenda developed by the International Union of Soil Scientists lists as the first two agenda items (1) assessment of status and trends of soil degradation at the global scale and (2) definition of impact indicators and tools for monitoring and evaluation (Hurni et al., 2006). There is clearly a need for international standards to measure

B.N. Moebius-Clune (✉)
Department of Crop and Soil Sciences, Cornell University,
Ithaca, NY 14853, USA
e-mail: bnm5@cornell.edu

SQ. These could be useful for agricultural research and extension agencies, non-governmental organizations, governments, and farmers to better understand, implement, and monitor sustainable soil management practices.

Soil Quality, Its Assessment, and Indicators

Soil quality includes an inherent and a dynamic component (Carter, 2002). The former is an expression of the soil-forming factors, documented by soil surveys as expressed by land capability classification. Dynamic SQ, however, refers to the condition of soil that is changeable in a short period of time largely due to human impact and management (Carter, 2002). The SQ concept encompasses the chemical, physical, and biological soil characteristics needed to support healthy plant growth, maintain environmental quality, and promote human and other animal health (Doran and Parkin, 1994). With farmer and lay audiences, the term "soil health" is often preferred when referring to this dynamic SQ concept as it suggests a holistic approach to soil management (Idowu et al., 2008).

New regulations have catalyzed a proliferation of various indicators and "environmental report cards" for assessing vulnerability and improvement toward sustainability (Riley, 2001). Indicator suitability can be judged by several criteria, such as relevance, accessibility to users, and measurability (Nambiar et al., 2001). Criteria and thresholds for relevant indicators must then be set by which to assess performance level relative to a standard (Manhoudt et al., 2005).

SQ cannot be measured directly, but soil properties that are sensitive to changes in management can be used as indicators (Larson and Pierce, 1991). Methods for measuring individual indicators and minimum data sets (Dexter, 2004) and for calculating indices from groups of indicators (Andrews et al., 2004) are being developed for SQ monitoring over time and for evaluating the integrated sustainability of agricultural management practices. However, such tests must be inexpensive and dependent on minimal infrastructure if they are to be widely adopted beyond the research domain, especially in the developing regions such as Africa.

Limited experience exists with the use of such methods, other than for standard agricultural soil tests. Such tests have provided farmers and consultants around the world with relevant information for nutrient and lime input management. In a more holistic SQ paradigm, integrative assessment of the three SQ domains (physical, biological, and chemical) would be accomplished by SQ indicators that represent soil processes relevant to soil functions and provide information that is useful for practical soil management. Our approach identifies soil constraints and aids in the selection of management solutions (Idowu et al., 2008). The interpretation of CSHT results requires professional judgment that takes into account the land use objectives and resource availability to devise locally appropriate strategies.

The objective of this paper is to (1) discuss the methods of the selection of key SQ indicators, as implemented through the CSHT, (2) highlight the utility of the test through the results from example cases based in New York State, United States, and (3) discuss the potential for applications of internationally applicable SQ standards using the CSHT as the starting framework.

Methods – Cornell Soil Health Test Development

Approach

The CSHT was developed through a triage process for potential SQ indicators and streamlining of methodologies. The new SQ test was envisioned to provide critical quantitative information that would allow for better management and protection of agricultural soil resources. Specifically, the test was developed for the following reasons: (1) improved soil inventory assessment by adding evaluation of dynamic SQ to the inherent SQ reported in soil surveys; (2) quantifying soil degradation or aggradation from management; (3) targeting management practices to address measured soil constraints; (4) education through addressing site-specific SQ and soil management issues; and (5) land valuation to facilitate monetary rewards for good land management.

Thirty-nine potential SQ indicators were evaluated (Table 1). The suitability of the soil properties as such

Table 1 Thirty-nine soil health indicators evaluated for the Cornell Soil Health Test in 2003–2005

Physical indicators	Biological indicators	Chemical indicators
Bulk density	Root health assessment	pH
Macro-porosity	Organic matter content	Phosphorus
Meso-porosity	Beneficial nematode population	Potassium
Micro-porosity	Parasitic nematode population	Magnesium
Available water capacity	Potentially mineralizable nitrogen	Calcium
Residual porosity	Decomposition rate	Iron
Penetration resistance at 10 kPa	Particulate organic matter	Aluminum
Saturated hydraulic conductivity	Active carbon	Manganese
Dry aggregate size (<0.25 mm)	Weed seed bank	Zinc
Dry aggregate size (0.25–2 mm)	Microbial respiration rate	Copper
Dry aggregate size (2–8 mm)	Soil Proteins	
Wet aggregate stability (0.25–2 mm)		
Wet aggregate stability (2–8 mm)		
Surface hardness (penetrometer)		
Subsurface hardness (penetrometer)		
Field infiltrability		

was evaluated through samples from (i) long-term, replicated research experiments related to tillage, rotation, harvest type, and cover cropping studies; and (ii) commercial farms (including grain, dairy, vegetable, and fruit operations on a number of soil types) that provided real-world perspective under the range of soil management conditions in New York State.

Sampling and Analysis

For all management units, two undisturbed soil core samples were collected from 5 to 66 mm depth using stainless steel rings (61 mm height, 72 mm ID, 1.5 mm wall thickness). Disturbed samples were collected from 5 to 150 mm depth using trowels. All samples were stored at 2°C until analysis.

Physical tests were mostly based on standard methodology, as discussed by Moebius et al. (2007), except for wet aggregate stability which involved the application of simulated rainfall of known energy (Ogden et al., 1997) to aggregates on sieves. Biological tests also mostly involved established methods: decomposition rate was based on loss of filter paper volume over 3 weeks of incubation on soil. The active carbon test involved a $KMnO_4$ oxidation procedure based on work by Weil et al. (2003). The root health assessment involved a bioassay method where sampled soil is planted with snap bean seeds and root damage is rated based on root morphological features (Abawi and Widmer, 2000).

Analysis of the chemical indicators was based on the standard soil fertility test offered by the Cornell Nutrient Analysis Laboratory. Available nutrients are extracted with Morgan's solution, a sodium acetate/acetic acid solution, buffered at pH 4.8. The extraction slurry is filtered and analyzed for K, Ca, Mg, Fe, Al, Mn, and Zn on an inductively coupled plasma spectrometer (ICP) and plant-available PO_4-P is measured using an automated rapid flow analyzer. Using a standard pH meter, pH is determined from a 1:1 soil:water mix.

Indicator Selection

The general criteria used for physical and biological indicator selection into the test included the following: (1) sensitivity to management, i.e., frequency of significant treatment effects in controlled experiments and directional consistency of these effects; (2) precision of measurement method, i.e., coefficients of variability; (3) relevance to important functional soil processes such as aeration, water infiltration/transmission, water retention, root proliferation, nitrogen mineralization, and development of root diseases; and (4) ease and cost of sampling and analysis (Moebius et al., 2007).

Many soil physical properties were rejected as suitable indicators due to the requirement for undisturbed samples or due to high variability. For example, although it is widely regarded as an important physical indicator, bulk density was not included because of the impractical need for undisturbed core samples (Moebius et al., 2007) and generally strong correlations with other physical indicators in the test. The use of ring samplers for bulk density proved to be a serious obstacle with field practitioners and technicians. Therefore, the reliability of the results was questionable due to frequent improper sampling, especially with soils containing coarse fragments. Many soil biological indicators were rejected due to the high cost of analysis, often associated with labor intensity and high variability. The seven soil chemical indicators adopted in the integrated SQ test are part of well-established standard soil nutrient analysis tests that are widely used at reasonable cost in NYS.

Selected Test Indicators

Table 2 shows the physical, biological, and chemical indicators that have been selected for the soil health test (Idowu et al., 2008). These are indicators of critical soil processes (e.g., aeration, infiltration, water and nutrient retention, root proliferation, N mineralization, toxicity prevention, and pest suppression), which in turn relate to soil functions such as plant production, landscape water partitioning and filtration, and habitat support. The standard soil health test thereby evaluates the soil's ability to accommodate ecosystem functioning within landscapes.

The indicators are measured based on a composited disturbed sample, which we recommend to be obtained from two locations nested within five sites on a management unit (Gugino et al., 2007). The test includes penetrometer measurements as the only in-field assessment. Soil texture is an integrative property and provides the basis for result interpretation through scoring functions. Root health assessment is an integrative biological measurement related to overall pressure from soil-borne disease organisms (Abawi and Widmer, 2000). The minor elements of the chemical analysis were grouped to prevent a bias of the soil health assessment in favor of chemical quality. Based on an economic analysis (Moebius et al., 2007), the standard test was offered for less than US $50 in NYS in 2006. In countries with lower wages, and with careful selection of a subset of indicators, these costs could be significantly reduced.

Most indicators were shown to have significant within-season variability (Moebius et al., 2007), and soil management practices can be a confounding influence for soil physical and biological indicators. Thus, samples should be collected at an appropriate and consistent time to be established regionally. In NYS, early spring sampling prior to tillage is best, due to favorable soil water conditions (near-field capacity) and relatively uniform biological conditions following over-wintering.

Data Interpretation and Scoring Curves

Effective use of soil health test results requires the development of an interpretive framework for the measured data. The general approach of Andrews et al. (2004) was applied for this purpose. Different scoring functions for the three main textural classes, sand, silt, and clay, were developed for all soil indicators to rate test results. Scoring functions were defined in the simple linear-plateau framework in 2006. Three types of scoring functions were considered, "more is better," "less is better," and "optimum" (Fig. 1). The critical high and low cutoff values were developed based on the frequency distribution of data throughout NYS. Test results with values less than the 25th percentile were given scores of 1, and greater than the 75th percentile were given scores of 10. This approach was evaluated relative to literature reports and in some cases, minor modifications were made. Scoring curves for currently available indicators are reported in Gugino et al. (2007).

Soil Health Test Report

A standard soil health test report was designed for practitioner audiences and facilitates both integrative assessment and targeted identification of soil constraints. This is accomplished through the combined use of quantitative measured values and ratings (Fig. 2) and color coding (not shown here; see Gugino et al., 2007 for sample reports in color). The physical, biological, and chemical indicators are grouped (by blue, green, and yellow colors, respectively, in reports provided in the United States). For each indicator, the

Table 2 Soil quality indicators included in the standard Cornell Soil Health Test and associated processes

Soil indicator	Soil process
Physical	
Soil texture	All
Aggregate stability	Aeration, infiltration, shallow rooting, crusting
Available water capacity	Water retention
Surface hardness	Rooting at the plow layer
Subsurface hardness	Rooting at depth, internal drainage
Biological	
Organic matter content	Energy/C storage, water, and nutrient retention
Active carbon content	Organic material to support biological functions
Potentially mineralizable nitrogen	Ability to supply N
Root rot rating	Soil-borne pest pressure
Chemical standard	
pH	Toxicity, nutrient availability
Extractable P	P availability, environmental loss potential
Extractable K	K availability
Minor element contents	Micronutrient availability, elemental imbalances, toxicity

Fig. 1 Models of scoring curves used for the interpretation of measured values of soil quality indicators in 2006

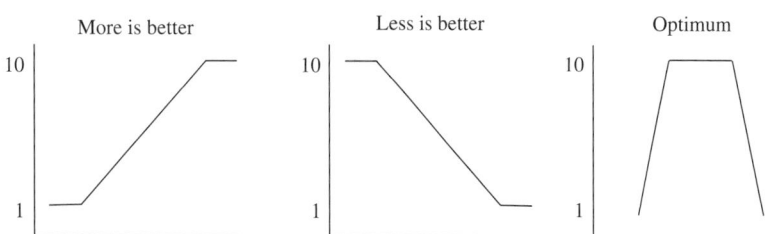

measured value, as well as the associated rating from its scoring function, is reported. The latter is interpreted in that ratings of less than 3 are considered constraints (and thus are coded in red), ratings greater than 8 are considered optimal (and thus receive a green code). Those in between receive an yellow code. This provides for an intuitive overview of the test report. If results are coded red, the associated soil constraints are additionally listed (Figs. 2 and 3). Finally, the percentile rating is shown for each indicator, based on the sample's ranking in the database of soil indicator measurements (Fig. 2). An overall soil health score is provided at the bottom of the report, which is standardized to a scale from 1 to 100. It is noted that the interpretation of the test results is generalized for agricultural systems and may require alternative interpretation in other cases. Hence, we recommend that the reports are interpreted by professional consultants and include consideration of site-specific information.

Soil management recommendations were developed to address specific soil management constraints in NYS agricultural systems (Gugino et al., 2007), which

may be partly applicable to other climates and soil types. A training manual was developed to explain the basic approaches to soil health assessment, the reporting and interpretation of the results, and the suggested management approaches. The most up-to-date version of this manual can be accessed and downloaded from the Cornell Soil Health website at http://soilhealth.cals.cornell.edu.

Results and Discussion

Case Study 1: Two Vegetable Production Scenarios

Figure 2 shows the test reports for two very different scenarios of vegetable production. Figure 2a reports data of a farm near Geneva, NY, on a glacial till-derived Honeoye Lima silt loam. This farm has been used for production of processing vegetables (cabbage,

838

Fig. 2 (**a**) Conventionally managed vegetable farm. (**b**) No-till vegetable garden on organically managed dairy farm
Note: Ratings of less than 3 indicate constraints; ratings greater than 8 indicate optimal functioning

a)

INDICATORS		VALUE	RATING	CONSTRAINT	PERCENTILE RATING*
PHYSICAL	Aggregate Stability (%)	17.6	1.0	aeration, infiltration, rooting	
	Available Water Capacity (m/m)	0.17	2.0	water retention	
	Surface Hardness (psi)	178	4.0		
	Subsurface Hardness (psi)	290	3.0		
BIOLOGICAL	Organic Matter (%)	2.3	1.0	energy storage, C sequestration, water retention	
	ActiveCarbon (ppm)	575	3.0		
	Potentially Mineralizable Nitrogen (µgN/gdwsoil/week)	5.1	3.0		
	Root Health Rating (1-9)	5.6	5.0		
CHEMICAL	pH (see CNAL Report)	7.2	10.0		
	Extractable Phosphorus (see CNAL Report)	9.8	10.0		
	Extractable Potassium (see CNAL Report)	53	7.5		
	Minor Elements (see CNAL Report)		10.0		50th Percentile →BETTER
OVERALL QUALITY SCORE (OUT OF 100)			LOW		49.6

b)

INDICATORS		VALUE	RATING	CONSTRAINT	PERCENTILE RATING*
PHYSICAL	Aggregate Stability (%)	22.1	3.0		
	Available Water Capacity (m/m)	0.32	10.0		
	Surface Hardness (psi)	40	10.0		
	Subsurface Hardness (psi)	145	10.0		
BIOLOGICAL	Organic Matter (%)	4.5	10.0		
	Active Carbon (ppm)	1011	10.0		
	Potentially Mineralizable Nitrogen (µgN/gdwsoil/week)	11.8	10.0		
	Root Health Rating (1-9)	2.4	9.0		
CHEMICAL	pH (see CNAL Report)	6.7	10.0		
	Extractable Phosphorus (see CNAL Report)	37.2	3.0		
	Extractable Potassium (see CNAL Report)	92	10.0		
	Minor Elements (see CNAL Report)		10.0		50th Percentile →BETTER
OVERALL QUALITY SCORE (OUT OF 100)			VERY HIGH		87.5

Fig. 3 Tillage management comparison of soil quality reports of (**a**) PT and (**b**) NT on the same farm in a Honeoye Lima silt loam
Note: Ratings of less than 3 indicate constraints; ratings greater than 8 indicate optimal functioning

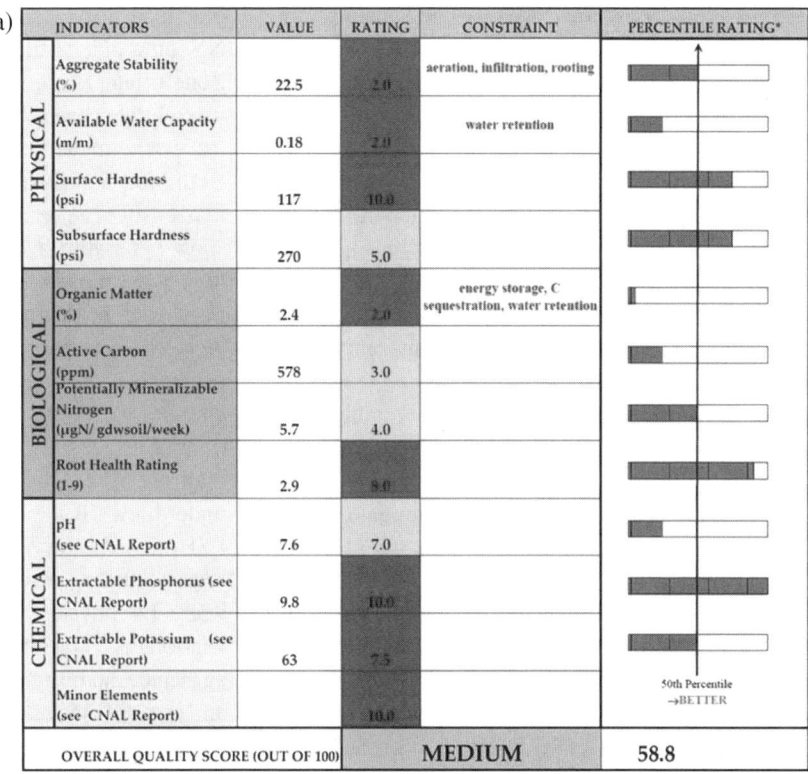

beets, sweet corn, snap beans, etc.) using intensive, conventional (moldboard plow) tillage. Figure 2b reports data of the organic vegetable garden that is part of an organic dairy near Keeseville, NY, on a Nellis/Amenia gravelly sandy loam. The garden is being managed without tillage and with large manure applications. Both test reports show generally favorable results for chemical indicators, with high rating scores (7.5 or above). Only P in the vegetable garden is suboptimally high, likely a result of high manure applications, which could lead to environmental P loading.

For the conventionally managed vegetable operation, the remaining indicators have low scores and therefore show evidence of low physical and biological SQ. Very unfavorable results for aggregate stability, available water capacity, and organic matter content (1, 2, and 1, respectively) are evidence of soil degradation from long-term intensive tillage and limited use of soil-building crops. Low to intermediate scores for active carbon, PNM, and root health (3, 2, and 5, respectively) indicate that the soil is biologically degraded and unbalanced. Scores of 3–4 for soil hardness indicate a mild soil compaction problem. The overall score of 49.5 signifies considerable opportunity for targeted improvement.

Biological and physical SQ of the vegetable garden, by contrast, are high, likely due to careful, concentrated management using ample organic matter additions, crop rotations, and no tillage. All indicators except aggregate stability show ratings of 9 or above. Low aggregation is common in sandy soils. Nevertheless, stability is greater in the vegetable garden than in the conventionally managed operation on a silt loam. Furthermore, high aggregation in a sandy soil is not as essential as with finer textured soils, as in a sandy soil aeration, infiltration, shallow rooting, and crusting are not generally limiting. The overall score of 87.5 signifies that only minor changes in current management may be advisable.

Case Study 2: Comparison of Tillage Management

Figure 3 shows the test reports for two tillage management styles: (a) plow tillage and (b) zone tillage,

practiced side by side on a research farm near Aurora, NY, on a glacial till-derived Honeoye Lima silt loam. Zone tillage is a conservation tillage system that limits soil disturbance to the area of the planting row and leaves the areas between the crop rows undisturbed. Both plow and zone tillage treatments have been under maize–soybean rotation since 1992. Both test reports show generally favorable results for chemical indicators, with mostly high rating scores (7.0 or above). Both root health ratings are also high (8 and 9 for plow till and zone till, respectively), likely because the bean root assay is mostly sensitive to vegetable diseases which are uncommon in maize–soybean rotations. Surface hardness is better under plow till (10) than no till (7), likely due to loosening after traffic under plow till.

However, the remaining indicators have lower scores under plow tillage, showing evidence of degraded physical and biological SQ for the conventionally plowed fields. Especially low scores for aggregate stability (2), available water capacity (2), and organic matter (2), and intermediate scores for subsurface hardness (5), active carbon (3), and potentially mineralizable nitrogen (4) are evidence of soil degradation from long-term intensive tillage and lacking use of soil-building crops or organic matter additions. The overall score of 58.8 for plow till as compared to 81.7 for no till signifies that no till is better able to maintain physical and biological SQ, and considerable opportunity for improvement exists in this plow till system.

These reports exemplify the need for broader assessment of SQ. Based on traditional soil testing methodology, i.e., the chemical indicators, all soils appeared to be of good quality. This is commonly the case, as most NYS farmers are diligent about submitting soil samples for nutrient analysis and subsequently correcting the deficiencies. Chemical constraints are readily remedied by application of inorganic chemicals, which generally provides instant results. In contrast, the lack of routine tests for soil physical and biological indicators has resulted in inadequate attention to these facets of the soil, especially in larger scale, conventional operations. Moreover, enhancing the physical and biological quality of soils generally requires a longer term commitment to soil management through practices such as conservation tillage, improved rotations, cover cropping, and organic amendments, as discussed in Gugino et al. (2007). The soil health test therefore identifies a broader set of constraints and provides farmers with information that allows for holistic soil management.

Applications in Africa

There is great potential to adapt the CSHT for international use by carrying out case studies to determine its utility. One such study took place at the Kakamega and Nandi Forest Margins in western Kenya during July and August of 2007. This study (1) evaluated the CSHT's ability to measure long-term trends in soil degradation status and specific constraints, as well as aggradation due to short-term organic matter additions; (2) evaluated soil reflectance using visible/near-infrared and mid-infrared reflectance spectroscopy (VNIRRS) as a method for rapid and very inexpensive SQ assessment; and (3) evaluated the CSHT's relationship with maize yield, using a chronosequence of farms converted from primary forest between 0 and 100+ years ago.

Further development of SQ indicators has significance to farmers, communities, and applied researchers, and may be adaptable to other tropical conditions worldwide. Standardized SQ tests and management recommendations could be provided on a for fee basis to larger commercial farmers, to help them manage specific constraints. For example, compaction and erosion problems are common in the sugarcane industry in Kenya, and better management could help cut down the costs of tillage and fertilizer application ($525/ha, Odipo, 2007, Process Chemist, Personal communication), while preventing the non-point source nutrient pollution of Lake Victoria (Ochala, 2007, Biochemist, Lake Basin Development Program for Lake Victoria, Personal communication).

Standardized SQ tests could also be subsidized or provided at no cost by locally active agricultural non-profits, international research organizations, governments, and universities that have access to microloans and development grants, and a stake in improving environmental quality and food security. The availability of such tests, and development programs formed around their use, can motivate and empower innovative farmers and communities to experiment with soil management strategies. Subsistence farmers who are experimenting with raised beds, organic methods, water-harvesting strategies, and other methods are expressing interest in learning about their soils' constraints and alternative management strategies (Mwoshi, 2007, Subsistence Farmer, Personal Communication). Self-designed innovations are more likely to take advantage of locally available resources and practices and to be widely adopted via information sharing and demonstrations within farmer-to-farmer networks.

The simplicity of the proposed SQ tests, in conjunction with their low cost and infrastructure requirements, makes them excellent tools for numerous low-budget extension and NGO-based experiments established in collaboration with local farmers, and based on the environmental and economic needs of, and resources available to communities. Additionally, the new SQ test may have global implications by establishing a standard for widespread assessment of soil degradation and calling attention to the need to internationally coordinate soil protection measures. Inexpensive analysis will allow for widespread assessment, monitoring, and evaluation of SQ across farms, regions, and countries. Standard monitoring raises awareness and can lead to environmental policy regulations based on measurable criteria, as has been the case with the establishment of water and air quality standards.

Conclusions

Soil quality management requires an integrated approach that recognizes the physical, biological, and chemical processes in soils. The development of an inexpensive integrated SQ test was seen as a priority to allow widespread soil monitoring and better management decisions. The CSHT developed in NYS is a significant step forward from conventional soil tests, which focus exclusively on chemical indicators. The use of a holistic test that provides information about the three aspects of soils, physical, biological, and chemical, is a more meaningful approach to monitoring SQ and provides farmers, consultants, and agencies with a tool to identify soil constraints and target management practices or remediation strategies. This tool has great potential as a basic framework from which to establish international SQ standards that similarly address soil quality issues.

Acknowledgments We acknowledge support from the USDA Northeast Sustainable Agriculture Research and Education Program (USDA 2003-3860-12985), the Northern New York Agricultural Development Program, USDA-Hatch funds, and graduate fellowship funds from the Saltonstall family and the National Science Foundation (GK-12, DGE0231913).

References

Abawi GS, Widmer TL (2000) Impact of soil health management practices on soilborne pathogens, nematodes and root diseases of vegetable crops. Appl Soil Ecol 15:37–47

Andrews SS, Karlen DL, Cambardella CA (2004) The soil management assessment framework: a quantitative soil quality evaluation method. Soil Sci Soc Am J 68:1945–1962

Carter MR (2002) Soil quality for sustainable land management: organic matter and aggregation interactions that maintain soil functions. Agron J 94:38–47

Dexter AR (2004) Soil physical quality – Part I. Theory, effects of soil texture, density, and organic matter, and effects on root growth. Geoderma 120:201–214

Doran JW, Parkin TB (1994) Defining and assessing soil quality. In: Doran JW, Coleman DC, Bezdicek DF, Stewart BAE (eds) Defining soil quality for a sustainable environment. Soil Science Society of America, Madison, WI, pp 3–21

Eswaran H, Almaraz R, van den Berg E, Reich P (2005) An assessment of the soil resources of Africa in relation to productivity. World soil resources, soil survey division, USDA. Natural Resources Conservation Service, Washington, DC

FAO (2005) Land degradation assessment, TERRASTAT CD-ROM, FAO land and water digital media series #20

Gugino BK, Idowu OJ, Schindelbeck RR, van Es HM, Wolfe DW, Moebius BN, Thies JE, Abawi GS (2007) Cornell soil health assessment training manual. Cornell University, Geneva, NY

Hurni H, Giger M, Meyer K (eds) (2006) Soils on the global agenda. Developing international mechanisms for sustainable land management. Prepared with the support of an international group of specialists of the IASUS Working Group of the International Union of Soil Sciences (IUSS). Centre for Development and Environment, Bern

Idowu OJ, van Es HM, Abawi GS, Wolfe DW, Ball JI, Gugino BK, Moebius BN, Schindelbeck RR, Bilgili AV (2008) Farmer-oriented assessment of soil quality using field, laboratory, and VNIR spectroscopy methods. Plant Soil 307: 243–253

Larson WE, Pierce FJ (1991) Conservation and enhancement of soil quality. In: Evaluation for sustainable land management in the developing world, vol 2. International Board for Soil Research and Management, Proceeding No. 12, Jatujak Thailand, Bangkok, Thailand, pp 175–203

Manhoudt AGE, de Haes HAU, de Snoo GR (2005) An indicator of plant species richness of semi-natural habitats and crops on arable farms. Agric Ecosyst Environ 109:166–174

Moebius BN, van Es HM, Schindelbeck RR, Idowu JO, Thies JE, Clune DJ (2007) Evaluation of laboratory-measured soil properties as indicators of soil physical quality. Soil Sci 172:895–912

Nambiar KKM, Gupta AP, Fu QL, Li S (2001) Biophysical, chemical and socio-economic indicators for assessing agricultural sustainability in the Chinese coastal zone. Agric Ecosyst Environ 87:209–214

Ogden CB, van Es HM, Schindelbeck RR (1997) Miniature rain simulator for field measurement of soil infiltration. Soil Sci Soc Am J 61:1041–1043

Riley J (2001) The indicator explosion: local needs and international challenges – Preface. Agric Ecosyst Environ 87: 119–120

Weil RR, Islam KR, Stine MA, Gruver JB, Samson-Liebig SE (2003) Estimating active carbon for soil quality assessment: a simplified method for laboratory and field use. Am J Altern Agric 18:3–17

Winder J (2003) Soil quality monitoring programs: a literature review. Alberta environmentally sustainable agriculture soil quality monitoring program, Edmonton, AB

Increasing Productivity Through Maize–Legume Intercropping in Central Kenya

M. Mucheru-Muna, D.N. Mugendi, P. Pypers, J. Mugwe, B. Vanlauwe, R. Merckx, and J.B. Kung'u

Abstract Declining land productivity is a major problem facing smallholder farmers in Kenya today. This decline results from a reduction in soil fertility caused by continuous cultivation without adequate addition of external inputs. Improved agronomic measures integrating grain legumes into maize cropping systems can enhance overall system's productivity. Trials were established in two sites in central Kenya (Mukuuni and Machang'a) to evaluate contribution of various legumes (beans, cowpea, and groundnut) and plant spacing to overall productivity of the intercropping system. The conventional spacing (a legume row alternating a cereal row) was compared to managing beneficial interactions in legume intercrop (MBILI) spacing (two legume rows alternating two cereal rows), both with and without phosphorus (P). In Mukuuni (more fertile), neither legumes nor maize responded to P application; legume yield was increased by on average 100% and maize yields more than doubled by P application in Machang'a. In Mukuuni, groundnut production was poor (<500 kg ha^{-1}); in Machang'a, highest legume yields were obtained with cowpea. In both sites, legume yields tended to be higher in the conventional intercrop, irrespective of legume species or P application, though not consistently significant in all seasons. In contrast, in Machang'a, maize yields were generally highest when planted using MBILI spacing, provided P was applied. Without P application, higher yields were observed for maize in the conventional intercrop, but only when intercropped with beans. Maize yields were significantly higher with conventional intercrop when intercropped with groundnut, while in MBILI spacing, highest yields were observed for maize intercropped with beans. In Mukuuni, benefit–cost ratio (BCR) was higher in treatments without P and in the MBILIs. In Machang'a, BCR was not significantly different between the MBILI and the conventional intercrop. Return to labour was higher in the MBILI in Mukuuni.

Keywords Benefit–cost ratio · Conventional intercrop · Grain legumes · MBILI · P fertilizer

Introduction

Soil fertility decline is a major problem contributing to the low maize grain yield in sub-Saharan African smallholder farms today. Maize is a major staple food in this region where 95% of the production is consumed by humans (McCann, 2005). In large parts of the African sub-continent, smallholder agricultural production and food security were reduced to a catastrophic level (Kumwenda, 1998; Sanchez, 2002). Low soil fertility, limited resources, nutrient mining, and terminal droughts are the main factors limiting maize productivity in sub-Saharan Africa. Meru South district is an overly populated area with more than 700 persons per km^{-2} (Government of Kenya, 2001). The high population has led to the replacement of traditional systems of shifting cultivation with continuous cultivation, leading to soil nutrient depletion which already exceeds 30 kg N per year (Smaling, 1993). As a result of the growing population pressure, people have moved from the high-potential areas to marginal

M. Mucheru-Muna (✉)
Department of Environmental Sciences, School of Environmental Studies, Kenyatta University, Nairobi, Kenya
e-mail: moniquechiku@yahoo.com

A. Bationo et al. (eds.), *Innovations as Key to the Green Revolution in Africa*,
DOI 10.1007/978-90-481-2543-2_84, © Springer Science+Business Media B.V. 2011

ones like Mbeere district. This leads to the expansion of cultivation onto marginal soil types and the reduction of fallow periods, and results in systematic soil degradation and yield decline.

The need for soil fertility replenishment in Africa has been recognized for decades and a number of technologies have been promoted. Farmers with limited resources can afford technologies only if they are feasible in terms of labour, land, and/or capital investments (Barrett et al., 2002), and for many of the proposed technologies, this is not the case (Sanchez and Jama, 2002). One possibility to alleviate soil nutrient depletion is to use mineral fertilizers. As the costs of mineral fertilizers in Kenya are about two to six times higher than in Europe, North America, and Asia, the use of fertilizers has generally been restricted to a few farms with high income (Sanchez, 2002). Kihanda (1996) reported that less than 25% of maize growers in the central highlands of Kenya use mineral fertilizers. The situation is further aggravated by the fact that even the farmers using mineral fertilizers hardly use the recommended rates (60 kg N ha^{-1}) in the area, with most of them applying less than 20 kg N ha^{-1} (Adiel, 2004). As a result, soil fertility and hence productivity has continued to decline drastically.

Intercropping of legumes and cereals has been suggested as a way of improving soil fertility. It is a very common practice in Africa where 98% of cowpeas (*Vigna unguiculata)* are grown in mixed cropping systems (Gutierrez et al., 1975). Maize and beans (*Phaseolus vulgaris*) intercrops are predominant in East Africa, while maize intercropped with cowpeas and groundnuts is common in southern Africa. The importance of intercrops is widely recognized by farmers. This arises from the stabilizing effect of crops on food security; enhanced use of efficiency of land, water, and labour; and risk aversion in case of crop failure (Chivenge et al., 2000; Mwale, 2000). Hence, intercropping of cereals and legumes could be an option as a means of better utilization of available resources like soil N, moisture, and light.

Legumes also have other beneficial effects. The soil structure and the water holding capacity are improved by added organic matter and root penetration (Hulugalle et al., 1986), which positively influence the soil fertility. Legume intercrops are also a source of plant N through atmospheric fixation that can offer a practical complement to mineral fertilizers (Jerenyama et al., 2000) and reduce competition for N from cereals

component (Fujita et al., 1992). Legumes also contribute to the economy of intercropping systems either by transferring N to the cereal crop during the growing period (Ofori et al., 1987) or as residual N that is available to the subsequent crop (Papastylianou, 1988). A yield advantage in species mixture may occur when component crops differ in their use of growth resources in such a way that when they are grown together they are able to complement each other and so make better overall use of resources than when separately grown (Willey, 1979). Furthermore, additional labour input associated with cultivating grain legumes is minimal and seed costs are low compared to other green manure or agroforestry crops (Sakala et al., 2003).

With this background, an experiment integrating grain legumes with maize was set in Meru South and Mbeere districts to evaluate their effects on maize yield improvement. Two systems of intercrops were used: the conventional method which includes one row of maize alternated by one row of legume and the MBILI (managing beneficial interactions in legume intercrop) method where two rows of cereal are alternated with two rows of legume. The two intercrops allow the same number of maize plants in an area. The advantage of the MBILI method is that more light penetration especially for the legume component takes place which reduces the competition for this growth resource.

Materials and Methods

The Study Area

The study was conducted in Meru South and Mbeere districts, Kenya. In Meru South, the experiment was in Mukuuni (00°23'30.3''S; 37°39'33.7''E) which is located in upper midland 3 with an altitude of approximately 1287 m above sea level. The soils are Humic Nitisols (Jaetzold and Schmidt, 1983), which are deep, well weathered with moderate- to high-inherent fertility (Table 1).

The area is characterized by rapid population growth and low soil fertility (Government of Kenya, 2001). The rainfall distribution is bimodal, with the long rains (LR) lasting from March to June and short rains (SR) from October to December. The mean annual rainfall in the area is 1200 mm (Jaetzold

Table 1 Physico-chemical properties of the soil in Mukuuni, Meru South district, Kenya

Soil parameters	
pH in water	5.9
Total N	0.24%
Total soil organic carbon	2.72%
Exchangeable P	20.95 ppm
Exchangeable K	2.65 cmol kg^{-1}
Exchangeable Ca	4.44 cmol kg^{-1}
Exchangeable Mg	1.44 cmol kg^{-1}
CEC	26.0 cmol kg^{-1}
Clay	67%
Sand	31%
Silt	2%

et al., 2006). Rainfall for the four seasons in which the experiment was conducted is presented in Fig. 1.

The main staple food crop is maize (*Zea mays L.*) which is commonly intercropped with beans (*P. vulgaris L*). Other food crops are Irish potatoes, bananas, sweet potatoes, vegetables, and fruits that are mainly grown for subsistence consumption. The main cash crops include coffee, tobacco, and tea. Livestock production is a major enterprise, especially dairy cattle of improved breeds. Other livestock are sheep, goats, and poultry.

In Mbeere district, the experiment was conducted in Machang'a (00°47′26.8″S; 37°39′45.3″E) with an altitude of approximately 1028 m above sea level and annual mean temperature of about 23°C. The soils are sandy–clay–loam, blackish grey, or reddish brown, classified as the Ferralic Arenosols (Jaetzold and Schmidt, 1983). They are shallow (about 1 m deep) and lose their organic matter, including nutrient-rich aggregates within 3–4 years of cultivation without adequate internal/external organic material inputs and soil protection from water erosion (Jaetzold and Schmidt, 1983; Warren et al., 1998; Micheni et al., 2004). Table 2 shows the soil characteristics of the soils in Machang'a.

The major cropping enterprises are maize and beans. Other food crops are cowpea (*V. unguiculata*), millet (*Eleusine coracana*), sorghum (*Sorghum bicolor* L.), green grams (*Vigna radiata*), and fruits (pawpaws (*Carica papaya L*) and mangoes (*Mangifera indica*)). Livestock (cows, goats, sheep, and poultry) production is a major enterprise and the farmers mainly keep the local breeds. Bee keeping is also a major enterprise in the area. Farmers plant food crops and keep livestock with a high staple and economic value as they do not grow any "cash crop". The crops they grow are therefore used for self-supply and sold on the market.

Machang'a is located between the marginal cotton (LM4) and the main cotton (LM3) agro-ecological zones (Jaetzold and Schmidt, 1983). The rainfall is bimodal with LR from March to June and SR from October to December. The total rainfall is, however, variable, with a mean annual rainfall of 900 mm (Government of Kenya, 1997). Total rainfall per season during the study period ranged between 209 and 731 mm. Rainfall for the four seasons in which the experiment was conducted is presented in Fig. 2.

Experimental Layout and Management

The experiment was established in October 2004 and laid out as a randomized complete block design (RCBD) with plot sizes measuring 6 m × 4.5 m replicated thrice. The maize varieties H513 and Katumani were intercropped with the legumes – beans, cowpea, and groundnuts – in conventional intercrop and MBILI.

In the conventional intercropping system, maize was planted at a spacing of 0.75 and 0.5 m inter- and intra-row, respectively, in Meru South district and at a spacing of 0.90 and 0.6 m inter- and intra-row, respectively, in Mbeere district. The legumes were planted between the maize rows. In the modified intercropping system, designated MBILI according to the *kiswahili* word for "two", two 50-cm maize rows separated by a 1-m strip reserved for the legume planted at 0.33 m apart were installed (Tungani et al., 2002). In this way, the same maize populations are maintained and grown with a wider selection of higher value pulses such as groundnuts, green gram, and soybean. Treatments for determining the maize response to mineral N fertilizer applied at the rates of 30, 60, and 90 kg N ha^{-1} were also included. Calcium ammonium nitrate (CAN) was the source of mineral N fertilizer. One-third of the fertilizer rate was applied 4 weeks after planting and two-thirds were applied 8 weeks after planting. P was applied as triple superphosphate (TSP) in some plots at the rate of 60 kg P ha^{-1} which is the recommended rate for an optimum crop production in the area (FURP, 1987). Other agronomic procedures for maize production were appropriately followed after

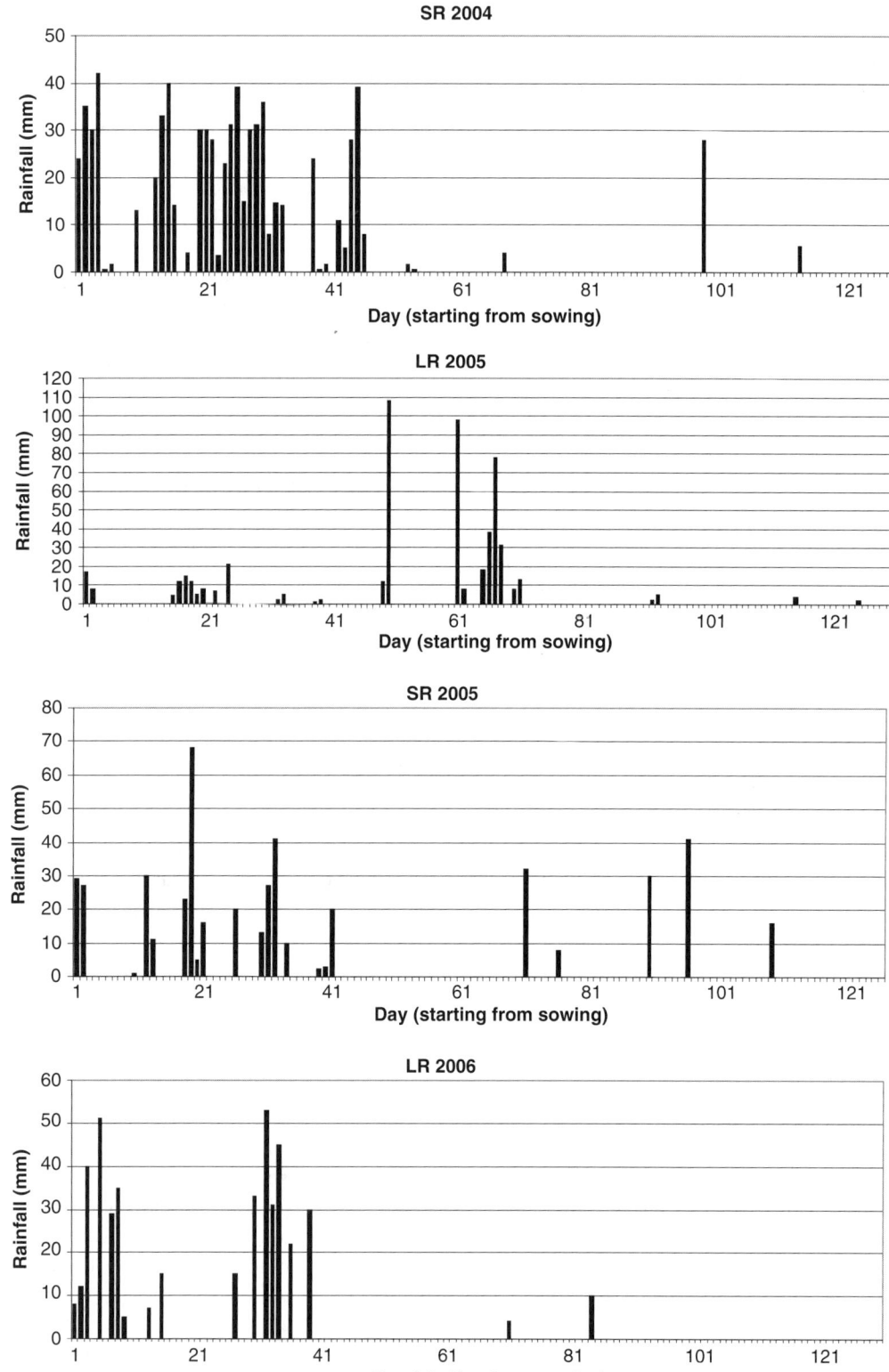

Fig. 1 Rainfall distribution from 2004 to 2006 in Mukuuni site, Meru South district, Kenya

Table 2 Physico-chemical properties of the soil characterization in Machang'a site, Mbeere district, Kenya

Soil parameters	
pH in water	5.0
Total N	0.04%
Total soil organic carbon	0.3%
Exchangeable P	6.28 ppm
Exchangeable K	0.45 cmol kg^{-1}
Exchangeable Ca	1.1 cmol kg^{-1}
Exchangeable Mg	0.2 cmol kg^{-1}
CEC	5.2 cmol kg^{-1}
Clay	7%
Sand	78%
Silt	15%

planting. At maturity, the maize and legumes were harvested and the fresh weight of both grain and stover was taken. The grain was air dried and the weight taken and expressed on dry weight basis. Afterwards, the legume stovers were incorporated into the soil during planting.

Economic Analysis

The information used for economic analysis in this study was collected at the specific time of each activity during the season. Farmers and agro-input retailers were interviewed. The time for each activity was measured by using a timer and the labour was valued at the local wage of USD 1.47 (Ksh 100) per man day equivalent to eight working hours. Maize stover was used as cattle feed in the area (with a market value of USD 22.1 per tonne) and was thus accounted for as a benefit in addition to the maize grain. The economic analysis was done using the farm gate prices of the various inputs (Table 3).

Root Length Density

Root sampling in the two intercropping systems was done during the active (growth) physiological podding stage of the legume and maize plants. Soil core root sampling technique was adapted in this study for root length and density quantification (Anderson and Ingram, 1993). In all cropping systems, stratification was done on an area basis (within rows vs. between

rows). Roots placed on a sieve were cleaned from soil particles by water, separated from debris and other dead roots, dried with serviettes, and randomly spread on a 1-cm grid (photocopied on an acetate folio for overhead projection) of 28 cm × 18 cm. All intersections of roots and grid lines were counted and the horizontal (H) and vertical (V) grid lines were added to give the total (N) according to the Newman line intersect method (Tennant, 1975). The equation used for the estimation of the root length was as follows:

$$\text{Root length}\,(R) = \text{number of intercepts}\,(N)$$
$$\times \text{ length conversion factor}$$

where conversion factor is 0.3928 for 0.5-cm grid, 0.7857 for 1-cm grid, 1.5714 for 2-cm grid, and 2.3571 for 3-cm grid squares.

Root length density, defined as root length per unit volume of soil (Newman, 1966), was obtained by calculation using the following formula:

$$\text{Root length density}\,(\text{cm cm}^{-3}) = \frac{\text{Root length}\,(\text{cm})}{\text{Soil volume}}$$

Statistical Analysis

The data were subjected to analysis of variance using GenStat programme and the means separated using Tukey's test and t-test at $p < 0.05$.

Results and Discussion

Maize Grain Yield

There was a wide variability in maize grain yields between Mukuuni and Machang'a sites across the four seasons. The yields were significantly higher in Mukuuni compared to Machang'a during the four seasons (t-test, $p < 0.001$). There was a significant ($p < 0.001$) effect of seasons on maize grain yield. The mean yields were highest during the LR 05 season followed by SR 05, SR 04, and LR 06 seasons (4.6, 2.4, 1.7, and 1.1 t ha^{-1}, respectively) (Table 4).

In Mukuuni site, maize/beans MBILI with P, maize/groundnut conventional intercrop without P,

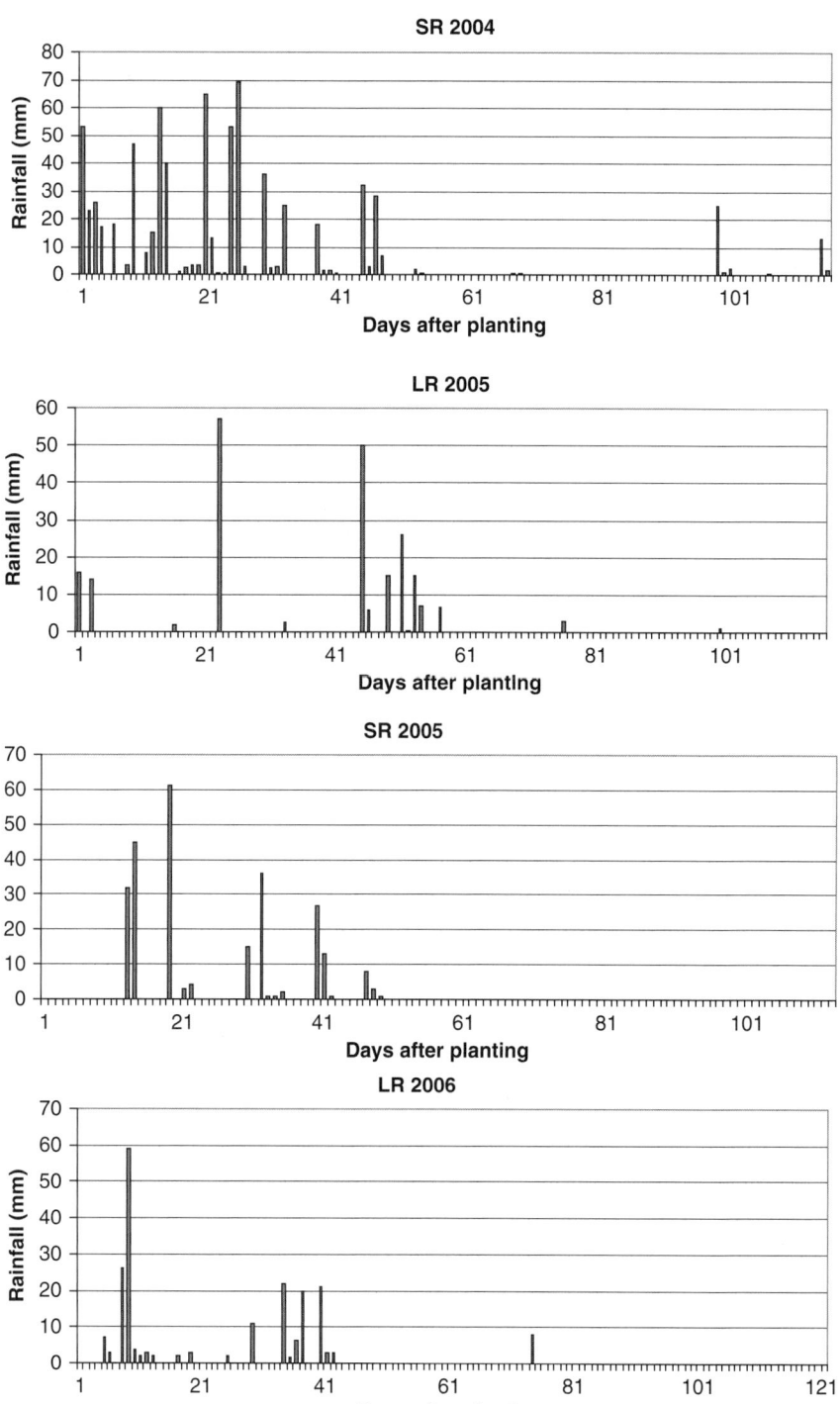

Fig. 2 Rainfall distribution from 2004 to 2006 in Machang'a site, Mbeere district, Kenya

maize/beans MBILI without P, and maize/beans conventional intercrop with P recorded the highest maize grain yields during the SR 04, LR 05, SR 05, and LR 06 seasons (2.2, 6.5, 3.9, and 1.9 t ha^{-1}, respectively).

Maize/cowpea MBILI without P, maize/beans conventional intercrop without P, maize/cowpea conventional without P, and maize/cowpea MBILI without P gave the lowest maize grain yields during the SR 04, LR 05,

Table 3 Parameters used to calculate the economic returns for the different nutrient replenishment technologies

Parameter	Actual values
Price of CAN (26:0:0)	1.69 USD kg^{-1} N
Price of TSP (0:46:0)	1.05 USD kg^{-1} N
Labour cost	0.18 USD hr^{-1}
Price of beans	0.66 USD kg^{-1}
Price of cowpea	0.59 USD kg^{-1}
Price of ground nuts	2.2 USD kg^{-1}
Price of maize	0.24 USD kg^{-1}
Price of stover	0.022 USD kg^{-1}

Exchange rate: Ksh 68 = 1 USD (September 2007)

SR 05, and LR 06 seasons (1.2, 3.0, 1.8, and 0.9 t ha^{-1}, respectively).

On average, in Mukuuni site, maize grain yields from plots with P treatments were significantly higher than those without P treatments ($p < 0.05$ and $p < 0.01$, respectively), during the SR 05 and LR 06 seasons (Table 4). For instance, the maize/cowpea MBILI treatments recorded 23 and 123% higher maize grain yields with P than without P during the SR 05 and LR 06 seasons, respectively.

There were no significant differences between maize grain yields in the MBILI and the conventional intercrop in Mukuuni during the SR 04 season. However, during the LR 05 and LR 06 seasons in treatments with and without P, in the maize/cowpea and the maize/groundnut intercrops, maize grain yields were significantly higher in the conventional intercrop compared to the MBILI. During the LR 05 season in the treatments with P, for instance, maize grain yield was 31 and 29% higher in the conventional intercrop compared to the MBILI in the maize/cowpea and maize/groundnut intercrops, respectively. Similarly in the treatments without P, the maize grain yield was 109 and 40% higher in the conventional intercrop compared to the MBILI in the maize/groundnut and maize/cowpea intercrops, respectively.

In the maize/bean intercrop during the LR 05 season, maize grain yields in the MBILI were significantly higher than those in the conventional intercrop in treatments with P and without P and during the LR 06 season the maize grain yields were only significantly higher in the MBILI in the treatments without P. During the LR 05 season, maize grain yield was 52 and 20% higher in the MBILI compared to the conventional intercrop in the treatments without P and with P, respectively, while in the LR 06 season, maize grain

yield was higher by 110% in the MBILI than in the conventional intercrop in the treatments without P in the maize/bean intercrop.

During the SR 05 season, treatments with P gave significantly higher maize grain yields in the MBILI compared to the conventional intercrop. For instance, the maize grain yields were 60, 49, and 34% higher in the maize/bean, maize/groundnut, and maize/cowpea MBILIs compared to the conventional intercrops. The treatments without P also recorded higher maize grain yields in the maize/bean and maize/cowpea MBILIs though they were not significant.

There was a significant ($p < 0.001$) effect of seasons on maize grain yield in Machang'a. The mean maize grain yields were highest during the LR 05 season followed by SR 05, LR 06, and SR 04 seasons (0.8, 0.6, 0.4, and 0.3 t ha^{-1}, respectively) (Table 5). During the LR 05 season, the maize grain yield ranged between 0.22 and 0.93 t ha^{-1}, while during the SR 04 season, the maize grain yield ranged between 0.08 and 0.64 t ha^{-1} (Table 5).

In Machang'a, maize/groundnut MBILI with P gave the highest maize grain yields during the SR 04, LR 05, and SR 05 seasons, while maize/beans MBILI with P gave the highest maize grain yield during the LR 06 season (0.64, 1.52, 1.17, and 0.90 t ha^{-1}, respectively). The lowest maize grain yields were reported in the maize/cowpea conventional intercrop without P during all the seasons (0.08, 0.22, 0.03, and 0.06 t ha^{-1}). In Machang'a, the maize grain yields generally declined from the LR 05 through to the LR 06 season.

Treatments with P performed significantly better than did the treatments without P across all the four seasons except in the maize/bean conventional intercrop in Machang'a (Table 5). For instance, during the LR 05 season, maize grain yield increased by 347 and 323% in the maize/groundnut MBILI and the maize/cowpea conventional intercrop; on the other hand, during the LR 06 season, maize grain yield in the maize/beans MBILI and maize/groundnut MBILI increased by 1400 and 800%, respectively, in the treatments with P compared to those without P.

MBILIs generally performed better than the conventional intercrops in all treatments with P fertilizer in Machang'a. For instance, during the LR 05 season, maize grain yields in the maize/groundnut, maize/beans, and maize/cowpea MBILIs were 111, 30, and 3% higher, respectively, than those in the

Table 4 Maize and legume yields (t ha^{-1}) in different intercrop systems during the SR 04, LR 05, SR 05, and LR 06 seasons at Mukuuni site in Meru South district

Treatment	SR 04				LR 05				SR 05				LR 06			
	-P		+P		-P		+P		-P		+P		-P		+P	
	C	M	C	M	C	M	C	M	C	M	C	M	C	M	C	M
Maize yield																
Beans	1.78	1.93	1.91	2.24	3.04	4.63	4.06	4.86	2.05	3.85	2.17	3.47	0.92	1.93	1.94	1.79
Cowpea	1.30	1.22	1.55	1.46	5.11	3.65	5.09	3.90	1.80	2.00	1.83	2.45	0.85	0.22	0.73	0.49
Groundnut	2.01	1.49	1.63	1.51	6.50	3.11	6.01	4.65	2.38	1.88	1.81	2.69	1.55	1.00	1.31	0.72
Control	1.86		2.02		3.01		3.53		2.12		3.05		1.35		1.18	
SED (P level)	n.s.															
SED (spacing)	n.s.															
SED (intercrop)	n.s.															
SED (P level × spacing)	n.s.				n.s.											
SED (P level × intercrop)	n.s.				n.s.											
SED (spacing × intercrop)	n.s.				0.52***											
SED (P level × spacing × intercrop)	n.s.				n.s.				0.39*				0.24**			
Legume yield																
Beans	0.21	0.21	0.09	0.27	0.51	0.26	0.44	0.40	0.63	0.43	0.74	0.45	0.24	0.29	0.29	0.24
Cowpea	0.10	0.13	0.12	0.09	0.14	0.15	0.19	0.14	0.41	0.36	0.36	0.36	0.42	0.36	0.54	0.26
Groundnut	0.09	0.11	0.11	0.13	0.02	0.03	0.03	0.02	0.21	0.16	0.21	0.14	0.51	0.27	0.51	0.27
SED (P level)	0.06				0.10**				0.10**				0.11			
SED (intercrop)																
SED (P level × intercrop)																

Key: C, Conventional intercrop; M, MBILI; n.s., not significantly different; *, **, *** significant at $p < 0.05$, $p < 0.01$, and $p < 0.001$, respectively; SED, standard error of means

Table 5 Maize and legume yields (t ha^{-1}) in different intercrop systems during the SR 04, LR 05, SR 05, and LR 06 seasons at Machang'a site

Treatment	SR 04 -P C	SR 04 -P M	SR 04 +P C	SR 04 +P M	LR 05 -P C	LR 05 -P M	LR 05 +P C	LR 05 +P M	SR 05 -P C	SR 05 -P M	SR 05 +P C	SR 05 +P M	LR 06 -P C	LR 06 -P M	LR 06 +P C	LR 06 +P M
Maize yield																
Beans	0.27	0.18	0.34	0.60	0.90	0.46	0.89	1.16	0.73	0.23	0.24	1.05	0.43	0.06	0.46	0.90
Cowpea	0.08	0.42	0.17	0.51	0.22	0.69	0.93	0.96	0.03	0.60	0.63	1.13	0.06	0.19	0.59	0.43
Groundnut	0.21	0.27	0.37	0.64	0.40	0.34	0.72	1.52	0.14	0.24	0.81	1.17	0.06	0.07	0.43	0.63
Control	0.64		1.52		0.46		1.03		0.33		0.57		0.26		0.61	
SED (P level)	0.10*															
SED (spacing)	0.10*															
SED (intercrop)	n.s.															
SED (P level × spacing)	n.s.															
SED (P level × intercrop)	n.s.															
SED (spacing × intercrop)	n.s.															
SED (P level × spacing × intercrop)	n.s.				0.26*				0.25*				0.13**			
Legume yield																
Beans	0.52	0.24	0.79	0.52	0.08	0.10	0.20	0.21	015	0.16	0.26	0.16	0.09	0.02	0.26	0.10
Cowpea	0.67	0.78	1.60	1.14	0.52	0.49	0.81	0.76	0.29	0.38	0.67	0.48	0.38	0.67	0.59	0.70
Groundnut	0.68	0.46	0.79	0.54	0.38	0.38	0.35	0.44	0.30	0.31	0.30	0.24	0.20	0.14	0.16	0.20
SED (P level)																
SED (intercrop)																
SED (P level × intercrop)	0.26**				0.14***				0.11**				0.09***			

Key: *, **, *** significant at $p < 0.05$, $p < 0.01$, and $p < 0.001$, respectively; C, conventional intercrop; M, MBILI; n.s., not significantly different; SED, standard error of means conventional intercrop. In the treatments without P, MBILI also performed better than the conventional intercrop except in the maize/bean intercrop.

Legume Grain Yield

The grain legume yields in Mukuuni were different from those in Machang'a sites across the four seasons. During the SR 04 and LR 05 seasons, the legume grain yields were significantly higher in Machang'a than in Mukuuni site (t-test, $p < 0.001$, and $p = 0.026$, respectively). On the other hand, during the SR 05 and LR 06 seasons, the legume grain yields were higher in Mukuuni site though not significant (t-test, $p = 0.360$, and $p = 0.460$, respectively). There was a significant ($p < 0.001$) effect of seasons on legume grain yield in Mukuuni. The mean legume grain yields were highest during the SR 05 season followed by LR 06, LR 05, and SR 04 seasons (0.37, 0.35, 0.19, and 0.14 t ha^{-1}, respectively) (Table 4). In Mukuuni site, there were no significant differences between the legume grain yields in the MBILI and the conventional intercrop in the maize/cowpea and maize/groundnut intercrops across the four seasons in treatments with and without P (Table 4).

In the maize/beans intercrop, legume grain yields in the MBILI were significantly higher than those in the conventional intercrop in treatments with P during the SR 04 season in Mukuuni (Table 4). The legume grain yield was 200% higher in the MBILI compared to the conventional intercrop. During the LR 05 and SR 05 seasons in the maize/beans intercrop, legume grain yields in the conventional intercrop were significantly higher than those in the MBILI with and without P.

There were no significant differences between the legume grain yield in the treatments with P and between that with P and without P treatments in the maize/cowpea and maize/groundnut intercrops in all the seasons in Mukuuni; however, the bean grain yields in the conventional intercrop were significantly higher in the treatments without P in the SR 04 season (Table 4). The bean grain yield in the conventional intercrop without P was 133% higher than that with P.

There was a significant ($p < 0.001$) effect of seasons on legume grain yield in Machang'a. The mean legume grain yields were highest during the SR 04 season followed by LR 05, SR 05, and LR 06 seasons (0.73, 0.39, 0.31, and 0.29 t ha^{-1}, respectively) (Table 5). Legume grain yields ranged between 00.24 and 1.6 t ha^{-1} during the SR 04 season, while the legume grain yield ranged between 0.09 and 0.70 t ha^{-1} during the LR 06 season (Table 5).

In Machang'a, there were no significant differences in legume grain yield between the MBILI and the conventional intercrop in the maize/beans and maize/groundnut intercrops in the first three seasons (Table 5). However, in most seasons the MBILI had higher legume grain yields compared to the conventional intercrop. For instance, during the SR 04 season, in the treatments with P, legume grain yields in the MBILI were higher than those in the conventional intercrop with 52, 46, and 40% for beans, groundnuts, and cowpea, respectively. In the treatments without P, legume grain yields in the conventional intercrop were higher than those in the MBILI with 31, 7, and 3% for cowpea, beans, and groundnuts, respectively.

However, during the LR 06 season, legume grain yields were significantly higher in the MBILI compared to the conventional intercrop. For instance, without P, bean grain yields were 350% higher in the MBILI compared to the conventional intercrop, while groundnut yields were 43% higher in the MBILI than in the conventional intercrop. With P, bean grain yields were 160% higher in the MBILI compared to the conventional intercrop, while groundnuts and cowpea were 25 and 19%, respectively, in the MBILI than in the conventional intercrop.

There were no significant differences in the legume grain yields between the treatments with P and without P in the maize/beans and maize/groundnut intercrops in the four seasons (Table 5). However, the treatments with P had generally higher legume grain yields compared to the treatments without P. For instance, during the LR 06 season, in the conventional intercrop, bean grain yields were 189% higher in treatments with P than without P, while in the MBILI, beans were 400% higher with P than without P.

Economic Returns

The results of the economic analysis indicate that there were significant differences between the treatments in terms of net benefit, benefit–cost ratio, and return to labour in Mukuuni at $p < 0.001$ during the SR 05 season (Table 6). Maize/bean MBILI without P recorded the highest net benefit followed closely by maize/bean MBILI with P with USD 1062 and USD 944, respectively. On the other hand, maize/cowpea conventional intercrop with P recorded the lowest net benefit (USD 479). The highest BCR was recorded by maize/bean MBILI without P followed closely by maize/bean conventional intercrop without P (4.8 and 3.5, respectively), while the lowest BCR (1.7) was recorded by maize/groundnut conventional intercrop with P. The highest return to labour was recorded by maize/cowpea conventional intercrop without P followed closely by maize/groundnut MBILI without P (8.1 and 7.7, respectively), while the lowest return to labour (3.8) was recorded in the maize/bean MBILI without P (Table 6).

In the conventional intercrop, the treatments without P had higher net benefits in the maize/groundnut intercrop. Net benefits were not significantly different between the treatments with P and without P in the MBILIs. The maize/bean intercrop, MBILI system without any application of P fertilizer had significantly higher net benefits compared to the conventional intercropping system, whereas in the maize/groundnut intercrop the conventional intercropping system had

higher net benefits compared to the MBILI system. In Mukuuni, the BCR was significantly higher in the treatments without P in the maize/bean MBILI and in the maize/cowpea and maize/groundnut conventional intercrops. In the treatments without P, BCR was significantly higher in the MBILI compared to the conventional intercrop in the maize/bean intercrop.

In the treatments without P, the return to labour in the MBILI was significantly higher in the maize/groundnut intercrop, while return to labour in the conventional intercrop was significantly higher in the maize/cowpea intercrop. There were no significant differences between the return to labour in the treatments with P in both the MBILI and the conventional intercrop. In the MBILIs, treatments with P had significantly higher return to labour in the maize/bean intercrop, while treatments without P had significantly higher return to labour in the maize/groundnut intercrop. In the conventional intercrop, in the maize/cowpea intercrop, the treatments without P had significantly higher return to labour than those with P.

In Machang'a, there were significant differences between the treatments in terms of net benefit, benefit–cost ratio, and return to labour at $p < 0.001$ during the SR 05 season (Table 7). Maize/groundnut

Table 6 Net benefit, benefit–cost ratio, and return to labour during the SR 05 season in Mukuuni, Meru South, Kenya

Treatment	Net benefit				BCR				Return to labour			
	−P		+P		−P		+P		−P		+P	
	C	M	C	M	C	M	C	M	C	M	C	M
Maize/bean intercrop	764	1062	795	944	3.5	4.80	2.86	3.34	5.14	3.76	6.39	7.31
Maize/cowpea intercrop	566	565	479	602	2.88	2.74	1.85	2.34	8.11	4.79	4.44	5.76
Maize/groundnut intercrop	817	598	613	644	2.76	2.02	1.71	1.78	5.22	7.74	3.88	3.96
Maize monocrop (control)	486		591		3.3		2.8		6.38		7.74	
SED (P level)	92.3*				0.39*				0.83*			

Key: * significant at $p < 0.001$; C, conventional intercrop; M, MBILI; SED, standard error of means

Table 7 Net benefit, benefit–cost ratio, and return to labour during the SR 05 season in Machang'a, Mbeere, Kenya

Treatment	Net benefit				BCR				Return to labour			
	−P		+P		−P		+P		−P		+P	
	C	M	C	M	C	M	C	M	C	M	C	M
Maize/bean intercrop	136	16	−218	−211	0.91	0.11	−0.47	0.39	1.81	0.23	−2.65	−2.82
Maize/cowpea intercrop	43	192	−249	−258	0.27	1.06	−0.31	−0.32	0.47	1.73	−2.18	−2.07
Maize/groundnut intercrop	479	526	−312	−308	2.14	2.23	−0.26	−0.31	4.48	4.55	−2.42	−2.51
Maize monocrop (control)	−180		−8		−0.56		−0.08		−0.17		−2.82	
SED (P level)	85.3*				0.45*				0.91*			

Key: * significant at $p < 0.001$; C, conventional intercrop; M, MBILI; SED, standard error of means

MBILI without P recorded the highest net bene-fit followed closely by maize/groundnut conventional intercrop without P (USD 526 and USD 479, respec-tively), while maize/groundnut conventional intercrop with P recorded the lowest net benefit (USD 312) (Table 7). The highest BCR was recorded in the maize/groundnut MBILI without P followed closely by maize/groundnut conventional intercrop without P (2.23 and 2.14, respectively), whereas the lowest BCR (–0.56) was recorded in the maize/bean conventional intercrop with P. Maize/groundnut MBILI without P recorded the highest return to labour followed closely by maize/groundnut conventional intercrop without P (4.55 and 4.48, respectively), while maize/bean MBILI with P and maize monocrop with P recorded the lowest return to labour of –2.82 (Table 7).

The net benefit was not significantly different between the MBILI and conventional intercrop in treatments both with P and without P. In both the MBILI and the conventional intercrop, the treatments without P recorded significantly higher net benefits than did those with P. The BCR was not significantly different between the MBILI and conventional inter-crop in treatments both with P and without P. In the MBILIs, treatments without P recorded significantly higher BCR in the maize/cowpea and maize/groundnut intercrops while in the conventional intercrops, treat-ments without P also recorded significantly higher BCR in the maize/bean and maize/groundnut inter-crops in Machang'a. Return to labour was not signifi-cantly different between the MBILI and the convention intercrop in treatments both with P and without P. In both the MBILI and the conventional intercrop, return to labour was significantly higher in the treatments without P compared to those with P.

Discussions

In general, the maize grain yields from the sites Mukuuni and Machang'a were very low. In Machang'a, the maize grain yields were less than 1 t ha^{-1} in most seasons. The consistently low yields in the intercrops indicate that grain legumes do not contribute to soil improvement in these intercropping systems. They use most or all of the nitrogen they fix, and the organic matter produced is negligible. Therefore, intercropping cereals with grain legumes without addition of other inputs will not be able to solve the problem of reduced soil fertility in the study area.

In Machang'a, there was a response to P during all the seasons, whereas in Mukuuni, the response was manifested from the SR 05 season (third season). The lack of response to P in Mukuuni at the beginning of the experiment could be as a result of the high avail-able P levels that were recorded in these soils (Table 1). These high P values could have masked the effect of the P that was applied in the beginning. But as cropping continued, the available P could have declined leading to more response as seasons progressed. In Machang'a, there were responses to P throughout the cropping sea-sons, which could have been a result of the soils being deficient in P.

The higher maize grain yields in the treatments that received P compared to those that did not receive P especially in Machang'a reflect the importance of adding P to the plants. P in particular is needed by plants at early stages of growth since it is involved in root development (Tisdale et al., 1985). Early adequate uptake of phosphorus by a maize crop results in high grain yield (Barry and Miller, 1989) as maize is quite sensitive to low phosphate supply particularly at early stages of growth (Ahn, 1993).

Maize/cowpea intercrop without P recorded consis-tently the lowest maize grain yields in both Machang'a and Mukuuni during the four seasons. This could be associated with the competition of cowpea with maize for soil water during the four seasons that had poor and unevenly distributed rainfall (Figs. 1 and 2). This corresponds to the results of Shumba et al. (1990), who reported that cowpea intercropping was advan-tageous with maize in seasons with adequate rainfall, but the cowpea competed strongly with the maize crop for soil water when rainfall was limited. Harder and Hoost (1990) observed that in mixed cropping of maize and cowpea, there was competition for N and P which resulted in nutrient deficiency and depressed yields. The low maize yields could also have been a result of P deficiency in the soils, leading to poor early root development of the maize.

The yields of the grain legumes were very low dur-ing the four cropping seasons both in Machang'a and in Mukuuni. For instance, the bean grain yields ranged from 0.02 to 0.79 t ha^{-1} in Machang'a and from 0.09 to 0.74 t ha^{-1} in Mukuuni. Cowpea grain yields were between 0.29 and 1.6 t ha^{-1} in Machang'a and from

0.09 to 0.54 t ha^{-1} in Mukuuni. The yields of groundnuts ranged from 0.14 to 0.79 t ha^{-1} in Machang'a, while they were between 0.02 and 0.51 t ha^{-1} in Mukuuni. The yields of the groundnuts were within the range of yields attained by smallholder farmers who harvest about 0.7 t ha^{-1} and often much less according to Freeman et al. (1999). The low yields could be the result of the lack of external inputs other than P in this study. This corresponds to findings by Kibunja et al. (2000), who reported that beans require external inputs to sustain high yields (over 1 t ha^{-1}) in low-fertility soils. This is contrary to what smallholder farmers practice as they hardly use any mineral fertilizers on the bean crop.

The yields of the legumes were generally low ranging from 0.02 to 0.74 t ha^{-1} in Mukuuni and from 0.02 to 1.6 t ha^{-1} in Machang'a across the four seasons. These results match with the traditional view which states that legumes are weak competitors compared with cereals (Brandsater and Netland, 1999; Ikerra et al., 1999; Fofana et al., 2004; Fan et al., 2006). The low legume grain yields in both intercropping systems could be the result of the shading effect imposed by the tall maize plants on the shorter legumes (Jensen, 1996). Reduction in photosynthesis, and roots are no longer adequately supplied with photosynthates resulting in a shortage of energy required for normal plant functioning (Wahua and Millar, 1978; Nambian et al., 1983). Similarly, competition for water and nutrients could also have contributed to the low yields. The low competitive capacity of legume compared to cereal has been ascribed to its small root system, shallow root distribution, and resulting low-competitive ability for mineral N. Roots of legumes have been reported to be confined to the upper soil layers, whereas cereals occupy both shallow and deeper soil layers (Hauggaard-Nielsen et al., 2001) and this was the case in this study.

The higher increase in maize grain yields in the MBILI, especially in Machang'a site, corresponds with results of Tungani et al. (2002), who reported that maize yields increased from 1.2 to 1.3 t ha^{-1} under MBILI system in unfertilized plots which is an increase of about 7.7%. The maize and legume grain yields increased in the MBILI with addition of P fertilizer. Tungani et al. (2002) also reported that with a modest diammonium phosphate (DAP) fertilizer (100 kg ha^{-1}), yields improved from 1.5 to 1.7 t ha^{-1} (an increase of 15%) in the MBILI.

Maize and legumes (especially cowpea and groundnuts) responded much better to P fertilizer in the MBILI compared to the conventional intercrop in most seasons in Machang'a. This concurs with findings by Tungani et al. (2002), who reported that groundnuts responded much better to fertilizer in MBILI system (139%) compared to the convectional system (64%). The higher yields in the MBILI could be attributed to the better agronomic conditions offered by the MBILI rearranged cereal–legume rows reported by Reddy et al. (1989) and Tungani et al. (2002). They obtained higher yields after two rows of cereal and observed that spatial arrangement of intercrops is an important management practice that improves radiation interception through increased ground cover by cereal and legume.

Some authors (Tungani et al., 2002; Thuita, 2006) have reported that MBILI performs better because of better root distribution. In Mukuuni site, in treatments with P, the maize grain yields were higher in the conventional intercrop compared to the MBILI in the maize/groundnut intercrop. On the other hand, the maize grain yields were higher in the MBILI compared to the conventional intercrop in the maize/bean intercrop. However, the root length density data in this study could not explain the differences in the performance of the maize in these two different intercrop systems as expected. The root length density of the MBILI and conventional intercrop for the groundnuts and beans showed that both the maize/groundnut and the maize/beans intercrops had higher root length densities in the conventional intercrops compared to the MBILI (Fig. 3). These results do not match with those of Thuita (2006), who observed that the crops of the MBILI system had higher root length densities compared to crops of the conventional intercropping system.

In the top 20 cm of the site with the maize/groundnut intercropping system, the root length density was about 71% each in the conventional intercrop and in the MBILI. In the top 20 cm in the maize/beans intercrop, the root density was about 79% in the conventional intercrop, while it was 66% in the MBILI. This indicates that most of the roots in both intercrops were in the range of 0–20 cm. Therefore, the root distribution in terms of soil depth could not explain the difference in the performance of the two intercrops.

The maize grain yields in the intercrops were higher than those in the control (maize monocrop) in some

Fig. 3 Root length densities at different soil depths of maize/bean and maize groundnut MBILI and conventional intercrops in Mukuuni, Meru South district

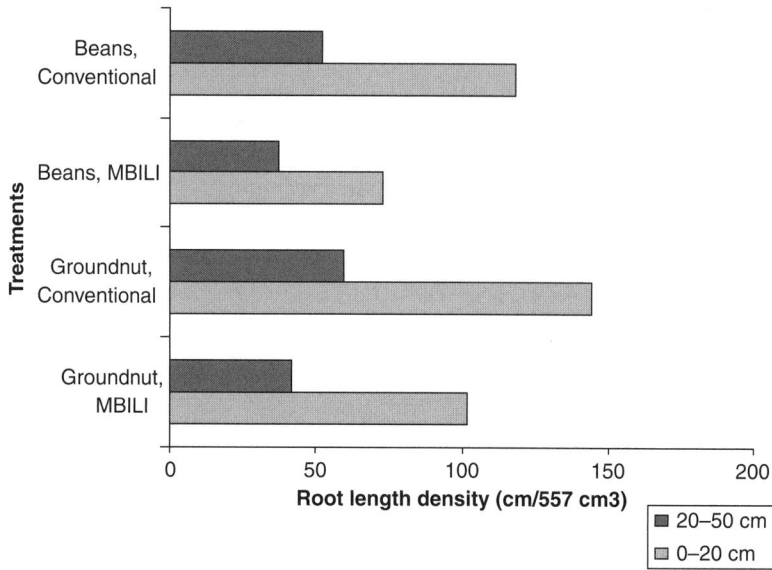

instances in both Mukuuni and Machang'a sites. These results correspond to the results of Giller (2001), who reported that there is evidence of soil fertility improvement in cropping systems that include legumes such as cowpea as shown by the yield of the succeeding maize crop. Mbaga and Friesen (2003) also reported an increase of 50% in the subsequent maize crop due to soil and moisture improvement when legumes were intercropped with maize.

In Machang'a, there were no significant differences in net benefit, BCR, and return to labour between the MBILI and the conventional intercrop in treatments both with and without P, indicating that there is no economic benefit for the MBILI in the system. In Mukuuni, net benefits were improved significantly when P was not applied in the different intercrops. For instance, beans gave higher net benefits and BCR when in the MBILI, while groundnut recorded higher net benefit in the conventional intercrop but, on the other hand, had a higher return to labour in the MBILI.

In both Machang'a and Mukuuni, treatments without P had higher net benefits, BCR, and return to labour in both the MBILI and the conventional intercrop. This indicates that the additional 60 kg of P fertilizer did not offer an economic benefit, though both, the legume and the maize grain yields, were shown to have been improved by the application of P fertilizer, especially in Machang'a. Groundnuts and bean intercrops did not especially give better economic return

with the addition of P in Mukuuni. Maize/groundnut intercrops without P recorded the highest net benefit in Machang'a; however, these results do not correspond to those of Lagat et al. (2005), who reported that the greatest economic return resulted from groundnut intercropped in the MBILI arrangement with addition of DAP, in their study in western Kenya. The high net benefit from the maize/groundnut intercrop could have been partly because the prices for groundnut were very high.

Conclusions

There were significant responses to P in Machang'a for both the maize and legumes throughout the four seasons. In Mukuuni, the P response was significant only from the third season for maize, but there was no significant response for the legumes. In Mukuuni, the conventional intercrops performed better in the maize/cowpea and maize/groundnut intercrops, while the MBILIs performed better in the maize/bean intercrops. In Machang'a, the MBILIs generally performed better for both maize and legumes than did the conventional intercrops in all treatments with and without P fertilizer except the maize/bean intercrop that was better in the conventional intercrop without P. Generally net benefits, BCR, and return to labour were highest when P fertilizer was not applied in both the MBILI and the conventional intercrop in Machang'a and Mukuuni. In Machang'a, there were no economic advantages of MBILI over the conventional intercrop; however, in Mukuuni, MBILI had more economic benefits compared to the conventional intercrop, especially in the maize/bean intercrop.

Acknowledgements The authors wish to thank Vlaamse Inter-Universitaire Raad (VLIR) for providing financial support for the field experimentation. They also appreciate the contribution and collaborative efforts by the Tropical Soil Biology and Fertility Institute of CIAT (TSBF-CIAT), Nairobi, Kenya; Agricultural Research Institute (KARI-Embu), Kenya; Forestry Research Institute (KEFRI-Muguga); and the Department of Environmental Sciences, Kenyatta University, in administering field activities.

References

Adiel RK (2004) Assessment of on-farm adoption potential of nutrient management strategies in Chuka division, Meru South, Kenya. MSc thesis, Kenyatta University, Kenya

Ahn PM (1993) Tropical soils and fertilizer use. Harlow Longman Scientific and Technical, UK, p 264

Anderson JM, Ingram JSL (1993) Tropical soil biology and fertility: a handbook of methods. CAB International, Wallingford, UK

Barrett CB, Place F, Aboud A, Brown DR (2002) The challenge of stimulating adoption of improved natural resource management practices in African agriculture. In: Barrett CB, Place F, Aboud AA (eds) Natural resource management in African agriculture. CAB International, Wallingford, CT, pp 1–21

Barry DAJ, Miller MH (1989) Phosphorus nutrition requirements of maize seedlings for maximum yield. Agron J 81:95–99

Brandsater LO, Netland J (1999) Winter annual legumes for use as cover crops in row crops in northern regions: I. Field experiments. Crop Sci 39(1369):1379

Chivenge P, Murwiri HK, Giller KE (2000) Efficacy of soil organic matter fractionation methods on soils of different texture under similar management. In: The biology and fertility of tropical soils. TSBF report 1997–1998

Fan F, Zhang F, Song Y, Sun J, Bao X, Guo T, Li L (2006) Nitrogen fixation of faba bean (*Vicia faba* L.) interacting with a non-legume in two contrasting intercropping systems. Plant Soil 283:275–286

Fertilizer Use Recommendation Project (FURP) (1987) Description of first priority trial site in the various districts. Final report, vol 24. Embu District, National Agricultural Research Laboratories, Nairobi

Fofana B, Breman H, Carsky RJ, Van Reuler H, Tamelokpo AF, Gnakpenou KD (2004) Using mucuna and P fertilizer to increase maize grain yield and N fertilizer use efficiency

in the coastal savanna of Togo. Nutr Cycling Agroecosyst 68(213):222

Freeman HA, Nigam SN, Kelley TG, Ntare BR, Subrahmanyam P, Boughton D (1999) The world groundnut economy: facts, trends and outlook. International Crops Research Institute for the Semi-Arid Tropics, Andhra Pradesh, India, 52pp

Fujita K, Ofosu-Budu KG, Ogata S (1992) Biological nitrogen fixation in mixed legume–cereal cropping systems. Plant Soil 141:155–175

Giller KE (2001) Nitrogen fixation in tropical cropping systems, 2nd edn. CAB International, Wallingford, CT, 423pp

Government of Kenya (1997) Mbeere district development plan, 1997–2001. Ministry of Planning and National Development, Nairobi

Government of Kenya (2001) The 1999 Kenya national census results. Ministry of Home Affairs, Nairobi

Gutierrez VM, Infante M, Pinchinot A (1975) Situación del cultivo de frijol en America Latina. Centro Internacional de Agricultura Tropical, Cali, Colombi

Harder R, Horst WJ (1990) Nitrogen and phosphorus use in maize sole cropping and maize/cowpea mixed cropping systems on an alfisol in northern Guinea savannah of Ghana. Biol Fertil soils 10:267–275

Hauggaard-Nielsen H, Ambus H, Jensen ES (2001) Temporal and spatial distribution of roots and competition for nitrogen in pea–barley intercrops: a field study employing 32 P technique. Plant Soil 236:63–74

Hulugalle NR, Lal R, Kuile CHH (1986) Amelioration of soil physical properties by mucuna following mechanized land clearing of a tropical rain forest. Soil Sci 141:219–224

Ikerra ST, Maghembe JA, Smithson PC, Buresh RJ (1999) Soil nitrogen dynamics and relationships with maize yields in a *Gliricidia* maize intercrop in Malawi. Plant Soil 211:155–164

Jaetzold R, Schmidt H (1983) Farm management handbook of Kenya. Natural conditions and farm information, vol 11/C. East Kenya. Ministry of Agriculture, Kenya

Jaetzold R, Schmidt H, Hornetz B, Shisanya CA (2006) Farm management handbook of Kenya. Natural conditions and farm information, 2nd edn, vol 11/C. Eastern Province. Ministry of Agriculture/GTZ, Nairobi

Jensen ES (1996) Grain yield, symbiotic N^2 fixation and interspecific competition for inorganic N in pea–barley intercrops. Plant Soil 182:25–38

Jerenyama P, Hestermann OB, Waddington SR, Hardwood RR (2000) Relay intercropping of sunnhemp and cowpea into a smallholder maize system in Zimbabwe. Agron J 92: 239–244

Kibunja CN, Gikonyo EW, Thuranira EG, Wamaitha J, Wamae DK, Nandwa SM (2000) Sustainability of long-term bean productivity using maize stover incorporation. Residual fertilizers and manure. 18th Soil Science Society of East Africa

Kihanda FM (1996) The role of farmyard manure in improving maize production in the sub-humid central highlands of central Kenya. PhD thesis, UK

Kumwenda AS (1998) Soil fertility research for maize-based farming systems in Malawi and Zimbabwe. In: Waddington SR, Murwira HK, Kumwenda JDT, Tagwira F (eds) Soil fertility research for maize-based farming systems in Malawi and Zimbabwe. Proceedings of the soil fertility net results and planning workshop, CIMMIT, Mutara, pp 263–269, 7–11 July 1997

Lagat M, Mukhwana E, Woomer PL (2005) Managing beneficial interactions in legume intercrops (MBILI). SACRED, Africa, Organic resources notebook

Mbaga TE, Friesen D (2003) Maize/legume systems for improved maize production in northern Tanzania. Afr Crop Sci Conf Proc 6:587–591

McCann JC (2005) Maize and grace. Africa's encounter with a new world crop 1500–2000. Harvard University Press, Cambridge, MA, 289pp

Micheni A, Kihanda FM, Irungu J (2004) Soil organic matter: the basis for improved crop production in arid and semi-arid climates of Eastern Kenya. In: Bationo A (ed) Managing nutrient cycles to sustain soil fertility in sub-Saharan Africa. Academy Science, Nairobi, pp 239–248

Mwale M (2000) Effect of application of *S. sesban* prunnings on maize yields. In: The biology and fertility of tropical soils. TSBF report 1997–1998

Nambian PT, Rao MR, Reddy MS, Floyd CN, Dart PJ, Willey RW (1983) Effect of intercropping on nodulation and N fixation by groundnut. Exp Agric 19:79–86

Newman EI (1966) A method of estimating the total length of root in a sample. J Appl Ecol 3:139–145

Ofori F, Pate JS, Stern WR (1987) Evaluation of N_2 fixation and nitrogen economy of a maize/cowpea intercrop system using ^{15}N dilution methods. Plants Soil 102:149–160

Papastylianou I (1988) The ^{15}N methodology in estimating N_2 fixation by vetch and pea grown in pure stand or in mixtures with oats. Plant Soil 107:183–188

Reddy SN, Reddy EVR, Reddy VM, Reddy MS, Reddy PP (1989) Row arrangement in groundnut/pigeon pea intercropping. Trop Agric J 66:309–312

Sakala WD, Kumwenda JDT, Saka AR (2003) The potential of green manures to increase soil fertility and maize yields in Malawi. Biol Agric Hort 21:121–130

Sanchez PA (2002) Soil fertility and hunger in Africa. Science 295:2019–2020

Sanchez PA, Jama BA (2002) Soil fertility replenishment takes off in east and southern Africa. In: Vanlauwe B, Diels J, Sanginga N, Merckx R (eds) Integrated plant nutrient management in sub-Saharan Africa. CAB International, Wallingford, CT, pp 23–45

Shumba EM, Dhliwayo HA, Mukok OZ (1990) The potential of maize–cowpea intercropping in low rainfall areas of Zimbabwe. Zimbabwe J Agric Res 29:81–85

Smaling E (1993) Soil nutrient depletion in sub-Saharan Africa. In: Van Reuler H, Prins W (eds) The role of plant nutrients for sustainable food crop production in sub-Saharan Africa. VKP, Leidschendam, pp 367–375

Tennant D (1975) A test of a modified line intersection method of estimating root length. J Ecol 63:995–1103

Thuita MN (2006) A study to better understand the 'MBILI' intercropping system in terms of grain yields, nutrient uptake and root distribution in relation to conventional intercrops in western Kenya. M.Phil. thesis, Department of Soil Science, Moi University

Tisdale SL, Werner LN, Beaton JD (1985) Soil fertility and fertilizers, 4th edn. Macmillan, New York, NY

Tungai JO, Mukhwana E, Woomer PL (2002) Mbili is number 1: a handbook for innovative maize–legume intercropping. SACRED Africa, Bungoma

Wahua TA, Millar DA (1978) Relative yield totals and yield components of intercropped sorghum and soyabeans. Agron J 70:287–291

Warren GP (1998) Effects of continuous manure application on grain yields at seven sites of trial, from 1993 to 1996. In: Kihanda FM, Warren GP (eds) Maintenance of soil fertility and organic matter in dryland areas. The Department of Soil Science, University of Reading occasional publication no. 3. The University of Reading, UK

Willey RW (1979) Intercropping – Its importance and research needs. Part 1. Competition and yield advantages. Field Crops Abstr 32:1–10

Contributions of Cowpea and Fallow to Soil Fertility Improvement in the Guinea Savannah of West Africa

B.V. Bado, F. Lompo, A. Bationo, Z. Segda, P.M. Sédogo, and M.P. Cescas

Abstract The effects of previous cowpea (*Vigna unguiculata*) and annual fallow on N recoveries, succeeding sorghum yields and soil properties were studied using a 5-year (1995–1999)-old field experiment at Kouaré (11°59′ North, 0°19′ West and 850 m altitude) in Burkina Faso. A 5 × 4 factorial design in a split plot arrangement with five rotation treatments and four fertilizer treatments was used. Total N uptake by succeeding sorghum increased from 26 kg N ha^{-1} in monocropping of sorghum to 31 and 48 kg N ha^{-1} when sorghum was rotated with fallow or cowpea, respectively. Nitrogen derived from fertilizer increased from 10% in monocropping of sorghum to 22 and 26% when sorghum was rotated with fallow or cowpea, respectively. While fallow did not increase N derived from soil, cowpea doubled the quantity of N derived from soil (Ndfs). Sorghum grain yields increased from 75 to 100% when sorghum was rotated with fallow or cowpea, respectively. All rotation treatments decreased soil organic C and N but soil organic C was highest in fallow–sorghum rotation. It was concluded that cowpea–sorghum rotation was more effective than fallow–sorghum rotation and five management options were suggested to improve traditional system productivity.

Keywords Crop rotations · Fallow · Fertilizer · Legume · Soil

B.V. Bado (✉)
Institute of Environment and Agricultural Research (INERA), BP 910, Bobo-Dioulasso, Burkina Faso
e-mail: V.Bado@cgiar.org

Introduction

Traditional agricultural systems in the Soudanian zone of West Africa are characterized by monocropping of cereals, such as pearl millet (*Pennisetum glaucum* L.), sorghum (*Sorghum bicolor*) or sometimes maize (*Zea mays*). Livestock are not well-integrated with agricultural activities. Crop residues are usually exported from the farm for household needs and animal feeding (Berger et al., 1987). Because of low financial resources and other socio-economic factors, chemical fertilizers are less used in traditional systems. Fertilizers are usually only applied on cash crops which can provide financial resources to buy agricultural inputs. The traditional bush-fallow system was used by farmers. New fields are installed on the most old bush fallow (10–15 years). The soil is cultivated for 3–5 years or until crop yields start to decrease, indicating the decline of soil fertility over time. So, the farmer leaves the old and poor soil for a new field under a new old bush fallow. In this system, soil fertility maintenance was based on the use of long-term bush-fallow period (10–15 years) followed by a short cropping period (3–5 years) (Steiner, 1991).

The traditional bush-fallow system worked well when population was low compared to land availabilities. However in recent years, increased population pressure has resulted in significant changes of the traditional bush-fallow systems. Long-term fallow systems are no longer practical because of high demand for lands. Long-term cultivation in lack of, or low applications of chemical fertilizers lead to soil fertility and crop yield decline (Pichot et al., 1981). The process of soil fertility decline has been widely explained by data of many long-term experiments, showing that soil organic carbon decline under long-term cultivation is the factor (Pichot et al., 1981; Pieri, 1989; Bado et al., 1997; Bationo and Mokwunye, 1991b). So, traditional long-term fallow was used by farmers as a means to replenish soil organic carbon.

Agronomic research should investigate other practices that can limit soil fertility decline. For example, short-term fallow of 1–3 years could be an alternative. Using the soil alternatively for one season for cereal and the second season free tillage under natural bush fallow could probably delay soil organic C decline compared to the continuous cultivation of cereal. Any technology to improve crop yield and soil fertility must take into account the socio-economic conditions of farmers who have limited resources to invest in agriculture. Alternative, cost-effective and socio-economically viable means of improving and maintaining soil fertility and productivity must therefore be developed and pilot-tested before providing recommendations to the farmers. These include increasing the efficiency of the little applied fertilizer and enhancing the nutrient input addition and recycling from other sources at the level of cropping systems.

Some N_2-fixing legume crops such as cowpea (*Vigna unguiculata*) are frequently used by farmers in their cropping systems. Legume crops serve as food as well as cash crops for farmers. They can provide the necessary financial resources to buy agricultural inputs, for instance fertilizers. After a long period of monocropping of cereals (5–10 years), legume crops (cowpea or groundnut) are frequently used on degraded soils. Farmers use N_2-fixing legume crops to replenish soil fertility after many years of monocropping of cereals. Beneficial effects of N_2-fixing legume crops on subsequent cereal yields have been reported for the Sahelian zone of West Africa (Bationo and Ntare, 2000; Bagayoko et al., 2000). The main effect of legumes on succeeding crops is commonly attributed to an increase in soil N fertility as a result of biological N_2 fixation (Chalk, 1998). In the savannahs of the West African Sahel, little information exists on the long-term effects of crop rotations with legumes or short-term fallow on N recoveries, soil organic C and soil fertility maintenance.

We hypothesized that traditional system productivity could be improved by an integrated management of cropping systems using short-term fallow, N_2-fixing legume crops such as cowpea and low-fertilizer inputs

of farmers. This research aimed to study different management options using short-term fallow or cowpea in cropping systems, combined with low quantities of chemical and organic fertilizers and local agro-mineral resources (phosphate rock and dolomite) as low-cost options to improve N recoveries, soil fertility and traditional system productivities. The purpose was to identify novel and socio-economically viable management options to improve soil fertility and smallholder farmer's systems.

Materials and Methods

Site, Climate and Soil

The study was undertaken at the agronomic research station of Kouaré (11°59′ North, 0°19′ West and 850 m altitude), located in the Soudanian savannah zone of Burkina Faso (WARDA, 2000). This agro-ecological zone has one rainy season per year, starting in May–June and ending in October. High variations of rainfall were observed both between years and within the years of experimentation (Table 1). The soil was an Alfisol of low fertility, representative of the study area. The main soil properties were weakly low acid (pH 5.5), sandy (75%) with low clay (11%) and organic carbon (0.50%) contents. Calcium, Mg, K and ECE were low (Table 4).

Experiment Description

The experiment was initiated in 1994 on an 8-year native grass fallow. The experiment was conducted for 5 years (from 1995 to 1999). In general, planting occurred in June and harvesting in October. A factorial 3×4 design in a split plot arrangement with four replications was used. Three rotation treatments (cowpea–sorghum, fallow–sorghum and sorghum–sorghum) and four fertilization treatments (chemical nitrogen (N), phosphorous (P), and potassium (K) fertilizer; NPK+dolomite; P+manure; and control without any fertilization) were used. Crop rotations were randomized in the main plots and fertilizer treatments were randomized in the sub-plots. The main plot size was 8 m × 4 m (32 m² per plot). Each main plot

Table 1 Monthly and seasonal rainfall (mm) at Kouaré during the period of experimentation (1995–1999)

| Years | Months | | | | | | Total |
	May	June	July	August	September	October	
1995	27	69	150	372	138	50	806
1996	113	173	92	210	165	26	779
1997	97	90	121	141	164	57	670
1998	89	171	222	244	178	90	994
1999	91	89	151	178	228	34	771
Mean	83	118	147	229	175	51	804

was split into four sub-plots of 2 m × 4 m (8 m^2 per sub-plot).

Improved varieties of sorghum (*S. bicolor*) (ICSV-1049) and cowpea (*V. unguiculata*) (KVX-61-1), recommended by the national agronomic research institute (INERA), were used at the recommended planting density of 62,500 and 125,000 plants ha^{-1} for sorghum and cowpea, respectively. The short-term fallow was a natural fallow where plots were kept free of cultivation under natural grasses for 1 year. Any fertilizer is applied. At the start of the next season, all the biomass of grasses from the fallow was incorporated in the soil. Chemical NPK fertilizers were applied each season at sowing at 14, 10, 11 and 6 kg ha^{-1} of N, P, K and S, respectively, only on the two crops (cowpea and sorghum). A complementary dose of 23 kg N ha^{-1} was applied on sorghum 40 after sowing in the form of urea. Three tons per hectare of air-dried cattle manure was annually applied at sowing to the manure treatments. Cattle manure contained 1.8, 18.40, 0.31 and 0.16% of N, C, P and K, respectively. Annual applications of 1 t ha^{-1} of dolomite (249 and 114 kg ha^{-1} of Ca and Mg, respectively) were used. All fertilizers were broadcasted and incorporated manually with traditional hoe. Soil of the experiment was ploughed each year at 15 cm depth at the start of the season. Like farmers' practices, the main aboveground parts of residues (stems and leaves) of sorghum (stover) and cowpea (haulms) were exported during harvesting. Aboveground parts and remaining parts of crop residues were incorporated with ploughing. Manual seedling of at least 5 cm depth was used for the two crops. Manual weeding with traditional hoe was used two or three times for weed control. Any herbicide or irrigation was used.

The two crops were harvested at maturity. Grains were air-dried for 5 weeks and yields were calculated at 15% of humidity. Stover and haulms were air-dried for 8 weeks and weighed for yield calculation.

Isotopic Experiment

The effects of cowpea and fallow on N recoveries from soil and fertilizer by sorghum were studied after 5 years of cultivation (in 1999). The isotopic dilution methodology with ^{15}N was used (Varvel and Peterson, 1990); 14 kg N ha^{-1} was applied as urea (5% atomic excess in ^{15}N) 40 days after sowing (second application of N) in microplots of 3.2 m × 1.6 m (5.12 m^2) delineated in three replications of the chemical NPK fertilizer treatments of the three cropping systems. To avoid lateral percolations and equal access of plants to ^{15}N microplots were isolated by sheet metals driven in at 15 cm below the soil. To assure homogeneous distribution, the ^{15}N fertilizer was diluted in water before application in microplots.

The total aboveground biomass of the two plants was harvested at physiological maturity. All plant samples were dried at 60°C for 72 h, ground and ^{15}N atomic excess was determined by mass spectrometry at the IAEA Seibersdorf Laboratory. The indirect method was used for N recovery studies. Labelled urea with ^{15}N was used as tracer for N recovery assessment. Fertilizer nitrogen use efficiency (FNUE) was calculated using the percentage of N derived from fertilizer (Ndff) and the total N applied by fertilization treatment (Chalk, 1998) (equation 1).

$$\text{FNUE} = \left(\text{Ndff kg N ha}^{-1} / \text{N applied kg ha}^{-1} \right) \times 100$$

(1)

Soil Analysis

Soil samples were taken from the top 20 cm depth in the experimental plots in 1994 (first year) and 1999 (last year) for laboratory analysis to assess changes in soil properties. Soil pH was measured in water and 1 N KCl using a 2:1 solution to soil ratio and exchangeable acidity was measured. Organic carbon was measured by the wet chemical digestion procedure of Walkley and Black (1934). Total nitrogen was determined by the Kjeldahl procedure. Calcium and Mg were determined by atomic absorption, while K and Na were determined using flame photometry. Effective cation exchange capacity of the soil (ECEC) was calculated by the total exchange bases. Available phosphorous was determined using Bray I method (Fixen and Grove, 1990).

Statistical Methods

Statistical analyses were conducted using software package SYSTAT for Windows (version 8.0). Effects of rotations, fertilizer treatments and season factors on crop yields, N uptake and soil properties were analysed using a multi-factor comparison with the general linear model (GLM) procedure. Treatment means were compared with Fisher's least significant difference (LSD) test (Gomez and Gomez, 1983). Data of original soil were not included in the analysis.

Results and Discussion

Crop Yields

High variations of both total rainfall and its distribution were observed during the 5 years (Table 1). Total rainfall varied from 670 mm (year 1997) to 994 mm (year 1998). But significant correlations were not observed between sorghum or cowpea yields and total rainfall. This indicated that other factors also affected crop yields over the years.

Cowpea

Cowpea was only used in cowpea–sorghum rotation. Thus, only the effects of fertilizer and cropping season can be evaluated. Cowpea grain and haulm yields were affected by fertilizer applications ($p < 0.05$) and cropping season ($p < 0.01$) but interaction was not

Table 2 Sorghum and cowpea yields (kg ha^{-1}) as affected by fertilization and cropping systems during 4 years (1996–1999)

Fertilization	Sorghum		Cowpea	
	Grain	Stover	Grain	Haulms
NPK + dolomite	914[a]	4489[a]	611[ab]	1228[a]
P + manure	746[bc]	3769[c]	649[a]	1284[a]
NPK	798[b]	4205[ab]	510[c]	1028[a]
Control	346[d]	1611[d]	265[d]	380[b]
Cropping systems				
Cowpea–sorghum	938[a]	4281[a]	na	na
Fallow–sorghum	819[b]	3493[b]	na	na
Sorghum–sorghum	467[c]	2912[c]	na	na

Values affected by the same letter in the same column are not significantly different at $p < 0.05$, according to Fisher's test
Na, Not applicable

observed between the two factors (data not shown). Compared to control (without fertilizer) fertilizer increased cowpea haulm yields and significant differences were not observed between fertilizer treatments (Table 2). However, significant differences were observed between fertilizers on cowpea grain yields. Compared to mineral NPK fertilizer alone, cowpea grain yields increased from 20% when dolomite was associated with NPK fertilizer. The combined application of P fertilizer with manure produced higher grain yields than mineral NPK fertilizer, meaning that chemical N and K applications were not necessary for cowpea when manure was used.

Sorghum

The sorghum grain and stover yields were affected by fertilization treatments ($p < 0.001$), rotations ($p < 0.01$) and cropping season ($p < 0.001$). But interactions were not observed between the three factors. Otherwise, the three factors affected sorghum yields without significant influences within them. For example, rotations did not influence sorghum responses to fertilizer and vice versa.

A global analysis of data of the 5 years indicated that all fertilizer treatments improved sorghum grain and stover yields (Table 2). Chemical NPK fertilizer improved sorghum grain yields but higher yields were obtained when chemical NPK fertilizer was associated with dolomite. The application of chemical P fertilizer (in lack of N and K) associated with manure was as effective on sorghum grain yield as chemical NPK fertilizer alone.

Sorghum produced less than 500 kg ha^{-1} of grain in monocropping. But sorghum produced highest grain yields when fallow or cowpea was used in the cropping system (fallow–sorghum or cowpea–sorghum rotation). The succeeding sorghum grain yields increased from 75 to 100% when it was rotated with fallow and cowpea, respectively. However, significant differences were not observed between the effects of cowpea and fallow on sorghum grain yields during the first 2 years. The effects of rotations increased over time and cowpea was most effective on sorghum yield increase during the last 3 years (Fig. 1). A general trend of sorghum yield increases was observed over the first 3 years of cultivation in all fertilizer and rotation treatments (Fig. 1). However, sorghum yields of the control (without fertilizer) and control rotation (sorghum–sorghum) only increased during the second season and remained constant during the last 3 years. Yield increases over the years with fertilizer applications can be explained by the effects of

fertilizers on soil fertility replenishment over time. The slight yield increase in control plots only during the second year was probably a seasonal effect.

Nitrogen Recoveries

Nitrogen uptake by sorghum and N derived from soil and fertilizer were affected by rotations ($p < 0.05$) (Table 3). Compared to monocropping of sorghum, total N uptake increased when sorghum was rotated with fallow or cowpea. However, sorghum absorbed more N in cowpea–sorghum rotation. Compared to monocropping of sorghum, N derived from fertilizer (FNUE) increased when sorghum was rotated with fallow or cowpea, but difference was not observed between the two rotations. Soil of cowpea–sorghum rotation increased N derived from soil (Ndfs) and sorghum absorbed two times more N from soil when it was rotated with cowpea.

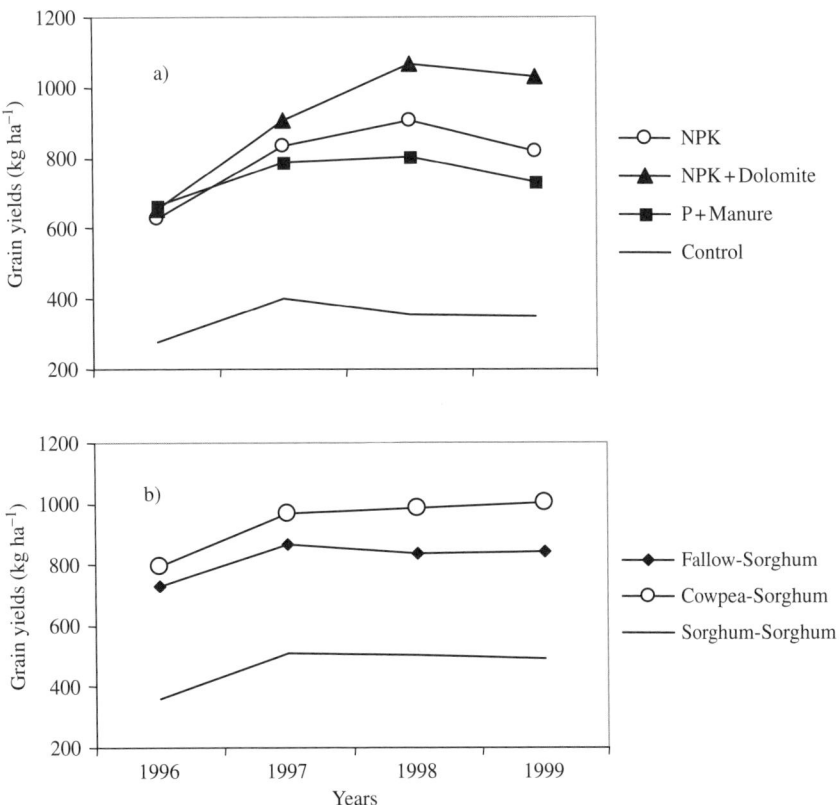

Fig. 1 Effects of (**a**) fertilizer applications and (**b**) crop rotations on succeeding sorghum grain yields for 4 years (1996–1999)

Table 3 Effects of cowpea and fallow on total N uptake by succeeding sorghum, fertilizer N use efficiencies (FNUE) and N derived from soil (Ndfs) and fertilizer (Ndff) in 1999 after 5 years of cultivation

Cropping systems	Total N uptake (kg N ha^{-1})	FNUE (%)	Ndff (kg ha^{-1})	Ndfs (kg ha^{-1})
Fallow–sorghum	31^b	26^a	10^a	21^b
Cowpea–sorghum	48^a	22^{ab}	8^{ab}	40^a
Sorghum–sorghum	26^c	17^c	6^{bc}	20^{bc}

Values affected by the same letter in the same column are not significantly different at $p < 0.05$, according to Fisher's test

Soil Properties

After 5 years of cultivation (1994–1999), fertilization and rotation effects on soil properties were examined with reference to the original soil (Table 4). Soil organic carbon was significantly affected by crop rotation ($p < 0.05$), but fertilization did not affect soil organic C. The highest concentration of organic C was observed in soils of fallow–sorghum rotation and lowest quantities were observed in monocropping of sorghum and cowpea–sorghum rotation.

Soil total N was affected by fertilization and rotations ($p < 0.05$) and interaction was not observed between the two factors. As observed with organic C, the highest quantity of N was observed in soils of fallow–sorghum rotation and lowest quantities were observed in monocropping of sorghum and cowpea–sorghum rotation. Compared to original soil, all rotations and fertilizer treatments decreased soil N. All fertilizer treatments increased soil N and highest quantities of N were observed in manure containing treatment (PK + manure). Dolomite increased soil pH and maintained soil bases (Ca^{++} and Mg^{++}), ECEC and Al^{3+} saturation at the same levels as the original soil.

Soil pH and base saturation were significantly greater with combined application of P and manure than chemical NPK fertilizer alone. Chemical NPK fertilizer increased Al^{3+}. But manure application decreased Al^{3+} saturation.

Good responses to fertilizer applications are likely due to the original low fertility of the soil (Pichot et al., 1981; Pieri, 1989; Sédogo et al., 1991) and yield increase over years indicated the effects of fertilizers on soil fertility replenishment by long-time application of fertilizers (Bationo and Mokwunye, 1991b; Bado et al., 1997). Our data showed that fertilizer application improved soil properties. Dolomite and manure improved soil properties. By supplying Ca^{2+} and Mg^{2+} that reduced Al^{3+} saturation and soil acidity, dolomite improved chemical fertilizer efficiency, leading to crop yield increases (Bado et al., 1993).

Yield improvement by manure applications is due to N effect as observed by soil total N increases and effect of manure on nutrient availability. Phosphorous deficiencies and high P fixation capacities of soils for P has been pointed out as an important limiting factor in West Africa (Bationo and Mokwunye, 1991a). Because of the low clay contents, cation exchange

Table 4 Some soil (0–20 cm layer) properties as affected by fertilizers and crop rotations after 5 years of cultivation (1995–1999)

	pH (KCl)	C. org (%)	Total N (kg N ha^{-1})	K+ (cmol kg^{-1})	Ca++ (cmol kg^{-1})	Mg++ (cmol kg^{-1})	ECEC (cmol kg^{-1})	Bases (%)	Al (%)
Cropping systems									
Fallow–sorghum	5.3	0.39^a	360^a	0.11	1.43^a	0.44	2.10	95	3
Cowpea–sorghum	5.2	0.29^b	294^b	0.09	1.09^b	0.53	1.84	94	3
Sorghum–sorghum	5.3	0.29^b	335^b	0.14	1.07^b	0.35	1.68	94	3
Fertilization									
P+manure	5.2^b	0.32	318^a	0.10	1.15	0.33	1.72	93^b	4^b
NPK	5.0^c	0.33	317^a	0.10	1.12	0.27	1.70	88^c	6^a
NPK+dolomite	5.6^a	0.32	270^b	0.10	1.38	0.62	2.17	99^a	0^d
Control	5.3^b	0.32	237^c	0.15	1.13	0.54	1.91	97^a	2^c
Original soil	5.5	0.50	427	0.13	1.59	0.51	2.30	99	0

EA, Exchange acidity; ECEC, Effective Cation Exchange Capacity
Values affected by the same letter in the same column are not significantly different at $p < 0.05$, according to Fisher's test

capacities and nutrient availabilities are mainly related to soil organic C (Bationo and Mokwunye, 1991b). So, efficiency of P +manure treatment can be explained by its ability to supply chemical P, organic N and improvement of P availability with manure application.

The positive effects of the two rotations on succeeding sorghum yield increases can be explained by the N effect of cowpea and fallow and soil fertility improvement through the recycling of crop residues in cowpea–sorghum and fallow–sorghum rotations as observed by high N uptake and the improvement of fertilizer N use efficiency by succeeding sorghum. Both cowpea and fallow recycled residues in the soil. Despite the exportation of legume fodders from the field, the remaining residues and the belowground parts of cowpea can contribute to soil organic matter and thereby to soil N, leading to soil mineral N increase through mineralization of organic N. The cowpea residues supplied easily decomposable organic N resulting in increased levels of soil mineral N (Barrios et al., 1998; Bagayoko et al., 2000; Bationo and Ntare, 2000; Bloem and Barnard, 2001, Bado et al., 2006). Our results indicated increases of soil total N and high total N uptake values when sorghum was rotated with fallow or cowpea as an increase of N supply from sources, the fertilizer and the soil (Bado et al., 2006). So, N_2-fixing legumes such as cowpea improve N availability in cropping systems as an indirect effect of N from biological N fixation on succeeding non-fixing crop (Chalk, 1998). By increasing fertilizer N use efficiency (Varvel and Peterson, 1990; Mvondo Awono, 1997; Mahadev-Pramanick and Pramanick, 2000) and N derived from soil, soil of cowpea–sorghum rotation improved N nutrition and succeeding sorghum yields. Beneficial effects of the inclusion of N_2-fixing legume crops on the production of succeeding non-fixing crops have been reported in many works (Peoples et al., 1995; Wani et al., 1995; Chalk, 1998; Bationo and Ntare, 2000). Peoples and Crasswell (1992) and Kouyaté et al. (2000) reported that legume–cereal rotations can increase cereal yields from 50 to 350%.

With residues of fallow, low N and low-quality residues are recycled in soil. So, less mineral N is released for sorghum. But in the monocropping of sorghum, residues are mainly exported for traditional systems. Less and poor residues remained in the soil and could not improve soil N. The decomposition of low-quality residues of sorghum probably induced mineral N immobilization, limiting N uptake.

Conclusions

These results indicate that soil fertility can be improved by integrated management of sorghum rotation with cowpea or short-term fallow, manure, dolomite and chemical fertilizers. Five management options could be recommended.

Option 1: Cowpea–sorghum rotation with application of the recommended rate of chemical P fertilizer with 3 t ha^{-1} of manure on cowpea and NPK with 1 t ha^{-1} of dolomite on sorghum. The farmer will achieve high yields of cowpea and sorghum. Moreover, farmers will save money by decreasing chemical application of N and K fertilizer with good cash crop production (cowpea) which will increase farmer's financial capacities to invest in agricultural inputs.

Option 2: Cowpea–sorghum is used with application of recommended doses of chemical NPK fertilizer on sorghum and P + manure on cowpea. This is an alternative option where dolomite is not available. It has the same socio-economic advantages of option 1.

Considering that local rock phosphate can be used on legumes as source of P and the abilities of legumes on P solubilization, the first two options are promising as sustainable and low-cost options for traditional systems improvement.

Option 3: Cowpea–sorghum is used with application of recommended doses of chemical NPK fertilizer (for each crop) and 1 t ha^{-1} of dolomite on the two crops. This is an improvement of fertilizer recommendations of the national agricultural institute (INERA) by the addition of low quantity of dolomite for soil acidity correction.

Option 4: Fallow–sorghum is used with application of recommended dose of chemical NPK with 1 t ha^{-1} of dolomite on sorghum. Fallow and dolomite are used for sustainable management of soil fertility.

Option 5: As an alternative of option 4 when dolomite is not available, fallow–sorghum can be used with application of recommended dose of chemical NPK on sorghum.

However, cowpea has been found to be most effective than short-term fallow systems. Moreover, due to high pressure on cultivable lands induced by population increase, the fallow–sorghum systems should be less sustainable than legume–sorghum systems.

Acknowledgements This study has been funded by the "Institut de l'Environnement et de Recherche Agricole" (INERA), International Crops Research Institute for the Semi-Arid Tropics (ICRISAT), Tropical Soil Biology and Fertility Institute (TSBF, institute of CIAT) and the International Atomic Energy Agency (IAEA) under research contract No. BKF-10952 of the FAO/IAEA Co-ordinated Research Project on Tropical Acid Soils with the support of Africa Rice Center (WARDA) for manuscript preparation. The authors are grateful to Dr. S. Koala (ICRISAT, Sahelian Centre), Dr. N. Sanginga (TSBF), Dr. F. Zapata and the staff of the FAO/IAEA Agriculture and Biotechnology Laboratory for providing ^{15}N labelled fertilizer and sample analysis.

References

Bado BV, Dakyo D, N'dayegamiye A, Cescas MP (1993) Effets de la dolomie sur la production et les propriétés chimiques d'un sol ferrallitique. Agrosol VI(2):22–24

Bado BV, Bationo A, Cescas MP (2006) Assessment of cowpea and groundnut contributions to soil fertility and succeeding sorghum yields in the Guinean Savannah Zone of Burkina Faso (West Africa). Biol Fertil Soil 43:171–176

Bado BV, Sedogo MP, Cescas MP, Lompo F, Bationo A (1997) Effet à long terme des fumures sur le sol et les rendements du maïs au Burkina Faso. Cah Agric 6:571–575

Bagayoko M, Buerkert A, Lung G, Bationo A, Römheld V (2000) Cereal/legume rotation effects on cereal growth in Sudano-Sahelian West Africa: soil mineral nitrogen, mycorrhizae and nematodes. Plant Soil 218:103–116

Barrios E, Kwesiga F, Buresh RJ, Sprent JJ, Coe R (1998) Relating pre-season soil nitrogen to maize yield in tree legume-maize rotations. Soil Sci Soc Am J 62(6): 1604–1609

Bationo A, Mokwunye AU (1991a) Alleviating soil fertility constraints to increased crop production in West Africa: the experience of the Sahel. In: Mokwunye AU (ed) Alleviating soil fertility constraints to increased crop production in West Africa. Kluwer, Dordrecht, pp 217–225

Bationo A, Mokwunye AU (1991b) Role of manure and crop residues in alleviating soil fertility constraints of crop production with special reference to the Sahelian and Sudanian zones of West Africa. Fert Res 29:117–125

Bationo A, Ntare BR (2000) Rotation and nitrogen fertilizer effects on pearl millet, cowpea and groundnut yield and soil chemical properties in a sandy soil in the semi-arid tropics, West Africa. J Agric Sci 134:277–284

Berger M, Belem PC, Dakouo D, Hien V (1987) Le maintien de la fertilité des sols dans l'Ouest du Burkina Faso et la nécessité de l'association agriculture-élévage. Cot Fib Trop XLII, Fasc 3:10–14

Bloem A, Barnard RO (2001) Effect of annual legumes on soil nitrogen and on subsequent yield of maize and grain sorghum. S Afr J Plant Soil 18:56–61

Chalk PM (1998) Dynamics of biologically fixed N in legume-cereal rotations: a review. Aust J Res 49:303–316

Fixen PE, Grove JH (1990) Testing soil for phosphorus. In: Westerman RL (ed) Soil testing and plant analysis. Soil Science Society of America, Madison, WI, pp 141–180

Gomez AK, Gomez AA (1983) Statistical procedures for agricultural research. Wiley, New York, NY

Kouyaté Z, Franzluebbers K, Juo-Anthony SR, Hossner-Lloyd R (2000) Tillage, crop residue, legume rotation and green manure effects on sorghum and millet yields in the semiarid tropics of Mali. Plant Soil 225:141–151

Mahadev-Pramanick MD, Pramanick M (2000) The effect of fertilizer and organic manure on rice-based cropping sequences in West Bengal. Trop Agric 77 (2):111–113

Mvondo Awono JP (1997) Fertilisation azotée du maïs-grain (*Zea mays* L.) en rotation avec une luzerne non dormante (*Medicago sativa* L. var. nitro). PhD thesis, Université Laval, Québec city, QC

Peoples MB, Crasswell ET (1992) Biological nitrogen fixation: investment, expectation and actual contribution to agriculture. Plant Soil 141:13–39

Peoples MP, Herridge DF, Ladha JK (1995) Biological nitrogen fixation: an efficient source of nitrogen for sustainable agricultural production. Plant Soil 174:3–28

Pichot J, Sedogo MP, Poulain JF (1981) Évolution de la fertilité d'un sol ferrugineux tropical sous l'influence des fumures minérales et organiques. Agron Trop 36:122–133

Pieri C (1989) Fertilité des terres de savane. Bilan de 30 années de recherche et de développement agricoles au sud du Sahara. Agridoc-International, Paris

Sédogo MP, Bado BV, Hien V, Lompo F (1991) Utilisation efficace des engrais azotés pour une augmentation de la production vivrière: L'expérience du Burkina Faso. In: Mokwunye U (ed) Alleviating soil fertility constraints to increased crop production in West Africa. Kluwer, Dordrecht, pp 115–123

Shumba EM (1990) Response of maize in rotation with cowpea to NPK fertilizer in a low rainfall area. Zimbabwe J Agric Res 28:39–45

Steiner KG (1991) Overcoming soil fertility constraints to crop production in West Africa: impact of traditional and improved cropping systems on soil fertility. In: Mokwunye AU (ed) Alleviating soil fertility constraints to improve crop production in West Africa. Kluwer, Dordrecht, pp 69–91

Varvel GE, Peterson TA (1990) Nitrogen fertilizer recovery by corn in monoculture and rotation systems. Agron J 82:935–938

Walkley A, Black JA (1934) An examination of the Detjareff method for determining soil organic matter and a proposed modification of the chromatic acid titration method. Soil Sci 37:29–38

Wani SP, Rupela OP, Lee KK (1995) Sustainable agriculture in the semi-arid tropics through biological nitrogen fixation in grain legumes. Plant Soil 174:29–49

WARDA (2000) Promising technologies for rice production in West and Central Africa. Available via DIALOG. http://www.warada.org/publications/varieties.pdf. Cited 28 Feb 2006